Graduate Texts in Mathematics 61

Graduate Texts in Mathematics

1 TAKEUTI/ZARING. Introduction to Axiomatic Set Theory. 2nd. ed.
2 OXTOBY. Measure and Category. 2nd ed.
3 SCHAEFFER. Topological Vector Spaces.
4 HILTON/STAMMBACH. A Course in Homological Algebra.
5 MAC LANE. Categories for the Working Mathematician.
6 HUGHES/PIPER. Projective Planes.
7 SERRE. A Course in Arithmetic.
8 TAKEUTI/ZARING. Axiometic Set Theory.
9 HUMPHREYS. Introduction to Lie Algebras and Representation Theory.
10 COHEN. A Course in Simple Homotopy Theory.
11 CONWAY. Functions of One Complex Variable. 2nd ed.
12 BEALS. Advanced Mathematical Analysis.
13 ANDERSON/FULLER. Rings and Categories of Modules.
14 GOLUBITSKY GUILEMIN. Stable Mappings and Their Singularities.
15 BERBERIAN. Lectures in Functional Analysis and Operator Theory.
16 WINTER. The Structure of Fields.
17 ROSENBLATT. Random Processes. 2nd ed.
18 HALMOS. Measure Theory.
19 HALMOS. A Hilbert Space Problem Book. 2nd ed., revised.
20 HUSEMOLLER. Fibre Bundles 2nd ed.
21 HUMPHREYS. Linear Algebraic Groups.
22 BARNES/MACK. An Algebraic Introduction to Mathematical Logic.
23 GREUB. Linear Algebra. 4th ed.
24 HOLMES. Geometric Functional Analysis and its Applications.
25 HEWITT/STROMBERG. Real and Abstract Analysis.
26 MANES. Algebraic Theories.
27 KELLEY. General Topology.
28 ZARISKI/SAMUEL. Commutative Algebra. Vol. I.
29 ZARISKI/SAMUEL. Commutative Algebra. Vol. II.
30 JACOBSON. Lectures in Abstract Algebra I. Basic Concepts.
31 JACOBSON. Lectures in Abstract Algebra II. Linear Algebra.
32 JACOBSON. Lectures in Abstract Algebra III. Theory of Fields and Galois Theory.
33 HIRSCH. Differential Topology.
34 SPITZER. Principles of Random Walk. 2nd ed.
35 WERMER. Banach Algebras and Several Complex Variables. 2nd ed.
36 KELLEY/NAMIOKA et al. Linear Topological Spaces.
37 MONK. Mathematical Logic.
38 GRAUERT/FRITZSCHE. Several Complex Variables.
39 ARVESON. An Invitation to C^*-Algebras.
40 KEMENY/SNELL/KNAPP. Denumerable Markov Chains. 2nd ed.
41 APOSTOL. Modular Functions and Dirichlet Series in Number Theory.
42 SERRE. Linear Representations of Finite Groups.
43 GILLMAN/JERISON. Rings of Continuous Functions.
44 KENDIG. Elementary Algebraic Geometry.
45 LOÈVE. Probability Theory I. 4th ed.
46 LOÈVE. Probability Theory II. 4th ed.
47 MOISE. Geometric Topology in Dimentions 2 and 3.

continued after index

George W. Whitehead

Elements of Homotopy Theory

Springer-Verlag
New York Berlin Heidelberg London
Paris Tokyo Hong Kong Barcelona

George W. Whitehead, Professor Emeritus
Department of Mathematics
Massachusetts Institute of Technology
Cambridge, Massachusetts 02139

AMS Subject Classifications: Primary: 55B-H. Secondary: 54 B ; 54 C 15, 20, 35; 54 D 99; 54 E 60; 57 F 05, 10, 15, 20, 25

Library of Congress Cataloging in Publication Data
Whitehead, George William, 1918-
Elements of homotopy theory.
(Graduate texts in mathematics; 61)
Bibliography: p.
Includes index.
1. Homotopy theory. I. Title. II. Series.
QA612.7.W45 514'.24 78-15029

Softcover reprint of the hardcover 1st edition 1978

a division of R. R. Donnelley & Sons Company.

9 8 7 6 5 4 3 2

ISBN 978-1-4612-6320-3 ISBN 978-1-4612-6318-0 (eBook)
DOI 10.1007/978-1-4612-6318-0

In memory of
Norman Earl Steenrod (1910-1971)
and
John Henry Constantine Whitehead (1904–1960)

Preface

As the title suggests, this book is concerned with the elementary portion of the subject of homotopy theory. It is assumed that the reader is familiar with the fundamental group and with singular homology theory, including the Universal Coefficient and Künneth Theorems. Some acquaintance with manifolds and Poincaré duality is desirable, but not essential.

Anyone who has taught a course in algebraic topology is familiar with the fact that a formidable amount of technical machinery must be introduced and mastered before the simplest applications can be made. This phenomenon is also observable in the more advanced parts of the subject. I have attempted to short-circuit it by making maximal use of elementary methods. This approach entails a leisurely exposition in which brevity and perhaps elegance are sacrificed in favor of concreteness and ease of application. It is my hope that this approach will make homotopy theory accessible to workers in a wide range of other subjects—subjects in which its impact is beginning to be felt.

It is a consequence of this approach that the order of development is to a certain extent historical. Indeed, if the order in which the results presented here does not strictly correspond to that in which they were discovered, it nevertheless does correspond to an order in which they *might* have been discovered had those of us who were working in the area been a little more perspicacious.

Except for the fundamental group, the subject of homotopy theory had its inception in the work of L. E. J. Brouwer, who was the first to define the degree of a map and prove its homotopy invariance. This work is by now standard in any beginning treatment of homology theory. More subtle is the fact that, for self-maps of the n-sphere, the homotopy class of a map is characterized by its degree. An easy argument shows that it is sufficient to prove that any map of degree zero is homotopic to a constant map. The book begins, after a few pages of generalities, with Whitney's beautiful elementary proof of this fact. It may seem out of place to include a detailed proof so early in an introductory chapter. I have done so for two reasons: firstly, in order to have the result ready for use at the appropriate time, without breaking the line of thought; secondly, to emphasize the point (if emphasis be needed) that algebraic topology does not consist solely of the juggling of categories, functors and the like, but has some genuine geometric content.

Most of the results of elementary homotopy theory are valid in an arbitrary category of topological spaces. If one wishes to penetrate further into the subject, one encounters difficulties due to the failure of such properties as the exponential law, relating cartesian products and function spaces, to be universally valid. It was Steenrod who observed that, if one remains within the category of compactly generated spaces (this entails alteration of the standard topologies on products and function spaces), these difficulties evaporate. For this reason we have elected to work within this category from the beginning.

A critical role in homotopy theory is played by the homotopy extension property. Equally critical is the "dual", the homotopy lifting property. This notion is intimately connected with that of fibration. In the literature various notions of fibrations have been considered, but the work of Hurewicz has led to the "correct" notion: a fibre map is simply a continuous map which has the homotopy lifting property for arbitrary spaces.

The first chapter of the present work expounds the notions of the last three paragraphs. In Chapter II, relative CW-complexes are introduced. These were introduced, in their absolute form, by J. H. C. Whitehead, and it is clear that they supply the proper framework within which to study homotopy theory, particularly obstruction theory.

Chapter III is a "fun" chapter. After presenting evidence of the desirability of studying homotopy theory in a category of spaces with base points, the "dual" notions of H-spaces and H′-spaces are introduced. A space X is an H-space if and only if the set $[Y, X]$ of homotopy classes of maps of Y into X admits a law of composition which is natural with respect to maps of the domain; the definition of H′-space is strictly dual. H-spaces are characterized by the property that the folding map $X \vee X \to X$ can be extended over $X \times X$, while H′-spaces are characterized by the compressibility of the diagonal map $X \to (X \times X, X \vee X)$. The most important H′-spaces are the spheres, and the set $[S^n, Y] = \pi_n(Y)$ has a natural group structure, which is abelian if $n \geqq 2$.

Chapter IV takes up the systematic study of the homotopy groups $\pi_n(Y)$. Relative groups are introduced, and an exact sequence for the homotopy groups of a pair is established. Homotopy groups are seen to behave in many respects like homology groups; this resemblance is pointed up by the Hurewicz map, a homomorphism $\rho : \pi_n(X) \to H_n(X)$. The Hurewicz Theorem, which asserts that ρ is an isomorphism if X is $(n-1)$-connected, is proved. Homotopy groups behave particularly well for fibrations, and this fact facilitates the calculation of the first few homotopy groups of the classical groups.

The fifth chapter is devoted to the homotopy properties of CW-complexes. The first half of the chapter is inspired by the work of J. H. C. Whitehead. The effect on the homotopy groups of the adjunction of a cell, or, more generally, the adjunction of a collection of cells of the same dimension, is considered. This allows one to construct a CW-complex with given homotopy groups. Moreover, if X is an arbitrary space, there is a CW-complex K and a map $f : K \to X$ which induces isomorphisms of the homotopy groups in all dimensions; i.e., f is a weak homotopy equivalence. Such a map is called a CW-approximation, and it induces isomorphisms of the homology groups as well. The device of CW-approximations allows one to replace the study of arbitrary spaces by that of CW-complexes.

The second part of Chapter V is concerned with obstruction theory. This powerful machinery, due to Eilenberg, is concerned with the *extension problem*: given a relative CW-complex (X, A) and a map $f : A \to Y$, does there exist an extension $g : X \to Y$ of f? This problem is attacked by a stepwise extension process: supposing that f has an extension g_n over the n-skeleton X_n of (X, A), one attempts to extend g_n over X_{n+1}. The attempt leads to an $(n+1)$-cochain c^{n+1} of (X, A) with coefficients in the group $\pi_n(Y)$. The fundamental property of the obstruction cochain c^{n+1} is that it is a cocycle whose cohomology class vanishes if and only if it is possible to alter g_n on the n-cells, without changing it on the $(n-1)$-skeleton, in such a way that the new map can be extended over X_{n+1}.

One can obtain further results by making simplifying assumptions on the spaces involved. One of the most important is the Hopf–Whitney Extension Theorem: if Y is $(n-1)$-connected and $\dim(X, A) \leqq n+1$, then the extension problem

$$\begin{array}{ccc} A & \xrightarrow{f} & Y \\ {\scriptstyle i}\downarrow & \nearrow & \\ X & & \end{array}$$

has a solution if and only if the algebraic problem

$$\begin{array}{ccc} H^n(A; \Pi) & \xleftarrow{f^*} & H^n(Y; \Pi) \\ {\scriptstyle i^*}\uparrow & \swarrow & \\ H^n(X; \Pi) & & \end{array}$$

has a solution $(\Pi = \pi_n(Y))$.

Another important application occurs when Y is an Eilenberg–Mac Lane space $K(\Pi, n)$, i.e., $\pi_i(Y) = 0$ for all $i \neq n$. In this case, if X is an arbitrary CW-complex, then $[X, Y]$ is in one-to-one correspondence with the group $H^n(X; \Pi)$. In other words, the functor $H^n(\ \ ; \Pi)$ is *representable*.

The problem of finding a cross-section of a fibration $p: X \to B$ whose base space is a connected CW-complex can be attacked by similar methods. If $f: B_n \to X$ is a cross-section over B_n, the problem of extending f over an $(n+1)$-cell E_α gives rise to an element $c^{n+1}(e_\alpha) \in \pi_n(F_\alpha)$, where F_α is the fibre $p^{-1}(x_\alpha)$ over some point $b_\alpha \in \dot{E}_\alpha$. Now if b_0 and b_1 are points of B, the fibres $F_i = p^{-1}(b_i)$ have isomorphic homotopy groups; but the isomorphism is not unique, but depends on the choice of a homotopy class of paths in B from b_0 to b_1. Thus c^{n+1} is not a cochain in the usual sense. The machinery necessary to handle this more general situation is provided by Steenrod's theory of *homology with local coefficients*. A system G of local coefficients in a space B assigns to each $b \in B$ an abelian group $G(b)$ and to each homotopy class ξ of paths joining b_0 and b_1 an isomorphism $G(\xi): G(b_1) \to G(b_0)$. These are required to satisfy certain conditions which can be most concisely expressed by the statement that G is a functor from the fundamental groupoid of B to the category of abelian groups. To each space B and each system G of local coefficients in B there are then associated homology groups $H_n(B; G)$ and cohomology groups $H^n(B; G)$. These have properties very like those of ordinary homology and cohomology groups, to which they reduce when the coefficient system G is simple. These new homology groups are studied in Chapter VI. An important theorem of Eilenberg asserts that if B has a universal covering space $\tilde{B}$, the groups $H^n(B; G)$ are isomorphic with the *equivariant* homology and cohomology groups of $\tilde{B}$ with ordinary coefficients in $G_0 = G(b_0)$.

Having set up the machinery of cohomology with local coefficients the appropriate obstruction theory can be set up without difficulty; the obstructions c^{n+1} are cochains with coefficients in the system $\pi_n(\mathscr{F})$ of homotopy groups of the fibres. Results parallel to those of obstruction theory can then be proved. As an application, one may consider the universal bundle for the orthogonal group $\mathbf{O}_n$, whose base space is the Grassmannian of n-planes in $\mathbf{R}^\infty$. There are associated bundles whose fibres are the Stiefel manifolds $\mathbf{V}_{n,k}$, and the primary obstructions to the existence of cross-sections in these bundles are the Whitney characteristic classes.

If $F \to X \to B$ is a fibration, the relationships among the homotopy groups of the three spaces are expressed by an exact sequence. The behavior of the homology groups is much more complicated. In Chapter VII we study the behavior of the homology groups in certain cases which, while they are very special, nevertheless include a number of very important examples. In the first instance we assume that B is the suspension of a space

W and establish an exact sequence, the generalized Wang sequence, which expresses certain important relations among the homology groups of F, X and W. When X is contractible this allows us to calculate the homology groups of F by an induction on the dimension. When the coefficient group is a field, the result can be expressed by the statement that $H_*(F)$ is the tensor algebra over the graded vector space $H_*(W)$. The way the tensor algebra over a module M is built up out of M has its geometric analogue in the reduced product of James. Indeed, if X is a space with base point e, one forms the reduced product $J(X)$ by starting with the space of finite sequences of points of X and identifying two sequences if one can be obtained from the other by a finite number of insertions and deletions of the base point. The natural imbedding of X in ΩSX then extends to a map of $J(X)$ into ΩSX which is a weak homotopy equivalence. In particular, if X is a CW-complex, then $J(X)$ is a CW-approximation to ΩSX.

The case when B is a sphere is of special interest because the classical groups admit fibrations over spheres. The Wang sequence then permits us to calculate the cohomology rings (in fact, the cohomology Hopf algebras) for the most important coefficient domains.

Another case of special interest is that for which the fibre F is a sphere. When the fibration is orientable there is a *Thom isomorphism* $H^q(B) \approx H^{q+n+1}(\hat{X}, X)$, where $\hat{X}$ is the mapping cylinder of p. This leads to the Gysin sequence relating the cohomology groups of B and of X.

While the homology groups of F, X and B do not fit together to form an exact sequence, they do so in a certain range of dimensions. Specifically, if F is $(m-1)$-connected and B is $(n-1)$-connected, then $p_* : H_q(X, F) \to H_q(B)$ is an isomorphism for $q < m + n$ and an epimorphism for $q = m + n$. From this fact the desired exact sequence is constructed just as in the case of homotopy groups. This result is due to Serre; an important application is the Homotopy Excision Theorem of Blakers and Massey. To appreciate this result, let us observe that the homotopy groups do not have the Excision Property; i.e., if $(X; A, B)$ is a (nice) triad and $X = A \cup B$, the homomorphism

$$i_* : \pi_q(B, A \cap B) \to \pi_q(X, A)$$

induced by the inclusion map i is not, in general, an isomorphism. However, if $(A, A \cap B)$ is m-connected and $(B, A \cap B)$ in n-connected, then i_* is an isomorphism for $q < m + n$ and an epimorphism for $q = m + n$. The fact that this result can be deduced from the Serre sequence is due to Namioka. As a special case we have the Freudenthal Suspension Theorem: the homomorphism $E : \pi_q(S^n) \to \pi_{q+1}(S^{n+1})$ induced by the suspension operation is an isomorphism for $q < 2n - 1$ and an epimorphism for $q = 2n - 1$.

In Chapter V it was shown that the cohomology functor $H^n(\ \ ; \Pi)$ has a natural representation as $[\ \ , K(\Pi, n)]$. In a similar way, the natural transformations $H^n(\ \ ; \Pi) \to H^q(\ \ ; G)$ correspond to homotopy classes of maps between their representing spaces, i.e., to $[K(\Pi, n), K(G, q)] \approx H^q(K(\Pi, n); G)$. These *cohomology operations* are the object of study in Chapter VIII. Because the suspension $H^{r+1}(SX; A) \to H^r(X; A)$ is an isomorphism, each operation $\theta: H^n(\ ; \Pi) \to H^q(\ ; G)$ determines a new operation $H^{n-1}(\ ; \Pi) \to H^{q-1}(\ ; G)$, called the *suspension* of θ. By the remarks above the suspension can be thought of as a homomorphism $\sigma^*: H^q(\Pi, n; G) \to H^{q-1}(\Pi, n-1; G)$. Interpreting this homomorphism in the context of the path fibration

$$K(\Pi, n-1) = \Omega K(\Pi, n) \to \mathbf{P}K(\Pi, n) \to K(\Pi, n),$$

we deduce from the Serre exact sequence that σ^* is an isomorphism for $q < 2n$ and a monomorphism for $q = 2n$. Indeed, the homomorphisms σ^* can be imbedded in an exact sequence, valid in dimensions through $3n$. The remaining groups in the sequence are cohomology groups of $K(\Pi, n) \wedge K(\Pi, n)$, and interpretation of the remaining homomorphisms in the sequence yields concrete results on the kernel and cokernel of σ^*.

Examples of cohomology operations are the mod 2 Steenrod squares. They are a sequence of stable cohomology operations Sq^i $(i = 0, 1, \ldots)$. These are characterized by a few very simple properties. More sophisticated properties are due to Cartan and to Adem. The former are proved in detail; as for the latter, only a few instances are proved. With the aid of these results it follows that the Hopf fibrations $S^{2n-1} \to S^n$ and their iterated suspensions are essential; moreover, certain composites (for example $S^{n+2} \to S^{n+1} \to S^n$) of iterated Hopf maps are also.

Chapter VIII concludes with the calculation of the Steenrod operations in the cohomology of the classical groups (and the first exceptional group $\mathbf{G}_2$).

If X is an arbitrary (0-connected) space and N a positive integer, one can imbed X in a space X^N in such a way that (X^N, X) is an $(N+1)$-connected relative CW-complex and $\pi_q(X^N) = 0$ for all $q > N$. The pair (X^N, X) is unique up to homotopy type (rel. X); and the inclusion map $X \subsetneq X^{N+1}$ can be extended to a map of X^{N+1} into X^N, which is homotopically equivalent to a fibration having an Eilenberg–MacLane space $K(\pi_{N+1}(X), N+1)$ as fibre. The space X^{N+1} can be constructed from X^N with the aid of a certain cohomology class $k^{N+2} \in H^{N+2}(X^N; \pi_{N+1}(X))$. The system $\{X^N, k^{N+2}\}$ is called a Postnikov system for X, and the space X is determined up to weak homotopy type by its Postnikov system. The Postnikov system of X can be used to give an alternative treatment of obstruction theory for maps into X. These questions are treated in Chapter IX.

In Chapter X we return to the study of H-spaces. However, further conditions are imposed, in that the group axioms are assumed to hold up to homotopy. For such a space X the set $[Y, X]$ is a group for every Y. This group need not be abelian. However, under reasonable conditions it is nilpotent, and its nilpotency class is intimately related to the Lusternik–Schnirelmann category of Y. Of importance in studying these groups is the Samelson product. If $f : Y \to X$ and $g : Z \to X$ are maps, then the commutator map

$$(y, z) \to (f(y)g(z))(f(y)^{-1}g(z)^{-1})$$

of $Y \times Z$ into X is nullhomotopic on $Y \vee Z$ and therefore determines a well-defined homotopy class of maps of $Y \wedge Z$ into X. When Y and Z are spheres, so is $Y \wedge Z$, and we obtain a bilinear pairing $\pi_p(X) \otimes \pi_q(X) \to \pi_{p+q}(X)$. This pairing is commutative (up to sign) but is not associative. Instead one has a kind of Jacobi identity with signs.

Suppose, in particular, that X is the loop space of a space W. Then the isomorphisms $\pi_{r-1}(X) \approx \pi_r(W)$ convert the Samelson product in X to a pairing $\pi_p(W) \otimes \pi_q(W) \to \pi_{p+q-1}(W)$. This pairing is called the Whitehead product after its inventor, J. H. C. Whitehead, and the algebraic properties already deduced for the Samelson product correspond to like properties for that of Whitehead. Chapter X then concludes with a discussion of the relation between the Whitehead product and other operations in homotopy groups.

Chapter XI is devoted to homotopy operations. These are quite analogous to the cohomology operations discussed earlier. Universal examples for operations in several variables are provided by clusters of spheres

$$\Sigma = \mathbf{S}^{n_1} \vee \cdots \vee \mathbf{S}^{n_k}.$$

Indeed, each element $\alpha \in \pi_n(\Sigma)$ determines an operation $\theta_\alpha : \pi_{n_1} \times \cdots \times \pi_{n_k} \to \pi_n$ as follows. If $\alpha_i \in \pi_{n_i}(X)$ is represented by a map $f_i : \mathbf{S}^{n_i} \to X$ $(i = 1, \ldots, k)$, then the maps f_i together determine a map $f : \Sigma \to X$. We then define $\theta_\alpha(\alpha_1, \ldots, \alpha_k) = f_*(\alpha)$. And the map $\alpha \to \theta_\alpha$ is easily seen to be a one-to-one correspondence between $\pi_n(\Sigma)$ and the set of all operations having the same domain and range as θ_α.

Thus it is of importance to study the homotopy groups of a cluster of spheres. This was done by Hilton, who proved the relation

$$\pi_n(\Sigma) \approx \bigoplus_{r=1}^{\infty} \pi_n(\mathbf{S}^{n_r}),$$

where $\{n_r\}$ is a sequence of integers tending to ∞. The inclusion $\pi_n(\mathbf{S}^{n_r}) \to \pi_n(\Sigma)$ is given by $\beta \to \alpha_r \circ \beta$, where $\alpha_r \in \pi_{n_r}(\Sigma)$ is an iterated Whitehead product of the homotopy classes ι_j of the inclusion maps $\mathbf{S}^{n_j} \subsetneq \Sigma$ $(j = 1, \ldots, k)$. Hilton's theorem was generalized by Milnor in that the spheres $\mathbf{S}^{n_i}$ were replaced by arbitrary suspensions $\mathbf{S}X_i$. Then Σ has to be replaced by $\mathbf{S}X$, where $X = X_1 \vee \cdots \vee X_k$. The Hilton–Milnor Theorem then asserts that if the spaces X_i are connected CW-complexes, then $J(X)$ has the same homotopy type as the (weak) cartesian product

$$\prod_{r=1}^{\infty} J(X_r),$$

where X_r is an itered reduced join of copies of $X_1, \ldots, X_k$. The isomorphism in question is induced by a certain collection of iterated Samelson products.

One consequence of the Hilton Theorem is an analysis of the algebraic properties of the composition operation. The map $(\alpha, \beta) \to \beta \circ \alpha$ $(\alpha \in \pi_n(\mathbf{S}^r)$, $\beta \in \pi_r(X))$ is clearly additive in α, but it is not, in general, additive in β. The universal example here is $\beta = \iota_1 + \iota_2$, where ι_1 and ι_2 are the homotopy classes of the inclusions $\mathbf{S}^r \to \mathbf{S}^r \vee \mathbf{S}^r$. Application of the Hilton Theorem and naturality show that, if $\beta_1, \beta_2 \in \pi_r(X)$, then

$$(\beta_1 + \beta_2) \circ \alpha = \beta_1 \circ \alpha + \beta_2 \circ \alpha + \sum_{j=0}^{\infty} w_j(\beta_1, \beta_2) \circ h_j(\alpha),$$

where $w_j(\beta_1, \beta_2)$ is a certain iterated Whitehead product and $h_j : \pi_n(\mathbf{S}^r) \to \pi_n(\mathbf{S}^{n_j})$ is a homomorphism, the jth *Hopf–Hilton homomorphism.*

The suspension operation induces a map of $[X, Y]$ into $[\mathbf{S}X, \mathbf{S}Y]$ for any spaces X, Y. We can iterate the procedure to obtain an infinite sequence

$$[X, Y] \to [\mathbf{S}X, \mathbf{S}Y] \to [\mathbf{S}^2X, \mathbf{S}^2Y] \to \cdots \to [\mathbf{S}^nX, \mathbf{S}^nY] \to \cdots$$

in which almost all of the sets involved are abelian groups and the maps homomorphisms. Thus we may form the direct limit

$$\{X, Y\} = \varinjlim_{n} [\mathbf{S}^nX, \mathbf{S}^nY];$$

it is an abelian group whose elements are called *S-maps* of X into Y. In particular, if $X = \mathbf{S}^n$, we obtain the nth *stable homotopy group* $\sigma_n(Y) = \{\mathbf{S}^n, Y\}$.

We have seen that the homotopy and homology groups have many properties in common. The resemblance between stable homotopy groups and homology groups is even closer. Indeed, upon defining relative groups in the appropriate way, we see that they satisfy all the Eilenberg–Steenrod axioms for homology theory, *except for the Dimension Axiom.*

Examination of the Eilenberg–Steenrod axioms reveals that the first six axioms have a very general character, while the seventh, the Dimension Axiom, is very specific. In fact, it plays a normative role, singling out standard homology theory from the plethora of theories which satisfy the first six. That it is given equal status with the others is no doubt due to the fact that very few interesting examples of non-standard theories were known. But the developments of the last fifteen or so years has revealed the existence of many such theories: besides stable homotopy, one has the various K-theories and bordism theories.

Motivated by these considerations, we devote the remainder of Chapter XII to a discussion of homology theories without the dimension axiom. The necessity of introducing relative groups being something of a nuisance, we avoid it by reformulating the axioms in terms of a category of spaces with base point, rather than a category of pairs. The two approaches to homology theory are compared and shown to be completely equivalent.

The book might well end at this point. However, having eschewed the use of the heavy machinery of modern homotopy, I owe the reader a sample of things to come. Therefore a final chapter is devoted to the Leray–Serre spectral sequence and its generalization to non-standard homology theories. If $F \to X \to B$ is a fibration whose base is a CW-complex, the filtration of B by its skeleta induces one of X by their counterimages. Consideration of the homology sequences of these subspaces of X and their interrelations gives rise, following Massey, to an exact couple; the latter, in turn gives rise to a spectral sequence leading from the homology of the base with coefficients in the homology of the fibre to the homology of the total space. Some applications are given and the book ends by demonstrating the power of the machinery with some qualitative results on the homology of fibre spaces and on homotopy groups.

As I have stated, this book has been a mere introduction to the subject of homotopy theory. The rapid development of the subject in recent years has been made possible by more powerful and sophisticated algebraic techniques. I plan to devote a second volume to these developments.

The results presented here are the work of many hands. Much of this work is due to others. But mathematics is not done in a vacuum, and each of us must recognize in his own work the influence of his predecessors. In my own case, two names stand out above all the rest: Norman Steenrod and J. H. C. Whitehead. And I wish to acknowledge my indebtedness to these two giants of our subject by dedicating this book to their memory.

I also wish to express my indebtedness to my friends and colleagues Edgar H. Brown, Jr., Nathan Jacobson, John C. Moore, James R. Munkres, Franklin P. Peterson, Dieter Puppe, and John G. Ratcliffe, for reading portions of the manuscript and/or cogent suggestions which have helped me over many sticky points. Thanks are also due to my students in several courses based on portions of the text, particularly to Wensor Ling and Peter Welcher, who detected a formidable number of typographical errors and infelicities of style.

Thanks are also due to Miss Ursula Ostneberg for her cooperation in dealing with the typing of one version after another of the manuscript, and for the fine job of typing she has done.

This book was begun during my sabbatical leave from M.I.T. in the spring term of 1973. I am grateful to Birkbeck College of London University for providing office space and a congenial environment.

There remains but one more acknowledgment to be made: to my wife, Kathleen B. Whitehead, not merely for typing the original version of the manuscript, but for her steady encouragement and support, but for which this book might never have been completed.

George W. Whitehead

Massachusetts Institute of Technology

June, 1978.

Contents

CHAPTER I

Introductory Notions

In this Chapter we set the stage for the developments to come. The introductory section is devoted to a general discussion of the most primitive notions of homotopy theory: extension and lifting problems. The notion of homotopy is introduced, and its connection with the above problems discussed. This leads to a formulation of fibrations and cofibrations, which have played such a fundamental role in the development of the subject.

Section 2 is devoted to a list of the standard notations which are used throughout the book. An important source of examples for us will be the classical groups: orthogonal, unitary and symplectic, and their coset spaces: Grassmann and Stiefel manifolds. As special cases of the latter appear the spheres and projective spaces.

Apart from the fundamental group, the oldest notion in homotopy theory is that of the degree of a mapping. This is due to Brouwer, who introduced the notion in [1] in 1912 and proved its homotopy invariance. This was sufficient for many important applications (invariance of domain, the existence of fixed points for maps of the disc, etc.) and it was not until 1926 that the converse was proved by Hopf [1]: two maps of $\mathbf{S}^n$ into itself having the same degree are homotopic. It is sufficient to prove that a map of degree zero is nullhomotopic, and a beautiful elementary proof of the latter statement was given by Whitney [2] in 1937. As this proof uses no machinery with which the reader is unfamiliar, and as it serves well as an introduction to the subject, we have devoted §3 to it.

In the elementary phase of the subject, there is no need to place any particular hypotheses on the spaces with which we are working, and we may as well operate in the category of all topological spaces. However, as more complex notions appear, more demands on the spaces become inevitable, and some attention has been paid in the literature to the discussion of a suitable category in which to study homotopy theory. In 1959 Milnor [1] proposed the category $\mathscr{W}$ of spaces having the homotopy type of CW-complexes. This category is completely satisfactory in many respects. One especially pleasant feature is that weak homotopy equivalences in $\mathscr{W}$ are, in fact, homotopy equivalences. On the other hand, it suffers from the disadvantage that in many cases the function space $F(X, Y)$ fails, in many cases, to belong to $\mathscr{W}$. In 1967 Steenrod [5] proposed the category $\mathscr{K}$ of compactly generated spaces, and we have found his arguments sufficiently cogent to impel us to adopt his suggestion. One feature of $\mathscr{K}$ is that certain spaces, such as products and function spaces, do not carry the usual topology (although they do in many important cases—in particular, $\mathbf{I} \times X$ has the usual topology, so that the notion of homotopy is unaffected). In §4 we enumerate the most important properties of the compactly generated category $\mathscr{K}$.

Much of homotopy theory has to do with pairs (X, A). In order for many standard constructions to work efficiently, it is necessary to make use of the homotopy extension property. It was Borsuk who first realized the importance of this notion, and many of his early papers were devoted to its study. Particularly significant was [1], written in 1937; one of the major results of this paper is a homotopy lifting theorem, the earliest one known to me. In the intervening years, several authors gave sufficient conditions, in the form of local smoothness conditions on (X, A), that an inclusion map $A \hookrightarrow X$ be a cofibration. Finally Steenrod proved in [5] the equivalence between the NDR condition and the absolute homotopy extension property. In §5 we summarize Steenrod's results.

The mapping cylinder of a continuous map $f: X \rightarrow Y$ has been a most fruitful notion. Introduced by J. H. C. Whitehead in 1939 in a combinatorial setting [1], it was studied in 1943 in a general context by Fox [2], who analysed the notions of retraction and deformation in terms of the behavior of the spaces involved *vis-à-vis* the mapping cylinder. These results are also given in §5.

Many spaces met with in homotopy theory (for example, CW-complexes and countable products) are built up as the unions of ascending sequences of topological spaces. Indeed, it is often the case that the union does not have an *a priori* given topology, and it is necessary to construct one from the topologies of the subspaces. In order that the space X should have desirable properties, it is sometimes necessary to impose restrictions on the topologies of the X_n. These restrictions are discussed, again following Steenrod [5], in §6.

The first definition of fibre space was given by Hurewicz and Steenrod [1] in 1941 (although the notion of fibre bundle appears in the work of Whitney as early as 1935 [1]). Realizing the importance of the homotopy lifting property (HLP), they gave conditions on a family of local cross-sections which enabled them to prove the HLP. In subsequent years a number of minor variants were proposed by various authors. But it was Serre [1] who took the bull by the horns in 1950 and defined a fibre space to be one which satisfied the HLP for maps of finite complexes. In 1955 the final step was taken by Hurewicz [2], who required the property in question to hold for maps of arbitrary spaces. As almost any map one would want to call a fibration has the Hurewicz property, we have chosen his notion as the basic one. And two theorems of Strøm [1, 2] reveal a beautiful and satisfying connection between fibrations and cofibrations.

Our final section is devoted to fibrations. The most common examples are given and the interplay between fibrations and cofibrations is exploited. Induced fibrations and fibre homotopy equivalence are studied. Finally, by analogy with the mapping cylinder construction, it is shown that every map is homotopically equivalent to a fibre map.

1 The Fundamental Problems: Extension, Homotopy, and Classification

A basic problem of topology is that of "factoring one continuous function through another". Specifically, given a diagram

$$\begin{array}{ccc} A & & \\ {\scriptstyle i}\big\downarrow & \searrow^{f} & \\ X & & Y \end{array}$$

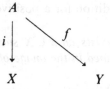

in some category $\mathscr{C}$ of topological spaces and continuous maps, can it be completed to a commutative diagram

$$\begin{array}{ccc} A & & \\ {\scriptstyle i}\big\downarrow & \searrow^{f} & \\ X & \xrightarrow[g]{} & Y \end{array}$$

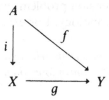

in the same category? This is the *right factorisation problem*, and is conveniently symbolised by the diagram

(1.1)
$$\begin{array}{ccc} A & & \\ {\scriptstyle i}\big\downarrow & \searrow^{f} & \\ X & \underset{g}{\dashrightarrow} & Y \end{array}$$

where the dashed arrow stands for a map whose existence is in question.

Dually, one has the *left factorisation problem*

$$\textbf{(1.2)}\qquad \begin{array}{ccc} Y & \overset{g}{\dashrightarrow} & X \\ & {\scriptstyle f}\searrow & \downarrow{\scriptstyle p} \\ & & B \end{array}$$

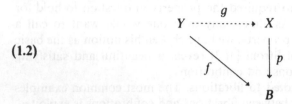

When the map i of (1.1) is one-to-one, the corresponding factorization problem is called an *extension problem*, and g an *extension* of f. When the map p of (1.2) maps X *upon* B, the corresponding factorization problem is called a *lifting problem*, and g a *lifting* of f.

A necessary condition for a positive solution of (1.1) is

(1.3) *For all* $a_1, a_2 \in A$, *if* $i(a_1) = i(a_2)$, *then* $f(a_1) = f(a_2)$.

If (1.3) holds, and if, in addition, i maps A upon X, then there is a unique function g such that $g \circ i = f$; and g is continuous provided that i is a *proclusion*, i.e., that X has the identification topology determined by i, so that a subset U of X is open in X if and only if $i^{-1}(U)$ is open in A.

Similarly, a necessary condition for a positive solution of (1.2) is

(1.4) *For each* $y \in Y$, *there exists an* $x \in X$ *such that* $f(y) = p(x)$; *in other words, the image of* f *is contained in the image of* p.

If (1.4) holds, and if, in addition, p is one-to-one, then there is a unique function g such that $p \circ g = f$; and g is continuous provided that p is an *inclusion*, so that we may regard X as a subspace of B.

Numerous examples of extension problems with positive solutions occur as basic theorems of general topology. For example,

(1.5) *If* A *is a dense subspace of the metric space* X, Y *is a complete metric space, and* $f: A \to Y$ *is uniformly continuous, then* f *has a uniformly continuous extension* $g: X \to Y$.

(1.6) (Tietze's Extension Theorem) *If* X *is normal,* A *is a closed subspace of* X, *and* J *is an interval of real numbers, then every map* $f: A \to J$ *has an extension* $g: X \to J$.

Negative solutions also occur. For example, the Brouwer fixed point theorem is equivalent to

(1.7) *The unit sphere* $\mathbf{S}^n$ *is not a retract of the unit disc* $\mathbf{E}^{n+1}$; i.e., *the identity map of* $\mathbf{S}^n$ *cannot be extended to a continuous map* $r: \mathbf{E}^{n+1} \to \mathbf{S}^n$.

This is proved with the aid of homology theory. In fact, if there is a commutative diagram

$$\begin{array}{ccc} \mathbf{S}^n & \xrightarrow{\quad 1 \quad} & \mathbf{S}_n \\ & {\scriptstyle i}\searrow \quad \nearrow{\scriptstyle r} & \\ & \mathbf{E}^{n+1} & \end{array}$$

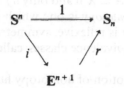

then application of the functor H_n yields a commutative diagram

$$\begin{array}{ccc} H_n(\mathbf{S}^n) & \xrightarrow{\quad H_n(1) = 1 \quad} & H_n(\mathbf{S}^n) \\ & {\scriptstyle H_n(i)}\searrow \quad \nearrow{\scriptstyle H_n(r)} & \\ & H_n(\mathbf{E}^{n+1}) & \end{array}$$

But $H_n(\mathbf{E}^{n+1}) = 0$, while $H_n(\mathbf{S}^n) \neq 0$, a contradiction.

Many negative results in extension theory are proved by an elaboration of this argument. In fact, let F be any functor defined on the category $\mathscr{C}$. Application of F to the diagram (1.1) yields a diagram

(1.8)

$$\begin{array}{ccc} F(A) & & \\ {\scriptstyle F(i)}\downarrow & \searrow{\scriptstyle F(f)} & \\ F(X) & \dashrightarrow & F(Y) \end{array}$$

which is a right factorization problem in the range category of F. Thus a necessary condition that (1.1) have a positive solution is that (1.8) have one. However, a positive solution for (1.8) need not imply one for (1.1); for if $\phi : F(X) \to F(Y)$ is a solution of (1.8), so that $\phi \circ F(i) = F(f)$, there need not exist $g : X \to Y$ such that $\phi = F(g)$. Moreover, even if such a map g does exist, the equality $F(g \circ i) = F(g) \circ F(i) = F(f)$ need not imply the equality $g \circ i = f$.

An important special case of the extension problem is the *homotopy problem*. Let $f_0, f_1 : X \to Y$ be maps, and let $\mathbf{I}$ be the closed interval $[0, 1]$. Then f_0 and f_1 are homotopic ($f_0 \simeq f_1$) if and only if the map of $\dot{\mathbf{I}} \times X$ into Y which sends (t, x) into $f_t(x)$ $(t = 0, 1)$ has an extension $f : \mathbf{I} \times X \to Y$. Such an extension is called a *homotopy* of f_0 to f_1, and determines, for each $t \in \mathbf{I}$, a map $f_t : X \to Y$, given by

$$f_t(x) = f(t, x) \qquad (t \in \mathbf{I},\ x \in X).$$

If A is a subspace of X and $f_0 \mid A = f_1 \mid A$, we say that f_0 and f_1 are *homotopic relative to* A ($f_0 \simeq f_1$ (rel. A)) if and only if there is a homotopy

$f: \mathbf{I} \times X \to Y$ of f_0 to f_1 satisfying the additional condition $f_t | A = f_0 | A$ for all $t \in I$. The homotopy f is *stationary at* $x \in X$ if and only if $f(t, x) = f(0, x)$ for all $t \in \mathbf{I}$; f is *stationary on* $A \subset X$ if and only if f is stationary at each point of A. Finally, a *stationary homotopy* is one which is stationary on X.

The relation of homotopy is reflexive, symmetric, and transitive, and the study of the sets $[X, Y]$ of equivalence classes, called *homotopy classes*, is the object of homotopy theory.

Let us consider how the notion of homotopy fits in with the factorization problems discussed above. For example, if

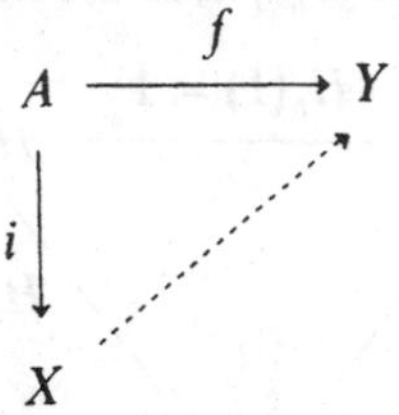

is an extension problem, and if $f' : A \to Y$ is homotopic to f, one may ask whether the existence of a solution for f implies that of one for f'. (It need not; let L_n be the line segment in $\mathbf{R}^2$ joining the points $(0, 1/n)$ and $(1, 0)$, $(n = 1, 2, \ldots)$, L_0 the line segment joining $(0, 0)$ and $(1, 0)$, $A = \bigcup_{n=0}^{\infty} L_n = Y$, X the unit square $[0, 1] \times [0, 1]$, f the identity map (Figure 1.1). It is an easy exercise to see that A is not a retract of X, so that f has no extension $g : X \to Y$. But A is contractible to the point $y_0 = (1, 0)$, and the constant map of A into the point y_0 has an extension).

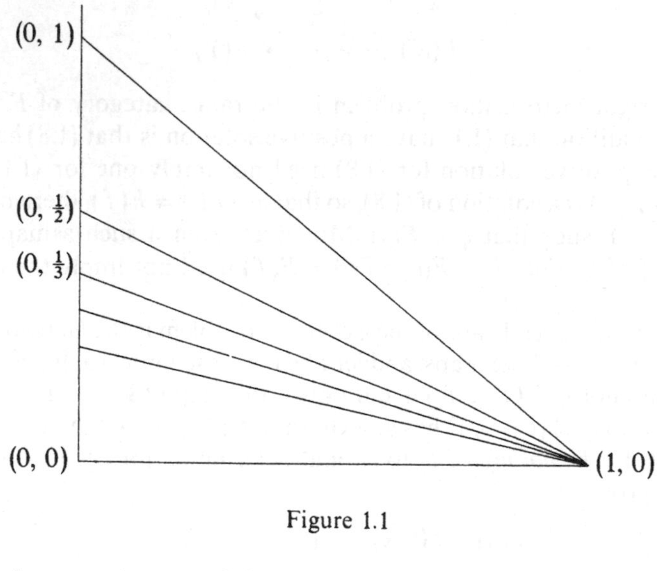

Figure 1.1

If two homotopic maps f, f' admit extensions g, g', we may ask whether the extensions need be homotopic. This suggests the *homotopy extension problem*: let $f\colon X \to Y$, and let $g\colon \mathbf{I} \times A \to Y$ be a homotopy of $f \mid A$ to a map $f'\colon A \to Y$. Then f and g define a map $h\colon 0 \times X \cup \mathbf{I} \times A \to Y$ and we may ask whether the extension problem

$$\begin{array}{ccc} 0 \times X \cup \mathbf{I} \times A & \xrightarrow{\ h\ } & Y \\ \big\downarrow & \nearrow & \\ \mathbf{I} \times X & & \end{array} \tag{1.9}$$

has a solution. When this is so, the extension is a homotopy of f to an extension of f'.

When the problem (1.9) has a solution for every space Y and map h, we say that the inclusion map $i\colon A \hookrightarrow X$ is a *cofibration*. Cofibrations are likely to be pathological unless i is a homeomorphism of A with a closed subset of X. In this case we say that i is a *closed cofibration*. This is equivalent to saying that (X, A) is an NDR-pair and is a notion of crucial importance in homotopy theory. We shall study this notion in §§5, 6 below.

Dually, let

$$\begin{array}{ccc} & & X \\ & \nearrow & \big\downarrow p \\ Y & \xrightarrow[\ f\ ]{} & B \end{array}$$

be a lifting problem. Again, if $f \simeq f'$, we may ask whether the existence of a solution for f implies that of one for f'. (Again, this need not be true: let X be the closure in $\mathbf{R}^2$ of the graph of the function $\sin(1/x)$ $(0 < x \le 1)$, $B = Y = [0, 1]$, p the vertical projection (Figure 1.2). Then the identity map $f\colon Y \to B$ cannot be lifted, but f is homotopic to a constant map, which can).

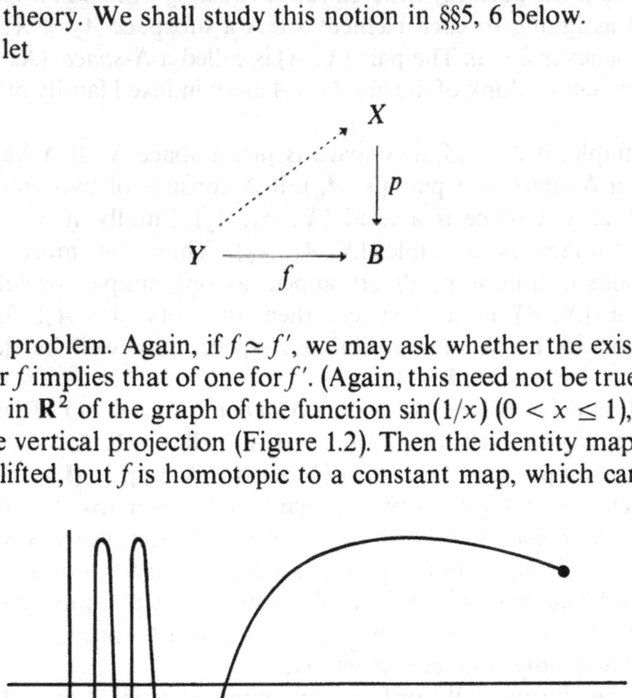

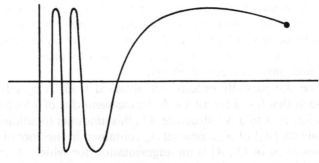

Figure 1.2

Again we may formulate the *homotopy lifting problem*

$$\begin{array}{ccc} 0 \times Y & \xrightarrow{f} & X \\ \cap & \nearrow & \downarrow p \\ \mathbf{I} \times Y & \xrightarrow[g]{} & B \end{array}$$

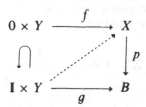

When this problem has a solution for every space Y and pair of maps f, g we say that $p: X \to B$ is a *fibration*. This notion is again of the utmost importance in homotopy theory; we shall study it in §7.

If (X, A) and (Y, B) are pairs, two maps $f_0, f_1 : (X, A) \to (Y, B)$ are homotopic as maps of pairs if and only if there is a homotopy of f_0 to f_1 which maps $\mathbf{I} \times A$ into B, and we may wish to study the set of homotopy classes $[X, A; Y, B]$ of such maps of pairs. Occasionally other more complicated configurations may arise (e.g. maps of triples, triads, etc.). These can be included under one heading, by formalizing the notion of configuration.

Let Λ be a finite partially ordered set. *A Λ-configuration in a space X* is a function A assigning to each element $\lambda \in \Lambda$ a subspace A_λ of X, such that $A_\lambda \subset A_\mu$ whenever $\lambda \leq \mu$. The pair (X, A) is called a *Λ-space*. (As usual, it is often convenient to think of the function A as an indexed family of subspaces of X).

For example, if $\Lambda = \varnothing$, a Λ-space is just a space X. If Λ has just one element 0, a Λ-space is a pair (X, A_0). If Λ consists of two incomparable elements 1, 2, a Λ-space is a triad $(X; A_1, A_2)$. Finally, if $\Lambda = \{1, 2\}$ and $2 < 1$, a Λ-space is a triple (X, A_1, A_2). Thus the most important configurations in homotopy theory appear as very simple special cases.

Again, if (X, A) is a Λ-space, then the sets $\{\mathbf{I} \times A_\lambda\}$ form a Λ-configuration in $\mathbf{I} \times X$; the resulting Λ-space may well be denoted by $\mathbf{I} \times (X, A)$ or by $(\mathbf{I} \times X, \mathbf{I} \times A)$.

If (X, A) and (Y, B) are Λ-spaces, a *Λ-map* $f: (X, A) \to (Y, B)$ is a continuous map $f: X \to Y$ such that $f(A_\lambda) \subset B_\lambda$ for all $\lambda \in \Lambda$. A *Λ-homotopy* between two Λ-maps $f_0, f_1 : (X, A) \to (Y, B)$ is a Λ-map $f: (\mathbf{I} \times X, \mathbf{I} \times A) \to (Y, B)$ which is also a homotopy between f_0 and f_1 as maps of X into Y. Two Λ-maps are *Λ-homotopic* if and only if there is a Λ-homotopy between them. The relation of being Λ-homotopic is an equivalence relation; denote the set of equivalence classes by $[X, A; Y, B]$ be the set of equivalence classes.

The *core* of a Λ-space is the set $\bigcap_{\lambda \in \Lambda} A_\lambda$, and a Λ-space (X, A) is said to be *jejune* if and only if its core is empty.

Let Λ^+ be the partially ordered set obtained from Λ by adjoining an element 0 such that $0 < \lambda$ for all $\lambda \in \Lambda$. An *augmentation* of a Λ-space (X, A) is an extension of A to a Λ^+ structure A^+; this amounts to adjoining to the family of subsets $\{A_\lambda\}$ of X, a new set A_0 contained in the core of (X, A). A *strict augmentation* of (X, A) is an augmentation for which A_0^+ is a single point x_0. Evidently (X, A) has a strict augmentation if and only if it is not jejune. It is convenient to use the notation (X, A, x_0) for a strictly augmented Λ-space.

2 Standard Notations and Conventions

$\mathbf{Z}$	is the ring of integers.
$\mathbf{Z}_m = \mathbf{Z}/m\mathbf{Z}$	is the ring of integers modulo m ($m = 2, 3, \ldots$).
$\mathbf{Z}_0$	is the field of rational numbers.
$\mathbf{R}$	is the real number system, with its usual topology.
$\mathbf{C}$	is the field of complex numbers.
$\mathbf{Q}$	is the algebra of quaternions.
$\mathbf{K}$	is the algebra of Cayley numbers.
$\mathbf{I}$	is the closed interval $[0, 1]$ of real numbers, with the relative topology, and $\dot{\mathbf{I}}$ is the subspace of $\mathbf{I}$ consisting of the two points 0, 1.

If V is a finite-dimensional vector space over $\mathbf{R}$, V is given the least topology making all linear functionals continuous; i.e., the sets $f^{-1}(U)$, where f ranges over all linear maps of V into $\mathbf{R}$, and U over all open subsets of $\mathbf{R}$, form a sub-basis for the topology of V.

If W is an arbitrary vector space over $\mathbf{R}$, W is given the weak topology determined by the family of all finite-dimensional subspaces, so that a subset C of W is closed if and only if $C \cap V$ is closed in V for every finite-dimensional subspace V of W.

$\mathbf{R}^\infty$ is the vector space of all sequences $x = \{x_i \mid i = 0, 1, \ldots,\}$ of real numbers which vanish from some point on, i.e., there exists a non-negative integer N (depending on x) such that $x_i = 0$ for all $i \geq N$. $\mathbf{R}^\infty$ is an inner-product space with inner product

$$x \cdot y = \sum_{i=0}^{\infty} x_i y_i$$

and norm

$$\|x\| = \sqrt{x \cdot x}.$$

If $\mathbf{e}_i$ is the ith unit vector, whose ith coordinate is 1 and all other coordinates are 0, then $\mathbf{e}_0, \mathbf{e}_1, \ldots$ is an orthonormal basis for $\mathbf{R}^\infty$.

$$\mathbf{S}^\infty = \{x \in \mathbf{R}^\infty \mid \|x\| = 1\}$$

$$\mathbf{E}^\infty = \{x \in \mathbf{R}^\infty \mid \|x\| \leq 1\}$$

$$\mathbf{R}^n = \{x \in \mathbf{R}^\infty \mid x_i = 0 \text{ for all } i \geq n\} \qquad (n \geq 0)$$

$$\mathbf{S}^{n-1} = \mathbf{S}^\infty \cap \mathbf{R}^n$$

$$\mathbf{E}^n = \mathbf{E}^\infty \cap \mathbf{R}^n.$$

(Warning: the inner product in $\mathbf{R}^\infty$ determines a metric but the topology we are using on $\mathbf{R}^\infty$ is not that determined by the metric; in fact, it is not even metrisable. However, the subspaces $\mathbf{R}^n$, $\mathbf{S}^{n-1}$, $\mathbf{E}^n$ have their customary topology, which is indeed induced by the metric.)

Δ^n is the convex hull of the set $\{\mathbf{e}_0, \ldots, \mathbf{e}_n\}$. A point x belongs to Δ^n if and only if

$$x_i \geq 0 \quad \text{for all } i,$$

$$x_i = 0 \quad \text{for all } i > n,$$

$$\Sigma x_i = 1.$$

The numbers $x_0, \ldots, x_n$ are *barycentric coordinates* of x.

$\dot{\Delta}^n = \{x \in \Delta^n \mid x_i = 0 \text{ for some } i, 0 \leq i \leq n\}$.

b_n is the *barycenter* $(1/(n+1)) \sum_{i=0}^n e_i$ of Δ^n.

$d_i^n : \Delta^{n-1} \to \Delta^n$ is the affine map sending $\mathbf{e}_j$ into $\mathbf{e}_j$ for $j < i$ and into $\mathbf{e}_{j+1}$ for $j \geq i$ $(i = 0, 1, \ldots, n)$.

$s_i^n : \Delta^{n+1} \to \Delta^n$ is the affine map sending $\mathbf{e}_j$ into $\mathbf{e}_j$ for $j \leq i$ and into $\mathbf{e}_{j-1}$ for $j > i$ $(i = 0, 1, \ldots, n)$.

(When no confusion can arise, d_i^n and s_i^n are often abbreviated to d_i and s_i, respectively).

A *singular* n-*simplex* in a space X is a map $u : \Delta^n \to X$. For such a map u, its ith face $\partial_i u$ and the ith degeneracy $\int_i u$ are defined by

$$\partial_i u = u \circ d_i^n \qquad (i = 0, 1, \ldots, n),$$

$$\int_i u = u \circ s_i^n \qquad (i = 0, 1, \ldots, n).$$

The totality of singular simplices in X, with the operations ∂_i, $\int_i$, is a semi-simplicial complex, the *total singular complex* $\mathfrak{S}(X)$ of X.

The *orthogonal* group $\mathbf{O}(n)$ is the set of all linear transformations $T : \mathbf{R}^n \to \mathbf{R}^n$ which preserve the inner product: $(Tx) \cdot (Ty) = x \cdot y$ for all $x, y \in \mathbf{R}^n$. The subgroup of $\mathbf{O}(n+1)$ leaving the last coordinate vector $\mathbf{e}_n$ fixed can be identified with $\mathbf{O}(n)$. Thus we have inclusions $\mathbf{O}(n) \subset \mathbf{O}(n+1)$ for all n; the *full orthogonal group* $\mathbf{O}$ is the union $\bigcup_{n=1}^\infty \mathbf{O}(n)$. The *rotation group* $\mathbf{O}^+(n)$ is the subgroup of $\mathbf{O}(n)$ consisting of transformations with determinant $+1$; and the *full rotation group* $\mathbf{O}^+$ is the union $\bigcup_{n=1}^\infty \mathbf{O}^+(n)$. It is usually convenient to think of $\mathbf{O}(n)$ as a group of transformations of the unit sphere $\mathbf{S}^{n-1}$. The groups $\mathbf{O}(n)$, $\mathbf{O}^+(n)$ are topologized as subspaces of the function space $\mathbf{F}(\mathbf{S}^{n-1}, \mathbf{S}^{n-1})$; equivalently, they may be regarded as subspaces of the Euclidean space of $n \times n$ matrices. Then $\{\mathbf{O}(n)\}$, $\{\mathbf{O}^+(n)\}$ are expanding sequences of spaces, and their unions $\mathbf{O}$ and $\mathbf{O}^+$ are given the direct limit topology as in §6.

The orthogonal complement of $\mathbf{R}^m$ in $\mathbf{R}^{m+n}$ is the subspace $\hat{\mathbf{R}}^n$ spanned by the last n basic vectors $\mathbf{e}_m, \ldots, \mathbf{e}_{m+n-1}$. The orthogonal group of $\hat{\mathbf{R}}^n$ can be identified with the subgroup $\hat{\mathbf{O}}(n)$ which fixes the *first m* basic vectors

$\mathbf{e}_0, \ldots, \mathbf{e}_{m-1}$. The subgroup $\mathbf{O}(n, m)$ of $\mathbf{O}(m+n)$ consisting of those transformations which send $\mathbf{R}^m$ into itself is the direct product $\hat{\mathbf{O}}(n) \times \mathbf{O}(m)$, and we may form the *Grassmann manifold*

$$\mathbf{G}_{n,m} = \mathbf{O}(m+n)/\mathbf{O}(n, m)$$

as well as the *Stiefel manifold*

$$\mathbf{V}_{m+n,m} = \mathbf{O}(m+n)/\hat{\mathbf{O}}(n).$$

The correspondences

$$\sigma \to \sigma(\mathbf{R}^m)$$

$$\sigma \to (\sigma(\mathbf{e}_0), \ldots, \sigma(\mathbf{e}_{m-1}))$$

identify $\mathbf{G}_{n,m}$ with the set of all m-planes through the origin in $\mathbf{R}^{m+n}$ and $\mathbf{V}_{m+n,m}$ with the set of all m-frames in $\mathbf{R}^{m+n}$, respectively.

We shall need to consider complex and quaternionic Euclidean spaces. It is convenient to do this in the following way. Let V be a real vector space, and let $I : V \to V$ be a linear transformation such that $I^2 = -E$, where E is the identity transformation of V. Then V is a complex vector space under the scalar product given by

$$(a + ib)x = ax + bI(x) \qquad (a, b \in R, x \in X).$$

Similarly, suppose that I, J, K are linear transformations such that

$$I^2 = J^2 = K^2 = -E,$$

$$IJ = -JI = K, \qquad JK = -KJ = I, \qquad KI = -IK = J.$$

Then V is a (left) vector space over the quaternion algebra $\mathbf{Q}$ under

$$(a + bi + cj + dk)x = a + bI(x) + cJ(x) + dK(x).$$

Let $\mathbf{C}^\infty$ be the vector space $\mathbf{R}^\infty$, with the complex structure defined by

$$I(\mathbf{e}_{2j}) = \mathbf{e}_{2j+1},$$

$$I(\mathbf{e}_{2j+1}) = -\mathbf{e}_{2j},$$

for all $j \geq 0$. Then the vectors $\mathbf{e}_0, \mathbf{e}_2, \ldots$ form a basis for $\mathbf{C}^\infty$, and the vectors $\mathbf{e}_0, \mathbf{e}_2, \ldots, \mathbf{e}_{2n-2}$ form a basis for a subspace $\mathbf{C}^n$, which is evidently $\mathbf{R}^{2n}$ with the complex structure defined by the restriction of I to $\mathbf{R}^{2n}$.

Similarly, we define $\mathbf{Q}^\infty$ to be the vector space $\mathbf{R}^\infty$, with the quaternionic structure defined by

$$J\mathbf{e}_{4i} = \mathbf{e}_{4i+2}, \qquad K\mathbf{e}_{4i} = \mathbf{e}_{4i+3};$$

$$J\mathbf{e}_{4i+1} = -\mathbf{e}_{4i+3}, \qquad K\mathbf{e}_{4i+1} = \mathbf{e}_{4i+2};$$

$$J\mathbf{e}_{4i+2} = -\mathbf{e}_{4i}, \qquad K\mathbf{e}_{4i+2} = -\mathbf{e}_{4i+1};$$

$$J\mathbf{e}_{4i+3} = \mathbf{e}_{4i+1}, \qquad K\mathbf{e}_{4i+3} = -\mathbf{e}_{4i};$$

and where I is defined as for $\mathbf{C}^\infty$. Then the vectors $\mathbf{e}_0, \mathbf{e}_4, \ldots$ form a basis for $\mathbf{Q}^\infty$, and the vectors $\mathbf{e}_0, \mathbf{e}_4, \ldots, \mathbf{e}_{4n-4}$ form a basis for a subspace $\mathbf{Q}^n$, which is evidently $\mathbf{R}^{4n}$ with the quaternionic structure defined by the restrictions of I, J, K to $\mathbf{R}^{4n}$.

The *unitary group* $\mathbf{U}(n)$ is defined to be the set of all linear transformations of $\mathbf{C}^n$ which belong to the orthogonal group $\mathbf{O}(2n)$; and the *unimodular unitary group* $\mathbf{U}^+(n)$ is the subgroup of $\mathbf{U}(n)$ consisting of all elements of determinant $+1$ (as complex-linear transformations; every member of $\mathbf{U}(n)$ has determinant $+1$ when considered as a real-linear transformation). Similarly, the *symplectic group* $\mathbf{Sp}(n)$ is the set of all linear transformations of $\mathbf{Q}^n$ which belong to $\mathbf{O}(4n)$; there is no quaternionic analogue of $\mathbf{U}^+(n)$. There are then inclusions

$$\mathbf{Sp}(n) \subset \mathbf{U}(2n),$$
$$\mathbf{U}(n) \subset \mathbf{O}(2n),$$

as well as inclusions

$$\mathbf{U}(n) \subset \mathbf{U}(n+1),$$
$$\mathbf{Sp}(n) \subset \mathbf{Sp}(n+1),$$

and we have the *full unitary and symplectic groups*, defined by

$$\mathbf{U} = \bigcup_n \mathbf{U}(n),$$
$$\mathbf{Sp} = \bigcup_n \mathbf{Sp}(n).$$

There are also complex and quaternionic analogues

$$\mathbf{V}_{n,k}(\mathbf{C}) = \mathbf{U}(n)/\hat{\mathbf{U}}(n-k),$$
$$\mathbf{V}_{n,k}(\mathbf{Q}) = \mathbf{Sp}(n)/\widehat{\mathbf{Sp}}(n-k),$$
$$\mathbf{G}_{k,l}(\mathbf{C}) = \mathbf{U}(k+l)/\mathbf{U}(k,l),$$
$$\mathbf{G}_{k,l}(\mathbf{Q}) = \mathbf{Sp}(k+l)/\mathbf{Sp}(k,l),$$

of the Stiefel and Grassman manifolds.

As special cases of the Grassman manifolds, we have the projective spaces

$$\mathbf{P}^n = \mathbf{P}^n(\mathbf{R}) = \mathbf{G}_{n,1},$$
$$\mathbf{P}^n(\mathbf{C}) = \mathbf{G}_{n,1}(\mathbf{C}),$$
$$\mathbf{P}^n(\mathbf{Q}) = \mathbf{G}_{n,1}(\mathbf{Q}),$$

and, as special cases of the Stiefel manifolds, the spheres

$$\mathbf{V}_{n+1,1} = \mathbf{S}^n,$$
$$\mathbf{V}_{n+1,1}(\mathbf{C}) = \mathbf{S}^{2n+1},$$
$$\mathbf{V}_{n+1,1}(\mathbf{Q}) = \mathbf{S}^{4n+3}.$$

Remark. For typographical convenience, we shall often write $\mathbf{O}_n$, $\mathbf{U}_n$, $\mathbf{Sp}_n$ instead of $\mathbf{O}(n)$, $\mathbf{U}(n)$, $\mathbf{Sp}(n)$, respectively.

3 Maps of the n-sphere into Itself

The notion of the *degree* of a map of the n-sphere $\mathbf{S}^n$ into itself ($n > 0$) is due to L. E. J. Brouwer. In modern terms, the degree of a map $f: \mathbf{S}^n \to \mathbf{S}^n$ is the unique integer $d(f)$ such that $f_*(u) = d(f) \cdot u$ for all u belonging to the infinite cyclic group $H_n(\mathbf{S}^n)$. Then the degree is a homotopy invariant of f. Clearly all nullhomotopic maps have degree zero. In this section we shall prove the converse: *every map of $\mathbf{S}^n$ into itself of degree zero is nullhomotopic.*

We shall need certain triangulations of $\mathbf{S}^n$. Let K be the simplicial complex consisting of all proper faces of the standard $(n+1)$-simplex $\Delta^{n+1} = \Delta$. All triangulations of $\mathbf{S}^n$ that we shall need are subdivisions of K (in fact, they may be taken to be iterated barycentric subdivisions).

Let us recall that, if L is any triangulation of $\mathbf{S}^n$, then L is a closed n-dimensional pseudomanifold, i.e.,

(3.1) *Every simplex of L is a face of an n-simplex of L.*

(3.2) *Every $(n-1)$-simplex of L is a face of exactly two n-simplices of L.*

(3.3) *L is strongly connected, in the sense that any two n-simplices of L can be joined by a finite sequence of n-simplices of L, each member of which has an $(n-1)$-dimensional face in common with the next.*

Moreover, L is orientable, i.e.,

(3.4) *The n-simplices of L can be so oriented that their sum is an n-cycle (called a fundamental cycle) whose homology class generates the infinite cyclic group $H_n(L)$.*

For the rest of this section, we shall make the following conventions about orientation. Choose a generator for $H_n(K)$. If L is any subdivision of K, there is a canonical isomorphism of $H_n(L)$ with $H_n(K)$. Use this isomorphism to determine a generator for $H_n(L)$. Finally, orient the simplices of L so that the sum of the positively oriented n-simplices is a cycle representing the chosen generator.

For each $x \in \Delta^{n+1}$, let $\lambda_0(x), \ldots, \lambda_{n+1}(x)$ be the barycentric coördinates of x. Let Δ_0^n be the face of Δ^{n+1} opposite to the vertex $\mathbf{e}_0$, so that $x \in \Delta_0^n$ if and only if $\lambda_0(x) = 0$. For $x \in \Delta_0^n$, let

$$\lambda(x) = (n+1)\min\{\lambda_1(x), \ldots, \lambda_{n+1}(x)\},$$

and let Δ^* be the set of all points $x \in \Delta_0^n$ with $\lambda(x) \geq 1/2$. Then Δ^* is an n-simplex interior to Δ_0^n with vertices $e_1^*, \ldots, e_{n+1}^*$, where

$$e_i^* = (1/2)(\mathbf{e}_i + b_*)$$

and

$$b_* = \frac{1}{n+1} \sum_{i=1}^{n+1} \mathbf{e}_i$$

is the barycenter of Δ_0^n, as well as of Δ^*.

Let r be the radial projection of $\Delta_0^n - \{b_*\}$ into its boundary $\dot{\Delta}_0^n$. We now define a map ϕ of $\dot{\Delta}^{n+1}$ into itself, as follows. Firstly, $\phi \mid \Delta^*$ is the simplicial map carrying $\mathbf{e}_i^*$ into $\mathbf{e}_i$ $(i = 1, \ldots, n+1)$. Secondly, ϕ maps the closure of $\dot{\Delta}^{n+1} - \Delta_0^n$ into $\mathbf{e}_0$. Finally for each $x \in \dot{\Delta}^*$, ϕ maps the line segment $[x, r(x)]$ linearly upon the line segment $[r(x), \mathbf{e}_0]$.

The map ϕ is homotopic to the identity under a deformation in which each point moves along a broken line segment. An explicit homotopy is given by

$$\Phi(x, t) = \begin{cases} (1-t)x + t\mathbf{e}_0 & (x \in \overline{\dot{\Delta}^{n+1} - \Delta_0^n}), \\ (1+t)x - tb_* & (x \in \Delta_0^n,\ t \leq (1+t)\lambda(x)), \\ \alpha(x - \lambda(x)b_*) + \beta\mathbf{e}_0 & (x \in \Delta_0^n,\ t \geq (1+t)\lambda(x)), \end{cases}$$

where

$$\alpha = \frac{(1-t) + \lambda(x)(1+t)}{1 - \lambda(x)},$$

$$\beta = t - (1+t)\lambda(x).$$

Let L be a subdivision of K. A map f of $\dot{\Delta}^{n+1}$ into itself is said to be *standard* (with respect to L) if and only if, for each n-simplex σ of L, either $f(\sigma) = \mathbf{e}_0$ or $f \mid \sigma = \phi \circ h_\sigma$, where $h_\sigma : \sigma \to \Delta_0^n$ is a nondegenerate simplicial map. For example, if g is a simplicial map of L into K, then $\phi \circ g$ is standard. However, not every standard map has this form. In fact, if f is a standard map, then f maps the $(n-1)$-skeleton L^{n-1} of L into the point $\mathbf{e}_0$; and it follows that, if σ is any n-simplex of K, one can alter f inside σ without changing it elsewhere in such a way that the resulting map is still standard. Thus standard maps are more flexible than simplicial maps; and this flexibility will be most useful in proving the main theorem of this section.

Let f be a standard map. Then, for each n-simplex σ of L on which f is not constant, the map h_σ is uniquely determined, and we shall say that f is positive or negative on σ according as h_σ preserves or reverses orientation. Let $p(f)$, $n(f)$ be the number of n-simplices on which f is positive or negative respectively. Then an easy argument shows that $d(f) = p(f) - n(f)$.

By the simplicial approximation theorem, any map of $\dot{\Delta}^{n+1}$ into itself is homotopic to a simplicial map of L into K, for some subdivision L of K.

Since ϕ is homotopic to the identity, any simplicial map is homotopic to a standard map. To prove the main result of this section, then, it suffices to prove

Theorem *Let L be a subdivision of K and let $f: \dot{\Delta}^{n+1} \to \dot{\Delta}^{n+1}$ be a standard map with respect to L such that $d(f) = 0$. Then f is nullhomotopic.*

The theorem is proved by induction on $p(f) = n(f)$. If $p(f) = 0$, then f is constant. Hence it suffices to show that, if $p(f) > 0$, then f is homotopic to a map f' such that $p(f') < p(f)$. This is accomplished with the aid of three Lemmas.

Lemma 1 *Let Δ be an n-simplex, and let $k : (\Delta, \dot{\Delta}) \to (\Delta, \dot{\Delta})$ be a simplicial map which permutes the vertices of Δ evenly. Then k is homotopic to the identity.*

Let M_n be the simplicial complex which is the union of two n-simplices Δ_R, Δ_L, whose intersection Δ is an $(n-1)$-dimensional face of each. Let Δ_0 be an $(n-2)$-dimensional face of Δ. Let e_R, e_L be the vertices of Δ_R, Δ_L respectively, opposite to Δ, and let e_0 be the vertex of Δ opposite to Δ_0. Then M_n is the join of Δ_0 with the subcomplex M_1 which is the union of the two 1-simplices $[e_R, e_0]$ and $[e_0, e_L]$. Let $\rho : \Delta_L \to \Delta_R$ be the simplicial map which is the identity on Δ_0 and sends e_L into e_0, e_0 into e_R. Let $\rho' : M_n \to M_n$ be the simplicial map which interchanges e_R and e_L and is the identity on Δ. Let $\dot{M}_n$ be the union of all the $(n-1)$-simplices of M_n except Δ. Finally, let Y be a topological space, and let $y_0 \in Y$.

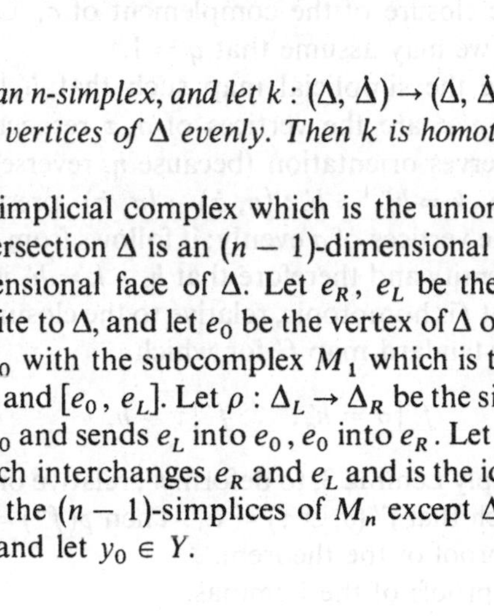

Figure 1.3

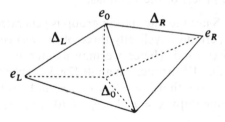

Lemma 2 *Let $h: (M_n, \dot{M}_n) \to (Y, y_0)$ be a map such that $h(\Delta_L) = y_0$. Let $h' : (M_n, \dot{M}_n) \to (Y, y_0)$ be the map such that*

$$h'(\Delta_R) = y_0,$$
$$h' | \Delta_L = h \circ \rho.$$

Then h is homotopic (rel. $\dot{M}_n$) to h'.

Lemma 3 *Let $h : (M_n, \dot{M}_n) \to (Y, y_0)$ be a map such that $h \circ \rho' = h$. Then h is homotopic (rel. $\dot{M}_n$) to the constant map of M_n into y_0.*

Before proving the Lemmas, we show how they imply the truth of the Theorem. Let f be a standard map with $p(f) > 0$. Let σ, τ be n-simplices such that f is positive on σ, negative on τ. Join σ and τ by a chain of n-simplices $\sigma_0 = \sigma, \sigma_1, \ldots, \sigma_{q-1}, \sigma_q = \tau$ as in (3.3). Discarding certain simplices from the head and tail of this chain, if necessary, we may assume that $f(\sigma_i) = \mathbf{e}_0$ $(i = 1, \ldots, q-1)$. It follows from Lemma 2 that, if $q > 0$, f is homotopic, relative to the closure of the complement of $\sigma_0 \cup \sigma_1$, to a map f_1 such that $f_1(\sigma_0) = \mathbf{e}_0$; since the map ρ has degree 1, f_1 is positive on σ_1. Applying Lemma 2 repeatedly, we find that f is homotopic to a standard map f' such that $f'(\sigma_0 \cup \cdots \cup \sigma_{q-2}) = \mathbf{e}_0$, f' maps σ_{q-1} positively and σ_q negatively, and f' agrees with f on the closure of the complement of $\sigma_0 \cup \cdots \cup \sigma_q$, so that $p(f') = p(f)$. Hence we may assume that $q = 1$.

Let $h'_\sigma : \sigma \to \Delta_0^n$ be the simplicial map such that $h'_\sigma | \sigma \cap \tau = h_\tau | \sigma \cap \tau$, $h'_\sigma(e) = h_\tau(e')$, where e, e' are the vertices of σ, τ respectively, opposite to $\sigma \cap \tau$. Then h'_σ preserves orientation (because h_τ reverses it) and therefore the simplicial map $k = h_\sigma^{-1} \circ h'_\sigma : (\sigma, \dot{\sigma}) \to (\sigma, \dot{\sigma})$ preserves orientation. Hence k permutes the vertices of σ evenly; it follows from Lemma 1, that k is homotopic to the identity and therefore that $h_\sigma \circ k = h'_\sigma$ is homotopic to h_σ. It follows in turn that f is homotopic, relative to the closure C of the complement of $\sigma \cup \tau$, to a standard map f' for which

$$f' | \sigma = h'_\sigma, \qquad f' | \tau = h_\tau .$$

Finally, we can apply Lemma 3, to deform f', relative once more to C, to a standard map f'' such that $f''(\sigma \cup \tau) = \mathbf{e}_0$. Then $p(f'') = p(f) - 1 < p(f)$. This completes the proof of the theorem.

We now give the proofs of the Lemmas.

Proof of Lemma 1. Since the alternating group is generated by 3-cycles, we may assume that k permutes cyclically three of the vertices of Δ and leaves the remaining vertices fixed. Hence we may assume $n = 2$. Let ρ_t be the rotation of the unit disc $\mathbf{E}^2$ through an angle $(2\pi/3)t$. Then there is a homeomorphism $h : \Delta \to \mathbf{E}^2$ such that $k = h^{-1} \circ \rho_1 \circ h$; and k is homotopic to the identity under the homotopy $k_t = h^{-1} \circ \rho_t \circ h$ $(0 \le t \le 1)$. □

Proof of Lemma 2. For $-1 \le t \le 1$, let $\alpha(t)$ be the point of M_1 given by

$$\alpha(t) = \begin{cases} (1+t)e_0 - te_R & (-1 \le t \le 0), \\ (1-t)e_0 + te_L & (\ \ 0 \le t \le 1); \end{cases}$$

as t increases from -1 to $+1$, $\alpha(t)$ moves along a broken line segment from e_R through e_0 to e_L.

Each point $x \in M_n$ can be represented in the form

$$x = (1-s)z + s\alpha(t)$$

with $z \in \Delta_0$, $-1 \le t \le 1$, $s \in \mathbf{I}$; note that $x \in \Delta_R$ if and only if $t \le 0$, $x \in \Delta_L$ if and only if $t \geq 0$. An explicit homotopy of h to h' is then given by

$$H(u, x) = \begin{cases} y_0 & (t + 1 \leq u \leq 1), \\ h((1 - s)z + s\alpha(t - u)) & (t \leq u \leq t + 1), \\ y_0 & (0 \leq u \leq t). \end{cases}$$

Continuity of H is easily established with the aid of the fact that, since the compact space M_n is an identification space of $M_1 \times \Delta_0 \times \mathbf{I}$, the product $\mathbf{I} \times M_n$ is an identification space of $\mathbf{I} \times (M_1 \times \Delta_0 \times \mathbf{I})$. □

Proof of Lemma 3. With the same notation as above, an explicit homotopy of h to a constant map is given by

$$H(u, x) = \begin{cases} h((1 - u)x + ue_L) & (x \in \Delta_L), \\ h((1 - u)x + ue_R) & (x \in \Delta_R). \end{cases}$$

Continuity of H is again easily established with the aid of the fact that, since the compact space M_n is an identification space of the disjoint union $\Delta_L \cup \Delta_R$, the product $\mathbf{I} \times M_n$ is an identification space of the disjoint union $\mathbf{I} \times (\Delta_R \cup \Delta_L)$. □

This completes the proofs of the three Lemmas and, accordingly, that of the Theorem. It should be observed that the same argument can be used to show that any map of degree $d > 0$ is homotopic to a standard map which maps each of d mutually disjoint simplices positively and carries the complement of their union into the base point $\mathbf{e}_0$. It is fairly evident that any two such standard maps are homotopic, and hence that any two maps of the same degree are homotopic. However, this will follow from what we have already proved by an easy argument to be given later (§4, Chapter IV).

4 Compactly Generated Spaces

Most of elementary homotopy theory can be carried out in a quite arbitrary category of (pointed) topological spaces. However, when one plunges more deeply into the subject, it becomes necessary to make certain constructions: product spaces $X \times Y$, function spaces $\mathbf{F}(X, Y)$, identification spaces, among others. If, as is fashionable nowadays, one wishes to work in a fixed category of spaces, it is desirable to verify that the category in question is closed under these operations. Moreover, it is desirable that certain natural relationships among these operations (for example, the "exponential law" $\mathbf{F}(X, \mathbf{F}(Y, Z)) = \mathbf{F}(X \times Y, Z))$ should hold in complete generality. For example, the category of all topological spaces is unsatisfactory because of the failure of the exponential law; on the other hand, the category of CW-complexes does not admit function spaces (nor does Milnor's category $\mathscr{W}$ of spaces having the homotopy type of a CW-complex).

A number of solutions to these difficulties have been proposed. Of these, perhaps the most satisfactory is Steenrod's proposal to use the category of compactly generated spaces.

In this book we shall consistently adopt Steenrod's solution. Accordingly, we shall assume throughout, unless explicitly stated to the contrary, that all spaces are compactly generated. Naturally, if in the course of an argument, a new space is constructed out of old ones, it will be necessary to prove that it too, is compactly generated. In this and subsequent sections, we shall list without proof a number of properties which will facilitate this process; for the proofs, the reader is referred to Steenrod's paper [5].

We recall that a space X is *compactly generated* if and only if X is a Hausdorff space and each subset A of X with the property that $A \cap C$ is closed for every compact subset C of X is itself closed. Let $\mathscr{K}$ be the category whose objects are all compactly generated spaces and whose morphisms are all continuous maps between such spaces. Thus $\mathscr{K}$ is a full subcategory of the category $\mathscr{T}$ of all topological spaces, as well as of the category $\mathscr{T}_2$ of all Hausdorff spaces.

(4.1) *Every locally compact Hausdorff space belongs to* $\mathscr{K}$.

(4.2) *Every Hausdorff space satisfying the first axiom of countability belongs to* $\mathscr{K}$.

In particular,

(4.3) *Every metrizable space belongs to* $\mathscr{K}$.

If X is a Hausdorff space, the *associated compactly generated space* is the space $k(X)$ defined as follows: $k(X)$ and X have the same underlying set, and a subset A of X is closed in $k(X)$ if and only if $A \cap C$ is closed in X for every compact subset C of X. For any function $f: X \to Y$, let $k(f)$ be the same function, regarded as a map of $k(X)$ into $k(Y)$.

(4.4) *The identity map* $k(X) \to X$ *is continuous.*

(4.5) $k(X)$ *is compactly generated.*

(4.6) *If* X *is compactly generated, then* $k(X) = X$.

(4.7) $k(X)$ *and* X *have the same compact sets.*

(4.8) *If* $f: X \to Y$ *is a function, then* $k(f)$ *is continuous if and only if* $f \mid C : C \to Y$ *is continuous for every compact set* $C \subset X$.

(4.9) *If X is compactly generated, then the operation of composition with the identity map $k(Y) \to Y$ is a one-to-one correspondence between all continuous maps of X into $k(Y)$ and all continuous maps of X into Y.*

(4.10) *$k(X)$ and X have the same total singular complexes, and therefore the same singular homology and cohomology groups.*

(4.11) *k is a functor: $\mathcal{T}_2 \to \mathcal{K}$, and k is the right adjoint of the inclusion functor $\mathcal{K} \to \mathcal{T}_2$.*

(This means that the sets $\mathcal{T}_2(X, Y)$ and $\mathcal{K}(X, k(Y))$ are in natural one-to-one correspondence, for any $X \in \mathcal{K}$, $Y \in \mathcal{T}_2$).

If X and Y are compactly generated, their Cartesian product Z in the customary topology need not be so. However, let $p : Z \to X$, $q : Z \to Y$ be the projections. Then $k(p) : k(Z) \to k(X) = X$ and $k(q) : k(Z) \to k(Y) = Y$.

(4.12) *$k(Z)$, with the morphisms $k(p)$, $k(q)$ is a product for X and Y in the category $\mathcal{K}$.*

(This means that the operation of composition with $k(p)$ and $k(q)$ establishes a one-to-one correspondence between the sets $\mathcal{K}(W, k(Z))$ and $\mathcal{K}(W, X) \times \mathcal{K}(W, Y)$ for any compactly generated space W).

Thus it is reasonable to define $X \times Y$ to be the retopologized Cartesian product $k(Z)$. We shall call this the *categorical product* (or simply *product*), reserving the term *Cartesian product* in its usual sense.

(4.13) *The categorical product is commutative and associative, up to natural homeomorphisms.*

It is often useful to know that in some circumstances the new product agrees with the old one, i.e., that the Cartesian product topology is compactly generated. This is assured by

(4.14) *If X is locally compact and Y compactly generated, their Cartesian product is compactly generated.*

Let **I** be the closed interval $[0, 1]$ of real numbers, in its usual topology. Then (4.14) implies that, if X is compactly generated, then $\mathbf{I} \times X$ has the usual topology. Thus the notion of homotopy is unchanged.

Let $f : X \to X'$, $g : Y \to Y'$ be maps in $\mathcal{K}$. If $p : X \times Y \to X$, $q : X \times Y \to Y$, $p' : X' \times Y' \to X'$ and $q' : X' \times Y' \to Y'$ are the projections, there is a unique map $f \times g : X \times Y \to X' \times Y'$ such that $p' \circ (f \times g) = f \circ p$, $q' \circ (f \times g) = g \circ q$. Thus the categorical product, like the Cartesian, is a functor.

Just as the notion of product requires some modification in order to remain within the category $\mathscr{K}$, so does the notion of subspace. For if X is compactly generated and A is a subset of X, then A need not be compactly generated in its relative topology. The solution to this difficulty is again to use the functor k. Thus we shall define a *subspace* of X to be a space of the form $k(A)$ for any subset A of X taken with the relative topology. The following statement shows that in many familiar situations the new notion of subspace coincides with the old.

(4.15) *If X is compactly generated, all closed subsets and all regular open subsets are compactly generated.*

(An open subset U is *regular* if and only if each point of U has a closed neighborhood which is contained in U).

In what follows, we shall assume that all subsets of a compactly generated space X have been retopologized in this way. A map $i : X \to Y$ in $\mathscr{K}$ will be called an *inclusion* if and only if i is a homeomorphism of X with the subspace $i(X)$ of Y.

(4.16) *If $i : X \to X'$ and $j : Y \to Y'$ are inclusions in $\mathscr{K}$ so is $i \times j : X \times Y \to X' \times Y'$.*

A map $f : X \to Y$ is called a *proclusion* if and only if f maps X *upon* Y and Y has the identification topology imposed by f, so that a subset U of Y is open if and only if $f^{-1}(U)$ is open in X.

(4.17) *If X is compactly generated, Y is a Hausdorff space, $p : X \to Y$ is a proclusion, then Y is compactly generated.*

(4.18) *If $f : X \to X'$ and $g : Y \to Y'$ are proclusions in $\mathscr{K}$, so is $f \times g : X \times Y \to X' \times Y'$.*

We now turn our attention to function spaces. Unfortunately even if X and Y are compactly generated, the function space $C(X, Y)$ of all continuous maps of X into Y, *with the compact-open topology*, need not be so. Thus it is natural to use the functor k to retopologize $C(X, Y)$, and we define $\mathbf{F}(X, Y) = k(C(X, Y))$. Important in this context is the evaluation map $\mathbf{e} : \mathbf{F}(X, Y) \times X \to Y$, defined by $\mathbf{e}(f, x) = f(x)$.

(4.19) *If X and Y are compactly generated, then the evaluation map $\mathbf{e} : \mathbf{F}(X, Y) \times X \to Y$ is continuous.*

(4.20) *If X, Y, and Z are compactly generated, then the spaces $\mathbf{F}(X, Y \times Z)$ and $\mathbf{F}(X, Y) \times \mathbf{F}(X, Z)$ are naturally homeomorphic.*

The homeomorphism in question is given by $f \to (p_1 \circ f, p_2 \circ f)$, where p_1, p_2 are the projections of the product $Y \times Z$ into its first and second factors, respectively.

(4.21) *If X, Y, and Z are compactly generated, then the spaces* $\mathbf{F}(X \times Y, Z)$ *and* $\mathbf{F}(Y, \mathbf{F}(X, Z))$ *are naturally homeomorphic.*

The homeomorphism in question sends a map $g : X \times Y \to Z$ into the map $\tilde{g} : Y \to \mathbf{F}(X, Z)$ defined by

$$\tilde{g}(y)(x) = g(x, y).$$

The maps g and $\tilde{g}$ are said to be *adjoint* to each other.

(4.22) *If X, Y, and Z are compactly generated, then the operation of composition is a continuous mapping of* $\mathbf{F}(Y, Z) \times \mathbf{F}(X, Y)$ *into* $\mathbf{F}(X, Z)$.

Again, it is easy to see that $\mathbf{F}$ defines a functor, contravariant in the first argument and covariant in the second. In particular, if $g : X' \to X$ and $h : Y \to Y'$ are maps, then the composition with g and h define continuous functions

$$\underline{g} = F(g, 1) : F(X, Y) \to F(X', Y),$$
$$\underline{h} = F(1, h) : F(X, Y) \to F(X, Y').$$

It follows that these functions induce operations on the sets of homotopy classes

$$\underline{g} : [X, Y] \to [X', Y],$$
$$\underline{h} : [X, Y] \to [X, Y'].$$

Moreover, because of (4.21)

(4.23) *The functors* $X \times \quad$ *and* $\mathbf{F}(X, \quad)$ *form an adjoint pair.*

5 NDR-pairs

We shall frequently have occasion to deal with the category $\mathscr{K}^2$ of all pairs (X, A), where X is compactly generated and A is a subspace of X, together with all maps between such pairs. Often it will be desirable that A be nicely imbedded in X, so that, for example, the homotopy extension property holds. Again we refer the reader to Steenrod's paper for the proofs.

Let X be compactly generated and let A be a subspace of X. Then (X, A) is said to be an NDR-*pair* if and only if there are continuous mappings $u : X \to \mathbf{I}$, $h : \mathbf{I} \times X \to X$, such that

(1) $A = u^{-1}(0)$;
(2) $h(0, x) = x$ for all $x \in X$;
(3) $h(t, x) = x$ for all $t \in \mathbf{I}$, $x \in A$;
(4) $h(1, x) \in A$ for all $x \in X$ such that $u(x) < 1$.

If, instead of (4), we assume

(4′) $h(1 \times X) \subset A$,

we say that (X, A) is a DR-*pair*. The pair (u, h) is said to *represent* (X, A) as an NDR-pair (DR-pair).

Let $\mathcal{K}_*^2$ be the full subcategory of $\mathcal{K}^2$ consisting of all NDR-pairs and all continuous maps between such pairs.

Note that, if (X, A) is an NDR-pair, then A is a closed G_δ-subset of X. Moreover, A is a retract of its neighborhood $U = \{x \in X \mid u(x) < 1\}$, and so a neighborhood retract of X. We shall refer to U as a *retractile neighborhood* of A.

(5.1) *If X is compactly generated and A is closed in X, the following four conditions are equivalent*:

(1) (X, A) *is an* NDR-*pair*:
(2) $(\mathbf{I} \times X, 0 \times X \cup \mathbf{I} \times A)$ *is a* DR-*pair*;
(3) $0 \times X \cup \mathbf{I} \times A$ *is a retract of* $\mathbf{I} \times X$;
(4) (X, A) *has the homotopy extension property with respect to arbitrary spaces (i.e., the inclusion map of A in X is a cofibration).*

(5.2) *If (X, A) and (Y, B) are* NDR-*pairs, then so are all the pairs which can be formed from the array*

$$X \times Y - X \times B \cup A \times Y \begin{array}{c} \nearrow X \times B \searrow \\ \\ \searrow A \times Y \nearrow \end{array} A \times B$$

In particular, the pair $(X, A) \times (Y, B) = (X \times Y, X \times B \cup A \times Y)$ is an NDR-*pair. Moreover, if (X, A) or (Y, B) is a* DR-*pair, so is $(X, A) \times (Y, B)$.*

(5.3) *If $B \subset A \subset X$ and (A, B) and (X, A) are* NDR-*pairs* (DR-*pairs*), *then (X, B) is an* NDR-*pair* (DR-*pair*).

A mapping $f: (X, A) \to (Y, B)$ is said to be a *relative homeomorphism* if and only if f is a proclusion of X on Y and $f \mid X - A$ is a homeomorphism of $X - A$ with $Y - B$. For example, let $(X, A) \in \mathcal{K}^2$ and let $h : A \to B$ be a map in $\mathcal{K}$; then the *adjunction space* $X \cup_h B$, obtained from the disjoint union

$X + B$ by identifying each $x \in A$ with $h(x)$, contains a homeomorphic copy of B, and the identification map $f: (X + B, A + B) \to (X \cup_h B, B)$ is a relative homeomorphism. In particular, if B is a point then $X \cup_h B = X/A$ is the identification space obtained from X by collapsing A to a point $*$ and the identification map is a relative homeomorphism $(X, A) \to (X/A, *)$.

(5.4) *Let* (X, A) *be an* NDR-*pair* (DR-*pair*), *and let* $f: (X, A) \to (Y, B)$ *be a relative homeomorphism. Then* (Y, B) *is an* NDR-*pair* (DR-*pair*). *In particular, let* (X, A) *be an* NDR-*pair, and let* $h: A \to B$ *be a map in* $\mathscr{K}$. *Then* $(X \cup_h B, B)$ *is an* NDR-*pair.*

An important special case of the above construction is the *mapping cylinder* $\mathbf{I}_f$ of a map $f: X \to Y$ in $\mathscr{K}$. This is the space $\mathbf{I} \times X \cup_h Y$, where $h: 1 \times X \to Y$ is defined by $h(1, x) = f(x)$. Let $\langle t, x \rangle$ be the image in $\mathbf{I}_f$ of the point $(t, x) \in \mathbf{I} \times X$ under the identification map. We may identify $x \in X$ with $\langle 0, x \rangle$ and $y \in Y$ with its image under the identification, thereby obtaining inclusions $i: X \subsetneq \mathbf{I}_f, j: Y \subsetneq \mathbf{I}_f$. The projection $\hat{f}: \mathbf{I}_f \to Y$, defined by

$$\hat{f}(\langle t, x \rangle) = f(x),$$
$$\hat{f}(y) = y$$

is a deformation retraction, and $\hat{f} \circ i = f$. Thus the map $f: X \to Y$ is homotopically equivalent to the inclusion map $i: X \subsetneq \mathbf{I}_f$. This useful device will often be exploited.

Similarly, the *mapping cone* of f is the space $\mathbf{T}_f = \mathbf{I}_f/X$.

If $f: X \to Y$ and $g: Y \to X$ are maps such that $g \circ f \simeq 1$, the identity map of X, we say that g is a *left homotopy inverse* of f, and f a *right homotopy inverse* of g. The map g is a *homotopy inverse* of f if and only if it is both a right and a left homotopy inverse of f, and f is said to be a *homotopy equivalence* if and only if it has a homotopy inverse. A necessary and sufficient condition for a map f to be a homotopy equivalence is that it have a left homotopy inverse g_L and a right homotopy inverse g_R. For then

$$g_L \simeq g_L \circ (f \circ g_R) = (g_L \circ f) \circ g_R \simeq g_R,$$
$$f \circ g_L \simeq f \circ g_R \simeq 1,$$

so that g_R (and likewise g_L) is a homotopy inverse of f. We also say that X *dominates* Y if and only if there is a map $f: X \to Y$ having a right homotopy inverse.

Of special importance is the case of an inclusion map. If $i: A \subsetneq X$ has a right homotopy inverse, we say that X is *deformable* into A.

(5.5) Theorem (Fox). *A map* $f: X \to Y$ *has a right homotopy inverse if and only if* $\mathbf{I}_f$ *is deformable into* X.

For if $g : Y \to X, f \circ g \simeq 1_Y$, let $q = g \circ \hat{f} : \mathbf{I}_f \to X$. Since $j \circ \hat{f} \simeq 1 : \mathbf{I}_f \to \mathbf{I}_f$ and $j \circ f \simeq i$,

$$i \circ q \simeq j \circ f \circ g \circ \hat{f} \simeq j \circ \hat{f} \simeq 1_{\mathbf{I}_f}.$$

Conversely, let $q : \mathbf{I}_f \to X$, $i \circ q \simeq 1_{\mathbf{I}_f}$ and let $g = q \circ j : Y \to X$. Then

$$f \circ g = f \circ q \circ j = \hat{f} \circ i \circ q \circ j \simeq \hat{f} \circ j = 1_Y. \qquad \square$$

We may also consider the condition that $i : A \hookrightarrow X$ have a left homotopy inverse. If, however, (X, A) is an NDR-pair, and $f : X \to A$ is a left homotopy inverse of i, so that $f \circ i \simeq 1$, then, by the homotopy extension property, f is homotopic to a map $r : X \to A$ such that $r \circ i = 1$. Such a map is called a *retraction*, and A a *retract* of X. Thus

(5.6) Theorem *If (X, A) is an* NDR-*pair, then the inclusion map $i : A \hookrightarrow X$ has a left homotopy inverse if and only if A is a retract of X.* $\square$

(5.7) Theorem *A map $f : X \to Y$ has a left homotopy inverse if and only if the inclusion map $i : Y \hookrightarrow \mathbf{I}_f$ has a left homotopy inverse.*

For if $g : Y \to X$, $g \circ f \simeq 1_X$, and $q = g \circ \hat{f}$, then $q \circ i = g \circ \hat{f} \circ i = g \circ f \simeq 1_X$. Conversely, if $q : \mathbf{I}_f \to X$, $q \circ i \simeq 1_X$, $g = q \circ j : Y \to X$, then $g \circ f = q \circ j \circ f \simeq q \circ i \simeq 1_X$. $\square$

(5.8) Corollary (Fox). *A map $f : X \to Y$ has a left homotopy inverse if and only if X is a retract of $\mathbf{I}_f$.*

For $(\mathbf{I}_f, X)$ is an NDR-pair, by (5.4). $\square$

Finally, we may consider when an inclusion $i : A \hookrightarrow X$ is a homotopy equivalence. As in the case of left homotopy equivalence, there is a useful notion, stronger for general pairs, but equivalent for NDR-pairs. Specifically, we shall say that A is a *deformation retract*[1] of X if and only if there is a homotopy $F : \mathbf{I} \times X \to X$, called a *retracting deformation*, such that

$$\begin{aligned} &F(0, x) = x && (x \in X), \\ &F(t, a) = a && (a \in A,\ t \in \mathbf{I}), \\ &F(1 \times X) \subset A. \end{aligned}$$

The end-value of such a homotopy is a retraction $r = F_1 : X \to A$, called a *deformation retraction*. Note that, if A is a deformation retract of X, then (X, A) is a DR-pair if and only if there is a map $h : X \to \mathbf{I}$ such that $A = u^{-1}(0)$.

[1] Some authors append the word *strong* for this notion; we omit it in accordance with the general principle of using the simplest language for the most important notions.

(5.9) Theorem *If (X, A) is an NDR-pair, then the inclusion map $i : A \subsetneq X$ is a homotopy equivalence if and only if A is a deformation retract of X.*

Let $f : X \to A$ be a homotopy inverse of i, so that there are homotopies $F : \mathbf{I} \times A \to A$, $G : \mathbf{I} \times X \to X$ of 1_A to $f \circ i$, 1_X to $i \circ f$, respectively. Since (X, A) is an NDR-pair, f is homotopic to a retraction $r : X \to A$, and therefore $i \circ r \simeq i \circ f \simeq 1_X$. Therefore we may assume that f is already a retraction, and G is a homotopy of 1_X to $i \circ f$.

Let $P = \mathbf{I} \times X$, $Q = \dot{\mathbf{I}} \times X \cup \mathbf{I} \times A$; since $(\mathbf{I}, \dot{\mathbf{I}})$ and (X, A) are NDR-pairs, so is their product (P, Q). Define $H_* : 0 \times P \cup \mathbf{I} \times Q \to X$ by

$$\begin{aligned} H_*(s, 0, x) &= x, \\ H_*(s, 1, x) &= G(1 - s, f(x)), \\ H_*(s, t, a) &= G((1 - s)t, a), \\ H_*(0, t, x) &= G(t, x). \end{aligned}$$

The verification that H_* is well-defined is trivial, except for the observation that the second and fourth lines agree at $(0, 1, x)$, i.e., that $G(1, f(x)) = G(1, x)$. The end value of the homotopy G is $G_1 = i \circ f$; and $f(x) \in A$ implies that $G(1, f(x)) = f(f(x)) = f(x) = G(1, x)$.

Since (P, Q) is an NDR-pair, H_* has an extension $H : \mathbf{I} \times P \to X$. Then the end value of H is the desired retracting deformation. □

(5.10) Corollary (Fox). *A map $f : X \to Y$ is a homotopy equivalence if and only if X is a deformation retract of $\mathbf{I}_f$.* □

Let (X, A) be a pair, $h : A \to B$, $Y = X \cup_h B$, and let $f : X \to Y$ be the identification map, $Z = \mathbf{I}_f$, $C = \mathbf{I}_h$. Then C is a subspace of Z, $C \cap X = A$.

(5.11) Theorem *If (X, A) is an NDR-pair, then $(Z, X \cup C)$ is a DR-pair.*

Let $P = \mathbf{I} \times X$, $Q = 0 \times X \cup \mathbf{I} \times A$; then (P, Q) is a DR-pair, and so, therefore, is $(P + B, Q + B)$. There is a commutative diagram

$$\begin{array}{ccc} (\mathbf{I} \times X) + X + B & \xrightarrow{\;l\;} & (\mathbf{I} \times X) + B \\ \Big\downarrow{\scriptstyle f_1} & & \Big\downarrow{\scriptstyle q} \\ (\mathbf{I} \times X) + Y & \xrightarrow[\;p\;]{} & Z \end{array}$$

in which $l \mid (\mathbf{I} \times X) + B$ is the identity, $l(x) = (1, x)$, $f_1 \mid \mathbf{I} \times X$ is the identity, $f_1 \mid X + B : X + B \to Y$ is the identification map, $p : (\mathbf{I} \times X) + Y \to Z$ is the identification map, and $q|\mathbf{I}+X = p|\mathbf{I}+X$, $q|B$ is the composite of the inclusions $B \subsetneq Y \subsetneq Z$. The maps p and f_1 are proclusions, and it follows that q is a proclusion. Moreover, $q^{-1}(X \cup C) = Q + B$, and q maps $(P + B) - (Q + B) = P - Q$ homeomorphically upon $Z = (X \cup C)$. Our conclusion then follows from (5.4). □

(5.12) Corollary *Suppose that $h : A \to B$ is a homotopy equivalence. Then $f: X \to X \cup_h B$ is also a homotopy equivalence.*

By Corollary (5.10), A is a deformation retract of C, and therefore (C, A) is a DR-pair. Hence $(X \cup C, X)$ is a DR-pair. But we have seen that $(Z, X \cup C)$ is a DR-pair. By (5.3), (Z, X) is a DR-pair. Again by Corollary (5.10), the map $f: X \to Y$ is a homotopy equivalence. □

An important special case is

(5.13) Corollary *If (X, A) is an* NDR-*pair and A is contractible, the identification map $p : X \to X/A$ is a homotopy equivalence.* □

We conclude this section with a variant of the definition of NDR-pair due to A. Strøm (who, however, does not demand that A be closed). This variant will be useful later.

(5.14) *If A is closed in X, then (X, A) is an* NDR-*pair if and only if there are maps $u : X \to \mathbf{I}$, $h : \mathbf{I} \times X \to X$ such that*

(1) $A \subset u^{-1}(0)$,
(2) $h(0, x) = x$ *for all* $x \in X$,
(3) $h(t, a) = a$ *for all* $(t, a) \in \mathbf{I} \times A$,
(4) $h(t, x) \in A$ *whenever* $t > u(x)$.

If (X, A) is an NDR-pair, then by (3) of (5.1), there is a retraction $r : \mathbf{I} \times X \to 0 \times X \cup \mathbf{I} \times A$. Define u and h by

$$u(x) = \sup_{t \in \mathbf{I}} |t - p_1 r(t, x)|,$$

$$h(t, x) = p_2 r(t, x),$$

where $p_1 : \mathbf{I} \times X \to \mathbf{I}$, $p_2 : \mathbf{I} \times X \to X$ are the projections.

Conversely, suppose u and h satisfy (1)–(4) above. Then a retraction $r : \mathbf{I} \times X \to 0 \times X \cup \mathbf{I} \times A$ is defined by

$$r(t, x) = \begin{cases} (0, h(t, x)) & \text{if } t \le u(x), \\ (t - u(x), h(t, x)) & \text{if } t \ge u(x). \end{cases}$$

□

6 Filtered Spaces

Let X be a space (not necessarily compactly generated), $\{A_\alpha \mid \alpha \in J\}$ a family of closed subsets whose union is X. We say that X has the *weak topology with respect to the* A_α if and only if it satisfies the following condition: a subset C of X, whose intersection with each of the sets A_α is closed, is itself closed.

For example, let X be a Hausdorff space. Then X is compactly generated if and only if it has the weak topology with respect to the collection of all *compact* subsets of X.

Let X be a set, and let $\{A_\alpha\}$ be a collection of topological spaces, each a subset of X. We shall say that $\{A_\alpha\}$ is a *coherent family* (of topological spaces) on X if and only if

(1) $X = \bigcup_\alpha A_\alpha$;
(2) $A_\alpha \cap A_\beta$ is a *closed* subset of A_α for every α, β;
(3) for every α, β, the topologies induced on $A_\alpha \cap A_\beta$ by A_α and A_β coincide.

Let $\{A_\alpha\}$ be a coherent family on X. Define a subset C of X to be closed if and only if $C \cap A_\alpha$ is closed in A_α for every α. Then

(1) X is a topological space (i.e., the complements of the closed sets form a topology on X);
(2) A_α is a subspace of X;
(3) A_α is closed in X;
(4) X has the weak topology with respect to the A_α.

For example, let X be the space of an infinite simplicial complex K. Then each simplex of K has a natural topology and K is a coherent family of topological spaces on X. And the weak topology with respect to K is one of the standard topologies on X.

If X has the weak topology with respect to a coherent family $\{A_\alpha\}$, then X need not be a Hausdorff space even if each A_α is. Therefore X may fail to be compactly generated, even if each A_α is, just for this reason. However,

(6.1) *If $\{A_\alpha\}$ is a coherent family of compactly generated spaces on X, and if X is a Hausdorff space (in the weak topology), then X is compactly generated.*

For suppose that C is a subset of X such that $C \cap K$ is closed for every compact set K. Then, for every compact subset K of A_α, $(C \cap A_\alpha) \cap K = C \cap K$ is closed in A_α; since A_α is compactly generated, $C \cap A_\alpha$ is closed in A_α. Since X has the weak topology with respect to $\{A_\alpha\}$, C is closed in X. □

Of special interest is the case of an *expanding sequence* of spaces. This is a sequence $\{X_n \mid n \geq 0\}$ of spaces such that X_n is a closed subspace of X_{n+1} for every n. If this is the case, $\{X_n\}$ is a coherent family on $X = \bigcup_{n=0}^{\infty} X_n$. The space X, with the weak topology defined by the family of subspaces X_n, is said to be a *filtered space, filtered by the* X_n, if and only if X is Hausdorff (and therefore compactly generated).

(6.2) *Let X have the weak topology with respect to the expanding sequence $\{X_n\}$. Then every compact subset C of X is contained in X_n for some n.*

(6.3) *Let X have the weak topology with respect to the expanding sequence $\{X_n\}$. Suppose that (X_{n+1}, X_n) is an* NDR*-pair for every n. Then X is compactly generated and (X, X_n) is an* NDR*-pair for every n.*

(6.4) *Let X have the weak topology with respect to the expanding sequence of subspaces $\{X_n\}$. Let $f: X \to Y$ be a proclusion, and suppose that X_n is saturated with respect to f. Then $Y_n = f(X_n)$ is an expanding sequence and Y has the weak topology with respect to the Y_n.*

(6.5) *Let X, Y be compactly generated spaces, filtered by $\{X_n\}$, $\{Y_n\}$, respectively. Then $X \times Y$ is filtered by $\{Z_n\}$, where*

$$Z_n = \bigcup_{i=0}^{n} X_i \times Y_{n-i}. \tag{6.6}$$

An NDR-*filtration* of X is a filtration $\{X_n\}$ of X such that (X_{n+1}, X_n) (and therefore (X, X_n)) is an NDR-pair for every n.

(6.7) *If $\{X_n\}$ and $\{Y_n\}$ are* NDR*-filtrations of X, Y respectively, and Z_n is defined by* (6.6), *then $\{Z_n\}$ is an* NDR*-filtration of $X \times Y$.*

A special case of interest is the product of countably many spaces X_i ($i = 1, 2, \ldots$), each having a *nondegenerate base point* $*$, i.e., $(X_i, *)$ is an NDR-pair. Let $X = \prod_{i=1}^{\infty} X_i$ be the subset of the Cartesian product of the X_i consisting of all x such that $x_i = *$ for almost all i. Let $Y_n = \{x \in X \mid x_i = *$ for all $i > n\}$. Then $Y_n \subset Y_{n+1}$, and Y_{n+1} is in one-to-one correspondence with $Y_n \times X_{n+1}$. Let us topologize the Y_n inductively by requiring that this correspondence be a homeomorphism. Then (Y_{n+1}, Y_n) is homeomorphic with $(X_n \times Y_n, \{*\} \times Y_n) = (X_n, \{*\}) \times Y_n$, and therefore is an NDR-pair. Then $\{Y_n\}$ is an expanding sequence of spaces, and therefore $X = \bigcup_{n=1}^{\infty} Y_n$ is a compactly generated space, and each of the pairs (X, X_n) is an NDR-pair. We shall refer to X as the *weak product* of the spaces X_n.

Fortunately, we shall never have to deal with the full Cartesian product of infinitely many factors, nor with uncountable weak products.

7 Fibrations

Let $p : X \to B$ be a map and Y a space in the category $\mathscr{K}$. A *homotopy lifting problem* for (p, Y) is symbolized by a commutative diagram

(7.1)
$$\begin{array}{ccc} Y & \xrightarrow{\ f\ } & X \\ {\scriptstyle i_0}\downarrow & \nearrow & \downarrow{\scriptstyle p} \\ \mathbf{I} \times Y & \xrightarrow[\ G\ ]{} & B \end{array}$$

where $i_0(y) = (0, y)$ for all $y \in Y$; and the maps $f : Y \to X$, $G : \mathbf{I} \times Y \to B$ are said to constitute the *data* for the problem in question. The map G is a homotopy of $p \circ f$; and a solution to the problem is a homotopy $F : \mathbf{I} \times Y \to X$ of f such that $p \circ F = G$; thus F *lifts* the homotopy G of $p \circ f$ to a homotopy of f.

Remark. Often it is convenient to prescribe the terminal, rather than the initial value, of the lifting of a homotopy. Thus, if $i_1 : Y \to \mathbf{I} \times Y$ is the map such that $i_1(y) = (1, y)$, two maps $f : Y \to X$, $G : \mathbf{I} \times Y \to B$ such that $G \circ i_1 = p \circ f$, are often referred to, loosely, as the data for a homotopy lifting problem.

The map p has the *homotopy lifting property with respect to* Y if and only if every problem (7.1) has a solution; and p is said to be a *fibration* (or *fibre map*) if and only if it has the homotopy lifting property (HLP) with respect to every space Y. If $p : X \to B$ is a fibration, the *fibre* over $b \in B$ is the set $F_b = p^{-1}(b)$. In Chapter IV we shall see that, if B is pathwise connected, then all the fibres F_b have the same homotopy type. The space B is called the *base space*, the space X the *total space*, of the fibration p; and we often say that X is a *fibre space over* B *with respect to* p.

We shall often use the locution

$$\text{“}F \xrightarrow{\ i\ } X \xrightarrow{\ p\ } B \text{ is a fibration”}$$

to mean that $p : X \to B$ is a fibration, that F is the fibre over some designated point of B, and that $i : F \subsetneq X$.

For any space Y, let $\mathbf{F}(Y) = \mathbf{F}(\mathbf{I}, Y)$ be the space of all paths in Y. For any map $p : X \to B$, let

(7.2)
$$\mathbf{I}^p = \{(x, u) \in X \times \mathbf{F}(B) \mid p(x) = u(0)\};$$

thus the points of $W = \mathbf{I}^p$ are the data for all homotopy lifting problems for (p, P), where P is a single point. A *connection for* p is a map $\lambda : W \to \mathbf{F}(X)$

with the following two properties:

(7.3) $$\lambda(x, y)(0) = x,$$

(7.4) $$p \circ \lambda(x, u) = u,$$

for all $(x, u) \in W$. Thus $\lambda(x, u)$ is a solution for the homotopy lifting problem having (x, u) as data; and a connection for p is a simultaneous solution of all homotopy lifting problems for (p, P). Moreover λ is a connection if and only if its adjoint $\bar{\lambda} : \mathbf{I} \times W \to X$ is a solution for the homotopy lifting problem

(7.5)

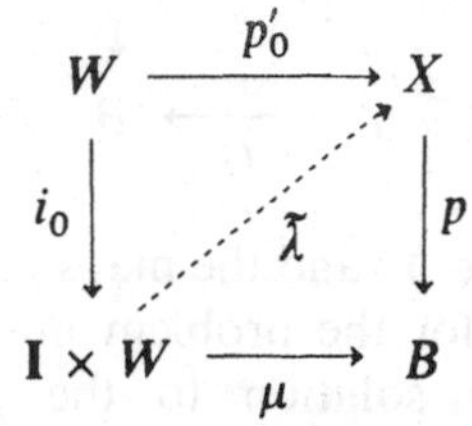

where $p'_0(x, u) = x$, $\mu(t, x, u) = u(t)$. Thus it is not surprising that

(7.6) *A map $p : X \to B$ is a fibration if and only if there exists a connection for p.*

Let λ be a connection, $\bar{\lambda}$ its adjoint, and let $f : Y \to X$, $G : \mathbf{I} \times Y \to B$ be the data for a homotopy lifting problem. The map $f : Y \to X$ and the adjoint $\tilde{G} : Y \to \mathbf{F}(B)$ of G define a map $\theta : Y \to W$, and $p'_0 \circ \theta = f$, $\mu \circ (1 \times \theta) = G$. Hence $\bar{\lambda} \circ (1 \times \theta)$ is the desired solution.

Conversely, if p is a fibration, then the homotopy lifting problem (7.5) has a solution, whose adjoint λ is a connection for p. □

Remark. The space $\mathbf{I}^p$, with the data (p'_0, μ) of (7.5), is a *universal example* for the homotopy lifting problem, in that it is a special case whose solution entails the solution of the problem in general. The method of the universal example is a powerful one, and, although we do not attempt to formalize it, the notion will recur again and again.

The next few theorems given some important examples of fibrations. Perhaps the simplest is given by

(7.7) Theorem *For any spaces B, F, the projection $p_2 : F \times B \to B$ is a fibration.*

In fact, if

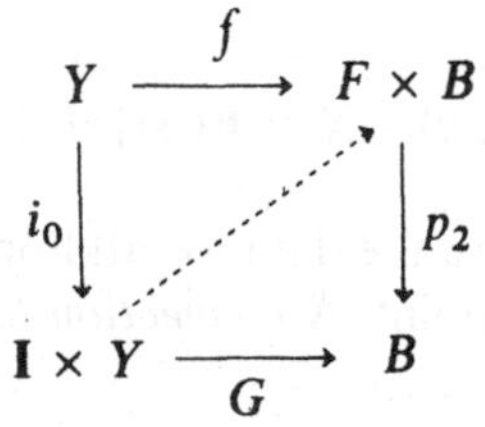

is a homotopy lifting problem, the map $H : \mathbf{I} \times Y \to F \times B$ defined by

$$H(t, y) = (p_1 f(y), G(t, y))$$

is a solution. □

The projection $p_2 : F \times B \to B$ is said to be a *trivial fibration.*

(7.8) Theorem *If (X, A) is an* NDR-*pair and if $i : A \subsetneq X$, then the restriction map*

$$\bar{i} : \mathbf{F}(X, Z) \to \mathbf{F}(A, Z)$$

is a fibration, for any space Z.

For the problem

$$\begin{array}{ccc} Y & \longrightarrow & \mathbf{F}(X, Z) \\ \downarrow & \nearrow & \downarrow \\ \mathbf{I} \times Y & \longrightarrow & \mathbf{F}(A, Z) \end{array}$$

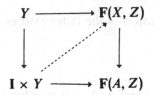

translates, with the aid of (4.23) and (4.13) to a problem

$$\begin{array}{ccc} X \times Y & \longrightarrow & Z \\ \downarrow & \nearrow & \uparrow \\ \mathbf{I} \times X \times Y & \longleftarrow & \mathbf{I} \times A \times Y \end{array}$$

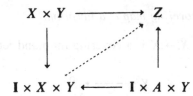

but the latter problem is a homotopy extension problem for the pair $(X \times Y, A \times Y) = (X, A) \times (Y, \varnothing)$. Since (X, A) and $(Y, \varnothing)$ are NDR-pairs, so is their product, by (5.2), and therefore the homotopy extension problem has a solution. Hence the original one does as well. □

(7.9) Corollary *The maps $p : \mathbf{F}(X) \to X \times X$, $p_i : \mathbf{F}(X) \to X$ defined by*

$$p(u) = (u(0), u(1)),$$
$$p_i(u) = u(i) \qquad (i = 1, 2)$$

are fibrations. □

(7.10) Theorem *If $p : X \to B$ is a fibration, then, for any space Z, $\bar{p} : \mathbf{F}(Z, X) \to \mathbf{F}(Z, B)$ is a fibration.*

For any homotopy lifting problem

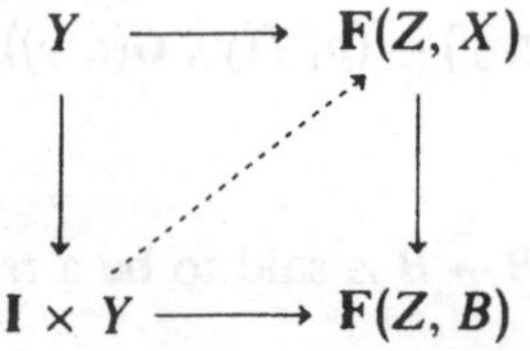

translates, with the aid of (4.23) and (4.13), into a homotopy lifting problem

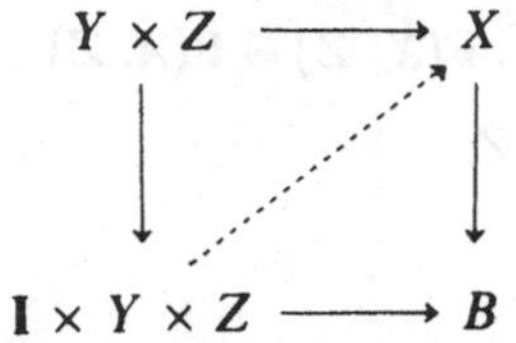

and the existence of a solution of the latter assures us of a solution of the former. □

The easy proof of

(7.11) Theorem *If* $p : X \to Y$ *and* $q : Y \to B$ *are fibrations, so is* $q \circ p : X \to B$.

is left to the reader.

(7.12) Theorem *Every covering map is a fibration.*

For suppose that $p : \tilde{X} \to X$ is a covering map and let

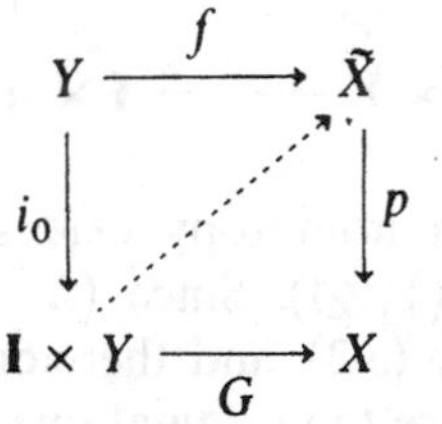

be a homotopy lifting problem. By the general theory of covering spaces, for each $y \in Y$ there is a unique path $u_y : \mathbf{I} \to \tilde{X}$ such that $u_y(0) = f(y)$ and $pu_y(t) = G(t, y)$. What has to be shown is that the map $(t, y) \to u_y(t)$ is a continuous map $F : \mathbf{I} \times Y \to \tilde{X}$.

Let $y \in Y$. For each $t \in \mathbf{I}$, there exist neighborhoods $U = U(t, y)$ of t, $V = V(t, y)$ of y such that $G(U \times V)$ is contained in some open set in X which is evenly covered by p. Let $\eta > 0$ be a Lebesgue number for the open covering $\{U(t, y) \mid t \in \mathbf{I}\}$ of $\mathbf{I}$, and let $0 = t_0 < t_1 < \cdots < t_n = 1$ be a partition with $t_i - t_{i-1} < \eta$ for $i = 1, \ldots, n$. Then, for each i, $[t_{i-1}, t_i] \subset U(t_i', y)$ for some t_i' and therefore $[t_{i-1}, t_i] \times V(t_i', y) \subset U(t_i', y) \times V(t_i', y)$ so that $G([t_{i-1}, t_i] \times V(t_i', y))$ is contained in an open set W_i evenly covered by p. Let $V(y) = \bigcap_{i=1}^{n} V(t_i', y)$.

We next define a continuous function $F_y : \mathbf{I} \times V(y) \to \tilde{X}$ such that (1) $F_y(0, z) = f(z)$ and (2) $pF_y(t, z) = G(t, z)$ for $z \in V(y), t \in \mathbf{I}$. Firstly, define F_y on $\{t_0\} \times V(y)$ by (1). Suppose that F_y has been defined on $[0, t_i] \times V(y)$. Let $p^{-1}(W_i)$ be represented as the union of disjoint open sets $\tilde{W}_\alpha$, each of which is mapped homeomorphically upon W_i by p, and let $q_\alpha = (p|\hat{W}_\alpha)^{-1} : W_i \to \tilde{W}_\alpha$. Then $V(y)$ is the union of the disjoint open sets

$$V_\alpha = \{z \in V(y) | F_y(t_i, z) \in \tilde{W}_\alpha\}$$

and therefore $[t_i, t_{i+1}] \times V(y)$ is the union of the disjoint open sets $[t_i, t_{i+1}] \times V_\alpha$. We can then extend F_y over $[0, t_{i+1}] \times V(y)$ by setting

$$F_y(t, z) = q_\alpha(G(t, z))$$

for $t \in [t_i, t_{i+1}]$, $z \in V_\alpha$. This completes the inductive construction of F_y.

Let $z \in V(y)$; then

$$u_z(0) = f(z) = F_y(0, z),$$
$$pu_z(t) = G(t, z) = pF_y(t, z);$$

by the uniqueness theorem for lifting of paths, $u_z(t) = F_y(t, z)$ for all $t \in \mathbf{I}$, $z \in V(y)$. Hence F and F_y agree on the open set $\mathbf{I} \times V(y)$. Since the F_y are continuous and the $V(y)$ cover Y, F is continuous. □

The next result is due to Hurewicz. The proof is long and complicated, involving intricate properties of paracompact spaces not needed elsewhere in this book. Therefore we shall omit the proof, referring the reader to [D, Chapter XX, §§3–4].

Theorem (Hurewicz). *Let $p : X \to B$ be a continuous map. Suppose that B is paracompact, and that there is an open covering $\mathfrak{V}$ of B such that, for each $V \in \mathfrak{V}$, $p|p^{-1}(V) : p^{-1}(V) \to V$ is a fibration. Then p is a fibration.*

The importance of Hurewicz's Theorem for us lies in its corollary:

(7.13) *Let $p : X \to B$ be the projection of a fibre bundle, and suppose that B is paracompact. Then p is a fibration.*

In fact, the coset decompositions of compact Lie groups modulo their closed subgroups give rise to fibre bundles, and these will serve as a valuable source of examples. The results on compact Lie groups which we shall need are described in Appendix A.

Another useful result is

(7.14) Theorem *Let $p : X \to B$ be a fibration, and let B_0 be a closed subspace of B, $X_0 = p^{-1}(B_0)$. If (B, B_0) is an* NDR*-pair, so is (X, X_0).*

We shall use the criterion of (5.14). Let $u : B \to \mathbf{I}$ and $h : \mathbf{I} \times B \to B$ be maps satisfying the conditions of (5.14). Then the identity map $1 : X \to X$

and the map $h \circ (1 \times p) : \mathbf{I} \times X \to B$ form the data for a homotopy lifting problem, and so there is a commutative diagram

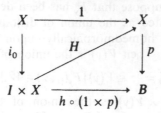

$$\begin{array}{ccc} X & \xrightarrow{\;1\;} & X \\ {\scriptstyle i_0}\downarrow & \overset{H}{\nearrow} & \downarrow{\scriptstyle p} \\ I \times X & \xrightarrow[h \circ (1 \times p)]{} & B \end{array}$$

Define $H' : \mathbf{I} \times X \to X$ by

$$H'(t, x) = \begin{cases} H(t, x) & \text{if } t \leq u(p(x)), \\ H(u(p(x)), x) & \text{if } t \geq u(p(x)). \end{cases}$$

Then it suffices to verify that $u \circ p : X \to \mathbf{I}$ and $H' : \mathbf{I} \times X \to X$ satisfy the conditions of (5.14). The others being trivial, we verify only the last one. Suppose that $t > u(p(x))$. Then

$$\begin{aligned} pH'(t, x) &= pH(u(p(x)), x) \\ &= h(u(p(x)), p(x)). \end{aligned}$$

For all $s > u(p(x))$, $h(s, p(x)) \in B_0$ by (4) for the pair (u, h). Since B_0 is closed, $h(u(p(x)), p(x)) \in B_0$, and therefore $H'(t, x) \in X_0$. □

In studying the homotopy lifting problem, it is natural to ask whether a partial lifting can be extended; i.e., whether a diagram (homotopy lifting extension problem)

$$\begin{array}{ccc} 0 \times Y \cup \mathbf{I} \times A & \longrightarrow & X \\ \cap\!\downarrow & \nearrow & \downarrow{\scriptstyle p} \\ \mathbf{I} \times Y & \longrightarrow & B \end{array}$$

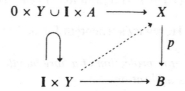

can always be completed. That some sort of hypothesis is necessary is clear, for if B is a point, the constant map of X into B is a fibration; but in this case the problem reduces to the homotopy extension problem for the pair (Y, A) with respect to X. Therefore it is natural to require that (Y, A) should be an NDR-pair. We shall show that this condition is sufficient for the problem to have a solution for an arbitrary fibration. We first prove

(7.15) Lemma *Let $p : X \to B$ be a fibration, and let (Y, A) be a* DR-*pair. Then every lifting extension problem*

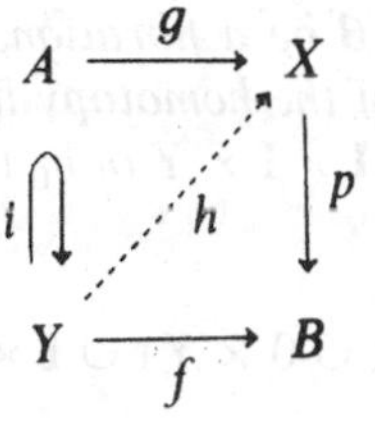

has a solution.

For let $u: Y \to \mathbf{I}$, $\phi: \mathbf{I} \times Y \to Y$, and $r: Y \to A$ be maps such that $A = u^{-1}(0)$, ϕ_1 is the identity map, $\phi_0 = i \circ r$, and ϕ is stationary on A. Define a map $\Phi: \mathbf{I} \times Y \to Y$ by

$$\Phi(t, y) = \begin{cases} \phi(t/u(y), y) & (t < u(y)), \\ \phi(1, y) & (t \geq u(y)). \end{cases}$$

Evidently the restriction of Φ to the open set $\mathbf{I} \times (Y - A)$ is continuous. Let $(t, y) \in \mathbf{I} \times A$, and let U be a neighborhood of y. Then $\mathbf{I} \times \{y\} \subset \phi^{-1}(U)$ and therefore there is a neighborhood V of y such that $\mathbf{I} \times V \subset \phi^{-1}(U)$. The set $\mathbf{I} \times V$ is a neighborhood of (t, y) in $\mathbf{I} \times Y$ and $\Phi(\mathbf{I} \times V) \subset \phi(\mathbf{I} \times V) \subset U$. This proves that Φ is continuous at each point of $\mathbf{I} \times A$, and therefore Φ is continuous.

The maps $f \circ \Phi: \mathbf{I} \times Y \to B$ and $g \circ r: Y \to X$ form the data for a homotopy lifting problem, and therefore there is a map $H: \mathbf{I} \times Y \to X$ such that $H_0 = g \circ r$ and $p \circ H = f \circ \Phi$. Then the map $h: Y \to X$ defined by

$$h(y) = H(u(y), y)$$

is the desired solution. □

(7.16) Theorem *Let $p: X \to B$ be a fibration, (Y, A) an* NDR*-pair. Then every homotopy lifting extension problem*

(7.17)

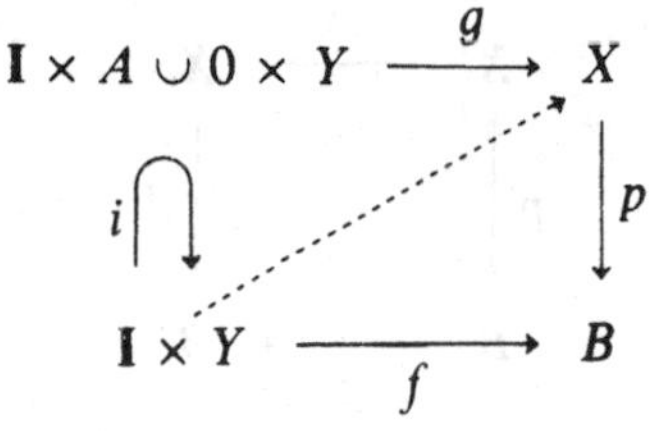

has a solution $h: \mathbf{I} \times Y \to X$.

For $(\mathbf{I} \times Y, \mathbf{I} \times A \cup 0 \times Y)$ is a DR-pair, by (2) of (5.1), and the result follows from Lemma (7.15). □

We now show to what extent the solution of a homotopy lifting extension problem is unique.

(7.18) Theorem *Let $p: X \to B$ be a fibration, (Y, A) an NDR-pair. Let h_0, $h_1: \mathbf{I} \times Y \to X$ be solutions of the homotopy lifting extension problem (7.17). Then there is a homotopy $h: \mathbf{I} \times \mathbf{I} \times Y$ of h_0 to h_1 (rel. $\mathbf{I} \times A \cup 0 \times Y$) such that $p \circ h$ is stationary.*

Define maps $G: \mathbf{I} \times (\mathbf{I} \times A \cup 0 \times Y) \cup \dot{\mathbf{I}} \times \mathbf{I} \times Y \to X, F: \mathbf{I} \times \mathbf{I} \times Y \to B$ by

$$\begin{aligned} G(s, t, y) &= g(t, y) \qquad (t, y) \in \mathbf{I} \times A \cup 0 \times Y, \\ G(0, t, y) &= h_0(t, y), \\ G(1, t, y) &= h_1(t, y), \\ F(s, t, y) &= f(t, y). \end{aligned}$$

Then (F, G) are the data for a lifting extension problem

$$\begin{array}{ccc} \mathbf{I} \times \mathbf{I} \times A \cup \mathbf{I} \times 0 \times Y \cup \dot{\mathbf{I}} \times \mathbf{I} \times Y & \xrightarrow{G} & X \\ \big\downarrow & \nearrow & \big\downarrow p \\ \mathbf{I} \times \mathbf{I} \times Y & \xrightarrow[F]{} & B \end{array}$$

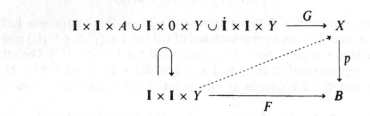

But $(\mathbf{I}, \dot{\mathbf{I}})$ and (Y, A) are NDR-pairs and $(\mathbf{I}, 0)$ a DR-pair. Hence $(\mathbf{I} \times \mathbf{I} \times Y, \mathbf{I} \times \mathbf{I} \times A \cup \mathbf{I} \times 0 \times Y \cup \dot{\mathbf{I}} \times \mathbf{I} \times Y) = (\mathbf{I}, \dot{\mathbf{I}}) \times (\mathbf{I}, 0) \times (Y, A)$ is a DR-pair. By Lemma (7.15) this problem has a solution $h: \mathbf{I} \times \mathbf{I} \times Y \to X$. □

Suppose that $p: X \to B$, $f: B' \to B$ are maps (not necessarily fibrations). Let $X' = \{(b', x) \in B' \times X \mid f(b') = p(x)\}$. Then the restrictions to X' of the projections of $B' \times X$ into its factors are maps $p': X' \to B, f': X' \to X$, and the diagram

(7.19)
$$\begin{array}{ccc} X' & \xrightarrow{f'} & X \\ {\scriptstyle p'}\big\downarrow & & \big\downarrow{\scriptstyle p} \\ B' & \xrightarrow[f]{} & B \end{array}$$

is commutative.

The diagram (7.19) has the following universal property:

(7.20) *Let $g: Y \to X$, $q: Y \to B'$ be maps such that $p \circ g = f \circ q$. Then there is a unique map $h: Y \to X'$ such that $f' \circ h = g$ and $p' \circ h = q$.*

(In other words, the problem

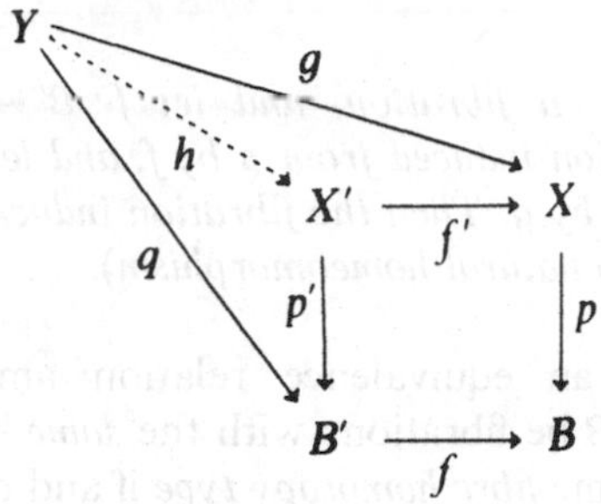

has a unique solution; thus the diagram (7.19) is a "pullback").

The maps q, g define a map of Y into $B' \times X$ whose image, because of the conditions we have imposed on q and g, lies in X', and therefore defines a map $h: Y \to X'$. Uniqueness follows from the fact that the map of X' into $B' \times X$ defined by p', f' is an inclusion. □

(7.21) Theorem *If $p: X \to B$ is a fibration, so is $p': X' \to B'$.*

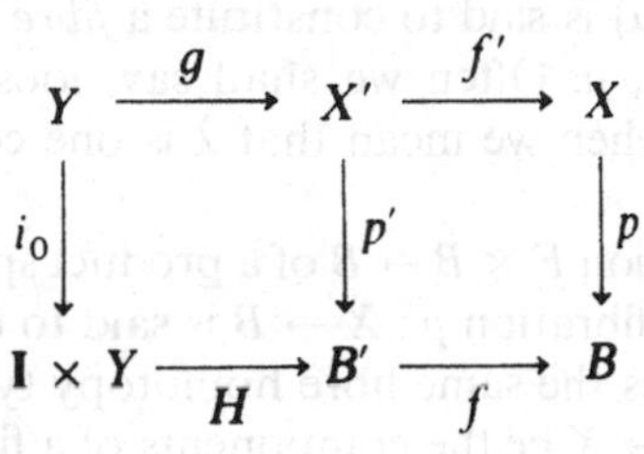

For if $q: Y \to X'$ and $H: \mathbf{I} \times Y \to B'$ form the data for a homotopy lifting problem for the map p', then $f' \circ g: Y \to X$ and $f \circ H: \mathbf{I} \times Y \to B$ form the data for a homotopy lifting problem for p, and the latter problem has a solution $G': \mathbf{I} \times Y \to X$. By (7.20), the maps $G': \mathbf{I} \times Y \to X$ and $H: \mathbf{I} \times Y \to B'$ induce a map $G: \mathbf{I} \times Y \to X'$, which is evidently a solution of the original problem. □

The fibration $p': X' \to B'$ is said to be *induced from $p: X \to B$ by f*. The construction of induced fibrations is a fruitful source of new examples. Let us observe that the map $f': X' \to X$ maps the fibre over $b' \in B'$ homeomorphically upon the fibre over $f(b') \in B$. Moreover, if $f: B' \to B$ is an inclusion map, then f' is an inclusion, mapping X homeomorphically upon $p^{-1}(B')$. Thus

(7.22) Corollary *If $p: X \to B$ is a fibration and $B' \subset B$, $X' = p^{-1}(B')$, then $p|X': X' \to B'$ is a fibration.* □

The notation of induced fibration also has a sort of transitivity property, expressed in

(7.23) *Let $p : X \to B$ be a fibration, and let $f : B' \to B$, $g : B'' \to B'$. Let $p' : X' \to B'$ be the fibration induced from p by f, and let $p'' : X'' \to B''$ be the fibration induced from p' by g. Then the fibration induced from p by the map $f \circ g : B'' \to B$ is p'' (up to natural homeomorphism).* □

We now introduce an equivalence relation among fibrations. Let $p : X \to B$ and $p' : X' \to B$ be fibrations with the *same* base space B. We say that p and p' have the same *fibre homotopy type* if and only if there are maps $\lambda : X \to X'$, $\mu : X' \to X$ such that $p' \circ \lambda = p$, $p \circ \mu = p'$, and homotopies $\Lambda : \mathbf{I} \times X \to X$, $\Lambda' : \mathbf{I} \times X' \to X'$ between $\mu \circ \lambda$ and the identity map of X on the one hand, and between $\lambda \circ \mu$ and the identity map of X' on the other, such that the homotopies $p \circ \Lambda$ and $p' \circ \Lambda'$ are stationary. Thus not only are λ and μ homotopy inverses, so that the total spaces of the fibrations in question have the same homotopy type, but their composites $\lambda \circ \mu$ and $\mu \circ \lambda$ are connected to the respective identity maps by "vertical" homotopies, in which the image of each point remains within the fibre containing it.

The pair of maps (λ, μ) is said to constitute a *fibre homotopy equivalence* whose *components* are λ, μ. Often we shall say, loosely, that λ is a fibre homotopy equivalence when we mean that λ is one component of such an equivalence.

We regard the projection $F \times B \to B$ of a product space into one factor as a trivial fibration; and a fibration $p : X \to B$ is said to be *fibre-homotopically trivial* if and only if it has the same fibre homotopy type as such a product.

Let $\lambda : X \to X'$, $\mu : X' \to X$ be the components of a fibre homotopy equivalence between $p : X \to B$ and $p' : X' \to B$. Let B_0 be a subspace of B, and let $X_0 = p^{-1}(B_0)$, $X'_0 = p^{-1}(B'_0)$. Then $\lambda(X_0) \subset X'_0$, $\mu(X'_0) \subset X_0$, and it is clear that

(7.24) *The maps $\lambda_0 = \lambda | X_0 : X_0 \to X'_0$ and $\mu_0 = \mu | X'_0 : X'_0 \to X_0$ are the components of a fibre homotopy equivalence between the fibrations $p | X_0 : X_0 \to B_0$ and $p' | X'_0 : X'_0 \to B_0$. In particular, the pairs (X, X_0) and (X', X'_0) have the same homotopy type.* □

A useful example of fibre-homotopically equivalent fibrations occurs in the following situation. Let $p : X \to B$ be a fibration, and let $f_0, f_1 : B' \to B$ be homotopic maps. Let $p'_t : X'_t \to B'$ be the fibration induced by f_t $(t = 0, 1)$. Then

(7.25) Theorem *The fibrations $p'_0 : X'_0 \to B'$ and $p'_1 : X'_1 \to B'$ have the same fibre homotopy type.*

Let $f : \mathbf{I} \times B' \to B$ be a homotopy of f_0 to f_1, so that $f \circ i_t = f_t$ $(t = 0, 1)$.

Then the maps $i'_0, i'_1 : B' \to \mathbf{I} \times B'$ are homotopic, and, from the transitivity property (7.23), it suffices to prove the following statement (in which we have dropped the primes for typographical reasons):

(7.26) *Let $p : X \to \mathbf{I} \times B$ be a fibration, and let $p_t : X_t \to B$ be the fibration induced by $i_t : B \to \mathbf{I} \times B$ $(t = 0, 1)$. Then p_0 and p_1 have the same fibre homotopy type.*

Since the prism $\mathbf{I} \times B$ retracts by deformation into either end, X retracts by deformation into X_0 and into X_1. Restricting these deformations to X_1, X_0 yields maps $X_1 \to X_0$, $X_0 \to X_1$ respectively, and these maps are the components of a fibre homotopy equivalence. The details follow.

We have commutative diagrams

$$\begin{array}{ccc} X_t & \xrightarrow{f_t} & X \\ {\scriptstyle p_t}\downarrow & & \downarrow{\scriptstyle p} \\ B & \xrightarrow[i_t]{} & \mathbf{I} \times B \end{array} \qquad (t = 0, 1).$$

The homotopy lifting problem

$$\begin{array}{ccc} X_0 & \xrightarrow{f_0} & X \\ {\scriptstyle i_0}\downarrow & H_0 & \downarrow{\scriptstyle p} \\ \mathbf{I} \times X_0 & \xrightarrow[1 \times p_0]{} & \mathbf{I} \times B \end{array}$$

has a solution $H_0 : \mathbf{I} \times X_0 \to X$, such that

$$\begin{aligned} H_0(0, x_0) &= f_0(x_0) \\ pH_0(s, x_0) &= (s, p_0(x_0)), \end{aligned} \qquad (s \in \mathbf{I},\ x_0 \in X_0).$$

Similarly, there is a map $H_1 : \mathbf{I} \times X_1 \to X$ such that

$$\begin{aligned} H_1(1, x_1) &= f_1(x_1), \\ pH_1(s, x_1) &= (s, p_1(x_1)), \end{aligned} \qquad (s \in \mathbf{I},\ x_1 \in X_1).$$

Define $\lambda : X_0 \to X_1$, $\mu : X_1 \to X_0$ by

$$\begin{aligned} f_1 \lambda(x_0) &= H_0(1, x_0), \\ f_0 \mu(x_1) &= H_1(0, x_1). \end{aligned}$$

Then $i_1 p_1 \lambda(x_0) = p f_1 \lambda(x_0) = pH_0(1, x_0) = (1, p_0(x_0)) = i_1 p_0(x_0)$, and therefore $p_1 \circ \lambda = p_0$. Similarly, $p_0 \circ \mu = p_1$.

Define maps $g : 1 \times \mathbf{I} \times X_0 \cup \mathbf{I} \times \dot{\mathbf{I}} \times X_0 \to X$, $K : \mathbf{I} \times \mathbf{I} \times X_0 \to \mathbf{I} \times B$ by

$$g(s, t, x_0) = \begin{cases} H_0(s, x_0) & (t = 0), \\ f_1 \lambda(x_0) & (s = 1), \\ H_1(s, \lambda(x_0)) & (t = 1); \end{cases}$$

$$K(s, t, x_0) = (s, p_0(x_0)).$$

The reader may verify that g is well defined and that K is an extension of $p \circ g$. By (7.15), g has an extension $G : \mathbf{I} \times \mathbf{I} \times X_0 \to X$ such that $p \circ G = K$. Then $pG(0, t, x_0) = K(0, t, x_0) = (0, p_0(x_0)) = i_0 p_0(x_0)$, and therefore $G | 0 \times \mathbf{I} \times X_0$ defines a map $G_0 : \mathbf{I} \times X_0 \to X_0$ and $p_0 G_0(t, x_0) = p_0(x_0)$. Moreover

$$G_0(0, x_0) = x_0,$$

$$G_0(1, x_0) = \mu\lambda(x_0),$$

so that G_0 is a vertical homotopy between the identity map and $\mu \circ \lambda$. Similarly, $\lambda \circ \mu$ is vertically homotopic to the identity of X_1. □

Remark. The fibre homotopy equivalence $\lambda : X_0 \to X_1$ has a large degree of arbitrariness. In fact, examination of the proof of (7.26) reveals that *any* map λ, having the property that there is a homotopy $H_0 : \mathbf{I} \times X_0 \to X$ of f_0 to $f_1 \circ \lambda$ such that $p \circ H_0 = 1 \times p_0 : \mathbf{I} \times X_0 \to \mathbf{I} \times B$, is a fibre homotopy equivalence.

(7.27) Corollary *Let $p : X \to B$ be a fibration, and suppose that B is contractible. Then p is fibre homotopically trivial. Therefore, for any subspace B_0 of B, the pairs $(X, p^{-1}(B_0))$ and $(F \times B, F \times B_0)$ have the same homotopy type.*

For the identity map of B is homotopic to a constant map and therefore p is fibre homotopically equivalent to the fibration induced by the constant map. But the latter fibration is trivial. □

A fibre homotopy equivalence $\lambda : F \times B \to X$ is called a *trivialization* of the fibration $p : X \to B$. If moreover, b_0 is a designated point of B, $F = p^{-1}(b_0)$, and $\lambda(y, b_0) = y$ for all $y \in F$, we shall call λ a *strong trivialization.*

It will be useful later to give a more or less explicit construction of a trivialization $\lambda : F \times B \to X$. Let $h : \mathbf{I} \times B \to B$ be a homotopy of the constant map to the identity map of B and let $H : \mathbf{I} \times F \times B \to B$ be the composite of h with $1 \times p_2$, where $p_2 : F \times B \to B$ is the projection, so that H is a homotopy of the constant map to the map p_2. Let $i : F \hookrightarrow X$, and let $p_1 : F \times B \to F$ be the projection. Then the homotopy H can be lifted to a homotopy $G : \mathbf{I} \times F \times B \to X$ of $i \circ p_1$ to a map $\lambda : F \times B \to X$ such that $p \circ \lambda = p_2$.

(7.28) Theorem *The map λ is a trivialization of p.*

Consider the diagram

$$\begin{array}{ccccc} X_t & \xrightarrow{f_t} & \bar{X} & \xrightarrow{\bar{h}} & X \\ \downarrow{\scriptstyle p_t} & & \downarrow{\scriptstyle \bar{p}} & & \downarrow{\scriptstyle p} \\ B & \xrightarrow[i_t]{} & \mathbf{I} \times B & \xrightarrow[h]{} & B \end{array}$$

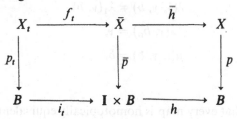

where $\bar{p} : \bar{X} \to \mathbf{I} \times B$ is the fibration induced by h and $p_t : X_t \to B$ the fibration induced from $\bar{p}$ by the inclusion i_t. We may identify X_0 with $F \times B$ and X_1 with X, so that the maps in the diagram are given by

$$p_0(y, b) = b, \quad p_1 = p,$$
$$f_0(y, b) = (0, b, y), \qquad f_1(x) = (1, p(x), x).$$

Define $H_0 : \mathbf{I} \times X_0 \to \bar{X}$ by

$$H_0(t, y, b) = (t, b, G(t, y, b)).$$

Then

$$H_0(0, y, b) = (0, b, i(y)) = f_0(y, b),$$
$$H_0(1, y, b) = (1, b, \lambda(y, b)) = f_1 \lambda(y, b).$$

By the Remark before Corollary (7.27), λ is a fibre homotopy equivalence. □

Remark. If we make the stronger assumption that $(B, \{b_0\})$ is a DR-pair, then λ is a strong trivialization. For then h can be chosen to be stationary at b_0, so that H is stationary on $F \times \{b_0\}$. The homotopy G can then be chosen to be stationary on $F \times \{b_0\}$. Moreover,

(7.29) Theorem *Let $p : X \to B$ be a fibration, $b_0 \in B$, and suppose that $(B, \{b_0\})$ is a DR-pair. Then any two strong trivializations $\lambda_0, \lambda_1 : F \times B \to X$ are vertically homotopic (rel. $F \times \{b_0\}$).*

For $(\mathbf{I} \times F \times B, \dot{\mathbf{I}} \times F \times B \cup \mathbf{I} \times F \times \{b_0\}) = (\mathbf{I}, \dot{\mathbf{I}}) \times (F, \varnothing) \times (B, \{b_0\})$ is a DR-pair, by (5.2); by Lemma (7.15), the lifting extension problem

$$\begin{array}{ccc} \dot{\mathbf{I}} \times F \times B \cup \mathbf{I} \times F \times \{b_0\} & \xrightarrow{\lambda} & X \\ \big\downarrow & \overset{\Lambda}{\nearrow} & \downarrow{\scriptstyle p} \\ \mathbf{I} \times F \times B & \xrightarrow[\mu]{} & B \end{array}$$

where

$$\lambda(0, y, b) = \lambda_0(y, b),$$
$$\lambda(1, y, b) = \lambda_1(y, b),$$
$$\lambda(y, b_0) = y,$$
$$\mu(t, y, b) = b,$$

has a solution Λ. □

We next show that every map is homotopically equivalent to a fibre map.

Let $f: X \to Y$, and consider the space $\mathbf{I}^f$ of (7.2). Note that $\mathbf{I}^f$ is the total space of the fibration $p_0' : \mathbf{I}^f \to X$ induced by f from the fibration $p_0 : \mathbf{F}(Y) \to Y$ of Corollary (7.9). Thus there is a commutative diagram

$$\begin{array}{ccc} \mathbf{I}^f & \xrightarrow{f'} & \mathbf{F}(Y) \\ {\scriptstyle p_0'}\downarrow & & \downarrow{\scriptstyle p_0} \\ X & \xrightarrow[f]{} & Y \end{array}$$

Define a map $p : \mathbf{I}^f \to Y$ by $p = p_1 \circ f'$. (Recall that $p_0, p_1 : \mathbf{F}(Y) \to Y$ are the maps defined by $p_t(u) = u(t)$ $(t = 0, 1)$; they are fibre maps, according to Corollary (7.9)).

The identity map of X and the map $x \to e_{f(x)}$ = the constant path at $f(x)$ define, by the universal property of (7.20), a map $\lambda: X \to \mathbf{I}^f$ such that $p_0' \circ \lambda = 1$ and $f'(\lambda(x)) = e_{f(x)}$.

(7.30) Theorem *The map $p : \mathbf{I}^f \to Y$ is a fibration, and p_0', λ are homotopy equivalences. Moreover,*

$$p_0' \circ \lambda = 1, \qquad \lambda \circ p_0 \simeq 1,$$
$$p \circ \lambda = f \qquad f \circ p_0' \simeq p.$$

In fact, let $g : P \to \mathbf{I}^f$, $H : \mathbf{I} \times P \to Y$ be maps such that $pg(z) = H(0, z)$ for all $z \in P$. Let $g' = f' \circ g$, $g'' = p_0' \circ g$ and define maps $G'' : \mathbf{I} \times P \to X$, $G_0' : (\dot{\mathbf{I}} \times \mathbf{I} \cup \mathbf{I} \times 0) \times P \to Y$ by

$$G''(t, z) = g''(z)$$

$$G_0'(s, t, z) = \begin{cases} fg''(z) & (s = 0), \\ H(t, z) & (s = 1), \\ g'(z)(s) & (t = 0) \end{cases}$$

and extend G_0' to a map $\bar{G}' : \mathbf{I} \times \mathbf{I} \times P \to Y$ (this is possible since $(\mathbf{I} \times P,$

$\dot{\mathbf{I}} \times P$) is an NDR-pair). Let $G' : \mathbf{I} \times P \to \mathbf{F}(Y)$ be the adjoint of $\bar{G}'$, so that $G'(t, z)(s) = \bar{G}'(s, t, z)$. Then $p_0 G'(t, z) = \bar{G}'(0, t, z) = fg''(z) = fG''(t, z)$, so that G', G'' determine a map $G : \mathbf{I} \times P \to \mathbf{I}^f$. Then $G(0, z) = g(z)$ and $p \circ G = H$, so that G is the required map. □

(7.31) Theorem *If $f : X \to Y$ is already a fibration, then the maps $p : \mathbf{I}^f \to Y$ and $f : X \to Y$ have the same fibre homotopy type.*

We have already defined a map $\lambda : X \to \mathbf{I}^f$ such that $p \circ \lambda = f$. By the relation $f \circ p_0' \simeq p$ and the homotopy lifting property, there is a map $L : \mathbf{I} \times \mathbf{I}^f \to X$ such that $L(0, x, u) = x$ $(= p_0'(x, u))$ and $fL(t, x, u) = u(t)$ for $t \in \mathbf{I}$, $(x, u) \in \mathbf{I}^f$. Define $\mu : \mathbf{I}^f \to X$ by $\mu(x, u) = L(1, x, u)$.

Define $\Lambda(t, x) = L(t, \lambda(x))$; then Λ is a vertical homotopy between the identity map and $\mu \circ \lambda$. The definition of $\Lambda' : \mathbf{I} \times \mathbf{I}^f \to \mathbf{I}^f$ is slightly more difficult. For each $t \in \mathbf{I}$, $u : \mathbf{I} \to Y$, let $\omega(t, u)$ be the path such that

$$\omega(t, u)(s) = u(s + t - st)$$

for all $s \in \mathbf{I}$, so that $\omega(t, u)$ is a path from $u(t)$ to $u(1)$, $\omega(0, u) = u$, and $\omega(1, u) = e_{u(1)}$. Then Λ' is defined by

$$\Lambda'(t, x, u) = (L(t, x, u), \omega(t, u)),$$

and Λ' is the desired vertical homotopy between the identity and $\lambda \circ \mu$.

□

The process of replacing the map $f : X \to Y$ by the homotopically equivalent fibration $p : \mathbf{I}^f \to Y$ is, in some sense, analogous to that of replacing f by the inclusion map of X into the mapping cylinder of f; the latter is a cofibration, rather than a fibration. Pursuing this analogy further, we may consider the fibre $\mathbf{T}^f$ of p over a designated point of Y. We shall call $\mathbf{T}^f$ the *mapping fibre of f* (resisting firmly the temptation to call $\mathbf{I}^f$ and $\mathbf{T}^f$ the mapping cocylinder and cocone of f!).

For future reference, let us consider the diagram

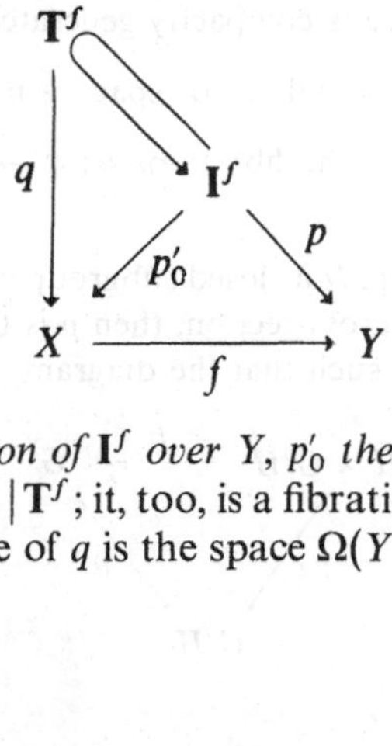

We shall call p the *fibration of* $\mathbf{I}^f$ *over* Y, p_0' *the fibration of* $\mathbf{I}^f$ *over* X. The map q is defined by $q = p_0' \mid \mathbf{T}^f$; it, too, is a fibration, called the *fibration of* $\mathbf{T}^f$ over X. Note that the fibre of q is the space $\Omega(Y)$ of all *loops* (paths starting and ending at x_0) in Y.

The above diagram is analogous to one involving the mapping cone and cylinder of a map $g : Y \to X$:

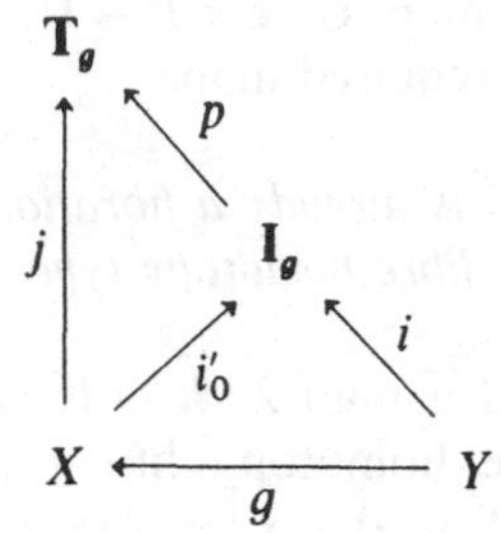

Here i, i'_0 and j are inclusions and $p : \mathbf{I}_g \to \mathbf{T}_g$ is the collapsing map.

Exercises

1. Let $\mathbf{S}^n \vee \mathbf{S}^n$ be the subspace $\mathbf{S}^n \times \mathbf{e}_0 \cup \mathbf{e}_0 \times \mathbf{S}^n$ of $\mathbf{S}^n \times \mathbf{S}^n$. Let $\mathbf{E}^n_+(\mathbf{E}^n_-)$ be the set of all $x = (x_0, \ldots, x_n) \in \mathbf{S}^n$ such that $x_n \geq 0$ $(x_n \leq 0)$. Let $\psi : (\mathbf{S}^n; \mathbf{E}^n_+, \mathbf{E}^n_-) \to (\mathbf{S}^n \vee \mathbf{S}^n; \mathbf{S}^n \times \mathbf{e}_0, \mathbf{e}_0 \times \mathbf{S}^n)$ be a map such that the maps $\psi_+, \psi_- : \mathbf{S}^n \to \mathbf{S}^n$ defined by
$$\psi_+ | \mathbf{E}^n_+ = \psi | \mathbf{E}^n_+$$
$$\psi_- | \mathbf{E}^n_- = \psi | \mathbf{E}^n_-$$
$$\psi_+(\mathbf{E}^n_-) = \psi_-(\mathbf{E}^n_+) = (\mathbf{e}_0, \mathbf{e}_0)$$
have degree 1. Let $h : \mathbf{S}^n \vee \mathbf{S}^n \to \mathbf{S}^n$ be a map such that the maps f, g defined by
$$f(x) = h(x, \mathbf{e}_0),$$
$$g(x) = h(\mathbf{e}_0, x),$$
have degree r, s respectively. Prove that the map $h \circ \psi : \mathbf{S}^n \to \mathbf{S}^n$ has degree $r + s$.
2. Prove that two maps of $\mathbf{S}^n$ into itself, having the same degree, are homotopic.
3. Let C be a compact subset of a vector space V. Prove that C is contained in some finite-dimensional subspace W.
4. Prove that every vector space is compactly generated.
5. Prove that no infinite-dimensional vector space is metrizable.
6. Let $f \simeq g : X \to Y$. Prove that the fibrations $p : \mathbf{I}^f \to Y$, $q : \mathbf{I}^g \to Y$ have the same fibre homotopy type.
7. Let G be a compact Lie group, H a closed subgroup of G. Prove that, if the natural fibration $p : G \to G/H$ has a cross-section, then p is trivial (i.e., there is a homeomorphism $h : H \times G/H \to G$ such that the diagram

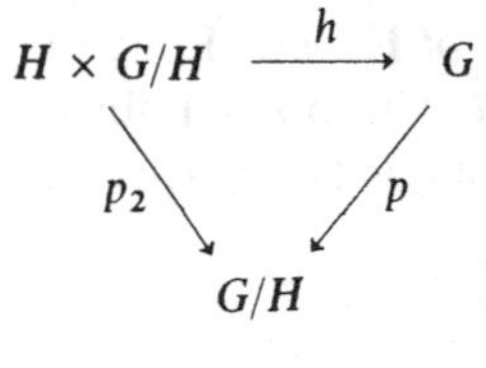

is commutative).

8. If (X, A) is an NDR pair and Y is contractible, then

   (i) every map $f: A \to Y$ has an extension $g: X \to Y$;
   (ii) if $f_0, f_1: X \to Y$ are maps such that $f_0 | A = f_1 | A$, then $f_0 \simeq f_1$ (rel. A).

9. If $p: X \to B$ is a fibration, Y is contractible, and B is pathwise connected, then

   (i) every map $f: Y \to B$ can be lifted to a map $g: Y \to X$;
   (ii) if $(Y, \{*\})$ is a DR-pair and $f_0, f_1: Y \to X$ are maps such that $p \circ f_0 = p \circ f_1$ and $f_0(*) = f_1(*)$, there is a homotopy $F: I \times Y \to X$ of f_0 to f_1 (rel. $\{*\}$) such that $p \circ F$ is stationary.

CHAPTER II

CW-complexes

The category of simplicial complexes probably provides the most suitable setting in which to study elementary algebraic topology. This is due to the fact that the simplicity of its structure allows explicit and elegant descriptions of the basic operations of the theory in combinatorial terms, including an algorithm for calculating homology groups. On the other hand, this very simplicity implies a concomitant rigidity, which has certain disadvantages, both theoretical (the Cartesian product of two simplicial complexes, in its natural decomposition, is not simplicial) and practical (the number of simplices needed to triangulate even a very simple space may be enormous).

One may think of a simplicial complex as being built up by (possible transfinite) iteration of a single process, viz. that of adjoining a cell. In this case, the process is particularly simple; the boundary of each new simplex Δ is already present as a subcomplex, constituting a kind of "hole", and adjunction of the simplex just fills in the hole. Thus the new complex is the mapping cone of the inclusion map i of $\dot{\Delta}$ into the smaller complex. Now the homotopy type of a mapping cone depends only on the homotopy class of the map in question. Thus we may replace the map i by any homotopic map i' without changing the homotopy type; and, while i is an inclusion, i' need not be. This suggests the idea of building up a space by a succession of cell-adjunctions, but by arbitrary continuous maps rather than the special ones which occur in the simplicial case.

The idea of attaching cells with singular boundaries is an old one; as early as 1921 Veblen [V] remarked that any group G can be realized as the fundamental group of a 2-dimensional complex, which he built up out of a given presentation of G by starting with a cluster of circles, one for each generator, and adjoining a 2-cell for each relation. The attaching maps in this case, however, have rather simple singularities, and the idea of using arbitrary continuous maps did not appear until J. H. C. Whitehead introduced CW-complexes in his 1949 lecture on Combinatorial Homotopy [5], thereby creating one of the most important building blocks of the subject of homotopy theory.

Whitehead first presented a CW-complex as a collection of cells, but it seems more appropriate to construct it skeleton by skeleton, each new skeleton being obtained from the preceding by the simultaneous adjunction of a family of cells. This is the procedure we adopt in §1, where the most elementary properties of CW-complexes are presented. For a more thorough treatment of the subject the reader is referred to the recent book by Lundell and Weingram [L–W].

In §2 we show how to associate to each (relative) CW-complex (X, A) a chain complex $\Gamma(X, A)$ whose homology groups are the singular homology groups of (X, A); the nth chain group $\Gamma_n(X, A)$ is a free abelian group having one basis element for each n-cell. The idea of the construction is due to Eilenberg and Steenrod [E–S] in their proof that their axioms characterize homology theory on compact triangulable spaces. In this Section we also discuss cohomology and show how the cross products in singular homology and cohomology fit into this setting.

Cellular maps play a role in the theory of CW-complexes analogous to that played by simplicial maps between simplicial complexes. The main result of §4 is the Cellular Approximation Theorem, which asserts that every map between CW-complexes is homotopic to a cellular map. The construction of the approximation is made by induction over the skeleta; the main tool in the proof is the theorem that a map of an n-dimensional relative CW-complex into an n-connected pair is compressible. The necessary geometrical groundwork is laid in §3.

As the chain groups of a relative CW-complex (X, A) are free abelian groups, the boundary operator is described by a sequence of matrices of integers, the *incidence matrices* of (X, A). Moreover, a cellular map of (X, A) into (Y, B) is described by a sequence of integral matrices. In §5 we discuss the calculation of these matrices. Although these calculations are not effective, the incidence matrices of a *regular cell complex* can be effectively calculated (when the complex is finite) and these calculations are discussed in §6. Regular cell complexes constitute a compromise between simplicial and CW-complexes. On the one hand, they are very close to simplicial complexes (in fact, only one barycentric subdivision away), so that they share many of the advantages of the latter. On the other hand, they lack the rigidity of the simplicial ones; for example, the Cartesian product of regular cell complexes has a natural structure as a regular cell complex. A detailed treatment of regular cell complexes is found in the book of Cooke and Finney [C–F].

Finally, in §7, we discuss cup products and the cohomology ring and calculate the latter for some standard examples.

1 Construction of CW-complexes

A CW-complex is a filtered space in which each stage is obtained from the preceding by a uniform procedure, that of n-cellular extension. We first describe the basic properties of this operation; the standard properties of CW-complexes in general then follow easily.

Let Δ^J be the topological sum of a collection $\{\Delta_\alpha^n \mid \alpha \in J\}$ of copies of the standard n-simplex Δ^n; we may think of Δ^J as the Cartesian product of Δ^n with the indexing set J, considered as a discrete topological space. Let $\dot{\Delta}^J$ be the subspace of Δ^J which is the union of the boundaries of the simplices Δ_α^n, so that $\dot{\Delta}^J$ is the topological sum of the $(n-1)$-spheres $\dot{\Delta}_\alpha^n$.

Let (X, A) be a pair with A compactly generated. We shall say that X is *an n-cellular extension of A* if and only if there is a relative homeomorphism $h : (\Delta^J + A, \dot{\Delta}^J + A) \to (X, A)$. For such a map h, let $f = h|\dot{\Delta}^J : \dot{\Delta}^J \to A$, so that X is homeomorphic with the adjunction space $\Delta^J \cup_f A$. The sets $E_\alpha^n = h(\Delta_\alpha^n)$ are called the *n-cells* of (X, A), and may be described, without reference to h, as the closures of the components of $X - A$. The sets $\operatorname{Int} E_\alpha^n = E_\alpha^n - A$ are called *open n-cells*; let $\dot{E}_\alpha^n = E_\alpha^n - \operatorname{Int} E_\alpha^n = E_\alpha^n \cap A$; $\dot{E}_\alpha^n$ is called the *boundary* of E_α^n. The map $h_\alpha = h|\Delta_\alpha^n : (\Delta_\alpha^n, \dot{\Delta}_\alpha^n) \to (E_\alpha^n, \dot{E}_\alpha^n)$ is called a *characteristic map*, and $f_\alpha = f \mid \dot{\Delta}_\alpha^n$ an *attaching map*, for the cell E_α^n. The map f is called a *simultaneous attaching map*.

The map h_α is a relative homeomorphism; we shall also use the term *characteristic map* somewhat loosely to mean any relative homeomorphism $h'_\alpha : (E, \dot{E}) \to (E_\alpha, \dot{E}_\alpha)$, where E is a closed n-cell.

It follows from §§4–6 of Chapter I that

(1.1) *If X is an n-cellular extension of A, then X is compactly generated, A is closed in X, and (X, A) is an* NDR-*pair. Moreover, a subset C of X is closed if and only if $C \cap A$ is closed and $C \cap E_\alpha$ is closed for every α; and any compact subset of X is contained in the union of A with finitely many cells of (X, A).* □

(1.2) Lemma *Let X be an n-cellular extension of A, A_0 a closed subset of A, and X_0 the union of A_0 with a sub-collection of the cells of (X, A), the boundary of each of which is contained in A_0. Then X_0 is an n-cellular extension of A_0, and is closed in X.*

We have $X_0 = A_0 \cup \bigcup_{\beta \in J_0} E_\beta$, where $J_0 \subset J$. Let $\Delta_0 = \bigcup_{\beta \in J_0} \Delta_\beta$, and let $h_0 = h|\Delta_0 + A_0 : \Delta_0 + A_0 \to X_0$. It suffices to verify that, if C is any subset of X_0 such that $h_0^{-1}(C)$ is closed, then C is closed in X. Since h is a proclusion, it suffices to verify that $h^{-1}(C)$ is closed. But

$$h^{-1}(C) \cap A = C \cap A = C \cap X_0 \cap A = C \cap A_0 = h_0^{-1}(C) \cap A_0;$$

if $\beta \in J_0$,

$$h^{-1}(C) \cap E_\beta = h_0^{-1}(C) \cap E_\beta;$$

if $\beta \notin J_0$,

$$h^{-1}(C) \cap E_\beta = h^{-1}(C) \cap \dot{E}_\beta = f_\beta^{-1}(C) = f_\beta^{-1}(C \cap A_0);$$

and each of these sets is closed. $\square$

The following Lemma will be useful later in constructing homotopies of maps of (X, A).

(1.3) Lemma *Let X be an n-cellular extension of A, with characteristic maps $h_\alpha : (\Delta_\alpha, \dot{\Delta}_\alpha) \to (X, A)$. Let $f : (X, A) \to (Y, B)$ be a map and, for each α, let $g_\alpha : (\mathbf{I} \times \Delta_\alpha, \mathbf{I} \times \dot{\Delta}_\alpha) \to (Y, B)$ be a homotopy of $f \circ h_\alpha$ (rel. $\dot{\Delta}_\alpha$). Then there is a homotopy $g : (\mathbf{I} \times X, \mathbf{I} \times A) \to (Y, B)$ of f (rel. A) such that $g \circ (1 \times h_\alpha) = g_\alpha$.*

Let $G : \mathbf{I} \times (\Delta^J + A) \to Y$ be the map such that

$$G | \mathbf{I} \times \Delta_\alpha = g_\alpha,$$

$$G(t, a) = f(a) \qquad (t \in \mathbf{I}, a \in A).$$

The map $1 \times h : \mathbf{I} \times (\Delta^J + A) \to \mathbf{I} \times X$ is a proclusion, and it is trivial to verify that, if $(1 \times h)(t, x) = (1 \times h)(t', x')$, then $G(t, x) = G(t', x')$. Hence the map $g = G \circ (1 \times h)^{-1}$ is continuous, and evidently has the desired properties. $\square$

(1.4) Lemma *Let Y be the union of two closed subspaces X, B, and suppose that X is an n-cellular extension of $A = X \cap B$. Then Y is an n-cellular extension of B with the same n-cells.*

For if $h : (\Delta^J + A, \dot{\Delta}^J + A) \to (X, A)$ is a relative homeomorphism, there is a commutative diagram

$$\begin{array}{ccc} \Delta^J + A + B & \xrightarrow{h_1} & \Delta^J + B \\ \downarrow k_1 & & \downarrow h_2 \\ X + B & \xrightarrow[k_2]{} & X \cup B \end{array}$$

in which $h_1 | \Delta^J + B$ is the identity map, $h_1 | A : A \subsetneq B$, $h_2 | \Delta^J = h | \Delta^J : \Delta^J \to X$, $h_2 | B : B \subsetneq X \cup B$, $k_1 | \Delta^J + A = h$, $k_1 | B$ is the identity map of B, and $k_2 | X$, $k_2 | B$ are inclusions. But k_1 and k_2 are proclusions and therefore h_2 is a proclusion. As h_2 maps $\Delta^J - \dot{\Delta}^J$ homeomorphically upon $Y - B = X - A$, it is a relative homeomorphism. $\square$

(1.5) Lemma *Let Y be the union of two closed subspaces X, B, and suppose that both X and B are n-cellular extensions of $A = X \cap B$. Then Y is an n-cellular extension of A.*

The easy proof is left to the reader. □

We are now ready to define the notion of relative CW-complex. Let (X, A) be a pair with A compactly generated. A *CW-decomposition* of (X, A) is a filtration $\{X_n\}$ of X such that

(1) $A \subset X_0$,
(2) for each $n \geq 0$, X_n is an n-cellular extension of X_{n-1} (it is convenient to agree that $X_n = A$ for $n < 0$).

If $\{X_n\}$ is a CW-decomposition of (X, A), we say that (X, A) is a *relative* CW-*complex* (with respect to the filtration $\{X_n\}$); if $A = \varnothing$, we say that X is a CW-*complex*. The n-cells of (X_n, X_{n-1}) are called *n-cells* of (X, A). The set X_n is called the *n-skeleton* of (X, A). We shall say that (X, A) is finite (countable) if and only if it has only finitely (countably) many cells.

The following properties are easily deduced from the results of Chapter I.

(1.6) *Let $\{X_n\}$ be a* CW-*decomposition of (X, A). Then*

(1) *X is compactly generated;*
(2) *X_n is closed in X for all n;*
(3) *(X, X_{n-1}) and (X_n, X_{n-1}) are* NDR-*pairs; in particular, (X, A) is an* NDR-*pair;*
(4) *X has the weak topology with respect to the collection of skeleta $\{X_n\}$;*
(5) *A subset C of X is closed if and only if $C \cap A$ is closed and $C \cap E_\alpha^n$ is closed for every n-cell E_α^n of (X, A);*
(6) *Every compact subset of X is contained in the union of A with finitely many cells of (X, A);*
(7) *If (Y, B) is a relative* CW-*complex with skeleta Y_n, then $(X, A) \times (Y, B) = (X \times Y, A \times Y \cup X \times B)$ is a relative* CW-*complex with skeleta*

$$Z_n = \bigcup_{i=0}^{n} X_i \times Y_{n-i}.$$ □

If (X, A) is a relative CW-complex, then the interiors of the cells of (X, A) and the set A are mutually disjoint and cover X. If $x \in X - A$, the unique n-cell E_α^n such that $x \in \operatorname{Int} E_\alpha^n$ is called the *carrier cell* of x. Moreover, the dimension n of an n-cell E_α^n is characterized by the property that E_α^n has a dense open set consisting of points having a neighborhood homeomorphic with Euclidean n-space $\mathbf{R}^n$.

The sets $\operatorname{Int} E_\alpha^n$ are called *open cells* of (X, A). The terminology is a bit misleading, because $\operatorname{Int} E_\alpha^n$ need not be an open set in X (although it is open in X_n).

The following examples are readily verified to be (relative) CW-complexes.

Example 1. If K is a simplicial complex and L a subcomplex of K, then $(|K|, |L|)$ is a relative CW-complex with n-skeleton $|K_n \cup L|$.

Example 2. The n-sphere is a CW-complex with one 0-cell E^0 and one n-cell $\mathbf{S}^n$ with $\dot{\mathbf{S}}^n = E^0$.

Example 3. The unit interval is a CW-complex with two 0-cells $\{0\}$, $\{1\}$ and one 1-cell $\mathbf{I}$, with $\dot{\mathbf{I}} = \{0, 1\}$.

Example 4. If (X, A) is a relative CW-complex, then $(\mathbf{I} \times X, \mathbf{I} \times A)$ is a relative CW-complex with n-cells

$$\begin{array}{ll} 0 \times E_\alpha^n,\ 1 \times E_\alpha^n & (E_\alpha^n \text{ an } n\text{-cell of } (X, A)), \\ \mathbf{I} \times E_\beta^{n-1} & (E_\beta^{n-1} \text{ an } (n-1)\text{-cell of } (X, A)). \end{array}$$

Example 5. If (X, A) is a relative CW-complex, then $Y = X/A$ is a CW-complex with $Y_n = X_n/A$ for $n \geq 0$.

Example 6. If (X, A) is a relative CW-complex and n a non-negative integer, then (X, X_n) and (X_n, A) are also relative CW-complexes.

If X is an n-cellular extension of A, and A is a CW-complex, it is tempting to conclude that X is a CW-complex. However, it is necessary to resist this temptation; for suppose that A is an n-sphere $(n \geq 2)$ and f is a map of $\dot{\Delta}^n$ upon the whole of A, then $X = \Delta^n \cup_f A$ cannot be a CW-complex. Therefore some condition must be imposed on the attaching maps.

(1.7) Lemma *If X is an n-cellular extension of the* CW*-complex A, and if the image of the simultaneous attaching map $f: \dot{\Delta}^J \to A$ is contained in the $(n-1)$-skeleton A_{n-1} of A, then X is a* CW*-complex with skeleta*

$$X_q = \begin{cases} A_q & (q < n), \\ A_q \cup_f \Delta^J & (q \geq n). \end{cases}$$

It follows from Lemmas (1.4) and (1.5) that X_q is a q-cellular extension of X_{q-1} for each q. We leave it to the reader to verify that X has the weak topology with respect to the X_q. □

Let (Y, B) be a relative CW-complex. A *subcomplex* of (Y, B) is a pair (X, A) such that

(1) X is a subspace of Y;
(2) A is a closed subspace of B;
(3) X is the union of A with a sub-collection of the open cells of Y, the boundary of each of which is contained in X.

Let (X, A) be a subcomplex of (Y, B), and let $X_n = X \cap Y_n$. It follows from Lemma (1.2), by induction on n, that X_n is an n-cellular extension of X_{n-1}, and that X_n is closed in Y_n, and therefore in Y. If C is a subset of X whose intersection with X_n is closed for every n, then

$$C \cap Y_n = C \cap X \cap Y_n = C \cap X_n$$

is closed, and therefore C is closed. Hence $\{X_n\}$ is a CW-decomposition of (X, A). We have proved

(1.8) *If (X, A) is a subcomplex of (Y, B), then (X, A) is a relative* CW-*complex with n-skeleton $X_n = X \cap Y_n$. Moreover, X is a closed subspace of Y.* □

EXAMPLE 1. If (X, A) is a relative CW-complex, then (X_n, A) is a subcomplex of (X, A).

EXAMPLE 2. If $\{(X_\alpha, A_\alpha)\}$ is a family of subcomplexes of (X, A), then $(\cap X_\alpha, \cap A_\alpha)$ is also a subcomplex of (X, A).

EXAMPLE 3. If $\{(X_\alpha, A_\alpha)\}$ is a family of subcomplexes of (X, A) and if $\cup A_\alpha$ is closed, then $(\cup X_\alpha, \cup A_\alpha)$ is a subcomplex of (X, A).

EXAMPLE 4. If (X, A) is a relative CW-complex, (X', A') is a subcomplex of (X_{n-1}, A), and if $\{E^n_\beta\}$ is any collection of n-cells of (X, A), each of whose boundaries is contained in X', then $(X' \cup \bigcup_\beta E^n_\beta, A')$ is a subcomplex of (X, A).

EXAMPLE 5. In Lemma (1.7) A is a subcomplex of X.

We next prove a strengthened form of (6) of (1.6).

(1.9) *If (X, A) is a relative* CW-*complex and C is a compact subset of X, then there is a finite subcomplex (X', A') such that $C \subset X'$.*

It follows from (6) of (1.6) that $C \subset X_n$ for some n. Therefore it suffices to prove, by induction on n, that, if C is any compact subset of (X_n, A), then $C \subset X'$ for some finite subcomplex (X', A'). This is trivial for $n = -1$. By Part (6) of (1.6), C is contained in the union of A with a finite collection $\mathscr{F}$ of cells of (X_n, A). Let C' be the union of $C \cap X_{n-1}$ with all the cells of $\mathscr{F}$ of dimension $< n$; then C' is compact. If (X', A') is a finite subcomplex of (X_{n-1}, A) such that $C' \subset X'$, if $E^n_1, \ldots, E^n_r$ are the n-cells of $\mathscr{F}$, and if $X'' = X' \cup \bigcup_{i=1}^r E^n_i$, then by Example 4, (X'', A') is a finite subcomplex of (X_n, A) and $C \subset X''$. This completes the inductive proof. □

The reader should not be misled, by analogy with the case of simplicial complexes, into the belief that the (closed) cells of a CW-complex X, or their boundaries, are subcomplexes of X. However, we have

(1.10) *If E is an n-cell of the relative* CW-*complex* (X, A), *there is a finite subcomplex* (X', A') *whose only cell of dimension not less than n is E.*

In fact, $\dot{E}$ is compact, so that by (1.9) there is a finite subcomplex (X'_0, A') of (X^{n-1}, A) such that $\dot{E} \subset X'_0$. According to Example 4, $(E \cup X'_0, A')$ is a subcomplex having the desired property. □

If K is a simplicial complex and $\tilde{K}$ any covering space of K, then $\tilde{K}$ is a simplicial complex whose simplices lie nicely over those of K. For CW-complexes, much the same is true. As a first step, we prove

(1.11) Lemma *Let A be a locally pathwise connected Hausdorff space, X an n-cellular extension of A. Let* $p : \tilde{X} \to X$ *be a covering map, and let* $\tilde{A} = p^{-1}(A)$. *Then* $\tilde{X}$ *is an n-cellular extension of* $\tilde{A}$, *and the restriction of p to each cell of* $(\tilde{X}, \tilde{A})$ *is a relative homeomorphism upon a cell of* (X, A).

(The example of the exponential map of the real line upon the circle, considered as a 1-cellular extension of a point, shows that the restriction of p to each cell need not be a homeomorphism).

Clearly $\tilde{X}$ is locally pathwise connected. Let $h_\alpha : \Delta^n \to X$ be the characteristic maps for the cells of (X, A) and let $x_\alpha = h_\alpha(\mathbf{e}_0)$. Let $p^{-1}(x_\alpha) = \{\tilde{x}_{\alpha\beta}\}$, and let $\tilde{h}_{\alpha\beta} : \Delta^n \to \tilde{X}$ be the map such that

$$p \circ \tilde{h}_{\alpha\beta} = h_\alpha,$$

$$\tilde{h}_{\alpha\beta}(\mathbf{e}_0) = \tilde{x}_{\alpha\beta}.$$

Let $\tilde{U}$ be a subset of $\tilde{X}$ such that $\tilde{U} \cap \tilde{A}$ and each of the sets $\tilde{h}_{\alpha\beta}^{-1}(\tilde{U})$ are open. We shall prove that $\tilde{U}$ is open. The space $\tilde{X}$ has a basis $\mathscr{B}$ consisting of open sets B, each of which is mapped homeomorphically by p upon an open subset of X, and we may assume that $\tilde{U} \subset B$ for some $B \in \mathscr{B}$. It suffices, then, to prove that $U = p(\tilde{U})$ is open. But the readily verified equalities

$$h_\alpha^{-1}(U) = \bigcup_\beta \tilde{h}_{\alpha\beta}^{-1}(\tilde{U}),$$

$$U \cap A = p(\tilde{U} \cap \tilde{A}),$$

and the fact that $p | \tilde{A} : \tilde{A} \to A$ is a covering map and therefore open, show that U is indeed open.

Let E_α^n, $\tilde{E}_{\alpha\beta}^n$ be the images of the maps h_α, $\tilde{h}_{\alpha\beta}$, respectively. As the restriction of p to the compact Hausdorff space $\tilde{E}_{\alpha\beta}^n$ maps it continuously upon the Hausdorff space E_α^n, it is a proclusion; and since p is a one-to-one map of the open set Int $\tilde{E}_{\alpha\beta}^n$ upon the open set Int E_α^n, $p | \tilde{E}_{\alpha\beta}^n$ is a relative homeomorphism. Finally, the equality $p \circ \tilde{h}_{\alpha\beta} = h_\alpha$ shows that $\tilde{h}_{\alpha\beta}$ maps Int Δ^n homeomorphically upon Int $\tilde{E}_{\alpha\beta}^n$, and this completes the proof that $\tilde{X}$ is an n-cellular extension of $\tilde{A}$ and therewith that of the Lemma. □

(1.12) Theorem *Let X be a* CW-*complex with skeleta* X_n, *and let* $p : \tilde{X} \to X$ *be a covering map. Then* $\tilde{X}$ *is a* CW-*complex with skeleta* $\tilde{X}_n = p^{-1}(X_n)$, *and* *the restriction of p to each cell of $\tilde{X}$ is a relative homeomorphism upon a cell of X.*

By Lemma (1.11), each of the sets $\tilde{X}_n$ is an n-cellular extension of $\tilde{X}_{n-1}$. It remains, therefore, to verify that $\tilde{X}$ has the weak topology. Accordingly, let $\tilde{U}$ be a subset of $\tilde{X}$ whose intersection with each of the sets $\tilde{X}_n$ is relatively open. As in the proof of Lemma (1.11), we may assume that $\tilde{U}$ is contained in an open set which is mapped homeomorphically by p, and it suffices, as before, to prove that $U = p(\tilde{U})$ is open. But $p(\tilde{U} \cap \tilde{X}_n) = p(\tilde{U} \cap p^{-1}(X_n)) = p(\tilde{U}) \cap X_n = U \cap X_n$, and the fact that $\tilde{U} \cap \tilde{X}_n$ is relatively open implies that $U \cap X_n$ is relatively open. Since this is true for every n, U is indeed open. □

Now suppose that X is a CW-complex with skeleta X_n and A is a subcomplex, with skeleta $A_n = X_n \cap A$. Then we have seen in Lemma (1.2) that (X, A) is a relative CW-complex with skeleta $X_n^* = X_n \cup A$. We also have $X_n = (X_n^* - A) \cup A_n$; and the boundary of every n-cell of (X, A) is contained in X_{n-1}. Conversely, we have

(1.13) Lemma *Let (X, A) be a relative CW-complex with skeleta X_n^*, and suppose that A is a CW-complex with skeleta A_n. Let $X_n = (X_n^* - A) \cup A_n$, and suppose that the boundary of each n-cell of (X, A) is contained in X_{n-1}. Then X is a CW-complex with skeleta X_n, and A is a subcomplex of X.*

Note that $X_n = (X_n^* - X_{n-1}^*) \cup X_{n-1} \cup A_n$. By Lemma (1.2), $(X_n^* - X_{n-1}^*) \cup X_{n-1}$ is an n-cellular extension of X_{n-1}. By Lemma (1.4), $X_{n-1} \cup A_n$ is an n-cellular extension of X_{n-1}. By Lemma (1.5), X_n is an n-cellular extension of X_{n-1}.

Suppose that $C \cap X_n$ is closed for every n. Now A_n is closed in A, and A is closed in X; thus $C \cap A_n = (C \cap X_n) \cap A_n$ is closed. Since A has the weak topology with respect to the A_n, $C \cap A$ is closed. Also $C \cap X_n^* = C \cap (X_n \cup A) = (C \cap X_n) \cup (C \cap A)$ is closed for every n. Hence C is closed. Therefore X has the weak topology with respect to the X_n, and so is a CW-complex. That A is a subcomplex of X is clear. □

2 Homology Theory of CW-complexes

We shall be concerned exclusively with singular homology and cohomology groups of a pair (X, A) with coefficients in an abelian group G; these are denoted, as usual, by

$$H_n(X, A; G) \quad \text{and} \quad H^n(X, A; G)$$

respectively; often these will be abbreviated to

$$H_n(X, A) \quad \text{and} \quad H^n(X, A)$$

when the coefficient group is the additive group of integers, or when it is obvious from the context.

We are going to show how the homology (and cohomology) of a relative CW-complex is determined from its structure as a complex. Our treatment is comparable to that in Eilenberg–Steenrod [E–S] for simplicial complexes. We shall be dealing extensively with excisions, and it will be convenient to have available certain results which are slightly stronger forms of analogous results which play an important role in Chapter III of [E–S].

In much of our work we have to deal with inclusion maps and the homomorphisms of algebraic structures (homotopy groups, homology groups, etc.) which they induce. We shall consistently refer to such homomorphisms as *injections*. The reader is warned that some authors use this word differently, referring to any one-to-one function as an injection. However, our usage is traditional in algebraic topology, and takes historical precedence over the other. We state, then for emphasis; *an injection need be neither a monomorphism, nor an epimorphism*!

As many arguments in homology theory make use of intricate diagrams involving boundary homomorphisms on the one hand and injection homomorphisms on the other, it will be useful to make the

Fundamental notational convention *In any diagram involving homology groups, the symbol*

$$H_n(X, A) \to H_{n-1}(A, B)$$

denotes the boundary operator of the homology sequence of the triple (X, A, B), *and the symbol*

$$H_n(X, A) \to H_n(X', A')$$

(*where* $A \subset X \cap A'$, $X \cup A' \subset X$) *denotes the injection. This convention remains in force unless a stipulation to the contrary is expressly made.*

The analogous convention for other theories (cohomology, homotopy, etc.) will likewise be made.

Furthermore, (although we do not make a formal stipulation to this effect), the letters i, j, k, l, with or without subscripts, will usually denote injections.

A triad $(X; A, B)$ is said to be *proper* if and only if the injections $H_n(A, A \cap B) \to H_n(A \cup B, B)$ (equivalently, if and only if the injections $H_n(B, A \cap B) \to H_n(A \cup B, A)$) are isomorphisms for all n. The triad $(X; A, B)$ is an NDR-*triad* if and only if $X = A \cup B$, A and B are closed, and one of the pairs $(A, A \cap B)$, $(B, A \cap B)$ is an NDR-pair.

(2.1) Lemma *Let* (X, A) *be an* NDR-*pair, and let* $p: X \to X/A$ *be the natural proclusion. Then*

$$p_* : H_n(X, A) \to H_n(X/A, *)$$

is an isomorphism.

For let $\mathbf{T}X$ be the cone over X; then $X \cup \mathbf{T}A$ is a subspace of $\mathbf{T}X$, and $(X \cup \mathbf{T}A, \mathbf{T}A)$ is an NDR-pair. By Corollary (5.13) of Chapter I, the proclusion $q: X \cup \mathbf{T}A \to X \cup \mathbf{T}A/\mathbf{T}A$ is a homotopy equivalence. Let $h: X/A \to X \cup \mathbf{T}A/\mathbf{T}A$ be the natural homeomorphism. Consider the commutative diagram

$$\begin{array}{ccc}
H_n(X, A) & \xrightarrow{p_*} & H_n(X/A, *) \\
\downarrow{\scriptstyle i_1} & & \downarrow{\scriptstyle h_*} \\
H_n(X \cup \mathbf{T}A, \mathbf{T}A) & \xrightarrow{q_1} & H_n(X \cup \mathbf{T}A/\mathbf{T}A, *) \\
\uparrow{\scriptstyle i_2} & \nearrow{\scriptstyle q_2} & \\
H_n(X \cup \mathbf{T}A, *) & &
\end{array}$$

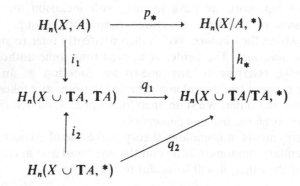

in which q_1 and q_2 are the homomorphisms induced by q. Then A is a deformation retract of a neighborhood of itself in $\mathbf{T}A$, and therefore i_1 is an isomorphism. That i_2 is an isomorphism follows from exactness of the homology sequence of the triple $(X \cup \mathbf{T}A, \mathbf{T}A, *)$ and the contractibility of $\mathbf{T}A$. Since q is a homotopy equivalence, q_2 is an isomorphism, and therefore q_1 is an isomorphism. But h_* is an isomorphism, and it follows that p_* is an isomorphism. □

(2.2) Theorem *Let* $(X; A, B)$ *be an* NDR-*triad. Then* $(X; A, B)$ *is proper.*

We may assume $(A, A \cap B)$ is an NDR-pair, and therefore (X, B) is also. Let $p: (A, A \cap B) \to (A/A \cap B, *)$ and $q: (X, B) \to (X/B, *)$ be the identification maps and let $i: (A, A \cap B) \subsetneq (X, B)$. The induced map $h: (A/A \cap B, *) \to (X/B, *)$ is a homeomorphism, and the diagram

$$\begin{array}{ccc}
H_n(A, A \cap B) & \xrightarrow{p_*} & H_n(A/A \cap B, *) \\
\downarrow{\scriptstyle i_*} & & \downarrow{\scriptstyle h_*} \\
H_n(X, B) & \xrightarrow{q_*} & H_n(X/B, *)
\end{array}$$

is commutative. Because $(A, A \cap B)$ and (X, B) are NDR-pairs, p_* and q_* are isomorphisms. Since h_* is an isomorphism, it follows that i_* is an isomorphism. □

(2.3) Theorem (Map Excision Theorem). *Let $f : (X, A) \to (Y, B)$ be a relative homeomorphism of* NDR-*pairs. Then $f_* : H_n(X, A) \approx H_n(Y, B)$ for all n.*

Again, let $p : (X, A) \to (X/A, *)$ and $q : (X, B) \to (Y/B, *)$ be the natural proclusions. Then f induces a map $g : (X/A, *) \to (Y/B, *)$ such that $g \circ p = q \circ f$. Since q and f are proclusions, g is a proclusion; since g is a one-to-one continuous map of X/A upon Y/B, it is a homeomorphism. Commutativity of the diagram

$$\begin{array}{ccc} H_n(X, A) & \xrightarrow{p_*} & H_n(X/A, *) \\ \downarrow{\scriptstyle f_*} & & \downarrow{\scriptstyle g_*} \\ H_n(Y, B) & \xrightarrow{q_*} & H_n(Y/B, *) \end{array}$$

and the fact that p_*, q_*, and g_* are isomorphisms imply that f_* is an isomorphism. □

Another useful property of NDR-triads is

(2.4) Lemma *Let $(X; A_1, A_2)$ be an* NDR *triad, $(K; L_1, L_2)$ a compact triangulable triad, $f : (K; L_1, L_2) \to (X; A_1, A_2)$. Then there are subspaces M_1, M_2 of K such that*

(1) *the triad $(K; M_1, M_2)$ is triangulable;*
(2) $K = M_1 \cup M_2$;
(3) $M_i \supset L_i$ $(i = 1, 2)$;
(4) *there is a homotopy*

$$F : (\mathbf{I} \times K; \mathbf{I} \times L_1, \mathbf{I} \times L_2) \to (X; A_1, A_2)$$

of f (rel. $L_1 \cap L_2$) to a map $f' : (K; M_1, M_2) \to (X; A_1, A_2)$.

Let $h_0 : \mathbf{I} \times A_1 \to A_1$, $u_0 : A_1 \to \mathbf{I}$ represent $(A_1, A_1 \cap A_2)$ as an NDR-pair, and let $U_0 = u_0^{-1}([0, 1))$. Define $h : \mathbf{I} \times X \to X$, $u : X \to \mathbf{I}$ by

$$h \mid \mathbf{I} \times A_1 = h_0,$$
$$h(t, x) = x \qquad (x \in A_2),$$
$$u_0 \mid A_1 = u,$$
$$u_0(x) = 0 \qquad (x \in A_2);$$

then h, u represent (X, A_2) as an NDR-pair and $U = u^{-1}([0, 1)) = A_2 \cup U_0$ is a neighborhood of A_2 in X. The sets $A_1 - A_2$, U form an open covering of X; let $\delta > 0$ be a Lebesque number for the open covering $\{f^{-1}(A_1 - A_2), f^{-1}(U)\}$ of K. Choose a triangulation of $(K; L_1, L_2)$ of mesh $< \delta$ and let M_1 (M_2) be the union of all simplices σ of the triangulation such that $f(\sigma) \subset A_1$ ($f(\sigma) \subset U$). Then the first three conditions are satisfied.

Let $F = h \circ (1 \times f) : \mathbf{I} \times K \to X$. Then $F(\mathbf{I} \times L_i) \subset A_i$ $(i = 1, 2)$ and $f'(M_i) \subset A_i$. □

Remark. Let $\{K_n\}$ be a sequence of subdivisions of K such that $\lim_{n\to\infty} \text{mesh } K_n = 0$. Then the triangulation of (1) can be chosen to be K_q for some q.

(2.5) Theorem (van Kampen Theorem). *Let $(X; A, B)$ be an* NDR-*triad, and suppose that A and B are* 1-*connected and $A \cap B$ is* 0-*connected. Then X is* 1-*connected.*

We shall suppose that the base point $*$ for the fundamental groups of all the subspaces X, A, B, $A \cap B$ is in $A \cap B$. Let $w : (\mathbf{I}, \dot{\mathbf{I}}) \to (X, *)$ be a loop in X. By Lemma (2.4) there is a partition $0 = t_0 < t_1 < \cdots < t_m = 1$ of $\mathbf{I}$ and a path $w' \simeq w$ such that, for each i, either $w'([t_{i-1}, t_i]) \subset A$ or $w'([t_{i-1}, t_i]) \subset B$, i.e., w is homotopic to a product $w'_1 \cdots w'_m$ such that, for each i, w'_i is either a path in A or a path in B. (We may ignore the parenthesization because multiplication of paths is associative up to homotopy.) We may assume that, if the image of w'_i lies in A (B) then the image of w'_{i+1} lies in B (A). Hence the end points of each of the paths w'_i lie in $A \cap B$. Let v_i be a path in $A \cap B$ from $*$ to the initial point of w'_i $(i = 1, \ldots, m - 1)$. Then w is homotopic to the product

$$(w'_1 v_1^{-1})(v_1 w'_1 v_2^{-1}) \cdots (v_{m-1} w'_m)$$

and each of the parenthesized paths is a *loop* in either A or B. Since A and B are 1-connected, these loops are all nullhomotopic and therefore w is also. □

Remark. What we have called the "van Kampen theorem" is only a special case of the general van Kampen theorem, which expresses $\pi_1(X)$ in terms of $\pi_1(A)$, $\pi_1(B)$ and the injections $\pi_1(A \cap B) \to \pi_1(A)$, $\pi_1(A \cap B) \to \pi_1(B)$. The latter theorem is also true for NDR-triads. As we shall not need the stronger theorem, we have relegated it to the Exercises.

As a corollary to Theorem (2.2), we have the *Direct Sum Theorem*:

(2.6) *Let $X_1, \ldots, X_r, A$, be closed subsets of a space X such that*

(1) $X = X_1 \cup \cdots \cup X_r \cup A$;
(2) $X_\alpha \cap X_\beta \subset A$ *for all* $\alpha \neq \beta$;
(3) $(X_\alpha, X_\alpha \cap A)$ *is an* NDR-*pair for each* α.

Then[1]

(1) *the injections*

$$i_\alpha : H_n(X_\alpha, X_\alpha \cap A) \to H_n(X, A)$$

represent $H_n(X, A)$ *as a direct sum;*

(2) *let* $X_\alpha^* = X - X_\alpha$; *then the injections*

$$j_\alpha : H_n(X, A) \to H_n(X, X_\alpha^*)$$

represent $H_n(X, A)$ *as a direct product;*

(3) *the injection* $k_\alpha = j_\alpha \circ i_\alpha : H_n(X_\alpha, X_\alpha \cap A) \to H_n(X, X_\alpha^*)$ *is an isomorphism, and the representations* $\{i_\alpha\}$, $\{j_\alpha\}$ *are weakly dual.* □

We shall need to extend (2.6) to the case of certain infinite collections of sets. The desired extension is

(2.7) (General Direct Sum Theorem). *Let* $\{X_\alpha \mid \alpha \in J\}$ *be a collection of closed subsets of a space* X, *and let* A *be a closed subset of* X. *Suppose that*

(1) $X = A \cup \bigcup_{\alpha \in J} X_\alpha$;

(2) $X_\alpha \cap X_\beta \subset A$ *for* $\alpha \neq \beta$;

(3) $(X_\alpha, X_\alpha \cap A)$ *is an* NDR-*pair for each* α;

(4) X *has the weak topology with respect to the collection of sets* $\{A, X_\alpha\}$.

Then[1]

(1) *the injections*

$$i_\alpha : H_n(X_\alpha, X_\alpha \cap A) \to H_n(X, A)$$

represent $H_n(X, A)$ *as a direct sum;*

(2) *let* $X_\alpha^* = X - X_\alpha$; *then the injections*

$$j_\alpha : H_n(X, A) \to H_n(X, X_\alpha^*)$$

represent $H_n(X, A)$ *as a weak direct product;*

(3) *the injection* $j_\alpha \circ i_\alpha : H_n(X_\alpha, X_\alpha \cap A) \to H_n(X, X_\alpha^*)$ *is an isomorphism, and the representations* $\{i_\alpha\}$, $\{j_\alpha\}$ *are weakly dual.*

The sets $X_\alpha - A$ are open and mutually disjoint; hence any compact set C is contained in the union of A with finitely many of the X_α. The theorem now follows from the fact that singular homology has compact supports. □

(A different argument, based on the *additivity* of singular homology, is given in Theorem (6.9) of Chapter XII.)

[1] cf. Appendix B, §1.

Let (X, A) be a relative CW-complex with skeleta X_n. Let $\{E_\alpha^n \mid \alpha \in J\}$ be the n-cells of (X, A), with characteristic maps

$$h_\alpha : (\Delta^n, \dot{\Delta}^n) \to (E_\alpha^n, \dot{E}_\alpha^n).$$

Let $X_\alpha = E_\alpha^n \cup X_{n-1}$. The group $H_n(\Delta^n, \dot{\Delta}^n)$ is an infinite cyclic group generated by the homology class δ_n of the identity map $\Delta^n \to \Delta^n$. Let $e_\alpha^n \in H_n(E_\alpha^n, \dot{E}_\alpha^n)$ be the image of δ_n under the homomorphism induced by h_α; we shall also denote by e_α^n the images of e_α^n under the injections

$$H_n(E_\alpha^n, \dot{E}_\alpha^n) \to H_n(X_\alpha, X_{n-1}),$$
$$H_n(E_\alpha^n, \dot{E}_\alpha^n) \to H_n(X_n, X_{n-1}).$$

It follows from (2.7) that

(2.8) *The injections $H_q(E_\alpha^n, \dot{E}_\alpha^n) \to H_q(X_n, X_{n-1})$ represent the group $H_q(X_n, X_{n-1})$ as a direct sum of the groups $H_q(E_\alpha^n, \dot{E}_\alpha^n)$.* □

On the other hand, it follows from Theorem (2.2) that

(2.9) *The injections $H_q(E_\alpha^n, \dot{E}_\alpha^n) \to H_q(X_\alpha, X_{n-1})$ are isomorphisms,* □

and from Theorem (2.3) that

(2.10) *The homomorphisms $h_{\alpha *} : H_q(\Delta^n, \dot{\Delta}^n) \to H_q(E_\alpha^n, \dot{E}_\alpha^n)$ are isomorphisms.* □

We summarize the results of (2.8) to (2.10) in

(2.11) Theorem *If $q \neq n$, $H_q(X_n, X_{n-1}) = 0$. The group $H_n(X_n, X_{n-1})$ is a free abelian group, for which the elements e_α^n constitute a basis.* □

(2.12) Corollary *If $p > q \geq r$, then $H_p(X_q, X_r) = 0$. In particular, if $p > q$, then $H_p(X_q, A) = 0$.* □

(2.13) Corollary *If $p \leq r \leq q$, then $H_p(X_q, X_r) = 0$.* □

Since singular homology has compact supports and X has the weak topology with respect to the skeleta X_q, the group $H_p(X, X_r)$ is the direct limit of the groups $H_p(X_q, X_r)$ under the injections $H_p(X_q, X_r) \to H_p(X_{q+1}, X_r)$. Thus

(2.14) Corollary *If $p \leq r$, then $H_p(X, X_r) = 0$.* □

(2.15) Corollary *If $q > p$, the injection*

$$H_p(X_q, A) \to H_p(X, A)$$

is an isomorphism. Moreover, the injection

$$H_p(X_p, A) \to H_p(X, A)$$

is an epimorphism whose kernel is the image of the boundary operator

$$H_{p+1}(X_{p+1}, X_p) \to H_p(X_p, A)$$

of the homology sequence of the triple (X_{p+1}, X_p, A). □

Let $\Gamma_n(X, A) = H_n(X_n, X_{n-1})$, and let

$$\partial_n : \Gamma_n(X, A) \to \Gamma_{n-1}(X, A)$$

be the boundary operator of the homology sequence of the triple (X_n, X_{n-1}, X_{n-2}). The composite $\partial_n \circ \partial_{n+1}$ can be factored as follows:

$$\begin{aligned} H_{n+1}(X_{n+1}, X_n) \to H_n(X_n, A) \to H_n(X_n, X_{n-1}) \\ \to H_{n-1}(X_{n-1}, A) \to H_{n-1}(X_{n-1}, X_{n-2}). \end{aligned}$$

The middle part of this array is a portion of the homology sequence of the triple (X_n, X_{n-1}, A). Hence $\partial_n \circ \partial_{n+1} = 0$, so that

(2.16) *The graded group* $\{\Gamma_n(X, A)\}$ *is a free chain complex* $\Gamma(X, A)$ *with respect to the endomorphisms* $\{\partial_n\}$. □

If G is an abelian group, let

$$\Gamma_n(X, A; G) = H_n(X_n, X_{n-1}; G);$$

these are the components of a chain complex $\Gamma(X, A; G)$ with respect to the boundary operator ∂_n of the homology sequence (with coefficients in G) of the triple (X_n, X_{n-1}, X_{n-2}). By the universal coefficient theorem, there are isomorphisms $G \otimes \Gamma_n(X, A) \to \Gamma_n(X, A; G)$ making a commutative diagram

$$\begin{array}{ccc} G \otimes H_n(X_n, X_{n-1}) & \xrightarrow{1 \otimes \partial_n} & G \otimes H_{n-1}(X_{n-1}, X_{n-2}) \\ \downarrow & & \downarrow \\ H_n(X_n, X_{n-1}; G) & \xrightarrow[\partial_n]{} & H_{n-1}(X_{n-1}, X_{n-2}; G) \end{array}$$

Thus

(2.17) *The complexes* $G \otimes \Gamma(X, A)$ *and* $\Gamma(X, A; G)$ *are isomorphic.* □

Similarly, let

$$\Gamma^n(X, A; G) = H^n(X_n, X_{n-1}; G),$$

and let $\delta_n : \Gamma^n(X, A; G) \to \Gamma^{n+1}(X, A; G)$ be the coboundary operator of the cohomology sequence of the triple (X_{n+1}, X_n, X_{n-1}); then the $\Gamma^n(X, A; G)$ are the components of a graded cochain complex $\Gamma^*(X, A; G)$. Applying the universal coefficient theorem for cohomology, we again obtain a commutative diagram

$$\begin{array}{ccc} H^n(X_n, X_{n-1}; G) & \xrightarrow{(-1)^n \delta_n} & H^{n+1}(X_{n+1}, X_n; G) \\ \downarrow & & \downarrow \\ \operatorname{Hom}(H_n(X_n, X_{n-1}), G) & \xrightarrow[\operatorname{Hom}(\partial_{n+1}, 1)]{} & \operatorname{Hom}(H_{n+1}(X_{n+1}, X_n), G) \end{array}$$

in which the vertical arrows represent isomorphisms. Thus

(2.18) *The cochain complexes* $\Gamma^*(X, A; G)$ *and* $\operatorname{Hom}(\Gamma(X, A), G)$ *are isomorphic.* □

Remark. If C is a graded chain complex, $C^* = \operatorname{Hom}(C, G)$, it is traditional to define the coboundary operator $\delta : C^* \to C^*$ by the formula

$$\delta f(c) = f(\partial c)$$

for $f \in C^q = \operatorname{Hom}(C_q, G)$, $c \in C_{q+1}$. With this definition, however, certain peculiar signs crop up. For example, the formula for the coboundary of a cross product becomes

$$\delta(u \times v) = (-1)^q \, \delta u \times v + u \times \delta v$$

($u \in C^p(K)$, $v \in C^q(L)$). For this reason and other reasons, we shall *not* use the traditional definition, but introduce a sign, so that

$$\delta f(c) = (-1)^q f(\partial c)$$

for f, c as above. (This is in keeping with the philosophy that, in a graded universe, one should always introduce a sign $(-1)^{pq}$ when interchanging two symbols of degrees p, q).

We come at last to the main result of this section.

(2.19) Theorem *Let* (X, A) *be a relative* CW-*complex. Then*

$$H_n(X, A; G) \approx H_n(\Gamma(X, A; G))$$

and

$$H^n(X, A; G) \approx H^n(\Gamma^*(X, A; G))$$

for any coefficient group G.

Because of (2.17), (2.18) and the universal coefficient theorems, it suffices to prove this for the case of integral homology. Consider the diagram

$$\text{(2.20)}\quad \begin{array}{ccccccccc} H_{n+1}(X_{n+1}, X_n) & \xrightarrow{d} & H_n(X_n, A) & \xrightarrow{i} & H_n(X_n, X_{n-1}) & \xrightarrow{d'} & H_{n-1}(X_{n-1}, A) & \xrightarrow{i'} & H_{n-1}(X_{n-1}, X_{n-2}) \\ & & \downarrow{\scriptstyle j} & & & & & & \\ & & H_n(X, A) & & & & & & \end{array}$$

in which i, i', j are injections and d, d' boundary operators. By Corollary (2.12), $H_n(X_{n-1}, A) = 0$; by exactness of the homology sequence of the triple (X_n, X_{n-1}, A), i is a monomorphism; similarly, i' is a monomorphism. Now the cycles and boundaries of the chain-complex $\Gamma(X, A)$ are

$$Z_n(\Gamma(X, A)) = \operatorname{Ker}(i' \circ d') = \operatorname{Ker} d' = \operatorname{Im} i,$$

$$B_n(\Gamma(X, A)) = \operatorname{Im}(i \circ d),$$

and these are isomorphic under i^{-1} with

$$H_n(X_n, A),$$

$$\operatorname{Im} d,$$

respectively. Hence

$$\begin{aligned} H_n(\Gamma(X, A)) &\approx H_n(X_n, A)/\operatorname{Im} d \\ &= H_n(X_n, A)/\operatorname{Ker} j \approx H_n(X, A) \end{aligned}$$

by Corollary (2.15). □

Now suppose that (X, A) and (Y, B) are relative CW-complexes with skeleta X_n, Y_n respectively. We have seen in (7) of (1.6) that their product

$$(Z, C) = (X, A) \times (Y, B) = (X \times Y, X \times B \cup A \times Y)$$

is a relative CW-complex with skeleta

$$Z_n = \bigcup_{i=0}^{n} (X_i \times Y_{n-i}).$$

The n-cells of (Z, C) are thus of the form $E_\alpha^p \times E_\beta^q$, where E_α^p is a p-cell of (X, A), E_β^q a q-cell of (Y, B) and $p + q = n$; and $(E_\alpha^p \times E_\beta^q)^\bullet = E_\alpha^p \times \dot{E}_\beta^q \cup \dot{E}_\alpha^p \times E_\beta^q$. Moreover, if $f: (\Delta^p, \dot{\Delta}^p) \to (E_\alpha^p, \dot{E}_\alpha^p)$ and $g: (\Delta^q, \dot{\Delta}^q) \to (E_\beta^q, \dot{E}_\beta^q)$ are characteristic maps, then

$$f \times g: (\Delta^p, \dot{\Delta}^p) \times (\Delta^q, \dot{\Delta}^q) \to (E_\alpha^p, \dot{E}_\alpha^p) \times (E_\beta^q, \dot{E}_\beta^q)$$

is a characteristic map for the product cell.

We now recall some properties of the cross product in singular homology.

If (X, A) and (Y, B) are NDR-pairs, $u \in H_p(X, A)$ $v \in H_q(Y, B)$, their cross product

$$u \times v \in H_{p+q}((X, A) \times (Y, B))$$

is defined using the Eilenberg–Zilber map of the tensor product of the singular complexes $\mathfrak{S}(X)$, $\mathfrak{S}(Y)$ of X, Y, respectively, into $\mathfrak{S}(X \times Y)$.

In particular, if (X, A) and (Y, B) are relative CW-complexes, the cross product is a pairing

$$\begin{aligned}\Gamma_p(X, A) \otimes \Gamma_q(Y, B) &= H_p(X_p, X_{p-1}) \otimes H_q(Y_q, Y_{q-1}) \\ &\to H_{p+q}(X_p \times Y_q, X_p \times Y_{q-1} \cup X_{p-1} \times Y_q);\end{aligned}$$

the latter group injects into $H_{p+q}(Z_{p+q}, Z_{p+q-1}) = \Gamma_{p+q}((X, A) \times (Y, B))$; we shall denote the image of $u \otimes v$ by $u \times v \in \Gamma_{p+q}((X, A) \times (Y, B))$. If δ_p, δ_q are generators of $H_p(\Delta^p, \dot{\Delta}^p)$, $H_q(\Delta^q, \dot{\Delta}^q)$, respectively, then $\delta_p \times \delta_q$ generates $H_{p+q}((\Delta^p, \dot{\Delta}^p) \times (\Delta^q, \dot{\Delta}^q))$, and therefore

$$(f \times g)_*(\delta_p \times \delta_q) = f_*(\delta_p) \times g_*(\delta_q) = e_\alpha^p \times e_\beta^q$$

generates the group $H_{p+q}((E_\alpha^p, \dot{E}_\alpha^p) \times (E_\beta^q, \dot{E}_\beta^q))$. Thus

(2.21) *The cross product induces an isomorphism of the graded groups* $\Gamma(X, A) \otimes \Gamma(Y, B)$ and $\Gamma((X, A) \times (Y, B))$. □

In fact, we have

(2.22) Theorem *The chain complexes* $\Gamma(X, A) \otimes \Gamma(Y, B)$ *and* $\Gamma((X, A) \times (Y, B))$ *are isomorphic under the operation of cross product.*

To prove this, we must calculate the effect of the boundary operator on a cross product. This follows from a more general result on cross products which we now explain.

Let (X, A, A') and (Y, B, B') be NDR-triples, and let $u \in H_p(X, A)$, $v \in H_q(Y, B)$. One then has elements $\partial u \in H_{p-1}(A, A')$, $\partial v \in H_{q-1}(B, B')$, where the symbol "$\partial$" denotes indiscriminately the boundary operator of the appropriate homology sequence. One can then form the cross products

$$\partial u \times v \in H_{p+q-1}(A \times Y, A \times B \cup A' \times Y),$$

$$u \times \partial v \in H_{p+q-1}(X \times B, X \times B' \cup A \times B).$$

These elements lie in different groups; to compare them, it is convenient to inject into the group

(2.23) $\quad H_{p+q-1}(X \times B \cup A \times Y, X \times B' \cup A \times B \cup A' \times Y);$

let us continue to denote their images under these injections by the same symbols. The boundary operator of the homology sequence of the triple

$$(X \times Y, X \times B \cup A \times Y, X \times B' \cup A \times B \cup A' \times Y)$$

also sends $u \times v$ into an element $\partial(u \times v)$ of the group (2.23).

(2.24) Lemma *In the group* (2.23), *the relation*

(2.25) $$\partial(u \times v) = \partial u \times v + (-1)^p u \times \partial v$$

holds.

This is proved by selecting singular relative cycles representing u and v, using the boundary formula in $\mathfrak{S}(X) \otimes \mathfrak{S}(Y)$, and applying the Eilenberg–Zilber map. □

Now apply this result to the triples (X_p, X_{p-1}, X_{p-2}) and (Y_q, Y_{q-1}, Y_{q-2}) to obtain

(2.26) Corollary *If* $u \in H_p(X_p, X_{p-1})$, $v \in H_q(Y_q, Y_{q-1})$, *then, in the group*

$$H_{p+q-1}(X_p \times Y_{q-1} \cup X_{p-1} \times Y_q, X_p \times Y_{q-2} \cup X_{p-1} \times Y_{q-1} \cup X_{p-2} \times Y_q),$$

the relation (2.25) *holds.* □

Injecting the latter group into the group

$$\Gamma_{p+q-1}((X, A) \times (Y, B)) = H_{p+q-1}(Z_{p+q-1}, Z_{p+q-2}),$$

we have

(2.27) Corollary *If* $u \in \Gamma_p(X, A)$, $v \in \Gamma_q(Y, B)$, *then the relation* (2.25) *holds in* $\Gamma_{p+q-1}((X, A) \times (Y, B))$. □

But this is precisely what is needed to prove Theorem (2.22).

Remark 1. Because of Corollary (2.27), the cross product of chains induces a cross product pairing in homology:

$$H_p(\Gamma(X, A)) \otimes H_q(\Gamma(Y, B)) \to H_{p+q}(\Gamma((X, A) \times (Y, B))).$$

We also have the cross product pairing in singular homology:

$$H_p(X, A) \otimes H_q(Y, B) \to H_{p+q}((X, A) \times (Y, B)).$$

Because the cross product of chains was defined in terms of singular homology, it is not unreasonable to expect that the two homology cross products agree. Specifically, we have, for each relative CW-complex (X, A), a uniquely defined isomorphism

$$\theta = \theta_{(X, A)} : H_*(\Gamma(X, A)) \approx H_*(X, A);$$

and we claim that, for $u \in H_p(\Gamma(X, A))$, $v \in H_q(\Gamma(Y, B))$,

$$\theta(u \times v) = \theta(u) \times \theta(v).$$

The easy proof of this fact is relegated to the Exercises.

Remark 2. Our discussion of cross product goes through without essential change for cross products with arbitrary coefficients. Specifically, if a pairing $G \otimes H \to K$ of abelian groups is given, there are cross products of chains

$$\Gamma_p(X, A; G) \otimes \Gamma_q(Y, B; H) \to \Gamma_{p+q}((X, A) \times (Y, B); K),$$

giving rise to cross products in homology, and these agree with those of singular theory.

Let us apply the above considerations to the product of the unit interval with a relative CW-complex (X, A). Now $\mathbf{I}$ has two 0-cells, $\{0\}$, $\{1\}$, and one 1-cell $\mathbf{i}$; and

$$\partial \mathbf{i} = \{1\} - \{0\}.$$

The cross products with these cells give a representation of $\Gamma_n(\mathbf{I} \times X, \mathbf{I} \times A)$ as a direct sum of two copies of $\Gamma_n(X, A)$ and one of $\Gamma_{n-1}(X, A)$; and one has, for $c \in \Gamma_n(X, A)$,

(2.28)
$$\begin{aligned} \partial(1 \times c) &= 1 \times \partial c, \\ \partial(0 \times c) &= 0 \times \partial c, \\ \partial(\mathbf{i} \times c) &= 1 \times c - 0 \times c - \mathbf{i} \times \partial c. \end{aligned}$$

Let us also consider the product

$$(\mathbf{I}, \dot{\mathbf{I}}) \times (X, A) = (\mathbf{I} \times X, \mathbf{I} \times A \cup \dot{\mathbf{I}} \times X) = (X^*, A^*);$$

this time $\Gamma_{n+1}((\mathbf{I}, \dot{\mathbf{I}}) \times (X, A)) \approx \Gamma_n(X, A)$. In fact, the map $c \to \mathbf{i} \times c$ is an isomorphic chain map, of degree -1, of $\Gamma(X, A)$ with $\Gamma(X^*, A^*)$. Hence

(2.29) Theorem *The cross product with the generator* $\mathbf{i}$ *of* $H_1(\mathbf{I}, \dot{\mathbf{I}}; \mathbf{Z})$ *induces an isomorphism*

$$\mathbf{i}\times : H_q(X, A; G) \approx H_{q+1}(X^*, A^*; G)$$

for any coefficient group G. □

The discussion of cross products in cohomology is considerably more intricate than the corresponding one in homology. Indeed, the cross product of chains is the composite

$$\begin{aligned} &C_p(\mathfrak{S}(X)/\mathfrak{S}(A)) \otimes C_q(\mathfrak{S}(Y)/\mathfrak{S}(B)) \\ &\qquad \subsetneq C_{p+q}(\mathfrak{S}(X) \otimes \mathfrak{S}(Y)/\mathfrak{S}(X) \otimes \mathfrak{S}(B) + \mathfrak{S}(A) \otimes \mathfrak{S}(Y)) \\ &\qquad \to C_{p+q}(\mathfrak{S}(X \times Y)/\mathfrak{S}(X \times B) + \mathfrak{S}(A \times Y)) \\ &\qquad \to C_{p+q}(\mathfrak{S}(X \times Y)/\mathfrak{S}(X \times B \cup A \times Y)); \end{aligned}$$

the second map is induced by the Eilenberg–Zilber map of $\mathfrak{S}(X) \otimes \mathfrak{S}(Y)$ into $\mathfrak{S}(X \times Y)$ and the third by the inclusion $\mathfrak{S}(X \times B) + \mathfrak{S}(A \times Y) \subsetneq \mathfrak{S}(X \times B \cup A \times Y)$. On the other hand, if we attempt to copy this procedure for cochains, we obtain the composite

$$\begin{aligned} &C^p(\mathfrak{S}(X)/\mathfrak{S}(A)) \otimes C^q(\mathfrak{S}(Y)/\mathfrak{S}(B)) \\ &\quad \hookrightarrow C^{p+q}(\mathfrak{S}(X) \otimes \mathfrak{S}(Y)/\mathfrak{S}(X) \otimes \mathfrak{S}(B) + \mathfrak{S}(A) \otimes \mathfrak{S}(Y)) \\ &\quad \rightarrow C^{p+q}(\mathfrak{S}(X \times Y)/\mathfrak{S}(X \times B) + \mathfrak{S}(A \times Y)) \\ &\quad \leftarrow C^{p+q}(\mathfrak{S}(X \times Y)/\mathfrak{S}(X \times B + A \times Y)); \end{aligned}$$

the second map is induced by the Eilenberg–Zilber map of $\mathfrak{S}(X \times Y)$ into $\mathfrak{S}(X) \otimes \mathfrak{S}(Y)$; the third map, induced by the inclusion $\mathfrak{S}(X \times B) + \mathfrak{S}(A \times Y) \hookrightarrow \mathfrak{S}(X \times B \cup A \times Y)$ goes in the wrong direction. If (X, A) and (Y, B) are NDR-pairs, the latter inclusion is a chain equivalence. Therefore there is no difficulty in defining a pairing

$$H^p(X, A) \otimes H^q(Y, B) \rightarrow H^{p+q}(X \times Y, X \times B \cup A \times Y)$$

with good properties; but calculations with the cohomology cross product at the cochain level are more complicated for this reason. A more detailed account of the cohomology cross product has been given by Steenrod and Rothenburg. Unfortunately, their paper was never published although its principal results were announced in [1].

Let $G \otimes H \rightarrow K$ be a pairing of abelian groups; we shall be concerned with cross-product pairings,

$$H^p(\quad ; G) \otimes H^q(\quad ; H) \rightarrow H^{p+q}(\quad ; K).$$

To simplify writing, however, we shall suppress the coefficient group from the notation.

A basic property of the cross product in cohomology is a coboundary formula dual to (2.25). Let (X, A, A') and (Y, B, B') be NDR-triples. Then there is a commutative diagram (Figure 2.1), in which the homomorphisms k_i are isomorphisms, by the Excision Theorem. It is a standard result (see, for example, [Sp, Theorem 5.6.6]) that, if $u \in H^p(A, A')$, $v \in H^q(B, B')$, then

$$\begin{aligned} \delta_1 k_1^{-1}(u \times v) &= \delta u \times v \in H^{p+q+1}((X, A) \times (B, B')), \\ \delta_2 k_2^{-1}(u \times v) &= (-1)^p u \times \delta v \in H^{p+q+1}((A, A') \times (Y, B)). \end{aligned}$$

We shall prove

(2.30) *Let $u \in H^p(A, A')$, $v \in H^q(B, B')$. Then*

$$\delta_0 k_0^{-1}(u \times v) = j_1 k_3^{-1}(\delta u \times v) + (-1)^p j_2 k_4^{-1}(u \times \delta v).$$

By the Direct Sum Theorem, the homomorphisms j_1, j_2 represent the group $H^{n+1}(X \times B \cup A \times Y, X \times B' \cup A \times B \cup A' \times Y)$ as a direct sum, while the homomorphisms l_1, l_2 form a weakly dual representation as a

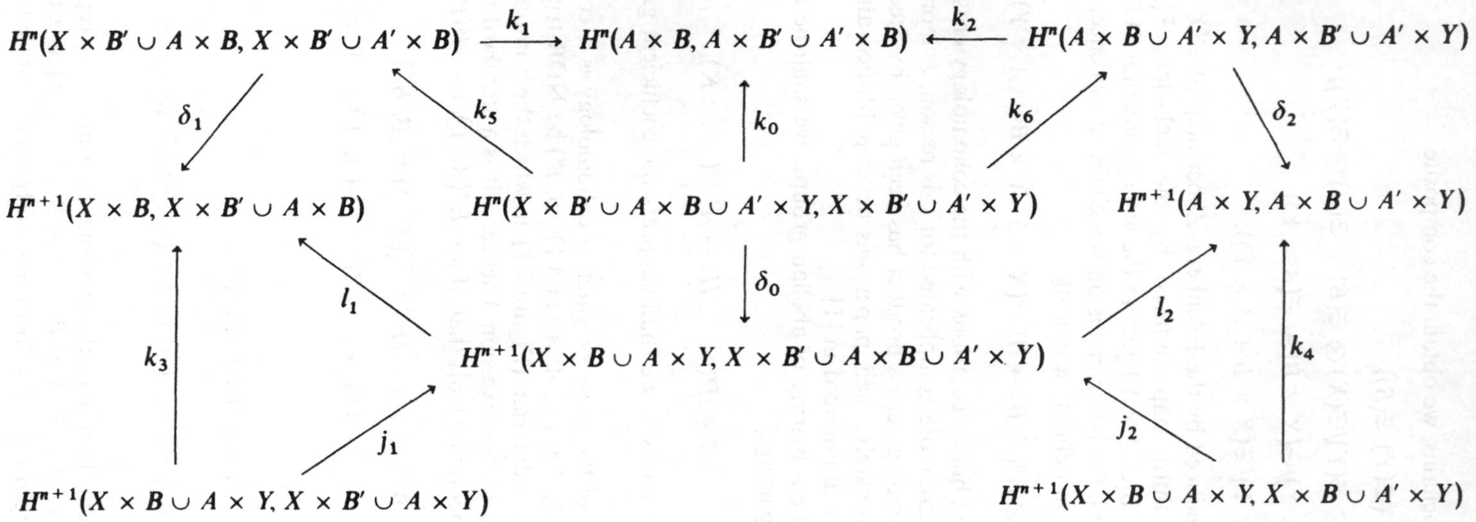
$H^n(X \times B' \cup A \times B, X \times B' \cup A' \times B)$
k_1
$H^n(A \times B, A \times B' \cup A' \times B)$
k_2
$H^n(A \times B \cup A' \times Y, A \times B' \cup A' \times Y)$
δ_1
k_5
k_0
k_6
δ_2
$H^{n+1}(X \times B, X \times B' \cup A \times B)$
$H^n(X \times B' \cup A \times B \cup A' \times Y, X \times B' \cup A' \times Y)$
$H^{n+1}(A \times Y, A \times B \cup A' \times Y)$
l_1
δ_0
l_2
k_3
$H^{n+1}(X \times B \cup A \times Y, X \times B' \cup A \times B \cup A' \times Y)$
k_4
j_1
j_2
$H^{n+1}(X \times B \cup A \times Y, X \times B' \cup A \times Y)$
$H^{n+1}(X \times B \cup A \times Y, X \times B \cup A' \times Y)$

Figure 2.1

direct product. Commutativity of Figure 2.1 implies that

$$l_1 \delta_0 k_0^{-1}(u \times v) = \delta_1 k_5 k_0^{-1}(u \times v) = \delta_1 k_1^{-1}(u \times v) = \delta u \times v,$$
$$l_2 \delta_0 k_0^{-1}(u \times v) = \delta_2 k_6 k_0^{-1}(u \times v) = \delta_2 k_2^{-1}(u \times v) = (-1)^p u \times \delta v,$$

and (2.30) follows immediately. □

Let (X, A), (Y, B) be relative CW-complexes, so that $(Z, C) = (X, A) \times (Y, B)$ is also a relative CW-complex. Let $K_{p,q} = X_p \times Y_q$, $L_{p,q} = X_p \times Y_{q-1} \cup X_{p-1} \times Y_q$. Then

$$Z_n = \bigcup_{p+q=n} K_{p,q}$$

and it follows from the Direct Sum Theorem (2.6) that

(2.31) *The injections $H_n(K_{p,q}, L_{p,q}) \to H_n(Z_n, Z_{n-1})$ $(p + q = n)$ form a representation of the latter group as a direct sum.*

The dual result in cohomology asserts

(2.32) *The injections $H^n(Z_n, Z_{n-1}) \to H^n(K_{p,q}, L_{p,q})$ $(p + q = n)$ form a representation of the former group as a direct product.* □

Therefore, there are monomorphisms

$$i_{p,q} : H^n(K_{p,q}, L_{p,q}) \to H^n(Z_n, Z_{n-1})$$

forming the representation, dual to that of (2.31), of the latter group as a direct sum. In fact, $i_{p,q}$ is the composite of the injection

$$H^n(Z_n, K'_{p,q}) \to H^n(Z_n, Z_{n-1})$$

with the inverse of the injection

$$H^n(Z_n, K'_{p,q}) \to H^n(K_{p,q}, L_{p,q}),$$

which is an isomorphism by the Excision Theorem ($K'_{p,q}$ is the union of the set $K_{r,s}$ for all $(r, s) \neq (p, q)$ with $r + s = n$).

Let $u \in \Gamma^p(X, A) = H^p(X_p, X_{p-1})$, $v \in \Gamma^q(Y, B)$. Their cross product in singular cohomology lies in the group

$$H^{p+q}(K_{p,q}, L_{p,q}) = H^{p+q}((X_p, X_{p-1}) \times (Y_p, Y_{p-1}))$$

The image of this cross-product in $\Gamma^{p+q}(Z, C)$ under the map $i_{p,q}$ will also be denoted by $u \times v$.

(2.33) Theorem *Let $u \in \Gamma^p(X, A)$, $v \in \Gamma^q(Y, B)$. Then, in the group $\Gamma^{p+q+1}(Z, C)$, the relation*

$$\delta(u \times v) = \delta u \times v + (-1)^p u \times \delta v$$

holds.

The proof is a matter of (sufficiently patient) diagram chasing. □

Thus, as in the case of homology, the cross product of cochains induces a pairing of cohomology groups, and the cross products so defined correspond, under the isomorphisms of Theorem (2.19), to the cross products in singular cohomology.

We conclude this section with a companion result to Theorem (2.29). Let $\mathbf{i}^* \in H^1(\mathbf{I}, \dot{\mathbf{I}}; \mathbf{Z})$ be the generator such that $\langle \mathbf{i}^*, \mathbf{i} \rangle = +1$. Then

(2.34) Theorem *The cross product with* $\mathbf{i}^*$ *induces an isomorphism*

$$\mathbf{i}^* \times : H^q(X, A; G) \approx H^{q+1}(X^*, A^*; G)$$

for any coefficient group G. □

3 Compression Theorems

In order to pave the way for the proof of the Cellular Approximation Theorem in §4, we introduce the concept of *n*-connected pair, and prove a series of results which are needed, not only for the aforementioned proof, but in many other contexts.

A pair (Y, B) (not necessarily a relative CW-complex) is said to be *n-connected* if and only if, for every relative CW-complex (X, A) with $\dim(X, A) \leq n$, any map $f: (X, A) \to (Y, B)$ is homotopic (rel. A) to a map of X into B. (Such a map is said to be *compressible*, and the homotopy a *compression*).

(3.1) Lemma *A necessary and sufficient condition that* (Y, B) *be n-connected is that every map* $f: (\Delta^q, \dot{\Delta}^q) \to (Y, B)$ *be compressible* $(q = 0, 1, \ldots, n)$.

Since $(\Delta^q, \dot{\Delta}^q)$ admits a CW-decomposition, the necessity is clear. The sufficiency is proved inductively. Accordingly, suppose that $q < n$ and $f(X_{q-1}) \subset B$. It suffices to show that $f \mid X_q : (X_q, X_{q-1}) \to (Y, B)$ is compressible. For the homotopy extension property for the pair (X, X_q) ensures that any compression of $f \mid X_q$ can be extended to a homotopy (rel. X_{q-1}) of f to a map g such that $g(X_q) \subset B$.

Now X_q is a q-cellular extension of X_{q-1}; let $h_\alpha : \Delta^q \to X_q$ be the characteristic maps. Since $q < n$, $f \circ h_\alpha : (\Delta^q, \dot{\Delta}^q) \to (Y, B)$ is compressible. If $g_\alpha : (\mathbf{I} \times \Delta^q, \mathbf{I} \times \dot{\Delta}^q) \to (Y, B)$ is a compression, it follows from Lemma (1.3) that there is a homotopy $g : (\mathbf{I} \times X_q, \mathbf{I} \times X_{q-1}) \to (Y, B)$ of $f \mid X_q$ such that $g \circ (1 \times h_\alpha) = g_\alpha$; and g is the desired compression. □

(3.2) Lemma *Let* (X, A) *be an* NDR-*pair, and let* $f: (\mathbf{I} \times X, \mathbf{I} \times A) \to (Y, B)$

be a homotopy of $f_0 : (X, A) \to (Y, B)$ *to a map* f_1 *such that* $f_1(X) \subset B$. *Then* f_0 *is compressible.*

Define $g : \mathbf{I} \times A \cup 0 \times X \to B$ by

$$g(s, a) = f(1 - s, a),$$
$$g(0, x) = f(1, x).$$

Since (X, A) is an NDR-pair, g has an extension $G : \mathbf{I} \times X \to B$.
Define

$$h : (0 \times \mathbf{I} \times X \cup \mathbf{I} \times \mathbf{I} \times A \cup \mathbf{I} \times \dot{\mathbf{I}} \times X, \mathbf{I} \times \mathbf{I} \times A \cup \mathbf{I} \times 1 \times X) \to (Y, B)$$

by

$$h(s, 0, x) = f_0(x),$$
$$h(s, 1, x) = G(s, x),$$
$$h(0, t, x) = f(t, x),$$
$$h(s, t, a) = f((1 - s)t, a).$$

Since $(\mathbf{I}, \dot{\mathbf{I}})$ and (X, A) are NDR-pairs, their product $(\mathbf{I} \times X, \dot{\mathbf{I}} \times X \cup \mathbf{I} \times A)$ is an NDR-pair. Hence h has an extension $H : \mathbf{I} \times \mathbf{I} \times X \to Y$. Let $f'(t, x) = h(1, t, x)$. Then $f' : \mathbf{I} \times X \to Y$, $f'(1 \times X) = h(1 \times 1 \times X) = G(1 \times X) \subset B$, $f'(t, a) = h(1, t, a) = f(0, a)$, and $f'(0, x) = h(1, 0, x) = f_0(x)$. Hence f' is a compression of f_0. □

(3.3) Lemma *Let* (Y, B) *be n-connected, and let* (X, A) *be a relative* CW-*complex with* $\dim(X, A) \le n$. *Then every map*

$$f: (0 \times X \cup \mathbf{I} \times A, 1 \times A) \to (Y, B)$$

has an extension

$$F : (\mathbf{I} \times X, 1 \times X) \to (Y, B).$$

Since (X, A) is an NDR-pair, f has an extension $f_1 : (\mathbf{I} \times X, 1 \times A) \to (Y, B)$. Since $\dim(X, A) \le n$, the map $f_1 | 1 \times X : (1 \times X, 1 \times A) \to (Y, B)$ is compressible. Hence there is a map

$$f_2 : (0 \times \mathbf{I} \times X \cup \mathbf{I} \times 1 \times X, 1 \times 1 \times X \cup \mathbf{I} \times 1 \times A) \to (Y, B)$$

such that

$$f_2(0, t, x) = f_1(t, x) \qquad ((t, x) \in \mathbf{I} \times X),$$
$$f_2(s, 1, a) = f(1, a) \qquad (s \in \mathbf{I}, a \in A).$$

Extend f_2 to a map

$$f_3 : (0 \times \mathbf{I} \times X \cup 1 \times \mathbf{I} \times X \cup \mathbf{I} \times 1 \times X \cup \mathbf{I} \times \mathbf{I} \times A, 1 \times \mathbf{I} \times X \cup \mathbf{I} \times \mathbf{I} \times A) \to (Y, B)$$

by setting

$$f_3(s, t, a) = \begin{cases} f(s+t, a) & (0 \le s \le 1-t,\ a \in A), \\ f(1, a) & (1-t \le s \le 1,\ a \in A), \end{cases}$$

$$f_3(1, t, x) = f_2(1, 1, x) \qquad ((t, x) \in \mathbf{I} \times X).$$

Since $(\mathbf{I}, \dot{\mathbf{I}})$ is an NDR-pair and $(\mathbf{I}, 1)$ a DR-pair, their product $(\mathbf{I} \times \mathbf{I}, \dot{\mathbf{I}} \times \mathbf{I} \cup \mathbf{I} \times 1)$ is a DR-pair; as (X, A) is also an NDR-pair, the product

$$(\mathbf{I} \times \mathbf{I} \times X, \dot{\mathbf{I}} \times \mathbf{I} \times X \cup \mathbf{I} \times 1 \times X \cup \mathbf{I} \times \mathbf{I} \times A),$$

of the latter two pairs is a DR-pair, and therefore f_3 has an extension

$$f_4 : (\mathbf{I} \times \mathbf{I} \times X, \mathbf{I} \times \mathbf{I} \times A \cup 1 \times \mathbf{I} \times X) \to (Y, B).$$

Define F by

$$F(s, x) = f_4(s, 0, x) \qquad ((s, x) \in \mathbf{I} \times X);$$

F is the desired extension. □

(3.4) Lemma *Suppose that (Y, B, B') is a triple such that both (Y, B) and (B, B') are n-connected. Then (Y, B') is n-connected.*

In fact, if (X, A) is a relative CW-complex with $\dim(X, A) \le n$ and $f : (X, A) \to (Y, B')$ is a map, then the n-connectedness of (Y, B) implies that f is homotopic (rel. A) to a map g such that $g(X) \subset B$. Again, the fact that (B, B') is n-connected implies that $g : (X, A) \to (B, B')$ is homotopic (rel. A) to a map h such that $h(X) \subset B'$. □

(3.5) Corollary *Let $\{Y_q\}$ be a filtration of a space Y such that each of the pairs (Y_{q+1}, Y_q) is n-connected $(q \ge 0)$. Then (Y, Y_0) is n-connected.*

For let $k \le n$, and let $f: (\Delta^k, \dot{\Delta}^k) \to (Y, Y_0)$ be a map. Since Δ^k is compact, there exists q such that $f(\Delta^k) \subset Y_q$. It follows by induction from Lemma (3.4) that the pair (Y_q, Y_0) is n-connected. Hence $f: (\Delta^k, \dot{\Delta}^k) \to (Y_q, Y_0)$ is compressible and therefore $f: (\Delta^k, \dot{\Delta}^k) \to (Y, Y_0)$ is also. □

(3.6) Lemma *The pair $(\Delta^n, \dot{\Delta}^n)$ is $(n-1)$-connected.*

Let $f: (\Delta^q, \dot{\Delta}^q) \to (\Delta^n, \dot{\Delta}^n)$, $q < n$. By the Simplicial Approximation Theorem, there is a subdivision (K, L) of $(\Delta^q, \dot{\Delta}^q)$ and a simplicial map $g : (K, L) \to (\Delta^n, \dot{\Delta}^n)$ such that g is homotopic to f. Since $q < n$, the image of g is contained in the $(n-1)$-skeleton $\dot{\Delta}^n$ of Δ^n. By Lemma (3.2), f is compressible. □

(3.7) Corollary *Let $b \in \operatorname{Int} \Delta^n$. Then the pair $(\Delta^n, \Delta^n - \{b\})$ is $(n-1)$-connected.*

For $\dot{\Delta}^n$ is a deformation retract of $\Delta^n - \{b\}$. □

(3.8) Corollary *The pair* (Int Δ^n, Int $\Delta^n - \{b\}$) *is* $(n-1)$*-connected.*

For if $f: (\Delta^q, \dot{\Delta}^q) \to (\text{Int } \Delta^n, \text{Int } \Delta^n - \{b\})$ and $q < n$, then there is a closed n-simplex Δ^n_* such that $b \in \text{Int } \Delta^n_*$, and $f(\Delta^q) \subset \Delta^n_* \subset \text{Int } \Delta^n$. Then $f: (\Delta^q, \dot{\Delta}^q) \to (\Delta^n_*, \Delta^n_* - \{b\})$ is compressible in Δ^n_* and therefore in Int Δ^n. □

(3.9) Theorem *Let X be an n-cellular extension of A. Then (X, A) is $(n-1)$-connected.*

Let $q < n$, $f: (\Delta^q, \dot{\Delta}^q) \to (X, A)$. By (1.1), there are finitely many cells $E^n_1, \ldots, E^n_r$ of (X, A) such that $f(\Delta^q) \subset A \cup \bigcup_{i=1}^r E^n_i = X'$. If (X', A) is $(n-1)$-connected, then f is compressible in X' and therefore in X. Hence we may assume $X' = X$. Let $X_k = A \cup \bigcup_{i=1}^k E^n_i$, so that $A = X_0 \subset X_1 \subset \cdots \subset X_r = X$, and, for each k, X_k is an n-cellular extension of X_{k-1} with just one n-cell E^n_k. If (X_k, X_{k-1}) is $(n-1)$-connected, so is (X, A), by Corollary (3.5). Therefore we may assume that (X, A) has exactly one cell E^n, with characteristic map $h: (\Delta^n, \dot{\Delta}^n) \to (X, A)$.

Let $\mathbf{b}_n$ be the barycenter of Δ^n, and let $U = \text{Int } E^n$, $V = A \cup h(\Delta^n - \{\mathbf{b}_n\})$. Then $\{U, V\}$ is an open covering of X, and A is a deformation retract of V. Let $q < n$, $f: (\Delta^q, \dot{\Delta}^q) \to (X, A)$. Let $\eta > 0$ be a Lebesgue number for the open covering $\{f^{-1}(U), f^{-1}(V)\}$ of Δ^q, and choose a simplicial subdivision K of Δ^q of mesh $< \eta$. Let L be the union of those simplices of K which are contained in $f^{-1}(U)$, M the union of those which are contained in $f^{-1}(V)$. Then L and M are subcomplexes of K, and $\dot{\Delta}^q \subset M$. Now $f|L: (L, L \cap M) \to (U, U - \{h(\mathbf{b}_n)\})$ and so $h^{-1} \circ (f|L): (L, L \cap M) \to (\text{Int } \Delta^n, \text{Int } \Delta^n - \{\mathbf{b}_n\})$. By Corollary (3.8), the latter pair is $(n-1)$-connected, and therefore $h^{-1} \circ (f|L)$ is compressible. Hence $f|L$ is compressible. Because $f(M) \subset V$, the map $f: (K, M) \to (X, V)$ is compressible; since $\dot{\Delta}^q \subset M$, $f: (\Delta^q, \dot{\Delta}^q) \to (X, V)$ is compressible. Since A is a deformation retract of V, $f: (\Delta^q, \dot{\Delta}^q) \to (X, A)$ is compressible. □

(3.10) Corollary *If (X, A) is a relative* CW*-complex, then (X_q, X_n) and (X, X_n) are n-connected for all $q \geq n$.*

By Theorem (3.9), the pair (X_q, X_{q-1}) is $(q-1)$-connected, and therefore n-connected, if $q > n$. The result now follows from Corollary (3.5). □

A pair (Y, B) is said to be ∞-connected if and only if it is n-connected for every positive integer n. For such a pair, every map $f: (X, A) \to (Y, B)$ of a *finite-dimensional* relative CW-complex is compressible. But more is true, viz:

(3.11) Theorem *If (Y, B) is ∞-connected and (X, A) is an arbitrary relative* CW-*complex, then every map $f: (X, A) \to (Y, B)$ is compressible.*

We construct a sequence of maps $f_n : (0 \times X_n \cup \mathbf{I} \times X_{n-1}, \mathbf{I} \times X_{n-1}) \to (Y, B)$ with the following properties:

(1) $f_n(0, x) = f(x)$ $(x \in K_n)$;
(2) f_{n+1} is an extension of f_n;
(3) $f_n(t, a) = f(a)$ $(t \in \mathbf{I}, a \in A)$.

First, define $f_{-1} : \mathbf{I} \times A \to B$ by (3). Assume that f_n has been defined and satisfies (1)–(3) for all $n \leq N$. By Lemma (3.3), the map f_N has an extension

$$f'_{N+1} : (\mathbf{I} \times X_N, 1 \times X_N) \to (Y, B).$$

Extend f'_{N+1} to $f_{N+1} : 0 \times X_{N+1} \cup \mathbf{I} \times X_N \to Y$ by

$$f_{N+1}(0, x) = f(x).$$

Then f_n satisfy (1)–(3) for all $n \leq N + 1$.
Let $g : \mathbf{I} \times X \to Y$ be the map such that

$$g | \mathbf{I} \times X_n = f_{n+1} | \mathbf{I} \times X_n;$$

because of (1)–(3), g is well-defined, $g(0, x) = f(x)$, and $g(t, a) = f(a)$ for $a \in A$. Since $g(1 \times X_n) = f_{n+1}(1 \times X_n) \subset B$, $g(1 \times X) \subset B$, and therefore g is a compression of f. □

(3.12) Corollary *If the relative* CW-*complex (X, A) is ∞-connected, then A is a deformation retract of X.*

For the identity map $1 : (X, A) \to (X, A)$ is compressible. □

We shall need a "relative" version of Theorem (3.11). A map $f: (X; A_1, A_2) \to (Y; B_1, B_2)$ is said to be *right compressible* if and only if the map $f_1 : (X, A_1) \to (Y, B_1)$ defined by f is homotopic (rel. A_2) to a map $g : (X, A_1) \to (B_2, B_1 \cap B_2)$.

(3.13) Theorem *Suppose that $(Y; B_1, B_2)$ is a triad such that the pairs $(Y, B_1 \cup B_2)$, $(B_1, B_1 \cap B_2)$, and $(B_1 \cup B_2, B_2)$ are ∞-connected. If $(X; A_1, A_2)$ is a triad such that the pairs $(A_1, A_1 \cap A_2)$ and $(X, A_1 \cup A_2)$ are relative* CW-*complexes, then every map $f: (X; A_1, A_2) \to (Y; B_1, B_2)$ is right compressible.*

The desired compression is carried out in three steps.
Step I. Since $(Y, B_1 \cup B_2)$ is ∞-connected, the map $f : (X, A_1 \cup A_2) \to (Y, B_1 \cup B_2)$ is compressible. Thus there is a homotopy $F_1 : I \times X \to Y$

such that

$$F_1(0, x) = f(x) \qquad (x \in X),$$
$$F_1(t, a) = f(a) \qquad (a \in A_1 \cup A_2),$$
$$F_1(1, x) \in B_1 \cup B_2.$$

The map f_1 defined by

$$f_1(x) = F_1(1, x)$$

sends $(X; A_1, A_2)$ into $(B_1 \cup B_2; B_1, B_2)$.

Step II. The map $f_1 | A_1 : (A_1, A_1 \cap A_2) \to (B_1, B_1 \cap B_2)$ is compressible, since $(B_1, B_1 \cap B_2)$ is ∞-connected. Thus there is a map $F_2' : \mathbf{I} \times A_1 \to B_1$ such that

$$F_2'(0, a_1) = f_1(a_1) = f(a_1) \qquad (a_1 \in A_1),$$
$$F_2'(t, a_0) = f_1(a_0) = f(a_0) \qquad (a_0 \in A_1 \cap A_2),$$
$$F_2'(1, a_1) \in B_1 \cap B_2 \qquad (a_1 \in A_1).$$

Extend F_2' to a map $F_2'' : 0 \times X \cup \mathbf{I} \times (A_1 \cup A_2) \to B_1 \cup B_2$ by

$$F_2''(0, x) = f_1(x) \qquad (x \in X),$$
$$F_2''(t, a_2) = f_1(a_2) = f(a_2) \qquad (a_2 \in A_2).$$

Since $(X, A_1 \cup A_2)$ is an NDR-pair, F_2'' can be extended to a map $F_2 : \mathbf{I} \times X \to B_1 \cup B_2$.

Define $f_2 : (X; A_1, A_2) \to (B_1 \cup B_2; B_1 \cap B_2, B_2)$ by

$$f_2(x) = F_2(1, x).$$

Step III. Since $(B_1 \cup B_2, \ B_2)$ is ∞-connected, the map $f_2 : (X, A_1 \cup A_2) \to (B_1 \cup B_2, B_2)$ is compressible. Hence there is a map $F_3 : \mathbf{I} \times X \to B_1 \cup B_2$ such that

$$F_3(0, x) = f_2(x) \qquad (x \in X),$$
$$F_3(t, a) = f_2(a) \qquad (a \in A_1 \cup A_2),$$
$$F_3(1, x) \in B_2 \qquad (x \in X).$$

Define $f_3 : \mathbf{I} \times X \to B_2$ by $f_3(x) = F_3(1, x)$.

The three homotopies F_1, F_2, F_3 can be put end to end to define a homotopy F of f to f_3. Examination of the properties of the F_i reveals that

$$F(\mathbf{I} \times A_1) \subset B_1,$$

$$F \text{ is stationary on } A_2,$$

$$f_3(X) \subset B_2;$$

hence $f_3(A_1) \subset B_1 \cap B_2$, and therefore F is a right compression of f. □

(3.14) Corollary *Let $(X; A_1, A_2)$ be a triad such that $(A_1, A_1 \cap A_2)$ and $(X, A_1 \cup A_2)$ are relative CW-complexes and each of the pairs $(A_1, A_1 \cap A_2)$, $(A_1 \cup A_2, A_2)$ and $(X, A_1 \cup A_2)$ is ∞-connected. Then the pair $(A_2, A_1 \cap A_2)$ is a deformation retract of the pair (X, A_1).*

For the identity map of $(X; A_1, A_2)$ is compressible. □

4 Cellular Maps

If $f: (X, A) \to (Y, B)$ is a continuous map, it is desirable to be able to calculate the induced homomorphism $f_* : H_q(X, A) \to H_q(Y, B)$. If (X, A) and (Y, B) are simplicial pairs, and f is a simplicial map, then f induces a chain map between the chain complexes of (X, A) and (Y, B), and the homomorphism of homology groups induced by the latter is just f_* (up to the identifications of the homology groups of the pairs in question with those of their chain complexes). If, however, f is not simplicial, the Simplicial Approximation Theorem assures us of the existence of a simplicial map homotopic to f, to which the above machinery can be applied. The use of the Simplicial Approximation Theorem, however, has the disadvantage, technical as well as aesthetic, that it necessitates the use of a subdivision of the original triangulation. And, while the process of (say) barycentric subdivision is nicely adapted to the simplicial theory, it becomes awkward and messy in the general case.

Suppose, then, that (X, A) and (Y, B) are relative CW-complexes. A map $f: (X, A) \to (Y, B)$ is said to be *cellular* if and only if $f(X_n) \subset Y_n$ for every n. Such a map induces homomorphisms

$$f_\# : H_n(X_n, X_{n-1}) \to H_n(Y_n, Y_{n-1}),$$
$$f_\natural : H_n(X_n, A) \to H_n(Y_n, B),$$

as well as

$$f_* : H_n(X, A) \to H_n(Y, B).$$

These map the diagram of (2.20) for (X, A) into that for (Y, B). We therefore have

(4.1) Theorem *If $f: (X, A) \to (Y, B)$ is a cellular map, then the homomorphisms $f_\# : H_n(X_n, X_{n-1}) \to H_n(Y_n, Y_{n-1})$ are the components of a chain map $\Gamma(f) = f_\# : \Gamma(X, A) \to \Gamma(Y, B)$ and the homomorphism of homology groups induced by $f_\#$ coincides, up to the natural isomorphism of Theorem* (2.19), *with $f_* : H_*(X, A) \to H_*(Y, B)$.* □

In categorical terms, this can be expressed as follows. Let $\mathscr{W}$ be the category whose objects are CW-complexes and whose morphisms are cellular maps, $\mathscr{C}_*$ the category of graded chain complexes and chain maps of degree 0, $\mathscr{A}_*$ the category of graded abelian groups (and homomorphisms of degree 0). One has homology functors $H : \mathscr{W} \to \mathscr{A}_*$, $H : \mathscr{C}_* \to \mathscr{A}_*$. Moreover, the construction Γ determines a functor $\Gamma : \mathscr{W} \to \mathscr{C}_*$, and the functors H, $H \circ \Gamma : \mathscr{W} \to \mathscr{A}_*$ are naturally equivalent.

We also have a notion of cellular homotopy. Let $f: (\mathbf{I} \times X, \mathbf{I} \times A) \to (Y, B)$ be a map, so that f is a homotopy between the maps $f_0, f_1 : (X, A) \to (Y, B)$ defined by

$$f_t(x) = f(t, x) \qquad (t = 0, 1; x \in X).$$

If f is cellular, then f_0 and f_1 are cellular, and we refer to f as a *cellular homotopy* between f_0 and f_1.

(4.2) Theorem *Let f be a cellular homotopy between the cellular maps f_0, $f_1 : (X, A) \to (Y, B)$. Then the chain maps*

$$\Gamma(f_i) : \Gamma(X, A) \to \Gamma(Y, B) \qquad (i = 0, 1)$$

are chain homotopic. □

The use of cellular maps often allows us to construct new CW-complexes out of old ones. The following result is often useful.

(4.3) Theorem *Let X, Y be* CW*-complexes, A a subcomplex of X, $f: A \to Y$ a cellular map. Then the adjunction space $X \cup_f Y$ is a* CW*-complex having Y as a subcomplex.* (Cf. Lemma (1.7) and the remarks preceding it).

The proclusion $h: (X + Y, A + Y) \to (X \cup_f Y, Y)$ being a relative homeomorphism, it follows that the latter pair is a relative CW-complex with skeleta $W_n^* = X_n \cup_{f_n} Y$, where $f_n = f \mid A_n : A_n \to Y$. It follows from Lemma (1.13) that $W = X \cup_f Y$ is a CW-complex with skeleta $(W_n^* - Y) \cup Y_n = h(X_n - A_n) \cup Y_n = X_n \cup_{f_n} Y_n$, and that Y is a subcomplex of W. □

(4.4) Corollary *The mapping cylinder $\mathbf{I}_f$, and the mapping cone $\mathbf{T}_f$, of a cellular map $f: X \to Y$, are* CW*-complexes; X and Y are disjoint subcomplexes of $\mathbf{I}_f$, and Y is a subcomplex of $\mathbf{T}_f$.* □

The main goal of this section is to prove

(4.5) Theorem (Cellular Approximation Theorem). *Let (X, A), (Y, B) be relative* CW*-complexes, and let $f: (X, A) \to (Y, B)$ be a continuous map. Then f is homotopic* (rel. A) *to a cellular map h.*

We shall construct, inductively, a sequence of maps

$$g_p : \mathbf{I} \times X_p \to Y$$

with the following properties:

(1) $g_p(0, x) = f(x)$ $\quad (x \in X_p)$,
(2) $g_p(t, a) = f(a)$ $\quad ((t, a) \in \mathbf{I} \times A)$,
(3) $g_p | \mathbf{I} \times X_{p-1} = g_{p-1}$,
(4) $g_p(1 \times X_p) \subset Y_p$.

Once this has been done, the function $g : \mathbf{I} \times X \to Y$ defined by

$$g | \mathbf{I} \times X_p = g_p$$

is well-defined, because of (3), and continuous because $\mathbf{I} \times X$ has the weak topology with respect to the subsets $\mathbf{I} \times X_p$; and g is a homotopy (rel. A) between f and the map h defined by

$$h(x) = g(1, x).$$

The map h is cellular, by (4).

It remains to define the g_p. We begin by setting $g_{-1}(t, x) = f(x)$ for $x \in X_{-1} = A$, as required by (2). Suppose that g_p has been defined for $p < n$ and satisfies (1)–(4). By Lemma (3.3), the map $g'_n : (0 \times X_n \cup \mathbf{I} \times X_{n-1}, 1 \times X_{n-1}) \to (Y, Y_{n-1}) \subset (Y, Y_n)$ defined by

$$g'_n(0, x) = f(x),$$
$$g'_n(t, x) = g_{n-1}(t, x) \quad ((t, x) \in \mathbf{I} \times X_{n-1}),$$

has an extension

$$g_n : (\mathbf{I} \times X_n, 1 \times X_n) \to (Y, Y_n),$$

and it is evident that (1)–(4) hold with $p = n$. □

(4.6) Corollary *Let (X, A), (Y, B) be relative* CW*-complexes and let (X', A') be a subcomplex of (X, A). Let $f: (X, A) \to (Y, B)$ be a map such that $f | X' : (X', A') \to (Y, B)$ is cellular. Then f is homotopic* (rel. X') *to a cellular map $h : (X, A) \to (Y, B)$.*

Just apply the Cellular Approximation Theorem to the relative CW-complex $(X, X' \cup A)$. □

(4.7) Corollary *Let $f_0, f_1 : (X, A) \to (Y, B)$ be homotopic cellular maps. Then there is a cellular homotopy $h : (\mathbf{I} \times X, \mathbf{I} \times A) \to (Y, B)$ of f_0 to f_1.*

Let $f: (\mathbf{I} \times X, \mathbf{I} \times A) \to (Y, B)$ be any homotopy of f_0 to f_1. Apply Corollary (4.6) to the relative CW-complex $(\mathbf{I} \times X, \mathbf{I} \times A)$ and its subcomplex $(\dot{\mathbf{I}} \times X \cup \mathbf{I} \times A, \mathbf{I} \times A)$. □

5 Local Calculations

We have shown that, if (X, A) is a relative CW-complex, then its homology (and cohomology) is that of the free chain complex $\Gamma(X, A)$ for which

$$\Gamma_p(X, A) = H_p(X_p, X_{p-1}).$$

If $h_\alpha : (\Delta^p, \dot{\Delta}^p) \to (E_\alpha, \dot{E}_\alpha)$ are characteristic maps for the p-cells of (X, A) $(\alpha \in J_p)$, then the homology classes $e_\alpha \in H_p(X_p, X_{p-1})$ of the singular relative cycles h_α form a basis for $\Gamma_p(X, A)$. Thus, for each $\alpha \in J_p$,

$$\partial e_\alpha = \sum_{\beta \in J_{p-1}} [e_\alpha : e_\beta] e_\beta,$$

where $[e_\alpha : e_\beta]$ are integers, almost all zero; they are called the *incidence numbers* for the pairs (E_α, E_β). In this section we shall show how to calculate them.

Similarly, suppose that $f: (X, A) \to (Y, B)$ is a cellular map between relative CW-complexes. Then f maps the pair (X_p, X_{p-1}) into the pair (Y_p, Y_{p-1}), and the homomorphisms of $H_p(X_p, X_{p-1})$ into $H_p(Y_p, Y_{p-1})$ are the components of a chain map $f_\# : \Gamma(X, A) \to \Gamma(Y, B)$. If $\{E_\beta \mid \beta \in K_p\}$ are the p-cells of (Y, B), then, for each $\alpha \in J_p$, there are integers $f_{\alpha\beta}$, almost all zero, such that

$$f_\#(e_\alpha) = \sum_{\beta \in K_p} f_{\alpha\beta} e_\beta.$$

We shall also show in this section how to calculate the $f_{\alpha\beta}$.

Let us consider the second problem first. If E_α is a p-cell of (X, A), then $E_\alpha / \dot{E}_\alpha$ is a p-sphere, oriented by the image s_α of e_α under the homomorphism induced by the collapsing map. Similarly, if E_β is a p-cell of (Y, B) and $E_\beta^* = \overline{Y_p - E_\beta}$, then Y_p / E_β^* is homeomorphic with the p-sphere $E_\beta / \dot{E}_\beta$; the latter is oriented by the image s_β of e_β. The composite map

$$(E_\alpha, \dot{E}_\alpha) \subsetneq (X_p, X_{p-1}) \xrightarrow{\quad f \quad} (Y_p, Y_{p-1}) \subsetneq (Y_p, E_\beta^*)$$

induces a map

$$\bar{f}: E_\alpha / \dot{E}_\alpha \to E_\beta / \dot{E}_\beta.$$

We have

$$\bar{f}_*(s_\alpha) = c_{\alpha\beta} s_\beta,$$

where $c_{\alpha\beta}$ is an integer. Note that $c_{\alpha\beta} = 0$ if the image of $\bar{f}$ is a proper subset of the sphere $E_\beta / \dot{E}_\beta$, i.e., $f(E_\alpha) \not\supset E_\beta$. As the compact set $f(E_\alpha)$ is contained in the union of Y_{p-1} with finitely many cells E_β, almost all the integers $c_{\alpha\beta}$ are zero.

(5.1) Theorem *The integers $f_{\alpha\beta}$ and $c_{\alpha\beta}$ coincide.*

For there is a commutative diagram

$$\begin{array}{ccccccc}
H_p(E_\alpha, \dot{E}_\alpha) & \xrightarrow{i_\alpha} & H_p(X_p, X_{p-1}) & \xrightarrow{f_\#} & H_p(Y_p, Y_{p-1}) & \xrightarrow{j_\beta} & H_p(Y_p, E_\beta^*) \\
\Big\downarrow q_\alpha & & & & & & \Big\downarrow q_\beta \\
H_p(E_\alpha/\dot{E}_\alpha) & & & \xrightarrow[\bar{f}_*]{} & & & H_p(E_\beta/\dot{E}_\beta)
\end{array}$$

where q_α and q_β are induced by the collapsing maps. By the General Direct Sum Theorem (2.7),

$$j_\beta f_\# i_\alpha(e_\alpha) = f_{\alpha\beta} k_\beta(e_\beta),$$

where $k_\beta : H_p(E_\beta, \dot{E}_\beta) \to H_p(Y_p, Y_{p-1})$ is the injection. But

$$\begin{aligned} c_{\alpha\beta} s_\beta &= \bar{f}_*(s_\alpha) = \bar{f}_* q_\alpha(e_\alpha) \\ &= q_\beta j_\beta f_\# i_\alpha(e_\alpha) = f_{\alpha\beta} q_\beta k_\beta(e_\beta) = f_{\alpha\beta} s_\beta; \end{aligned}$$

since s_β has infinite order, $c_{\alpha\beta} = f_{\alpha\beta}$. □

Let us return to the first problem. This time there is a commutative diagram

(5.2)

$$\begin{array}{ccccccc}
H_p(\Delta^p, \dot{\Delta}^p) & \xrightarrow{h_{\alpha *}} & H_p(E_\alpha, \dot{E}_\alpha) & \xrightarrow{i_\alpha} & H_p(X_p, X_{p-1}) & & \\
\Big\downarrow \partial_1 & & \Big\downarrow \partial_2 & & \Big\downarrow j_\beta \circ \partial & & \\
H_{p-1}(\dot{\Delta}^p) & \xrightarrow[h'_\alpha]{} & H_{p-1}(\dot{E}_\alpha) & \xrightarrow[i'_\alpha]{} & H_{p-1}(X_{p-1}, E_\beta^*) & \xrightarrow[q_\beta]{} & H_{p-1}(E_\beta/\dot{E}_\beta)
\end{array}$$

and the bottom line is induced by a map of the oriented $(p-1)$-sphere $\dot{\Delta}^p$ into the oriented sphere $E_\beta/\dot{E}_\beta$; let $a_{\alpha\beta}$ be the degree of this map. Again, $a_{\alpha\beta} = 0$ unless $\dot{E}_\alpha \supset E_\beta$, and this can be true for only a finite number of β.

Again we appeal to the General Direct Sum Theorem to prove

(5.3) Theorem *The integers $a_{\alpha\beta}$ and $[e_\alpha : e_\beta]$ coincide.*

In fact, by the said Theorem,

$$j_\beta \partial i_\alpha(e_\alpha) = [e_\alpha : e_\beta] k_\beta(e_\beta),$$

and our result is obtained by chasing the diagram (5.2). □

6 Regular Cell Complexes

While CW-complexes are adapted very well to homotopy theory, there are occasional problems caused by the fact that the cells may not be homeomorphic with Δ^n. Consider, for example, the problem of finding a chain map $K \to K \times K$ approximating the diagonal map. Of course, the existence of such a map is guaranteed by the Cellular Approximation Theorem. But if one wants a map with good local properties (e.g., the image of a cell E of K should be contained in $E \times E$), one is forced into a cell-by-cell construction—the image of ∂E is a cycle z lying in $E \times E$, and if the latter space is acyclic, z bounds a chain which we may define to be the image of E (cf. the construction of Steenrod reduced powers in [St$_2$]).

For this reason it seems desirable to introduce a class of complexes lying between the simplicial and the CW-complexes. A CW-complex K is called a *regular cell complex* if and only if

(1) each n-cell of K is homeomorphic with Δ^n;
(2) if E is an n-cell of K, then $\dot{E}$ is the union of finitely many $(n-1)$-cells of K.

Remark. Condition (2) is actually superfluous, and is merely included for convenience. To see that this is the case, as well as to verify the remaining statements made without proof in this section, see [C–F] (especially p. 229 ff.).

Let K be a regular cell complex. An $(n-1)$-cell contained in the boundary of an n-cell E is said to be an $(n-1)$-face of E. More generally, if E and E' are cells of K, E' is said to be a k-face of E if and only if there is a sequence $E' = E_k \subset E_{k+1} \subset \cdots \subset E_n = E$ of cells of K such that E_p is a p-face of E_{p+1} for each p.

(6.1) *Let E be a cell of the regular cell complex K. Then the sets $E \cap K_n$ form a CW-decomposition of E as a subcomplex of K. Similarly, the sets $\dot{E} \cap K_n$ form a CW-decomposition of $\dot{E}$ as a subcomplex of K. Each of these subcomplexes is a regular cell complex.* □

(6.2) *Let K be a regular cell decomposition of* $\mathbf{S}^n$. *Then*

(1) *every cell of K is a face of an n-cell of K;*
(2) *every $(n-1)$-cell of K is a face of exactly two n-cells of K;*
(3) *if E and E' are n-cells of K, there is a sequence $E = E_0, E_1, \ldots, E_k = E'$ of n-cells of K such that, for each i, E_i and E_{i+1} have an $(n-1)$-face in common.*

To prove (1), let E be a principal cell (i.e., one which is not a face of any other cell). If $x \in \operatorname{Int} E$ then the local homology group $H_k(E|x) =$

$H_k(E, E - \{x\})$ is non-trivial ($k = \dim E$). But Int E is an open subset of $\mathbf{S}^n$ and therefore the injection $H_k(E|x) \to H_k(\mathbf{S}^n|x)$ is an isomorphism. Since $\mathbf{S}^n$ is an n-manifold, the latter group is non-zero only if $k = n$.

To prove (2), let E be an $(n-1)$-cell which is a face of exactly k n-cells, and let $x \in \operatorname{Int} E$. Then $H_n(\mathbf{S}^n|x)$ is a free abelian group of rank $k-1$. Again, since $\mathbf{S}^n$ is an n-manifold, we must have $k = 2$.

Property (3) follows from the fact that $\mathbf{S}^n$ cannot be separated by any closed set of dimension less than $n - 1$. □

Thus, the above three properties express the fact that $\mathbf{S}^n$ is an n-dimensional (pseudo-) manifold.

Let E be an n-cell of K, F an $(n-1)$-face of E. Let F' be the closure of $\dot{E} - F$, so that $\dot{E} = F \cup F'$, $F \cap F' = \dot{F}$. Then $(F, \dot{F})$ is an NDR-pair, and, if $x \in \operatorname{Int} F$, then $(F - \{x\}, \dot{F})$ is a DR-pair. Hence $(\dot{E}, F')$ is an NDR-pair and $(\dot{E} - \{x\}, F')$ a DR-pair. But $\dot{E} - \{x\}$ is contractible, and therefore F' is contractible. It follows that (E, F') is a DR-pair. Thus the boundary operator

$$\partial : H_n(E, \dot{E}) \to H_{n-1}(\dot{E}, F')$$

is an isomorphism. Moreover, the injection

$$k : H_{n-1}(F, \dot{F}) \to H_{n-1}(\dot{E}, F')$$

is an isomorphism, by the Excision Theorem. Thus

$$\sigma(E, F) = k^{-1} \circ \partial : H_n(E, \dot{E}) \to H_{n-1}(F, \dot{F})$$

is an isomorphism, called the *incidence isomorphism* of the pair (E, F).

An *orientation* of the n-cell E is a generator e of the infinite cyclic group $H_n(E, \dot{E})$; the pair (E, e) is called an *oriented n-cell*. An *orientation* of the CW-complex K is a function assigning to each cell E of K an orientation of E; and K is said to be *oriented* when a specific orientation has been chosen. The orientation assigned to a cell will be called the *preferred orientation*. We agree that the preferred orientation of a vertex E is the homology class of the point E.

Let E, F be cells of the regular cell-complex K, of dimensions n, $n-1$, respectively, and let e, f be orientations of these cells. If $F \not\subset E$, let $[e:f] = 0$. If $F \subset E$, the incidence isomorphism carries e into $[e:f]f$, where $[e:f] = \pm 1$. The number $[e:f]$ is called the *incidence number* of the oriented cells, e, f.

(6.3) Theorem *Let K be an oriented regular cell-complex. Let E be an n-cell of K with preferred orientation e, and let F_α $(\alpha \in J_{p-1})$ be the $(n-1)$-cells of K with preferred orientations f_α. Then, in the chain complex $\Gamma(K)$, the relation*

$$\partial e = \sum_\alpha [e : f_\alpha] f_\alpha$$

holds.

Let $F_\alpha^* = K_{n-2} \cup \bigcup_{\beta \neq \alpha} F_\beta$. Consider the commutative diagram

$$
(6.4)\qquad
\begin{array}{ccccc}
H_n(E, \dot{E}) & \xrightarrow{\;i\;} & H_n(K_n, K_{n-1}) & & \\
 & & \big\downarrow \partial_1 & \searrow^{\partial_2} & \\
H_{n-1}(F_\alpha, \dot{F}_\alpha) & \xrightarrow[i_\alpha]{} & H_{n-1}(K_{n-1}, K_{n-2}) & \xrightarrow[j_\alpha]{} & H_{n-1}(K_{n-1}, F_\alpha^*)
\end{array}
$$

and note that the injection $k_\alpha = j_\alpha \circ i_\alpha$ is an isomorphism. By (2.7), the i_α represent $\Gamma_{n-1}(K) = H_{n-1}(K_{n-1}, K_{n-2})$ as a direct sum; the dual representation as a weak direct product is given by $p_\alpha = k_\alpha^{-1} \circ j_\alpha$. Thus we must prove that $p_\alpha\, \partial_1\, i(e) = 0$ if $F_\alpha \not\subset E$, while $p_\alpha\, \partial_1\, i(e) = [e : f_\alpha] f_\alpha$, i.e., $p_\alpha \circ \partial_1 \circ i$ is the incidence isomorphism, if $F_\alpha \subset E$.

Suppose $F_\alpha \not\subset E$; then $\dot{E} \subset F_\alpha^*$, and the diagram

$$
\begin{array}{ccccc}
H_n(E, \dot{E}) & \xrightarrow{\;i\;} & H_n(K_n, K_{n-1}) & & \\
\big\downarrow & & \big\downarrow \partial_1 & \searrow^{\partial_2} & \\
H_{n-1}(\dot{E}) & \longrightarrow & H_{n-1}(K_{n-1}, K_{n-2}) & & \\
\big\downarrow & & & \searrow^{j_\alpha} & \\
0 = H_{n-1}(F_\alpha^*, F_\alpha^*) & & \longrightarrow & & H_{n-1}(K_{n-1}, F_\alpha^*)
\end{array}
$$

is commutative, and therefore $0 = \partial_2 \circ i = j_\alpha \circ \partial_1 \circ i$, which implies that $p_\alpha \circ \partial \circ i = 0$.

On the other hand, if $F_\alpha \subset E$, then $F'_\alpha = \dot{E} \cap F_\alpha^*$ and there is a commutative diagram

$$
(6.5)\qquad
\begin{array}{ccc}
H_n(E, \dot{E}) & \xrightarrow{\;i\;} & H_n(K_n, K_{n-1}) \\
\big\downarrow \partial & & \big\downarrow \partial_2 \\
H_{n-1}(\dot{E}, \dot{E} \cap F_\alpha^*) & \xrightarrow{\;k_2\;} & H_{n-1}(K_{n-1}, F_\alpha^*) \\
\big\uparrow k & \nearrow_{k_\alpha} & \\
H_{n-1}(F_\alpha, \dot{F}_\alpha) & &
\end{array}
$$

while k_2 and $k_\alpha = k_2 \circ k$ are isomorphisms, by the Excision Theorem. By commutativity of the diagrams (6.4) and (6.5),

$$\begin{aligned} p_\alpha \circ \partial_1 \circ i = k_\alpha^{-1} \circ j_\alpha \circ \partial_1 \circ i = k_\alpha^{-1} \circ \partial_2 \circ i \\ = k_\alpha^{-1} \circ k_2 \circ \partial = k^{-1} \circ \partial \end{aligned}$$

is the incidence isomorphism $\sigma(E, F_\alpha)$. □

Applying the Universal Coefficient Theorem, we obtain the analogous result for cohomology.

(6.3*) Theorem *The cohomology operator* $\delta : \Gamma^{p-1}(K; G) \to \Gamma^p(K; G)$ *is given by*

$$\delta c(e) = (-1)^{p-1} \sum_\alpha [e : f_\alpha] c(f_\alpha)$$

for any oriented p-cell e of K. □

An *incidence system* on the regular cell complex K is a function which assigns to each ordered pair (E, F), (E a p-cell of K, F a $(p-1)$-cell of K) a number $[E : F] = 0$ or ± 1, in such a way that

(1) $[E : F] = 0$ if $F \not\subset E$;
(2) $[E : F] = \pm 1$ if $F \subset E$;
(3) let G be a $(q-2)$-dimensional face of the q-cell E ($q \geq 2$), and let F_1, F_2 be the $(q-1)$-dimensional faces of E which contain G. Then
$$[E : F_1][F_1 : G] + [E : F_2][F_2 : G] = 0;$$
(4) let E be a 1-cell of K, F_1 and F_2 its 0-dimensional faces. Then
$$[E : F_1] + [E : F_2] = 0.$$

Let K be oriented, and if E, F are faces of dimensions p, $p-1$, respectively, let $[E : F] = [e : f]$, where e and f are the preferred orientations.

(6.6) Theorem *The function* [:] *so defined is an incidence system on K.*

We have seen that the first two conditions are fulfilled. The other two follow immediately from

(6.7) Theorem *Let E be an n-cell, and let* F_1 *and* F_2 *be* $(n-1)$*-faces of E which have a common* $(n-2)$*-face G. Then*

(1) *if* $n \geq 2$,
$$\sigma(F_1, G) \circ \sigma(E, F_1) + \sigma(F_2, G) \circ \sigma(E, F_2) = 0;$$
(2) *if* $n = 1$, *P is a space with just one point, and* $p_i : F_i \to P$ *is the unique map, then*
$$H_0(p_1) \circ \sigma(E, F_1) + H_0(p_2) \circ \sigma(E, F_2) = 0.$$

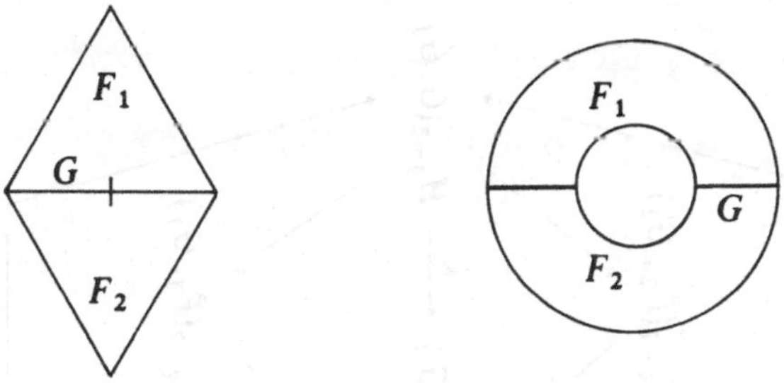

Figure 2.2

Let $F_0 = F_1 \cup F_2$, $F_i' = \overline{\dot{E} - F_i}$, $F_0' = \dot{E} - \{\text{Int } F_1 \cup \text{Int } F_2 \cup \text{Int } G\}$, $G_i' = \overline{\dot{F}_i - G}$, $G' = \overline{\dot{F}_1 \cup \dot{F}_2 - G}$ $(i = 1, 2)$ (cf. Figure 2.2). Then there is a commutative diagram (Figure 2.3), in which the homomorphisms labelled "k" are isomorphisms, by the Excision Theorem. Figure 2.3 can be consolidated to a simpler diagram (Figure 2.4), which is readily verified to satisfy the hypotheses of the Hexagonal Lemma (note, for example, that l_1 is the composite of the injection $H_{q-1}(F_1 \cup \dot{F}_2, \dot{F}_1 \cup \dot{F}_2) \to H_{q-1}(F_0, \dot{F}_1 \cup \dot{F}_2)$ with the isomorphic injection $H_{q-1}(F_1, \dot{F}_1) \to H_{q-1}(F_1 \cup \dot{F}_2, \dot{F}_1 \cup \dot{F}_2)$). But $k_4^{-1} \circ (k_2^{-1} \circ \partial_2) = \sigma(E, F_2)$ and $k_3^{-1} \circ (k_1^{-1} \circ \partial_1) = \sigma(E, F_1)$, and the conclusion of the Hexagonal Lemma is Property (1).

The proof of Property (2) is a simpler version of that of Property (1), and is left to the reader. □

We have seen that each orientation of K determines an incidence system. The converse is also true, viz.:

(6.8) Theorem *Let $[\ :\ ]$ be an incidence system on the regular cell complex K. Then K can be oriented so that, if E and F are cells of dimensions q, $q - 1$, respectively, and if e, f are their preferred orientations, then $[E : F] = [e, f]$.*

Because of Theorem (6.6) it suffices to prove the following statement:

(6.9) *Let $[\ :\ ]$, $[\ :\ ]'$ be incidence systems on K. Then there are functions ε_p, assigning the value ± 1 to each p-cell of K, such that, for every E, F,*

(6.10) $$[E : F]' = \varepsilon_p(E)\varepsilon_{p-1}(F)[E : F].$$

Let $\varepsilon_0(E) = 1$ for every 0-cell E. Suppose that ε_p has been defined and satisfies (6.9) for all $p \leq n$. Let E be an $(n + 1)$-cell of K. By (1) of (6.2), E has a face F of dimension n. Let

$$\eta_{n+1}(E, F) = \varepsilon_n(F)[E : F]'[E : F].$$

Let F' be another n-face of E; we shall show that $\eta_{n+1}(E, F') = \eta_{n+1}(E, F)$.

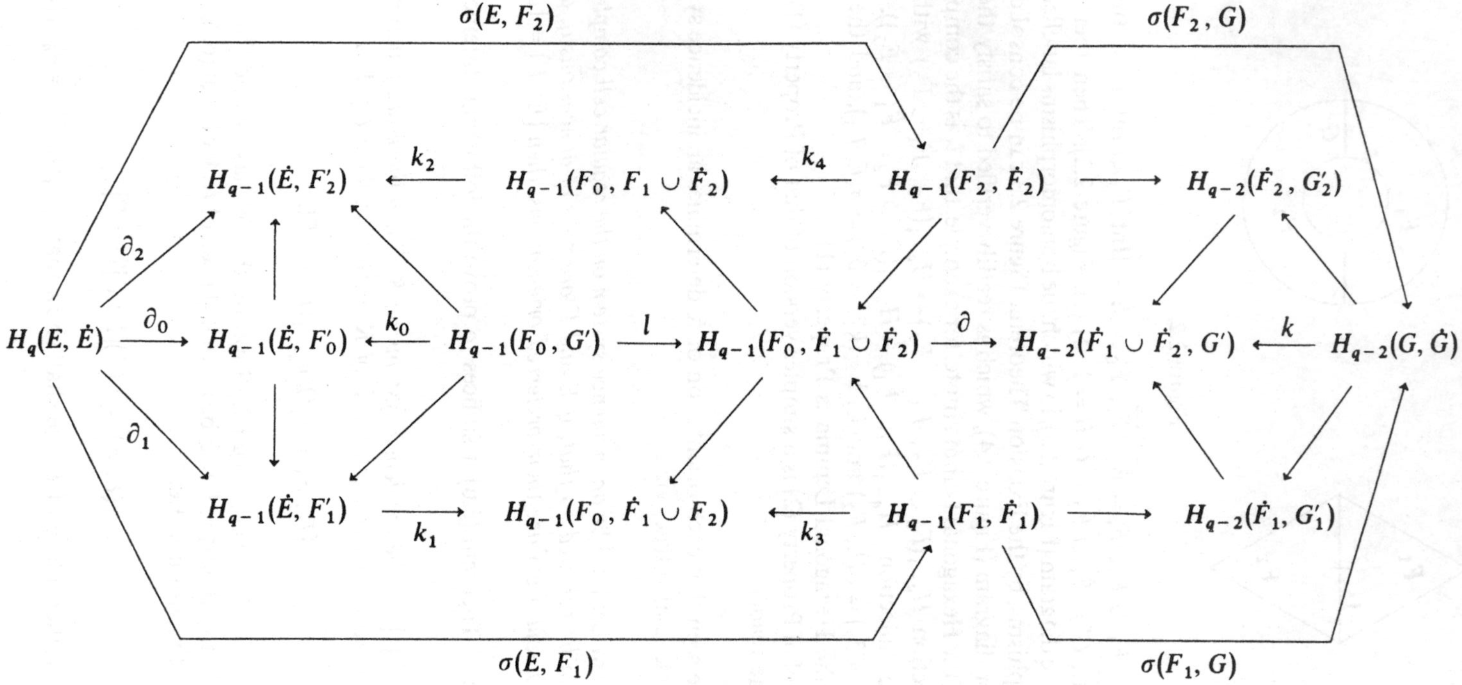
σ(E, F₂)
σ(F₂, G)
Hq(E, Ė)
∂₀
∂₁
∂₂
Hq−1(Ė, F′₂)
Hq−1(Ė, F′₀)
Hq−1(Ė, F′₁)
k₂
k₀
k₁
Hq−1(F₀, F₁ ∪ Ḟ₂)
Hq−1(F₀, G′)
Hq−1(F₀, Ḟ₁ ∪ F₂)
l
k₄
k₃
Hq−1(F₀, Ḟ₁ ∪ Ḟ₂)
Hq−1(F₂, Ḟ₂)
Hq−1(F₁, Ḟ₁)
∂
Hq−2(Ḟ₂, G′₂)
Hq−2(Ḟ₁ ∪ Ḟ₂, G′)
Hq−2(Ḟ₁, G′₁)
k
Hq−2(G, Ġ)
σ(E, F₁)
σ(F₁, G)

Figure 2.3

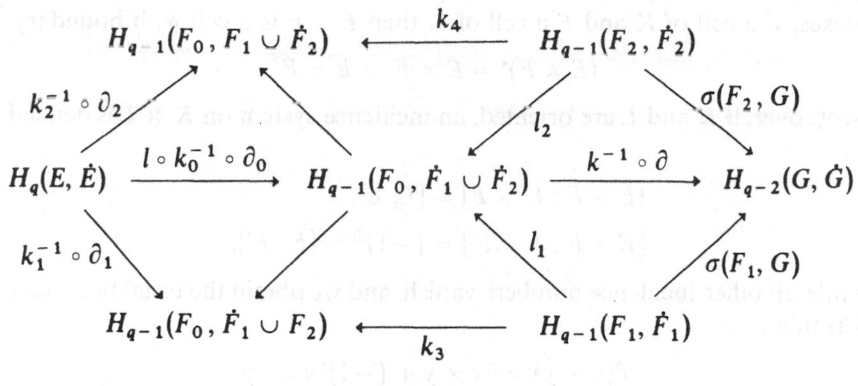

Figure 2.4

Because of (3) of (6.2), we may assume that F and F' have an $(n-1)$-face G in common. Then

$$\eta_{n+1}(E, F') = \varepsilon_n(F')[E : F']'[E : F'].$$

By Property (3) for the incidence system [:], we have

$$[E : F'] = -[F' : G][E : F][F : G]$$

and similarly,

$$[E : F']' = -[F' : G]'[E : F]'[F : G]'.$$

By induction hypothesis,

$$[F' : G]' = \varepsilon_n(F')\varepsilon_{n-1}(G)[F' : G]$$
$$[F : G]' = \varepsilon_n(F)\varepsilon_{n-1}(G)[F : G],$$

so that

$$\begin{aligned}[E : F']' &= -[E : F]'\varepsilon_n(F)\varepsilon_n(F')[F' : G][F : G] \\ &= \varepsilon_n(F)\varepsilon_n(F')[E : F]'[E : F'][E : F].\end{aligned}$$

Thus

$$\eta_{n+1}(E, F') = \varepsilon_n(F)[E : F]'[E : F] = \eta_{n+1}(E, F).$$

Define $\varepsilon_{n+1}(E)$ to be the common value of $\eta_{n+1}(E, F)$ for all n-faces F of E. Then (6.10) holds whenever F is a face of E. If F is not a face of E, then $[E : F] = [E : F]' = 0$ by Property (1). Therefore (6.10) holds in all cases.

The above argument needs to be modified slightly, using Property (4) instead of Property (3), in the case $n = 1$. The details are left to the reader. □

One advantage of regular cell complexes over simplicial complexes is that they behave well under products. Indeed, if K and L are regular cell complexes, E a cell of K and F a cell of L, then $E \times F$ is a cell with boundary

$$(E \times F)^{\bullet} = \dot{E} \times F \cup E \times \dot{F}.$$

Moreover, if K and L are oriented, an incidence system on $K \times L$ is defined by

$$[E \times F : E' \times F] = [E, \dot{E}],$$
$$[E \times F : E \times F'] = (-1)^{\dim E}[F : F'],$$

while all other incidence numbers vanish, and we obtain the usual boundary formula

$$\partial(x \times y) = \partial x \times y + (-1)^p x \times \partial y$$

for $x \in \Gamma_p(K)$, $y \in \Gamma_q(L)$.

7 Products and the Cohomology Ring

In §2 we studied cross-products in the homology and cohomology of a relative CW-complex, and showed how they could be expressed in terms of the structure of these complexes. Now these are external products, and one has corresponding internal products, defined with the aid of the diagonal map. Specifically, if X is a space, $u \in H^p(X; G)$, $v \in H^q(X; H)$, and if a pairing $G \otimes H \to K$ is given, the cup product $u \smile v \in H^{p+q}(X; K)$ is defined by

$$u \smile v = \Delta^*(u \times v),$$

where $\Delta^* : H^{p+q}(X \times X; K) \to H^{p+q}(X; K)$ is the homomorphism induced by the diagonal map. More generally, if $(X; A, B)$ is a proper triad, $u \in H^p(X, A; G)$, $v \in H^q(X, B; H)$, then $u \times v \in H^{p+q}((X, A) \times (X, B); K)$ and

$$\Delta : (X, A \cup B) \to (X \times X, A \times X \cup X \times B)$$

so that we may define

$$u \smile v = \Delta^*(u \times v) \in H^{p+q}(X, A \cup B; K).$$

Suppose, for example, that X is a CW-complex. Then we can calculate $u \times v$ in terms of the structure of X as a complex. In principle, then, we can calculate $u \smile v$, provided that we can calculate Δ^*. If the diagonal map $\Delta : X \to X \times X$ were cellular, this could be done. But Δ is almost never cellular; therefore one must first use the Cellular Approximation Theorem to find a cellular map $D : X \to X \times X$ homotopic to Δ, and then use the cochain map induced by D. When X is a simplicial complex, there is an explicit approximation D available, and this gives the Alexander–Čech–Whitney formula for calculating cup products. Unfortunately, no such simple procedure works for a general CW-complex, and this renders the calculation of cup products difficult. In this section we shall use certain devices (mostly Poincaré duality) to calculate the cohomology rings of some useful spaces.

EXAMPLE 1 (Real Projective Space). We give a CW-decomposition of the infinite-dimensional sphere $\mathbf{S}^\infty$ which induces a CW-decomposition of $\mathbf{S}^n$ for every n. Let $\tau : \mathbf{S}^\infty \to \mathbf{S}^\infty$ be the antipodal map, so that $\tau(x) = -x$ for all $x \in \mathbf{S}^n$.

Let $\mathbf{E}_+^n = \{x \in \mathbf{S}^n \mid x_n \geq 0\}$, $\mathbf{E}_-^n = \tau\mathbf{E}_+^n = \{x \in \mathbf{S}^n \mid x_n \leq 0\}$. Then $\mathbf{E}_+^n \cap \mathbf{E}_-^n = \mathbf{S}^{n-1} = \mathbf{E}_+^{n-1} \cup \mathbf{E}_-^{n-1}$, the cells $\{\mathbf{E}_+^n, \mathbf{E}_-^n\}$ give a CW-decomposition of $\mathbf{S}^\infty$, and the map τ is cellular. We may regard the chain groups as modules over the group ring of the cyclic group of order two generated by τ. Since τ maps $(\mathbf{E}_+^n, \dot{\mathbf{E}}_+)$ homeomorphically upon $(\mathbf{E}_-^n, \dot{\mathbf{E}}_-^n)$, it follows that if e^n is an orientation of $\mathbf{E}_+^n$, then τe^n is an orientation of $\mathbf{E}_-^n$.

We claim: *Orientations e^n of the cells $\mathbf{E}_+^n$ can be found so that*

(7.1)
$$\partial e^n = \begin{cases} (1-\tau)e^{n-1} & (n \text{ odd}), \\ (1+\tau)e^{n-1} & (n \text{ even}). \end{cases}$$

In fact, let e^0 be the homology class of the point $\mathbf{E}_+^0$; then $\partial : H_1(\mathbf{E}_+^1, \mathbf{S}^0) \approx \tilde{H}_0(\mathbf{S}^0)$, and the homology class of the cycle $e^0 - \tau e^0 = (1-\tau)e^0$ generates the latter group. Hence there exists $e^1 \in H^1(\mathbf{E}_+^1, \mathbf{S}^0)$ such that $\partial e^1 = (1-\tau)e^0$.

Suppose that e^n have been defined and satisfy (7.1) for all $n < r$, $(r \geq 2)$. Suppose that r is even. Then

$$\partial e^{r-1} = (1-\tau)e^{r-2},$$

$$\partial \tau e^{r-1} = \tau \partial e^{r-1} = \tau(1-\tau)e^{r-2} = (\tau - 1)e^{r-2},$$

so that $(1+\tau)e^{r-1}$ is a cycle whose homology class manifestly generates $H_{r-1}(\mathbf{S}^{r-1})$. Since $\partial : H_r(\mathbf{E}_+^r, \mathbf{S}^{r-1}) \approx H_{r-1}(\mathbf{S}^{r-1})$, there exists e^r such that $\partial e^r = (1+\tau)e^{r-1}$, and e^r generates $H_r(\mathbf{E}_+^r, \mathbf{S}^{r-1})$. The case r odd is proved similarly.

Let $p : \mathbf{S}^\infty \to \mathbf{P}^\infty$ be the covering map, so that τ generates the group of covering translations. Then the cells $\bar{\mathbf{E}}^n = p(\mathbf{E}_+^n)$ give a CW-decomposition of $\mathbf{P}^\infty$, and the elements $\bar{e}^n = p(e^n)$ are orientations of these cells. As $p \circ \tau = p$, the relations (7.1) imply

(7.2)
$$\partial \bar{e}^n = \begin{cases} 0 & (n \text{ odd}), \\ 2\bar{e}^{n-1} & (n \text{ even}). \end{cases}$$

This determines the homology groups of $\mathbf{P}^\infty$, and, as the cells $\bar{\mathbf{E}}^r$ with $r \leq n$ form a CW-decomposition of $\mathbf{P}^n$, those of $\mathbf{P}^n$ as well. We shall be interested primarily in the mod 2 homology and cohomology; in fact, it follows from (7.2) that

(7.3) *The mod 2 homology and cohomology of real projective space* $\mathbf{P}^n$ $(n \leq \infty)$ *is given by*

$$H_q(\mathbf{P}^n; \mathbf{Z}_2) \approx H^q(\mathbf{P}^n; \mathbf{Z}_2) \approx \begin{cases} \mathbf{Z}_2 & (0 \leq q < n+1), \\ 0 & (q > n). \end{cases}$$

Moreover, for $m < n \leq \infty$, *the injections*

$$H_q(\mathbf{P}^m; \mathbf{Z}_2) \to H_q(\mathbf{P}^n; \mathbf{Z}_2)$$

$$H^q(\mathbf{P}^n; \mathbf{Z}_2) \to H^q(\mathbf{P}^m; \mathbf{Z}_2)$$

are isomorphisms for all $q \leq m$. □

To determine the cohomology ring of $\mathbf{P}^n$, we shall use the fact that $\mathbf{P}^n$ is a manifold for $n < \infty$ and therefore satisfies Poincaré duality. Let $u = u_n$ be the non-zero element of $H^1(\mathbf{P}^n; \mathbf{Z}_2)$. We shall prove

(7.4) Theorem *The cohomology algebra* $H^*(\mathbf{P}^\infty; \mathbf{Z}_2)$ *is the polynomial algebra* $\mathbf{Z}_2[u]$. *If* $n < \infty$, $H^*(\mathbf{P}^n; \mathbf{Z}_2)$ *is the truncated polynomial algebra* $\mathbf{Z}_2[u]/(u^{n+1})$.

Because of the last sentence in (7.3), it suffices to prove that, if $r \leq n$, then in $H^r(\mathbf{P}^n; \mathbf{Z}_2)$ we have $u_n^n \neq 0$. This is clearly true for $n = 1$. Suppose that $u_{n-1}^{n-1} \neq 0$. As the injection i^* maps u_n into u_{n-1}, we have $i^* u_n^{n-1} = u_{n-1}^{n-1} \neq 0$ and therefore $u_n^{n-1} \neq 0$. By Poincaré duality there is an element $v \in H^1(\mathbf{P}^n; \mathbf{Z}_2)$ such that $v \smile u_n^{n-1} \neq 0$. Then $v = u_n$ and therefore $u_n^n \neq 0$. □

Example 2 (Complex Projective Space). Recall from §2, Chapter I that we may regard $\mathbf{S}^\infty$ as the unit sphere in complex Euclidean space $\mathbf{C}^\infty$, and $\mathbf{S}^{2n+1}$ as the unit sphere in $\mathbf{C}^{n+1}$. The operation of scalar multiplication by complex numbers of absolute value 1 defines a free action of $\mathbf{S}^1$ on $\mathbf{S}^{2n+1}$; the quotient space is complex projective space $\mathbf{P}^n(\mathbf{C})$, and the natural map $p: \mathbf{S}^{2n+1} \to \mathbf{P}^n(\mathbf{C})$, is a fibration, called the *Hopf fibration.* Thus the points of $\mathbf{P}^n(\mathbf{C})$ may be described by *homogeneous coordinates* $z_0, \ldots, z_n$ with $\sum_{i=0}^n |z_i|^2 = 1$, and two $(n+1)$-tuples represent the same point if and only if they are proportional (the proportionality factor is necessarily a complex number of absolute value 1). Let $[z_0, \ldots, z_n]$ be the point of $\mathbf{P}^n(\mathbf{C})$ having $z_0, \ldots, z_n$ as its coordinates; i.e., $p(z_0, \ldots, z_n) = [z_0, \ldots, z_n]$.

Let $\mathbf{E}_*^{2n}$ be the set of all points $(z_0, \ldots, z_n) \in \mathbf{S}^{2n+1}$ such that z_n is real and non-negative. Then $\mathbf{E}_*^{2n}$ is a cell with boundary $\mathbf{S}^{2n-1}$, and $p: (\mathbf{E}_*^{2n}, \mathbf{S}^{2n-1}) \to (\mathbf{P}^n(\mathbf{C}), \mathbf{P}^{n-1}(\mathbf{C}))$ is a relative homeomorphism, while $p \mid \mathbf{S}^{2n-1}$ is just the Hopf fibration $\mathbf{S}^{2n-1} \to \mathbf{P}^{n-1}(\mathbf{C})$. Hence $\mathbf{P}^n(\mathbf{C})$ is a $2n$-cellular extension of $\mathbf{P}^{n-1}(\mathbf{C})$ and it follows by induction that $\mathbf{P}^n(\mathbf{C})$ is a CW-complex with cells $\mathbf{E}^0, \mathbf{E}^2, \ldots, \mathbf{E}^{2n}$, where $\mathbf{E}^{2n} = p(\mathbf{E}_*^{2n})$. As there are no cells of odd dimension, the boundary operator is zero, and we have

(7.5) *The integral homology and cohomology of complex projective space* $\mathbf{P}^n(\mathbf{C})$ $(n \leq \infty)$ *are given by*

$$H_{2q}(\mathbf{P}^n(\mathbf{C})) \approx H^{2q}(\mathbf{P}^n(\mathbf{C})) \approx \begin{cases} \mathbf{Z} & (0 \leq q < n+1) \\ 0 & (q > n); \end{cases}$$

$$H_{2q+1}(\mathbf{P}^n(\mathbf{C})) \approx H^{2q+1}(\mathbf{P}^n(\mathbf{C})) = 0$$

Moreover, for $m < n \leq \infty$, *the injections*

$$H_q(\mathbf{P}^m(\mathbf{C})) \to H_q(\mathbf{P}^n(\mathbf{C}))$$
$$H^q(\mathbf{P}^n(\mathbf{C})) \to H^q(\mathbf{P}^m(\mathbf{C}))$$

are isomorphisms for all $q \leq 2m$. □

Again, since $\mathbf{P}^n(\mathbf{C})$ is a manifold (orientable, of course!) we can use Poincaré duality to calculate its integral cohomology ring. We shall need the following version of Poincaré duality:

Let M be a compact orientable m-manifold and let $z \in H_m(M)$ *be an orientation of M. Let* $x_1, \ldots, x_r$ *be a basis for the free part of* $H^q(M)$ *(i.e., their residue classes modulo the torsion subgroup form a basis for the free abelian quotient), and let* $y_1, \ldots, y_s$ *be a basis for the free part of* $H^{m-q}(M)$. *Then* $r = s$, *and the integral matrix* $\langle x_i \smile y_j, z \rangle$ *has determinant* ± 1.

The calculation of the cohomology ring now parallels that for the real case, and we content ourselves with stating the result.

(7.6) Theorem *Let u be a generator of the infinite cyclic group* $H^2(\mathbf{P}^n(\mathbf{C}))$ $(1 \leq n \leq \infty)$. *Then* $H^*(\mathbf{P}^\infty(\mathbf{C}))$ *is the polynomial ring* $\mathbf{Z}[u]$, *while* $H^*(\mathbf{P}^n(\mathbf{C}))$ *is the truncated polynomial ring* $\mathbf{Z}[u]/(u^{n+1})$. □

Example 3 (The infinite-dimensional lens spaces $\mathbf{L}^\infty(m)$). Let m be a positive integer; then the cyclic group of order m is a subgroup of the circle group $\mathbf{S}^1$, and therefore acts on complex Euclidean space $\mathbf{C}^\infty$ as well as on the unit sphere $\mathbf{S}^\infty$. Specifically, let ω be a primitive mth root of unity, and define $\tau : \mathbf{S}^\infty \to \mathbf{S}^\infty$ by $\tau(z_0, z_1, \ldots) = (\omega z_0, \omega z_1, \ldots)$. Then the group Γ generated by τ is strongly discontinuous, and the quotient space $\mathbf{S}^\infty/\Gamma = \mathbf{L}^\infty(m)$ is covered by $\mathbf{S}^\infty$; moreover, Γ acts on $\mathbf{S}^{2k+1}$ for every k, and $\mathbf{S}^{2k+1}/\Gamma = \mathbf{L}^{2k+1}(m)$ is a *lens space*. (Warning: these do not exhaust the lens spaces which have appeared in the literature; for example, if we define $\tau(z_0, z_1, \ldots) = (\omega z_0, \omega^{i_1} z_1, \ldots, \omega^{i_n} z_n, \ldots)$, where $i_1, i_2, \ldots$ is a more or less arbitrary sequence of integers relatively prime to m, we obtain other examples (cf. Exercise 3, below).)

We now describe a CW-decomposition of $\mathbf{S}^\infty$, invariant under τ, and inducing a CW-decomposition of the $\mathbf{L}^{2k+1}(m)$. Let

$$E^{2k} = \{z \in \mathbf{S}^{2k+1} \mid z_k \geq 0\},$$

$$E^{2k+1} = \left\{z \in \mathbf{S}^{2k+1} \,\middle|\, 0 \leq \arg z_k \leq \frac{2\pi}{m}\right\}.$$

Then it is not hard to see that the cells E^n and their images $\tau^i E^n$ under powers of τ give a CW-decomposition of $\mathbf{S}^\infty$, those of dimension $\leq 2k+1$ giving one of $\mathbf{S}^{2k+1}$. Moreover τ is a cellular map, and the induced chain map makes the chain groups into modules over the group ring of Γ. Moreover, by an argument not unlike that for the real projective space, we can find orientations e_n for the cells in such a way that

$$\partial e_{2k} = \Sigma e_{2k-1},$$

$$\partial e_{2k+1} = \Delta e_{2k},$$

where $\Sigma = 1 + \tau + \cdots + \tau^{m-1}$, $\Delta = 1 - \tau$ are elements of the group ring. The images of the cells E^n under the covering map $p : \mathbf{S}^\infty \to \mathbf{L}^\infty(m)$ give a CW-decomposition of $\mathbf{L}^\infty(m)$, and the map p is cellular. Thus, if $e_n^* = p(e_n)$, we have

$$\partial e_{2k}^* = m e_{2k-1}^*,$$

$$\partial e_{2k+1}^* = 0.$$

(7.7) *The integral homology groups of the lens space* $\mathbf{L}^{2k+1}(m)$ $(k \leq \infty)$ *are given by*

$$H_0(\mathbf{L}^{2k+1}(m)) = \mathbf{Z},$$

$$H_{2q}(\mathbf{L}^{2k+1}(m)) = 0 \qquad (q > 0),$$

$$H_{2q+1}(\mathbf{L}^{2k+1}(m)) = \mathbf{Z}_m \qquad (0 \leq q < k),$$

$$H_{2k+1}(\mathbf{L}^{2k+1}(m)) = \mathbf{Z},$$

$$H_{2q+1}(\mathbf{L}^{2k+1}(m)) = 0 \qquad (q > k).$$

Moreover, for $k < l \leq \infty$, *the injection*

$$H_q(\mathbf{L}^{2k+1}(m)) \to H_q(\mathbf{L}^{2l+1}(m))$$

is an isomorphism for $q < 2k+1$ *and an epimorphism for* $q = 2k+1$. □

Suppose that m is a prime p. Then we can apply the universal coefficient theorem to obtain the homology and cohomology groups with Z_p coefficients, with the following result:

(7.8) *If* p *is a prime, the* mod p *homology and cohomology groups of the lens space* $\mathbf{L}^{2k+1}(p)$ $(k \leq \infty)$ *are given by*

$$H_q(\mathbf{L}^{2k+1}(p); \mathbf{Z}_p) \approx H^q(\mathbf{L}^{2k+1}(p); \mathbf{Z}_p) \approx \mathbf{Z}_p \qquad (0 \leq q < 2k+2).$$

For $k < l \leq \infty$ the injections

$$H_q(\mathbf{L}^{2k+1}(p); \mathbf{Z}_p) \to H_q(\mathbf{L}^{2l+1}(p); \mathbf{Z}_p)$$

$$H^q(\mathbf{L}^{2l+1}(p); \mathbf{Z}_p) \to H^q(\mathbf{L}^{2k+1}(p); \mathbf{Z}_p)$$

are isomorphisms for $0 \leq q \leq 2k+1$. Finally, the Bocksteins

$$\beta_p : H_{2q}(\mathbf{L}^{2k+1}(p); \mathbf{Z}_p) \to H_{2q-1}(\mathbf{L}^{2k+1}(p); \mathbf{Z}_p)$$

$$\beta_p^* : H^{2q-1}(\mathbf{L}^{2k+1}(p); \mathbf{Z}_p) \to H^{2q}(\mathbf{L}^{2k+1}(p); \mathbf{Z}_p)$$

are isomorphisms for $0 < q < k$. □

If $p = 2$, the lens spaces in question are real projective spaces, and we have determined their cohomology rings in Theorem (7.4). Therefore we shall assume p odd. Let u be a generator of $H^1(\mathbf{L}^{2k+1}(p); \mathbf{Z}_p)$, and let $v = \beta_p^* u$, so that v generates $H^2(\mathbf{L}^{2k+1}(p); \mathbf{Z}_p)$. By the commutation rule for the cup product, we have

$$u \smile u = -u \smile u;$$

since p is odd, this implies that $u \smile u = 0$. Thus u generates an exterior algebra $\Lambda(u)$. By Poincaré duality in the orientable manifold $\mathbf{L}^{2k+1}(p)$, we deduce, as in the case of $\mathbf{P}^\infty$, that

$$u \smile : H^{2k}(\mathbf{L}^{2k+1}(p); \mathbf{Z}_p) \to H^{2k+1}(\mathbf{L}^{2k+1}(p); \mathbf{Z}_p)$$

and

$$v \smile : H^{2k-1}(\mathbf{L}^{2k+1}(p); \mathbf{Z}_p) \to H^{2k+1}(\mathbf{L}^{2k+1}(p); \mathbf{Z}_p)$$

are non-trivial, and the determination of the cohomology rings of the lens spaces is now easily completed, with the following result:

(7.9) Theorem *If p is an odd prime, the* mod *p cohomology ring of $\mathbf{L}^\infty(p)$ is the tensor product*

$$H^*(\mathbf{L}^\infty(p); \mathbf{Z}_p) = \Lambda(u) \otimes \mathbf{Z}_p[v]$$

of the exterior algebra generated by an element $u \in H^1(\mathbf{L}^\infty(p); \mathbf{Z}_p)$ and the polynomial algebra generated by $v = \beta_p^* u \in H^2(\mathbf{L}^\infty(p); \mathbf{Z}_p)$. If $k < \infty$, the mod *p cohomology ring is the tensor product*

$$H^*(\mathbf{L}^{2k+1}(p); \mathbf{Z}_p) = \Lambda(u) \otimes \{\mathbf{Z}_p[v]/(v)^{k+1}\}$$

of the exterior algebra $\Lambda(u)$ and the truncated polynomial ring generated by v. □

Exercises

1. Let X be a Hausdorff space, A a closed subspace of X. $K_n = \{E_\alpha^n \mid \alpha \in J_n\}$ a collection of closed subspaces of X $(n \geq 0)$, $X_n = A \cup \bigcup_{p \leq n} \bigcup_{\alpha \in J_p} E_\alpha^p$,

$\dot{E}^n_\alpha = E^n_\alpha \cap X_{n-1}$, $\operatorname{Int} E^n_\alpha = E^n_\alpha - X_{n-1}$. Suppose that

(i) The sets $\operatorname{Int} E^n_\alpha$ are mutually disjoint, and

$$X - A = \bigcup_n \bigcup_{\alpha \in J_n} \operatorname{Int} E^n_\alpha;$$

(ii) For each α, n, there is a relative homeomorphism

$$f^n_\alpha : (\Delta^n, \dot{\Delta}^n) \to (E^n_\alpha, \dot{E}^n_\alpha);$$

(iii) For each α, n, E^n_α meets only finitely many of the sets $\operatorname{Int} E^q_\beta$;
(iv) A subset C of X is closed if and only if $C \cap A$ and $C \cap E^n_\alpha$ are closed for each α, n.

Prove that the $\{X_n\}$ form a CW-decomposition of (X, A), and, conversely, if (X, A) is a relative CW-complex, its cells satisfy the above conditions.

2. Let $(X; A, B)$ be a triad. Prove that, if the injection $k_1 : H_q(A, A \cap B) \to H_q(X, B)$ is a monomorphism for $q = n$ and an epimorphism for $q = n + 1$, then the injection $k_2 : H_q(B, A \cap B) \to H_q(X, A)$ has the same properties. Deduce that k_1 is an isomorphism for all q if and only if k_2 is.

3. Let $\mathbf{Z}_p$ act on $\mathbf{S}^{2k+1}$ by

$$\tau(z_0, \ldots, z_k) = (\omega z_0, \omega^{i_1} z_1, \ldots, \omega^{i_k} z_k),$$

where $I = (i_1, \ldots, i_k)$ is a sequence of integers, each relatively prime to p, and ω is a primitive pth root of unity. Let $\mathbf{L}^{2k+1}_I(p)$ be the orbit space. Calculate the integral homology groups and the mod p cohomology ring of $\mathbf{L}^{2k+1}_I(p)$.

4. Prove the statement made in Remark 1, §2.

5. (Milnor [1]). Let X be a space. Let S_q be the set of all singular q-simplices in X, considered as a discrete space, and let $Y_q = \Delta^q \times S_q$. Let Y be the topological sum of the Y_q for all $q \geq 0$. An equivalence relation $\sim$ in Y is generated by the *elementary equivalences*

$$(d^q_i(x), u) \sim (x, \partial_i u) \qquad (u \in S_q,\ x \in \Delta^{q-1},\ 0 \leq i \leq q),$$

$$(s^q_i(x), u) \sim (x, \textstyle\int_i u) \qquad (u \in S_q,\ x \in \Delta^{q+1},\ 0 \leq i \leq q).$$

Let W be the quotient space. Prove that

(i) Y and W can be given the structure of CW-complexes in such a way that the quotient map $p : Y \to W$ is cellular;
(ii) The map of $f: Y \to X$ defined by

$$f(x, u) = u(x) \qquad (u \in S_q,\ x \in \Delta^q)$$

induces a map $g : W \to X$ with $g \circ p = f$;
(iii) the map g is a homology equivalence, i.e., $g_* : H_q(W) \approx H_q(W)$ for all q.

What can be said about the homology groups of Y?

The space W is called the *geometric realization* of the total singular complex $\mathfrak{S}(X)$.

6. Prove the van Kampen Theorem, in the following form. Let $(X; A_1, A_2)$ be an

NDR-triad, and suppose that $X = A_1 \cup A_2$ and that $A_0 = A_1 \cap A_2$ is 0-connected. Let

$$\Pi_k = \pi_1(A_k) \qquad (k = 0, 1, 2),$$
$$\Pi = \pi_1(X),$$

and let

$$i : \Pi_0 \to \Pi$$
$$i_k : \Pi_0 \to \Pi_k \qquad (k = 1, 2),$$
$$j_k : \Pi_k \to \Pi \qquad (k = 1, 2),$$

be the injections. Let Π^* be the free product of Π_1 and Π_2, and let $j_k^* : \Pi_k \to \Pi^*$ be the inclusion. Let $j : \Pi^* \to \Pi$ be the homomorphism such that $j \circ j_k^* = j_k$. Define a map $\lambda : \Pi_0 \to \Pi^*$ by

$$\lambda(x) = j_1^* i_1(x)^{-1} \cdot j_2^* i_2(x).$$

Then j is an epimorphism and Ker j is the smallest normal subgroup of Π^* containing Im λ.

7. A pair (X, A) is 0-connected if and only if each path component of X contains a point of A.

8. A pair (X, A) is 1-connected if and only if the following two conditions are satisfied:

   (1) each path component of X contains exactly one path component of A;
   (2) for each $a \in A$, the injection $\pi_1(A, a) \to \pi_1(X, a)$ is an epimorphism.

9. Prove that, if (X, A) is m-connected and (Y, B) is n-connected, then $(X, A) \times (Y, B)$ is $(m + n + 1)$-connected. (Hint: first prove that if K is a finite simplicial complex of dimension r and $-1 \le p \le r$, then there is a subdivision K_1 of K and subcomplexes P, Q, P_0 and Q_0 of K_1 such that $K = P \cup Q$, $|P_0|$ is a deformation retract of $|P|$, $|Q_0|$ is a deformation retract of $|Q|$, dim $P_0 \le p$, and dim $Q_0 \le r - p - 1$).

CHAPTER III

Generalities on Homotopy Classes of Mappings

The set $[X, Y]$ of homotopy classes of maps between two compactly generated spaces X, Y has no particular algebraic structure. This Chapter is devoted to the study of conditions on one or both spaces in order that $[X, Y]$ support additional structure of interest. Guided by the fact that $\pi_1(X, x_0) = [S^1, y_0; X, x_0]$ is a group, while $[S^1, X]$ is in one-to-one correspondence with the set of all conjugacy classes in $\pi_1(X, x_0)$, and the latter set has no algebraic structure of interest, we discuss in §1 the way in which $[X, x_0; Y, y_0]$ depends on the base points. It turns out that under reasonable conditions the sets $[X, x_0; Y_0, y_0]$ and $[X, x_0; Y, y_1]$ are isomorphic. However, there is an isomorphism between them for every homotopy class of paths in Y from y_1 to y_0. In particular, the group $\pi_1(Y, y_0)$ operates on $[X, x_0; Y, y_0]$, and $[X, Y]$ can be identified with the quotient of the latter set under the action of the group. This action for the case $X = \mathbf{S}^n$ was first studied by Eilenberg [1] in 1939; it, and an analogous action of $\pi_1(B, y_0)$ on the set $[X, A, x_0; B, y_0]$, are discussed in §1.

The discussion of §1 suggests as the primary objects of study the sets $[X, x_0; Y, y_0]$, where the base point of X is non-degenerate (i.e., $(X, \{x_0\})$ is an NDR-pair), while no condition need be imposed on the base point in Y. Spaces with non-degenerate base point form a full subcategory $\mathscr{K}_*$ of the category $\mathscr{K}_0$ of spaces with base point, and many constructions in $\mathscr{K}_*$ will have interest for us and are discussed in §2: reduced joins, cones, suspensions. On the other hand, others (path and loop spaces) are valid in the larger category $\mathscr{K}_0$.

We next address ourselves to the problem of defining a natural binary operation in $[X, Y]$ for X, Y in $\mathscr{K}_*$. The question divides naturally into two parts:

(1) Given Y, does there exist a natural product in $[X, Y]$ for all X?
(2) Given X, does there exist a natural product in $[X, Y]$ for all Y?

When the answer to the first is affirmative, we call Y an H-space; when the answer to the second is affirmative, we call X an H'-space. In each case, there is a universal example. For the first question, it is $Y \times Y$, and Y is an H-space if and only if Y admits a continuous multiplication with unit, i.e., a map $\mu : Y \times Y \to Y$ such that $\mu | Y \vee Y$ is the folding map which identifies each copy of Y in $Y \vee Y$ with the space Y. For the second question, the universal example is $X \vee X$, and X is an H'-space if and only if there is a map $\theta : X \to X \vee X$ which is homotopic in $X \times X$ to the diagonal map. Examples of H-spaces are topological groups and loop spaces; examples of H'-spaces are suspensions. The two types of spaces and their mutual relationships are explored in §§3–5.

If $f: X \to Y$ is a map, we may form the mapping cone $\mathbf{T}_f$, and there is a natural inclusion $j : Y \subsetneq \mathbf{T}_f$. We may then form the mapping cone of j and obtain an inclusion $k : \mathbf{T}_f \subsetneq \mathbf{T}_j$. And this process can be iterated indefinitely. However, the space $\mathbf{T}_j$ has the same homotopy type as the suspension $\mathbf{S}X$ of X, and the space $\mathbf{T}_k$ has the homotopy type of $\mathbf{S}Y$ in such a way that the inclusion map of $\mathbf{T}_j$ in $\mathbf{T}_k$ corresponds to the suspension $\mathbf{S}f$ of the original map f. In this way we obtain an infinite sequence

$$X \xrightarrow{f} Y \xrightarrow{j} \mathbf{T}_f \xrightarrow{q} \mathbf{S}X \xrightarrow{\mathbf{S}f} \mathbf{S}Y$$
$$\xrightarrow{\mathbf{S}j} \mathbf{ST}_f \xrightarrow{\mathbf{S}q} \mathbf{S}^2X \to \cdots.$$

Given another space W, we may apply the contravariant functor $[\ \ , W]$ to the above sequence, to obtain a sequence

$$\cdots \to [\mathbf{S}^{n+1}X, W] \to [\mathbf{S}^n\mathbf{T}_f, W] \to [\mathbf{S}^nY, W] \to [\mathbf{S}^nX, W] \to \cdots$$
$$\to [\mathbf{T}_f, W] \to [Y, W] \to [X, W].$$

Except for the last few terms, this is an exact sequence of abelian groups and homomorphisms, and is due to Barratt [1] who introduced it in 1955 under the name *track group sequence* (for any positive integer n, the set $[\mathbf{S}^nZ, W]$ is a group, called a *track group*). A few years later, a careful study of the track group sequence was made by Puppe [1].

The above construction can be dualized; given a map $f: Y \to X$, we may iterate the process of forming the mapping fibre to obtain a sequence homotopically equivalent to the sequence

$$\cdots \to \Omega^{n+1}X \to \Omega^n\mathbf{T}^f \to \Omega^nY \to \Omega^nX \to \cdots \to \mathbf{T}^f \to Y \to X,$$

applying the covariant functor $[Z,\ \ ]$, we obtain an infinite exact sequence

$$\cdots \to [Z, \Omega^{n+1}X] \to [Z, \Omega^n\mathbf{T}^f] \to [Z, \Omega^nY] \to [Z, \Omega^nX] \to \cdots$$
$$\to [Z, \mathbf{T}^f] \to [Z, Y] \to [Z, X],$$

again consisting, except for the last few terms, of abelian groups and homomorphisms. These sequences are discussed in §6.

The condition that a space be an H- or an H′-space has important consequences for the cohomology. For example, the homology groups of an H-space X form a graded ring, the Pontryagin ring of X. This fact was exploited by Pontryagin [1] in 1939, when he calculated the homology groups of the classical groups. Section 7 is devoted to a discussion of these questions.

If X is any space, its homology groups with coefficients in a field form a coalgebra. In addition, when X is an H-space, the Pontryagin product makes $H_*(X)$ into a Hopf algebra. This fact was exploited by Hopf [6] in 1941 when he proved that a compact Lie group has the same rational cohomology ring as a product of spheres of odd dimension. This had been conjectured by Elie Cartan [1] in 1929 and verified for the classical groups by Pontryagin [1] and Brauer [1]. Further implications of the Hopf algebra structure were found by Samelson [1], Leray [1] and Borel [1]. §8 we discuss these questions, citing without proof the algebraic properties of Hopf algebras which are involved.

1 Homotopy and the Fundamental Group

In studying the homotopy classes of maps of a space X into a space Y, one may first fix base points $x_0 \in X$, $y_0 \in Y$, and attempt to classify the maps of the *pair* (X, x_0) into the pair (Y, y_0) under homotopies which leave x_0 at y_0 throughout. Having accomplished this, one may then study the effect of changing the base point. Thus one is led to the notion of *free homotopy*.

Specifically, let (X, x_0) be a space with *nondegenerate base point* (i.e., $(X, \{x_0\})$ is an NDR-pair). Let $f_0, f_1 : X \to Y$ be maps and let $u : \mathbf{I} \to Y$ be a path in Y. We shall say that f_0 is *freely homotopic to* f_1 *along* u ($f_0 \underset{u}{\simeq} f_1$) if and only if there is a homotopy $f : \mathbf{I} \times X \to Y$ of f_0 to f_1 such that $f(t, x_0) = u(t)$ for all t; thus $f_0(x_0)$ is the initial point $u(0)$, and $f_1(x_0)$ is the terminal point $u(1)$, of u.

The following properties of free homotopy are immediate:

(1.1) *If* $f : X \to Y$ *is a map, and* $e : \mathbf{I} \to Y$ *is the constant map of* $\mathbf{I}$ *into* $f(x_0)$, *then* $f \underset{e}{\simeq} f$. □

(1.2) *If* $f_0, f_1 : X \to Y$ *and* $u : \mathbf{I} \to Y$ *are maps such that* $f_0 \underset{u}{\simeq} f_1$, *and if* $v : \mathbf{I} \to Y$ *is the path inverse to* u, *then* $f_1 \underset{v}{\simeq} f_0$. □

(1.3) If $f_0, f_1, f_2 : X \to Y$ and $uv : \mathbf{I} \to Y$ *are maps such that* $f_0 \underset{u}{\simeq} f_1$, $f_1 \underset{v}{\simeq} f_2$, *and if* $w : \mathbf{I} \to Y$ *is the product of* u *and* v, *then* $f_0 \underset{w}{\simeq} f_2$. □

The homotopy extension property easily yields

(1.4) *If $f_0 : X \to Y$ and $u : \mathbf{I} \to Y$ are maps such that $f(x_0) = u(0)$, then there exists $f_1 : X \to Y$ such that $f_0 \underset{u}{\simeq} f_1$.* □

Since $(\mathbf{I}, \dot{\mathbf{I}})$ and (X, x_0) are NDR-pairs so is their product $(\mathbf{I} \times X, \dot{\mathbf{I}} \times X \cup \mathbf{I} \times x_0)$. This is used to prove

(1.5) *If $f_0, f_1 : X \to Y$ and $u, v : \mathbf{I} \to Y$ are maps such that $f_0 \underset{u}{\simeq} f_1$ and $u \simeq v$ (rel. $\dot{\mathbf{I}}$) then $f_0 \underset{v}{\simeq} f_1$.*

In fact, let $f : \mathbf{I} \times X \to Y$ be a free homotopy of f_0 to f_1 along u, and let $g : \mathbf{I} \times \mathbf{I} \to Y$ be a homotopy of u to v (rel. $\dot{\mathbf{I}}$). Then the map of the subspace $0 \times \mathbf{I} \times X \cup \mathbf{I} \times \dot{\mathbf{I}} \times X \cup \mathbf{I} \times \mathbf{I} \times x_0$ into Y defined by

$$h(t, s, x) = \begin{cases} f(s, x) & (t = 0), \\ g(t, s) & (x = x_0), \\ f_s(x) & (s = 0, 1), \end{cases}$$

has an extension $H : \mathbf{I} \times \mathbf{I} \times X \to Y$. Let $f'(t, x) = H(1, t, x)$; then $f' : \mathbf{I} \times X \to Y$ is a free homotopy of f_0 to f_1 along v. □

(1.6) *If $f_0, f_1 : (X, x_0) \to (Y, y_0)$, $g_0, g_1 : (X, x_0) \to (Y, y_0)$, $u : (\mathbf{I}; 0, 1) \to (Y; y_0, y_1)$, and if $f_0 \simeq f_1$ (rel. x_0), $g_0 \simeq g_1$ (rel. x_0), $f_0 \underset{u}{\simeq} g_0$, then $f_1 \underset{u}{\simeq} g_1$.*

This follows from (1.5) with the aid of (1.2) and (1.3). □

Let $f : (X, x_0) \to (Y, y_0)$, $u : (\mathbf{I}; 0, 1) \to (Y; y_1, y_0)$. By (1.4), there exists $g : (X, x_0) \to (Y, y_1)$ such that $g \underset{u}{\simeq} f$. It follows from (1.5) and (1.6) that the homotopy class of g depends only on the homotopy classes α, ξ of f, u respectively. Let $\tau_\xi(\alpha)$ be the homotopy class of g.

(1.7) *If $\alpha \in [(X, x_0), (Y, y_0)]$, $\xi \in \pi_1(Y; y_1, y_0)$, $\eta \in \pi_1(Y; y_2, y_1)$, then $\tau_{\eta\xi}(\alpha) = \tau_\eta(\tau_\xi(\alpha))$. If $\xi \in \pi_1(Y; y_0, y_0)$ is the homotopy class of the constant path, then $\tau_\xi(\alpha) = \alpha$.* □

(1.7) has a categorical interpretation. The *fundamental groupoid* $\Pi_1(Y)$ is a category whose objects are the points of Y and whose morphisms are homotopy classes of paths with fixed end points. For each $y \in Y$, let $M(y) = [(X, x_0), (Y, y)]$. $M(y)$ is a set with distinguished base point, viz. the homotopy class of the constant map of X into y. For each $\xi : y_0 \to y_1$ (i.e., ξ is a homotopy class of paths from y_1 to y_0), define $M(\xi) : M(y_0) \to M(y_1)$ by

$$M(\xi)(\alpha) = \tau_\xi(\alpha).$$

(1.8) *M is a (covariant) functor from the category* $\Pi_1(Y)$ *to the category* $\mathscr{M}_0$ *of pointed sets.* □

In particular,

(1.9) *If* y_0 *and* y_1 *belong to the same path component of* Y, *then* $M(y_0)$ *and* $M(y_1)$ *are in one-to-one correspondence.* □

(1.10) *The fundamental group* $\pi_1(Y, y_0)$ *operates on the set* $M(y_0)$. □

Finally, we have

(1.11) *If* Y *is* 0*-connected, then* $[X, Y]$ *is the quotient of* $M(y_0)$ *under the action of* $\pi_1(Y, y_0)$. □

We now give a different interpretation to the operations of $\pi_1(Y, y_0)$ on the set $M(y_0)$. Suppose that Y is semilocally 1-connected, and so has a universal covering space $\tilde{Y}$; let $p : \tilde{Y} \to Y$ be the covering map, and choose $\tilde{y}_0 \in p^{-1}(y_0)$. Suppose further that X is 1-connected. Then it is proved in the theory of covering spaces that

(1.12) *The operation* $\underline{p} : [X, x_0; \tilde{Y}, \tilde{y}_0] \to [X, x_0; Y, y_0]$ *defined by composition with p is a one-to-one correspondence.* □

Moreover, it follows from (1.11) that

(1.13) *The injection* $[X, x_0; \tilde{Y}, \tilde{y}_0] \to [X, \tilde{Y}]$ *is a one-to-one correspondence.* □

Let Π be the group of covering translations of $\tilde{Y}$ over Y; the elements of Π are self-homeomorphisms $h : \tilde{Y} \to \tilde{Y}$ such that $p \circ h = p$. Again by covering space theory, we have

(1.14) *The group* Π *is isomorphic with* $\pi_1(Y, y_0)$. □

An isomorphism ϕ is defined as follows. Let $h \in \Pi$, and let $u : \mathbf{I} \to \tilde{Y}$ be a path from $\tilde{y}_0$ to $h(\tilde{y}_0)$. Then $\phi(h)$ is the homotopy class of the loop $p \circ h : (\mathbf{I}, \dot{\mathbf{I}}) \to (Y, y_0)$.

Finally, the group Π operates on $[X, \tilde{Y}]$ by composition. Identifying Π with $\pi_1(Y, y_0)$ by ϕ and $[X, \tilde{Y}]$ with $[X, x_0; Y, y_0]$ by the one-to-one correspondences of (1.12), (1.13), we obtain two *a priori* different operations of Π on $[X, \tilde{Y}]$. In fact

(1.15) *The two operations described above are identical. In other words, for each covering translation* $h \in \Pi$, *the diagram*

$$\begin{array}{ccccc}
[X, \tilde{Y}] & \longleftarrow & [X, x_0; \tilde{Y}, \tilde{y}_0] & \xrightarrow{\underline{p}} & [X, x_0; Y, y_0] \\
\Big\downarrow{\scriptstyle \underline{h}} & & & & \Big\downarrow{\scriptstyle \tau_{\phi(h)}} \\
[X, \tilde{Y}] & \longleftarrow & [X, x_0; \tilde{Y}, \tilde{y}_0] & \xrightarrow[\underline{p}]{} & [X, x_0; Y, y_0]
\end{array}$$

is commutative.

In fact, let $f : (X, x_0) \to (\tilde{Y}, \tilde{y}_0)$, and let u be a path in $\tilde{Y}$ from $\tilde{y}_0$ to $h(\tilde{y}_0)$. Then there is a map $f' : (X, x_0) \to (\tilde{Y}, \tilde{y}_0)$ such that $f' \underset{u}{\simeq} h \circ f$. Then $p \circ f' \underset{p \circ u}{\simeq} p \circ h \circ f = p \circ f$, and the commutativity of the diagram follows. □

If we consider maps of pairs $(X, A) \to (Y, B)$ with $A \neq \varnothing$, the above discussion goes through with minor changes provided that both $(A, \{x_0\})$ and (X, A) are NDR-pairs; the paths along which the base point $a_0 \in A$ are deformed are required to lie in B. Thus to each homotopy class α of maps: $(X, A, x_0) \to (Y, B, y_0)$ and to each element $\xi \in \pi_1(B; y_1, y_0)$, there is associated an element $\tau'_\xi(\alpha) \in [(X, A, x_0), (Y, B, y_1)]$; if f, u are representatives of α, ξ, respectively, then a map $g : (X, A, x_0) \to (Y, B, y_0)$ represents $\tau'_\xi(\alpha)$ if and only if there is a homotopy $h : (\mathbf{I} \times X, \mathbf{I} \times A) \to (Y, B)$ of g to f such that $h(t, x_0) = u(t)$.

(1.16) *If $\alpha \in [(X, A, x_0), (Y, B, y_0)]$, $\xi \in \pi_1(B; y_1, y_0)$, $\eta \in \pi_1(B; y_2, y_1)$, then $\tau'_{\eta\xi}(\alpha) = \tau'_\eta(\tau'_\xi(\alpha))$. If $\xi \in \pi_1(B, y_0)$ is the homotopy class of the constant path, then $\tau'_\xi(\alpha) = \alpha$.* □

As before, these constructions determine a (covariant) functor $M : \Pi_1(B) \to \mathcal{M}_0$ with $M(y) = [(X, A, x_0), (Y, B, y)]$, and we have

(1.17) *If y_0 and y_1 belong to the same path component of B, then the sets $M(y_0)$ and $M(y_1)$ are in one-to-one correspondence.* □

(1.18) *The fundamental group $\pi_1(B, y_0)$ operates on $M(y_0)$.* □

(1.19) *If B is 0-connected, then $[(X, A), (Y, B)]$ is the quotient of $M(y_0)$ by the action of $\pi_1(B, y_0)$.* □

Again, suppose that B and Y are semilocally 1-connected spaces, and that the injection $\pi_1(B, y_0) \to \pi_1(Y, y_0)$ is an isomorphism. If $p : \tilde{Y} \to Y$ is a universal covering map, and $\tilde{B} = p^{-1}(B)$, then $p \mid \tilde{B} : \tilde{B} \to B$ is a universal covering map. Moreover, every covering translation h maps $\tilde{B}$ into $\tilde{B}$. The group Π of covering translations is isomorphic with $\pi_1(Y, y_0)$ and therefore with $\pi_1(B, y_0)$. Suppose X is 1-connected. Then $\underline{p}: [X, A, x_0; \tilde{Y}, \tilde{B}, \tilde{y}_0] \to [X, A, x_0; Y, B, y_0]$ is a one-to-one correspondence. Because of (1.19) the injection

$$[X, A, x_0; \tilde{Y}, \tilde{B}, \tilde{y}_0] \to [X, A; \tilde{Y}, \tilde{B}]$$

is a one-to-one correspondence. Thus, as in the absolute case, there are two ways of operating with the group Π on the set $[X, A; \tilde{Y}, \tilde{B}]$, and these two ways coincide.

(1.20) *The diagram*

$$\begin{array}{ccccc}
[X, A; \tilde{Y}, \tilde{B}] & \longleftarrow & [X, A, x_0; \tilde{Y}, \tilde{B}, \tilde{y}_0] & \xrightarrow{\underline{p}} & [X, A, x_0; Y, B, y_0] \\
\Big\downarrow{\scriptstyle \underline{h}} & & & & \Big\downarrow{\scriptstyle \tau'_{\phi(h)}} \\
[X, A; \tilde{Y}, \tilde{B}] & \longleftarrow & [X, A, x_0; \tilde{Y}, \tilde{B}, \tilde{y}_0] & \xrightarrow[\underline{p}]{} & [X, A, x_0; Y, B, y_0]
\end{array}$$

is commutative. □

The extension of these ideas to maps of more complicated configurations is left to the reader (cf. the remarks at the end of §1 of Chapter I).

2 Spaces with Base Points

The reader may have been led by the considerations of §1 to the belief that it will be convenient to work in some category of spaces with prescribed base points. He should, however, have observed that, in the study of the action of $\pi_1(Y, y_0)$ on the set $[X, x_0; Y, y_0]$, it was appropriate, indeed essential, to impose the condition of nondegeneracy on the base point of X, while no such condition on the base point of Y was needed. Thus it would not be possible to work in the category $\mathscr{K}_0$ of all compactly generated spaces with base points; and to work in the full subcategory $\mathscr{K}_*$ consisting of all spaces with nondegenerate base points would amount to the imposition of inessential and totally irrelevant hypotheses with the sole intent of forcing the theory to fit into a given formal system. On the other hand, it would be perverse to refuse to recognize the convenience that category theory has to offer in its clarity and economy of statement of results in many diverse branches of mathematics. Thus the attitude taken in this book is that, while the notions of category theory form a convenient language in which to state many theorems, they do not constitute a universal framework into which all of mathematics must, willy-nilly, be set. And we intend to steer a course between the Scylla of blind adherence to category theory despite its disadvantages, and the Charybdis of an equally blind refusal to accept its many conveniences.

Occasionally it will be necessary to deal with (compactly generated) spaces without a prescribed base point (called *free spaces*) and maps and homotopies of maps between such spaces (called *free maps* and *free homotopies*, respectively). To handle these we shall adopt the device of adjoining an external base point. Specifically, if X is a free space, let $X^+ = X + P$ be the topological sum of X and a space P consisting of a single point $*$, the base point of X^+; and if $f: X \to Y$ is a free map, $f^+ : X^+ \to Y^+$ is the extension of f which preserves the base point. Free homotopies are handled in a similar way. The assignment $X \to X^+, f \to f^+$ is a faithful functor from the category $\mathscr{K}$ to the category $\mathscr{K}_*$; thus we may regard $\mathscr{K}$ as a subcategory of $\mathscr{K}_*$. Usually we shall drop the plus except when necessary for emphasis.

If (X, x_0) and (Y, y_0) are spaces in $\mathscr{K}_0$, it is natural to choose (x_0, y_0) as the base point of $X \times Y$, for the projections $p_1 : X \times Y \to X$, $p_2 : X \times Y \to Y$ are maps in $\mathscr{K}_0$, and it is easy to see that $(X \times Y, (x_0, y_0))$, with the maps p_1, p_2, is the categorical product of (X, x_0) and (Y, y_0). The same remarks apply to the category $\mathscr{K}_*$, because of (5.2) of Chapter I.

This is not the case for the sum, for the disjoint union $X + Y$ does not have a natural base point. However, the space $X \vee Y$ obtained from $X + Y$ by identifying x_0 with y_0, together with the natural inclusions $j_1 : X \to X \vee Y, j_2 : Y \to X \vee Y$, is the sum of X and Y in the category $\mathscr{K}_0$. Again, the same results apply to $\mathscr{K}_*$.

As is always the case in a pointed category, there is a natural map k of the sum $X \vee Y$ into the product $X \times Y$; this map is characterized by the properties

(1) $p_1 \circ k \circ j_1$ is the identity map of X,
(2) $p_2 \circ k \circ j_2$ is the identity map of Y,
(3) $p_1 \circ k \circ j_2$ and $p_2 \circ k \circ j_1$ are constant maps.

The map k is, in fact, an inclusion, mapping $X \vee Y$ upon the subspace $X \times \{y_0\} \cup \{x_0\} \times Y$ of $X \times Y$, and we shall normally identify $X \vee Y$ with this subspace.

An important construction in the category $\mathscr{K}_*$ is the *reduced join* (or smash product) of two spaces X, Y. This is the quotient space $X \wedge Y = X \times Y / X \vee Y$. For $x \in X$, $y \in Y$, let $x \wedge y$ be the image of (x, y) under the proclusion $X \times Y \to X \wedge Y$. Thus $x \wedge y_0 = x_0 \wedge y = x_0 \wedge y_0$, the base-point of $X \wedge Y$.

(N.B.: This construction can be made in $\mathscr{K}_0$, but does not have many useful properties there).

The following properties are readily verified with the aid of the considerations of §4 of Chapter I.

(2.1) *The reduced join is commutative and associative, up to natural homeomorphism.* □

(2.2) *The operation of reduced join is distributive over that of addition; i.e., the spaces $X \wedge (Y \vee Z)$ and $(X \wedge Y) \vee (X \wedge Z)$ are naturally homeomorphic.* □

(2.3) *If (X, A) is an* NDR-*pair and $Y \in \mathscr{K}_*$, the spaces $(X/A) \wedge Y$ and $X \wedge Y/A \wedge Y$ are naturally homeomorphic.* □

It may be of interest to examine the reduced join when one or both factors are free spaces. In fact, one has

(2.4) *If $X \in \mathscr{K}_*$, $Y \in \mathscr{K}$, then $X \wedge Y^+ = X \times Y/\{x_0\} \times Y$.* □

(2.5) *If X, $Y \in \mathscr{K}$, then $X^+ \wedge Y^+ = (X \times Y)^+$.* □

Let $f : X \to X'$, $g : Y \to Y'$ be maps in $\mathscr{K}_*$. Then the map $f \times g : X \times Y \to X' \times Y'$ sends $X \vee Y$ into $X' \vee Y'$, thereby inducing

$$f \vee g : X \vee Y \to X' \vee Y',$$

as well as

$$f \wedge g : X \wedge Y \to X' \wedge Y',$$

and it is clear that

(2.6) *The reduced join and the sum are functors on $\mathscr{K}_* \times \mathscr{K}_*$ to $\mathscr{K}_*$.* □

We next turn our attention to function spaces, returning momentarily to the category $\mathscr{K}$. If (X, A), (Y, B) are pairs, with X and Y compactly generated and A and B closed, then the set $\mathbf{F}(X, A; Y, B)$ of all maps of (X, A) into (Y, B) is a subset of $\mathbf{F}(X, Y)$ which we topologize as in §4 of Chapter I. In particular, if B is a point, the operation of composition with the proclusion $p : X \to X/A$ induces a map $\bar{p} : \mathbf{F}(X/A, *; Y, *) \to \mathbf{F}(X, A; Y, *)$.

(2.7) *The map $\bar{p}$ is a homeomorphism.*

For the proof see Steenrod [5, 5.11]. □

We now return to the category $\mathscr{K}_0$. If X, $Y \in \mathscr{K}_0$, the function space $\mathbf{F}(X, Y)$ is constructed as above, and we choose the constant map of X into y_0 as the base point. Thus the construction of a function space determines a functor $\mathbf{F} : \mathscr{K}_0 \times \mathscr{K}_0 \to \mathscr{K}_0$, contravariant in the first argument and covariant in the second. We then have the analogues, in the category $\mathscr{K}_0$, of (4.19), (4.21), (4.22), and (4.23) of Chapter I.

(2.8) *The evaluation map $\mathbf{e} : \mathbf{F}(X, Y) \wedge X \to Y$ is continuous.* □

(2.9) *The function spaces* $\mathbf{F}(X \wedge Y, Z)$ *and* $\mathbf{F}(X, \mathbf{F}(Y, Z))$ *are naturally homeomorphic.* □

(2.10) *The operation of composition induces a continuous map of* $\mathbf{F}(Y, Z) \wedge \mathbf{F}(X, Y)$ *into* $\mathbf{F}(X, Z)$. □

(2.11) *The functors* $\wedge Y$ *and* $\mathbf{F}(Y, \quad)$ *form an adjoint pair.* □

The notion of reduced join is useful in handling homotopies of maps between spaces with base point. In fact, if $f_0, f_1 : X \to Y$ are maps in $\mathscr{K}_0$, and the base point x_0 of X is nondegenerate, any base point preserving homotopy of f_0 to f_1 is a map $f : \mathbf{I} \times X \to Y$ which sends $\mathbf{I} \times x_0$ into y_0. If we regard $\mathbf{I}$ as a free space in the category $\mathscr{K}_*$, we see that, by (2.4), $\mathbf{I} \wedge X$ is the quotient $\mathbf{I} \times X/\mathbf{I} \times \{x_0\}$; thus f induces a map of $\mathbf{I} \wedge X$ into Y. Conversely, a map $f : \mathbf{I} \wedge X \to Y$ induces a homotopy between the maps $f_0, f_1 : X \to Y$ defined by

$$f_t(x) = f(t \wedge x) \qquad (t = 0, 1, x \in X).$$

The space $\mathbf{I} \wedge X$ is the (reduced) *prism* over X, and the maps $x \to t \wedge x$ $(t = 0, 1)$ are inclusions, mapping X homeomorphically upon the subspaces $0 \wedge X$, $1 \wedge X$ of $\mathbf{I} \wedge X$. The prism defined in this way differs from the usual prism $\mathbf{I} \times X$ in that the subspace $\mathbf{I} \times \{x_0\}$ of the latter has been shrunk to a point. Since $\mathbf{I} \times \{x_0\}$ is contractible and $(\mathbf{I} \times X, \mathbf{I} \times \{x_0\})$ is an NDR-pair, this process does not affect the homotopy type. (This last observation is not really necessary, since both $\mathbf{I} \times X$ and $\mathbf{I} \wedge X$ have copies of X as deformation retracts!).

It is also convenient to consider the unit interval as a space with 0 as base point; to distinguish this pointed space from the free space $\mathbf{I}$, we define $\mathbf{T}$ to be the pair $(\mathbf{I}, 0)$. Let also $\dot{\mathbf{T}} = (\dot{\mathbf{I}}, 0)$. The (reduced) *cone* over a space X is then defined to be the space $\mathbf{T}X = \mathbf{T} \wedge X$. In this case the map $x \to 1 \wedge x$ is an inclusion, mapping X homeomorphically upon the subspace $\dot{\mathbf{T}} \wedge X = 1 \wedge X$ of $\mathbf{T}X$, which we shall normally identify with X, while $0 \wedge X$ is the base point. The space $\mathbf{T}X$ is contractible; it differs from the usual cone in that the contractible subspace of the latter which is the cone over the base point has been shrunk to a point. Moreover, it follows from (2.3) that $\mathbf{T}X = \mathbf{I} \wedge X/0 \wedge X$. Therefore, by the above remarks,

(2.12) *A map* $f : X \to Y$ *is nullhomotopic if and only if* f *has an extension* $g : \mathbf{T}X \to Y$. □

Let $\mathbf{S} = \mathbf{I}/\dot{\mathbf{I}}$; then the (reduced) *suspension* of X is the space $\mathbf{S}X = \mathbf{S} \wedge X$. Because of (2.3), $\mathbf{S}X = \mathbf{I} \wedge X/\dot{\mathbf{I}} \wedge X = \mathbf{T}X/X$. Again, it differs from the usual suspension in that an appropriate contractible subset has been shrunk to a point.

Define subsets

$$\mathbf{T}_+ X = \{\bar{t} \wedge x \mid 0 \le t \le 1/2\},$$

$$\mathbf{T}_- X = \{\bar{t} \wedge x \mid 1/2 \le t \le 1\},$$

of $\mathbf{S}X$ (here $\bar{t}$ is the image of $t \in \mathbf{I}$ under the identification map $\varpi : (\mathbf{I}, \dot{\mathbf{I}}) \to (\mathbf{S}, *)$). Then $\mathbf{T}_+ X$ and $\mathbf{T}_- X$ are homeomorphic copies of the cone $\mathbf{T}X$, and $\mathbf{T}_+ X \cap \mathbf{T}_- X$ is homeomorphic with X.

The operations of forming the reduced cone and reduced suspension determine functors, also denoted by $\mathbf{T}$ and $\mathbf{S}$, respectively; if $f: X \to Y$, then $\mathbf{T}f: \mathbf{T}X \to \mathbf{T}Y$ and $\mathbf{S}f: \mathbf{S}X \to \mathbf{S}Y$ are defined by

$$\mathbf{T}f = 1 \wedge f, \qquad \mathbf{S}f = 1 \wedge f,$$

where the symbol "1" denotes the appropriate identity map.

There are parallel constructions involving function spaces. A *free path* in X is just a map of $\mathbf{I}$ into X; a (based) *path* in X a map of $\mathbf{T}$ into X (i.e., one that starts at x_0); and a *loop* in X is a map of $\mathbf{S}$ into X. Thus we have the three function spaces

$$\mathbf{F}X = \mathbf{F}(\mathbf{I}, X), \quad \text{the space of free paths in } X,$$

$$\mathbf{P}X = \mathbf{F}(\mathbf{T}, X), \quad \text{the space of paths in } X,$$

$$\Omega X = \mathbf{F}(\mathbf{S}, X), \quad \text{the space of loops in } X.$$

The maps $u \to u(0)$, $u \to u(1)$ are, by Corollary (7.9) of Chapter I, fibrations, mapping $\mathbf{F}X$ into X, with fibre $\mathbf{P}X$. Similarly, the map $u \to u(1)$ is a fibration, mapping $\mathbf{P}X$ into X, with fibre ΩX.

These constructions, like the suspension and cone functors, do not go outside the category $\mathscr{K}_*$. That is

(2.13) If $X \in \mathscr{K}_*$, *the spaces* $\mathbf{F}X$, $\mathbf{P}X$ *and* ΩX *belong to* $\mathscr{K}_*$.

We prove this for $\mathbf{F}X$; the proofs for the other two are entirely similar. Let (u, h) be a representation of $(X, \{*\})$ as an NDR-pair. Define $\bar{u} : \mathbf{F}X \to \mathbf{I}$, $\bar{h} : \mathbf{I} \times \mathbf{F}X \to \mathbf{F}X$ by

$$\bar{u}(f) = \sup_{s \in I} u(f(s)).$$

$$\bar{h}(t, f)(s) = h(t, f(s)).$$

Then $(\bar{u}, \bar{h})$ is a representation of $(\mathbf{F}X, \{e\})$ as an NDR-pair. □

Finally, the (reduced) *mapping cylinder* $\mathbf{I}_f$ and *mapping cone* $\mathbf{T}_f$ of a map $f: X \to Y$ are the spaces obtained from $(\mathbf{I} \wedge X) \vee Y$, $(\mathbf{T} \wedge X) \vee Y$, respectively, by identifying the point $1 \wedge x$ with $f(x)$ for each $x \in X$. There are homeomorphic imbeddings $X \vee Y \to \mathbf{I}_f$, $Y \to \mathbf{T}_f$, and Y is a deformation retract of $\mathbf{I}_f$, while $\mathbf{I}_f / X = \mathbf{T}_f$.

Since the construction of reduced join and function space are functors, it follows that the above constructions, too are functorial. In particular,

(2.14) *The functors* $\mathbf{S}$ *and* Ω *form an adjoint pair; in fact, the spaces* $\mathbf{F}(\mathbf{S}X, Y)$ *and* $\mathbf{F}(X, \Omega Y)$ *are naturally homeomorphic.* □

One can iterate the suspension functor n times; it is natural to denote this functor by $\mathbf{S}^n$. In particular, one can suspend the space $\mathbf{S}$ $n-1$ times, and it is natural to denote this space by $\mathbf{S}^n$. On the other hand, in §2 of Chapter I, we have used the same symbol $\mathbf{S}^n$ to denote the unit sphere in $\mathbf{R}^{n+1}$. This involves no real contradiction, for the spaces involved are homeomorphic. But it is desirable to identify these spaces by a particular homeomorphism, which we now construct.

To avoid confusion between the two spaces, let us temporarily denote the unit sphere in $\mathbf{R}^{n+1}$ by $\mathbf{S}_0^n$. We shall take the unit vector $\mathbf{e}_0$ as the base point for each of the spheres $\mathbf{S}_0^n$.

The map $t \to (\cos 2\pi t, \sin 2\pi t)$ is a relative homeomorphism of $(\mathbf{I}, \dot{\mathbf{I}})$ with $(\mathbf{S}_0^1, \mathbf{e}_0)$, and therefore induces a map $h_1 : \mathbf{S} \to \mathbf{S}_0^1$. Suppose we have defined a homeomorphism $h_n : \mathbf{S}^n \to \mathbf{S}_0^n$. Then the map

$$(x, t) \to (\cos^2 \pi t)\mathbf{e}_0 + (\sin^2 \pi t)h_n(x) + \sqrt{\frac{1 - h_n(x) \cdot \mathbf{e}_0}{2}}(\sin 2\pi t)\mathbf{e}_{n+1}$$

is readily verified to be a relative homeomorphism of $(\mathbf{S}^n \times \mathbf{I}, \mathbf{S}^n \times \dot{\mathbf{I}} \cup \{*\} \times \mathbf{I}) = (\mathbf{S}^n, \{*\}) \times (\mathbf{I}, \dot{\mathbf{I}})$ with $(\mathbf{S}_0^{n+1}, \mathbf{e}_0)$, and therefore to induce a homeomorphism h_{n+1} of $\mathbf{S}^n \wedge (\mathbf{I}/\dot{\mathbf{I}}) = \mathbf{S}^n \wedge \mathbf{S} = \mathbf{S}^{n+1}$ with $\mathbf{S}_0^{n+1}$.

In a similar way we shall want to identify $\mathbf{TS}^n$ with $\mathbf{E}^{n+1}$. This time the map

$$(t, x) \to \left(\sin^2 \frac{\pi}{2} t\right)h_n(x) + \left(\cos^2 \frac{\pi}{2} t\right)\mathbf{e}_0$$

is a relative homeomorphism of

$$(\mathbf{I} \times \mathbf{S}^n, 0 \times \mathbf{S}^n \cup \mathbf{I} \times \{*\}) = (\mathbf{I}, 0) \times (\mathbf{S}^n, \{*\})$$

with $(\mathbf{E}^{n+1}, \mathbf{e}_0)$, and therefore induces a homeomorphism h'_{n+1} of $\mathbf{T} \wedge \mathbf{S}^n$ with $\mathbf{E}^{n+1}$.

It is somewhat inconvenient that the operation of multiplication of paths is not strictly associative, nor is the constant path a true unit. We can surmount this difficulty by using the notion of *measured path*, i.e., a mapping of a closed interval of variable length into the space. More precisely, a measured path in X is an ordered pair (r, f), with r a non-negative real number and $f: [0, r] \to X$; we say that (r, f) *starts* at $f(0)$ and *ends* at $f(r)$. Such a function has a canonical extension (still denoted by f) over the non-negative real axis $\mathbf{R}^+$, such that

$$f(t) = f(r)$$

for all $t \geq r$. Thus we may (and shall) regard the space $\mathbf{F}^*(X)$ as a subspace of the product $\mathbf{R}^+ \times \mathbf{F}(\mathbf{R}^+, X)$. (Recall that, as is appropriate when dealing with the compactly generated category, products, function spaces and subspaces are topologized as in Chapter I). A map $p : \mathbf{F}^*(X) \to X \times X$ is defined by

$$p(r, f) = (f(0), f(r)).$$

(2.15) Theorem *The map p is a fibration.*

Continuity of p follows from that of the evaluation map ((4.19) of Chapter I). Let

$$\begin{array}{ccc} Y & \xrightarrow{\ \phi\ } & \mathbf{F}^*(X) \\ \downarrow & \nearrow & \downarrow p \\ \mathbf{I} \times Y & \xrightarrow[\ \psi\ ]{} & X \times X \end{array}$$

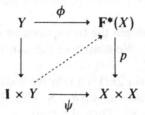

be a homotopy lifting problem. The first component of ϕ is a map $\lambda : Y \to \mathbf{R}^+$, the adjoint of the second a map $\hat{\phi} : \mathbf{R}^+ \times Y \to X$. The components of ψ are maps $\psi_i : Y \to X$ $(i = 1, 2)$. These satisfy the conditions

$$\begin{aligned} \psi_1(0, y) &= \hat{\phi}(0, y), \\ \psi_2(0, y) &= \hat{\phi}(\lambda(y), y) \\ &= \hat{\phi}(s, y) \qquad (s \geq \lambda(y)). \end{aligned}$$

We are required to find maps

$$\Lambda : \mathbf{I} \times Y \to \mathbf{R}^+,$$

$$\hat{\Phi} : \mathbf{R}^+ \times \mathbf{I} \times Y \to X$$

such that

$$\begin{aligned} \Lambda(0, y) &= \lambda(y), \\ \hat{\Phi}(s, 0, y) &= \hat{\phi}(s, y), \\ \hat{\Phi}(0, t, y) &= \psi_1(t, y), \\ \psi_2(t, y) &= \hat{\Phi}(\Lambda(t, y), t, y) \\ &= \hat{\Phi}(s, t, y) \qquad (s \geq \Lambda(t, y)). \end{aligned}$$

The desired lifting $\Phi : \mathbf{I} \times Y \to \mathbf{F}^*(X)$ is then defined by

$$\Phi(t, y) = (\Lambda(t, y), \hat{\Phi}(\cdot, t, y)).$$

Let

$$\Lambda(t, y) = \begin{cases} \lambda(y) + (1 - \lambda(y))t & \text{if } \lambda(y) \le 1, \\ \lambda(y) & \text{if } \lambda(y) \ge 1. \end{cases}$$

For $y \in Y$, let

$$P(y) = \{(s, t) \in \mathbf{R}^+ \times \mathbf{I} \mid 0 \le s \le \Lambda(t, y)\},$$

$$Q(y) = \{(s, t) \in P(y) \mid s = 0 \text{ or } t = 0 \text{ or } s = \Lambda(t, y)$$

(see Figure 3.1), and let $\rho_y : P(y) \to Q(y)$ be the radial projection from the point $(\frac{1}{2}, 2) \in \mathbf{R}^+ \times \mathbf{R}^+$. Let

$$P = \mathbf{R}^+ \times \mathbf{I} \times Y,$$

$$Q = \{(s, t, y) \in P \mid s = 0 \text{ or } t = 0 \text{ or } s \ge \Lambda(t, y)\},$$

and let $\rho : P \to Q$ be the function such that

$$\rho(s, t, y) = \begin{cases} (\rho_y(s, t), y) & \text{if } s \le \Lambda(t, y), \\ (s, t, y) & \text{if } s \ge \Lambda(t, y). \end{cases}$$

In order to prove the continuity of ρ, we need explicit formulae for the functions ρ_y.

Case I ($\lambda = \lambda(y) \le 1$). Here

(1) if $0 \le s \le \frac{1}{4}t$,

$$\rho_y(s, t) = \left(0, \frac{t - 4s}{1 - 2s}\right);$$

(2) if $\frac{1}{4}t \le s \le \lambda + \frac{1}{4}(1 - 2\lambda)t$,

$$\rho_y(s, t) = \left(\frac{4s - t}{2(2 - t)}, 0\right);$$

(3) if $\lambda + \frac{1}{4}(1 - 2\lambda)t \le s \le \lambda + t(1 - \lambda)$,

$$\rho_y(s, t) = (\sigma, \tau),$$

where

$$\sigma = \frac{(4s - t)(1 - \lambda) - \lambda(1 - 2s)}{2(2 - t)(1 - \lambda) - (1 - 2s)},$$

$$\tau = \frac{(4s - t) - 2\lambda(2 - t)}{2(2 - t)(1 - \lambda) - (1 - 2s)}.$$

Case II ($\lambda \ge 1$). If (1) $0 \le s \le \frac{1}{4}t$ or (2) $\frac{1}{4}t \le s \le \lambda + \frac{1}{4}(1 - 2\lambda)t$, then $\rho_y(s, t)$ is defined by the same formula as above. However

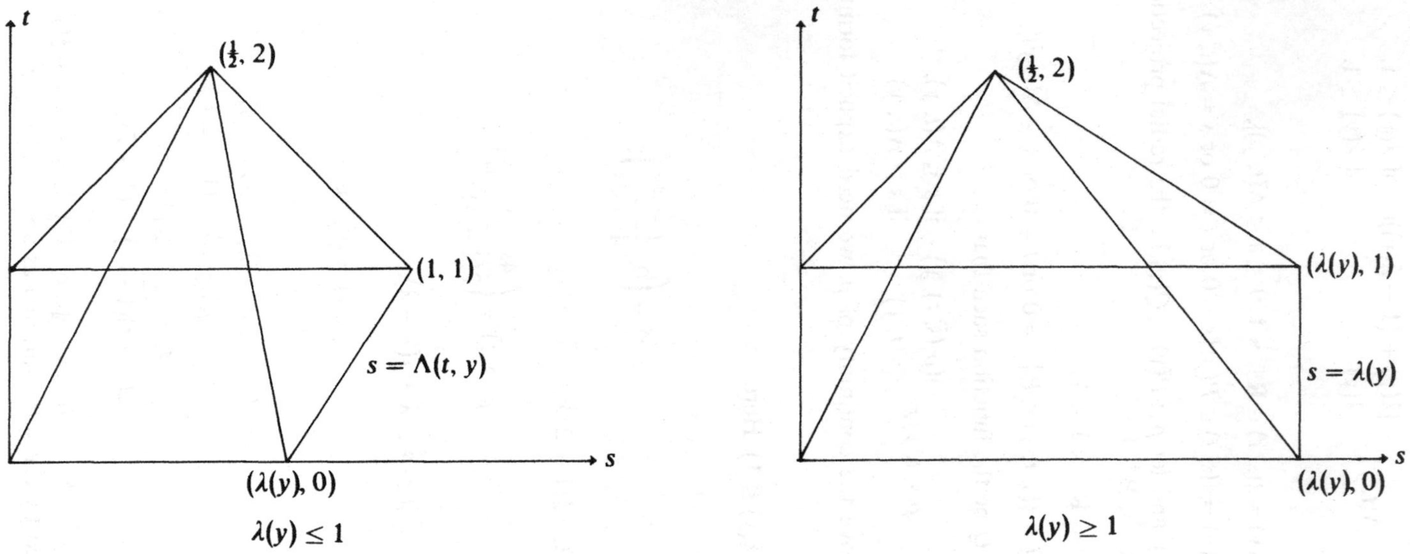

Figure 3.1

(3) if $\lambda + \frac{1}{4}(1 - 2\lambda)t \leq s \leq \lambda$, then

$$\rho_y(s, t) = \left(\lambda, \frac{4s - t - 2(2 - t)\lambda}{2s - 1}\right).$$

We leave it to the reader to verify that the functions ρ_y are consistently defined. Continuity of ρ then follows from the continuity of the expressions occurring in the formulae (1)–(3), above. The only possible snag is the vanishing of the denominators of the appropriate fractions. In (1), $s \leq \frac{1}{4}t \leq \frac{1}{4}$, so that $1 - 2s \geq \frac{1}{2}$; in (2) $t \leq 1$ so that $2 - t \geq 1$. In (3) of Case II, $s \geq (1 - t/2) + t/4 \geq 1 - t/2 \geq \frac{1}{2}$. Finally, in (3) of Case I,

$$\begin{aligned} 2(2 - t)(1 - \lambda) - (1 - 2s) &= 2s - 2t(1 - \lambda) + (3 - 4\lambda) \\ &\geq 2[\lambda + \tfrac{1}{4}(1 - 2\lambda)t] - 2t(1 - \lambda) + (3 - 4\lambda) \\ &= (3 - \tfrac{3}{2}t) + \lambda(t - 2) = (2 - t)(\tfrac{3}{2} - \lambda) \\ &\geq \tfrac{1}{2}(2 - t) \geq \tfrac{1}{2}. \end{aligned}$$

The map $\rho : P \to Q$ is a retraction. The function $\Psi : Q \to X$ defined by

$$\begin{aligned} &\Psi(s, 0, y) = \hat{\phi}(s, y), \\ &\Psi(0, t, y) = \psi_1(t, y), \\ &\Psi(s, t, y) = \psi_2(t, y) \qquad (s \geq \Lambda(t, y)) \end{aligned}$$

is then continuous and thus has the continuous extension $\hat{\Phi} = \Psi \circ \rho : P \to X$. It is clear that the functions Λ, $\hat{\Phi}$ have the desired properties. □

Let

$$\begin{aligned} \mathbf{P}^*(X) &= \{(r, f) \in \mathbf{F}^*(X) \mid f(0) = *\}, \\ \mathbf{P}'^*(X) &= \{(r, f) \in \mathbf{F}^*(X) \mid f(r) = *\}, \\ \mathbf{\Omega}^*(X) &= \mathbf{P}^*(X) \cap \mathbf{P}'^*(X). \end{aligned}$$

The space $\mathbf{\Omega}^*(X)$ is called the space of *measured loops* in X.

Define maps $q^* : \mathbf{P}^*(X) \to X$, $q'^* : \mathbf{P}'^*(X) \to X$ by

$$\begin{aligned} q(r, f) &= f(r), \\ q'(r, f) &= f(0). \end{aligned}$$

It follows easily from Theorem (2.15) that

(2.16) Theorem *The maps q and q' are fibrations with the same fibre $\mathbf{\Omega}^*(X)$.* □

An important property of the new path spaces is

(2.17) *The spaces $\mathbf{P}^*(X)$ and $\mathbf{P}'^*(X)$ are contractible.*

The spaces in question being homeomorphic, we shall prove that $\mathbf{P}'^*(X)$ is contractible. An explicit contraction is given by

$$\phi(t; r, f) = (tr, f \mid [0, tr]).$$

To prove ϕ continuous, it suffices to prove that the first component of ϕ is a continuous map $\phi_1 : \mathbf{I} \times \mathbf{P}'^*X \to \mathbf{R}^+$, and that the adjoint of the second component is a continuous map $\bar{\phi}_2 : \mathbf{R}^+ \times \mathbf{I} \times \mathbf{P}'^*X \to X$. The map ψ_1 is the composite

$$\mathbf{I} \times \mathbf{P}'^* \subsetneq \mathbf{I} \times \mathbf{R}^+ \times \mathbf{F} \xrightarrow{\mu \times 1} \mathbf{R}^+ \times \mathbf{F} \xrightarrow{p_1} \mathbf{R}^+,$$

where μ is the operation of multiplication of real numbers and p_1 the projection on the first factor. The maps in question being continuous, so is their composite ϕ_1.

Let

$$X_1 = \{(s, t; r, f) \in \mathbf{R}^+ \times \mathbf{I} \times \mathbf{P}'^* \mid s \le tr\},$$
$$X_2 = \{(s, t; r, f) \in \mathbf{R}^+ \times \mathbf{I} \times \mathbf{P}'^* \mid s \ge tr\}.$$

Then X_1 and X_2 are closed subsets of $\mathbf{R}^+ \times \mathbf{I} \times \mathbf{P}'^*$ whose union is $\mathbf{R}^+ \times \mathbf{I} \times \mathbf{P}'^*$, and it suffices to show that $\bar{\phi}_2 \mid X_1$ and $\bar{\phi}_2 \mid X_2$ are continuous. But $\bar{\phi}_2 \mid X_1$ is the composite

$$X \subsetneq \mathbf{R}^+ \times \mathbf{I} \times \mathbf{P}'^* \subsetneq \mathbf{R}^+ \times \mathbf{I} \times \mathbf{F} \xrightarrow{p_1 \times 1} \mathbf{R}^+ \times \mathbf{F} \xrightarrow{\mathbf{e}} X,$$

where p_1 is the projection of $\mathbf{R}^+ \times \mathbf{I}$ into its first factor $\mathbf{R}^+$ and $\mathbf{e}$ is the evaluation map. Therefore $\bar{\phi}_2 \mid X_1$ is continuous. Similarly $\bar{\phi}_2 \mid X_2$ is continuous. □

We can now compare the new path spaces with the old. Let us recall that the map $q : \mathbf{P}(X) \to X$ defined by $q(f) = f(1)$ is a fibration with fibre $\Omega(X)$, as is the map $q' : \mathbf{P}'(X) \to X$ defined by $q'(f) = f(0)$.

(2.18) *The fibrations $q : \mathbf{P}(X) \to X$ and $q^* : \mathbf{P}^*(X) \to X$, as well as the fibrations $q' : \mathbf{P}'(X) \to X$ and $q'^* : \mathbf{P}'^*(X) \to X$, have the same fibre homotopy type.*

For any map $v : [0, r] \to X$ and positive real number s, define $v_s : [0, r/s] \to X$ by

$$v_s(t) = v(st).$$

Define $h : \mathbf{P}(X) \to \mathbf{P}^*(X)$, $h' : \mathbf{P}^*(X) \to \mathbf{P}(X)$ by

$$h(u) = (1, u);$$

$$h'(r, v) = \begin{cases} v_r & \text{if } r > 0, \\ e_* & \text{if } r = 0, \end{cases}$$

where e_* is the constant path. The map h is obviously continuous. The adjoint $\tilde{h} : \mathbf{I} \times \mathbf{P}^*(X) \to X$ is given by

$$\tilde{h}'(t, r, v) = v(rt);$$

it is the composite

$$\mathbf{I} \times \mathbf{P}^*(X) \hookrightarrow \mathbf{I} \times \mathbf{R}^+ \times \mathbf{F}(\mathbf{R}^+, X) \xrightarrow{\mu \times 1} \mathbf{R}^+ \times \mathbf{F}(\mathbf{R}^+, X) \xrightarrow{\mathbf{e}} X,$$

where μ is the operation of multiplication of real numbers and $\mathbf{e}$ the evaluation map. It follows that h' is continuous.

The composite $h' \circ h$ is the identity map of $P(X)$. The composite $h \circ h'$ is homotopic to the identity under the map

$$H(t, r, v) = (\theta, v_\theta) \quad \text{if } t < 1 \text{ or } r > 0,$$
$$H(1, 0, v) = (0, v),$$

where $\theta = 1 - t + tr$. We leave it to the reader to verify the continuity of H. It is then clear that h, h' and H define a fibre homotopy equivalence. □

Remark: (2.18) also follows from Exercise 2.

(2.19) Corollary *The space $\Omega(X)$ and $\Omega^*(X)$ have the same homotopy type.* □

We can now define the product of measured paths. Let M be the subspace of $\mathbf{F}^*(X) \times \mathbf{F}^*(X)$ consisting of all pairs $(r, f; s, g)$ such that $f(r) = g(0)$; then a function $\mu : M \to \mathbf{F}^*(X)$ is defined by $\mu(r, f; s, g) = (r + s, h)$, where

$$h(t) = \begin{cases} f(t) & (0 \le t \le r), \\ g(t - r) & (t \ge r). \end{cases}$$

(2.20) Theorem *The function $\mu : M \to \mathbf{F}^*(X)$ is continuous.*

We must prove that the first component of μ is a continuous map of M into $\mathbf{R}^+$, and that the adjoint of its second component is a continuous map of $\mathbf{R}^+ \times M \to X$. These maps are given by

$$\mu_1(r, f; s, x) = r + s;$$

$$\tilde{\mu}_2(t; r, f; s, g) = \begin{cases} f(t) & (0 \le t \le r), \\ g(t - r) & (t \ge r). \end{cases}$$

The map μ_1 is the composite

$$M \hookrightarrow \mathbf{F}^* \times \mathbf{F}^* \hookrightarrow \mathbf{R}^+ \times \mathbf{F} \times \mathbf{R}^+ \times \mathbf{F} \xrightarrow{p_1 \times p_1} \mathbf{R}^+ \times \mathbf{R}^+ \xrightarrow{\mu_0} \mathbf{R}^+,$$

where $\mathbf{F}^* = \mathbf{F}^*(X)$, $\mathbf{F} = \mathbf{F}(R^+, X)$, $p_1 : \mathbf{R}^+ \times \mathbf{F} \to \mathbf{R}^+$ is the projection on the first factor, and $\mu_0 : \mathbf{R}^+ \times \mathbf{R}^+ \to \mathbf{R}^+$ is the operation of multiplication of real numbers. As each of the component maps is continuous, μ_1 is continuous.

To prove continuity of $\tilde{\mu}_2$, let

$$X_1 = \{(t; r, f; s, g) \in \mathbf{R}^+ \times M \mid t \le r\},$$
$$X_2 = \{(t; r, f; s, g) \in \mathbf{R}^+ \times M \mid t \ge r\};$$

then X_1 and X_2 are closed subspaces of $\mathbf{R}^+ \times M$ whose union is M, and it suffices to prove the continuity of $\tilde{\mu}_2 | X_1$ and $\tilde{\mu}_2 | X_2$. For example, $\tilde{\mu}_2 | X_2$ is the composite

$$X_2 \hookrightarrow \mathbf{R}^+ \times M \hookrightarrow \mathbf{R}^+ \times \mathbf{R}^+ \times \mathbf{F} \times \mathbf{R}^+ \times \mathbf{F} \xrightarrow{p} \mathbf{R}^+ \times \mathbf{R}^+ \times \mathbf{F}$$
$$\xrightarrow{\sigma \times 1} \mathbf{R}^+ \times \mathbf{F} \xrightarrow{\mathbf{e}} X,$$

where p is the projection $(t; r, f; s, g) \to (t, r, g)$, $\sigma(t, r) = |r - t|$, and i is the evaluation map. □

(2.21) Corollary *The product μ determines continuous functions*

$$\mathbf{P}'^*(X) \times \mathbf{P}^*(X) \to \mathbf{F}^*(X),$$
$$\mathbf{P}'^*(X) \times \Omega^*(X) \to \mathbf{P}'^*(X),$$
$$\Omega^*(X) \times \mathbf{P}^*(X) \to \mathbf{P}^*(X),$$
$$\Omega^*(X) \times \Omega^*(X) \to \Omega^*(X).$$

□

Unlike the usual path-multiplication, the new product is strictly associative. Moreover, for any point $x \in X$, the pair $\varepsilon_x = (0, e_x)$ is a strict unit. More precisely:

(2.22) Theorem *Let $\alpha, \beta, \gamma \in \mathbf{F}^*(X)$. If $\alpha\beta$ is defined and $(\alpha\beta)\gamma$ is defined, then $\beta\gamma$ and $\alpha(\beta\gamma)$ are defined, and $(\alpha\beta)\gamma = \alpha(\beta\gamma)$. If $\alpha \in F^*(X)$ starts at x and ends at y then $\varepsilon_x \alpha = \alpha = \alpha\varepsilon_y$.* □

In replacing the loop space ΩX by the space $\Omega^* X$ of measured loops, we have gained, in the sense that the new space has nicer algebraic properties. On the other hand, we have lost the adjointness property

$$\mathbf{F}(\mathbf{S}X, Y) = \mathbf{F}(W, \Omega Y) \neq \mathbf{F}(X, \Omega^* Y);$$

though it is still true up to homotopy:

$$[\mathbf{S}X, Y] \approx [X, \Omega Y] \approx [X, \Omega^* Y].$$

In fact, the maps

$$\mathbf{F}(1, h) : \mathbf{F}(X, \Omega Y) \to \mathbf{F}(X, \Omega^* Y),$$
$$\mathbf{F}(1, h') : \mathbf{F}(X, \Omega^* Y) \to \mathbf{F}(X, \Omega Y)$$

are homotopy equivalences, inducing isomorphisms

$$\underline{h} : [X, \Omega Y] \approx [X, \Omega^* Y],$$
$$\underline{h}' : [X, \Omega^* Y] \approx [X, \Omega Y].$$

If $f : X \to \Omega^* Y$, and if $\bar{f} : \mathbf{S}X \to Y$ is the map corresponding to $h' \circ f$ under the adjointness relation

$$\mathbf{F}(X, \Omega Y) \approx \mathbf{F}(\mathbf{S}X, Y),$$

we shall refer to f and $\bar{f}$ as adjoints of each other.

3 Groups of Homotopy Classes

In Part A of Chapter I it was suggested (rightly, as it turns out) that the set of homotopy classes of maps of $\mathbf{S}^n$ into itself form a group, isomorphic with the additive group of integers. It is natural to ask how special this phenomenon is; in other words, under what circumstances can the set $[X, Y]$ be given a group structure? As stated, the question is meaningless, for any set can be given some group structure. Evidently, some naturality properties must be imposed.

Let us first investigate the possibility of defining a binary operation in $[X, Y]$. In view of the conventions we have just made, the set $[X, Y]$ has a distinguished element, the homotopy class $*$ of the constant map. We shall consider only operations for which $*$ is a two-sided identity. We shall also assume that our operations are natural; in other words, if $f : X' \to X$ and $g : Y \to Y'$ are maps, then the operations

$$\bar{f} : [X, Y] \to [X', Y]$$

and

$$\underline{g} : [X, Y] \to [X, Y']$$

induced by composition with f, g, respectively, are homomorphisms.

It will be seen from the discussion that follows that there is no natural product defined in $[X, Y]$ for every X, Y. This being so, the following two questions suggest themselves:

(1) Given Y, does there exist a natural product defined in $[X, Y]$ for all X?
(2) Given X, does there exist a natural product defined in $[X, Y]$ for all Y?

Examples of both kinds come immediately to mind. For example, if Y is a topological group and $f, g : X \to Y$ we can define a product $f \cdot g : X \to Y$ by the formula

$$(f \cdot g)(x) = f(x) \cdot g(x);$$

this product is compatible with homotopy and induces a natural group operation in $[X, Y]$. On the other hand, if $X = \mathbf{S}^1$, then the usual group structure in $[X, Y] = \pi_1(Y)$ is a natural group operation.

We shall say that Y is an H-*space* if and only if (1) holds, and that X is an H′-*space* if and only if (2) holds. In the next two sections we shall study H-spaces and H′-spaces, respectively. When Y is an H-space, we shall write the composition multiplicatively; when X is an H′-space we shall write it additively (except in the case $X = \mathbf{S}$, i.e., $[X, Y] = \pi_1(Y)$). In the next two sections, we shall be working in the category $\mathscr{K}_*$, so that all spaces under discussion will be assumed to have non-degenerate base-points. The results of these sections are mainly due to Copeland [2]; see also [Hi].

4 H-spaces

For any space Y, let p_1, p_2 be the projections of $Y \times Y$ on the first and second factors, respectively. Let $i_1, i_2 : Y \to Y \times Y$ be the inclusions, defined by

(4.1) $$i_1(y) = (y, y_0), \qquad i_2(y) = (y_0, y) \qquad (y \in Y).$$

Then $p_1 \circ i_1 = p_2 \circ i_2 = 1$, the identity map of Y, while $p_1 \circ i_2 = p_2 \circ i_1 = *$.

Suppose that Y is an H-space. Then we can multiply the homotopy classes of p_1 and p_2 to obtain an element $[\mu] = [p_1] \cdot [p_2]$ for some $\mu : Y \times Y \to Y$. By the naturality of the product we have

(4.2) $$[\mu] \circ [i_1] = ([p_1] \cdot [p_2]) \circ [i_1] = [p_1 \circ i_1] \cdot [p_2 \circ i_1] = [1] \cdot [*] = [1],$$

and similarly

(4.3) $$[\mu] \circ [i_2] = [1].$$

If X is any space, let $\Delta : X \to X \times X$ be the diagonal map, given by

(4.4) $$\Delta(x) = (x, x);$$

then, if $f_1, f_2 : X \to Y$, we have

(4.5) $$p_i \circ (f_1 \times f_2) \circ \Delta = f_i \qquad (i = 1, 2).$$

Again by naturality, we have

$$\begin{aligned}[\mu] \circ [f_1 \times f_2] \circ [\Delta] &= ([p_1] \cdot [p_2]) \circ [f_1 \times f_2] \circ [\Delta] \\ &= ([p_1] \circ [f_1 \times f_2] \circ [\Delta] \cdot ([p_2] \circ [f_1 \times f_2] \circ [\Delta]) \\ &= [f_1] \cdot [f_2].\end{aligned}$$

The conditions (4.2), (4.3) tell us that $\mu \circ i_1 \simeq 1 \simeq \mu \circ i_2$, and therefore $\mu \circ k \simeq \nabla$, where $k : Y \vee Y \to Y \times Y$ is the inclusion and $\nabla : Y \vee Y \to Y$ is the *folding map*, defined by

(4.6) $$\nabla(y, y_0) = \nabla(y_0, y) = y.$$

Now $(Y \times Y, Y \vee Y)$ is an NDR-pair, by (5.2) of Chapter I, and therefore there exists $\mu' \simeq \mu$ such that $\mu' \circ k = \nabla$.

Conversely, suppose that Y is a space and $\mu : Y \times Y \to Y$ a map such that $\mu \circ i_1 \simeq 1 \simeq \mu \circ i_2$. (Such a map μ is called a *product* in Y). Then, given f_1, $f_2 : X \to Y$, we may define $f_1 \cdot f_2 = \mu \circ (f_1 \times f_2) \circ \Delta$. The composition so defined is compatible with homotopy, and induces a natural product in $[X, Y]$. Thus Y is an H-space; moreover, $p_1 \cdot p_2 = \mu \circ (p_1 \times p_2) \circ \Delta = \mu \circ 1 = \mu$.

Summarizing, we have

(4.7) Theorem *A space Y is an* H-*space if and only if there is a map $\mu : Y \times Y \to Y$ such that $\mu \circ i_1 = \mu \circ i_2 = 1$. The map μ then satisfies the condition $[\mu] = [p_1] \cdot [p_2]$. Moreover, if $f_1, f_2 : X \to Y$ are maps, then $[f_1] \cdot [f_2]$ is the homotopy class of the composite*

$$X \xrightarrow{\Delta} X \times X \xrightarrow{f_1 \times f_2} Y \times Y \xrightarrow{\mu} Y. \qquad \square$$

The multiplication we have defined in $[X, Y]$ when Y is an H-space need not be associative. In fact, if $f_1, f_2, f_3 : X \to Y$ then $(f_1 \cdot f_2) \cdot f_3$ and $f_1 \cdot (f_2 \cdot f_3)$ are the homotopy classes of the composites

$$\textbf{(4.8)} \quad X \xrightarrow{\Delta_3} X \times X \times X \xrightarrow{f_1 \times f_2 \times f_3} Y \times Y \times Y \xrightarrow{\mu \times 1} Y \times Y \xrightarrow{\mu} Y,$$

$$\textbf{(4.9)} \quad X \xrightarrow{\Delta_3} X \times X \times X \xrightarrow{f_1 \times f_2 \times f_3} Y \times Y \times Y \xrightarrow{1 \times \mu} Y \times Y \xrightarrow{\mu} Y,$$

where $\Delta_3 = (\Delta \times 1) \circ \Delta = (1 \times \Delta) \circ \Delta : X \to X \times X \times X$ is the diagonal map. Hence the condition

$$\textbf{(4.10)} \qquad \mu \circ (\mu \times 1) \simeq \mu \circ (1 \times \mu)$$

is sufficient for associativity. It is also necessary; for if $X = Y \times Y \times Y$ and f_1, f_2, f_3 are the projections of $Y \times Y \times Y$ into Y, then $(f_1 \times f_2 \times f_3) \circ \Delta_3$ is the identity map.

We shall say that the H-space Y is *homotopy associative* if and only if (4.10) holds in $[Y \times Y \times Y, Y]$. When the maps in (4.10) are actually equal, we say that Y is *strictly associative.* Then

(4.11) Theorem *The sets $[X, Y]$ have a monoid structure natural with respect to X if and only if Y is a homotopy associative* H-*space.* $\square$

If Y is a homotopy associative H-space, it is natural to ask when the monoids $[X, Y]$ are groups. If this is so, then the homotopy class of the identity map $1 \in [Y, Y]$ has an inverse $[j]$, so that the composites

(4.12) $$Y \xrightarrow{\Delta} Y \times Y \xrightarrow{j \times 1} Y \times Y \xrightarrow{\mu} Y,$$

(4.13) $$Y \xrightarrow{\Delta} Y \times Y \xrightarrow{1 \times j} Y \times Y \xrightarrow{\mu} Y$$

are nullhomotopic. Conversely, suppose that the composites (4.12) and (4.13) are nullhomotopic. Then the commutativity of the diagram

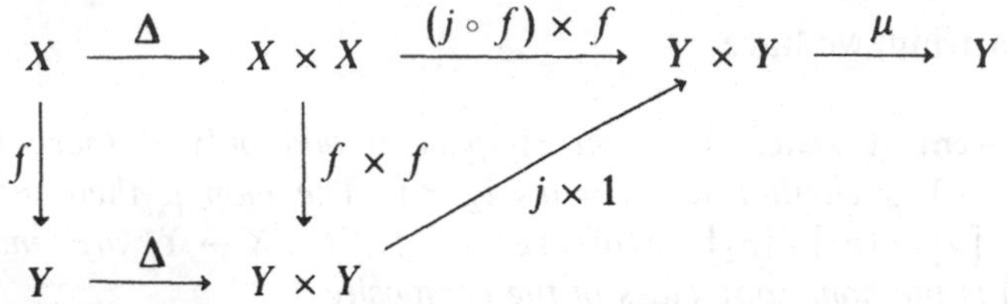

shows that $(j \circ f) \cdot f \simeq *$. Similarly, $f \cdot (j \circ f) \simeq *$.

A *group-like space* is a homotopy associative H-space Y for which there exists a map j as above. In other words, Y satisfies the group axioms up to homotopy.

Remark. In order that Y be group-like, it suffices that either one of the composites (4.12), (4.13) be nullhomotopic. For then every element of the monoid $[X, Y]$ has a one-sided inverse, and this is well-known, in the presence of a unit element, to imply that $[X, Y]$ is a group.

(4.14) Theorem *The sets $[X, Y]$ have group structures natural with respect to X if and only if Y is a group-like space.* □

Let us investigate more closely the question of the existence of inverses. Let $\phi : Y \times Y \to Y \times Y$ be the *shear map*, given by

(4.15) $$\phi(x, y) = (x, xy).$$

If Y is a topological group, then ϕ is a homeomorphism with

$$\phi^{-1}(u, v) = (u, u^{-1}v),$$

and therefore

$$j = p_2 \circ \phi^{-1} \circ i_1.$$

Suppose now that ϕ is a *homotopy equivalence* with homotopy inverse ψ, and define $j \in [Y, Y]$ by

(4.16) $$j = p_2 \circ \psi \circ i_1.$$

Then

$$p_1 \circ \phi = p_1, \qquad p_2 \circ \phi = \mu$$

and therefore

$$p_1 \simeq p_1 \circ \phi \circ \psi = p_1 \circ \psi, \qquad p_2 \simeq p_2 \circ \phi \circ \psi = \mu \circ \psi.$$

In particular,

$$p_1 \circ \psi \circ i_1 \simeq p_1 \circ i_1 = 1.$$

Hence

$$\begin{aligned} \mu \circ (1 \times j) \circ \Delta &= \mu \circ (p_1 \circ \psi \circ i_1 \times p_2 \circ \psi \circ i_1) \circ \Delta \\ &= \mu \circ (p_1 \times p_2) \circ (\psi \circ i_1 \times \psi \circ i_1) \circ \Delta \\ &= \mu \circ (p_1 \times p_2) \circ \Delta \circ \psi \circ i_1 \\ &= \mu \circ \psi \circ i_1 \simeq p_2 \circ i_1 = *; \end{aligned}$$

thus j is a right inverse of the identity map. It follows from the argument above that every element of $[X, Y]$ has a left inverse, and hence that $[X, Y]$ is a group. Conversely, if Y is group-like, the map $(u, v) \to (u, j(u)v)$ is clearly a homotopy inverse of the shear map ϕ. Thus

(4.17) *If Y is a homotopy-associative* H-*space, then Y is group-like if and only if the shear map* (4.15) *is a homotopy equivalence.* □

We now give some examples of H-spaces. Further examples will occur in §5 below.

EXAMPLE 1. Every topological group is an H-space.

EXAMPLE 2. The unit spheres in the division algebras **C**, **Q**, **K** are H-spaces. In fact $\mathbf{S}^1$ and $\mathbf{S}^3$ are topological groups, but the product $\mathbf{S}^7$ is not associative. Later we shall see that it is not even homotopy associative.

An important property of H-spaces is given by

(4.18) Theorem *If Y is an* H-*space, then $\pi_1(Y)$ operates trivially on $[X, Y]$ for any space X.*

For let $u : (\mathbf{I}, \dot{\mathbf{I}}) \to (Y, e)$ and $f : (X, *) \to (Y, e)$ be maps. Then the map $g : \mathbf{I} \times X \to Y$ defined by

$$g(t, x) = u(t) \cdot f(x)$$

is a free homotopy of f to itself along u (we are assuming, as we may, that the base point of Y is a unit element, and not just one up to homotopy), and therefore $\tau_\xi(\alpha) = \alpha$, where ξ, α are the elements of $\pi_1(Y)$, $[X, Y]$ represented by u, f, respectively. □

Theorem (4.18) has a relative form too. An NDR-pair (Y, B) is said to be an H-*pair* if and only if Y and B are H-spaces and the restriction of the product in Y to $B \times B$ is homotopic (and therefore may be assumed equal) to the product in B.

(4.19) Theorem *If* (Y, B) *is an* H-*pair, then* $\pi_1(B)$ *operates trivially on* $[X, A; Y, B]$ *for any* NDR-*pair* (X, A).

The proof is an easy modification of that of Theorem (4.18). □

Let Y, Y' be H-spaces with products μ, μ', respectively. A map $f: Y \to Y'$ is said to be an H-*map* if and only if the diagram

$$\begin{array}{ccc} Y \times Y & \xrightarrow{f \times f} & Y' \times Y' \\ \mu \downarrow & & \downarrow \mu' \\ Y & \xrightarrow[f]{} & Y \end{array}$$

is homotopy commutative. Clearly

(4.20) *If* Y, Y' *are* H-*spaces and* $f: Y \to Y'$ *an* H-*map, then* $\underline{f}: [X, Y] \to [X, Y']$ *is a homomorphism for any space* X. □

(4.21) *If* (Y, B) *is an* NDR-*pair of* H-*spaces, then* (Y, B) *is an* H-*pair if and only if the inclusion map of* B *into* Y *is an* H-*map.* □

An important example of an H-map is given by

(4.22) *The map* $h \mid \Omega(X): \Omega(X) \to \Omega^*(X)$ *of the proof of* (2.18) *is an* H-*map.* □

If Y is an H-space, the map $[X, Y] \times [X, Y] \to [X, Y]$ may be thought of as an *internal product*. There is a corresponding *external product* $[X_1, Y] \times [X_2, Y] \to [X_1 \times X_2, Y]$, defined by $(f_1, f_2) \to f_1 \otimes f_2$, where $f_1 \otimes f_2$ is the composite

$$X_1 \times X_2 \xrightarrow{f_1 \times f_2} Y \times Y \xrightarrow{\mu} Y.$$

(These are comparable with the cup and cross products in cohomology theory). The two kinds of products determine each other as follows: if $f_1, f_2 : X \to Y$, then

$$f_1 \cdot f_2 = (f_1 \otimes f_2) \circ \Delta,$$

while, if $f_1 : X_1 \to Y, f_2 : X_2 \to Y$, and if $p_i : X_1 \times X_2 \to X_i$ is the projection on the ith factor, then

$$f_1 \otimes f_2 = (f_1 \circ p_1) \cdot (f_2 \circ p_2).$$

5 H′-spaces

Let $j_1, j_2 : X \to X \vee X$ be the inclusions, representing $X \vee X$ as the sum of two copies of X, so that $k \circ j_\alpha = i_\alpha \quad (\alpha = 1, 2)$, and let $q_\alpha = p_\alpha \circ k : X \vee X \to X$. Then $q_\alpha \circ j_\alpha = 1 \quad (\alpha = 1, 2)$, while $q_1 \circ j_2 = q_2 \circ j_1 = *$.

Suppose that X is an H′-space. Then we can add the homotopy classes of j_1 and j_2 to obtain an element $[\theta] = [j_1] + [j_2]$ for some map $\theta : X \to X \vee X$. By the naturality of the sum, we have

(5.1) $$[q_1] \circ [\theta] = [q_1] \circ ([j_1] + [j_2]) = [q_1 \circ j_1] + [q_1 \circ j_2] = [1] + [*] = [1]$$

and similarly

(5.2) $$[q_2] \circ [\theta] = [1].$$

The conditions (5.1) and (5.2) are equivalent to

(5.3) $$k \circ \theta \simeq \Delta$$

and a map θ satisfying (5.3) is called a *coproduct* in X. Unlike the case of H-spaces, we cannot replace the homotopy in (5.3) by an equality (for that would imply that the image of Δ is contained in the image $X \vee X$ of k, which is never the case if X is nondegenerate).

With θ as above, let $f_1, f_2 : X \to Y$. Then

(5.4) $$\nabla \circ (f_1 \vee f_2) \circ j_\alpha = f_\alpha \qquad (\alpha = 1, 2)$$

and therefore

$$\begin{aligned}[\nabla] \circ [f_1 \vee f_2] \circ [\theta] &= [\nabla] \circ [f_1 \vee f_2] \circ ([j_1] + [j_2]) \\ &= ([\nabla] \circ [f_1 \vee f_2] \circ [j_1]) + ([\nabla] \circ [f_1 \vee f_2] \circ [j_2]) \\ &= [f_1] + [f_2].\end{aligned}$$

Conversely, let X be a space and let $\theta : X \to X \vee X$ be a map such that $q_1 \circ \theta \simeq 1 \simeq q_2 \circ \theta$. Then, given $f_1, f_2 : X \to Y$, we may define $f_1 + f_2 = \nabla \circ (f_1 \vee f_2) \circ \theta$. The sum so defined is compatible with homotopy and induces a natural operation in $[X, Y]$. Thus X is an H′-space, and

$$j_1 + j_2 = \nabla \circ (j_1 \vee j_2) \circ \theta = \theta.$$

Summarizing, we have

(5.5) Theorem *A space X is an* H′*-space if and only if there is a map* $\theta : X \to X \vee X$ *such that* $q_1 \circ \theta \simeq 1 \simeq q_2 \circ \theta$. *The map* θ *then satisfies the condition* $[\theta] = [j_1] + [j_2]$. *Moreover, if* $f_1, f_2 : X \to Y$, *then* $[f_1] + [f_2]$ *is the homotopy class of the composite*

$$X \xrightarrow{\theta} X \vee X \xrightarrow{f_1 \vee f_2} Y \vee Y \xrightarrow{\nabla} Y. \qquad \square$$

Let $f_1, f_2, f_3 : X \to Y$. Then $(f_1 + f_2) + f_3$ and $f_1 + (f_2 + f_3)$ are the homotopy classes of the composites

$$X \xrightarrow{\theta} X \vee X \xrightarrow{\theta \vee 1}$$
$$X \vee X \vee X \xrightarrow{f_1 \vee f_2 \vee f_3} Y \vee Y \vee Y \xrightarrow{\nabla_3} Y,$$

$$X \xrightarrow{\theta} X \vee X \xrightarrow{1 \vee \theta}$$
$$X \vee X \vee X \xrightarrow{f_1 \vee f_2 \vee f_3} Y \vee Y \vee Y \xrightarrow{\nabla_3} Y,$$

where $\nabla_3 = \nabla \circ (\nabla \vee 1) = \nabla \circ (1 \vee \nabla)$ is the 3-fold folding map. As in the case of H-spaces, we see that the sum in $[X, Y]$ is associative if and only if $(\theta \vee 1) \circ \theta \simeq (1 \vee \theta) \circ \theta$, in which case we say that the H′-space X is *homotopy associative.* Thus

(5.6) Theorem *The sets* $[X, Y]$ *have a monoid structure natural with respect to* Y *if and only if* X *is a homotopy associative* H′*-space.* $\square$

Finally, we say that a homotopy associative H′-space X is cogroup-like if and only if the homotopy class of the identity map $1 \in [X, X]$ has an inverse $[j]$, so that the composites

(5.7) $$X \xrightarrow{\theta} X \vee X \xrightarrow{j \vee 1} X \vee X \xrightarrow{\nabla} X$$

(5.8) $$X \xrightarrow{\theta} X \vee X \xrightarrow{1 \vee j} X \vee X \xrightarrow{\nabla} X$$

are null homotopic.

(5.9) Theorem *The sets* $[X, Y]$ *have group structures natural with respect to* Y *if and only if* X *is cogroup-like.* $\square$

To clarify the question of the existence of inverses in a homotopy-associative H′-space X, consider the *coshear map* $\chi : X \vee X \to X \vee X$, defined by

(5.10) $$\chi \circ j_1 = j_1, \qquad \chi \circ j_2 = \theta.$$

(5.11) Theorem *If* X *is a homotopy associative* H′*-space, then* X *is cogroup-like if and only if the coshear map* $\chi : X \vee X \to X \vee X$ *is a homotopy equivalence.*

For if $\omega : X \vee X \to X \vee X$ is a homotopy inverse of χ, then

$$(5.12) \qquad j_1 \simeq \omega \circ \chi \circ j_1 = \omega \circ j_1, \qquad j_2 \simeq \omega \circ \chi \circ j_2 = \omega \circ \theta,$$

$$q_1 \circ \omega \circ j_1 \simeq q_1 \circ j_1 = 1.$$

and therefore, if we define

$$j = q_1 \circ \omega \circ j_2,$$

then

$$\begin{aligned}\nabla \circ (1 \vee j) \circ \theta &= \nabla \circ (q_1 \circ \omega \circ j_1 \vee q_1 \circ \omega \circ j_2) \circ \theta \\ &= \nabla \circ (q_1 \circ \omega \vee q_1 \circ \omega) \circ (j_1 \vee j_2) \circ \theta \\ &= q_1 \circ \omega \circ \nabla \circ (j_1 \vee j_2) \circ \theta \\ &= q_1 \circ \omega \circ \theta \simeq q_1 \circ j_2 = *.\end{aligned}$$

As before the converse is easy. □

For our first example, we observe that the group structure in $\pi_1(Y) = [\mathbf{S}, Y]$ is natural, and therefore **S** *is an* H′*-space.* It is left to the reader to verify that the map $\theta : \mathbf{S} \to \mathbf{S} \vee \mathbf{S}$ defined by

$$(5.13) \qquad \theta(\bar{t}) = \begin{cases} (\overline{2t}, *) & (0 \le t \le \frac{1}{2}), \\ (*, \overline{2t - 1}) & (\frac{1}{2} \le t \le 1), \end{cases}$$

where $\bar{t}$ is the image of the number $t \in \mathbf{I}$ under the identification map $\mathbf{I} \to \mathbf{S}$, is a coproduct in **S**. Moreover, **S** is cogroup-like, since $\pi_1(Y)$ is a group.

(5.14) Theorem *If X is an* H′*-space, then $X \wedge Y$ is an* H′*-space for any space Y. Moreover, $X \wedge Y$ is homotopy associative if X is homotopy associative, and cogroup-like if X is cogroup-like.*

In fact, we may use the one-to-one correspondence

$$[X \wedge Y, Z] \approx [X, \mathbf{F}(Y, Z)]$$

induced by the homeomorphism

$$\mathbf{F}(X \wedge Y, Z) \approx \mathbf{F}(X, \mathbf{F}(Y, Z))$$

of (2.9), to transfer the given operation in $[X, \mathbf{F}(Y, Z)]$ to one in $[X \wedge Y, Z]$; and the new operation is associative or a group operation if the old one is. □

(5.15) Corollary *For any space Y, the suspension* **S***Y of Y is a cogroup-like space. In particular,* $\mathbf{S}^n$ *is a cogroup-like space* $(n \ge 1)$. □

(5.16) Theorem *If X is an* H′*-space, then* $\mathbf{F}(X, Y)$ *is an* H*-space for every* *space Y. Moreover,* $\mathbf{F}(X, Y)$ *is homotopy associative if X is homotopy associative, and group-like if X is cogroup-like.*

Again, we use the one-to-one correspondence

$$[W, \mathbf{F}(X, Y)] \approx [W \wedge X, Y],$$

together with the commutative law $W \wedge X = X \wedge W$ for the reduced join, to transfer the operation in $[X \wedge W, Y]$ given by Theorem (5.14) to one in $[W, \mathbf{F}(X, Y)]$. □

(5.17) Corollary *For any space Y, the loop space* ΩY *is group-like.* □

(5.18) Theorem *If Y is an* H-*space, then* $\mathbf{F}(X, Y)$ *is an* H-*space for any space X. Moreover,* $\mathbf{F}(X, Y)$ *is homotopy associative if Y is homotopy associative, and group-like if Y is group-like.*

This time we use the one-to-one correspondence

$$[W, \mathbf{F}(X, Y)] \approx [W \wedge X, Y]. \qquad \square$$

(5.19) Theorem *Let X be an* H′-*space. Then for any spaces Y, Z, the projections* $\underline{p_1}$, $\underline{p_2}$ *induce an isomorphism of* $[X, Y \times Z]$ *with the direct product* $[X, Y] \times [X, Z]$.

Let $h : X \to Y \times Z$; the map τ in question carries $[h]$ into the pair $([p_1 \circ h], (p_2 \circ h])$; that τ is a homomorphism follows from the fact that $\underline{p_1}$ and $\underline{p_2}$ are homomorphisms. To prove τ an isomorphism, define $\sigma : [X, Y] \times [X, Z] \to [X, Y \times Z]$ by $\sigma([f], [g]) = [(f \times g) \circ \Delta]$, and verify that $\sigma = \tau^{-1}$. □

(5.20) Theorem *Let Y be an* H-*space. Then, for any spaces W, X, the injections* $\bar{j}_1$, $\bar{j}_2$ *induce an isomorphism of* $[W \vee X, Y]$ *with the direct product* $[W, Y] \times [X, Y]$.

The proof of this is parallel to that of Theorem (5.19), and is left to the reader. □

Now suppose that X is an H′-space, Y an H-space. Then the set $[X, Y]$ has, *a priori*, two structures, one derived from the coproduct in X, the other from the product in Y. In fact

(5.21) Theorem *If X is an* H′-*space and Y an* H-*space, the two structures in* $[X, Y]$ *coincide, and the operation in* $[X, Y]$ *is commutative and associative.*

Let $f_1, f_2 : X \to Y$. Then the diagram

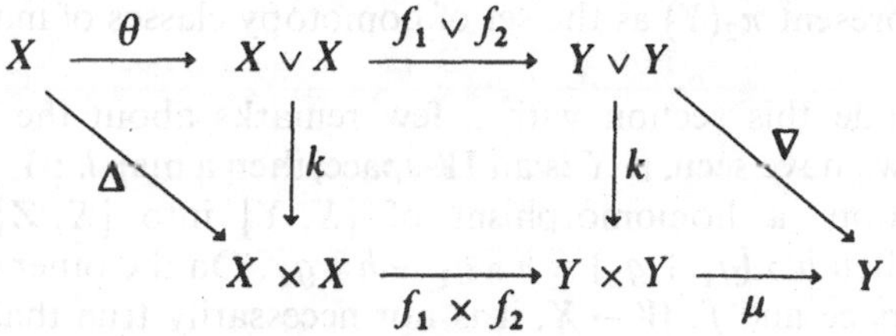

is homotopy-commutative. But $\nabla \circ (f_1 \vee f_2) \circ \theta$ represents $[f_1] + [f_2]$, while $\mu \circ (f_1 \times f_2) \circ \Delta$ represents $[f_1] \cdot [f_2]$. Hence $[f_1] + [f_2] = [f_1] \cdot [f_2]$.

Because of naturality, composition with the homotopy class of μ is a homomorphism $\mu_* : [X, Y \times Y] \to [X, Y]$. Composition of μ_* with the isomorphism $\sigma : [X, Y] \times [X, Y] \to [X, Y \times Y]$ of Theorem (5.19) is then a homomorphism of $[X, Y] \times [X, Y]$ into $[X, Y]$. Now $\mu_* \circ \sigma([f], [g]) = [\mu] \circ [f \times g] \circ \Delta = [f] \cdot [g]$. But it is well known (and trivial) that a multiplicative system G with unit, for which the operation is a homomorphism of $G \times G$ into G, is commutative and associative. □

(5.22) Corollary *For any spaces X, Y, the three sets $[\mathbf{S}^2X, Y]$, $[\mathbf{S}X, \Omega Y]$, and $[X, \Omega^2 Y]$ are isomorphic abelian groups.* □

We have seen that $\mathbf{S}^n$ is a cogroup-like space ($n \geq 1$), and accordingly $\pi_n(Y) = [\mathbf{S}^n, Y]$ is a group, the nth *homotopy group* of Y. By Corollary (5.22), with $X = \mathbf{S}^{n-2}$, we have □

(5.23) Corollary *If $n \geq 2$, the homotopy group $\pi_n(Y)$ is abelian.* □

(5.24) Corollary *If Y is an* H*-space, then $\pi_1(Y)$ is abelian.* □

A direct argument that $\pi_n(Y)$ is abelian is suggested by Figure 3.2, in

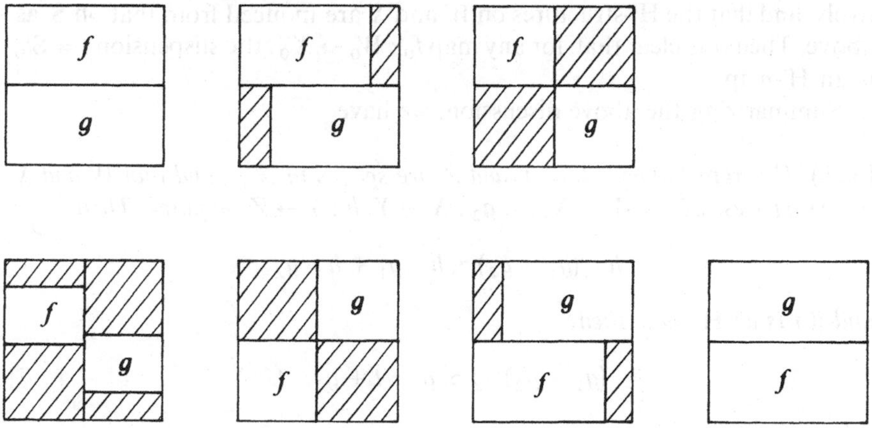

Figure 3.2

which we represent $\pi_2(Y)$ as the set of homotopy classes of maps of $(\mathbf{I}^2, \dot{\mathbf{I}}^2)$ into (Y, y_0).

We conclude this section with a few remarks about the composition product. As we have seen, if X is an H'-space, then a map $h : Y \to Z$ induces, by composition, a homomorphism of $[X, Y]$ into $[X, Z]$, i.e., if g_1, $g_2 : X \to Y$, then $h \circ (g_1 + g_2) \simeq h \circ g_1 + h \circ g_2$. On the other hand, if W is also an H'-space and $f: W \to X$, it is not necessarily true that the map of $[X, Y]$ into $[W, Y]$, induced by composition with f, be a homomorphism. However, this will be the case if f is an H'-map, in the sense that if $\theta : W \to W \vee W$ and $\phi : X \to X \vee X$ are the coproducts in W and X, respectively, then the diagram

$$\begin{array}{ccc} W & \xrightarrow{\theta} & W \vee W \\ {\scriptstyle f}\downarrow & & \downarrow{\scriptstyle f \vee f} \\ X & \xrightarrow[\phi]{} & X \vee X \end{array}$$

is homotopy commutative. For then, if $g_1, g_2 : X \to Y$, the diagram

$$\begin{array}{ccccccc} W & \xrightarrow{\theta} & W \vee W & & & & \\ {\scriptstyle f}\downarrow & & \downarrow{\scriptstyle f \vee f} & \searrow^{g_1 \circ f \vee g_2 \circ f} & & & \\ X & \xrightarrow[\phi]{} & X \vee X & \xrightarrow[g_1 \vee g_2]{} & Y \vee Y & \xrightarrow[\nabla]{} & Y \end{array}$$

is also homotopy commutative.

Suppose that W and X are the suspensions of spaces W_0 and X_0, respectively, and that the H'-structures on W and X are induced from that on $\mathbf{S}$, as above. Then it is clear that, for any map $f_0 : W_0 \to X_0$, the suspension $f = \mathbf{S}f_0$ is an H'-map.

Summarizing the above discussion, we have

(5.25) Theorem *Suppose that Y and Z are spaces in $\mathscr{K}_*$, and that W and X are* H'*-spaces. Let $f: W \to X$, $g_1, g_2 : X \to Y$, $h : Y \to Z$ be maps. Then*

$$h \circ (g_1 + g_2) \simeq h \circ g_1 + h \circ g_2;$$

and if f is an H'*-map, then*

$$(g_1 + g_2) \circ f \simeq g_1 \circ f + g_2 \circ f. \qquad \square$$

(5.26) Corollary *If W_0, X_0, Y, Z are spaces in $\mathscr{K}_*$ and $f_0 : W_0 \to X_0$,*

$g_1, g_2 : \mathbf{S}X_0 \to Y, h : Y \to Z$, *then*

$$h \circ (g_1 + g_2) \simeq h \circ g_1 + h \circ g_2,$$
$$(g_1 + g_2) \circ \mathbf{S}f_0 \simeq g_1 \circ \mathbf{S}f_0 + g_2 \circ \mathbf{S}f_0. \qquad \square$$

Finally, we observe that if X is an H′-space, there is an external product

$$[X, Y_1] \times [X, Y_2] \to [X, Y_1 \vee Y_2],$$

defined by $(f_1, f_2) \to f_1 \oplus f_2$, where $f_1 \oplus f_2$ is the composite

$$X \xrightarrow{\theta} X \vee X \xrightarrow{f_1 \vee f_2} Y_1 \vee Y_2.$$

As in the case of H-spaces, the external and internal products determine each other: if $f_1, f_2 : X \to Y$, then $f_1 + f_2$ is the composite

$$X \xrightarrow{f_1 \oplus f_2} Y \vee Y \xrightarrow{\nabla} Y;$$

and if $f_i : X \to Y_i$ $(i = 1, 2)$, then

$$f_1 \oplus f_2 = (j_1 \circ f_1) + (j_2 \circ f_2),$$

where $j_i : Y_i \to Y_1 \vee Y_2$ is the inclusion.

6 Exact Sequences of Mapping Functors

A diagram

(6.1)
$$X \xrightarrow{f} Y \xrightarrow{g} Z$$

in $\mathscr{K}_0$ is said to be *left exact* if and only if, for every space W in $\mathscr{K}_*$, the diagram

$$[W, X] \xrightarrow{f} [W, Y] \xrightarrow{g} [W, Z]$$

is exact in the category $\mathscr{M}_0$, of sets with base point, i.e., $\operatorname{Im} \underline{f} = \operatorname{Ker} \underline{g} = \{\alpha \in [W, Y] \mid g(\alpha) = *\}$. Similarly, the diagram (6.1) is *right exact* if and only if, for every space W in $\mathscr{K}_0$, the diagram

$$[Z, W] \xrightarrow{\overline{g}} [Y, W] \xrightarrow{\overline{f}} [X, W]$$

is exact in $\mathscr{M}_0$.

For example,

(6.2) *Every fibration*

$$F \xrightarrow{i} X \xrightarrow{p} B$$

in $\mathscr{K}_0$ is left exact.

For suppose that $g : (Y, *) \to (X, *)$ is a map such that $p \circ g \simeq *$ (rel. $*$). By the homotopy lifting extension property (Theorem (7.16) of Chapter I), g is homotopic (rel. $*$) to a map of Y into F. Hence Ker $\underline{p} \subset$ Im $\underline{i}$; the opposite inclusion is trivial. □

Dually, we have

(6.3) *If (X, A) is an* NDR*-pair, $i : A \subsetneq X$, and if $p : X \to X/A$ is the identification map, then the sequence*

$$A \xrightarrow{\ i\ } X \xrightarrow{\ p\ } X/A$$

is right exact. □

A sequence of spaces and maps

(6.4) $$\cdots \to X_{n+1} \xrightarrow{f_{n+1}} X_n \xrightarrow{f_n} X_{n-1} \to \cdots$$

(which may terminate in either direction) is left (right) exact if and only if each of its sections

$$X_{n+1} \longrightarrow X_n \longrightarrow X_{n-1}$$

is left (right) exact.

Let $f : X \to Y$ be an arbitrary map. Then there is a homotopy commutative diagram

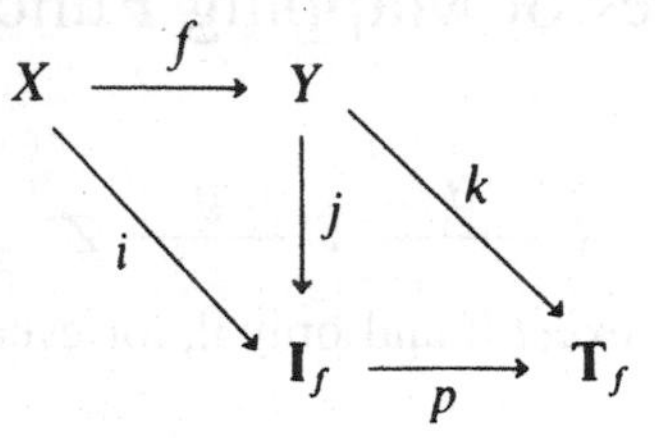

in which $\mathbf{I}_f$ ($\mathbf{T}_f$) is the mapping cylinder (cone) of f, i, j and k are inclusions and p is the identification map. Moreover, $(\mathbf{I}_f, X)$ is an NDR-pair and j is a homotopy equivalence. We have seen that the sequence

$$X \longrightarrow \mathbf{I}_f \longrightarrow \mathbf{T}_f$$

is right exact; it follows that the sequence

$$X \xrightarrow{\ f\ } Y \xrightarrow{\ k\ } \mathbf{T}_f$$

is right exact.

Iterating the above procedure, we find:

(6.5) *Let $f : X \to Y$ be a map in $\mathscr{K}_*$. Then there is an infinite right exact* *sequence*

(6.6) $$X^0 \xrightarrow{f^0} X^1 \xrightarrow{f^1} X^2 \to \cdots \to X^n \xrightarrow{f^n} X^{n+1} \to \cdots$$

such that $f^0 = f$ *and* X^n *is the mapping cone of* f^{n-2} $(n \geq 2)$. □

On the other hand, we may use Theorem (7.30) of Chapter I to replace an arbitrary map f by a homotopically equivalent fibre map, obtaining a homotopy commutative diagram

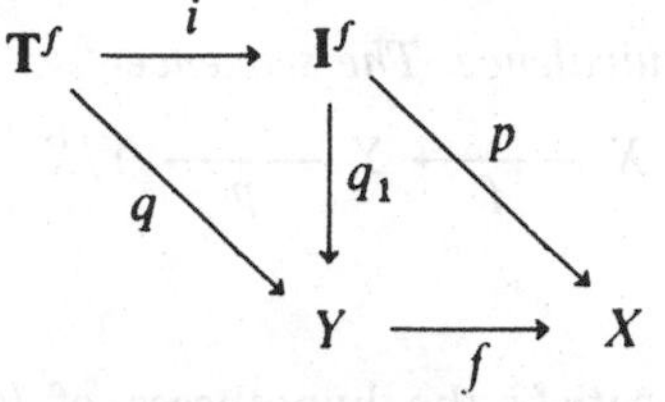

in which p, q, and q_1 are fibrations, i the inclusion map, and q_1 is a homotopy equivalence. Since the sequence

$$\mathbf{T}^f \xrightarrow{i} \mathbf{I}^f \xrightarrow{p} X$$

is left exact, so is the sequence

$$\mathbf{T}^f \xrightarrow{q} Y \xrightarrow{f} X.$$

Again, we may iterate the above procedure to obtain

(6.7) *Let* $f: Y \to X$ *be a map in* $\mathcal{K}_*$. *Then there is an infinite left exact sequence*

(6.8) $$\cdots \to X_{n+1} \xrightarrow{f_n} X_n \to \cdots \to X_1 \xrightarrow{f_0} X_0$$

such that $f_0 = f$ and $X_n = \mathbf{T}^{f_{n-2}}$ $(n \geq 2)$. □

The constructions of the sequences (6.6) and (6.8) are canonical. Our objective in this section is to study them in detail.

Let us begin by examining the first few terms of the sequence (6.6). They are

$$X \xrightarrow{f} Y \xrightarrow{k} \mathbf{T}_f \xrightarrow{l} \mathbf{T}_k,$$

where l is the inclusion map.

Suppose that (Y, X) is an NDR-pair and f is the inclusion map. Then $\mathbf{T}_f$ is the subspace $Y \cup \mathbf{T}X$ of the cone $\mathbf{T}Y$, and $(\mathbf{T}_f, \mathbf{T}X)$ is an NDR-pair. By Corollary (5.13) of Chapter I, the identification map of $\mathbf{T}_f$ into $\mathbf{T}_f/\mathbf{T}X$ is a homotopy equivalence. But $\mathbf{T}_f/\mathbf{T}X$ is naturally homeomorphic with $Y/Y \cap \mathbf{T}X = Y/X$. Hence

(6.9) *Let (Y, X) be an NDR-pair, $f : X \hookrightarrow Y$, $k : Y \hookrightarrow \mathbf{T}_f$, and let $p : Y \to Y/X$, $p_1 : \mathbf{T}_f \to Y/X$ be the identification maps. Then there is a commutative diagram*

$$\begin{array}{ccccc} X & \xrightarrow{f} & Y & \xrightarrow{k} & \mathbf{T}_f \\ & & & \searrow^{p} & \downarrow{\scriptstyle p_1} \\ & & & & Y/X \end{array}$$

and p_1 is a homotopy equivalence. The sequence

$$X \xrightarrow[f]{} Y \xrightarrow[p]{} Y/X$$

is right exact. □

Even if f does not satisfy the hypotheses of (6.9), the map k does. Moreover, the quotient space $\mathbf{T}_f/Y = Y \cup_f \mathbf{T}X/Y$ is naturally homeomorphic with $\mathbf{S}X$. Thus

(6.10) *There is a commutative diagram*

$$\begin{array}{ccccc} Y & \xrightarrow{k} & \mathbf{T}_f & \xrightarrow{l} & \mathbf{T}_k \\ & & & \searrow^{q} & \downarrow{\scriptstyle q_1} \\ & & & & \mathbf{S}X \end{array}$$

such that q_1 is a homotopy equivalence and the sequence

$$Y \xrightarrow{k} \mathbf{T}_f \xrightarrow{q} \mathbf{S}X$$

is right exact. □

The next theorem clarifies the structure of the sequence (6.6).

(6.11) Theorem *Let $f: X \to Y$ be a map in $\mathscr{K}_*$. Then the sequence*

(6.12)

$$X \xrightarrow{f} Y \xrightarrow{k} \mathbf{T}_f \xrightarrow{q} \mathbf{S}X \to \cdots$$
$$\to \mathbf{S}^{n-1}\mathbf{T}_f \xrightarrow{\mathbf{S}^{n-1}q} \mathbf{S}^n X \xrightarrow{\mathbf{S}^n f} \mathbf{S}^n Y \xrightarrow{\mathbf{S}^n k} \mathbf{S}^n \mathbf{T}_f \xrightarrow{\mathbf{S}^n q} \mathbf{S}^{n+1} X \to \cdots$$

is right exact.

We first explain the relationship between the sequences (6.6) and (6.12). In order to formulate this, we shall need the *antisuspension* operator $-\mathbf{S}$.

This operator assigns to each map $f: X \to Y$ the map $-\mathbf{S}f: \mathbf{S}X \to \mathbf{S}Y$ defined by

$$(-\mathbf{S}f)(\bar{t} \wedge x) = \overline{1 - t} \wedge f(x)$$

for $t \in I$, $x \in X$. Let us observe that the iterate $(-\mathbf{S}) \circ (-\mathbf{S})$ is $\mathbf{S}^2$. Moreover, the homotopy class of $-\mathbf{S}f$ in the group $[\mathbf{S}X, \mathbf{S}Y]$ is minus that of $\mathbf{S}f$.

Let us modify the sequence (6.12) by replacing the maps $\mathbf{S}^n f, \mathbf{S}^n k, \mathbf{S}^n q$ by their negatives for all odd n. The resulting sequence

$$\textbf{(6.13)}\quad X \xrightarrow{f} Y \xrightarrow{k} \mathbf{T}_f \xrightarrow{q} \mathbf{S}X \xrightarrow{-\mathbf{S}f}$$
$$\mathbf{S}Y \xrightarrow{-\mathbf{S}k} \mathbf{ST}_f \xrightarrow{-\mathbf{S}q} \mathbf{S}^2X \xrightarrow{\mathbf{S}^2 f} \mathbf{S}^2 Y \to \cdots$$

is right exact if and only if (6.12) is. And we shall show that (6.13) is right exact by proving

(6.14) Theorem *The sequences* (6.6) *and* (6.13) *are homotopically equivalent.*

This means that there are homotopy equivalences $h^n : X^n \to Y^n$, Y^n being the nth term of (6.13), such that the diagrams

$$\begin{array}{ccc} X^n & \xrightarrow{f^n} & X^{n+1} \\ \big\downarrow{\scriptstyle h^n} & & \big\downarrow{\scriptstyle h^{n+1}} \\ Y^n & \xrightarrow[g_n]{} & Y^{n+1} \end{array}$$

are homotopy commutative.

Let us now scrutinize more carefully the construction of (6.10). The map q_1 is the composite $r_1^{-1} \circ \bar{q}$, where

$$\bar{q} : \mathbf{T}_k \to \mathbf{T}_k/\mathbf{T}Y$$

is the identification map and

$$r_1 : \mathbf{S}X = \mathbf{T}X/X \to \mathbf{T}_k/\mathbf{T}Y$$

is the homeomorphism induced by the composite

$$(\mathbf{T}X, X) \to (\mathbf{T}_f, Y) = (\mathbf{T}_f, \mathbf{T}_f \cap \mathbf{T}Y) \subsetneq (\mathbf{T}_f \cup \mathbf{T}Y, \mathbf{T}Y) = (\mathbf{T}_k, \mathbf{T}Y).$$

Let us apply the same construction with f replaced by $k : Y \to \mathbf{T}_f$. We obtain a commutative diagram

$$\begin{array}{ccccc} \mathbf{T}_f & \xrightarrow{l} & \mathbf{T}_k & \xrightarrow{m} & \mathbf{T}_l \\ & & & \searrow^{q'} & \downarrow{\scriptstyle q_2} \\ & & & & \mathbf{S}Y \end{array}$$

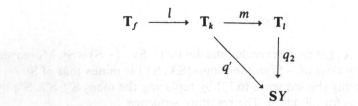

analogous to that of (6.10). The map q_2 is the composite $r_2^{-1} \circ \tilde{q}$, where

$$\tilde{q} : \mathbf{T}_l \to \mathbf{T}_l/\mathbf{T}\mathbf{T}_f$$

is the identification map and

$$r_2 : \mathbf{S}Y = \mathbf{T}Y/Y \to \mathbf{T}_l/\mathbf{T}\mathbf{T}_f$$

is the homeomorphism induced by the composite

$$(\mathbf{T}Y, Y) \to (\mathbf{T}_k, \mathbf{T}_f) = (\mathbf{T}_k, \mathbf{T}_k \cap \mathbf{T}\mathbf{T}_f) \subsetneq (\mathbf{T}\mathbf{T}_f \cup \mathbf{T}_k, \mathbf{T}\mathbf{T}_f) = (\mathbf{T}_l, \mathbf{T}\mathbf{T}_f).$$

The map q' is the composite $r_2'^{-1} \circ q_2'$, where

$$q_2' : \mathbf{T}_k \to \mathbf{T}_k/\mathbf{T}_f$$

is the identification map and the homeomorphism

$$r_2' : \mathbf{S}Y = \mathbf{T}Y/Y \to \mathbf{T}_k/\mathbf{T}_f$$

is induced by the inclusion

$$(\mathbf{T}Y, Y) \to (\mathbf{T}_k, \mathbf{T}_f).$$

The next Lemma is the crucial step in the proof of Theorem (6.14).

(6.15) Lemma *The diagram*

$$\begin{array}{ccc} \mathbf{T}_k & \xrightarrow{m = f^3} & \mathbf{T}_l \\ {\scriptstyle q_1}\downarrow & & \downarrow{\scriptstyle q_2} \\ \mathbf{S}X & \xrightarrow[-\mathbf{S}f]{} & \mathbf{S}Y \end{array}$$

is homotopy commutative.

Let us adjoin to the diagram whose commutativity is to be proved the map $q' : \mathbf{T}_k \to \mathbf{S}Y$; we obtain a new diagram

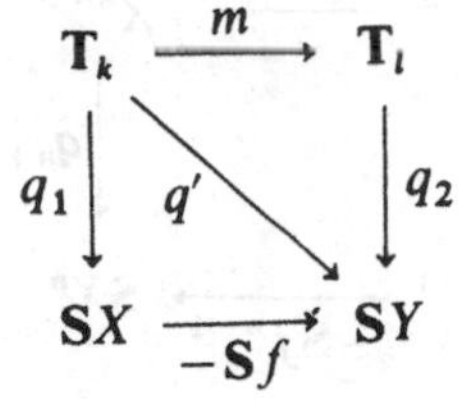

in which the right-hand triangle is already commutative. Therefore it suffices to prove that the left-hand triangle is also commutative. Note that q_1 maps $\mathbf{T}Y$ into the base point and $q_1 | \mathbf{T}_f : Y \cup_f \mathbf{T}X = \mathbf{T}_f \to \mathbf{S}X$ is defined by the maps

$$\begin{cases} y \to *, \\ t \wedge x \to \bar{t} \wedge x; \end{cases}$$

while q' maps $\mathbf{T}_f$ into the base point and

$$q'(t \wedge y) = \bar{t} \wedge y$$

for $t \in T$, $y \in Y$.

Let us consider the map $\psi : \mathbf{T}_k \to \mathbf{S}Y$ defined as follows. The restriction of ψ to $\mathbf{T}Y$ is the natural homeomorphism with $\mathbf{T}_+ Y : \psi(t \wedge y) = \overline{\tfrac{1}{2}t} \wedge y$ for $t \in \mathbf{T}$, $y \in Y$. On the other hand, the restriction of ψ to $\mathbf{T}_f = Y \cup_f \mathbf{T}X$ is defined by the maps

$$\begin{cases} y \to \overline{\tfrac{1}{2}t} \wedge y, \\ t \wedge x \to \overline{(1 - \tfrac{1}{2}t)} \wedge f(x); \end{cases}$$

note that they are consistent with the identifications $1 \wedge y = f(x)$. Contracting the cone $\mathbf{T}_+ Y$ to a point, we obtain a homotopy of ψ to a map ψ_1 such that $\psi_1(\mathbf{T}Y) = *$, while $\psi_1 | \mathbf{T}_f$ is defined by the maps

$$\begin{cases} y \to *, \\ t \wedge x \to \overline{1 - t} \wedge f(x); \end{cases}$$

this map is clearly $(-\mathbf{S}f) \circ q_1$. On the other hand, contraction of $\mathbf{T}_- Y$ to a point yields a homotopy of ψ to the map ψ_2 such that $\psi_2(\mathbf{T}_f) = *$, while

$$\psi_2(t \wedge y) = \bar{t} \wedge y;$$

this map is, equally clearly, q'. □

Iterating the construction of q_1 gives us a sequence of maps $q_k : X^{k+2} \to \mathbf{S}X^{k-1}$ $(k = 1, 2, 3, \ldots)$ (cf. Figure 3.3), and it follows from Lemma (6.13) that

(6.16) *The diagrams*

$$\begin{array}{ccc} X^{n+2} & \xrightarrow{f^{n+2}} & X^{n+3} \\ {\scriptstyle q_n}\downarrow & & \downarrow{\scriptstyle q_{n+1}} \\ \mathbf{S}X^{n-1} & \xrightarrow[-\mathbf{S}f^{n-1}]{} & \mathbf{S}X^{n} \end{array}$$

are homotopy commutative.

Let us examine the diagram of Figure 3.3. The top row is the sequence (6.6). Each successive row is formed from the preceding one by applying the anti-suspension operator and shifting the resulting sequence three units to the right. The third entry $\mathbf{S}^i X_2$ in the $(i + 1)$st row is mapped into the leading entry $\mathbf{S}^{i+1} X_0$ by $\mathbf{S}^i q$ or $-\mathbf{S}^i q$ according as i is even or odd. Finally, the maps $\mathbf{S}^i q_n : \mathbf{S}^i X^{n+2} \to \mathbf{S}^{i+1} X^{n-1}$ complete the diagram.

According to (6.10), the diagram

$$\begin{array}{ccc} X^{2} & \xrightarrow{f^{2}} & X^{3} \\ & {\scriptstyle q}\searrow & \downarrow{\scriptstyle q_1} \\ & & \mathbf{S}X^{0} \end{array}$$

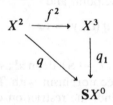

is homotopy commutative. According to (6.16), each of the top row of squares in Figure 3.3 is homotopy commutative. It follows by induction that

(6.17) *The diagram of Figure* 3.3 *is homotopy commutative.* □

The sequence which runs along the bottom edge of Figure 3.3 is precisely (6.13). Therefore the truth of Theorem (6.14) follows from the commutativity of Figure 2, which we have just established. □

From the right exactness of the sequence (6.12) we conclude

(6.18) *For any map $f: X \to Y$ in $\mathcal{K}_*$ and any space $W \in \mathcal{K}_*$, there is an exact sequence*

(6.19)

$$\cdots \to [\mathbf{S}^{n+1}X, W] \longrightarrow [\mathbf{S}^n\mathbf{T}_f, W] \longrightarrow [\mathbf{S}^n Y, W] \longrightarrow [\mathbf{S}^n X, W]$$

$$\to \cdots \to [\mathbf{S}X, W] \xrightarrow{\bar{q}} [\mathbf{T}_f, W] \xrightarrow{\bar{k}} [Y, W] \xrightarrow{\bar{f}} [X, W].$$

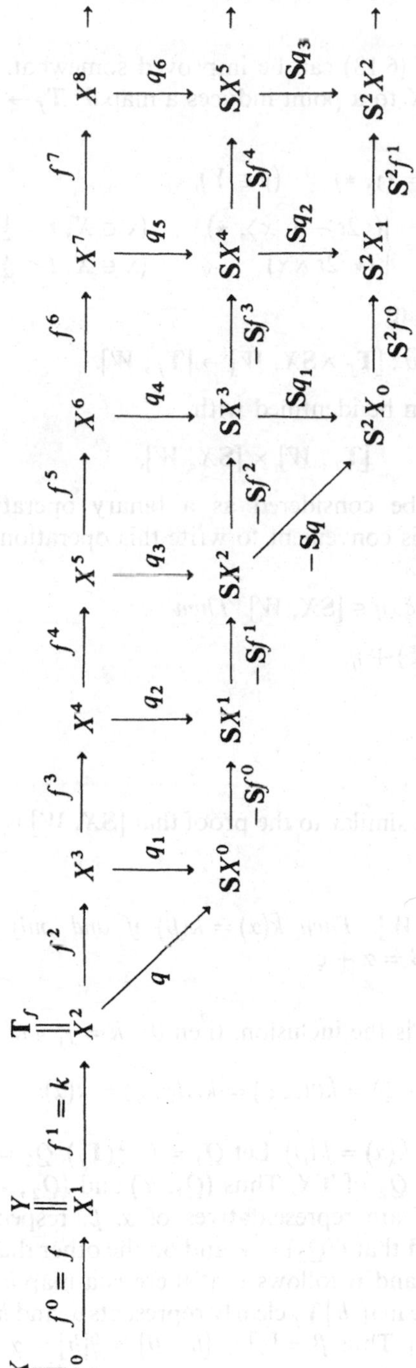

Figure 3.3

The terms of the sequence (6.19), *except possibly for the last three, are groups and the maps of the sequence, except for the last three, are homomorphisms.* □

The last sentence of (6.18) can be improved somewhat. Collapsing the mid-section $\frac{1}{2} \wedge X$ of $\mathbf{T}X$ to a point induces a map $\theta : \mathbf{T}_f \to \mathbf{T}_f \wedge \mathbf{S}X$; more precisely, θ is given by

$$\theta(y) = (y, *) \qquad (y \in Y),$$

$$\theta(\langle t, x \rangle) = \begin{cases} (\langle 2t - 1, x \rangle, *) & (x \in X, t \geq \frac{1}{2}), \\ (*, \overline{2t} \wedge x) & (x \in X, t \leq \frac{1}{2}). \end{cases}$$

The map θ induces a map

$$\bar{\theta} : [\mathbf{T}_f \wedge \mathbf{S}X, W] \to [\mathbf{T}_f, W].$$

But the domain of $\bar{\theta}$ can be identified with

$$[\mathbf{T}_f, W] \times [\mathbf{S}X, W],$$

and therefore $\bar{\theta}$ can be considered as a binary operation $[\mathbf{T}_f, W] \times [\mathbf{S}X, W] \to [\mathbf{T}_f, W]$. It is convenient to write this operation as a sum.

(6.20) *Let* $\alpha \in [\mathbf{T}_f, W]$, $\xi, \eta \in [\mathbf{S}X, W]$. *Then*

(1) $\alpha + (\xi + \eta) = (\alpha + \xi) + \eta$,
(2) $\alpha + 0 = \alpha$,
(3) $0 + \xi = \bar{q}(\xi)$,
(4) $\bar{q}(\xi) + \eta = \bar{q}(\xi + \eta)$.

The proof of (6.20) is similar to the proof that $[\mathbf{S}X, W]$ is a group, and is left to the reader. □

(6.21) *Let* $\alpha, \beta \in [\mathbf{T}_f, W]$. *Then* $\bar{k}(\alpha) = \bar{k}(\beta)$ *if and only if there exists* $\xi \in [\mathbf{S}X, W]$ *such that* $\beta = \alpha + \xi$.

If $j_1 : \mathbf{T}_f \to \mathbf{T}_f \wedge \mathbf{S}X$ is the inclusion, then $\theta \circ k = j_1 \circ k$. It follows that

$$\bar{k}(\alpha + \xi) = \bar{k}\bar{\theta}(\alpha, \xi) = \bar{k}\bar{j}_1(\alpha, \xi) = \bar{k}(\alpha).$$

Conversely, suppose $\bar{k}(\alpha) = \bar{k}(\beta)$. Let $Q_1 = \theta^{-1}(\mathbf{T}_f)$, $Q_2 = \theta^{-1}(\mathbf{S}X)$; then Q_1 is a copy of $\mathbf{I}_f$ and Q_2 of $\mathbf{T}X$. Thus (Q_1, Y) and $(Q_2, *)$ are DR-pairs. Hence, if $f, g : \mathbf{T}_f \to W$ are representatives of α, β, respectively, we may assume on the one hand that $f(Q_2) = *$, and on the other that $f \mid Q_1 = g \mid Q_1$. Then $g(Q_1 \cap Q_2) = *$, and it follows that there is a map $h : \mathbf{T}_f \vee \mathbf{S}X \to W$ such that $h \circ \theta = g$. The map $h \mid \mathbf{T}_f$ clearly represents α, and $h \mid \mathbf{S}X$ represents an element $\xi \in [\mathbf{S}X, W]$. Thus $\beta = [g] = [h \circ \theta] = \bar{\theta}[h] = \alpha + \xi$. □

Suppose, in particular, that (Y, X) is an NDR-pair, $f: X \hookrightarrow Y$. Then the homotopy commutative diagram

$$\begin{array}{ccccccc} X & \overset{f}{\hookrightarrow} & Y & \xrightarrow{k} & \mathbf{T}_f & \xrightarrow{q} & \mathbf{S}X \\ & & & \searrow^{p} & \downarrow{\scriptstyle p_1} & \nearrow^{h} & \\ & & & & Y/X & & \end{array}$$

of (6.9) can be completed; we may take $h = q \circ p_1'$, where $p_1' : Y/X \to \mathbf{T}_f$ is a homotopy inverse of p_1. The map $h : Y/X \to \mathbf{S}X$ is called a *connecting map* for the pair (Y, X). For this important special case Theorem (6.11) takes the form

(6.22) Theorem *Let* (Y, X) *be an NDR-pair,* $f: X \hookrightarrow Y$, $p: Y \to Y/X$ *the identification map,* $h: Y/X \to \mathbf{S}X$ *a connecting map. Then the sequence*

$$X \xrightarrow{f} Y \xrightarrow{p} Y/X \xrightarrow{h} \mathbf{S}X \to \cdots$$
$$\to \mathbf{S}^{n-1}(Y/X) \xrightarrow{\mathbf{S}^{n-1}h} \mathbf{S}^n X \xrightarrow{\mathbf{S}^n f}$$
$$\mathbf{S}^n Y \xrightarrow{\mathbf{S}^n p} \mathbf{S}^n(Y/X) \xrightarrow{\mathbf{S}^n h} \mathbf{S}^{n+1}X \to \cdots$$

is homotopy-equivalent to (6.6) *and therefore is right exact.* □

Let us now turn our attention to the "dual" sequence (6.8). We shall first suppose that $f: Y \to X$ is a fibration, with fibre F. By Theorem (7.31) of Chapter I, the maps $p : \mathbf{I}^f \to X$ and $f : Y \to X$ have the same fibre homotopy type, and, in particular, their fibres $\mathbf{T}^f$ and F have the same homotopy type. In fact, the map $z \to (z, *)$ of F into $\mathbf{T}^f$ is a homotopy equivalence j_1, and if $q : \mathbf{T}^f \to Y$ is the fibration of $\mathbf{T}^f$ over Y, then $q \circ j_1$ is the inclusion map $j : F \hookrightarrow Y$. Thus

(6.9*) *Let* $f: Y \to X$ *be a fibration, with fibre* F *and inclusion* $j: F \hookrightarrow Y$. *Let* $i_1 : F \to \mathbf{T}^f$, $q: \mathbf{T}^f \to Y$ *be as above. Then the diagram*

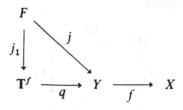

$$\begin{array}{ccccc} F & & & & \\ \downarrow{\scriptstyle j_1} & \searrow^{j} & & & \\ \mathbf{T}^f & \xrightarrow[q]{} & Y & \xrightarrow[f]{} & X \end{array}$$

is commutative, and j_1 is a homotopy equivalence. The sequence

$$F \xrightarrow{j} Y \xrightarrow{f} X$$

is exact. □

Even though f may not satisfy the hypotheses of (6.9*), the map q does. Moreover, the fibre of q is the space of pairs (y, u) such that $y = *$ and $u(0) = u(1) = *$, and this is homeomorphic with ΩX. Thus

(6.10*) *There is a commutative diagram*

$$\begin{array}{ccccc} \Omega X & & & & \\ {\scriptstyle k_1}\downarrow & \searrow^{k} & & & \\ \mathbf{T}^q & \xrightarrow[r]{} & \mathbf{T}^f & \xrightarrow[q]{} & Y \end{array}$$

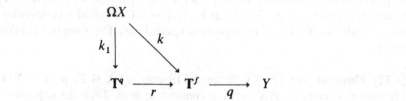

such that k_1 is a homotopy equivalence and the sequence

$$\Omega X \xrightarrow[k]{} \mathbf{T}^f \xrightarrow[q]{} Y$$

is left exact. □

Iterating the above construction, we obtain a commutative diagram

$$\begin{array}{ccccc} \Omega Y & & & & \\ {\scriptstyle k_2}\downarrow & \searrow^{k'} & & & \\ \mathbf{T}^r & \xrightarrow[s]{} & \mathbf{T}^q & \xrightarrow[r]{} & \mathbf{T}^f \end{array}$$

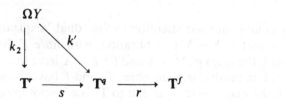

We shall need the anti loop operator $-\Omega$; if $f: X \to Y$, then $-\Omega f: \Omega X \to \Omega Y$ is defined by

$$(-\Omega f)(u)(t) = f(u(1 - t)).$$

and the homotopy class of $-\Omega f$ in $[\Omega X, \Omega Y]$ is minus that of f.

(6.15*) Lemma *The diagram*

$$\begin{array}{ccc} \Omega Y & \xrightarrow{-\Omega f} & \Omega X \\ {\scriptstyle k_2}\downarrow & & \downarrow{\scriptstyle k_1} \\ \mathbf{T}^r & \xrightarrow[s]{} & \mathbf{T}^q \end{array}$$

is homotopy commutative.

As in the proof of Lemma (6.15), it suffices to show that the diagram

$$\begin{array}{ccc} \Omega Y & \xrightarrow{-\Omega f} & \Omega X \\ & {}_{k'}\searrow & \downarrow k_1 \\ & & \mathbf{T}^q \end{array}$$

is homotopy commutative. Let us observe that, if $v \in \Omega Y$, then

$$k'(v) = (*, e_X, v) \in Y \times \mathbf{F}(X) \times \mathbf{F}(Y),$$

where e_X is the constant map of $\mathbf{I}$ into the base point $*$ of X. On the other hand,

$$\begin{aligned} k_1(-\Omega f(v)) &= k_1(f \circ v^{-1}) \\ &= (*, f \circ v^{-1}, e_Y). \end{aligned}$$

A homotopy between these maps is given by

$$H(t, v) = (v(t), u_t, w_t),$$

where $u_t : \mathbf{I} \to X$ is given by

$$u_t(s) = \begin{cases} * & \text{if } t \leq s, \\ f(v(t-s)) & \text{if } t \geq s, \end{cases}$$

and $w_t : \mathbf{I} \to Y$ is given by

$$w_t(s) = \begin{cases} * & \text{if } s + t \geq 1, \\ v(s+t) & \text{if } s + t \leq 1. \end{cases}$$ □

We can now prove, in much the same way as (6.11)

(6.11*) Theorem *Let $f: X \to Y$ be a map in $\mathscr{K}_*$. Then the sequence*

(6.12*) $$\cdots \to \Omega^{n+1}X \xrightarrow{\Omega^n k} \Omega^n\mathbf{T}^f \xrightarrow{\Omega^n q} \Omega^n Y \xrightarrow{\Omega^n f} \Omega^n X \xrightarrow{\Omega^{n-1}k}$$
$$\Omega^{n-1}\mathbf{T}^f \to \cdots \to \Omega X \xrightarrow{k} \mathbf{T}^f \xrightarrow{q} Y \xrightarrow{f} X$$

is left exact. □

(6.18*) *For any map $f: X \to Y$ in $\mathscr{K}_*$ and any space $Z \in \mathscr{K}_*$, there is an exact sequence*

(6.19*) $$\cdots \to [Z, \Omega^{n+1}X] \to [Z, \Omega^n \mathbf{T}^f] \to [Z, \Omega^n Y] \to [Z, \Omega^n X]$$

$$\to \cdots \to [Z, \Omega X] \xrightarrow{\underline{k}} [Z, \mathbf{T}^f] \xrightarrow{\underline{q}} [Z, Y] \xrightarrow{\underline{f}} [Z, X]. \quad \square$$

By analogy with the map $\theta : \mathbf{T}_f \to \mathbf{T}_f \wedge \mathbf{S}X$ constructed above, we can construct a map $\psi : \mathbf{T}^f \times \Omega X \to \mathbf{T}^f$. The map ψ is defined as follows: let $(y, u) \in \mathbf{T}^f$, so that $y \in Y$, $u : \mathbf{I} \to X$, $f(y) = u(0)$, and $u(1) = *$; then if $v \in \Omega X$,

$$\psi(y, u; v) = (y, uv).$$

For any space Z, we can identify $[Z, \mathbf{T}^f \times \Omega X]$ with $[Z, \mathbf{T}^f] \times [Z, \Omega X]$. Hence ψ defines an operation

$$\underline{\psi} : [Z, \mathbf{T}^f] \times [Z, \Omega X] \to [Z, \mathbf{T}^f];$$

we shall write this operation as a product.

(6.20*) *Let* $\alpha \in [Z, \mathbf{T}^f]$, $\xi, \eta \in [Z, \Omega X]$. *Then*

(1) $\alpha(\xi\eta) = (\alpha\xi)\eta$;
(2) $\alpha 1 = \alpha$;
(3) $1\xi = \underline{k}(\xi)$;
(4) $\underline{k}(\xi)\eta = \underline{k}(\xi\eta)$. $\square$

Again, the proof is left to the reader.

(6.21*) *Let* $\alpha, \beta \in [Z, \mathbf{T}_f]$. *Then* $\underline{q}(\alpha) = \underline{q}(\beta)$ *if and only if there exists* $\xi \in [Z, \Omega X]$ *such that* $\beta = \alpha\xi$.

The map q sends $(y, u) \in \mathbf{T}^f$ into y; hence $q \circ \psi = q \circ p_1$ where $p_1 : \mathbf{T}^f \times \Omega X \to \mathbf{T}^f$ is the projection on the first factor. Thus

$$\underline{q}(\alpha\xi) = \underline{q}\underline{\psi}(\alpha, \xi) = \underline{q}\underline{p}_1(\alpha, \xi) = \underline{q}(\alpha).$$

Conversely, suppose that $\underline{q}(\alpha) = \underline{q}(\beta)$. If $f, g : Z \to \mathbf{T}^f$ are maps representing α, β, respectively, then $q \circ f \simeq q \circ g$. By the homotopy lifting property, g is homotopic to a map g' such that $q \circ g' = q \circ f$. If $z \in Z$, we have

$$f(z) = (h(z), u(z)),$$
$$g'(z) = (h(z), v(z)),$$

where $u(z), v(z)$ are paths in X beginning at the same point $fh(z)$ and ending at the base point $*$. Let $w(z) = u(z)^{-1}v(z)$, so that $w : Z \to \Omega X$. Then

$$\begin{aligned}\psi(f(z), w(z)) &= (h(z), u(z)w(z)) \\ &= (h(z), u(z)(u(z)^{-1}v(z))),\end{aligned}$$

and this map is clearly homotopic to g'. Thus, if $\xi \in [Z, \Omega X]$ is the element represented by w, we have $\alpha\xi = \beta$. □

Suppose that $f: Y \to X$ is a fibration, with fibre F. Then the diagram

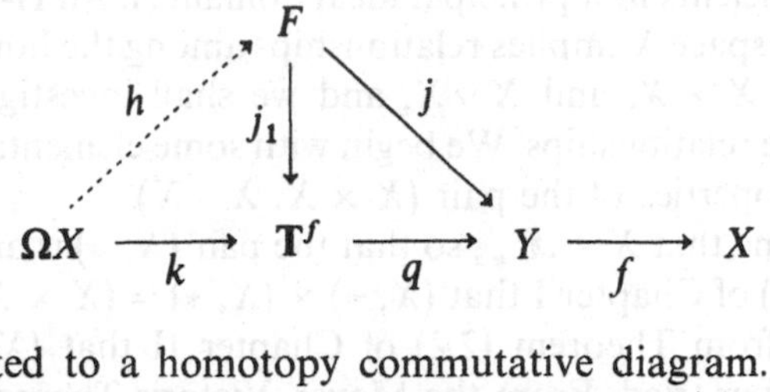

can be completed to a homotopy commutative diagram. For example, if $\lambda : \mathbf{I}^f \to \mathbf{F}(Y)$ is a connection for f, then, we may take $h(u) = \lambda(*, u)(1)$ for all $u \in \Omega(X)$. The map h is called a *connecting map* for the fibration f, and we have

(6.22*) Theorem *Let $f: Y \to X$ be a fibration with fibre F and inclusion map $j: F \subseteq Y$. Let $h : \Omega X \to F$ be a onnecting map for f. Then the sequence*

$$\cdots \to \Omega^{n+1}X \xrightarrow{\Omega^n h} \Omega^n F \xrightarrow{\Omega^n j} \Omega^n Y \xrightarrow{\Omega^n f} \Omega^n X \to \cdots$$

$$\to \Omega X \xrightarrow{h} F \xrightarrow{j} Y \xrightarrow{f} X$$

is left exact. □

Remark. We have referred parenthetically to a sort of duality, some aspects of which are exemplified by the following table:

Fibration	Cofibration
Loop space $\Omega(X)$	Suspension $\mathbf{S}(X)$
Path space $\mathbf{P}(X)$	Cone $\mathbf{T}(X)$
$\mathbf{I}^f$	Mapping cylinder $\mathbf{I}_f$
$\mathbf{T}^f$	Mapping cone $\mathbf{T}_f$

Moreover many diagrams (like those of (6.9) and (6.10)) appear to have dual diagrams (like those of (6.9*) and (6.10*)). It does not, however, appear that there is a formal (categorical) duality underlying these phenomena, comparable, for example, with the character theory for locally compact abelian groups. Nevertheless, the examples we have are so striking, and so basic, as to motivate a systematic treatment of duality phenomena in homotopy theory. Such an exposition has been given by Hilton [Hi]. In the present work we shall, therefore, abstain from attempting a systematic treatment of duality, but shall content ourselves with pointing out such examples as may naturally crop up.

7 Homology Properties of H-spaces and H′-spaces

Throughout this section we shall consider homology and cohomology groups with coefficients in a principal ideal domain A. An H-structure or an H′-structure on a space X implies relationships among the homology groups of the spaces X, $X \times X$, and $X \vee X$, and we shall investigate some consequences of these relationships. We begin with some elementary remarks on the homology properties of the pair $(X \times X, X \vee X)$.

We shall assume that $X \in \mathscr{K}_*$, so that the pair $(X, *)$ is an NDR-pair. It follows from (5.2) of Chapter I that $(X, *) \times (X, *) = (X \times X, X \vee X)$ is an NDR-pair, and from Theorem (2.2) of Chapter II that $(X \vee X; X \times \{*\}, \{*\} \times X)$ is a proper triad. From the Mayer–Vietoris Theorem we deduce

(7.1) *The injections* $j_{1*}, j_{2*} : H_q(X) \to H_q(X \vee X)$ *represent the latter group as a direct sum. The dual representation as a direct product is given by* q_{1*}, $q_{2*} : H_q(X \vee X) \to H_q(X)$. □

Let us consider the homology sequence

$$\cdots \to H_q(X \vee X) \xrightarrow{k_*} H_q(X \times X) \xrightarrow{l_*} H_q(X \times X, X \vee X) \xrightarrow{\partial_*} \cdots$$

of the pair $(X \times X, X \vee X)$. If $p_1, p_2 : X \times X \to X$ are the projections, and $\lambda = j_{1*} \circ p_{1*} + j_{2*} \circ p_{2*} : H_q(X \times X) \to H_q(X \vee X)$, then

$$\lambda \circ k_* \circ j_{1*} = \lambda \circ i_{1*} = j_{1*} \circ p_{1*} \circ i_{1*} + j_{2*} \circ p_{2*} \circ i_{1*} = j_{1*}$$

(because $p_1 \circ i_1$ is the identity and $p_2 \circ i_1$ the constant map into the basepoint) and similarly

$$\lambda \circ k_* \circ j_{2*} = j_{2*}.$$

It follows from (7.1) that $\lambda \circ k_*$ is the identity. Moreover, $\operatorname{Ker} \lambda = \operatorname{Ker} p_{1*} \cap \operatorname{Ker} p_{2*}$. Therefore

(7.2) Theorem *The injection* $k_* : H_q(X \vee X) \to H_q(X \times X)$ *is a split monomorphism, so that*

$$H_q(X \times X) \approx H_q(X \vee X) \oplus H_q(X \times X, X \vee X)$$

for all q. Moreover, the injection l_* *maps the subgroup* $\operatorname{Ker} p_{1*} \cap \operatorname{Ker} p_{2*}$ *isomorphically upon* $H_q(X \times X, X \vee X)$. □

Thus, if $\beta : H_q(X \times X, X \vee X) \to H_q(X \times X)$ is the inverse of the restriction of l_* to $\operatorname{Ker} p_{1*} \cap \operatorname{Ker} p_{2*}$, we have

(7.3) Corollary *The homomorphisms*

$$i_{1*}, i_{2*} : H_q(X) \to H_q(X \times X)$$
$$\beta : H_q(X \times X, X \vee X) \to H_q(X \times X)$$

represent the group $H_q(X \times X)$ *as a direct sum*

$$H_q(X \times X) \approx H_q(X) \oplus H_q(X) \oplus H_q(X \times X, X \vee X).$$

The dual representation as a product is given by

$$p_{1*}, p_{2*} : H_q(X \times X) \to H_q(X),$$
$$l_* : H_q(X \times X) \to H_q(X \times X, X \vee X).$$

For each $u \in H_q(X \times X)$, *we have*

$$u = i_{1*}p_{1*}u + i_{2*}p_{2*}u + \beta l_* u. \qquad \square$$

An element $x \in H_q(X)$ is said to be *primitive* if and only if $l_* \Delta_* x = 0$ (equivalently, $\Delta_* x = i_{1*} x + i_{2*} x$). Let $M_q(X)$ be the set of all primitive homology classes; it is a submodule of $H_q(X)$.

Remark. Suppose that $H_*(X)$ is torsion-free, so that $H_*(X \times X) \approx H_*(X) \otimes H_*(X)$ under the isomorphism defined by the cross product. Moreover, $i_{1*}(x) = x \times 1$ and $i_{2*}(x) = 1 \times x$ for all $x \in H_q(X)$. When this is the case, $H_*(X)$ is a *coalgebra*, (cf. §8, below), and an element of $H_q(X)$ is primitive if and only if it is a primitive element of the coalgebra $H_*(X)$.

Let x be a primitive element of $H_q(X)$, and let $f: X \to Y$ be a map. The calculation

$$\begin{aligned} \Delta_* f_* x &= (f \times f)_* \Delta_* x = (f \times f)_*(i_{1*}x + i_{2*}x) \\ &= i_{1*} f_* x + i_{2*} f_* x \end{aligned}$$

proves

(7.4) *If* $f: X \to Y$, *then* $f_* M_q(X) \subset M_q(Y)$. $\square$

Suppose that $H_q(X) = 0$ for all $q < n$. By the relative Künneth Theorem, $H_q(X \times X, X \vee X) = 0$ for all $q < 2n$. Therefore

(7.5) *If* $H_q(X) = 0$ *for all* $q < n$, *then* $M_q(X) = H_q(X)$ *for all* $q < 2n$. $\square$

In particular,

(7.6) *Each element of* $H_n(\mathbf{S}^n)$ *is primitive* $(n > 0)$. $\square$

A homology class $x \in H_q(X)$ is said to be *spherical* if and only if there is a map $f: \mathbf{S}^q \to X$ such that $x \in f_* H_q(\mathbf{S}^q)$. Let $\Sigma_q(X)$ be the set of all spherical homology classes in $H_q(X)$. It follows from (7.4) and (7.6) that

(7.7) *Every spherical homology class is primitive* (i.e., $\Sigma_q(X) \subset M_q(X)$). □

Suppose that X is an H′-space with coproduct θ. If $x \in H_q(X)$, then $\Delta_* x = k_* \theta_* x$, and therefore $l_* \Delta_* x = 0$. Thus

(7.8) *If X is an* H′*-space, then every element of $H_q(X)$ is primitive.* □

(7.9) *If X is an* H′*-space with coproduct θ, and $x \in H_q(X)$, then* $\theta_* x = j_{1*} x + j_{2*} x$.

For $q_1 \circ \theta = q_2 \circ \theta = 1$ and therefore $x = q_{1*} \theta_* x = q_{2*} \theta_* x$, and our result follows from (7.1). □

This is used to prove

(7.10) Theorem *If X is an* H′*-space, $f, g : X \to Y$, and if $x \in H_q(X)$, then*

$$(f+g)_* x = f_* x + g_* x.$$

For

$$\begin{aligned}(f+g)_* x &= \nabla_*(f \vee g)_* \theta_* x = \nabla_*(f \vee g)_*(j_{1*} x + j_{2*} x)\\ &= \nabla_*(j_{1*} f_* x + j_{2*} g_* x) = f_* x + g_* x.\end{aligned}$$ □

Thus the correspondence $(f, x) \to f_* x$ induces a pairing

$$[X, Y] \otimes H_*(X) \to H_*(Y).$$

Unfortunately, (cf. (7.16) below), no analogous result holds for the case where Y is an H-space.

(7.11) Corollary *For any space X, the set $\Sigma_q(X)$ is a submodule of $H_q(X)$.* □

We next suppose that X is an H-space. It will sometimes be convenient here to treat X as a free space, ignoring the role of the base point. To avoid confusion, let X_* be the resulting free space. The reduced homology group $\tilde{H}_q(X_*)$ is defined by Eilenberg–Steenrod [E–S, p. 18] to be the kernel of the homomorphism $H_q(X_*) \to H_q(*)$ induced by the unique map of X_* into the base point $*$. The composite

$$\tilde{H}_q(X_*) \to H_q(X_*) \to H_q(X_*, *) = H_q(X)$$

being an isomorphism, we may consider $H_q(X)$ as a subgroup of $H_q(X_*)$. (Since $\tilde{H}_q(X_*) = H_q(X_*)$ except when $q = 0$, this may seem a bit pedantic; but we have in mind applications in later chapters involving homology theories not satisfying the dimension axiom).

Let $\mu : X_* \times X_* \to X_*$ be the product; then μ induces

$$\mu_* : H_n(X_* \times X_*) \to H_n(X_*).$$

The homology cross-product is a pairing

$$H_p(X_*) \otimes H_q(X_*) \to H_{p+q}(X_* \times X_*).$$

Combining these, we obtain a pairing

$$H_p(X_*) \otimes H_q(X_*) \to H_{p+q}(X_*).$$

Thus $H_*(X_*)$ is a graded algebra over A, the *Pontryagin algebra* of X. The map $X_* \to *$ is evidently an H-map, and it follows that the induced homomorphism $H_*(X_*) \to H_*(*)$ is a homomorphism of algebras. Thus $\tilde{H}_*(X_*)$ is an ideal in $H_*(X_*)$, and so we may consider $H_*(X)$ as an algebra in its own right, the *reduced Pontryagin* algebra of X. Clearly $H_*(X_*)$ is the direct sum of $\tilde{H}_*(X_*)$ and the free cyclic module generated by the homology class e of the base point. As e is the unit element of $H_*(X_*)$, the essential structure of the latter is already contained in $H_*(X)$. Thus

(7.12) Theorem *If X is an* H-*space, then $H_*(X_*)$ is a graded algebra with unit element e, and $H_*(X)$ is an ideal in $H_*(X_*)$, complementary to the free cyclic module Ae.* □

It may be useful to make explicit the product in the ideal $H_*(X)$. We have defined above a split monomorphism

$$\beta : H_q(X \times X, X \vee X) \to H_q(X \times X);$$

composing β with $\mu_* : H_q(X \times X) \to H_q(X)$, we obtain a homomorphism

$$\tau_* : H_q(X \times X, X \vee X) \to H_q(X).$$

An easy calculation, using Corollary (7.3), yields the formula

(7.13) $$\tau_* \circ l_* = \mu_* - p_{1*} - p_{2*};$$

since l_* is an epimorphism, this characterizes τ_*. The elements of the image of τ_* are called *reductive*.

The cross-product is a pairing $H_p(X) \otimes H_r(X) \to H_q(X \times X, X \vee X)$ $(p + r = q)$. We leave it to the reader to verify that, if $u \in H_p(X)$, $v \in H_r(X)$, then $\tau_*(u \times v) = u \cdot v$, the Pontryagin product in the ideal $H_*(X)$. Thus

(7.14) *Every decomposable element of $H_*(X)$ is reductive. Conversely, if $H_*(X)$ is torsion-free, then every reductive element is decomposable.* □

Suppose further that $H_*(X_*)$ is torsion-free. Then the cross product pairing is an isomorphism

$$H_*(X_*) \otimes H_*(X_*) \to H_*(X_* \times X_*);$$

thus the homomorphism Δ_* may be regarded as a map

$$H_*(X_*) \to H_*(X_*) \otimes H_*(X_*),$$

and this map, together with the algebra structure described above, makes $H_*(X_*)$ into a *Hopf algebra* over A (cf. §8, below). The product in $H_*(X_*)$ need be neither commutative nor associative; however, commutativity and associativity properties of the diagonal map carry over to the coproduct. Thus

(7.15) Theorem *If X is an* H-*space and $H_*(X_*)$ is a torsion-free A-module (in particular, if A is a field), then $H_*(X_*)$ is a cocommutative and coassociative Hopf algebra over A.* □

As we remarked above, the analogue of Theorem (7.10) for H-spaces is false. The best we can do in this direction is

(7.16) Theorem *Let Y be an* H-*space, $f, g : X \to Y$, and let $x \in H_q(X)$ be primitive. Then*

$$(f \cdot g)_*(x) = f_*(x) + g_*(x).$$

In fact, $f \cdot g$ is the composite

$$X \xrightarrow{\Delta} X \times X \xrightarrow{f \times g} Y \times Y \xrightarrow{\mu} Y$$

and so

$$\begin{aligned}(f \cdot g)_* x &= \mu_*(f \times g)_* \Delta_* x = \mu_*(f \times g)_*(i_{1*}x + i_{2*}x) \\ &= \mu_*(i_{1*}f_*x + i_{2*}g_*x) = f_*x + g_*x.\end{aligned}$$ □

Remark. Let W be an H-space, $Y = W \times W$, so that Y is also an H-space. Then $\mu \circ (i_1 \times i_2) \circ \Delta = \Delta$, and therefore $[\Delta] = [i_1] \cdot [i_2]$. But $\Delta_*(x) = i_{1*}(x) + i_{2*}(x)$ if and only if x is primitive. Thus the hypothesis of primitivity in Theorem (7.16) is essential.

The above results dualize to cohomology.

(7.1*) *The projections $q_1^*, q_2^* : H^r(X) \to H^r(X \vee X)$ represent the latter group as a direct sum. The dual representation as a direct product is given by j_1^*, $j_2^* : H^r(X \vee X) \to H^r(X)$.* □

Consider the cohomology sequence

$$\cdots \xrightarrow{\delta^*} H^r(X \times X, X \vee X) \xrightarrow{l^*} H^r(X \times X) \xrightarrow{k^*} H^r(X \vee X) \to \cdots$$

of the pair $(X \times X, X \vee X)$, and let

$$\lambda^* = p_1^* \circ j_1^* + p_2^* \circ j_2^* : H^r(X \vee X) \to H^r(X \times X).$$

Now

$$j_1^* \circ k^* \circ \lambda^* = i_1^* \circ \lambda^* = j_1^*,$$
$$j_2^* \circ k^* \circ \lambda^* = i_2^* \circ \lambda^* = j_2^*,$$

and it follows from (7.1)* that $k^* \circ \lambda^*$ is the identity. Moreover, $\operatorname{Im} \lambda^* = \operatorname{Im} p_1^* + \operatorname{Im} p_2^*$. Thus

(7.2*) Theorem *The injection $k^* : H^r(X \times X) \to H^r(X \vee X)$ is a split epimorphism, so that*

$$H^r(X \times X) \approx H^r(X \vee X) \oplus H^r(X \times X, X \vee X)$$

for all r. Moreover, the injection l^* induces an isomorphism of $H^r(X \times X, X \vee X)$ with the quotient group $H^r(X \times X)/\operatorname{Im} p_1^* + \operatorname{Im} p_2^*$. □

Composition of the inverse of the latter isomorphism with the natural projection of $H^r(X \times X)$ into its quotient group $H^r(X \times X)/\operatorname{Im} p_1^* + \operatorname{Im} p_2^*$ is a homomorphism $\beta^* : H^r(X \times X) \to H^r(X \times X, X \vee X)$, and we have

(7.3*) Corollary *The homomorphisms*

$$p_1^*, p_2^* : H^r(X) \to H^r(X \times X)$$
$$l^* : H^r(X \times X, X \vee X) \to H^r(X \times X)$$

represent the group $H^r(X \times X)$ as a direct sum

$$H^r(X \times X) \approx H^r(X) \oplus H^r(X) \oplus H^r(X \times X, X \vee X).$$

The dual representation as a product is given by

$$i_1^*, i_2^* : H^r(X \times X) \to H^r(X)$$
$$\beta^* : H^r(X \times X) \to H^r(X \times X, X \vee X).$$

For each $u \in H^r(X \times X)$, we have

$$u = p_1^* i_1^* u + p_2^* i_2^* u + l^* \beta^* u.$$ □

An element $x \in H^r(X)$ is *reductive* if and only if $x = \Delta^* l^* y$ for some $y \in H^r(X \times X, X \vee X)$. Let $\Delta^r(X)$ be the set of all reductive elements of $H^r(X)$; it is a submodule of $H^r(X)$. Now $H^*(X_*)$ is an algebra over A, the *cohomology algebra* of X_*, under the cup product, and $H^*(X)$ is an ideal; $H^*(X_*)$ is the direct sum of $H^*(X)$ with the free cyclic module generated by the unit element. If $u \in H^p(X)$, $v \in H^q(X)$, then

$$u \times v \in H^{p+q}(X \times X, X \vee X),$$

and $\Delta^* l^*(u \times v) = uv$, their product in $H^*(X)$. Thus, if $D^*(X)$ is the submodule of $H^*(X)$ spanned by all such products (i.e., $D^*(X)$ is the square of the ideal $H^*(X)$), then $D^r(X) \subset \Delta^r(X)$ for every r. The elements of $D^r(X)$ are said to be *decomposable.*

We leave it to the reader to verify

(7.4*) *If $f: X \to Y$, then $f^*\Delta^r(Y) \subset \Delta^r(X)$ and $f^*D^r(Y) \subset D^r(X)$.* □

(7.5*) *If $H_q(X) = 0$ for all $q < n$, then $\Delta^r(X) = D^r(X) = 0$ for all $q < 2n$.* □

Suppose that $H_*(X)$ is free of finite type (i.e., $H_r(X)$ is a free module of finite rank for each r). Then the cup product map is an isomorphism

$$H^*(X_*) \otimes H^*(X_*) \to H^*(X_* \times X_*)$$

which maps $H^*(X) \otimes H^*(X)$ upon $H^*(X \times X, X \vee X)$. Thus $D^*(X) = \Delta^*(X)$ in this case.

Suppose that X is an H'-space with coproduct θ, $y \in H^q(X \times X, X \vee X)$. Then $\Delta^* l^* y = \theta^* k^* l^* y = 0$, and therefore

(7.8*) *If X is a H'-space, then $\Delta^r(X) = 0$ for all r. In particular, all cup products in $H^*(X)$ vanish.* □

(7.10*) Theorem *If X is an H'-space, $f, g : X \to Y$, and if $y \in H^r(Y)$, then*

$$(f + g)^* y = f^* y + g^* y.$$

For

$$\begin{aligned}(f + g)^* y &= \theta^*(f \vee g)^* \nabla^* y = \theta^*(f \vee g)^*(q_1^* y + q_2^* y) \\ &= \theta^*(q_1^* f^* y + q_2^* g^* y) = f^* y + g^* y.\end{aligned}$$ □

Now suppose that X is an H-space with product μ. Then $\mu^* : H^r(X) \to H^r(X \times X)$. We shall say that $x \in H^r(X)$ is *primitive* if and only if $\beta^* \mu^* x = 0$. By Corollary (7.3*), this is so if and only if $\mu^* x = p_1^* x + p_2^* x$.

The injection $l^* : H^r(X \times X, X \vee X) \to H^r(X \times X)$ is a monomorphism, and the image of the homomorphism

$$v^* = \mu^* - p_1^* - p_2^* : H^r(X) \to H^r(X \times X)$$

belongs to the submodule $\operatorname{Ker} i_1^* \cap \operatorname{Ker} i_2^* = \operatorname{Im} l^*$. Thus there is a unique homomorphism $\tau^* : H^r(X) \to H^r(X \times X, X \vee X)$ such that $l^* \circ \tau^* = v^*$. Evidently an element $x \in H^r(X)$ is primitive if and only if $\tau^* x = 0$.

Suppose further that $H_*(X)$ is free and of finite type (i.e., $H_r(X)$ is a free module of finite rank for each r). Then we may regard μ^* as a homomorphism of $H^*(X_*)$ into $H^*(X_*) \otimes H^*(X_*)$. As such, it is readily verified that

(7.15*) Theorem *If X is an H-space and $H_*(X)$ is free of finite type, then $H^*(X_*)$ is a commutative and associative Hopf algebra over A.* □

When this is the case, an element is primitive if and only if it is a primitive element of the coalgebra $H^*(X_*)$.

Again the analogue of Theorem (7.10*) for H-spaces fails, and we can only prove

(7.16*) Theorem *If Y is an H-space, $f, g : X \to Y$, and if $y \in H^r(Y)$ is primitive, then*

$$(f \cdot g)^* y = f^* y + g^* y.$$ □

We conclude this section by using the Kronecker index to relate the notions we have considered above in homology and cohomology.

(7.17) Theorem *If $x \in H_r(X)$ is primitive, then x is orthogonal to $\Delta^r(X)$. Conversely, if $H_*(X)$ is free of finite type, and x is orthogonal to $\Delta^r(X)$, then x is primitive.*

Suppose x is primitive. Then, for all $y \in H^r(X \times X, X \vee X)$, we have

$$\langle \Delta^* l^* y, x \rangle = \langle y, l_* \Delta_* x \rangle = 0$$

and x is orthogonal to $\Delta^r(X) = \mathrm{Im}(\Delta^* \circ l^*)$. Conversely, suppose that x is orthogonal to $\Delta^r(X)$. Then $\langle y, l_* \Delta_* x \rangle = 0$ for all $y \in H^r(X \times X, X \vee X)$, and the hypothesis that $H_*(X)$ is free of finite type allows us to deduce that $l_* \Delta_* x = 0$, so that x is primitive. □

(7.18) Theorem *If X is an H-space and $H_*(X)$ is free of finite type, then $H_*(X_*)$ and $H^*(X_*)$ are dual Hopf algebras.*

Beweis klar! □

8 Hopf Algebras

Let X be a 0-connected H-space, A a principal ideal domain, and let us assume throughout this section that

(8.1) *For each q, the homology module $H_q(X; A)$ is a free module of finite rank.*

When this is so, we shall say that X has *finite type*. It follows from the Universal Coefficient Theorem that

(8.2) *The modules $H_q(X; A)$, $H^q(X; A)$ are dual A-modules.* □

It follows from the Künneth Theorem that

(8.3) *The homology and cohomology cross products induce isomorphisms*

$$H_*(X; A) \otimes H_*(X; A) \approx H_*(X \times X; A),$$

$$H^*(X; A) \otimes H^*(X; A) \approx H^*(X \times X; A). \qquad \square$$

In this section we shall use these isomorphisms to identify $H_*(X \times X; A)$ and $H^*(X \times X; A)$ as the appropriate tensor products. Thus the diagonal map $\Delta : X \to X \times X$ and the product $\mu : X \times X \to X$ may be considered to induce homomorphisms

$$\Delta_* : H_*(X) \to H_*(X) \otimes H_*(X),$$

$$\mu_* : H_*(X) \otimes H_*(X) \to H_*(X),$$

as well as homomorphisms

$$\mu^* : H^*(X) \to H^*(X) \otimes H^*(X),$$

$$\Delta^* : H^*(X) \otimes H^*(X) \to H^*(X).$$

(Here, and for the remainder of this section, we shall suppress the name of the coefficient domain A, writing $H_*(X)$, $H^*(X)$, instead of $H_*(X; A)$, $H^*(X; A)$).

(8.4) Theorem *Under the homomorphisms μ_*, Δ_*, the module $H_*(X)$ is a Hopf algebra over A. Under the homomorphisms Δ^*, μ^*, the module $H^*(X)$ is a Hopf algebra over A. Moreover, $H_*(X)$ and $H^*(X)$ are dual Hopf algebras.* $\square$

Remark. The standard reference for the theory of Hopf algebras is Milnor–Moore [1]. It should be remarked, however, that our terminology differs somewhat from theirs, in that we do not assume any commutative or associative laws in the definition of a Hopf algebra. Thus a Hopf algebra over A is a graded A-module H, together with maps

$$\phi : H \otimes H \to H, \qquad \eta : A \to H$$

$$\psi : H \to H \otimes H, \qquad \varepsilon : H \to A$$

such that the following diagrams are commutative:

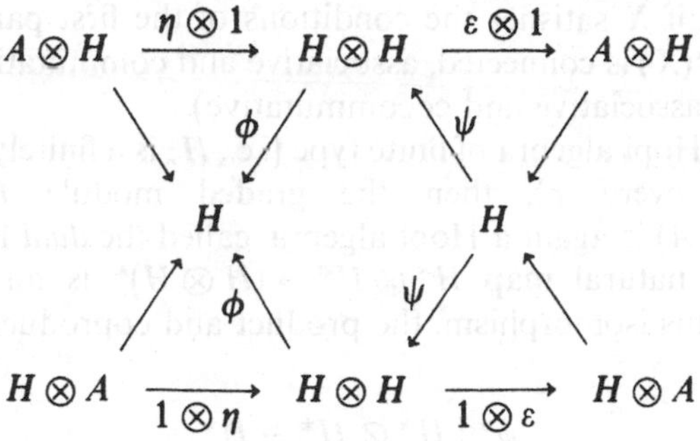

$$\begin{array}{ccccc} H\otimes H & \xrightarrow{\phi} & H & \xrightarrow{\psi} & H\otimes H \\ \Big\downarrow{\scriptstyle \psi\otimes\psi} & & & & \Big\uparrow{\scriptstyle \phi\otimes\phi} \\ H\otimes H\otimes H\otimes H & & \xrightarrow[1\otimes\tau\otimes 1]{} & & H\otimes H\otimes H\otimes H \end{array}$$

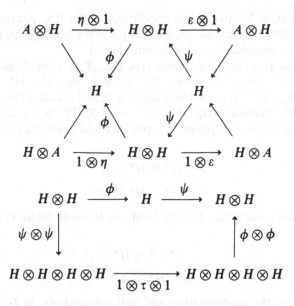

where the unlabelled arrows decide the obvious isomorphisms and where $\tau : H \otimes H \to H \otimes H$ is the twisting map defined by

$$\tau(x \otimes y) = (-1)^{pq} y \otimes x$$

for $x \in H_p$, $y \in H_q$.

Thus H is an algebra with unit with respect to the maps ϕ, η and H is a coalgebra with augmentation (counit) with respect to ψ, ε, and the map $\phi : H \otimes H \to H$ is a homomorphism of coalgebras, while $\psi : H \to H \otimes H$ is a homomorphism of algebras ($H \otimes H$ inheriting the obvious structure as algebra (coalgebra) from the corresponding structure on H).

We shall say that the Hopf algebra is *associative* (*commutative*) if and only if its underlying algebra has the property in question. (The term "commutative" is used in the graded sense: H is commutative if and only if the diagram

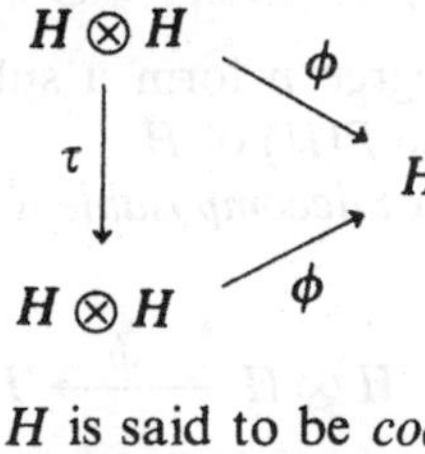

is commutative). Similarly, H is said to be *coassociative* (*cocommutative*) if and only if its underlying coalgebra has the requisite property. H is said to be *biassociative* (*bicommutative*) if and only if it is both associative and coassociative (commutative and cocommutative). Finally, H is said to be *connected* if and only if η maps A *upon* H_0 (equivalently, $\varepsilon \,|\, H_0 : H_0 \approx A$). Let us observe that, if X satisfies the conditions of the first paragraph of this section, then $H^*(X)$ is connected, associative and commutative (and $H_*(X)$ is connected, coassociative and cocommutative).

If H is a free Hopf algebra of finite type (i.e., H_p is a finitely generated free A-module for every p), then the graded module H^* such that $H^p = \mathrm{Hom}(H_p, A)$ is again a Hopf algebra, called the *dual* Hopf algebra of H. In fact, the natural map $H^* \otimes H^* \to (H \otimes H)^*$ is an isomorphism; making use of this isomorphism, the product and coproduct in H^* are the duals

$$\psi^* : H^* \otimes H^* \to H^*$$

$$\phi^* : H^* \to H^* \otimes H^*$$

of the coproduct and product, respectively, in H, while the unit and augmentation maps

$$\varepsilon^* : A \to H^*,$$

$$\eta^* : H^* \to A,$$

are the duals of the augmentation and unit, respectively, in H.

Let us observe that, since $\varepsilon \circ \eta = 1 : A \to H$, the composite

$$\mathrm{Ker}\,\varepsilon \underset{i}{\hookrightarrow} H \xrightarrow[p]{} \mathrm{Cok}\,\eta,$$

where p is the natural projection, is an isomorphism. Thus it is convenient to identify them, so that $\bar{H} = \mathrm{Ker}\,\varepsilon = \mathrm{Cok}\,\eta$ may be regarded (at our convenience) as either a submodule or a quotient module of H. If H is connected, then $\bar{H}_0 = 0$, $\bar{H}_n = H_n$ for all $n > 0$.

An element $x \in \bar{H}$ is said to be *primitive* if and only if x belongs to the kernel of the composite map

$$\bar{H} \xrightarrow{i} H \xrightarrow{\psi} H \otimes H \xrightarrow{p \otimes p} \bar{H} \otimes \bar{H}.$$

Note that, if H is connected, then $x \in H_n$ is primitive if and only if

$$\psi(x) = x \otimes 1 + 1 \otimes x.$$

The primitive elements of degree n form a submodule P_n of H_n; the P_n determine a graded submodule $P(H)$ of $\bar{H}$.

An element $x \in \bar{H}$ is said to be *decomposable* if and only if x belongs to the image of the composite map

$$\bar{H} \otimes \bar{H} \xrightarrow{i \otimes i} H \otimes H \xrightarrow{\phi} H \xrightarrow{p} \bar{H}$$

If H is connected, then an element $x \in \bar{H}_n = H_n$ is decomposable if and only if x has the form

$$x = \Sigma y_i z_i,$$

where $y_i \in H_{p_i}$, $z_i \in H_{n-p_i}$, $0 < p_i < n$. The decomposable elements of degree

n form a submodule D_n of H_n, and the D_n determine a graded module $D(H)$ of $\bar{H}$. The elements of the quotient module $Q(H) = \bar{H}/D(H)$ are termed, by abuse of language, *indecomposable elements* of H.

A useful observation is the following:

(8.5) *Let H be connected, and let K be a graded submodule of $\bar{H}$ such that $K + D(H) = \bar{H}$. Let B be a basis for K. Then B generates H as an algebra with unit.* □

Let H be free of finite type. Then Cok ε^* is the dual space of Ker ε (and Ker η^* of Cok η), so that we may regard the module $\overline{H^*}$ as the dual space $\bar{H}^*$ of $\bar{H}$. Moreover $i^* : H^* \to \text{Cok } \varepsilon^*$ and $p^* : \text{Ker } \eta^* \to H^*$, play the same roles for H^* that p and i play for H. It follows that

(8.6) *The module $D(H^*)$ is the annihilator of $P(H)$, so that $P(H)$ and $Q(H^*)$ (as well as $Q(H)$ and $P(H^*)$) are dual graded modules.* □

Let H be associative. An element $x \in \bar{H}$ is said to have *height* h if and only if $x^h = 0$ but $x^{h-1} \neq 0$; if no such h exists, we say that x has *infinite height*.

(8.7) Theorem *Let H be a connected, associative and commutative Hopf algebra over a field A, and let $x \in H_n$ be a primitive element. Let p ($= 0$ or a prime) be the characteristic of A. Then*

(i) *if n is odd and $p \neq 2$, x has height 2;*
(ii) *if n is even and $p = 0$, then x has infinite height;*
(iii) *if n is even and p is odd, then the height of x is ∞ or a power of p;*
(iv) *if $p = 2$, then the height of x is ∞ or a power of 2.*

Since H is commutative, $x^2 = (-1)^n x^2$. Thus, if n is odd, $2x^2 = 0$ and therefore $x^2 = 0$ unless $p = 2$. Suppose, then that n is even or that $p = 2$. Since x is primitive, $\psi(x) = x \otimes 1 + 1 \otimes x$, and therefore

$$\psi(x^h) = \psi(x)^h = (x \otimes 1 + 1 \otimes x)^h;$$

we can expand the latter by the binomial theorem to obtain

$$\textbf{(8.8)} \quad \psi(x^h) = \sum_{i=0}^{h} \binom{h}{i} x^i \otimes x^{h-i} = x^h \otimes 1 + 1 \otimes x^h + \sum_{i=1}^{h-1} \binom{h}{i} x^i \otimes x^{h-i}.$$

Suppose that x has finite height h, so that $x^{h-1} \neq 0$, $x^h = 0$. If $p = 0$, the left-hand side of (8.8) is zero, while the right-hand side is non-zero, a contradiction. If, on the other hand, $p \neq 0$, then each of the mod p binomial coefficients $\binom{h}{i}$ must vanish ($i = 1, \ldots, p - 1$), and this is well-known to imply that h is a power of p. □

An important consequence is

(8.9) Corollary *If $p = 0$ and H has finite dimension, then H contains no primitive elements of even degree.* □

Let us call the Hopf algebra H *monogenic* if and only if

(1) H is connected;
(2) H is commutative and associative;
(3) as an algebra, H is generated by a single element $x \in H_n$.

The element x is necessarily primitive, so that the coproduct ψ is determined by (8.8), while the structure of H as an algebra is determined by Theorem (8.7). In fact, we have

(8.10) Theorem *Let H be a monogenic Hopf algebra, generated by an element $x \in H_n$, over a field A of characteristic p. Then*

(i) *if n is odd and $p \neq 2$, H is the exterior algebra $\Lambda(x)$;*
(ii) *if n is even and $p = 0$, H is the polynomial algebra $A[x]$;*
(iii) *if n is even and p is an odd prime, either $H \approx A[x]$ or H is a truncated polynomial algebra $A[x]/(x^{p^e})$;*
(iv) *if $p = 2$, then $H \approx A[x]$ or $H \approx A[x]/(x^{2^e})$.*

In all cases, the coproduct in H is given by (8.8). □

The importance of the notion of monogenicity is shown by

(8.11) Theorem (Borel). *Let H be a connected, associative and commutative Hopf algebra of finite type over a perfect field A. Then H is isomorphic, as an algebra, with a tensor product of monogenic Hopf algebras.* □

(8.12) Corollary *Let X be a 0-connected H-space having finite type and let A be a field of characteristic p. Then the cohomology ring $H^*(X; A)$ is a tensor product $\bigotimes_{i=1}^{\infty} B_i$, where each B_i is generated by one element x_i of degree n_i. If n_i is odd and $p \neq 2$, $B_i = \Lambda(x_i)$; if n_i is even and $p = 0$, $B_i = A[x_i]$; if n_i is even and p is an odd prime, then $B_i = A[x_i]$ or $A[x_i]/(x_i^{p^{e_i}})$; and if $p = 2$, then $B_i = A[x_i]$ or $A[x_i]/(x_i^{2^{e_i}})$.* □

(Note that the hypothesis that the field A be perfect is not needed here. For $H^*(X; A) \approx H^*(X; \mathbf{Z}_p) \otimes A$ and we can first apply Theorem (7.11) for the perfect field $\mathbf{Z}_p$, and obtain the general result by forming the tensor product with A).

(8.13) Corollary (Hopf). *If X is a 0-connected H-space, A is a field of characteristic zero, and if $H^*(X; A)$ is finite-dimensional, then $H^*(X; A)$ is an exterior algebra $\Lambda(x_1, \ldots, x_l)$ generated by elements x_i of odd degree.* □

This result of Hopf was the first general result on the homology of H-spaces. It was conjectured by Elie Cartan [1] and verified for the classical groups, by different methods, by Pontryagin [1] and Brauer [1].

As the theorem of Borel is a purely algebraic result we shall refrain from proving it here, referring the reader to Milnor–Moore [1]. The same statement applies to

(8.14) Theorem (Samelson–Leray). *Let H be a connected, biassociative, strictly commutative Hopf algebra over a field A. Suppose that H is generated, as an algebra, by elements of odd degree. Let $\{x_\alpha\}$ be a basis for the space $P(H)$ of primitive elements of H. Then H is the exterior algebra $\Lambda(\{x_\alpha\})$.* □

Remark 1. Because of the hypotheses on H, there is a unique homomorphism of $\Lambda\{x_\alpha\}$ into H which sends each element x_α into x_α. Theorem (8.14) states that this homomorphism is, in fact, an isomorphism.

Remark 2. An algebra is said to be *strictly commutative* if and only if it is commutative and, in addition, $x^2 = 0$ for every element x of odd degree. Because the commutative law implies that $2x^2 = 0$, this additional condition has force only when the coefficient field has characteristic two.

(8.15) Corollary *If X is a 0-connected homotopy associative* H*-space, A is a field of characteristic zero, and $H^*(X; A)$ is finite-dimensional, then $H^*(X; A)$ is an exterior algebra $\Lambda(x_1, \ldots, x_l)$, where the x_i are primitive elements of odd degree.* □

(8.16) Corollary *If X is a connected compact Lie group, and A a field of characteristic zero, then $H^*(X; A)$ is an exterior algebra generated by primitive elements of odd degree.* □

Exercises

1. Let (X, A) be an NDR-pair (in the category $\mathscr{K}_*$), and suppose that X is contractible. Prove that the pairs (X, A) and $(\mathbf{T}A, A)$ have the same homotopy type, and therefore X/A has the same homotopy type as $\mathbf{S}A$.
2. Let $p : X \to B$ be a fibration, and suppose that X is contractible. Show that p has the same fibre homotopy type as $p_1 : \mathbf{P}(B) \to B$, and therefore the fibre of p has the same homotopy type as $\Omega(B)$.
3. Prove that, if n is even, $\mathbf{S}^n$ is not an H-space.
4. Prove that $\mathbf{P}^2(\mathbf{C})$ is not an H′-space.
5. Let X be an H′-space. Find a coproduct in $X \wedge Y$ corresponding to the natural operation of Theorem (5.14).
6. Let X be an H′-space. Find a product in $\mathbf{F}(X, Y)$ corresponding to the natural operation of Theorem (5.16).

7. Let Y be an H-space. Find a product in $\mathbf{F}(X, Y)$ corresponding to the natural operation of Theorem (5.18).

8. Let X be an H′-space. Prove that $[\mathbf{T}X, X; Y, B]$ has a natural composition. If X is homotopy associative (group-like), then $[\mathbf{T}X, X; Y, B]$ is a monoid (group).

9. If X is an H′-space, the restriction map induces a homomorphism

$$\partial_* : [\mathbf{T}X, X; Y, B] \to [X, B].$$

Moreover, there is an exact sequence

$$[\mathbf{S}X, B] \to [\mathbf{S}X, Y] \to [\mathbf{T}X, X; Y, B] \to [X, B] \to [X, Y].$$

10. Show that there is a natural group structure in $[X, Y]$ where X ranges over all CW-complexes of dimension $\leq 2n$ and Y over all n-connected spaces.

11. Let X be an orientable n-manifold, and suppose that a fundamental class $z \in H_n(X)$ is spherical. Prove that X is a homology sphere (i.e., $H_q(X) = 0$ for $0 < q < n$).

CHAPTER IV

Homotopy Groups

In Chapter III we saw that, if X is any space with base point, then $\pi_n(X) = [\mathbf{S}^n, X]$ is a group for any positive integer n. In fact, π_n is a functor from the category $\mathscr{K}_0$ to the category of groups if $n = 1$, abelian groups if $n > 1$. In certain respects, they resemble the homology groups, and one of the objectives of this chapter is to pursue this analogy and see where it may lead.

One point of similarity is the existence of relative groups. The nth relative homotopy group of a pair (X, A) (with base point $* \in A$) is the set $\pi_n(X, A) = [\mathbf{E}^n, \mathbf{S}^{n-1}; X, A]$; it is a group for $n = 2$ and abelian for $n \geq 3$. Moreover, π_n is a functor. The restriction map sending $[\mathbf{E}^n, \mathbf{S}^{n-1}; X, A]$ into $[\mathbf{S}^{n-1}, A]$ is a homomorphism $\partial_* : \pi_n(X, A) \to \pi_{n-1}(A)$, and we may form the homotopy sequence of the pair. Like the homology sequence, it is exact. These elementary properties are developed in §§1-2.

The fundamental group owes its existence to Poincaré [1]. In 1932 Čech [1] suggested how to define higher homotopy groups, but he did not pursue the notion and it was Hurewicz who first studied them, as homotopy groups of the appropriate function spaces, in 1935–1936 [1]. In 1939 Eilenberg [1] showed how the fundamental group operates on the higher homotopy groups. In the case of a pair, (X, A), the group $\pi_1(A)$ operates on $\pi_n(X, A)$, as we saw in §1 of Chapter III; in fact, $\pi_1(A)$ is a group of operators on the entire homotopy sequence, as we see in §3.

The connection between homology and homotopy is made explicit in §4, with the introduction of the Hurewicz map. This is a homomorphism $\rho : \pi_n(X) \to H_n(X)$; if $\alpha \in \pi_n(X)$ is represented by a map $f : \mathbf{S}^n \to X$, then $\rho(\alpha) = f_*(s)$, where s generates the infinite cyclic group $H_n(\mathbf{S}^n)$. The Hurewicz map can be relativized, and gives rise to a map of the homotopy sequence of a pair into its homology sequence.

In the next two Sections we develop machinery needed for the proof of the Hurewicz Theorem: if X is $(n-1)$-connected, then $\rho : \pi_n(X) \to H_n(X)$ is an isomorphism. The tools developed are firstly, the Homotopy Addition Theorem, a formula analogous to the boundary formula in simplicial homology, and secondly, a sequence of functors leading to groups intermediate between the homology and homotopy groups. These are due to Eilenberg [3] in the absolute case and to Blakers [1] in the relative.

The Hurewicz Theorem and its relative counterpart are proved in §7. Sufficient conditions for the converse are developed, and a useful theorem of J. H. C. Whitehead [3] is proved: a map $f: X \to Y$ which induces isomorphisms of the homotopy groups also induces isomorphisms of the homology groups.

While homology groups behave very well under cofibrations, homotopy groups do not. For fibrations the situation is reversed. There is an exact sequence relating the homotopy groups of the fibre, total space and base space of a fibration. Moreover, the fibres over two different points of the (pathwise connected) base are connected by homotopy equivalences depending on the choice of a homotopy class of paths joining the two points. Thus the fundamental group of the base acts on the homology groups (and in many cases, on the homotopy groups) of the fibre. These results are developed in §8. The last two sections apply the theory of this Chapter to certain special fibrations and yield information on the homotopy groups of certain compact Lie groups and their coset spaces.

1 Relative Homotopy Groups

We have defined the higher homotopy groups $\pi_n(X) = [\mathbf{S}^n, X]$. They are groups for $n \geq 1$, and even abelian groups if $n \geq 2$. It is often convenient to allow n to be zero; in this case $\pi_0(X) = [\mathbf{S}^0, X]$ is in one-to-one correspondence with the set of path-components of X. It has no natural group structure, but is to be regarded simply as a set with base point.

The homotopy groups are functors, and therefore it is natural to study the properties of the homomorphism $\pi_n(f) = f_* : \pi_n(X) \to \pi_n(Y)$ induced by a map $f: X \to Y$. Of special interest is the case when f is an inclusion. Even when this is so, f_* need not be a monomorphism; for example, if X is a space with $\pi_n(X) \neq 0$, we may consider the inclusion map $i : X \subsetneq \mathbf{T}X$. Since $\mathbf{T}X$ is contractible, $\pi_n(\mathbf{T}X) = 0$ and therefore $i_* = 0$ is not a monomorphism.

The method we shall use for studying this question is one that occurs again and again in algebraic topology. We describe first an example which will be familiar to the reader from his study of elementary topology. Suppose that K is a simplicial complex, L a subcomplex of K, $\alpha \in H_q(L)$, and $i_*(\alpha) = 0$. (Again $i_* : H_q(L) \to H_q(K)$ is the injection). If $z \in Z_q(L)$ is a cycle representing α, then the condition $i_*(\alpha) = 0$ implies that $z \sim 0$ in K; i.e., there is a chain $c \in C_{q+1}(K)$ such that $\partial c = z$. Thus c is a relative cycle of K

modulo L, and the relative homology group $H_{q+1}(K, L)$ is composed of equivalence classes of such relative cycles. And the reader will recall that the blanks in the diagram

$$\cdots H_{q+1}(L) \xrightarrow{i_*} H_{q+1}(K) \cdots H_q(L) \xrightarrow{i_*} H_q(K) \cdots$$

can be filled in by inserting the appropriate relative groups to obtain the homology sequence of (K, L) whose exactness is so crucial in homology theory.

This suggests the desirability of defining relative homotopy groups $\pi_{n+1}(X, A)$. Again, let $\alpha \in \pi_n(A)$, $i_*(\alpha) = 0$, where $i_* : \pi_n(A) \to \pi_n(X)$ is the injection; and let $f : \mathbf{S}^n \to A$ be a map representing α. (Let us recall that, in §2 of Chapter III, we established explicit identifications of the n-fold reduced join $\mathbf{S} \wedge \cdots \wedge \mathbf{S}$ with the unit sphere $\mathbf{S}^n$ in $\mathbf{R}^{n+1}$, and of $\mathbf{TS}^n$ with the unit disc $\mathbf{E}^{n+1}$). Then f, considered as a map of $\mathbf{S}^n$ into X, is nullhomotopic, so that there is a map $g : (\mathbf{E}^{n+1}, \mathbf{S}^n) \to (X, A)$ such that $g | \mathbf{S}^n = f$. This suggests that we define $\pi_{n+1}(X, A)$, for all $n \geq 0$, to be the set of homotopy classes of maps of $(\mathbf{E}^{n+1}, \mathbf{S}^n)$ into (X, A).

This defines $\pi_{n+1}(X, A)$ as a set. In order to impose a group structure on this set, we could relativize the discussion of §5 of Chapter III as in Exercise 8 of the same Chapter. However, the following device makes it unnecessary to do so. Let $\Omega^{n+1}(X, A)$ be the function space of maps of $(\mathbf{E}^{n+1}, \mathbf{S}^n)$ into (X, A), so that $\pi_{n+1}(X, A)$ may be regarded as the set of path-components of $\Omega^{n+1}(X, A)$. If $n \geq 1$, $\Omega^{n+1}(X, A)$ is a subspace of the function space $\mathbf{F}(\mathbf{T} \wedge \mathbf{S}^n, X) = \mathbf{F}(\mathbf{T} \wedge \mathbf{S}^1 \wedge \mathbf{S}^{n-1}, X) \approx \mathbf{F}(\mathbf{S}^1 \wedge \mathbf{T} \wedge \mathbf{S}^{n-1}, X)$ which is homeomorphic by (2.9) of Chapter III with $\mathbf{F}(\mathbf{S}^1, \mathbf{F}(\mathbf{E}^n, X)) = \Omega\mathbf{F}(\mathbf{E}^n, X)$. Under the homeomorphism the subspace $\Omega^{n+1}(X, A)$ obviously corresponds to $\Omega(\Omega^n(X, A))$. Thus

(1.1) *If $n \geq 1$, $\pi_{n+1}(X, A)$ is in one-to-one correspondence with $\pi_1(\Omega^n(X, A))$.* □

We shall use the above one-to-one correspondence to impose a group structure on $\pi_{n+1}(X, A)$ for $n \geq 1$. On the other hand, $\pi_1(X, A)$ is to be considered merely as a set with base point.

If $n \geq 2$, then $\Omega^n(X, A)$ is homeomorphic with the H-space $\Omega(\Omega^{n-1}(X, A))$. Hence

(1.2) *If $n \geq 2$, $\pi_{n+1}(X, A)$ is an abelian group.* □

It is clear that π_{n+1} is a functor from the category of pairs (of compactly generated spaces with base point) into the category of

$$\begin{cases} \text{sets with base point} & \text{if } n = 0, \\ \text{groups} & \text{if } n = 1, \\ \text{abelian groups} & \text{if } n \geq 2. \end{cases}$$

If $A = \{*\}$, the base point of X, then $\Omega^{n+1}(X, A)$ is homeomorphic with $\mathbf{F}(\mathbf{T} \wedge \mathbf{S}^n/\mathbf{S}^n, X) = \mathbf{F}(\mathbf{S}^1 \wedge \mathbf{S}^n, X) = \mathbf{F}(\mathbf{S}^{n+1}, X) = \Omega^{n+1}(X)$ and this induces an isomorphism $\pi_{n+1}(X, \{*\}) \approx \pi_{n+1}(X)$.

At this point it seems appropriate to pause in order to make the group operations in $\pi_n(X)$ and $\pi_{n+1}(X, A)$ more explicit. Let $\varpi : (\mathbf{I}, \dot{\mathbf{I}}) \to (\mathbf{S}^1, *)$ be the identification map; then $\varpi^n = \varpi \times \cdots \times \varpi : \mathbf{I}^n \to \mathbf{S}^1 \times \cdots \times \mathbf{S}^1$. Composition of ϖ^n with the natural proclusion $\mathbf{S}^1 \times \cdots \times \mathbf{S}^1 \to \mathbf{S}^1 \wedge \cdots \wedge \mathbf{S}^1 = \mathbf{S}^n$ yields a relative homeomorphism $\varpi_n : (\mathbf{I}^n, \dot{\mathbf{I}}^n) \to (\mathbf{S}^n, *)$. We may use this map ϖ_n to identify $\Omega^n(X)$ with the function space $\bar{\Omega}^n(X) = F(\mathbf{I}^n, \dot{\mathbf{I}}^n; X, *)$, and $\pi_n(X) = [\mathbf{S}^n, X]$ with $[\mathbf{I}^n, \dot{\mathbf{I}}^n; X, *]$. If $f : \mathbf{S}^n \to X$ represents $\alpha \in \pi_n(X)$, we shall say that $f \circ \varpi_n : (\mathbf{I}^n, \dot{\mathbf{I}}^n) \to (X, *)$ *represents* α. It then follows from the definitions of the coproduct in $\mathbf{S}^1$ given in (5.13) of Chapter III that

(1.3) *If $f, g : (\mathbf{I}^n, \dot{\mathbf{I}}^n) \to (X, *)$ represent $\alpha, \beta \in \pi_n(X)$, respectively, then $\alpha + \beta$ is represented by the map: $f + g : (\mathbf{I}^n, \dot{\mathbf{I}}^n) \to (X, *)$ given by*

$$(f+g)(t_1, \ldots, t_n) = \begin{cases} f(2t_1, t_2, \ldots, t_n) & (0 \le t_1 \le \frac{1}{2}), \\ g(2t_1 - 1, t_2, \ldots, t_n) & (\frac{1}{2} \le t_1 \le 1). \end{cases}$$

Moreover, the operation $(f, g) \to f + g$ makes $\bar{\Omega}^n(X)$ into an H-*space.* □

If $1 < k \le n$, we can compose f and g in a different way, to obtain a new map h, given by

(1.4)

$$h(t_1, \ldots, t_n) = \begin{cases} f(t_1, \ldots, t_{k-1}, 2t_k, t_{k+1}, \ldots, t_n) & (0 \le t_k \le \frac{1}{2}), \\ g(t_1, \ldots, t_{k-1}, 2t_k - 1, t_{k+1}, \ldots, t_n) & (\frac{1}{2} \le t_k \le 1). \end{cases}$$

We shall say that h is obtained from f and g by "adding along the kth coordinate." It is intuitively evident that h represents the sum $\alpha + \beta$ of the elements represented by f and g. However, to give a direct argument by writing down explicit formulae would be very complicated and ugly. Instead, we may use the results of §5 of Chapter III as follows. The natural homeomorphism $\mathbf{F}(\mathbf{S}^{k-1} \wedge \mathbf{S}^{n-k+1}, X)$ with $\mathbf{F}(\mathbf{S}^{k-1}, \mathbf{F}(\mathbf{S}^{n-k+1}, X))$ induces a homeomorphism of $\bar{\Omega}^n(X)$ with $\bar{\Omega}^{k-1}(\bar{\Omega}^{n-k+1}(X))$, as well as an isomorphism of $\pi_n(X)$ with $\pi_{k-1}(\bar{\Omega}^{n-k+1}(X))$. The latter sethas two compositions, the former coming from the coproduct in $\mathbf{S}^{k-1}$, the latter from the above H-structure in $\bar{\Omega}^{n-k+1}(X)$. By Theorem (5.21) of Chapter III, these two compositions coincide. But these two operations correspond to those obtained by adding along the first and the kth coordinates, respectively. Thus

(1.5) *If $f, g : (\mathbf{I}^n, \dot{\mathbf{I}}^n) \to (X, *)$ represent $\alpha, \beta \in \pi_n(X)$, respectively, and if $h : (\mathbf{I}^n, \dot{\mathbf{I}}^n) \to (X, *)$ is defined by* (1.4), *then h represents $\alpha + \beta$.* □

We can treat relative homotopy groups in a similar manner. Composition of the proclusion $1 \times \varpi^n : \mathbf{I} \times \mathbf{I}^n \to \mathbf{T} \times \mathbf{S}^1 \times \cdots \times \mathbf{S}^1$ with the proclusion $\mathbf{T} \times \mathbf{S}^1 \times \cdots \times \mathbf{S}^1 \to \mathbf{T} \wedge \mathbf{S}^n$ is a relative homeomorphism $\varpi'_{n+1} : (\mathbf{I}^{n+1},$

$\dot{\mathbf{I}}^{n+1}) \to (\mathbf{E}^{n+1}, \mathbf{S}^n)$ whose restriction to $\dot{\mathbf{I}}^{n+1}$ maps the subset $J^n = 0 \times \mathbf{I}^n \cup \mathbf{I} \times \dot{\mathbf{I}}^n$ into the base point and is a relative homeomorphism of $(1 \times \mathbf{I}^n, 1 \times \dot{\mathbf{I}}^n)$ upon $(\mathbf{S}^n, *)$ (in fact, $\varpi'_{n+1}(1, x) = \varpi_n(x)$ for $x \in \mathbf{I}^n$). Composition with ω'_{n+1} maps the function space $\Omega^{n+1}(X, A)$ homeomorphically upon the function space $\bar{\Omega}^{n+1}(X, A) = \mathbf{F}(\mathbf{I}^{n+1}, \dot{\mathbf{I}}^{n+1}, J^n; X, A, *)$. Again, if $f: (\mathbf{T} \wedge \mathbf{S}^n, \mathbf{S}^n) \to (X, A)$ represents $\alpha \in \pi_{n+1}(X, A)$, we shall say that $f \circ \varpi'_{n+1}$ *represents* α.

We can repeat essentially the same argument as in the case of the absolute homotopy groups to obtain

(1.6) *Let $f, g : (\mathbf{I}^n, \dot{\mathbf{I}}^n, J^{n-1}) \to (X, A, *)$ represent $\alpha, \beta \in \pi_n(X, A)$, respectively. Let $1 < k \leq n$, and define h by* (1.4). *Then $h : (\mathbf{I}^n, \dot{\mathbf{I}}^n, J^{n-1}) \to (X, A, *)$ represents $\alpha + \beta \in \pi_n(X, A)$.* □

(Note that, if $k = 1$, the formula (1.4) does not define a continuous function. In particular, if $n = 1$, our procedure does not define a composition in $\pi_n(X, A)$).

Remark. Although $\pi_0(X, A)$ is not defined, it is sometimes convenient to use the phrase "$\pi_0(X, A) = 0$" to mean that (X, A) is 0-connected, or equivalently, that every path-component of X meets A.

2 The Homotopy Sequence

Just as in homology theory, we can define a boundary operator on relative homotopy groups. In fact, if $f: (\mathbf{E}^{n+1}, \mathbf{S}^n) \to (X, A)$, the homotopy class of $f \mid \mathbf{S}^n : \mathbf{S}^n \to A$ depends only on that of f, so that we may define $\partial_{n+1} = \partial_{n+1}(X, A) : \pi_{n+1}(X, A) \to \pi_n(A)$ by $\partial_{n+1}(\alpha) = [f \mid \mathbf{S}^n]$.

(2.1) *The map $\partial_{n+1}(X, A) : \pi_{n+1}(X, A) \to \pi_n(A)$ is a homomorphism, if $n > 0$.*

To see this, we represent $\pi_{n+1}(X, A)$ and $\pi_n(A)$ as $\pi_1(\Omega^n(X, A))$ and $\pi_1(\Omega^{n-1}(A))$, respectively. The restriction map $r : \Omega^n(X, A) \to \Omega^{n-1}(A)$, which associates to each map $f : (\mathbf{E}^n, \mathbf{S}^{n-1}) \to (X, A)$ its restriction $r(f) = f \mid \mathbf{S}^{n-1} : \mathbf{S}^{n-1} \to A$, is continuous, by (4.22) of Chapter I; moreover, $\partial_{n+1}(X, A) = \pi_1(r)$, the homomorphism of fundamental groups induced by r. □

(This can also be seen to follow from (1.5) and (1.6)).

Let R be the functor which assigns to each pair (X, A) the space A, and to each map $f: (X, A) \to (Y, B)$ its restriction $f \mid A : A \to B$. It is then immediately clear that

(2.2) *The operation ∂_{n+1} is a natural transformation of the functor π_{n+1} into the functor $\pi_n \circ R$.* □

We can now set up the homotopy sequence of a pair (X, A). Let $i: A \hookrightarrow X$, $j: (X, *) \hookrightarrow (X, A)$; then we have injections

$$i_* = \pi_n(i): \pi_n(A) \to \pi_n(X),$$
$$j_* = \pi_n(j): \pi_n(X) \to \pi_n(X, A),$$

as well as

$$\partial_* = \partial_{n+1}(X, A): \pi_{n+1}(X, A) \to \pi_n(A).$$

Thus we have a sequence

$$\cdots \to \pi_{n+1}(X, A) \xrightarrow{\partial_*} \pi_n(A) \xrightarrow{i_*} \pi_n(X) \xrightarrow{j_*} \pi_n(X, A) \to \cdots$$

(2.3)

$$\to \pi_1(X) \xrightarrow{j_*} \pi_1(X, A) \xrightarrow{\partial_*} \pi_0(A) \xrightarrow{i_*} \pi_0(X),$$

called the *homotopy sequence of the pair* (X, A).

(2.4) Theorem *The homotopy sequence* (2.3) *of the pair* (X, A) *is exact.*

(Note that the last three terms in the sequence are not groups, but merely sets with base point, and the last three maps are just base point preserving functions. Moreover, $\pi_0(X, A)$ is not defined. Even in these cases, exactness has the usual meaning: the image of each map is the "kernel" of the next, i.e., the counterimage of the base point. In fact, a little more is true; cf. Exercise 2 below).

The proof of Theorem (2.4) is divided into six parts.

(1) $i_* \circ \partial_* = 0$. For if $f: (\mathbf{E}^{n+1}, \mathbf{S}^n) \to (X, A)$ represents $\alpha \in \pi_{n+1}(X, A)$, then $f|\mathbf{S}^n: \mathbf{S}^n \to A$ represents $\partial_*(\alpha)$, and $i \circ f|\mathbf{S}^n: \mathbf{S}^n \to X$ represents $i_*\partial_*(\alpha)$. But $f: \mathbf{TS}^n \to X$ is a nullhomotopy of $i \circ f|\mathbf{S}^n$.

(2) $j_* \circ i_* = 0$. For if $f: \mathbf{S}^n = \mathbf{S}^1 \wedge \mathbf{S}^{n-1} \to A$ represents $\alpha \in \pi_n(A)$, then $f \circ (\varpi \wedge 1): (\mathbf{TS}^{n-1}, \mathbf{S}^{n-1}) \to (A, *)$ represents α, and

$$f' = j \circ i \circ f \circ (\varpi \wedge 1): (\mathbf{TS}^{n-1}, \mathbf{S}^{n-1}) \to (X, A)$$

represents $j_* i_*(\alpha)$. But $f'(\mathbf{TS}^{n-1}) \subset A$, and the contractibility of the cone $\mathbf{TS}^{n-1}$ implies that f' is nullhomotopic.

(3) $\partial_* \circ j_* = 0$. For if $f: \mathbf{S}^n = \mathbf{S}^1 \wedge \mathbf{S}^{n-1} \to X$ represents $\alpha \in \pi_n(X)$, then $f' = j \circ f \circ (\varpi \wedge 1): (\mathbf{TS}^{n-1}, \mathbf{S}^{n-1}) \to (X, A)$ represents $j_*(\alpha)$, and $f'|\mathbf{S}^{n-1}$ represents $\partial_* j_*(\alpha)$. But $f'(\mathbf{S}^{n-1}) = *$.

(1′) Kernel $i_* \subset$ Image ∂_*. For if $f: \mathbf{S}^n \to A$ is a map representing $\alpha \in$ Kernel i_*, and $g: \mathbf{TS}^n \to X$ is a nullhomotopy of $i \circ f$, then the map $g: (\mathbf{TS}^n, \mathbf{S}^n) \to (X, A)$ represents an element $\beta \in \pi_{n+1}(X, A)$ and $\partial_*(\beta) = \alpha$.

(2′) Kernel $j_* \subset$ Image i_*. For if $f: (\mathbf{E}^n, \mathbf{S}^{n-1}) \to (X, *)$ represents an element $\alpha \in$ Kernel j_*, then $j \circ f$ is nullhomotopic. By Lemma (3.2) of Chapter II, $j \circ f$ is compressible; i.e., f is homotopic (rel. $\mathbf{S}^{n-1}$) to a map of $\mathbf{E}^n$ into A. The latter map represents an element $\beta \in \pi_n(A)$ such that $i_*(\beta) = \alpha$.

(3′) Kernel $\partial_* \subset$ Image j_*. For if $f: (\mathbf{E}^n, \mathbf{S}^{n-1}) \to (X, A)$ represents $\alpha \in$ Kernel ∂_*, then $f | \mathbf{S}^{n-1} : \mathbf{S}^{n-1} \to A$ is nullhomotopic. By the homotopy extension property, f is homotopic to a map $f' : \mathbf{E}^n \to X$ such that $f(\mathbf{S}^{n-1}) = *$. Then f' represents an element $\beta \in \pi_n(X)$ such that $j_*(\beta) = \alpha$. □

(2.5) Corollary *For any space X,* $\pi_n(X, X) = 0$ *for all* $n \geq 1$. □

This follows formally from Theorem (2.4), even if $n = 1$! (The astute reader will, of course, have recognized that Corollary (2.5) was used at one point in the proof of Theorem (2.4); in fact, the proof of Corollary (2.5) was essentially given there).

Let (X, A, B) be a triple in $\mathscr{K}_0$. Then we have injections

$$k_* : \pi_n(A, B) \to \pi_n(X, B),$$
$$l_* : \pi_n(X, B) \to \pi_n(X, A),$$

as well as a boundary operator

$$\partial_* : \pi_{n+1}(X, A) \to \pi_n(A, B)$$

which is defined as the composite

$$\pi_{n+1}(X, A) \to \pi_n(A) \to \pi_n(A, B)$$

where the first map is the boundary operator of the homotopy sequence of the pair (X, A), while the second is the injection. We then have a commutative diagram (Figure 4.1) consisting of four sequences arranged along sinusoidal curves. Those numbered 1, 2, 3 are the homotopy sequences of the pairs (A, B), (X, B), (X, A) respectively; the fourth is the homotopy sequence of the triple (X, A, B). That $l_* \circ k_* = 0$ follows from commutativity of the diagram

$$\begin{array}{ccc} \pi_n(A, B) & \longrightarrow & \pi_n(X, B) \\ \downarrow & & \downarrow \\ \pi_n(A, A) & \longrightarrow & \pi_n(X, A) \end{array}$$

of injections and Corollary (2.5). Routine diagram-chasing, as in the analogous situation in homology theory completes the proof of

(2.6) Theorem *The homotopy sequence of a triple* (X, A, B) *is exact.* □

Let us recall that in §2 of Chapter II we established our Fundamental Notational Convention. As we promised the reader there, the analogous convention will be applied to homotopy groups.

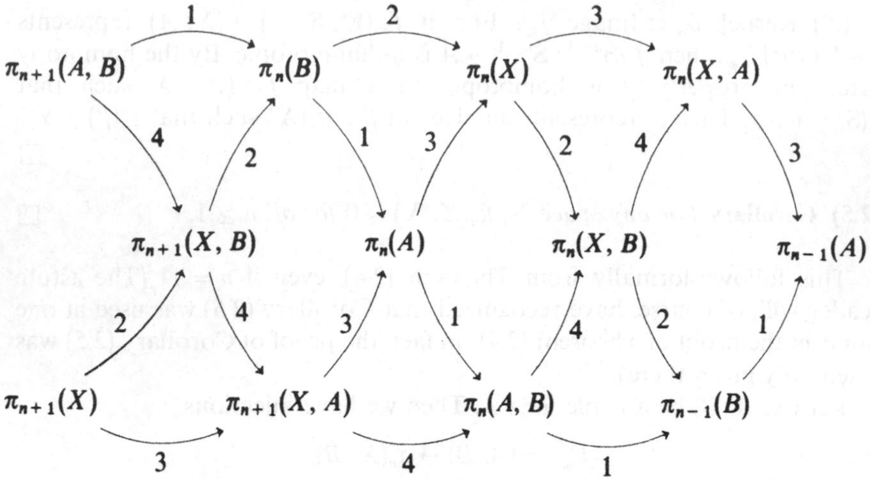

Figure 4.1

3 The Operations of the Fundamental Group on the Homotopy Sequence

We have seen in Chapter III how the fundamental group $\pi_1(A)$ operates on the homotopy groups $\pi_n(A)$. Moreover, $\pi_1(A)$ operates on the relative homotopy groups $\pi_n(X, A)$. Finally, the injection $\pi_1(A) \to \pi_1(X)$, together with the operation of $\pi_1(X)$ on $\pi_n(X)$, defines an operation of $\pi_1(A)$ on $\pi_n(X)$. Thus $\pi_1(A)$ operates on each of the terms of the homotopy sequence of (X, A).

(3.1) Theorem *If (X, A) is a pair, the fundamental group $\pi_1(A)$ operates on the homotopy sequence of (X, A); i.e., the maps i_*, j_*, ∂_* of the sequence are operator homomorphisms.*

In diagrammatic terms, this means that, for each $\xi \in \pi_1(A)$, the diagram

$$\begin{array}{ccccccccc}
\cdots \to \pi_{n+1}(X, A) & \xrightarrow{\partial_*} & \pi_n(A) & \xrightarrow{i_*} & \pi_n(X) & \xrightarrow{j_*} & \pi_n(X, A) \to \cdots \\
\downarrow{\scriptstyle \tau'_\xi} & & \downarrow{\scriptstyle \tau_\xi} & & \downarrow{\scriptstyle \tau_\eta} & & \downarrow{\scriptstyle \tau'_\xi} \\
\cdots \to \pi_{n+1}(X, A) & \xrightarrow{\partial_*} & \pi_n(A) & \xrightarrow{i_*} & \pi_n(X) & \xrightarrow{j_*} & \pi_n(X, A) \to \cdots
\end{array}$$

(where $\eta = i_*(\xi)$) is commutative.

The easy proof of the theorem is left to the reader. □

The operations of $\pi_1(A)$ on $\pi_n(A)$ and $\pi_n(X, A)$ are consistent with the group operations in the latter sets. Specifically,

(3.2) Theorem *If $\xi \in \pi_1(A)$, then $\tau_\xi : \pi_n(A) \to \pi_n(A)$ and $\tau'_\xi : \pi_{n+1}(X, A) \to \pi_{n+1}(X, A)$ are homomorphisms whenever $n \geq 1$.*

We shall prove this for the relative groups; the proof for the absolute groups is entirely similar. Let $\xi \in \pi_1(A)$, and let $u : (\mathbf{I}, \dot{\mathbf{I}}) \to (A, *)$ be a path representing ξ. What has to be shown is that, if $f_0 \underset{u}{\simeq} f_1$, $g_0 \underset{u}{\simeq} g_1$, then $f_0 + g_0 \underset{u}{\simeq} f_1 + g_1$. We shall use the representation of $\pi_{n+1}(X, A)$ given at the end of §1. Thus f_0, f_1, g_0, g_1 are maps of $(\mathbf{I}^{n+1}, \dot{\mathbf{I}}^{n+1}, J^n)$ into $(X, A, *)$ and there are homotopies $f_t, g_t : (\mathbf{I}^{n+1}, \mathbf{I}^{n+1}) \to (X, A)$ of f_0 to f_1 and g_0 to g_1 such that $f_t(J^n) = g_t(J^n) = u(t)$ for $0 \leq t \leq 1$. Define $h_t : (\mathbf{I}^{n+1}, \dot{\mathbf{I}}^{n+1}) \to (X, A)$ by Formula (1.4) with $k = 1$, say. Then h_t is well-defined and is the desired free homotopy of $f_0 + g_0$ to $f_1 + g_1$ along u. □

Let $\omega_n(X)$ be the subgroup of $\pi_n(X)$ generated by all elements of the form $\alpha - \tau_\xi(\alpha)$ with $\xi \in \pi_1(X)$, $\alpha \in \pi_n(X)$, and let $\pi_n^*(X)$ be the quotient group $\pi_n(X)/\omega_n(X)$. Note that, if $n = 1$, $\omega_n(X)$ is just the commutator subgroup of $\pi_n(X)$.

Similarly, let $\omega'_n(X, A)$ be the subgroup of $\pi_n(X, A)$ generated by all elements of the form $\alpha - \tau'_\xi(\alpha)$ with $\xi \in \pi_1(A)$, and let $\pi_n^\dagger(X, A)$ be the quotient group $\pi_n(X, A)/\omega'_n(X, A)$.

(3.3) Lemma *Let $\alpha, \beta \in \pi_2(X, A)$, $\xi = \partial_* \beta \in \pi_1(A)$. Then $\tau'_\xi(\alpha) = \beta + \alpha - \beta$.*

We prove this by the method of the universal example. Indeed, let $h : (\mathbf{E}^2 \vee \mathbf{E}^2, \mathbf{S}^1 \vee \mathbf{S}^1) \to (X, A)$ be a map such that $h \circ j_1 : (\mathbf{E}^2, \mathbf{S}^1) \to (X, A)$ represents α and $h \circ j_2 : (\mathbf{E}^2, \mathbf{S}^1) \to (X, A)$ represents β. Then the elements ι_1, $\iota_2 \in \pi_2(\mathbf{E}^2 \vee \mathbf{E}^2, \mathbf{S}^1 \vee \mathbf{S}^1)$ represented by $j_1, j_2 : (\mathbf{E}^2, \mathbf{S}^1) \to (\mathbf{E}^2 \vee \mathbf{E}^2, \mathbf{S}^1 \vee \mathbf{S}^1)$ have the property that $h_*(\iota_1) = \alpha$ and $h_*(\iota_2) = \beta$. Then, if $\lambda = \partial_* \iota_2$, we have $h_*(\lambda) = \xi \in \pi_1(A)$, and

$$
\begin{aligned}
h_* \tau'_\lambda(\iota_1) &= \tau'_{h_*(\lambda)}(h_*(\iota_1)) \\
&= \tau'_\xi(\alpha).
\end{aligned}
$$

If Lemma (3.3) holds for the pair of elements $\iota_1, \iota_2 \in \pi_2(\mathbf{E}^2 \vee \mathbf{E}^2, \mathbf{S}^1 \vee \mathbf{S}^1)$, we have $\tau'_\lambda(\iota_1) = \iota_2 + \iota_1 - \iota_2$ and therefore

$$
\begin{aligned}
h_* \tau'_\lambda(\iota_1) &= h_*(\iota_2 + \iota_1 - \iota_2) \\
&= h_* \iota_2 + h_* \iota_1 - h_* \iota_2 \\
&= \beta + \alpha - \beta.
\end{aligned}
$$

Therefore it suffices to prove the special case. In this case,

$$\begin{aligned}\partial_* \tau'_\lambda(\iota_1) &= \tau_\lambda(\partial_* \iota_1) && \text{by Theorem (3.1),}\\ &= \lambda(\partial_* \iota_1)\lambda^{-1} \\ &= (\partial_* \iota_2)(\partial_* \iota_1)(\partial_* \iota_2)^{-1} && \text{by definition of } \lambda,\\ &= \partial_*(\iota_2 + \iota_1 - \iota_2) && \text{since } \partial_* \text{ is a homomorphism.}\end{aligned}$$

But $\mathbf{E}^2 \vee \mathbf{E}^2$ is contractible, and it follows from the exactness of the homotopy sequence of $(\mathbf{E}^2 \vee \mathbf{E}^2, \mathbf{S}^1 \vee \mathbf{S}^1)$ that

$$\partial_* : \pi_2(\mathbf{E}^2 \vee \mathbf{E}^2, \mathbf{S}^1 \vee \mathbf{S}^1) \to \pi_1(\mathbf{S}^1 \vee \mathbf{S}^1)$$

is an isomorphism. Hence

$$\tau'_\lambda(\iota_1) = \iota_2 + \iota_1 - \iota_2 . \qquad \square$$

(3.4) Corollary *The group $\pi_2^\dagger(X, A)$ is abelian.*

For $\omega'(X, A)$ contains all elements of the form

$$\alpha - \tau'_\xi(\alpha) = \alpha - \beta - \alpha + \beta$$

($\xi = \partial_* \beta$ as above) and these generate the commutator subgroup. $\square$

(3.5) Corollary *The image of the injection $\pi_2(X) \to \pi_2(X, A)$ is contained in the center of $\pi_2(X, A)$.* $\square$

A space A, or a pair (X, A), is said to be *n-simple* if and only if $\pi_1(A)$ operates trivially on $\pi_n(A)$, $\pi_n(X, A)$, respectively. A space or pair which is n-simple for every n is said to be *simple.* Theorems (4.18) and (4.19) of Chapter III then imply

(3.6) *Every* H-*space or* H-*pair is simple.* $\square$

4 The Hurewicz Map

Like the homology groups, the homotopy groups are functors from the category of pairs to that of abelian groups. The resemblance is enhanced by the fact that the homotopy sequence of a pair, like the homology sequence, is exact. A more precise connection between them, due to W. Hurewicz, is a homomorphism $\rho : \pi_n(X, A) \to H_n(X, A)$, defined as follows: if $f: (\mathbf{E}^n, \mathbf{S}^{n-1}) \to (X, A)$ represents $\alpha \in \pi_n(X, A)$ and $e \in H_n(\mathbf{E}^n, \mathbf{S}^{n-1})$ is a generator, then $\rho(\alpha) = f_*(e) \in H_n(X, A)$. That $\rho(\alpha)$ is well-defined follows from the Homotopy Axiom for homology theory).

(4.1) Theorem *The Hurewicz map $\rho : \pi_n(X, A) \to H_n(X, A)$ is a homomorphism if $n > 1$ or $n = 1$ and $A = \{*\}$.*

Let $f, g : (\mathbf{E}^n, \mathbf{S}^{n-1}) \to (X, A)$ represent $\alpha, \beta \in \pi_n(X, A)$, respectively. Then $\alpha + \beta$ is represented (cf. Exercise 8 of Chapter III) by the composite

$$(\mathbf{E}^n, \mathbf{S}^{n-1}) \xrightarrow{\mathbf{T}\theta} (\mathbf{E}^n \vee \mathbf{E}^n, \mathbf{S}^{n-1} \vee \mathbf{S}^{n-1}) \xrightarrow{f \vee g}$$

$$(X \vee X, A \vee A) \xrightarrow{\nabla} (X, A)$$

(Let $\theta : \mathbf{S}^{n-1} \to \mathbf{S}^{n-1} \vee \mathbf{S}^{n-1}$ be the coproduct which was defined in §5 of Chapter III. Then $\mathbf{T}\theta$ maps $\mathbf{TS}^{n-1} = \mathbf{E}^n$ into $\mathbf{T}(\mathbf{S}^{n-1} \vee \mathbf{S}^{n-1}) = \mathbf{E}^n \vee \mathbf{E}^n$ and its restriction to $\mathbf{S}^{n-1}$ is the map θ).

It suffices, by (7.1) of Chapter III, to show that $(\mathbf{T}\theta)_*(e) = j_{1*}(e) + j_{2*}(e)$, where $j_1, j_2 : (\mathbf{E}^n, \mathbf{S}^{n-1}) \to (\mathbf{E}^n \vee \mathbf{E}^n, \mathbf{S}^{n-1} \vee \mathbf{S}^{n-1})$ are the inclusions. For then

(4.2)
$$\begin{aligned}\rho(\alpha + \beta) &= \nabla_*(f \vee g)_*(j_{1*}(e) + j_{2*}(e)) \\ &= \nabla_*(j_{1*}\, f_*(e) + j_{2*}\, g_*(e)) = f_*(e) + g_*(e) \\ &= \rho(\alpha) + \rho(\beta).\end{aligned}$$

The diagram[1]

(4.3)
$$\begin{array}{ccc} H_n(\mathbf{E}^n, \mathbf{S}^{n-1}) & \xrightarrow{(T\theta)_*} & H_n(\mathbf{E}^n \vee \mathbf{E}^n, \mathbf{S}^{n-1} \vee \mathbf{S}^{n-1}) \\ \downarrow \partial_* & & \downarrow \partial_* \\ H_{n-1}(\mathbf{S}^{n-1}) & \xrightarrow{\theta_*} & H_{n-1}(\mathbf{S}^{n-1} \vee \mathbf{S}^{n-1}), \end{array}$$

as well as the diagrams obtained from it by replacing θ by j_1 and by j_2, is commutative, and the homomorphisms ∂_* are isomorphisms (by exactness of the homology sequences of the appropriate pairs and the fact that $\mathbf{E}^n$ and $\mathbf{E}^n \vee \mathbf{E}^n$ are contractible). Thus it suffices to show that

(4.4)
$$\theta_* = j_{1*} + j_{2*}.$$

Now j_{1*} and j_{2*} represent $H_{n-1}(\mathbf{S}^{n-1} \vee \mathbf{S}^{n-1})$ as a direct sum, while $q_{1*}, q_{2*} : H_{n-1}(\mathbf{S}^{n-1} \vee \mathbf{S}^{n-1}) \to H_{n-1}(\mathbf{S}^{n-1})$ is the dual representation of the same group as a product. The relation (4.4) then follows from the relations $q_1 \circ \theta \simeq 1 \simeq q_2 \circ \theta$ of (5.1), (5.2) of Chapter III.

This proves the theorem in the relative case; that in the absolute case follows by a slight simplification of the same argument. □

(4.5) Corollary *The Hurewicz map $\rho : \pi_n(S^n) \to H_n(S^n)$ is an isomorphism for $n \geq 1$.*

[1] In this, as in all other diagrams involving homology groups of spaces with base point, $H_q(X)$ is an abbreviation for $H_q(X, *)$.

It was proved in §3 of Chapter I that every map of $\mathbf{S}^n$ into itself of degree 0 is nullhomotopic. Now, if $f: \mathbf{S}^n \to \mathbf{S}^n$ represents $\alpha \in \pi_n(S^n)$, then $\rho(\alpha) = f_*(s) = d(f) \cdot s$ for a generator s of $H_n(\mathbf{S}^n)$. Thus ρ is a monomorphism. But if f is the identity map, then $d(f) = 1$ and hence ρ is a epimorphism as well. □

(4.6) Corollary *The Hurewicz map* $\rho: \pi_n(\mathbf{E}^n, \mathbf{S}^{n-1}) \to H_n(\mathbf{E}^n, \mathbf{S}^{n-1})$ *is an isomorphism for* $n \geq 2$. □

The definition of the Hurewicz map depends on the choice of a generator ε_n of the homology group $H_n(\mathbf{T}\mathbf{S}^{n-1}, \mathbf{S}^{n-1})$. As the group in question is infinite cyclic, this determines ρ up to a sign. It is desirable, for two reasons, to make this choice explicit.

The first reason is that we may wish to compare the homotopy and homology sequences of a pair (X, A). In particular, there is a diagram

(4.7)
$$\begin{array}{ccc} \pi_n(X, A) & \xrightarrow{\partial_*} & \pi_{n-1}(A) \\ \rho \downarrow & & \downarrow \rho \\ H_n(X, A) & \xrightarrow[\partial_*]{} & H_{n-1}(A) \end{array}$$

which is commutative up to sign for arbitrary choices of the generators ε_n, ε_{n-1}. To make it strictly commutative entails relating the choices in consecutive dimensions.

The second reason is that in certain situations it may be convenient to use other spaces than the standard n-cell $\mathbf{E}^n$ to define elements of homotopy groups. Indeed if E is an n-cell with boundary $\dot{E}, f: (E, \dot{E}) \to (X, A)$ a map, and $h: (\mathbf{E}^n, \mathbf{S}^{n-1}) \to (E, \dot{E})$ a homeomorphism, then $f \circ h$ represents an element $\alpha \in \pi_n(X, A)$, and it is convenient to say that f represents α. The element depends on the choice of the homeomorphism h. Now the group $\pi_n(E, \dot{E})$ is, by Corollary (4.6), an infinite cyclic group, and h represents a generator; thus α is determined up to a sign. If, however, $\mathbf{E}^n$ has been given a standard orientation, and if E is oriented, we may normalize the choice of h by requiring that it preserve orientation.

Let us recall, then, that if E is an n-cell, the boundary $\dot{E}$ of E is uniquely characterized, by Brouwer's theorem of invariance of domain, as the image of $\dot{\Delta}^n$ under any homeomorphism of Δ^n with E. An *orientation* of E is just a generator ε of the infinite cyclic group $H_n(E, \dot{E})$; the pair (E, ε) is called an *oriented n-cell.* If S is an n-sphere (with base point), an *orientation* σ of S is a generator of the infinite cyclic group $H_n(S, *)$; the pair (S, σ) is called an *oriented n-sphere*). If E is an n-cell with base point $* \in \dot{E}$, then $\dot{E}$ is an $(n-1)$-sphere, and the isomorphism $\partial_*: H_n(E, \dot{E}) \to H_{n-1}(\dot{E}, *)$ carries an orientation ε of E into an orientation $\partial_* \varepsilon$ of $\dot{E}$; we say that ε and $\partial_* \varepsilon$ are *coherent*. Let E be an n-cell, E' an $(n-1)$-cell contained in $\dot{E}$; we shall say that E' is a *face* of E if and only if the pairs $(\Delta^n, \Delta_0^{n-1})$ and (E, E') are homeomorphic. It follows that, if E'' is the closure of $\dot{E} - E'$, then E'' is also a cell with $E' \cap E'' = \dot{E}' = \dot{E}''$, the injection $j_* : H_{n-1}(E', \dot{E}') \to H_{n-1}(\dot{E}, E'')$ is an isomorphism, and the composite of j_*^{-1} with the boundary operator

$$\partial'_* : H_n(E, \dot{E}) \to H_{n-1}(\dot{E}, E'')$$

of the homology sequence of the triple $(E, \dot{E}, E'')$ is an isomorphism, called the *incidence isomorphism* $[E : E']$. We then say that orientations ε, ε' of E, E' are *coherent* if and only if $[E : E']\varepsilon = \varepsilon'$; and when this is so we also say that the orientations $\partial_* \varepsilon$ of $\dot{E}$ and ε' are *coherent*.

We now choose definite orientations of many of the cells and spheres to be encountered later.

$\mathbf{S}^0 = \dot{\mathbf{T}} = \{0, 1\}$ is oriented by the homology class $\mathbf{s}_0$ of the point 1.

$\mathbf{T}$ is oriented by the element $\mathbf{t}$ such that $\partial_* \mathbf{t} = \mathbf{s}_0$.

$\mathbf{S}^1$ is oriented by $\mathbf{s} = \varpi_*(\mathbf{t})$, where $\varpi : (\mathbf{T}, \dot{\mathbf{T}}) \to (\mathbf{S}^1, *)$ is the identification map.

$\mathbf{S}^n = \mathbf{S}^1 \wedge \cdots \wedge \mathbf{S}^1$ is oriented by $\mathbf{s}_n = \mathbf{s} \wedge \cdots \wedge \mathbf{s}$.

$\mathbf{E}^n = \mathbf{T} \wedge \mathbf{S}^{n-1}$ is oriented by $\varepsilon_n = \mathbf{t} \wedge \mathbf{s}_{n-1}$.

Δ^0 is oriented by the homology class δ_0 of the point $\mathbf{e}_0$.

$\Delta_0^{n-1} \subset \dot{\Delta}^n$ is oriented so that the map $d_0^n : \Delta^{n-1} \to \Delta_0^{n-1} \subset \Delta^n$ is orientation-preserving, and Δ^n is oriented coherently with Δ_0^{n-1}.

Clearly $(\mathbf{TS}^{n-1})^\bullet = \mathbf{S}^{n-1}$, and $\mathbf{TS}^{n-1}$ and $\mathbf{S}^{n-1}$ are coherently oriented. It then follows that

(4.8) *The diagram* (4.7) *is commutative.*

Thus ρ is a map of the homotopy sequence of (X, A) into its homology sequence.

(4.9) *The homomorphism* $\rho : \pi_n(X) \to H_n(X)$ *annihilates the subgroup* $\omega_n(X)$, *and so induces a homomorphism*

$$\bar{\rho} : \pi_n^*(X) \to H_n(X).$$

(4.10) *The homomorphism* $\rho : \pi_n(X, A) \to H_n(X, A)$ *annihilates the subgroup* $\omega'_n(X, A)$, *and so induces a homomorphism*

$$\tilde{\rho} : \pi_n^\dagger(X, A) \to H_n(X, A).$$

The proofs of these two statements are similar; we prove the first one. Let $\alpha \in \pi_n(X)$, $\xi \in \pi_1(X)$, and let $a : \mathbf{S}^n \to X$ and $a' : \mathbf{S}^n \to X$ be representatives of α, $\tau_\xi(\alpha)$, respectively. Then a and a' are homotopic as free maps, and therefore

$$\rho(\alpha) = a_*(\mathbf{s}_n) = a'_*(\mathbf{s}_n) = \rho(\tau_\xi(\alpha)), \qquad \rho(\alpha - \tau_\xi(\alpha)) = 0.$$

5 The Eilenberg and Blakers Homology Groups

In this section we introduce a sequence of groups, intermediate between the homotopy and the homology groups. Those groups are singular groups, based on singular simplices which are subjected to certain auxiliary conditions. These groups, as the names suggest, were defined by Eilenberg [3] in the absolute case and by Blakers [1] in the relative.

Let (X, A) be a pair with base point $*$. For simplicity, we shall assume that X and A are 0-connected. Let $\mathfrak{S}^{(n)}(X, A)$ be the subcomplex of the total singular complex $\mathfrak{S}(X)$ of X consisting of all singular simplices $u : \Delta^q \to X$ such that u maps the vertices of Δ^q into $*$ and the n-skeleton of Δ^q into A. It is clear that $\mathfrak{S}^{(n)}(X, A)$ is indeed a subcomplex. Moreover,

$$\mathfrak{S}(X) \supset \mathfrak{S}^{(0)}(X, A) \supset \mathfrak{S}^{(1)}(X, A) \supset \cdots,$$
$$\bigcap \mathfrak{S}^{(n)}(X, A) = \mathfrak{S}^{(0)}(A, *),$$

while $\mathfrak{S}^{(n)}(X, A)$ and $\mathfrak{S}^{(0)}(A, *)$ have the same q-simplices for all $q \leq n$. It is also useful to observe that, if $q > n$, a singular q-simplex belongs to $\mathfrak{S}^{(n)}(X, A)$ if and only if all of its faces do.

(5.1) Theorem *If* (X, A) *is n-connected, the inclusion map of the pair* $(\mathfrak{S}^{(n)}(X, A), \mathfrak{S}^{(0)}(A, *))$ *into the pair* $(\mathfrak{S}(X), \mathfrak{S}(A))$ *is a chain equivalence of pairs.*

This is proved by constructing a chain deformation retraction. Specifically, we construct, inductively for each singular simplex $u : \Delta^q \to X$ a singular prism $Pu : \mathbf{I} \times \Delta^q \to X$ with the following properties

(1) $(Pu) \circ i_0 = u$,
(2) $(Pu) \circ i_1 \in \mathfrak{S}^{(n)}(X, A)$,
(3) $P(u \circ d_i) = (Pu) \circ (1 \times d_i)$,
(4) Pu is stationary if $u \in \mathfrak{S}^{(n)}(X, A)$.
(5) if $u \in \mathfrak{S}(A)$ then $Pu(\mathbf{I} \times \Delta^q) \subset A$.

It follows as usual that, if we define $\alpha : \mathfrak{S}(X) \to \mathfrak{S}^{(n)}(X, A)$ by $\alpha u = (Pu) \circ i_1$, then α defines a chain retraction and P a chain homotopy between α and the identity.

We define P inductively, as follows. If $q = 0$, then, since X is 0-connected, we can choose a path $\lambda_u : \mathbf{I} \to X$ from $u(\mathbf{e}_0)$ to $*$; if $u(\mathbf{e}_0) = *$, we take λ_u to be the constant path. Then define

$$Pu(t, \mathbf{e}_0) = \lambda_u(t).$$

Suppose that Pv has been defined, and satisfies (1)–(5), for all simplices v of dimension $< q$. Let $u : \Delta^q \to X$ be a q-simplex. Then the maps

$P(u \circ d_i) : \mathbf{I} \times \Delta^{q-1} \to X$ "fit together" with u, in the sense that there is a map

$$\phi : 0 \times \Delta^q \cup \mathbf{I} \times \dot{\Delta}^q \to X$$

such that

$$\phi(0, x) = u(x) \qquad (x \in \Delta^q),$$
$$\phi(t, d_i x) = P(u \circ d_i)(t, x) \qquad (t \in \mathbf{I},\ x \in \Delta^{q-1}).$$

If $q \leq n$, $\phi(1 \times \dot{\Delta}^q) \subset A$. Since $(\Delta^q, \dot{\Delta}^q)$ is an NDR-pair, ϕ has an extension $Pu : \mathbf{I} \times \Delta^q \to X$. If $q \leq n$, it follows from Lemma (3.3) of Chapter II that we can choose the extension Pu so that $Pu(1 \times \Delta^q) \subset A$. If each of the maps $P(u \circ d_i)$ is stationary, then $\phi \mid \mathbf{I} \times \dot{\Delta}^q$ is stationary, and we can choose Pu to be stationary. The map P is then readily verified to satisfy conditions (1)–(5) in dimension q, and the induction is complete. □

Since X is 0-connected, we may take $A = *$ and conclude:

(5.2) Corollary *The inclusion $\mathfrak{S}^{(0)}(X, *) \subsetneq \mathfrak{S}(X)$ is a chain equivalence.*

□

If A is any subspace of X, then $\mathfrak{S}^{(0)}(X, A) = \mathfrak{S}^{(0)}(X, *)$. Hence

(5.3) Corollary *If A is any subspace of X, then the inclusion $\mathfrak{S}^{(0)}(X, A) \subsetneq \mathfrak{S}(X)$ is a chain equivalence.* □

The *Eilenberg homology groups of* X are the groups $H_n^{(q)}(X) = H_n(\mathfrak{S}^{(q)}(X, *), \mathfrak{S}(*))$. The *Blakers homology groups of* (X, A) are the groups

$$H_n^{(q)}(X, A) = H_n(\mathfrak{S}^{(q)}(X, A), \mathfrak{S}^{(0)}(A, *)).$$

The groups $H_n^{(q)}(X, A)$ are zero if $n \leq q$. Moreover, there are injections

(5.4) $\quad H_n^{(n-1)}(X, A) \to H_n^{(n-2)}(X, A) \to \cdots \to H_n^{(0)}(X, A) \to H_n(X, A).$

If $f : (X, A) \to (Y, B)$ then the chain map of $\mathfrak{S}(X)$ into $\mathfrak{S}(Y)$ induced by f carries $\mathfrak{S}^{(q)}(X, A)$ into $\mathfrak{S}^{(q)}(Y, B)$, thereby inducing homomorphisms $f_*^{(q)} : H_n^{(q)}(X, A) \to H_n^{(q)}(Y, B)$. Thus $H_n^{(q)}$, like H_n, is a functor. Moreover, the injections of (5.4) are natural transformations.

Since the groups $H_n^{(q)}(X, A)$ are the relative homology groups of the pair $(\mathfrak{S}^{(q)}(X, A), \mathfrak{S}^{(0)}(A, *))$, there are boundary operators mapping $H_n^{(q)}(X, A)$ into $H_{n-1}^{(0)}(A)$. Since the latter group injects isomorphically, by Corollary (5.2), into $H_{n-1}(A)$, we may regard these boundary operators as homomorphisms

$$\partial_*^{(q)} : H_n^{(q)}(X, A) \to H_{n-1}(A).$$

Clearly $\partial_*^{(q)}$ is natural transformation of functors. Moreover the diagram

$$\text{(5.5)}\qquad \begin{array}{ccc} H_n^{(q)}(X, A) & \xrightarrow{i_*} & H_n(X, A) \\ & \partial_*^{(q)} \searrow \quad \swarrow \partial_* & \\ & H_{n-1}(A) & \end{array}$$

where i_* is the injection, is commutative.

Let us study the relationship between the group $\pi_n(X, A)$ and the Blakers homology groups. We have seen that the Hurewicz map $\rho : \pi_n(X, A) \to H_n(X, A)$ annihilates the subgroup $\omega_n'(X, A)$ and thereby induces a homomorphism $\tilde{\rho} : \pi_n^\dagger(X, A) \to H_n(X, A)$. We now prove that $\tilde{\rho}$ factors through $H_n^{(n-1)}(X, A)$, i.e.,

(5.6) Lemma *Let (X, A) be a pair with A and X 0-connected. If $n \geq 2$, there is a homomorphism $\tilde{\rho}' : \pi_n^\dagger(X, A) \to H_n^{(n-1)}(X, A)$ such that $i_* \circ \tilde{\rho}' = \tilde{\rho}$.*

The proof breaks up into three steps.

Step I. Construction of $\rho' : \pi_n(X, A) \to H_n^{(n-1)}(X, A)$ with $i_* \circ \rho' = \rho$.
Step II. Proof that ρ' is a homomorphism.
Step III. Proof that ρ' annihilates $\omega_n'(X, A)$.

Step I. Because $(\mathbf{E}^n, \mathbf{S}^{n-1})$ is $(n-1)$-connected, the injection $i_* : H_n^{(n-1)}(\mathbf{E}^n, \mathbf{S}^{n-1}) \to H_n(\mathbf{E}^n, \mathbf{S}^{n-1})$ is an isomorphism. Let $\varepsilon' = i_*^{-1}(\varepsilon_n)$. If $\alpha \in \pi_n(X, A)$ is represented by $f: (\mathbf{E}^n, \mathbf{S}^{n-1}) \to (X, A)$, define $\rho'(\alpha) = f_*^{(n-1)}(\varepsilon')$. The fact that $i_* \circ \rho' = \rho$ then follows from the naturality of i_*, i.e., commutativity of the diagram

$$\text{(5.7)}\qquad \begin{array}{ccc} H_n^{(n-1)}(\mathbf{E}^n, \mathbf{S}^{n-1}) & \xrightarrow{f_*^{(n-1)}} & H_n^{(n-1)}(X, A) \\ \downarrow i_* & & \downarrow i_* \\ H_n(\mathbf{E}^n, \mathbf{S}^{n-1}) & \xrightarrow[f_*]{} & H_n(X, A) \end{array}$$

Step II. Recall the diagram (4.3), and observe that there is a comparable commutative diagram

$$\text{(5.8)}\qquad \begin{array}{ccc} H_n^{(n-1)}(\mathbf{E}^n, \mathbf{S}^{n-1}) & \xrightarrow{(T\theta)_*^{(n-1)}} & H_n^{(n-1)}(\mathbf{E}^n \vee \mathbf{E}^n, \mathbf{S}^{n-1} \vee \mathbf{S}^{n-1}) \\ \downarrow \partial_*^{(n-1)} & & \downarrow \partial_*^{(n-1)} \\ H_{n-1}(\mathbf{S}^{n-1}) & \xrightarrow[\theta_*]{} & H_{n-1}(\mathbf{S}^{n-1} \vee \mathbf{S}^{n-1}) \end{array}$$

Moreover, because of the commutativity of the diagram (5.5) there is a map of the diagram (5.8) into the diagram (4.3) which is the identity on the bottom row, while the groups in the top row are mapped by the injections i_*. The latter are isomorphisms, by Theorem (5.1); since the homomorphisms ∂_* are isomorphisms, so are the homomorphisms $\partial_*^{(n-1)}$. It follows that $(\mathbf{T}\theta)_*^{(n-1)} = j_{1*}^{(n-1)} + j_{2*}^{(n-1)}$, and the proof that ρ' is a homomorphism now follows from the formula (4.2) by substituting ρ' for ρ and $\nabla_*^{(n-1)}, \ldots$ for $\nabla_*, \ldots$.

Step III. It suffices to show that, if $f, f' : (\mathbf{E}^n, \mathbf{S}^{n-1}) \to (X, A)$ are maps representing $\alpha, \alpha' \in \pi_n(X, A)$ and if f and f' are freely homotopic via a loop in A, then $\rho'(\alpha) = \rho'(\alpha')$, i.e., $f_*^{(n-1)}(e') = f_*'^{(n-1)}(e')$. Since $n \geq 2$, this follows from

(5.9) Lemma *Let $f, f' : (Y, B) \to (X, A)$ and suppose that f and f' are freely homotopic via a loop in A. Then, for all $q \geq 1$, $f_*^{(q)} = f_*'^{(q)} : H_n^{(q)}(Y, B) \to H_n^{(q)}(X, A)$.*

In fact, if $F : (\mathbf{I} \times Y, \mathbf{I} \times B) \to (X, A)$ is such a free homotopy, and if, for each singular n-simplex $u : \Delta^n \to Y$, $Pu : \mathbf{I} \times \Delta^n \to X$ is the map given by $Pu = F \circ (1 \times u)$, then the operation P induces a chain homotopy between the chain maps $f_\#, f'_\# : \mathfrak{S}(Y) \to \mathfrak{S}(X)$ defined by f, f'. What is to be shown is that P sends chains of $\mathfrak{S}^{(q)}(Y, B)$ into chains of $\mathfrak{S}^{(q)}(X, A)$; in other words, if $u \in \mathfrak{S}^{(q)}(Y, B)$, so that all faces of Δ^n of dimension $\leq q$ are mapped into B by u, then all faces of $\mathbf{I} \times \Delta^n$, in the standard subdivision, of dimension $\leq q$, are mapped into A by Pu. It is left to the reader to verify that, since $q \geq 1$, this is the case. □

We are going to prove that the map $\tilde{\rho}'$ is an isomorphism. In fact, let C_q be the group of q-chains of $\mathfrak{S}^{(n-1)}(X, A)$, A_q the intersection of C_q with the chain group of A. Then

$$H_n^{(n-1)}(X, A) = C_n/(A_n + \partial C_{n+1}).$$

The group C_n is freely generated by all singular n-simplices $u : \Delta^n \to X$ such that $u(\dot{\Delta}^n) \subset A$ and $j(e_i) = *$ for $i = 0, \ldots, n$. Since $\pi_n^\dagger(X, A)$ is abelian (even if $n = 2$, according to Corollary (3.4)), there is a unique homomorphism $\sigma : C_n \to \pi_n^\dagger(X, A)$ such that, for each such u, $\sigma(u)$ is the image in $\pi_n^\dagger(X, A)$ of the element of $\pi_n(X, A)$ represented by u. Evidently $\sigma(A_n) = 0$. If $\sigma(\partial C_{n+1}) = 0$, then σ induces a homomorphism $\sigma' : H_n^{(n-1)}(X, A) \to \pi_n^\dagger(X, A)$. An easy argument shows

(5.10) Lemma *If $\sigma(\partial C_{n+1}) = 0$, then $\tilde{\rho}'$ is an isomorphism whose inverse is σ'.*

The proof that σ annihilates ∂C_{n+1}, while intuitively clear, presents formidable difficulties in practice. The argument depends on a geometric proposition known as the Homotopy Addition Theorem. A direct proof of this theorem has been given by Hu [1] in the case that (X, A) is n-simple. In $[W_1]$ I have given a direct proof in the general case, but using singular cubes instead of simplices.

The Eilenberg groups can be treated in a similar, but easier, fashion.

(5.11) Lemma *Let X be a 0-connected space in $\mathscr{K}_0$. If $n \geq 1$, there is a homomorphism $\bar{\rho}' : \pi_n^*(X) \to H_n^{(n-1)}(X)$ such that $i_* \circ \bar{\rho}' = \bar{\rho}$.*

In fact, Lemma (5.11) follows formally from Lemma (5.6), except for the case $n = 1$. However, it is not difficult to prove Lemma (5.11) directly for all $n \geq 1$. The details are left to the reader. □

We shall need, however,

(5.12) Lemma *The homomorphism $\bar{\rho}'$ is an isomorphism if $n = 1$.*

As in the relative case, let C_q be the group of q-chains of $\mathfrak{F}^{(0)}(X, *)$ and define $\sigma : C_1 \to \pi_1^*(X)$ in the same way. What has to be proved is that $\sigma(\partial C_2) = 0$. Let $u : \Delta^2 \to X$ be a singular simplex with vertices at the base point $*$ of X. Then $\sigma(\partial u)$ is the image in $\pi_1^*(X)$ of the product $\xi = \xi_2 \xi_0 \xi_1^{-1}$, where ξ_i is the element of $\pi_1(X)$ represented by $u \circ d_i$ $(i = 0, 1, 2)$. But $\xi = u_*(\eta)$, where $\eta = \eta_2 \eta_0 \eta_1^{-1}$ and η_i is the homotopy class of the rectilinear path in Δ^2 from $d_i(e_0)$ to $d_i(e_1)$. Since Δ^2 is contractible, $\eta = 1$ and therefore $\xi = 1$, $\sigma(\partial u) = 0$. □

6 The Homotopy Addition Theorem

Let (X, A) be a pair, and let $f: (\dot{\Delta}^{n+1}, \mathbf{e}_0) \to (X, *)$ be a map which carries each $(n-1)$-dimensional face of Δ^{n+1} into A. $(n \geq 1)$. Then f represents an element $\alpha \in \pi_n(X)$, and the map $f \circ d_i : (\Delta^n, \dot{\Delta}^n, \mathbf{e}_0) \to (X, A, fd_i(\mathbf{e}_0))$ represents an element $\alpha_i \in \pi_n(X, A, fd_i(\mathbf{e}_0))$. If $i > 0$, $d_i(\mathbf{e}_0) = \mathbf{e}_0$, $fd_i(\mathbf{e}_0) = *$. On the other hand, if $i = 0$, $fd_i(\mathbf{e}_0)$ may not be the base-point $*$; but if ξ is the homotopy class of the rectilinear path from $\mathbf{e}_0$ to $\mathbf{e}_1$, then $\tau'_{f_*\xi}(\alpha_0) \in \pi_n(X, A, *)$. Let $j_* : \pi_n(X, *) \to \pi_n(X, A, *)$ be the injection.

The homotopy addition theorem is a formula relating the element $j_*(\alpha)$ with the elements $\alpha_0' = \tau'_{f_*\xi}(\alpha_0), \alpha_1, \ldots, \alpha_{n+1}$. It will be used in the proof of the Hurewicz theorem, and it appears implicitly in obstruction theory in the fact that the obstruction cochain is a cocycle. The form taken by this formula differs slightly, according as $n = 1$, when it relates certain homotopy classes of paths, $n = 2$, when the group $\pi_n(X, A, *)$ is not necessarily abelian, or $n > 2$, when $\pi_n(X, A, *)$ is an abelian group. These distinctions in low dimensions also crop up in the proof.

(6.1) Theorem (Homotopy Addition Theorem). *The elements $j_*\alpha$, α'_0, α_i described above are related by:*

$$j_*\alpha = \begin{cases} \alpha_2\alpha_0\alpha_1^{-1} \quad (= \alpha'_0\alpha_2\alpha_1^{-1}) & \text{if } n = 1; \\ \alpha'_0 + \alpha_2 - \alpha_1 - \alpha_3 & \text{if } n = 2; \\ \alpha'_0 + \sum_{i=1}^{n+1} (-1)^i\alpha_i & \text{if } n \geq 3. \end{cases}$$

The truth of the formula for $n = 1$ is evident (note that, in this case, $\alpha'_0 = \alpha_2\alpha_0\alpha_2^{-1}$).

Let us observe that, because of the functorial properties of relative homotopy groups, the homotopy addition theorem is natural; i.e., if it holds for a pair (X, A) and if $g : (X, A) \to (Y, B)$ is a (base point preserving) map, then it holds in (Y, B). This allows us to use the method of the *universal example*; if we can prove the homotopy addition theorem in the special case $X = \Delta^{n+1}$, A is the union of the $(n-1)$-dimensional faces of Δ^{n+1}, and f is the identity map, then it will follow in general.

Let K be the simplicial complex consisting of all the faces of Δ^{n+1}. Then $K_n = \dot{\Delta}^{n+1}$, and the formula to be proved holds in the group $\pi_n(K_n, K_{n-1})$. Moreover, the elements $\alpha_i \in \pi_n(K_n, K_{n-1})$ are the images, under the homomorphisms induced by the characteristic maps for the n-simplices of K, of the generator of the infinite cyclic group $\pi_n(\Delta^n, \dot{\Delta}^n)$ represented by the identity map. And the element α is the image under $\partial_* : \pi_{n+1}(K_{n+1}, K_n) \to \pi_n(K_n)$ of the generator of the group $\pi_{n+1}(K_{n+1}, K_n)$ represented by the identity map.

This should remind the reader of our discussion of the homology theory of a CW-complex. In fact, the proof that the groups $\Gamma_n(X, A) = H_n(X_n, X_{n-1})$ of a CW-pair (X, A) form a chain complex does not use the excision theorem. Accordingly, the groups $\pi_n(X_n, X_{n-1})$, with the operators $\partial = j_* \circ \partial_* : \pi_{n+1}(X_{n+1}, X_n) \to \pi_n(X_n, X_{n-1})$, form a chain complex (with some reservations concerning the behavior in low dimensions). And the homotopy addition theorem is just a calculation of the boundary operator of that chain complex in a special case.

In view of these observations, the homotopy addition theorem follows from

(6.2) Theorem *The image of the generator α of $\pi_{n+1}(K_{n+1}, K_n)$ represented by the identity map of Δ^{n+1} under the boundary operator $\partial : \pi_{n+1}(K_{n+1}, K_n) \to \pi_n(K_n, K_{n-1})$ is given by*

(6.3)
$$\partial\alpha = \begin{cases} \alpha'_0\alpha_2\alpha_1^{-1} & \text{if } n = 1; \\ \alpha'_0 + \alpha_2 - \alpha_1 - \alpha_3 & \text{if } n = 2; \\ \alpha'_0 + \sum_{i=1}^{n+1} (-1)^i\alpha_i & \text{if } n \geq 3. \end{cases}$$

As remarked above, the theorem is true if $n = 1$. Suppose that $n \geq 2$. Then there is a commutative diagram

$$\begin{array}{ccccc} \pi_{n+1}(K_{n+1}, K_n) & \xrightarrow{\partial} & \pi_n(K_n, K_{n-1}) & \xrightarrow{\partial} & \pi_{n-1}(K_{n-1}, K_{n-2}) \\ \downarrow \rho & & \downarrow \rho & & \\ H_{n+1}(K_{n+1}, K_n) & \xrightarrow{\partial} & H_n(K_n, K_{n-1}) & & \end{array} \tag{6.4}$$

where ρ is the Hurewicz map. Now $\partial : H_{n+1}(K_{n+1}, K_n) \to H_n(K_n, K_{n-1})$ is a monomorphism (because K is acyclic) and $\rho : \pi_{n+1}(K_{n+1}, K_n) \to H_{n+1}(K_{n+1}, K_n)$ is an isomorphism (by Corollary (4.6)) and therefore $\rho \circ \partial = \partial \circ \rho : \pi_{n+1}(K_{n+1}, K_n) \to H_n(K_n, K_{n-1})$ is a monomorphism. Let $\beta_n \in \pi_n(K_n, K_{n-1})$ be the right-hand side of the formula (6.3). We shall prove

(6.5) $$\partial\beta_n = 0 \qquad (n \geq 2).$$

We prove (6.5) by induction on n. We shall need to calculate in the group $\pi_{n-1}(K_{n-1}, K_{n-2})$ and accordingly must introduce some further notations. For each pair (i, j) with $0 \leq i < j \leq n+1$, let α_{ij} be the element of $\pi_{n-1}(K_{n-1}, K_{n-2})$ represented by the map $d_i \circ d_j : \Delta^{n-1} \to \Delta^{n+1}$. Let ξ_{ij} be the homotopy class of the rectilinear path from e_i to e_j, and let $\sigma_{ij} = \tau'_{\xi_{ij}}$ be the corresponding operation on relative homotopy groups.

Suppose $n = 2$. Then

$$\partial\alpha_0 = \xi_{12}\xi_{23}\xi_{13}^{-1}$$

and therefore

$$\partial\alpha'_0 = \partial(\sigma_{01}\alpha_0) = \sigma_{01}(\partial\alpha_0) = \sigma_{01}(\xi_{12}\xi_{23}\xi_{13}^{-1}) = \xi_{01}\xi_{12}\xi_{23}\xi_{13}^{-1}\xi_{01}^{-1};$$

while

$$\partial\alpha_2 = \xi_{01}\xi_{13}\xi_{03}^{-1},$$
$$\partial\alpha_1 = \xi_{02}\xi_{23}\xi_{03}^{-1},$$
$$\partial\alpha_3 = \xi_{01}\xi_{12}\xi_{02}^{-1}.$$

Then

$$\partial\beta_2 = \xi_{01}\xi_{12}\xi_{23}\xi_{13}^{-1}\xi_{01}^{-1}\xi_{01}\xi_{13}\xi_{03}^{-1}\xi_{03}\xi_{23}^{-1}\xi_{02}^{-1}\xi_{02}\xi_{12}^{-1}\xi_{01}^{-1} = 1.$$

Suppose $n = 3$. We then have, by induction hypothesis

$$\partial\alpha_0 = \sigma_{12}(\alpha_{01}) + \alpha_{03} - \alpha_{02} - \alpha_{04}$$

and therefore

$$\begin{aligned} \partial(\sigma_{01}\alpha_0) &= \sigma_{01}\,\partial\alpha_0 \\ &= \sigma_{01}\sigma_{12}(\alpha_{01}) + \sigma_{01}(\alpha_{03}) - \sigma_{01}(\alpha_{02}) - \sigma_{01}(\alpha_{04}). \end{aligned}$$

It is tempting to put $\sigma_{01} \circ \sigma_{12} = \sigma_{02}$, because $\xi_{01}\xi_{12} = \xi_{02}$, but this relation, while true in K_q for $q \geq 2$, is not true in K_1. As we are working in $\pi_2(K_2, K_1)$, we must be more careful. In fact

$$\begin{aligned}\sigma_{01}\sigma_{12}(\alpha_{01}) &= \sigma_{01}\sigma_{12}\sigma_{02}^{-1}(\sigma_{02}\alpha_{01}) \\ &= \tau'_{\partial\alpha_{34}}(\sigma_{02}(\alpha_{01})) \\ &= \alpha_{34} + \sigma_{02}(\alpha_{01}) - \alpha_{34}\end{aligned}$$

by Lemma (3.3). Hence

$$\partial\sigma_{01}(\alpha_0) = \alpha_{34} + \sigma_{02}(\alpha_{01}) - \alpha_{34} + \sigma_{01}(\alpha_{03}) - \sigma_{01}(\alpha_{02}) - \sigma_{01}(\alpha_{04}),$$

while

$$\begin{aligned}\partial\alpha_1 &= \sigma_{02}(\alpha_{01}) + \alpha_{13} - \alpha_{12} - \alpha_{14}, \\ \partial\alpha_2 &= \sigma_{01}(\alpha_{02}) + \alpha_{23} - \alpha_{12} - \alpha_{24}, \\ \partial\alpha_3 &= \sigma_{01}(\alpha_{03}) + \alpha_{23} - \alpha_{13} - \alpha_{34}, \\ \partial\alpha_4 &= \sigma_{01}(\alpha_{04}) + \alpha_{24} - \alpha_{14} - \alpha_{34}.\end{aligned}$$

The group $\pi_2(K_2, K_1)$ is not abelian, and if we add the terms in the order given, there appears to be little cancellation. However, it should be remembered that the operator ∂ factors as $j_* \circ \partial_*$, and the image of j_* is contained in the center of $\pi_2(K_2, K_1)$; thus each of the terms in the required sum is in the center. Hence we may start the calculation as follows:

$$\begin{aligned}\partial\{\sigma_{01}(\alpha_0) - \alpha_1\} &= \{\alpha_{34} + \sigma_{02}(\alpha_{01}) - \alpha_{34} + \sigma_{01}(\alpha_{03}) - \sigma_{01}(\alpha_{02}) \\ &\quad - \sigma_{01}(\alpha_{04})\} + \{\alpha_{14} + \alpha_{12} - \alpha_{13} - \sigma_{02}(\alpha_{01})\} \\ &= \alpha_{34} + \{\alpha_{14} + \alpha_{12} - \alpha_{13} - \sigma_{02}(\alpha_{01})\} + \sigma_{02}(\alpha_{01}) \\ &\quad - \alpha_{34} + \sigma_{01}(\alpha_{03}) - \sigma_{01}(\alpha_{02}) - \sigma_{01}(\alpha_{04}) \\ &= \alpha_{34} + \alpha_{14} + \alpha_{12} - \alpha_{13} - \alpha_{34} + \sigma_{01}(\alpha_{03}) \\ &\quad - \sigma_{01}(\alpha_{02}) - \sigma_{01}(\alpha_{04}).\end{aligned}$$

Proceeding similarly with each of the remaining terms, we find that everything eventually cancels, so that $\partial\beta_3 = 0$.

The proof that $\partial\beta_n = 0$ for $n \geq 4$ is much simpler, because $\sigma_{01} \circ \sigma_{12} = \sigma_{02}$ and because $\pi_{n-1}(K_{n-1}, K_{n-2})$ is abelian. The argument to prove that $\partial\beta_n = 0$ is then not much more difficult than the standard argument that $\partial \circ \partial = 0$ in the chain-complex of a simplicial complex, and may safely be left to the reader. □

7 The Hurewicz Theorems

These were the first non-trivial theorems about homotopy groups. The absolute theorem was proved by Hurewicz in his pioneering series of papers [1]. The relative theorem was known to Hurewicz and had become folklore by the time Hu published his exposition [1] of relative homotopy theory in 1947.

(7.1) Theorem (Absolute Hurewicz Theorem). *Let X be an $(n-1)$-connected space $(n \geq 1)$. Then*

$$\bar{\rho} : \pi_n^*(X) \approx H_n(X)$$

(7.2) Theorem (Relative Hurewicz Theorem). *Let (X, A) be an $(n-1)$-connected pair $(n \geq 2)$ such that A and X are 0-connected. Then*

$$\tilde{\rho} : \pi_n^\dagger(X, A) \approx H_n(X, A).$$

We have seen that the homomorphism $\tilde{\rho}$ factors as the composite

$$\pi_n^\dagger(X, A) \xrightarrow{\tilde{\rho}'} H_n^{(n-1)}(X, A) \xrightarrow{i_*} H_n(X, A).$$

Moreover, it follows from Theorem (5.1) that, if (X, A) is $(n-1)$-connected, then the injection i_* is an isomorphism. Therefore the Relative Hurewicz Theorem is equivalent to the proposition

H(n) *The homomorphism $\tilde{\rho}' : \pi_n^\dagger(X, A) \to H_n^{(n-1)}(X, A)$ is an isomorphism.*

We have also seen that the relative theorem implies the absolute one if $n \geq 2$. If $n = 1$, the homomorphism $\bar{\rho}$ factors as the composite

$$\pi_1^*(X) \xrightarrow{\bar{\rho}'} H_1^{(0)}(X, *) \xrightarrow{i_*} H_1(X, *).$$

By Corollary (5.2), the injection i_* is an isomorphism. By Lemma (5.12), $\bar{\rho}'$ is an isomorphism. Hence $\bar{\rho}$ is an isomorphism and the absolute theorem is true if $n = 1$.

We have also remarked that the fact that $\tilde{\rho}'$ is an isomorphism is related to the Homotopy Addition Theorem. Let $A(n)$ be the conclusion of Theorem (6.2) for the given value of n. We shall prove the Relative Hurewicz Theorem and the Homotopy Addition Theorem by simultaneous induction, i.e.,

(7.3) Theorem *For each $n \geq 2$, the implications*

$$H(n-1) \Rightarrow A(n),$$
$$A(n) \Rightarrow H(n)$$

hold.

(N.B.: We agree that $H(1)$ is true).

To prove the first implication, let us recall that K is the simplicial complex consisting of Δ^{n+1}, together with all of its faces, and that β_n is the right-hand side of (6.3).

We first prove

(7.4) Lemma *The space K_{n-1} is $(n-2)$-connected.*

By Corollary (3.10) of Chapter II, the pair (K, K_{n-1}) is $(n-1)$-connected and therefore $\pi_j(K, K_{n-1}) = 0$ for all $j \le n-1$. Since K is contractible, $\pi_i(K) = 0$ for all i. By exactness of the homotopy sequence

$$\pi_{i+1}(K, K_{n-1}) \to \pi_i(K_{n-1}) \to \pi_i(K)$$

of the pair (K, K_{n-1}), $\pi_i(K_{n-1}) = 0$ for $i+1 \le n-1$, i.e., $i \le n-2$, and therefore K_{n-1} is $(n-2)$-connected. □

(7.5) Lemma *The injection $j : \pi_{n-1}(K_{n-1}) \to \pi_{n-1}(K_{n-1}, K_{n-2})$ is a monomorphism.*

By exactness, it suffices to prove that the injection $\pi_{n-1}(K_{n-2}) \to \pi_{n-1}(K_{n-1})$ is zero. If $n = 2$, $\pi_{n-1}(K_{n-2}) = 0$. Suppose $n > 2$; then there is a commutative diagram

$$\begin{array}{ccc}
\pi_{n-1}(K_{n-2}) & \xrightarrow{i_1} & \pi_{n-1}(K_{n-1}) \\
\downarrow \rho & & \downarrow \rho \\
H_{n-1}(K_{n-2}) & \xrightarrow[i_2]{} & H_{n-1}(K_{n-1}).
\end{array}$$

Since $H_{n-1}(K_{n-2}) = 0$, we have $0 = i_2 \circ \rho = \rho \circ i_1$. But ρ is an isomorphism, by $H(n-1)$, and therefore $i_1 = 0$.

(7.6) Lemma *The boundary homomorphism*

$$\partial_1 : \pi_n(K_{n+1}, K_{n-1}) \to \pi_{n-1}(K_{n-1})$$

is an isomorphism.

This follows by exactness from the contractibility of $K = K_{n+1}$.

We now prove $A(n)$. Consider the composite

$$\pi_{n+1}(K_{n+1}, K_n) \xrightarrow{\partial} \pi_n(K_n, K_{n-1}) \xrightarrow{i} \pi_n(K_{n+1}, K_{n-1})$$
$$\xrightarrow[\partial_1]{} \pi_{n-1}(K_{n-1}) \xrightarrow[j]{} \pi_{n-1}(K_{n-1}, K_{n-2})$$

By (6.5), $j\partial_1 i\beta_n = 0$. But ∂_1 and j are monomorphisms and therefore $i\beta_n = 0$.

By exactness, $\beta_n = \partial\alpha'$ for some $\alpha' \in \pi_{n+1}(K_{n+1}, K_n)$. Consider the commutative diagram

$$\begin{array}{ccc} \pi_{n+1}(K_{n+1}, K_n) & \xrightarrow{\partial} & \pi_n(K_n, K_{n-1}) \\ \downarrow \rho & & \downarrow \rho \\ H_{n+1}(K_{n+1}, K_n) & \xrightarrow[\partial]{} & H_n(K_n, K_{n-1}) \end{array}$$

and observe that $\rho(\beta_n) = \partial\rho(\alpha)$ (this fact is just a statement of the standard formula for the boundary operator in a simplicial complex). And we saw in §6 that the composite $\rho \circ \partial = \partial \circ \rho$ is a monomorphism. But $\partial\rho(\alpha') = \rho\partial\alpha' = \rho(\beta_n) = \partial\rho(\alpha)$, so that $\alpha = \alpha'$, $\beta_n = \partial\alpha$. This proves the first implication.

To prove the second, it suffices, by Lemma (5.10), to show that the homomorphism $\sigma : H_n^{(n-1)}(X, A) \to \pi_n^\dagger(X, A)$ annihilates ∂C_{n+1}. It suffices, then, to show that, if $u : \Delta^{n+1} \to X$ belongs to C_{n+1}, then $\sigma(\partial u) = 0$. Let α, α_0', α_i be the elements specified in the statement of Theorem (6.1). Then $j_* \alpha = 0$ because the map $f = u | \dot{\Delta}^{n+1}$ has the extension $u : \Delta^{n+1} \to X$. Let $\bar{\alpha}_i$, $\bar{\alpha}_0'$ be the images of the elements α_i, α_0' in $\pi_n^\dagger(X, A)$. Then

$$\sigma(\partial u) = \sum_{i=0}^{n+1} (-1)^i \bar{\alpha}_i .$$

By the Homotopy Addition Theorem,

$$0 = \bar{\alpha}_0' + \sum_{i=1}^{n+1} (-1)^i \bar{\alpha}_i .$$

But $\bar{\alpha}_0' = \bar{\alpha}_0$, so that $\sigma(\partial u) = 0$. This completes the proof of the second implication and therewith that of Theorem (7.3).

(7.7) Corollary *Let X be $(n-1)$-connected $(n \geq 2)$. Then $H_i(X) = 0$ for all $i < n$, and $\rho : \pi_n(X) \approx H_n(X)$.*

(7.8) Corollary *Let X be 1-connected, and suppose that $H_i(X) = 0$ for all $i < n$, $(n \geq 2)$. Then X is $(n-1)$-connected, and $\rho : \pi_n(X) \approx H_n(X)$.*

This partial converse to the (absolute) Hurewicz theorem is proved by induction on n. The hypothesis that X be 1-connected cannot be dropped, as the numerous known examples of acyclic spaces with non-trivial fundamental groups evince.

(7.9) Corollary *Let (X, A) be an $(n-1)$-connected pair $(n \geq 2)$, and suppose that A and X are both 0-connected. Then $H_i(X, A) = 0$ for all $i < n$, and $\tilde{\rho} : \pi_n^\dagger(X, A) \approx H_n(X, A)$.* □

(7.10) Corollary *Let (X, A) be an $(n-1)$-connected pair $(n \geq 2)$, and suppose that A is 1-connected. Then $H_i(X, A) = 0$ for all $i < n$, and $\rho : \pi_n(X, A) \approx H_n(X, A)$.* □

(7.11) Corollary *Let (X, A) be a 1-connected pair, and let A be 1-connected. Suppose that $H_i(X, A) = 0$ for all $i < n$ $(n \geq 2)$. Then (X, A) is $(n-1)$-connected, and $\rho : \pi_n(X, A) \approx H_n(X, A)$.* □

The relative homology and homotopy groups were designed to study the homomorphisms of (absolute) homology and homotopy groups induced by an inclusion map. If $f: X \to Y$ is an arbitrary continuous map, we can use the device of the mapping cylinder $\mathbf{I}_f$ to reduce the study of f_* to that of a suitable injection.

We shall say that a map $f: X \to Y$ is *n-connected* if and only if the pair $(\mathbf{I}_f, X)$ is n-connected.

(7.12) Lemma *Let X, Y be 0-connected spaces with base points x_0, y_0. Then a map $f: (X, x_0) \to (Y, y_0)$ is n-connected if and only if $f_* : \pi_q(X, x_0) \to \pi_q(Y, y_0)$ is an isomorphism for all $q < n$ and an epimorphism for $q = n$.*

This is an easy consequence of the exactness of the homotopy sequence of the pair $(\mathbf{I}_f, X)$ and Lemmas (3.1), (3.2) of Chapter II. □

The next theorem is due to J. H. C. Whitehead.

(7.13) Theorem (Whitehead Theorem). *Let $f: X \to Y$ be an n-connected map between 0-connected spaces. Then $f_* : H_q(X) \to H_q(Y)$ is an isomorphism for all $q < n$ and an epimorphism for $q = n$. Conversely, suppose that X and Y are 1-connected and that $f_* : H_n(X) \to H_n(Y)$ is an isomorphism for all $q < n$ and an epimorphism for $q = n$. Then f is n-connected.*

The above conditions on the homology groups are equivalent to the condition: $H_q(\mathbf{I}_f, X) = 0$ for all $q \leq n$. The Whitehead theorem now follows easily from Lemma (7.12) and Corollaries (7.10), (7.11). □

(7.14) Corollary *If f is n-connected, then, for any coefficient group G, $f_* : H_q(X, G) \to H_q(Y, G)$ is an isomorphism for all $q < n$ and an epimorphism for $q = n$.*

For we have seen that $H_q(\mathbf{I}_f, X) = 0$ for all $q \leq n$. By the Universal Coefficient Theorem, $H_q(\mathbf{I}_f, X; G) = 0$ for $q \leq n$ and any abelian group G, and the conclusion follows. □

A map which is n-connected for all n is called a *weak homotopy equivalence*. A map $f: X \to Y$ such that $f_* : H_q(X) \approx H_q(Y)$ for all q is called a *homology equivalence*. The following theorem is also sometimes called the Whitehead theorem.

(7.15) Theorem *If $f: X \to Y$ is a weak homotopy equivalence, and X and Y are 0-connected, then f is a homology equivalence. Conversely, if f is a homology equivalence and both X and Y are 1-connected, then f is a weak homotopy equivalence.* □

(7.16) Theorem *A map $f: X \to Y$ is n-connected if and only if*

(1) *for every* CW-*complex K with* $\dim K < n$, $\underline{f}: [K, X] \to [K, Y]$ *is a one-to-one correspondence;*
(2) *for every* CW-*complex K, with* $\dim K = n$, $\underline{f}$ *maps* $[K, X]$ *upon* $[K, Y]$.

Let $\mathbf{I}_f$ be the mapping cylinder of f, $i: X \hookrightarrow \mathbf{I}_f$, $Y \hookrightarrow \mathbf{I}_f$, and let $p: \mathbf{I}_f \to Y$ be the projection. Let f be n-connected, so that $(\mathbf{I}_f, X)$ is n-connected, and let K be a CW-complex of dimension $\leq n$, $h: K \to Y$. Then $j \circ h: K \to (\mathbf{I}_f, X)$ is compressible, i.e., $j \circ h \simeq i \circ g$ for some map $g: K \to X$. Then $h = p \circ j \circ h \simeq p \circ i \circ g = f \circ g$, and therefore $\underline{f}[K, X] = [K, Y]$. Let K be a CW-complex of dimension $< n$, and let $g_0, g_1: K \to X$ be maps such that $f \circ g_0 \simeq f \circ g_1$. The map $p: \mathbf{I}_f \to Y$ is a homotopy equivalence, and $p \circ j = f$; it follows that $i \circ g_0 \simeq i \circ g_1$. Let $f: \mathbf{I} \times K \to \mathbf{I}_f$ be a homotopy of $i \circ g_0$ to $i \circ g_1$. Since $(\mathbf{I}_f, X)$ is n-connected and $\dim K < n$, $\dim(\mathbf{I} \times K) \leq n$ and therefore f is compressible to a map $g: \mathbf{I} \times K \to X$. But g is a homotopy of g_0 to g_1. Hence $\underline{f}$ is one-to-one.

Conversely, suppose that conditions (1) and (2) are satisfied. Since $p: \mathbf{I}_f \to Y$ is a homotopy equivalence, $p \circ i = f$, it follows that the conditions (1′), (2′) obtained from (1) and (2) by replacing Y by $\mathbf{I}_f$ and f by the inclusion $i: X \hookrightarrow \mathbf{I}_f$ are satisfied. By Lemma (3.1) of Chapter II, it suffices to prove that every map $g: (\Delta^q, \dot{\Delta}^q) \to (\mathbf{I}_f, X)$ is compressible. The map $g_0 = g \mid \dot{\Delta}^q: \dot{\Delta}^q \to X$ is nullhomotopic in $\mathbf{I}_f$; by (1′), g_0 is nullhomotopic in X and therefore there is a map $g': \Delta^q \to X$ extending g_0. The map $h: (\mathbf{I} \times \Delta^q)^\bullet \to \mathbf{I}_f$ defined by

$$h(0, y) = g(y), h(1, y) = g'(y) \qquad (y \in \Delta^q)$$

$$h(t, y) = g_0(y) \qquad (t \in \mathbf{I}, y \in \dot{\Delta}^q)$$

is homotopic, because of (2′), to a map $h': (\mathbf{I} \times \Delta^q)^\bullet \to X$. By Lemma (3.2) of Chapter II, applied to the pair $((\mathbf{I} \times \Delta^q)^\bullet, 1 \times \Delta^q \cup \mathbf{I} \times \dot{\Delta}^q)$, the map $h: ((\mathbf{I} \times \Delta^q)^\bullet, 1 \times \Delta^q \cup \mathbf{I} \times \dot{\Delta}^q) \to (\mathbf{I}_f, X)$ is compressible. Therefore the composite of h with the map $i_0: (\Delta^q, \dot{\Delta}^q) \to (\mathbf{I} \times \Delta^q, 1 \times \Delta^q \cup \mathbf{I} \times \dot{\Delta}^q)$, sending y into $(0, y)$, is also compressible. But the latter map is our original map g. □

(7.17) Theorem *A necessary and sufficient condition that a map $f: X \to Y$ be a weak homotopy equivalence is that $\underline{f}: [K, X] \to [K, Y]$ be a one-to-one correspondence for every* CW-*complex K.*

The sufficiency is an immediate consequence of Theorem (7.16). So is the necessity, when K is finite-dimensional. To prove the necessity in general, we use Theorem (3.11) of Chapter II. Suppose that f is a weak homotopy equivalence. Then the pair $(\mathbf{I}_f, X)$ is ∞-connected. Let $h: K \to Y$; then $j \circ h: K \to (\mathbf{I}_f, X)$ is compressible, i.e., $j \circ h \simeq i \circ g$ for a map $g: K \to X$. Thus $f \circ g = p \circ i \circ g \simeq p \circ j \circ h = h$. Hence f is an epimorphism.

Let $g_0, g_1 : K \to X$ be maps such that $f \circ g_0 \simeq f \circ g_1$. Then $p \circ i \circ g_0 \simeq p \circ i \circ g_1$; as p is a homotopy equivalence, $i \circ g_0 \simeq i \circ g_1$. Let G be a homotopy of $i \circ g_0$ to $i \circ g_1$. Then G is compressible, i.e., G is homotopic (rel. $\dot{\mathbf{I}} \times K$) to a map $G' : \mathbf{I} \times K \to X$. But G' is a homotopy of g_0 to g_1. Therefore f is a monomorphism. □

We conclude this section with a number of examples relevant to the preceding discussion.

Example 1. There are 0-connected spaces X, Y such that $\pi_n(X) \approx \pi_n(Y)$ for all n, but X and Y do not have isomorphic homology groups. (Therefore there is no map $f: X \to Y$ inducing isomorphisms of the homotopy groups).

For this example, we take $X = \mathbf{P}^m \times \mathbf{S}^n$, $Y = \mathbf{S}^m \times \mathbf{P}^n$ with $m > n > 1$. Then $\pi_1(X) \approx \pi_1(Y) \approx \mathbf{Z}_2$, and the universal covering spaces of X and Y are both $\mathbf{S}^m \times \mathbf{S}^n$, so that the higher homotopy groups of X and Y are isomorphic, by (8.10) below. On the other hand, the Künneth theorem assures us that $H_{m+n}(X) \approx H_m(\mathbf{P}^m) \otimes H_n(\mathbf{S}^n) \approx H_m(\mathbf{P}^m)$, $H_{m+n}(Y) \approx H_n(\mathbf{P}^n)$. Therefore, if m is even and n odd, say, we have $H_{m+n}(X) = 0 \neq H_{m+n}(Y)$.

Example 2. There are 1-connected spaces X, Y such that $H_n(X) \approx H_n(Y)$ for all n, but X and Y do not have isomorphic homotopy groups. (Again, there can be no map $f: X \to Y$ inducing isomorphisms of the homology groups).

For this example, take $X = \mathbf{S}^2 \vee \mathbf{S}^4$, $Y = \mathbf{P}^2(\mathbf{C})$. Then X and Y do indeed have isomorphic homology groups ($H_0 \approx H_2 \approx H_4 \approx \mathbf{Z}$, $H_q = 0$ otherwise). On the other hand, $\mathbf{S}^4$ is a retract of X so that $\pi_4(\mathbf{S}^4) = \mathbf{Z}$ is a retract of $\pi_4(X)$, while $\pi_4(Y) \approx \pi_4(\mathbf{S}^5) = 0$, by (8.13) below.

Example 3. The reader may have wondered whether a map $f: X \to Y$ is a weak homotopy equivalence if $f_* : \pi_1(X) \to \pi_1(Y)$ as well as $f_* : H_q(X) \to H_q(Y)$ are isomorphisms for all q. The following example shows that this is not the case.

Let $X = \mathbf{S}^1 \vee \mathbf{S}^2$, and let $\xi \in \pi_1(X)$, $\alpha \in \pi_2(X)$ be the images of generators of $\pi_1(\mathbf{S}^1)$, $\pi_2(\mathbf{S}^2)$ under the injections. Let $\beta = 2\alpha - \tau_\xi(\alpha) \in \pi_2(X)$, and let $h : \mathbf{S}^2 \to X$ be a map representing β. Let $Y = X \cup_h \mathbf{E}^3$ be the mapping cone of h, so that Y is a CW-complex and X a subcomplex. The boundary of the 3-cell in Y is the image of β under the Hurewicz map $\rho : \pi_2(X) \to H_2(X)$. As $\rho(\tau_\xi(\alpha)) = \rho(\alpha)$, $\rho(\beta) = \rho(\alpha)$ is a generator of $H_2(X)$. An easy calculation now shows that Y is a homology 1-sphere, and the injection $i_* : H_q(\mathbf{S}^1) \to H_q(Y)$ is an isomorphism for all q.

Now, (Y, X) is 2-connected, so that the injection $\pi_1(X) \to \pi_1(Y)$ is an isomorphism. As the injection $\pi_1(\mathbf{S}^1) \to \pi_1(X)$ is also an isomorphism, $i_* : \pi_1(\mathbf{S}^1) \approx \pi_1(Y)$.

On the other hand, the universal covering space $\tilde{X}$ of X is the real line $\mathbf{R}$, with a copy of $\mathbf{S}^2$ attached at each integral point ($\tilde{X}$ is an infinite string of balloons!), and a generator of the group of covering translations translates each sphere into the next. Thus $\pi_2(X) \approx \pi_2(\tilde{X}) \approx H_2(\tilde{X})$ is the free abelian group with basis $\alpha_i = \tau_\xi^i(\alpha)$ for all $i \in Z$. The sequence

$$\pi_3(Y, X) \xrightarrow{\partial_*} \pi_2(X) \xrightarrow{i_*} \pi_2(Y) \to 0$$

is exact, and it follows from a result to be proved later (Chapter V, Corollary (1.4)) that the image of ∂_* is the operator subgroup of $\pi_2(X)$ generated by β. Thus $\pi_2(Y)$ is generated by $\{\bar{\alpha}_i\}$, subject to the relations $2\bar{\alpha}_i - \bar{\alpha}_{i+1} = 0$; this is just the group of dyadic rationals, and is not zero. As $\pi_2(\mathbf{S}^1) = 0$, the injection $i_* : \pi_2(\mathbf{S}^1) \to \pi_2(Y)$ is not an isomorphism.

Remark. The same example shows that, unlike the homology groups, the homotopy groups of a finite complex need not be finitely generated.

EXAMPLE 4. There is no relative form of the Whitehead theorem. That is, a map $f: (X, A) \to (Y, B)$ with $f_* : \pi_q(X, A) \approx \pi_q(Y, B)$ for all q, need not induce isomorphisms of the relative homology groups. In fact, let $X = Y \times A, B = \{*\}$. Then f is a fibration, with fibre A, so that, by Theorem (8.5) below, f_* maps $\pi_q(X, A)$ isomorphically upon $\pi_q(Y, B)$ for all q. But it follows from the Künneth theorem that the kernel of $f_* : H_n(X, A) \to H_n(Y, B)$ is isomorphic with $H_n(Y \times A, Y \vee A) = H_n((Y, *) \times (A, *))$. As this group is, in general, non-zero, f_* is not an isomorphism in homology.

EXAMPLE 5. Neither is there a relative form of the converse of the Whitehead theorem. That is, a map $f: (X, A) \to (Y, B)$ with $f_* : H_q(X, A) \approx H_q(Y, B)$ for all q, need not map the relative homotopy groups isomorphically. For an example, let $Y = \mathbf{S}^2$, and let X, B be hemispheres intersecting in an equator $\mathbf{S}^1$. That f_* is an isomorphism in homology follows from the Excision Theorem. Now

$$\pi_3(X, A) \approx \pi_2(A) = \pi_2(\mathbf{S}^1) = 0$$

since X is contractible, and

$$\mathbf{Z} = \pi_3(\mathbf{S}^2) = \pi_3(Y) \approx \pi_3(Y, B)$$

since B is contractible.

While the relative form of the Whitehead theorem fails in general, we can ensure its conclusion by strengthening the hypotheses. Specifically,

(7.18) Theorem *Let (X, A) and (Y, B) be 0-connected pairs such that A and B are also 0-connected. Let $f: (X, A) \to (Y, B)$ be a map such that f_* maps the homotopy groups of two of the three pairs $(X, *)$, (X, A), $(A, *)$ isomorphically. Then f_* maps the homotopy sequence, as well as the homology sequence with arbitrary coefficients, of (X, A) isomorphically upon those of (Y, B).*

That f_* maps the homotopy groups of the third pair isomorphically follows from the Five-Lemma. By Corollary (7.14), $f_* : H_q(A) \to H_q(B)$ and $f_* : H_q(X) \to H_q(Y)$ are isomorphisms for all q. Again by the Five-Lemma, $f_* : H_q(X, A) \approx H_q(Y, B)$ for all q.

8 Homotopy Relations in Fibre Spaces

For most of this section we shall be working in the category of free spaces and maps. We first examine the influence of the fundamental groupoid of the base space of a fibration on the fibres.

Let $p : X \to B$ be a fibration, and let $u : \mathbf{I} \to B$ be a path in B from b_0 to b_1. Let $F_t = p^{-1}(b_t)$, and let $i_t : F_t \subsetneq X$ $(t = 0, 1)$. We shall describe a map $h : F_1 \to F_0$ as *u-admissible* if and only if there is a homotopy $H : \mathbf{I} \times F_1 \to X$ of h to i_1 such that $pH(t, y) = u(t)$ for all $(t, y) \in \mathbf{I} \times F_1$. (Such a homotopy is said to *lie over* the path u.)

Clearly,

(8.1) *If $u : \mathbf{I} \to B$ is the constant map of $\mathbf{I}$ into b_0, then the identity map of F_0 is u-admissible.* □

(8.2) *Let $u : \mathbf{I} \to B$, $v : \mathbf{I} \to B$ be paths such that $u(0) = v(1) = b_1$, and let $w : \mathbf{I} \to B$ be their product. If $h : F_1 \to F_0$ is u-admissible and $k : F_2 \to F_1$ is v-admissible, then $h \circ k : F_2 \to F_0$ is w-admissible.*

For let $H : \mathbf{I} \times F_1 \to X$, $K : \mathbf{I} \times F_2 \to X$ be homotopies of h to i_1 and of k to i_2, which lie over u and v respectively. Then the map $L : \mathbf{I} \times F_2 \to X$ defined by

$$L(x, t) = \begin{cases} H(2t, k(x)) & (t \le \frac{1}{2}), \\ K(2t - 1, x) & (t \ge \frac{1}{2}), \end{cases}$$

lies over w and deforms $h \circ k$ into i_2. □

More difficult to prove is

(8.3) *Let u_0, $u_1 : \mathbf{I} \to B$ be maps which are homotopic (rel. $\dot{\mathbf{I}}$), and let h_0, $h_1 : F_1 \to F_0$ be maps such that h_t is u_t-admissible $(t = 0, 1)$. Then $h_0 \simeq h_1$.*

For let $F : \mathbf{I} \times \mathbf{I} \to B$ be a homotopy of u_0 to u_1 (rel. $\dot{\mathbf{I}}$), and let $H_s : \mathbf{I} \times F_1 \to X$ be a homotopy of h_s to i_1 lying over u_s $(s = 0, 1)$. Define maps $H : (\dot{\mathbf{I}} \times \mathbf{I} \cup \mathbf{I} \times 1) \times F_1 \to X$, $G : \mathbf{I} \times \mathbf{I} \times F_1 \to B$ by

$$H(s, t, y) = \begin{cases} H_0(t, y) & (s = 0), \\ y & (t = 1), \\ H_1(t, y) & (s = 1); \end{cases}$$

$$G(s, t, y) = F(s, t).$$

Then G is an extension of $p \circ H$. The pair

$$(\mathbf{I} \times \mathbf{I} \times F_1, (\dot{\mathbf{I}} \times \mathbf{I} \cup \mathbf{I} \times 1) \times F_1) = (\mathbf{I}, \dot{\mathbf{I}}) \times (I, 1) \times (F_1, \emptyset)$$

is a product of NDR-pairs, one of which is a DR-pair, and therefore is itself a DR-pair. By Lemma (7.15) of Chapter I, H has an extension $H' : \mathbf{I} \times \mathbf{I} \times F_1 \to X$ such that $p \circ H' = G$. Then $pH'(s, 0, y) = G(s, 0, y) = b_0$ and therefore there is a map $h : \mathbf{I} \times F_1 \to F_0$ such that $i_0 \circ h(s, 0, y) = H'(s, 0, y)$. The map h is the desired homotopy of h_0 to h_1. □

It follows from (8.1)–(8.3) that for each homotopy class ξ of paths in B, there is a unique homotopy class of maps of the fibre over the terminal point of ξ into the fibre over the initial point of ξ; and that under this correspondence multiplication of paths corresponds to composition of maps. In categorical terms, this means that the fibration determines a functor F from the fundamental groupoid of B to the category of spaces and homotopy classes of maps.

(8.4) Corollary *If $p : X \to B$ is a fibration and B is 0-connected, then any two fibres have the same homotopy type.* □

In particular, the fundamental group $\pi_1(B, b_0)$ operates on the homology groups of the fibre $F_0 = p^{-1}(b_0)$. If $\xi \in \pi_1(B, b_0)$, and if $h : F_0 \to F_0$ is u-admissible for a representative path u of ξ, let $\theta_\xi = h_* : H_q(F_0) \to H_q(F_0)$. It follows from (8.1)–(8.3) that the correspondence $\xi \to \theta_\xi$ defines an action of $\pi_1(B, b_0)$ on $H_q(F_0)$; i.e.,

(1) *if ξ is the identity, then θ_ξ is the identity;*
(2) *if $\xi, \eta \in \pi_1(B, b_0)$, then $\theta_\xi \circ \theta_\eta = \theta_{\xi\eta}$.*

As for the homotopy groups, there is the difficulty that we cannot control the behavior of the base point under u-admissible maps. Therefore we can only say that $\pi_1(B, b_0)$ operates on the groups $\pi_q^*(F_0)$. Of course, if F_0 is q-simple, then $\pi_1(B, b_0)$ does indeed operate on $\pi_q(F_0)$.

We now return to the category $\mathscr{K}_0$ of spaces with base point. By a fibration in $\mathscr{K}$ we shall mean a map in $\mathscr{K}_0$ which becomes a fibration in $\mathscr{K}$ if we ignore the base point. If $p : X \to B$ is such a fibration, then the base point x_0 of X lies in the *fibre* $F = p^{-1}(b_0)$, and we shall agree to consider F as a space with base point x_0.

(8.5) Theorem *Let $p: X \to B$ be a fibration with fibre F. Then $p_*: \pi_n(X, F) \to \pi_n(B, \{b_0\}) = \pi_n(B)$ is an isomorphism for all $n \geq 1$.*

We first show that p_* is an epimorphism. Let $f: (\mathbf{E}^n, \mathbf{S}^{n-1}) \to (B, *)$ be a map. Since $\mathbf{E}^n$ is contractible, the constant map of $\mathbf{E}^n$ into the base point $*$ of B is homotopic (rel. $*$) to f, considered as a map $(\mathbf{E}^n, *) \to (B, *)$. By the homotopy lifting extension property (Theorem (7.16) of Chapter I), the constant map of $\mathbf{E}^n$ into the base point of X is homotopic (rel. $*$) to a map $g: \mathbf{E}^n \to X$ such that $p \circ g = f$. Then $pg(\mathbf{S}^{n-1}) = *$ and therefore $g: (\mathbf{E}^n, \mathbf{S}^{n-1}, *) \to (X, F, *)$ represents an element $\alpha \in \pi_n(X, F)$ whose image under p_* is the element of $\pi_n(B)$ represented by f.

To show that p_* is a monomorphism, it suffices, if $n > 1$, to show that Kernel $p_* = \{0\}$. Let $f: (\mathbf{E}^n, \mathbf{S}^{n-1}, *) \to (X, F, *)$ and suppose that $p \circ f: (\mathbf{E}^n, \mathbf{S}^{n-1}) \to (B, *)$ is nullhomotopic. Again by Theorem (7.16) of Chapter I, f is compressible. By Corollary (2.5), f is nullhomotopic.

If $n = 1$, the proof that p_* is a monomorphism fails, since $\pi_1(X, F)$ is not, *a priori*, a group. In this case, let $f_0, f_1: (\mathbf{I}, \dot{\mathbf{I}}, 0) \to (X, F, *)$ and suppose $p \circ f_1 \simeq p \circ f_0$ (rel. $\dot{\mathbf{I}}$). Let $g: \mathbf{I} \to X$ be the product of the inverse of the path f_0 with f_1. Then $p \circ g$ is nullhomotopic (rel. $\dot{\mathbf{I}}$), and therefore g is homotopic (rel. $\dot{\mathbf{I}}$) to a path h in F. Hence f_1 is homotopic (rel. $\dot{\mathbf{I}}$) to the product $f_0 \cdot h$. Since $h(\mathbf{I}) \subset F$, $f_0 \cdot h: (\mathbf{I}, \dot{\mathbf{I}}, 0) \to (X, F, *)$ is homotopic (rel. $\dot{\mathbf{I}}$) to f_0. Hence f_0 and f_1 represent the same element of $\pi_1(X, F)$. □

The composite of $\partial_*: \pi_n(X, F) \to \pi_{n-1}(F)$ with the inverse of the isomorphism $p_*: \pi_n(X, F) \approx \pi_n(B)$ is a homomorphism $\Delta_*: \pi_n(B) \to \pi_{n-1}(F)$, and we have

(8.6) Corollary *Let $p: X \to B$ be a fibration with fibre F. Then there is an exact sequence*

$$\text{(8.7)} \quad \cdots \to \pi_n(F) \xrightarrow{i_*} \pi_n(X) \xrightarrow{p_*} \pi_n(B) \xrightarrow{\Delta_*} \pi_{n-1}(F) \xrightarrow{i_*} \pi_{n-1}(X) \to \cdots \to \pi_1(B) \xrightarrow{\Delta_*} \pi_0(F) \xrightarrow{i_*} \pi_0(X) \xrightarrow{p_*} \pi_0(B). \quad \square$$

(8.8) Corollary *Let B_0 be a subspace of B, $X_0 = p^{-1}(B_0)$. Then $p_0 = p \mid X_0 : X_0 \to B_0$ is a fibration, and $p_*: \pi_n(X, X_0) \to \pi_n(B, B_0)$ is an isomorphism for all $n \geq 1$.*

That p_0 is a fibration was pointed out in Corollary (7.22) of Chapter I. The homomorphisms induced by p map the homotopy sequence of the triple

(X, X_0, F) into that of $(B, B_0, *)$ and the homomorphisms of $\pi_q(X_0, F)$ into $\pi_q(B_0)$ and of $\pi_q(X, F)$ into $\pi_q(B)$ are isomorphisms. By the Five-Lemma, so are the homomorphisms $p_* : \pi_n(X, X_0) \to \pi_n(B, B_0)$.

(N.B.—The argument needs a mild modification if $n = 1$ or 2, like that needed in the proof of Theorem (8.5) itself—this is left to the reader). □

The sequence (8.7) is called the *homotopy sequence of* the fibration.

(8.9) Corollary *Let $f : X \to Y$ be an arbitrary map, $F = \mathbf{T}^f$ the mapping fibre of f, $q : F \to X$ the fibration over X. Then there is an exact sequence*

$$\cdots \to \pi_{q+1}(Y) \xrightarrow{\Delta_*} \pi_q(F) \xrightarrow{q_*} \pi_q(X) \xrightarrow{f_*}$$

$$\pi_q(Y) \longrightarrow \pi_{q-1}(F) \to \cdots$$

called the fibre homotopy sequence of f. □

(8.10) Corollary *If $p : \tilde{B} \to B$ is a covering map, then $p_* : \pi_q(\tilde{B}) \approx \pi_q(B)$ is an isomorphism for all $q \geq 2$.* □

Since $\mathbf{R}$ covers $\mathbf{S}^1$ and $\mathbf{R}$ is contractible, we have in particular

(8.11) Corollary *The homotopy groups $\pi_q(\mathbf{S}^1)$ vanish for all $q \geq 2$.* □

(8.12) Theorem *Let $p : X \to B$ be a fibration whose fibre F is contractible in X. Then the homotopy sequence (8.7) breaks up into a family of splittable short exact sequences*

$$0 \to \pi_n(X) \xrightarrow{p_*} \pi_n(B) \xrightarrow{\Delta_*} \pi_{n-1}(F) \to 0.$$

In particular, $\pi_n(B) \approx \pi_n(X) \oplus \pi_{n-1}(F)$ for all $n \geq 2$.

Since F is contractible in X, the injection $\pi_n(F) \to \pi_n(X)$ is zero for every n. Let $h : (\mathbf{T} \wedge F, F) \to (X, F)$ be a nullhomotopy of the inclusion map $i : F \subsetneq X$. Then $\mathbf{T} \wedge F$ is contractible, so that $\partial_* : \pi_n(\mathbf{T} \wedge F, F) \approx \pi_{n-1}(F)$ for all n. The composite

$$\pi_{n-1}(F) \xrightarrow{\partial_*^{-1}} \pi_n(\mathbf{T} \wedge F, F) \xrightarrow{h_*} \pi_n(X, F) \xrightarrow{p_*} \pi_n(B)$$

is the desired left inverse of Δ_*. □

We can apply Theorem (8.12) to the Hopf fibrations

$$\mathbf{S}^{2n+1} \to \mathbf{P}^n(\mathbf{C}) \qquad (\text{fibre } \mathbf{S}^1),$$

$$\mathbf{S}^{4n+3} \to \mathbf{P}^n(\mathbf{Q}) \qquad (\text{fibre } \mathbf{S}^3),$$

$$\mathbf{S}^{15} \to \mathbf{S}^8 \qquad (\text{fibre } \mathbf{S}^7);$$

in each case the fibre is a proper subspace of a sphere, and so contractible in the total space. Hence

(8.13) Corollary *The homotopy groups of complex and quaternionic projective space are given by*

$$\pi_q(\mathbf{P}^n(\mathbf{C})) \approx \pi_q(\mathbf{S}^{2n+1}) \oplus \pi_{q-1}(\mathbf{S}^1) \qquad (\approx \pi_q(\mathbf{S}^{2n+1}) \text{ for } q \geq 3),$$
$$\pi_q(\mathbf{P}^n(\mathbf{Q})) \approx \pi_q(\mathbf{S}^{4n+3}) \oplus \pi_{q-1}(\mathbf{S}^3).$$

In particular,

$$\pi_q(\mathbf{S}^2) \approx \pi_q(\mathbf{S}^3) \qquad (q \geq 3),$$
$$\pi_q(\mathbf{S}^4) \approx \pi_q(\mathbf{S}^7) \oplus \pi_{q-1}(\mathbf{S}^3),$$
$$\pi_q(\mathbf{S}^8) \approx \pi_q(\mathbf{S}^{15}) \oplus \pi_{q-1}(\mathbf{S}^7).$$
□

Note that $\pi_3(\mathbf{S}^2) \approx \pi_3(\mathbf{S}^3) \approx \mathbf{Z}$. Thus the homotopy groups, unlike the homology groups, need not vanish in dimensions greater than that of the space.

(Warning: the singular homology groups of a (bad) space X may also fail to vanish in dimensions greater than that of X. An example of this phenomenon has been given by Barratt and Milnor [1]. The construction uses the Hopf map $\mathbf{S}^3 \to \mathbf{S}^2$ in an essential way).

Let $p : X \to B$ be a fibration. A *cross-section* of p is a map $\lambda : B \to X$ such that $p \circ \lambda$ is the identity map of B. Because of the homotopy lifting property, p has a cross-section if and only if there is a map $\lambda' : B \to X$ such that $p \circ \lambda'$ is homotopic to the identity.

(8.14) Theorem *If the fibration $p : X \to B$ has a cross-section then the homotopy sequence breaks up into a family of short exact sequences*

$$0 \to \pi_q(F) \xrightarrow{i_*} \pi_q(X) \xrightarrow{p_*} \pi_q(B) \to 0.$$

Each of these sequences splits, and therefore

$$\pi_q(X) \approx \pi_q(F) \oplus \pi_q(B)$$

for all $q \geq 2$, while $\pi_1(X)$ is a semi-direct product of $\pi_1(F)$ by $\pi_1(B)$.

In fact, if $\lambda : B \to X$ is a cross-section, then $p_* \circ \lambda_*$ is the identity and therefore $p_* : \pi_q(X) \to \pi_q(B)$ is a split epimorphism. By exactness of the sequence (8.7), $\Delta_* = 0$ and i_* is a monomorphism, and the result follows.

□

Next, let $p : X \to B$ be a fibration with fibre F, and let B_0 be a subspace of B, $X_0 = p^{-1}(B_0)$. Let $\hat{X}$ be the mapping cylinder of p, $\hat{p} : \hat{X} \to B$ the projection, and let $\hat{X}_0 = \hat{p}^{-1}(B_0)$, so that $\hat{X}_0$ is the mapping cylinder of $p \mid X_0 : X_0 \to B_0$. We are going to relate the homotopy sequences of the pairs

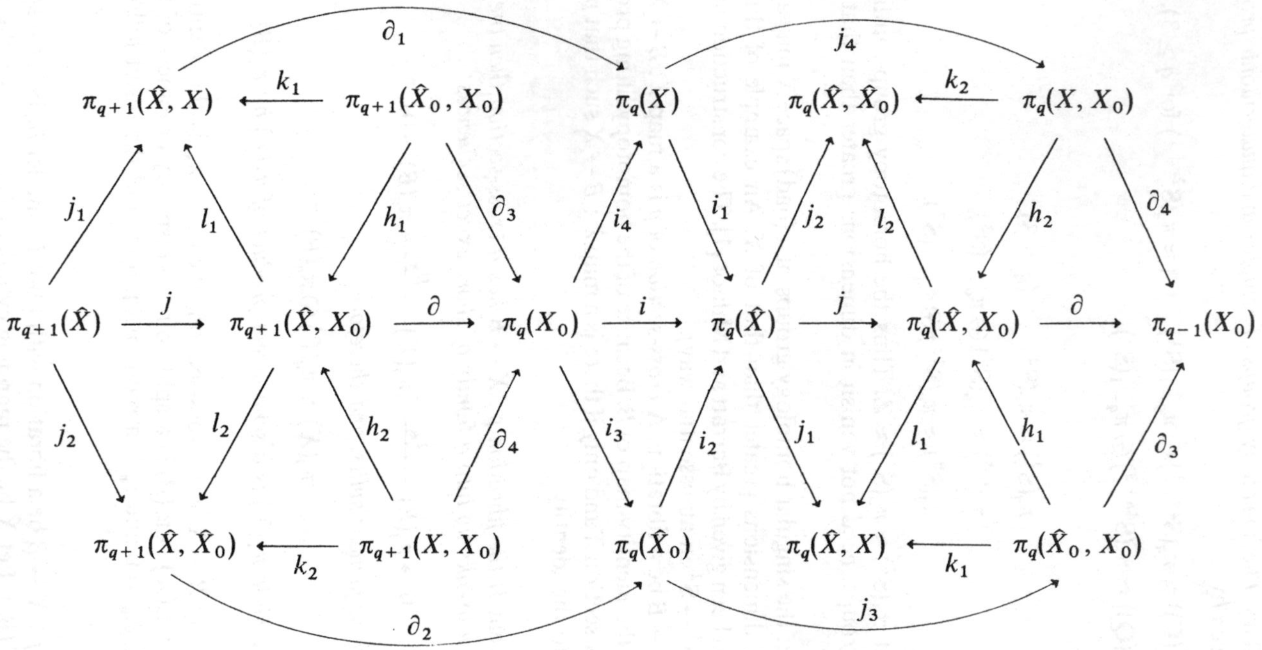
$\pi_{q+1}(\hat{X}, X)$
$\pi_{q+1}(\hat{X}_0, X_0)$
$\pi_q(X)$
$\pi_q(\hat{X}, \hat{X}_0)$
$\pi_q(X, X_0)$
$\pi_{q+1}(\hat{X})$
$\pi_{q+1}(\hat{X}, X_0)$
$\pi_q(X_0)$
$\pi_q(\hat{X})$
$\pi_q(\hat{X}, X_0)$
$\pi_{q-1}(X_0)$
$\pi_{q+1}(\hat{X}, \hat{X}_0)$
$\pi_{q+1}(X, X_0)$
$\pi_q(\hat{X}_0)$
$\pi_q(\hat{X}, X)$
$\pi_q(\hat{X}_0, X_0)$
∂_1 j_4 k_1 k_2 j_1 l_1 h_1 ∂_3 i_4 i_1 j_2 l_2 h_2 ∂_4
j ∂ i j ∂
j_2 l_2 h_2 ∂_4 i_3 i_2 j_1 l_1 h_1 ∂_3
k_2 k_1 ∂_2 j_3

Figure 4.2

$(\hat{X}, X)$, $(\hat{X}, \hat{X}_0)$, $(\hat{X}_0, X_0)$, (X, X_0) and $(\hat{X}, X_0)$. In fact, we have a diagram (Figure 2) containing all of the above homotopy sequences. The remaining maps in the diagram are injections.

(8.15) Lemma *If k_2 is a monomorphism in dimension q and an epimorphism in dimension $q + 1$, then so is k_1.*

The proof is an exercise in diagram-chasing, a discipline with which the reader is presumably familiar! □

(8.16) Corollary *If k_2 is an isomorphism for all q, so is k_1.* □

In fact,

(8.17) Lemma *The homomorphisms k_1 and k_2 are isomorphisms for all q.*

To see this, consider the commutative diagram

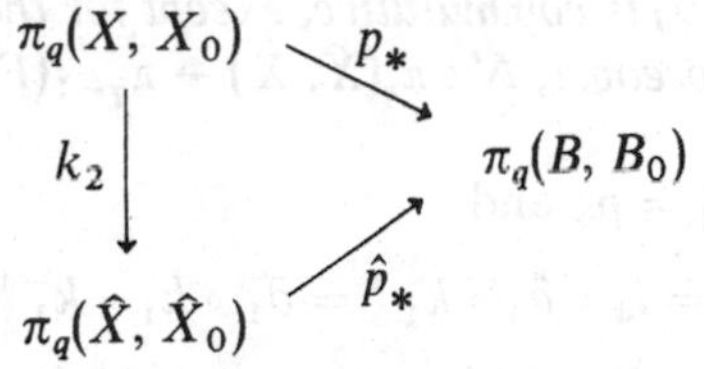

and observe that p_* is an isomorphism by Corollary (8.8) and $\hat{p}_*$ is an isomorphism because $\hat{p}$ is a homotopy equivalence. Therefore k_2 is an isomorphism. □

(8.18) Lemma *The subdiagram*

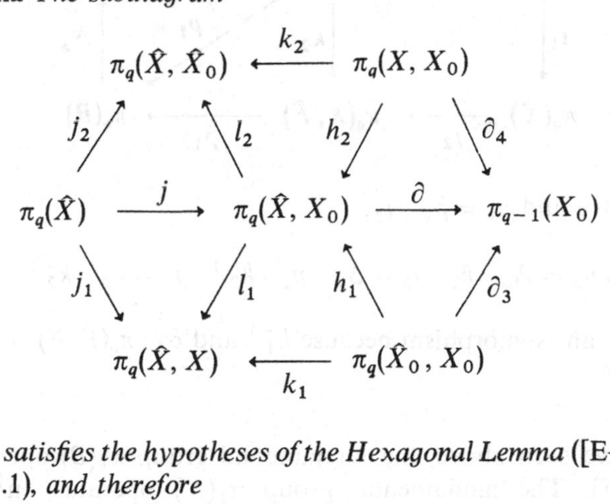

of Figure 4.2 satisfies the hypotheses of the Hexagonal Lemma ([E–S], *Chapter I, Lemma* 15.1), *and therefore*

$$\partial_3 \circ k_1^{-1} \circ j_1 = -\partial_4 \circ k_2^{-1} \circ j_2 : \pi_q(\hat{X}) \to \pi_{q-1}(X_0).$$

Commutativity of each triangle in the diagram is clear. Moreover, (∂, j), (l_2, h_1), and (l_1, h_2) are pairs of successive homomorphisms in the appropriate homotopy sequences, so that the kernel of the first member of each pair is the image of the second. By Lemma (8.17), k_1 and k_2 are isomorphisms. $\square$

Let us now apply the above discussion to the special case $B_0 = \{*\}$, $X_0 = F$; $\hat{X}_0 = \hat{F}$ is the cone over F. Let $\Delta' = \partial_3 \circ k_1^{-1} : \pi_q(\hat{X}, X) \to \pi_{q-1}(F)$. Then we have a diagram

(8.19)
$$\begin{array}{ccccccccc}
\cdots \to \pi_q(X) & \xrightarrow{i_1} & \pi_q(\hat{X}) & \xrightarrow{j_1} & \pi_q(\hat{X}, X) & \xrightarrow{\partial_1} & \pi_{q-1}(X) \to \cdots \\
\downarrow{\scriptstyle 1} & & \downarrow{\scriptstyle \hat{p}_*} & & \downarrow{\scriptstyle \Delta'} & & \downarrow{\scriptstyle 1} \\
\cdots \to \pi_q(X) & \xrightarrow[p_*]{} & \pi_q(B) & \xrightarrow[\Delta_*]{} & \pi_{q-1}(F) & \xrightarrow[i_4]{} & \pi_{q-1}(X) \to \cdots
\end{array}$$

(8.20) *The diagram* (8.19) *is commutative, except for the middle square, which is anti-commutative. Moreover,* $\Delta' : \pi_q(\hat{X}, X) \to \pi_{q-1}(F)$ *is an isomorphism.*

The relations $\hat{p}_* \circ i_1 = p_*$ and

$$i_4 \circ \Delta' = i_4 \circ \partial_3 \circ k_1^{-1} = \partial_1 \circ k_1 \circ k_1^{-1} = \partial_1$$

are clear. Now $\Delta' \circ j_1 = \partial_3 \circ k_1^{-1} \circ j_1 = -\partial_4 \circ k_2^{-1} \circ j_2$, and it remains to show that $\partial_4 \circ k_2^{-1} \circ j_2 = \Delta_* \circ \hat{p}_*$. But the diagram

$$\begin{array}{ccccc}
\pi_q(X) & \xrightarrow{j_4} & \pi_q(X, F) & \xrightarrow{\partial_4} & \pi_{q-1}(F) \\
\downarrow{\scriptstyle i_1} & & \downarrow{\scriptstyle k_2} & \searrow{\scriptstyle p_1} & \uparrow{\scriptstyle \Delta_*} \\
\pi_q(\hat{X}) & \xrightarrow[j_2]{} & \pi_q(\hat{X}, \hat{F}) & \xrightarrow[\hat{p}_1]{} & \pi_q(B)
\end{array}$$

is commutative and $\hat{p}_* = \hat{p}_1 \circ j_2$,

$$\Delta_* \circ \hat{p}_* = \Delta_* \circ \hat{p}_1 \circ j_2 = \Delta_* \circ p_1 \circ k_2^{-1} \circ j_2 = \partial_4 \circ k_2^{-1} \circ j_2.$$

Finally, Δ' is an isomorphism because k_1^{-1} and $\partial_3 : \pi_q(\hat{F}, F) \to \pi_{q-1}(F)$ are isomorphisms. $\square$

As we have seen above, the fundamental group $\pi_1(B)$ operates on the groups $\pi_q^*(F)$. The fundamental group $\pi_1(X)$ operates on the group

$\pi_{q+1}(\hat{X}, X)$, and, through the homomorphism $p_* : \pi_1(X) \to \pi_1(B)$, on the groups $\pi_q^*(F)$ as well.

(8.21) Theorem *If F is 0-connected, the composite*

$$\pi_{q+1}(\hat{X}, X) \xrightarrow{\Delta'} \pi_q(F) \longrightarrow \pi_q^*(F)$$

is an operator homomorphism.

It will simplify the proof slightly if we assume that F has a non-degenerate base point $*$. We leave to the reader the task of making the necessary modifications to handle the general case.

Let $\alpha \in \pi_{q+1}(\hat{X}, X)$, $\xi \in \pi_1(X)$, and let $\Delta^* : \pi_{q+1}(\hat{X}, X) \to \pi_q^*(F)$ be the above composite. Let $f: (\mathbf{E}^{q+1}, \mathbf{S}^q, *) \to (\hat{F}, F, *)$ be a map representing $k_1^{-1}(\alpha)$, and let $u: (\mathbf{I}, \dot{\mathbf{I}}) \to (X, *)$ be a representative of ξ, so that $v = p \circ u: (\mathbf{I}, \dot{\mathbf{I}}) \to (B, *)$ represents $p_*(\xi)$. Let $h: \mathbf{I} \times F \to X$ be a map such that

$$h(1, y) = y,$$

$$h(t, *) = u(t),$$

$$ph(t, y) = v(t)$$

for all $t \in \mathbf{I}$, $y \in F$; then the initial value of h is a v-admissible map $h_0 : F \to F$. The map h has an extension $\hat{h}: (\mathbf{I} \times \hat{F}, \mathbf{I} \times F) \to (\hat{X}, X)$, defined by

$$\hat{h}(t, \langle s, y\rangle) = \langle s, h(t, y)\rangle$$

(this is well defined when $x = 1$ because $\langle 1, h(t, y)\rangle = ph(t, y) = v(t)$ is independent of y). The continuity of $\hat{h}$ follows from (4.18) of Chapter I.

The map $\hat{h}_0 \circ f : (\mathbf{E}^{q+1}, \mathbf{S}^q, *) \to (\hat{F}, F, *)$ represents an element $\beta \in \pi_{q+1}(\hat{F}, F)$. The map $\hat{h} \circ (1 \times f): (\mathbf{I} \times \mathbf{E}^{q+1}, \mathbf{I} \times \mathbf{S}^q) \to (\hat{X}, X)$ is a free homotopy along u of $\hat{h}_0 \circ f$ to $\hat{h}_1 \circ f = f$. Hence $k_1(\beta) = \tau'_\xi(\alpha)$. The restriction $h_0 \circ (f \,|\, \mathbf{S}^q): \mathbf{S}^q \to F$ of $\hat{h}_0 \circ f$ thus represents $\partial_3 \beta = \partial_3 k_1^{-1} \tau'_\xi(\alpha) = \Delta' \tau'_\xi(\alpha) \in \pi_q(F)$ and also $\Delta^* \tau'_\xi(\alpha) \in \pi_q^*(F)$.

On the other hand, $f \,|\, \mathbf{S}^q$ represents $\Delta'(\alpha)$, and also $\Delta^*(\alpha)$. Hence $h_0 \circ (f \,|\, \mathbf{S}^q)$ represents $\theta_\xi(\Delta^*(\alpha))$. □

(8.22) Corollary *If F is 0-connected, the composite*

$$\pi_{q+1}(\hat{X}, X) \xrightarrow{\Delta'} \pi_q(F) \xrightarrow{\rho} H_q(F)$$

is an operator homomorphism. □

9 Fibrations in Which the Base or Fibre is a Sphere

These arise quite often in homotopy theory. For example, the classical groups admit fibre maps

$$\mathbf{O}_{n+1} \to \mathbf{S}^n,$$
$$\mathbf{U}_n \to \mathbf{S}^{2n-1},$$
$$\mathbf{Sp}_n \to \mathbf{S}^{4n-1}.$$

On the other hand, the bundle X of unit tangent vectors of a Riemannian n-manifold M admits a fibration $p : X \to M$ with $\mathbf{S}^{n-1}$ as fibre.

We shall give here the most elementary facts about such fibrations. They will be used in Chapter V and elsewhere.

Suppose first that $p : X \to \mathbf{S}^n$ is a fibration with fibre F. Then exactness of the homotopy sequence

$$\cdots \to \pi_{q+1}(\mathbf{S}^n) \xrightarrow{\Delta_*} \pi_q(F) \xrightarrow{i_*} \pi_q(X) \xrightarrow{p_*} \pi_q(\mathbf{S}^n) \to \cdots$$

implies

(9.1) *The injection $i_* : \pi_q(F) \to \pi_q(X)$ is an isomorphism for $q < n - 1$ and an epimorphism for $q = n - 1$.* □

Let $\iota_n \in \pi_n(\mathbf{S}^n)$ be the homotopy class of the identity map, so that $\pi_n(\mathbf{S}^n)$ is the infinite cyclic group generated by ι_n. The element $\omega_{n-1} = \Delta_*(\iota_n) \in \pi_{n-1}(F)$ is called the *characteristic element* of the fibration, and we have

(9.2) *The kernel of the injection $i_* : \pi_{n-1}(F) \to \pi_{n-1}(X)$ is the cyclic subgroup generated by ω_{n-1}.* □

Let us try to describe the characteristic element. Since $p_* : \pi_n(X, F) \to \pi_n(\mathbf{S}^n)$ is an isomorphism, there is a map $f: (\mathbf{E}^n, \mathbf{S}^{n-1}) \to (X, F)$ such that $p \circ f$ represents ι_n. The map f is called a *sectional element for* p, and we have

(9.3) *If $f: (\mathbf{E}^n, \mathbf{S}^{n-1}) \to (X, F)$ is a sectional element, then $f \,|\, \mathbf{S}^{n-1} : \mathbf{S}^{n-1} \to F$ represents the characteristic element ω_{n-1}.* □

The sectional element $f: (\mathbf{E}^n, \mathbf{S}^{n-1}) \to (X, F)$ will be called *regular* if and only if $p \circ f: (\mathbf{E}^n, \mathbf{S}^{n-1}) \to (\mathbf{S}^n, *)$ is the identification map.

It is sometimes important to know whether the fibration $p : X \to \mathbf{S}^n$ has a cross-section. If this is so, then the homomorphism $\Delta_* : \pi_{q+1}(\mathbf{S}^n) \to \pi_q(F)$ is trivial, and, in particular, $\omega_{n-1} = \Delta_*(\iota_n) = 0$. Conversely, suppose that $\omega_{n-1} = 0$. Then exactness of the homotopy sequence implies that

$p_* : \pi_n(X) \to \pi_n(\mathbf{S}^n)$ is an epimorphism. Let $\alpha \in \pi_n(X)$ be an element such that $p_*(\alpha) = \iota_n$. If $\lambda' : \mathbf{S}^n \to X$ is a representative of α, then $p \circ \lambda'$ is homotopic to the identity map of $\mathbf{S}^n$, and therefore λ' is homotopic to a map $\lambda : \mathbf{S}^n \to X$ such that $p \circ \lambda = 1$, i.e., λ is a cross-section. Thus

(9.4) *A necessary and sufficient condition that a fibration $p : X \to \mathbf{S}^n$ have a cross-section is that its characteristic element vanish.* □

We can apply the Hurewicz theorem to study the behavior of the homology groups. In fact, the Hurewicz theorem implies that $H_q(X, F) = 0$ for all $q < n$, and that $H_n(X, F)$ is an infinite cyclic group. (The fundamental group of F operates trivially on $\pi_n(X, F)$. Why?). Thus

(9.5) *The injection $i_* : H_q(F) \to H_q(X)$ is an isomorphism for $q < n - 1$ and an epimorphism for $q = n - 1$. The kernel of the injection $i_* : H_{n-1}(F) \to H_{n-1}(X)$ is generated by the image $\rho(\omega_{n-1})$ of the characteristic element under the Hurewicz map.* □

Now suppose that $p : X \to B$ is a fibration whose fibre F has the homotopy type of $\mathbf{S}^n$. We shall assume $n \geq 1$; this ensures that F is 0-connected and rules out the case of a two-sheeted covering. Such a fibration is said to be *n-spherical.*

Exactness of the homotopy sequence of p implies

(9.6) *The projection $p_* : \pi_q(X) \to \pi_q(B)$ is an isomorphism for $q < n$ and an epimorphism for $q = n$.* □

As in §8, let $\hat{X}$ be the mapping cylinder of p, $\hat{p} : \hat{X} \to B$ the projection, $\hat{F} = \hat{p}^{-1}(*)$ the cone over the fibre F. By (8.20) the homomorphism $\Delta' : \pi_{q+1}(\hat{X}, X) \to \pi_q(F)$ is an isomorphism, and therefore

(9.7) *The pair $(\hat{X}, X)$ is n-connected, and $\pi_{n+1}(\hat{X}, X)$ is an infinite cyclic group.* □

To study the homology groups, consider the commutative diagram

$$\begin{array}{ccccc}
\pi_{n+1}(\hat{X}, X) & \xleftarrow{k_1} & \pi_{n+1}(\hat{F}, F) & \xrightarrow{\partial_3} & \pi_n(F) \\
\downarrow{\scriptstyle \rho_1} & & \downarrow{\scriptstyle \rho_2} & & \downarrow{\scriptstyle \rho} \\
H_{n+1}(\hat{X}, X) & \xleftarrow[\bar{k}_1]{} & H_{n+1}(\hat{F}, F) & \xrightarrow[\bar{\partial}_3]{} & H_n(F)
\end{array}$$

and note that ∂_3 and $\bar{\partial}_3$ are isomorphisms (since $\hat{F}$ is contractible), ρ and ρ_2 are isomorphisms and ρ_1 an epimorphism (by the Hurewicz theorem), and

k_1 an isomorphism (by Lemma (8.17)). Hence $\bar{k}_1$ is an epimorphism, and ρ_1 and $\bar{k}_1$ have isomorphic kernels. It follows from Theorem (8.21) that

(9.8) *The injection $\bar{k}_1$ is an epimorphism, and the kernel of $\bar{k}_1$ is the subgroup generated by all elements of the form $\bar{\partial}_3^{-1}(\alpha - \theta_\xi(\alpha))$ with $\alpha \in H_n(F)$, $\xi \in \pi_1(B)$. Thus $\bar{k}_1$ is an isomorphism if and only if $\pi_1(B)$ operates trivially on $H_n(F)$.* □

10 Elementary Homotopy Theory of Lie Groups and Their Coset Spaces

In this section we shall apply the theory developed so far to obtain some elementary results on the homotopy groups of the spaces mentioned in the title. Our principal tool will be the homotopy sequence of a fibration.

I. The Orthogonal Groups

The map $p_n : \mathbf{O}_{n+1}^+ \to \mathbf{S}^n$, defined by $p_n(\tau) = \tau(\mathbf{e}_n)$ is a fibration with fibre $\mathbf{O}_n^+$. The sequence

$$\cdots \to \pi_{q+1}(\mathbf{S}^n) \xrightarrow{\Delta_*} \pi_q(\mathbf{O}_n^+) \longrightarrow \pi_q(\mathbf{O}_{n+1}^+) \longrightarrow \pi_q(\mathbf{S}^n) \to \cdots$$

being exact, we deduce

(10.1) *The injection $\pi_q(\mathbf{O}_n^+) \to \pi_q(\mathbf{O}_{n+1}^+)$ is an isomorphism for $q < n - 1$ and an epimorphism for $q = n - 1$ (i.e., the pair $(\mathbf{O}_{n+1}^+, \mathbf{O}_n^+)$ is $(n-1)$-connected).* □

Repeated application of this result yields

(10.2) *The pair $(\mathbf{O}_{n+k}^+, \mathbf{O}_n^+)$ is $(n-1)$-connected.* □

(10.3) *The pair $(\mathbf{O}^+, \mathbf{O}_n^+)$ is $(n-1)$-connected.* □

The kernel of the injection $\pi_{n-1}(\mathbf{O}_n^+) \to \pi_{n-1}(\mathbf{O}_{n+1}^+)$ is the cyclic group generated by the element $\omega_{n-1} = \Delta_*(\iota_n)$, where $\iota_n \in \pi_n(\mathbf{S}^n)$ is the homotopy class of the identity map. We next give an explicit description of the element ω_{n-1}.

For each $x \in \mathbf{S}^n$, let $f(x)$ be the reflection of $\mathbf{R}^{n+1}$ about the hyperplane orthogonal to x; thus

$$f(x)(y) = y - 2(x \cdot y)x.$$

Then $f: \mathbf{S}^n \to \mathbf{O}_{n+1}$ is continuous; considered as a map of $\mathbf{S}^n$ into the function space $\mathbf{F}(\mathbf{S}^n, \mathbf{S}^n)$, f has an adjoint $\bar{f}: \mathbf{S}^n \times \mathbf{S}^n \to \mathbf{S}^n$.

(10.4) Lemma *The map $\bar{f}: \mathbf{S}^n \times \mathbf{S}^n \to \mathbf{S}^n$ has type $(1-(-1)^n, -1)$; i.e., $\bar{f}\,|\,\mathbf{S}^n \times y$ has degree $1-(-1)^n$ and $\bar{f}\,|\,x \times \mathbf{S}^n$ has degree -1, for each x, $y \in \mathbf{S}^n$.*

The second part is clear, for $f(x)$ is a reflection in a subspace of codimension 1. Since $\mathbf{S}^n$ is pathwise connected, it suffices to prove the first statement for a particular y, say $y = -\mathbf{e}_n$. Let $g: \mathbf{S}^n \to \mathbf{S}^n$ be the map in question, so that

$$g(x) = f(x)(-\mathbf{e}_n) = 2(x \cdot \mathbf{e}_n)x - \mathbf{e}_n.$$

Note that $g(-x) = g(x)$, i.e., $g \circ a = g$, where a is the antipodal map. Since a has degree $(-1)^{n+1}$, we have $(-1)^{n+1} \cdot d = d$, where d is the degree of g. Therefore $d = 0$ if n is even.

Suppose that n is odd. The map g carries the equator $S^{n-1} = \{x \in S^n \mid x \cdot \mathbf{e}_n = 0\}$ into $-\mathbf{e}_n$; let g_+, g_- be maps such that

$$g_+ \,|\, \mathbf{E}^n_+ = g \,|\, \mathbf{E}^n_+, \quad g_+(\mathbf{E}^n_-) = -\mathbf{e}_n;$$

$$g_- \,|\, \mathbf{E}^n_- = g \,|\, \mathbf{E}^n_-, \quad g_-(\mathbf{E}^n_+) = -\mathbf{e}_n.$$

The element of $\pi_n(\mathbf{S}^n, -\mathbf{e}_n)$ represented by g is the sum of the elements represented by g_+ and g_-, and so $d = \deg g_+ + \deg g_-$. But $g_- = g_+ \circ a$, so that $d = (1 + (-1)^{n+1})\deg g_+ = 2 \deg g_+$. It is easy to see that for no $x \in \mathbf{S}^n$ is $g_+(x) = -x$, and it follows that g_+ is homotopic to the identity, so that $\deg g_+ = +1$. □

It will be useful to give the matrix representation for the map f. For this, note that, if $x \in \mathbf{S}^n$, then

$$\begin{aligned} f(x)(\mathbf{e}_i) &= \mathbf{e}_i - 2(x \cdot \mathbf{e}_i)x = \mathbf{e}_i - 2x_i x \\ &= \mathbf{e}_i - 2x_i \sum_{j=0}^{n} x_j \mathbf{e}_j \\ &= \sum_{j=0}^{n} (\delta_{ij} - 2x_i x_j)\mathbf{e}_j; \end{aligned}$$

thus the matrix of $f(x)$ is

$$\begin{pmatrix} 1-2x_0^2 & -2x_0x_1 & \cdots & -2x_0x_n \\ -2x_0x_1 & 1-2x_1^2 & \cdots & -2x_1x_n \\ & & \cdots & \\ -2x_0x_n & -2x_1x_n & \cdots & 1-2x_n^2 \end{pmatrix}.$$

Define $f_n: \mathbf{S}^n \to \mathbf{O}^+_{n+1}$ by $f_n(x) = f(x)f(\mathbf{e}_0)$. Since the matrix of $f(\mathbf{e}_0)$ is diag$\{-1, 1, \ldots, 1\}$, the matrix of $f_n(x)$ is obtained from that of $f(x)$ by changing the sign of the first column. The map $f_n|\mathbf{E}^n_+$ sends $(\mathbf{E}^n_+, \mathbf{S}^{n-1})$ into $(\mathbf{O}^+_{n+1}, \mathbf{O}^+_n)$, and

$$\begin{aligned} p_n f_n(x) &= f_n(x)(\mathbf{e}_n) = f(x)(f(\mathbf{e}_0)(\mathbf{e}_n)) \\ &= f(x)(\mathbf{e}_n) = -g_+(x) \text{ for } x \in \mathbf{E}^n_+. \end{aligned}$$

Extending $p_n \circ f_n$ over all of $\mathbf{S}^n$ by sending $\mathbf{E}^n_-$ into $\mathbf{e}_n$, we see that the resulting map is $a \circ g_+$ and therefore has degree $(-1)^{n+1}$. Hence $p_n \circ f_n$ represents $(-1)^{n+1}\iota_n$. Moreover, $f_n|\mathbf{S}^{n-1} = f_{n-1}$. Thus

(10.5) Theorem *The map* $f_{n-1} : \mathbf{S}^{n-1} \to \mathbf{O}^+_n$ *represents the element* $(-1)^{n+1}\omega_{n-1} \in \pi_{n-1}(\mathbf{O}^+_n)$. □

It follows from Lemma (10.4) that the map $p_{n-1} \circ f_{n-1} : \mathbf{S}^{n-1} \to \mathbf{S}^{n-1}$ has degree 2 if n is even. Hence

(10.6) Corollary *If n is even then ω_{n-1} generates an infinite cyclic subgroup of* $\pi_{n-1}(\mathbf{O}^+_n)$. □

Let us use this result to calculate some homotopy groups of the rotation groups. The group $\mathbf{O}^+_2$ is a circle, and the map $p_1 : \mathbf{O}^+_2 \to \mathbf{S}^1$ is a homeomorphism. We have the exact sequence

$$\begin{aligned} \to \pi_2(\mathbf{O}^+_2) \longrightarrow \pi_2(\mathbf{O}^+_3) \xrightarrow{p_*} \pi_2(\mathbf{S}^2) \xrightarrow{\Delta_*} \\ \pi_1(\mathbf{O}^+_2) \longrightarrow \pi_1(\mathbf{O}^+_3) \to 0 \end{aligned}$$

and $\pi_2(\mathbf{O}^+_2) = \pi_2(\mathbf{S}^1) = 0$, while Δ_* maps the infinite cyclic group $\pi_2(\mathbf{S}^2)$ isomorphically on the subgroup of $\pi_1(\mathbf{O}^+_2)$ generated by $2\omega_1$. Hence $p_* = 0$, and therefore $\pi_2(\mathbf{O}^+_3) = 0$, $\pi_1(\mathbf{O}^+_3) = \mathbf{Z}_2$. Applying (9.2) and (10.1), we have

$$\begin{aligned} &\pi_1(\mathbf{O}^+_2) = \mathbf{Z}, \\ &\pi_1(\mathbf{O}^+_n) \approx \pi_1(\mathbf{O}^+) = \mathbf{Z}_2 \qquad (n \geq 3), \\ &\pi_2(\mathbf{O}^+_n) = 0 \qquad (n \geq 2). \end{aligned}$$

(The fact that the second homotopy group of a compact Lie group always vanishes was proved by Elie Cartan [Ca]. Cartan's proof depends heavily on the structure theory of compact Lie groups. A different argument, using only homological properties, has been given by W. Browder [1]).

A little more difficult is the calculation of the third homotopy group. We have seen that $\mathbf{O}^+_2$ is a circle. We next show that $\mathbf{O}^+_3$ is homeomorphic with the real projective 3-space $\mathbf{P}^3$. For each unit quaternion $x \in \mathbf{S}^3$, let λ_x, ρ_x be the operations of left, right translation by x, respectively. Then λ_x and ρ_x are orthogonal transformations and the maps sending x into λ_x, ρ_x respectively are continuous maps $\lambda, \rho : \mathbf{S}^3 \to \mathbf{O}^+_4$. Because of the associative law, we have the relations

$$\lambda_x \circ \lambda_y = \lambda_{xy},$$

$$\rho_x \circ \rho_y = \rho_{yx},$$

$$\lambda_x \circ \rho_y = \rho_y \circ \lambda_x.$$

Evidently $\tau_x = \lambda_x \circ \rho_x^{-1}$ is the inner automorphism by x; it belongs to the subgroup of $\mathbf{O}_4^+$ leaving the identity e fixed, and this subgroup is isomorphic with $\mathbf{O}_3^+$. Thus we obtain a map $\tau : \mathbf{S}^3 \to \mathbf{O}_3^+$. Now $\tau_x = \tau_y$ if and only if $\tau_{x^{-1}y} = E$, and this is true if and only if $x^{-1}y$ belongs to the center of $\mathbf{Q}$. As $\|x\| = \|y\| = 1$, this is true if and only if $x = \pm y$. Thus τ induces a one-to-one map $\tau : \mathbf{P}^3 \to \mathbf{O}_3^+$. The image of τ is closed (because $\mathbf{P}^3$ is compact) and open (by Brouwer's theorem of invariance of domain) and therefore τ is a homeomorphism of $\mathbf{P}^3$ with $\mathbf{O}_3^+$.

The space $\mathbf{O}_4^+$ is homeomorphic with $\mathbf{O}_3^+ \times \mathbf{S}^3$; in fact, the projection $\alpha \to \alpha(e)$ of $\mathbf{O}_4^+$ into $\mathbf{S}^3$ is a fibration with fibre $\mathbf{O}_3^+$. This fibration has the cross-section $\rho : \mathbf{S}^3 \to \mathbf{O}_4^+$, and the map $(x, \alpha) \to \rho_x \circ \alpha$ is a homeomorphism of $\mathbf{S}^3 \times \mathbf{O}_3^+$ with $\mathbf{O}_4^+$.

Thus

$$\pi_3(\mathbf{O}_2^+) = \pi_3(\mathbf{S}^1) = 0,$$

$$\pi_3(\mathbf{O}_3^+) = \pi_3(\mathbf{P}^3) \approx \pi_3(\mathbf{S}^3) = \mathbf{Z},$$

$$\pi_3(\mathbf{O}_4^+) \approx \pi_3(\mathbf{S}^3 \times \mathbf{O}_3^+) = \pi_3(\mathbf{S}^3) \oplus \pi_3(\mathbf{O}_3^+) = \mathbf{Z} \oplus \mathbf{Z}$$

and the injection $\pi_3(\mathbf{O}_4^+) \to \pi_3(\mathbf{O}_5^+)$ is an epimorphism whose kernel is generated by ω_3.

We next verify by direct calculation that

$$\lambda_x \circ \rho_x = f_3(x).$$

In fact, the matrices of λ_x and ρ_x are

$$\begin{pmatrix} x_0 & -x_1 & -x_2 & -x_3 \\ x_1 & x_0 & -x_3 & x_2 \\ x_2 & x_3 & x_0 & -x_1 \\ x_3 & -x_2 & x_1 & x_0 \end{pmatrix}, \quad \begin{pmatrix} x_0 & -x_1 & -x_2 & -x_3 \\ x_1 & x_0 & x_3 & -x_2 \\ x_2 & -x_3 & x_0 & x_1 \\ x_3 & x_2 & -x_1 & x_0 \end{pmatrix}$$

and their product is the matrix

$$\begin{pmatrix} 2x_0^2 - 1 & -2x_0x_1 & -2x_0x_2 & -2x_0x_3 \\ 2x_0x_1 & 1 - 2x_1^2 & -2x_1x_2 & -2x_1x_3 \\ 2x_0x_2 & -2x_1x_2 & 1 - 2x_2^2 & -2x_2x_3 \\ 2x_0x_3 & -2x_1x_3 & -2x_2x_3 & 1 - 2x_3^2 \end{pmatrix}$$

which we have seen to be the matrix of $f_3(x)$.

This being so, we have

$$f_3(x) = \lambda_x \circ \rho_x^{-1} \circ \rho_x^2 = \tau_x \circ \rho_x^2.$$

It follows from Theorem (5.21) of Chapter III that if $\xi \in \pi_3(\mathbf{O}_3^+)$ is the element represented by $\tau : \mathbf{S}^3 \to \mathbf{O}_3^+$ and if $\eta \in \pi_3(\mathbf{O}_4^+)$ is the element represented by $\rho : \mathbf{S}^3 \to \mathbf{O}_4^+$, then

$$\omega_3 = i_*(\xi) + 2\eta.$$

Since $\pi_3(\mathbf{O}_3^+)$ is generated by ξ, it follows from the representation of $\mathbf{O}_4^+$ as a product given above that $\pi_3(\mathbf{O}_4^+)$ is the free abelian group generated by $i_*(\xi)$ and η. Hence

(10.7) Theorem *The first three homotopy groups of the rotation groups are given by the following table*

	$\mathbf{O}_2^+$	$\mathbf{O}_3^+$	$\mathbf{O}_4^+$	$\mathbf{O}_n^+$ $(n \geq 5)$
π_1	$\mathbf{Z}$	$\mathbf{Z}_2$	$\mathbf{Z}_2$	$\mathbf{Z}_2$
π_2	0	0	0	0
π_3	0	$\mathbf{Z}$	$\mathbf{Z}+\mathbf{Z}$	$\mathbf{Z}$

□

If n is odd, then $2\omega_{n-1} = \Delta_*(2\iota_n) = \Delta_* p_{n*}(\omega_n) = 0$. Thus ω_{n-1} generates a cyclic group of order at most two. If $\omega_{n-1} = 0$, then $p_{n*} : \pi_n(\mathbf{O}_{n+1}^+) \to \pi_n(\mathbf{S}^n)$ is an epimorphism, and it follows that the fibration p_n has a cross-section. In fact, the fibration is trivial; for if $\lambda : \mathbf{S}^n \to \mathbf{O}_{n+1}^+$ is a cross-section, the map $(x, \tau) \to \lambda(x) \cdot \tau$ is a homeomorphism of $\mathbf{S}^n \times \mathbf{O}_n^+$ with $\mathbf{O}_{n+1}^+$ and $p_n(x, \tau) = p_n(\lambda(x) \cdot \tau) = p_n(\lambda(x)) = x$.

The fibration p_n has a cross-section if $n = 1, 3, 7$ (and only in these cases, as we shall prove later). In fact, let us regard $\mathbf{S}^n$ as the unit sphere in the algebra **C**, **Q**, or **K**, according as $n = 1, 3, 7$, and let $\lambda(x)$ be the operation of left multiplication by $-x\mathbf{e}_n$. Then $p_n \lambda(x) = (-x\mathbf{e}_n)\mathbf{e}_n = x(-\mathbf{e}_n^2) = x$ (recall that the Cayley algebra is alternative so that the use of the associative law is justified even if $n = 7$)!

Summarizing, we have

(10.8) Theorem *If n is odd, then ω_{n-1} has order at most two, and $\omega_{n-1} = 0$ if and only if the fibration $p_n : \mathbf{O}_{n+1}^+ \to \mathbf{S}^n$ has a cross-section. In particular, $\omega_{n-1} = 0$ if $n = 1, 3,$ or 7.* □

Suppose that n is odd. Then $p_{n-1} \circ f_{n-1}$ has degree 0, and so is nullhomotopic. Since p_{n-1} is a fibration, f_{n-1} is compressible into $\mathbf{O}_{n-1}^+$. Suppose f_{n-1} is compressible into $\mathbf{O}_{n-k}^+$; by the homotopy extension property, $f_n | \mathbf{E}_+^n : (\mathbf{E}_+^n, \mathbf{S}^{n-1}) \to (\mathbf{O}_{n+1}^+, \mathbf{O}_n^+)$ is homotopic to a map $g : (\mathbf{E}_+^n, \mathbf{S}^{n-1}) \to (\mathbf{O}_{n+1}^+, \mathbf{O}_{n-k}^+)$.

Consider the commutative diagram

$$\begin{array}{ccc} \pi_n(\mathbf{O}_{n+1}^+, \mathbf{O}_{n-k}^+) & \xrightarrow{i_*} & \pi_n(\mathbf{O}_{n+1}^+, \mathbf{O}_n^+) \\ \downarrow p'_* & & \downarrow p_* \\ \pi_n(\mathbf{V}_{n+1,k+1}) & \xrightarrow[q_*]{} & \pi_n(\mathbf{S}^n) \end{array}$$

where i_* is the injection and the remaining homomorphisms are induced by the appropriate fibre maps; p'_* and p_* are isomorphisms, by Theorem (8.5). Thus q_* is an epimorphism if and only if i_* is. But q_* is an epimorphism if and only if the fibration $q : \mathbf{V}_{n+1,k+1} \to \mathbf{S}^n$ has a cross-section. Such a cross-section attaches continuously to each point $x \in \mathbf{S}^n$ a $(k+1)$-frame $(x_0, \ldots, x_{k-1}, x)$; the vectors $x_0, \ldots, x_{k-1}$ are orthogonal to each other and to x, so that their translates with origin x are tangent to $\mathbf{S}^n$ there. Thus q has a cross-section if and only if $\mathbf{S}^n$ admits a family of k mutually orthogonal fields of unit tangent vectors.

We have seen

(10.9) Theorem *The following conditions are equivalent*:

(1) f_{n-1} *is compressible into* $\mathbf{O}_{n-k}^+$;
(2) *the injection* $\pi_n(\mathbf{O}_{n+1}^+, \mathbf{O}_{n-k}^+) \to \pi_n(\mathbf{O}_{n+1}^+, \mathbf{O}_n^+)$ *is an epimorphism*;
(3) *the fibration* $q : \mathbf{V}_{n+1,k+1} \to \mathbf{S}^n$ *has a cross-section*;
(4) *the n-sphere* $\mathbf{S}^n$ *admits k mutually orthogonal fields of unit tangent vectors.* □

The problem of determining, as a function of n, the maximum k for which these conditions hold is known as the *vector field problem.*

Suppose that n is even. Then we have seen that $p_{n-1} \circ f_{n-1}$ has degree ± 2, and it follows that f_{n-1} is not compressible into $\mathbf{O}_{n-2}^+$. Thus $\mathbf{S}^n$ does not admit a field of unit tangent vectors. This also follows from the general result of Hopf: *a closed manifold M admits a continuous vector field without zeroes if and only if the Euler characteristic* $\chi(M) = 0$.

On the other hand, if n is odd, $\mathbf{S}^n$ admits a 1-field. For we may think of $\mathbf{S}^{2k-1}$ as the unit sphere in complex n-space $\mathbf{C}^n$. If $x \in \mathbf{S}^{2k-1}$, then the vectors x and ix are linearly independent (and even orthonormal) over the real field. Similarly, we can see that $\mathbf{S}^{4n-1}$ admits a 3-field and $\mathbf{S}^{8n-1}$ a 7-field.

Another result along these lines is

(10.10) *If* $\mathbf{S}^{n-1}$ *admits a k-field, then* $\mathbf{S}^{rn-1}$ *admits a k-field for every positive integer r.*

Suppose that v is a vector field on $\mathbf{S}^{n-1}$. Then v has an extension $\bar{v}$ over $\mathbf{E}^n$, defined by

$$\bar{v}(tx) = tv(x) \qquad (t \in \mathbf{I},\ x \in \mathbf{S}^{n-1}).$$

If $x \in \mathbf{S}^{rn-1}$, we can write $x = (x_1, \ldots, x_r)$ with $x_i \in \mathbf{E}^n$ and $\sum_{i=0}^{r} \|x_i\|^2 = 1$. If $v_1, \ldots, v_k$ is a k-field on $\mathbf{S}^{n-1}$, define w_i by

$$w_i(x_1, \ldots, x_n) = (\bar{v}_i(x_1), \ldots, \bar{v}_i(x_r));$$

then $(w_1, \ldots, w_k)$ is the required k-field on $\mathbf{S}^{rn-1}$. □

Let us examine the map f_n more carefully. The map $f: \mathbf{S}^n \to \mathbf{O}_{n+1}$ has the property that $f(-x) = f(x)$; accordingly f induces a map $g: \mathbf{P}^n \to \mathbf{O}_{n+1}$. Suppose that $f(x) = f(y)$; it follows from the definition of f that the linear subspaces spanned by x and y have the same orthogonal complement, the fixed set of the reflection $f(x) = f(y)$. Hence y is a scalar multiple of x; as both are unit vectors, $y = \pm x$. Therefore g is a homeomorphic imbedding of $\mathbf{P}^n$ in $\mathbf{O}_{n+1}$. It follows that f_n induces an imbedding g_n of $\mathbf{P}^n$ in $\mathbf{O}_{n+1}^+$. From the properties of f which we have obtained above, we can deduce

(10.11) Theorem *The map $f_n : \mathbf{S}^n \to \mathbf{O}_{n+1}^+$ identifies antipodal points of $\mathbf{S}^n$ and so induces a map $g_n : \mathbf{P}^n \to \mathbf{O}_{n+1}^+$. The maps g_n are related by*

$$g_n | \mathbf{P}^{n-1} = g_{n-1},$$

and therefore they define a map $g_\infty : \mathbf{P}^\infty \to \mathbf{O}^+$. The map g_n is an imbedding of $\mathbf{P}^n$ in $\mathbf{O}_{n+1}^+$. The map $p_n \circ g_n : (\mathbf{P}^n, \mathbf{P}^{n-1}) \to (\mathbf{S}^n, \mathbf{e}_n)$ is a relative homeomorphism. □

11. The Real Stiefel Manifolds

Consider the fibration $\mathbf{O}_n^+ \to \mathbf{V}_{n,k}$ with fibre $\mathbf{O}_{n-k}^+$, and its homotopy sequence

$$\cdots \to \pi_q(\mathbf{O}_{n-k}^+) \xrightarrow{i} \pi_q(\mathbf{O}_n^+) \longrightarrow \pi_q(\mathbf{V}_{n,k}) \longrightarrow$$
$$\pi_{q-1}(\mathbf{O}_{n-k}^+) \xrightarrow{i'} \pi_{q-1}(\mathbf{O}_n^+) \to \cdots$$

By (10.2), the injection i is an epimorphism and i' a monomorphism if $q \le n - k - 1$. Hence

(10.12) *The Stiefel manifold $\mathbf{V}_{n,k}$ is $(n-k-1)$-connected.* □

We next determine the first non-trivial group $\pi_{n-k}(\mathbf{V}_{n,k})$. Let $r = n - k$; then we have the fibrations $\mathbf{V}_{k+r,k} \to \mathbf{V}_{k+r,1} = \mathbf{S}^{k+r-1}$, the fibre being $\mathbf{V}_{k+r-1,k-1}$, and therefore the exact sequences

$$\pi_{r+1}(\mathbf{S}^{k+r-1}) \to \pi_r(\mathbf{V}_{k+r-1,k-1}) \to \pi_r(\mathbf{V}_{k+r,k}) \to \pi_r(\mathbf{S}^{k+r-1}).$$

If $k > 2$, the extreme groups are both zero, and therefore the intermediate groups are isomorphic. For $k = 2$, the sequence becomes

$$\pi_{r+1}(\mathbf{S}^{r+1}) \xrightarrow{\Delta_*} \pi_r(\mathbf{V}_{r+1,1}) \longrightarrow \pi_r(\mathbf{V}_{r+2,2}) \to 0.$$

To calculate Δ_*, note that the fibrations $\mathbf{O}^+_{r+2} \to \mathbf{S}^{r+1}$ and $\mathbf{V}_{r+2,2} \to \mathbf{S}^{r+1}$ are related by a commutative diagram

$$\begin{array}{ccccc}
\mathbf{O}^+_{r+1} & \longrightarrow & \mathbf{O}^+_{r+2} & & \\
\big\downarrow & & \big\downarrow & \searrow & \\
 & & & & \mathbf{S}^{r+1} \\
 & & & \nearrow & \\
\mathbf{V}_{r+1,1} & \longrightarrow & \mathbf{V}_{r+2,2} & &
\end{array}$$

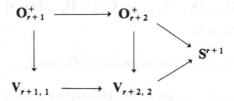

and so there is a commutative diagram

$$\begin{array}{ccccc}
 & \overset{\Delta_*}{\nearrow} & \pi_r(\mathbf{O}^+_{r+1}) & \longrightarrow & \pi_r(\mathbf{O}^+_{r+2}) \\
\pi_{r+1}(\mathbf{S}^{r+1}) & & \big\downarrow p_* & & \big\downarrow \\
 & \underset{\Delta_*}{\searrow} & \pi_r(\mathbf{V}_{r+1,1}) & \longrightarrow & \pi_r(\mathbf{V}_{r+2,2})
\end{array}$$

relating their homotopy sequences. But we have seen in Lemma (10.4) that $p_* \Delta_*(\iota_{r+1}) = 0$ or $2\iota_r$ according as r is even or odd. We have proved

(10.13) Theorem *If r is even, then $\pi_r(\mathbf{V}_{r+k,k})$ is infinite cyclic for all $k \geq 1$. If r is odd, then $\pi_r(\mathbf{V}_{r+1,1})$ is infinite cyclic and $\pi_r(\mathbf{V}_{r+k,k})$ is cyclic of order two for all $k \geq 2$.* □

A case of special interest is $\mathbf{V}_{r+2,2}$, which we may identify with the tangent bundle of $\mathbf{S}^{r+1}$; $\mathbf{V}_{r+2,2}$ is an orientable manifold of dimension $2r + 1$, and we have just seen that it is $(r-1)$-connected. By the Hurewicz theorem, $H_r(\mathbf{V}_{r+2,2}) = \mathbf{Z}$ or $\mathbf{Z}_2$ according as r is even or odd. By Poincaré duality, we have

(10.14) Corollary *If r is even,*

$$H_q(\mathbf{V}_{r+2,2}) = \begin{cases} \mathbf{Z} & \text{if } q = 0, r, r+1, \text{ or } 2r+1; \\ 0 & \text{otherwise.} \end{cases}$$

If r is odd,

$$H_q(\mathbf{V}_{r+2,2}) = \begin{cases} \mathbf{Z} & \text{if } q = 0 \text{ or } 2r+1; \\ \mathbf{Z}_2 & \text{if } q = r; \\ 0 & \text{otherwise.} \end{cases}$$

□

III. The Grassmann Manifolds

The manifold $\mathbf{G}_{k,l}$ is the quotient $\mathbf{O}_{k+l}/\mathbf{O}(k, l)$, where $\mathbf{O}(k, l)$ is the subgroup of $\mathbf{O}_{k+l}$ consisting of all transformations which carry $\mathbf{R}^l$ into itself. Evidently $\mathbf{G}_{k,l}$ is homeomorphic with $\mathbf{G}_{l,k}$; therefore we shall assume $l \leq k$.

The Grassmann and Stiefel manifolds are related by a fibration

$$\mathbf{O}_{k+l}/\mathbf{O}_k = \mathbf{V}_{k+l,l} \to \mathbf{G}_{k,l} = \mathbf{O}_{k+l}/\mathbf{O}(k, l)$$

with fibre $\mathbf{O}(k, l)/\mathbf{O}_k \approx \mathbf{O}_l$.

(10.15) Lemma *If $k \geq l$, the inclusion map of $\mathbf{O}_l$ into $\mathbf{V}_{k+l,l}$ is nullhomotopic.*

The space $\mathbf{V}_{k+l,l}$ can be represented as the space of all $(k + l) \times l$ matrices with orthonormal columns. Under this representation, $\mathbf{O}_l$ is represented by those matrices of the form

$$\begin{pmatrix} X \\ 0 \end{pmatrix},$$

where 0 is a $k \times l$ matrix of zeroes. Let $F(t, X)$ be the matrix

$$\begin{pmatrix} cX \\ sE \\ 0 \end{pmatrix}$$

where $c = \cos \frac{1}{2}\pi t$, $s = \sin \frac{1}{2}\pi t$, E is the $l \times l$ identity matrix, and 0 is a $(k - l) \times l$ matrix of zeroes. Then F is the desired nullhomotopy. □

It follows from Lemma (10.15) that the homotopy sequence of the fibration in question breaks up into a family of split short exact sequences. Hence

(10.16) Theorem *If $k \geq l$, then*

$$\pi_q(\mathbf{G}_{k,l}) \approx \pi_q(\mathbf{V}_{k+l,l}) \oplus \pi_{q-1}(\mathbf{O}_l)$$

for all q. □

IV. The complex case

This time we have the fibration

$$\mathbf{U}_n \to \mathbf{S}^{2n-1}$$

with fibre $\mathbf{U}_{n-1}$. We then have, in analogy with (10.1)–(10.3);

(10.17) *The injection $\pi_q(\mathbf{U}_{n-1}) \to \pi_q(\mathbf{U}_n)$ is an isomorphism for $q < 2n - 2$ and an epimorphism for $q = 2n - 2$,* i.e., *the pair $(\mathbf{U}_n, \mathbf{U}_{n-1})$ is $(2n - 2)$-connected.* □

(10.18) *The pair* $(\mathbf{U}_{n+k}, \mathbf{U}_{n-1})$ *is* $(2n-2)$*-connected for each* $k \geq 0$. □

(10.19) *The pair* $(\mathbf{U}, \mathbf{U}_{n-1})$ *is* $(2n-2)$*-connected.* □

We now define a map of $\mathbf{SP}^n(\mathbf{C})$ into $\mathbf{U}_{n+1}$ which plays a role analogous to that played by the mapping $g_n : \mathbf{P}^n \to \mathbf{O}_{n+1}^+$ in the case of the orthogonal group. It is convenient to think of points of $\mathbf{C}^{n+1}$ as column vectors $\binom{x}{\xi}$ where $x \in \mathbf{C}$ and ξ is an $(n \times 1)$ complex matrix, and the elements of $\mathbf{U}_{n+1}$ as matrices of the form

$$\begin{pmatrix} x & \eta^* \\ \xi & A \end{pmatrix}$$

where $x \in \mathbf{C}$, ξ and η are $n \times 1$ complex matrices, A is an $n \times n$ complex matrix, and the asterisk denotes conjugate transposition. Let $u \in \mathbf{S}^1$, $X = \binom{x}{\xi} \in \mathbf{S}^{2n+1}$, and let $\tilde{F}_{n+1}(u, X)$ be the matrix

$$\begin{pmatrix} u + (1-u)x\bar{x} & x(1-u)\xi^* \\ -\bar{x}(1-\bar{u})\xi & E - (1-\bar{u})\xi\xi^* \end{pmatrix}.$$

One verifies immediately that the matrix $\tilde{F}_{n+1}(u, X)$ is unitary and $\tilde{F}_{n+1} : \mathbf{S}^1 \times \mathbf{S}^{2n+1} \to \mathbf{U}_{n+1}$ is continuous. Moreover, $\tilde{F}_{n+1}(\mathbf{S}^1 \vee \mathbf{S}^{2n+1})$ is the identity matrix, and $\tilde{F}_{n+1}(u, vX) = \tilde{F}_{n+1}(u, X)$ for any $v \in \mathbf{S}^1$. Therefore $\tilde{F}_{n+1}$ induces a map $F_{n+1} : \mathbf{S}^{2n+2} \to \mathbf{U}_{n+1}$, as well as a map $G_{n+1} : \mathbf{SP}^n(\mathbf{C}) \to \mathbf{U}_{n+1}$. If $X \in \mathbf{S}^{2n-1}$, so that the last coordinate of ξ is zero, then the matrix $\tilde{F}_{n+1}(u, X)$ belongs to $\mathbf{U}_n$, the subgroup of $\mathbf{U}_{n+1}$ leaving the last basis vector $\mathbf{e}_{2n}$ fixed; in fact $\tilde{F}_{n+1}(u, X) = \tilde{F}_n(u, X)$. Therefore, if $p_{n+1} : \mathbf{U}_{n+1} \to \mathbf{S}^{2n+1}$ is the usual fibre map, then $p_{n+1} \circ G_{n+1}$ maps $\mathbf{SP}^{n-1}(\mathbf{C})$ into the point $\mathbf{e}_{2n}$.

(10.20) Lemma *The map* $p_{n+1} \circ G_{n+1} : (\mathbf{SP}^n(\mathbf{C}), \mathbf{SP}^{n-1}(\mathbf{C})) \to (\mathbf{S}^{2n+1}, \mathbf{e}_{2n})$ *is a relative homeomorphism.*

It is sufficient, because all the spaces involved are compact Hausdorff spaces, to prove that $p_{n+1} \circ G_{n+1}$ is one-to-one on the complement of $\mathbf{SP}^n(\mathbf{C})$. Suppose, then, that $p_{n+1}\tilde{F}_{n+1}(u, X) = p_{n+1}\tilde{F}_{n+1}(v, Y)$ and that $u \neq 1 \neq v$, $X = \binom{x}{\xi}$, $Y = \binom{y}{\eta}$, $\xi_n \neq 0 \neq \eta_n$. We may further assume that x and y are real and non-negative. Then

$$\begin{aligned} x(1-u)\bar{\xi}_n &= y(1-v)\bar{\eta}_n, \\ \mathbf{e}_{2n} - (1-\bar{u})\bar{\xi}_n\xi &= \mathbf{e}_{2n} - (1-\bar{v})\bar{\eta}_n\eta. \end{aligned}$$

Then $\eta = c\xi$, where

$$c = \frac{(1-\bar{u})\bar{\xi}_n}{(1-\bar{v})\bar{\eta}_n};$$

in particular, $\eta_n = c\xi_n$, and therefore

$$\eta_n \bar{\eta}_n = \frac{1 - \bar{u}}{1 - \bar{v}} \bar{\xi}_n \xi_n$$

so that the complex number $(1 - \bar{u})/(1 - \bar{v})$ is real. Thus $\bar{u}$ and $\bar{v}$ are two points of the unit circle lying on the same line through the point 1; as $\bar{u} \neq 1 \neq \bar{v}$, this implies $\bar{u} = \bar{v}$, and therefore $x\bar{\xi}_n = y\bar{\eta}_n$, $y = (\bar{\xi}_n/\bar{\eta}_n)x = cx$. Therefore $Y = cX$. We have proved that $u = v$, $Y = cX$, so that $u \wedge X$ and $v \wedge Y$ correspond to the same point of $\mathbf{S}^1 \wedge \mathbf{P}^n(\mathbf{C})$. □

Now consider the homotopy sequence

$$\cdots \to \pi_{2n+1}(\mathbf{S}^{2n+1}) \xrightarrow{\Delta_*} \pi_{2n}(\mathbf{U}_n) \longrightarrow \pi_{2n}(\mathbf{U}_{n+1}) \to \cdots$$

and let $\omega_{n+1} = \Delta_*(\iota_{2n+1})$.

(10.21) *The map* $F_n : \mathbf{S}^{2n} \to \mathbf{U}_n$ *represents the element* ω_n.

Recall that $(\mathbf{P}^n(\mathbf{C}), \mathbf{P}^{n-1}(\mathbf{C}))$ is a relative CW-complex with one $2n$-cell, whose attaching map is the Hopf fibre map $h : \mathbf{S}^{2n-1} \to \mathbf{P}^{n-1}(\mathbf{C})$, and therefore $(\mathbf{SP}^n(\mathbf{C}), \mathbf{SP}^{n-1}(\mathbf{C}))$ is a relative CW-complex with one $(2n+1)$-cell whose attaching map is the suspension of h. We have a commutative diagram

$$\begin{array}{ccc}
\pi_{2n+1}(\mathbf{E}^{2n+1}, \mathbf{S}^{2n}) & \xrightarrow{\partial_*} & \pi_{2n}(\mathbf{S}^{2n}) \\
\downarrow & & \downarrow {\scriptstyle Sh_*} \\
\pi_{2n+1}(\mathbf{SP}^n(\mathbf{C}), \mathbf{SP}^{n-1}(\mathbf{C})) & \xrightarrow{\partial_*} & \pi_{2n}(\mathbf{SP}^{n-1}(\mathbf{C})) \\
{\scriptstyle G_{n+1*}} \downarrow & & {\scriptstyle G_{n*}} \downarrow \\
\pi_{2n+1}(\mathbf{U}_{n+1} \mathbf{U}_n) & \xrightarrow{\partial_*} & \pi_{2n}(\mathbf{U}_n) \\
{\scriptstyle p_*} \downarrow & \nearrow {\scriptstyle \Delta_*} & \\
\pi_{2n+1}(\mathbf{S}^{2n+1}) & &
\end{array}$$

Since $p \circ G_{n+1}$ and the attaching map for the $(2n+1)$-cell are relative homeomorphisms, their composite is a relative homeomorphism $(\mathbf{E}^{2n+1}, \mathbf{S}^{2n}) \to (\mathbf{S}^{2n+1}, \mathbf{e}_{2n})$ representing the generator ι_{2n+1}. By commutativity, the element $\Delta_*(\iota_{2n+1})$ is represented by the map $G_n \circ h = F_n$. □

Remark. Later we shall need to know that the image of the map G_n is contained in $\mathbf{U}_{n+1}^+$, i.e.,

(10.22) *For any* $u \in \mathbf{S}^1$, $X \in \mathbf{S}^{2n+1}$,

$$\det \tilde{F}_{n+1}(u, X) = 1.$$

Let

$$H(u, X) = \begin{pmatrix} 1 & 0 \\ -\bar{u}\xi & xE \end{pmatrix} \tilde{F}_{n+1}(u, X) \begin{pmatrix} x & 0 \\ \xi & E \end{pmatrix}.$$

A little calculation reveals that

$$H(u, X) = \begin{pmatrix} x & x(1-u)\xi^* \\ 0 & xE \end{pmatrix}$$

so that $\det H(u, x) = x^{n+1}$. But

$$\det\begin{pmatrix} 1 & 0 \\ -\bar{u}\xi & xE \end{pmatrix} = x^n,$$

$$\det\begin{pmatrix} x & 0 \\ \xi & E \end{pmatrix} = x,$$

and therefore, if $x \neq 0$,

$$\det \tilde{F}_{n+1}(u, X) = 1.$$

As the set of points (u, X) with $x \neq 0$ is dense and $\tilde{F}_{n+1}$ is continuous, $\det \tilde{F}_{n+1}(u, X) = 1$ everywhere. □

Exercises

1. Prove Theorem (2.6) (paying strict attention to low-dimensional cases).
2. Let (X, A) be a pair, and let

$$\pi_1(A) \xrightarrow{i_*} \pi_1(X) \xrightarrow{j_*} \pi_1(X, A) \xrightarrow{\partial_*} \pi_0(A) \xrightarrow{i_*} \pi_0(X)$$

be a portion of the homotopy sequence of (X, A).
   (i) If $\alpha, \beta \in \pi_1(X)$ and $j_*(\alpha) = j_*(\beta)$, then there exists $\gamma \in \pi_1(A)$ such that $\beta = \alpha \cdot i_*(\gamma)$;
   (ii) Show how multiplication of paths can be used to define an operation of $\pi_1(X)$ on $\pi_1(X, A)$;
   (iii) Use the operation defined in (ii) to show that if $\alpha, \beta \in \pi_1(X, A)$ and $\partial_*(\alpha) = \partial_*(\beta)$, then β can be obtained from α by operating with ξ for some $\xi \in \pi_1(X)$.

3. (Wall [1]). Suppose given a commutative sinusoidal diagram like that of Figure 4.3. Suppose, moreover, that the last three of the sequences

$$\cdots \to C_{n+1} \xrightarrow{\partial} D_n \xrightarrow{i} X_n \xrightarrow{j} C_n \to \cdots$$

$$\cdots \to D_{n+1} \xrightarrow{\partial_1} A_n \xrightarrow{i_1} Y_n \xrightarrow{j_1} D_n \to \cdots$$

$$\cdots \to X_{n+1} \xrightarrow{\partial_2} A_n \xrightarrow{i_2} B_n \xrightarrow{j_2} X_n \to \cdots$$

$$\cdots \to C_{n+1} \xrightarrow{\partial_3} Y_n \xrightarrow{i_3} B_n \xrightarrow{j_3} C_n \to \cdots$$

are exact. Then $\operatorname{Ker} \partial = \operatorname{Im} j$, $\operatorname{Ker} i = \operatorname{Im} \partial$, and

$$\operatorname{Ker} j/\operatorname{Ker} j \cap \operatorname{Im} i \approx \operatorname{Im} i/\operatorname{Ker} j \cap \operatorname{Im} i.$$

Thus the top sequence is exact if either $\operatorname{Im} i \subset \operatorname{Ker} j$ or $\operatorname{Ker} j \subset \operatorname{Im} i$.

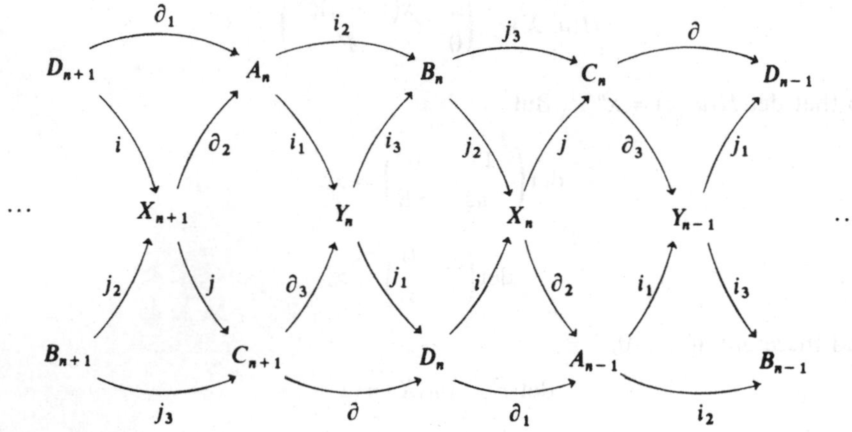

Figure 4.3

4. Let X be a semi-locally 1-connected space, $\tilde{X}$ its universal covering space. Prove that $H_n^{(1)}(X) \approx H_n(\tilde{X})$ for all n.

5. Let $p : X \to B$ be a fibration with fibre F, and let $\alpha, \beta \in \pi_1(B)$ be elements such that $\Delta_*(\alpha) = \Delta_*(\beta) \in \pi_0(F)$. Then there exists $\xi \in \pi_1(X)$ such that $\beta = \alpha \cdot p_*(\xi)$.

6. Let $\{X_n \mid n = 1, 2, \ldots\}$ be a sequence of spaces with base points, $X = \prod_{n=1}^{\infty} X_n$. Prove that $\pi_q(X) \approx \bigoplus_{n=1}^{\infty} \pi_q(X_n)$ for all $q > 1$. What happens if $q = 1$?

7. (Milnor [1]). Let W be the geometric realization of the total singular complex $\mathfrak{S}(X)$ of a space X (Exercise 5, Chapter II). Prove that the map $g : W \to X$ is a weak homotopy equivalence.

CHAPTER V

Homotopy Theory of CW-complexes

In Chapter II we proved that if X is an n-cellular extension of the pathwise connected space A, then the pair (X, A) is $(n-1)$-connected, so that $\pi_i(X, A) = 0$ for all $i < n$. The next step is the determination of $\pi_n(X, A)$. By the results of §2 of Chapter II, $H_n(X, A)$ is a free abelian group with one basis element for each n-cell of (X, A). We have seen that the Hurewicz map $\rho : \pi_n(X, A) \to H_n(X, A)$ is an epimorphism whose kernel is generated by all elements of the form $\alpha - \tau'_\xi(\alpha)$ with $\alpha \in \pi_n(X, A)$, $\xi \in \pi_1(A)$. If $\Pi = \pi_1(A)$ operates trivially on $\pi_n(X, A)$, then ρ is an isomorphism. But this condition is not easy to verify *a priori*.

J. H. C. Whitehead addressed himself to this problem in a series of papers [1, 2, 4] written between 1939 and 1949, culminating in [6] with a complete solution of the problem. If $n \geq 3$, the answer is very simple—$\pi_n(X, A)$ is a *free* $\mathbf{Z}(\Pi)$-module with one basis element for each n-cell. For $n = 2$, the situation is considerably more complicated; the group $\pi_2(X, A)$ being non-abelian, it is not a module at all over $\mathbf{Z}(\Pi)$; however, it is a crossed module, and Whitehead proved in [6] that it is a free crossed module with one basis element for each 2-cell.

In §1 we prove Whitehead's theorem for $n \geq 3$. We shall not need the full result for $n = 2$, and, as the algebra involved is of considerable complexity, we shall content ourselves with proving that $\pi_2(X, A)$ is generated, as an additive group, by the elements $\tau'_\xi(\varepsilon_\alpha)$, where the ε_α are in one-to-one correspondence with the 2-cells of (X, A) and ξ ranges over $\pi_1(A)$.

A consequence of these results is that the injection $\pi_i(A) \to \pi_i(X)$ is an isomorphism for $i < n - 1$ and an epimorphism if $i = n - 1$; in the latter case the kernel is the operator subgroup of $\pi_{n-1}(A)$ generated by the attaching maps for the n-cells of (X, A). This result allows us to construct a CW-complex with given homotopy groups. This, and a few related results, are proved in §2; the ideas again are due to J. H. C. Whitehead [5].

One consequence of the results of §2 is that, if X is an arbitrary space, there is a CW-complex K and a weak homotopy equivalence $f: K \to X$. The complex K is then uniquely determined up to homotopy type. By the Whitehead Theorem, the map f induces isomorphisms $f_* : H_q(K) \approx H_q(X)$ for all q. Thus, from the point of view of algebraic topology, we may as well replace X by K. This is the method of CW-approximation; generalized to the relative case, it is expounded in §3.

A 0-connected space is said to be *aspherical* if and only if its homotopy groups vanish in all dimensions greater than 1. Examples of such spaces are very common in mathematics; for example if G is a strongly discontinuous group acting on a convex open set D in R^n, then the orbit space D/G is aspherical. In particular, every closed surface, except for the sphere and the projective plane, is aspherical. In §4 these spaces are discussed, and it is shown, that an aspherical CW-complex X is determined up to homotopy type by its fundamental group Π; and if K is an arbitrary CW-complex, then $[K, X]$ is in one-to-one correspondence with the conjugacy classes of homomorphisms of $\pi_1(K)$ into Π.

Some of the results of §4 were proved by a stepwise extension process. A map of the n-skeleton K_n of K into a space X was given, and it was required to extend the map over the $(n + 1)$-skeleton. And the vanishing of the homotopy group $\pi_n(K)$ allowed the extension to be carried out over each cell separately, and therefore (because K_{n+1} has the weak topology) over all of K_{n+1}. This method of *obstruction theory* has considerably wider range of application; it was developed in 1940 by Eilenberg [2].

The principle of the method is this: suppose that (X, A) is a relative CW-complex, Y an n-simple space, and $f: X_n \to Y$ is a map. If $h_\alpha : \Delta^{n+1} \to X_n$ is the attaching map for an $(n+1)$-cell E_α, then $f \circ h_\alpha : \Delta^{n+1} \to Y$ represents an element $c(e_\alpha) \in \pi_n(Y)$. The function c determines a homomorphism of the $(n + 1)^{\text{st}}$ chain group $\Gamma_{n+1}(X, A)$ into $\pi_n(Y)$, i.e., a cochain $c^{n+1}(f) \in \Gamma^{n+1}(X, A; \pi_n(Y))$.

The cochain $c^{n+1}(f)$ is called the *obstruction* to extending f; it is a *cocycle*, whose vanishing is necessary and sufficient for the map f to be extendible over X_{n+1}. Moreover, f is a coboundary if and only if $f \mid X_{n-1}$ can be extended over X_{n+1}. In practice, we are usually given a map $f_0 : A \to Y$ and are seeking necessary and sufficient conditions for f_0 to be extendible over X. In order to solve this, we need to know how the obstructions $c^{n+1}(f)$, $c^{n+1}(f')$ of two different extensions of f_0 over X_n are related. We are a long way from doing this.

The above obstruction theory is presented in §5. In §6 we make the simplifying assumption that the space Y is $(n - 1)$-connected. If $f: A \to Y$ is an arbitrary map, then, by the above theory, f can be extended over X_n and any two extensions over X_n are homotopic on X_{n-1}. If $g : X_n \to Y$ is any extension of f, the cohomology class of the obstruction $c^{n+1}(g)$ depends only on f; it is an element $\gamma^{n+1}(f) \in H^{n+1}(X, A; \pi_n(Y))$, whose vanishing is necessary and sufficient that f be extendible over X_{n+1}.

The class $\gamma^{n+1}(f)$ can be given a description in terms of a certain "universal class" $\iota^n(Y) \in H^n(Y; \pi_n(Y))$ and standard operations in cohomology theory. By the Hurewicz Theorem $\rho : \pi_n(Y) \approx H_n(Y)$, and the class $\iota^n(Y)$ is determined by the condition

$$\langle \iota^n(Y), z \rangle = \rho^{-1}(z)$$

for all $z \in H_n(Y)$. The class $\gamma^{n+1}(f)$ is then, up to sign, the image of $\iota^n(Y)$ under the composite

$$H^n(Y; \pi_n(Y)) \xrightarrow{f^*} H^n(A; \pi_n(Y)) \xrightarrow{\delta^*} H^{n+1}(X, A; \pi_n(Y)).$$

Thus $\gamma^{n+1}(f) = 0$ if and only if there is a class $u \in H^n(X; \pi_n(Y))$ which is mapped into $f^*(\iota^n(Y))$ by the injection.

Suppose that $\dim (X, A) \leq n + 1$. Then f can be extended over X if and only if $f^*\iota^n(Y)$ belongs to the image of the injection. This is a modern form of Hopf's extension theorem, proved in [4] in 1933 for the case $Y = S^n$. The formulation in terms of cohomology was given by Whitney [3] in 1937.

In [2] Eilenberg gave a set of conditions sufficient to ensure that the vanishing of $\gamma^{n+1}(f)$ implies the extendibility of f over all of X. These conditions are satisfied, in particular, when $\pi_n(Y)$ is the only non-vanishing homotopy group of Y. The corresponding homotopy classification theorem asserts that, for such a space Y, the map $f \to f^*\iota^n(Y)$ induces a one-to-one correspondence between $[X, Y]$ and $H^n(X; \pi_n(Y))$ for any CW-complex X.

Spaces with only one non-zero homotopy group were first investigated by Eilenberg and Mac Lane [1] in 1945. They are of enormous importance in homotopy theory, not only because of the above-mentioned classification theorem, but also because, as we shall see in Chapter IX, every CW-complex can be built up, up to homotopy type, out of these "Eilenberg–Mac Lane" spaces. In §7 we investigate some of their elementary properties.

The classification theorem has another important consequence. A cohomology operation is a natural transformation of functors

$$\theta : H^n(\quad; \Pi) \to H^q(\quad; G).$$

They and their higher-order generalizations have turned out to be powerful tools in studying obstruction theory. Because of the classification theorem, such an operation is completely determined by its effect on the fundamental class $\mathbf{b}_n = \iota^n(K(\Pi, n))$. This observation was made in 1953 by Serre [2]. Cohomology operations are introduced in §8 and a few examples given. A deeper study is made in Chapter VIII.

1 The Effect on the Homotopy Groups of a Cellular Extension

Let A be a space, which we shall assume to be 0-connected, and let X be an n-cellular extension of A, with cells $\{E_\alpha \mid \alpha \in J\}$ $(n \geq 2)$ and characteristic maps $h_\alpha : (\Delta^n, \dot{\Delta}^n) \to (X, A)$. We have seen that the pair (X, A) is

$(n-1)$-connected, and therefore the Hurewicz map sends $\pi_n^\dagger(X, A)$ isomorphically upon $H_n(X, A)$. In §2 of Chapter II we have determined the group $H_n(X, A)$; it is a free abelian group with one basis element e_α for each cell E_α; the element e_α is the image under the homomorphism induced by h_α of a generator δ_n of the infinite cyclic group $H_n(\Delta^n, \dot{\Delta}^n)$. If we know *a priori* that $\pi_1(A)$ operates trivially on $\pi_n(X, A)$, (for example, if A is 1-connected), we can deduce the structure of the latter group; it is a free abelian group with one basis element ε_α for each $\alpha \in J$; the element ε_α is the image under $\pi_n(h_\alpha)$ of a generator ε_n of the infinite cyclic group $\pi_n(\Delta^n, \dot{\Delta}^n)$.

If A is not 1-connected, the problem becomes more difficult. Suppose $n > 2$; then the pair (X, A) is 2-connected, so that the injection $\pi_1(A) \to \pi_1(X)$ is an isomorphism. Let $p : \tilde{X} \to X$ be a universal covering map; then, if $\tilde{A} = p^{-1}(A)$, the map $p \mid \tilde{A} : \tilde{A} \to A$ is also a universal covering map. Let Π be the group of covering translations of $\tilde{X}$, so that $\Pi \approx \pi_1(X) \approx \pi_1(A)$. Then we have isomorphisms

$$\rho : \pi_n(\tilde{X}, \tilde{A}) \approx H_n(\tilde{X}, \tilde{A})$$

(because $\tilde{A}$ is 1-connected and $(\tilde{X}, \tilde{A})$ $(n-1)$-connected); and

$$p_* : \pi_n(\tilde{X}, \tilde{A}) \approx \pi_n(X, A)$$

(because of Corollary (8.10) of Chapter IV, or by covering space theory). The group Π operates on $H_n(\tilde{X}, \tilde{A})$, and it follows from (1.20) of Chapter III that the composite $p_* \circ \rho^{-1} : H_n(\tilde{X}, \tilde{A}) \to \pi_n(X, A)$ is an operator isomorphism.

Since $H_n(\tilde{X}, \tilde{A})$ is an abelian group on which Π operates, it is naturally a $\mathbf{Z}(\Pi)$-module, where $\mathbf{Z}(\Pi)$ is the integral group ring of Π. The structure of this module is easily determined; for each cell E_α of (X, A), choose a cell $\tilde{E}_\alpha$ of $(\tilde{X}, \tilde{A})$ such that $p \circ \tilde{h}_\alpha = h_\alpha$, where h_α, $\tilde{h}_\alpha$ are characteristic maps for these cells. Then the map $h \to h(\tilde{E}_\alpha)$ establishes a one-to-one correspondence between Π and the set of cells of $(\tilde{X}, \tilde{A})$ which lie over E_α. An additive basis for $H_n(\tilde{X}, \tilde{A})$ is thus formed by the elements $h_*(\tilde{e}_\alpha)$ as α ranges over the indexing set J and h over the group Π. Therefore, as a module, $H_n(\tilde{X}, \tilde{A})$ is free with the $\tilde{e}_\alpha$, $\alpha \in J$, as a basis.

Interpreting our result in the base space X, we have

(1.1) Theorem *If X is an n-cellular extension of the* 0*-connected space A, and $n \geq 3$, then $\pi_n(X, A)$ is a free module over the integral group ring $\mathbf{Z}(\pi_1(A))$. If $\{E_\alpha \mid \alpha \in J\}$ are the n-cells of (X, A), with characteristic maps $h_\alpha : (\Delta^n, \dot{\Delta}^n) \to (X, A)$, and if, for each $\alpha \in J$, u_α is a homotopy class of paths in A from the base point e_0 to $h_\alpha(e_0)$, and if $\varepsilon_\alpha = h_{\alpha*}(\varepsilon_n) \in \pi_n(X, A, h_\alpha(e_0))$, then the elements $\varepsilon'_\alpha = \tau'_{u_\alpha}(\varepsilon_\alpha)$ form a module basis for $\pi_n(X, A)$.* □

It will simplify the discussion below and elsewhere if we make the following definition. Let $\alpha \in \pi_n(X, A, x_0)$, $\beta \in \pi_n(X, A, x_1)$; then α is a *translate* of β if and only if there exists a homotopy class ξ of paths in A from x_0 to x_1 such that $\tau'_\xi(\beta) = \alpha$.

(1.2) Corollary *If $n > 2$, the elements ε'_α and their translates generate the group $\pi_n(X, A)$.* □

The case $n = 2$ is much more difficult. In the first place, the injection $\pi_1(A) \to \pi_1(X)$ is only an epimorphism, and not an isomorphism, and the covering space argument breaks down (cf., however, the Remark at the end of this section). Secondly, the group $\pi_2(X, A)$ is, in general, non-abelian, so that it cannot be regarded as a module over the group ring $\mathbf{Z}(\Pi)$. However, it is a *crossed module* over $\mathbf{Z}(\Pi)$ and it is proved by J. H. C. Whitehead in [6] that it is a *free* crossed module having the ε'_α as basis. It follows that Corollary (1.2) holds even if $n = 2$.

As the proof of the analogue of Theorem (1.1) for the case $n = 2$ is quite complicated, as it involves notions like that of crossed module which are not needed elsewhere in this book, and as we shall not need the complete result, we shall not give a proof here. We shall, however, need the fact that Corollary (1.2) remains valid, and accordingly, we proceed to prove

(1.3) Theorem *The elements ε'_α generate $\pi_n(X, A)$, as a group with operators.*

We show successively that the theorem is true

(1) when $(X, A) = (K_n, K_{n-1})$ for some simplicial subdivision K of Δ^n;
(2) when (X, A) has just one n-cell;
(3) when (X, A) has only a finite number of n-cells;
(4) in general.

In Case (1), suppose first that $n = 2$; then $\partial_* : \pi_2(X, A) \to \pi_1(A)$ is an isomorphism, and we must show that $\pi_1(A)$ is generated, as a normal subgroup, by the elements $\partial_* \varepsilon'_\alpha$.

The elements $\partial_* \varepsilon'_\alpha$ depend on certain choices—on the choice of the characteristic map h_α and of the homotopy class of paths u_α. But it is clear that different choices for these elements has the effect of replacing $\partial_* \varepsilon'_\alpha$ by a conjugate, and this does not affect the normal subgroup generated by these elements.

Recall that, if K is any connected simplicial complex with base vertex $*$, then $\pi_1(K) \approx \pi_1(K_2)$ can be calculated as follows. Order the vertices of K and choose a maximal tree T in K. For each vertex e of K, let ξ_e be the unique homotopy class of paths in T from $*$ to e. For each 1-simplex σ of $K - T$ with vertices $a < b$, let $\eta_{a,b}$ be the unique homotopy class of paths in σ from a to b. The elements $\zeta_{a,b} = \xi_a \eta_{a,b} \xi_b^{-1}$, considered as homotopy classes of paths in K_1, form a set of free generators for $\pi_1(K_1)$. To find the kernel R of the epimorphic injection $\pi_1(K_1) \to \pi_1(K_2)$, for each 2-simplex α of K with vertices $a < b < c$, let $\omega_\alpha = \zeta_{a,b}\zeta_{b,c}\zeta_{a,c}^{-1}$ (we agree that $\zeta_{x,y} = 1$ if $x < y$ are the vertices of a 1-simplex in T). Then R is generated, as a normal subgroup, by the elements ω_α. In our case K is 1-connected, so that $\pi_1(K_1)$ is generated, as a normal subgroup, by the elements ω_α. If we choose h_α to be the simplicial map of Δ^2 into α which sends $\mathbf{e}_0, \mathbf{e}_1, \mathbf{e}_2$ into a, b, c respectively, and if we choose $u_\alpha = \xi_a$, we see that $\omega_\alpha = \partial_* \varepsilon'_\alpha$. This proves our contention when $n = 2$.

If $n > 2$, the argument is much simpler. For then K_{n-1} is 1-connected, so that the Hurewicz map $\rho: \pi_n(K_n, K_{n-1}) \to H_n(K_n, K_{n-1})$ is an isomorphism. As $\rho(\varepsilon'_\alpha) = e_\alpha$ and the elements e_α generate $H_n(K_n, K_{n-1})$, the ε'_α generate $\pi_n(K_n, K_{n-1})$.

In order to handle Case (2), we shall need

(1.4) Lemma *Let X be an n-cellular extension of A, having just one n-cell E with characteristic map $h: (\Delta^n, \dot{\Delta}^n) \to (X, A)$. Let $f: (\Delta^n, \dot{\Delta}^n, \mathbf{e}_0) \to (X, A, *)$ be a map. Then there is a subdivision K of Δ^n and a map $f_2 \simeq f$ such that, for each n-simplex σ of K, either $f_2(\sigma) \subset A$ or $f_2 | \sigma = h \circ f_\sigma$ for some map $f_\sigma: (\sigma, \dot{\sigma}) \to (\Delta^n, \dot{\Delta}^n)$.*

As in the proof of Theorem (3.9) of Chapter II, let $U = \operatorname{Int} E^n$, $V = A \cup h(\Delta^n - \{b_n\})$, where b_n is the barycenter of Δ^n. Let K be a simplicial subdivision of Δ^n so fine that each simplex of K is mapped by f into either U or V; let L be the union of those mapped into U, M the union of those mapped into V, so that L and M are subcomplexes of K and $\dot{\Delta}^n \subset M$, $L \cap M \subset \operatorname{Int} \Delta^n$. Then $f(L) \subset h(\operatorname{Int} \Delta^n)$ and h maps $\operatorname{Int} \Delta^n$ homeomorphically; thus $h^{-1} \circ f | L$ is continuous, and we may regard it as a map $g: (L, L \cap M) \to (\Delta^n, \Delta^n - \{b_n\})$.

By Corollary (3.7) of Chapter II, the pair $(\Delta^n, \Delta^n - \{b_n\})$ is $(n-1)$-connected. Hence, if L' is the union of $L \cap M$ with the $(n-1)$-skeleton of L, the map $g | L': (L', L \cap M) \to (\Delta^n, \Delta^n - \{b_n\})$ is compressible. By the homotopy extension property for the pair (L, L'), g is homotopic (rel. $L \cap M$) to a map $g': (L, L') \to (\Delta^n, \Delta^n - \{b_n\})$. Now $\dot{\Delta}^n$ is a deformation retract of $\Delta^n - \{b_n\}$; thus g', and therefore g, is homotopic to a map $g'': (L, L') \to (\Delta^n, \dot{\Delta}^n)$. Composing a homotopy of g to g'' with h, we obtain a homotopy k_t of $f | L: (L, L \cap M) \to (X, V)$ to the map $h \circ g''$. Combining k_t with the stationary homotopy of $f | \dot{\Delta}^n$, and using the homotopy extension property for the pair $(K, L \cup \dot{\Delta}^n)$, we obtain a homotopy of f to a map $f_1: (K, \dot{\Delta}^n) \to (X, A)$ such that $f_1(M, (L \cap M) \cup \dot{\Delta}^n) \subset (V, A)$ and $f_1 | L = h \circ g''$. Since A is a deformation retract of V, the map $f_1 | M: (M, (L \cap M) \cup \dot{\Delta}^n) \to (V, A)$ is compressible. Hence f_1 is homotopic (rel. $L \cup \dot{\Delta}^n$) to a map $f_2: (\Delta^n, \dot{\Delta}^n) \to (X, A)$ with the desired property.

Let ι be the element of $\pi_n(K_n, K_{n-1}, \mathbf{e}_0)$ represented by the identity map of Δ^n. For each n-simplex σ of K, let ξ_σ be a homotopy class of paths in K_n from $\mathbf{e}_0$ to a vertex e_σ of σ, and let $\iota_\sigma \in \pi_n(K_n, K_{n-1}, \mathbf{e}_0)$ be the element represented by a characteristic map for σ. By Part (1) above, ι is a sum of terms, each of which is, up to sign, a translate of ι_σ.

Now $\beta = f_*(\iota) = f_{2*}(\iota)$ is a sum of translates of $\pm f_{2*}(\iota_\sigma)$. If $f_2(\sigma) \subset A$, then

$f_{2*}(\iota_\sigma) = 0$. Otherwise $f_2|\sigma = h \circ f_\sigma$ with $f_\sigma : (\sigma, \dot{\sigma}) \to (\Delta^n, \dot{\Delta}^n)$. Then $f_{\sigma*}(\iota_\sigma)$ is a translate of $n_\sigma \varepsilon_n$ for some integer n_σ; thus $f_{2*}(\iota_\sigma)$ is a translate of $n_\sigma h_*(\varepsilon_n)$ and therefore of $n_\sigma \varepsilon$. Hence β is a sum of integral multiples of translates of ε'. This proves our contention in Case (2).

Now suppose (3) holds, so that there are only finitely many cells $E_1, \ldots, E_r$. The case $r = 1$ having been treated in Case (2), we may assume that $r > 1$ and that the desired result holds for the pair (X', A), where $X' = E_2 \cup \cdots \cup E_r \cup A$. There is an exact sequence

$$\textbf{(1.4)} \qquad \pi_n(X', A) \xrightarrow{\ i\ } \pi_n(X, A) \xrightarrow{\ j\ } \pi_n(X, X') \to 0$$

(recall that (X', A) is $(n-1)$-connected!) where i, j are the injections. The first two groups in (1.4) have operators in $\pi_1(A)$, while the third has operators in $\pi_1(X')$. Since the injection $\pi_1(A) \to \pi_1(X')$ is an epimorphism and i, j are operator homomorphisms, we may regard (1.4) as an exact sequence of groups with operators in $\pi_1(A)$. As $\pi_n(X', A)$ has a set of generators which are mapped by i into $\varepsilon_2', \ldots, \varepsilon_r'$, and as $\pi_n(X, X')$ has one generator $j(\varepsilon_1')$, it is a matter of elementary algebra to verify that $\pi_n(X, A)$ is generated by $\varepsilon_1', \ldots, \varepsilon_r'$.

The general case (4) now follows by a direct limit argument of a type we have used several times before. □

(1.5) Corollary *The injection $\pi_{n-1}(A) \to \pi_{n-1}(X)$ is an epimorphism. Its kernel is the operator subgroup generated by the elements $\partial_*(\varepsilon_\alpha')$.* □

Note that the elements $\partial_*(\varepsilon_\alpha')$ are just (translates of) the homotopy classes of the attaching maps for the cells of (X, A).

Remark. It should be observed that the covering space argument used in the proof of Theorem (1.1) is valid, even when $n = 2$, provided that the injection $\pi_1(A) \to \pi_1(X)$ is an isomorphism; note that then the injection $\pi_2(X) \to \pi_2(X, A)$ is an epimorphism, so that $\pi_2(X, A)$ is abelian. By Corollary (1.4), this is so if and only if the attaching maps for the 2-cells are all trivial. This means that (X, A) has the same homotopy type as $(A \vee \Sigma, A)$, where Σ is a cluster of 2-spheres.

Suppose, more generally, that $X = A \vee \Sigma^r$, where Σ^r is a cluster of r-spheres. Then A is a retract of X, and therefore the homotopy exact sequence of (X, A) breaks up into a family of short exact sequences

$$0 \to \pi_r(A) \to \pi_r(X) \to \pi_r(X, A) \to 0$$

of $\pi_1(A)$-modules. In fact, these sequences split; for if $f: X \to A$ is a retraction, then $f_* : \pi_r(X) \to \pi_r(A)$ is a left inverse of the injection $\pi_r(A) \to \pi_r(X)$. Since (X, A) is $(n-1)$-connected, we have:

(1.6) Theorem *If A is 0-connected and $X = A \vee \Sigma^n$, where $\Sigma^n = \bigvee_{\alpha \in J} \mathbf{S}_\alpha^n$ is a cluster of n-spheres $(n \geq 2)$, then the injection $\pi_r(A) \to \pi_r(X)$ is an isomorphism* *for* $r < n$. *Moreover, there is an isomorphism*

$$\pi_n(X) \approx \pi_n(A) \oplus \pi_n(X, A)$$

of $\pi_1(A)$*-modules, and the module* $\pi_n(X, A)$ *is free. The images of the generators of the groups* $\pi_n(\mathbf{S}^n_\alpha)$ *under the injections*

$$\pi_n(\mathbf{S}^n_\alpha) \to \pi_n(\Sigma^n) \to \pi_n(X)$$

form a basis for a submodule of $\pi_n(X)$ *which is mapped isomorphically upon* $\pi_n(X, A)$ *by the injection.* □

2 Spaces with Prescribed Homotopy Groups

It is natural to ask to what extent the homotopy groups of a space can be prescribed. Specifically, if that $\pi_1, \pi_2, \ldots$ is a sequence of groups, what conditions must be imposed in order that there exist a space X and isomorphisms $\pi_i(X) \to \pi_i$ for all i? Of course, the groups π_i must be abelian if $i \geq 2$. Since $\pi_1(X)$ operates on $\pi_i(X)$, it is natural to require that π_1 operate on π_i for each i and that the above isomorphisms be operator isomorphisms. (N.B.—The group π_1 operates on itself by inner automorphisms, so that any isomorphism is an operator isomorphism if $i = 1$).

It turns out that no further conditions are necessary. In fact, we have

(2.1) Theorem *Let* π_1 *be a group, and, for each* $q \geq 2$, *let* π_q *be an abelian group on which* π_1 *operates. Then there is a connected* CW*-complex* X *and a family of isomorphisms* $\phi_q : \pi_q(X) \to \pi_q$ *such that, for all* $\xi \in \pi_1(X)$, $\alpha \in \pi_q(X)$, *we have*

$$\phi_q(\tau_\xi(\alpha)) = \phi_1(\xi) \cdot \phi_q(\alpha).$$

We shall construct the space X skeleton by skeleton. Two kinds of processes are involved; the first is that of attaching spheres so as to create new generators; the second, of attaching cells to create new relations.

To begin with, let A_1 be a set of generators for the group π_1, and let $X_1 = \bigvee_{\alpha \in A_1} \mathbf{S}^1_\alpha$ be a cluster of circles, indexed by A_1. Then $\pi_1(X_1)$ is a free group; the images $\sigma_\alpha \in \pi_1(X_1)$ of generators of the groups $\pi_1(\mathbf{S}^1_\alpha)$ form a basis, and a homomorphism $\psi_1 : \pi_1(X_1) \to \pi_1$ is defined by $\psi_1(\sigma_\alpha) = \alpha$ for each $\alpha \in A_1$. Since A_1 generates π_1, ψ_1 is an epimorphism.

Suppose that X_n is an n-dimensional CW-complex ($n \geq 1$) and that $\phi'_i : \pi_i(X_n) \to \pi_i$, $(1 \leq i \leq n-1)$ are operator isomorphisms, and $\psi_n : \pi_n(X_n) \to \pi_n$ is an operator epimorphism. Let B_n be a set of generators for Ker ψ_n as a $\pi_1(X_n)$-module (if $n = 1$, assume instead that B_n generates Ker ψ_n as a normal subgroup of the free group $\pi_1(X_1)$). Let $\Sigma_n = \bigvee_{\beta \in B_n} \mathbf{S}^n_\beta$ be a cluster of n-spheres indexed by B_n, and let $h_n : \Sigma_n \to X_n$ be a map such that $h_n | \mathbf{S}^n_\beta$ represents $\beta \in \pi_n(X_n)$. Let X'_{n+1} be the mapping cone of h_n. By

Corollary (1.4), the injection $\pi_n(X_n) \to \pi_n(X'_{n+1})$ is a epimorphism whose kernel is the operator subgroup generated by B_n; thus this injection induces an isomorphism $\phi'_n : \pi_n(X'_{n+1}) \to \pi_n$. Moreover, if $i < n$, the injection $\pi_i(X_n) \to \pi_i(X'_{n+1})$ is an isomorphism, and the ϕ'_i determine isomorphisms, which we continue to denote by $\phi'_i : \pi_i(X'_{n+1}) \approx \pi_i$.

Let A_{n+1} be a set of generators for the module π_{n+1}. Let $\Sigma'_{n+1} = \bigvee_{\alpha \in A_{n+1}} \mathbf{S}_\alpha^{n+1}$ be a cluster of $(n+1)$-spheres, indexed by A_{n+1}, and let $X_{n+1} = X'_{n+1} \vee \Sigma'_{n+1}$. It follows from Theorem (1.6) that

(1) *the injection* $\pi_i(X'_{n+1}) \to \pi_i(X_{n+1})$ *is an isomorphism for all* $i \leq n$;
(2) *the injection* $i : \pi_{n+1}(X'_{n+1}) \to \pi_{n+1}(X_{n+1})$ *is a monomorphism.*
(3) $\pi_{n+1}(X_{n+1})$ *is the direct sum of the image of i with the submodule F freely generated by the elements* σ_α *represented by the inclusion maps* $\mathbf{S}_\alpha^{n+1} \hookrightarrow X_{n+1}$.

Hence the operator isomorphisms $\phi'_i : \pi_i(X'_{n+1}) \approx \pi_i$ induce operator isomorphisms $\phi_i : \pi_i(X_{n+1}) \approx \pi_i$ for each $i \leq n$. Let $\theta : \pi_{n+1}(X'_{n+1}) \to \pi_{n+1}$ be an arbitrary operator homomorphism (for example, we may take $\theta = 0$). Then there is an operator homomorphism $\psi_{n+1} : \pi_{n+1}(X_{n+1}) \to \pi_{n+1}$ defined by

$$\psi_{n+1} \circ i = \theta,$$
$$\psi_{n+1}(\sigma_\alpha) = \alpha;$$

ψ_{n+1} is an epimorphism because A_{n+1} generates the module π_{n+1}.

This completes the inductive construction of an expanding sequence of spaces $\{X_n\}$. Their union X is a CW-complex having X_n as its n-skeleton. Since the injections $\pi_n(X_{n+1}) \to \pi_n(X)$ are isomorphisms, the isomorphisms $\phi_n : \pi_n(X_{n+1}) \to \pi_n$ induce isomorphisms $\pi_n(X) \approx \pi_n$ which evidently have the required properties. □

We can modify the details of the above proof to obtain other useful results. For example, suppose that Y is a 0-connected space and $\pi_i = \pi_i(Y)$. Then there is a map $f_1 : X_1 \to Y$ such that $f_1 | \mathbf{S}_\alpha^1$ represents the element $\alpha \in \pi_1 = \pi_1(Y)$. Suppose that $f_n : X_n \to Y$ is a map such that

$$\phi'_i = \pi_i(f_n) : \pi_i(X_n) \to \pi_i \qquad (i < n),$$
$$\psi_n = \pi_n(f_n) : \pi_n(X_n) \to \pi_n.$$

Then, for each $\beta \in B_n$, $f_n \circ h_n | \mathbf{S}_\beta^n$ represents $f_{n*}(\beta) = \psi_n(\beta) = 0$. Hence $f_n \circ h_n$ is nullhomotopic, and therefore $f_n : X_n \to Y$ has an extension $f'_{n+1} : X'_{n+1} \to Y$, and

$$\phi'_i = \pi_i(f'_{n+1}) : \pi_i(X'_{n+1}) \to \pi_i \qquad (i \leq n).$$

Now extend f'_{n+1} to a map $f_{n+1} : X_{n+1} \to Y$ by requiring that $f_{n+1} | \mathbf{S}_\alpha^{n+1}$ represent $\alpha \in \pi_{n+1} = \pi_{n+1}(Y)$. Then

$$\phi_i = \pi_i(f_{n+1}) : \pi_i(X_{n+1}) \to \pi_i \qquad (i \leq n);$$

and, if we choose the homomorphism $\theta : \pi_{n+1}(X'_{n+1}) \to \pi_{n+1}$ to be $\pi_{n+1}(f'_{n+1})$, then

$$\psi_{n+1} = \pi_{n+1}(f_{n+1}) : \pi_{n+1}(X_{n+1}) \to \pi_{n+1}.$$

Each map f_{n+1} being an extension of the preceding, there is a map $f : X \to Y$ such that $f \mid X_n = f_n$ for every n. Therefore we have

(2.2) Theorem *Let Y be a 0-connected space. Then there is a connected* CW-*complex X and a map $f : X \to Y$ such that $f_* : \pi_n(X) \approx \pi_n(Y)$ for every n.*

□

Remark. Suppose that Y is m-connected ($m \geq 1$). Then the groups π_i are zero for all $i \leq m$, and we can choose X so that its m-skeleton is a single point.

This line of reasoning can be carried a little further, to obtain

(2.3) Theorem *Let A, Y be 0-connected spaces, $f_0 : A \to Y$, and let n be a non-negative integer. Then there is a relative* CW-*complex (X, A), having no cells of dimension $\leq n$, and an extension $f : X \to Y$ of f_0, such that*

(1) *the injection $\pi_q(A) \to \pi_q(X)$ is an isomorphism for $q < n$ and an epimorphism for $q = n$;*
(2) *the homomorphism $\pi_q(f) : \pi_q(X) \to \pi_q(Y)$ is a monomorphism for $q = n$ and an isomorphism for $q > n$.*

Let us first define the groups π_q by

$$\pi_q = \begin{cases} \pi_q(A) & \text{if } q < n, \\ \operatorname{Im} \pi_n(f_0) : \pi_n(A) \to \pi_n(Y) & \text{if } q = n, \\ \pi_q(Y) & \text{if } q > n. \end{cases}$$

Let $X_n = A$, $f_n = f_0 : X_n \to Y$, and let

$$\phi'_q : \pi_q(X_n) \to \pi_q$$

be the identity map ($q < n$), while

$$\psi_n : \pi_n(X_n) \to \pi_n$$

is the homomorphism induced by $\pi_n(f_0)$. Then ϕ'_q is an isomorphism for all $q < n$ and ψ_n is an epimorphism. The argument used to prove Theorem (2.2) can now be repeated almost *verbatim* and yields the desired conclusion.

□

Taking Y to be a point, we obtain

(2.4) Corollary *Let A be a 0-connected space, n a non-negative integer. Then there is a relative* CW-*complex (X, A) such that*

(1) *the injections $\pi_i(A) \to \pi_i(X)$ are isomorphisms for all $i < n$;*
(2) *$\pi_i(X) = 0$ for all $i \geq n$.*

The space X can be obtained from A by adjoining cells of dimensions $\geq n+1$. □

A final application is a relative CW-approximation theorem. Let us say that a map $f: (X, X_0) \to (Y, Y_0)$ is a *weak homotopy equivalence* if and only if the maps $f_1: X \to Y$ and $f_0: X_0 \to Y_0$ defined by f are weak homotopy equivalences. It follows from the Five-Lemma (with the usual modification in low dimensions), that

(2.5) *If $f: (X, X_0) \to (Y, Y_0)$ is a weak homotopy equivalence, then $f_*: \pi_q(X, X_0) \to \pi_q(Y, Y_0)$ is an isomorphism for all $q \geq 1$.* □

Our application is

(2.6) Theorem *Let (X, A) be an arbitrary pair, $g: L \to A$ a CW-approximation. Then there is a CW-pair (K, L) and a weak homotopy equivalence $f: (K, L) \to (X, A)$ extending g. If (X, A) is m-connected, the complex K can be so chosen that its m-skeleton is contained in L.*

By Theorem (3.2) below we can find a CW-complex M and a weak homotopy equivalence $h: M \to X$. Let $i: A \to X$ be the inclusion. Since $h: M \to X$ is a weak homotopy equivalence, there is a map $g_1: L \to M$ such that $h \circ g_1 \simeq i \circ g$. By the cellular approximation theorem, we may assume g_1 to be cellular. Let K be the mapping cylinder of g_1. Then (K, L) is a CW-pair, and, if $p: K \to M$ is the projection, $f_1 = h \circ p$, then $f_1 | L \simeq i \circ g$. Since (K, L) has the homotopy extension property, f_1 is homotopic to a map f such that $f | L = i \circ g$. Then $f(L) \subset A$ and f is the desired map.

Suppose that (X, A) is m-connected. Let us recall from the proof of Theorems (2.1) and (2.2) that the construction of a weak homotopy equivalence $h: M \to X$ is an inductive one; if M_q is a q-dimensional CW-complex and $h_q: M_q \to X$ a q-connected map, then there is a $(q+1)$-dimensional CW-complex M_{q+1}, which is a $(q+1)$-cellular extension of M_q, and a $(q+1)$-connected map $h_{q+1}: M_{q+1} \to X$ extending h_q. In our case, the map $i \circ g: L \to X$ is m-connected and we may start the induction with this map. Thus L is obtained from M by adjoining cells of dimensions $\geq m+1$, and so L and M have the same m-skeleton. □

3 Weak Homotopy Equivalence and CW-approximation

Let us recall from §7 of Chapter IV that a map $f: X \to Y$ is a weak homotopy equivalence if and only if, for every CW-complex K, the induced map

$$\underline{f}: [K, X] \to [K, Y]$$

is a one-to-one correspondence. Taking K to be a single point, it follows that f induces a one-to-one correspondence between the path-components of X and those of Y. Hence, if A is a path-component of X and B is the path-component of Y containing $f(A)$, the restriction of f to A induces a one-to-one correspondence between $[K, A]$ and $[K, B]$ for any connected CW-complex K. It follows easily that $f \mid A: A \to B$ is a weak homotopy equivalence.

Conversely, suppose that a map $f: X \to Y$ induces a one-to-one correspondence between the path-components of X and those of Y, and that the restriction of f to each path-component A of X is a weak homotopy equivalence of A with the corresponding path-component of Y. Then it is easy to see that f is a weak homotopy equivalence.

It follows from these remarks and Lemma (7.12) of Chapter IV that

(3.1) Theorem *A map $f: X \to Y$ is a weak homotopy equivalence if and only if, for each $x \in X$, $f_* : \pi_q(X, x) \to \pi_q(Y, f(x))$ is an isomorphism for all q.* □

It follows from Theorems (3.1) and (2.2) that

(3.2) Theorem *For any space Y, there exists a* CW-*complex X and a weak homotopy equivalence $f: X \to Y$. If Y is m-connected, we may assume that the m-skeleton of X is a single point.* □

The following is easily verified.

(3.3) *Let $f: X \to Y$, $g: Y \to Z$. If any two of the three maps f, g, and $g \circ f: X \to Z$ are weak homotopy equivalences, so is the third.* □

A space X is *weakly contractible* if and only if $\pi_q(X) = 0$ for all q. If P is a space consisting of a single point, then X is weakly contractible if and only if the unique map of X into P is a weak homotopy equivalence. Thus

(3.4) *If X is weakly contractible, then every map of a* CW-*complex K into X is nullhomotopic.* □

A homotopy equivalence is a weak homotopy equivalence. The converse is false, as the example of a circle, into which a one-sided "sin $1/x$" singularity has been introduced, shows (Figure 5.1). (The inclusion map of a point into this space X is a weak homotopy equivalence, which is not a homotopy equivalence because the first Čech cohomology group of X does not vanish). However,

(3.5) Theorem *Let X, Y be* CW-*complexes, $f : X \to Y$ a weak homotopy equivalence. Then f is a homotopy equivalence.*

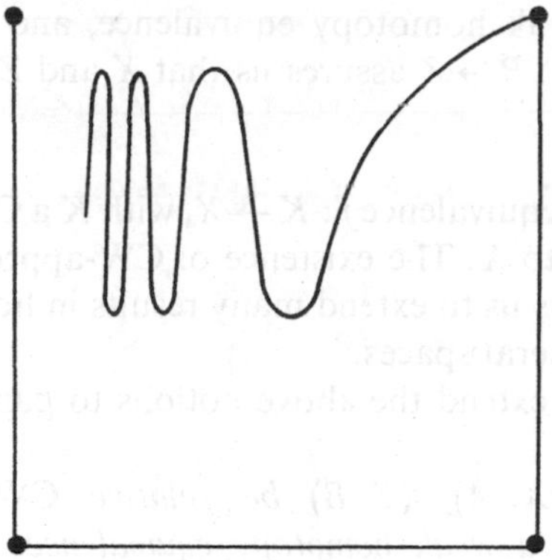

Figure 5.1

For $\underline{f} : [Y, X] \to [Y, Y]$ is a one-to-one correspondence, and therefore there exists a map $f' : Y \to X$ such that $f \circ f'$ is homotopic to the identity map of Y. Moreover, $\underline{f} : [X, X] \to [X, Y]$ is a one-to-one correspondence and $f \circ f' \circ f \simeq 1_Y \circ f = f = f \circ 1_X$; therefore $f' \circ f \simeq 1_X$. Hence f' is a homotopy inverse of f. □

(3.6) Corollary *If X is an m-connected* CW*-complex, then there is a* CW*-complex K, of the same homotopy type as X, whose m-skeleton is a single point.* □

It is tempting to define X and Y to have the same weak homotopy type if and only if there is a weak homotopy equivalence $f : X \to Y$. However, this relation is not symmetric. Hence we must modify this definition a little if we wish to obtain an equivalence relation. Accordingly, we say that *X and Y have the same weak homotopy type* if and only if there is a space P and weak homotopy equivalences $f: P \to X$, $g : P \to Y$.

Remark. We may assume P is a CW-complex. For, by Theorem (3.2), there is a CW-complex K and a weak homotopy equivalence $h : K \to P$. By (3.3), $f \circ h : K \to X$ and $g \circ h : K \to Y$ are weak homotopy equivalences.

(3.7) Theorem *Having the same weak homotopy type is an equivalence relation.*

Only transitivity presents any problem. Accordingly, suppose $f: P \to X$, $g : P \to Y$, $f' : Q \to Y$, $g' : Q \to Z$ are weak homotopy equivalences, P and Q being CW-complexes. Then $\underline{f'} : [P, Q] \to [P, Y]$ is a one-to-one correspondence, and therefore there exists $k : P \to Q$ such that $f' \circ k \simeq g$. Since g and f' are weak homotopy equivalences, so is $k : P \to Q$, by (3.3). Again by (3.3),

$g' \circ k : P \to Z$ is a weak homotopy equivalence, and the existence of the maps $f: P \to X$, $g' \circ k : P \to Z$ assures us that X and Z have the same weak homotopy type. □

A weak homotopy equivalence $f: K \to X$, with K a CW-complex, is called a CW-*approximation* to X. The existence of CW-approximations is a most useful tool, as it allows us to extend many results in homotopy theory from CW-complexes to general spaces.

It will be useful to extend the above notions to pairs. In the first place

(3.8) Theorem *Let* (X, A), (Y, B) *be relative* CW-*complexes and let* $f: (X, A) \to (Y, B)$ *be a weak homotopy equivalence. Suppose further that* $f \mid A : A \to B$ *is a homotopy equivalence. Then* $f: (X, A) \to (Y, B)$ *is a homotopy equivalence of pairs.*

We may assume that f is cellular. Let (Z, C) be the mapping cylinder of f. Since $f \mid A : A \to B$ is a homotopy equivalence, the inclusion map $i_A : A \to C$ is a homotopy equivalence; since, moreover, (C, A) is an NDR-pair, A is a deformation retract of C. Therefore the pair (X, A) is a deformation retract of the pair $(X \cup C, C)$. The projection $p : (Z, C) \to (Y, B)$ induces isomorphisms

$$p_* : \pi_q(Z, C) \approx \pi_q(Y, B),$$

and the inclusion $i : (X, A) \to (Z, C)$ induces homomorphisms

$$i_* : \pi_q(X, A) \to \pi_q(Z, C),$$

such that $p_* \circ i_*$ is the isomorphism f_*. Hence the injection i_* is an isomorphism for all q. Now i_* is the composite of the injections

$$i_{1*} : \pi_q(X, A) \to \pi_q(X \cup C, C),$$
$$i_{2*} : \pi_q(X \cup C, C) \to \pi_q(Z, C).$$

But we have seen that (X, A) is a deformation retract of $(X \cup C, C)$, so that i_{1*} is an isomorphism for all q. Hence i_{2*} is an isomorphism, and therefore $\pi_q(Z, X \cup C) = 0$, for all q. Since $(Z, X \cup C)$ is a relative CW-complex, it follows from Corollary (3.12) of Chapter II that $X \cup C$ is a deformation retract of Z. Hence the pair $(X \cup C, C)$ is a deformation retract of (Z, C), and therefore (X, A) is also a deformation retract of (Z, C). But this implies that the inclusion map $(X, A) \subsetneq (Z, C)$ is a homotopy equivalence. Finally, because the projection $p : (Z, C) \to (Y, B)$ is a homotopy equivalence, so is $f = p \mid (X, A)$. □

(3.9) Corollary *If* (X, A) *and* (Y, B) *are* CW-*pairs and* $f: (X, A) \to (Y, B)$ *a weak homotopy equivalence, then* f *is a homotopy equivalence.* □

(3.10) Theorem *A map $f: (X, A) \to (Y, B)$ is a weak homotopy equivalence if and only if, for every CW-pair (K, L), the map $\underline{f} : [K, L; X, A] \to [K, L; Y, B]$ is a one-to-one correspondence.*

Suppose first that f is a weak homotopy equivalence, and let (Z, C) be its mapping cylinder. Then the projection $p: (Z, C) \to (Y, B)$ is a homotopy equivalence, and therefore the inclusion map $i: (X, A) \subsetneq (Z, C)$ is a weak homotopy equivalence. In particular, the pair (C, A) is ∞-connected.

(3.11) Lemma *Let $(W; P, Q)$ be an NDR-triad, and suppose that the pair $(P, P \cap Q)$ is ∞-connected. Then the pair (W, Q) is ∞-connected.*

It suffices, by Lemma (3.1) of Chapter II, to prove that every map $g: (\Delta^q, \dot{\Delta}^q) \to (W, Q)$ is compressible. By Lemma (2.4) of Chapter II, there is a triangulation $(K; L_1, L_2)$ of Δ^q such that $K = L_1 \cup L_2$, $\dot{\Delta}^q \subset L_2$, and g is homotopic to a map $g' : (K; L_1, L_2) \to (W; P, Q)$. Since $(P, P \cap Q)$ is ∞-connected, $g' | L_1 : (L_1, L_1 \cap L_2) \to (P, P \cap Q)$ is compressible, and therefore $g' : (K, L_2) \to (W, Q)$ is compressible. In particular, $g' : (\Delta^q, \dot{\Delta}^q) \to (W, Q)$ is compressible. □

It follows from Lemma (3.11) that the pair $(X \cup C, X)$ is ∞-connected. Since (Z, X) is ∞-connected, it follows with the aid of Lemma (3.2) of Chapter II that $(Z, X \cup C)$ is ∞-connected. Therefore the triad $(Z; C, X)$ satisfies the conditions of Theorem (3.13) of Chapter II, and hence

(3.12) *If $(K; L_1, L_2)$ is any CW-triad, then every map $f: (K; L_1, L_2) \to (Z; C, X)$ is right compressible.* □

We can now prove Theorem (3.10). Let (K, L) be a CW-pair, $g: (K, L) \to (Y, B)$. Then $j \circ g : (K; L, \varnothing) \to (Z; C, X)$ is right compressible, and therefore there is a map $g' : (K, L) \to (X, A)$ such that $i \circ g' \simeq j \circ g$. Hence $f \circ g' = p \circ i \circ g' \simeq p \circ j \circ g = g$, so that $\underline{f}$ is an epimorphism.

Let $g_0, g_1 : (K, L) \to (X, A)$ and suppose that $f \circ g_0 \simeq f \circ g_1$. Then $p \circ i \circ g_0 \simeq p \circ i \circ g_1$; as p is a homotopy equivalence, $i \circ g_0 \simeq i \circ g_1$. Let $G: (I \times K; I \times L, \dot{I} \times K) \to (Z; C, X)$ be a homotopy of $i \circ g_0$ to $i \circ g_1$. Again G is right compressible, to a map $G' : (I \times K, I \times L) \to (X, A)$, and G' is a homotopy of g_0 to g_1.

Conversely, suppose that $\underline{f} : [K, L; X, A] \to [K, L; Y, B]$ is a one-to-one correspondence for every CW-pair (K, L). In particular, we may take $L = \varnothing$, so that $[K, L; X, A] = [K, X]$, $[K, L; Y, B] = [K, Y]$; thus $\underline{f} : [K, X] \to [K, Y]$ is an isomorphism for every CW-complex K. By Theorem (7.17) of Chapter IV, $f : X \to Y$ is a weak homotopy equivalence. Again, we may take $K = \mathbf{I} \times L$, observing that

$$[\mathbf{I} \times L, 0 \times L; X, A] \approx [L, A]$$

under the map $g \to g|0 \times L : L \to A$. Hence $\underline{f|A} : [L, A] \approx [L, B]$ for every CW-complex L, and, again by Theorem (7.17) of Chapter IV, $f|A$ is a weak homotopy equivalence. Hence $f: (X, A) \to (Y, B)$ is a weak homotopy equivalence. □

A weak homotopy equivalence $f: (K, L) \to (X, A)$, with (K, L) a CW-pair, is called a *CW-approximation.*

(3.13) Theorem *Let $f: (K, L) \to (X, A)$, $f': (K', L') \to (X, A)$ be CW-approximations. Then there is a map $g: (K, L) \to (K', L')$ such that $f' \circ g \simeq f$. The map g is a homotopy equivalence, and g is unique up to homotopy.*

By Theorem (3.10), $\underline{f}' : [K, L; K', L'] \to [K, L; X, A]$ is a one-to-one correspondence, and therefore there is a map $g: (K, L) \to (K', L')$, unique up to homotopy, such that $f' \circ g \simeq f$. Similarly, there is a map $g': (K', L') \to (K, L)$ such that $f \circ g' \simeq f'$. Then $f' \circ g \circ g' \simeq f \circ g' \simeq f'$; since $\underline{f}'$ is one-to-one, $g \circ g' \simeq 1$. Similarly, $g' \circ g \simeq 1$, so that g is a homotopy equivalence with homotopy inverse g'. □

Since the condition for a weak homotopy equivalence can be expressed in terms of the behavior of the homotopy groups, it is natural to ask: suppose that $f: X \to Y$ is a map such that $\underline{f}: [K, X] \to [K, Y]$ is a one-to-one correspondence for every *finite* complex K. Is f a weak homotopy equivalence? The reader is invited to ponder on this question, awaiting the answer which we shall give in §4.

4 Aspherical Spaces

A space X is *aspherical* if and only if it is 0-connected and $\pi_q(X) = 0$ for all $q \geq 2$. If X has a universal covering space $\tilde{X}$, then asphericity of X is equivalent to the weak contractibility of $\tilde{X}$. Thus

(4.1) *If G is a strongly discontinuous group acting on $\mathbf{R}^n$, then the orbit space $\mathbf{R}^n/G$ is aspherical.*

In particular,

(4.2) *Every closed surface, except for the sphere and the projective plane, is aspherical.* □

Thus aspherical spaces are quite common in mathematics.

Let K be a connected CW-complex, X an aspherical space. We now show how to classify the maps of K into X. As in §1 of Chapter III, we first fix base points $k_0 \in K$, $x_0 \in X$, and classify the maps of (K, k_0) into (X, x_0). Such a map f induces a homomorphism $f_* : \pi_1(K) \to \pi_1(X)$, and homotopic maps induce the same homomorphism.

(4.3) Theorem *If K is a connected CW-complex, X an aspherical space, then the correspondence $f \to f_*$ induces a one-to-one correspondence between $[K, k_0; X, x_0]$ and $\mathrm{Hom}(\pi_1(K), \pi_1(X))$.*

By Corollary (3.6), we may assume that the 0-skeleton of K is a single point. We may also assume that the attaching maps for the r-cells of K carry the vertices of Δ^r into the base point k_0. Let $\eta : \pi_1(K) \to \pi_1(X)$. For each 1-cell E_α^1 of X, with characteristic map $h_\alpha : (\Delta^1, \dot{\Delta}^1) \to (K, k_0)$, the map h_α represents an element $\xi_\alpha \in \pi_1(K)$. Let $f_\alpha : (\Delta^1, \dot{\Delta}^1) \to (X, x_0)$ be a map representing the element $\eta(\xi_\alpha) \in \pi_1(X)$. Then there is a map $f_1 : (K_1, k_0) \to (X, x_0)$ such that $f_1 \circ h_\alpha = f_\alpha$. It follows that, if $i : \pi_1(K_1) \to \pi_1(K)$ is the injection, then $f_{1*} = \eta \circ i$.

Suppose that f_1 has been extended to a map $f_r : K_r \to X$. If E_β^{r+1} is an $(r+1)$-cell of K, with characteristic map $h_\beta : (\Delta^{r+1}, \dot{\Delta}^{r+1}) \to (K_{r+1}, K_r)$, then $f_r \circ h_\beta \,|\, \dot{\Delta}^{r+1}$ represents an element $c_\beta \in \pi_r(X, x_0)$. Suppose that $c_\beta = 0$. Then the map $f_r \circ h_\beta \,|\, \dot{\Delta}^{r+1}$ has an extension $g_\beta : \Delta^{r+1} \to X$, and $f_r \,|\, \dot{E}_\beta^{r+1}$ has the extension $g_\beta \circ h_\beta^{-1} : E_\beta^{r+1} \to X$. Therefore, if all the elements c_β are 0, there is a map $f_{r+1} : K_{r+1} \to X$ extending f_r.

If $r > 1$, then $c_\beta = 0$ because $\pi_r(X) = 0$. Suppose that $r = 1$. Then $h_\beta | \dot{\Delta}^2 : \dot{\Delta}^2 \to K_1$ represents an element $\xi \in \pi_1(K_1)$, and $i(\xi) = 0$, because $h_\beta | \dot{\Delta}^2$ has the extension $h_\beta : \Delta^2 \to K$. Then $c_\beta = f_{1*}(\xi) = \eta(i(\xi)) = 0$. Hence $c_\beta = 0$ in all cases.

The maps $f_r : K_r \to X$ fit together to define a map $f : (K, k_0) \to (X, x_0)$, and $f_* \circ i = f_{1*} = \eta \circ i$. Since the injection $i : \pi_1(K_1) \to \pi_1(K)$ is an epimorphism, we have $f_* = \eta$.

Now suppose that $f_0, f_1 : (K, k_0) \to (X, x_0)$ are maps such that $f_{0*} = f_{1*} : \pi_1(K) \to \pi_1(X)$. Let L_r be the subcomplex $0 \times K \cup 1 \times K \cap \mathbf{I} \times K_{r-1}$ of $\mathbf{I} \times K$. Define $F_1 : L_1 \to X$ by

$$\begin{aligned} F_1(0, x) &= f_0(x) \qquad (x \in K); \\ F_1(1, x) &= f_1(x) \qquad (x \in K); \\ F_1(t, k_0) &= x_0 \qquad ((t, k_0) \in \mathbf{I} \times K_{r-1}). \end{aligned}$$

For each 1-cell of E_α^1 of K, its characteristic map h_α represents an element $\xi_\alpha \in \pi_1(K)$, and $f_{0*}(\xi_\alpha) = f_{1*}(\xi_\alpha)$. Hence there is a homotopy F_α of $f_0 \circ h_\alpha$ to $f_1 \circ h_\alpha$ (rel. $\dot{\Delta}_1$). By Lemma (1.3) of Chapter II, there is a homotopy $F_2 : (\mathbf{I} \times K_1, \mathbf{I} \times K_0) \to (X, x_0)$ of $f_0 | K_1$ to $f_1 | K_1$ such that $F_2 \circ (1 \times h_\alpha) = F_\alpha$. Extend F_2 over L_2 by $F_2(t, x) = f_t(x)$ for $x \in K$, $t = 0, 1$. Then F_2 is an extension of F_1.

Suppose that $r \geq 2$, and $F_r : L_r \to X$ is an extension of F_1. For each r-cell E_γ^r of K, the map

$$F_r \circ (1 \times h_\gamma) \,|\, (\mathbf{I} \times \Delta^r)^\bullet$$

represents an element of $\pi_r(X) = 0$. Hence this map has an extension $F_\gamma : \mathbf{I} \times \Delta^r \to X$. Again by Lemma (1.3) of Chapter II, there is a map $F_{r+1} : \mathbf{I} \times K_r \to X$ such that $F_{r+1} \circ (1 \times h_\gamma) = F_\gamma$. The map F_{r+1} extends to a map $F_{r+1} : L_{r+1} \to X$ which clearly extends F_r.

As before, the maps F_r fit together to define a map $F : \mathbf{I} \times K \to X$ which is an extension of F_1 and therefore a homotopy of f_0 to f_1 (rel. k_0). This completes the proof. □

As in §1 of Chapter III, we obtain

(4.4) Corollary *The free homotopy classes of free maps of K into X are in one-to-one correspondence with the conjugacy classes of homomorphisms of $\pi_1(K)$ into $\pi_1(X)$.*

For if $f_0, f_1 : (K, k_0) \to (X, x_0)$, and if $f_0 \underset{u}{\simeq} f_1$ where u is a loop in (X, x_0), then, for all $\xi \in \pi_1(K)$, $f_{1*}(\xi) = [u]^{-1} f_{0*}(\xi)[u]$. □

(4.5) Corollary *Two aspherical spaces having isomorphic fundamental groups have the same weak homotopy type.*

We may assume the two spaces are CW-complexes K, L. Let $\eta : \pi_1(K) \to \pi_1(L)$ be an isomorphism. Then there is a map $f : (K, k_0) \to (L, l_0)$ such that $f_* = \eta$. Moreover, there is a map $g : (L, l_0) \to (K, k_0)$ such that $g_* = \eta^{-1}$. Then $(f \circ g)_* = f_* \circ g_*$ is the identity and therefore $f \circ g$ is homotopic (rel. l_0) to the identity map of L. Similarly, $g \circ f$ is homotopic (rel. k_0) to the identity map of K. □

(4.6) Corollary *If X and Y are aspherical spaces with isomorphic fundamental groups, then $H_q(X; G) \approx H_q(Y; G)$ and $H^q(X; G) \approx H^q(Y; G)$ for any group G.*

This follows from the Whitehead Theorem (7.14) of Chapter IV. It states that the homology and cohomology of an aspherical space depend only on its fundamental group (and on the coefficient group). We now make this dependence explicit.

Let X be an aspherical CW-complex, $\tilde{X}$ its universal covering space, Π the group of covering translations, so that $\Pi \approx \pi_1(X)$. Now $\tilde{X}$ is a CW-complex, and each covering translation is a cellular map; thus Π operates on the chain groups $\Gamma_q(\tilde{X})$, and, as we have seen in §1, $\Gamma_q(\tilde{X})$ is a free $\mathbf{Z}(\Pi)$-module. Since $\tilde{X}$ is acyclic, the $\Gamma_q(\tilde{X})$ form a free resolution of the additive group of integers considered as a $\mathbf{Z}(\Pi)$-module with trivial action. Hence if

G is any right $\mathbf{Z}(\Pi)$-module, we may form the complex $G \otimes_{\mathbf{Z}(\Pi)} \Gamma(\tilde{X})$; its homology groups are known in homological algebra as $\operatorname{Tor}_q^{\mathbf{Z}(\Pi)}(G, Z)$. Similarly, if G is a left $\mathbf{Z}(\Pi)$-module, the homology groups of the cochain complex $\operatorname{Hom}_{\mathbf{Z}(\Pi)}(\Gamma(\tilde{X}), G)$ are $\operatorname{Ext}^q_{\mathbf{Z}(\Pi)}(Z, G)$. These groups are also known under the names $H_q(\Pi; G)$, $H^q(\Pi; G)$, respectively.

Suppose that Π operates trivially on G. Then $G \otimes_{\mathbf{Z}(\Pi)} \Gamma(\tilde{X})$ is isomorphic with $G \otimes \Gamma(X)$ and $\operatorname{Hom}_{\mathbf{Z}(\Pi)}(\Gamma(\tilde{X}), G)$ with $\operatorname{Hom}(\Gamma(X), G)$, and we have

(4.7) Corollary *If X is an aspherical space with fundamental group Π, then, for any abelian group G,*

$$H_q(X; G) \approx \operatorname{Tor}_q^{\mathbf{Z}(\Pi)}(G, \mathbf{Z}),$$
$$H^q(X; G) \approx \operatorname{Ext}^q_{\mathbf{Z}(\Pi)}(\mathbf{Z}, G). \qquad \square$$

If Π acts non-trivially on G, the above results continue to be valid if the ordinary homology and cohomology groups are replaced by those with local coefficients, as we shall see in Chapter VI.

From a historical point of view, the statement of Corollary (4.7) puts the cart before the horse. In fact, Hurewicz [1, IV] proved Corollary (4.6) in 1935, long before the invention of homological algebra; and it was in the effort to make this dependence explicit which led Eilenberg and Mac Lane in the late forties to the homology of groups. Finally, the homology of groups was one of the main ideas which led to the creation of the subject of homological algebra.

We now show that the question propounded at the end of §3 must be answered in the negative. Let G be the group of all permutations of the integers which leave almost all integers fixed, and let H be the subgroup of G consisting of all permutations leaving every negative integer fixed. We shall need two Lemmas.

Lemma 1 *Every finitely generated subgroup of G is conjugate to a subgroup of H.*

For if F is a finite subset of G, there exist integers m, n such that each $\sigma \in F$ fixes all integers not belonging to the closed interval $[-m, n]$. If $m \leq 0$, $F \subset H$. Suppose $m > 0$. Then there is a permutation τ such that

$$\tau(x) = \begin{cases} x & \text{if } x < -m \text{ or } x > m + n, \\ x + m & \text{if } -m \leq x \leq n, \\ x - m - n - 1 & \text{if } n + 1 \leq x \leq m + n. \end{cases}$$

Then $\tau \in G$, and, if $x < 0$, $\sigma \in F$, then either $x = \tau^{-1}(x) < -m$ or $\tau^{-1}(x) > n$. Hence $\sigma\tau^{-1}(x) = \tau^{-1}(x)$, $\tau\sigma\tau^{-1}(x) = x$. Thus $\tau F \tau^{-1} \subset H$. $\square$

Lemma 2 *Let K be a finitely generated subgroup of H, and let ρ be an element of G such that $\rho K \rho^{-1} \subset H$. Then there exists $\alpha \in H$ such that $\alpha^{-1}\rho$ commutes with every element of K.*

For let $\sigma_1, \ldots, \sigma_r$ be a set of generators for K, $\tau_i = \rho\sigma_i\rho^{-1}$. Then there exist non-negative integers $m \leq n$ such that $\sigma_i(x) = \tau_i(x) = x$ for all i unless $m \leq x \leq n$, and, in addition, $\rho(x) = x$ unless $x \leq n$. Let $S = [m, n] \cap \rho^{-1}[m, n]$. Then there is a permutation α such that

$$\alpha(x) = \begin{cases} x & \text{if } x \notin [m, n], \\ \rho(x) & \text{if } x \in S. \end{cases}$$

We next observe that, *if $x \notin S$, then $\sigma_i(x) = x$ for all i*. In fact, if $\sigma_i(x) \neq x$, then $x \in [m, n]$ and $\rho(x) \neq \rho\sigma_i(x) = \tau_i\rho(x)$, and therefore $\rho(x) \in [m, n]$, so that $x \in S$. Similarly, *if $x \notin \rho(S)$, then $\tau_i(x) = x$ for all i*.

Note also that $x \in S$ implies $\sigma_i(x) \in S$. For $x \in [m, n]$, and therefore $\sigma_i(x) \in [m, n]$; while $\rho(x) \in [m, n]$, and $\rho\sigma_i(x) = \tau_i\rho(x) \in [m, n]$, so that $\sigma_i(x) \in S$.

The element $\alpha \in H$ has the desired property. For if $x \notin [m, n]$, then $\alpha\sigma_i(x) = \alpha(x)$ and $\tau_i\alpha(x) = \alpha(x)$. If $x \in S$, then $\sigma_i(x) \in S$ and therefore $\alpha\sigma_i(x) = \rho\sigma_i(x) = \tau_i\rho(x) = \tau_i\alpha(x)$. Finally, if $x \in [m, n] - S$, then $\sigma_i(x) = x$ and therefore $\alpha\sigma_i(x) = \alpha(x)$. Now $\alpha(x) \notin \rho(S)$; for if $\rho^{-1}\alpha(x) \in S$, then $\alpha\rho^{-1}\alpha(x) = \rho\rho^{-1}\alpha(x) = \alpha(x)$, $x = \rho^{-1}\alpha(x) \in S$, a contradiction. Thus $\tau_i\alpha(x) = \alpha(x)$. Therefore $\alpha\sigma_i(x) = \tau_i\alpha(x)$ for every x,

$$\alpha\sigma_i = \tau_i\alpha = \rho\sigma_i\rho^{-1}\alpha$$

so that $\alpha^{-1}\rho$ commutes with every σ_i. □

The Lemmas having been proved, we take X and Y to be aspherical CW-complexes with fundamental groups H, G respectively, and $f: X \to Y$ a map inducing the inclusion map of H into G. We may assume that f itself is an inclusion. Then f is not a homotopy equivalence.

However, let K be a finite complex with fundamental group P, so that P is finitely generated. By Corollary (4.4), $[K, X]$ is in one-to-one correspondence with the conjugacy classes of homomorphisms of P into H, and similarly for $[K, Y]$; and $\underline{f}: [K, X] \to [K, Y]$ is induced by the inclusion map $H \hookrightarrow G$. By Lemma 1, every homomorphism of P into G is conjugate to a homomorphism of P into H. By Lemma 2, two homomorphisms of P into H which are conjugate in G, are already conjugate in H. Hence $\underline{f}: [K, X] \to [K, Y]$ is a one-to-one correspondence.

5 Obstruction Theory

In the proof of Theorem (4.3), we were confronted with the problem of extending a map from a subcomplex of a CW-complex K to the whole complex. This was carried out one skeleton at a time. In extending from one skeleton to the next, we considered each cell separately; the map was already defined on the boundary of the cell, and composition of this map with the attaching map for the cell gave a map of a sphere into the range space, representing a certain element of one of the homotopy groups of the latter space. And the vanishing of this element was necessary and sufficient for the existence of an extension over the cell in question.

In this section we shall formalize this procedure, and use it as a tool for studying the extension problem in general. Let (X, A) be a relative CW-complex, $f: A \to Y$. Then f can be extended over X_0, by defining $f(x)$ arbitrarily for any vertex x of X_0. If E is a 1-cell with characteristic map $h: (\Delta^1, \dot{\Delta}^1) \to (X_1, X_0)$, then $y_0 = fh(\mathbf{e}_0)$ and $y_1 = fh(\mathbf{e}_1)$ are points of Y, and f can be extended over E if and only if they belong to the same path-component of Y. Hence, if Y is 0-connected any map of A into Y can be extended to a map of X_1 into Y. Assuming that f has been extended to a map of X_n into Y, we wish to study the problem of extending from X_n to X_{n+1}.

Accordingly, throughout the rest of this section, we shall assume that (X, A) is a relative CW-complex, n a positive integer, $f: X_n \to Y$ a map. We shall further assume, in order to avoid any considerations involving base points, that the space Y is n-simple, so that any map of an oriented n-sphere into Y represents a well defined element of $\pi_n(Y)$. Because of this hypothesis, the group $\pi_n(Y)$ is abelian, even if $n = 1$.

Let $\{E_\alpha^{n+1} \mid \alpha \in J\}$ be the $(n+1)$-cells of (X, A), with characteristic maps $h_\alpha: (\Delta^{n+1}, \dot{\Delta}^{n+1}) \to (X_{n+1}, X_n)$. Then $f \circ h_\alpha | \dot{\Delta}^{n+1}: \dot{\Delta}^{n+1} \to X_n$ represents an element $c_\alpha \in \pi_n(Y)$. The assignment $e_\alpha \to c_\alpha$ is a function from a basis of the chain group $\Gamma_{n+1}(X, A)$ to the abelian group $\pi_n(Y)$; this function extends to a homomorphism of $\Gamma_{n+1}(X, A)$ into $\pi_n(Y)$, or, what is the same thing, a cochain

$$c^{n+1} = c^{n+1}(f) \in \Gamma^{n+1}(X, A; \pi_n(Y)).$$

This can be thought of as a "local" description of c^{n+1}. We can also give a "global" description, due to J. H. C. Whitehead [8], as follows. Consider the diagram

$$\Gamma_{n+1}(X, A) = H_{n+1}(X_{n+1}, X_n) \xleftarrow{\ \rho\ } \pi_{n+1}(X_{n+1}, X_n) \xrightarrow{\ \partial_*\ } \pi_n(X_n) \xrightarrow{\ f_*\ } \pi_n(Y),$$

where ρ is the Hurewicz map and ∂_* the boundary operator of the homotopy sequence of the pair (X_{n+1}, X_n). By the Relative Hurewicz Theorem, ρ is an epimorphism with kernel $\omega'_{n+1}(X_{n+1}, X_n)$. By Theorem (3.1) of Chapter IV, ∂_* maps $\omega'_{n+1}(X_{n+1}, X_n)$ into $\omega_n(X_n)$. Since $f_*(\omega_n(X_n)) \subset \omega_n(Y) = 0$, we see that $f_* \circ \partial_*$ annihilates the kernel of ρ, and therefore $f_* \circ \partial_* \circ \rho^{-1}$ is a well defined homomorphism, which is readily verified to be the homomorphism c^{n+1} constructed above. The cochain $c^{n+1}(f)$ is called the *obstruction to extending f*.

The following properties of the obstruction cochain are immediate:

(5.1) *For each $(n+1)$-cell E_α^{n+1} of (X, A), $f \mid \dot{E}_\alpha^{n+1}$ can be extended over E_α^{n+1} if and only if $c^{n+1}(e_\alpha^{n+1}) = 0$.* □

(5.2) *The map $f: X_n \to Y$ can be extended over X_{n+1} if and only if $c^{n+1}(f) = 0$.* □

(5.3) *If (X', A') is a relative* CW-*complex and $g: (X', A') \to (X, A)$ a cellular map, then $c^{n+1}(f \circ (g \mid X'_n)) = g^{\#} c^{n+1}(f)$.* □

(5.4) *If Y' is an n-simple space and $h: Y \to Y'$, then $c^{n+1}(h \circ f) = h_* \circ c^{n+1}(f)$.* □

(5.5) *If $f_0 \simeq f_1 : X_n \to Y$, then $c^{n+1}(f_0) = c^{n+1}(f_1)$.* □

Less obvious is

(5.6) Theorem *The obstruction cochain $c^{n+1}(f)$ is a cocycle.*

Consider the commutative diagram

$$
\begin{array}{ccccccc}
H_{n+2}(X_{n+2}, X_{n+1}) & \xleftarrow{\rho_1} & \pi_{n+2}(X_{n+2}, X_{n+1}) & & & & \\
\downarrow{\scriptstyle \partial_1} & & \downarrow{\scriptstyle \partial_2} & & & & \\
H_{n+1}(X_{n+1}) & \xleftarrow{\rho} & \pi_{n+1}(X_{n+1}) & & & & \\
\downarrow{\scriptstyle i_1} & & \downarrow{\scriptstyle i_2} & & & & \\
H_{n+1}(X_{n+1}, X_n) & \xleftarrow[\rho_2]{} & \pi_{n+1}(X_{n+1}, X_n) & \xrightarrow[\partial_3]{} & \pi_n(X_n) & \xrightarrow[f_*]{} & \pi_n(Y)
\end{array}
$$

in which the homomorphisms denoted by ρ are the Hurewicz maps, those denoted by ∂ are boundary operators of the appropriate homology or homotopy sequences, and those denoted by i are injections. Then

$$
\begin{aligned}
(-1)^{n+1}(\delta c^{n+1}) \circ \rho_1 &= (c^{n+1} \circ i_1 \circ \partial_1) \circ \rho_1 \\
&= f_* \circ \partial_3 \circ \rho_2^{-1} \circ i_1 \circ \partial_1 \circ \rho_1 \\
&= f_* \circ \partial_3 \circ \rho_2^{-1} \circ \rho_2 \circ i_2 \circ \partial_2 \\
&= f_* \circ \partial_3 \circ i_2 \circ \partial_2;
\end{aligned}
$$

but i_2 and ∂_3 are consecutive homomorphisms in the homotopy sequence of the pair (X_{n+1}, X_n), and therefore $\partial_3 \circ i_2 = 0$. Hence $(\delta c^{n+1}) \circ \rho_1 = 0$; since ρ_1 is an epimorphism, $\delta c^{n+1} = 0$. □

We shall also need to study the obstructions to extending homotopies.

Let $(\hat{X}, \hat{A}) = \mathbf{I} \times (X, A)$; then $(\hat{X}, \hat{A})$ is a relative CW-complex with $\hat{X}_n = \mathbf{I} \times X_{n-1} \cup \dot{\mathbf{I}} \times X_n$. A map $F : \hat{X}_n \to Y$ consists of a pair of maps f_0, $f_1 : X_n \to Y$, together with a homotopy $G : \mathbf{I} \times X_{n-1} \to Y$ between $f_0 \mid X_{n-1}$ and $f_1 \mid X_{n-1}$. As usual, we regard $\mathbf{I}$ as a CW-complex with two 0-cells, $\{0\}$ and $\{1\}$, and a 1-cell $\mathbf{i}$ with $\partial \mathbf{i} = \{1\} - \{0\}$. The *difference cochain of* (f_0, f_1) *with respect to* G is then defined to be the cochain $d^n = d^n(f_0, G, f_1) = d^n(F) \in \Gamma^n(X, A; \pi_n(Y))$ such that

$$d^n(c) = (-1)^n c^{n+1}(F)(\mathbf{i} \times c)$$

for all $c \in \Gamma_n(X, A)$. An important special case is that in which f_0 and f_1 agree on X_{n-1} and the homotopy G is stationary; in this case we shall abbreviate $d^n(f_0, G, f_1)$ to $d^n(f_0, f_1)$.

Remark. Like the obstruction, the difference cochain has "local" description. If E_α^n is an n-cell of (X, A) with characteristic map h_α, then $\mathbf{I} \times \Delta^n$ is an oriented $(n+1)$-cell whose boundary $(\mathbf{I} \times \Delta^n)^\bullet$ is an oriented n-sphere, and the composite of the map F with the restriction of the map $1 \times h_\alpha$ to this sphere represents $(-1)^n$ times the value of the cochain $d^n(F)$ on the oriented cell e_α^n.

The properties of the difference cochain analogous to (5.1)–(5.5) are equally evident:

(5.1′) *For each n-cell E_α^n of (X, A), $F \mid (\mathbf{I} \times E_\alpha^n)^\bullet$ can be extended over $\mathbf{I} \times E_\alpha^n$ if and only if $d^n(e_\alpha^n) = 0$.* □

(5.2′) *There is a homotopy of f_0 to f_1 extending G if and only if $d^n = 0$.* □

(5.3′) *If (X', A') is a relative CW-complex, $g : (X', A') \to (X, A)$ a cellular map, and if $g' : \hat{X}'_n \to \hat{X}_n$ is the restriction of $1 \times g$, then $d^n(F \circ g') = g^\# d^n(F)$.* □

(5.4′) *If Y' is an n-simple space and $h : Y \to Y'$, then $h_* \circ d^n(F) = d^n(h \circ F)$.* □

(5.5′) *If $F \simeq F' : \hat{X}_n \to Y$, then $d^n(F) = d^n(F')$.* □

The difference cochain is not a cocycle. However, we have a useful coboundary formula:

(5.6′) Theorem *The coboundary of the difference cochain is given by*

$$\delta d^n(f_0, G, f_1) = c^{n+1}(f_1) - c^{n+1}(f_0).$$

In fact, if $c \in \Gamma_{n+1}(X, A)$

$$\text{(5.7)} \qquad \delta d^n(c) = (-1)^n d^n(\partial c) = c^{n+1}(F)(\mathbf{i} \times \partial c).$$

But $c^{n+1}(F)$ is a cocycle, by Theorem (5.6). Thus

(5.8)
$$\begin{aligned} 0 &= (-1)^{n+1}\delta c^{n+1}(F)(\mathbf{i} \times c) = c^{n+1}(F)(\partial(\mathbf{i} \times c)) \\ &= c^{n+1}(F)(1 \times c - 0 \times c - \mathbf{i} \times \partial c) \end{aligned}$$

But clearly

(5.9)
$$c^{n+1}(F)(t \times c) = c^{n+1}(f_t)(c) \qquad (t = 0, 1)$$

and the coboundary formula follows from (5.7)–(5.9). □

Before stating further properties of the difference cochain, we state two Lemmas which will be needed here and elsewhere.

For the first one, let K be a CW-complex whose cells are: one 0-cell $*$, the base point of K; one $(n-1)$-cell S, so that $\dot{S} = *$ and S is an $(n-1)$-sphere; and two n-cells E_0, E_1 with $\dot{E}_0 = \dot{E}_1 = S$. Choose orientations e_0, e_1, s for the n- and $(n-1)$-cells so that $\partial e_0 = \partial e_1 = s$. Then K is an n-sphere and $e_0 - e_1$ is an orientation of K. (If $n = 1$, this description must be modified slightly; there are two 0-cells, $*$ and E^0, and S is the 0-sphere $\{*\} \cup E^0$).

(5.10) Lemma *Let $f: (E_1, *) \to (X, *)$ and let $\alpha \in \pi_n(X)$. Then f has an extension $g: (K, *) \to (X, *)$ representing α.*

For let $g_0 : (K, *) \to (X, *)$ be any representative of α. Since E_1 is contractible, f and $g_0 | E_1$ are homotopic to the constant map of E_1 into X; hence $g_0 | E_1 \simeq f$. By the homotopy extension property, g_0 is homotopic to a map g such that $g | E_1 = f$. □

For the second Lemma, let us enlarge K to a complex L by adjoining still another n-cell E_2 with $\dot{E}_2 = S$, and let e_2 be an orientation of E_2 with $\partial e_2 = s$. Then $K_0 = E_1 \cup E_2$, $K_1 = E_0 \cup E_2$, and $K_2 = E_0 \cup E_1$ are spheres; and $s_0 = e_1 - e_2$, $s_1 = e_0 - e_2$, $s_2 = e_0 - e_1$ are orientations of these spheres.

(5.11) Lemma *Let $f: (L, *) \to (X, *)$, and let $\alpha_i \in \pi_n(X)$ be the element represented by $f | K_i$ with respect to the orientation s_i. Then $\alpha_1 = \alpha_2 + \alpha_0$.*

We shall use the "method of the universal example," observing that if the theorem is true in the special case $L = X, f =$ the identity map, then its truth in general follows from the functorial property of the homotopy groups. Hence we may assume that we are dealing with the special case.

If $n = 1$, the characteristic maps for the one-cells can be regarded as paths in X from the base point $*$ to the other point E^0 of S; let ξ_i be the homotopy class of the path corresponding to the 1-cell E_i. Then $\alpha_0 = \xi_1 \xi_2^{-1}$, $\alpha_1 = \xi_0 \xi_2^{-1}$, $\alpha_2 = \xi_0 \xi_1^{-1}$, so that $\alpha_2 \alpha_0 = \xi_0 \xi_1^{-1} \xi_1 \xi_2^{-1} = \xi_0 \xi_2^{-1} = \alpha_1$. (As always, when dealing with π_1, we have switched from additive to multiplicative notation!).

Now suppose $n \geq 2$. Since L is $(n-1)$-connected, the Hurewicz map $\rho : \pi_n(L) \to H_n(L)$ is an isomorphism. Since $H_n(L_{n-1}) = 0$, the injection $i : H_n(L) \to H_n(L_n, L_{n-1})$ is a monomorphism. Therefore $i \circ \rho : \pi_n(L) \to H_n(L_n, L_{n-1})$ is a monomorphism. But

$$i\rho(\alpha_t) = s_t \qquad (t = 0, 1, 2)$$

and $s_1 = s_2 + s_0$. Hence $\alpha_1 = \alpha_2 + \alpha_0$. □

(5.12) *Let $F_0 : 0 \times X_n \cup \mathbf{I} \times X_{n-1} \to Y$ be a map and let $d \in \Gamma^n(X, A; \pi_n(Y))$. Then F_0 has an extension $F : \hat{X}_n \to Y$ such that $d^n(F) = d$.*

We use the local description of the difference cochain; and we shall use Lemma (5.10) with $K = (\mathbf{I} \times \Delta^n)^\bullet$, $E_0 = 0 \times \Delta^n \cup \mathbf{I} \times \dot{\Delta}^{n-1}$, $E_1 = 1 \times \Delta^n$. For each n-cell of E_α^n of (X, A) with characteristic map h, the composite of F_0 with the restriction of $1 \times h_\alpha$ to E_0 has, by Lemma (5.10), an extension $F_\alpha : K \to X$ representing $(-1)^n d(e_\alpha)$. The maps $F_\alpha \circ (1 \times h_\alpha^{-1}) | (\mathbf{I} \times E_\alpha^n)^\bullet$ are then well-defined and fit together to give the required extension F. □

(5.13) *Let $F', F'' : \hat{X}_n \to Y$ be maps such that $F'(1, x) = F''(0, x)$ for all $x \in X_n$, and let $F : \hat{X}_n \to Y$ be the map such that, for $(t, x) \in \hat{X}_n$*

$$F(t, x) = \begin{cases} F'(2t, x) & (0 \leq t \leq \frac{1}{2}), \\ F''(2t-1, x) & (\frac{1}{2} \leq t \leq 1). \end{cases}$$

Then $d^n(F) = d^n(F') + d^n(F'')$.

Again we use the local description of the difference cochain, and we shall use Lemma (5.11) with $L = (\mathbf{I} \times \Delta^n)^\bullet \cup (\frac{1}{2} \times \Delta^n)$, $E_0 = ([0, \frac{1}{2}] \times \Delta^n)^\bullet$, $E_2 = ([\frac{1}{2}, 1] \times \Delta^n)^\bullet$, $E_1 = \frac{1}{2} \times \Delta^n$. Let $\tilde{X}_n = \hat{X}_n \cup (\frac{1}{2} \times X_n)$; then the map $F : \hat{X}_n \to Y$ has an extension $H : \tilde{X}_n \to Y$ defined by putting $H(\frac{1}{2}, x) = F'(1, x) = F''(0, x)$. Again, let E_α^n be an n-cell of (X, A) with characteristic map h_α, and let $\tilde{h}_\alpha : L \to \tilde{X}_n$ be the restriction of $(1 \times h_\alpha)$ to L. Then it is clear that the elements α_i are given by

$$\alpha_0 = (-1)^n d^n(F'')(e^n),$$
$$\alpha_2 = (-1)^n d^n(F')(e^n),$$
$$\alpha_1 = (-1)^n d^n(F)(e^n),$$

and application of Lemma (5.11) completes the proof. □

(5.14) Theorem *Let $f : X_n \to Y$. Then $f | X_{n-1}$ can be extended over X_{n+1} if and only if $c^{n+1}(f) \sim 0$.*

For suppose that $f \mid X_{n-1}$ has an extension $f' : X_{n+1} \to Y$. Define $F : \hat{X}_n \to Y$ by

$$F(0, x) = f'(x), \qquad F(1, x) = f(x) \qquad (x \in X_n)$$

$$F(t, x) = f(x) = f'(x) \qquad (x \in X_{n-1}, t \in \mathbf{I}).$$

Then, by Theorem (5.6′),

$$\delta d^n(F) = c^{n+1}(f) - c^{n+1}(f' \mid X_n)$$

But $c^{n+1}(f' \mid X_n) = 0$, by (5.2). Hence $c^{n+1}(f)$ is a coboundary.

Conversely, suppose that $c^{n+1}(f) = \delta d$ with $d \in \Gamma^n(X, A; \pi_n(Y))$. By (5.12), there is a map $F : \hat{X}_n \to Y$ such that

$$F(0, x) = f(x) \qquad (x \in X_n),$$

$$F(t, x) = f(x) \qquad (x \in X_{n-1}),$$

and $d^n(F) = -d$. Define $f' : X_n \to Y$ by

$$f'(x) = F(1, x).$$

By Theorem (5.6′),

$$c^{n+1}(f) = \delta d = -\delta d^n(F) = c^{n+1}(f) - c^{n+1}(f').$$

Hence $c^{n+1}(f') = 0$; by (5.2), f' can be extended over X_{n+1}. But $f' \mid X_{n-1} = f \mid X_{n-1}$, and so $f \mid X_{n-1}$ can be extended over X_{n+1}. □

Suppose we are given a map $f: A \to Y$. We wish to determine whether f can be extended over X. If $f_n : X_n \to Y$ is an extension over X_n, then (5.2) gives a necessary and sufficient condition that f_n be extended over X_{n+1}. But there are many different extensions f_n, and we may consider the set $\mathcal{O}'_{n+1}(f)$ of obstructions of all extensions of f over X_n. It follows from Theorem (5.6′) (cf. the proof of Theorem (5.14)) that, if two extensions agree on X_{n-1}, their obstructions are cohomologous; and from Theorem (5.14) that any cocycle cohomologous to an obstruction is itself an obstruction. Thus the set $\mathcal{O}'_{n+1}(f)$ is a union of cohomology classes; thus it may be regarded as a subset $\mathcal{O}_{n+1}(f)$ of the cohomology group $H^{n+1}(X, A; \pi_n(Y))$; $\mathcal{O}_{n+1}(f)$ is called the $(n + 1)$-*dimensional obstruction set* of f. And we have the formal statement: a map $f: A \to Y$ can be extended over X_{n+1} if and only if $0 \in \mathcal{O}_{n+1}(f)$. Of course, this formal statement is worthless unless we can give some kind of concrete description of $\mathcal{O}_{n+1}(f)$.

Theorem (5.14) may be regarded as the first of a sequence of theorems, of the form: "$f \mid X_{n-r}$ can be extended over X_{n+1} if and only if ...". But such an infinite set of theorems is out of sight at this time. We shall return to this subject later.

In the next section we shall make simplifying assumptions on (X, A) and Y which allow us to give a solution to the problem in certain very important special cases. Before doing so, however, we shall formulate a homotopy theorem analogous to the extension theorem (5.14).

Let $f_0, f_1 : X \to Y$ be maps and let $G : \mathbf{I} \times A \to Y$ be a homotopy of $f_0 \mid A$ to $f_1 \mid A$. In our discussion of the difference cochain, we were concerned with maps of X_n into Y, and we considered the product complex $\mathbf{I} \times (X, A)$. Here, on the other hand, we are dealing with maps of X into Y, and it is appropriate to consider instead the product complex

$$\begin{aligned}(X^*, A^*) &= (\mathbf{I}, \dot{\mathbf{I}}) \times (X, A)\\ &= (\mathbf{I} \times X, \mathbf{I} \times A \cup \dot{\mathbf{I}} \times X)\end{aligned}$$

whose cells are exactly those of the form $\mathbf{I} \times E^n_\alpha$ for E^n_α an n-cell of (X, A). Then $X^*_n = \mathbf{I} \times X_{n-1} \cap \dot{\mathbf{I}} \times X$. In this case, the map $c \to \mathbf{i} \times c$ is a chain map, of degree 1, which is an isomorphism between the chain complexes of (X, A) and (X^*, A^*). And, if $\mathbf{i}^*$ is the integral 1-cochain of $(\mathbf{I}, \dot{\mathbf{I}})$ whose value on the 1-cell $\mathbf{i}$ is $+1$, then the correspondence $c \to \mathbf{i}^* \times c$ is an isomorphism, of degree $+1$, between their cochain complexes. Thus, in the present case, the difference cochain is a *cocycle* which corresponds to the obstruction cocycle under this isomorphism. We can then apply the extension theory to this special case.

(5.15) Theorem *Let $f_0, f_1 : X \to Y$, be maps such that $f_0 \mid X_{n-1} = f_1 \mid X_{n-1}$. Then*

(1) *$d^n(f_0, f_1)$ is an n-cocycle of (X, A) with coefficients in $\pi_n(Y)$;*
(2) *$d^n(f_0, f_1) = 0$ if and only if $f_0 \mid X_n \simeq f_1 \mid X_n$ (rel. X_{n-1});*
(3) *$d^n(f_0, f_1) \backsim 0$ if and only if $f_0 \mid X_n \simeq f_1 \mid X_n$ (rel. X_{n-2}).*

The proof involves no new ideas and will be left to the reader. □

6 Homotopy Extension and Classification Theorems

Let (X, A) be a relative CW-complex, Y an $(n-1)$-connected space ($n \geq 1$; if $n = 1$, we assume $\pi_1(Y)$ to be abelian, so that Y is 1-simple; if $n > 1$, Y is automatically n-simple). Finally, let $f: A \to Y$ be a map.

(6.1) Theorem *The map $f: A \to Y$ can be extended to a map $g : X_n \to Y$. If $g_0, g_1 : X_n \to Y$ are extensions of f, then $g_0 \mid X_{n-1} \simeq g_1 \mid X_{n-1}$ (rel. A), and $c^{n+1}(g_0) \backsim c^{n+1}(g_1)$.*

We have seen that f can be extended over X_1. Suppose that $h : X_r \to Y$ is an extension of f $(1 \leq r \leq n-1)$. Then $c^{r+1}(h) \in \Gamma^{r+1}(X, A; \pi_r(Y))$ and

$\pi_r(Y) = 0$ because Y is $(n-1)$-connected. Hence h can be extended over X_{r+1}. This gives an inductive proof of the first statement.

Let $g_0, g_1 : X_n \to Y$ be extensions of f. Then we may apply the result already proved, to the relative CW-complex $(\mathbf{I}, \dot{\mathbf{I}}) \times (X_n, A)$, and the map of $\mathbf{I} \times A \cup \dot{\mathbf{I}} \times X_n$ defined by g_0, g_1 and the stationary homotopy of $g_0 | A = f$, to deduce the existence of an extension of this map to a map F of the n-skeleton $\mathbf{I} \times X_{n-1} \cup \dot{\mathbf{I}} \times X_n = \hat{X}_n$ into Y. But this extension just defines a homotopy (rel. A) between $g_0 | X_{n-1}$ and $g_1 | X_{n-1}$. Applying Theorem (5.6′) to the map F, we find that $c^{n+1}(g_0) \backsim c^{n+1}(g_1)$. □

It follows from Theorem (6.1) that there is a uniquely defined cohomology class $\gamma^{n+1}(f) \in H^{n+1}(X, A; \pi_n(Y))$; it is the cohomology class of $c^{n+1}(g)$ for any extension $g : X_n \to Y$ of f, and is called the *primary obstruction* to extending f.

We next give a description of $\gamma^{n+1}(f)$ in terms of standard operations of homology theory and a suitable universal class. This class, the *characteristic class* $\iota^n(Y) \in H^n(Y; \pi_n(Y))$ of the $(n-1)$-connected space Y, is defined as follows. By the Universal Coefficient Theorem and the fact that $H_{n-1}(Y)$ is free (it is zero if $n > 1$ and infinite cyclic if $n = 1$), there is a natural isomorphism

$$H^n(Y; \pi_n(Y)) \approx \operatorname{Hom}(H_n(Y), \pi_n(Y)).$$

The class $\iota^n(Y)$ is then defined to be that class which corresponds under the above isomorphism, to the inverse of the Hurewicz isomorphism $\rho : \pi_n(Y) \approx H_n(Y)$.

Suppose, for example, that Y is a CW-complex. Since Y is $(n-1)$-connected, we may assume that Y_{n-1} is a single point $*$. If E_α^n is an n-cell of Y with characteristic map h_α, then $h_\alpha : (\Delta^n, \dot{\Delta}^n) \to (Y, *)$ represents an element $u_\alpha \in \pi_n(Y)$. The cochain $u \in \Gamma^n(Y; \pi_n(Y))$ such that $u(e_\alpha) = u_\alpha$ for all u is then a cocycle whose cohomology class is $\iota^n(Y)$.

Now consider the diagram

$$H^n(Y; \Pi) \xrightarrow{f^*} H^n(A; \Pi) \xrightarrow{\delta^*} H^{n+1}(X, A; \Pi)$$

where $\Pi = \pi_n(Y)$ and δ^* is the coboundary operator. We now prove

(6.2) Theorem *The primary obstruction $\gamma^{n+1}(f)$ to extending f is given by*

$$\gamma^{n+1}(f) = (-1)^n \delta^* f^* \iota^n(Y).$$

Let $g : X_n \to Y$ be an extension of f. Then there is a commutative diagram

$$
\begin{array}{ccccc}
H^n(Y;\Pi) & \xrightarrow{f^*} & H^n(A;\Pi) & \xrightarrow{\delta^*} & H^{n+1}(X,A;\Pi) \\
& \searrow^{g^*} & \uparrow{\scriptstyle i_1} & & \uparrow{\scriptstyle i_2} \\
& & H^n(X_n;\Pi) & \xrightarrow{\delta_1} & H^{n+1}(X,X_n;\Pi) \\
& & & \searrow^{\delta_2} & \downarrow{\scriptstyle j_1} \\
& & & & H^{n+1}(X_{n+1},X_n;\Pi)=\Gamma^{n+1}(X,A;\Pi)
\end{array}
$$

Now j_1 is a monomorphism and its image is the group $Z^{n+1}(X, A; \Pi)$ of cocycles. Moreover, $i_2 \circ j_1^{-1} : Z^{n+1}(X, A; \Pi) \to H^{n+1}(X, A; \Pi)$ is the natural projection from the group of cocycles to the cohomology group.

We claim that

(6.3) $$\delta_2 g^* \iota^n(Y) = (-1)^n c^{n+1}(g).$$

If this is the case, we have

$$
\begin{aligned}
(-1)^n \gamma^{n+1}(f) &= (-1)^n i_2 j_1^{-1} c^{n+1}(g) \\
&= i_2 j_1^{-1} \delta_2 g^* \iota^n(Y) \\
&= i_2 j_1^{-1} j_1 \delta_1 g^* \iota^n(Y) = i_2 \delta_1 g^* \iota^n(Y) \\
&= \delta^* i_1 g^* \iota^n(Y) = \delta^* f^* \iota^n(Y).
\end{aligned}
$$

To prove (6.3), consider the commutative diagram (Figure 5.2).

$$
\begin{array}{ccccc}
H^n(Y;\Pi) & \xrightarrow{g^*} & H^n(X_n;\Pi) & \xrightarrow{\delta_2} & H^{n+1}(X_{n+1},X_n;\Pi) \\
\downarrow{\scriptstyle \beta_1} & & \downarrow{\scriptstyle \beta_2} & & \downarrow{\scriptstyle \beta_3} \\
\mathrm{Hom}(H_n(Y),\Pi) & \xrightarrow{g_1^*} & \mathrm{Hom}(H_n(X_n),\Pi) & \xrightarrow{\delta_3} & \mathrm{Hom}(H_{n+1}(X_{n+1},X_n),\Pi) \\
\downarrow{\scriptstyle \rho_1^*} & & \downarrow{\scriptstyle \rho_2^*} & & \downarrow{\scriptstyle \rho_3^*} \\
\mathrm{Hom}(\pi_n(Y),\Pi) & \xrightarrow{g_2^*} & \mathrm{Hom}(\pi_n(X_n),\Pi) & \xrightarrow{\delta_4} & \mathrm{Hom}(\pi_{n+1}(X_{n+1},X_n),\Pi)
\end{array}
$$

Figure 5.2

The bottom half of Figure 5.2 is obtained by applying the contravariant functor Hom(, Π) to the commutative diagram

$$\begin{array}{ccccc}
H_n(Y) & \xleftarrow{g_1} & H_n(X_n) & \xleftarrow{(-1)^n\partial_3} & H_{n+1}(X_{n+1}, X_n) \\
\uparrow{\scriptstyle \rho_1} & & \uparrow{\scriptstyle \rho_2} & & \uparrow{\scriptstyle \rho_3} \\
\pi_n(Y) & \xleftarrow[g_2]{} & \pi_n(X_n) & \xleftarrow[(-1)^n\partial_4]{} & \pi_{n+1}(X_{n+1}, X_n)
\end{array}$$

in which the ρ_i are the Hurewicz maps, and g_1 and g_2 are induced by g. The homomorphisms β_i are those given by the Universal Coefficient Theorem. Now $\iota^n(Y)$ is defined by the relation

$$\beta_1 \iota^n(Y) = \rho^{-1}$$

and therefore

$$\rho_1^* \beta_1 \iota^n(Y) = 1,$$

the identity homomorphism of $\pi_n(Y) = \Pi$. Hence

$$\delta_4 g_2^* \rho_1^* \beta_1 \iota^n(Y) = \delta_4 g_2^*(1) = (-1)^{n+1} g_2 \circ \partial_4.$$

The obstruction $c^{n+1}(g)$ is defined by the condition

$$\beta_3 c^{n+1}(g) = g_2 \circ \partial_4 \circ \rho_3^{-1},$$

i.e.,

$$\rho_3^* \beta_3 c^{n+1}(g) = g_2 \circ \partial_4.$$

Hence

$$\begin{aligned}
(-1)^n \rho_3^* \beta_3 c^{n+1}(g) &= \delta_4 g_2^* \rho_1^* \beta_1 \iota^n(Y) \\
&= \rho_3^* \beta_3 \delta_2 g^* \iota^n(Y),
\end{aligned}$$

and the fact that β_3 is an isomorphism and ρ_3^* a monomorphism (since ρ_3 is an epimorphism) allows us to deduce (6.3). □

(6.4) Corollary *A map $f\colon A \to Y$ can be extended over X_{n+1} if and only if $\gamma^{n+1}(f) = \delta^* f^* \iota^n(Y) = 0$.* □

The following consequence is due to H. Hopf, although its formulation in cohomological terms is due to H. Whitney:

(6.5) Corollary (Hopf–Whitney Extension Theorem). *If $\dim(X, A) \leq n + 1$, a map $f\colon A \to Y$ can be extended over X if and only if $\delta^* f^* \iota^n(Y) = 0$.* □

We now replace the condition "$\dim(X, A) \leq n + 1$" by a sequence of conditions on (X, A) and on Y which guarantee that there are no further obstructions to extending over all of X. This generalization is due to S. Eilenberg.

(6.6) Theorem (Eilenberg Extension Theorem). *Suppose that Y is q-simple and that $H^{q+1}(X, A; \pi_q(Y)) = 0$ whenever $n + 1 \leq q < \dim(X, A)$. Then $f: A \to Y$ can be extended over X if and only if $\gamma^{n+1}(f) = 0$.*

For we have seen that f has an extension $f_{n+1} : X_{n+1} \to Y$. Then we can define inductively a sequence of maps $f_q : X_q \to Y$ such that $f_{q+1} | X_{q-1} = f_q | X_{q-1}$ for all $q \geq n + 1$. For suppose that $f_q : X_q \to Y$ is an extension of f. Then $c^{q+1}(f_q) \in H^{q+1}(X, A; \pi_q(Y)) = 0$ and therefore $f_q | X_{q-1}$ has an extension $f_{q+1} : X_{q+1} \to Y$. A map $f: X \to Y$ is then defined by $f | X_q = f_{q+1} | X_q$ for all $q \geq n + 1$. □

Remark. By exactness of the cohomology sequence of the pair (X, A), the condition $\delta^* f^* \iota^n(Y) = 0$ is equivalent to the condition: "there exists $u \in H^n(X; \Pi)$ such that $i^* u = f^* \iota^n(Y)$" where $i^* : H^n(X; \Pi) \to H^n(A; \Pi)$ is the injection. Consider the diagram

(6.7)

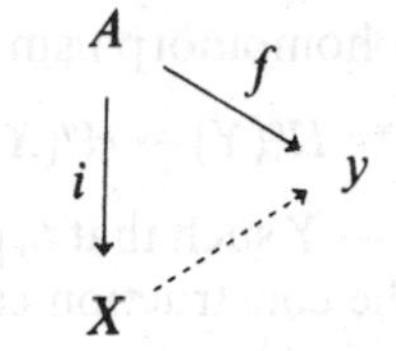

and apply the functor $H^n(\ ; \Pi)$ to obtain the diagram

(6.8)

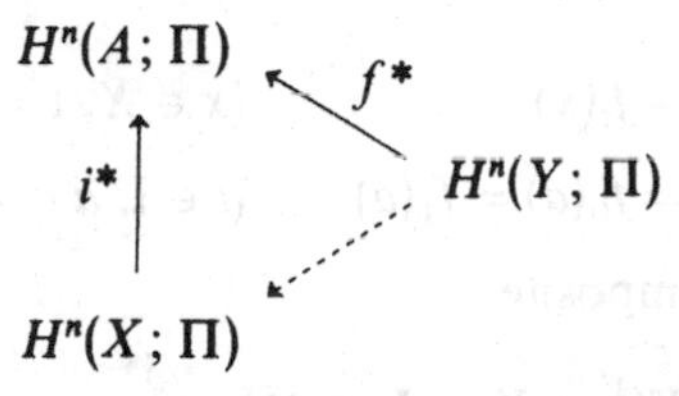

(recall the remarks in §1, Chapter I). The Eilenberg Extension Theorem tells us that, in the case at hand, the diagram (6.7) can be filled in if and only if the diagram (6.8) can. For if $g \circ i = f$, then $i^* \circ g^* = f^*$. But if $\phi : H^n(Y; \Pi) \to H^n(X; \Pi)$ is a homomorphism such that $i^* \circ \phi = f^*$, then $f^* \iota^n(Y) = i^* \phi \iota^n(Y)$, and so the extension exists.

(6.9) Corollary *If Y is an n-simple space and $\pi_i(Y) = 0$ for all $i \neq n$, then a map $f: A \to Y$ can be extended over X if and only if $\delta^* f^* \iota^n(Y) = 0$.* □

A space Y satisfying the hypothesis of the Corollary is called an *Eilenberg–Mac Lane space of type* (Π, n). We shall return to the study of these spaces in §7; they are of crucial importance for the further developments in homotopy theory. (An Eilenberg–Mac Lane space of type $(\Pi, 1)$ is just an aspherical space with abelian fundamental group).

We now apply the above considerations to the homotopy problem. Let $f_0, f_1 : X \to Y$ be two maps such that $f_0 | A = f_1 | A$. With the aid of the stationary homotopy between $f_0 | A$ and $f_1 | A$, they define a map $F : A^* = \dot{\mathbf{I}} \times X \cup \mathbf{I} \times A \to Y$, and we have the cohomology class $\gamma^{n+1}(F) \in H^{n+1}(X^*, A^*; \Pi)$. In §5 we saw that the cross product with $\mathbf{i}^* \in Z^1(\mathbf{I}, \dot{\mathbf{I}}; \mathbf{Z})$ induced an isomorphism, of degree $+1$, of the chain complex of (X, A) with that of (X^*, A^*), and therefore an isomorphism

$$\mathbf{i}^* \times : H^n(X, A; \Pi) \approx H^{n+1}(X^*, A^*; \Pi).$$

Let $\delta^n(f_0, f_1) \in H^n(X, A; \Pi)$ be the class defined by

$$\textbf{(6.10)} \qquad (-1)^n \mathbf{i}^* \times \delta^n(f_0, f_1) = \gamma^{n+1}(F);$$

$\delta^n(f_0, f_1)$ is the cohomology class of $d^n(F_n)$ for any extension $F_n : X_n^* \to Y$ of F. It is called the *primary obstruction to deforming f_0 into f_1*.

A description of $\delta^n(f_0, f_1)$ parallel to our description of $\gamma^{n+1}(f)$ can now be given. Instead of the homomorphism $f^* : H^q(Y) \to H^q(A)$ induced by a map $f : A \to Y$, we shall need a homomorphism

$$(f_0, f_1)^* : H^q(Y) \to H^q(X, A)$$

induced by two maps $f_0, f_1 : X \to Y$ such that $f_0 | A = f_1 | A$. We shall assume that (X, A) is a CW-pair, but the construction can be made for an arbitrary pair (X, A) (cf. Exercises 2, 3 below).

To construct the homomorphism $(f_0, f_1)^*$, let $F : \dot{\mathbf{I}} \times X \cup \mathbf{I} \times A \to Y$ be the map such that

$$F(t, x) = f_t(x) \qquad (x \in X, t = 0, 1),$$
$$F(t, a) = f_0(a) = f_1(a) \qquad (t \in \mathbf{I}, a \in A).$$

Then $(f_0, f_1)^*$ is the composite

$$H^q(Y) \xrightarrow{F^*} H^q(\dot{\mathbf{I}} \times X \cup \mathbf{I} \times A) \xrightarrow{\delta^*}$$
$$H^{q+1}(\mathbf{I} \times X, \dot{\mathbf{I}} \times X \cup \mathbf{I} \times A) \xrightarrow{(\mathbf{i}^* \times)^{-1}} H^q(X, A)$$

where δ^* is the coboundary operator and $\mathbf{i}^* \times$ the isomorphism of Chapter II, Theorem (2.34).

(6.11) Theorem *The operation $(f_0, f_1)^*$ has the following properties*:

(1) $(f_0, f_0)^* = 0$;
(2) $(f_0, f_1)^* + (f_1, f_2)^* = (f_0, f_2)^*$;
(3) $(f_1, f_0)^* = -(f_0, f_1)^*$;
(4) *If $j^* : H^q(X, A) \to H^q(X)$ is the injection, then*

$$j^* \circ (f_0, f_1)^* = f_1^* - f_0^*;$$

(5) *If $g : Y \to Y'$, then*

$$(g \circ f_0, g \circ f_1)^* = (f_0, f_1)^* \circ g^*;$$

(6) *If* $h:(X',A')\to(X,A)$ *is cellular, then*

$$(f_0\circ h, f_1\circ h)^* = h^*\circ(f_0,f_1)^*.$$

To prove (1), observe that, if $f_0=f_1$, the map $F:\dot{\mathbf{I}}\times X\cup\mathbf{I}\times A\to Y$ has a (stationary) extension $F':\mathbf{I}\times X\to Y$, and therefore $F^*=j^*\circ F'^*$, where $h^*:H^q(\mathbf{I}\times X)\to H^q(\dot{\mathbf{I}}\times X\cup\mathbf{I}\times A)$ is the injection. Hence, if $u\in H^q(Y)$,

$$\mathbf{i}^*\times(f_0,f_0)^*u=\delta^*F^*u=\delta^*j^*F'^*u=0,$$

since $\delta^*\circ j^*=0$. Since $\mathbf{i}^*\times$ is an isomorphism, $(f_0,f_0)^*u=0$.

To prove (2), let $I_0=[0,\frac{1}{2}]$, $I_1=[\frac{1}{2},1]$, and observe that by the Direct Sum Theorem, the appropriate injections induce an isomorphism

$$\textbf{(6.12)}\quad H^{q+1}(\mathbf{I}\times X,\dot{\mathbf{I}}\times X\cup\{\tfrac{1}{2}\}\times X\cup\mathbf{I}\times A)\approx$$
$$H^{q+1}(I_0\times X,\dot{I}_0\times X\cup I_0\times A)\times H^{q+1}(I_1\times X,\dot{I}_1\times X\cup I_1\times A).$$

The three maps $f_i:X\to Y$ induce a map

$$G:\dot{\mathbf{I}}\times X\cup\{\tfrac{1}{2}\}\times X\cup\mathbf{I}\times A\to Y$$

by

$$G(t,x)=f_{2t}(x)\qquad (t=0,\tfrac{1}{2},1;\ x\in X),$$
$$G(t,a)=f_0(a)=f_1(a)=f_2(a)\qquad (t\in\mathbf{I},\ a\in A).$$

Clearly, if $u\in H^q(Y)$, then $\delta^*G^*(u)$ corresponds, under the isomorphism (6.12) to the pair of elements $(i_0^*\times(f_0,f_1)^*u,\ i_1^*\times(f_1,f_2)^*u)$. Moreover, if

$$j^*:H^{q+1}(\mathbf{I}\times X,\mathbf{I}\times X\cup\{\tfrac{1}{2}\}\times X\cup\mathbf{I}\times A)\to$$
$$H^{q+1}(\mathbf{I}\times X,\dot{\mathbf{I}}\times X\cup\mathbf{I}\times A)$$

is the injection, $j^*\delta^*G^*(u)=\delta^*(G|\mathbf{I}\times X)^*u=\mathbf{i}^*\times(f_0,f_2)^*u$.

Let us make the isomorphism (6.12) more explicit. Consider the diagram

$$\begin{array}{ccc}
H^{q+1}(\mathbf{I}\times X,I_1\times X\cup\dot{I}\times X\cup\mathbf{I}\times A) & \xrightarrow{k_0^*} & H^{q+1}(I_0\times X,\dot{I}_0\times X\cup I_0\times A)\\
 & \searrow^{j_0^*}\qquad\qquad {}^{l_0^*}\swarrow & \\
\textbf{(6.13)}\qquad & H^{q+1}(\mathbf{I}\times X,\dot{\mathbf{I}}\times X\cup\{\tfrac{1}{2}\}\times X\cup\mathbf{I}\times A) & \\
 & \swarrow^{j_1^*}\qquad\qquad {}^{l_1^*}\searrow & \\
H^{q+1}(\mathbf{I}\times X,I_0\times X\cup\dot{\mathbf{I}}\times X\cup\mathbf{I}\times A) & \xrightarrow[k_1^*]{} & H^{q+1}(I_1\times X,\dot{I}_1\times X\cup I_1\times A)
\end{array}$$

in which k_0^* and k_1^* are isomorphisms by the Excision Theorem. The isomorphism (6.12) is induced by l_0^* and l_1^*, while j_0^*, j_1^* represent the middle group as a direct sum. In fact, the representations (l_0^*, l_1^*) and (j_0^*, j_1) are weakly dual to each other.

Let us consider first the special case $(X, A) = (P, \emptyset)$, where P is a single point. The group $H^1(I_t, \dot{I}_t)$ is an infinite cyclic group with generator i_t^* $(t = 0, 1)$, and therefore $H^1(\mathbf{I}, \dot{\mathbf{I}} \cup \{\frac{1}{2}\})$ is a free abelian group of rank 2, with basis

$$\tilde{i}_t^* = j_0^* k_0^{*-1}(i_t^*).$$

Moreover, the injection $H^1(\mathbf{I}, \dot{\mathbf{I}} \cup \{\frac{1}{2}\}) \to H^1(\mathbf{I}, \dot{\mathbf{I}})$ maps each of the two elements $\tilde{i}_0^*, \tilde{i}_1^*$ upon the generator $i^* \in H^1(\mathbf{I}, \dot{\mathbf{I}})$.

Since each of the injections of (6.13) is induced by a map of the form $j \times 1$, where j is an inclusion and 1 the identity map of (X, A), it follows that

$$\begin{aligned} \delta^* G^* u &= \tilde{i}_0^* \times (f_0, f_1)^* u + \tilde{i}_1^* \times (f_1, f_2)^* u, \\ j^* \delta^* G^* i &= i^* \times (f_0, f_1)^* u + i^* \times (f_1, f_2)^* u \\ &= i^* \times \{(f_0, f_1)^* u + (f_1, f_2)^* u\}. \end{aligned}$$

But we have seen that

$$j^* \delta^* G^* u = i^* \times (f_0, f_2)^* u,$$

and (2) follows from the fact that $\mathbf{i}^* \times$ is an isomorphism. Clearly (3) follows from (1) and (2).

To prove (4), note that the injection maps $F^*u \in H^q(\dot{\mathbf{I}} \times X \cup \mathbf{I} \times A)$ into $0^* \times f_0^* u + 1^* \times f_1^* u \in H^q(\dot{\mathbf{I}} \times X)$, where 0^* and 1^* are the cohomology classes dual to the homology classes of the points $0, 1 \in H_0(\dot{I})$. But $\delta^*(1^*) = -\delta^*(0^*) = \mathbf{i}^* \in H^1(\mathbf{I}, \dot{\mathbf{I}})$, and therefore δ^* maps the above class into $\mathbf{i}^* \times (f_1^* u - f_0^* u)$. Commutativity of the diagram

$$\begin{array}{ccccc}
H^q(Y) & & & & \\
\downarrow{\scriptstyle F^*} & & & & \\
H^q(\dot{\mathbf{I}} \times X \cup \mathbf{I} \times A) & \xrightarrow{\delta^*} & H^{q+1}(\mathbf{I} \times X, \dot{\mathbf{I}} \times X \cup \mathbf{I} \times A) & \xleftarrow{i^* \times} & H^q(X, A) \\
\downarrow{\scriptstyle j^*} & & \downarrow{\scriptstyle j^*} & & \downarrow{\scriptstyle j^*} \\
H^q(\dot{\mathbf{I}} \times X) & \xrightarrow[\delta^*]{} & H^{q+1}(\mathbf{I} \times X, \dot{\mathbf{I}} \times X) & \xrightarrow[i^* \times]{} & H^q(X)
\end{array}$$

completes the proof.

The proofs of (5) and (6) are easy and are left to the reader. □

It follows from the definition of $(f_0, f_1)^*$, (6.10), and Theorem (6.2) that

(6.14) $$\delta^n(f_0, f_1) = (-1)^n (f_0, f_1)^* \iota^n(Y).$$

We now have

(6.15) Theorem (Eilenberg Homotopy Theorem). *Suppose that Y is q-simple and that $H^q(X, A; \pi_q(Y)) = 0$ for all q such that $n+1 \leq q < 1 + \dim(X, A)$. Then $f_0 \simeq f_1$* (rel. A) *if and only if $(f_0, f_1)^* \iota^n(Y) = 0$.* □

(6.16) Corollary *The above conclusion holds if either*

(a) $\dim(X, A) \leq n$,

or

(b) *Y is an Eilenberg–Mac Lane space of type* (Π, n). □

Combining the above results, we obtain

(6.17) Theorem (Eilenberg Classification Theorem). *Suppose that*

(1) *Y is q-simple for $n+1 \leq q < 1 + \dim(X, A)$,*
(2) *$H^q(X, A; \pi_q(Y)) = 0$ for $n+1 \leq q < 1 + \dim(X, A)$,*
(3) *$H^{q+1}(X, A; \pi_q(Y)) = 0$ for $n+1 \leq q < \dim(X, A)$.*

Let $f_0 : X \to Y$ be a map. Then the correspondence $f \to (f_0, f)^* \iota^n(Y)$ induces a one-to-one correspondence between the homotopy classes (rel. A) *of extensions of $f_0 | A$ and the group $H^n(X, A; \Pi)$.*

Suppose that $(f_0, f)^* \iota^n(Y) = (f_0, g)^* \iota^n(Y)$. Then by Theorem (6.11), part (2), $(f, g)^* \iota^n(Y) = 0$. By Theorem (6.15), $f \simeq g$ (rel. A).

Let $z \in H^n(X, A; \Pi)$, and let $d \in Z^n(X, A; \Pi)$ be a representative cocycle. Let $F_0 : 0 \times X_n \cup \mathbf{I} \times X_{n-1} \to Y$ be the map such that

$$F_0(0, x) = f_0(x) \qquad (x \in X_n),$$

$$F_0(t, x) = f_0(x) \qquad (x \in X_{n-1}).$$

By (5.12), F_0 has an extension $F : \hat{X}_n \to Y$ such that $d^n(F) = d$. Let $f(x) = F(1, x)$ for $x \in X_n$. Then, by Theorem (5.6′),

$$0 = \delta d = \delta d^n(F_0) = c^{n+1}(f) - c^{n+1}(f_0 | X_n)$$
$$= c^{n+1}(f)$$

because $f_0 | X_n$ has the extension f_0. Therefore f has an extension over X_{n+1}. Because of (3), f_n can be extended to a map $f_1 : X \to Y$. Since $d^n(F_0) = d$, we have $z = \delta^n(f_0, f_1) = (-1)^n (f_0, f_1)^* \iota^n(Y)$. Hence every class can be realized. □

In particular, we may take $A = \varnothing$, f_0 a constant map of X into Y, to obtain

(6.18) Corollary *If $A = \varnothing$ and the hypotheses* (1)–(3) *of Theorem* (6.17) *are*

satisfied, then the correspondence $f \to f^*\iota^n(Y)$ *induces a one-to-one correspondence between* $[X, Y]$ *and* $H^n(X; \Pi)$. □

(6.19) Corollary (Hopf–Whitney Classification Theorem). *The homotopy classes of maps of an n-dimensional* CW-*complex* X *into an* $(n-1)$*-connected n-simple space* Y *are in one-to-one correspondence with the group* $H^n(X; \pi_n(Y))$. □

(6.20) Corollary *Let* Π *be an abelian group, n a positive integer, and let* Y *be an Eilenberg-Mac Lane space of type* (Π, n). *Then, for any* CW-*complex* X, $[X, Y]$ *is in one-to-one correspondence with* $H^n(X; \Pi)$. □

7 Eilenberg–Mac Lane Spaces

If Π is a group and n a positive integer, *an Eilenberg–Mac Lane space of type* (Π, n) is a space X whose homotopy groups vanish in all dimensions except n, while $\pi_n(X) \approx \Pi$. (Of course, Π has to be abelian if $n > 1$). We shall often use the notation $K(\Pi, n)$ for a CW-complex which is an Eilenberg–Mac Lane space of type (Π, n) (more precisely, $K(\Pi, n)$ is a pair (X, η) where X is a space such that $\pi_i(X) = 0$ for $i \neq n$ and η is an isomorphism of Π with $\pi_n(X)$).

(7.1) Theorem *Let n be a positive integer,* Π *a group (abelian if* $n > 1$*). Then there exists an Eilenberg–Mac Lane space (even a* CW-*complex of type* (Π, n) *and any two such spaces have the same weak homotopy type.*

Existence follows from the general existence theorem (2.1). Uniqueness follows from the more general

(7.2) Theorem *Let* Π, Π' *be abelian groups, n a positive integer, and let* $\eta : \Pi \to \Pi'$ *be a homomorphism. Then there is a unique homotopy class of maps* $f: K(\Pi, n) \to K(\Pi', n)$ *such that* $\pi_n(f) = \eta : \Pi \to \Pi'$.

Remark. If Π' is not required to be abelian, existence still holds, but uniqueness holds only up to conjugacy; cf. Corollary (4.4). This is enough to prove the uniqueness part of Theorem (7.1) when $n = 1$ and Π is non-abelian.

Let $X = K(\Pi, n)$, $X' = K(\Pi', n)$. By Corollary (6.20), the map $f \to f^*\iota^n(X')$ induces a one-to-one correspondence between $[X, X']$ and $H^n(X; \Pi')$. Now $H^n(X; \Pi') \approx \operatorname{Hom}(H_n(X), \Pi')$; let $f: X \to X'$ be the map such that $f^*\iota^n(X')$ is the cohomology class corresponding to the homomorphism $\eta \circ \rho^{-1} : H_n(X) \to \Pi'$. Composition with $H_n(f) : H_n(X) \to H_n(X')$ defines a homomorphism of $\operatorname{Hom}(H_n(X'), \Pi')$ into $\operatorname{Hom}(H_n(X), \Pi')$ which is part of a commutative diagram

$$\begin{array}{ccc} H^n(X';\Pi') & \xrightarrow{f^*} & H^n(X;\Pi') \\ \downarrow & & \downarrow \\ \operatorname{Hom}(H_n(X'),\Pi') & \longrightarrow & \operatorname{Hom}(H_n(X),\Pi') \end{array} \tag{7.3}$$

Since $\iota^n(X')$ corresponds to ρ^{-1}, $f^*\iota^n(X')$ corresponds to $\rho^{-1} \circ H_n(f)$, so that $\eta \circ \rho^{-1} = \rho^{-1} \circ H_n(f)$. But commutativity of the diagram

$$\begin{array}{ccc} \pi_n(X) = \Pi & \xrightarrow{\pi_n(f)} & \Pi' = \pi_n(X') \\ \rho\downarrow & & \downarrow\rho \\ H_n(X) & \xrightarrow[H_n(f)]{} & H_n(X') \end{array} \tag{7.4}$$

ensures that $\pi_n(f) \circ \rho^{-1} = \rho^{-1} \circ H_n(f) = \eta \circ \rho^{-1}$ and therefore $\pi_n(f) = \eta$. If f, f' are maps with $\pi_n(f) = \pi_n(f')$, then $H_n(f) = H_n(f')$ by commutativity of (7.4) and $f^* = f'^* : H^n(X'; \Pi') \to H^n(X; \Pi')$ by commutativity of (7.3). Hence $f^*\iota^n(X') = f'^*\iota^n(X')$ and therefore $f \simeq f'$ by the uniqueness part of Corollary (6.20). □

(7.5) Corollary *The homology and cohomology groups of an Eilenberg–Mac Lane space of type* (Π, n) *depend only on* Π *and* n, *and on the coefficient group.* □

They are usually denoted by

$$H_q(\Pi, n; G), \quad H^q(\Pi, n; G)$$

respectively.

Some examples of Eilenberg–Mac Lane spaces are at hand. For example

$$K(\mathbf{Z}, 1) = \mathbf{S}^1,$$

$$K(\mathbf{Z}_m, 1) = \mathbf{L}^\infty(m),$$

$$K(\mathbf{Z}, 2) = \mathbf{P}^\infty(\mathbf{C}).$$

In particular

$$K(\mathbf{Z}_2, 1) = \mathbf{P}^\infty(\mathbf{R}).$$

Hence we have

(7.6) Theorem *The groups $H_q(\Pi, n)$ are given in the following cases:*

(1) $n = 1, \Pi = Z$

$$H_q(\mathbf{Z}, 1) = \begin{cases} \mathbf{Z} & (q = 0, 1), \\ 0 & (q \geq 2). \end{cases}$$

(2) $n = 1, \Pi = \mathbf{Z}_m$

$$H_q(\mathbf{Z}_m, 1) = \begin{cases} \mathbf{Z} & (q = 0), \\ 0 & (q \text{ even} > 0), \\ \mathbf{Z}_m & (q \text{ odd}). \end{cases}$$

(3) $n = 2, \Pi = \mathbf{Z}$

$$H_q(Z, 2) = \begin{cases} \mathbf{Z} & (q \text{ even}), \\ 0 & (q \text{ odd}). \end{cases}$$ □

We can determine more of the groups $H_q(\Pi, n)$ with the aid of a "Künneth Theorem";

(7.7) Theorem *If Π, Π' are abelian groups, n a positive integer, then*

$$H_q(\Pi \oplus \Pi', n) \approx \bigoplus_{r+s=q} H_r(\Pi, n) \otimes H_s(\Pi', n)$$
$$\oplus \bigoplus_{r+s=q-1} \operatorname{Tor}\{H_r(\Pi, n), H_s(\Pi', n)\}.$$

This follows from the Künneth theorem for the homology of a product space and the fact that $K(\Pi \oplus \Pi', n) = K(\Pi, n) \times K(\Pi', n)$. □

In particular, we can compute $H_q(\Pi, 1)$ for any finitely generated abelian group.

We also have

(7.8) Theorem *If $n \geq 1$, $H_n(\Pi, n) \approx \Pi/[\Pi, \Pi]$. If $n \geq 2$, then $H_{n+1}(\Pi, n) = 0$.*

Since $K(\Pi, n)$ is $(n - 1)$-connected, the first result follows from the Hurewicz theorem. To prove the second, let

$$0 \to R \to F \to \Pi \to 0$$

be a short exact sequence with F a free abelian group. Then R is also free abelian; let B, A be bases for R, F respectively. We construct $X^* = K(\Pi, n)$ as follows. Let X_{n-1} be a single point $*$, and let $X_n = \bigvee_{a \in A} \mathbf{S}_a^n$ be a cluster of n-spheres. Then $F \approx H_n(X_n) \approx \pi_n(X_n)$ by the Hurewicz theorem (here we use the fact that $n > 1$). Let $Y = \bigvee_{b \in B} \mathbf{S}_b^n$, so that $H_n(Y) \approx R$, and let $f: Y \to X_n$ be a map such that $f_*: H_n(Y) \to H_n(X_n)$ is the imbedding of R into F. Let $X = X_{n+1}$ be the mapping cone of f. Now use Corollary (2.4) to imbed X in a space X^* so as to kill all the homotopy groups in dimensions $> n$, by adjoining cells of dimensions $\geq n+2$. Then $X^* = K(\Pi, n)$, and (X^*, X) is $(n+1)$-connected. By the Hurewicz theorem, $H_{n+1}(X^*, X) = 0$ and therefore the injection $H_{n+1}(X) \rightarrow H_{n+1}(X^*)$ is an epimorphism. So it suffices to show that $H_{n+1}(X) = 0$.

The chain complex of X reduces to

$$\Gamma_{n+1} = H_{n+1}(X, X_n) \xrightarrow{\partial} H_n(X_n) = \Gamma_n.$$

Now $H_{n+1}(X, X_n)$ is a free abelian group with one basis element for each cell; and ∂ maps each basis element of $H_{n+1}(X, X_n)$ into the corresponding basis element of R. Thus ∂ is a monomorphism, and therefore $H_{n+1}(X) = 0$. □

It is not true that $H_2(\Pi, 1) = 0$, even when Π is abelian. (A non-abelian example is given by any closed orientable surface of positive genus). In fact, the "Künneth Theorem" (7.7) gives

$$H_2(\mathbf{Z}_2 + \mathbf{Z}_2, 1) \approx H_1(\mathbf{Z}_2, 1) \otimes H_1(\mathbf{Z}_2, 1) \approx \mathbf{Z}_2 \otimes \mathbf{Z}_2 \approx \mathbf{Z}_2.$$

The knowledge of the homology groups of $K(\Pi, n)$ can be useful when there are "gaps" in the homotopy of a space. Specifically,

(7.9) Theorem *Let X be an $(n-1)$-connected space, and suppose that $\pi_i(X) = 0$ for $n < i < q$. Then*

$$H_i(X) \approx H_i(\Pi, n) \qquad (i < q),$$

$$H_q(X)/\Sigma_q(X) \approx H_q(\Pi, n),$$

where $\Sigma_q(X)$ is the image of the Hurewicz map $\rho : \pi_q(X) \rightarrow H_q(X)$.

(The elements of $\Sigma_q(X)$ are called *spherical homology classes*, or sometimes, loosely, *spherical cycles*).

Again we use Corollary (2.4) to imbed X in an Eilenberg–Mac Lane space X^* by attaching cells of dimensions $\geq q+1$. Then (X^*, X) is q-connected; by the Hurewicz theorem, $H_i(X^*, X) = 0$ for all $i \leq q$. Hence the injection $H_i(X) \rightarrow H_i(X^*)$ is an isomorphism for all $i < q$, and there is a commutative diagram

$$\begin{array}{ccccc}
\pi_{q+1}(X^*, X) & \xrightarrow{\partial_*} & \pi_q(X) & \longrightarrow & \pi_q(X^*) = 0 \\
\downarrow \rho & & \downarrow \rho & & \downarrow \rho \\
H_{q+1}(X^*, X) & \xrightarrow[\partial_*]{} & H_q(X) & \xrightarrow[i_*]{} & H_q(X^*) \rightarrow 0
\end{array}$$

Now $\rho : \pi_{q+1}(X^*, X) \to H_{q+1}(X^*, X)$ is an epimorphism, by the Hurewicz theorem, and $\partial_* : \pi_{q+1}(X^*, X) \to \pi_q(X)$ is an epimorphism, since $\pi_q(X^*) = 0$; therefore

$$\operatorname{Ker} i_* = \operatorname{Im} \partial_* = \operatorname{Im}(\partial_* \circ \rho) = \operatorname{Im}(\rho \circ \partial_*) = \operatorname{Im} \rho = \Sigma_q(X),$$
$$H_q(X^*) \approx H_q(X)/\operatorname{Ker} i_* = H_q(X)/\Sigma_q(X). \quad \square$$

(7.10) Corollary *If X is an $(n-1)$-connected space $(n \geq 2)$, then the Hurewicz map $\rho : \pi_{n+1}(X) \to H_{n+1}(X)$ is an epimorphism.*

Just take $q = n + 1$ and apply Theorems (7.8) and (7.9). □

This result is due to Fox [3], and independently to Hopf [8].

The fact that the homotopy classes of maps of a CW-complex X into $K(\Pi, n)$ have a group structure (which is evidently natural) may lead the reader, in view of the argument of Chapter III, §4, to suspect that $K(\Pi, n)$ might be an H-space. That this is indeed the case can be shown by observing that we could have replaced the category of spaces used there by a quite general category; and, if we had used the category of CW-complexes and continuous maps, the desired conclusion would have followed. However, there are at least two other ways of proving

(7.11) Theorem *If Π is an abelian group and n is a positive integer, then $K(\Pi, n)$ is an* H*-space.*

The first proof is prompted by the observation that, if X is any space, then ΩX is an H-space, and $\pi_i(\Omega X) \approx \pi_{i+1}(X)$. Thus, if $X = K(\Pi, n+1)$, then ΩX has the correct homotopy groups. Of course, ΩX is not a CW-complex (although it has the homotopy type of one, according to a theorem of Milnor). However, we have

(7.12) Lemma *If $f: X \to Y$ is a weak homotopy equivalence, X is a* CW-*complex, and Y is an* H-*space, then X admits an* H-*structure for which f is an* H-*map.*

For $f \times f: X \times X \to Y \times Y$ is then also a weak homotopy equivalence, and if $\mu : Y \times Y \to Y$ is a product in Y, then composition with $f: X \to Y$ induces a one-to-one correspondence between $[X \times X, X]$ and $[X \times X, Y]$. Therefore there is a map $\lambda : X \times X \to X$ such that the diagram

$$\begin{array}{ccc} X \times X & \xrightarrow{\lambda} & X \\ {\scriptstyle f \times f}\downarrow & & \downarrow{\scriptstyle f} \\ Y \times Y & \xrightarrow[\mu]{} & Y \end{array}$$

is homotopy commutative. If $\alpha = 1, 2$, we have

$$f \circ \lambda \circ i_\alpha \simeq \mu \circ (f \times f) \circ i_\alpha = \mu \circ i_\alpha \circ f \simeq f$$

and therefore $\lambda \circ i_\alpha \simeq 1$. Hence λ is a product in X. □

By Theorem (2.2), there is a weak homotopy equivalence $g: K(\Pi, n) \to \Omega K(\Pi, n+1)$, and it follows from the Lemma that $K(\Pi, n)$ is an H-space. □

We now give a second proof, and, in fact, we prove a little more.

(7.13) Theorem *If Π is an abelian group and n a positive integer, then $K(\Pi, n)$ has an* H-*structure which is unique up to homotopy.*

Let $X = K(\Pi, n)$, and consider the folding map $\nabla: X \vee X \to X$. The primary obstruction to extending ∇ lies in the group $G = H^{n+1}(X \times X, X \vee X; \Pi)$. If $n > 1$, then $H_n(X \times X, X \vee X)$ and $H_{n+1}(X \times X, X \vee X)$ vanish, by the Künneth Theorem for pairs; therefore $G = 0$. Suppose that $n = 1$; since X is 0-connected, we may assume that X_0 is a single point. Then the 2-cells of the relative CW-complex $(X \times X, X \vee X)$ are products $E_\alpha^1 \times E_\beta^1$ of 1-cells of X. Let us take the liberty of denoting by the same symbols α, β, the elements of $\pi_1(X)$ represented by characteristic maps for these cells. Then the attaching map for $E_\alpha^1 \times E_\beta^1$ is the commutator $[i_{1*}(\alpha), i_{2*}(\beta)]$, whence the element $c^2(e_\alpha^1 \times e_\beta^1)$ is $\nabla_*[i_{1*}(\alpha), i_{2*}(\beta)] = [\alpha, \beta] = 1$ because Π is abelian. Hence the primary obstruction vanishes in all cases.

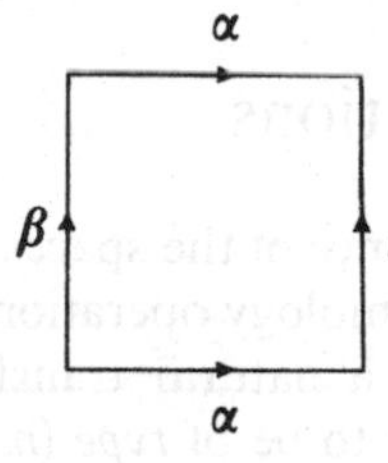

Figure 5.3

If $q > n$, then $\pi_q(X) = 0$ and therefore $H^{q+1}(X \times X, X \vee X; \pi_q(X)) = 0$. By Theorem (6.6), ∇ can be extended to a map $\mu: X \times X \to X$, which is a product in X.

If $\mu_0, \mu_1: X \times X \to X$ are products in X, then

$$(\mu_0, \mu_1)^* \iota^n(X) \in H^n(X \times X, X \vee X; \Pi) = 0$$

by the Künneth Theorem. As $H^q(X \times X, X \vee X; \pi_q(X)) = 0$ for all $q > n$, it follows from Theorem (6.15) that $\mu_0 \simeq \mu_1$ (rel. $X \vee X$). □

Since $K(\Pi, n)$ is an H-space, the set $[X, K(\Pi, n)]$ has a natural composition. In fact, $[X, K(\Pi, n)]$ is an abelian group. Moreover

(7.14) Theorem *The one-to-one correspondence of Corollary* (6.17) *is an isomorphism* $[X, K(\Pi, n)] \approx H^n(X; \Pi)$.

Let $f, g : X \to K(\Pi, n)$, and let $\mu : K \times K \to K$ be the product in $K = K(\Pi, n)$. Let h be the composite

$$X \xrightarrow{\Delta} X \times X \xrightarrow{f \times g} K \times K \xrightarrow{\mu} K.$$

What we must prove is the $h^*(\mathbf{b}_n) = f^*(\mathbf{b}_n) + g^*(\mathbf{b}_n)$, where $\mathbf{b}_n = \iota^n(K)$ is the characteristic class.

Since $H_i(K) = 0$ for $0 \le i < n$, the groups $H_q(K \times K, K \vee K)$ vanish in dimensions $< 2n$. By Corollary (7.3*) of Chapter III, the element $\mathbf{b}_n$ is primitive:

$$\mu^*\mathbf{b}_n = p_1^*\mathbf{b}_n + p_2^*\mathbf{b}_n$$

where $p_1, p_2 : K \times K \to K$ are the projections. Now $p_1 \circ (f \times g) = f \circ p_1$, $p_2 \circ (f \times g) = g \circ p_2$, and therefore

$$\begin{aligned}(f \times g)^*\mu^*\mathbf{b}_n &= (f \times g)^*p_1^*\mathbf{b}_n + (f \times g)^*p_2^*\mathbf{b}_n \\ &= p_1^* f^*\mathbf{b}_n + p_2^* g^*\mathbf{b}_n.\end{aligned}$$

But $p_1 \circ \Delta = p_2 \circ \Delta = 1$, and therefore

$$\begin{aligned}h^*\mathbf{b}_n &= \Delta^*(f \times g)^*\mu^*\mathbf{b}_n = \Delta^*p_1^* f^*\mathbf{b}_n + \Delta^*p_2^* g^*\mathbf{b}_n \\ &= f^*\mathbf{b}_n + g^*\mathbf{b}_n.\end{aligned}$$

□

8 Cohomology Operations

Another reason for the importance of the spaces $K(\Pi, n)$ is their connection (discovered by Serre) with cohomology operations. By a (primary) cohomology operation we shall mean a natural transformation $\theta : H^n(\ \ ; \Pi) \to H^q(\ \ ; G)$ of functors; θ is said to be of *type* $(n, q; \Pi, G)$. We shall assume that n and q are positive. For the case when one of them is zero, the reader is referred to the Exercises.

Examples of cohomology operations are numerous; we shall list only a few.

EXAMPLE 1 (Coefficient group homomorphism). Let $f : \Pi \to G$ be a homomorphism of abelian groups. If X is a CW-complex, composition with f is a chain map of the cochain complex $\operatorname{Hom}(\Gamma(X), \Pi)$ into $\operatorname{Hom}(\Gamma(X), G)$, and therefore induces a homomorphism

$$f_* : H^n(X; \Pi) \to H^n(X; G),$$

and it is clear that f_* is a cohomology operation of type $(n, n; \Pi, G)$, the *coefficient group homomorphism* induced by f.

EXAMPLE 2 (The Bockstein operator). Let

$$0 \to G \to E \to \Pi \to 0 \tag{8.1}$$

be a short exact sequence of abelian groups. If X is a CW-complex, its cochain complexes with respect to the above groups form a short exact sequence

$$0 \to \operatorname{Hom}(\Gamma(X), G) \to \operatorname{Hom}(\Gamma(X), E) \to \operatorname{Hom}(\Gamma(X), \Pi) \to 0$$

and the connecting homomorphism β of the resulting homology exact sequence

$$\cdots \to H^n(X; G) \to H^n(X; E) \to H^n(X; \Pi) \xrightarrow{\ \beta\ } H^{n+1}(X; G) \to \cdots$$

is a cohomology operation of type $(n, n+1; \Pi, G)$, called the *Bockstein operator* associated with the short exact sequence (8.1).

EXAMPLE 3 (The cup square). Let Π, G be abelian groups and suppose there is given a pairing $\Pi \otimes \Pi \to G$. If $u \in H^n(X; \Pi)$, this pairing allows us to construct the cup square $Sq^n(u) = u \smile u \in H^{2n}(X; G)$. Then Sq^n is a cohomology operation of type $(n, 2n; \Pi, G)$. Let us observe that this operation may not be additive; i.e., it is not true that $Sq^n : H^n(X; \Pi) \to H^{2n}(X; G)$ is a homomorphism for every X. (Just let $\Pi = G = \mathbf{Z}$ with the pairing given by multiplication, and take $X = \mathbf{P}^2(\mathbf{C})$, $n = 2$).

EXAMPLE 4. Let Π, G be abelian groups and let $u \in H^q(\Pi, n; G)$. Then if X is a CW-complex and $x \in H^n(X; \Pi)$, there is, by Corollary (6.20), a unique homotopy class of maps $f: X \to K(\Pi, n)$ such that $f^*\mathbf{b}_n = x$, where $\mathbf{b}_n = \iota^n(K(\Pi, n))$ is the characteristic class. Let $\theta_u(x) = f^*(u) \in H^q(X; G)$. Then $\theta_u : H^n(X, \Pi) \to H^q(X; G)$ is a cohomology operation of type $(n, q; \Pi, G)$.

Before embarking on a detailed study of cohomology operations, it behooves us to make precise the categories in which we are working. It is customary to consider the cohomology groups as functors with values in the category $\mathcal{A}$ of Abelian groups. However, Example 3 makes it clear that we shall have to consider mappings between abelian groups which are not homomorphisms. In fact, there is no *a priori* reason to make any restriction whatever on the maps involved. Therefore, in the present context, we shall take the range category $\bar{\mathcal{A}}$ to be the one whose objects are abelian groups and whose morphisms are arbitrary functions.

As for the domain, it is clearly desirable that it should contain the category $\mathscr{C}$ of CW-complexes and cellular maps. However, it is easy to show that any natural transformation defined over $\mathscr{C}$ has a unique extension over the whole category $\mathscr{K}$ of (compactly generated) spaces. In fact, if θ is such a transformation, X an arbitrary space, and $x \in H^n(X; \Pi)$, we can choose a weak homotopy equivalence $f: K \to X$ with K a CW-complex. Then $\theta(f^*x)$ is defined, and therefore we may define

$$\theta'(x) = f^{*-1}\theta(f^*x) \in H^q(X; G)$$

and a simple argument using the techniques of §3 shows that $\theta'(x)$ is uniquely defined, that θ' is a natural transformation over the larger category extending θ, and that θ' is unique.

These observations allow us to eat our cake and have it too. On the one hand, in a general discussion of cohomology operations, we may assume them everywhere defined; on the other, if we wish to construct a specific operation, we need only construct it on CW-complexes.

We shall now show that all cohomology operations are given by the construction of Example 4. More precisely,

(8.2) Theorem (Serre). *The correspondence $u \to \theta_u$ of Example 4 is a one-to-one correspondence between $H^q(\Pi, n; G)$ and the set of all cohomology operations of type $(n, q; \Pi, G)$. If θ is a cohomology operation, the element of $H^q(\Pi, n; G)$ to which it corresponds is $\theta(\mathbf{b}_n)$.*

In fact, let $u_\theta = \theta(\mathbf{b}_n)$. We must show that

(1) $\theta_{u_\theta} = \theta$,
(2) $u_{\theta_u} = u$.

If θ is a cohomology operation, $x \in H^n(X; \Pi)$ and if $f: X \to K(\Pi, n)$ is the map such that $f^*\mathbf{b}_n = x$, then

$$\theta_u(x) = f^*(u_\theta) = f^*\theta(\mathbf{b}_n) = \theta f^*(\mathbf{b}_n) = \theta(x),$$

so that $\theta_{u_\theta} = \theta$.

On the other hand, if $u \in H^q(\Pi, n; G)$, then

$$u_{\theta_u} = \theta_u(\mathbf{b}_n);$$

now the identity map 1 of $K(\Pi, n)$ has the property that $1^*\mathbf{b}_n = \mathbf{b}_n$, and therefore

$$\theta_u(\mathbf{b}_n) = 1^*(u) = u$$

so that $u_{\theta_u} = u$. □

It is natural to ask what conditions on an element $u \in H^q(\Pi, n; G)$ are equivalent to the additivity of the corresponding operation. The answer is given by

(8.3) Theorem *Let $u \in H^q(\Pi, n; G)$, and let $\theta = \theta_u$ be the corresponding operation. Then θ is additive if and only if u is primitive.*

(To understand this statement, recall that $K = K(\Pi, n)$ is an H-space with multiplication $\mu : K \times K \to K$, and that, if $p_1, p_2 : K \times K \to K$ are the projections on the first and second factors, respectively, then u is primitive if and only if

$$\mu^*u = p_1^*u + p_2^*u).$$

We have seen in the proof of Theorem (7.14) that the element $\mathbf{b}_n$ is primitive, and that, if $f: X \to K$, $g : X \to K$, and if $h = \mu \circ (f \times g) \circ \Delta$, then $h^*\mathbf{b}_n = f^*\mathbf{b}_n + g^*\mathbf{b}_n$.

Suppose that θ is additive, and let $u = u_\theta = \theta(\mathbf{b}_n)$. Then

$$\begin{aligned} \mu^*u = \mu^*\theta(\mathbf{b}_n) &= \theta\mu^*(\mathbf{b}_n) = \theta(p_1^*\mathbf{b}_n + p_2^*\mathbf{b}_n) \\ &= p_1^*\theta(\mathbf{b}_n) + p_2^*\theta(\mathbf{b}_n) = p_1^*u + p_2^*u, \end{aligned}$$

so that u is primitive.

Conversely, suppose that u is primitive and let $\theta = \theta_u$. Let $x, y \in H^n(X; \Pi)$ and let $f, g: X \to K(\Pi, n)$ be the corresponding maps. Let $h = \mu \circ (f \times g) \circ \Delta$. Then $h^*\mathbf{b}_n = f^*\mathbf{b}_n + g^*\mathbf{b}_n = x + y$, while

$$\begin{aligned} \theta(x + y) &= \theta h^*\mathbf{b}_n = h^*\theta(\mathbf{b}_n) = h^*(u) \\ &= \Delta^*(f \times g)^*\mu^*u = \Delta^*(f \times g)^*(p_1^*u + p_2^*u) \\ &= \Delta^*(p_1^*f^*u + p_2^*g^*u) = f^*u + g^*u \\ &= \theta(x) + \theta(y), \end{aligned}$$

so that θ is additive. □

EXERCISES

1. (Barratt and Whitehead [1]). Let

$$\begin{array}{ccccccccccc}
\cdots \to A_{n+2} & \xrightarrow{\alpha_{n+2}} & A_{n+1} & \xrightarrow{\alpha_{n+1}} & A_n & \xrightarrow{\alpha_n} & A_{n-1} & \xrightarrow{\alpha_{n-1}} & A_{n-2} \to \cdots \\
\downarrow{\scriptstyle \phi_{n+2}} & & \downarrow{\scriptstyle \phi_{n+1}} & & \downarrow{\scriptstyle \phi_n} & & \downarrow{\scriptstyle \phi_{n-1}} & & \downarrow{\scriptstyle \phi_{n-2}} \\
\cdots \to B_{n+2} & \xrightarrow[\beta_{n+2}]{} & B_{n+1} & \xrightarrow[\beta_{n+1}]{} & B_n & \xrightarrow[\beta_n]{} & B_{n-1} & \xrightarrow[\beta_{n-1}]{} & B_{n-2} \to \cdots
\end{array}$$

be a commutative diagram with exact rows. Suppose that ϕ_n is an isomorphism whenever $n \equiv 0 \pmod 3$.

(i) Deduce the existence of an exact sequence

$$\cdots \to A_{3n+2} \to A_{3n+1} \oplus B_{3n+2} \to B_{3n+1} \to A_{3n-1} \to A_{3n-2} \oplus B_{3n-1} \to \cdots$$

(ii) Use (i) to derive the Mayer–Vietoris sequence of a proper triad.

(iii) Let $p : X \to B$ be a fibration, $f : B' \to B$ a map, $p' : X' \to B'$ the induced fibration. Show that there is an exact sequence

$$\cdots \to \pi_{n+1}(X') \to \pi_{n+1}(B') \oplus \pi_{n+1}(X) \to \pi_{n+1}(B) \to$$
$$\pi_n(X') \to \pi_n(B') \oplus \pi_n(X) \to \cdots$$

2. Let (X, A) be an arbitrary pair, and let H^* be cohomology with coefficients in an arbitrary group G. Show that there are isomorphisms

$$H^n(X, A) \to H^n(0 \times X, 0 \times A) \to H^n(\mathbf{I} \times A \cup \dot{\mathbf{I}} \times X, \mathbf{I} \times A \cup 1 \times X) \to$$
$$H^{n+1}(\mathbf{I} \times X, \mathbf{I} \times A \cup \dot{\mathbf{I}} \times X)$$

such that the composite reduces, when (X, A) is a relative CW-complex, to the isomorphism $\mathbf{i}^* \times : H^n(X, A) \to H^{n+1}(X^*, A^*)$ of (2.34) of Chapter II.

3. (Steenrod [2]). Using the result of #2, define $(f_0, f_1)^* : H^n(Y) \to H^n(X, A)$ for any two maps $f_0, f_1 : X \to Y$ such that $f_0 | A = f_1 | A$ ((X, A) being an arbitrary pair) and prove its principal properties.

4. (Hopf [7]). Let Π be a group, F a free group, $p : F \to \Pi$ an epimorphism with kernel R. Prove that

$$H_2(\Pi, 1) \approx R \cap [F, F]/[F, R].$$

5. (Miller [1]). Let Π be an abelian group. Then

$$H_2(\Pi, 1) \approx \Pi \otimes \Pi/D,$$

where D is the "diagonal subgroup," generated by all elements of the form $x \otimes x$, $x \in \Pi$.

6. Note that $H^n(\Pi, n; G) \approx \operatorname{Hom}(H_n(\Pi, n), G) \approx \operatorname{Hom}(\Pi, G)$. Show that if $u \in H^n(\Pi, n; G)$ corresponds to the homomorphism $f : \Pi \to G$, then the operation θ_u is just the coefficient group homomorphism f_*. Thus the cohomology operations of type $(n, n; \Pi, G)$ are just the coefficient group homomorphisms induced by homomorphisms of Π into G.

7. Note that $H^{n+1}(\Pi, n; G) \approx \operatorname{Ext}(H_n(\Pi, n), G) \approx \operatorname{Ext}(\Pi, G)$. Show that, if $u \in H^{n+1}(\Pi, n; G)$ corresponds to the element $\varepsilon \in \operatorname{Ext}(\Pi, G)$ which is the equivalence class of a short exact sequence

$$0 \to G \to E \to \Pi \to 0,$$

then the operation θ_u is just the associated Bockstein operator. Thus the cohomology operations of type $(n, n+1; \Pi, G)$ are just the Bockstein operators associated with arbitrary short exact sequences as above.

8. Classify all operations of type $(n, q; \Pi, G)$ when n or q is zero.

CHAPTER VI

Homology with Local Coefficients

Let $p : X \to B$ be a fibration whose base space is a CW-complex. One may attempt to find a cross-section to p by a stepwise extension process like the one of the last chapter. If $f: B_n \to X$ is a cross-section over the n-skeleton, the problem of extending f reduces to a family of local problems: for each $(n+1)$-cell E_α of B, the induced fibration over Δ^{n+1} is fibre homotopically trivial. Its total space may thus be represented as a product $\Delta^{n+1} \times F_\alpha$, where F_α is the fibre over some point b_α of E_α. The cross-section thus defines a map of Δ^{n+1} into F_α, representing an element $c(E_\alpha) \in \pi_n(F_\alpha)$ whose vanishing is necessary and sufficient for the cross-section to be extendible over E_α. The groups $\pi_n(F_\alpha)$, for different cells E_α, are all isomorphic; thus it is tempting to regard the function c as a cochain of B with coefficients in the abstract group $\pi_n(F)$. The difficulty is that the isomorphisms $\pi_n(F_\alpha) \approx \pi_n(F_\beta)$ are not uniquely defined; they depend on the choice of a homotopy class of paths joining b_α and b_β.

This difficulty was surmounted in 1943 by Steenrod [1], who showed how the whole apparatus of classical homology theory could be extended to handle the above problem. A bundle of abelian groups G in a pathwise connected space X assigns to each point $x \in X$ a group G_x and to each element $\xi \in \pi_1(X; x, y)$ an isomorphism $G(\xi) : G_y \to G_x$, satisfying certain conditions which are best expressed by the statement that G is a functor from the fundamental groupoid of X to the category $\mathcal{A}$ of abelian groups. Steenrod showed how to associate to a pair (X, A) and a bundle G of abelian groups in X a family of abelian groups $H_q(X, A; G)$, called the *homology groups of* (X, A) *with local coefficients in* G. These have many properties in common with ordinary homology groups (to which they reduce when the coefficient bundle G is simple), and they and the corresponding cohomology groups form the proper setting for the study of the cross-section problem.

In §1 bundles of groups are introduced and various examples studied. In §2 homology and cohomology groups with local coefficients are introduced and properties analogous to the Eilenberg–Steenrod axioms are formulated. In §3 low-dimensional cases are discussed, and it is shown, following Eilenberg [4] that the local homology and cohomology groups of a space X are isomorphic with the equivariant groups of its universal covering space $\tilde{X}$ under the action of the group of covering translations. In §4 a discussion of the homology groups of a relative CW-complex, paralleling that given in §2 of Chapter II for ordinary homology groups, is given. In §§5, 6 the cross-section problem for fibrations is treated, and results parallel to those of §§5, 6 of Chapter V are obtained.

Applications to the theory of characteristic classes are given in §7. Let $p: X \to B$ be the projection of a fibre bundle having $\mathbf{R}^n$ as fibre and $\mathbf{O}(n)$ as structural group. The group $\mathbf{O}(n)$ operates on the Stiefel manifold $\mathbf{V}_{n,k}(0 \leq k \leq n)$, and there is an associated bundle ξ_k over B with fibre $\mathbf{V}_{n,k}$; the total space of the latter bundle is the set of all $(k+1)$-tuples $(b; x_1, \ldots, x_k)$ such that $b \in B$ and $(x_1, \ldots, x_k)$ is an orthonormal k-frame in the Euclidean space $p^{-1}(b)$. The fibre $\mathbf{V}_{n,k}$ is $(n-k-1)$-connected, and the primary obstruction to a cross-section of ξ_k is a cohomology class $W_{n-k+1}(\xi) \in H^{n-k+1}(B; \pi_{n-k}(\mathscr{F}_k))$; the group $\pi_{n-k}(\mathbf{V}_{n,k})$ is either $\mathbf{Z}$ or $\mathbf{Z}_2$, and the local coefficient system is, of course, simple in the latter case, but may be twisted in the former.

There is a classifying space for orthogonal $\mathbf{R}^n$-bundles; it is the Grassmannian space $\mathbf{G}(n)$ of n-planes in $\mathbf{R}^\infty$, and there is a bundle η_n over $\mathbf{G}(n)$ such that the correspondence $f \to f^*\eta_n$ induces a one-to-one correspondence between $[B, \mathbf{G}(n)]$ and the set of equivalence classes of bundles over B. Then the classes $W_r(n) = W_r(\eta_n)$ are defined and if $\xi = f^*\eta_n$, we have the relation

$$f^*W_r(n) = W_r(\xi).$$

The classes $W_r(n)$ are called the *universal Whitney classes*, while the $W_r(\xi)$ are the *Whitney classes of* ξ. Reducing the coefficient groups mod 2, we obtain classes

$$w_r(n) \in H^r(\mathbf{G}(n); \mathbf{Z}_2),$$
$$w_r(\xi) \in H^r(B; \mathbf{Z}_2);$$

these are called the *universal Stiefel–Whitney classes* and the *Stiefel–Whitney classes of* ξ, respectively.

In 1936 Stiefel [1] inaugurated the study of characteristic classes. He considered the tangent bundle of a manifold and defined characteristic homology classes; these were carried by the loci of singularities of fields of k-frames defined on the manifold, and corresponded by Poincaré duality to the Whitney cohomology classes. About the same time Whitney [1] defined characteristic invariants for sphere bundles over a complex K; with the discovery of cohomology he was able [4] to reformulate these invariants as cohomology classes (in the orientable case; in the non-orientable case he reduced the coefficients mod 2). Later [5] he treated the twisted case by the device of passing to a "locally isomorphic complex". It remained for Steenrod [1] to give the formulation in terms of local coefficients.

For a thorough treatment of characteristic classes the reader is referred to the book by Milnor and Stasheff [M–S].

There are many other applications of homology with local coefficients which we have chosen to omit. Two examples are:

(1) obstruction theory for maps into non-simple spaces (Olum [1]);
(2) Poincaré duality for non-orientable manifolds (Steenrod [1]).

1 Bundles of Groups

Let us recall that the *fundamental groupoid* of a space B is the category $\Pi_1(B)$ whose objects are the points of B and whose morphisms: $b_1 \to b_2$ constitute the set $\pi_1(B; b_2, b_1)$ of homotopy classes of paths in B from b_2 to b_1. By a *bundle of groups* in B we shall mean simply a functor G from the category $\Pi_1(B)$ to the category $\mathscr{G}$ of groups. Similarly, a *bundle of abelian groups* (or *local coefficient system*) is a functor from $\Pi_1(B)$ to the category $\mathscr{A}$ of abelian groups. Such a functor assigns to each $b \in B$ a group (an abelian group) $G(b)$ and to each homotopy class $\xi \in \pi_1(B; b_1, b_2)$ a homomorphism $G(\xi): G(b_2) \to G(b_1)$; these are required to satisfy

(1) *if* $\xi \in \pi_1(B, b) = \pi_1(B; b, b)$ *is the identity, then* $G(\xi): G(b) \to G(b)$ *is the identity*;
(2) *if* $\xi \in \pi_1(B; b_1, b_2)$, $\eta \in \pi_1(B; b_2, b_3)$, *then*

$$G(\xi\eta) = G(\xi) \circ G(\eta): G(b_3) \to G(b_1).$$

It then follows that

(3) *if* $\xi \in \pi_1(B; b_1, b_2)$, *then*

$$G(\xi^{-1}) = G(\xi)^{-1}: G(b_1) \to G(b_2),$$

so that the homomorphisms $G(\xi)$ are isomorphisms.

We have already encountered a number of examples of bundles of groups.

EXAMPLE 1. The homotopy groups $\pi_n(B, b)$, together with the operations τ_ξ (Chapter III, §1, and Chapter IV, §3) form a bundle $\Pi_n(B)$ in B.

EXAMPLE 2. If $B \subset A$, the relative homotopy groups $\pi_n(A, B)$, with the operations τ'_ξ (Chapter III, §1, and Chapter IV, §3) form a bundle $\Pi_n(A, B)$ in B.

EXAMPLE 3. If $p: X \to B$ is a fibration, the homology groups $H_n(F_b)$,

$\mathscr{F} = \{F_b\} = \{p^{-1}(b)\}$, with the operations of Chapter IV, §8, form a bundle $H_n(\mathscr{F})$ in B.

EXAMPLE 4. If $p : X \to B$ is a fibration, the groups $\pi_n^*(F_b)$, with the operations of Chapter IV, §8, form a bundle $\pi_n^*(\mathscr{F})$ in B. If the fibres are n-simple, then the homotopy groups $\pi_n(F_b)$ form a bundle $\pi_n(\mathscr{F})$ in B.

In connection with the subject of Poincaré duality, the following example is of crucial importance.

EXAMPLE 5. Let X be a connected n-manifold, and for each $x \in X$, let $G(x)$ be the local homology group

$$H_n(X \mid x) = H_n(X, X - \{x\}).$$

Let us observe that, the injection $H_n(U, U - \{x\}) \to H_n(X, X - \{x\})$ is an isomorphism, for any neighborhood U of x, by the Excision Theorem. In particular, we may take U to be an open n-cell, so that there is a homeomorphism $u : U \to \mathbf{R}^n$. The group $G(x) \approx H_n(\mathbf{R}^n, \mathbf{R}^n - \{x\})$ is then infinite cyclic; a generator of $G(x)$ is called a *local orientation of X about* x.

In order to make $\{G(x)\}$ into a local coefficient system, we first need a few geometric lemmas.

(1.1) Lemma *If V is a (closed) n-cell, x and y interior points of V, then there is a homeomorphism $h : V \to V$ such that $h \mid \dot{V}$ is the identity map and $h(x) = y$.*

We may assume that V is the unit disc $\|x\| \leq 1$ in $\mathbf{R}^n$. The homeomorphism h is then defined by the property: h maps each line segment $[x, z]$ linearly upon $[y, z]$ for every $z \in \dot{V}$ (see Figure 6.1). □

(1.2) Lemma *Let X be an n-manifold, U a connected open subset of X, $x, y \in U$. Then there is a homeomorphism $h : X \to X$ such that $h \mid X - U$ is the identity map and $h(y) = x$.*

Define a relation $\sim$ in U by $x \sim y$ if and only if there is a homeomorphism h as above. Then $\sim$ is an equivalence relation, and it follows from Lemma (1.1) that each equivalence class is open. Since U is connected, there is only one equivalence class. □

(1.3) Lemma *Let X be a connected n-manifold, $x, y \in X$. Then there is an n-cell $E \subset X$ such that both x and y are interior to E.*

We may assume $x \neq y$. Since X is locally connected, the components of $X - \{x\}$ are open; let W be that which contains y, W' the union of the remaining ones. Let E be an n-cell, such that $x \in U = \operatorname{Int} E$. Then

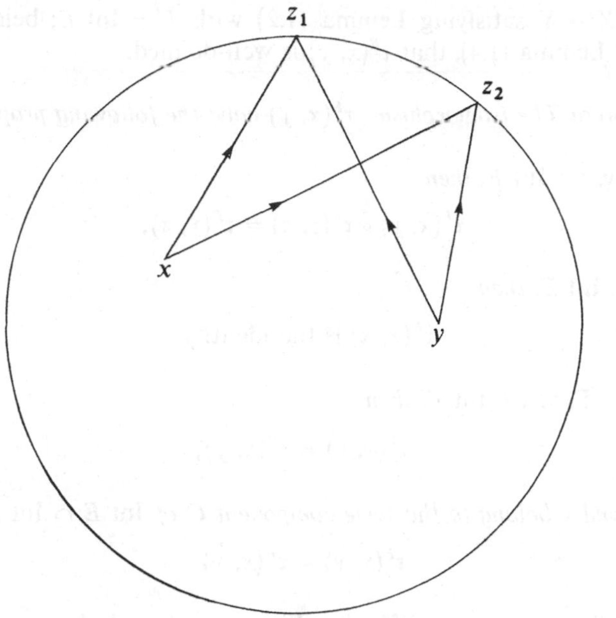

Figure 6.1

$U \cap W \neq \emptyset$; for if not, then $X = (W' \cup U) \cup W$ is a separation of X. Let $z \in U \cap W$; by Lemma (1.2), there is a homeomorphism $h: X \to X$ such that $h|X - W$ is the identity and $h(z) = y$. Then $h(E)$ is an n-cell whose interior $h(U)$ contains $h(x) = x$ and $h(z) = y$. □

(1.4) Lemma *Let X be an n-manifold, E an n-cell contained in X, $x \in \text{Int } E$. Let $h_0, h_1 : X \to X$ be homeomorphisms such that $h_0|X - E = h_1|X - E$ is the identity and $h_0(x) = h_1(x)$. Then there is a homotopy $h : \mathbf{I} \times X \to X$ of h_0 to h_1, (rel. $\{x\} \cup (X - E)$) such that, for each $t \in \mathbf{I}$, $h_t : X \to X$ is a homeomorphism i.e., h is an isotopy of h_0 to h_1.*

By Alexander's Lemma [1], there is an isotopy $h' : I \times E \to E$ of $h_1^{-1} \circ h_0|E$ to the identity (rel. $\{x\} \cup \dot{E}$). The isotopy h is then defined by

$$h_t|X - E \text{ is the identity,}$$
$$h_t|E = h_1 \circ h'_t. \quad \square$$

We can now construct the local coefficient system $\{G(x)\}$ in the connected n-manifold X. If E is an n-cell contained in X, we may define an isomorphism

$$\tau^E(x, y) : H_n(X, X - \{y\}) \to H_n(X, X - \{x\}),$$

for each pair of points $x, y \in \text{Int } E$, by $\tau^E(x, y) = h_*$ for any homeomorphism $h: X \to X$ satisfying Lemma (1.2) with $U = \text{Int } E$; being assured, because of Lemma (1.4), that $\tau^E(x, y)$ is well-defined.

(1.5) Theorem *The isomorphisms* $\tau^E(x, y)$ *have the following properties*:

(1.6) *If* $x, y, z \in \text{Int } E$, *then*

$$\tau^E(x, y) \circ \tau^E(y, z) = \tau^E(x, z);$$

(1.7) *If* $x \in \text{Int } E$, *then*

$$\tau^E(x, x) \text{ is the identity;}$$

(1.8) *If* $E \subset F$, $x, y \in \text{Int } E$, *then*

$$\tau^E(x, y) = \tau^F(x, y);$$

(1.9) *If* x *and* y *belong to the same component* C *of* $\text{Int } E \cap \text{Int } F$, *then*

$$\tau^E(x, y) = \tau^F(x, y).$$

Only (1.9) presents any difficulty. To prove it, apply Lemma (1.3) to the connected manifold C to find an n-cell $G \subset C$ and containing x and y in its interior. Then, by (1.8), we have

$$\tau^E(x, y) = \tau^G(x, y) = \tau^F(x, y). \qquad \square$$

Let $\mathfrak{E}$ be the set consisting of the interiors of all n-cells contained in X, so that $\mathfrak{E}$ is an open covering of X. Let $u: \mathbf{I} \to X$ be a path in X from x to y; then the sets $u^{-1}(U)$, $U \in \mathfrak{E}$, form an open covering $u^*\mathfrak{E}$ of the compactum $\mathbf{I}$. Let $\eta > 0$ be a Lebesgue number for this covering, and let $\Pi: 0 = t_0 < t_1 < \cdots < t_n = 1$ be a partition of $\mathbf{I}$ with $t_i - t_{i-1} < \eta$ for $i = 1, \ldots, m$. Then there are n-cells $E_1, \ldots, E_m$ in X such that $u([t_{i-1}, t_i]) \subset \text{Int } E_i$ for $i = 1, \ldots, m$. (A partition Π with the latter property is said to be *adapted* to u). Let

$$\tau_i = \tau^{E_i}(u(t_{i-1}), u(t_i)),$$
$$\phi(\Pi) = \tau_1 \circ \cdots \circ \tau_m : G(y) \to G(x).$$

It follows from (1.9) that the isomorphism τ_i is independent of the n-cell E_i, and therefore $\phi(\Pi)$ does not depend on the choice of the n-cells $E_1, \ldots, E_m$. We next prove that $\phi(\Pi)$ is independent of the partition Π. In fact, if Π' is another partition adapted to u, then Π and Π' have a common refinement Π'' which is also adapted to u, and it suffices to prove that $\phi(\Pi) = \phi(\Pi')$ whenever Π' is a refinement of Π. It suffices, in turn, to consider the case in which Π' is obtained from Π by inserting one subdivision point $t_* \in (t_{i-1}, t_i)$. We then have

$$\phi(\Pi) = \rho \circ \tau_i \circ \sigma$$
$$\phi(\Pi') = \rho \circ \tau' \circ \tau'' \circ \sigma$$

where

$$\rho = \tau_1 \circ \cdots \circ \tau_{i-1}, \qquad \sigma = \tau_{i+1} \circ \cdots \circ \tau_m,$$
$$\tau' = \tau^{E'}(u(t_{i-1}), u(t_*)),$$
$$\tau'' = \tau^{E''}(u(t_*), u(t_i)),$$

where E' and E'' are n-cells containing $u([t_{i-1}, t_*])$, $u([t_*, t_i])$, respectively, in their interiors. Because of (1.8) and (1.9), we may assume $E' = E'' = E_i$. But then (1.6) assures us that $\tau' \circ \tau'' = \tau_i$ and therefore $\phi(\Pi) = \phi(\Pi')$.

We have just seen that the isomorphism $\phi_u = \phi(\Pi)$ depends only on the path u, and not on the partition Π used in its definition. Let u_1, u_2 be paths which can be multiplied, so that $u_1(1) = u_2(0)$, and let $u = u_1 u_2$ be their product. If

$$\Pi_1 : 0 = s_0 < s_1 < \cdots < s_p = 1,$$
$$\Pi_2 : 0 = t_0 < t_1 < \cdots < t_q = 1$$

are partitions adapted to u_1, u_2 respectively, then the partition

$$\Pi : 0 = \tfrac{1}{2}s_0 < \tfrac{1}{2}s_1 < \cdots < \tfrac{1}{2}s_p < \tfrac{1}{2}(1 + t_1) < \cdots < \tfrac{1}{2}(1 + t_q) = 1$$

is adapted to u. Clearly $\phi(\Pi) = \phi(\Pi_1) \circ \phi(\Pi_2)$, and therefore $\phi_u = \phi_{u_1} \circ \phi_{u_2}$. Moreover, if u is a constant path, ϕ_u is the identity.

We next show that ϕ_u depends only on the homotopy class ξ of u, so that we may define $G(\xi) = \phi_u$ for any $u \in \xi \in \pi_1(X; x, y)$. In fact, let $w : \mathbf{I}^2 \to X$ be a homotopy (rel. $\dot{\mathbf{I}}$) between paths u, v from x to y, so that

$$w(0, t) = u(t), \qquad w(1, t) = v(t),$$
$$w(s, 0) = x, \qquad w(s, 1) = y.$$

Let $\eta > 0$ be a Lebesgue number for the open covering $w^*\mathfrak{E}$ of $\mathbf{I}^2$, and choose partitions

$$0 = s_0 < s_1 < \cdots < s_p = 1,$$
$$0 = t_0 < t_1 < \cdots < t_q = 1,$$

so that each of the rectangles $R_{ij} = [s_{i-1}, s_i] \times [t_{j-1}, t_j]$ has diameter $< \eta$, (see Figure 6.2), and therefore $u(R_{ij})$ is contained in the interior of some n-cell E_{ij}. Let u_i be the path defined by $u_i(t) = u(s_i, t)$; then $u_0 = u$, $u_1 = v$, and it suffices to show that $\phi_{u_{i-1}} = \phi_u$ $(i = 1, \ldots, p)$. Therefore we may as well assume $p = 1$, $u = u_0$, $v = u_1$.

Let

$$x_i = u(t_i),\ y_i = v(t_i),$$
$$\sigma_i = \tau^{E_i}(x_{i-1}, x_i),$$
$$\tau_i = \tau^{E_i}(y_{i-1}, y_i),$$
$$\rho_i = \tau^{E_i}(x_i, y_i);$$

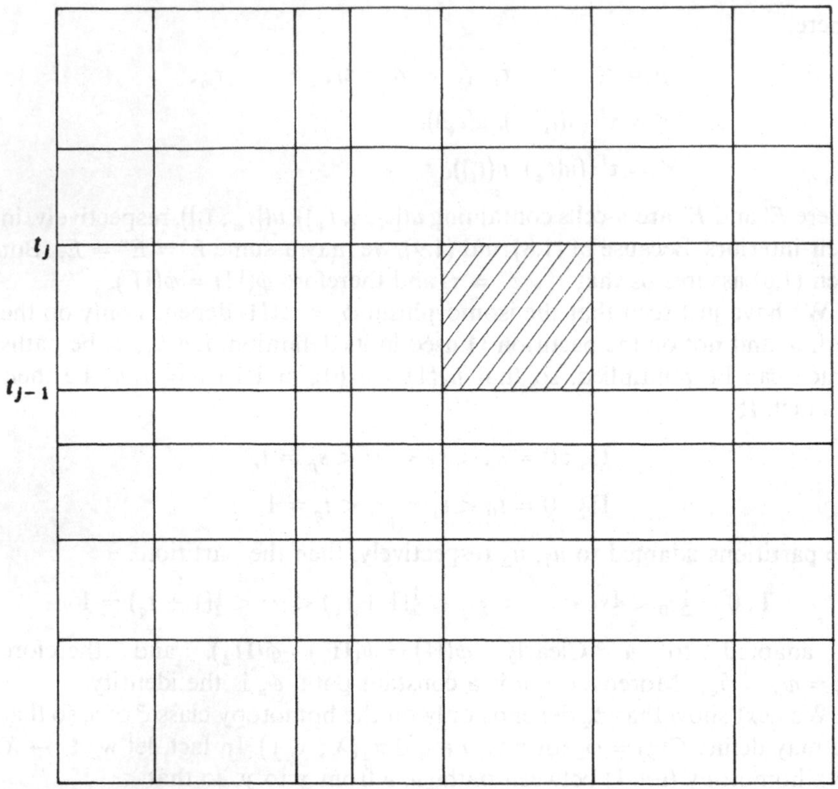

Figure 6.2

note that, because of (1.9),

(1.10) $$\rho_i = \tau^{E_{i+1}}(x_i, y_i).$$

Two applications of (1.6) yield

$$\begin{aligned}\sigma_i &= \tau^{E_i}(x_{i-1}, y_{i-1}) \circ \tau^{E_i}(y_{i-1}, x_i)\\ &= \tau^{E_i}(x_{i-1}, y_{i-1}) \circ \tau^{E_i}(y_{i-1}, y_i) \circ \tau^{E_i}(y_i, x_i)\\ &= \rho_{i-1} \circ \tau_i \circ \rho_i^{-1} \quad \text{by (1.6) and (1.10).}\end{aligned}$$

Hence

$$\begin{aligned}\tau_u &= \sigma_1 \circ \sigma_2 \circ \cdots \circ \sigma_q\\ &= (\rho_0 \circ \tau_1 \circ \rho_1^{-1}) \circ (\rho_1 \circ \tau_2 \circ \rho_2^{-1}) \circ \cdots \circ (\rho_{q-1} \circ \tau_q \circ \rho_q^{-1})\\ &= \rho_0 \circ (\tau_1 \circ \cdots \circ \tau_q) \circ \rho_q^{-1}\\ &= \rho_0 \circ \tau_v \circ \rho_q^{-1} = \tau_v,\end{aligned}$$

since ρ_0 and ρ_q are identity maps, by (1.7).

Since $\phi_{uv} = \phi_u \circ \phi_v$, it follows that $G(\xi\eta) = G(\xi) \circ G(\eta)$ whenever the relevant products are defined. Since, moreover, $G(\varepsilon)$ is the identity whenever ε is the homotopy class of a constant path, the conditions for a bundle of abelian groups are satisfied. The bundle G so defined is called the *orientation bundle* of the manifold X.

If G is a group, G determines a *constant* bundle $\underline{G}$ with

$$\underline{G}(b) = G,$$
$$\underline{G}(\xi) = \text{the identity map.}$$

A bundle isomorphic (*v. infra*) with a constant bundle is said to be *simple*. This is the case if and only if the morphism $G(\xi)$ is independent of ξ.

Let G, H be bundles of groups in B. By a *homomorphism* $\Phi : G \to H$ we shall mean simply a natural transformation of functors. Thus Φ determines, for each $b \in B$, a homomorphism $\Phi(b) : G(b) \to H(b)$ such that the diagram is commutative for any $\xi \in \pi_1(B; b_1, b_2)$

$$\begin{array}{ccc} G(b_2) & \xrightarrow{G(\xi)} & G(b_1) \\ {\scriptstyle \Phi(b_2)}\downarrow & & \downarrow{\scriptstyle \Phi(b_1)} \\ H(b_2) & \xrightarrow[H(\xi)]{} & H(b_1) \end{array}$$

We say that Φ is an isomorphism (monomorphism, epimorphism) if and only if each of the homomorphisms $\Phi(b)$ is an isomorphism (monomorphism, epimorphism).

If G is a bundle of groups in B, and $b_0 \in B$, then $\pi_1(B, b_0)$ operates on $G(b_0)$. Conversely, if $\pi_1(B, b_0)$ operates on a group G_0, and if B is 0-connected, then there is a bundle of groups G, unique up to isomorphism, such that $G_{b_0} = G_0$ and G induces the given operation. More precisely,

(1.11) Theorem *Let B be 0-connected, and let G, G' be bundles of groups in B, and let $\phi : G_{b_0} \to G'_{b_0}$ be an operator isomorphism. Then there is a unique isomorphism $\Phi : G \to G'$ such that $\Phi(b_0) = \phi$.*

(1.12) Theorem *Let B be 0-connected, and let G_0 be a group on which $\pi_1(B, b_0)$ operates. Then there exists a bundle of groups G in B such that $G_{b_0} = G_0$ and which induces the given operation of $\pi_1(B, b_0)$ on G_0.*

For each $b \in B$, choose an element $\xi(b) \in \pi_1(B; b_0, b)$; we may assume that $\xi(b_0) = 1$, the identity of $\pi_1(B, b_0)$. To prove Theorem (1.11), define $\Phi(b) : G(b) \to G'(b)$ by

$$\Phi(b) = G'(\xi(b))^{-1} \circ \phi \circ G(\xi(b)).$$

Let $\eta \in \pi_1(B; b_1, b_2)$. Then

$$\begin{aligned}\Phi(b_1) \circ G(\eta) &= G'(\xi(b_1))^{-1} \circ \phi \circ G(\xi(b_1)) \circ G(\eta)\\ &= G'(\xi(b_1))^{-1} \circ \phi \circ G(\xi(b_1)\eta),\\ G'(\eta) \circ \Phi(b_2) &= G'(\eta) \circ G'(\xi(b_2))^{-1} \circ \phi \circ G(\xi(b_2))\\ &= G'(\eta \cdot \xi(b_2)^{-1}) \circ \phi \circ G(\xi(b_2)).\end{aligned}$$

Let $\alpha = \xi(b_1)\eta\xi(b_2)^{-1}$, so that $\alpha \in \pi_1(B, b_0)$. Because ϕ is an operator homomorphism, we have $\phi \circ G(\alpha) = G'(\alpha) \circ \phi$. Thus

$$\begin{aligned}\Phi(b_1) \circ G(\eta) &= G'(\xi(b_1))^{-1} \circ \phi \circ G(\alpha\xi(b_2))\\ &= G'(\xi(b_1))^{-1} \circ \phi \circ G(\alpha) \circ G(\xi(b_2))\\ &= G'(\xi(b_1))^{-1} \circ G'(\alpha) \circ \phi \circ G(\xi(b_2))\\ &= G'(\xi(b_1)^{-1}\alpha) \circ \phi \circ G(\xi(b_2))\\ &= G'(\eta\xi(b_2)^{-1}) \circ \phi \circ G(\xi(b_2)) = G'(\eta) \circ \Phi(b_2),\end{aligned}$$

so that Φ is a natural transformation. And

$$\begin{aligned}\Phi(b_0) &= G'(\xi(b_0))^{-1} \circ \phi \circ G(\xi(b_0))\\ &= G'(1)^{-1} \circ \phi \circ G(1) = \phi.\end{aligned}$$

Uniqueness follows from the fact that, if $b \in B$ and ξ is any homotopy class of paths from b_0 to b, then the diagram

$$\begin{array}{ccc} G(b) & \xrightarrow{G(\xi)} & G(b_0)\\ \Phi(b)\downarrow & & \downarrow\phi\\ G'(b) & \xrightarrow[G'(\xi)]{} & G'(b_0)\end{array}$$

is commutative, so that

$$\Phi(b) = G'(\xi)^{-1} \circ \phi \circ G(\xi).$$

To prove Theorem (1.12), let $G(b) = G$ for each $b \in B$. If $\eta \in \pi_1(B; b_1, b_2)$, then $\xi(b_1)\eta\xi(b_2)^{-1} \in \pi_1(B, b_0)$, and the effect of operating with this element on G_0 is an automorphism $G(\eta): G(b_2) = G_0 \to G_0 = G(b_1)$. If $b_1 = b_2$ and $\eta = 1 \in \pi_1(B, b_1)$, then $\xi(b_1)\eta\xi(b_1)^{-1} = 1$ and therefore $G(\eta) = 1$. If, on the other hand, $\zeta \in \pi_1(B; b_2, b_3)$, then

$$(\xi(b_1)\eta\xi(b_2)^{-1}) \cdot (\xi(b_2)\zeta\xi(b_3)^{-1}) = \xi(b_1)\eta\zeta\xi(b_3)^{-1}$$

and therefore $G(\eta\zeta) = G(\eta) \circ G(\zeta)$. Thus G is a bundle of groups in B, and if $b_1 = b_2 = b_0$, $\eta \in \pi_1(B, b_0)$, then $\xi(b_1)\eta\xi(b_2)^{-1} = \eta$, and therefore the operation $G(\eta)$ of $\pi_1(B, b_0)$ on $G_0 = G(b_0)$ induced by G coincides with the given operation. □

Let $f: A \to B$ be a map. Then composition with f induces a functor[1] $F : \Pi_1(A) \to \Pi_1(B)$. If G is a bundle of groups in B, so that $G : \Pi_1(B) \to \mathcal{G}$, then $G \circ F : \Pi_1(A) \to \mathcal{G}$ is a bundle of groups in A. Let $f^*(G) = G \circ F$; we call it the bundle over A *induced by the map* $f: A \to B$. Evidently

(1.13) *If* $f: A \to B$ *and* $g : B \to C$, *and if* H *is a bundle of groups in* C, *then* $(g \circ f)^*(H) = f^*(g^*(H))$. □

If A is a subspace of B, $i : A \subsetneq B$, it is often convenient to write $G|A$ instead of i^*G.

Now suppose that B is 0-connected and semilocally 1-connected, and let $b_0 \in B$. Let G be a bundle of groups in B, and let K be the set of all $\xi \in \pi_1(B, b_0)$ such that $G(\xi)$ is the identity map of $G(b_0)$. Then K is a subgroup of $\pi_1(B, b_0)$; by the general theory of covering spaces, there is a covering map $p : \tilde{B} \to B$ and a point $\tilde{b}_0 \in p^{-1}(b_0)$ such that $p_* \pi_1(\tilde{B}, \tilde{b}_0) = K$. Clearly

(1.14) *The bundle* $p^*(G)$ *is simple.* □

Finally, let $p : X \to B$ be a fibration with n-simple fibres, and let $f : B' \to B$. Let $p' : X' \to B'$ be the induced fibration. Then

(1.15) *The systems* $\pi_n(\mathcal{F})$, $\pi_n(\mathcal{F}')$ *of local coefficients are related by*

$$\pi_n(\mathcal{F}') = f^*\pi_n(\mathcal{F}).$$ □

2 Homology with Local Coefficients

Let X be a space, G a bundle of abelian groups in X. Let $C_q(X; G)$ be the set of all functions c with the following properties:

(1) for every singular q-simplex $u : \Delta^q \to X$, $c(u)$ is defined and belongs to the group $G(u(\mathbf{e}_0))$;
(2) the set of singular simplices u such that $c(u) \neq 0$ is finite.

The elements of $C_q(X; G)$ are called *singular q-chains with coefficients in G*. Such a singular chain is called *elementary* if and only if $c(u) \neq 0$ for at most one singular simplex u, and it is then convenient to write $c = c(u) \cdot u$. Each

[1] The category-minded reader may already have observed that Π_1 is a functor from the category $\mathcal{K}$ to the category of groupoids. A morphism in this latter category is just a functor. In particular, $F = \Pi_1(f)$.

$c \in C_q(X; G)$ is a finite sum of elementary chains, so that c can be expressed as a finite sum

$$c = \sum_{i=1}^{m} g_i u_i,$$

where u_i are singular simplices and $g_i \in G(u_i(\mathbf{e}_0))$.

We are going to make the graded group $C_*(X; G)$ into a chain-complex. In order to do so, it suffices to define the boundary of an elementary q-chain c, and to prove that $\partial\partial c = 0$. Let us first observe that, if $u : \Delta^q \to X$, and if $u_i = \partial_i u$, then

$$u_i(\mathbf{e}_0) = \begin{cases} u(\mathbf{e}_0) & \text{if } i > 0, \\ u(\mathbf{e}_1) & \text{if } i = 0. \end{cases}$$

If $g \in G(u(\mathbf{e}_0))$, we should like to define $\partial(gu) = \sum_{i=0}^{q} (-1)^i g \partial_i u$, but are prevented from doing so by the fact that, because $g \notin G(u_0(\mathbf{e}_0))$, the right-hand side of the above relation is not a singular $(q-1)$-chain with coefficients in G. We can correct for this anomaly, however, by observing that the path $t \to u((1-t)\mathbf{e}_1 + t\mathbf{e}_0)$ joins $u(\mathbf{e}_1)$ to $u(\mathbf{e}_0)$, and therefore its homotopy class σ_u induces an isomorphism $G(\sigma_u) : G(u(\mathbf{e}_0)) \to G(u(\mathbf{e}_1))$. Thus we may define

$$\partial(gu) = G(\sigma_u)(g) \cdot \partial_0 u + \sum_{i=1}^{q} (-1)^i g \cdot \partial_i u.$$

To prove that $\partial\partial c = 0$, we should first observe that

$$\sigma_{\partial_i u} = \sigma_u \quad \text{if } i > 1,$$

while

$$\sigma_{\partial_1 u} = \sigma_{\partial_0 u} \sigma_u;$$

the latter equality is due to the fact that each side is the homotopy class of the composite of u with a path in Δ^q from $\mathbf{e}_2$ to $\mathbf{e}_0$, and the fact that Δ^q is simply connected. With these facts in mind, the details of the calculation of $\partial\partial(gu)$ are readily carried out.

The graded group $C_*(X; G)$ is, then a chain complex, and its homology groups

$$H_q(X; G) = H_q(C_*(X; G))$$

are called the *homology groups of X with coefficients in the bundle G*. They are also referred to as homology groups with *local coefficients in G*.

(2.1) *If the space X is 0-connected and G is simple, then $H_q(X; G) \approx H_q(X; G_0)$, where $G_0 = G(x_0)$ for some $x_0 \in X$.* □

Thus the homology groups with local coefficients are a true generalization of the ordinary homology groups.

Let A be a subspace of X, $i : A \subsetneq X$. Let $c = \sum g_j u_j \in C_q(A; i^*G) = C_q(A; G \mid A)$, so that $u_j : \Delta^q \to A$ and

$$g_j \in (i^*G)(u_j(\mathbf{e}_0)) = G((i \circ u_j)(\mathbf{e}_0)) = G(u_j'(\mathbf{e}_0)),$$

where $u_j' = i \circ u_j$. Then

$$i_\#(c) = \sum g_j u_j' \in C_q(X; G),$$

and $i_\# : C_q(A; G \mid A) \to C_q(X; G)$ is a monomorphism, so that we may regard $C_*(A; G \mid A)$ as a subgroup of $C_*(X; G)$, which is even a subcomplex. Accordingly, we may define

$$C_q(X, A; G) = \operatorname{Cok} i_\# : C_q(A; G \mid A) \to C_q(X; G),$$

and we have a short exact sequence

$$0 \to C_*(A; G \mid A) \to C_*(X; G) \to C_*(X, A; G) \to 0,$$

of chain complexes. Accordingly, we may define

$$H_q(X, A; G) = H_q(C_*(X, A; G)),$$

and obtain a long exact sequence

$$\cdots \to H_{q+1}(X, A; G) \xrightarrow{\partial_*} H_q(A; G \mid A) \xrightarrow{i_*}$$
$$H_q(X; G) \xrightarrow{j_*} H_q(X, A; G) \to \cdots.$$

Let G and H be systems of local coefficients on X, and let $\Phi : G \to H$ be a homomorphism. Then Φ induces a homomorphism $\Phi_\# : C_*(X; G) \to C_*(X; H)$;

$$\Phi_\#(\sum g_i u_i) = \sum \Phi(u_i(\mathbf{e}_0))(g_i) u_i.$$

Clearly $\Phi_\#$ is a chain map. Moreover, if $A \subset X$, $\Phi_\#$ maps $C_*(A; G \mid A)$ into $C_*(A; H \mid A)$. Thus Φ induces

$$\Phi_* : H_q(X, A; G) \to H_q(X, A; H).$$

Let $f: (X, A) \to (Y, B)$, and let G be a system of local coefficients in Y. A slight modification of the above argument shows that f induces a chain map $f_\# : C_*(X; f^*G) \to C_*(Y; G)$ which sends $C_*(A, f^*G \mid A)$ into $C_*(B, G \mid B)$. If, moreover, F is a system of local coefficients in X and $\Phi : F \to f^*G$ a homomorphism, we may compose the chain maps $f_\#$, $\Phi_\#$, to obtain a chain map of $C_*(X; F)$ into $C_*(Y; G)$. The latter map, in turn, carries $C_*(A; F \mid A)$ into $C_*(B; G \mid B)$ and so induces a chain map of $C_*(X, A; F)$ into $C_*(Y, B; G)$. In turn, the last-mentioned chain map induces a homomorphism of $H_q(X, A; G)$ into $H_q(Y, B; H)$.

The above discussion suggests that the proper setting for homology with local coefficients is the category $\mathscr{L}$ whose objects are triples $(X, A; G)$ with (X, A) a pair in $\mathscr{K}$ and G a system of local coefficients in X. A morphism

$\phi : (X, A; G) \to (Y, B; H)$ in $\mathscr{L}$ is a pair (ϕ_1, ϕ_2) such that

$$\phi_1 : (X, A) \to (Y, B)$$

is a continuous map and

$$\phi_2 : G \to \phi_1^* H$$

is a homomorphism. If $\psi : (Y, B; H) \to (Z, C; K)$, then $\psi \circ \phi = \omega$, where $\omega_1 = \psi_1 \circ \phi_1$ and $\omega_2 = (\phi_1^* \psi_2) \circ \phi_2$ and $\phi_1^* \psi_2 : \phi_1^* H \to \phi_1^* \psi_1^* K$ is the homomorphism such that

$$\phi_1^* \psi_2(x) = \psi_2(\phi_1(x)) : (\phi_1^* H)(x) = H(\phi_1(x)) \to (\phi_1^* \psi_1^* K)(x) = K(\omega_1(x)).$$

In order to deal with homotopies, we need the notion of the *prism* over an object in $\mathscr{L}$. If $X \in \mathscr{K}$, there are maps

$$X \underset{i_1}{\overset{i_0}{\rightrightarrows}} \mathbf{I} \times X \xrightarrow{\ p\ } X$$

defined by $i_t(x) = (t, x)$, $p(t, x) = x$. The prism $\mathbf{I} \times (X, A; G)$ over an object $(X, A; G)$ of $\mathscr{L}$ is the object

$$(\mathbf{I} \times X, \mathbf{I} \times A; p^* G).$$

If 1 is the identity map of G into $i_t^* p^* G = G$, then $(i_t, 1)$ is a map

$$i_t : (X, A; G) \to \mathbf{I} \times (X, A; G) \qquad (t = 0, 1).$$

If $\phi, \psi : (X, A; G) \to (Y, B; H)$ are maps in $\mathscr{L}$, *a homotopy of ϕ to ψ* is a map $\lambda : \mathbf{I} \times (X, A; G) \to (Y, B; H)$ such that $\lambda \circ i_0 = \phi$, $\lambda \circ i_1 = \psi$. Thus $\lambda_1 : (\mathbf{I} \times X, \mathbf{I} \times A) \to (Y, B)$ is a homotopy of ϕ_1 to ψ_1, and $\lambda_2 : p^* G \to \lambda_1^* H$ is a homomorphism such that $i_0^* \lambda_2 = \phi_2 : G \to \phi^* H$ and $i_1^* \lambda_2 = \psi_2 : G \to \psi^* H$. We say that *$\phi$ is homotopic to ψ* ($\phi \simeq \psi$) if and only if there is a homotopy λ of ϕ to ψ.

The reader is invited to verify:

(2.2) *Homotopy between morphisms in $\mathscr{L}$ is an equivalence relation.* □

(2.3) *If $\phi \simeq \psi : (X, A; F) \to (Y, B; G)$ and $\phi' \simeq \psi' : (Y, B; G) \to (Z, C; H)$, then $\phi' \circ \phi \simeq \psi' \circ \psi : (X, A; F) \to (Z, C; H)$.* □

The *shift operator* is the functor $R : \mathscr{L} \to \mathscr{L}$, defined for objects in $\mathscr{L}$ by

$$R(X, A; G) = (A, \varnothing; G \mid A),$$

and for morphisms $\phi : (X, A; G) \to (Y, B; K)$ by $R(\phi) = \psi$, where $\psi_1 : A \to B$ is the restriction of $\phi_1 : (X, A) \to (Y, B)$, $i : A \hookrightarrow X$, $j : B \hookrightarrow Y$, so that $j \circ \psi_1 = \phi_1 \circ i$, and

$$\psi_2 = i^* \phi_2 : G \mid A = i^* G \to i^* \phi_1^* H = \psi_1^* j^* H = \psi_1^* (H \mid B).$$

We can now state a list of properties of the homology groups with local coefficients, analogous to those formulated by Eilenberg and Steenrod for ordinary homology groups. In the first place, we are given a sequence of functors $H_q: \mathscr{L} \to \mathscr{A}$. Secondly, we are given a natural transformation $\partial_q: H_q \to H_{q-1} \circ R$. If $(X_1, A_1; G_1)$ and $(X_2, A_2; G_2)$ are objects in $\mathscr{L}$ such that X_1 is a subspace of X_2 and A_1 a subspace of A_2, and if $G_2 | X_1 = G_1$, then the inclusion map $(X_1, A_1) \subsetneq (X_2, A_2)$, together with the identity map $G_1 \to G_2 \mid X_1$, define a morphism $k: (X_1, A_1; G_1) \to (X_2, A_2; G_2)$, which will also be called an *inclusion*; moreover, the homomorphism $H_q(k): H_q(X_1, A_1; G_1) \to H_q(X_2, A_2; G_2)$ will be called the *injection.*

We can now formulate the analogues of the Eilenberg–Steenrod axioms for homology with local coefficients.

(2.4) Theorem *The functors H_q and natural transformations ∂_q have the following properties:*

(2.5) (Exactness). *If $(X, A; G)$ is an object in $\mathscr{L}$, i and j are appropriate inclusions, then the sequence*

$$\cdots \to H_{q+1}(X, A; G) \xrightarrow{\partial_{q+1}(X, A; G)} H_q(A; G|A) \xrightarrow{H_q(i)} H_q(X; G) \xrightarrow{H_q(j)} H_q(X, A; G) \to \cdots$$

is exact.

(2.6) (Homotopy). *If $\phi_0, \phi_1: (X, A; G) \to (Y, B; H)$ are homotopic morphisms in $\mathscr{L}$, then*

$$H_q(\phi_0) = H_q(\phi_1): H_q(X, A; G) \to H_q(Y, B; H)$$

for all q.

(2.7) (Excision). *Let $(X; X_1, X_2)$ be a triad in $\mathscr{K}$ such that $X = \text{Int } X_1 \cup \text{Int } X_2$, and let G be a system of local coefficients in X. Then the injection $H_q(X_1, X_1 \cap X_2; G|X_1) \to H_q(X, X_2; G)$ is an isomorphism for all q.*

(2.8) (Dimension). *If X is a space consisting of a single point $*$, then $H_q(X; G) = 0$ for all $q \neq 0$ and $H_0(X; G) \approx G(*)$.*

(2.9) (Additivity). *Let X be the union of a family of mutually disjoint open sets X_α, and let A be a subspace of X, $A_\alpha = A \cap X_\alpha$. Let G be a system of local coefficients in X, $G_\alpha = G|X_\alpha$. Then the injections*

$$H_n(X_\alpha, A_\alpha; G_\alpha) \to H_n(X, A; G)$$

represent the latter group as a direct sum.

The proofs of these analogues of the Eilenberg–Steenrod axioms, and of the additivity property, are not difficult, and are left to the reader. □

In a similar way we can define *singular cohomology groups* $H^n(X, A; G)$. Let $C^n(X; G)$ be the set of all functions c which assign, to each singular simplex $u : \Delta^q \to X$, an element $c(u) \in G(u(e_0))$. The set $C^n(X; G)$ is an abelian group under addition of functional values. The coboundary operator $\delta : C^n(X; G) \to C^{n+1}(X; G)$ is defined by

$$(-1)^n\, \delta c(u) = G(\sigma_u)^{-1} c(\partial_0 u) + \sum_{i=1}^{n+1} (-1)^i c(\partial_i u)$$

for each singular simplex $u : \Delta^{n+1} \to X$. Then δ is a homomorphism and $\delta \circ \delta = 0$, so that $C^*(X; G)$ is a cochain complex. If A is a subspace of X, the restriction map $i^\#$ is a cochain map of $C^*(X; G)$ *upon* $C^*(A; G \mid A)$; thus the kernel of $i^\#$ is a subcomplex $C^*(X, A; G)$ and we define its cohomology groups to be the singular cohomology groups $H^n(X, A; G)$ of the pair (X, A) with coefficients in the local system G.

Let $\mathscr{L}^*$ be the category whose objects are triples $(X, A; G)$ with (X, A) a pair in $\mathscr{K}$ and G a system of local coefficients in X, and such that a morphism $\phi : (X, A; G) \to (Y, B; H)$ in $\mathscr{L}^*$ is a pair (ϕ_1, ϕ_2), where $\phi_1 : (X, A) \to (Y, B)$ is a continuous map and $\phi_2 : \phi_1^* H \to G$ is a homomorphism of local coefficient systems in X. The product of two morphisms $\phi : (X, A; G) \to (Y, B; H)$ and $\psi : (Y, B; H) \to (Z, C; K)$ is the morphism $\omega = \psi \circ \phi$ such that

$$\omega_1 = \psi_1 \circ \phi_1 : (X, A) \to (Z, C),$$

while

$$\omega_2(x) = \phi_2(x) \circ \psi_2(\phi_1(x)) : \phi_1^* \psi_1^* K \to G.$$

(Note that, while the categories $\mathscr{L}$ and $\mathscr{L}^*$ have the same objects, their morphisms are different).

It is then routine that a morphism $\phi : (X, A; G) \to (Y, B; H)$ induces a cochain map

$$\phi^\# : C^*(Y, B; H) \to C^*(X, A; G);$$

for a singular simplex $u : \Delta^n \to X$, $c \in C^n(Y, B; H)$, we have

$$\phi^\# c(u) = \phi_2(u(e_0))(c(\phi_1(u))).$$

The cochain map $\phi^\#$ in turn induces a homomorphism

$$\phi^* = H^n(\phi) : H^n(Y, B; H) \to H^n(X, A; G).$$

Thus the H^n become *contravariant* functors: $\mathscr{L}^* \to \mathscr{A}$. In a similar way, we have coboundary operators; these are natural transformations

$$\delta_n^* = \delta_n(X, A; G) : H^n(A; G \mid A) \to H^{n+1}(X, A; G).$$

If (X_1, A_1) and (X_2, A_2) are pairs in $\mathcal{K}$ such that A_1 is a subspace of A_2 and X_1 of X_2, and if G is a system of local coefficients in X_2, then the inclusion map $(X_1, A_1) \subseteq (X_2, A_2)$, together with the identity map of $G|X_1$, form a morphism $k : (X_1, A_1; G|X_1) \to (X_2, A_2; G)$ in $\mathcal{L}^*$, called an *inclusion.* The induced homomorphism

$$k^* : H^n(X_2, A_2; G) \to H^n(X_1, A_1; G|X_1)$$

is called an *injection.*

The prism over an object $(X, A; G)$ in $\mathcal{L}^*$ is the object $(\mathbf{I} \times X, \mathbf{I} \times A; p^*G)$, and the maps $i_0, i_1 : (X, A) \to (\mathbf{I} \times X, \mathbf{I} \times A)$, together with the identity maps of $i_t^* p^* G = G$ into G, define morphisms $i_0, i_1 : (X, A; G) \to (\mathbf{I} \times X, \mathbf{I} \times A; p^*G)$ in $\mathcal{L}^*$. As in the case of homology, it is obvious how to define the notion of homotopy between morphisms in $\mathcal{L}^*$, and to prove

(2.2*) *Homotopy between morphisms in $\mathcal{L}^*$ is an equivalence relation.* □

(2.3*) *If $\phi \simeq \psi : (X, A; F) \to (Y, B; G)$ and $\phi' \simeq \psi' : (Y, B, G) \to (Z, C; H)$, then $\phi' \circ \phi \simeq \psi' \circ \psi : (X, A; F) \to (Z, C; H)$.* □

We can now formulate the Eilenberg–Steenrod axioms for cohomology with local coefficients.

(2.4*) Theorem *The functors H^q and natural transformations δ_q have the following properties:*

(2.5*) (Exactness). *If $(X, A; G)$ is an object in $\mathcal{L}^*$, i and j are the appropriate inclusions, then the sequence*

$$\cdots \to H^{q-1}(A; G|A) \xrightarrow{\delta_{q-1}(X, A; G)} H^q(X, A; G) \xrightarrow{H^q(j)} H^q(X; G) \xrightarrow{H^q(i)} H^q(A; G|A) \to \cdots$$

is exact.

(2.6*) (Homotopy). *If $\phi_0, \phi_1 : (X, A; G) \to (Y, B; H)$ are homotopic morphisms in $\mathcal{L}^*$, then*

$$H^q(\phi_0) = H^q(\phi_1) : H^q(Y, B; H) \to H^q(X, A; G)$$

for all q.

(2.7*) (Excision). *Let $(X; X_1, X_2)$ be a triad in $\mathcal{K}$ such that $X = \operatorname{Int} X_1 \cup \operatorname{Int} X_2$, and let G be a system of local coefficients in X. Then the injection $H^q(X, X_2; G) \to H^q(X_1, X_1 \cap X_2; G|X_1)$ is an isomorphism for all q.*

(2.8*) (Dimension). *If X is a space consisting of a single point $*$, then $H^q(X; G) = 0$ for all $q \neq 0$ and $H^0(X; G) \approx G(*)$.*

(2.9*) (Additivity). *Let X be the union of a family of mutually disjoint subspaces X_α, A a subspace of X, $A_\alpha = A \cap X_\alpha$. Let G be a system of local coefficients in X, $G_\alpha = G \mid X_\alpha$. Then the injections*

$$H^n(X, A; G) \to H^n(X_\alpha, A_\alpha; G_\alpha)$$

represent the former group as a direct product. □

Remark. When we are dealing with various subsets A of a fixed space X, it is often convenient to abbreviate $H_n(A_1, A_2; G \mid A_1)$ and $H^n(A_1, A_2; G \mid A_1)$ to $H_n(A_1, A_2; G)$ and $H^n(A_1, A_2; G)$ for any system G of local coefficients in X. Thus, for example, the additivity properties (2.9) and (2.9)* can be written

$$H_n(X, A; G) \approx \bigoplus H_n(X_\alpha, A_\alpha; G),$$

$$H^n(X, A; G) \approx \Pi H^n(X_\alpha, A_\alpha; G).$$

Let (X, A) be a relative CW-complex, and let $\{(X_n, A) \mid n = 0, 1, \ldots\}$ be an expanding sequence of subcomplexes with union X. Examples can be given to show that, even in the case of ordinary cohomology theory, it need not be true that $H^q(X, A)$ is the inverse limit of the system $H^q(X_n, A)$.

For example, let $\{A_n\}$ be a sequence of circles, $f_n : A_n \to A_{n+1}$ a map of degree 2, B_n the mapping cylinder of f_n; we may assume that $B_n \cap B_m = \varnothing$ if $|m - n| > 1$, while $B_n \cap B_{n+1} = A_{n+1}$. Let $X = \bigcup_{k=1}^{\infty} B_k$, $X_n = \bigcup_{k=1}^{n} B_k$. Then A_{n+1} is a deformation retract of X_n, so that $H_1(X_n) \approx \mathbf{Z}$, $H_2(X_n) = 0$, while the injection $H_1(X_n) \to H_1(X_{n+1})$ maps a generator of the former group into twice a generator of the latter. Now $H_q(X)$ is the direct limit of the sequence of groups $H_q(X_n)$ under the injections. Thus $H_1(X)$ is the group J_2 of dyadic rationals (i.e., rational numbers whose denominators are powers of 2), while $H_2(X) = 0$.

By the Universal Coefficient Theorem,

$$\begin{aligned} H^2(X; \mathbf{Z}) &\approx \operatorname{Hom}(H_2(X), \mathbf{Z}) \oplus \operatorname{Ext}(H_1(X), \mathbf{Z}) \\ &\approx \operatorname{Ext}(J_2, \mathbf{Z}), \end{aligned}$$

while $H^2(X_n; \mathbf{Z}) = 0$ for every n. Thus it remains to show that the group $\operatorname{Ext}(J_2; \mathbf{Z})$ is non-zero; in fact, it is uncountable.

To see this, let $\mathbf{Z}_0$ be the additive group of rationals, so that there is a short exact sequence

$$0 \to \mathbf{Z} \to \mathbf{Z}_0 \to \mathbf{Z}_0/\mathbf{Z} \to 0,$$

and therefore an exact sequence

$$\begin{aligned} 0 &\to \operatorname{Hom}(J_2, \mathbf{Z}) \to \operatorname{Hom}(J_2, \mathbf{Z}_0) \to \operatorname{Hom}(J_2, \mathbf{Z}_0/\mathbf{Z}) \\ &\to \operatorname{Ext}(J_2, \mathbf{Z}) \to \operatorname{Ext}(J_2; \mathbf{Z}_0) \to \operatorname{Ext}(J_2, \mathbf{Z}_0/\mathbf{Z}) \to 0. \end{aligned}$$

Now $\text{Hom}(J_2, \mathbf{Z}) = 0$ since $\mathbf{Z}$ has no 2-divisible subgroups, and $\text{Ext}(J_2, \mathbf{Z}_0) = \text{Ext}(J_2, \mathbf{Z}_0/\mathbf{Z}) = 0$ because $\mathbf{Z}_0$ and $\mathbf{Z}_0/\mathbf{Z}$ are divisible abelian groups and therefore injective $\mathbf{Z}$-modules. Thus we are concerned with the short exact sequence

$$0 \to \text{Hom}(J_2, \mathbf{Z}_0) \to \text{Hom}(J_2, \mathbf{Z}_0/\mathbf{Z}) \to \text{Ext}(J_2, \mathbf{Z}) \to 0.$$

The group $\text{Hom}(J_2, \mathbf{Z}_0)$ is isomorphic with $\mathbf{Z}_0$, and therefore countable; for if a is any rational number there is a unique homomorphism $f: J_2 \to \mathbf{Z}_0$ such that $f_2(1) = a$. On the other hand, the homomorphisms of J_2 into $\mathbf{Z}_0/\mathbf{Z}$ are in one-to-one correspondence with the sequences $(a_0, a_1, a_2, \ldots)$ of elements of the latter group satisfying the conditions $2a_{i+1} = a_i$ $(i = 0, 1, 2, \ldots)$. For each choice of elements $a_0, \ldots, a_k$ satisfying the above conditions, there are exactly two ways of choosing a_{k+1} so that the conditions are still satisfied. It follows that the set of all such sequences is uncountable, so that $\text{Hom}(J_2, \mathbf{Z}_0/\mathbf{Z})$ is uncountable. Hence $\text{Ext}(J_2, \mathbf{Z})$ is uncountable, too.

Milnor [3] has shown how the relationship between $H^q(X, A; G)$ and $\underleftarrow{\lim}\, H^q(X_n, A; G)$ can be made explicit. To accomplish this, one needs the notion of the first derived functor $\underleftarrow{\lim}^1$ of the inverse limit functor. Specifically, let

$$G_1 \xleftarrow{f_1} G_2 \xleftarrow{f_2} G_3 \xleftarrow{f_3} G_4 \leftarrow \cdots$$

be an inverse system of abelian groups and homomorphisms. Let $G = \prod_{n=1}^{\infty} G_n$, and define an endomorphism $d : G \to G$ by

$$d(x_1, x_2, \ldots) = (x_1 - f_1(x_2), x_2 - f_2(x_3), \ldots).$$

Then $\text{Ker}\, d = \underleftarrow{\lim}\, G_n$, and we define

$$\underleftarrow{\lim}^1 G_n = \text{Cok}\, d.$$

Milnor's theorem, generalized to the case of local coefficients, asserts:

(2.10*) Theorem *Let (X, A) be a relative* CW*-complex, and let $\{(X_n, A) \mid n = 0, 1, 2, \ldots\}$ be an expanding sequence of subcomplexes such that $\bigcup_{n=0}^{\infty} X_n = X$. Let G be a system of local coefficients in X. Then there is an exact sequence*

$$0 \longrightarrow \underleftarrow{\lim}^1 H^{q-1}(X_n, A; G) \xrightarrow{\alpha} H^q(X, A; G) \xrightarrow{\beta}$$

$$\underleftarrow{\lim}\, H^q(X_n, A; G) \longrightarrow 0,$$

in which β is induced by the injections $H^q(X, A; G) \to H^q(X_n, A; G)$.

Milnor's proof applies without essential change; we shall merely sketch it. Let $\mathbf{R}^+$ be the set of non-negative real numbers, given the structure of a CW-complex whose vertices are the non-negative integers and whose 1-cells are the closed intervals $[n, n+1]$. Let

$$L = \bigcup_{n=0}^{\infty} X_n \times [n, n+1],$$

so that $(L, A \times \mathbf{R}^+)$ is a subcomplex of the CW-complex $(X \times \mathbf{R}^+, A \times \mathbf{R}^+)$. The restriction to $(L, A \times \mathbf{R}^+)$ of the projection of $(X \times \mathbf{R}^+, A \times \mathbf{R}^+)$ on the first factor is a homotopy equivalence $p : (L, A \times \mathbf{R}^+) \to (X, A)$. Let

$$L_1 = \bigcup_{i \geq 0} X_{2i} \times [2i, 2i+1],$$

$$L_2 = \bigcup_{i \geq 0} X_{2i+1} \times [2i+1, 2i+2],$$

$$A_i = L_i \cap (A \times \mathbf{R}^+) \qquad (i = 1, 2).$$

Then there is an exact Mayer–Vietoris sequence

(2.11)

$$\cdots \to \xrightarrow{\lambda_{q-1}} H^{q-1}(L_1 \cap L_2, A_1 \cap A_2; \bar{G}) \to H^q(L, A \times \mathbf{R}^+; \bar{G}) \to$$

$$H^q(L_1, A_1; \bar{G}) \oplus H^q(L_2, A_2; \bar{G}) \xrightarrow{\lambda_q} H^q(L_1 \cap L_2, A_1 \cap A_2; \bar{G}) \to \cdots$$

with local coefficients in $\bar{G} = p^*G$. But the Additivity Property implies that

$$H^q(L_i, A_i; \bar{G}) \approx \prod_{k=0}^{\infty} H^q(X_{2k+i}, A; G)$$

so that

$$H^q(L_1, A_1; \bar{G}) \oplus H^q(L_2, A_2; \bar{G}) \approx \prod_{n=0}^{\infty} H^q(X_n, A; G),$$

while

$$H^q(L_1 \cap L_2, A_1 \cap A_2; G) \approx \prod_{n=0}^{\infty} H^q(X_n, A; G).$$

Under these isomorphisms the homomorphism λ_q induced by the appropriate injections corresponds to the endomorphism d constructed above, in the definition of $\underleftarrow{\lim}^1$. But the exact sequence (2.11) induces a short exact sequence

$$0 \to \operatorname{Cok} \lambda_{q-1} \to H^q(L, A \times \mathbf{R}^+; \bar{G}) \to \operatorname{Ker} \lambda_q \to 0.$$

Since the middle group can be identified with $H^q(X, A; G)$ and the two end groups with the appropriate $\underleftarrow{\lim}^1$, $\underleftarrow{\lim}$, the desired result follows. □

3 Computations and Examples

(3.1) Theorem *Let X be a space, G a system of local coefficients in X, $\{X_\alpha \mid \alpha \in J\}$ the path-components of X, $G_\alpha = G \mid X_\alpha$. Then the injections*

$$H_n(X_\alpha; G_\alpha) \to H_n(X; G)$$

represent the latter group as a direct sum.

In fact, since the image $u(\Delta^q)$ of a singular simplex $u : \Delta^q \to X$ is pathwise connected, $u(\Delta^q) \subset X_\alpha$ for some α, and the chain complex $C_*(X; G)$ decomposes as the direct sum of the subcomplexes $C_*(X_\alpha; G_\alpha \mid X_\alpha)$. □

Because of Theorem (3.1) we may often assume that the spaces with which we are dealing are pathwise connected.

Let X be a 0-connected space with base point x_0, and let G be a system of local coefficients in X, $G_0 = G(x_0)$. Then G_0 is a $\pi_1(X)$-module; let H_0 be the subgroup of G_0 generated by all elements of the form $x - \xi x$ with $x \in G_0$, $\xi \in \pi_1(X)$, and let H_0^* be the subgroup of G_0 consisting of all elements x such that $\xi x = x$ for all $x \in \pi_1(X)$.

(3.2) Theorem *The 0-dimensional homology and cohomology groups of a pathwise-connected space X with coefficients in the local system G are given by*

$$H_0(X; G) \approx G_0/H_0,$$
$$H^0(X; G) \approx H_0^*.$$

Define a homomorphism $\phi : G_0 \to C_0(X; G) = Z_0(X; G)$ by

$$\phi(g) = g \cdot x_0.$$

If $\xi \in \pi_1(X, x_0)$, $u : (\Delta^1, \dot{\Delta}^1) \to (X, x_0)$ is a representative path, then

$$\begin{aligned}\partial(gu) &= G(\xi)(g)u(1) - gu(0)\\ &= (G(\xi)g - g)x_0,\end{aligned}$$

so that $\phi(H_0) \subset B_0(X; G)$ and ϕ induces

$$\bar{\phi} : G_0/H_0 \to H_0(X; G)$$

The easy proof that $\bar{\phi}$ is an isomorphism is left to the reader, as is the calculation of $H^0(X; G)$. □

Before stating our next results, we need some notions from algebra. Let Π be a group, and let G be an abelian group on which Π operates. A *crossed homomorphism* of Π into G is a function $f : \Pi \to G$ such that $f(\alpha\beta) = f(\alpha) + \alpha f(\beta)$ for all $\alpha, \beta \in \Pi$. The set of all functions from Π to G is an abelian group under the operation of pointwise addition. And it is trivial to verify that the set of all crossed homomorphisms of Π into G is a subgroup $Q(\Pi, G)$.

A *principal homomorphism* of Π into G is a function f such that

$$f(\alpha) = \alpha g - g$$

for some $g \in G$ and all $\alpha \in \Pi$. Every principal homomorphism is a crossed homomorphism, and the principal homomorphisms form a subgroup $P(\Pi, G)$ of $Q(\Pi, G)$.

(3.3) Theorem *Let X be a 0-connected space, $x_0 \in X$, and let G be a system of local coefficients in X, $G_0 = G(x_0)$. Then the local coefficient system G determines an action of $\pi_1(X)$ on G_0, and, with respect to this action,*

$$H^1(X; G) \approx Q(\pi_1(X), G_0)/P(\pi_1(X), G_0).$$

It will simplify the proof to consider only singular simplices $u : \Delta^q \to X$ with the property that u maps each vertex of Δ^q into the base point $*$ (cf. Exercise 8, below).

Let $\Pi = \pi_1(X)$, and let $\phi \in Q(\Pi, G_0)$. Let $\hat{\phi}$ be the 1-cochain such that, for every singular 1-simplex $u : (\Delta^1, \dot{\Delta}^1) \to (X, *)$

$$\hat{\phi}(u) = \phi(\sigma_u^{-1}).$$

The function $\hat{\phi}$ is a cocycle. For let $w : \Delta^2 \to X$ be a singular 2-simplex, and let $\xi_i = \sigma_{\partial_i w}^{-1}$ $(i = 0, 1, 2)$. Then $\xi_1 = \xi_2 \xi_0$ (cf. Figure 6.3), and therefore

$$\phi(\xi_1) = \phi(\xi_2) + \xi_2 \phi(\xi_0),$$
$$\hat{\phi}(\partial_1 w) = \hat{\phi}(\partial_2 w) + \xi_2 \hat{\phi}(\partial_0 w).$$

But $\xi_2 = \sigma_w^{-1}$, and so

$$-\delta\hat{\phi}(w) = \xi_2 \hat{\phi}(\partial_0 w) - \hat{\phi}(\partial_1 w) + \hat{\phi}(\partial_2 w) = 0.$$

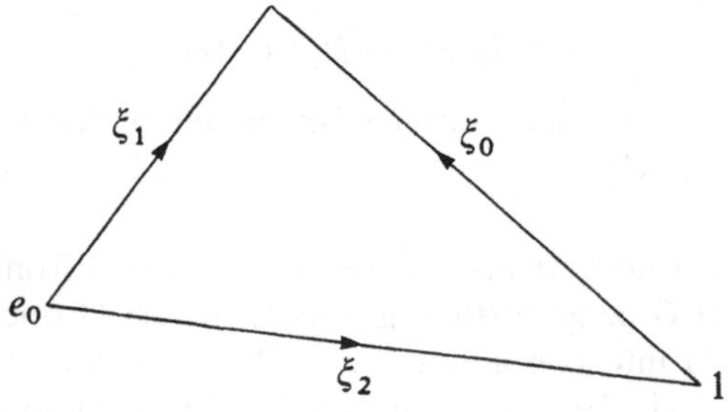

Figure 6.3

The correspondence $\phi \to \hat{\phi}$ is evidently a homomorphism (in fact, a monomorphism)

$$\theta : Q(\Pi, G_0) \to Z^1(X; G).$$

If ϕ is a principal homomorphism, so that $\phi(\xi) = \xi g - g$ with $g \in G_0$, then g may be regarded as a 0-cochain, and

$$\hat{\phi}(u) = \phi(\sigma_n^{-1}) = \sigma_u^{-1} g - g = \delta g(u),$$

so that $\hat{\phi} = \delta g$ is a coboundary. Conversely, if $\hat{\phi} = \delta g$, then

$$\phi(\sigma_u^{-1}) = \hat{\phi}(u) = \delta g(u) = \sigma_u^{-1} g - g$$

and therefore

$$\phi(\xi) = \xi g - g$$

for any $\xi \in \Pi$. Hence ϕ is a principal homomorphism.

It follows that θ induces a monomorphism

$$\bar{\theta} : Q(\Pi, G_0)/P(\Pi, G_0) \to H^1(X; G),$$

and the proof of the theorem will be complete when we have shown that $\bar{\theta}$ is an epimorphism.

Let $f \in Z^1(X; G)$. Let $u : (\Delta^1, \dot{\Delta}^1) \to (X, *)$ represent an element $\xi \in \Pi$ (i.e., $\sigma_u^{-1} = \xi$); then $f(u) \in G_0$. Let v be another representative of ξ, and $h : \mathbf{I} \times \Delta^1 \to X$ a homotopy of u to v (rel. $\dot{\Delta}^1$). Let $p : \mathbf{I} \times \Delta^1 \to \Delta^2$ be the map given by

$$p(s, (1 - t)\mathbf{e}_0 + t\mathbf{e}_1) = (1 - t)\mathbf{e}_0 + t(1 - s)\mathbf{e}_1 + st\mathbf{e}_2 .$$

Then the reader may verify that $w = h \circ p^{-1} : \Delta^2 \to X$ is well-defined and that $\partial_0 w = *$, $\partial_1 w = v$, $\partial_2 w = u$. Since f is a cocycle,

$$\begin{aligned} 0 = -\delta f(w) &= \xi f(*) - f(v) + f(u) \\ &= f(u) - f(v). \end{aligned}$$

Therefore $f(u)$ depends only on the homotopy class of u and hence there is a function $\phi : \Pi \to G_0$ such that $\phi(\xi) = f(u)$ for any 1-simplex u such that $\sigma_u^{-1} = \xi$. Again the reader may verify that ϕ is a crossed homomorphism and that $\hat{\phi} = f$. □

Let X be a space on which a group Π acts (on the left, say). Then Π acts on the singular chain complex $C_*(X)$; if $\xi \in \Pi$ and $u : \Delta^q \to X$ is a singular simplex, we may define $\xi u : \Delta^q \to X$ by

$$\xi u(t) = \xi \cdot u(t) \qquad (t \in \Delta^n).$$

The action of Π on the singular simplices is extended by linearity to an action on $C_*(X)$, so that the latter group becomes a Π-module. Moreover, it is clear that the boundary operator of $C_*(X)$ commutes with the action, so that $C_*(X)$ is a complex of Π-modules. Thus Π operates on $H_*(X)$.

If G is an abelian group on which Π operates on the right, we may form the complex $G \otimes_\Pi C_*(X)$; it is the quotient of the tensor product $G \otimes C_*(X)$ by the subgroup $Q(G, X)$ generated by all elements of the form $g\xi \otimes c - g \otimes \xi c$ ($g \in G$, $\xi \in \Pi$, $c \in C_*(X)$). As the boundary operator of $G \otimes C_*(X)$ maps $Q(G, X)$ into itself, it induces an endomorphism ∂ of $G \otimes_\Pi C_*(X)$. The latter group is a chain complex under ∂, and its homology groups

$$E_q(X; G) = H_q(G \otimes_\Pi C_*(X))$$

are called the *equivariant homology groups of* X with coefficients in the Π-module G. The projection

$$p : G \otimes C_*(X) \to G \otimes_\Pi C_*(X)$$

is a chain map, inducing a homomorphism

$$p_* : H_q(X; G) \to E_q(X; G).$$

If, on the other hand, G is a left Π-module, the group of operator homomorphisms

$$\mathrm{Hom}^\Pi(C_*(X), G)$$

is a subcomplex of the cochain complex

$$C^*(X; G) = \mathrm{Hom}\{C_*(X), G\}.$$

Accordingly we may define its homology groups to be the equivariant cohomology groups $E^q(X; G)$, and there is an injection homomorphism

$$i_* : E^q(X; G) \to H^q(X; G).$$

Now let X be a 0-connected and semilocally 1-connected space with base point x_0, and let G be a system of local coefficients in X. Let $p : \tilde{X} \to X$ be the universal covering of X, $\tilde{x}_0 \in p^{-1}(x_0)$ the base point of $\tilde{X}$, Π the group of covering translations. There is an isomorphism $h : \pi_1(X, x_0) \to \Pi$ such that, if $\xi \in \pi_1(X, x_0)$ and $u : \mathbf{I} \to \tilde{X}$ is a path from $\tilde{x}_0$ to $h_\xi(\tilde{x}_0)$, then $p \circ u : (\mathbf{I}, \dot{\mathbf{I}}) \to (X, x_0)$ is a representative of ξ.

The group $\pi_1(X, x_0)$ acts on G_0 (on the left). We convert this to a right action of Π on G_0 by

$$g \cdot h_\xi = G(\xi)^{-1}(g).$$

(3.4) Theorem (Eilenberg). *The homology groups $H_q(X; G)$ with respect to the system G of local coefficients are isomorphic with the equivariant homology groups $E_q(\tilde{X}; G_0)$.*

We shall exhibit an isomorphism

$$\tilde{p} : G_0 \otimes_\Pi C_*(\tilde{X}) \to C_*(X; G).$$

We first observe that, since $\tilde{X}$ is 1-connected, there is, for each $y \in \tilde{X}$, a unique homotopy class $\xi(y)$ of paths from y to $\tilde{x}_0$. Let $g_0 \in G_0$, $w : \Delta^q \to \tilde{X}$, $u = p \circ w$, $y = w(\mathbf{e}_0)$, $x = u(\mathbf{e}_0) = p(y_0)$, and define $\tilde{p}_0 : G_0 \otimes C_*(\tilde{X}) \to C_*(X; G)$ by

$$\tilde{p}_0(g_0 \otimes w) = g \cdot u,$$

where

$$g = G(\underline{p}(\xi(y)))(g_0) \in G(x).$$

(See Figure 6.4). Let $\eta \in \pi_1(X, x_0)$. Then $\underline{h}_\eta(\xi(y)) \in \pi_1(\tilde{X}; h_\eta(y), h_\eta(\tilde{x}_0))$, and $\xi(h_\eta(\tilde{x}_0)) \in \pi_1(\tilde{X}; h_\eta(\tilde{x}_0), \tilde{x}_0)$. Therefore their product is defined and equal to $\xi(h_\eta(y))$, and hence

$$\tilde{p}_0(g_0 \otimes h_\eta \circ w) = g' \cdot u,$$

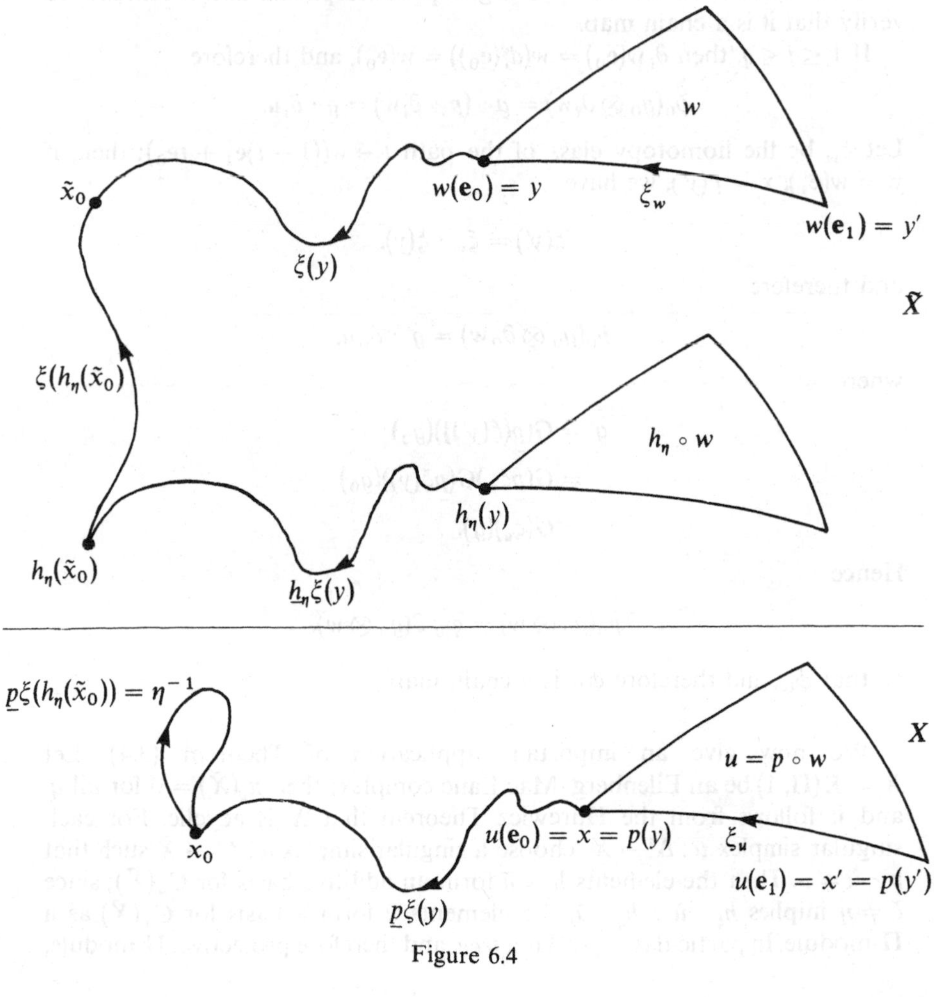

Figure 6.4

where

$$\begin{aligned} g' &= G(\underline{p}\xi(h_\eta(y)))(g_0) \\ &= G(\underline{p}h_\eta \xi(y))G(\underline{p}\xi(h_\eta(\tilde{x}))(g_0) \\ &= G(\underline{p}\xi(y))(G(\eta^{-1}))(g_0) \\ &= G(\underline{p}\xi(y))(g_0 \eta). \end{aligned}$$

But

$$\tilde{p}_0(g_0 \eta \otimes w) = g'u,$$

and therefore $\tilde{p}_0 Q(G, X) = 0$. Thus $\tilde{p}_0$ induces a homomorphism

$$\tilde{p} : G_0 \otimes_\Pi C_*(\tilde{X}) \to C_*(X; G).$$

It is not difficult to see that $\tilde{p}$ is a group isomorphism, and it remains to verify that it is a chain map.

If $1 \le i \le q$, then $\partial_i w(\mathbf{e}_0) = w(d_i^q(\mathbf{e}_0)) = w(\mathbf{e}_0)$, and therefore

$$\tilde{p}_0(g_0 \otimes \partial_i w) = g \cdot (p \circ \partial_i w) = g \cdot \partial_i u.$$

Let ξ_w be the homotopy class of the path $t \to w((1-t)\mathbf{e}_1 + t\mathbf{e}_0)$; then, if $y' = w(\mathbf{e}_1)$, $x' = p(y')$, we have

$$\xi(y') = \xi_w \cdot \xi(y),$$

and therefore

$$\tilde{p}_0(g_0 \otimes \partial_0 w) = g' \cdot \partial_0 u,$$

where

$$\begin{aligned} g' &= G(\underline{p}(\xi(y')))(g_0) \\ &= G(\underline{p}\xi_w)G(\underline{p}\xi(y))(g_0) \\ &= G(\xi_u)(g). \end{aligned}$$

Hence

$$\partial\phi_0(g_0 \otimes w) = \phi_0\, \partial(g_0 \otimes w),$$

so that ϕ_0, and therefore ϕ_1, is a chain map. □

We now give an important application of Theorem (3.4). Let $X = K(\Pi, 1)$ be an Eilenberg–Mac Lane complex; then $\pi_q(\tilde{X}) = 0$ for all q, and it follows from the Hurewicz Theorem that $\tilde{X}$ is acyclic. For each singular simplex $u : \Delta^q \to X$, choose a singular simplex $\tilde{u} : \Delta^q \to \tilde{X}$ such that $p \circ \tilde{u} = u$. Then the elements $h_\xi \circ \tilde{u}$ form an additive basis for $C_*(\tilde{X})$; since $\xi \ne \eta$ implies $h_\xi \circ \tilde{u} \ne h_\eta \circ \tilde{u}$, the elements $\tilde{u}$ form a basis for $C_*(\tilde{X})$ as a Π-module. In particular, $C_*(\tilde{X})$ is a *free*, and therefore projective, Π-module, so that $C_*(\tilde{X})$ is a projective resolution of the group $\mathbf{Z}$, considered as a Π-module with trivial action. Hence

$$E_q(\tilde{X}; G_0) = \mathrm{Tor}_q^{\Pi}(\mathbf{Z}, G_0)$$

and therefore

(3.5) Theorem *If* Π *is a group and* G *a system of local coefficients in* $K(\Pi, 1)$, *then*

$$H_q(K(\Pi, 1), G) \approx \mathrm{Tor}_q^{\Pi}(G_0, \mathbf{Z}). \qquad \square$$

This is the generalization to local coefficients of Corollary (4.7) of Chapter V.

The corresponding results for cohomology also hold; their proofs involve little that is new, and may safely be left to the reader.

(3.4*) Theorem *The cohomology groups* $H^q(X; G)$ *of a space* X *with respect to the system* G *of local coefficients are isomorphic with the equivariant cohomology groups* $E^q(\tilde{X}; G_0)$ *of its covering space.* $\square$

(3.5*) Theorem *If* Π *is a group, and* G *a system of local coefficients in* $K(\Pi, 1)$, *then*

$$H^q(K(\Pi, 1); G) \approx \mathrm{Ext}_{\Pi}^q(\mathbf{Z}; G_0). \qquad \square$$

4 Local Coefficients in CW-complexes

In this section we shall show how the results of §2 of Chapter II can be modified to take local coefficients into account.

Let (X, A) be a relative CW-complex, G a system of local coefficients in X. As we shall be dealing with various pairs $(X_1, A_1) \subset (X, A)$, it will be convenient, in accordance with the Remark after (2.9*), to abbreviate $H_q(X_1, A_1; G \mid X_1)$ to $H_q(X_1, A_1; G)$. As our discussion will be based, for the most part, on Theorem (2.4), the arguments follow very closely those of §2, Chapter II. Therefore our treatment will be rather sketchy, attention being paid to the points of difference.

Let X_n be the n-skeleton of (X, A), and let $\{E_\alpha^n\}$ be the n-cells of (X, A), with characteristic maps $h_\alpha : (\Delta^n, \dot{\Delta}^n) \to (X_n, X_{n-1})$. Let $G_\alpha = h_\alpha^* G$; since Δ^n is 1-connected, the system G_α is simple. This means that the groups at different points of Δ^n are connected by *uniquely defined* isomorphisms; i.e., they form a *transitive system* of groups in the sense of Eilenberg–Steenrod [E–S, p. 17]. Let $\bar{G}_\alpha$ be the group defined by this transitive system (i.e., its direct limit). Then there are uniquely defined isomorphisms

$$H_q(\Delta^n, \dot{\Delta}^n; G_\alpha) \approx H_q(\Delta^n, \dot{\Delta}^n; \bar{G}_\alpha).$$

In particular, $H^q(\Delta^n, \dot{\Delta}^n; G_\alpha) = 0$ for all $q \neq n$, while

$$H_n(\Delta^n, \dot{\Delta}^n; G_\alpha) \approx \bar{G}_\alpha.$$

If $g \in \bar{G}_\alpha$, let $g\delta_n$ be the element of $H_q(\Delta^n, \dot{\Delta}^n; G_\alpha)$ which corresponds to it under the above isomorphisms. Let $z_\alpha = h_\alpha(\mathbf{e}_0)$; then the map h_α induces a well-defined isomorphism $\bar{G}_\alpha \to G(z_\alpha)$. If $g \in G(z_\alpha)$ and g' is the corresponding element of $\bar{G}_\alpha$, let ge_α^n be the image in $H_n(E_\alpha^n, \dot{E}_\alpha^n; G)$ (or in $H_n(X_n, X_{n-1}; G)$) of $g'\delta_n$ under the homomorphism induced by h_α.

(4.1) Theorem *The homomorphisms* $H_q(\Delta^n, \dot{\Delta}^n; \bar{G}_\alpha) \to H_q(X_n, X_{n-1}; G)$ *induced by the characteristic maps* $h_\alpha : (\Delta^n, \dot{\Delta}^n) \to (X_n, X_{n-1})$ *represent the latter group as a direct sum. If* $q \neq n$, *then* $H_q(X_n, X_{n-1}; G) = 0$. *The group* $H_n(X_n, X_{n-1}; G)$ *is isomorphic with the group of all formal sums* $\sum_\alpha g_\alpha e_\alpha^n$ *with* $g_\alpha \in G(z_\alpha)$ *and* $g_\alpha = 0$ *for almost all* α.

Let U be the subset of X_n obtained by removing the point $x_\alpha = h_\alpha(b_n)$ for each α. Then X_{n-1} is a deformation retract of U, and therefore the injection

$$i_1 : H_q(X_n, X_{n-1}; G) \to H_q(X_n, U; G)$$

is an isomorphism. Let $V = X_n - X_{n-1}$, $W = V \cap U$; then X_n is the union of the relatively open sets U, V, and therefore the injection

$$i_2 : H_q(V, W; G) \to H_q(X_n, U; G)$$

is an isomorphism. Finally, let $V_\alpha = \operatorname{Int} E_\alpha^n$, $W_\alpha = V_\alpha \cap U$. By the Additivity Theorem, the injections

$$i_\alpha : H_q(V_\alpha, W_\alpha; G) \to H_q(V, W; G)$$

represent the latter group as a direct sum. The homomorphisms

$$H_q(\Delta^n, \dot{\Delta}^n; G_\alpha) \to H_q(X_n, X_{n-1}; G),$$
$$H_q(\Delta^n, \Delta^n - \{b_n\}; G_\alpha) \to H_q(X_n, U; G),$$
$$H_q(\operatorname{Int} \Delta^n, \operatorname{Int} \Delta^n - \{b_n\}; G_\alpha) \to H_q(V, W; G)$$

induced by h_α define homomorphisms

$$h_1 : \bigoplus_\alpha H_q(\Delta^n, \dot{\Delta}^n; G_\alpha) \to H_q(X_n, X_{n-1}; G),$$

$$h_2 : \bigoplus_\alpha H_q(\Delta^n, \Delta^n - \{b_n\}; G_\alpha) \to H_q(X_n, U; G),$$

$$h_3 : \bigoplus_\alpha H_q(\operatorname{Int} \Delta^n, \operatorname{Int} \Delta^n - \{b_n\}; G_\alpha) \to H_q(V, W; G).$$

The map $h_\alpha \,|\, \text{Int}\,\Delta^n : (\text{Int}\,\Delta^n, \text{Int}\,\Delta^n - \{b_n\}) \to (V_\alpha, W_\alpha)$ is a homeomorphism, inducing an isomorphism

$$H_q(\text{Int}\,\Delta^n, \text{Int}\,\Delta^n - \{b_n\}; G_\alpha) \approx H_n(V_\alpha, W_\alpha; G);$$

the direct sum of the latter isomorphisms is an isomorphism

$$h_4 : \bigoplus_\alpha H_n(\text{Int}\,\Delta^n, \text{Int}\,\Delta^n - \{b_n\}; G_\alpha) \approx \bigoplus_\alpha H_n(V_\alpha, W_\alpha; G).$$

There is a commutative diagram (Figure 6.5), in which j_1 and j_2 are the direct sums of the appropriate injections and i is the homomorphism induced by the i_α. Since j_1, j_2, i_1, i_2, i, and h_4 are isomorphisms, so is h_1. This proves the first statement of Theorem (4.1), and the remaining ones follow immediately.

$$\begin{array}{ccccccc}
\bigoplus_\alpha H_q(\text{Int}\,\Delta^n, \text{Int}\,\Delta^n - \{b_n\}; G_\alpha) & & \xrightarrow{j_2} & \bigoplus_\alpha H_q(\Delta^n, \Delta^n - \{b_n\}; G_\alpha) & \xleftarrow{j_1} & \bigoplus_\alpha H_q(\Delta^n, \dot{\Delta}^n; G_\alpha) \\
\swarrow h_4 \quad \searrow h_3 & & & \downarrow h_2 & & \downarrow h_1 \\
\bigoplus_\alpha H_q(V_\alpha, W_\alpha; G) & \xrightarrow[i]{} & H_q(V, W; G) \xrightarrow[i_2]{} & H_q(X_n, U; G) & \xleftarrow[i_1]{} & H_q(X_n, X_{n-1}; G)
\end{array}$$

Figure 6.5

(4.2) Corollary *If* $q \le n < m$, *then* $H_q(X_m, X_n; G) = 0$. *If* $q > m > n$, *then* $H_q(X_m, X_n; G) = 0$.

This follows as in §2 of Chapter II. □

(4.3) Corollary *If* $q \le n$, *then* $H_q(X, X_n; G) = 0$.

This follows from the fact that singular homology with local coefficients, like ordinary singular theory, has compact carriers. Alternatively, it can be deduced from the axioms, including additivity, as in Milnor [3]. □

Continuing with the argument suggested by §2 of Chapter II, let us define

$$\Gamma_n(X, A; G) = H_n(X_n, X_{n-1}; G)$$

and observe that the composite

$$H_{n+1}(X_{n+1}, X_n; G) \xrightarrow{\partial_{n+1}} H_n(X_n, X_{n-1}; G) \xrightarrow{\partial_n} H_{n-1}(X_{n-1}, X_{n-2}; G)$$

is zero. Thus $\{\Gamma_n(X, A; G);\ \partial_n : \Gamma_n(X, A; G) \to \Gamma_{n-1}(X, A; G)\}$ is a graded chain complex $\Gamma_*(X, A; G)$, and

(4.4) Theorem *The homology groups* $H_n(\Gamma_*(X, A; G))$ *are isomorphic with the singular homology groups* $H_n(X, A; G)$.

The calculation of the boundary operator is subject to the same difficulties as in the case of ordinary homology theory, with the additional complication due to the fact that the cells E_α^n may not be simply connected. Let us illustrate this with an example.

Let $(X, A) = (\mathbf{P}^2, \varnothing)$, where $\mathbf{P}^2 = \mathbf{P}^2(\mathbf{R})$ is the real projective plane. As in §7 of Chapter II, $\mathbf{P}^2$ has a CW-decomposition with one 2-cell $\bar{E}^2$, one 1-cell $\bar{E}^1$, and one 0-cell $\bar{E}^0 = *$, the base point for $\pi_1(\mathbf{P}^2) = \mathbf{Z}_2$. A characteristic map $h_1 : (\Delta^1, \dot{\Delta}^1) \to (\bar{E}^1, \bar{E}^0)$ is a relative homeomorphism. There is a characteristic map $h_2 : (\Delta^2, \dot{\Delta}^2) \to (\bar{E}^2, \bar{E}^1)$ such that $\partial_2 h_2 = \partial_0 h_2 = h_1$, while $\partial_1 h_2 : \Delta^1 \to \bar{E}^0$ is the constant map. (See Figure 6.6). The map h_1 represents the non-zero element ξ of $\pi_1(\mathbf{P}^2)$. And we see that

$$\sigma_{h_1} = \xi = \sigma_{h_2}.$$

Thus

$$\partial(gh_1) = (\xi \cdot g)h_0 - g \cdot h_0,$$
$$\partial(gh_2) = (\xi \cdot g)h_1 - g \cdot * + g \cdot h_1.$$

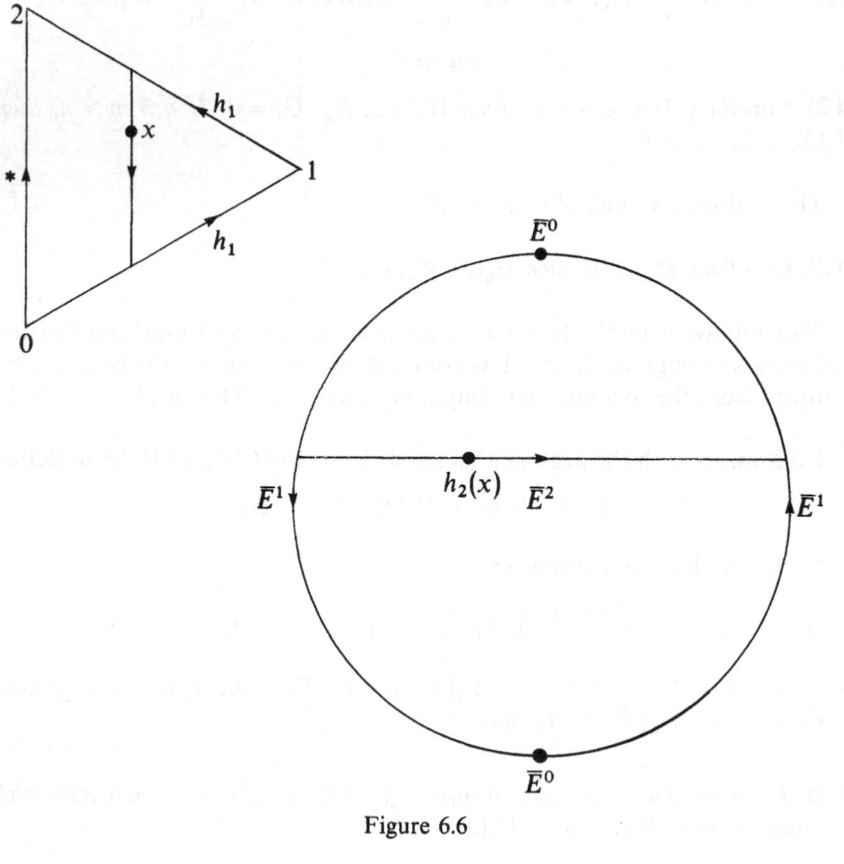

Figure 6.6

The chain groups $\Gamma_i(\mathbf{P}^2; G)$ are generated by the homology classes $1 \cdot \bar{e}_i$ of the singular chains h_i, while the homology class of the constant simplex $*$ is zero. Thus, in $\Gamma_*(\mathbf{P}^2; G)$, we have

$$\partial(g\bar{e}_2) = ((\xi \cdot g) + g)\bar{e}_1,$$

$$\partial(g\bar{e}_1) = ((\xi \cdot g) - g)\bar{e}_0.$$

Suppose, for example that G is the simple system $\mathbf{Z}$ of integers. Then $\xi \cdot g = g$, and we have

$$\partial(g\bar{e}_2) = 2g\bar{e}_1,$$

$$\partial(g\bar{e}_1) = 0,$$

and deduce

$$H_0(\mathbf{P}^2; \mathbf{Z}) = \mathbf{Z}, \quad H_1(\mathbf{P}^2; \mathbf{Z}) = \mathbf{Z}_2, \quad H_2(\mathbf{P}^2; \mathbf{Z}) = 0.$$

Suppose, on the other hand, that G is the system $\mathscr{Z}$ of *twisted integers*, the fundamental group operating non-trivially, so that $\xi \cdot g = -g$. Then

$$\partial(g\bar{e}_2) = 0,$$

$$\partial(g\bar{e}_1) = -2g\bar{e}_0,$$

and accordingly

$$H_0(\mathbf{P}^2; \mathscr{Z}) = \mathbf{Z}_2, \quad H_1(\mathbf{P}^2; \mathscr{Z}) = 0, \quad H_2(\mathbf{P}^2; \mathscr{Z}) = \mathbf{Z}.$$

We conclude our discussion of homology by describing the chain complex $\Gamma(X; G)$ when $X = |K|$ is the space of an ordered simplicial complex. Let $x_0 < x_1 < \cdots < x_p$ be the vertices of a simplex σ of K. Then we may define the characteristic map $h_\sigma : \Delta^p \to |K|$ by $h_\sigma = |f_\sigma|$, where f_σ is the simplicial map which preserves the order of the vertices. We also take the point z_σ to be the leading vertex x_0 of σ. Then

$$z_{\partial_i \sigma} = z_\sigma = x_0 \quad \text{if } i > 0,$$

but

$$z_{\partial_0 \sigma} = x_1.$$

We shall calculate the boundary operator by the method of the universal example. That is, we shall first derive the desired formula for the case $X = \Delta^q$, h_σ is the identity map. When this is so, and G is an abelian group, we have

$$\text{(4.5)} \qquad \partial(g \cdot h_\sigma) = \sum_{i=0}^{p} (-1)^i g \cdot \partial_i h_\sigma .$$

If, instead, we assume that G is a system of local coefficients then G is simple, so that

$$\text{(4.6)} \qquad \Gamma_*(X; G) \approx \Gamma_*(X, G_x)$$

for any point $x \in X$. Thus we must modify (4.5) by assuming that $g \in G(x_0) = G(z_\alpha)$. Then $g \in G(z_{\partial i \sigma})$ if $i > 0$, but $g \notin G(z_{\partial_0 \sigma})$. The isomorphism (4.6) requires that we replace the coefficient $g \in G(x_0)$ of $\partial_0 h_\sigma$ by $G(\xi)(g) \in G(x_1)$ for any path in Δ^p from x_1 to x_0 (they are all homotopic, since Δ^p is 1-connected). We may as well take ξ to be the homotopy class ξ_σ of the rectilinear path. Then (4.5) has to be rewritten

$$\partial(g \cdot h_\sigma) = G(\xi_\sigma)(g) \cdot \partial_0 h_\sigma + \sum_{i=1}^{p} (-1)^i g \cdot \partial_i h_\sigma. \tag{4.7}$$

Applying the argument of the universal example now gives the desired result in general.

Let us summarize what we have found. For an ordered simplicial complex K and system G of local coefficients in $|K|$, let us define an ordered q-cell of G to be a sequence $x = \langle x_0, \ldots, x_q \rangle$ of vertices of K, all of which belong to a simplex of K, and such that $x_0 < x_1 < \cdots < x_q$. The ith *face* $\partial_i x$ of x is the sequence $\langle x_0, \ldots, \hat{x}_i, \ldots, x_q \rangle$. The *leading vertex* $e_0(x)$ of x is x_0; its *leading edge* is $\langle x_0, x_1 \rangle$, and ξ_x is the homotopy class of the path $t \to (1-t)x_1 + tx_0$. The group $\Gamma_q(K; G)$ is the additive group of all finite formal sums $\sum_{i=1}^m g_i x_i$, where the x_i are ordered q-cells of K and $g_i \in G(e_0(x_i))$. And the boundary operator $\partial : \Gamma_q(K; G) \to \Gamma_{q-1}(K; G)$ is defined by

$$\partial(gx) = G(\xi_x)(g) \cdot \partial_0 x + \sum_{i=1}^{q} (-1)^i g \cdot \partial_i x.$$

Under the above definition the graded group $\Gamma_*(K; G)$ becomes a chain complex, and

(4.8) Theorem *If K is an ordered simplicial complex and G a system of local coefficients in $|K|$, then the homology groups $H_q(|K|; G)$ are isomorphic with the homology groups $H_q(\Gamma_*(K; G))$.* □

In Chapter II we derived our results on cohomology from those on homology with the aid of the Universal Coefficient Theorem. This recourse not being available to us now, we must use more direct methods. The key to the argument is the use, paralleling Milnor's in [3], of the Additivity Theorem.

(4.1*) Theorem *The homomorphisms $H^q(X_n, X_{n-1}; G) \to H^q(\Delta^n, \dot{\Delta}^n; \bar{G}_\alpha)$ induced by the characteristic maps $h_\alpha : (\Delta^n, \dot{\Delta}^n) \to (X_n, X_{n-1})$ represent the former group as a direct product. If $q \neq n$, then $H^q(X_n, X_{n-1}; G) = 0$. The group $H^n(X_n, X_{n-1}; G)$ is isomorphic with the group of all functions c which assign to each n-cell E_α^n of (X, A) an element $c(e_\alpha^n) \in G(z_\alpha)$.*

The proof is entirely parallel to that of Theorem (4.1). □

(4.2*) Corollary *If $q \le n < m$, then $H^q(X_m, X_n; G) = 0$. If $q > m > n$, then $H^q(X_m, X_n; G) = 0$.* □

More difficult is the proof of

(4.3*) Theorem *If $q \le n$, then $H^q(X, X_n; G) = 0$.*

To prove Theorem (4.3*), we need Theorem (2.10*). We have seen that $H^q(X_{n+1}, X_n; G) = 0$ if $q \le n$; by exactness, the injection

$$H^q(X_{n+1}, A; G) \to H^q(X_n, A; G)$$

is an isomorphism if $n > q$, and a monomorphism if $n = q$. Similarly, the injection

$$H^{q-1}(X_{n+1}, A; G) \to H^{q-1}(X_n, A; G)$$

is an isomorphism if $n \ge q$. From the second statement it follows that

$$\underleftarrow{\lim}^1 H^{q-1}(X_n, A; G) = 0,$$

from the first that the natural homomorphism

$$\underleftarrow{\lim}\, H^q(X_m, A; G) \to H^q(X_n, A; G)$$

is an isomorphism if $n > q$ and a monomorphism if $n = q$. Because the $\underleftarrow{\lim}^1$ term vanishes, Theorem (2.10*) implies that the injection

$$H^q(X, A; G) \to H^q(X_n, A; G)$$

has the same properties; by exactness, $H^q(X, X_n; G) = 0$ for $n \ge q$. □

Let $\Gamma^n(X, A; G) = H^n(X_n, X_{n-1}; G)$, and let

$$\delta : \Gamma^n(X, A; G) \to \Gamma^{n+1}(X, A; G)$$

be the coboundary operator of the cohomology sequence of the triple (X_{n+1}, X_n, X_{n-1}). Then $\delta \circ \delta = 0$, so that $\Gamma^*(X, A; G)$ is a cochain complex, and it now follows, exactly as in §2 of Chapter II, that

(4.4*) Theorem *The homology groups of the cochain complex $\Gamma^*(X, A; G)$ are isomorphic with the singular cohomology groups $H^q(X, A; G)$.* □

Let us conclude this section by relating the complexes $\Gamma_*(X, A; G)$ and $\Gamma^*(X, A; G)$ to the universal covering space $\tilde{X}$ of X. Let X be connected and semilocally 1-connected, so that X has a universal covering $p : \tilde{X} \to X$. Let $\tilde{A} = p^{-1}(A)$; then $p | \tilde{A} : \tilde{A} \to A$ is also a covering map. By Theorem (1.12) of Chapter II, $(\tilde{X}, \tilde{A})$ is a relative CW-complex and $p : (\tilde{X}, \tilde{A}) \to (X, A)$ is a cellular map. Moreover, each covering translation $h : (\tilde{X}, \tilde{A}) \to (\tilde{X}, \tilde{A})$ is cellular. Hence the group Π operates on the chain-complex $\Gamma_*(\tilde{X}, \tilde{A})$; thus, if G_0 is a right (left) Π-module, we can form the chain complexes

$$G \otimes_\Pi \Gamma_*(\tilde{X}, \tilde{A}),$$
$$\mathrm{Hom}^\Pi(\Gamma_*(\tilde{X}, \tilde{A}), G),$$

and the corresponding equivariant homology and cohomology groups

$$E_q(\Gamma_*(\tilde{X}, \tilde{A}); G_0),$$
$$E^q(\Gamma_*(\tilde{X}, \tilde{A}); G_0).$$

(4.9) Theorem *Let G be a system of local coefficients in X. Then there are isomorphisms*

$$G_0 \otimes_\Pi \Gamma_*(\tilde{X}, \tilde{A})) \approx \Gamma_*(X, A; G),$$
$$\mathrm{Hom}^\Pi(\Gamma_*(\tilde{X}, \tilde{A}), G_0) \approx \Gamma^*(X, A; G),$$

and therefore

$$H_q(X, A; G) \approx E_q(\Gamma_*(\tilde{X}, \tilde{A}); G_0),$$
$$H^q(X, A; G) \approx E^q(\Gamma_*(\tilde{X}, \tilde{A}); G_0).$$

(4.10) Corollary *The groups*

$$E_q(\Gamma_*(\tilde{X}, \tilde{A}); G_0) \quad \text{and} \quad E_q(\tilde{X}, \tilde{A}; G_0),$$

as well as the groups

$$E^q(\Gamma_*(\tilde{X}, \tilde{A}); G_0) \quad \text{and} \quad E^q(\tilde{X}, \tilde{A}; G_0)$$

are isomorphic.

The restriction $p|\tilde{X}_n : (\tilde{X}_n, \tilde{X}_{n-1}) \to (X_n, X_{n-1})$ induces a homomorphism $p_* : H_n(\tilde{X}_n, \tilde{X}_{n-1}; p^*(G|X_n)) \to H_n(X_n, X_{n-1}; G|X_n)$. Now

$$p^*(G|X_n) = p^*G|\tilde{X}_n;$$

since $\tilde{X}$ is 1-connected, p^*G is simple; thus

$$\begin{aligned} H_n(\tilde{X}_n, \tilde{X}_{n-1}; p^*(G|X_n)) &\approx H_n(\tilde{X}_n, \tilde{X}_{n-1}; G_0) \\ &\approx G_0 \otimes H_n(\tilde{X}_n, \tilde{X}_{n-1}) \\ &= G_0 \otimes \Gamma_n(\tilde{X}, \tilde{A}). \end{aligned}$$

Hence p_* becomes a homomorphism

$$p_* : G_0 \otimes \Gamma_n(\tilde{X}, \tilde{A}) \to \Gamma_n(X, A; G),$$

which is evidently a chain map.

The latter homomorphism can be made explicit, much as in §3. Let $\tilde{E}^n_\beta$ be a n-cell of (X, A), with characteristic map $\tilde{h}_\beta : (\Delta^n, \dot{\Delta}^n) \to (\tilde{X}_n, \tilde{X}_{n-1})$. Then $p \circ \tilde{h}_\beta = h_\alpha$, the characteristic map for the cell $E^n_\alpha = p(\tilde{E}^n_\beta)$. Let ξ be the (unique) homotopy class of paths in $\tilde{X}$ joining the point $\tilde{z}_\beta = \tilde{h}_\beta(\mathbf{e}_0)$ to the base point $\tilde{x}_0$. Then $G(p\xi): G_0 = G(x_0) \to G(z_\alpha)$; let $g = G(p\xi)(g_0)$. Then $p_*(g_0 \otimes \tilde{e}_\beta) = ge_\alpha$. By the argument of §2, we now see that p_* annihilates the kernel of the natural epimorphism $G_0 \otimes \Gamma_n(\tilde{X}, \tilde{A}) \to G_0 \otimes_\Pi \Gamma_n(\tilde{X}, \tilde{A})$ and induces an isomorphism

$$\bar{p}_* : G_0 \otimes_\Pi \Gamma_n(\tilde{X}, \tilde{A}) \to \Gamma_n(X, A; G).$$

Similarly,

$$\begin{aligned} p^* : H^n(X_n, X_{n-1}; G|X_n)| &\to H^n(\tilde{X}_n, \tilde{X}_{n-1}; p^*(G|X_n)) \\ &= H^n(\tilde{X}_n, \tilde{X}_{n-1}; p^*G|\tilde{X}_n) \\ &\approx H^n(\tilde{X}_n, \tilde{X}_{n-1}; G_0) \\ &\approx \operatorname{Hom}(H_n(\tilde{X}_n, \tilde{X}_{n-1}), G_0); \end{aligned}$$

p^* is a cochain map, and we may verify without difficulty that it is a monomorphism whose image is the subgroup

$$\operatorname{Hom}^\Pi(H_n(\tilde{X}_n, \tilde{X}_{n-1}), G_0)$$

of

$$\operatorname{Hom}(H_n(\tilde{X}_n, \tilde{X}_{n-1}); G_0). \quad \square$$

Suppose that $n \geq 3$; then the injection $\pi_1(\tilde{X}_{n-1}) \to \pi_1(\tilde{X})$ is an isomorphism, and therefore $\tilde{X}_{n-1}$ is 1-connected. By the relative Hurewicz Theorem,

$$\rho : \pi_n(\tilde{X}_n, \tilde{X}_{n-1}) \approx H_n(\tilde{X}_n, \tilde{X}_{n-1}).$$

On the other hand, the fibre map $p : \tilde{X} \to X$ induces isomorphisms

$$p_* : \pi_n(\tilde{X}_n, \tilde{X}_{n-1}) \approx \pi_n(X_n, X_{n-1}).$$

Moreover, the diagram

$$\begin{array}{ccccc} H_{n+1}(\tilde{X}_{n+1}, \tilde{X}_n) & \xleftarrow{\rho} & \pi_{n+1}(\tilde{X}_{n+1}, \tilde{X}_n) & \xrightarrow{p_*} & \pi_{n+1}(X_{n+1}, X_n) \\ \downarrow{\scriptstyle \partial_1} & & \downarrow{\scriptstyle \partial_2} & & \downarrow{\scriptstyle \partial_3} \\ H_n(\tilde{X}_n, \tilde{X}_{n-1}) & \xleftarrow[\rho]{} & \pi_n(\tilde{X}_n, \tilde{X}_{n-1}) & \xrightarrow[p_*]{} & \pi_n(X_n, X_{n-1}) \end{array}$$

is commutative.

From this we have immediately

(4.11) Theorem *If $n \geq 3$, there are isomorphisms*

$$\theta_n : \operatorname{Hom}^\Pi(\pi_n(X_n, X_{n-1}), G_0) \to \Gamma^n(X, A; G)$$

such that the diagram

$$\begin{array}{ccc} \mathrm{Hom}^{\Pi}(\pi_n(X_n, X_{n-1}), G_0) & \xrightarrow{\theta_n} & \Gamma^n(X, A; G) \\ \Big\downarrow \mathrm{Hom}^{\Pi}(\partial_3, \ ) & & \Big\downarrow (-1)^n\delta \\ \mathrm{Hom}^{\Pi}(\pi_{n+1}(X_{n+1}, X_n); G_0) & \xrightarrow[\theta_{n+1}]{} & \Gamma^{n+1}(X, A; G) \end{array}$$

is commutative. □

The case $n = 2$ is a bit more delicate. In this case, $\pi_1(\tilde{X}_{n-1})$ is not, in general, zero; rather, it is isomorphic with the kernel of the injection

$$i_* : \pi_1(X_1) \to \pi_1(X).$$

The Hurewicz map $\rho : \pi_2(\tilde{X}_2, \tilde{X}_1) \to H_2(\tilde{X}_2, \tilde{X}_1)$ is no longer an isomorphism; however, it is an epimorphism whose kernel is the subgroup $\omega_2'(\tilde{X}_2, \tilde{X}_1)$ generated by all elements of the form $\alpha - \tau_\xi'(\alpha)$ with $\alpha \in \pi_2(\tilde{X}_2, \tilde{X}_1)$ and $\xi \in \pi_1(\tilde{X}_1)$. But $\pi_1(\tilde{X}_2) = 0$ and therefore $\xi = \partial_* \beta$ for some $\beta \in \pi_2(\tilde{X}_2, \tilde{X}_1)$. By Lemma (3.3) of Chapter IV, $\alpha - \tau_\xi'(\alpha)$ is just the commutator $\alpha - \beta - \alpha + \beta$ of α and β, so that $\omega_2'(\tilde{X}_2, \tilde{X}_1)$ is the commutator subgroup of $\pi_2(\tilde{X}_2, \tilde{X}_1)$. The projection $p_* : \pi_2(\tilde{X}_2, \tilde{X}_1) \to \pi_2(X_2, X_1)$ is, however, still an isomorphism, inducing an isomorphism

$$p_*' : \pi_2^\S(\tilde{X}_2, \tilde{X}_1) \approx \pi_2^\S(X_2, X_1)$$

between their commutator quotients.

The group $\pi_2^\S(X_2, X_1)$ is a Π-module, and the composite

$$\partial_3^\S : \pi_3(X_3, X_2) \to \pi_2(X_2, X_1) \to \pi_2^\S(X_2, X_1)$$

is an operator homomorphism. Hence

(4.12) Theorem *There is an isomorphism*

$$\theta_2 : \mathrm{Hom}^{\Pi}(\pi_2^\S(X_2, X_1), G_0) \to \Gamma^2(X, A; G)$$

such that the diagram

$$\begin{array}{ccc} \mathrm{Hom}^{\Pi}(\pi_2^\S(X_2, X_1), G_0) & \xrightarrow{\theta_2} & \Gamma^2(X, A; G) \\ \Big\downarrow \mathrm{Hom}(\partial_3^\S, \ ) & & \Big\downarrow \delta \\ \mathrm{Hom}^{\Pi}(\pi_3(X_3, X_2), G_0) & \xrightarrow[\theta_3]{} & \Gamma^3(X, A; G) \end{array}$$

is commutative. □

5 Obstruction Theory in Fibre Spaces

Let $p : X \to B$ be a fibration. In this and the following section we shall be concerned with the problem of finding a cross-section to B, and of classifying the cross-sections up to vertical homotopy (a homotopy $f : \mathbf{I} \times Y \to X$ between two maps $f_0, f_1 : Y \to X$ is said to be *vertical* if and only if $p \circ f : \mathbf{I} \times Y \to B$ is stationary).

The problem of finding a cross-section is a special case of the lifting problem: given $f : Y \to B$, is there a map $h : Y \to X$ such that $p \circ h = f$? However, the general problem can be reduced to the special case; for let $q : W \to Y$ be the fibration induced by f, so that there is a commutative diagram

(5.1)
$$\begin{array}{ccc} W & \xrightarrow{g} & X \\ {\scriptstyle q}\downarrow & & \downarrow{\scriptstyle p} \\ Y & \xrightarrow[f]{} & B \end{array}$$

If $h : Y \to X$ is a lifting of f, then the maps $1 : Y \to Y$, $h : Y \to X$ have the property that $f \circ 1 = p \circ h$; by the universal property of the diagram (5.1), there is a unique map $k : Y \to W$ such that $q \circ k = 1$, $g \circ k = h$, and thus k is a cross-section for g. Conversely, if $k : Y \to W$ is a cross-section for q, then $g \circ k : Y \to X$ is a lifting of f. Thus the liftings of f are in one-to-one correspondence with the cross-sections of the fibration induced by f. Similarly, two liftings of f are vertically homotopic if and only if the corresponding cross-sections are.

Finally, the problem of finding a vertical homotopy between two liftings can be viewed as a *relative cross-section problem*, viz., that of extending a cross-section already defined on a subspace. Therefore we shall address ourselves to the latter problem.

Despite the fact that liftings are no more general than cross-sections, it will allow for greater flexibility and neater statements of the results to study the problem from the point of view of liftings.

Accordingly, let $p : X \to B$ be a fibration with 0-connected base space B and fibre F. Let (K, L) be a connected relative CW-complex, and let $\phi : K \to B$, $f : L \to X$ be maps such that $p \circ f = \phi | L$. Under these circumstances we shall say that f is a *partial lifting* of ϕ. The *lifting extension problem* is that of finding a map $g : K \to X$ such that $p \circ f = \phi$ and $g | L = f$. Following the methods of §§5, 6 of Chapter V as closely as we may, we shall attack the problem by a stepwise extension process. The main point of difference is that, while in ordinary obstruction theory the coefficients for the obstructions lie in fixed groups (the homotopy groups of the range space), in the present case the coefficients form a local system.

Since B is 0-connected, p maps X *upon* B. Therefore, if e is any vertex of (K, L), there is a point $x \in p^{-1}\phi(e)$. Thus $f: L \to X$ can be extended to a partial lifting $g_0 : K_0 \to X$. Let E_α^1 be a 1-cell of (K, L) with characteristic map $h_\alpha : (\Delta^1, \dot{\Delta}^1) \to (E_\alpha^1, \dot{E}_\alpha^1)$. Since Δ^1 is contractible, the fibration $q_\alpha : X_\alpha \to \Delta^1$ induced by the map $\phi \circ h_\alpha$ is fibre homotopically trivial, and the space X_α, having the same homotopy type as $\Delta^1 \times F$, and therefore as F, is 0-connected. Therefore the partial cross-section of q_α, defined by the map $g_0 \circ h_\alpha | \dot{\Delta}^1 : \dot{\Delta}^1 \to X_\alpha$, has an extension $k_\alpha : \Delta^1 \to X_\alpha$. The map k_α need not be a cross-section; however, the map $q_\alpha \cdot k_\alpha$ coincides with the identity map on $\dot{\Delta}^1$. Since Δ^1 is 1-connected, $q_\alpha \circ k_\alpha \simeq 1$ (rel. $\dot{\Delta}^1$). By the homotopy lifting extension property (Theorem 7.16 of Chapter I), k_α is homotopic (rel. $\dot{\Delta}^1$) to a cross-section $k'_\alpha : \Delta^1 \to X_\alpha$. The map k'_α defines, in turn, a lifting $l_\alpha : \Delta^1 \to X$ of $\phi \circ h_\alpha$ extending the partial lifting $g_0 \circ h_\alpha | \dot{\Delta}^1$. Then $f_\alpha = l_\alpha \circ h_\alpha^{-1} : E_\alpha^1 \to X$ is a partial lifting defined over E_α^1 and extending the map $g_0 | \dot{E}_\alpha^1$. The partial liftings f_α, for all the 1-cells of (K, L), fit together to define a partial lifting $g_1 : L \to X$ extending g_0.

We have proved

(5.2) Theorem *If $p : X \to B$ is a fibration with* 0*-connected base space and fibre, (K, L) is a relative* CW*-complex, $\phi : K \to B$, then any partial lifting $f: L \to X$ of ϕ can be extended to a partial lifting $g_1 : K_1 \to X$.* □

Suppose that $g : K_n \to X$ is a partial lifting of $\phi (n \geq 1)$, and assume that the fibre F is n-simple, so that the homotopy groups $\{\pi_n(F_b)\}$ form a system $\pi_n(\mathscr{F})$ of local coefficients in B, and therefore $\phi^*\pi_n(\mathscr{F})$ is a system of local coefficients in K. It will simplify the notation and should not cause undue confusion to write $\pi_n(\mathscr{F})$ instead of $\phi^*\pi_n(\mathscr{F})$.

Let E_α^{n+1} be a cell of (K, L) with characteristic map $h_\alpha : (\Delta^{n+1}, \dot{\Delta}^{n+1}) \to (E_\alpha^{n+1}, \dot{E}_\alpha^{n+1})$; since K is connected, we may assume $h_\alpha(\mathbf{e}_0) = *$. The map $g \circ h_\alpha | \dot{\Delta}^{n+1}$ defines a partial cross-section $k_\alpha : \dot{\Delta}^{n+1} \to W_\alpha$ of the fibration $q_\alpha : W_\alpha \to \Delta^{n+1}$ induced by $\phi \circ h_\alpha$. Since Δ^{n+1} is contractible, q_α is fibre homotopically trivial, and therefore W_α has the same homotopy type as the fibre $q_\alpha^{-1}(\mathbf{e}_0) = F_\alpha$. Hence k_α represents a uniquely defined element $\bar{c}^{n+1}(e_\alpha) \in \pi_n(F_\alpha)$. The function $\bar{c}^{n+1}$ so defined is a cochain

$$\bar{c}^{n+1} = \bar{c}^{n+1}(g) \in \Gamma^{n+1}(K, L; \pi_n(\mathscr{F})),$$

called the *obstruction* to extending the partial lifting g.

As in §5 of Chapter V, we shall give a global description of the obstruction equivalent to the preceding local definition. In order to do so, we shall make use of Theorems (4.11) and (4.12).

Let $(\hat{X}, \hat{F})$ be the mapping cylinder of $p : (X, F) \to (B, *)$. In §8 of Chapter IV, we introduced an isomorphism

$$\Delta' : \pi_{n+1}(\hat{X}, X) \to \pi_n(F);$$

it is the composite of the boundary operator of the homotopy sequence of the pair $(\hat{F}, F)$ with the inverse of the (isomorphic) injection $\pi_{n+1}(\hat{F}, F) \to \pi_{n+1}(\hat{X}, X)$. In Theorem (8.21) of Chapter IV, it was proved that Δ' is an operator isomorphism, $\pi_1(X)$ operating on $\pi_n(F)$ via the homomorphism $p_* : \pi_1(X) \to \pi_1(B)$. The sequence

$$\pi_1(F) \xrightarrow{i_*} \pi_1(X) \xrightarrow{p_*} \pi_1(B) \longrightarrow 0$$

is exact, since F is 0-connected. The kernel of p_* operates trivially on $\pi_n(F)$, and therefore on $\pi_{n+1}(\hat{X}, X)$. Therefore $\pi_{n+1}(\hat{X}, X)$ is a Π-module and Δ' an isomorphism of Π-modules ($\Pi = \pi_1(B)$).

(5.3) Lemma *Every partial lifting $g : K_n \to X$ of f can be extended to a map $\hat{g} : (K_{n+1}, K_n) \to (\hat{X}, X)$ such that $\hat{p} \circ \hat{g} : K_{n+1} \to B$ is homotopic* (rel. K_n) *to $\phi \mid K_{n+1}$. If $n \geq 2$, the homomorphism*

$$\Delta' \circ \hat{g}_* : \pi_{n+1}(K_{n+1}, K_n) \to \pi_n(F)$$

corresponds, under the isomorphism θ_{n+1} of Theorem (4.11) *to the cochain $\bar{c}^{n+1}(g)$. If $n = 1$, then $\Delta' \circ \hat{g}_*$ annihilates the commutator subgroup of $\pi_2(K_2, K_1)$ and corresponds, under the isomorphism θ_2 of Theorem* (4.12), *to $\bar{c}^2(g)$.*

Let E_α^{n+1} be a cell of (K, L), with characteristic map $h_\alpha : (\Delta^{n+1}, \dot{\Delta}^{n+1}) \to (K_{n+1}, K_n)$. Define $\hat{g}_\alpha : (\Delta^{n+1}, \dot{\Delta}^{n+1}) \to (\hat{X}, X)$ by

$$\hat{g}_\alpha((1-t)b_{n+1} + tz) = \begin{cases} \phi h_\alpha((1-2t)b_{n+1} + 2tz) & (t \leq \frac{1}{2},\ z \in \dot{\Delta}^{n+1}), \\ \langle 2(1-t),\ gh_\alpha(z)\rangle & (t \geq \frac{1}{2},\ z \in \dot{\Delta}^{n+1}); \end{cases}$$

then $\hat{p} \circ \hat{g}_\alpha = \phi \circ h_\alpha \circ d$ for a map $d : (\Delta^{n+1}, \dot{\Delta}^{n+1}) \to (\Delta^{n+1}, \dot{\Delta}^{n+1})$ such that $d \mid \dot{\Delta}^{n+1} = 1$. Since Δ^{n+1} is convex, d is homotopic (rel. $\dot{\Delta}^{n+1}$) to the identity map of Δ^{n+1}. The maps $\hat{g}_\alpha \circ h_\alpha^{-1}$ fit together to yield a map $\hat{g} : K_{n+1} \to \hat{X}$ such that $\hat{g} \circ h_\alpha = \hat{g}_\alpha$; and $\hat{p} \circ \hat{g} \circ h_\alpha = \hat{p} \circ \hat{g}_\alpha = \phi \circ h_\alpha \circ d \simeq \phi \circ h_\alpha$ (rel. $\dot{\Delta}^{n+1}$). These homotopies therefore fit together to yield a homotopy of $\hat{p} \circ \hat{g}$ to $\phi \mid K_{n+1}$ (rel. K_n).

Let us now recall (Theorem (1.3) of Chapter V) that the group $\pi_{n+1}(K_{n+1}, K_n)$ is generated, as a $\pi_1(B_n)$-module, by the homotopy classes ε_α of the characteristic maps h_α. Therefore, in order to calculate the homomorphism $\Delta' \circ \hat{g}_*$, it suffices to calculate the elements $\Delta'\hat{g}_*(\varepsilon_\alpha) = \Delta'(\eta_\alpha)$, where η_α is the homotopy class of $\hat{g}_\alpha$. If $\hat{q}_\alpha : \hat{W}_\alpha \to \Delta^{n+1}$ is the map induced from $\hat{p} : \hat{X} \to B$ by the map $\phi \circ h_\alpha$, then W_α is a subspace of $\hat{W}_\alpha$, $\hat{q}_\alpha \mid W_\alpha = q_\alpha : W_\alpha \to \Delta^{n+1}$, and $\hat{W}_\alpha$ can be identified with the mapping cylinder of q_α. The maps $\hat{g}_\alpha : \Delta^{n+1} \to \hat{X}$, $d : \Delta^{n+1} \to \Delta^{n+1}$ define a map $\hat{k}_\alpha : \Delta^{n+1} \to \hat{W}_\alpha$ such that $\hat{k}_\alpha \mid \dot{\Delta}^{n+1} = k_\alpha : \dot{\Delta}^{n+1} \to W_\alpha$.

There is a commutative diagram

$$
\text{(5.4)}\qquad
\begin{array}{ccccc}
\pi_{n+1}(\Delta^{n+1}, \dot{\Delta}^{n+1}) & \xrightarrow{\hat{k}_{\alpha *}} & \pi_{n+1}(\hat{W}_\alpha, W_\alpha) & \xleftarrow{k_1} & \pi_{n+1}(\hat{F}, F) \\
\downarrow \partial_0 & & \downarrow \partial_1 & & \downarrow \partial_3 \\
\pi_n(\dot{\Delta}^{n+1}) & \xrightarrow[k_{\alpha *}]{} & \pi_n(W_\alpha) & \xleftarrow[i_4]{} & \pi_n(F)
\end{array}
$$

in which the right-hand square is a subdiagram of Figure 4.2 for the fibration $q_\alpha : W_\alpha \to \Delta^{n+1}$, and the left-hand square is a subdiagram of the map of the homotopy sequence of $(\Delta^{n+1}, \dot{\Delta}^{n+1})$ into that of $(\hat{W}_\alpha, W_\alpha)$ induced by the map $\hat{k}_\alpha$. In §8 of Chapter IV we saw that k_1 and ∂_3 are isomorphisms; since Δ^{n+1} is contractible, so is $\hat{W}_\alpha$, so that i_4 and ∂_1 are isomorphisms.

Now $\Delta'(\eta_\alpha) = \partial_3 k_1^{-1}(\eta_\alpha)$, and therefore

$$i_4 \Delta'(\eta_\alpha) = i_4 \partial_3 k_1^{-1}(\eta_\alpha) = \partial_1 k_1 k_1^{-1}(\eta_\alpha) = \partial_1(\eta_\alpha)$$

is represented by the map $k_\alpha : \dot{\Delta}^{n+1} \to W_\alpha$. But the element $i_4 \bar{c}^{n+1}(e_\alpha)$ is also represented by k_α. Hence

$$\Delta' \hat{q}_*(\varepsilon_\alpha) = \Delta'(\eta_\alpha) = \bar{c}^{n+1}(e_\alpha),$$

as desired.

If $n = 1$, the group $\pi_n(F)$ is abelian because of our hypothesis that F is 1-simple. Therefore $\Delta' \circ \hat{g}_*$ must annihilate the commutator subgroup. □

(5.5) Theorem *The obstruction cochain $\bar{c}^{n+1}(g)$ has the following properties:*

(1) *The map g can be extended to a partial lifting over $K_n \cup E_\alpha^{n+1}$ if and only if $\bar{c}^{n+1}(e_\alpha) = 0$;*
(2) *The map g can be extended to a partial lifting over K_{n+1} if and only if $\bar{c}^{n+1}(g) = 0$;*
(3) *If $\psi : (K', L') \to (K, L)$ is a cellular map and $g' = g \circ \psi \mid K'_n : K'_n \to X$, then*

$$\bar{c}^{n+1}(g') = \psi^{\#} \bar{c}^{n+1}(g);$$

(4) *If g_0, g_1 are partial liftings over K_n which are vertically homotopic* (rel. A), *then $\bar{c}^{n+1}(g_0) = \bar{c}^{n+1}(g_1)$;*
(5) *The cochain $\bar{c}^{n+1}(g)$ is a cocycle of (K, L) with coefficients in the local system $\phi^* \pi_n(\mathscr{F})$.*

To prove Property 1, we use the local definition. If $g \mid \dot{E}_\alpha^{n+1}$ has an extension to a partial lifting $g' : E_\alpha^{n+1} \to X$, then $g_\alpha = g \circ h_\alpha \mid \dot{\Delta}^{n+1}$ has the extension $g' \circ h_\alpha : \Delta^{n+1} \to X$, and therefore k_α has an extension $k'_\alpha : \Delta^{n+1} \to W_\alpha$, so that $\bar{c}^{n+1}(e_\alpha) = 0$. Conversely, suppose that k_α has an extension

$k'_\alpha : \Delta^{n+1} \to W_\alpha$. This means that there is a commutative diagram

$$\begin{array}{ccc} \Delta^{n+1} & \xrightarrow{\;g'_\alpha\;} & X \\ {\scriptstyle d}\big\downarrow & & \big\downarrow{\scriptstyle p} \\ \Delta^{n+1} & \xrightarrow[\;\phi \circ h_\alpha\;]{} & B \end{array}$$

such that $g'_\alpha \,|\, \dot{\Delta}^{n+1} = g_\alpha$ and $d \,|\, \dot{\Delta}^{n+1} =$ the identity map. The map d being homotopic (rel. $\dot{\Delta}^{n+1}$) to the identity, the map $p \circ g'_\alpha = \phi \circ h_\alpha \circ d$ is homotopic (rel. $\dot{\Delta}^{n+1}$) to $\phi \circ h_\alpha$. By the homotopy lifting extension property (Theorem (7.16) of Chapter I), g'_α is homotopic (rel. $\dot{\Delta}^{n+1}$) to a map g''_α such that $p \circ g_\alpha = \phi \circ h_\alpha$. Then $g''_\alpha : \Delta^{n+1} \to X$ is an extension of $g_\alpha = g \circ h_\alpha$, and therefore $g''_\alpha \circ h_\alpha^{-1}$ is an extension of $f \,|\, \dot{E}_\alpha^{n+1}$ to a partial lifting over E_α.

Property 2 follows from Property 1 by patching together the extensions over each cell.

To prove (3), we use the global definition. The map ψ induces an operator homomorphism $\psi_* : \pi_{n+1}(K'_{n+1}, K'_n) \to \pi_{n+1}(K_{n+1}, K_n)$. Let

$$\eta : \pi_{n+1}(K_{n+1}, K_n) \to \pi_n(F)$$

be the homomorphism corresponding, under the isomorphism θ_{n+1} to the cochain $\bar{c}^{n+1}(f)$; then $\eta \circ \psi_*$ corresponds to $\psi^{\#} \bar{c}^{n+1}(f)$. But the homomorphism corresponding to $\bar{c}^{n+1}(f')$ is clearly $\eta \circ \psi_*$.

If g_0 and g_1 are vertically homotopic, then, for each cell E_α^{n+1}, the cross-section of $q_\alpha : W_\alpha \to \Delta^{n+1}$ induced by $g_0 \circ h_\alpha \,|\, \dot{\Delta}^{n+1}$ and $g_1 \circ h_\alpha \,|\, \dot{\Delta}^{n+1}$ are vertically homotopic. In particular, they are homotopic and therefore the elements assigned to e_α by the cochains $\bar{c}^{n+1}(f_0)$ and $\bar{c}^{n+1}(f_1)$ are equal. Since this is true for every cell, $\bar{c}^{n+1}(f_0) = \bar{c}^{n+1}(f_1)$.

To prove Property 5, we use the global definition. Let $\hat{j} : B \hookrightarrow \hat{X}$; then $\hat{j}$ is a homotopy equivalence and $\hat{p} \circ \hat{j} = 1$, $\hat{j} \circ \hat{p} \simeq 1$. The map $\hat{g} : K_{n+1} \to \hat{X}$ of Lemma (5.3) has the property that $\hat{p} \circ \hat{g} \simeq \phi \,|\, K_{n+1}$ (rel. K_n), and therefore

$$\hat{g} \simeq \hat{j} \circ \hat{p} \circ \hat{g} \simeq \hat{j} \circ \phi \,|\, K_{n+1} (\text{rel. } K_n).$$

Since (K_{n+2}, K_{n+1}) is an NDR-pair, $\hat{j} \circ \phi \,|\, K_{n+2}$ is homotopic to a map $\hat{g}_1 : K_{n+2} \to \hat{X}$ such that $\hat{g}_1 \,|\, K_{n+1} = \hat{g}$. The diagram

$$\begin{array}{ccccc} \pi_{n+2}(K_{n+2}, K_{n+1}) & \xrightarrow{\;\hat{g}_{1*}\;} & \pi_{n+2}(\hat{X}, \hat{X}) = 0 & & \\ {\scriptstyle \partial_*}\big\downarrow & & \big\downarrow{\scriptstyle \partial'_*} & & \\ \pi_{n+2}(K_{n+1}, K_n) & \xrightarrow[\;\hat{g}_*\;]{} & \pi_{n+1}(\hat{X}, X) & \xrightarrow[\;\Delta'\;]{} & \pi_n(F) \end{array}$$

is commutative, and therefore $\Delta' \circ \hat{g}_* \circ \partial_* = 0$. But $\Delta' \circ \hat{g}_* \circ \partial_*$ is the homomorphism corresponding, by Theorems (4.11) and (4.12), to $\delta\bar{c}^{n+1}(g)$.

We can apply the above results to study the existence of a vertical homotopy between two partial liftings. In fact, let $g_0, g_1 : K_n \to X$ be partial liftings of ϕ, and let $g' : \mathbf{I} \times K_{n-1} \to X$ be a vertical homotopy (rel. A) between $g_0 | K_{n-1}$ and $g_1 | K_{n-1}$. These maps fit together to define a partial lifting $g : \dot{\mathbf{I}} \times K_n \cup \mathbf{I} \times K_{n-1} \to X$ of $\phi \circ p_2$, where $p_2 : \mathbf{I} \times K \to K$ is the projection on the second factor. The *difference cochain* of g_0, g_1 *with respect to* g' is the cochain

$$\bar{d}^n = \bar{d}^n(g) = \bar{d}^n(g_0, g_1, g') \in C^n(K, L; \pi_n(\mathscr{F}))$$

such that $\bar{d}^n(e_\alpha)$ is $(-1)^n$ times the value of the cochain $\bar{c}^{n+1}(G)$ on $\mathbf{i} \times e_\alpha$. If the homotopy g' is stationary, so that $g_0 | K_{n-1} = g_1 | K_{n-1}$, we write $\bar{d}^n(g_0, g_1)$ instead of $\bar{d}^n(g_0, g_1, g')$.

(5.6) Theorem *The difference cochain has the following properties*:

(1) *The map* $g' : \mathbf{I} \times K_{n-1} \to X$ *can be extended to a vertical homotopy* $\bar{g}_\alpha : \mathbf{I} \times (K_{n-1} \cup E_\alpha^n)$ *between the restrictions of* g_0 *and* g_1 *to* $K_{n-1} \cup E_\alpha^n$ *if and only if* $\bar{d}^n(e_\alpha) = 0$;
(2) *The map* g' *can be extended to a vertical homotopy* $\bar{g} : \mathbf{I} \times K_n \to X$ *between* g_0 *and* g_1 *if and only if* $\bar{d}^n(g) = 0$;
(3) *The coboundary of* $\bar{d}^n(g)$ *is given by*

$$\delta\bar{d}^n(h) = \bar{c}^{n+1}(g_1) - \bar{c}^{n+1}(g_0);$$

(4) *If* $g_0 : K_n \to X$ *is a partial lifting of* ϕ, $g' : \mathbf{I} \times K_{n-1} \to X$ *is a vertical homotopy of* $g_0 | K_{n-1}$ *to a map* $f'_1 : K_{n-1} \to X$, *and* $d \in C^n(K, L; \pi_n(\mathscr{F}))$, *then* f'_1 *can be extended to a partial lifting* $g_1 : K_n \to X$ *of* ϕ *such that* $\bar{d}^n(g_0, g_1, g') = d$;
(5) *Let* $g_i : K_n \to X$ *be a partial lifting of* ϕ $(i = 0, 1, 2)$, *and let* $g'_{01} : \mathbf{I} \times K_{n-1} \to X$, $g'_{12} : \mathbf{I} \times K_{n-1} \to X$ *be vertical homotopies between the restrictions of* g_0 *and* g_1 *and of* g_1 *and* g_2, *respectively, to* K_{n-1}. *Let* $g'_{02} : \mathbf{I} \times K_{n-1} \to X$ *be the homotopy of* g_0 *to* g_2 *defined by* g'_{01} *and* g'_{02}. *Then*

$$\bar{d}^n(g_0, g_2, g'_{02}) = \bar{d}^n(g_0, g_1, g'_{01}) + \bar{d}^n(g_1, g_2, g'_{12});$$

(6) *Let* $\psi : (K', L') \to (K, L)$ *be a cellular map, and let* $g : \dot{\mathbf{I}} \times K_n \cup \mathbf{I} \times K_{n-1} \to X$ *be a partial lifting of* $\phi \circ p_2$. *Then*

$$\bar{d}^n(g \circ \psi | \dot{\mathbf{I}} \times K_n \cup \mathbf{I} \times K_{n-1}) = \psi^\# \bar{d}^n(g).$$

The reader who has mastered the proof of Theorem (5.5), as well as the results of §§5, 6 of Chapter V, should have no difficulty with the proof of Theorem (5.6). □

(5.7) Corollary *The cochain $\bar{c}^{n+1}(g)$ is a coboundary if and only if the map $g \mid K_{n-1}$ can be extended to a partial lifting $g_1 : K_{n+1} \to X$ of ϕ.* □

We conclude this section by making explicit the cochains $\bar{c}^{n+1}(g), \bar{d}^n(g)$ in the special case of a trivial fibration. Suppose, then, that $X = B \times F$, $p(b, y) = b$; then the local coefficient system $\pi_n(\mathscr{F})$ is simple and can be identified with the group $\pi_n(F)$. If $\phi : K \to B$, a partial lifting $g : K' \to B$ defined on a subspace K' of K, is given by $g(x) = (\phi(x), \psi(x))$ for a map $\psi : K' \to F$. In particular, if $g : K_n \to B$ and E_α^{n+1} is an $(n+1)$-cell, then the projection of $\Delta^{n+1} \times F$ its second factor is a homotopy inverse of the inclusion $F \subsetneq \Delta^{n+1} \times F$. Hence $\bar{c}^{n+1}(e_\alpha)$ is the element of $\pi_n(F)$ represented by the map $\psi \circ h_\alpha \mid \dot{\Delta}^{n+1} : \dot{\Delta}^{n+1} \to F$; but this is just the value at e_α of the obstruction $c^{n+1}(\psi)$. Thus

(5.8) Theorem *If $p : B \times F \to B$ is a trivial fibration, and a partial lifting $g : K_n \to B \times F$ of $\phi : K \to B$ is determined by a map $\psi : K_n \to F$, then*

$$\bar{c}^{n+1}(g) = c^{n+1}(\psi) \in Z^{n+1}(K, L; \pi_n(F)).$$ □

Similarly, we can prove

(5.9) Theorem *If $p : B \times F \to B$ is a trivial fibration and a partial lifting $g : \dot{\mathbf{I}} \times K_n \cup \mathbf{I} \times K_{n-1} \to B \times F$ of $\phi \circ p_2 : \mathbf{I} \times K \to B$ is determined by a map $\psi : \dot{\mathbf{I}} \times K_n \cup \mathbf{I} \times K_{n-1} \to F$, then*

$$\bar{d}^n(g) = d^n(\psi) \in C^n(K, L; \pi_n(F)).$$ □

6 The Primary Obstruction to a Lifting

Let $p : X \to B$ be a fibration with 0-connected base and $(n-1)$-connected fibre F ($n \geq 1$; if $n = 1$, we assume that F is 1-simple). Let (K, L) be a connected relative CW-complex, $\phi : K \to B$, $f : L \to X$, $p \circ f = \phi \mid L$. The following statement is an immediate consequence of Theorems (5.5) and (5.6).

(6.1) Theorem *The map f can be extended to a partial lifting $g : K_n \to X$ of ϕ. If g_0 and g_1 are two such partial liftings, then $g_0 \mid K_{n-1}$ and $g_1 \mid K_{n-1}$ are vertically homotopic* (rel. L).

Let $g_0, g_1 : K_n \to X$ be partial liftings of ϕ, and let $g' : \mathbf{I} \times K_{n-1} \to X$ be a vertical homotopy between their restrictions to K_{n-1}. Then the cochain

$$\bar{d}^n = \bar{d}^n(g_0, g_1, g') \in C^n(K, L; \pi_n(\mathscr{F}))$$

is defined, and

$$\delta \bar{d}^n = \bar{c}^{n+1}(g_1) - \bar{c}^{n+1}(g_0)$$

by Theorem (5.6). Therefore the obstruction cocycles $\bar{c}^{n+1}(g)$, for all possible partial liftings g extending f, lie in a single cohomology class

$$\bar{\gamma}^{n+1} = \bar{\gamma}^{n+1}(f) \in H^{n+1}(K, L; \phi^*\pi_n(\mathscr{F})).$$

The class $\bar{\gamma}^{n+1}$ is called the *primary obstruction to extending* f; if $A = \varnothing$, then $\bar{\gamma}^{n+1} \in H^{n+1}(K; \pi_n(\mathscr{F}))$ is called the *primary obstruction to lifting* ϕ.

(It is often convenient to assume $A \neq \varnothing$. This is not a serious restriction; for we may choose a vertex $*$ of K and define the lifting g by picking a point of the fibre F over the base point $* \in B$).

In the special case of a trivial fibration, it follows from Theorem (5.8) that

(6.2) Theorem *If $p: B \times F \to B$ is a trivial fibration and a partial lifting $f: L \to B \times F$ of a map $\phi: (K, L) \to (B, A)$ is determined by a map $\psi: L \to F$, then*

$$\bar{\gamma}^{n+1}(f) = \gamma^{n+1}(\psi) \in H^{n+1}(K, L; \pi_n(F)). \qquad \square$$

The following properties of $\bar{\gamma}^{n+1}(f)$ are immediate consequences of Theorem (5.5).

(6.3) Theorem *The primary obstruction $\bar{\gamma}^{n+1}(f)$ has the following properties:*

(1) *The map f can be extended to a partial lifting $g: K_{n+1} \to X$ of ϕ if and only if $\bar{\gamma}^{n+1}(f) = 0$;*

(2) *If $\psi: (K', L') \to (K, L)$ is a cellular map and $f: L \to X$ is a partial lifting of $\phi: K \to L$, then $f \circ \psi \mid L': L' \to X$ is a partial lifting of $\phi \circ \psi$, and*

$$\bar{\gamma}^{n+1}(f \circ \psi' \mid L) = \psi^*\bar{\gamma}^{n+1}(f) \in H^{n+1}(K', L'; \psi^*\phi^*\pi_n(\mathscr{F}));$$

(3) *If $f_0, f_1: L \to X$ are partial liftings of ϕ which are vertically homotopic, then*

$$\bar{\gamma}^{n+1}(f_0) = \bar{\gamma}^{n+1}(f_1). \qquad \square$$

Let $g_0, g_1: K \to X$ be liftings of ϕ which agree on L. Then there is a vertical homotopy $g': \mathbf{I} \times K_{n-1} \to X$ (rel. L) between $g_0 \mid K_{n-1}$ and $g_1 \mid K_{n-1}$. Because $g'_0 = g_0 \mid K_n$ and $g'_1 = g_1 \mid K_n$ have extensions over K, their obstructions vanish, and therefore, by Theorem (5.6), their difference cochain is a cocycle $\bar{d}^n$. In fact, $\bar{d}^n$ corresponds, under the isomorphism

$$\Gamma^n(K, L; G) \to \Gamma^{n+1}(\mathbf{I} \times K, \dot{\mathbf{I}} \times K \cup \mathbf{I} \times L; p_2^*G)$$

to the obstruction to extending the partial lifting g of $\phi \circ p_2$ defined by the maps g_0, g_1, and g'. It therefore follows that the cohomology class $\bar{\delta}^n(g_0, g_1)$ of $\bar{d}^n(g)$ depends only on g_0 and g_1, and not on the homotopy g';

$$\bar{\delta}^n(g_0, g_1) \in H^n(K, L; \pi_n(\mathscr{F})).$$

The class $\bar{\delta}^n(g_0, g_1)$ is called the *primary difference* of the liftings g_0, g_1.

Again, we deduce from Theorem (5.9):

(6.4) Theorem *If $p: B \times F \to B$ is a trivial fibration, and if liftings g_0, $g_1: K \to B \times F$ of a map $\phi: K \to B$ are determined by maps $\psi_0, \psi_1: K \to F$ such that $\psi_0 | L = \psi_1 | L$, then*

$$\bar{\delta}^n(g_0, g_1) = \delta^n(\psi_0, \psi_1) \in H^n(K, L; \pi_n(F)). \qquad \square$$

And from Theorem (5.6):

(6.5) Theorem *The primary difference $\bar{\delta}^n(g_0, g_1)$ has the following properties:*

(1) *The maps $g_0 | K_n$, $g_1 | K_n$ are vertically homotopic* (rel. L) *if and only if* $\bar{\delta}^n(g_0, g_1) = 0$;

(2) *If $g_0, g_1, g_2: K \to X$ are liftings of ϕ agreeing on L, then*

$$\bar{\delta}^n(g_0, g_2) = \bar{\delta}^n(g_0, g_1) + \bar{\delta}^n(g_1, g_2);$$

(3) *If $\psi: (K', L') \to (K, L)$ is a cellular map, then*

$$\bar{\delta}^n(g_0 \circ \psi, g_1 \circ \psi) = \psi^* \bar{\delta}^n(g_0, g_1) \in H^n(K', L'; \psi^* \phi^* \pi_n(\mathscr{F})). \qquad \square$$

We can now show how these classes can be defined for an arbitrary lifting extension problem

$$\begin{array}{ccc} Q & \xrightarrow{\ f\ } & X \\ \cap\big\downarrow & \nearrow & \big\downarrow p \\ P & \xrightarrow{\ \phi\ } & B \end{array}$$

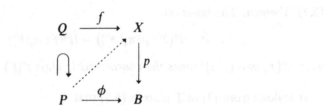

((P, Q) not necessarily a CW-pair). Let (K, L) be a CW-pair, $\psi: (K, L) \to (P, Q)$ a weak homotopy equivalence. Then $f \circ \psi | L: L \to X$ is a partial lifting of $\phi \circ \psi: K \to B$, so that

$$\bar{\gamma}^{n+1}(f \circ \psi | L) \in H^{n+1}(K, L; \psi^* \phi^* \pi_n(\mathscr{F}))$$

is defined. If $\psi': (K', L') \to (P, Q)$ is another weak homotopy equivalence, then, by Theorem (3.13) of Chapter V, there is a map $\omega: (K, L) \to (K', L')$, which we may assume to be cellular, such that $\psi' \circ \omega \simeq \psi$. Then

$$\bar{\gamma}^{n+1}(f \circ \psi' | L') \in H^{n+1}(K', L'; \psi'^* \phi^* \pi_n(\mathscr{F})),$$

and it follows from Theorem (5.5) that

$$\textbf{(6.6)} \qquad \begin{aligned} \omega^* \bar{\gamma}^{n+1}(f \circ \psi' | L') &= \bar{\gamma}^{n+1}(f \circ \psi' \circ \omega | L) \\ &= \bar{\gamma}^{n+1}(f \circ \psi | L) \end{aligned}$$

$$\in H^{n+1}(K, L; \omega^* \psi'^* \phi^* \pi_n(\mathscr{F})) = H^{n+1}(K, L; \psi^* \phi^* \pi_n(\mathscr{F})).$$

Therefore we may define

$$\bar{\gamma}^{n+1}(f) = \psi^{*-1}\bar{\gamma}^{n+1}(f \circ \psi | L) \in H^{n+1}(P, Q; \phi^*\pi_n(\mathscr{F})),$$

and an easy calculation using (6.6) shows that $\bar{\gamma}^{n+1}(f)$ is well-defined.

Similarly, let $g_0, g_1 : P \to X$ be liftings of ϕ which agree on Q. Then, for any CW-pair (K, L) and weak homotopy equivalence $\psi : (K, L) \to (P, Q)$, $g_0 \circ \psi$ and $g_1 \circ \psi$ are liftings of $\phi \circ \psi$ which agree on L, and we may define

$$\bar{\delta}^n(g_0, g_1) = \psi^{*-1}\bar{\delta}^n(g_0 \circ \psi, g_1 \circ \psi) \in H^n(P, Q; \phi^*\pi_n(\mathscr{F})),$$

verifying as before that the element so defined is independent of the CW-approximation ψ.

We next introduce a new cohomology class, associated to a fibration $p : X \to B$ with a cross-section $f : B \to X$. The identity map $1 : X \to X$ and the maps $f \circ p : X \to X$ are liftings of $p : X \to B$, and therefore we may define

$$\bar{\varepsilon}^n(f) = \bar{\delta}^n(1, f \circ p) \in H^n(X; p^*\pi_n(\mathscr{F})).$$

It will be useful to explore the connection of $\bar{\varepsilon}^n(f)$ with other cohomology classes related to the fibration p.

Let $i : F \subsetneq X$; since $p \circ i$ is the constant map, the local coefficient system i^*p^*G in F is simple for any local coefficient system G in B.

(6.7) Theorem *The injection*

$$i^* : H^n(X; p^*\pi_n(\mathscr{F})) \to H^n(F; \pi_n(F))$$

maps $\bar{\varepsilon}^n(f)$ into $(-1)^n$ times the characteristic class $\iota^n(F)$.

It follows from (3) of Theorem (6.5) that

$$\begin{aligned} i^*\bar{\varepsilon}^n(f) &= i^*\bar{\delta}^n(1, f \circ p) \\ &= \bar{\delta}^n(i, f \circ p \circ i) = \bar{\delta}^n(i, *), \end{aligned}$$

where $*$ is the constant map. (Here i and $*$ are to be regarded as liftings of the constant map of F into B). The fibration induced by the latter map is, of course, trivial, and therefore

$$\begin{aligned} \bar{\delta}^n(i, *) &= \delta^n(1, *) \quad \text{by Theorem (6.4)} \\ &= (-1)^n\iota^n(F) \quad \text{by (6.11) and (6.14) of Chapter V.} \end{aligned}$$ □

(6.8) Theorem *Let $f_0, f_1 : B \to X$ be cross-sections of the fibration $p : X \to B$. Then*

$$p^*\bar{\delta}^n(f_0, f_1) = \bar{\varepsilon}^n(f_1) - \bar{\varepsilon}^n(f_0).$$

In fact,

$$\begin{aligned}\bar{\varepsilon}^n(f_1) - \bar{\varepsilon}^n(f_0) &= \bar{\delta}^n(1, f_1 \circ p) - \bar{\delta}^n(1, f_0 \circ p) \\ &= \bar{\delta}^n(f_0 \circ p, f_1 \circ p) \quad \text{by (2) of Theorem (6.5),} \\ &= p^*\bar{\delta}^n(f_0, f_1) \text{ by (3) of Theorem (6.5).}\end{aligned}$$ □

Let us next consider the problem of extending a cross-section. Let A be a subspace of B, $W = p^{-1}(A)$, $p_0 = p \mid W : W \to A$, and let $f: A \to W$ be a partial cross-section. Then

$$\bar{\gamma}^{n+1}(f) \in H^{n+1}(B, A; \pi_n(\mathscr{F})),$$

$$\bar{\varepsilon}^n(f) \in H^n(W; p_0^*\pi_n(\mathscr{F})).$$

(6.9) Theorem *The cohomology classes $\bar{\gamma}^{n+1}(f)$, $\bar{\varepsilon}^n(f)$ are connected by the relation*

$$p^*\bar{\gamma}^{n+1}(f) = \delta^*\bar{\varepsilon}^n(f) \in H^{n+1}(X, W; p^*\pi_n(\mathscr{F})).$$

We may assume that (B, A) is a CW-pair; let $\psi : (K, L) \to (X, W)$ be a weak homotopy equivalence. The map $p \circ \psi$ is homotopic to a cellular map; by the homotopy lifting property, ψ is homotopic to a map ψ' such that $p \circ \psi'$ is cellular. Therefore we may assume that $p \circ \psi$ is already cellular.

The map $f: A \to W$ has an extension $g : B_n \to X$, and we may consider the obstruction $\bar{c}^{n+1}(g) \in Z^{n+1}(B, A; \pi_n(\mathscr{F}))$. Then $\psi^*p^*\bar{\gamma}^{n+1}(f)$ is represented by the cochain $(p \circ \psi)^{\#}\bar{c}^{n+1}(g)$. Let $\lambda : \mathbf{I} \times K_{n-1} \subset \dot{\mathbf{I}} \times K_n \to X$ be a map such that

$$\lambda(0, x) = \psi(x),$$

$$\lambda(1, x) = (f \circ p \circ \psi)(x).$$

Thus $\lambda \mid \mathbf{I} \times K_{n-1}$ is a vertical homotopy between the restrictions of the above two maps to K_{n-1}. Then

$$\bar{d}^n(\lambda) \in C^n(K; (p \circ \psi)^*\pi_n(\mathscr{F}))$$

is a cochain whose restriction to L represents $\psi_0^*\bar{\varepsilon}^n(f)$. Hence its coboundary $\delta\bar{d}^n(\lambda) \in Z^{n+1}(K, L; (p \circ \psi)^*\pi_n(\mathscr{F}))$ represents $\delta^*\psi_0^*\bar{\varepsilon}^n(f) = \psi^*\delta^*\bar{\varepsilon}^n(f)$. But

$$\delta\bar{d}^n(\lambda) = \bar{c}^{n+1}(g \circ p \circ \psi \mid K_n) - \bar{c}^{n+1}(\psi \mid K_n),$$

by (3) of Theorem (5.6). The second term on the right is zero because $\psi \mid K_n$ has the extension ψ; the first is equal, by (3) of Theorem (5.5), to $(p \circ \psi)^{\#}\bar{c}^{n+1}(g)$. Thus

$$\psi^*\delta^*\bar{\varepsilon}^n(f) = (p \circ \psi)^*\bar{\gamma}^{n+1}(f) = \psi^*p^*\bar{\gamma}^{n+1}(f).$$

But ψ^* is an isomorphism, so that

$$\delta^*\bar{\varepsilon}^n(f) = p^*\bar{\gamma}^{n+1}(f).$$ □

Suppose, in particular, that $A = \{*\}$, the base point of B. Then $W = F$, and $\bar{\varepsilon}^n(f) = (-1)^n \iota^n(F) \in H^n(F; \pi_n(F))$. Thus

(6.10) Corollary *The primary obstruction $\bar{\gamma}^{n+1} \in H^{n+1}(B, \{*\}; \pi_n(\mathscr{F}))$ and the characteristic class $\iota^n(F) \in H^n(F; \pi_n(F))$ are connected by the relation*

$$p^*\bar{\gamma}^{n+1} = (-1)^n \delta^* \iota^n(F) \in H^{n+1}(X, F; p^*\pi_n(\mathscr{F})). \qquad \square$$

It can be shown that $p^* : H^{n+1}(B, \{*\}; G) \to H^{n+1}(X, F; p^*G)$ is a monomorphism. Therefore Corollary (6.10) contains a characterization of $\bar{\gamma}^{n+1}$.

We now have

(6.11) Theorem (Extension Theorem). *Suppose that F is q-simple and that $H^{q+1}(K, L; \pi_q(\mathscr{F})) = 0$ whenever $n + 1 \leq q < \dim(K, L)$. Then a partial lifting $f : L \to X$ of a map $\phi : K \to B$ can be extended to a lifting $g : K \to X$ of ϕ if and only if $\bar{\gamma}^{n+1}(f) = 0$.* $\square$

(6.12) Theorem (Homotopy Theorem). *Suppose that F is q-simple and that $H^q(K, L; \pi_q(\mathscr{F})) = 0$ for all q such that $n + 1 \leq q < 1 + \dim(K, L)$. Then two liftings $f_0, f_1 : K \to X$ of $\phi : K \to B$ which agree on L are vertically homotopic* (rel. L) *if and only if $\bar{\delta}^n(f_0, f_1) = 0$.* $\square$

(6.13) Theorem (Classification Theorem). *Suppose that*

(1) *F is q-simple for $n + 1 \leq q < 1 + \dim(K, L)$;*
(2) *$H^q(K, L; \pi_q(\mathscr{F})) = 0$ for $n + 1 \leq q < 1 + \dim(K, L)$;*
(3) *$H^{q+1}(K, L; \pi_q(\mathscr{F})) = 0$ for $n + 1 \leq q < \dim(K, L)$.*

Let $f_0 : K \to X$ be a lifting of $\phi : K \to B$. Then the correspondence $f \to \bar{\delta}^n(f_0, f)$ is a one-to-one correspondence between the set of vertical homotopy classes (rel. L) *of liftings of ϕ which agree with f_0 on L and the group $H^n(K, L; \pi_n(\mathscr{F}))$.* $\square$

(6.14) Corollary *If $F = K(\Pi, n)$ or if $\dim(K, L) \leq n + 1$, then a partial lifting $f : L \to X$ of a map $\phi : K \to B$ can be extended to a lifting of ϕ if and only if $\bar{\gamma}^{n+1}(f) = 0$.* $\square$

(6.15) Corollary *If $F = K(\Pi, n)$ or if $\dim(K, L) \leq n$, then two liftings of $\phi : K \to B$ which agree on L are vertically homotopic* (rel. L) *if and only if $\bar{\delta}^n(f_0, f_1) = 0$.* $\square$

(6.16) Corollary *If $F = K(\Pi, n)$ or if $\dim(K, L) \leq n$, and if $\phi : K \to B$ can be lifted to a map of K into X, then the vertical homotopy classes* (rel. L) *of liftings of ϕ are in $1:1$ correspondence with $H^n(K, L; \pi_n(\mathscr{F}))$.* $\square$

We conclude this section by discussing the behavior of the obstructions in certain composite fibrations. Let

$$\begin{array}{c} F'' \longrightarrow X' \xrightarrow{\ q\ } X, \\ F \longrightarrow X \xrightarrow{\ p\ } B \end{array}$$

be fibrations; then $p' = p \circ q : X' \to B$ is a fibration, by Theorem (7.11) of Chapter I. If F' is the fibre of p', then the restriction of q to F' is a fibration

(6.17) $$F'' \longrightarrow F' \longrightarrow F.$$

Let us assume that F is $(m-1)$-connected and F'' is $(n-1)$-connected $(n \leq m)$. Exactness of the homotopy sequence of the fibration (6.17) implies that F' is $(n-1)$-connected. There are local coefficient systems $\pi_n(\mathscr{F}')$ over B and $\pi_n(\mathscr{F}'')$ over X, associated with the fibrations $p' : X' \to B$ and $q : X' \to X$, respectively. For each $x \in X$, we have an inclusion $F''_x \subset F'_{p(x)}$, and it is easy to see that

(6.18) *The injections $\pi_n(F''_x) \to \pi_n(F'_{p(x)})$ define a homomorphism $\rho : \pi_n(\mathscr{F}'') \to p^*\pi_n(\mathscr{F}')$.* □

Now consider the obstructions

$$\bar{\gamma}_1 \in H^{n+1}(B; \pi_n(\mathscr{F}')),$$
$$\bar{\gamma}_2 \in H^{n+1}(X; \pi_n(\mathscr{F}'')),$$

to cross-sections of the fibrations $p' : X' \to B$, $q : X' \to X$, respectively. Consider the diagram

$$\begin{array}{ccc} H^{n+1}(B; \pi_n(\mathscr{F}')) & \overset{p^*}{\searrow} & \\ & & H^{n+1}(X; p^*\pi_n(\mathscr{F}')) \\ H^{n+1}(X; \pi_n(\mathscr{F}'')) & \underset{\rho_*}{\nearrow} & \end{array}$$

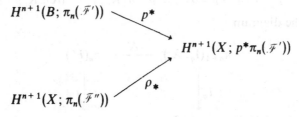

where p^* is induced by the map $p : X \to B$ (more precisely, by the pair $(p, 1)$, where 1 is the identity map of $p^*\pi_n(\mathscr{F}')$) and ρ_* by the homomorphism ρ (more precisely, by the pair $(1, \rho)$, where 1 is the identity map of X).

(6.19) Theorem *Under the above circumstances, the obstructions $\bar{\gamma}_1$, $\bar{\gamma}_2$ are related by*

$$p^*\bar{\gamma}_1 = \rho_* \bar{\gamma}_2 \in H^{n+1}(X; p^*\pi_n(\mathscr{F}')).$$

In fact, let

$$\bar{\gamma} \in H^{n+1}(X; p^*\pi_n(\mathscr{F}'))$$

be the obstruction to lifting p typified by the diagram

$$\begin{array}{ccc} & & X' \\ & \nearrow & \big\downarrow p' \\ X & \xrightarrow[p]{} & B \end{array}$$

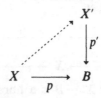

We shall show that $p^*\bar{\gamma}_1 = \rho_* \bar{\gamma}_2 = \bar{\gamma}$.

That $p^*\bar{\gamma}_1 = \bar{\gamma}$ follows from (2) of Theorem (6.3). To prove the second relation, let $h : K \to X$ be a CW-approximation. By Theorem (6.1), there is a map $f: K_n \to X'$ such that $q \circ f = h | K_n$. Let $\hat{q} : \mathbf{I}_q \to X$ be the projection of the mapping cylinder on its base. By Lemma (5.3), f has an extension $\hat{f}: (K_{n+1}, K_n) \to (\mathbf{I}_q, X')$ such that $\hat{q} \circ \hat{f} \simeq h | K_{n+1} : K_{n+1} \to X$ (rel. K_n). Let $\hat{p}' : \mathbf{I}_{p'} \to B$ be the projection. Then there is a map $l: \mathbf{I}_q \to \mathbf{I}_{p'}$ such that

$$l(\langle t, x' \rangle) = \langle t, x' \rangle \qquad (t \in I, x' \in X')$$

and the diagram

$$\begin{array}{ccc} \mathbf{I}_q & \xrightarrow{l} & \mathbf{I}_{p'} \\ \hat{q} \big\downarrow & & \big\downarrow \hat{p}' \\ X & \xrightarrow[p]{} & B \end{array}$$

is commutative. Then $l \circ \hat{f}: (K_{n+1}, K_n) \to (\mathbf{I}_{p'}, X')$ and

$$\hat{p}' \circ l \circ \hat{f} = p \circ \hat{q} \circ \hat{f} \simeq p \circ h | K_{n+1} \qquad \text{(rel. } K_n\text{)}.$$

Moreover the diagram

$$\begin{array}{ccc} \pi_{n+1}(\mathbf{I}_q, X') & \xrightarrow{\Delta'} & \pi_n(F'') \\ l_* \big\downarrow & & \big\downarrow i \\ \pi_{q+1}(\mathbf{I}_{p'}, X') & \xrightarrow[\Delta']{} & \pi_n(F') \end{array}$$

is commutative, i being the injection. Hence

$$i \circ \Delta' \circ \hat{f}_* = \Delta' \circ l_* \circ \hat{f}_* : \pi_{n+1}(K_{n+1}, K_n) \to \pi_n(F').$$

But

$$\Delta' \circ f_* : \pi_{n+1}(K_{n+1}, K_n) \to \pi_n(F'')$$

and

$$\Delta' \circ (l \circ \hat{f})_* : \pi_{n+1}(K_{n+1}, K_n) \to \pi_n(F')$$

are the homomorphisms corresponding to the obstruction cocycles $\bar{c}^{n+1}(h)$, $\bar{c}^{n+1}(p \circ h)$ for the fibrations $q : X' \to X$, $p' : X' \to B$, respectively. It follows that $\rho_* \bar{\gamma}_2 = \bar{\gamma}$, as desired. □

7 Characteristic Classes of Vector Bundles

Let $p : X \to B$ be the projection of a fibre bundle ξ having $\mathbf{R}^n$ as fibre and the orthogonal group $\mathbf{O}(n)$ as structural group (such a bundle will be called an *orthogonal vector bundle*). Then $\mathbf{O}(n)$ acts on the Stiefel manifold $\mathbf{V}_{n,k}(0 \le k \le n)$, and there is an *associated bundle* $p_k : X_k \to B$ having $\mathbf{V}_{n,k}$ as fibre; the points of X_k are the $(k+1)$-tuples $(b; x_1, \ldots, x_k)$, where $b \in B$ and $(x_1, \ldots, x_k)$ is an orthonormal k-frame in the Euclidean space $p^{-1}(b)$. Moreover, $p_k(b; x_1, \ldots, x_k) = b$. In particular, we may identify $\mathbf{V}_{n,n}$ with $\mathbf{O}(n)$; then $p_n : X_n \to B$ is the (projection of) the *principal* associated bundle of ξ. Moreover, X_1 is the space of *unit* vectors in X.

The map $(b; x_1, \ldots, x_k) \to (b; x_1, \ldots, x_{k-1})$ is a fibre bundle $q_k : X_k \to X_{k-1}$, and the fibre of q_k is $\mathbf{S}^{n-k}$. Thus there is a tower of fibrations

$$X_n \xrightarrow{q_n} X_{n-1} \longrightarrow \cdots \longrightarrow X_k \xrightarrow{q_k} X_{k-1} \longrightarrow \cdots \xrightarrow{q_1} X_0 = B,$$

and $p_k = q_1 \circ \cdots \circ q_k$.

As a special case, we may take $X_n = \mathbf{V}(n)$, the space of n-frames in $\mathbf{R}^\infty$, $B = \mathbf{G}(n)$ the Grassmannian space of n-dimensional subspaces of $\mathbf{R}^\infty$ (cf. §2 of Appendix A), and we have the tower of fibrations

$$\mathbf{V}(n) = \mathbf{V}_n(n) \xrightarrow{\mathbf{q}_n(n)} \mathbf{V}_{n-1}(n) \xrightarrow{\mathbf{q}_{n-1}(n)} \cdots \longrightarrow \mathbf{V}_1(n) \xrightarrow{\mathbf{q}_1(n)} \mathbf{V}_0(n) = \mathbf{G}(n).$$

The space $\mathbf{G}(n)$ is a classifying space for $\mathbf{O}(n)$ (Theorem (2.1) of Appendix A), and therefore there is a unique homotopy class of maps $f: B \to \mathbf{G}(n)$ such that $f^*(\eta_n) = \xi$, where η_n is the $\mathbf{R}^n$-bundle associated with the principal bundle $\mathbf{V}(n) \to \mathbf{G}(n)$. It follows easily that

(7.1) Theorem *There is a commutative diagram*

$$\begin{array}{ccccccccccc}
X_n & \xrightarrow{q_n} & X_{n-1} & \to\cdots\to & X_k & \xrightarrow{q_k} & X_{k-1} & \to\cdots\to & X_1 & \xrightarrow{q_1} & B \\
\downarrow{\scriptstyle f_n} & & \downarrow{\scriptstyle f_{n-1}} & & \downarrow{\scriptstyle f_k} & & \downarrow{\scriptstyle f_{k-1}} & & \downarrow{\scriptstyle f_1} & & \downarrow{\scriptstyle f} \\
\mathbf{V}(n) & \xrightarrow[\mathbf{q}_n(n)]{} & \mathbf{V}_{n-1}(n) & \to\cdots\to & \mathbf{V}_k(n) & \xrightarrow[\mathbf{q}_k(n)]{} & \mathbf{V}_{k-1}(n) & \to\cdots\to & \mathbf{V}_1(n) & \xrightarrow[\mathbf{q}_1(n)]{} & \mathbf{G}(n)
\end{array}$$

The fibration $q_k : X_k \to X_{k-1}$ *is induced from the fibration* $\mathbf{q}_k(n) : \mathbf{V}_k(n) \to \mathbf{V}_{k-1}(n)$ *by the map* f_{k-1}, *and the fibration* $p_k : X_k \to B$ *is induced from* $\mathbf{p}_k(n) = \mathbf{q}_1(n) \circ \cdots \circ \mathbf{q}_k(n)$ *by the map* f. □

Consider the fibration $p_k : X_k \to B$. The fibre $\mathbf{V}_{n,k}$ of p_k is $(n-k-1)$-connected, by (10.12) of Chapter IV. By Theorem (10.13) of the same chapter, the first non-vanishing homotopy group $\pi_{n-k}(\mathbf{V}_{n,k})$ is cyclic of order two if $n-k$ is odd and $k \geq 2$, and is infinite cyclic otherwise. In the former case, the local coefficient system $\pi_{n-k}(\mathscr{F}_k)$ is simple. In the latter case, the system $\pi_{n-k}(\mathscr{F}_k)$ is *simple*, when $\pi_1(B)$ operates trivially, or *twisted*, when the action is non-trivial. Because of (1.15), it suffices to determine the action of $\pi_1(B)$ in the special case $B = \mathbf{G}(n)$. By Theorem (10.16) of Chapter IV, with $l = n$ and $k \to \infty$, $\pi_1(\mathbf{G}(n)) \approx \pi_0(\mathbf{O}(n)) = \mathbf{Z}_2$. If α is the non-zero element of $\pi_1(\mathbf{G}(n))$, we have

(7.2) Theorem *The action of* $\pi_1(\mathbf{G}(n))$ *on* $\pi_{n-k}(\mathbf{V}_{n,k})$ *is determined by*

$$\theta_\alpha(x) = -x \qquad x \in \pi_{n-k}(\mathbf{V}_{n,k}).$$

Thus the local coefficient system $\pi_{n-k}(\mathscr{F}_k)$ *over* $\mathbf{G}(n)$ *is twisted if and only if* $n-k$ *is even or* $k = 1$.

As $\mathbf{G}(n)$ is the direct limit of its subspaces $\mathbf{G}_{q,n}$, it suffices to consider the case $B = \mathbf{G}_{q,n} = \mathbf{O}(q+n)/\hat{\mathbf{O}}(q) \times \mathbf{O}(n)$, $X_n = \mathbf{V}_{n+q,n} = \mathbf{O}(q+n)/\hat{\mathbf{O}}(q)$, so that $X_k = \mathbf{O}(q+n)/\hat{\mathbf{O}}(q) \times \mathbf{O}(n-k)$, for large values of q. Let $\tilde{u} : \mathbf{I} \to \mathbf{O}(q+n)$ be a path such that $\tilde{u}(1)$ is the identity matrix, while $\tilde{u}(0) = \operatorname{diag}\{J(q), J(n)\}$, where $J(n)$ is an improper orthogonal matrix. Then the projection of $\tilde{u}$ into B is a loop $u : \mathbf{I} \to B$ which represents the element α. Let $F_k = \mathbf{V}_{n,k}$ be the fibre of p_k, $i : F_k \subsetneq X_k$. Then the map

$$h : \mathbf{I} \times F_k \to X_k$$

defined by

$$h(t, y) = \tilde{u}(t) \cdot i(y)$$

lies over u and ends at the inclusion map i. Hence its initial value $h_0 : F_k \to X_k$ has the form $h_0 = i \circ l$ for a u-admissible map $l : F_k \to F_k$; and $l_* = \theta_\alpha : \pi_q(F) \to \pi_q(F)$ for any q.

Now $F = \mathbf{V}_{n,k} \approx \mathbf{O}(n)/\mathbf{O}(n-k) \approx \mathbf{O}(q) \times \mathbf{O}(n)/\mathbf{O}(q) \times \mathbf{O}(n-k)$ can be represented as the space of $n \times k$ matrices with orthonormal columns. The orthogonal group $\mathbf{O}(n)$ can be represented as the group of $n \times n$ orthogonal matrices, and the operation of $\mathbf{O}(n)$ on F by left multiplication. In terms of this representation, a generator β of $\pi_{n-k}(F)$ is represented by the map $g : \mathbf{S}^{n-k} \to F$, where

$$g(x) = \begin{pmatrix} x & 0 \\ 0 & E_{k-1} \end{pmatrix},$$

$x \in \mathbf{S}^{n-k}$ is a unit column vector with $n-k+1$ components, E_{k-1} is the identity matrix of order $k-1$, and the 0's are matrices of zeroes of the appropriate size. We may assume that $J(n) = \operatorname{diag}\{-1, E_{n-1}\}$; then

$$J(n)g(x) = g(-x)$$

and therefore $\theta_\alpha(\beta) = -\beta$. □

(7.3) Corollary *Let $f: B \to \mathbf{G}(n)$. Then the local coefficient system $\pi_{n-k}(\mathscr{F}_k)$ over B determined by the induced fibration is simple if and only if either*

(1) *$n-k$ is odd and $k \geq 2$*

or

(2) *$f_* : \pi_1(B) \to \pi_1(\mathbf{G}(n))$ is trivial.* □

We have now determined the coefficient groups for the primary obstructions to the cross-sections $p_{n-k+1} : X_{n-k+1} \to B$. The obstructions themselves are cohomology classes

$$W_k(\xi) \in H^k(B; \pi_{k-1}(\mathscr{F}_{n-k+1})) \qquad (k = 2, 3, \ldots, n).$$

The class $W_1(\xi)$ is not defined by the above procedure, since the fibre $\mathbf{V}_{n,n} = \mathbf{O}(n)$ is not pathwise connected. It is convenient to define $W_1(\xi)$ instead by means of Theorem (3.3). The local coefficient system $\mathscr{Z}$ for $W_1(\xi)$ is defined by the following action of $\pi_1(B)$ on the additive group $\mathbf{Z}$ of integers: let $f: B \to \mathbf{G}(n)$ be the classifying map; then an element $\lambda \in \pi_1(B)$ operates trivially or not according as $f_*(\lambda) = 1$ or $f_*(\lambda)$ is the non-zero element $\alpha \in \pi_1(\mathbf{G}(n))$. The reader may then verify that the function $\phi : \pi_1(B) \to \mathbf{Z}$ defined by

$$\phi(\lambda) = \begin{cases} 0 & \text{if } f_*(\lambda) = 1, \\ 1 & \text{if } f_*(\lambda) = \alpha, \end{cases}$$

is a crossed homomorphism. The class $W_1(\xi)$ is then defined to be the element of $H^1(B; \mathscr{Z})$ associated to ϕ by the isomorphism of Theorem (3.3).

Finally, it is convenient to define $W_0(\xi) \in H^0(B; \mathbf{Z})$ to be the unit element of the integral cohomology ring of B.

The classes $W_k(\xi)$ $(k = 0, 1, 2, \ldots, n)$ are called the *Whitney characteristic classes* of the vector bundle ξ. They have the following naturality property:

(7.4) Theorem *Let $g : B' \to B$ be a continuous map. Then $W_k(g^*\xi) = g^*W_k(\xi)$ $(k = 0, 1, \ldots, n)$.*

This follows from (2) of Theorem (6.3) if $2 \leq k \leq n$. The remaining cases $k = 0, 1$ are easily settled. □

In particular, we may consider the universal example $B = \mathbf{G}(n)$ and the bundle η_n. The resulting classes $W_k(n)$ are called the *universal Whitney classes.*

(7.5) Corollary *Let $f: B \to \mathbf{G}(n)$ be a classifying map for the vector bundle ξ. Then*

$$W_k(\xi) = f^* W_k(n). \qquad \square$$

Since cohomology with twisted integer coefficients is relatively unfamiliar, it is often useful to reduce the coefficient groups mod 2. The resulting local coefficient systems are all simple, and the resulting characteristic classes

$$w_k(\xi) \in H^k(B; \mathbf{Z}_2)$$

are called the *Stiefel–Whitney classes of* ξ, and the classes

$$w_k(n) \in H^k(G(n); \mathbf{Z}_2)$$

the *universal Stiefel–Whitney classes.* Of course,

(7.6) Theorem *Let $g: B' \to B$ be a continuous map. Then $w_k(g^*\xi) = g^* w_k(\xi)$ $(k = 0, 1, \ldots, n)$.* $\square$

(7.7) Corollary *Let $f: B \to \mathbf{G}(n)$ be a classifying map for the vector bundle ξ. Then*

$$w_k(\xi) = f^* w_k(n). \qquad \square$$

Let us now compare the above characteristic classes for different values of n. We may begin with the observation that, the projection $\mathbf{p}_n(n): \mathbf{V}(n) \to \mathbf{G}(n)$ being a principal fibration with group $\mathbf{O}(n)$, the space $\mathbf{G}(n)$ is naturally homeomorphic with the quotient space $\mathbf{V}(n)/\mathbf{O}(n)$. Through the monomorphism $\mathbf{O}(n-k) \to \hat{\mathbf{O}}(n-k) \subset \mathbf{O}(n)$, the group $\mathbf{O}(n-k)$ acts on $\mathbf{V}(n)$, and the quotient space $\mathbf{V}(n)/\mathbf{O}(n-k)$ is easily seen to be the space $\mathbf{V}_k(n)$.

Let $i: \mathbf{R}^\infty \to \mathbf{R}^\infty$ be the map which sends $\mathbf{e}_j$ into $\mathbf{e}_{j+1}$ $(j = 0, 1, 2, \ldots)$. The map i is an isometric imbedding and induces an imbedding of $\mathbf{V}(n-1)$ in $\mathbf{V}(n)$: if $(x_1, \ldots, x_{n-1})$ is an orthonormal $(n-1)$-frame in $\mathbf{R}^\infty$, then $(\mathbf{e}_0, i(x_1), \ldots, i(x_{n-1}))$ is an orthonormal n-frame in $\mathbf{R}^\infty$. This imbedding is equivariant with respect to the action of $\mathbf{O}(n-1)$ (where $\mathbf{O}(n-1)$ is considered as the subgroup of $\mathbf{O}(n)$ fixing the first unit vector), and therefore there is a commutative diagram (Figure 6.7). The spaces $\mathbf{V}_{k-1}(n-1)$ and $\mathbf{V}_k(n)$ are classifying spaces for the group $\mathbf{O}(n-k)$ and it follows as in Theorem (2.3) of Appendix A that

(7.8) Theorem *The map $i_k: \mathbf{V}_{k-1}(n-1) \to \mathbf{V}_k(n)$ is a homotopy equivalence.* $\square$

(7.9) Corollary *The space $\mathbf{V}_k(n)$ has the same homotopy type as $G(n-k)$.* $\square$

The map $j = \mathbf{q}_1(n) \circ i_1$ is easily seen to be the canonical imbedding, associating with each $(n-1)$-plane π in $\mathbf{R}^\infty$ the n-plane $\mathbf{R}\mathbf{e}_0 \oplus i(\pi)$.

$$\begin{array}{ccc}
\mathbf{V}(n-1) = \mathbf{V}_{n-1}(n-1) & \xrightarrow{i_n} & \mathbf{V}_n(n) = \mathbf{V}(n)\\
\downarrow & & \downarrow\\
\cdots & & \cdots\\
\downarrow & & \downarrow\\
\mathbf{V}(n-1)/\mathbf{O}(n-k-1) = \mathbf{V}_k(n-1) & \xrightarrow{i_{k+1}} & \mathbf{V}_{k+1}(n) = \mathbf{V}(n)/\mathbf{O}(n-k-1)\\
\mathbf{q}_k(n-1)\Big\downarrow & & \Big\downarrow \mathbf{q}_{k+1}(n)\\
\mathbf{V}(n-1)/\mathbf{O}(n-k) = \mathbf{V}_{k-1}(n-1) & \xrightarrow{i_k} & \mathbf{V}_k(n) = \mathbf{V}(n)/\mathbf{O}(n-k)\\
\downarrow & & \downarrow\\
\cdots & & \cdots\\
\downarrow & & \downarrow\\
\mathbf{G}(n-1) = \mathbf{V}(n-1)/\mathbf{O}(n-1) = \mathbf{V}_0(n-1) & \xrightarrow{i_1} & \mathbf{V}_1(n) = \mathbf{V}(n)/\mathbf{O}(n-1)\\
& \searrow^{j} & \Big\downarrow \mathbf{q}_1(n)\\
& & \mathbf{V}_0(n) = \mathbf{V}(n)/\mathbf{O}(n) = \mathbf{G}(n)
\end{array}$$

Figure 6.7

It is likewise easy to see that

(7.10) *The fibration* $\mathbf{q}_k(n-1)$ *is induced by the map* i_k *from the fibration* $\mathbf{q}_{k+1}(n)$. □

Let us now consider the commutative diagram

$$\begin{array}{ccc}
\mathbf{V}_{n-k-1}(n-1) & \xrightarrow{i_{n-k}} & \mathbf{V}_{n-k}(n)\\
q_1 = \mathbf{p}_{k-1}(n-1)\Big\downarrow & & \Big\downarrow \mathbf{p}'_k(n) = q\\
\mathbf{G}(n-1) & \xrightarrow{i_1} & \mathbf{V}_1(n)\\
& \searrow^{j} & \Big\downarrow \mathbf{q}_1(n) = p\\
& & \mathbf{G}(n)
\end{array}$$

where $\mathbf{p}'_k(n) = \mathbf{q}_2(n) \circ \cdots \circ \mathbf{q}_k(n)$. It follows from (7.10) that

(7.11) *The fibration q_1 is induced by i_1 from the fibration q.* □

The local coefficient systems defined by the fibrations in the above diagram are

$$\pi_k(\mathscr{F}'') \quad \text{for } q,$$
$$\pi_k(\mathscr{F}') \quad \text{for } p' = p \circ q,$$
$$i_1^*\pi_k(\mathscr{F}'') \quad \text{for } q_1.$$

In (6.18) we saw that the injections define a homomorphism $\rho : \pi_k(\mathscr{F}'') \to p^*\pi_k(\mathscr{F}')$. Similarly, the injections define a homomorphism $\sigma = i_1^*\rho : i_1^*\pi_k(\mathscr{F}'') \to j^*\pi_k(\mathscr{F}')$. The map $i_1 : \mathbf{G}(n-1) \to \mathbf{V}_1(n)$, together with the identity map of $i_1^*\pi_k(\mathscr{F}'')$, define a morphism

$$\bar{i}_1 : (\mathbf{G}(n-1), i_1^*\pi_k(\mathscr{F}'')) \to (\mathbf{V}_1(n), \pi_k(\mathscr{F}''))$$

in the category $\mathscr{L}^*$. The identity map of $\mathbf{G}(n-1)$, together with the homomorphism σ, defines a morphism

$$\tilde{\sigma} : (\mathbf{G}(n-1), j^*\pi_k(\mathscr{F}')) \to (\mathbf{G}(n-1), i_1^*\pi_k(\mathscr{F}'')).$$

The map i_1, together with the identity map of $j^*\pi_k(\mathscr{F}')$, defines a morphism

$$\hat{i}_1 : (\mathbf{G}(n-1), j^*\pi_k(\mathscr{F}')) \to (\mathbf{V}_1(n), p^*\pi_k(\mathscr{F}')).$$

The identity map of $\mathbf{V}_1(n)$, together with the homomorphism ρ, defines a morphism

$$\tilde{\rho} : (\mathbf{V}_1(n), p^*\pi_k(\mathscr{F}')) \to (\mathbf{V}_1(n), \pi_k(\mathscr{F}'')).$$

An easy calculation assures us that

(7.12) $\quad \tilde{\rho} \circ \hat{i}_1 = \bar{i}_1 \circ \tilde{\sigma} : (\mathbf{G}(n-1), j^*\pi_k(\mathscr{F}')) \to (\mathbf{V}_1(n), \pi_k(\mathscr{F}''))$.

On the other hand, the map $p : \mathbf{V}_1(n) \to \mathbf{G}(n)$, together with the identity map of $\pi_k(\mathscr{F}')$, defines a morphism $\tilde{p} : (\mathbf{V}_1(n), p^*\pi_k(\mathscr{F}')) \to (\mathbf{G}(n), \pi_k(\mathscr{F}'))$. And the map $j : \mathbf{G}(n-1) \to \mathbf{G}(n)$, with the identity map of $\pi_k(\mathscr{F}')$, defines a morphism $\tilde{j}: (\mathbf{G}(n-1), j^*\pi_k(\mathscr{F}')) \to (\mathbf{G}(n), \pi_k(\mathscr{F}'))$, and we have

(7.13) $\quad \tilde{p} \circ \hat{i}_1 = \tilde{j}: (\mathbf{G}(n-1), j^*\pi_k(\mathscr{F}')) \to (\mathbf{G}(n), \pi_k(\mathscr{F}'))$.

Let $W_{k+1}^* \in H^{k+1}(\mathbf{V}_1(n), \pi_k(\mathscr{F}''))$ be the primary obstruction to a cross-section of q. Because of (7.11),

$$\bar{i}_1^* W_{k+1}^* = W_{k+1}(n-1) \in H^{k+1}(\mathbf{G}(n-1), i_1^*\pi_k(\mathscr{F}'')).$$

By Theorem (6.19),

(7.14) $\quad \tilde{p}^* W_{k+1}(n) = \tilde{\rho}^* W_{k+1}^* \in H^{k+1}(\mathbf{V}_1(n), p^*\pi_k(\mathscr{F}'))$.

Applying the homomorphism $\hat{i}_1^*$ to both sides of (7.14) and using (7.12), we find that

$$\text{(7.15)}\qquad \begin{aligned} \tilde{j}^* W_{k+1}(n) &= \hat{i}_1^* \tilde{p}^* W_{k+1}(n) \\ &= \hat{i}_1^* \tilde{\rho}^* W_{k+1}^* \\ &= \tilde{\sigma}^* \tilde{i}_1^* W_{k+1}^* \\ &= \tilde{\sigma}^* W_{k+1}(n-1). \end{aligned}$$

The relation (7.15) appears somewhat indigestible as it stands. However, if $n - k > 2$ or $n - k = 2$ and n is even, then the injection

$$\pi_k(\mathbf{V}_{n-1,\, n-k-1}) \to \pi_k(\mathbf{V}_{n,\, n-k})$$

is an isomorphism, by Theorem (10.13) of Chapter IV. It follows that σ is an isomorphism of local coefficient systems (integers mod 2 or twisted integers, as the case may be). Identifying these isomorphic systems, we then have

(7.16) Theorem *If $k < n - 2$ or $k = n - 2$ is even, the injection j^* maps $W_{k+1}(n)$ into $W_{k+1}(n-1)$.* □

We should observe that we have used the definition of the Whitney classes as obstructions. Therefore the proof of Theorem (7.16) will not be complete until we have studied the cases $k = -1$ and $k = 0$. But these present no difficulty and are left to the reader.

Suppose that $k = n - 2$ is odd. Then the injection

$$\mathbf{Z} = \pi_{n-2}(\mathbf{V}_{n-1,\,1}) \to \pi_{n-2}(\mathbf{V}_{n,\,2}) = \mathbf{Z}_2$$

is an epimorphism. In this case the homomorphism σ maps the system $i_1^* \pi_{n-2}(\mathscr{F}'')$ of twisted integers into the simple system $j^* \pi_{n-2}(\mathscr{F}')$ of integers modulo two. Reducing the coefficient groups for $W_{n-1}(n-1)$ mod 2, we obtain the Stiefel–Whitney class $w_{n-1}(n-1)$, and

(7.17) Theorem *If n is odd, the injection j^* maps $W_{n-1}(n)$ into $w_{n-1}(n-1)$.* □

Finally, if $k = n - 1$, the class $W_n(n-1) = 0$, so that

(7.18) Theorem *The injection j^* maps $W_n(n)$ into zero.* □

Theorems (7.16)–(7.18) cover the behavior of the Whitney classes. If we reduce mod 2, we obtain the Stiefel–Whitney classes, and the corresponding result is

(7.19) Theorem *The injection j^* maps $w_k(n)$ into $w_k(n-1)$ $(k = 0, 1, \ldots, n-1)$, while $j^* w_k(n) = 0$.* □

It is customary to abbreviate $W_k(n)$, $w_k(n)$ to W_k, w_k, respectively. Because of Theorems (7.16)–(7.19) this should cause little confusion.

Exercises

1. Let $f_1 : X \to Y$ be a homotopy equivalence in $\mathscr{K}$, and let Q be a system of local coefficients in Y. Then f_1 extends to a homotopy equivalence $f : (X; f_1^* Q) \to (Y; Q)$ in $\mathscr{L}$.

2. Let K, L be ordered simplicial complexes, $\phi : K \to L$ an order-preserving simplicial map. Let G, H be local coefficient systems on $|K|$, $|L|$, respectively, and let $\psi : G \to |\phi|^* H$ be a homomorphism. Show that ϕ, ψ induce a chain map of $\Gamma_*(K; G)$ into $\Gamma_*(L; H)$ whose induced homomorphism coincides, up to the isomorphism of Theorem (4.8), with the homomorphism
$$(\phi, \psi)_* : H_*(|K|; G) \to H_*(|L|; H)$$
induced by the morphism (ϕ, ψ) of the category $\mathscr{L}$.

3. Let K be an ordered simplicial complex, K' its first barycentric subdivision. Let $\phi : K' \to K$ be the simplicial map such that, for each simplex σ of K, $\phi(b_\sigma)$ is the last vertex of σ. Prove that $\phi_* : H_q(K'; G) \to H_q(K; G)$ is an isomorphism for every local coefficient system G in $|K| = |K'|$.

4. Prove the analogues of the results of Exercises 2 and 3 for cohomology with local coefficients.

5. Let $p : X \to B$ be a fibration with fibre F, and suppose that B and F are 1-connected. Show that the local coefficient system $\pi_q(\Omega\mathscr{F})$ for the fibration $\Omega p : \Omega X \to \Omega B$ is simple.

6. Let $p : X \to B$ be a fibration with fibre F, and suppose that B is 1-connected and F $(n-1)$-connected $(n \geq 2)$. Let (K, L) be a relative CW-complex and let $f : L \to \Omega X$ be a partial lifting of $\phi : K \to \Omega B$. Let $\bar{f} : \mathrm{S}L \to X$ and $\bar{\phi} : \mathrm{S}K \to B$ be the adjoints of f, ϕ, respectively. Let $\alpha^* : H^{n+1}(\mathrm{S}K, \mathrm{S}L; \pi_n(F)) \to H^n(K, L; \pi_{n-1}(\Omega F))$ be the composite of the suspension operator
$$H^{n+1}(\mathrm{S}K, \mathrm{S}L; \pi_n(F)) \approx H^n(K, L; \pi_n(F))$$
with the isomorphism
$$H^n(K, L; \pi_n(F)) \approx H^n(K, L; \pi_{n-1}(\Omega F))$$
induced by the isomorphism
$$\Delta_* : \pi_n(F) \to \pi_{n-1}(\Omega F)$$
of the homotopy sequence of the fibration $p : P'(F) \to F$. Show that $\alpha^* \gamma^{n+1}(\bar{f}) = \gamma^n(f)$.

7. Let $p : X \to B$ be a fibration with fibre F, and suppose that B is 1-connected and F $(n-1)$-connected $(n \geqq 2)$. Let $\bar{\gamma}^{n+1} \in H^{n+1}(B; \pi_n(F))$, $\bar{\gamma}^n \in H^n(\Omega B; \pi_{n-1}(\Omega F))$ be the primary obstruction to cross-sections of p, Ωp, respectively. Let β^* be the composite of the cohomology suspension
$$\sigma^* : H^{n+1}(B; \pi_n(F)) \to H^n(\Omega B; \pi_n(F))$$

of Chapter VIII with the homomorphism

$$H^n(\Omega B; \pi_n(F)) \to H^n(\Omega B; \pi_{n-1}(\Omega F))$$

induced by the coefficient group isomorphism Δ_*. Prove that $\beta^* \bar{\gamma}^{n+1} = \bar{\gamma}^n$.

8. Prove that, if X is a 0-connected space, $x_0 \in X$, if G is a bundle of abelian groups in X, and $C_q^{(0)}(X; G)$ is the set of all singular q-chains $\sum g_i u_i$, where $u_i : \Delta^q \to X$ are singular simplices, all of whose vertices are at x_0, then the graded group $C_*^{(0)}(X; G)$ is a subcomplex of $C_*(X; G)$, and the injection $H_q(C_*^{(0)}(X; G)) \to H_q(C_*(X; G))$ is an isomorphism.

CHAPTER VII

Homology of Fibre Spaces: Elementary Theory

The relations among the homotopy groups of the fibre F, total space X and base space B of a fibration are rather simple, as we saw in §8 of Chapter IV. The behavior of the homology groups is much more complicated. In the simplest case, that of a trivial fibration, the relationship is given by the Künneth Theorem. The general case will be treated in Chapter XIII with the aid of the complicated machinery of spectral sequences. In this Chapter we shall treat by more elementary methods certain important special cases. There are two reasons for this. The first is the hope that the geometrical considerations of this Chapter will help motivate the spectral sequence. The second is that the present route leads quickly to certain applications we have in mind, e.g., the homology of the classical groups.

In §1, we assume that the base space B is the suspension of a space W. The decomposition of the base into two copies of the contractible space TW induces a decomposition of the total space into two subspaces, each of which has the homotopy type of $TW \times F$, and therefore of F. Their intersection then has the homotopy type of $W \times F$, and a consideration of the relationship between these subspaces leads to an exact sequence which makes the relationships among the homology groups of the three spaces F, X and B reasonably perspicuous. The case $B = \mathbf{S}^n$ is of especial interest, and the resulting exact sequence was found by Wang [1] in 1949.

Another special case of importance is that of the path space fibration over $\mathbf{S}W$. The adjoint of the identity map of $\mathbf{S}W$ is an imbedding of W in the fibre $\Omega\mathbf{S}W$. Here the process of iterated multiplication in the H-space $\Omega\mathbf{S}W$ leads to a sequence of maps of $W^n = W \times \cdots \times W$ into $\Omega\mathbf{S}W$. The spaces W^n can be pasted together, by identifying $(w_1, \ldots, w_{n+1})$ with $(w_1, \ldots, \hat{w}_i, \ldots, w_{n+1})$ whenever w_i is the base point, to produce a space $J(W)$, called the *reduced product*, together with a map of $J(W)$ into $\Omega\mathbf{S}W$. This construction was discovered in 1955 by James [1], who proved that the map in question is a weak homotopy equivalence. A comparable construction was found about the same time by Toda [1]. Both were anticipated by many years by Morse [Mo], who used it as his calculation of the homology of the space of rectifiable paths on a sphere. The reduced product is discussed in §2, while §3 is devoted to some further properties of the Wang sequence. In §4 the fibrations

$$\mathbf{O}_n^+ \to \mathbf{S}^{n-1}$$

$$\mathbf{U}_n \to \mathbf{S}^{2n-1}$$

$$\mathbf{Sp}_n \to \mathbf{S}^{4n-1}$$

are used to calculate the cohomology rings of the classical groups (with $\mathbf{Z}_2$ coefficients for $\mathbf{O}_n^+$ and general coefficients for the others).

Another special case of interest is that in which the fibre is a sphere. If we further assume the fibration to be orientable, we obtain the Thom Isomorphism Theorem [1] $H^p(B) \approx H^{p+n+1}(\hat{X}, X)$, where $\hat{X}$ is the mapping cylinder of p. This leads to an exact sequence due to Gysin [1] in 1941, relating the homology groups of the total space with those of the base, which in many cases allow us to make explicit calculations. These results are discussed in §5.

As remarked before, there is a simple relationship among the homotopy groups of the three spaces F, X, B; it is an exact sequence

$$\cdots \to \pi_r(F) \to \pi_r(X) \to \pi_r(B) \to \pi_{r-1}(F) \to \cdots.$$

For homology groups there is nothing so simple; however, if B is $(m-1)$-connected and F is $(n-1)$-connected, there is a parallel result in the form of an exact sequence relating the homology groups in dimensions $\leq m+n$. This was discovered by Serre [1] in 1950; an important consequence is that, if B is $(m-1)$-connected, then $H_q(\Omega B) \approx H_{q+1}(B)$ for $q \leq 2m-2$. Section 6 is devoted to this result.

The homology and homotopy groups have many points of resemblance. One important difference is that, while the Excision Property is satisfied by the homology groups, it fails for the homotopy groups. In 1951 Blakers and Massey [1] introduced the homotopy groups of a triad, which were designed to measure the extent to which the Excision Property fails. Their main result was that, under reasonable assumptions about the spaces involved, if $X = A \cup B$ and $(A, A \cap B)$ is $(m-1)$-connected, $(B, A \cap B)$ is $(n-1)$-connected, then the injections

$$\pi_i(B, A \cap B) \to \pi_i(X, A),$$

$$\pi_i(A, A \cap B) \to \pi_i(X, B),$$

are isomorphisms for $i < m+n-2$ and epimorphisms for $i = m+n-2$. A consequence is the Suspension Theorem: if X is $(n-1)$-connected then $\pi_i(X) \approx \pi_{i+1}(\mathrm{S}X)$ for $i < 2n-1$. This theorem was proved by Freudenthal in 1937 [1] in the case $X = \mathbf{S}^n$ and is one of the landmarks of homotopy theory. The Blakers–Massey Theorem is proved in §7, following Namioka [1], as a consequence of the results of §6.

1 Fibrations over a Suspension

Let W be a space with non-degenerate base point $*$, $B = \mathbf{S}W$, and let $p: X \to B$ be a fibration with fibre F. As usual, if $t \in \mathbf{I}$, $\bar{t} = \varpi(t)$ is the point of $\mathbf{S} = \mathbf{S}^1$ to which it corresponds under the identification map ϖ. Then B is the union $\mathbf{T}_+ W \cup \mathbf{T}_- W$ of two copies of the cone over W;

$$\mathbf{T}_+ W = \{\bar{t} \wedge w \mid t \leq \tfrac{1}{2}\},$$
$$\mathbf{T}_- W = \{\bar{t} \wedge w \mid t \geq \tfrac{1}{2}\},$$

and $\mathbf{T}_+ W \cap \mathbf{T}_- W$ is a copy of W.

Let $X_+ = p^{-1}(\mathbf{T}_+ W)$, $X_- = p^{-1}(\mathbf{T}_- W)$, $X_0 = X_+ \cap X_- = p^{-1}(W)$. Since $(\mathbf{T}_\pm W, W)$ and $(\mathbf{S}W, \mathbf{T}_\pm W)$ are NDR-pairs, so are the pairs $(X_\pm, X_0)$ and $(X, X_\pm)$, by Theorem (7.14) of Chapter I. Since $\mathbf{T}_- W$ is contractible, the fibre F is a deformation retract of X_-, and therefore the injection

$$i_1 : H_q(X, F) \to H_q(X, X_-)$$

is an isomorphism. By Theorem (2.2) of Chapter II, the triad $(X; X_+, X_-)$ is proper, so that the injection

$$i_2 : H_q(X_+, X_0) \to H_q(X, X_-)$$

is an isomorphism.

As $\mathbf{T}_+ W$ is homeomorphic with $\mathbf{T}W$ under the map $\bar{t} \wedge w \to 2t \wedge w$, we may regard X_+ as a fibre space over $\mathbf{T}W$. Since the latter is contractible, the projection $p_1 : \mathbf{T}W \times F \to \mathbf{T}W$ is homotopic to the constant map. Hence the projection $p_2 : \mathbf{T}W \times F \to F$ is homotopic in X_+ to a map $h: \mathbf{T}W \times F \to X_+$ such that $p \circ h = p_1$. By the remarks preceding Corollary (7.27) of Chapter I, $H: (\mathbf{T}W \times F, W \times F) \to (X_+, X_0)$ is a fibre homotopy equivalence, and by that same Corollary,

$$h_* : H_q(\mathbf{T}W \times F, W \times F) \approx H_q(X_+, X_0).$$

The above homeomorphism of $\mathbf{T}W$ with $\mathbf{T}_+ W$ is homotopic in $B = \mathbf{S}W$ to the proclusion $\varpi' = \varpi \wedge 1 : (\mathbf{T}W, W) \to (\mathbf{S}W, *)$; under this homotopy, we may assume the points of W remain in $\mathbf{T}_- W$. Hence $h: (\mathbf{T}W \times F, W \times F) \to (X_+, X_0)$ is homotopic in (X, X_-) to a map $\phi: (\mathbf{T}W \times F, W \times F) \to (X, F)$. Then $i_1 \circ \phi_* = i_2 \circ h_*$, and the fact that i_1, i_2 and h_* are isomorphisms implies that $\phi_* : H_q(\mathbf{T}W \times F, W \times F) \to H_q(X, F)$ is an isomorphism.

The above discussion holds for homology with arbitrary coefficients, and the dual results hold for cohomology. We summarize this discussion in

(1.1) Theorem *Let $p : X \to \mathbf{S}W$ be a fibration with fibre F. Then there is a map $\phi : (\mathbf{T}W \times F, W \times F) \to (X, F)$ such that the diagram*

$$\begin{array}{ccc} (\mathbf{T}W \times F, W \times F) & \xrightarrow{\ \phi\ } & (X, F) \\ \downarrow{\scriptstyle p_1} & & \downarrow{\scriptstyle p} \\ (\mathbf{T}W, W) & \xrightarrow[\ \varpi'\ ]{} & (\mathbf{S}W, *) \end{array}$$

is commutative and such that, for any coefficient group G,

$$\phi_* : H_q(\mathbf{T}W \times F, W \times F; G) \to H_q(X, F; G)$$

and

$$\phi^* : H^q(X, F; G) \to H^q(\mathbf{T}W \times F, W \times F; G)$$

are isomorphisms. Moreover, $\phi \mid \{*\} \times F : F \to F$ is homotopic to the identity. □

The map ϕ is called a *structural map*, and $\psi = \phi \mid W \times F : W \times F \to F$ a *characteristic map*, for the fibration p.

(1.2) Corollary *There are exact sequences*

(1.3) $$\cdots \to H_q(F; G) \to H_q(X; G) \to H_q(\mathbf{T}W \times F, W \times F; G) \to H_{q-1}(F; G) \to H_{q-1}(X; G) \to \cdots$$

(1.4) $$\cdots \to H^{q-1}(X; G) \to H^{q-1}(F; G) \to H^q(\mathbf{T}W \times F, W \times F; G) \to H^q(X; G) \to H^q(F; G) \to \cdots$$

□

Since $\mathbf{T}W$ is contractible, the subspace $F = \{*\} \times F$ is a deformation retract of $\mathbf{T}W \times F$, and, by exactness of the homology sequence of the triple $(\mathbf{T}W \times F, W \times F, F)$, we have isomorphisms

$$\partial_* : H_q(\mathbf{T}W \times F, W \times F) \approx H_{q-1}(W \times F, F).$$

Let us consider W as a space with base point and F as a free space; then the quotient $W \times F/F$ is just the reduced join $W \wedge F$; then $H_{q-1}(W \times F, F) \approx$

$H_{q-1}(W \wedge F)$, by Lemma (2.1) of Chapter II. Hence the sequences (1.3), (1.4) become

(1.5) $$\cdots \to H_{q+1}(F; G) \to H_{q+1}(X; G) \to H_q(W \wedge F; G) \xrightarrow{\tilde{\psi}} H_q(F; G) \to H_q(X; G) \to \cdots$$

(1.6) $$\cdots \to H^q(X; G) \to H^q(F; G) \xrightarrow{\tilde{\psi}^*} H^q(W \wedge F; G) \to H^{q+1}(X; G) \to H^{q+1}(F; G) \to \cdots$$

Let us make the maps $\tilde{\psi}$, $\tilde{\psi}^*$ explicit. The homology sequence

$$\cdots \to H_q(F) \to H_q(W \times F) \to H_q(W \times F, F) \to \cdots$$

of the pair $(W \times F, F)$ with coefficients in G breaks up (because F is a retract of $W \times F$) into a family of (split) short exact sequences

$$0 \to H_q(F) \xrightarrow{i_*} H_q(W \times F) \xrightarrow{p_*} H_q(W \wedge F) \to 0.$$

Let $p_2 : W \times F \to F$ be the projection on the second factor; then $p_2 \circ i = 1 \simeq \psi \circ i$, and therefore $\psi_* \circ i_* = p_{2*} \circ i_* : H_q(F) \to H_q(F)$. Thus $\psi_* - p_{2*}$ annihilates the image of i_* and so induces a homomorphism

$$\tilde{\psi} : H_q(W \wedge F; G) \to H_q(F; G)$$

such that

(1.7) $$\tilde{\psi} \circ p_* = \psi_* - p_{2*}.$$

Similarly, there is a homomorphism

$$\tilde{\psi}^* : H^q(F; G) \to H^q(W \wedge F; G)$$

such that $p^* \circ \tilde{\psi}^* = \psi^* - p_2^* : H^q(F; G) \to H^q(W \times F; G)$.

Consider the diagram

(1.8) $$\begin{array}{ccc} H_{q+1}(\mathbf{T}W \times F, W \times F) & \xrightarrow{\phi_*} & H_{q+1}(X, F) \\ \downarrow \partial_* & & \downarrow \partial_* \\ H_q(W \times F) & \xrightarrow{\psi_* - p_{2*}} & H_q(F) \\ \downarrow p_* & \nearrow \tilde{\psi} & \\ H_q(W \wedge F) & & \end{array}$$

Since ψ is the restriction of ϕ to a map of $W \times F$ into F, we have

$$\psi_* \circ \partial_* = \partial_* \circ \phi_*.$$

Now $p_{2*} \circ \partial_* = 0$ because the diagram

$$\begin{array}{ccc} H_{q+1}(\mathbf{T}W \times F, W \times F) & \xrightarrow{p_{2*}} & H_{q+1}(F, F) = 0 \\ \downarrow \partial_* & & \downarrow \partial_* \\ H_q(W \times F) & \xrightarrow[p_{2*}]{} & H_q(F) \end{array}$$

is commutative, and therefore the top square in the diagram (1.8) is commutative. By (1.7), the whole diagram (1.8) is commutative.

The sequences (1.3), (1.4) (or (1.5), (1.6)) are called the *Wang sequences* for the fibration p; they were discovered by H. C. Wang in the case that W is a sphere. As this case is extremely important, we single it out for special mention:

(1.9) Corollary *If $p : X \to \mathbf{S}^n$ is a fibration with fibre F ($n > 1$), there are exact sequences*

$$\text{(1.10)} \quad \cdots \to H_q(F; G) \to H_q(X; G) \to H_{q-n}(F; G) \xrightarrow{\theta_*} H_{q-1}(F; G) \to H_{q-1}(X; G) \to \cdots$$

$$\text{(1.11)} \quad \cdots \to H^{q-1}(X; G) \to H^{q-1}(F; G) \xrightarrow{\theta^*} H^{q-n}(F; G) \to H^q(X; G) \to H^q(F; G) \to \cdots \qquad \square$$

These are obtained from (1.3) and (1.4) after observing that by the Künneth theorem, the cross product operations induce isomorphisms

$$H_n(\mathbf{T}\mathbf{S}^{n-1}, \mathbf{S}^{n-1}) \otimes H_{q-n}(F; G) \approx H_q(\mathbf{T}\mathbf{S}^{n-1} \times F, \mathbf{S}^{n-1} \times F; G),$$

$$H^n(\mathbf{T}\mathbf{S}^{n-1}, \mathbf{S}^{n-1}) \otimes H^{q-n}(F; G) \approx H^q(\mathbf{T}\mathbf{S}^{n-1} \times F, \mathbf{S}^{n-1} \times F; G).$$

Suppose, in particular, that G is (the additive group of) a commutative ring with unit. Then we have

(1.12) Theorem *The map $\theta^* : H^*(F; G) \to H^*(F; G)$ of* (1.11) *is a derivation, of degree $n - 1$, of the graded ring $H^*(F; G)$.*

Let $\mathbf{e}$ be the canonical generator of the infinite cyclic group $H^n(\mathbf{E}^n, \mathbf{S}^{n-1})$, and let $\mathbf{s} = \delta^{*-1}\mathbf{e} \in H^{n-1}(\mathbf{S}^{n-1})$, so that the Kronecker index $\langle \mathbf{e}, \varepsilon_n \rangle = 1$. (See the discussion of orientation in §4 of Chapter IV). Then $\delta^*\mathbf{s} = \mathbf{e}$, and there is a commutative diagram

$$\begin{array}{ccc} H^{q-1}(F) & \xrightarrow{\psi^*} & H^{q-1}(\mathbf{S}^{n-1} \times F) \\ \downarrow \delta^* & & \downarrow \delta^* \\ H^q(X, F) & \xrightarrow[\phi^*]{} & H^q(\mathbf{E}^n \times F, \mathbf{S}^{n-1} \times F) \end{array}$$

and θ^* is defined by the condition

$$\textbf{(1.13)} \qquad \phi^*\delta^*x = \mathbf{e} \times \theta^*(x).$$

The unit element 1 of $H^*(\mathbf{S}^{n-1})$ generates the infinite cyclic group $H^0(\mathbf{S}^{n-1})$, and the maps $u \to 1 \times u$, $v \to \mathbf{s} \times v$ represent $H^{q-1}(\mathbf{S}^{n-1} \times F)$ as the direct sum $H^{q-1}(F) \oplus H^{q-n}(F)$, and

$$\textbf{(1.14)} \qquad \begin{aligned} \delta^*(1 \times u) &= 0, \\ \delta^*(\mathbf{s} \times v) &= \delta^*\mathbf{s} \times v = \mathbf{e} \times v. \end{aligned}$$

If $x \in H^{q-1}(F)$, we have $\psi^*x = 1 \times u + \mathbf{s} \times v$, for some $u \in H^{q-1}(F)$, $v \in H^{q-n}(F)$. Now the last sentence of Theorem (1.1) implies that $u = x$, and (1.13), (1.14) imply that $v = \theta^*(x)$. Thus

$$\psi^*(x) = 1 \times x + \mathbf{s} \times \theta^*(x).$$

Now ψ^* is a ring homomorphism. But $x \in H^{q-1}(F)$, $y \in H^{r-1}(F)$ imply

$$\begin{aligned} \psi^*(xy) &= 1 \times xy + \mathbf{s} \times \theta^*(xy) \\ \psi^*(x)\psi^*(y) &= \{1 \times x + \mathbf{s} \times \theta^*(x)\}\{1 \times y + \mathbf{s} \times \theta^*(y)\} \\ &= 1 \times xy + \mathbf{s} \times \theta^*(x)y + (-1)^{(n-1)(q-1)}\mathbf{s} \times x\theta^*(y), \end{aligned}$$

so that

$$\theta^*(xy) = \theta^*(x)y + (-1)^{(n-1)\dim x}x\theta^*(y). \qquad \square$$

Another special case of importance arises when the fibre F *acts on the fibration*; i.e., when there is a map $\mu : X \times F \to X$ such that

$\mu \circ i_1 \simeq 1$, the identity map of X,

$\mu \circ i_2 \simeq i : F \hookrightarrow X$,

$p \circ \mu = p \circ p_1$, where $p_1 : X \times F \to X$ is the projection.

In particular, the restriction of μ to $F \times F$ is a multiplication $\mu_0 : F \times F \to F$, making F into an H-space. It is convenient to think of μ as a kind of multiplication, and to write $x \cdot y$ instead of $\mu(x, y)$.

N.B.: We have defined what might appropriately be called a *right* action. Sometimes it is more convenient to deal with *left* actions instead; these are maps of $F \times X \to X$ with the appropriate properties.

We shall say that the action μ is *associative* if and only if $x \cdot (y_1 \cdot y_2) = (x \cdot y_1) \cdot y_2$ for all $y_1, y_2 \in F$, $x \in X$; and μ is *homotopy associative* if and only if there is a homotopy h of $\mu \circ (1 \times \mu_0)$ to $\mu \circ (\mu \times 1) : X \times F \times F \to X$ such that $p \circ h$ is stationary.

There are two important examples of actions.

EXAMPLE 1. Let $q : \mathbf{P}^*(B) \to B$ be the fibration of Theorem (2.16), Chapter III; the fibre of q is the space $\Omega^*(B)$ of measured loops, and the operation of multiplication of measured paths defines an associative left action of $\Omega^*(B)$ on the fibration q.

EXAMPLE 2. Let G be a compact Lie group, H a closed subgroup of G, and let G/H be the space of *left* cosets, $p : G \to G/H$ the natural map, defined by $p(x) = xH$. Then the restriction of the group operation in G to $G \times H$ defines a right action of G on p; and this action is, of course, associative.

Suppose that F acts on the fibration $p : X \to \mathbf{S}W$. Let $h_0 : \mathbf{T}W \to X_+$ be a map such that $p \circ h_0$ is the identity map of $\mathbf{T}W$. The map $\mu : X \times F \to X$ sends $X_+ \times F$ into X_+, and composition of the latter map with $h_0 \times 1 : \mathbf{T}W \times F \to X_+ \times F$ defines a map $h : \mathbf{T}W \times F \to X_+$. Similarly, there is a map $\phi_0 : \mathbf{T}W \to X$ such that $p \circ \phi_0 = \varpi \wedge 1$. Again, let $\phi = \mu \circ (\phi_0 \times 1) : \mathbf{T}W \times F \to X$; then these maps h, ϕ will serve in the proof of Theorem (1.1). Hence

(1.15) Theorem *If $p : X \to \mathbf{S}W$ is a fibration with fibre F, and if F acts on the fibration p, then there is a map $\phi_0 : (\mathbf{T}W, W) \to (X, F)$ such that the map $\phi = \mu \circ (\phi_0 \times 1) : (\mathbf{T}W \times F, W \times F) \to (X, F)$ is a structural map; and if $\psi_0 : W \to F$ is the restriction of ϕ_0, then $\psi = \mu_0 \circ (\psi_0 \times 1) : W \times F \to F$ is a characteristic map.* □

Let us return to Example 1, the fibration $\mathbf{P}^*\mathbf{S}W \to \mathbf{S}W$, with fibre $F = \Omega^*\mathbf{S}W$. In this case, the total space is contractible, and the Wang sequence (1.5) reduces to a family of isomorphisms

(1.16) $$H_q(W \wedge F) \approx H_q(F),$$

the cases $q = 0, 1$ being possible exceptions. Now X is 0-connected, and therefore the injection $H_0(F) \to H_0(X)$ is an epimorphism. If W is 0-connected, then $\mathbf{S}W$ is 1-connected, by the van Kampen Theorem, and therefore $F = \Omega^*\mathbf{S}W$ is 0-connected, so that the injection $H_0(F) \to H_0(X)$ is an isomorphism. Hence (1.16) holds for all $q \geq 0$. This allows a recursive calculation of the homology groups of F. For, by the Künneth Theorem

$$\begin{aligned} H_q(F) &\approx H_q(W \wedge F) \\ &\approx \bigoplus_{r+s=q} H_r(W, *) \otimes H_s(F) \oplus \bigoplus_{r+s=q-1} \operatorname{Tor}\{H_r(W, *), H_s(F)\}. \end{aligned}$$

Since W is 0-connected, $H_r(W, *) = 0$ for $r \leq 0$, and therefore the right-hand side of the above formula involves only the homology groups of F in dimensions less than q. To give an explicit formula would be rather complicated in general. However, let us suppose that the coefficient group is a field (or else that the coefficient ring G is a principal ideal domain and $H_*(W; G)$ has no torsion). Let $p_k(Y)$ be the kth Betti number of the space Y and assume $p_k(Y) < \infty$ for all k; then the *Poincaré series* of Y is the formal power series $Y(t) = \sum_{k=0}^{\infty} p_k(Y)t^k$. Now the Künneth Theorem implies that the Poincaré series of a product space is the product of the Poincaré series of its factors. The Poincaré series of $(W, *)$ is $\tilde{W}(t) = W(t) - 1$. Thus

$$F(t)\tilde{W}(t) = F(t),$$

$$F(t) = \frac{1}{1 - \tilde{W}(t)} = 1 + \tilde{W}(t) + \tilde{W}(t)^2 + \cdots$$

If W is a sphere $\mathbf{S}^n$, then $\tilde{W}(t) = t^n$. Thus

(1.17) Corollary *The homology groups of the loop space* $\Omega^{n+1} = \Omega(\mathbf{S}^{n+1})$ *are given by*

$$\begin{aligned} H_{qn}(\Omega^{n+1}) &\approx \mathbf{Z} \qquad (q = 0, 1, 2, \ldots) \\ H_q(\Omega^{n+1}) &= 0 \qquad \textit{otherwise}. \end{aligned}$$

□

This result for $\Omega(\mathbf{S}^{n+1})$ (or rather, for the space of *rectifiable* loops in $\mathbf{S}^{n+1}$) was first proved by Marston Morse.

We have determined the homology *groups* of the space $F = \Omega^* SW$. But there is additional structure, deriving from the fact that F is an H-space. In fact $H_*(F)$ is a graded algebra, the Pontryagin algebra, over the coefficient ring G; the product in $H_*(F)$ is defined by $u \cdot v = \mu_{0*}(u \times v)$, where $\mu_0 : F \times F \to F$ is the product in F. Let M be the G-module $H_*(W, *)$.

Remark. By Corollary (2.19) of Chapter III, $\Omega(SW)$ and $\Omega^*(SW)$ have isomorphic homology and cohomology groups, and even isomorphic cohomology rings. That they have isomorphic Pontryagin rings follows from the observation in (4.22) of Chapter III that the homotopy equivalence h is an H-map.

(1.18) Theorem *If W is a 0-connected space with non-degenerate base point $*$, and if G is a field, or else G is a principal ideal domain and $M = H_*(W, *; G)$*

is torsion-free, then the Pontryagin algebra $H_*(\Omega SW)$ is isomorphic with the tensor algebra $T(M)$.

Let us recall [*B*] that the additive group of $T(M)$ is the direct sum

$$\bigoplus_{n=0}^{\infty} M_n,$$

where $M_0 = G$ and $M_{n+1} = M \otimes M_n$; $T(M)$ is an associative algebra with unit, and the multiplication in $T(M)$ is uniquely determined by those properties and the formula

(1.19) $$x \cdot u = x \otimes u$$

for $x \in M = M_1$ and $u \in M_n$.

By the Künneth theorem,

$$H_*(W \wedge F) \approx H_*(W, *) \otimes H_*(F) = M \otimes H_*(F),$$

and so $\bar{\psi}$ may be regarded as a map of $M \otimes H_*(F)$ into $H_*(F)$. Let $M'_0 = H_0(F)$, and define $M'_n \subset H_*(F)$ inductively by

$$\bar{\psi}(M \otimes M'_n) = M'_{n+1}.$$

We have seen that F is 0-connected, so that the augmentation $\varepsilon : H_0(F; G) \to G$ is an isomorphism. Let $f_0 : M_0 = G \to M'_0 = H_0(F; G)$ be the inverse of ε. If $f_n : M_n \to M'_n$ is an isomorphism, the composite

$$M_{n+1} = M \otimes M_n \xrightarrow{1 \otimes f_n} M \otimes M'_n \xrightarrow{\bar{\psi}} H_*(F)$$

is a monomorphism with image M'_{n+1}, and therefore an isomorphism $f_{n+1} : M_{n+1} \to M'_{n+1}$. The $f_n : M_n \to M'_n$ induce a homomorphism $f : M \to H_*(F)$. We shall show

(1) f is an epimorphism;
(2) f is a monomorphism;
(3) f is a homomorphism of algebras.

Let $P_n = \sum_{q=0}^{n} M_q$, $P'_n = \sum_{q=0}^{n} M'_q$. Then $P'_0 = M'_0 = H_0(F)$. We shall prove that $H_n(F) \subset P'_n$ for all n; it follows that f is an epimorphism. In fact, suppose that $H_q(F) \subset P'_q$ for all $q \leq n$, and let $x \in H_{n+1}(F)$. The fact that $\bar{\psi}$ is an epimorphism implies that $x = \bar{\psi}(u)$ for some $u \in M \otimes H_*(F)$. Now M has no elements of dimension ≤ 0, and therefore the components of u all have dimension $\leq n$; by the induction hypothesis, $u \in M \otimes P'_n$, and therefore

$$x = \bar{\psi}(u) \in \sum_{q \leq n} \bar{\psi}(M \otimes M'_q) = \sum_{q \leq n} M'_{q+1} = P'_{n+1}.$$

Now $f \mid P_0 = f_0 : P_0 = M_0 \to H_0(F) \subset H_*(F)$ is a monomorphism. Suppose that $f \mid P_n : P_n \to H_*(F)$ is a monomorphism. Then $(1 \otimes f) \mid M \otimes P_n : M \otimes P_n \approx M \otimes P'_n$, and, as $\bar{\psi} \mid M \otimes P'_n : M \otimes P'_n \to H_*(F)$ is a monomorphism, their composite is a monomorphism $g : M \otimes P_n \to H_*(F)$.

The restriction of this monomorphism to $M \otimes M_q$ being f_{q+1}, it follows that $g = f \mid M \otimes P_n$. But P_{n+1} is the direct sum $M_0 \oplus (M \otimes P_n)$ and $f(M_0) \subset H_0(F)$, while $f(M \otimes P_n) \subset H_+(F)$, the ideal of elements of positive dimension, and therefore $f \mid P_{n+1}$ is also a monomorphism. Since this is true for all n, f is a monomorphism.

To prove that f is a homomorphism of algebras, we shall use Theorem (1.15). There is a map $\psi_0 : W \to F$ such that $\psi = \mu_0 \circ (\psi_0 \times 1)$, where $\mu_0 : F \times F \to F$ is the operation of multiplication of measured paths.

(1.20) Lemma *The map $\bar{\psi} : M \otimes H_*(F) \to H_*(F)$ is given by*

$$\bar{\psi}(u \otimes v) = \bar{\psi}_0(u) \cdot v$$

for all $u \in M$, $v \in H_*(F)$.

(The homomorphism $\bar{\psi}_0 : H_*(W, *) \to H_*(F)$ is defined in a similar way to $\bar{\psi}$; the diagram

$$\begin{array}{ccc} H_q(W, *) & \xrightarrow{\bar{\psi}_0} & H_q(F) \\ \uparrow & \nearrow_{\psi_{0*} - c_*} & \\ H_q(W) & & \end{array}$$

is required to be commutative, where c is the constant map).

The proof of the Lemma is an exercise in diagram-chasing, which is left to the reader with the hint that $p_2 \simeq \mu_0 \circ (c \times 1)$.

It is now easy to complete the proof of Theorem (1.18). It suffices, in view of the remarks we have made above on the structure of the tensor algebra, and, in particular, of (1.19) to prove that, if $u \in M$, $v \in M_n$, then $f(u \otimes v) = f(u) \cdot f(v)$, i.e.,

$$f_{n+1}(u \otimes v) = f_1(u) \cdot f_n(v)$$

But

$$\begin{aligned} f_{n+1}(u \otimes v) &= \bar{\psi}(1 \otimes f_n)(u \otimes v) \\ &= \bar{\psi}(u \otimes f_n(v)) \\ &= \bar{\psi}_0(u) \cdot f_n(v); \end{aligned}$$

while

$$f_1(u) = f_1(u \otimes 1) = \bar{\psi}(u \otimes f_0(1)) = \bar{\psi}(u \otimes 1) = \bar{\psi}_0(u). \qquad \square$$

(1.21) Corollary *The Pontryagin ring $H_*(\Omega^{n+1})$ is the polynomial ring* **Z**[u], *where u generates the infinite cyclic group $H_n(\Omega^{n+1})$.* □

Note that $H_*(\Omega SW)$ need not be commutative as a graded ring (i.e., it is not true that $uv = (-1)^{pq}vu$ for $u \in H_p$, $v \in H_q$).

We shall not calculate the cohomology ring of ΩSW in general. However, the special case $W = S^n$ will be very important for us, and will illustrate the complications that may be expected to arise in general.

In calculating $H^*(\Omega^{n+1})$, let us first assume that n is even. We shall use Corollary (1.9) and Theorem (1.12). Now $H^q(\Omega^{n+1}) = 0$ unless $q = 0 (\text{mod } n)$, while $\theta^* : H^{qn}(\Omega^{n+1}) \approx H^{(q-1)n}(\Omega^{n+1})$ for all $q \geq 1$. Moreover, Ω^{n+1} is 0-connected, and so $H^0(\Omega^{n+1})$ is the infinite cyclic group generated by the unit element 1. Define a sequence $\{z_q\}$ inductively by

$$z_0 = 1$$
$$\theta^* z_k = z_{k-1} \qquad (k \geq 1).$$

We now claim that

(1.22) $$k!\, z_k = z_1^k$$

for every positive integer k. In fact, θ^* is a derivation of even degree n, and therefore $\theta^*(z_1^k) = k z_1^{k-1} \theta^*(z_1) = k z_1^{k-1}$ for all k. If (1.22) holds for a certain integer k, then

$$\begin{aligned} \theta^*(z_1^{k+1}) &= (k+1) z_1^k = (k+1) k!\, z_k = (k+1)!\, z_k \\ &= \theta^*((k+1)!\, z_{k+1}); \end{aligned}$$

since θ^* is an isomorphism, $z_1^{k+1} = (k+1)!\, z_{k+1}$. As (1.22) is patently true for $k = 1$, its truth for all k follows.

This determines the structure of $H^*(\Omega^{n+1})$ in the case that n is even. We shall call it the *divided polynomial* ring $P^*(z)$ generated by the sequence $z = (z_1, z_2, \ldots)$.

Suppose now that n is odd, and choose the elements z_k as before. By the commutative law for the cup-product, $z_1^2 = -z_1^2$, and therefore $z_1^2 = 0$ since $H^{2n}(\Omega^{n+1})$ has no torsion. Therefore (1.22) cannot hold for all k. Instead, we have the somewhat more complicated relations

(1.23) $$\begin{aligned} z_1 z_{2k} &= z_{2k+1}, \\ z_1 z_{2k+1} &= 0, \\ z_2^k &= k!\, z_{2k}. \end{aligned}$$

In fact, the relations (1.23) hold for $k = 0$. If they hold for all $j < k$, then

$$\begin{aligned} \theta^*(z_1 z_{2k}) &= \theta^*(z_1) z_{2k} - z_1 \theta^*(z_{2k}) \\ &= z_{2k} - z_1 z_{2k-1} = z_{2k} = \theta^*(z_{2k+1}) \end{aligned}$$

and therefore $z_1 z_{2k} = z_{2k+1}$. Also

$$z_1 z_{2k+1} = z_1(z_1 z_{2k}) = z_1^2 z_{2k} = 0.$$

Finally, note that $(\theta^*)^2 = \theta^* \circ \theta^*$ is a derivation of even degree, which maps z_{2k} into z_{2k-2}, and the third formula of (1.23) now follows by a calculation similar to that by which (1.21) was proved.

We can now explicate the structure of $H^*(\Omega^{n+1})$ for n odd. The elements $z_2, z_4, \ldots$ generate a divided polynomial algebra isomorphic with $H^*(\Omega^{2n+1})$, and the element z_1 generates an exterior algebra isomorphic with $H^*(S^n)$. Finally, the whole algebra is the tensor product of these two subalgebras.

We have proved

(1.24) Theorem *If n is even, $H^*(\Omega^{n+1})$ is the divided polynomial algebra generated by a sequence of elements $\{z_k\}$ such that*

$$z_1^k = k!\, z_k$$

for all $k \geq 0$. If n is odd, $H^*(\Omega^{n+1}) \approx H^*(S^n) \otimes H^*(\Omega^{2n+1})$. □

2 The James Reduced Products

The considerations of the preceding section suggest the possibility of constructing a model for the loop space of a suspension through approximation by iterated products. Such a procedure was first carried out by Marston Morse for the sphere, and, in full generality by I. M. James.

Let W be a space with nondegenerate base point e. Let $J(W)$ be the free monoid with unit element generated by W. Intuitively the points of $J(W)$ are formal products of elements of W, two such products being equal if they become equal after deleting all occurrences of the base point e. More precisely, the points of $J(W)$ are equivalence classes of finite sequences of points of W under the equivalence relation defined as follows. An *elementary equivalence* is the operation of replacing the sequence $(w_1, \ldots, w_q)$ by the sequence $(w_1, \ldots, w_{i-1}, e, w_i, \ldots, w_q)$ for some $i (1 \leq i \leq q+1)$; or the inverse of such an operation. Two sequences $(w_1, \ldots, w_q)$ and $(w_1', \ldots, w_q')$ are *equivalent* if and only if one can be obtained from the other by a finite sequence of elementary equivalences. The product of two sequences $(w_1, \ldots, w_p)$ and $(w_1', \ldots, w_q')$ is the sequence $(w_1, \ldots, w_p, w_1', \ldots, w_q')$ obtained by juxtaposition. It is evident that an elementary equivalence on either factor induces one on the product, and therefore the operation of multiplying two sequences induces an operation in $J(W)$, making it into a monoid. Each element of W, considered as a sequence of length 1, determines an element of $J(W)$, and this correspondence is a one-to-one map of W into $J(W)$ which sends the base point e of W into the unit element of $J(W)$. Then W generates $J(W)$ as a monoid (in fact, if $(w_1, \ldots, w_q)$ is a finite sequence of elements of W, the element of $J(W)$ it determines is just the product $w_1 \cdots w_q$). Moreover, $J(W)$ is free in the customary sense; any function $f: W \to M$ of W into a monoid M with unit has a unique extension to a homomorphism $\hat{f}: J(W) \to M$.

We proceed to impose a topology on $J(W)$. Let W^m be the product of m copies of W, and let $i_k : W^{m-1} \to W^m$ be the map defined by

$$i_k(w_1, \ldots, w_{m-1}) = (w_1, \ldots, w_{k-1}, e, w_k, \ldots, w_{m-1}) \qquad (k = 1, \ldots, m).$$

Let W_*^{m-1} be the union of the images of the i_k; it is the set of all points of W^m with at least one coordinate equal to e.

(The reader is reminded here of the fact that the topology we are imposing on the product of two or more spaces is not the customary one, and that some properties which are obvious for the customary topology need verification here. As most of these verifications are extremely easy, we shall leave them to him).

Let $J^m(W)$ be the set of all elements of $J(W)$ which are products of at most m factors. Let $\pi_m : W^m \to J^m(W)$ be the natural map, defined by

$$\pi_m(w_1, \ldots, w_m) = w_1 \cdots w_m,$$

and topologize $J^m(W)$ be requiring that π_m be a proclusion.

(2.1) Lemma *The space $J^m(W)$ is a Hausdorff space.*

For let $w = w_1 \cdots w_m$, $w' = w'_1 \cdots w'_m$ be distinct points of $J^m(W)$. Then we may write $w = x_1 \cdots x_p$, $w' = x'_1 \cdots x'_q$ where $x_i, x'_j \in W - \{e\}$, and the sequences $(x_1, \ldots, x_p)$, $(x'_1, \ldots, x'_q)$ are distinct. Choose neighborhoods U of e, V_i of x_i, V'_j of x'_j such that

$$U \cap V_i = U \cap V'_j = \emptyset,$$
$$V_i \cap V'_j = \emptyset \quad \text{if } x_i \neq x'_j.$$

Let

$$P = V_1 \times \cdots \times V_p \times \underbrace{U \times \cdots \times U}_{m-p \text{ factors}},$$

$$P' = V'_1 \times \cdots \times V'_q \times \underbrace{U \times \cdots \times U}_{m-q \text{ factors}},$$

and let Q be the union of all the sets which can be obtained from P by permuting the factors but leaving the order of the factors $V_1, \ldots, V_p$ unchanged, and define Q' in a similar way from P'. Then Q is open in the cartesian product topology, and *a fortiori* Q is open in W^m. Similarly, Q' is open in W^m; since both are saturated, under π_m, their images in $J^m(W)$ are open and contain w, w' respectively. It remains only to prove that $Q \cap Q' = \emptyset$, a task which we cheerfully leave to the reader! □

It follows from Lemma (2.1) and from (4.17) of Chapter I that the space $J^m(W)$ is compactly generated; thus our construction does not take us outside the category $\mathscr{K}$.

(2.2) Lemma *The map $J^{m-1}(W) \subset J^m(W)$ is a homeomorphism of $J^{m-1}(W)$ with a closed subspace of $J^m(W)$, and the proclusion $\pi_m : (W^m, W_*^{m-1}) \to (J^m(W), J^{m-1}(W))$ is a relative homeomorphism. The latter pair is an NDR-pair.*

Let J_*^{m-1} be the image of W_*^{m-1} under the proclusion π_m; then $W_*^{m-1} = \pi_m^{-1}(J_*^{m-1})$ is closed in W^m, and $\pi_m | W_*^{m-1} : W_*^{m-1} \to J_*^{m-1}$ is a proclusion; hence J_*^{m-1} is closed and $\pi_m : (W^m, W_*^{m-1}) \to (J^m(X), J_*^{m-1})$ is a relative homeomorphism. By (5.4) of Chapter I, the latter pair is an NDR-pair. As $J^{m-1}(X)$ and J_*^{m-1} coincide as sets, it remains only to verify that they have the same topology.

Let $\pi_* = \pi_m | W_*^{m-1}$; then the diagram

$$\begin{array}{ccc} W^{m-1} & \xrightarrow{\ i_k\ } & W_*^{m-1} \\ \Big\downarrow{\scriptstyle \pi_{m-1}} & & \Big\downarrow{\scriptstyle \pi_*} \\ J^{m-1}(X) & = & J_*^{m-1} \end{array}$$

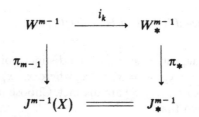

is commutative. If a set A is closed in $J^{m-1}(X)$ then $\pi_{m-1}^{-1}(A)$ is closed in W^{m-1} and therefore $i_k \pi_{m-1}^{-1}(A)$ is closed in W_*^{m-1}. But then $\pi_*^{-1}(A) = \bigcup_{k=1}^m i_k \pi_{m-1}^{-1}(A)$ is closed. This proves that A is closed in J_*^{m-1}. Conversely, if A is closed in J_*^{m-1}, then $\pi_*^{-1}(A)$ is closed and therefore $\pi_{m-1}^{-1}(A) = i_k^{-1}\pi_*^{-1}(A)$ is closed because i_k is continuous. Hence A is closed in $J^{m-1}(X)$. □

It follows from Lemma (2.2) that the spaces $J^m(W)$ form an expanding sequence of spaces in the sense of §6 of Chapter I. Giving $J(W)$ the topology of the union, we see from (6.3) of Chapter I that $J(W)$ is a filtered space under the NDR-filtration $\{J^m(W)\}$.

The following remark follows easily from the above observation.

(2.3) *Let $f_m : W^m \to X$ be a sequence of maps such that $f_m \circ i_k = f_{m-1}$ for $k = 1, \ldots, m$. Then there is a map $f : J(W) \to X$ such that $(f | J^m(W)) \circ \pi_m = f_m$ for $m = 1, 2, 3, \ldots$.* □

Commutativity of the diagram

$$\begin{array}{ccc} W^m \times W^n & \longrightarrow & W^{m+n} \\ \Big\downarrow \pi_m \times \pi_n & & \Big\downarrow \pi_{m+n} \\ J^m(W) \times J^n(W) & \longrightarrow & J^{m+m}(W) \end{array}$$

and the fact that $\pi_m \times \pi_n$ is a proclusion, by (4.18) of Chapter I, allows us to conclude that the map of $J^m(W) \times J^n(W)$ into $J^{m+n}(W)$ defined by the product in $J(W)$ is continuous. By (6.7) of Chapter I, $J(W) \times J(W)$ is filtered by the spaces $\bigcup_{m=0}^{n} J^m(W) \times J^{n-m}(W)$, and it follows that

(2.4) Theorem *The space $J(W)$ is a strictly associative* H-*space.* □

(2.5) Theorem *The* H-*space $J(W)$ is freely generated by W in the sense that, for any strictly associative* H-*space X with unit e and any continuous map $f: (W, e) \to (X, e)$, there is a continuous homomorphism $\hat{f}: J(W) \to X$ extending f.*

For let $f_m : W^m \to X$ be the map such that

$$f_m(w_1, \ldots, w_m) = f(w_1) \cdots f(w_m).$$

Clearly $f_m \circ i_k = f_{m-1}$; by (2.3), these induce $\hat{f}: J(W) \to X$, such that

$$\hat{f}(w_1 \cdots w_m) = f_m(w_1, \ldots, w_m)$$

for all $w_i \in W$. Clearly $\hat{f}$ is a homomorphism, which is unique since W generates $J(W)$ as a monoid. □

The map $\hat{f}$ is called the *canonical extension* of f.

In particular, suppose that $f: W \to W'$ is a map. Then the composite

$$W \xrightarrow{\;f\;} W' \overset{i}{\hookrightarrow} J(W')$$

has the canonical extension

$$J(f) = \widehat{i \circ f}: J(W) \to J(W').$$

Clearly J determines a functor from $\mathscr{K}_*$ to the category of strictly associative H-spaces and continuous homomorphisms.

The identity map of $\mathbf{S}W$ is adjoint to the map $\lambda_0 : W \to \Omega \mathbf{S} W$ defined by

$$\lambda_0(w)(t) = t \wedge w$$

for $t \in I$, $w \in W$. Unfortunately, the H-space $\Omega \mathbf{S} W$ is not strictly associative. Instead, we shall consider the space $\Omega^* \mathbf{S} W$ and the imbedding $\lambda : W \to \Omega^* \mathbf{S} W$ defined as follows. Let $u : W \to \mathbf{I}$, $k : \mathbf{I} \times W \to W$, represent $(W, *)$ as an NDR-pair. Then

$$\lambda(w) = (u(w), v),$$

where $v : [0, u(w)] \to (\mathbf{S}W, *)$ is given by

$$v(t) = \begin{cases} (t/u(w)) \wedge w & \text{if } w \neq *, \\ * & \text{if } w = *. \end{cases}$$

The reader may verify that λ is continuous, and that, if $h' : \Omega^* \mathbf{S} W \to \Omega \mathbf{S} W$ is the homotopy equivalence of (2.18), Chapter III, then $h' \circ \lambda = \lambda_0$. Thus λ and λ_0 correspond under the isomorphism

$$\underline{h}' : [W, \Omega^* \mathbf{S} W] \approx [W, \Omega \mathbf{S} W].$$

By theorem (2.5), the map λ extends to a map

$$\hat{\lambda} : J(W) \to \Omega^* \mathbf{S} W.$$

An important result of I. M. James is

(2.6) Theorem *The map $\hat{\lambda} : J(W) \to \Omega^* \mathbf{S} W$ is a weak homotopy equivalence.*

(In fact, D. Puppe [2] has shown that under rather weak hypotheses it is a homotopy equivalence).

We first show that $\hat{\lambda}$ is a homology equivalence. Let us abbreviate $J(W)$, $J^m(W)$ to J, J^m, respectively. Define a map $\phi : W \times J \to J$ as the composite

$$W \times J \hookrightarrow J \times J \xrightarrow{\mu} J,$$

μ being the product in J. As in the case of $\Omega \mathbf{S} W$ (cf. the construction of $\bar{\psi}$ in §1), the homomorphism $\phi_* - p_{2*} : H_*(W \times J) \to H_*(J)$ annihilates the image of the injection $H_*(e \times J) \to H_*(W \times J)$ and so induces a homomorphism

$$\bar{\phi}_* : H_*(W \times J, e \times J) \to H_*(J).$$

(Here, as in §1, J is being considered as a free space).

(2.7) Lemma *The homomorphism $\bar{\phi}_*$ is an isomorphism.*

The map ϕ carries $W \times J^m$ into J^{m+1} for every m, and there are commutative diagrams

$$\begin{array}{ccc}
(W, e) \times (W^m, W^{m-1}_*) & \longrightarrow & (W^{m+1}, W^m_*) \\
\downarrow & & \downarrow \\
(W, e) \times (J^m, J^{m-1}) & \longrightarrow & (J^{m+1}, J^m)
\end{array}$$

in which the vertical arrows represent relative homeomorphisms, the upper horizontal arrow the identity map, and the lower horizontal the map of pairs induced by the restriction of ϕ. It follows that ϕ induces isomorphisms

$$\phi_m : H_*((W, e) \times (J^m, J^{m-1})) \approx H_*(J^{m+1}, J^m)$$

for all m.

In the homology sequence of the triple

$$(W \times J^m, W \times J^{m-1} \cup e \times J^m, e \times J^m)$$

the group

$$H_*(W \times J^{m-1} \cup e \times J^m, e \times J^m)$$

is isomorphic with the group

$$H_*(W \times J^{m-1}, e \times J^{m-1}),$$

by the Excision Theorem. Replacing the former group by the latter, we obtain the left-hand column of the diagram (Figure 7.1) the right-hand column of which is the homology sequence of the pair (J^{m+1}, J^m). The homomorphisms $\tilde{\phi}_m$ are defined in a similar way to the homomorphism $\tilde{\phi}_*$; the composite of the injection $H_*(J^m) \to H_*(J^{m+1})$ with

$$p_{2*} : H_*(W \times J^m) \to H_*(J^m)$$

agrees, on the image of the injection $H_*(e \times J^m) \to H_*(W \times J^m)$, with the homomorphism of $H_*(W \times J^m)$ into $H_*(J^{m+1})$ induced by the restriction of ϕ to $W \times J^m$; their difference thereby induces a homomorphism $\tilde{\phi}_m : H_*(W \times J^m, e \times J^m) \to H_*(J^{m+1})$.

(2.8) Lemma *The diagram in Figure* 1 *is commutative.*

That the first and fourth squares are commutative is evident. The commutativity of the other two is easily proved by diagram-chasing, with the aid of the observation that $p_{2*} : H_*(W \times J^m, W \times J^{m-1} \cup e \times J^m) \to H_*(J^{m+1}, J^m)$ is zero, because it factors through the group $H_*(J^m, J^m) = 0$.

We have seen that ϕ_m is an isomorphism for all m. By induction on m and the Five-Lemma we deduce that $\tilde{\phi}_m$ is an isomorphism for every m. Finally,

$$\begin{array}{ccc}
H_*(W \times J^{m-1}, e \times J^{m-1}) & \xrightarrow{\bar{\phi}_{m-1}} & H_*(J^m) \\
\downarrow & (1) & \downarrow \\
H_*(W \times J^m, e \times J^m) & \xrightarrow{\bar{\phi}_m} & H_*(J^{m+1}) \\
\downarrow & (2) & \downarrow \\
H_*(W \times J^m, W \times J^{m-1} \cup e \times J^m) & \xrightarrow{\phi_m} & H_*(J^{m+1}, J^m) \\
\downarrow & (3) & \downarrow \\
H_*(W \times J^{m-1}, e \times J^{m-1}) & \xrightarrow{\bar{\phi}_{m-1}} & H_*(J^m) \\
\downarrow & (4) & \downarrow \\
H_*(W \times J^m, e \times J^m) & \xrightarrow{\bar{\phi}_m} & H_*(J^{m+1})
\end{array}$$

Figure 7.1

$\bar{\phi}_*$ is an isomorphism because $H_*(W \times J, J)$ and $H_*(J)$ are the direct limits of the groups $H_*(W \times J^m, J^m)$ and $H_*(J^{m+1})$, respectively, and because of the commutativity of the top and bottom squares in Figure 1. □

The fact that $\hat{\lambda}$ is a homology equivalence can now be proved by an induction on the dimension. In fact, suppose that $\hat{\lambda}_* : H_q(J(W)) \approx H_q(\Omega^*SW)$ for all $q < n$. Then the diagram

$$\begin{array}{ccc}
H_n(W \times J(W), e \times J(W)) & \xrightarrow{(1 \times \hat{\lambda})_*} & H_n(W \times \Omega^*SW, e \times \Omega^*SW) \\
\downarrow \bar{\phi}_* & & \downarrow \bar{\psi}_* \\
H_n(J(W)) & \xrightarrow[\hat{\lambda}_*]{} & H_m(\Omega^*SW)
\end{array}$$

is commutative. The Künneth relation for the homology groups of the pairs in the top line involves (because W is 0-connected) only the homology groups of $J(W)$ and of Ω^*SW in dimensions less than n. Hence $(1 \times \hat{\lambda})_*$ is an isomorphism. As $\bar{\phi}_*$ and $\bar{\psi}_*$ are isomorphisms, so is $\hat{\lambda}_*$.

That $\hat{\lambda}$ is a weak homotopy equivalence follows immediately from

(2.9) Lemma *Let X and Y be 0-connected* H*-spaces, and let $f: X \to Y$ be a homology equivalence which is also an* H*-map. Then f is a weak homotopy equivalence.*

By Theorem (3.2) of Chapter V, there are CW-approximations $\phi: K \to X$, $\psi: L \to Y$. By Lemma (7.12) of Chapter V, K and L admit H-structures such that ϕ and ψ are H-maps. Since ψ is a weak homotopy equivalence, there is a map $g: K \to L$, unique up to homotopy, such that $\psi \circ g \simeq f \circ \phi$. Then g is a homology equivalence and a suitable diagram chase shows that g is an H-map. Moreover, g is a weak homotopy equivalence if and only if h is. Hence it suffices to prove Lemma (2.9) under the additional assumption that X and Y are themselves CW-complexes. We may further assume that f is cellular.

Now $(\mathbf{I}_f, X)$ is a CW-pair; we shall show that it is an H-pair. Then Theorems (4.18) and (4.19) of Chapter III assure us that X and $\mathbf{I}_f$ are q-simple for all q and that $\pi_1(X)$ operates trivially on the relative homotopy groups of $(\mathbf{I}_f, X)$. In particular, X and Y are 1-simple, so that $\pi_1(X)$ and $\pi_1(Y)$ are isomorphic with $H_1(X)$ and $H_1(Y)$, respectively, and therefore $f_*: \pi_1(X) \approx \pi_1(Y)$. It follows that $(\mathbf{I}_f, X)$ is 1-connected. Then induction and the Hurewicz theorem allow us to conclude that $\pi_q(\mathbf{I}_f, X) = 0$ for all q.

To prove $(\mathbf{I}_f, X)$ an H-pair, we must show that the map of $X \times X \cup \mathbf{I}_f \vee \mathbf{I}_f$ into $\mathbf{I}_f$ which restricts to the folding map on $\mathbf{I}_f \vee \mathbf{I}_f$ and to the product in X on $X \times X$, extends to a map of $\mathbf{I}_f \times \mathbf{I}_f$ into $\mathbf{I}_f$. Consider the homology sequence of the triple $(\mathbf{I}_f \times \mathbf{I}_f,\ X \times X \cup \mathbf{I}_f \vee \mathbf{I}_f,\ X \times X)$. Since f is a homology equivalence, so is $f \times f$, by the Künneth Theorem, and therefore $H_q(\mathbf{I}_f \times \mathbf{I}_f, X \times X) = 0$ for all q. By the Excision Theorem, the injection

$$H_q(\mathbf{I}_f \vee \mathbf{I}_f, X \vee X) \to H_q(X \times X \cup \mathbf{I}_f \vee \mathbf{I}_f, X \times X)$$

is an isomorphism. Since f is a homology equivalence, so is $f \vee f$, by the Direct Sum Theorem, and therefore $H_q(\mathbf{I}_f \vee \mathbf{I}_f,\ X \vee X) = 0$ for all q. By the exactness of the homology sequence of the above triple, $H_q(\mathbf{I}_f \times \mathbf{I}_f,\ X \times X \cup \mathbf{I}_f \vee \mathbf{I}_f) = 0$ for all q. By the Universal Coefficient Theorem the cohomology groups of the latter pair vanish with arbitrary coefficients. As $\mathbf{I}_f$ is q-simple for all q, we conclude from the Eilenberg extension theorem that the desired extension exists.

Let W, X be spaces with nondegenerate base points and let

$$f: (J^m(W), J^{m-1}(W)) \to (X, *)$$

We shall construct an extension $g: J(W) \to J(X)$, called the *combinatorial extension of f*. By (2.3), it suffices to construct a sequence of maps $f_n: W^n \to J(X)(n = 1, 2, \ldots)$ such that

$$f_n \circ i_k = f_{n-1} \qquad (k = 1, \ldots, n),$$

$$f_m = f \circ \pi_m$$

(and therefore f_n is the constant map for all $n < m$).

Let $n \geq m$, and let P_n be the set of all strictly increasing m-termed subsequences of $(1, \ldots, n)$, ordered lexicographically from the right, so that $\alpha < \beta$ if and only if there exists $j(1 \leq j \leq m)$ such that $\alpha_i = \beta_i$ for $i > j$ and $\alpha_j < \beta_j$. Let $\alpha_1, \ldots, \alpha_N$ be the $N = \binom{n}{m}$ elements of P_n, arranged in increasing order. For each $r(1 \leq r \leq N)$, define $g_r : W^n \to J^m(W)$ by

$$g_r(a_1, \ldots, a_n) = \pi_m(a_{\alpha_1}, \ldots, a_{\alpha_m}),$$

where $\alpha = \alpha_r$. Then a map $f_n : W^n \to J^N(X) \subset J(X)$ is defined by

$$f_n(x) = \pi_N(fg_1(x), \ldots, fg_N(x)).$$

If $m = n$, then $N = 1$, $g_1 = \pi_m$, and therefore $f_m = \pi_1 \circ f \circ \pi_m = f \circ \pi_m$. Suppose $n > m$, and let k be an integer, $1 \leq k \leq n$. The map $(1, \ldots, n-1) \to (1, \ldots, k-1, k+1, \ldots, n)$ induces an one-to-one map of P_{n-1} into P_n which is readily seen to be order-preserving; the image of this map is the set of all sequences α which omit the value k. The remaining sequences α_r have the property that $g_r i_k(x) = e$ for all $x \in W^{n-1}$. Therefore, if $N' = \binom{n-1}{m}$, the sequence

$$(fg_1 i_k(x), \ldots, fg_N i_k(x))$$

differs from the sequence

$$(fg_1(x), \ldots, fg_{N'}(x))$$

by the interpolation of $N - N'$ e's; thus

$$f_n i_k(x) = f_{n-1}(x).$$

The combinatorial extension $g : J(W) \to J(X)$ of f is defined by the condition

$$(g \,|\, J^n(W)) \circ \pi_n = f_n \qquad (n = 1, 2, 3 \ldots).$$

In particular, let $W^{(n)}$ be the n-fold reduced join $W \wedge \cdots \wedge W$; then the natural projection

$$h_n : W^n \to W^{(n)}$$

maps W_n^{n-1} into the base point, and so induces a map f_n of the pair $(J^m(W), J^{m-1}(W))$ into $(W^{(n)}, *)$. Let

$$g_n : J(W) \to J(W^{(n)})$$

be the combinatorial extension of f_n. Let $X = \bigvee_{n=1}^{\infty} W^{(n)}$, and let

$i_n: W^{(n)} \to X$ be the inclusion, $i'_n = J(i_n): J(W^{(n)}) \to J(X)$. If $x \in J^m(W)$, let

$$G_m(x) = \prod_{n=1}^{m} i'_n(g_n(x)).$$

If $x \in J^{m-1}(W)$, then $g_m(x) = e$; hence

$$G_m | J^{m-1}(W) = G_{m-1}.$$

Therefore the maps G_m together define a map

$$G: J(W) \to J(X).$$

Let $\tilde{G}: \mathbf{S}J(W) \to \mathbf{S}X$ be the adjoint of the composite map

$$J(W) \xrightarrow{G} J(X) \xrightarrow{\hat{\lambda}} \Omega^*\mathbf{S}X.$$

(2.10) Theorem *If W is 0-connected, the map $\tilde{G}$ is a weak homotopy equivalence.*

Since W is 0-connected, so are $J(W)$ and $X = \bigvee_{n=1}^{\infty} W^{(n)}$, and therefore $\mathbf{S}J(W)$ and $\mathbf{S}X$ are 1-connected. Therefore it suffices to prove that $\tilde{G}$ is a homology equivalence. Let $X_m = \bigvee_{n=1}^{m} W^{(n)}$; then it is clear that

$$\tilde{G}(\mathbf{S}J^m(W)) \subset \mathbf{S}X_m$$

for every m. Therefore $\tilde{G}$ induces

$$\tilde{G}_m: \mathbf{S}J^m(W)/\mathbf{S}J^{m-1}(W) \to \mathbf{S}X_m/\mathbf{S}X_{m-1}.$$

But $\mathbf{S}J^m(W)/\mathbf{S}J^{m-1}(W) = \mathbf{S}(J^m(W)/J^{m-1}(W))$ is homeomorphic with $\mathbf{S}W^{(m)}$, as is $\mathbf{S}X_m/\mathbf{S}X_{m-1}$. Moreover, $g_m | J^m(W)$ is the relative homeomorphism $f_m: (J^m(W), J^{m-1}(W)) \to (W^{(m)}, *)$, and it follows that $\tilde{G}_m$ is the natural homeomorphism. By induction and the Five-Lemma, it follows that

$$\tilde{G} | \mathbf{S}J^m(W): \mathbf{S}J^m(W) \to \mathbf{S}X_m$$

is a homology equivalence. But the homology groups of the filtered spaces $\mathbf{S}J(W)$ and $\mathbf{S}X$ are the direct limits of those of the subspaces $\mathbf{S}J^m(W)$ and $\mathbf{S}X_m$, respectively, and therefore $\tilde{G}$ is a homology equivalence. □

Of particular interest is the

(2.11) Corollary *If $n > 0$, $\tilde{G}: \mathbf{S}J(\mathbf{S}^n) \to \bigvee_{k=1}^{\infty} \mathbf{S}^{nk+1}$ is a homotopy equivalence, so that $\mathbf{S}\Omega\mathbf{S}^{n+1}$ has the same weak homotopy type as $\bigvee_{k=1}^{\infty} \mathbf{S}^{nk+1}$.* □

Finally, let $p_n: X \to W^{(n)}$ be the projection on its nth summand. In view of the natural isomorphism $\pi_q(J(Y)) \approx \pi_q(\Omega\mathbf{S}Y) \approx \pi_{q+1}(\mathbf{S}Y)$, the map

$J(p_n) \circ G : J(W) \to J(W^{(n)})$ induces homomorphisms

$$j_n : \pi_{q+1}(\mathbf{S}W) \to \pi_{q+1}(\mathbf{S}W^{(n)})$$

for every positive integer n. If $\alpha \in \pi_{q+1}(\mathbf{S}W)$, the element $j_n(\alpha)$ is called the nth *Hopf–James invariant* of α. In particular, if $W = \mathbf{S}^m$, we have

$$j_n : \pi_{q+1}(\mathbf{S}^{m+1}) \to \pi_{q+1}(\mathbf{S}^{mn+1}).$$

3 Further Properties of the Wang Sequence

Let $p : X \to \mathbf{S}^n$ be a fibration with fibre F. In §1 we established the Wang cohomology sequence

$$\cdots \to H^{q-1}(X) \xrightarrow{i^*} H^{q-1}(F) \xrightarrow{\theta^*} H^{q-n}(F) \xrightarrow{\alpha^*} H^q(X) \xrightarrow{i^*} H^q(F) \to \cdots$$

with coefficients in a commutative ring A, where the injection i^* is a ring homomorphism and θ^* is a derivation. The following property of α^* will be useful for us:

(3.1) Theorem *For* $x \in H^{q-n}(F)$, $y \in H^r(X)$, *we have*

$$\alpha^*(x \smile i^*(y)) = \alpha^*(x) \smile y.$$

To see this, recall that α^* is defined by commutativity of the diagram

$$\begin{array}{ccc} H^q(X, F) & \xrightarrow{j^*} & H^q(X) \\ \phi^* \downarrow & & \uparrow \alpha^* \\ H^q(\mathbf{E}^n \times F, \mathbf{S}^{n-1} \times F) & \xleftarrow{\mathbf{e}\,\times} & H^{q-n}(F) \end{array}$$

where ϕ is a structural map and $\mathbf{e}$ is an orientation of $\mathbf{E}^n$. Thus $\alpha^*(x) = j^*(z)$, where $z \in H^q(X, F)$ is the element such that $\phi^*(z) = \mathbf{e} \times x$.

There is a pairing: $H^q(X, F) \otimes H^r(X) \to H^{q+r}(X, F)$, defined by the cup-product; and this pairing has the property

$$j^*(z \smile y) = j^*(z) \smile y = \alpha^*(x) \smile y.$$

Let us calculate $\phi^*(z \smile y)$. The naturality of the cup-product gives

$$\phi^*(z \smile y) = \phi^*(z) \smile \phi_0^*(y),$$

where $\phi_0^* : H^r(X) \to H^r(\mathbf{E}^n \times F)$ is the homomorphism of the absolute groups induced by the structural map ϕ. It follows from the last sentence of

Theorem (1.1) that

$$\phi_0^*(y) = 1 \times i^*(y),$$

and therefore

$$\begin{aligned}\phi^*(z \smile y) &= \phi^*(z) \smile \phi_0^*(y)\\ &= (\mathbf{e} \times x) \smile (1 \times i^*(y))\\ &= \mathbf{e} \times (x \smile i^*(y))\end{aligned}$$

By definition of α^*, $\alpha^*(x \smile i^*(y)) = \alpha^*(x) \smile y$. □

(3.2) Corollary *If $x \in H^p(X)$, $y \in H^{q-n}(F)$, then*

$$\alpha^*(i^*(x) \smile y) = (-1)^{np} x \smile \alpha^*(y).$$ □

(3.3) Corollary *The image of α^* is an ideal in $H^*(X)$, the product of any two of whose elements is zero.* □

If x is the unit element $1 \in H^0(F)$ of the cohomology ring, the element $u = \alpha^*(1)$ belongs to $H^n(X)$, and we have

(3.4) Corollary *The endomorphism $d^* = \alpha^* \circ i^*$ of $H^*(X)$ is determined by*

$$d^*(y) = u \smile y$$

for all $y \in H^*(X)$. □

We can identify the element u. For recall that we have a commutative diagram

$$\begin{array}{ccc}(\mathbf{E}^n \times F, \mathbf{S}^{n-1} \times F) & \xrightarrow{\ \phi\ } & (X, F)\\ \Big\downarrow p_1 & & \Big\downarrow p\\ (\mathbf{E}^n, \mathbf{S}^{n-1}) & \xrightarrow[\ q\ ]{} & \mathbf{S}^n\end{array}$$

where p_1 is the projection on the first factor. Let $\mathbf{s}^n$ be the generator of $H^n(\mathbf{S}^n)$ such that $q^*\mathbf{s}^n = \mathbf{e}$. Then $\phi^*p^*\mathbf{s}^n = p_1^*q^*\mathbf{s}^n = p_1^*\mathbf{e} = \mathbf{e} \times 1$. By definition of α^*, $\alpha^*(1) = j^*p^*\mathbf{s}^n$; this element is the image of $\mathbf{s}^n$ under the homomorphism of $H^n(\mathbf{S}^n)$ into $H^n(X)$ induced by p. If we continue to denote this homomorphism by p^*, we have

(3.5) *The relation $\alpha^*(1) = p^*(\mathbf{s}^n)$ holds in $H^n(X)$.* □

Let us now suppose that F acts on the fibration p by $\mu : X \times F \to X$. Then the map $p \circ p_1 : X \times F \to \mathbf{S}^n$ is a fibration with fibre $F \times F$; and the diagram

$$\begin{array}{ccc} X \times F & \xrightarrow{\mu} & X \\ & {\scriptstyle p \circ p_1}\searrow \quad \swarrow{\scriptstyle p} & \\ & \mathbf{S}^n & \end{array}$$

is commutative. Hence

$$\mu^*(u) = \mu^* p^* \mathbf{s}^n = (p \circ p_1)^* \mathbf{s}^n = p_1^* p^* \mathbf{s}^n = p_1^* u = u \times 1.$$

Thus

(3.6) *The homomorphism* $\mu^* : H^*(X) \to H^*(X \times F)$ *sends* u *into* $u \times 1$. □

An important special case will be needed in §4. Suppose that the map $\mu : X \times F \to X$ has an extension $\mu_1 : X \times X \to X$ making (X, F) into an H-pair. In this case we shall say that *p is an* H-*fibration* (with respect to the given structure of (X, F) as an H-pair). For example, if X is a compact Lie group and F a closed subgroup, then $p : X \to X/F$ is an H-fibration.

(3.7) Theorem *If* $p : X \to \mathbf{S}^n$ *is an* H-*fibration and the coefficient ring is an integral domain, then the element* $u \in H^n(X)$ *is primitive.*

The exactness of the Wang sequence implies that the injection $i^* : H^q(X) \to H^q(F)$ is an isomorphism for $q < n - 1$, and the sequence

$$0 \longrightarrow H^{n-1}(X) \xrightarrow{i^*} H^{n-1}(F) \xrightarrow{\theta^*} H^0(F)$$

is exact. Since $H^0(F)$ is free and A is a principal ideal domain, the image of θ^* is free, and therefore $i^* : H^{n-1}(X) \to H^{n-1}(F)$ is a split monomorphism. It follows from the Künneth Theorem that $(1 \wedge i)^* : H^n(X \wedge X) \to H^n(X \wedge F)$ is a monomorphism. There is a commutative diagram

$$\begin{array}{ccccc} & & H^n(X \times X) & \xrightarrow{k_1^*} & H^n(X \wedge X) \\ & {\scriptstyle \mu_1^*}\nearrow & \downarrow {\scriptstyle (1 \times i)^*} & & \downarrow {\scriptstyle (1 \wedge i)^*} \\ H^n(X) & & & & \\ & {\scriptstyle \mu^*}\searrow & & & \\ & & H^n(X \times F) & \xrightarrow[k_2^*]{} & H^n(X \wedge F) \end{array}$$

in which the homomorphisms k_i^* are injections. By (3.6), $k_2^* \mu^*(u) = 0$; since $(1 \wedge i)^*$ is a monomorphism, $k_1^* \mu_1^*(u) = 0$, and so u is primitive. □

We shall also need to study the Wang sequence in homology. In doing so, we shall assume that the coefficients lie in a commutative ring A, that F acts on the fibration, and that the action is homotopy associative.

The action $\mu : (X \times F, F \times F) \to (X, F)$ gives rise to a commutative diagram

$$\begin{array}{ccccccccccc} \cdots \to H_q(F \times F) & \to & H_q(X \times F) & \to & H_q(X \times F, F \times F) & \to & H_{q-1}(F \times F) & \to & H_{q-1}(X \times F) \to \cdots \\ \downarrow \mu_0 & & \downarrow \mu_1 & & \downarrow \mu_2 & & \downarrow \mu_0 & & \downarrow \mu_1 \\ \cdots \to H_q(F) & \longrightarrow & H_q(X) & \longrightarrow & H_q(X, F) & \longrightarrow & H_{q-1}(F) & \longrightarrow & H_{q-1}(X) \to \cdots \end{array}$$

Moreover, if $u \in H_r(F)$ is a fixed element, then the cross-product with u gives rise to a commutative diagram

$$\begin{array}{ccccccccccc} \cdots \to H_{q-r}(F) & \longrightarrow & H_{q-r}(X) & \longrightarrow & H_{q-r}(X, F) & \longrightarrow & H_{q-r-1}(F) & \longrightarrow & H_{q-r-1}(X) \to \cdots \\ \downarrow \times u & & \downarrow \times u & & \downarrow \times u & & \downarrow \times u & & \downarrow \times u \\ \cdots \to H_q(F \times F) & \to & H_q(X \times F) & \to & H_q(X \times F, F \times F) & \to & H_{q-1}(F \times F) & \to & H_{q-1}(X \times F) \to \cdots \end{array}$$

Combining these gives a commutative diagram

$$\begin{array}{ccccccccccc} \cdots \to H_p(F) & \longrightarrow & H_p(X) & \to & H_p(X, F) & \longrightarrow & H_{p-1}(F) & \longrightarrow & H_{p-1}(X) \to \cdots \\ \downarrow \cdot u & & \downarrow \cdot u & & \downarrow \cdot u & & \downarrow \cdot u & & \downarrow \cdot u \\ \cdots \to H_{p+r}(F) & \to & H_{p+r}(X) & \to & H_{p+r}(X, F) & \to & H_{p+r-1}(F) & \to & H_{p+r-1}(X) \to \cdots \end{array}$$

Now the Wang sequence is obtained from the homology sequence of the pair (X, F) by replacing the group $H_r(X, F)$ by the isomorphic group $H_{r-n}(F)$. The homology class u operates on each term of the resulting sequence, and we obtain a diagram

(3.8)

$$\begin{array}{ccccccccccc} \cdots \to H_p(F) & \xrightarrow{i_*} & H_p(X) & \xrightarrow{\alpha_*} & H_{p-n}(F) & \xrightarrow{\theta_*} & H_{p-1}(F) & \xrightarrow{i_*} & H_{p-1}(X) \to \cdots \\ \downarrow \cdot u & & \downarrow \cdot u & & \downarrow \cdot u & & \downarrow \cdot u & & \downarrow \cdot u \\ \cdots \to H_{p+r}(F) & \xrightarrow[i_*]{} & H_{p+r}(X) & \xrightarrow[\alpha_*]{} & H_{p+r-n}(F) & \xrightarrow[\theta_*]{} & H_{p+r-1}(F) & \xrightarrow[i_*]{} & H_{p+r-1}(X) \to \cdots \end{array}$$

(3.9) Theorem *If $p : X \to \mathbf{S}^n$ is a fibration, with fibre F, on which F acts, and if $w = \rho(\omega) \in H_{n-1}(F)$ is the image of the characteristic element under the Hurewicz map, then*

$$\theta_*(x) = w \cdot x$$

for all $x \in H_{p-n}(F)$. *If the action is homotopy associative, then the diagram* (3.8) *is commutative.*

Let $\phi : (\mathbf{E}^n \times F,\ \mathbf{S}^{n-1} \times F) \to (X, F)$ be a structural map. Then the diagram

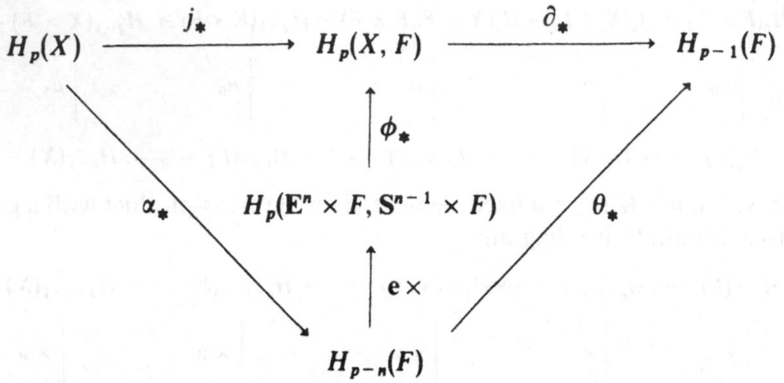

is commutative, by the definitions of α_*, θ_*. In our case, the map ϕ has the form $\mu \circ (\psi \times 1)$ for a map $\psi : (\mathbf{E}^n, \mathbf{S}^{n-1}) \to (X, F)$. Then

$$\begin{aligned}\theta_*(x) &= \partial_* \phi_*(\mathbf{e} \times x) = \partial_* \mu_*(\psi \times 1)_*(\mathbf{e} \times x)\\ &= \partial_* \mu_*(\psi_* \mathbf{e} \times x) = \mu_* \partial_*(\psi_* \mathbf{e} \times x) = \mu_*(w \times x) = w \cdot x\end{aligned}$$

where $w = \partial_* \psi_*(\mathbf{e}) \in H_{n-1}(F)$. Since $p \circ \psi$ represents the element $\iota_n \in \pi_n(\mathbf{S}^n)$, $w = \rho(\omega)$. This proves the first statement.

Suppose that the action is homotopy associative. Then

$$\theta_*(x \cdot u) = w \cdot (x \cdot u) = (w \cdot x) \cdot u = \theta_*(x) \cdot u.$$

On the other hand,

$$\begin{aligned} j_*(x \cdot u) = \phi_*(\mathbf{e} \times \alpha_*(x \cdot u)) &= \mu_*(\psi \times 1)_*(\mathbf{e} \times \alpha_*(x \cdot u))\\ &= \mu_*(\psi_*(\mathbf{e}) \times \alpha_*(x \cdot u))\\ &= \psi_*(\mathbf{e}) \cdot \alpha_*(x \cdot u),\end{aligned}$$

while

$$j_*(x) \cdot u = (\psi_*(\mathbf{e}) \cdot \alpha_*(x)) \cdot u = \psi_*(e) \cdot (\alpha_*(x) \cdot u)$$

Since $j_*(x \cdot u) = j_*(x) \cdot u$, the elements $\alpha_*(x \cdot u)$ and $\alpha_*(x) \cdot u$ are mapped into the same element by the isomorphism $\psi_*(e)\ \cdot$, and hence they are equal. □

We conclude this section by remarking that the Wang sequence is, in a certain sense, natural.

(3.10) Theorem *Suppose that*

$$\begin{array}{ccc} X & \xrightarrow{f} & X' \\ & {\scriptstyle p}\searrow \quad \swarrow{\scriptstyle p'} & \\ & \mathbf{S}^n & \end{array}$$

is a commutative diagram, and that p and p′ are fibrations with fibres F and F′, respectively. Then $f\colon (X, F) \to (X', F')$ *induces a map*

$$\begin{array}{ccccccc}
\cdots\to H_p(F) & \xrightarrow{i_*} & H_p(X) & \xrightarrow{\alpha_*} & H_{p-n}(F) & \xrightarrow{\theta_*} & H_{p-1}(F)\to\cdots \\
\downarrow f_0 & & \downarrow f_1 & & \downarrow f_2 & & \downarrow f_0 \\
\cdots\to H_p(F') & \xrightarrow[i'_*]{} & H_p(X') & \xrightarrow[\alpha'_*]{} & H_{p-n}(F') & \xrightarrow[\theta'_*]{} & H_{p-1}(F')\to\cdots
\end{array} \tag{3.11}$$

of the Wang sequence of p into that of p′ (i.e., *the diagram* (3.11) *is commutative*). *Moreover, there is a dual commutative diagram*

$$\begin{array}{ccccccc}
\cdots\to H^{q-1}(F') & \xrightarrow{\theta'^*} & H^{q-n}(F') & \xrightarrow{\alpha'^*} & H^q(X') & \xrightarrow{i'^*} & H^q(F')\to\cdots \\
\downarrow f_0^* & & \downarrow f_2^* & & \downarrow f_1^* & & \downarrow f_0^* \\
\cdots\to H^{q-1}(F) & \xrightarrow[\theta^*]{} & H^{q-n}(F) & \xrightarrow[\alpha^*]{} & H^q(X) & \xrightarrow[i^*]{} & H^q(F)\to\cdots
\end{array} \tag{3.11*}$$

The proofs are routine diagram-chasing, and are left to the reader. □

4 Homology of the Classical Groups

We shall study the homology properties of the classical groups by making use of the inclusions

$$\mathbf{O}_1^+ \subset \mathbf{O}_2^+ \subset \cdots \subset \mathbf{O}_n^+ \subset \mathbf{O}_{n+1}^+ \subset \cdots$$

$$\mathbf{U}_1 \subset \mathbf{U}_2 \subset \cdots \subset \mathbf{U}_{n-1} \subset \mathbf{U}_n \subset \cdots$$

$$\mathbf{Sp}_1 \subset \mathbf{Sp}_2 \subset \cdots \subset \mathbf{Sp}_{n-1} \subset \mathbf{Sp}_n \subset \cdots$$

and the observation that the coset spaces $\mathbf{O}_{n+1}^+/\mathbf{O}_n^+$, $\mathbf{U}_n/\mathbf{U}_{n-1}$, $\mathbf{Sp}_n/\mathbf{Sp}_{n-1}$ are spheres of dimension n, $2n-1$, $4n-1$, respectively. Thus we can make use of the Wang sequences (1.10) and (1.11). We have written $\mathbf{O}_n^+$, etc., instead of $\mathbf{O}^+(n)$ for typographical convenience.

As the unitary and symplectic groups are easier to handle, we shall study them first. The following results are not unexpected, in view of §8 of Chapter III.

(4.1) Theorem *The Hopf algebra $H^*(\mathbf{U}_n; A)$ with coefficients in an integral domain A is the exterior algebra*

$$\Lambda(x_1, \ldots, x_n),$$

where x_i is a primitive element of $H^{2i-1}(\mathbf{U}_n; A)$.

(4.2) Theorem *The Hopf algebra $H^*(\mathbf{Sp}_n; A)$ with coefficients in an integral domain A is the exterior algebra*

$$\Lambda(x_1, \ldots, x_n),$$

where x_i is a primitive element of $H^{4i-1}(\mathbf{Sp}_n; A)$.

We shall prove Theorem (4.1); the proof of Theorem (4.2) differs only in notation.

The group $\mathbf{U}_1$ is just the multiplicative group $\mathbf{S}^1$ of complex numbers of absolute value 1, and Theorem (4.1) is true in this case. Suppose that $H^*(\mathbf{U}_{n-1})$ is as described. Consider the Wang sequence

$$\cdots \to H^{q-1}(\mathbf{U}_{n-1}) \xrightarrow{\theta^*} H^{q-2n+1}(\mathbf{U}_{n-1}) \to H^q(\mathbf{U}_n) \xrightarrow{i^*}$$
$$H^q(\mathbf{U}_{n-1}) \xrightarrow{\theta^*} H^{q-2n+2}(\mathbf{U}_{n-1}) \to \cdots$$

Now θ^* has degree $-2n+2$, and therefore θ^* annihilates the generators $x_1, \ldots, x_{n-1}$ of $H^*(\mathbf{U}_{n-1})$. As θ^* is a derivation, θ^* vanishes identically, and the Wang sequence breaks up into a family of short exact sequences

$$\text{(4.3)} \quad 0 \to H^{q-2n+1}(\mathbf{U}_{n-1}) \xrightarrow{\alpha^*} H^q(\mathbf{U}_n) \xrightarrow{i^*} H^q(\mathbf{U}_{n-1}) \to 0.$$

Since $H^q(\mathbf{U}_{n-1})$ is a free A-module for all q, it follows that $H^q(\mathbf{U}_n)$ is also free. Moreover, the injection $i^*: H^q(\mathbf{U}_n) \to H^q(\mathbf{U}_{n-1})$ is an isomorphism for $q < 2n-1$, by (10.17) of Chapter IV. Let us denote the counterimages of the generators $x_1, \ldots, x_{n-1}$ of $H^*(\mathbf{U}_{n-1})$ by the same symbols $x_1, \ldots, x_{n-1}$. Let $x_n = \alpha^*(1) \in H^{2n-1}(\mathbf{U}_n)$. The monomials $x_I = x_{i_1} \cdots x_{i_k}$ $(i_k < \cdots < i_1 < n)$ form a module basis for $H^*(\mathbf{U}_{n-1})$, and, by the exactness of the sequence (4.3), the monomials x_I, $\alpha^*(x_J)$ form a module basis for $H^*(U_n)$. But

$$\alpha^*(x_J) = \alpha^*(x_{j_1} \cdots x_{j_k}) = \alpha^* i^*(x_{j_1} \cdots x_{j_k})$$
$$= x_n x_{j_1} \cdots x_{j_k} = x_n x_J,$$

and it follows that $H^*(\mathbf{U}_n)$ is the exterior algebra generated by $x_1, \ldots, x_n$. By hypothesis, the elements $i^*x_j (j < n)$ are primitive in $\mathbf{U}_{n-1}$; since i^* is an isomorphism in dimensions $\leq j$, so is $i^* \otimes i^*$, and therefore the x_j are primitive. By Theorem (3.7), x_n is primitive. □

Note. If A has characteristic $\neq 2$, the elements $x_i \in H^{2i-1}(\mathbf{U}_n; A)$ have square zero because of commutativity. In particular, this is true for $A = \mathbf{Z}$; moreover, $H^*(\mathbf{U}_n; \mathbf{Z})$ is free of finite rank. Hence, for any A, $H^*(\mathbf{U}_n; A)$ is isomorphic as an algebra with $H^*(U_n; \mathbf{Z}) \otimes A$, we may take the generators to be of the form $x_i \otimes 1$, where the x_i are the generators of $H^*(\mathbf{U}_n; \mathbf{Z})$; as $x_i^2 = 0$, $(x_i \otimes 1)^2 = 0$.

Since the x_i are primitive, the structure of $H^*(\mathbf{U}_n; A)$ and $H^*(\mathbf{Sp}_n; A)$ as Hopf algebras are determined. By duality, we can obtain the structure of the homology Hopf algebra, and this tells us the structure of the Pontryagin algebras. The results are:

(4.1$_*$) Theorem *The Hopf algebra $H_*(\mathbf{U}_n; A)$ with coefficients in an integral domain is the exterior algebra*

$$\Lambda(x'_1, \ldots, x'_n),$$

where x'_i is a primitive element of $H_{2i-1}(\mathbf{U}_n; A)$. In particular, the Pontryagin algebra of $\mathbf{U}_n$ is commutative.

(4.2$_*$) Theorem *The Hopf algebra $H_*(\mathbf{Sp}_n; A)$ with coefficients in an integral domain A is the exterior algebra*

$$\Lambda(x'_1, \ldots, x'_n),$$

where x'_i is a primitive element of $H_{4i-1}(\mathbf{Sp}_n; A)$. In particular, the Pontryagin algebra of $\mathbf{Sp}_n$ is commutative.

Again, we prove this in the unitary case. The elements $x_I = x_{i_1} \cdots x_{i_k}$, as I ranges over all sequences of positive integers $(i_1, \ldots, i_k)$ such that $i_1 < \cdots < i_k \leq n$, form a module basis for $H^*(\mathbf{U}_n; A)$. Let $\{x'_I\}$ be the dual basis.

If I, J are two such sequences having no term in common, let $I + J$ be the sequence obtained by arranging $i_1, \ldots, i_r, j_1, \ldots, j_s$ in increasing order. We then have

$$\mu^*(x_K) = \sum_{I+J=K} \eta(I, J) x_I \times x_J,$$

where $\eta(I, J)$ is the sign of the permutation

$$\begin{pmatrix} k_1, \ldots, k_r, k_{r+1}, \ldots, k_n \\ i_1, \ldots, i_r, j_1, \ldots, j_s \end{pmatrix}$$

Let P, Q be fixed sequences. We shall calculate $x'_P \cdot x'_Q$ by finding its Kronecker index with each basis element x_K. Now

$$\begin{aligned} \langle x_K, x'_P \cdot x'_Q \rangle &= \langle x_K, \mu_*(x'_P \times x'_Q) \rangle \\ &= \langle \mu^* x_K, x'_P \times x'_Q \rangle \\ &= \sum_{I+J=K} \eta(I, J) \langle x_I \times x_J, x'_P \times x'_Q \rangle. \end{aligned}$$

There is at most one non-zero term in this sum, that for which $I = P$ and $J = Q$. If P and Q have an element in common, or if they have no element in common, but $P + Q \neq K$, this cannot happen, and so $\langle x_K, x'_P \cdot x'_Q \rangle = 0$. If $P + Q = K$, then

$$\begin{aligned}\langle x_K, x'_P \cdot x'_Q \rangle &= \eta(P, Q)\langle x_P \times x_Q, x'_P \times x'_Q \rangle \\ &= (-1)^{pq}\eta(P, Q)\end{aligned}$$

where p, q are the lengths of the sequences P, Q respectively. Therefore

(4.4)

$$x'_P \cdot x'_Q = \begin{cases} (-1)^{pq}\eta(P, Q)x'_{P+Q} & \text{if } P \text{ and } Q \text{ have no common element,} \\ 0 & \text{otherwise.} \end{cases}$$

Interchanging P and Q in (4.4) yields

$$\begin{aligned} x'_Q \cdot x'_P &= (-1)^{pq}\eta(Q, P)x'_{P+Q} \\ &= \eta(Q, P)\eta(P, Q)x'_P \cdot x'_Q \quad \text{by (4.4)} \\ &= (-1)^{pq}x'_P \cdot x'_Q \end{aligned}$$

and therefore the algebra is commutative. Moreover, $x_i'^2 = 0$ for each i. To prove that $H_*(\mathrm{U}_n; A)$ is an exterior algebra, it remains to show that the monomials $x'_{i_1} \cdots x'_{i_k}$ with $i_1 > \cdots > i_k$ form a basis. For this it suffices to show that, if $I = (i_1, \ldots, i_k)$ then

(4.5) $$x'_I = x'_{i_k} \cdots x'_{i_1}.$$

This being manifestly true for $k = 1$, let $J = (i_2, \ldots, i_k)$ and assume that $x'_J = x'_{i_k} \cdots x'_{i_2}$. Then by (4.4),

$$\begin{aligned} x'_J \cdot x'_{i_1} &= (-1)^{k-1}\eta(J, \{i_1\})x'_I \\ &= (-1)^{k-1}(-1)^{k-1}x'_I = x'_I, \end{aligned}$$

and this completes the inductive proof of (4.5).

It remains to prove x'_i primitive. Now

$$\begin{aligned} \langle x_I \times x_J, \Delta_* x'_i \rangle &= \langle \Delta^*(x_I \times x_J), x'_i \rangle \\ &= \langle x_I \cdot x_J, x'_i \rangle. \end{aligned}$$

If I, J have a common element, $x_I \cdot x_J = 0$; otherwise, $x_I \cdot x_J = x_{I+J}$. Therefore, if either I or J has length > 1 or both of them have length 1, then $\langle x_I \times x_J, \Delta_* x'_i \rangle = 0$. The only non-zero Kronecker indices are

$$\langle x_i \times 1, \Delta_* x'_i \rangle = \langle x_i, x'_i \rangle = 1$$
$$\langle 1 \times x_i, \Delta_* x'_i \rangle = \langle x_i, x'_i \rangle = 1$$

and therefore

$$\Delta_* x_i' = x_i' \times 1 + 1 \times x_i',$$

i.e., x_i' is primitive. This completes the proof of Theorem $(4.1)_*$. □

We shall also need the corresponding results for the unitary unimodular group $\mathbf{U}_n^+$. Perhaps the simplest procedure is to repeat the inductive calculations we have made for $\mathbf{U}_n$, but starting with $\mathbf{U}_2^+ = \mathbf{S}^3$. We then find:

(4.1$^+$) Theorem *The cohomology ring $H^*(\mathbf{U}_n^+; A)$ is an exterior algebra generated by primitive elements $u_i \in H^{2i-1}(\mathbf{U}_n^+; A)$ $(i = 2, 3, \ldots, n)$.* □

(4.1$^+_*$) Theorem *The Pontryagin ring $H_*(\mathbf{U}_n^+; A)$ is an exterior algebra generated by primitive elements $u_i' \in H_{2i-1}(\mathbf{U}_n^+; A)$ $(i = 2, 3, \ldots, n)$.* □

In §10 of Chapter IV we constructed an imbedding $G_{n+1} : \mathbf{SP}^n(\mathbf{C}) \to \mathbf{U}_{n+1}^+$ for every positive integer n; this map has the following properties:

(i) If $p : \mathbf{U}_{n+1}^+ \to \mathbf{S}^{2n+1}$ is the restriction of $p_{n+1} : \mathbf{U}_{n+1} \to \mathbf{S}^{2n+1}$, then $p \circ G_{n+1} : (\mathbf{SP}^n(\mathbf{C}), \mathbf{SP}^{n-1}(\mathbf{C})) \to (\mathbf{S}^{2n+1}, *)$ is a relative homeomorphism;
(ii) $G_{n+1} | \mathbf{SP}^{n-1}(\mathbf{C}) = G_n : \mathbf{SP}^{n-1}(\mathbf{C}) \to \mathbf{U}_n^+$.

We now observe:

(4.6) *The injection $G_{n+1}^* : H^*(\mathbf{U}_{n+1}^+) \to H^*(\mathbf{SP}^n(\mathbf{C}))$ is an epimorphism; in fact, G_{n+1}^* maps the space P^* of primitive elements of $H^*(\mathbf{U}_{n+1}^+)$ isomorphically upon $H^*(\mathbf{SP}^n(\mathbf{C}))$.*

We have seen that the elements $u_2, \ldots, u_{n+1}$ are primitive, and it follows by an argument similar to that used in the case of the orthogonal group that they form a basis for P^* as an A-module. Therefore it remains only to verify that $G_{n+1}^*(u_j)$ is a generator of $H^{2j-1}(\mathbf{SP}^n(\mathbf{C}))$ $(j = 2, 3, \ldots, n+1)$.

This is proved by induction on n. Let us recall that the generators u_i of $H^*(\mathbf{U}_{n+1}^+)$ are mapped by the injection into the generators with the same names for $H^*(\mathbf{U}_n^+)$ $(i = 2, \ldots, n)$. Moreover, the remaining generator x_{n+1} is defined to be $\alpha^*(1)$ $(= p^*(\mathbf{S}^{2n+1})$, by (3.5)). It follows from (ii), above, that the diagram

$$\begin{array}{ccc} H^{2i-1}(\mathbf{U}_{n+1}^+) & \longrightarrow & H^{2i-1}(\mathbf{U}_n^+) \\ \downarrow G_{n+1}^* & & \downarrow G_n^* \\ H^{2i-1}(\mathbf{SP}^n(\mathbf{C})) & \longrightarrow & H^{2i-1}(\mathbf{SP}^{n-1}(\mathbf{C})), \end{array}$$

whose horizontal arrows denote injections, is commutative. Moreover the injections in question are isomorphisms for $i \le n$. Therefore it suffices, because of the inductive hypothesis, to prove that the last statement of (4.6) is true for $i = n + 1$. But the diagram

$$\begin{array}{ccc} H^{2n+1}(\mathbf{S}^{2n+1}) & \xrightarrow{(p \circ G_{n+1})^*} & H^{2n+1}(\mathbf{SP}^n(\mathbf{C}), \mathbf{SP}^{n-1}(\mathbf{C})) \\ \downarrow p^* & & \downarrow j^* \\ H^{2n+1}(\mathbf{U}_{n+1}^+) & \xrightarrow[G_{n+1}^*]{} & H^{2n+1}(\mathbf{SP}^n(\mathbf{C})), \end{array}$$

in which the injection j^* is an isomorphism, is commutative. As $(p \circ G_{n+1})^*$ is also an isomorphism, by (*i*), it follows that G_{n+1}^* is an epimorphism. □

The situation for the orthogonal group is more complicated because of the presence of 2-torsion in the integral homology. Accordingly, we shall discuss only the mod 2 homology, reserving the case of other coefficients for Volume II.

The discussion of the mod 2 cohomology of $\mathbf{O}_n^+$ resembles that of $\mathbf{U}_n$. There are two important points of difference. Firstly, the generators may have even degree. Secondly, the commutative law does not guarantee that $x^2 = 0$ if x has odd degree.

What was actually proved about $H^*(\mathbf{U}_n)$ was that the monomials $x_{i_1} \cdots x_{i_k}$ with $i_1 > i_2 > \cdots > i_k$ form an additive basis. That the x_i generate an exterior algebra then follows from the commutative law (in fact, a special argument was needed in the mod 2 case).

Let A be a (graded) algebra over $\mathbf{Z}_2$, and let $(x_1, x_2, \ldots)$ be a sequence (finite or infinite) of homogeneous elements of A. We shall say that the x_i form a *simple system of generators for* A if and only if the monomials $x_{i_1} \cdots x_{i_k}$ with $i_1 > \cdots > i_k$ form an additive basis for A. It follows that x_i^2 is a linear combination of such monomials; and a knowledge of these elements determines the structure of A as an algebra.

In studying the homology and cohomology of $\mathbf{O}_{n+1}^+$, we shall need to make use of the imbedding $g_n : \mathbf{P}^n \to \mathbf{O}_{n+1}^+$ of Theorem (10.11) of Chapter IV. We shall formulate and prove the results on homology and cohomology by simultaneous induction.

(4.7) Theorem *The cohomology algebra $H^*(\mathbf{O}_{n+1}^+; \mathbf{Z}_2)$ has a simple system of primitive generators $x_1, \ldots, x_n$, such that*

(4.8)
$$x_i^2 = \begin{cases} x_{2i} & \text{if } 2i \le n, \\ 0 & \text{if } 2i > n. \end{cases}$$

The elements $x_1, \ldots, x_n$ form a basis for the space M^n of primitive elements of $H^*(\mathbf{O}_{n+1}^+; \mathbf{Z}_2)$. The homomorphism g_n^* maps M^n isomorphically upon $H^*(\mathbf{P}^n; \mathbf{Z}_2)$.

(4.7$_*$) Theorem *The Pontryagin algebra $H_*(\mathbf{O}_{n+1}^+; \mathbf{Z}_2)$ is an exterior algebra $\Lambda(x_1', \ldots, x_n')$, where x_i' is a homogeneous element of degree i. The elements x_i' with i odd are primitive and form a basis for the space M_n of primitive elements.*

If $n = 1$, $\mathbf{O}_{n+1}^+ = \mathbf{S}^1$ and Theorems (4.7) and (4.7$_*$) are true in this case. Assume that both theorems are true for $\mathbf{O}_n^+$. Consider first the Wang sequence for cohomology:

$$\cdots \to H^{q-1}(\mathbf{O}_n^+) \xrightarrow{\theta^*} H^{q-n}(\mathbf{O}_n^+) \xrightarrow{\alpha^*}$$
$$H^q(\mathbf{O}_{n+1}^+) \xrightarrow{i^*} H^q(\mathbf{O}_n^+) \xrightarrow{\theta^*} \cdots$$

For $i < n - 1$, $\theta^* x_i \in H^{i-n+1}(\mathbf{O}_n^+) = 0$. We next show that $\theta^* x_{n-1} = 0$. If not, then $\theta^* x_{n-1} = 1$, and, by (3.5), $p_n^* \mathbf{s}^n = \alpha^*(1) = \alpha^* \theta^* x_{n-1} = 0$. The map $p_n \circ g_n : (\mathbf{P}^n, \mathbf{P}^{n-1}) \to (\mathbf{S}^n, \mathbf{e}_n)$ is a relative homeomorphism and the injection $H^n(\mathbf{P}^n, \mathbf{P}^{n-1}) \to H^n(\mathbf{P}^n)$ is an isomorphism. It follows that $g_n^* \circ p_n^* : H^n(\mathbf{S}^n) \to H^n(\mathbf{P}^n)$ is an isomorphism and $g_n^* p_n^* \mathbf{s}^n = 0$, a contradiction.

The derivation θ^* annihilates a set of generators for $H^*(\mathbf{O}_n^+)$, and therefore $\theta^* = 0$. Thus the Wang sequence breaks up into a family of short exact sequences

$$0 \to H^{q-n}(\mathbf{O}_n^+) \xrightarrow{\alpha^*} H^q(\mathbf{O}_{n+1}^+) \xrightarrow{i^*} H^q(\mathbf{O}_n^+) \to 0.$$

Now i^* is an isomorphism for $q < n$; let us denote the counter-images under i^* of the generators x_q of $H^*(\mathbf{O}_n^+)$ by the same symbols x_q ($q = 1, \ldots, n - 1$). Let $x_n = \alpha^*(1)$. Then it follows, just as in the unitary case, that $x_1, \ldots, x_n$ form a simple system of generators for $H^*(\mathbf{O}_{n+1}^+)$. We have seen above that $g_n^*(x_n)$ is the non-zero element u^n of $H^n(\mathbf{P}^n)$, and it follows by induction hypothesis and the fact that $g_n | \mathbf{P}^{n-1} = g_{n-1}$ that $g_n^* x_q = u^q$ for $q = 1, \ldots, n - 1$. Moreover, the elements $x_1, \ldots, x_{n-1}$ are clearly primitive, and x_n is primitive by Theorem (3.7).

As the x_i form a simple system of generators for $H^*(\mathbf{O}_{n+1}^+)$, the elements $x_I = x_{i_1} \cdots x_{i_k}$, as I ranges over all sequences $i_1 > \cdots > i_k$, form an additive basis. Let $\{x_I'\}$ be the dual basis for $H_*(\mathbf{O}_{n+1}^+)$. Then the proof of Theorem (4.1$_*$) can be repeated almost *verbatim*, up to the calculation of the coproduct, to show that the latter algebra is the exterior algebra generated by the elements x_i' ($i = 1, \ldots, n$), and $x_I' = x_{i_1}' \cdots x_{i_k}'$. We can now exploit the duality between the Hopf algebras $H^n = H^*(\mathbf{O}_{n+1}^+)$, $H_n = H_*(\mathbf{O}_{n+1}^+)$. By (8.6) of Chapter III, M^n is the dual space of $Q_n = H_n/D_n$ and M_n the dual space of $Q^n = H^n/D^n$, where D_n and D^n are the spaces of decomposable elements. Since H_n is an exterior algebra, D_n is spanned by the unit and the x_I where I has length ≥ 2; hence Q_n has one basis element of dimension q corresponding to the indecomposable element x_q' ($q = 1, \ldots, n$), so that M^n has one basis element in each dimension q ($q = 1, \ldots, n$). Hence the primitive elements x_q form a basis for M^n, which is mapped isomorphically by g_n^* on $H^*(\mathbf{P}^n)$.

The element x_q^2 is primitive and $g_n^*(x_q^2) = u^{2q}$. Hence $x_q^2 = 0$ if $2q > n$ and $x_q^2 = x_{2q}$ if $2q \leq n$. This completes the proof of Theorem (4.7). Moreover, it follows easily that $x_I = x_{i_1} \cdots x_{i_k}$ is decomposable if either $k > 1$ or $k = 1$ and i_1 is even; and if i is odd then x_i is indecomposable. It follows that the x_i' with i odd form a basis for M_n. □

In a similar way, we can calculate the mod 2 cohomology ring of the Stiefel manifolds $V_{n,m}$. The result is given by

(4.9) Theorem *Let* $p : \mathbf{O}_n^+ \to \mathbf{V}_{n,m} = \mathbf{O}_n^+/\mathbf{O}_{n-m}^+$ *be the natural fibration. Then* $p^* : H^*(\mathbf{V}_{n,m}; \mathbf{Z}_2) \to H^*(\mathbf{O}_n^+; \mathbf{Z}_2)$ *is a monomorphism, and the image of* p^* *is the subalgebra of* $H^*(\mathbf{O}_n^+; \mathbf{Z}_2)$ *generated by* $x_{n-m}, \ldots, x_{n-1}$.

(4.10) Corollary *The algebra* $H^*(\mathbf{V}_{n,m}; \mathbf{Z}_2)$ *has a simple system of generators* $(x_{n-m}, \ldots, x_{n-1})$.

Theorem (4.9) is proved by induction on n. If $n = m + 1$, $\mathbf{V}_{n,m} = \mathbf{S}^{n-1}$ is a sphere, and we have seen in the proof of Theorem (4.7) that the theorem is true in this case.

Assume that the cohomology of $\mathbf{V}_{n,m-1}$ is as stated. Then there is a commutative diagram

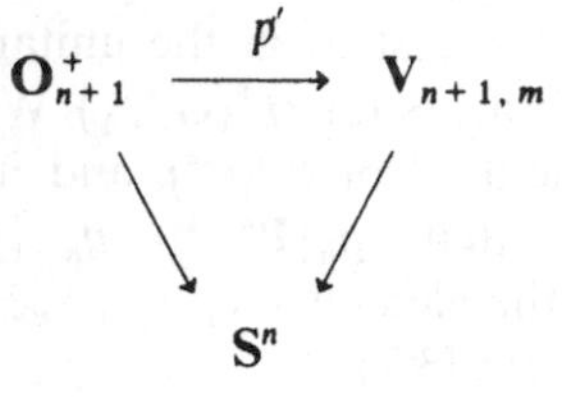

of fibrations, and the restriction of p' to the subgroup $\mathbf{O}_n^+$ is the fibration $p : \mathbf{O}_n^+ \to \mathbf{V}_{n,m-1}$. By Theorem (3.10), there is a commutative diagram (Figure 7.2) and, by the induction hypothesis, the homomorphisms p^* are all monomorphisms. We have seen that $\theta^* = 0$; therefore $0 = \theta^* \circ p^* = p^* \circ \theta'^*$, and the fact that p^* is a monomorphism implies that $\theta'^* = 0$. Thus the Wang sequence of the fibration $\mathbf{V}_{n+1,m} \to \mathbf{S}^n$, like that for the fibration $\mathbf{O}_{n+1}^+ \to \mathbf{S}^n$, breaks up into a family of short exact sequences, and we can now apply the Five-Lemma to deduce that p'^* is a monomorphism. By induction hypothesis, the image of $p^* : H^*(\mathbf{V}_{n,m-1}) \to H^*(\mathbf{O}_n^+)$ is generated by $x_{n-m}, \ldots, x_{n-1}$. But $x_n = \alpha^*(1) = \alpha^* p^*(1) = p'^* \alpha'^*(1)$ and it follows that the image of p'^* contains the subalgebra generated by $x_{n-m}, \ldots, x_n$, and it is a simple matter to deduce that equality does, in fact, hold. □

The cohomology rings of the complex and quaternionic Stiefel manifolds can be calculated in the same way.

$$
\begin{array}{ccccccccccc}
\cdots \to H^{q-1}(\mathbf{V}_{n,m-1}) & \xrightarrow{\theta'^*} & H^{q-n}(\mathbf{V}_{n,m-1}) & \xrightarrow{\alpha'^*} & H^q(\mathbf{V}_{n+1,m}) & \xrightarrow{i'^*} & H^q(\mathbf{V}_{n,m-1}) & \xrightarrow{\theta'^*} & H^{q-n+1}(\mathbf{V}_{n,m-1}) \to \cdots \\
\downarrow p^* & & \downarrow p^* & & \downarrow p'^* & & \downarrow p^* & & \downarrow p^* \\
\cdots \to H^{q-1}(\mathbf{O}_n^+) & \xrightarrow[\theta^*]{} & H^{q-n}(\mathbf{O}_n^+) & \xrightarrow[\alpha^*]{} & H^q(\mathbf{O}_{n+1}^+) & \xrightarrow[i^*]{} & H^q(\mathbf{O}_n^+) & \xrightarrow[\theta^*]{} & H^{q-n+1}(\mathbf{O}_n^+) \to \cdots
\end{array}
$$

Figure 7.2

5 Fibrations Having a Sphere as Fibre

Let $p : X \to B$ be an n-spherical fibration, and let A be a principal ideal domain. We shall be concerned with the homology and cohomology of the spaces involved with coefficients in an A-module M.

Let $\hat{X}$ be the mapping cylinder of p, $\hat{p} : \hat{X} \to B$ the projection. A *Thom class for p* is an element $u \in H^{n+1}(\hat{X}, X; A)$ whose image under the injection generates the free module $H^{n+1}(\hat{F}, F; A)$. The fibration p is said to be *A-orientable* if and only if a Thom class exists.

In §9 of Chapter IV we saw that $H_q(\hat{X}, X; \mathbf{Z}) = 0$ if $q \leq n$, while the injection $k_* : H_{n+1}(\hat{F}, F; \mathbf{Z}) \to H_{n+1}(\hat{X}, X; \mathbf{Z})$ is an epimorphism; it is an isomorphism if and only if $\pi_1(B)$ operates trivially on $H_n(F; \mathbf{Z})$.

There is a commutative diagram

$$
\begin{array}{ccccc}
H^{n+1}(\hat{X}, X; A) & \xrightarrow{k^*} & H^{n+1}(\hat{F}, F; A) & \xleftarrow{(-1)^n\delta^*} & H^n(F; A) \\
\downarrow & & \downarrow & & \downarrow \\
\mathrm{Hom}(H_{n+1}(\hat{X}, X; \mathbf{Z}), A) & & \downarrow & & \mathrm{Hom}(H_n(F; \mathbf{Z}), A) \\
& \searrow^{\mathrm{Hom}(k_*, 1)} & \downarrow & \swarrow^{\mathrm{Hom}(\partial_*, 1)} & \\
& & \mathrm{Hom}(H_{n+1}(\hat{F}, F; \mathbf{Z}), A) & &
\end{array}
$$

in which the vertical arrows denote the isomorphisms given by the universal coefficient theorem. Therefore

(5.1) *The injection k^* is a monomorphism; it is an isomorphism if and only if $\pi_1(B)$ operates trivially on* $\mathrm{Hom}(H_n(F; \mathbf{Z}), A)$.

(5.2) Theorem *If p is* **Z***-orientable, it is A-orientable for any principal ideal domain A. If p is not* **Z***-orientable, it is A-orientable if and only if A has characteristic two.*

In fact the group $H_n(F; \mathbf{Z})$ is infinite cyclic, and thus has only two automorphisms, viz. the identity and the reversal of sign; an element of $\pi_1(B)$ is accordingly said to *preserve* or to *reverse* orientation. The group $\operatorname{Hom}(H_n(F; \mathbf{Z}), A)$ is isomorphic with A under the correspondence $f \to f(1)$, and the operations of an element $\xi \in \pi_1(B)$ on A are given by

$$a \to a \quad \text{if } \xi \text{ preserves orientation,}$$

$$a \to -a \quad \text{if } \xi \text{ reverses orientation.}$$

Our result now follows easily.

Let B_0 be a closed subspace of B, $X_0 = p^{-1}(B_0)$, $\hat{X}_0 = \hat{p}^{-1}(B_0)$. The cup and cap products give rise to pairings

$$H^{n+1}(\hat{X}, X; A) \otimes H^p(\hat{X}, \hat{X}_0; M) \to H^{p+n+1}(\hat{X}, \hat{X}_0 \cup X; M),$$

$$H_p(\hat{X}, \hat{X}_0 \cup X; M) \otimes H^{n+1}(\hat{X}, X; A) \to H_{p-n-1}(\hat{X}, \hat{X}_0; M).$$

In particular, each element $u \in H^{n+1}(\hat{X}, X; A)$ determines homomorphisms

$$u \smile : H^p(\hat{X}, \hat{X}_0; M) \to H^{p+n+1}(\hat{X}, \hat{X}_0 \cup X; M),$$

$$\frown u : H_p(\hat{X}, \hat{X}_0 \cup X; M) \to H_{p-n-1}(\hat{X}, \hat{X}_0; M).$$

The principal object of this section is to prove

(5.3) Theorem (Thom Isomorphism Theorem). *If $p : X \to B$ is an A-orientable spherical fibration with Thom class $u \in H^{n+1}(\hat{X}, X; A)$, then, for any subspace B_0 of B and any coefficient module M, the homomorphisms $u \smile$ and $\frown u$ are isomorphisms.*

We shall reduce the proof of Theorem (5.3), step by step, to one special case, for which the proof is transparent.

We first observe that if u is a Thom class for the fibration $p : X \to B$ and if B_0 is a closed subspace of B, then the image u_0 of u under the injection $H^{n+1}(\hat{X}, X; A) \to H^{n+1}(\hat{X}_0, X_0; A)$ is a Thom class for the fibration $p_0 = p \mid X_0 : X_0 \to B_0$. Moreover,

(5.4) Lemma *If any two of the three homomorphisms*

$$\frown u_0 : H_p(\hat{X}_0, X_0; M) \to H_{p-n-1}(\hat{X}_0; M),$$

$$\frown u : H_p(\hat{X}, X; M) \to H_{p-n-1}(\hat{X}; M),$$

$$\frown u : H_p(\hat{X}, \hat{X}_0 \cup X; M) \to H_{p-n-1}(\hat{X}, \hat{X}_0; M)$$

are isomorphisms for all p, so is the third. Dually, if any two of the three homomorphisms

$$u_0 \smile : H^p(\hat{X}_0; M) \to H^{p+n+1}(\hat{X}_0, X_0; M),$$

$$u \smile : H^p(\hat{X}; M) \to H^{p+n+1}(\hat{X}, X; M),$$

$$u \smile : H^p(\hat{X}, \hat{X}_0; M) \to H^{p+n+1}(\hat{X}, \hat{X}_0 \cup X; M)$$

are isomorphisms for all p, so is the third.

$$
\begin{array}{ccccccccc}
H_p(\hat{X}_0, X_0; M) & & & & H_{p-1}(\hat{X}_0, X_0; M) & & & & \\
\downarrow{\scriptstyle k} & \overset{l}{\searrow} & & & \downarrow{\scriptstyle k} & \overset{l}{\searrow} & & & \\
H_p(\hat{X}_0 \cup X, X; M) & \xrightarrow{i} & H_p(\hat{X}, X; M) & \xrightarrow{j} & H_p(\hat{X}, \hat{X}_0 \cup X; M) & \xrightarrow{\partial} & H_{p-1}(\hat{X}_0 \cup X, X; M) & \xrightarrow{i} & H_{p-1}(\hat{X}, X; M) \\
\downarrow{\scriptstyle \phi} & & \downarrow{\scriptstyle \frown u} & & \downarrow{\scriptstyle \frown u} & & \downarrow{\scriptstyle \phi} & & \downarrow{\scriptstyle \frown u} \\
H_{p-n-1}(\hat{X}_0; M) & \xrightarrow{i'} & H_{p-n-1}(\hat{X}; M) & \xrightarrow{j'} & H_{p-n-1}(\hat{X}, \hat{X}_0; M) & \xrightarrow{\partial'} & H_{p-n-2}(\hat{X}_0; M) & \xrightarrow{i'} & H_{p-n-2}(\hat{X}; M)
\end{array}
$$

Figure 7.3

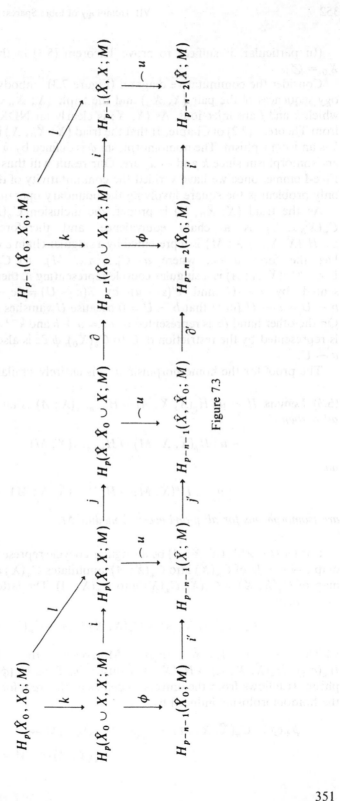

(In particular, it suffices to prove Theorem (5.3) in the absolute case $X_0 = \varnothing$).

Consider the commutative diagram (Figure 7.3) embodying the homology sequences of the pair $(\hat{X}, \hat{X}_0)$ and the triple $(\hat{X}; \hat{X}_0 \cup X, X)$, and in which k and l are injections. As $(\hat{X}, X)$ is clearly an NDR-pair, it follows from Theorem (2.2) of Chapter II that the triad $(\hat{X}; \hat{X}_0, X)$ is proper, and so k is an isomorphism. The homomorphism ϕ is defined by $\phi \circ k = \frown u_0$; it is an isomorphism since k and $\frown u_0$ are. Our result will thus follow from the Five-Lemma, once we have verified the commutativity of the diagram. The only problem is the square involving the boundary operators.

As the triad $(\hat{X}; \hat{X}_0, X)$ is proper, the inclusion $C_*(\hat{X}_0) + C_*(X)$ in $C_*(\hat{X}_0 \cup X)$ is a chain equivalence, and therefore any element $z \in H_p(\hat{X}, \hat{X}_0 \cup X; M)$ is represented by a singular chain c whose boundary has the form $a + b$, where $a \in C_{p-1}(\hat{X}_0;\ M)$, $b \in C_{p-1}(X;\ M)$. If $U \in C^{n+1}(\hat{X}, X; A)$ is a singular cocycle representing u, then $z \frown u$ is represented by $c \frown U$ and $\partial'(z \frown u)$ by $\partial(c \frown U) = \partial c \frown U = a \frown U + b \frown U = a \frown U$ (note that $b \frown U = 0$ because U vanishes on chains of X). On the other hand ∂z is represented by $\partial c = a + b$ and $k^{-1}\ \partial c$ by a. Since u_0 is represented by the restriction of U to $C_*(\hat{X}_0)$, $\phi\ \partial z$ is also represented by $a \frown U$.

The proof for the homomorphism $u \smile$ is entirely similar. □

(5.5) Lemma *If* $\frown u : H_p(\hat{X}, X; A) \to H_{p-n-1}(\hat{X}; A)$ *is an isomorphism for all* p, *then*

$$\frown u : H_p(\hat{X}, X; M) \to H_{p-n-1}(\hat{X}; M)$$

and

$$u \smile : H^p(\hat{X}; M) \to H^{p+n+1}(\hat{X}, X; M)$$

are isomorphisms for all p *and every* A*-module* M.

For let $U \in Z^{n+1}(\hat{X}, X; A)$ be a singular cocycle representing u; then the map $c \to c \frown U$ of $C_*(\hat{X})$ into $C_*(\hat{X}; A)$ annihilates $C_*(X)$ and so induces a map of $C_*(\hat{X}, X) = C_*(\hat{X})/C_*(X)$ into $C_*(\hat{X}; A)$. The latter determines in turn a map

$$\phi_A : C_*(\hat{X}, X; A) = C_*(\hat{X}, X) \otimes A \to C_*(\hat{X}; A)$$

which is a chain map. Moreover, the homomorphism $H_*(\phi_A) : H_*(\hat{X}, X; A) \to H_*(\hat{X}; A)$ is just $\frown u$. Thus $H_*(\phi_A)$ is an isomorphism. It follows from the universal coefficient theorem for A-modules that the homomorphisms induced by

$$\phi_A \otimes 1 : C_*(\hat{X}, X; M) = C_*(\hat{X}, X; A) \otimes_A M \to$$
$$C_*(\hat{X}; A) \otimes_A M = C_*(\hat{X}; M)$$

and

$$\mathrm{Hom}(\phi_A, 1): C^*(\hat{X}; M) = \mathrm{Hom}_A(C_*(\hat{X}; A), M)$$
$$\to \mathrm{Hom}_A(C_*(\hat{X}, X; A), M) = C_*(\hat{X}, X; M)$$

induce homology isomorphisms. But the latter are $\frown u$ and $u \smile$, respectively. □

Let $f: B' \to B$ be a map, and let $p': W \to B'$ be the induced fibration, $\hat{W}$ its mapping cylinder. Then there is a commutative diagram

$$\text{(5.6)} \qquad \begin{array}{ccc} W & \xrightarrow{f'} & X \\ {\scriptstyle p'}\downarrow & & \downarrow{\scriptstyle p} \\ B' & \xrightarrow[f]{} & B \end{array}$$

inducing, in turn, a commutative diagram

$$\text{(5.7)} \qquad \begin{array}{ccc} (\hat{W}, W) & \xrightarrow{\hat{f}'} & (\hat{X}, X) \\ {\scriptstyle \hat{p}'}\downarrow & & \downarrow{\scriptstyle \hat{p}} \\ (B', B') & \xrightarrow[f]{} & (B, B) \end{array}$$

Let $v = \hat{f}'^* u \in H^{n+1}(\hat{W}, W; A)$.

(5.8) Lemma *The class v is a Thom class for the fibration $p': W \to B'$. Moreover, the diagram*

$$\text{(5.9)} \qquad \begin{array}{ccc} H_p(\hat{W}, W; A) & \xrightarrow{\hat{f}'_*} & H_p(\hat{X}, X; A) \\ {\scriptstyle \frown v}\downarrow & & \downarrow{\scriptstyle \frown u} \\ H_{p-n-1}(\hat{W}; A) & \xrightarrow[\hat{f}'_*]{} & H_{p-n-1}(\hat{X}; A) \end{array}$$

is commutative.

Since the fibre of p' can be identified with the fibre F of p, there is a commutative diagram

$$\begin{array}{ccc} & F & \\ {}^{i'}\swarrow & & \searrow^{i} \\ W & \xrightarrow[f']{} & X \end{array}$$

and therefore a commutative diagram

$$\begin{array}{ccc} & (\hat{F}, F) & \\ {}^{\hat{\imath}'}\nearrow & & \nwarrow^{\hat{\imath}} \\ (\hat{W}, W) & \xrightarrow[\hat{f}']{} & (\hat{X}, X) \end{array}$$

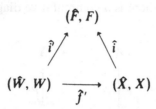

and the fact that v is a Thom class follows immediately. The commutativity of the diagram (5.9) is a consequence of the functional properties of the cap product.

(5.10) Corollary *If f is a weak homotopy equivalence, and $\frown v$ is an isomorphism, then $\frown u$ is an isomorphism.*

The homotopy sequences of the fibrations p', p are related by a commutative diagram

$$\begin{array}{ccccccccc} \cdots \to \pi_{n+1}(B') & \xrightarrow{\partial} & \pi_n(F) & \xrightarrow{i} & \pi_n(W) & \xrightarrow{j} & \pi_n(B') \to \cdots \\ \downarrow f_* & & \downarrow 1 & & \downarrow f'_* & & \downarrow f_* \\ \cdots \to \pi_{n+1}(B) & \xrightarrow[\partial]{} & \pi_n(F) & \xrightarrow[i]{} & \pi_n(X) & \xrightarrow[j]{} & \pi_n(B) \to \cdots \end{array}$$

and it follows from the Five-Lemma that $f' : W \to X$ is a weak homotopy equivalence. Since the diagram (5.7) is commutative and $\hat{p}' : \hat{W} \to B'$ and $\hat{p} : \hat{X} \to B$ are homotopy equivalences, $\hat{f}' : \hat{W} \to \hat{X}$ is a weak homotopy equivalence. By the Whitehead theorem,

$$f'_* : H_*(W) \to H_*(X)$$

and

$$\hat{f}'_* : H_*(\hat{W}) \to H_*(\hat{X})$$

are isomorphisms. Since $\hat{f}' \mid W : W \to X$ is just the map f', it follows from the Five-Lemma that

$$\hat{f}'_* : H_*(\hat{W}, W) \to H_*(\hat{X}, X)$$

is an isomorphism, and from the Universal Coefficient Theorem that

$$\hat{f}'_* : H_*(\hat{W}, W; A) \to H_*(\hat{X}, X; A)$$

and

$$\hat{f}'_* : H_*(\hat{W}; A) \to H_*(\hat{X}; A)$$

are isomorphisms. The Corollary follows from these facts and the commutativity of the diagram (5.9). □

Since B has a CW-approximation $f: K \to B$, it follows from the Corollary that it suffices to prove Theorem (5.3) for the case that B is itself a CW-complex, and even a simplicial complex. That we may even assume B to be a finite complex follows from

(5.11) Lemma *Let B be a CW-complex, and let $\{B_\alpha \mid \alpha \in J\}$ be the family of its finite subcomplexes. Let $(\hat{X}_\alpha, X_\alpha) = (\hat{p}^{-1}(\hat{B}_\alpha), p^{-1}(B_\alpha))$, and let $u_\alpha \in H^{n+1}(\hat{X}_\alpha, X_\alpha; A)$ be the image of u under the injection. If, for every α,*

$$\frown u_\alpha : H_p(\hat{X}_\alpha, X_\alpha; A) \to H_{p-n-1}(\hat{X}_\alpha; A)$$

is an isomorphism for all p, then so is

$$\frown u : H_p(\hat{X}, X; A) \to H_{p-n-1}(\hat{X}; A).$$

This follows from naturality properties of the cap product and the fact that $\frown u$ is the direct limit of the isomorphisms $\frown u_\alpha$.

The above considerations have allowed us to reduce the proof of Theorem (5.3) to the case where B is a finite simplicial complex. We shall treat this case by induction on the number of simplices in B. Accordingly, we may assume that $B = B_0 \cup B_1$, where B_1 is a principal r-simplex of B, B_0 is the union of the remaining simplicies of B, so that $B_0 \cap B_1 = B_{01}$ is the boundary of B_1 and that our theorem is true for B_0. Because of Lemma (5.4) it now suffices to prove that

$$\frown u : H_p(\hat{X}, \hat{X}_0 \cup X; A) \to H_{p-n-1}(\hat{X}, \hat{X}_0; A)$$

is an isomorphism. There is a commutative diagram

$$\begin{array}{ccc} H_p(\hat{X}_1, \hat{X}_{01} \cup X_1; A) & \xrightarrow{\frown u_1} & H_{p-n-1}(\hat{X}_1, \hat{X}_{01}; A) \\ \downarrow & & \downarrow \\ H_p(\hat{X}, \hat{X}_0 \cup X; A) & \xrightarrow[\frown u]{} & H_{p-n-1}(\hat{X}, \hat{X}_0; A) \end{array}$$

where the vertical arrows denote injections; these are isomorphisms, because the triads $(\hat{X}; \hat{X}_0 \cup X, \hat{X}_1)$ and $(\hat{X}; \hat{X}_0, \hat{X}_1)$ are proper. Therefore it suffices to prove that $\frown u_1$ is an isomorphism. But this last argument has reduced the problem to that of proving Theorem (5.3) for the special case in which B is an r-simplex with boundary B_0.

When this is the case, the base-space B is contractible and so the fibration is fibre homotopically trivial. By Corollary (7.27) of Chapter I, there is a fibre homotopy equivalence $h : (B \times F, B_0 \times F) \to (X, X_0)$, and this induces a homotopy equivalence

$$\hat{h} : (B \times \hat{F}; B_0 \times \hat{F}, B \times F) \to (\hat{X}; \hat{X}_0, X)$$

The groups

$$H_p(B \times \hat{F}, B_0 \times \hat{F} \cup B \times F; A),$$
$$H^{n+1}(B \times \hat{F}, B \times F; A),$$
$$H_{p-n-1}(B \times \hat{F}, B_0 \times \hat{F}; A),$$

are isomorphic with the groups

$$H_{p-n-1}(B, B_0; A) \otimes_A H_{n+1}(\hat{F}, F; A),$$
$$H^0(B; A) \otimes_A H^{n+1}(\hat{F}, F; A),$$
$$H_{p-n-1}(B, B_0; A) \otimes_A H_0(\hat{F}; A),$$

respectively, these isomorphisms being given by the appropriate cross products. The class u corresponds under the second of these isomorphisms to $1 \otimes w^*$, where w^* generates the free cyclic module $H^{n+1}(\hat{F}, F; A)$. Let $w \in H_{n+1}(\hat{F}, F; A)$ be the homology class such that $\langle w^*, w \rangle = 1$. The first and third of our groups are isomorphic with $H_{p-n-1}(B, B_0; A)$ under the correspondences $x \to x \times w$, $x \to x \times 1$, respectively. As

$$(x \times w) \frown (1 \times w^*) = \pm (x \frown 1) \times (w \frown w^*) = \pm x \times 1$$

it follows that $\frown u$ is indeed an isomorphism, and our proof is complete.

We can now use the information given by the Thom isomorphism to relate the homology groups of (X, X_0) with those of (B, B_0). In fact, let us consider the homology sequence of the triad $(\hat{X}; \hat{X}_0, X)$, i.e., the sequence obtained from the homology sequence of the triple $(\hat{X}, \hat{X}_0 \cup X, \hat{X}_0)$ by replacing the groups $H_p(\hat{X}_0 \cup \hat{X}, \hat{X}_0)$ by the isomorphic groups $H_p(X, X_0)$. Since $\hat{p} : (X, X_0) \to (B, B_0)$ is a homotopy equivalence, we may replace the groups $H_p(\hat{X}, \hat{X}_0)$ by $H_p(B, B_0)$. Finally, we can use the Thom isomorphism to replace $H_p(\hat{X}, \hat{X}_0 \cup X)$ by $H_{p-n-1}(B, B_0)$. We can make the dual constructions in cohomology. Thus we obtain

(5.12) Theorem (Gysin Theorem). *Let A be a principal ideal domain, and let $p : X \to B$ be an A-orientable n-spherical fibration. Then, for any closed sub-*

space B_0 of B, there are exact sequences

(5.13) $$\cdots \to H_{p-n}(B, B_0; M) \xrightarrow{\beta} H_p(X, X_0; M) \xrightarrow{p_*} H_p(B, B_0; M) \xrightarrow{\gamma} H_{p-n-1}(B, B_0; M) \to \cdots$$

(5.14) $$\cdots \to H^{p-n-1}(B, B_0; M) \xrightarrow{\gamma^*} H^p(B, B_0; M) \xrightarrow{p^*} H^p(X, X_0; M) \xrightarrow{\beta^*} H^{p-n}(B, B_0; M) \to \cdots$$

Moreover,

$$\gamma(x) = x \frown w \qquad (x \in H_p(B, B_0; M)),$$
$$\gamma^*(x) = w \smile x \qquad (x \in H^{n-p-1}(B, B_0; M)),$$

where $w \in H^{n+1}(B; A)$ is $(-1)^{n+1}$ times the image of the primary obstruction $\bar{\gamma}^{n+1} \in H^{n+1}(B; Z)$ under the (coefficient group) homomorphism induced by the map $n \to n \cdot 1$ of $\mathbf{Z}$ into A.

Let $j^* : H^{n+1}(\hat{X}, X; A) \to H^{n+1}(\hat{X}; A)$ be the injection, and let $w = \hat{p}^{*-1} j^*(u) \in H^{n+1}(B; A)$. The statement $\gamma(x) = x \frown w$ is a consequence of the commutativity of the diagram

$$\begin{array}{ccc}
H_p(\hat{X}, \hat{X}_0) & \xrightarrow{i_*} & H_p(\hat{X}, \hat{X}_0 \cup X) \\
 & \searrow^{\frown\, \hat{p}^* w} & \downarrow{\frown\, u} \\
\downarrow{\hat{p}_*} & & H_{p-n-1}(\hat{X}, \hat{X}_0) \\
 & & \downarrow{\hat{p}_*} \\
H_p(B, B_0) & \xrightarrow[\frown\, w]{} & H_{p-n-1}(B, B_0)
\end{array}$$

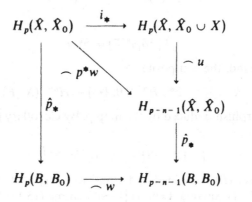

where i_* is the injection (we have omitted mention of the coefficient module, for simplicity's sake). Commutativity of each part of the diagram is a consequence of the naturality properties of the cap product. The proof that $\gamma^*(x) = w \smile x$ is similar.

It remains to prove the last assertion. It will suffice to do so in the special case $A = \mathbf{Z}$; the general case then follows by a simple diagram chase.

Consider the diagram

$$
\text{(5.15)}\qquad
\begin{array}{ccccccc}
 & & H^{q+1}(\hat{F}, F) & \xleftarrow{k_1} & H^{q+1}(\hat{X}, X) & & \\
 & \delta_1 \nearrow & & \nwarrow l_1 \quad h_1 \swarrow & & \searrow j_1 & \\
H^q(F) & & \xrightarrow{\delta} & H^{q+1}(\hat{X}, F) & \xrightarrow{j} & & H^{q+1}(\hat{X}) \\
 & \delta_2 \searrow & & \swarrow l_2 \quad h_2 \nwarrow & & \nearrow j_2 & \\
 & & H^{q+1}(X, F) & \xleftarrow[k_2]{} & H^{q+1}(\hat{X}, \hat{F}) & &
\end{array}
$$

whose maps are injections or the appropriate coboundary operators; the coefficient ring $\mathbf{Z}$ is understood. The homomorphism k_1 is an isomorphism for $q < n$ (because both groups involved are zero) and for $q = n$ (because the fibration is orientable). It follows as in Exercise 2 of Chapter II that k_2 is a monomorphism for $q = n$ (and an isomorphism for $q < n$). Thus the hypotheses of the Hexagonal Lemma fail to be satisfied by the diagram only in that k_2 may not be an epimorphism.

The Thom class $u \in H^{n+1}(\hat{X}, X)$ is determined only up to sign. Choice of such a class determines a generator $\delta_1^{-1} k_1(u)$ of $H^n(F)$; this class in turn determines an isomorphism of $\pi_n(F)$ with $\mathbf{Z}$, and we shall use this isomorphism to identify the coefficient groups in (5.15) with $\pi_n(F)$. It follows that

$$k_1^{-1} \delta_1 \iota^n(F) = u \in H^{n+1}(\hat{X}, X),$$

and therefore

$$j_1 k_1^{-1} \delta_1 \iota^n(F) = \hat{p}^* w.$$

On the other hand, the composite

$$k_2 \circ j_2^{-1} \circ \hat{p}^* : H^{n+1}(B, \{*\}) \to H^{n+1}(X, F)$$

is the homomorphism induced by the map p. By Corollary (6.10) of Chapter VI,

$$(-1)^n \delta_2 \iota^n(F) = k_2 j_2^{-1} \hat{p}^* \bar{\gamma}^{n+1},$$

so that $j_2 k_2^{-1} \delta_2 \iota^n(F)$ is defined and equal to $(-1)^n \hat{p}^* \bar{\gamma}^{n+1}$. The argument used to prove the Hexagonal Lemma [E-S], Lemma (15.1) of Chapter I) then applies and we have

$$(-1)^n \hat{p}^* \bar{\gamma}^{n+1} = -\hat{p}^* w.$$

Since $\hat{p}^*$ is an isomorphism, $\bar{\gamma}^{n+1} = (-1)^{n+1} w$. □

The homomorphism β^* is often referred to as "integration over the fibre." In fact, suppose that X and B are differential manifolds and p a differentiable fibration whose fibre is the n-dimensional sphere $\mathbf{S}^n$. Then the real cohomology of each of these manifolds can be calculated, by the de Rham theorem, from its algebra of differential forms. If ω is a p-form on X, then a process of "partial integration" along each fibre yields a $(p - n)$-form, which is easily seen to have the form $p^*\theta$, where θ is a form on B. And the correspondence $\omega \to \theta$ induces a homomorphism of $H^p(X; \mathbf{R})$ into $H^{p-n}(B; \mathbf{R})$, which can be shown to be our homomorphism β^*.

The sequences (5.13), (5.14) are called the *Gysin sequences* of the fibration p.

The homomorphism β^* has certain multiplicative properties analogous to those of the homomorphism α^* of §3. These, and more sophisticated properties of β^*, have been used by W. S. Massey in his study of the cohomology ring of a sphere-bundle [3].

(5.16) Theorem *If* $x \in H^p(X, X_0; A)$ $y \in H^q(B, B_0; A)$, *then*

$$\beta^*(x \smile p^*y) = \beta^*x \smile y \in H^{p+q-n}(B, B_0; A).$$

For there is a commutative diagram

$$\begin{array}{ccccc}
H^p(\hat{X}, \hat{X}_0) & \xrightarrow{j^*} & H^p(\hat{X}_0 \cup X, \hat{X}_0) & \xrightarrow{\delta^*} & H^{p+1}(\hat{X}, \hat{X}_0 \cup X) \\
\uparrow \hat{p}^* & & \downarrow i^* & & \uparrow u\smile \\
& & & & H^{p-n}(\hat{X}, \hat{X}_0) \\
& & & & \uparrow \hat{p}^* \\
H^p(B, B_0) & \xrightarrow[p^*]{} & H^p(X, X_0) & \xrightarrow[\beta^*]{} & H^{p-n}(B, B_0)
\end{array}$$

We make use of the following fact; if $x_1 \in H^p(\hat{X}_0 \cup X, \hat{X}_0)$, $y_1 \in H^q(\hat{X}, \hat{X}_0)$, then $\delta^*(x_1 \smile j^*y_1) = \delta^*x_1 \smile y_1$. Let $x_1 = i^{*-1}x$, $y_1 = \hat{p}^*y$. Then

$$x \smile p^*y = i^*x_1 \smile i^*j^*\hat{p}^*y = i^*(x_1 \smile j^*y_1),$$

and therefore

$$\begin{aligned} u \smile \hat{p}^*\beta^*(x \smile p^*y) &= u \smile \hat{p}^*\beta^*i^*(x_1 \smile j^*y_1) \\ &= \delta^*(x_1 \smile j^*y_1) = \delta^*x_1 \smile y_1, \end{aligned}$$

while

$$\begin{aligned} u \smile \hat{p}^*(\beta^*x \smile y) &= u \smile \hat{p}^*\beta^*i^*x_1 \smile \hat{p}^*y \\ &= \delta^*x_1 \smile y_1. \end{aligned}$$

Since $\hat{p}^*$ and $u \smile$ are isomorphisms,

$$\beta^*(x \smile p^*y) = \beta^*x \smile y. \qquad \square$$

We conclude this section with two applications of the Gysin sequence. The first one is to the exceptional Lie group $\mathbf{G}_2$ (see Appendix A, §5). There is a fibration $p: \mathbf{G}_2 \to \mathbf{V}_{7,2}$, the fibre being $\mathbf{S}^3$. Since $\mathbf{V}_{7,2}$ is 1-connected ((10.12) of Chapter IV), it follows from (5.1) that the fibration is orientable, so that there is a Gysin sequence

$$H^{q-4}(\mathbf{V}_{7,2}) \xrightarrow{\gamma^*} H^q(\mathbf{V}_{7,2}) \xrightarrow{p^*} H^q(\mathbf{G}_2) \xrightarrow{\beta^*} H^{q-3}(\mathbf{V}_{7,2})$$

for any coefficient domain A. Let us first take $A = \mathbf{Z}$, and recall from (10.14) of Chapter IV that $H_0(\mathbf{V}_{7,2}) \approx H_{11}(\mathbf{V}_{7,2}) \approx \mathbf{Z}$, $H_5(\mathbf{V}_{7,2}) \approx \mathbf{Z}_2$, and therefore

$$H^0(\mathbf{V}_{7,2}) = \mathbf{Z}, \quad \text{generated by the unit element } 1;$$
$$H^{11}(\mathbf{V}_{7,2}) = \mathbf{Z}, \quad \text{generated by a class } z,$$
$$H^6(\mathbf{V}_{7,2}) = \mathbf{Z}_2, \quad \text{generated by a class } h,$$

while all other homology and cohomology groups vanish. From the Gysin sequence we deduce

(5.17) Theorem *The integral cohomology groups $H^q(\mathbf{G}_2; \mathbf{Z})$ are given by*

$$H^0(\mathbf{G}_2) = \mathbf{Z}, \quad \textit{generated by } 1;$$
$$H^3(\mathbf{G}_2) = \mathbf{Z}, \quad \textit{generated by } x,\ \beta^*(x) = 1;$$
$$H^6(\mathbf{G}_2) = \mathbf{Z}_2, \quad \textit{generated by } p^*(h);$$
$$H^9(\mathbf{G}_2) = \mathbf{Z}_2, \quad \textit{generated by } x \smile p^*(h);$$
$$H^{11}(\mathbf{G}_2) = \mathbf{Z}, \quad \textit{generated by } p^*(z);$$
$$H^{14}(\mathbf{G}_2) = \mathbf{Z}, \quad \textit{generated by } x \smile p^*z;$$

while $H^q(\mathbf{G}_2) = 0$ for all other values of q.

Only the calculation of H^{14} needs further argument. In fact, the Wang sequence gives us a generator $z \in H^{14}(G_2)$ such that $\beta^* w = x$; and the element $w = x \smile p^*z$ satisfies this condition, by Theorem (5.16). □

We have not determined the ring structure, having left open the question of whether $x \smile x = p^*h$. We shall resolve this question in Chapter VIII, with the aid of the Steenrod squares.

From our knowledge of the integral homology, or from Theorem (4.9), we have (for $\mathbf{Z}_2$ coefficients):

$$H^0(\mathbf{V}_{7,2}) \approx \mathbf{Z}_2, \quad \text{generated by } 1;$$
$$H^5(\mathbf{V}_{7,2}) \approx \mathbf{Z}_2, \quad \text{generated by } y_5;$$
$$H^6(\mathbf{V}_{7,2}) \approx \mathbf{Z}_2, \quad \text{generated by } y_6;$$
$$H^{11}(\mathbf{V}_{7,2}) \approx \mathbf{Z}_2, \quad \text{generated by } y_5 \smile y_6.$$

Let $\tilde{x}$ be the image of a class x under reduction mod 2. Then

$$\tilde{h} = y_6 = \delta^* y_5,$$

where δ^* is the Bockstein operator: $H^5(X; \mathbf{Z}_2) \to H^6(X; \mathbf{Z}_2)$, and

$$\widetilde{p^* z} = y_5 \smile y_6 .$$

We can now determine the mod 2 groups of $\mathbf{G}_2$.

(5.18) Theorem *The* mod 2 *cohomology groups of* $\mathbf{G}_2$ *are given by*

$$H^0(\mathbf{G}_2) = \mathbf{Z}_2, \quad \textit{generated by } \tilde{1} = 1;$$
$$H^3(\mathbf{G}_2) = \mathbf{Z}_2, \quad \textit{generated by } \tilde{x};$$
$$H^5(\mathbf{G}_2) = \mathbf{Z}_2, \quad \textit{generated by } p^*(y_5);$$
$$H^6(\mathbf{G}_2) = \mathbf{Z}_2, \quad \textit{generated by } p^*(y_6);$$
$$H^8(\mathbf{G}_2) = \mathbf{Z}_2, \quad \textit{generated by } \tilde{x} \smile p^*(y_5);$$
$$H^9(\mathbf{G}_2) = \mathbf{Z}_2, \quad \textit{generated by } \tilde{x} \smile p^*(y_6);$$
$$H^{11}(\mathbf{G}_2) = \mathbf{Z}_2, \quad \textit{generated by } p^*(y_5 \smile y_6);$$
$$H_{14}(\mathbf{G}_2) = \mathbf{Z}_2, \quad \textit{generated by } \tilde{x} \smile p^*(y_5 \smile y_6),$$

while $H^q(\mathbf{G}_2) = 0$ *for all other values of* q.

This follows by examination of the mod 2 Gysin sequence and from Theorem (5.16), as before. □

Again, we have left open the question of determining $\tilde{x}^2$. Note, however, that

(5.19) Corollary *The cohomology ring* $H^*(\mathbf{G}_2; \mathbf{Z}_2)$ *has a simple system of generators* $\tilde{x}$, p^*y_5, p^*y_6. *The elements* $\tilde{x}$, $p^*(y_5)$ *are primitive.* □

Finally, we have

(5.20) Theorem *If* A *is a field of characteristic* p, *where* $p = 0$ *or an odd prime, then* $H^*(\mathbf{G}_2; A)$ *is an exterior algebra on primitive generators* x_3, x_{11}.

For our second application of the Gysin sequence, we shall calculate the mod 2 cohomology ring of the Grassmannian $\mathbf{G}(n)$. Let us recall from §7 of Chapter VI that there is a commutative diagram

(5.21)

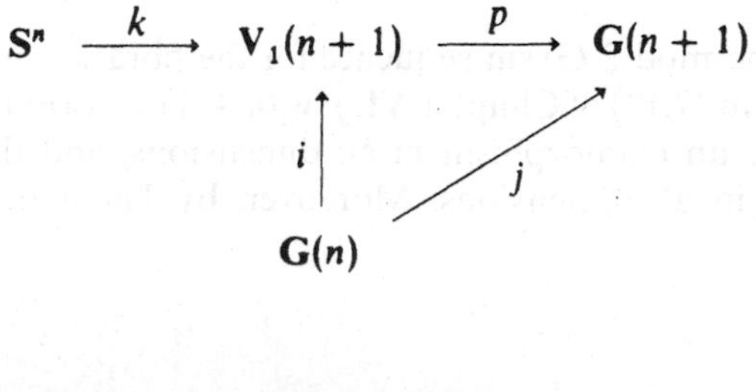

such that p is a fibration, i, j and k are inclusions, i is a homotopy equivalence, and k is the inclusion of the fibre in the total space $\mathbf{V}_1(n+1) = \mathbf{V}(n+1)/\mathbf{O}(n)$. We shall prove

(5.22) Theorem *The* mod 2 *cohomology ring* $H^*(\mathbf{G}(n); \mathbf{Z}_2)$ *is the polynomial ring* $\mathbf{Z}_2[w_1, w_2, \ldots, w_n]$ *in the universal Stiefel–Whitney classes* $w_k = w_k(n)$.

This is proved by induction on n. The space $\mathbf{G}(1)$ is the space of all 1-dimensional subspaces of $\mathbf{R}^\infty$, i.e.,

$$\mathbf{G}(1) = \mathbf{P}^\infty(\mathbf{R}),$$

so that $H^*(\mathbf{G}(1); \mathbf{Z}_2) = \mathbf{Z}_2[w]$ for a certain class w. If we can show that $w_1 \neq 0$, it will follow that $w = w_1$ and Theorem (5.22) is true for $n = 1$.

Let us recall that W_1 belongs to the group

$$Q(\mathbf{Z}_2, \mathbf{Z})/P(\mathbf{Z}_2, \mathbf{Z})$$

of crossed homomorphisms modulo principal homomorphisms, the non-zero element $a \in \mathbf{Z}_2$ operating on $\mathbf{Z}$ by change of sign. For each integer m, there is a unique crossed homomorphism $f_m : \mathbf{Z}_2 \to \mathbf{Z}$, defined by

$$f_m(1) = 0, \qquad f_m(a) = m;$$

and the crossed homomorphism f_m is principal if and only if m is even ($f_{2k}(a) = 2k = k - a \cdot k$). Thus

$$H^1(\mathbf{G}(1), \mathscr{Z}) \approx \mathbf{Z}_2,$$

and W_1 is the non-zero element of this group.

Now reduce mod 2. Then every crossed homomorphism is mapped into an ordinary homomorphism and every principal homomorphism into zero. Thus reduction mod 2 induces an isomorphism

$$H^1(\mathbf{G}(1); \mathscr{Z}) \approx H^1(\mathbf{G}(1); \mathbf{Z}_2)$$

which carries the non-zero element W_1 into w_1. Hence $w_1 \neq 0$.

Remark. Cf. our calculation of $H^q(\mathbf{P}^2(\mathbf{R}); \mathscr{Z})$ in §4 of Chapter VI.

The proof of Theorem (5.22) can now be completed by induction. Assume $H^*(\mathbf{G}(n))$ is as stated, and consider the sequence

$$\cdots \to H^{q-n-1}(\mathbf{G}(n+1)) \xrightarrow{\gamma^*} H^q(\mathbf{G}(n+1)) \xrightarrow{j^*}$$

$$H^q(\mathbf{G}(n)) \xrightarrow{\beta^*} H^{q-n}(\mathbf{G}(n+1)) \xrightarrow{\gamma^*} H^{q+1}(\mathbf{G}(n+1)) \to \cdots$$

obtained from the mod 2 Gysin sequence for the fibration p with the aid of (5.21). By Theorem (7.19) of Chapter VI, $j^*w_i(n+1) = w_i(n)$ $(i = 1, \ldots, n)$. It follows that j^* is an epimorphism in all dimensions, and therefore γ^* is a monomorphism in all dimensions. Moreover, by Theorem (5.12) $\gamma^*(x) =$

$w \smile x$, where $w = w_{n+1}(n+1)$ is the reduction mod 2 of the primary obstruction $\bar{\gamma}^{n+1}$ to a cross-section of p. An easy argument by induction on q now proves the inductive step. □

6 The Homology Sequence of a Fibration

We have seen that there is an exact sequence connecting the homotopy groups of the fibre, the total space, and the base space, of a fibration. This is a consequence of the fact (Theorem (8.5) of Chapter IV) that the projection induces an isomorphism $p_* : \pi_q(X, F) \approx \pi_q(B)$ in all dimensions.

It is easy to see that the corresponding theorem in homology is false. Consider, for example, the Hopf fibration $p : \mathbf{S}^{2n+1} \to \mathbf{P}^n(\mathbf{C})$, with fibre $\mathbf{S}^1$. Then $H_q(\mathbf{S}^{2n+1}, \mathbf{S}^1) = 0$ for $2 < q < 2n+1$, while $H_q(\mathbf{P}^n(\mathbf{C})) \approx \mathbf{Z}$ for all even values of q in that range. Again, consider the trivial fibration $p : B \times F \to B$, for which

$$\begin{aligned} H_r(B \times F, F) &= H_r((B, b_0) \times F) \\ &\approx \bigoplus_{p+q=r} H_p(B, b_0) \otimes H_q(F) \oplus \bigoplus_{p+q=r-1} \operatorname{Tor}(H_p(B, b_0), H_q(F)). \end{aligned}$$

This group has a direct summand isomorphic with $H_r(B, b_0)$, but is, in general, considerably larger. Suppose, however, that B and F are homologically $(m-1)$-connected and $(n-1)$-connected, respectively, i.e., $\tilde{H}_p(B) = 0$ for all $p < m$ and $\tilde{H}_q(F) = 0$ for all $q < n$. Then, if $r < m + n$, the only non-vanishing term in the above direct sum decomposition is $H_r(B, b_0) \otimes H_0(F) \approx H_r(B, b_0)$, and it is not hard to verify that $p_* : H_r(B \times F, F) \to H_r(B, b_0)$ is indeed an isomorphism for all $r < m + n$.

The purpose of this section is to show that the latter phenomenon, in a relative form, generalizes to more or less arbitrary fibrations. We shall prove a result which is just strong enough for the applications to be made in Chapter VIII; for stronger results, the reader is referred to Chapter XIII, Theorem (7.10).

(6.1) Theorem (Serre). *Let $p : X \to B$ be a fibration with fibre F, and suppose that $\tilde{H}_q(F) = 0$ for all $q < n$ $(n \geq 1)$. Let B_0 be a subspace of B, $X_0 = p^{-1}(B_0)$, and suppose that the pair (B, B_0) is $(m-1)$-connected $(m \geq 1)$. Then $p_* : H_r(X, X_0) \to H_r(B, B_0)$ is an isomorphism for $r < m + n$ and an epimorphism for $r = m + n$.*

Because of Theorem (2.6) of Chapter V, we may assume that (B, B_0) is a relative CW-complex having no cells of dimension $\leq m - 1$.

(6.2) Lemma *If p is fibre homotopically trivial, the conclusion of Theorem* (6.1) *holds.*

We may assume that p is trivial; $X = B \times F$, p is the projection on the first factor. The Künneth theorem gives

$$\begin{aligned} &H_r(B \times F, B_0 \times F) \\ &\quad \approx \bigoplus_{p+q=r} H_p(B, B_0) \otimes H_q(F) \oplus \bigoplus_{p+q=r-1} \operatorname{Tor}(H_p(B, B_0), H_q(F)). \end{aligned}$$

If $p < m$, $H_p(B, B_0) = 0$; if $r < m + n$ and $p \geq m$, $p + q = r$, then $q < n$ and so $H_q(F) = 0$ unless $q = 0$. In the same way, all the terms involving the torsion product vanish (even if $q = 0$). Thus

$$H_r(B \times F, B_0 \times F) \approx H_r(B, B_0) \otimes H_0(F) \approx H_r(B, B_0)$$

for all $r < m + n$. The isomorphism $H_r(B, B_0) \to H_r(B \times F, B_0 \times F)$ is given by $u \to u \times 1$, where $1 \in H_0(F)$ is the homology class of a point. As $p_*(u \times 1) = u$, p_* is the inverse of the Künneth isomorphism.

If $r = m + n$, there is an extra term $H_m(B, B_0) \otimes H_n(F)$. However, the fact that $p_*(u \times 1) = u$ shows that p_* is an epimorphism. Thus our result is best possible. □

Now suppose that $B = B_0 \cup_f \mathbf{E}^r$, $r \geq m$, for some map $f: \dot{\mathbf{E}}^r \to B_0$. Let $U = B - B_0$, $V = B - \{x_0\}$ for some $x_0 \in B - B_0$, so that U and V form an open covering of B. Let $U^* = p^{-1}(U)$, $V^* = p^{-1}(V)$, so that U^* and V^* are open sets covering X. Then there is a commutative diagram

$$\begin{array}{ccccc} H_r(X, X_0) & \xrightarrow{i_1} & H_r(X, V^*) & \xleftarrow{i_2} & H_r(U^*, U^* \cap V^*) \\ \downarrow{\scriptstyle p_1} & & \downarrow{\scriptstyle p_2} & & \downarrow{\scriptstyle p_3} \\ H_r(B, B_0) & \xrightarrow[j_1]{} & H_r(B, V) & \xleftarrow[j_2]{} & H_r(U, U \cap V) \end{array}$$

in which the horizontal arrows represent injections and the vertical ones the homomorphisms induced by p. The homomorphisms i_2 and j_2 are isomorphisms, by the Excision Theorem. Now B_0 is a deformation retract of V and therefore X_0 is a deformation retract of V^*; hence i_1 and j_1 are isomorphisms. Therefore, to show p_1 an isomorphism, it suffices to prove that p_3 is. But U is contractible and so the fibration $p \mid U^*$ is fibre homotopically trivial, and our result follows from Lemma (6.2).

Next, suppose that B is a q-cellular extension of B_0, with cells $\{E_\alpha^q\}$, $q \geq m$. By the General Direct Sum Theorem (Theorem (2.7), (Chapter II), the injections

$$H_r(B_0 \cup E_\alpha^q, B_0) \to H_r(B, B_0)$$

$$H_r(X_0 \cup p^{-1}(E_\alpha^q), X_0) \to H_r(X, X_0)$$

represent the groups $H_r(B, B_0)$, $H_r(X, X_0)$ as direct sums, and our result now follows immediately from the case of only one cell.

The general result now follows by induction and the Five-Lemma. For $X_{m-1} = X_0$ and therefore

$$p_* : H_r(X_m, X_0) \to H_r(B_m, B_0)$$

has the required properties, by the previous step. Suppose $q > m$. Then there is a commutative diagram

$$\begin{array}{ccccccccc} H_{r+1}(X_q, X_{q-1}) & \to & H_r(X_{q-1}, X_0) & \to & H_r(X_q, X_0) & \to & H_r(X_q, X_{q-1}) & \to & H_{r-1}(X_{q-1}, X_0) \\ \downarrow p_1 & & \downarrow p_2 & & \downarrow p_3 & & \downarrow p_4 & & \downarrow p_5 \\ H_{r+1}(B_q, B_{q-1}) & \to & H_r(B_{q-1}, B_0) & \to & H_r(B_q, B_0) & \to & H_r(B_q, B_{q-1}) & \to & H_{r-1}(B_{q-1}, B_0) \end{array}$$

The pair (X_q, X_{q-1}) is $(q-1)$-connected, and therefore p_4 is an isomorphism for $r < q + n$ and, in particular, for $r \leq m + n$, while p_1 is an isomorphism for $r < m + n$ and an epimorphism for $r = m + n$. If p_2 is an isomorphism for $r < m + n$ and an epimorphism for $r = m + n$, then p_5 is an isomorphism for $r \leq m + n$. By the Five-Lemma, p_3 has the requisite properties.

Theorem (6.1) now follows from the fact that $p_* : H_r(X, X_0) \to H_r(B, B_0)$ is the direct limit of the homomorphisms $p_* : H_r(X_q, X_0) \to H_r(B_q, B_0)$. □

As a Corollary, we have the *Serre exact sequence*:

(6.3) Corollary *If $p : X \to B$ is a fibration with fibre F, B is $(m-1)$-connected, and $\tilde{H}_q(F) = 0$ for all $q < n$, then there is an exact sequence*

$$H_{m+n-1}(F) \to H_{m+n-1}(X) \to H_{m+n-1}(B) \to \cdots$$

$$\to H_q(F) \xrightarrow{i_*} H_q(X) \xrightarrow{p_*} H_q(B) \xrightarrow{\Delta_*} H_{q-1}(F) \to \cdots \quad \square$$

In particular, we may suppose that $X = \mathbf{P}(B)$ and p is the path fibration: $p(u) = u(1)$ for $u \in X$. Thus $F = \Omega(B)$ is $(m-2)$-connected. By the Hurewicz Theorem, $\tilde{H}_q(F) = 0$ for $q \leq m - 2$. Therefore

(6.4) Corollary *If B is $(m-1)$-connected, then $H_{q-1}(\Omega B) \approx H_q(B)$ for $q \leq 2m - 2$.* □

We can improve Corollary (6.4) a little. For an arbitrary fibration $p : X \to B$, the composite

$$H_q(F) \overset{\partial_*^{-1}}{\rightsquigarrow} H_{q+1}(X, F) \xrightarrow{p_*} H_q(B)$$

is an additive relation (cf. Appendix B), called the *homology suspension*. If $X = \mathbf{P}(B)$, then ∂_* is an isomorphism, and therefore $p_* \circ \partial_*^{-1}$ is a homomorphism $\sigma_*: H_{q-1}(F) \to H_q(B)$. We then have

(6.5) Corollary *If B is $(m-1)$-connected, the suspension $\sigma_*: H_{q-1}(\Omega B) \to H_q(B)$ is an isomorphism for $q \leq 2m-2$ and an epimorphism for $q = 2m-1$.* □

For more detailed results on the homology suspension, the reader is referred to Chapter VIII.

7 The Blakers–Massey Homotopy Excision Theorem

We have seen in §2 of Chapter II that, if $(X; A, B)$ is a triad with $X = A \cup B$, then, under reasonable conditions, the injection $H_q(B, A \cap B) \to H_q(X, A)$ is an isomorphism for all q. On the other hand the corresponding statement for homotopy groups is false, as we have seen in Example 5, §7 of Chapter IV. However, Blakers and Massey [1] have shown that the homotopy excision theorem does hold in a range of dimensions. In this section, we shall show, following Namioka [1], how a slightly weakened, but nevertheless extremely useful, form of the Blakers–Massey theorem follows easily from the results of §6.

Let $(X; A, B)$ be a triad and let m, n be integers ≥ 2. We shall suppose

(7.1) *The spaces X, A, B, and $C = A \cap B$ are 1-connected;*

(7.2) *The pair (X, A) is $(m-1)$-connected and the pair (X, B) is $(n-1)$-connected.*

(7.3) *The injection*

$$H_q(B, C) \to H_q(X, A)$$

is an isomorphism for $q < m + n - 2$ and an epimorphism for $q = m + n - 2$.

We shall prove

(7.4) Theorem (Blakers–Massey). *Under the hypotheses* (7.1)–(7.3), *the injection*

$$\pi_q(B, C) \to \pi_q(X, A)$$

is an isomorphism for $q < m + n - 2$ and an epimorphism for $q = m + n - 2$.

Let P be the space of all paths in X which end in B. The map $\lambda: B \to P$ which sends each point $b \in B$ into the constant map of $\mathbf{I}$ into b is a homotopy equivalence. The map $u \to u(0)$ is a fibration $p: P \to X$, and if F is the fibre of p we have $\pi_q(F) \approx \pi_{q+1}(X, B)$ for all q. In particular, F is $(n-2)$-connected; by the Hurewicz theorem, $\tilde{H}_q(F) = 0$ for $q \leq n-2$. Let $P_0 = p^{-1}(A)$. Then it follows from Theorem (6.1) that

(7.5) *The homomorphism $p_* : H_q(P, P_0) \to H_q(X, A)$ is an isomorphism for $q < m + n - 1$ (and an epimorphism for $q = m + n - 1$).* □

The composite map $p \circ \lambda$ is the inclusion map $j : (B, C) \to (X, A)$. By (7.3), $p_* \circ \lambda_* = j_* : H_q(B, C) \to H_q(X, A)$ is an isomorphism for $g < m + n - 2$, and an epimorphism for $q = m + n - 2$. Hence

(7.6) *The homomorphism $\lambda_* : H_q(B, C) \to H_q(P, P_0)$ is an isomorphism for $q < m + n - 2$ and an epimorphism for $q = m + n - 2$.* □

Since $\lambda : B \to P$ is a homotopy equivalence, the homomorphism $\lambda_* : H_q(B) \to H_q(P)$ is an isomorphism for all q. From the Five-Lemma we deduce

(7.7) *The homomorphism $(\lambda | C)_* : H_q(C) \to H_q(P_0)$ is an isomorphism for $q < m + n - 3$ and an epimorphism for $q = m + n - 3$.* □

At this point we should like to apply the Whitehead Theorem. Since C is 1-connected, by hypothesis, it suffices to show that P_0 is 1-connected. Since $1 \leq m + n - 3$ and $H_1(C) = 0$, it follows that $H_1(P_0) = 0$. There is a commutative diagram

$$\begin{array}{ccccc}
 & & \pi_2(P, P_0) & \xrightarrow{\partial_*} & \pi_1(P_0) & \longrightarrow & \pi_1(P) \\
 & & \downarrow{\scriptstyle p_*} & & \downarrow & & \\
\pi_2(X) & \longrightarrow & \pi_2(X, A) & \longrightarrow & \pi_1(A) & &
\end{array}$$

Since A is 1-connected, $\pi_2(X, A)$ is a quotient of the abelian group $\pi_2(X)$ and therefore itself abelian. But $p_* : \pi_2(P, P_0) \approx \pi_2(X, A)$, so that $\pi_2(P, P_0)$ is abelian. As $\lambda : B \to P$ is a homotopy equivalence, $\pi_1(P) \approx \pi_1(B) = 0$. Hence ∂_* is an epimorphism and therefore $\pi_1(P_0)$ is abelian. Since $H_1(P_0) = 0$, it follows from the Poincaré Theorem that $\pi_1(P_0) = 0$. Therefore

(7.8) *The homomorphism $(\lambda | C)_* : \pi_q(C) \to \pi_q(P_0)$ is an isomorphism for $q < m + n - 3$ and an epimorphism for $q = m + n - 3$.* □

Using the Five-Lemma and the fact that $\lambda_* : \pi_q(B) \approx \pi_q(P)$, we deduce

(7.9) *The homomorphism $\lambda_* : \pi_q(B, C) \to \pi_q(P, P_0)$ is an isomorphism for $q < m + n - 2$ and an epimorphism for $q = m + n - 2$.* □

But $p_* : \pi_q(P, P_0) \to \pi_q(X, A)$ is an isomorphism for all q. Since $p_* \circ \lambda_*$ is the injection $\pi_q(B, C) \to \pi_q(X, A)$, our result follows from (7.9). □

We now make some remarks on the hypotheses. In the first place, it follows from (7.1) that each of the pairs (A, C), (B, C), (X, A), and (X, B) is 1-connected; moreover, the first non-vanishing homotopy group of each of the above pairs is, by the Relative Hurewicz Theorem, isomorphic with its first non-vanishing homology group. Thus the condition

(7.2′) *The pair (B, C) is $(m - 1)$-connected and the pair (A, C) is $(n - 1)$-connected*

is equivalent, in the presence of (7.1) and (7.3), to (7.2).

We shall say that the triad $(X; A, B)$ is *r-connected* $(r \geq 1)$ if and only if the injection

$$\pi_q(B, C) \to \pi_q(X, A)$$

is an isomorphism for $q < r$ and an epimorphism for $q = r$. (This is equivalent to the analogous condition with A and B interchanged; cf. Exercise 2, Chapter II). Thus the Blakers–Massey Theorem can be formulated as follows:

(7.10) Theorem *If $(X; A, B)$ is a triad satisfying* (7.1) *and* (7.3), *as well as either* (7.2) *or* (7.2′), *then $(X; A, B)$ is $(m + n - 2)$-connected.* □

Instead of (7.3) we may consider

(7.3′) *The triad $(X; A, B)$ is an* NDR-*triad.*

By Theorem (2.2) of Chapter II, this condition implies that the triad $(X; A, B)$ is proper, and therefore (7.3) holds in all dimensions. Hence

(7.11) Theorem *If $(X; A, B)$ satisfies* (7.1), (7.2) *or* (7.2′), *and* (7.3′), *then $(X; A, B)$ is $(m + n - 2)$-connected.* □

An important special case of the Blakers–Massey Theorem is the following:

(7.12) Theorem *Let (X, A) be an $(m - 1)$-connected* NDR-*pair, and suppose that A is $(n - 1)$-connected $(m \geq 2, n \geq 2)$. Let $p : (X, A) \to (X/A, *)$ be the collapsing map. Then*

$$p_* : \pi_q(X, A) \to \pi_q(X/A)$$

is an isomorphism for $q < m + n - 1$ and an epimorphism for $q = m + n - 1$.

For the triad $(X \cup \mathbf{T}A; X, \mathbf{T}A)$ satisfies (7.3′) (here $\mathbf{T}A$ is the unreduced cone, i.e., A is treated as a free space). Since $n \geq 2$, A is 1-connected; since $m \geq 2$, X is 1-connected, while $\mathbf{T}A$ is contractible. By the van Kampen Theorem, $X \cup \mathbf{T}A$ is 1-connected. Finally $(\mathbf{T}A, A)$ is n-connected. By Theorem (7.11), the triad $(X \cup \mathbf{T}A; X, \mathbf{T}A)$ is $(m + n - 1)$-connected, and therefore the injection

$$\pi_q(X, A) \to \pi_q(X \cup \mathbf{T}A, \mathbf{T}A)$$

is an isomorphism for $q < m + n - 1$ and an epimorphism for $q = m + n - 1$. But, by Corollary (5.13) of Chapter I, the homomorphism

$$p_1 : \pi_q(X \cup \mathbf{T}A, \mathbf{T}A) \to \pi_q(X/A, *)$$

induced by p is an isomorphism for all q, and p_* is the composite of p_1 with the injection; therefore p_* has the desired properties.

Consider the special case $X = \mathbf{T}A$; then $(\mathbf{T}A, A)$ is n-connected, so that we may take $m = n + 1$. Then $\mathbf{T}A/A$ is the suspension of A, and the boundary operator $\partial_* : \pi_{q+1}(\mathbf{T}A, A) \to \pi_q(A)$ is an isomorphism for all q. Moreover, the composite $E = p_* \circ \partial_*^{-1} : \pi_q(A) \to \pi_{q+1}(\mathbf{S}A)$ is easily seen to be the homomorphism

$$\pi_q(A) = [\mathbf{S}^q, A] \to [\mathbf{S}^{q+1}, \mathbf{S}A] = \pi_{q+1}(\mathbf{S}A)$$

induced by the suspension functor. From Theorem (7.12) we deduce

(7.13) Theorem (Freudenthal Suspension Theorem). *If A is $(n - 1)$-connected $(n \geq 2)$, then $E : \pi_q(A) \to \pi_{q+1}(\mathbf{S}A)$ is an isomorphism for $q < 2n - 1$ and an epimorphism for $q = 2n - 1$.* □

Another important consequence of the Blakers–Massey theorem is

(7.14) Theorem (Homotopy Map Excision Theorem). *Let (X, A), (Y, B) be NDR-pairs, $f: (X, A) \to (Y, B)$ a map such that $f_* : H_q(X, A) \approx H_q(Y, B)$ for all q. Suppose that X, A, and B are 1-connected, (X, A) is m-connected, and $f \mid A : A \to B$ is n-connected. Then $f_* : \pi_q(X, A) \to \pi_q(Y, B)$ is an isomorphism for $q < m + n$ and an epimorphism for $q = m + n$.*

For let Z be the mapping cylinder of f, C the mapping cylinder of $f \mid A : A \to B$. There are commutative diagrams

$$\begin{array}{ccc} \pi_q(X, A) & \xrightarrow{\pi_q(i)} & \pi_q(X \cup C, C) \\ {\scriptstyle \pi_q(f)}\downarrow & & \downarrow{\scriptstyle \pi_q(j)} \\ \pi_q(Y, B) & \xrightarrow[\pi_q(k)]{} & \pi_q(Z, C) \end{array} \qquad \begin{array}{ccc} H_q(X, A) & \xrightarrow{H_q(i)} & H_q(X \cup C, C) \\ {\scriptstyle H_q(f)}\downarrow & & \downarrow{\scriptstyle H_q(j)} \\ H_q(Y, B) & \xrightarrow[H_q(k)]{} & H_q(Z, C) \end{array}$$

where i, j, and k are inclusion maps. Since $H_q(f)$, $H_q(i)$ and $H_q(k)$ are isomorphisms for all q, so is $H_q(j)$; by exactness of the homology sequence of the triple $(Z, X \cup C, C)$, the groups $H_q(Z, X \cup C)$ are zero for all q. But X and C are 1-connected and their intersection A is 1-connected; from the van Kampen and Hurewicz Theorems we deduce that $\pi_q(Z, X \cup C) = 0$, and therefore $\pi_q(j)$ is an isomorphism, for all q. On the other hand, we can apply the Blakers–Massey Theorem to deduce that the triad $(X \cup C; X, C)$ is $(m + n)$-connected, and therefore $\pi_q(i)$ is an isomorphism for $q < m + n$ and an epimorphism for $q = m + n$. Since $\pi_q(k)$ is an isomorphism, $\pi_q(f)$ has the desired properties. □

Exercises

1. Let W^n be the space of *free* loops in $\mathbf{S}^n$. Calculate the homology groups of W^n for n odd. What can you say when n is even?

2. Let $p : \mathbf{S}^7 \to \mathbf{S}^4$ be the Hopf fibration, $f = \mathbf{S}^2 \times \mathbf{S}^2 \to \mathbf{S}^4$ a map of degree 1, $q : X \to \mathbf{S}^2 \times \mathbf{S}^2$ the induced fibration, $q' = p_2 \circ q : X \to \mathbf{S}^2$.

   (a) Calculate the homology groups of X;
   (b) Show that q' has a cross-section, but is not fibre-homotopically trivial.

3. Let $M(\mathbf{Z}_d, n) = \mathbf{S}^n \cup_f \mathbf{E}^{n+1}$, where $f: \mathbf{S}^n \to \mathbf{S}^n$ has degree d; $M = M(\mathbf{Z}_d, n)$ is called a *Moore space* of type $(\mathbf{Z}_d, n)$. Calculate the homology groups $H_*(\Omega M; \mathbf{Z}_d)$. (Hint: show that $H_m(\Omega M; \mathbf{Z}_d)$ is a direct sum of f_m copies of $\mathbf{Z}_d$, and find a recursion formula for f_m).

4. Show that the maps g_n, $J(p_n) \circ g : J(W) \to J(W^{(n)})$ coincide.

5. Show that, if W is a CW-complex, then $J(W)$ is a CW-complex and each of the spaces $J_m(W)$ is a subcomplex of $J_{m+1}(W)$ as well as of $J(W)$.

6. Show that $H_*(\mathbf{O}_{n+1}^+; \mathbf{Z}_2)$ has a system of generators $y_1, \ldots, y_n$ such that

$$\Delta_* y_r = \sum_{p+q=r} y_p \times y_q$$

   (where $y_0 = 1$). The elements

$$y_{2s+1} + \sum_{j=1}^{s} y_j y_{2s-j+1}$$

   form a basis for the space M_n of primitive elements.

CHAPTER VIII

The Homology Suspension

The homology groups of the space $\Omega(B)$ of loops in a space B are related to those of B by homomorphism $\sigma_* : H_q(\Omega B) \to H_{q+1}(B)$, called the *homology suspension*. Let $\mathbf{P}'(B)$ be the space of paths in B which end at the base point; then the map $p : \mathbf{P}'(B) \to B$ defined by $p(u) = u(0)$ is a fibration with ΩB as fibre. The total space $\mathbf{P}'(B)$ being acyclic, the boundary operator

$$\partial_* : H_{q+1}(\mathbf{P}'(B), \Omega B) \to H_q(\Omega B)$$

is an isomorphism, and the map p induces

$$p_* : H_{q+1}(\mathbf{P}'(B), \Omega B) \to H_{q+1}(B);$$

the homomorphism σ_* is the composite $p_* \circ \partial_*^{-1}$.

Suppose that B is n-connected $(n \geq 1)$; then ΩB is $(n-1)$-connected, and we saw in Corollary (6.5) of Chapter VII that σ_* is an isomorphism for $q \leq 2n-1$ and an epimorphism for $q = 2n$.

In §§1, 2 we study the behavior of the homology suspension in the range $q \leq 3n$. This is accomplished by the following device. Let $E = \mathbf{F}(B)$ be the space of all free paths in B, and let $X = E_0 \cup E_1$ be the set of all paths which either start or end at the base point. Then E_0 and E_1 are contractible, $E_0 \cap E_1 = \Omega B$, and therefore the boundary operator

$$\Delta_* : H_{q+1}(X) \to H_q(\Omega B)$$

of the Mayer–Vietoris sequence of the triad $(X; E_0, E_1)$ is an isomorphism. Moreover, E has the same homotopy type as B, and the composite

$$H_q(\Omega B) \xrightarrow{\Delta_*^{-1}} H_{q+1}(X) \longrightarrow H_{q+1}(E)$$

of Δ_*^{-1} with the injection is equivalent to the homology suspension σ_*. Thus we can obtain results on σ_* by studying the relative homology groups

$H_{q+1}(E, X)$. We prove that, if $q \leq 3n$, there are isomorphisms

$$H_{q-1}(\Omega B \times \Omega B, \Omega B \vee \Omega B) \approx H_{q+1}(E, X) \approx H_{q+1}(B \times B, B \vee B).$$

Moreover the injection $H_{q+1}(E) \to H_{q+1}(E, X)$ corresponds, under the second isomorphism, to the homomorphism

$$d_* : H_{q+1}(B) \to H_{q+1}(B \times B, B \vee B),$$

induced by the diagonal map of B, while the boundary operator $H_{q+1}(E, X) \to H_q(X)$ corresponds, under the first isomorphism, to the homomorphism

$$\tau_* : H_{q-1}(\Omega B \times \Omega B, \Omega B \vee \Omega B) \to H_{q-1}(\Omega B)$$

induced by the product in the H-space ΩB. Dual results are obtained for cohomology.

This calculation of the relative groups was made by the present author [7] in 1955. Some of its consequences are exploited in §§3, 4. Perhaps the most important is related to cohomology operations. If

$$\theta : H^n(\quad; \Pi) \to H^q(\quad; G)$$

is a cohomology operation, the composite

$$H^{n-1}(X; \Pi) \xrightarrow{\Delta^*} H^n(SX; \Pi) \xrightarrow{\theta(X)} H^q(SX; G) \xrightarrow{\Delta^{*-1}} H^{q-1}(X; G)$$

(Δ^* is the coboundary operator of the Mayer–Vietoris sequence of the triad $(\mathbf{S}X; \mathbf{T}_+ X, \mathbf{T}_- X)$) is again a cohomology operation $\sigma^*\theta$, called the suspension of θ. If θ corresponds to the element $u \in H^q(\Pi, n; G)$, then $\sigma^*\theta$ corresponds to the image of u under the cohomology suspension σ^*, dual to the homology suspension described above. Now the operation $\sigma^*\theta$ is always additive, but not every additive operation has the form $\sigma^*\theta$. However, it follows from the dual of the homology suspension theorem that if the above operation θ is additive and $q \leq 3n - 1$, then θ does have the form $\sigma^*\phi$ for some operation ϕ.

Let Π, G be abelian groups, k a non-negative integer. Then there is a sequence

$$\cdots \leftarrow H^{k+n}(\Pi, n; G) \xleftarrow{\sigma^*} H^{k+n+1}(\Pi, n+1; G) \leftarrow \cdots$$

and each element of the inverse limit

$$A^k(\Pi; G) = \varprojlim_n H^{k+n}(\Pi, n; G)$$

determines a sequence of operations θ_n. Such a sequence is called a *stable operation* of type $(k; \Pi, G)$; thus the stable operations of the said type are in one-to-one correspondence with $A^k(\Pi; G)$. In fact, the homomorphism σ^* is an isomorphism for all sufficiently large n, so that $A^k(\Pi, G) \approx H^{k+n}(\Pi, n; G)$

whenever $n \geq k + 1$. Stable operations are discussed in §5, and a particular sequence of stable operations, the Steenrod squares, are introduced in §6. These were constructed in 1947 by Steenrod [2] as generalizations of the cup square and provided the essential tool for his proof of the classification theorem for maps of an $(n + 1)$-dimensional complex into $\mathbf{S}^n$.

One application of the Steenrod squares is to the Hopf fibre maps $\alpha \in \pi_{2n-1}(\mathbf{S}^n)$ $(n = 2, 4, 8)$. The mapping cone of α is the projective plane over $\mathbf{C}$, $\mathbf{Q}$, or $\mathbf{K}$, as the case may be, and the fact that the cup square is non-zero in $\mathbf{T}_\alpha$ implies that $\alpha \neq 0$. But more is true; the operation Sq^n is non-zero in the mapping cone of the k-fold suspension $E^k\alpha$, and therefore the latter element, too, is non-zero (in other words, the element $E^k\alpha$ is *detectable by* Sq^n). Thus we obtain non-zero stable elements η, ν, σ in the stable homotopy groups of spheres.

The Steenrod squares have additional properties, which makes them even more powerful tools. In §7 we prove the product formula, due to Henri Cartan [1], for the effect of the Sq^i on a cup product. With its aid one can calculate the Sq^i in the cohomology of the Stiefel manifolds. As a consequence Steenrod and J. H. C. Whitehead [1] proved that the n-sphere $\mathbf{S}^n$ is not parallelizable unless $n + 1$ is a power of 2.

As the Steenrod operations Sq^i are stable, they can be composed to yield iterated squares. Relations among these were obtained by Adem [1]. At this point we have not developed enough machinery to prove these results. However, in §8 we derive a few of them in order to illustrate their use. On the one hand, one can prove non-existence theorems, e.g., no element of $\pi_{d+n-1}(\mathbf{S}^n)$ can be detected by Sq^d unless d is a power of 2. On the other hand, one can prove existence theorems, e.g., the stable elements $\eta \circ \eta$, $\nu \circ \nu$, $\sigma \circ \sigma$, $\eta \circ \sigma$ are all non-zero.

1 The Homology Suspension

In this Chapter, homology and cohomology groups will have coefficients in a module M over a principal ideal domain Λ. Usually the coefficient module will be understood. Let us recall that, if B is a space with non-degenerate base point $*$, then the map $p : \mathbf{P}'(B) \to B$ defined by $p(u) = u(0)$ is a fibration, with fibre ΩB. Since $\mathbf{P}'(B)$ is contractible, the homomorphism $\partial_* : H_{q+1}(\mathbf{P}'(B), \Omega(B)) \to H_q(\Omega(B))$ is an isomorphism. On the other hand, the map p induces a homomorphism $p_* : H_{q+1}(\mathbf{P}'(B), \Omega(B)) \to H_{q+1}(B)$. The composite

$$\sigma_* = p_* \circ \partial_*^{-1} : H_q(\Omega(B)) \to H_{q+1}(B)$$

is a homomorphism, the *homology suspension*.

If we replace homology by homotopy, the map p_* is also an isomorphism, and therefore the homomorphism analogous to σ_* is an isomorphism. But we have seen that p_* is not, in general, an isomorphism in homology, and so neither is σ_*. In this chapter, we shall assume that B is n-connected, and see what can be said about σ_* for a certain range of dimensions.

It will be convenient to look at σ_* in a slightly different context. Let $E = \mathbf{F}(B)$, the space of all free paths in B, and let

$$E_0 = \mathbf{P}(B) = \{u \in E \mid u(0) = *\},$$
$$E_1 = \mathbf{P}'(B) = \{u \in E \mid u(1) = *\},$$
$$X = E_0 \cup E_1;$$

note that $F = \Omega(B) = E_0 \cap E_1$. The map $P_i : E \to B$ given by $P_i(u) = u(i)$ ($i = 0, 1$) is a fibration by Corollary (7.9) of Chapter I; the fibre of P_i is E_i, and P_i is a homotopy equivalence. The two maps P_0 and P_1 are homotopic; a homotopy between them is defined by

$$P_t(u) = u(t)$$

for all $t \in I$. The restrictions $P_0' = P_0 \mid E_1 : E_1 \to B$ and $P_1' = P_1 \mid E_0 : E_0 \to B$ are also fibrations with the same fibre F. Furthermore, there are fibrations $p : E \times E \to B \times B$, $p' : E \to B \times B$, defined by

$$p(u, v) = (u(0), v(1)),$$
$$p'(u) = (u(0), u(1));$$

their fibres are $E_0 \times E_1$ and F, respectively. Note that $p' = p \circ d$, where $d : E \to E \times E$ is the diagonal map, and that $X = p'^{-1}(B \vee B)$.

The pair $(B, *)$ being an NDR-pair, it follows from (2.13) of Chapter III and from Theorem (7.14) of Chapter I that

(1.1) *The pairs (F, e), (E_0, F), and (E_1, F) are* NDR-*pairs.* □

Therefore, by (5.2) of Chapter I,

(1.2) *The pairs*

$$(F \times F, F \vee F) = (F, e) \times (F, e),$$
$$(E_1 \times F, F \times F \cup E_1 \times e) = (E_1, F) \times (F, e),$$
$$(E_1 \times F, F \times F) = (E_1, F) \times (F, \varnothing)$$

are NDR-*pairs.* □

Let us now introduce the *Main Diagram* (Figure 1), which will exhibit the relationships among the homology groups of many of the pairs of subspaces

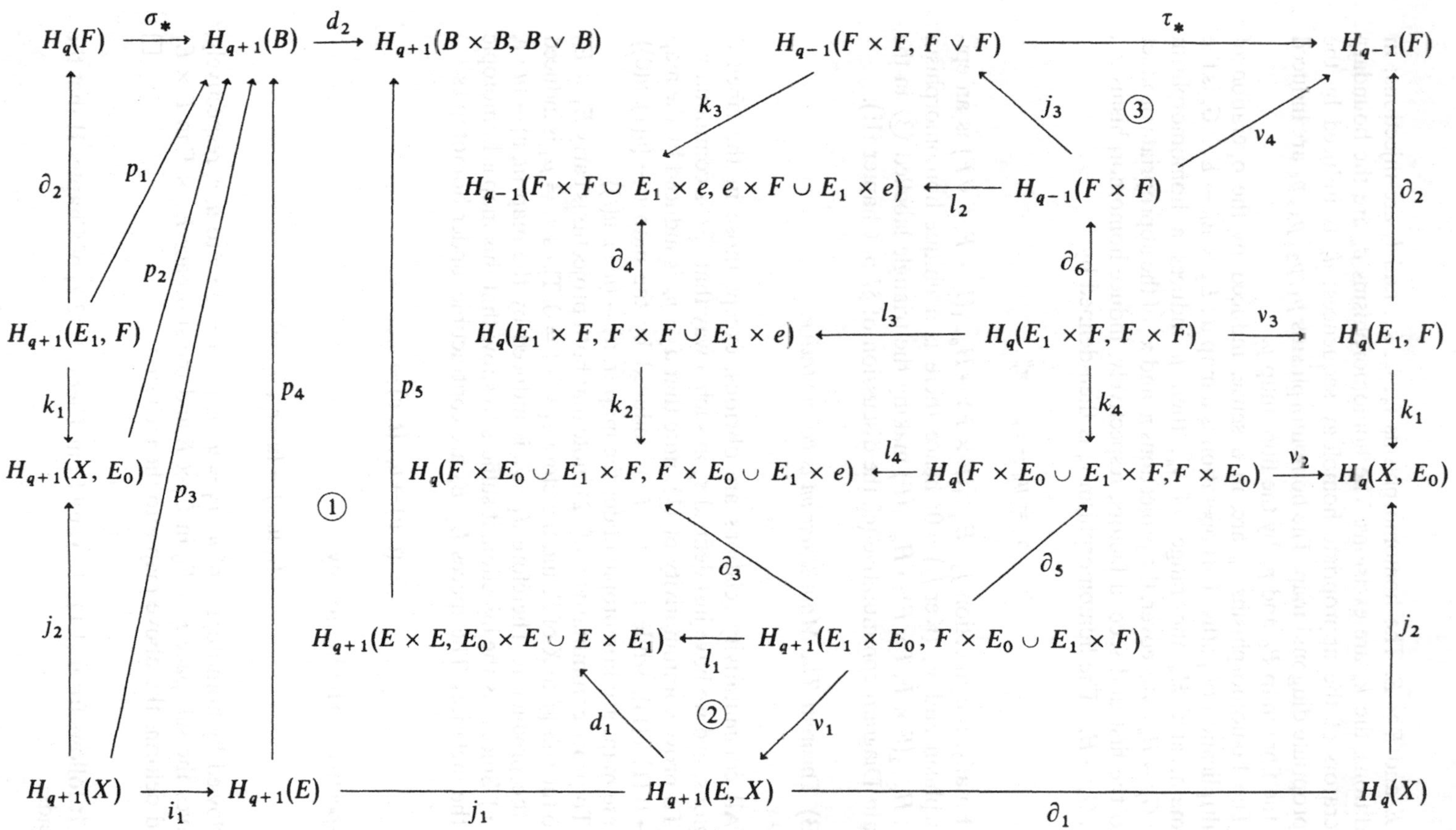

Figure 8.1 (Main diagram)

of E and $E \times E$. The homomorphisms i_n, j_n, k_n, and l_n are injections; in additions, the k_n are excisions. The homomorphisms ∂_n are the boundary operators of the appropriate homology sequences; d_n is induced by the appropriate diagonal map. The homomorphisms p_1, p_2, p_3, p_4 are induced by the fibre map P_0 and p_5 by the fibre map p.

The homomorphisms v_n are, in a sense, induced by the operation of multiplication of paths. This operation is a map $\mu : E_1 \times E_0 \to E$; if G_n is the domain, and H_n the range, of v_n, then μ induces a homomorphism $\mu_n : G_n \to H_n$. Moreover, the projections π' and π'' of the appropriate product into the first and second factors, respectively, induce homomorphisms π'_n, $\pi''_n : G_n \to H_n$. The homomorphism v_n is then defined by

$$v_n = \mu_n - \pi'_n - \pi''_n.$$

Finally, the injection $j_3 : H_{q-1}(F \times F) \to H_{q-1}(F \times F, F \vee F)$ is an epimorphism and v_4 (Ker $j_3) = 0$; hence there is a unique homomorphism $\tau_* : H_{q-1}(F \times F, F \vee F) \to H_{q-1}(F)$ making the triangle labelled ③ in the Main Diagram commutative (cf. the discussion of §7 of Chapter III).

(1.3) Theorem *The Main Diagram is commutative.*

All commutativity relations are obvious, except those in the labelled regions, and we have just defined τ_* in such a way that ③ is commutative.

To prove commutativity of ①, note that $d_2 \circ p_4$ is induced by the map $u \to (u(1), u(1))$, while $p_5 \circ d_1 \circ j_1$ is induced by the map $u \to (u(1), u(0))$. These maps are homotopic under the map $(t, u) \to (u(1), u(t))$.

To prove commutativity of ②, note that both projections carry $E_1 \times E_0$ into the subspace X of E, and therefore $\pi'_1 = \pi''_1 = 0$. Thus $v_1 = \mu_1$ is induced by the product μ. Therefore $d_1 \circ v_1$ is induced by the map $(u, v) \to (u \cdot v, u \cdot v)$. Since l_1 is the injection, it suffices to show that this map is homotopic to the inclusion. The spaces E_1, E_0 are contractible, under homotopies

$$(t, u) \to u_t, \ (t, v) \to v_t,$$

respectively. The homotopy

$$(t, u, v) \to (u \cdot v_t, u_t \cdot v),$$

followed by homotopies of $u \cdot v_1 = u \cdot e$, $u_1 \cdot v = e \cdot v$ to u, v, respectively, leaves the subspace $F \times E_0$ in $E_0 \times E$ and the subspace $E_1 \times F$ in $E \times E_1$ and deforms the above map into the inclusion. □

It follows from (1.1), (1.2) and from Theorem (2.2) of Chapter II that the triads

$$(X; E_0, E_1),$$
$$(F \times E_0 \cup E_1 \times F; E_1 \times F, F \times E_0 \cup E_1 \times e),$$
$$(F \times F \cup E_1 \times e; F \times F, e \times F \cup E_1 \times e),$$
$$(F \times E_0 \cup E_1 \times F; E_1 \times F, F \times E_0)$$

are proper, so that the excisions k_n are isomorphisms ($n = 1, 2, 3, 4$).

Since E_0 and E_1 are contractible, j_2 and ∂_2 are isomorphisms. Hence the homomorphism $\Delta_* = \partial_2 \circ k_1^{-1} \circ j_2$ is an isomorphism; it is, of course, the boundary operator of the Mayer–Vietoris sequence of the triad $(X; E_1, E_0)$. The fibre map $P_0 : E \to B$ is a homotopy equivalence, and hence its induced homomorphism $p_4 : H_{q+1}(E) \to H_{q+1}(B)$ is an isomorphism. Therefore the commutativity of the Main Diagram implies

(1.4) *The diagram*

$$\begin{array}{ccc} H_q(F) & \xrightarrow{\sigma_*} & H_{q+1}(B) \\ \uparrow{\scriptstyle \Delta_*} & & \uparrow{\scriptstyle p_4} \\ H_{q+1}(X) & \xrightarrow[i_1]{} & H_{q+1}(E) \end{array}$$

is commutative, so that σ_* is equivalent to the injection i_1.

Thus it behooves us to study the groups $H_{q+1}(E, X)$. On the one hand, the homomorphism $p_6 = p_5 \circ d_1$ is induced by the fibre map p'. The pairs (E_0, e) and (E_1, e) are DR-pairs; by (5.2) of Chapter I, the pairs

$$(E_1 \times E_0, F \times E_0 \cup E_1 \times e) = (E_1, F) \times (E_0, e)$$

and $(E_1 \times F, e \times F \cup E_1 \times e) = (E_1, e) \times (F, e)$ are DR-pairs and therefore have trivial homology. It follows from the exactness of the homology sequences of the appropriate triples that ∂_3 and ∂_4 are isomorphisms. Thus there is a homomorphism $\beta = v_1 \circ \partial_3^{-1} \circ k_2 \circ \partial_4^{-1} \circ k_3 : H_{q-1}(F \times F, F \vee F) \to H_{q+1}(E, X)$.

We can now formulate

(1.5) Theorem (Homology Suspension Theorem). *The diagram of Figure* 8.2 *is commutative, and the homomorphisms Δ_* and p_4 are isomorphisms. If B is n-connected ($n \geq 1$), then p_6 and β are isomorphisms for $q \leq 3n$.*

There is a dual result for cohomology.

(1.5*) Theorem *The diagram of Figure* 8.3 *is commutative, and the homomorphisms p_4^* and Δ^* are isomorphisms. If B is n-connected ($n \geq 1$), then p_6^* and β^* are isomorphisms for $q \leq 3n$.*

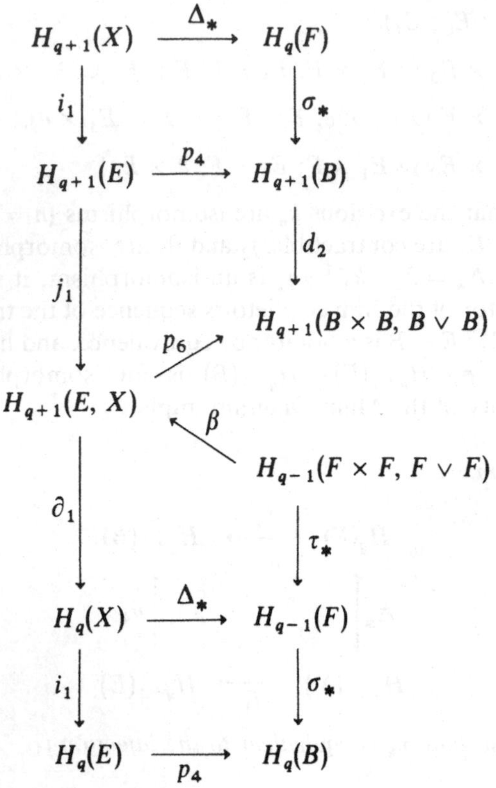

Figure 8.2

These results can be restated as follows:

(1.6) Corollary *If B is n-connected there are exact sequences*

$$H_{3n}(F) \xrightarrow{\sigma_*} H_{3n+1}(B) \to \cdots$$

$$\to H_q(F) \xrightarrow{\sigma_*} H_{q+1}(B) \to G_q \to H_{q-1}(F) \xrightarrow{\sigma_*} H_q(B) \to \cdots$$

$$H^{3n}(F) \xleftarrow{\sigma^*} H^{3n+1}(B) \leftarrow \cdots$$

$$\leftarrow H^q(F) \xleftarrow{\sigma^*} H^{q+1}(B) \leftarrow G^q \leftarrow H^{q-1}(F) \xleftarrow{\sigma^*} H^q(B) \leftarrow \cdots$$

where

$$G_q \approx H_{q-1}(F \wedge F) \approx H_{q+1}(B \wedge B),$$
$$G^q \approx H^{q-1}(F \wedge F) \approx H^{q+1}(B \wedge B).$$

(See Exercise 2 for a slightly improved result).

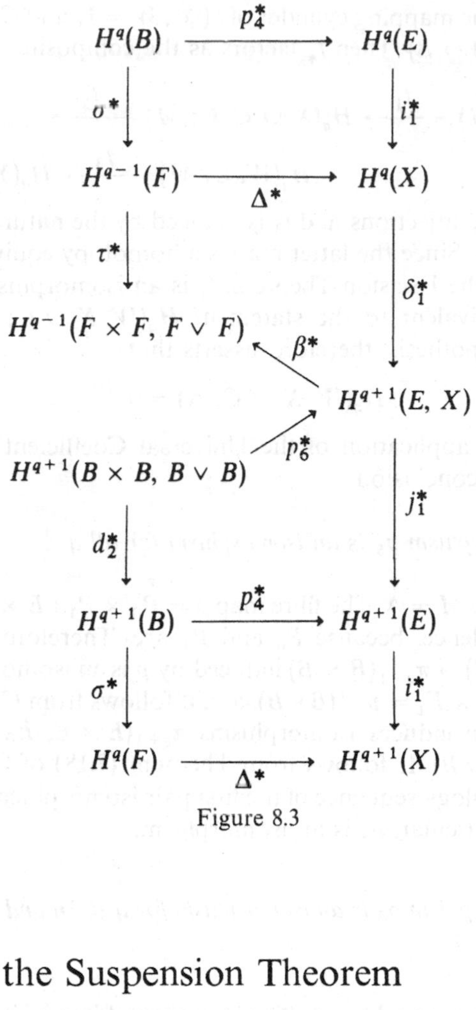

Figure 8.3

2 Proof of the Suspension Theorem

We shall prove Theorem (1.5); the proof is easily dualized to obtain Theorem (1.5*).

The commutativity of the diagram in Figure 8.2 is a consequence of that of the Main Diagram and of the relevant definitions. And we have already observed that Δ_* and p_4 are isomorphisms. It remains to prove that p_6 and β have the stated properties.

We first prove a result which allows us to reduce the question to the case when the coefficient group is Λ.

(2.1) Theorem *Let $f: (X, A) \to (Y, B)$, and suppose that $f_* : H_q(X, A; \Lambda) \to H_q(Y, B; \Lambda)$ is an isomorphism for all $q \le r$ and an epimorphism for $q = r + 1$. Then $f_* : H_q(X, A; M) \to H_q(Y, B; M)$ has the same properties for any coefficient module M.*

Let (W, C) be the mapping cylinder of f (i.e., $W = \mathbf{I}_f$ and C is the mapping cylinder of $f \mid A : A \to B$). Then f_* factors as the composite

$$H_q(X, A; M) \xrightarrow{f_1} H_q(X \cup C, C; M) \xrightarrow{f_2}$$
$$H_q(W, C; M) \xrightarrow{f_3} H_q(Y, B; M)$$

where f_1 and f_2 are injections and f_3 is induced by the natural projection of (W, C) into (Y, B). Since the latter map is a homotopy equivalence, f_3 is an isomorphism. By the Excision Theorem, f_1 is an isomorphism. The desired conclusion is equivalent to the statement $H_q(W, X \cup C; M) = 0$ for all $q \leq r + 1$. Our hypothesis, therefore, asserts that

$$H_q(W, X \cup C; \Lambda) = 0$$

for $q \leq r + 1$. An application of the Universal Coefficient Theorem then yields the desired conclusion. □

(2.2) *The homomorphism p_5 is an isomorphism for all q.*

We may assume $M = \Lambda$. The fibre map $p = P_0 \times P_1 : E \times E \to B \times B$ is a homotopy equivalence, because P_0 and P_1 are. Therefore the homomorphism $\pi_{q+1}(E \times E) \to \pi_{q+1}(B \times B)$ induced by p is an isomorphism for all q. Now $E_0 \times E \cup E \times E_1 = p^{-1}(B \vee B)$, and it follows from Corollary (8.8) of Chapter IV that p induces isomorphisms $\pi_{q+1}(E \times E, E_0 \times E \cup E \times E_1) \approx \pi_{q+1}(B \times B, B \vee B)$. It follows from Theorem (7.18) of Chapter IV that p_* maps the homology sequence of the first pair isomorphically upon that of the second; in particular, p_5 is an isomorphism. □

(2.3) *The homomorphism p_6 is an isomorphism for $q \leq 3n$ and an epimorphism for $q = 3n + 1$.*

Again, we may assume $M = \Lambda$. The homomorphism p_6 is induced by the fibre map p', and the fibre of p' is F. Since B is n-connected, $F = \Omega B$ is $(n - 1)$-connected. Moreover, the groups $H_{q+1}(B \times B, B \vee B)$ vanish for $q \leq 2n$, by the Künneth Theorem. Since B is 1-connected, so is $B \vee B$, by the van Kampen Theorem, and therefore $(B \times B, B \vee B)$ is $(2n + 1)$-connected. By the Hurewicz Theorem, $H_q(F) = 0$ for $q < n$. Therefore we can apply Theorem (6.1) of Chapter VII to deduce that p_6 is an isomorphism if $q + 1 < 2n + 2 + n$, i.e., $q \leq 3n$, and an epimorphism if $q = 3n + 1$. □

(2.4) Corollary *The homomorphism d_1 is an isomorphism for $q \leq 3n$.* □

(2.5) *The homomorphism l_1 is an isomorphism for $q \leq 3n$ and an epimorphism for $q = 3n + 1$.*

Once more, we may assume $M = \Lambda$. Consider the NDR-pairs $(E_1, F) \times (E_0, F)$ and $(E, E_0) \times (E, E_1)$. By the naturality of the Künneth Theorem for such pairs, there is a commutative diagram (Figure 8.4) whose columns are exact; the homomorphisms l_5 and l_6 are induced by the appropriate injections.

$$\begin{array}{ccc}
0 & & 0 \\
\downarrow & & \downarrow \\
\bigoplus_{i+j=q+1} H_i(E_1, F) \otimes H_j(E_0, F) & \xrightarrow{l_5} & \bigoplus_{i+j=q+1} H_i(E, E_0) \otimes H_j(E, E_1) \\
\downarrow & & \downarrow \\
H_{q+1}(E_1 \times E_0, E_1 \times F \cup F \times E_0) & \xrightarrow{l_1} & H_{q+1}(E \times E, E_0 \times E \cup E \times E_1) \\
\downarrow & & \downarrow \\
\bigoplus_{i+j=q} \operatorname{Tor}\{H_i(E_1, F), H_j(E_0, F)\} & \xrightarrow{l_6} & \bigoplus_{i+j=q} \operatorname{Tor}\{H_i(E, E_0), H_j(E, E_1)\} \\
\downarrow & & \downarrow \\
0 & & 0
\end{array}$$

Figure 8.4

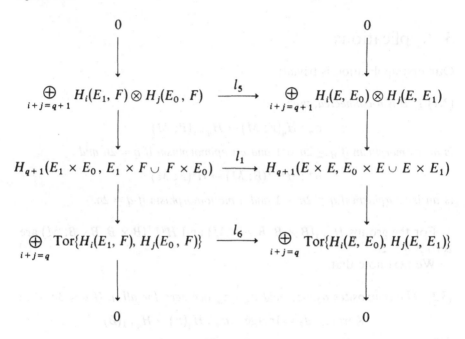

We have seen that p_6 is an isomorphism for $q \leq 3n$. Since $H_{q+1}(B \times B, B \vee B) = 0$ for $q \leq 2n$, we have $H_{q+1}(E, X) = 0$ for $q \leq 2n$. The injection $H_j(E_0, F) \to H_j(E, E_1)$ factors as the composite

$$H_j(E_0, F) \xrightarrow{k_1} H_j(X, E_1) \xrightarrow{l_7} H_j(E, E_1);$$

but we have seen that k_1 is an isomorphism for all j. Hence l_7 (and therefore $l_7 \circ k_1$) is an isomorphism for $j \leq 2n$ and an epimorphism for $j = 2n + 1$. Similarly, the injection $H_i(E_1, F) \to H_i(E, E_0)$ is an isomorphism for $i \leq 2n$ and an epimorphism for $i = 2n + 1$. Moreover, the groups $H_i(E_1, F)$, $H_i(E, E_0)$, $H_i(E_0, F)$, and $H_i(E, E_1)$ are zero for $i \leq n$. Thus l_5 is an isomorphism for $q \leq 3n$, and an epimorphism for $q = 3n + 1$, while l_6 is an isomorphism for $q \leq 3n + 1$. The desired property of l_1 now follows from the Five-Lemma. □

(2.6) Corollary *The homomorphism v_1 is an isomorphism for $q \leq 3n$.*

We can now complete the proof of the theorem. We have seen that ∂_3 and ∂_4, as well as k_2 and k_3, are isomorphisms. Hence the homomorphism $\beta' = \partial_3^{-1} \circ k_2 \circ \partial_4^{-1} \circ k_3$ is an isomorphism. Since $\beta = \nu_1 \circ \beta'$, it follows from Corollary (2.6) that β is an isomorphism for $q \leq 3n$. □

3 Applications

Our first application is trivial:

(3.1) *If B is n-connected, then*

$$\sigma_* : H_q(F; M) \to H_{q+1}(B; M)$$

is an isomorphism if $q \leq 2n - 1$ and an epimorphism if $q = 2n$ and

$$\sigma^* : H^{q+1}(B; M) \to H^q(F; M)$$

is an isomorphism if $q \leq 2n - 1$ and a monomorphism if $q = 2n$.

For the groups $H_{q+1}(B \times B, B \vee B; M)$ and $H^{q+1}(B \times B, B \vee B; M)$ are zero if $q \leq 2n$. □

We next note that

(3.2) *The composites $d_2 \circ \sigma_*$ and $\sigma_* \circ \tau_*$ are zero for all q. If $q \leq 3n$, then*

$$\text{Kernel} \quad d_2 = \text{Image} \quad \sigma_* : H_q(F) \to H_{q+1}(B)$$

and

$$\text{Image} \quad \tau_* = \text{Kernel} \quad \sigma_* : H_{q-1}(F) \to H_q(B).$$

Firstly,

$$d_2 \circ \sigma_* \circ \Delta_* = p_6 \circ j_1 \circ i_1 = 0$$

by commutativity of Figure 8.2; since Δ_* is an isomorphism, $d_2 \circ \sigma_* = 0$. Also

$$\begin{aligned} \sigma_* \circ \tau_* &= \sigma_* \circ \Delta_* \circ \partial_1 \circ \beta \\ &= p_4 \circ i_1 \circ \partial_1 \circ \beta = 0. \end{aligned}$$

If $q \leq 3n$,

$$\begin{aligned} \text{Ker}(d_2 \circ p_4) &= \text{Ker}(p_6 \circ j_1) = \text{Ker } j_1 \quad (\text{since } p_6 \text{ is an isomorphism}) \\ &= \text{Im } i_1; \end{aligned}$$

since p_4 is an isomorphism).

$$\begin{aligned} \text{Ker } d_2 &= \text{Im}(p_4 \circ i_1) = \text{Im}(\sigma_* \circ \Delta_*) \\ &= \text{Im } \sigma_* \quad (\text{since } \Delta_* \text{ is an isomorphism}). \end{aligned}$$

Again, if $q \leq 3n$,

$$\begin{aligned}\operatorname{Im} \tau_* &= \operatorname{Im}(\Delta_* \circ \partial_1 \circ \beta)\\ &= \operatorname{Im}(\Delta_* \circ \partial_1) \quad (\text{since } \beta \text{ is an isomorphism})\\ &= \Delta_*(\operatorname{Ker} i_1)\\ &= \Delta_* \operatorname{Ker}(p_4 \circ i_1) \quad (\text{since } p_4 \text{ is an isomorphism})\\ &= \Delta_* \operatorname{Ker}(\sigma_* \circ \Delta_*)\\ &= \operatorname{Ker} \sigma_* \quad (\text{since } \Delta_* \text{ is an isomorphism}).\end{aligned}$$

The elements of $\operatorname{Im} \sigma_*$ are said to be *transgressive.*

(3.3) Corollary *Every transgressive element of $H_{q+1}(B)$ is primitive. Conversely, if $q \leq 3n$, every primitive element of $H_{q+1}(B)$ is transgressive.* □

Let us recall from §7 of Chapter III that the elements of the image of τ_* are called *reductive*; moreover, if $M = \Lambda$, $u \in H_r(F)$, $v \in H_s(F)$, $r + s = q - 1$, then $u \times v$ belongs to $H_{q-1}(F \times F, F \vee F)$ and that $\tau_*(u \times v)$ is their Pontryagin product in the ideal $H_*(F) \subset H_*(F_*)$. Thus every decomposable element of $H_*(F)$ is reductive; the converse holds if $H_*(F)$ is torsion-free. (In fact, in order that each reductive element of $H_{q-1}(F)$ be decomposable, it suffices that $H_r(F)$ be torsion-free for all $r \leq q - n - 2$).

(3.4) Corollary *The suspension σ_* maps every reductive element of $H_{q-1}(F)$ into zero. Conversely, if $q \leq 3n$, every element of* Ker σ_* *is reductive. If $M = \Lambda$, $H_r(F)$ is torsion-free for all $r \leq q - n - 2$, and $q \leq 3n$, then every element of* Ker σ_* *is decomposable.* □

There are dual results for cohomology.

(3.2*) *The composites $\sigma^* \circ d_2^*$ and $\tau^* \circ \sigma^*$ are zero for all q. If $q \leq 3n$, then* $\operatorname{Im} d_2^* = \operatorname{Ker} \sigma^* : H^{q+1}(B) \to H^q(F)$ *and* $\operatorname{Ker} \tau^* = \operatorname{Im} \sigma^* : H^q(B) \to H^{q-1}(F)$. □

The elements of $\sigma^* H^q(B)$ are said to be *transgressive.*

(3.3*) *Every transgressive element of $H^{q-1}(F)$ is primitive. Conversely, if $q \leq 3n$, every primitive element of $H^{q-1}(F)$ is transgressive.* □

(3.4*) *The suspension σ^* maps every reductive element of $H^{q+1}(B)$ into zero. Conversely, if $q \leq 3n$, every element of* Ker σ^* *is reductive. If $M = \Lambda$ and $H_r(B)$ is a finitely-generated free module for all $r \leq q - n + 1$, and $q \leq 3n$, then every element of* Ker σ^* *is decomposable.*

Let us give an alternative description of the suspension homomorphism. Consider the triad $(\mathbf{S}F; \mathbf{T}_+ F, \mathbf{T}_- F)$. Since E_0 and E_1 are contractible, the inclusion maps of F into E_0 and E_1 have extensions $g_+ : (\mathbf{T}_+ F; F) \to (E_0, F)$, $g_- : (\mathbf{T}_- F, F) \to (E_1, F)$; the maps g_+, g_- agree on $\mathbf{T}_+ F \cap \mathbf{T}_- F = F$ and define a map

$$g : (\mathbf{S}F; \mathbf{T}_+ F, \mathbf{T}_- F) \to (X; E_0, E_1).$$

The map g induces a map of the Mayer–Vietoris sequence of the first triad into that of the second; and it follows from the contractibility of the triads $\mathbf{T}_+ F$, $\mathbf{T}_- F$, E_0, E_1 and the Five-Lemma that

(3.5) *The homomorphism $g_* : H_{q+1}(\mathbf{S}F) \to H_{q+1}(X)$ is an isomorphism for all q.*

The map g also induces homomorphisms $g_1 = p_3 \circ g_* : H_{q+1}(\mathbf{S}F) \to H_{q+1}(B)$, and we have

(3.6) *The composite*

$$H_q(F) \xrightarrow{\Delta_1^{-1}} H_{q+1}(\mathbf{S}F) \xrightarrow{g_1} H_{q+1}(B)$$

of g_1 with the inverse of the boundary operator Δ_1 of the Mayer–Vietoris sequence of the triad $(\mathbf{S}F; \mathbf{T}_+ F, \mathbf{T}_- F)$ is the homology suspension σ_*. The homomorphism g_1 is an isomorphism if $q \le 2n - 1$ and an epimorphism if $q = 2n$. □

The homomorphism g_1 can be described in still another way. The adjoint of the identity map $1 : \Omega B \to \Omega B$ is a map $h : \mathbf{S}F = \mathbf{S}\Omega B \to B$. It is not difficult (and left to the reader) to verify that the maps g and h are essentially the same; more precisely, for any $t \in I$, the map $(P_t | X) \circ g : \mathbf{S}F \to B$ is homotopic to h. The map h, or any map homotopic to it, will be called a *transfer of F into B.*

There is another operation which has a right to be called the homology suspension; it is the inverse of the boundary operator

$$s_* = \Delta_1^{-1} : H_q(Y) \to H_{q+1}(\mathbf{S}Y)$$

of the Mayer–Vietoris sequence of the triad $(\mathbf{S}Y; \mathbf{T}_+ Y, \mathbf{T}_- Y)$. Unlike σ_*, the homomorphism s_* is an isomorphism for all q. If W is another space, the maps

$$t \wedge (y \wedge w) \to (t \wedge y) \wedge w,$$
$$t \wedge (y \wedge w) \to y \wedge (t \wedge w),$$

are homeomorphisms of $\mathbf{S}(Y \wedge W)$ with $\mathbf{S}Y \wedge W$ and $Y \wedge \mathbf{S}W$, respectively.

Therefore there are isomorphisms

$$s_L : H_q(Y \wedge W) \to H_{q+1}(\mathbf{S}Y \wedge W),$$
$$s_R : H_q(Y \wedge W) \to H_{q+1}(Y \wedge \mathbf{S}W),$$

obtained by composing

$$s_* : H_q(Y \wedge W) \to H_{q+1}(\mathbf{S}(Y \wedge W))$$

with the isomorphisms induced by these homeomorphisms.

Let $Y = F$, $h : \mathbf{S}F \to B$ a transfer and let $\sigma_L : H_q(F \wedge W) \to H_{q+1}(B \wedge W)$ be the composite of s_L with the homomorphism induced by $h \wedge 1 : SF \wedge W \to B \wedge W$. Similarly, we may take $W = F$, and let $\sigma_R : H_q(Y \wedge F) \to H_{q+1}(Y \wedge B)$ be the composite of s_R with the homomorphism induced by $1 \wedge h : Y \wedge SF \to Y \wedge B$.

(3.7) Theorem *If B is n-connected and W is m-connected, then $\sigma_L : H_q(F \wedge W) \to H_{q+1}(B \wedge W)$ is an isomorphism if $q \leq m + 2n$ and an epimorphism if $q = m + 2n + 1$. If Y is m-connected, then $\sigma_R : H_q(Y \wedge F) \to H_{q+1}(Y \wedge B)$ is an isomorphism if $q \leq m + 2n$ and an epimorphism if $q = m + 2n + 1$.*

Since s_R and s_L are isomorphisms for all q, these results follow from (3.1) by applying the Künneth Theorem to the spaces $\mathbf{I}_h \wedge W, Y \wedge \mathbf{I}_h$, respectively.

Dually, one has the suspension homomorphism

$$s^* = \Delta^{*-1} : H^{q+1}(\mathbf{S}Y) \to H^q(Y),$$

and the homomorphisms

$$\sigma_L^* : H^{q+1}(B \wedge W) \to H^q(F \wedge W),$$
$$\sigma_R^* : H^{q+1}(Y \wedge B) \to H^q(Y \wedge F);$$

and

(3.7*) Theorem *If B is n-connected and W is m-connected, then $\sigma_L^* : H^{q+1}(B \wedge W) \to H^q(F \wedge W)$ is an isomorphism if $q \leq m + 2n$ and a monomorphism if $q = m + 2n + 1$. If Y is m-connected, then $\sigma_R^* : H^{q+1}(Y \wedge B) \to H^q(Y \wedge F)$ is an isomorphism if $q \leq m + 2n$ and a monomorphism if $q = m + 2n + 1$.* □

4 Cohomology Operations

Let us now suppose that $B = K(\Pi, n + 1)$ (and therefore $F = K(\Pi, n)$). Let us recall our discussion of cohomology operations in §8 of Chapter V. The groups $H^q(F; G)$ and $H^{q+1}(B; G)$ are in one-to-one correspondence with the sets of cohomology operations of types $(n, q; \Pi, G)$ and $(n+1, q+1; \Pi, G)$, respectively. Then the cohomology suspension corresponds to a transformation which associates to each cohomology operation of the second type, another one of the first. Let us attempt to determine how this transformation can be described in the context of cohomology operations.

Let, then, $\theta : H^{n+1}(\quad; \Pi) \to H^{q+1}(\quad; G)$ be an operation, and let X be a CW-complex. Then the suspension operator $s^* : H^{r+1}(\mathbf{S}X; M) \to H^r(X; M)$ is an isomorphism, for all r and every coefficient group M. Thus we can form the composite

$$H^n(X; \Pi) \xrightarrow{s^{*-1}} H^{n+1}(\mathbf{S}X; \Pi) \xrightarrow{\theta(\mathbf{S}X)} H^{q+1}(\mathbf{S}X; G) \xrightarrow{s^*} H^q(X; G).$$

Call this composite $\sigma^*\theta(X)$. Then it is clear that $\sigma^*\theta$ is a cohomology operation of type $(n, q; \Pi, G)$; we shall call $\sigma^*\theta$ the *suspension* of θ.

(4.1) Theorem *For each $u \in H^{q+1}(\Pi, n+1; G)$, and let θ_u be the corresponding operation. Then $\theta_{\sigma^*(u)} = \sigma^*\theta_u$.*

In other words, the two notions of suspension correspond.

Let us first remark that we have identified $\pi_n(F)$ and $\pi_{n+1}(B)$ with Π, and therefore with each other; i.e., there is implicitly defined an isomorphism of $\pi_n(F)$ with $\pi_{n+1}(B)$; let us choose this isomorphism to be the one which corresponds to the homology suspension under the Hurewicz map. It then follows that, if ι_F and ι_B are the fundamental classes of F and B, respectively, then $\sigma^*\iota_B = \iota_F$.

(4.2) Lemma *Let $K \in \mathscr{K}_*$, and let $f : K \to F$. Then there is a map $g : (\mathbf{S}K; \mathbf{T}_+K, \mathbf{T}_-K) \to (X; E_0, E_1)$ such that $f = g|K : K \to F$.*

Since E_0 is contractible and $\mathbf{T}_+K$ is a copy of the cone over K, the map $f : K \to F$ has an extension $g_+ : (\mathbf{T}_+K, K) \to (E_0, F)$. Similarly, f has an extension $g_- : (\mathbf{T}_-K, K) \to (E_1, F)$. The two maps agree on $\mathbf{T}_+K \cap \mathbf{T}_-K = K$ and so define a map $g : \mathbf{S}K \to E_0 \cup E_1 = X$. □

(4.3) Lemma *Let $g : (\mathbf{S}K; \mathbf{T}_+K, \mathbf{T}_-K) \to (X; E_0, E_1)$ be an extension of $f : K \to F$, and let $i : X \hookrightarrow E$, $g_0 = P_1 \circ i \circ g : \mathbf{S}K \to B$. Then, for any $u \in H^{q+1}(B; \Pi)$, we have*

$$s^* g_0^* u = f^* \sigma^* u.$$

For there is a commutative diagram

$$
\begin{array}{ccc}
H^q(K;\Pi) & \xleftarrow{\ f^*\ } & H^q(F;\Pi) \\
{\scriptstyle s^{*-1}=\Delta^*}\downarrow & & \downarrow{\scriptstyle \Delta^*} \\
H^{q+1}(SK;\Pi) & \xleftarrow{\ g^*\ } & H^{q+1}(X;\Pi) \\
{\scriptstyle g_0^*}\uparrow & & \uparrow{\scriptstyle i_1^*} \\
H^{q+1}(B;\Pi) & \xrightarrow[\ p_4^*\ ]{} & H^{q+1}(E;\Pi)
\end{array}
\tag{4.4}
$$

and

$$
\begin{aligned}
g_0^* u &= g^* i_1^* p_4^* u \\
&= g^* \Delta^* \Delta^{*-1} i_1^* p_4^* u \\
&= g^* \Delta^* \sigma^* u \\
&= \Delta^* f^* \sigma^* u = s^{*-1} f^* \sigma^* u. \qquad \square
\end{aligned}
$$

We can now prove Theorem (4.1). Let $f\colon X \to F$ be a map representing $x \in H^n(X; \Pi)$ (i.e., $f^*\iota_F = x$). Applying Lemma (4.3) to the class ι_B, we have

$$s^* g_0^* \iota_B = f^* \sigma^* \iota_B = f^* \iota_F = x,$$

so that g_0 represents $s^{*-1}x$. Now apply Lemma (4.3) again, this time to the class u, to obtain

$$
\begin{aligned}
\theta_u(s^{*-1}x) = g_0^* u &= s^{*-1} f^* \sigma^* u \\
&= s^{*-1} \theta_{\sigma^* u}(x), \\
\theta_{\sigma^* u}(x) = s^* \theta_u s^{*-1} x &= (\sigma^* \theta_u)(x). \qquad \square
\end{aligned}
$$

We have seen in §8 of Chapter V that a general cohomology operation need not be additive; a necessary and sufficient condition that an operation be additive is that the corresponding element of $H^q(\Pi, n; G)$ be primitive. It is easy to see that the suspension of any operation is additive. In fact, it follows from Theorem (8.3) of Chapter V and Corollary (3.3) above that

(4.5) Theorem *If θ is a cohomology operation of type $(n, q; \Pi, G)$, and if the corresponding element $u \in H^q(\Pi, n; G)$ is transgressive, then θ is additive. Conversely, if $q \leq 3n - 1$ and θ is additive, then u is transgressive.* □

From (3.1) we deduce

(4.6) Corollary *If* $q \leq 2n - 1$, *every cohomology operation of type* $(n, q; \Pi, G)$ *is additive.* □

We have formulated the notion of cohomology operation in the context of CW-complexes with base point. It is sometimes convenient to consider them in the context of free CW-complexes and pairs.

Suppose, then, that (X, A) is a free CW-pair (more generally, an NDR-pair). Then X/A is a space with nondegenerate base point. The subspace $X \cup \mathbf{T}A$ of $\mathbf{T}X$ has projections

$$p : X \cup \mathbf{T}A \to X \cup \mathbf{T}A/\mathbf{T}A = X/A,$$
$$q : X \cup \mathbf{T}A \to X \cup \mathbf{T}A/X = \mathbf{S}A;$$

the map p is a homotopy equivalence because $\mathbf{T}A$ is contractible. Let $p' : X/A \to X \cup \mathbf{T}A$ be a homotopy inverse of p. Then the composite $r = q \circ p' : X/A \to \mathbf{S}A$, or any map homotopic to it, is called a *connecting map* for the pair (X, A) (cf. §6 of Chapter III).

Let us make the construction for r a little more explicit. Since $\mathbf{T}_+ A$ and $\mathbf{T}_- A$ are contractible, the inclusion maps of A into $\mathbf{T}_+ A$ and $\mathbf{T}_- A$ extend to maps of (X, A) into $(\mathbf{T}_+, A)$ and of $(\mathbf{T}A, A)$ into $(\mathbf{T}_- A, A)$. These maps, in turn, fit together to define a map

$$g : (X \cup \mathbf{T}A; X, \mathbf{T}A) \to (\mathbf{S}A; \mathbf{T}_+ A, \mathbf{T}_- A).$$

Again using the contractibility of the cones, we find that there are homotopies of $g : X \cup \mathbf{T}A \to \mathbf{S}A$ to maps $g', g'' : X \cup \mathbf{T}A \to \mathbf{S}A$ such that $g'(\mathbf{T}A) = g''(X) = *$. Therefore there is a commutative diagram

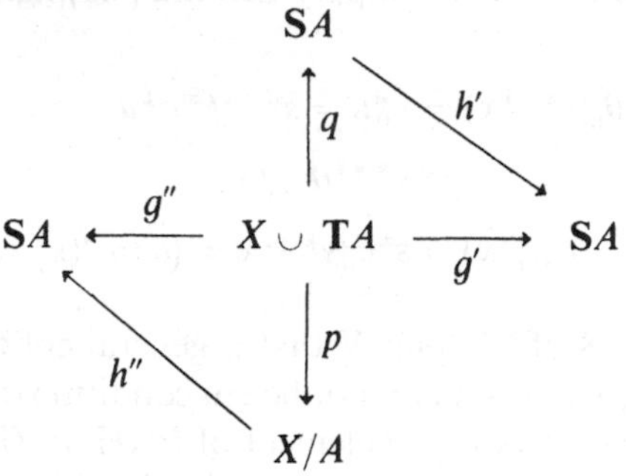

Clearly h' is homotopic to the identity map, so that $g' \simeq q$. Therefore

(4.7)
$$h'' \circ p \simeq g'' \simeq g' \simeq q,$$
$$h'' \simeq q \circ p' = r.$$

Let $p_0 = p|X : (X, A) \to (X/A, *)$, and let

$$\delta^* : H^{n-1}(A) \to H^n(X, A),$$
$$\Delta^* : H^{n-1}(A) \to H^n(\mathbf{S}A),$$

be the coboundary operators of the cohomology sequence of (X, A) and the Mayer–Vietoris sequence of the triad $(\mathbf{S}A; \mathbf{T}_+ A, \mathbf{T}_- A)$, respectively. (The coefficient group G is understood).

(4.8) Lemma *Let* (X, A) *be an* NDR-*pair. Then the diagram*

$$\begin{array}{ccc} H^{n-1}(A) & \xrightarrow{\delta^*} & H^n(X, A) \\ \Delta^* \downarrow & & \uparrow p_0^* \\ H^n(\mathbf{S}A) & \xrightarrow[r^*]{} & H^n(X/A) \end{array}$$

is anti-commutative.

For there is a commutative diagram

$$\begin{array}{ccccccc} & & & & H^n(X/A) & & \\ & & & \swarrow p_0^* & \downarrow p_1^* & \searrow p^* & \\ & & H^n(X, A) & \xleftarrow{k^*} & H^n(X \cup \mathbf{T}A, \mathbf{T}A) & \xrightarrow{i^*} & H^n(X \cup \mathbf{T}A) \\ & \nearrow \delta^* & \uparrow g_1^* & & \uparrow g_2^* & & \uparrow g_3^* \\ H^{n-1}(A) & \xrightarrow[\delta_1^*]{} & H^n(\mathbf{T}_+ A, A) & \xleftarrow[k_1^*]{} & H^n(\mathbf{S}A, \mathbf{T}_- A) & \xrightarrow[i_1^*]{} & H^n(\mathbf{S}A) \end{array} \tag{4.9}$$

in which δ^* and δ_1^* are coboundary operators, i^*, i_1^*, k^*, and k_1^* are injections, g_1^*, g_2^*, g_3^* are induced by the map g constructed above; p_0^* is induced by $p_0 : (X, A) \to (X/A, *)$, and p^*, p_1^* by $p : X \cup \mathbf{T}A \to X/A$. But we have seen that $g \simeq q$, whence $p^{*-1} \circ g_3^* = r^*$. Now $i_1^* \circ k_1^{*-1} \circ \delta_1^*$ is the coboundary operator of the Mayer Vietoris sequence of the triad $(\mathbf{S}A; \mathbf{T}_- A, \mathbf{T}_+ A)$, which is $-\Delta^*$. Commutativity of (4.9) now yields the desired conclusion. □

We can now describe the action of the suspension of an operation in terms of relative groups. First note that, if (X, A) is a free CW-pair, then the projection $p_0 : (X, A) \to (X/A, *)$ induces isomorphisms $p_0^* : H^r(X/A; H) \approx H^r(X, A; H)$ for every integer r and coefficient group H. Therefore we can make any cohomology operation act on the relative groups by demanding that the diagram

$$\begin{array}{ccc} H^n(X/A;\Pi) & \xrightarrow{\theta(X/A)} & H^q(X/A;G) \\ \downarrow p_0^* & & \downarrow p_0^* \\ H^n(X,A;\Pi) & \xrightarrow[\theta(X,A)]{} & H^q(X,A;G) \end{array}$$

be commutative. We then have:

(4.10) Theorem *Let* (X, A) *be a* CW-*pair, and let* θ *be a cohomology operation of type* $(n, q; \Pi, G)$. *Then the diagram*

$$\begin{array}{ccc} H^n(X,A;\Pi) & \xrightarrow{\theta} & H^q(X,A;G) \\ \uparrow \delta^* & & \uparrow \delta^* \\ H^{n-1}(A;\Pi) & \xrightarrow[\sigma^*\theta]{} & H^{q-1}(A;G) \end{array}$$

is commutative.

This follows readily from Lemma (4.8). Details are left to the reader. □

5 Stable Operations

For each integer r and pair of abelian groups Π, G, the groups $H_{n+r}(\Pi, n; G)$ are connected by homomorphisms

$$\sigma_* : H_{n+r}(\Pi, n; G) \to H_{n+r+1}(\Pi, n+1; G),$$

and so we may form the direct limit

$$A_r(\Pi; G) = \varinjlim_n H_{n+r}(\Pi, n; G).$$

As usual, we often abbreviate $A_r(\Pi; Z)$ to $A_r(\Pi)$.

Similarly, we may form the inverse limit

$$A^r(\Pi; G) = \varprojlim_n H^{n+r}(\Pi, n; G)$$

of the groups $H^{n+r}(\Pi, n; G)$ with respect to the homomorphisms

$$\sigma^* : H^{n+r+1}(\Pi, n+1; G) \to H^{n+r}(\Pi, n; G).$$

The above direct and inverse systems are trivial in the sense that, according to (3.1), the homomorphisms σ_* and σ^* are isomorphisms for all sufficiently large n. Thus the groups $H_{n+r}(\Pi, n; G)$ and $H^{n+r}(\Pi, n; G)$ are eventually constant, and we have

(5.1) *The natural homomorphism*

$$H_{n+r}(\Pi, n; G) \to A_r(\Pi; G)$$

is an isomorphism if $n \geq r+1$ and an epimorphism if $n = r$. Dually, the natural homomorphism

$$A^r(\Pi; G) \to H^{n+r}(\Pi, n; G)$$

is an isomorphism if $n \geq r+1$ and a monomorphism if $n = r$. □

The Universal Coefficient Theorem can then be applied to yield

(5.2) *There are splittable short exact sequences*

$$0 \to A_r(\Pi) \otimes G \to A_r(\Pi; G) \to \mathrm{Tor}(A_{r-1}(\Pi), G) \to 0,$$

$$0 \to \mathrm{Ext}(A_{r-1}(\Pi), G) \to A^r(\Pi; G) \to \mathrm{Hom}(A_r(\Pi), G) \to 0;$$

and if G is a field, then $A^r(\Pi; G)$ is the dual vector space of $A_r(\Pi; G)$. □

Just as elements of $H^{n+r}(\Pi, n; G)$ correspond to cohomology operations, so do elements of $A^r(\Pi; G)$ correspond to *stable operations.* Specifically, a *stable cohomology operation of degree r and type* (Π, G) is just a sequence of cohomology operations

$$\theta_n : H^n(\ \ ; \Pi) \to H^{n+r}(\ \ ; G)$$

with the property that $\sigma^*\theta_{n+1} = \theta_n$ for every n. If $\theta = \{\theta_n\}$ is such a stable operation and

$$u_n \in H^{n+r}(\Pi, n; G)$$

is the element corresponding to θ_n, then

$$\sigma^* u_{n+1} = u_n,$$

according to Theorem (4.1), and so the sequence $\{u_n\}$ is an element of the inverse limit group $A^r(\Pi; G)$. It is clear that

(5.3) *The set of all stable cohomology operations of degree r and type* (Π, G) *is in* 1 : 1 *correspondence with the group* $A^r(\Pi; G)$. □

It follows from Theorem (4.5) that

(5.4) *If $\theta = \{\theta_n\}$ is a stable cohomology operation, then each of the operations θ_n is additive.* □

Cohomology operations can sometimes be composed; the result is another cohomology operation. Specifically, let

$$\theta : H^n(\ \ ;\Pi) \to H^{n+r}(\ \ ;G),\ \phi : H^{n+r}(\ \ ;G) \to H^{n+r+s}(\ \ ;H)$$

be cohomology operations; then it is clear that, for every K, the map

$$H^n(K;\Pi) \xrightarrow{\theta(K)} H^{n+r}(K;G) \xrightarrow{\phi(K)} H^{n+r+s}(K;H)$$

defines a cohomology operation, which we may christen $\phi \circ \theta$. Clearly

(5.5) *If θ corresponds to $u \in H^{n+r}(\Pi, n; G)$, then $\phi \circ \theta$ corresponds to $\phi(u) \in H^{n+r+s}(\Pi, n; H)$.* □

The effect of suspension on the composition of operations is given by

(5.6) *Let θ and ϕ be cohomology operations of types $(n+r, n; \Pi, G)$, and $(n+r+s, n+r; G, H)$, respectively. Then*

$$\sigma^*(\phi \circ \theta) = (\sigma^*\phi) \circ (\sigma^*\theta).$$

For

$$(\sigma^*\phi)(X) = \Delta^{*-1} \circ \phi(\mathbf{S}X) \circ \Delta^*,$$
$$(\sigma^*\theta)(X) = \Delta^{*-1} \circ \theta(\mathbf{S}X) \circ \Delta^*,$$

and therefore

$$\begin{aligned}(\sigma^*\phi(X)) \circ (\sigma^*\theta(X)) &= \Delta^{*-1} \circ \phi(\mathbf{S}X) \circ \Delta^* \circ \Delta^{*-1} \circ \theta(\mathbf{S}X) \circ \Delta^* \\ &= \Delta^{*-1} \circ \phi(\mathbf{S}X) \circ \theta(\mathbf{S}X) \circ \Delta^* \\ &= \sigma^*(\phi \circ \theta)(X).\end{aligned}$$ □

Suppose now that θ and ϕ are stable operations:

$$\theta_n : H^n(\ \ ;\Pi) \to H^{n+r}(\ \ ;G),$$
$$\phi_n : H^m(\ \ ;G) \to H^{m+s}(\ \ ;G).$$

Then $\sigma^*(\phi_{n+r+1} \circ \theta_{n+1}) = \sigma^*(\phi_{n+r+1}) \circ \sigma^*(\theta_{n+1}) = \phi_{n+r} \circ \theta_n$. Therefore

(5.7) *The operations $\psi_n = \phi_{n+r} \circ \theta_n$ are the components of a stable operation $\psi = \phi \circ \theta$.* □

Evidently

(5.8) *Composition of stable operations is associative.* □

For any group Π, the fundamental class $\mathbf{b}_n$ of $K(\Pi, n)$ defines an operation, which is evidently the identity map of $H^n(\ \ ;\Pi)$. Moreover, we have seen that $\sigma^*\mathbf{b}_{n+1} = \mathbf{b}_n$; therefore the $\mathbf{b}_n$ are the components of a stable operation 1. Clearly

(5.9) *For any stable operation θ for which $\theta \circ 1$ $(1 \circ \theta)$ is defined, we have* $\theta \circ 1 = \theta$ $(1 \circ \theta = \theta)$. □

In particular, if Π is any abelian group, stable operations of type (Π, Π) can always be composed, and the stable operations of degree r and type (Π, Π) are the components $\mathfrak{O}_r(\Pi)$ of a graded ring $\mathfrak{O}(\Pi)$; $\mathfrak{O}(\Pi)$ is associative and $1 \in \mathfrak{O}_0(\Pi)$ is the unit element of this ring.

If Π is a field, then $\mathfrak{O}(\Pi)$ is a graded *algebra* over Π, the *algebra of stable operations of type* Π. In particular, when $\Pi = \mathbf{Z}_p$, $\mathfrak{O}(\Pi)$ is called the *Steenrod algebra* mod p.

We now give some examples of stable operations. Less trivial examples will be given in §6.

Example 1 (Coefficient group homomorphisms). Let Π, G be abelian groups, $f: \Pi \to G$ a homomorphism. Then the coefficient group homomorphism

$$f_*: H^n(\ ;\Pi) \to H^n(\ ;G),$$

as we saw in Example 1 of §8, Chapter V is a cohomology operation of type $(n, n; \Pi, G)$. Clearly f_* commutes with the coboundary operator of the Mayer-Vietoris sequence of a proper triad, and hence the f_*, for each value of n, are the components of a stable operation f_*.

Example 2 (Bockstein operators). This time the Bockstein operators

$$H^n(\ ;\Pi) \to H^{n+1}(\ ;G),$$

associated with a short exact sequence

$$0 \to G \to E \to \Pi \to 0,$$

commute with coboundary operators except for sign; the diagram

$$\begin{array}{ccc} H^n(A;\Pi) & \xrightarrow{\beta^*} & H^{n+1}(A;G) \\ \downarrow{\scriptstyle \delta^*} & & \downarrow{\scriptstyle \delta^*} \\ H^{n+1}(X, A;\Pi) & \xrightarrow[\beta^*]{} & H^{n+2}(X, A;G) \end{array}$$

is *anti-commutative*. Therefore the operators

$$(-1)^n\beta_n^*: H^n(\ ;\Pi) \to H^{n+1}(\ ;G)$$

are the components of a stable operation β^*.

6 The mod 2 Steenrod Algebra

We have seen that the cup square is an example of a cohomology operation. Its utility is apparent in the proof of

(6.1) Theorem *Let $d = 2, 4$, or 8, and let $\alpha \in \pi_{2d-1}(\mathbf{S}^d)$ be the homotopy class of the Hopf fibration. Then $\alpha \neq 0$.*

Let us recall that the mapping cone of α is the projective plane $\mathbf{P}^2(D)$, where $D = \mathbf{C}$, $\mathbf{Q}$, or $\mathbf{K}$ is the underlying division algebra. The space $\mathbf{P}^2(D)$ is a manifold whose mod 2 cohomology groups vanish, except that

$$H^0(\mathbf{P}^2(D)) \approx H^d(\mathbf{P}^2(D)) \approx H^{2d}(\mathbf{P}^2(D)) \approx \mathbf{Z}_2;$$

and it follows from Poincaré duality that, if u is the non-zero element of $H^d(\mathbf{P}^2(D))$, then $u \cup u \neq 0$. On the other hand, if $\alpha = 0$, then $\mathbf{P}^2(D)$ has the same homotopy type as $\mathbf{S}^d \vee \mathbf{S}^{2d}$, and all cup-products in $H^*(\mathbf{S}^d \vee \mathbf{S}^{2d})$ vanish. □

We may attempt to generalize this argument. If $\alpha \in \pi_n(\mathbf{S}^r)$, the mapping cone of α is a CW-complex with one cell in each of the dimensions $0, r, n+1$. If $n > r$, the boundary operator of its chain complex must be trivial, so that

$$H^r(\mathbf{T}_\alpha; G) \approx H^{n+1}(\mathbf{T}_\alpha; G) \approx G$$

for any coefficient group G. The proof of Theorem (6.1) generalizes to give

(6.2) Theorem *Let $\alpha \in \pi_n(\mathbf{S}^r)$, and suppose that there is a cohomology operation θ of type $(\Pi, r; G, n+1)$ such that $\theta : H^r(\mathbf{T}_\alpha; \Pi) \to H^{n+1}(\mathbf{T}_\alpha; G)$ is not zero. Then $\alpha \neq 0$.*

For if $\alpha = 0$, then $\mathbf{T}_\alpha$ has the homotopy type of $\mathbf{S}^r \vee \mathbf{S}^{n+1}$, and therefore $\mathbf{S}^r$ is a retract of $\mathbf{T}_\alpha$. If $f : \mathbf{T}_\alpha \to \mathbf{S}^r$ is a retraction, then

$$f^* : H^r(\mathbf{S}^r; \Pi) \approx H^r(\mathbf{T}_\alpha; \Pi).$$

But $H^{n+1}(\mathbf{S}^r; G) = 0$, and commutativity of the diagram

$$\begin{array}{ccc} H^r(\mathbf{T}_\alpha; \Pi) & \xrightarrow{\theta} & H^{n+1}(\mathbf{T}_\alpha; G) \\ \uparrow f^* & & \uparrow f^* \\ H^r(\mathbf{S}^r; \Pi) & \xrightarrow[\theta]{} & H^{n+1}(\mathbf{S}^r; G) \end{array}$$

shows that $\theta = 0$. □

Let us recall that the suspension functor induces a homomorphism $E : \pi_n(\mathbf{S}^r) \to \pi_{n+1}(\mathbf{S}^{r+1})$; iteration yields $E^k : \pi_n(\mathbf{S}^r) \to \pi_{n+k}(\mathbf{S}^{r+k})$.

(6.3) Theorem *Let $\alpha \in \pi_n(\mathbf{S}^r)$, and suppose that there is a cohomology operation θ of type $(\Pi, r+1; G, n+2)$ such that $\sigma^*\theta : H^r(\mathbf{T}_\alpha; \Pi) \to H^{n+1}(\mathbf{T}_\alpha; G)$ is non-zero. Then $E\alpha \neq 0$.*

For we may identify $\mathbf{T}_{E\alpha}$ with $\mathbf{ST}_\alpha$; thus there is a commutative diagram

$$\begin{array}{ccc} H^r(\mathbf{T}_\alpha; \Pi) & \xrightarrow{\sigma^*\theta} & H^{n+1}(\mathbf{T}_\alpha; G) \\ \uparrow s^* & & \uparrow s^* \\ H^{r+1}(\mathbf{T}_{E\alpha}; \Pi) & \xrightarrow[\theta]{} & H^{n+2}(\mathbf{T}_{E\alpha}; G); \end{array}$$

since s^* is an isomorphism $\theta \neq 0$, and the fact that $E\alpha \neq 0$ follows from Theorem (6.2). □

(6.4) Corollary *Let $\alpha \in \pi_n(\mathbf{S}^r)$, and suppose that there is a stable operation θ such that*

$$\theta : H^r(\mathbf{T}_\alpha; \Pi) \to H^{n+1}(\mathbf{T}_\alpha; G)$$

is non-zero. Then $E^k\alpha \neq 0$ for every k.

This suggests the question of determining whether the cup square operation is a component of a stable operation. But if $u, v \in H^n(X)$,

$$\begin{aligned}(u+v) \smile (u+v) &= u \smile u + u \smile v + v \smile u + v \smile v \\ &= u \smile u + [1 + (-1)^n]u \smile v + v \smile v;\end{aligned}$$

thus the cup square is additive if n is odd, but may fail to be additive if n is even. In the latter case, it cannot, because of Theorem (4.5), be the suspension of an operation. However, reduction mod 2 yields an additive operation even if n is even.

We can now prove[1]

(6.5) Theorem *There are unique stable cohomology operations Sq^i $(i = 0, 1, 2, \ldots)$ such that*

(1) *Sq^i has degree i;*
(2) *Sq^0 is the identity;*
(3) *$Sq^i x = x \smile x$ if $x \in H^i(K)$;*
(4) *$Sq^i x = 0$ if $x \in H^n(K)$, $i > n$.*

We begin by defining Sq^0 to be the identity. Suppose $n > 0$; then the element $x = \mathbf{b}_n \smile \mathbf{b}_n \in H^{2n}(\mathbf{Z}_2, n)$ is decomposable, and it follows from

[1] In this section the coefficient group for all homology and cohomology groups is $\mathbf{Z}_2$, and will be understood.

(3.4*) that $\sigma^* x = 0$. On the other hand b_n is primitive, and therefore

$$\begin{aligned}\mu^*(\mathbf{b}_n \smile \mathbf{b}_n) &= (\mu^*\mathbf{b}_n) \smile (\mu^*\mathbf{b}_n)\\ &= (\mathbf{b}_n \times 1 + 1 \times \mathbf{b}_n) \smile (\mathbf{b}_n \times 1 + 1 \times \mathbf{b}_n)\\ &= (\mathbf{b}_n \smile \mathbf{b}_n) \times 1 + \mathbf{b}_n \times \mathbf{b}_n + \mathbf{b}_n \times \mathbf{b}_n + 1 \times (\mathbf{b}_n \smile \mathbf{b}_n)\\ &= (\mathbf{b}_n \smile \mathbf{b}_n) \times 1 + 1 \times (\mathbf{b}_n \smile \mathbf{b}_n)\end{aligned}$$

since the coefficient field is $\mathbf{Z}_2$. Thus $\mathbf{b}_n \cup \mathbf{b}_n$ is primitive; by (3.3*), $\mathbf{b}_n \smile \mathbf{b}_n$ is transgressive. But

$$\sigma^* : H^{2n+1}(\mathbf{Z}_2, n+1) \to H^{2n}(\mathbf{Z}_2, n)$$

is a monomorphism, and

$$\sigma^* : H^{2n+r}(\mathbf{Z}_2, n+r) \to H^{2n+r-1}(\mathbf{Z}_2, n+r-1)$$

is an isomorphism if $r > 1$, by Theorem (3.1). Therefore there are unique elements

$$x_r \in H^{2n+r}(\mathbf{Z}_2, n+r)$$

such that $x_0 = \mathbf{b}_n \smile \mathbf{b}_n$ and $\sigma^* x_r = x_{r-1}$ (in particular, $x_r = 0$ for all $r < 0$), and the x_r define a stable operation Sq^n of degree n. It is clear that the Sq^n have the requisite properties, and uniqueness follows from the above monomorphic property of σ^*. □

(6.6) *The operation Sq^1 is the Bockstein operator associated with the exact sequence*

$$0 \to \mathbf{Z}_2 \to \mathbf{Z}_4 \to \mathbf{Z}_2 \to 0.$$

For we have seen (Theorem (7.8) of Chapter V) that $H_n(\mathbf{Z}_2, n; \mathbf{Z}) = \mathbf{Z}_2$ and $H_{n+1}(\mathbf{Z}_2, n; \mathbf{Z}) = 0$ for large n. Hence $H^{n+1}(\mathbf{Z}_2, n; \mathbf{Z}_2) \approx \operatorname{Ext}(\mathbf{Z}_2, \mathbf{Z}_2) \approx \mathbf{Z}_2$, and we have seen in Exercise 7 of Chapter V that the non-zero element of $H^{n+1}(\mathbf{Z}_2 n; \mathbf{Z}_2)$ is the Bockstein operator corresponding to the non-zero element of $\operatorname{Ext}(\mathbf{Z}_2, \mathbf{Z}_2)$, i.e., to the above short exact sequence. Thus the statement in question is true, provided that Sq^1 is not identically zero. But if $X = \mathbf{P}^2$ and $x \in H^1(X)$ is the non-zero element, then $Sq^1 x = x \smile x$ is the non-zero element of $H^2(X)$. □

A similar useful relation is

(6.7) *The operation Sq^1 is the composite*

$$H^q(X; \mathbf{Z}_2) \xrightarrow{\beta^*} H^{q+1}(X; \mathbf{Z}) \xrightarrow{\rho_2} H^{q+1}(X; \mathbf{Z}_2),$$

where β^* is the Bockstein operator associated with the short exact sequence

$$0 \longrightarrow \mathbf{Z} \xrightarrow{\times 2} \mathbf{Z} \longrightarrow \mathbf{Z}_2 \longrightarrow 0$$

and ρ_2 *is the homomorphism induced by the operation of reduction of the coefficients* mod 2.

The easy proof is left to the reader. □

7 The Cartan Product Formula

In order to calculate effectively with the Steenrod squares, additional information about their properties is needed. Such information is provided by the relation

(7.1)
$$Sq^k(x \smile y) = \sum_{i+j=k} Sq^i x \smile Sq^j y,$$

due to H. Cartan.

This relation is entirely equivalent to the relation

(7.2)
$$Sq^k(u \wedge v) = \sum_{i+j=k} Sq^i u \wedge Sq^j v,$$

for the $\wedge$-product pairing $H^*(X)$ with $H^*(Y)$ to $H^*(X \wedge Y)$.

Let $K_n = K(\mathbf{Z}_2, n)$, and let $\mathbf{b}_n \in H^n(K_n)$ be its fundamental class. The Cartan relation (7.2) follows by naturality from

(7.3)
$$Sq^k(\mathbf{b}_m \wedge \mathbf{b}_n) = \sum_{i+j=k} Sq^i\mathbf{b}_m \wedge Sq^j\mathbf{b}_n.$$

We shall prove

(7.4) Theorem *The relation* (7.3) *holds for all* k, m, *and* n, *and therefore the Cartan relations* (7.2) *hold for all* $u \in H^m(X)$, $v \in H^n(Y)$, *and the relations* (7.1) *for all* $x \in H^m(X)$, $y \in H^n(X)$.

We shall prove Theorem (7.4) by induction on $m + n$, observing that, if $m + n < k$, then both sides of (7.3) are zero. Suppose $m + n = k$. Then

$$\begin{aligned} Sq^k(\mathbf{b}_m \wedge \mathbf{b}_n) &= (\mathbf{b}_m \wedge \mathbf{b}_n) \smile (\mathbf{b}_m \wedge \mathbf{b}_n) \\ &= (\mathbf{b}_m \smile \mathbf{b}_m) \wedge (\mathbf{b}_n \smile \mathbf{b}_n), \end{aligned}$$

while all terms on the right hand side are zero except for

$$Sq^m\mathbf{b}_m \wedge Sq^n\mathbf{b}_n = (\mathbf{b}_n \smile \mathbf{b}_m) \wedge (\mathbf{b}_n \smile \mathbf{b}_n).$$

Therefore we shall assume that p and q are integers with $p + q > k$ and that (7.3) holds whenever $m + n = p + q - 1$. Let $h_1 : \mathbf{S}K_{p-1} \to K_p$, $h_2 : \mathbf{S}K_{q-1} \to K_q$ be transfers. Then the maps $h_1 \wedge 1 : \mathbf{S}K_{p-1} \wedge K_q \to K_p \wedge K_q$, $1 \wedge h_2 : K_p \wedge \mathbf{S}K_{q-1} \to K_p \wedge K_q$, induce a map

$$h : (\mathbf{S}K_{p-1} \wedge K_q) \vee (K_p \wedge \mathbf{S}K_{q-1}) \to K_p \wedge K_q.$$

(7.5) Lemma *The homomorphism*

$$h_* : H_r((SK_{p-1} \wedge K_q) \vee (K_p \wedge SK_{q-1})) \to H_r(K_p \wedge K_q)$$

is an epimorphism in mod 2 *homology if* $r < 2(p+q)$.

By the Künneth Theorem, it suffices to prove that, for each i, j with $i + j < 2p + 2q$, one of the homomorphisms

$$h_{1*} \otimes 1 : H_i(SK_{p-1}) \otimes H_j(K_q) \to H_i(K_p) \otimes H_j(K_q),$$
$$1 \otimes h_{2*} : H_i(K_p) \otimes H_j(SK_{q-1}) \to H_i(K_p) \otimes H_j(K_q),$$

is an epimorphism. If h_{1*} is not an epimorphism, it follows from (3.1) that $i \geq 2p$; but then $j < 2q$, and therefore h_{2*} is an epimorphism. □

(7.6) Corollary *If* $r < 2(p+q)$, *the homomorphism*

$$h^* : H^r(K_p \wedge K_q) \to H^r((SK_{p-1} \wedge K_q) \vee (K_p \wedge SK_{q-1}))$$

is a monomorphism in mod 2 *cohomology.* □

(7.7) Corollary *If* $r < 2(p+q)$, *the homomorphisms*

$$\sigma_L^* : H^r(K_p \wedge K_q) \to H^{r-1}(K_{p-1} \wedge K_q),$$
$$\sigma_R^* : H^r(K_p \wedge K_q) \to H^{r-1}(K_p \wedge K_{q-1})$$

induce a monomorphism

$$H^r(K_p \wedge K_q) \to H^{r-1}(K_{p-1} \wedge K_q) \oplus H^{r-1}(K_p \wedge K_{q-1})$$

in mod 2 *cohomology.* □

Moreover, the fact that the Sq^i are stable operations implies that they commute with s^*, s_R^* and s_L^*, and therefore with σ^*, σ_R^*, σ_L^*:

(7.8) *If* $u \in H^r(K_p \wedge K_q)$, *then*

$$\sigma_L^* Sq^i u = Sq^i \sigma_L^* u \in H^{r+i-1}(K_{p-1} \wedge K_q),$$
$$\sigma_R^* Sq^i u = Sq^i \sigma_R^* u \in H^{r+i-1}(K_p \wedge K_{q-1});$$

and if $u \in H^r(K_p)$, *then*

$$\sigma^* Sq^i u = Sq^i \sigma^* u \in H^{r+i-1}(K_{p-1}).$$ □

The inductive step in the proof of Theorem (7.4) now follows. For let

$$z = Sq^k(\mathbf{b}_p \wedge \mathbf{b}_q) - \sum_{i+j=k} Sq^i \mathbf{b}_p \wedge Sq^j \mathbf{b}_q.$$

Then

$$\begin{aligned}
\sigma_L^* z &= \sigma_L^* Sq^k(\mathbf{b}_p \wedge \mathbf{b}_q) - \sum_{i+j=k} \sigma_L^*(Sq^i\mathbf{b}_p \wedge Sq^j\mathbf{b}_q) \\
&= Sq^k\sigma_L^*(\mathbf{b}_p \wedge \mathbf{b}_q) - \sum_{i+j=k} \sigma^* Sq^i\mathbf{b}_p \wedge Sq^j\mathbf{b}_q \\
&= Sq^k(\sigma^*\mathbf{b}_p \wedge \mathbf{b}_q) - \sum_{i+j=k} Sq^i\sigma^*\mathbf{b}_p \wedge Sq^j\mathbf{b}_q \\
&= Sq^k(\mathbf{b}_{p-1} \wedge \mathbf{b}_q) - \sum_{i+j=k} Sq^i\mathbf{b}_{p-1} \wedge Sq^j\mathbf{b}_q \\
&= 0
\end{aligned}$$

by the induction hypothesis. Similarly, $\sigma_R^* z = 0$. By Corollary (7.7), $z = 0$. □

An important consequence of the Cartan formulas is

(7.9) Theorem *If X is an* H-*space, and $x \in H^q(X; \mathbf{Z}_2)$ is primitive, then $Sq^i u \in H^{q+i}(X; \mathbf{Z}_2)$ is also primitive.*

For if μ is the product in X, application of Sq^i to both sides of the formula

$$\mu^* x = x \times 1 + 1 \times x$$

yields

$$Sq^i\mu^* x = Sq^i x \times 1 + 1 \times Sq^i x,$$

the remaining terms vanishing because $Sq^j 1 = 0$ for $j > 0$. But

$$Sq^i\mu^* x = \mu^* Sq^i x$$

by the naturality of Sq^i. □

We now give some more consequences of the Cartan formulas by calculating the Sq^i in certain spaces of interest. We begin with $\mathbf{P}^\infty$.

(7.10) Theorem *Let u be the non-zero element of $H^1(\mathbf{P}^\infty; \mathbf{Z}_2)$; then*

$$Sq^i(u^j) = \binom{j}{i} u^{i+j}.$$

This is proved by induction on j. The cases $j = 0$, 1 are direct consequences of the properties of Sq^i listed in Theorem (6.5). Assume $Sq^i u^j$ is as stated for some $j \geq 1$ and all i. Then

$$\begin{aligned}
Sq^i(u^{j+1}) &= Sq^i(u^j \smile u) \\
&= \sum_{r+s=i} Sq^r u^j \smile Sq^s u \\
&= Sq^i u^j \smile u + Sq^{i-1} u^j \smile u^2 \qquad \text{(by the case } j = 1) \\
&= \left\{\binom{j}{i} + \binom{j-1}{i}\right\} u^{j+i+1} \\
&= \binom{j+1}{i} u^{j+i+1} \qquad \text{(by Pascal's triangle).}
\end{aligned}$$

□

(7.11) Corollary *Let u be the non-zero element of $H^1(\mathbf{P}^n; \mathbf{Z}_2)$. Then*

$$Sq^i u^j = \begin{cases} \binom{j}{i} u^{i+j} & \text{if } i + j \leq n, \\ 0 & \text{if } i + j > n. \end{cases}$$ □

Another application is to the cohomology of the Stiefel manifolds. Let us recall (Theorem (10) of Chapter VII) that $H^*(\mathbf{V}_{n+1,m+1}; \mathbf{Z}_2)$ has a simple system of generators $x_{n-m}, \ldots, x_n$ satisfying the condition

$$x_i^2 = \begin{cases} x_{2i} & \text{if } 2i \leq n, \\ 0 & \text{if } 2i > n. \end{cases}$$

We can now prove

(7.12) Theorem *The action of the Steenrod algebra in $H^*(\mathbf{V}_{n+1,m+1}; \mathbf{Z}_2)$ is given by*

$$Sq^i x_j = \begin{cases} \binom{j}{i} x_{i+j} & \text{if } i + j \leq n, \\ 0 & \text{if } i + j > n. \end{cases} \tag{7.13}$$

Because of Theorem (4.9) of Chapter VII, it suffices to prove this for $\mathbf{V}_{n+1,n} = \mathbf{O}_{n+1}^+$. By Theorem (4.7) of the same Chapter, the generators x_i have the additional properties:

(1) $x_1, \ldots, x_n$ form a basis for the space M^n of primitive elements;
(2) the homomorphism $g_n^* : H^*(\mathbf{O}_{n+1}^+) \to H^*(\mathbf{P}^n)$ maps M^n isomorphically upon $H^*(\mathbf{P}^n)$.

By Theorem (7.9) the space M^n is mapped into itself by the Sq^i. By (2) the action of Sq^i in M^n is determined by its action in $H^*(\mathbf{P}^n)$. This is given by Corollary (7.11). □

The Steenrod squares can now be applied to give a partial solution to the vector field problem.

(7.14) Theorem *Let $n + 1 = 2^k r$, where r is odd. Then the fibration*

$$\mathbf{V}_{n+1,2k+1} \to \mathbf{S}^n$$

does not admit a cross-section. Thus $\mathbf{S}^n$ does not admit a continuous field of 2^k-frames.

This follows from two Lemmas, one topological and one purely arithmetic.

(7.15) Lemma *If the fibration*

$$p : \mathbf{V}_{n+1,m+1} \to \mathbf{S}^n$$

has a cross section, then

$$\binom{n-i}{i} \equiv 0 \pmod 2 \quad \textit{for } i = 1, \ldots, m.$$

(7.16) Lemma *If $n + 1 = 2^k r$, with $r > 1$, then*

$$\binom{n-2^k}{2^i} \equiv 1 \qquad \pmod 2.$$

PROOF OF LEMMA (7.15). Let $\lambda : \mathbf{S}^n \to \mathbf{V}_{n+1,m+1}$ be a cross-section. Then $p \circ \lambda = 1$, so that $\lambda^* \circ p^* = 1$. If $1 \le i \le m$, then

$$Sq^i x_{n-i} = \binom{n-i}{i} x_n = \binom{n-i}{i} p^* \mathbf{s}^n,$$

and therefore

$$\binom{n-i}{i} \mathbf{s}^n = \binom{n-i}{i} \lambda^* p^* \mathbf{s}^n = \lambda^*(Sq^i x_{n-i})$$
$$= Sq^i \lambda^*(x_{n-i}) = 0$$

because $\lambda^* x_{n-i} \in H^{n-i}(\mathbf{S}^n) = 0$. Hence $\binom{n-i}{i} = 0$. □

To prove Lemma (7.16), we shall need to develop some properties of binomial coefficients.

(1) *Let $m = 2r + a$, $n = 2s + b$, where a and b are either 0 or 1. Then*

$$\binom{m}{n} \equiv \binom{r}{s} \cdot \binom{a}{b} \qquad \pmod 2$$

In the polynomial ring $Z_2[x]$, we have

$$(1 + x)^m = (1 + x)^{2r}(1 + x)^a = (1 + x^2)^r(1 + x)^a;$$

by the Binomial Theorem

$$\sum_{k=0}^{m} \binom{m}{k} x^k = \left\{ \sum_{i=0}^{k} \binom{r}{i} x^{2i} \right\} \cdot \left\{ \sum_{j=0}^{1} \binom{a}{j} x^j \right\};$$

when the expressions in braces are multiplied, each power of x appears in only one term. Therefore equating coefficients of like powers of x yields

$$\binom{m}{2s+b} = \binom{r}{s} \cdot \binom{a}{b},$$

as desired.

(2) *Let*

$$m = \sum_{i \geq 0} 2^i m_i,$$

$$n = \sum_{i \geq 0} 2^i n_i,$$

be the 2-adic expansions of m, n. Then

$$\binom{m}{n} \equiv \prod_{i \geq 0} \binom{m_i}{n_i} \quad \text{(mod 2)}.$$

This follows from (1) by induction.

(3) *The binomial coefficient* $\binom{m}{n}$ *is odd if and only if* $m_i \geq n_i$ *for each i* (i.e., *the 2-adic expansion of m "dominates" that of n*).

We can now prove Lemma (7.16). In fact, $r = 2s + 1$ with $s > 0$, and

$$\begin{aligned} n - 2^k &= 2^k(2s + 1) - 2^k - 1 \\ &= 2^{k+1}(s - 1) + 2^{k+1} - 1 \\ &= 1 + 2 + \cdots + 2^k + 2^{k+1}(s - 1); \end{aligned}$$

therefore $n_k = 1$, and our result follows from (3). □

Remark. This result is due to Steenrod and J. H. C. Whitehead. They also prove, using (3) that, if $1 \leq t < 2^k$, then $\binom{n-t}{t} \equiv 0 \pmod 2$. Therefore their argument cannot be improved. On the other hand, this result is not the best possible. In fact, a reformulation of an algebraic result of Hurewicz and Radon asserts that if

$$n + 1 = 2^k r \qquad (r \text{ odd}),$$

$$k = 4a + b \qquad (0 \leq b \leq 3,\ a \geq 0),$$

then S^n admits a q-field, where

$$q + 1 = 2^b + 8a;$$

and Adams has proved that this result is best possible [2].

Another application of the Steenrod process is

(7.17) Theorem *Let* $d = 2, 4,$ *or* 8, *and let* $\alpha \in \pi_{2d-1}(\mathbf{S}^d)$ *be the homotopy class of the Hopf map* $f: \mathbf{S}^{2d-1} \to \mathbf{S}^d$. *Then* $E^k\alpha \neq 0$ *for all k.*

In view of the remarks made earlier in this Chapter, this follows from Theorem (6.1), Corollary (6.4), and the existence of Sq^d. □

Thus we have established

(7.18) Corollary *There are non-zero elements*

$$\eta_n \in \pi_{n+1}(\mathbf{S}^n) \qquad (n \geq 2),$$
$$\nu_n \in \pi_{n+3}(\mathbf{S}^n) \qquad (n \geq 4),$$
$$\sigma_n \in \pi_{n+7}(\mathbf{S}^n) \qquad (n \geq 8). \qquad \square$$

8 Some Relations among the Steenrod Squares

As the Sq^i are stable operations, they can be composed to obtain *iterated squares*: if $I = (i_1, \ldots, i_k)$ is any sequence of positive integers, there is a corresponding stable operation

$$Sq^I = Sq^{i_1} Sq^{i_2} \cdots Sq^{i_k},$$

where composition of stable operations is denoted by juxtaposition.

We have seen in §5 that the stable operations in cohomology mod 2 form a graded algebra, the Steenrod algebra $\mathfrak{A} = \mathfrak{A}_2$. The following questions naturally present themselves:

(1) Do the Sq^i generate $\mathfrak{A}$ as an algebra; in other words, do the iterated squares Sq^I span $\mathfrak{A}$ as a $\mathbf{Z}_2$ vector space?
(2) Are the Sq^I linearly independent? If not, find a complete set of relations among the Sq^I.

The first question was settled in the affirmative by H. Cartan [1]. In fact, let us define a sequence $I = (i_1, \ldots, i_k)$ to be *admissible* if and only if $i_r \geq 2i_{r+1}$ for $r = 1, \ldots, k - 1$. Then Cartan proved that the admissible iterated squares form a *basis* for $\mathfrak{A}$ as a vector space over $\mathbf{Z}_2$.

It follows that the answer to the first question posed in (2) is negative; for example, a non-admissible iterated square, such as Sq^2Sq^2, must be expressible as a linear combination with $\mathbf{Z}_2$ coefficients, of admissible ones (in this case, $Sq^2Sq^2 = Sq^3Sq^1$).

Relations among the iterated squares were found by J. Adem [1]. These form a complete set in the following sense. Let $\tilde{\mathfrak{A}}$ be the free associative algebra generated by the Sq^i (so that the Sq^I for all sequences I form an additive basis for $\tilde{\mathfrak{A}}$). The kernel of the natural epimorphism $\pi : \tilde{\mathfrak{A}} \to \mathfrak{A}$ is a two-sided ideal $\mathscr{I}$, and the Adem relations determine a subset of $\mathscr{I}$ which spans $\mathscr{I}$ as an ideal. In other words, any linear relation among the iterated squares is a consequence of the Adem relations.

We shall defer the proofs of the theorems of Cartan and Adem to a later chapter, since they depend on more sophisticated relations among the homology groups of the fibre, total space, and base space of a fibration. However, we shall derive a few of the simpler of the Adem relations to illustrate the type of result which can be inferred from them.

(8.1) Theorem *The relation*

$$Sq^1Sq^n = \begin{cases} 0 & (n \text{ odd}) \\ Sq^{n+1} & (n \text{ even}) \end{cases}$$

holds for every positive integer n.

Let $K_n = K(\mathbf{Z}_2, n)$; then, by Theorem (7.8) of Chapter V,

$$H_n(K_n; \mathbf{Z}) = \mathbf{Z}_2, \qquad H_{n+1}(K_n; \mathbf{Z}) = 0$$

and therefore

$$H^n(K_n; \mathbf{Z}_2) \approx \operatorname{Hom}(H_n(K_n; \mathbf{Z}), \mathbf{Z}_2) \approx \mathbf{Z}_2,$$
$$H^{n+1}(K_n; \mathbf{Z}_2) \approx \operatorname{Ext}(H_n(K_n; \mathbf{Z}), \mathbf{Z}_2) \approx \mathbf{Z}_2.$$

Let $\mathbf{b}_n$ generate the former group; then we have seen in the proof of (6.6) that $Sq^1\mathbf{b}_n$ generates the latter.

We first observe that

$$Sq^1Sq^n(\mathbf{b}_n) = Sq^1(\mathbf{b}_n^2) = (Sq^1\mathbf{b}_n)\mathbf{b}_n + \mathbf{b}_n(Sq^1\mathbf{b}_n) = 0$$

by the Cartan formula and the commutative law for the cup product. Consider the following segment

$$H^{2n+2}(K_{n+1} \wedge K_{n+1}) \xrightarrow{d_2^*} H^{2n+2}(K_{n+1}) \xrightarrow{\sigma^*} H^{2n+1}(K_n)$$

of the suspension exact sequence of §1; we have taken $B = K_{n+1}$, $F = K_n$, and are using (3.2*). Since the Sq^i are stable operations, they commute with σ^*; hence

$$\sigma^*(Sq^1Sq^n\mathbf{b}_{n+1}) = Sq^1Sq^n\sigma_n^*\mathbf{b}_{n+1} = Sq^1Sq^n\mathbf{b}_n = 0,$$

and therefore $Sq^1Sq^n\mathbf{b}_{n+1} \in \operatorname{Im} d_2^*$. By the Künneth Theorem, $H^{2n+2}(K_{n+1} \wedge K_{n+1})$ is generated by $\mathbf{b}_{n+1} \wedge \mathbf{b}_{n+1}$, and $d_2^*(\mathbf{b}_{n+1} \wedge \mathbf{b}_{n+1}) = \mathbf{b}_{n+1}^2$. Thus $Sq^1Sq^n\mathbf{b}_{n+1} = \lambda\mathbf{b}_{n+1}^2$ for some $\lambda \in \mathbf{Z}_2$.

Next observe that $\sigma^* : H^{2n+3}(K_{n+2}) \to H^{2n+2}(K_{n+1})$ is a monomorphism, and $\sigma^*(Sq^{n+1}\mathbf{b}_{n+2}) = Sq^{n+1}\mathbf{b}_{n+1} = \mathbf{b}_{n+1}^2$. Since $H^{2n+3}(K_{n+2})$ is in the stable range, and

$$\begin{aligned} \sigma^*(Sq^1Sq^n\mathbf{b}_{n+2} &+ \lambda Sq^{n+1}\mathbf{b}_{n+2}) \\ &= Sq^1Sq^n\mathbf{b}_{n+1} + \lambda Sq^{n+1}\mathbf{b}_{n+1} \\ &= Sq^1Sq^n\mathbf{b}_{n+1} + \lambda\mathbf{b}_{n+1}^2 = 0, \end{aligned}$$

it follows that the stable operations Sq^1Sq^n and Sq^{n+1} satisfy

$$Sq^1Sq^n = \lambda Sq^{n+1}. \tag{8.2}$$

It remains to determine λ. To do so we shall apply both sides of (8.2) to a suitable cohomology class in a suitable space X. In fact, let $X = K_1 =$

$\mathbf{P}^{\infty}(\mathbf{R})$; then $H^q(X)$ is generated by $\mathbf{b}_1^q$ for every q, and

$$Sq^i(\mathbf{b}_1^q) = \binom{q}{i}\mathbf{b}_1^{q+1},$$

according to (7.10). Hence

$$\begin{aligned} Sq^1Sq^n(\mathbf{b}_1^{n+1}) &= \binom{n+1}{n}Sq^1(\mathbf{b}_1^{2n+1}) = (n+1)Sq^1(\mathbf{b}_1^{2n+1}) \\ &= (n+1)\mathbf{b}_1^{2n+2}; \end{aligned}$$

while

$$Sq^{n+1}(\mathbf{b}_1^{n+1}) = (\mathbf{b}_1^{n+1})^2 = \mathbf{b}_1^{2n+2},$$

and it follows that $\lambda = n+1$. □

Similar, but more complicated, is the proof of

(8.3) Theorem *If $n \geq 2$, then*

$$Sq^2Sq^n = Sq^{n+1}Sq^1 + \binom{n-1}{2}Sq^{n+2}.$$

Let $\theta = Sq^2Sq^n + Sq^{n+1}Sq^1$. Then

$$\begin{aligned} \theta(\mathbf{b}_n) &= Sq^2Sq^n(\mathbf{b}_n) + Sq^{n+1}(Sq^1\mathbf{b}_n) \\ &= Sq^2(\mathbf{b}_n^2) + (Sq^1\mathbf{b}_n)^2 \quad \text{by Theorem (6.5)} \\ &= (Sq^1\mathbf{b}_n)^2 + (Sq^1\mathbf{b}_n)^2 \quad \text{by the Cartan relation (7.1)} \\ &= 0. \end{aligned}$$

Hence $\sigma^*\theta(\mathbf{b}_{n+1}) = \theta(\sigma^*\mathbf{b}_{n+1}) = \theta(\mathbf{b}_n) = 0$; by exactness of the sequence

$$H^{2n+3}(K_{n+1} \wedge K_{n+1}) \xrightarrow{d_2^*} H^{2n+3}(K_{n+1}) \xrightarrow{\sigma^*} H^{2n+2}(K_n),$$

there is an element $x \in H^{2n+3}(K_{n+1} \wedge K_{n+1})$ such that $d_2^* x = \theta(\mathbf{b}_{n+1})$. By the Künneth Theorem, the group $H^{2n+3}(K_{n+1} \wedge K_{n+1})$ is generated by the elements $Sq^1\mathbf{b}_{n+1} \wedge \mathbf{b}_{n+1}$ and $\mathbf{b}_{n+1} \wedge Sq^1\mathbf{b}_{n+1}$, and each of these elements is mapped by d_2^* into $\mathbf{b}_{n+1}(Sq^1\mathbf{b}_{n+1})$. Thus

$$\theta(\mathbf{b}_{n+1}) = \lambda\mathbf{b}_{n+1}(Sq^1\mathbf{b}_{n+1})$$

for some $\lambda \in \mathbf{Z}_2$.

Now consider the sequence

$$\begin{aligned} H^{2n+4}(K_{n+2} \wedge K_{n+2}) \xrightarrow{d_2^*} H^{2n+4}(K_{n+2}) \xrightarrow{\sigma^*} \\ H^{2n+3}(K_{n+1}) \xrightarrow{\tau^*} H^{2n+3}(K_{n+1} \wedge K_{n+1}), \end{aligned}$$

and observe that

$$\tau^*(\mathbf{b}_{n+1}(Sq^1\mathbf{b}_{n+1})) = \mathbf{b}_{n+1} \wedge Sq^1\mathbf{b}_{n+1} + Sq^1\mathbf{b}_{n+1} \wedge \mathbf{b}_{n+1} \neq 0,$$

while

$$\tau^*\theta(\mathbf{b}_{n+1}) = \tau^*\theta(\sigma^*\mathbf{b}_{n+2}) = \tau^*\sigma^*\theta(\mathbf{b}_{n+2}) = 0.$$

It follows that $\lambda = 0$, $\theta(\mathbf{b}_{n+1}) = 0$. Since

$$\sigma^*\theta(\mathbf{b}_{n+2}) = \theta(\mathbf{b}_{n+1}) = 0,$$

$H^{2n+4}(K_{n+2} \wedge K_{n+2})$ is generated by $\mathbf{b}_{n+2} \wedge \mathbf{b}_{n+2}$, and $d_2^*(\mathbf{b}_{n+2} \wedge \mathbf{b}_{n+2}) = \mathbf{b}_{n+2}^2$, it follows that

$$\theta(\mathbf{b}_{n+2}) = \mu\mathbf{b}_{n+2}^2.$$

As in the proof of Theorem (8.1), we deduce that

(8.4)
$$\theta = \mu Sq^{n+2}$$

with $\mu \in \mathbf{Z}_2$, and it remains to determine μ.

This time we apply both sides of (8.4) to the class $\mathbf{b}_1^{n+2} \in H^{n+2}(K_1)$; of course,

$$\mu Sq^{n+2}(\mathbf{b}_1^{n+2}) = \mu(\mathbf{b}_1^{n+2})^2 = \mu\mathbf{b}_1^{2n+4}.$$

On the other hand

$$Sq^2Sq^n(\mathbf{b}_1^{n+2}) = \binom{n+2}{n}Sq^2(\mathbf{b}_1^{2n+2}) = \binom{n+2}{2}\binom{2n+2}{2}\mathbf{b}_1^{2n+4}$$
$$= \binom{n+2}{2}(n+1)\mathbf{b}_1^{2n+4};$$

$$Sq^{2n+1}Sq^1(\mathbf{b}_1^{n+2}) = (n+2)Sq^{2n+1}(\mathbf{b}_1^{n+3}) = (n+2)\binom{n+3}{n+1}\mathbf{b}_1^{2n+4}$$
$$= n\binom{n+3}{2}\mathbf{b}_1^{2n+4}.$$

Since the mod 2 binomial coefficient $\binom{n}{2}$ has period 4 in n, we need merely check that the relation

$$(n+1)\binom{n+2}{2} + n\binom{n+3}{2} = \binom{n-1}{2}$$

holds for four consecutive values of n. We leave this to the reader. □

A consequence of Theorems (8.1) and (8.3) is

(8.5) Corollary *If $n > 1$ is odd or $n > 2$, $n \equiv 2 \pmod 4$, then Sq^n is a decomposable element of $\mathfrak{A}$.*

In fact,

$$Sq^{2k+1} = Sq^1 Sq^{2k},$$
$$Sq^{4k+2} = Sq^2 Sq^{4k} + Sq^{4k+1} Sq^1. \qquad \square$$

Let us return to the considerations of §6. Let $\alpha \in \pi_n(\mathbf{S}^r)$, and let θ be a cohomology operation of type $(r, n+1; \Pi, G)$. We shall say that α is *detectable* by θ if and only if

$$\theta : H^r(\mathbf{T}_\alpha; \Pi) \to H^{n+1}(\mathbf{T}_\alpha; G)$$

is non-zero. For example, the elements η_n, ν_n, σ_n of Corollary (7.18) are detectable by Sq^2, Sq^4, Sq^8, respectively.

The cohomology operation θ is said to be *decomposable* if and only if there are operations ϕ_i, ψ_i of types $(r, m_i; \Pi, G_i)$, $(m_i, n+1; G_i, G)$ $(i = 1, \ldots, k)$ with $r < m_i < n+1$, such that

$$\theta = \sum_{i=1}^{k} \psi_i \circ \phi_i .$$

Clearly

(8.6) *If $\alpha \in \pi_n(\mathbf{S}^r)$ and θ is a decomposable operation of type $(r, n+1; \Pi, G)$, then α is not detectable by θ.*

For if $x \in H^r(\mathbf{T}_\alpha; \Pi)$, then

$$\phi_i(x) \in H^{m_i}(\mathbf{T}_\alpha; G_i) = 0$$

and therefore

$$\theta(x) = \sum_{i=1}^{k} \psi_i(\phi_i(x)) = \sum_{i=1}^{k} \psi_i(0) = 0. \qquad \square$$

(8.7) Corollary *If θ is a decomposable element of degree d of the Steenrod algebra $\mathfrak{A}$, then no element of $\pi_{k+n-1}(\mathbf{S}^n)$ is detectable by θ.* $\square$

(8.8) Corollary *If $\alpha \in \pi_{d+n-1}(\mathbf{S}^n)$ is detectable by Sq^d, then $d = 2$ or $d \equiv 0$* (mod 4). $\square$

In fact, Adem's relations can be used to show that Sq^i is decomposable unless i is a power of two. Combining Adem's results with those of Cartan, we can conclude that, if $\alpha \in \pi_{d+n-1}(\mathbf{S}^n)$ is detectable by *any* primary operation, then d must be a power of two. We shall return to these questions in Volume Two.

The Adem relations can also be used in a positive way, to show that certain maps are essential. For example,

(8.9) Theorem *For each integer $n \geq 2$, the element*

$$\eta_n \circ \eta_{n+1} \in \pi_{n+2}(\mathbf{S}^n)$$

is non-zero.

Let $f : \mathbf{S}^{n+1} \to \mathbf{S}^n$, $g : \mathbf{S}^{n+2} \to \mathbf{S}^{n+1}$ be maps representing η_n, η_{n+1}, respectively. The mapping cone of g is a CW-complex $K = \mathbf{S}^{n+1} \cup_g \mathbf{E}^{n+3}$; since η_{n+1} is detected by Sq^2, we have

$$Sq^2 : H^{n+1}(K) \approx H^{n+3}(K).$$

If $h : (\mathbf{E}^{n+3}, \mathbf{S}^{n+2}) \to (K, \mathbf{S}^{n+1})$ is a characteristic map for the $(n+3)$-cell, then $f \circ (h | \mathbf{S}^{n+2}) = f \circ g \simeq 0$, and therefore $f \circ g$ admits an extension $k_1 : \mathbf{E}^{n+3} \to \mathbf{S}^n$. It follows that $f : \mathbf{S}^{n+1} \to \mathbf{S}^n$ has the extension $k = k_1 \circ h^{-1} : K \to \mathbf{S}^n$. Let L be the mapping cone of k. Then $L = \mathbf{S}^n \cup \mathbf{E}^{n+2} \cup \mathbf{E}^{n+4}$ is a CW-complex, and the subcomplex $\mathbf{S}^n \cup \mathbf{E}^{n+2}$ is the mapping cone of f, while L/S^n is the suspension $\mathbf{S}K$. From the first fact, it follows that

$$Sq^2 : H^n(L) \approx H^{n+2}(L);$$

from the second that

$$Sq^2 : H^{n+2}(L) \approx H^{n+4}(L),$$

so that $Sq^2 Sq^2$ is non-zero in $H^*(L)$. But $Sq^2 Sq^2 = Sq^3 Sq^1$, by Theorem (8.3) and $Sq^3 Sq^1 = 0$ since $H^{n+1}(L) = 0$. This contradiction completes the proof. □

Similarly, using the more complicated Adem relations

$$Sq^4 Sq^4 = Sq^7 Sq^1 + Sq^6 Sq^2$$
$$Sq^8 Sq^8 = Sq^{15} Sq^1 + Sq^{14} Sq^2 + Sq^{12} Sq^4,$$

one can prove that $\nu_n \circ \nu_{n+3}$ and $\sigma_n \circ \sigma_{n+7}$ are non-zero whenever $n \geq 4$, $n \geq 8$, respectively. Again, we refer the reader to Volume II.

9 The Action of the Steenrod Algebra on the Cohomology of Some Compact Lie Groups

In §7 we calculated the action of the mod 2 Steenrod algebra on $H^*(\mathbf{O}_n^+; \mathbf{Z}_2)$ in order to deduce it on $H^*(\mathbf{V}_{n,m}; \mathbf{Z}_2)$. Let us recall the result.

(9.1) Theorem *The action of the Steenrod algebra on $H^*(\mathbf{O}_n^+; \mathbf{Z}_2)$ is* *determined by*

$$Sq^i x_j = \begin{cases} \binom{j}{i} x_{i+j} & \text{if } i+j<n, \\ 0 & \text{if } i+j \geq n. \end{cases}$$

We next turn to the unitary groups. Let us first calculate the action of $\mathfrak{A}$ on $H^*(\mathbf{P}^n(\mathbf{C}))$. As the injection $H^*(\mathbf{P}^\infty(\mathbf{C})) \to H^*(\mathbf{P}^n(\mathbf{C}))$ is an epimorphism, we may assume $n = \infty$. We have seen that $H^*(\mathbf{P}^\infty(\mathbf{C}))$ is the polynomial algebra $\mathbf{Z}_2[u]$, where u generates $H^2(\mathbf{P}^\infty(\mathbf{C}))$.

(9.2) Theorem *The action of* $\mathfrak{A}$ *on* $H^*(\mathbf{P}^\infty(\mathbf{C}))$ *is given by*

$$Sq^i(u^j) = 0 \quad \text{if } i \text{ is odd,}$$

$$Sq^{2i}(u^j) = \binom{j}{i} u^{i+j}.$$

This is proved by induction on j, after the observation that, since $H^q(\mathbf{P}^\infty(\mathbf{C})) = 0$ for all odd q, Sq^i must vanish for any odd i. Assume that $Sq^{2i}(u^j)$ is as stated. By the Cartan formula

$$\begin{aligned} Sq^{2i}(u^{j+1}) &= Sq^{2i}(u^j \smile u) \\ &= Sq^{2i}(u^j) \smile u + Sq^{2i-2}(u^j) \smile Sq^2 u \\ &= \binom{j}{i} u^{j+i+1} + \binom{j}{i-1} u^{j+i-1} \smile u^2 \\ &= \left\{\binom{j}{i} + \binom{j}{i-1}\right\} u^{j+i+1} = \binom{j+1}{i} u^{i+j+1} \end{aligned}$$

by Pascal's triangle. □

Since the Sq^i are stable operations, we may use (9.2) to calculate the action of $\mathfrak{A}$ in $H^*(\mathbf{SP}^\infty(\mathbf{C}))$. Let v_i be the generator of $H^{2i+1}(\mathbf{SP}^\infty(\mathbf{C}))$ which is mapped into u^i by the suspension operator σ^*.

(9.3) Corollary *The action of* $\mathfrak{A}$ *in* $H^*(\mathbf{SP}^\infty(\mathbf{C}))$ *is given by*

$$Sq^i v_j = 0 \quad \text{if } i \text{ is odd,}$$

$$Sq^{2i} v_j = \binom{j}{i} v_{j+i}. \qquad \square$$

We can now determine the action of $\mathfrak{A}$ on the cohomology of the unitary groups.

(9.4) Theorem *The action of* $\mathcal{A}$ *on* $H^*(\mathbf{U}_n^+)$ *is determined by*

$$Sq^i u_j = 0 \qquad \text{if } i \text{ is odd};$$

$$Sq^{2i} u_j = \begin{cases} \binom{j-1}{i} u_{i+j} & \text{if } i + j \leq n, \\ 0 & \text{if } i + j > n. \end{cases}$$

By (4.6) of Chapter VII, the restriction of G_{n+1}^* to the space P^* of primitive elements is an isomorphism of P^* with $H^*(\mathbf{SP}^n(\mathbf{C}))$. By Theorem (7.9), $Sq^i P^* \subset P^*$, and therefore the theorem follows from our calculation of the Sq^i in $H^*(\mathbf{SP}^n(\mathbf{C}))$, which is given by Corollary (9.3).

Similarly,

(9.5) Theorem *The action of* $\mathcal{A}$ *in* $H^*(\mathbf{U}_n)$ *is given by*

$$Sq^i x_j = 0 \qquad \text{if } i \text{ is odd};$$

$$Sq^{2i} x_j = \begin{cases} \binom{j-1}{i} x_{i+j} & \text{if } i + j \leq n, \\ 0 & \text{if } i + j > n. \end{cases}$$

The injection $k^* : H^*(\mathbf{U}_n) \to H^*(\mathbf{U}_n^+)$ is an epimorphism whose kernel is the ideal generated by x_1; as $\mathbf{U}_n^+$ is a subgroup of $\mathbf{U}_n$, $k^*P^*(\mathbf{U}_n) \subset P^*(\mathbf{U}_n^+)$ and it follows that $k^*(x_j) = u_j$. It follows that the asserted formulae hold except possibly for $j = 1$. But $\mathbf{U}_1$ is one-dimensional, so that

$$Sq^0 u_1 = u_1, \qquad Sq^1 u_1 = u_1^2 = 0,$$

$$Sq^i u_1 = 0 \qquad (i \geq 2),$$

and these agree with the stated results for $j = 1$. □

We can now complete the determination of the integral and mod 2 cohomology rings of $\mathbf{G}_2$, as well as the action of $\mathcal{A}$ on the latter.

(9.6) Theorem *The cohomology ring* $H^*(\mathbf{G}_2; \mathbf{Z}_2)$ *has a simple system of primitive generators* v_3, v_5, v_6. *The action of* $\mathcal{A}$ *on* $H^*(\mathbf{G}_2; \mathbf{Z}_2)$ *is determined by the table*

	v_3	v_5	v_6
Sq^1	0	v_6	0
Sq^2	v_5	0	0
Sq^3	v_6	0	0
Sq^4	0	0	0
Sq^5	0	0	0
Sq^6	0	0	0

(9.7) Corollary *As an algebra,* $H^*(\mathbf{G}_2; \mathbf{Z}_2)$ *is isomorphic with the tensor product*

$$\mathbf{Z}_2[v_3]/(v_3^4) \otimes \Lambda(Sq^2 v_3).$$

Remark. The above isomorphism is not an isomorphism of *algebras over* $\mathfrak{A}$, because of the relation $Sq^1 Sq^2 v_3 = Sq^3 v_3 = v_3^2$.

Let us recall (Theorem (5.18) of Chapter VII) that there are elements $v_3 = \tilde{x} \in H^3(\mathbf{G}_2)$, $v_5 = p^* y_5 \in H^5(\mathbf{G}_2)$, $v_6 = p^*(y_6) \in H^6(\mathbf{G}_2)$ which form a simple system of generators for $H^*(\mathbf{G}_2)$. Consider, moreover, the Wang sequence

$$H^{q-6}(\mathbf{U}_3^+) \xrightarrow{\alpha^*} H^q(\mathbf{G}_2) \xrightarrow{i^*} H^q(\mathbf{U}_3^+) \xrightarrow{\theta^*} H^{q-5}(\mathbf{U}_3^+)$$

associated with the fibration

$$\mathbf{U}_3^+ \longrightarrow \mathbf{G}_2 \longrightarrow \mathbf{S}^6,$$

and recall that $H^*(\mathbf{U}_3^+)$ is the exterior algebra on two primitive generators $u_2 \in H^3(\mathbf{U}_3^+)$, $u_3 \in H^5(\mathbf{U}_3^+)$, related by

$$Sq^2 u_2 = u_3.$$

Evidently $i^* v_3 = u_2$, and from exactness of the sequence

$$0 \to H^5(\mathbf{G}_2) \xrightarrow{i^*} H^5(\mathbf{U}_3^+) \xrightarrow{\theta^*} H^0(\mathbf{U}_3^+) \xrightarrow{\alpha^*} H^6(\mathbf{G}_2) \to 0$$

we deduce that $i^*(v_5) = u_3$, $\theta^*(u_3) = 0$, $\alpha^*(1) = v_6$. Then

$$i^* Sq^2 v_3 = Sq^2 i^* v_3 = Sq^2 u_2 = u_3 = i^* v_5,$$

and therefore

$$Sq^2 v_3 = v_5.$$

In $H^*(\mathbf{V}_{7,2})$ we have the relation

$$Sq^1 y_5 = y_6,$$

and therefore

$$v_6 = p^* y_6 = p^* Sq^1 y_5 = Sq^1 p^* y_5 = Sq^1 v_5.$$

By Theorem (8.1),

$$v_6 = Sq^1 v_5 = Sq^1 Sq^2 v_3 = Sq^3 v_3;$$

thus $v_3^2 = Sq^3 v_3 = v_6$, which settles the question left open in Theorem (5.18) of Chapter VII.

The element v_3 being primitive, it follows from Theorem (7.9) that v_5 and v_6 are primitive. We can now easily complete the table giving the $Sq^i v_j$:

$$\begin{aligned} Sq^2 v_6 = Sq^2(v_3^2) &= (Sq^1 v_3)^2 \quad \text{by the Cartan formula} \\ &= 0; \end{aligned}$$

$$Sq^3 v_5 = Sq^3 Sq^2 v_3 = Sq^1 Sq^2 Sq^2 v_3 = Sq^1 Sq^3 Sq^1 v_3$$
$$= 0 \quad \text{by the Adem relations in §8;}$$
$$Sq^3 v_6 = Sq^1 Sq^2 v_6 = 0 \qquad \text{(Adem);}$$
$$Sq^5 v_6 = Sq^2 Sq^3 v_6 + Sq^4 Sq^1 v_6 = 0 \qquad \text{(Adem).}$$

The remaining $Sq^i v_j$ lie in groups which are zero, except for $Sq^4 v_5$. But $Sq^4 v_5$ is primitive, by Theorem (7.9), while the non-zero element $v_3 v_6 = v_3^3$ satisfies

$$\mu^*(v_3^3) = \mu^*(v_3)^3 = (v_3 \times 1 + 1 \times v_3)^3$$
$$= v_3^3 \times 1 + v_3^2 \times v_3 + v_3 \times v_3^2 + 1 \times v_3^3$$

and accordingly is not primitive. □

Finally, we can complete the determination of the integral cohomology ring $H^*(\mathbf{G}_2; \mathbf{Z})$ (cf. Theorem (5.17) of Chapter VII).

(9.8) Theorem *The elements* $x \in H^3(\mathbf{G}_2; \mathbf{Z})$, $p^*(h) \in H^6(\mathbf{G}_2; \mathbf{Z})$ *are related by*

$$x^2 = p^*(h).$$

Thus $H^q(\mathbf{G}_2; \mathbf{Z})$ *is an infinite cyclic group generated by* 1, x, $p^*(z)$, $x \smile p^*(z)$ *for* $q = 0, 3, 11, 14$, *respectively, while* $H^6(\mathbf{G}_2; \mathbf{Z})$ *is a cyclic group of order* 2 *generated by* x^2, $H^9(\mathbf{G}_2; \mathbf{Z})$ *is a cyclic group of order two generated by* x^3, *and* $H^q(\mathbf{G}_2; \mathbf{Z}) = 0$ *for all other values of* q.

We need only observe that the coefficient homomorphism $H^*(\mathbf{G}_2; \mathbf{Z}) \to H^*(\mathbf{G}_2; \mathbf{Z}_2)$ induced by reduction mod 2 is a ring homomorphism and that x^2 is mapped into $v_3^2 = v_6 \neq 0$. Thus $x^2 \neq 0$ must be the non-zero element of $H^6(\mathbf{G}_2; \mathbf{Z})$. □

Exercises

1. In the notation of §1, let $q : X \to B$ be the map defined by $q(u) = u(\frac{1}{2})$. Prove that
   (a) q is a fibre map;
   (b) the fibre F_* of q is homeomorphic with the subspace $E_1 \times F \cup F \times E_0$ of $E_1 \times E_0$;
   (c) the boundary operator of the Mayer–Vietoris sequence of the triad $(F_*; E_1 \times F, F \times E_0)$ induces an isomorphism
   $$\gamma_* : H_{q+1}(F_*) \approx H_q(F \times F, F \vee F);$$
   (d) the diagram

$$\begin{array}{ccccc}
H_{q+1}(F_*) & \xrightarrow{i_*} & H_{q+1}(X) & \xrightarrow{j_*} & H_{q+1}(X, F_*) \\
\Big\downarrow \gamma_* & & \Big\downarrow \Delta_* & & \Big\downarrow q_* \\
H_q(F \times F, F \vee F) & \xrightarrow[\tau_*]{} & H_q(F) & \xrightarrow[\sigma_*]{} & H_{q+1}(B)
\end{array}$$

(where i_* and j_* are the injections) is commutative.

2. (Barcus and Meyer [1]). Suppose that B is n-connected. Prove that
   (a) F_* is $2n$-connected;
   (b) $q_* : H_r(X, F_*) \to H_r(B)$ is an isomorphism for $r \leq 3n + 1$;
   (c) there is a commutative diagram

$$\begin{array}{ccccccccccc}
H_{3n+1}(F_*) & \to & H_{3n+1}(X) & \to \cdots \to & H_{q+1}(F_*) & \to & H_{q+1}(X) & \to & H_{q+1}(B) & \longrightarrow & H_q(F_*) \to \cdots \\
\Big\downarrow \gamma_* & & \Big\downarrow \Delta_* & & \Big\downarrow \gamma_* & & \Big\downarrow \Delta_* & & \Big\downarrow 1 & & \Big\downarrow \gamma_* \\
H_{3n}(F \times F, F \vee F) & \to & H_{3n}(F) & \to \cdots \to & H_q(F \times F, F \vee F) & \to & H_q(F) & \to & H_{q+1}(B) & \to & H_{q-1}(F \times F, F \vee F) \to \cdots
\end{array}$$

where the top sequence is the homology sequence of the fibration q and the bottom one is that of Corollary (1.6), but extended one unit to the left.

3. Assuming that the above results can be dualized (as indeed they can!), show that the last sentence of Theorem (4.5) also holds for $q = 3n$.

4. (Steenrod [4]). Formulate the notion of *homology* operation, and prove that every homology operation is additive.

5. Prove that, if $p : X \to S^n$ is a fibration with fibre F and λ is any stable cohomology operation then the diagram

$$\begin{array}{ccccccc}
H^{q-1}(F; A) & \xrightarrow{\theta^*} & H^{q-n}(F; A) & \xrightarrow{\alpha^*} & H^q(X; A) & \xrightarrow{i^*} & H^q(F, A) \\
\Big\downarrow \lambda & & \Big\downarrow \lambda & & \Big\downarrow \lambda & & \Big\downarrow \lambda \\
H^{q+p-1}(F; B) & \xrightarrow[\theta^*]{} & H^{q+p-n}(F; B) & \xrightarrow[\alpha^*]{} & H^{q+p}(X; B) & \xrightarrow[i^*]{} & H^{q+p}(F; B)
\end{array}$$

whose top and bottom rows are the appropriate Wang sequences, is commutative.

6. Prove that there is a homomorphism

$$\sigma_2 : H_{q-1}(F \times F, F \vee F) \to H_{q+1}(B \times B, B \vee B)$$

such that the diagram

$$\begin{array}{ccccc}
\bigoplus_{r+s=q+1} H_{r-1}(F) \otimes H_{s-1}(F) & \to & H_{q-1}(F \times F, F \vee F) & \to & \bigoplus_{r+s=q} \operatorname{Tor}(H_{r-1}(F), H_{s-1}(F)) \\
\Big\downarrow \sigma_* \otimes \sigma_* & & \Big\downarrow \sigma_2 & & \Big\downarrow \operatorname{Tor}(\sigma_*, \sigma_*) \\
\bigoplus_{r+s=q+1} H_r(B) \otimes H_s(B) & \to & H_{q+1}(B \times B, B \vee B) & \to & \bigoplus_{r+s=q} \operatorname{Tor}(H_r(B), H_s(B))
\end{array}$$

is commutative and $p_6 \circ \beta = \sigma_2$.

7. (Adem [2]). Prove that $\nu_n \circ \eta_{n+3} \neq 0$ if $n = 4$ or 5, and that $\sigma_n \circ \eta_{n+7} \neq 0$ if $n = 8$ or 9.

8. Assuming the Adem relation

$$Sq^4 Sq^n = \binom{n-1}{4} Sq^{n+4} + \binom{n-2}{2} Sq^{n+3} Sq^1 + Sq^{n+2} Sq^2,$$

deduce that $\sigma_n \circ \eta_{n+7} \neq 0$ for every n, and that $\sigma_n \circ \nu_{n+7} \neq 0$ for $8 \leq n \leq 11$.

9. (Borel [3]). Prove that the cohomology rings $H^*(\mathbf{Spin}(7); \mathbf{Z})$ and $H^*(\mathbf{G}_2 \times \mathbf{S}^7; \mathbf{Z})$ are isomorphic.

10. Prove that $H^*(\mathbf{Spin}(7); \mathbf{Z}_2) \approx H^*(\mathbf{G}_2 \times \mathbf{S}^7; \mathbf{Z}_2)$ as an $\mathfrak{A}$-module.

11. Prove the corresponding statements for Spin(9):
    (a) the cohomology rings $H^*(\mathbf{Spin}(9); \mathbf{Z})$ and $H^*(\mathbf{Spin}(7) \times \mathbf{S}^{15}; \mathbf{Z})$ are isomorphic;
    (b) the $\mathfrak{A}$-modules $H^*(\mathbf{Spin}(9); \mathbf{Z}_2)$ and $H^*(\mathbf{Spin}(7) \times \mathbf{S}^{15}; \mathbf{Z}_2)$ are isomorphic.

CHAPTER IX

Postnikov Systems

In Chapter V we showed how to use the process of attaching cells to construct CW-complexes with desired properties. In this Chapter we shall exploit this process further, one of our aims being to show how any space can be built up, up to homotopy type, out of Eilenberg–MacLane spaces.

Obviously, given any 0-connected space X, we can kill its homotopy groups above a certain dimension N by attaching cells. We start with a family of maps of $\mathbf{S}^{N+1}$ into X representing a set of generators of $\pi_{N+1}(X)$. Attaching $(N+2)$-cells to X, we obtain a new space X' with the same homotopy groups as X through dimension N, but with $\pi_{N+1}(X') = 0$. In a similar way we kill $\pi_{N+2}(X')$ by attaching $(N+3)$-cells to obtain a space X''. Iterating this process indefinitely, we obtain a space $X^* \supset X$ with $\pi_i(X) \approx \pi_i(X^*)$ for $i \leq N$, $\pi_i(X^*) = 0$ for $i > N$.

Let us turn the inclusion map of X into X^* into a fibration by the method of §7, Chapter I. The fibre $\tilde{X}$ then admits a fibre map $p : \tilde{X} \to X$ and we have $\pi_i(\tilde{X}) = 0$ for $i \leq N$, while $p_* : \pi_i(\tilde{X}) \to \pi_i(X)$ for $i > N$.

The map p is called an N-connective fibre map. If $q : \bar{X} \to X$ is an $(N-1)$-connective fibre map, we can construct an N-connective fibration $q' : \tilde{X} \to \bar{X}$. The composite is then an N-connective fibration $p : \tilde{X} \to X$. Thus, starting with $N = 1$, we can construct a tower of fibrations

$$\cdots \to X_n \xrightarrow{p_n} X_{n-1} \to \cdots \to X_1 \xrightarrow{p_1} X_0 = X$$

such that $p_n : X_n \to X_{n-1}$ is an n-connective fibration and therefore so is $q_n = p_1 \circ \cdots \circ p_n$. The map p_1 has many of the properties of the universal covering map, and the construction of the tower can be thought of as a process of generalizing the universal covering space.

The map p_n was constructed from X_{n-1} with the aid of an auxiliary space X_{n-1}^*; the latter space is an Eilenberg–MacLane space $K_n = K(\pi_n(X), n)$.

The above process is essentially one of resolving X into Eilenberg–MacLane components. It was discovered independently by H. Cartan and Serre [1] and by the present author [5] in 1952, and was developed with a view to calculating the homotopy groups by studying the behavior of the homology groups of the tower and using the isomorphism $\pi_n(X) \approx \pi_n(X_{n-1}) \approx H_n(X_{n-1})$. This construction is expounded in §1.

For each n, let X^n be a space constructed from X by the process described in the second paragraph of this Chapter. Then there are essentially unique maps $f_{n+1} : X^{n+1} \to X^n$ such that $f_{n+1} | X$ is the inclusion of X in X^n. The problem of constructing a left inverse to f_{n+1} leads to an obstruction $\kappa^{n+2} \in H^{n+2}(\mathbf{T}_{f_{n+1}}, X^{n+1}; \pi_{n+1}(X))$ whose image

$$k^{n+2} \in H^{n+2}(X^n; \pi_{n+1}(X))$$

under the injection is called the $(n+2)^{nd}$ *Postnikov invariant* of X. These classes are introduced in §2 and their naturality properties formulated. In §3, it is shown that the pair (X^{n+1}, X) is determined up to homotopy equivalence by the pair (X^n, X) and the class k^{n+2}. In fact, the class k^{n+2} determines a map $h : X^n \to K(\pi_{n+1}(X), n+2) = K_n$, and we construct X^{n+1} as a CW-approximation to the total space W^{n+1} of the fibration induced by h from the path space fibration $P'(K_n) \to K_n$.

The above discussion has the effect of *analyzing* the space X. Let us consider the problem of *synthesizing* a space X from Postnikov data. Indeed, suppose that we are given

(1) a sequence $\Pi_1, \Pi_2, \ldots$ of abelian groups;
(2) a sequence $X^0, X^1, X^2, \ldots$ of CW-complexes;
(3) a sequence of elements $k^{n+1} \in H^{n+1}(X^{n-1}; \Pi_n)$ such that each of the complexes X^{n+1} is obtained from X^n and k^{n+2} by the process of the preceding paragraph.

We wish to determine whether there is a space X whose Postnikov system is given by the above data. Such a space X can be constructed as follows. Replace the maps $f_n : X^{n+1} \to X^n$ in turn by fibrations, obtaining a tower

$$\cdots \to Y^{n+1} \xrightarrow{p_n} Y^n \to \cdots \to Y^1 \to Y^0$$

of fibrations. Form the inverse limit Y of this sequence of spaces (N.B.: as usual, care must be taken to stay within the category $\mathscr{K}$). Finally, choose a CW-approximation $X \to Y$. This construction is dealt with in §4, where it is also proved that if the given system is the Postnikov system of a space X', then X and X' have the same homotopy type.

In §5 we discuss a few examples; in §6 it is indicated how to relativize the notions of this Chapter. And we conclude the Chapter with an alternative approach to obstruction theory, making use of Postnikov systems.

The ideas of this Chapter, as the name suggests, are due to Postnikov [1, 2] in the absolute case. The relativization was found in 1957 by Moore [1].

1 Connective Fibrations

Let $p : \tilde{X} \to X$ be the universal covering of a 0-connected space X. The map p then has the following properties:

(1.1) *p is a fibration;*

(1.2) *$\tilde{X}$ is 1-connected;*

(1.3) *$p_* : \pi_q(\tilde{X}) \approx \pi_q(X)$ for all $q \geq 2$.*

Of course, the universal covering has other beautiful and useful properties (for example, the isomorphism of $\pi_1(X)$ with the group of covering translations). These are, to some extent, counterbalanced by the fact that not every space has a universal covering space (in fact, a necessary and sufficient condition for its existence is that X be semilocally 1-connected). Moreover, it is not at all obvious how to generalize its construction to construct "higher covering spaces." However, if we relax our requirements, by demanding only that conditions (1.1)–(1.3) above should hold, it is very easy to make the appropriate construction, and also to find the generalization to higher dimensions.

Let X, then, be a 0-connected space. According to Corollary (2.4) of Chapter V, we can imbed X in a space X^* in such a way that

(1*) (X^*, X) is a relative CW-complex;
(2*) (X^*, X) is 1-connected;
(3*) $\pi_q(X^*) = 0$ for all $q \geq 2$.

(In fact, we may assume that (X^*, X) has no cells of dimension ≤ 2). Thus the injection $\pi_1(X) \to \pi_1(X^*)$ is an isomorphism, and X^* is an Eilenberg–MacLane space $K(\pi_1(X), 1)$.

Let $\tilde{X}$ be the mapping of the inclusion map $i : X \hookrightarrow X^*$, and let $p : \tilde{X} \to X$ be the fibration of $\tilde{X}$ over X. By Corollary (8.9) of Chapter IV, there is an exact sequence

$$\cdots \to \pi_{q+1}(X^*) \xrightarrow{\Delta_*} \pi_q(\tilde{X}) \xrightarrow{p_*} \pi_q(X) \xrightarrow{i_*} \pi_q(X^*) \to \cdots$$

If $q \geq 2$, $\pi_q(X^*) = \pi_{q+1}(X^*) = 0$, so that $p_* : \pi_q(\tilde{X}) \approx \pi_q(X)$. Since $\pi_2(X^*) = 0$ and $i_* : \pi_1(X) \approx \pi_1(X^*)$, we must have $\pi_1(\tilde{X}) = 0$. Since (X^*, X) is 1-connected and X is 0-connected, $\tilde{X}$ is 0-connected. Thus conditions (1*)–(3*) are satisfied.

It is now obvious how to generalize the above construction. Let us call a map $p : \tilde{X} \to X$ *N-connective* if and only if

(1) *$\tilde{X}$ is N-connected;*
(2) *$p_* : \pi_q(\tilde{X}) \approx \pi_q(X)$ for all $q \geq N + 1$.*

Furthermore, let us say that a space X is *N-anticonnected* if and only if $\pi_q(X) = 0$ for all $q \geq N$. A space X^* containing X is called an *N-anticonnected extension of X* if and only if X^* is N-anticonnected and (X^*, X) is N-connected; thus the injections

$$\pi_q(X) \to \pi_q(X^*)$$

are isomorphisms for all $q < N$. If, moreover, (X^*, X) is a relative CW-complex having no cells of dimension $\leq N$, we shall say that X^* is a *regular N-anticonnected* extension of X.

We can now repeat the preceding argument, assuming instead that X^* is a regular $(N + 1)$-anticonnected extension of X and that $\tilde{X}$ is the mapping fibre of the inclusion $i : X \hookrightarrow X^*$, to obtain

(1.4) Theorem *Let X be a 0-connected space, N a positive integer. Then there exists an N-connective fibration $p : \tilde{X} \to X$.* □

The process by which we constructed the N-connective fibration $p : \tilde{X} \to X$ is more or less unique. We first prove

(1.5) Theorem *Let X^* be a regular n-anticonnected extension of X, Y^* an m-anticonnected extension of Y, and let $f: X \to Y$. Then*

(1) *if $m \leq n$, f can be extended to a map $f^* : X^* \to Y^*$;*
(2) *if $m \leq n + 1$, any two extensions $f_0^*, f_1^* : X^* \to Y^*$ are homotopic* (rel. X).

Since (X^*, X) has no cells of dimension $\leq n$, f is already defined on the n-skeleton X_n^* of (X^*, X). Therefore the obstructions to extending f lie in the groups $H^{q+1}(X^*, X; \pi_q(Y^*))$ $(q \geq n)$ and these groups vanish, since $\pi_q(Y^*) = 0$, if $q \geq m$. If $m \leq n$, all the obstructions vanish and the extension exists.

If f_0^*, f_1^* are extensions of f, they already agree on the n-skeleton X_n^*. The obstructions to deforming f_0^* to f_1^* lie in the groups $H^q(X^*, X; \pi_q(Y^*))$ for $q \geq n + 1$, and these groups vanish, since $\pi_q(Y^*) = 0$, if $q \geq m$. If $m \leq n + 1$, all the obstructions vanish, and the two maps are homotopic (rel. X). □

(1.6) Corollary *Let X_1^* and X_2^* be regular $(n + 1)$-anticonnected extensions of X. Then the pairs (X_1^*, X) and (X_2^*, X) have the same homotopy type.*

By (1) of Theorem (1.5) with $m = n = N$, the identity map of X has extensions $f: (X_1^*, X) \to (X_2^*, X)$ and $g : (X_2^*, X) \to (X_1^*, X)$. By (2) of the same theorem, the maps $g \circ f: (X_1^*, X) \to (X_1^*, X)$ and $f \circ g : (X_2^*, X) \to (X_2^*, X)$ are homotopic to the appropriate identity maps (rel. X). □

(1.7) Corollary *Let X_1^* and X_2^* be regular $(N+1)$-anticonnected extensions of X, and let $p_i : \tilde{X}_i \to X$ be the N-connective fibrations constructed in the proof of Theorem (7.4). Then p_1 and p_2 have the same fibre homotopy type.*

Let $f: (X_1^*, X) \to (X_2^*, X)$, $g: (X_2^*, X) \to (X_1^*, X)$ be extensions of the identity map of X. Then composition with f induces a map $f_1 : \tilde{X}_1 \to X_2$, and $p_2 \circ f_1 = p_1 : \tilde{X}_1 \to X$; similarly composition with g defines a map $g_1 : \tilde{X}_2 \to \tilde{X}_1$, and $p_1 \circ g_1 = p_2 : \tilde{X}_2 \to X$. Let $F : \mathbf{I} \times X_1^* \to X_1^*$ be a homotopy of 1 to $g \circ f$ (rel. X); then the map $F_1 : \mathbf{I} \times \tilde{X}_1 \to \tilde{X}_1$ defined by

$$F_1(t, u)(s) = F(t, u(s))$$

is a vertical homotopy of the identity map of $\tilde{X}_1$ to $g_1 \circ f_1$. Similarly, a homotopy G of 1 to $f \circ g$ (rel. X) induces a vertical homotopy G_1 of the identity map of $\tilde{X}_2$ to $f_1 \circ g_1$. □

The process of constructing an N-connective fibration $p : \tilde{X} \to X$ can be broken up into several steps. In fact, let X_0 be a 0-connected space, and suppose we have constructed spaces X_r and r-connective fibrations $p_r : X_r \to X_{r-1} (r = 1, \ldots, N)$. Then the composition $q_N = p_1 \circ \cdots \circ p_N$ is an N-connective fibration. Applying the process of Theorem (1.4) to the space X_N, we construct a regular $(N+2)$-anticonnected extension X_N^* of X in order to construct the $(N+1)$-connective fibration $p_{N+1} : X_{N+1} \to X_N$. Note that X_N^* is an Eilenberg–MacLane space $K(\pi_{N+1}(X), N+1)$, and therefore the fibre of p_{N+1} is the space $\Omega X_N^* = K(\pi_{N+1}(X), N)$.

Summarizing, we have

(1.8) Theorem *If $X = X_0$ is a 0-connected space, there is a tower*

$$\cdots \to X_n \xrightarrow{p_n} X_{n-1} \to \cdots \to X_1 \xrightarrow{p_1} X_0 .$$

The maps p_n and $q_n = p_1 \circ p_2 \circ \cdots \circ p_n$ are n-connective fibrations. The fibre of p_n is an Eilenberg–MacLane space $K(\pi_n(X), n-1)$. □

The spaces X_n^* used in the above construction are Eilenberg–MacLane spaces: $X_n^* = K(\pi_n(X), n)$. Thus an arbitrary space X can be decomposed, in a sense, into a family of Eilenberg–MacLane spaces. It is natural to ask whether X has the same homotopy type as the weak product $\prod_{n=1}^{\infty} K(\pi_n(X), n)$. However, this is easily seen not to be the case. Suppose, for example, that X is a finite-dimensional real projective space $\mathbf{P}^n$ $(n \geq 2)$, so that $\pi_1(X) = \mathbf{Z}_2$. Then $H_q(X; \mathbf{Z}_2) = 0$ for all $q > n$. If X were a product, as above, then $\mathbf{P}^\infty = K(\mathbf{Z}_2, 1)$ would be a retract of X and the fact that $H_{n+1}(\mathbf{P}^\infty; \mathbf{Z}_2) \neq 0$ implies that $H_{n+1}(X; \mathbf{Z}_2) \neq 0$.

In this connection, the following theorem is frequently useful.

(1.9) Theorem (J. C. Moore). *A necessary and sufficient condition that a connected CW-complex K have the same homotopy type as a weak product of Eilenberg–MacLane spaces is that the Hurewicz map $\rho : \pi_n(X) \to H_n(X)$ have a left inverse $\lambda : H_n(X) \to \pi_n(X)$ for every positive integer n.*

Suppose first that $X = \prod_{k=1}^{\infty} K(G_k, n_k)$. For a given positive integer n, let $J = \{k \mid n_k = n\}$; and let $X' = \prod_{k \notin J} K(G_k, m_k)$. Then

$$\begin{aligned} X &= \prod_{k \in J} K(G_k, n) \times X' \\ &= K(G, n) \times X' \end{aligned}$$

where $G = \bigoplus_{k \in J} G_k$ and $\pi_n(X') = \bigoplus_{k \notin J} \pi_n(K(G_k, n_k)) = 0$. Thus $\pi_n(X) \approx G$. Let $p : X \to K(G, n)$ be the projection on the first factor, and let

$$u = p^* \iota^n(K(G, n)) \in H^n(X; G).$$

A homomorphism $\lambda : H_n(X) \to G = \pi_n(X)$ is then defined by

$$\lambda(z) = \langle u, z \rangle \qquad (z \in H_n(X)).$$

If $x \in G$, $\mathbf{b}_n = \iota^n(K(G, n))$, then

$$\begin{aligned} \lambda(\rho(x)) &= \langle u, \rho(x) \rangle = \langle p^* \mathbf{b}_n, \rho(x) \rangle \\ &= \langle \mathbf{b}_n, p_* \rho(x) \rangle = \langle \mathbf{b}_n, \rho p_*(x) \rangle \\ &= p_*(x). \end{aligned}$$

As $p_* : \pi_n(X) \to \pi_n(K(G, n)) = G$ is an isomorphism, this implies that ρ has a left inverse.

Conversely, suppose that $\lambda_n : H_n(X) \to \pi_n(X)$ is a left inverse of the Hurewicz map $(n \geq 1)$. For each n, choose a cohomology class $u_n \in H^n(X; \pi_n(X))$ such that the homomorphism of $H_n(X)$ into $\pi_n(X)$ defined by the Kronecker index with u_n is λ_n. Let $f_n : X \to K(\pi_n(X), n) = K_n$ be a map such that $f_n^* \mathbf{b}_n = u_n$; we may assume that f_n maps the $(n-1)$-skeleton X_{n-1} of X into the base point. It follows that the f_n define a map $f : X \to \prod_{n=1}^{\infty} K_n$ such that $p_n \circ f = f_n$, where p_n is the projection on the nth factor. If $\alpha \in \pi_n(X)$, then

$$\begin{aligned} \alpha = \lambda_n \rho(\alpha) &= \langle u_n, \rho(\alpha) \rangle = \langle f_n^* \mathbf{b}_n, \rho(\alpha) \rangle \\ &= \langle \mathbf{b}_n, f_{n*} \rho(\alpha) \rangle = \langle \mathbf{b}_n, \rho f_{n*}(\alpha) \rangle \\ &= f_{n*}(\alpha), \end{aligned}$$

so that $f_{n*} : \pi_n(X) \to \pi_n(K_n)$ is an isomorphism. As $f_{m*} : \pi_n(X) \to \pi_n(K_m)$ is zero for $m \neq n$, and as

$$\pi_n\left(\prod_{m=1}^{\infty} K_m\right) \approx \bigoplus_{m=1}^{\infty} \pi_n(K_m) \approx \pi_n(K_n)$$

it follows that $f_* : \pi_n(X) \to \pi_n(\prod_{m=1}^{\infty} K_m)$ is an isomorphism for all n, and therefore f is a weak homotopy equivalence. Since both spaces are CW-complexes, f is a homotopy equivalence, by Theorem (3.5) of Chapter V. □

2 The Postnikov Invariants of a Space

We saw in §1 how a space X can be resolved into simpler "constituents"—these are Eilenberg–MacLane spaces $K(\pi_n(X), n)$. We also saw that the constituents themselves do not suffice to determine the homotopy type of X. The further information necessary turns out to be provided by a sequence of cohomology classes, the *Postnikov invariants* of X, which we now describe.

Let X be a connected CW-complex. In addition, we shall assume that X is simple (i.e., n-simple for every n); thus $\pi_1(X)$ is abelian and operates trivially on the higher homotopy groups $\pi_n(X)$. Appealing once more to Theorem (2.3) of Chapter V, we construct a sequence of CW-complexes $X^0, X^1, X^2, \ldots$, each containing X as a subcomplex, such that X^{n-1} is a regular n-anticonnected extension of X ($n = 1, 2, \ldots$); in particular, X^0 is contractible. Let us call such a sequence $\{X^n\}$ a *resolving sequence* for X. A sequence of maps $f_n : X^n \to X^{n-1}$ is called a *bonding sequence*, and the f_n *bonding maps*, if and only if $f \circ i_n = i_{n-1}$, where $i_n : X \hookrightarrow X_n$. Finally, a *homotopy resolution* of X consists of a resolving sequence $\{X^n\}$, together with a bonding sequence $\{f_n\}$.

From Theorem (1.5) we conclude

(2.1) Theorem *If $\{X^n\}$ is a resolving sequence for X, then there exists a bonding sequence $\{f_n\}$. If $\{f_n\}$ and $\{f'_n\}$ are bonding sequences, then, for each n,* $f_n \simeq f'_n$ (rel. X). □

Let us suppose given a homotopy resolution for X. We may assume that the maps f_n are cellular. If $f = f_{n+1} : X^{n+1} \to X^n$ is one of the bonding maps, its mapping cylinder is a CW-complex containing $0 \times X^{n+1} \cup \mathbf{I} \times X \cup 1 \times X^n$ as a subcomplex. It is convenient to "flatten out" the prism $\mathbf{I} \times X$; more precisely, let $p : \mathbf{I} \times X \to X$ be the projection on the second factor, and let $\hat{X}^n$ be the adjunction space

$$\hat{X}^n = \mathbf{I}_f \cup_p X.$$

As the adjunction map p is a homotopy equivalence, the homotopy type of $\mathbf{I}_f$ is not altered by this process (Corollary 5.12, Chapter I); we shall call $\hat{X}^n$ the *relative mapping cylinder* of f. The space $\hat{X}^n$ is a CW-complex with subcomplexes isomorphic with X^n and X^{n+1}; these have been pasted together along their common subcomplex X.

Let $\langle t, x\rangle$ be the image of the point (t, x) in $\hat{X}^n$ under the compound identification $\mathbf{I} \times X^{n+1} \to \mathbf{I}_f \to \hat{X}^n$. Then we have the relations

$$\begin{aligned} \langle 0, y\rangle &= y & &(y \in X^{n+1}),\\ \langle t, x\rangle &= x & &(x \in X)\\ \langle 1, y\rangle &= f_{n+1}(y) & &(y \in X^{n+1}). \end{aligned}$$

We can also form the relative mapping cone $\check{X}^n = \hat{X}^n/X^{n+1}$. The inclusion $X^n \hookrightarrow \hat{X}^n$ gives rise to a mapping of X^n into $\check{X}^n$. The latter map, however, is not an inclusion, because the subspace $X \subset X^n$ has been collapsed. Instead, there is an inclusion $X^n/X \hookrightarrow \check{X}^n$.

From the commutativity of the diagram

$$\begin{array}{ccc} & \overset{\pi_q(i_{n+1})}{\nearrow} & \pi_q(X^{n+1}) \\ \pi_q(X) & & \downarrow \pi_q(f_{n+1}) \\ & \underset{\pi_q(i_n)}{\searrow} & \pi_q(X^n) \end{array}$$

we deduce

(2.2) Theorem *The homomorphism* $\pi_q(f_{n+1}) : \pi_q(X^{n+1}) \to \pi_q(X^n)$ *is an isomorphism for all* $q \neq n + 2$. *Hence*

$$\pi_q(\hat{X}^n, X^{n+1}) = 0 \qquad (q \neq n + 2),$$

while the composite

$$\pi_{n+2}(\hat{X}^n, X^{n+1}) \xrightarrow{\partial_*} \pi_{n+1}(X^{n+1}) \xrightarrow{\pi_{n+1}(i_{n+1})^{-1}} \pi_{n+1}(X)$$

is an isomorphism. □

Because of our hypothesis that X is $(n + 1)$-simple, the space X^{n+1} is $(n + 1)$-simple and the pair $(\hat{X}^n, X^{n+1})$ is $(n + 2)$-simple. Hence we can apply the Relative Hurewicz Theorem to obtain

(2.3) Corollary *The composite*

$$H_{n+2}(\hat{X}^n, X^{n+1}) \xrightarrow{\rho^{-1}} \pi_{n+2}(\hat{X}^n, X^{n+1}) \xrightarrow{\partial_*}$$
$$\pi_{n+1}(X^{n+1}) \xrightarrow{\pi_{n+1}(i_{n+1})^{-1}} \pi_{n+1}(X)$$

is an isomorphism $\kappa_{n+2} : H_{n+2}(\hat{X}^n, X^{n+1}) \to \pi_{n+1}(X)$. □

Since $H_{n+1}(\hat{X}^n, X^{n+1}) = 0$, the Universal Coefficient Theorem reduces to an isomorphism

$$H^{n+2}(\hat{X}^n, X^{n+1}; G) \approx \operatorname{Hom}\{H_{n+2}(\hat{X}^n, X^{n+1}), G\}.$$

Therefore the isomorphism κ_{n+2} corresponds to a cohomology class

$$k_1^{n+2} \in H^{n+2}(\hat{X}^n, X^{n+1}; \pi_{n+1}(X)),$$

whose image

$$k^{n+2}(X) = k^{n+2} \in H^{n+2}(X^n; \pi_{n+1}(X))$$

under the injection is called the $(n+2)^{nd}$ *Postnikov invariant* of X. The system $\{X^n, f_n, k^{n+2}\}$ is called a *Postnikov system* for X.

By exactness of the cohomology sequence of $(\hat{X}^n, X^{n+1})$, we have

(2.4) *The relation*

$$f_{n+1}^* k^{n+2} = 0$$

holds in the group $H^{n+2}(X^{n+1}; \pi_{n+1}(X))$. □

Since the inclusion maps $i_n : X \hookrightarrow X^n$, $i_{n+1} : X \hookrightarrow X^{n+1}$ are related by

$$f_{n+1} \circ i_{n+1} = i_n,$$

we have in turn

(2.5) *The injection*

$$i_n^* : H^{n+2}(X^n; \pi_{n+1}(X)) \to H^{n+2}(X; \pi_{n+1}(X))$$

sends the Postnikov invariant k^{n+2} *into zero.* □

The term "invariant" is used somewhat loosely here. In fact, k^{n+2} is a cohomology class of a space X^n, which has not been constructed in an invariant way. This difficulty, however, is not serious, for, as we shall show below, the construction of the space X^n can be made completely natural. The real difficulty arises from the fact that the space X may admit non-trivial self homotopy equivalences. (This sort of difficulty is a familiar one to the category theorist who has tried to construct a new category by identifying equivalent objects in an old one.)

Let us make explicit the sense in which the classes k^{n+2} may be regarded as invariants. Let $\{X^n, f_n\}$, $\{Y^n, g_n\}$ be homotopy resolutions for X, Y, respectively, and let $h : X \to Y$. By Theorem (1.5), the map h can be extended to a map $h_n : X^n \to Y^n$ for every n. Moreover, the maps $g_{n+1} \circ h_{n+1}$, $h_n \circ f_{n+1} : X^{n+1} \to Y^n$ are homotopic (rel. X). The map h_n induces a homomorphism $h_n^* : H^{n+2}(Y^n; G) \to H^{n+2}(X^n; G)$ for any coefficient group G. And the homomorphism of $\pi_{n+1}(X)$ into $\pi_{n+1}(Y)$ induced by h determines a (coefficient group) homomorphism

$$h_* : H^{n+2}(Z; \pi_{n+1}(X)) \to H^{n+2}(Z; \pi_{n+1}(Y))$$

for any space Z. We then have

(2.6) Theorem *The Postnikov invariants*

$$k^{n+2}(X) \in H^{n+2}(X^n; \pi_{n+1}(X)),$$
$$k^{n+2}(Y) \in H^{n+2}(Y^n; \pi_{n+1}(Y))$$

are related by

$$h_n^* k^{n+2}(Y) = h_* k^{n+2}(X) \in H^{n+2}(X^n; \pi_{n+1}(Y)).$$

A homotopy of $g_{n+1} \circ h_{n+1}$ to $h_n \circ f_{n+1}$ can be used to construct a map $\hat{G}: (\hat{X}^n, X^{n+1}) \to (\hat{Y}^n, Y^{n+1})$ whose restrictions to X^n, X^{n+1} are the maps h_n, h_{n+1} respectively. It suffices to prove that $\hat{G}^* k_1^{n+2}(X) = h_* k_1^{n+2}(Y)$. But the diagram

$$\begin{array}{ccccccc}
H_{n+2}(\hat{X}^n, X^{n+1}) & \xleftarrow{\rho} & \pi_{n+2}(\hat{X}^n, X^{n+1}) & \xrightarrow{\partial_*} & \pi_{n+1}(X^{n+1}) & \xleftarrow{\pi_{n+1}(i_{n+1})} & \pi_{n+1}(X) \\
\downarrow{\scriptstyle H_{n+2}(\hat{G})} & & \downarrow{\scriptstyle \pi_{n+2}(\hat{G})} & & \downarrow{\scriptstyle \pi_{n+1}(h_{n+1})} & & \downarrow{\scriptstyle \pi_{n+1}(h)} \\
H_{n+2}(\hat{Y}^n, Y^{n+1}) & \xleftarrow[\rho]{} & \pi_{n+2}(\hat{Y}^n, Y^{n+1}) & \xrightarrow[\partial_*]{} & \pi_{n+1}(Y^{n+1}) & \xleftarrow[\pi_{n+1}(j_{n+1})]{} & \pi_{n+1}(Y)
\end{array}$$

is commutative, and $k_1^{n+2}(X)$ is the cohomology class corresponding to the homomorphism

$$\pi_{n+1}(i_{n+1})^{-1} \circ \partial_* \circ \rho^{-1} \in \operatorname{Hom}(H_{n+2}(\hat{X}^n, X^{n+1}), \pi_{n+1}(X)),$$

and similarly for $k_1^{n+2}(Y)$. Their images under $\hat{G}^*$, h_* correspond to the composites of these homomorphisms with $H_{n+2}(\hat{G})$, $\pi_{n+1}(h)$, respectively, and the theorem follows from commutativity of the above diagram. □

We next show how the construction of the spaces X^{n+1} can be made natural. The idea is to use a "big enough" model. This is in accordance with the observation that large constructions are often best to give conceptual explanations and elegant proofs, while small constructions are better adapted to making specific computations.

Let X be a space and n a non-negative integer. Let $F_n = F_n(X)$ be the set of *all* continuous functions $\alpha: \dot{\Delta}^{n+1} \to X$; we shall consider F_n as a *discrete* topological space. The evaluation map $\mathbf{e} = \mathbf{e}_X$ sends $F_n \times \dot{\Delta}^{n+1}$ into X; let

$$Q_{n-1}(X) = X \cup_{\mathbf{e}} (F_n \times \Delta^{n+1}).$$

Let $f: X \to Y$; then composition with f is a map

$$\underline{f}: F_n(X) \to F_n(Y),$$

and $\mathbf{e}_Y \circ (\underline{f} \times 1) = f \circ \mathbf{e}_X$. Hence the maps $f: X \to Y$, $\underline{f} \times 1: F_n(X) \times \Delta^{n+1} \to F_n(Y) \times \Delta^{n+1}$ are compatible with the identification maps

$$X + (F_n(X) \times \Delta^{n+1}) \to Q_{n-1}(X),$$
$$Y + (F_n(Y) \times \Delta^{n+1}) \to Q_{n-1}(Y),$$

and therefore induce a map

$$Q_{n-1}(f) : Q_{n-1}(X) \to Q_{n-1}(Y)$$

Evidently we have defined a functor $Q_{n-1} : \mathscr{K} \to \mathscr{K}$. Moreover, the inclusion maps $j_{n-1}(X) : X \hookrightarrow Q_{n-1}(X)$ define a natural transformation j_{n-1} of the identity functor into Q_{n-1}.

(2.7) Theorem *The functor Q_{n-1} and the natural transformation j_{n-1} have the following properties:*

(1) *The pair $(Q_{n-1}(X), X)$ is a relative CW-complex (in fact, an $(n+1)$-cellular extension of X);*
(2) *The injection $\pi_i(X) \to \pi_i(Q_{n-1}(X))$ is an isomorphism for all $i < n$;*
(3) $\pi_n(Q_{n-1}(X)) = 0$.
(4) *If $f : X \to Y$ is an inclusion, so is*

$$Q_{n-1}(f) : Q_{n-1}(X) \to Q_{n-1}(Y).$$

The easy verification of these properties is left to the reader.

To construct a candidate for the space X^{n-1}, we simply iterate the above construction. Specifically, let

$$\begin{aligned} P_k^{n-1}(X) &= X && (k \le n), \\ P_{n+1}^{n-1}(X) &= Q_{n-1}(X), \\ P_k^{n-1}(X) &= Q_{k-2}(P_{k-1}^{n-1}(X)) && (k \ge n+2). \end{aligned}$$

This recursive definition gives rise to a relative CW-complex $(P^{n-1}(X), X)$ whose k-skeleton is $P_k^{n-1}(X)$. Moreover, if $f : X \to Y$, we can define $P_k^{n-1}(f)$ recursively by

$$\begin{aligned} P_k^{n-1}(f) &= f && (k \le n), \\ P_{n+1}^{n-1}(f) &= Q_{n-1}(f), \\ P_k^{n-1}(f) &= Q_{k-2}(P_{k-1}^{n-1}(f)) && (k \ge n+2). \end{aligned}$$

Clearly each of the maps $P_k^{n-1}(f)$ is an extension of the preceding, so that they define a map

$$P^{n-1}(f) : P^{n-1}(X) \to P^{n-1}(Y).$$

As before, we have defined a functor $P^{n-1} : \mathscr{K} \to \mathscr{K}$, and the inclusions $i_{n-1}(X) : X \hookrightarrow P^{n-1}(X)$ determine a natural transformation i_{n-1} of the identity functor into P^{n-1}, and we have

(2.8) Theorem *The functor P^{n-1} and the natural transformation i_{n-1} have the following properties:*

(1) *The pair $(P^{n-1}(X), X)$ is a relative CW-complex and*

$P^{n-1}(f):(P^{n-1}(X), X)\to(P^{n-1}(Y), Y)$ a cellular map for every $f:X\to Y$;
(2) *The space $P^{n-1}(X)$ is a regular n-anticonnected extension of X.*
(3) *If $f:X\to Y$ is an inclusion, so is $P^{n-1}(f):P^{n-1}(X)\to P^{n-1}(Y)$.* □

To study the relationship between $P^{n-1}(X)$ and $P^n(X)$, let us observe that, if $k\le n$, then

$$P^n_k(X)=X=P^{n-1}_k(X),$$

while

$$P^n_{n+1}(X)=X\subset Q_{n-1}(X)=P^{n-1}_{n+1}(X),$$
$$P^n_{n+2}(X)=Q_n(X)\subset Q_n(P^{n-1}_{n+1}(X))=P^{n-1}_{n+2}(X).$$

Assume that $P^n_{k-1}(X)$ is a subspace of $P^{n-1}_{k-1}(X)$, $k\ge n+3$. Then

$$P^n_k(X)=Q_{k-2}(P^n_{k-1}(X))\subset Q_{k-2}(P^{n-1}_{k-1}(X))=P^{n-1}_k(X).$$

Hence, by induction, $P^n_k(X)\subset P^{n-1}_k(X)$ for all k, and therefore

(2.9) *The relative* CW-*complex $(P^n(X), X)$ is a subcomplex of $(P^{n-1}(X), X)$. Moreover, the inclusions $P^n(X)\hookrightarrow P^{n-1}(X)$ define a natural transformation $f_n:P^n\to P^{n-1}$.* □

(2.10) Corollary *The sequence $\{P^n(X), f_n(X)\}$ is a homotopy resolution of X.* □

We shall refer to the above resolution as the *canonical homotopy resolution* of X.

3 Amplifying a Space by a Cohomology Class

Let X be a (connected) CW-complex, $u\in H^n(X;\Pi)$ $(n\ge 1)$. The pair (X, u) determines a new space $W(u)$, together with a map $q:W(u)\to X$, as follows. Let $K=K(\Pi, n)$; then, according to Corollary (6.20) of Chapter V, there is a unique homotopy class of maps

$$h:X\to K$$

such that $h^*\iota^n(K)=u$. Let $W(u)$ be the mapping fibre of h, $q:W(u)\to X$ its fibration over X. Then there is a commutative diagram

(3.1)
$$\begin{array}{ccc} W(u) & \xrightarrow{\;g\;} & \mathbf{P}'(K) \\ {\scriptstyle q}\downarrow & & \downarrow{\scriptstyle p} \\ X & \xrightarrow[\;h\;]{} & K \end{array}$$

The map $q: W(u) \to X$ (and, for the sake of brevity, the space $W(u)$) is called an *amplification* of X by k.

Remark 1. The map h is determined only up to homotopy. Thus (Exercise 6, Chapter I) the fibration q is determined only up to fibre homotopy type. In particular, the space $W(u)$ is determined only up to homotopy type.

Remark 2. The map q is a fibration whose fibre $\Omega(K)$ is an Eilenberg–MacLane space $K(\Pi, n-1)$.

By Corollary (8.9) of Chapter IV, there is an exact sequence

(3.2)
$$\cdots \to \pi_{r+1}(X) \xrightarrow{h_*} \pi_{r+1}(K) \xrightarrow{\Delta_*} \pi_r(W(u)) \xrightarrow{q_*} \pi_r(X) \to \cdots$$

Since $\pi_r(K) = 0$ for all $r \neq n$, we see that

$$q_* : \pi_r(W(u)) \approx \pi_r(X) \qquad (n-1 \neq r \neq n),$$
$$\pi_n(W(u)) = \operatorname{Ker} \tilde{u},$$

while there is a short exact sequence

$$0 \to \operatorname{Cok} \tilde{u} \to \pi_{n-1}(W(u)) \to \pi_{n-1}(X) \to 0,$$

where

$$\tilde{u} = \pi_n(h) : \pi_n(X) \to \pi_n(K(\Pi, n)) = \Pi.$$

The explicit calculation of the homomorphism $\tilde{u}$ is given by

(3.3) Theorem *For all* $\alpha \in \pi_n(X)$,
$$\tilde{u}(\alpha) = \langle u, \rho(\alpha) \rangle.$$

In fact, there is a commutative diagram

$$\begin{array}{ccc} \pi_n(X) & \xrightarrow{\tilde{u}} & \pi_n(K) = \Pi \\ \rho \downarrow & & \downarrow \rho \\ H_n(X) & \xrightarrow[h_*]{} & H_n(K), \end{array}$$

and

$$\begin{aligned} \langle u, \rho(\alpha) \rangle &= \langle h^* \iota^n(K), \rho(\alpha) \rangle = \langle \iota^n(K), h_* \rho(\alpha) \rangle \\ &= \langle \iota^n(K), \rho \tilde{u}(\alpha) \rangle \\ &= \tilde{u}(\alpha) \end{aligned}$$

by the defining property of the class $\iota^n(K)$. □

(3.4) Corollary *If X is $(n-1)$-anticonnected, then $W(u)$ is n-anticonnected, and*

$$q_* : \pi_r(W(u)) \approx \pi_r(X) \qquad (r \neq n-1),$$
$$\pi_{n-1}(W(u)) \approx \Pi. \qquad \square$$

Thus, in the passage from X to $W(u)$, the sequence of homotopy groups $(\pi_1(X), \ldots, \pi_{n-2}(X), 0, 0, \ldots)$ is altered by replacing the first written zero by Π).

We next show that each stage of a Postnikov system for a space X can be obtained, up to weak homotopy equivalence, by the above amplification process, from the preceding stage. Specifically,

(3.5) Theorem *Let $\{X^n, f_n, k^{n+2}\}$ be a Postnikov system for the connected simple* CW*-complex X. For each positive integer n, let $q_{n+1} : W^{n+1} \to X^n$ be the amplification of X^n by the cohomology class $k^{n+2} \in H^{n+2}(X^n; \pi_{n+1}(X))$. Then there is a weak homotopy equivalence $g_{n+1} : X^{n+1} \to W^{n+1}$ such that $q_{n+1} \circ g_{n+1} = f_{n+1}$.*

Let $\check{X}^n$ be the relative mapping cone of f_{n+1}, $\bar{p}_n : (\hat{X}^n, X^{n+1}) \to (\check{X}^n, *)$ the identification map, $j_n : X^n \to \check{X}^n$ the restriction of $\bar{p}_n$ to the subspace X^n of $\check{X}^n$. Since $\bar{p}_n$ is a relative homeomorphism,

$$\bar{p}_n^* : H^{n+2}(\check{X}^n; G) \approx H^{n+2}(\hat{X}^n, X^{n+1}; G)$$

for any coefficient group G. Let

$$h_1 : \check{X}^n \to K_n = K(\pi_{n+1}(X), n+2)$$

be a map such that

$$\bar{p}_n^* h_1^* \iota^{n+2}(K_n) = k_1^{n+2}.$$

Let $h = h_1 \circ j_n : X^n \to K_n$. Then

$$h^* \iota^{n+2}(K_n) = j_n^* h_1^* \iota^{n+2}(K_n)$$

is the image k^{n+2} of k_1^{n+2} under the injection. Thus there is a commutative diagram

$$\begin{array}{ccccc}
W^{n+1} & & \xrightarrow{\quad g \quad} & & \mathbf{P}'(K_n) \\
{\scriptstyle q_{n+1}}\downarrow & & & & \downarrow{\scriptstyle p} \\
X^n & \xrightarrow[j_n]{} & \check{X}^n & \xrightarrow[h_1]{} & K_n
\end{array}$$

The map $j_n \circ f_{n+1}$ is nullhomotopic; a nullhomotopy is given by

$$(t, y) \to \bar{p}_n(\langle 1-t, y \rangle).$$

Therefore the map $h \circ f_{n+1}$ is nullhomotopic via the homotopy

$$(t, y) \to h_1 p_n(\langle 1 - t, y \rangle).$$

The adjoint of the latter homotopy is a map $\bar{g} : X^{n+1} \to \mathbf{P}'(K_n)$ such that

$$\begin{aligned} p\bar{g}(y) = \bar{g}(y)(0) &= h_1 \bar{p}_n(\langle 1, y \rangle) = h_1 \bar{p}_n f_{n+1}(y) \\ &= h_1 j_n f_{n+1}(y) = hf_{n+1}(y). \end{aligned}$$

The pair of maps $(f_{n+1}, \bar{g})$ defines, in turn, a map $g_{n+1} : X^{n+1} \to W^{n+1}$ such that $q_{n+1} \circ g_{n+1} = f_{n+1}$ and $g \circ g_{n+1} = \bar{g}$.

To show that g_{n+1} is a weak homotopy equivalence, we first observe that, if $r \neq n + 1$, then, by Corollary (3.4),

$$\pi_r(q_{n+1}) : \pi_r(W^{n+1}) \approx \pi_r(X^r).$$

Since

$$\pi_r(f_{n+1}) : \pi_r(X^{n+1}) \to \pi_r(X^n)$$

is also an isomorphism in these dimensions, so is $\pi_r(g_{n+1})$. It remains to prove that $\pi_{n+1}(g_{n+1})$ is an isomorphism.

Define $\bar{g}_t(y) : \mathbf{I} \to K_n$, for each $y \in X^{n+1}$, $t \in \mathbf{I}$, by

$$\bar{g}_t(y)(s) = h_1 \bar{p}_n(\langle 1 - s + st, y \rangle);$$

then

(1) $\bar{g}_0(y) = \bar{g}(y)$;
(2) $\bar{g}_t(y)(1) = h_1 \bar{p}_n(\langle t, y \rangle)$;
(3) $\bar{g}_1(y) = e_{hf_{n+1}(y)}$;
(4) $\bar{g}_t(y)(0) = hf_{n+1}(y)$.

Because of these relations, the function defined by

$$\begin{aligned} h_{n+1}(\langle t, y \rangle) &= (f_{n+1}(y), \bar{g}_t(y)), \\ h_{n+1}(x) &= (x, e_{h(x)}) \qquad (x \in X), \end{aligned}$$

is a map $h_{n+1} : \hat{X}^n \to \bar{X}^n$ making the diagram

$$\begin{array}{ccccc} X^{n+1} & \hookrightarrow & \hat{X}^n & \xrightarrow{\bar{p}_n} & \check{X}^n \\ \big\downarrow{\scriptstyle g_{n+1}} & & \big\downarrow{\scriptstyle h_{n+1}} & & \big\downarrow \\ W^{n+1} & \hookrightarrow & \bar{X}^n & \xrightarrow[\bar{h}]{} & K_n \end{array}$$

commutative ($\bar{h} : \bar{X}^n \to K_n$ is the fibration of $\bar{X}^n = \mathbf{I}^h$ over K_n). Therefore the diagram

$$\begin{array}{ccccc}
\pi_{n+2}(K_n) & \xleftarrow{\pi_{n+2}(\bar{h})} & \pi_{n+2}(\bar{X}^n, W^{n+1}) & \xrightarrow{\partial'_*} & \pi_{n+1}(W^{n+1}) \\
\uparrow \pi_{n+2}(h_1) & & \uparrow \pi_{n+2}(h_{n+1}) & & \uparrow \pi_{n+1}(g_{n+1}) \\
\pi_{n+2}(\check{X}^n) & \xleftarrow[\pi_{n+2}(\bar{p}_n)]{} & \pi_{n+2}(\hat{X}^n, X^{n+1}) & \xrightarrow[\partial_*]{} & \pi_{n+1}(X^{n+1}) \\
\downarrow \rho & & \downarrow \rho & & \\
H_{n+2}(\check{X}^n) & \xleftarrow[H_{n+2}(\bar{p}_n)]{} & H_{n+2}(\hat{X}^n, X^{n+1}) & &
\end{array}$$

is commutative. Because $\bar{h}$ is a fibration, $\pi_{n+2}(\bar{h})$ is an isomorphism. Because $\bar{p}_n$ is a relative homomorphism, $H_{n+2}(\bar{p}_n)$ is an isomorphism. By the Hurewicz Theorems, the two homomorphisms ρ are isomorphisms, and therefore $\pi_{n+2}(\bar{p}_n)$ is an isomorphism. Because $\bar{X}^n$ and $\hat{X}^n$, having the same homotopy type as X^n, are $(n+1)$-anticonnected, ∂_* and ∂'_* are isomorphisms. We have seen that the homomorphism κ_{n+2} such that $\langle k_1^{n+2}, z\rangle = \kappa_{n+2}(z)$ is an isomorphism: $H_{n+2}(\hat{X}^n, X^{n+1}) \to \pi_{n+1}(X)$. But the map h_1 was so chosen that $k_1^{n+2} = \bar{p}_n^* h_1^* \iota^{n+2}(K_n)$. Hence

$$\begin{aligned}
\kappa_{n+2}(z) = \langle k_1^{n+2}, z\rangle &= \langle \bar{p}_n^* h_1^* \iota^{n+2}(K_n), z\rangle \\
&= \langle \iota^{n+2}(K_n), H_{n+2}(h_1) H_{n+2}(\bar{p}_n z\rangle \\
&= \rho^{-1} H_{n+2}(h_1) H_{n+2}(\bar{p}_n) z \\
&= \pi_{n+2}(h_1) \rho^{-1} H_{n+2}(\bar{p}_n) z.
\end{aligned}$$

It follows that $\pi_{n+2}(h_1)$ is an isomorphism, and so, therefore, are $\pi_{n+2}(h_{n+1})$ and $\pi_{n+1}(g_{n+1})$. □

Remark. The maps $g_{n+1} | X : X \to W^{n+1}$ is an inclusion, and it follows that

(3.6) *The map* $g_{n+1} : (X^{n+1}, X) \to (W^{n+1}, X)$ *is a weak homotopy equivalence of pairs.* □

4 Reconstruction of a Space from Its Postnikov System

Let X be a connected simple CW-complex. We have seen how to associate to X a sequence

$$\mathscr{P}(X) = \{\Pi_n, X^{n-1}, k^{n+1}, f_n \,|\, n \geq 1\},$$

where

(1) $\Pi_n(=\pi_n(X))$ is an abelian group;
(2) X^{n-1} is a CW-complex, and X^0 is contractible;
(3) $k^{n+1} \in H^{n+1}(X_{n-1}; \Pi_n)$;
(4) there is a commutative diagram

$$\begin{array}{ccc} X^n & \xrightarrow{g_n} & W(k^{n+1}) \\ & \searrow^{f_n} & \downarrow q_n \\ & & X^{n-1} \end{array}$$

such that g_n is a weak homotopy equivalence.

Let us study the question posed in the title to this section. More specifically, let us define a *Postnikov system* to be a system $\mathscr{P}$ satisfying (1)–(4) above.

From Corollary (3.4) and induction on n, we find

(4.1) Theorem *Let* $\mathscr{P} = \{\Pi_n, X^{n-1}, k^{n+1}, f_n\}$ *be a Postnikov system. Then*

(1) X^{n-1} *is connected and simple*;
(2) $\pi_i(X^n) \approx \Pi_i$ *for* $i = 1, \ldots, n$; $\pi_i(X^n) = 0$ *for* $i > n$;
(3) $\pi_i(f_n) : \pi_i(X^n) \to \pi_i(X^{n-1})$ *is an isomorphism for all* $i \neq n$.

It will be convenient to modify a Postnikov system by replacing each of the maps f_n in turn by an equivalent fibration. More precisely, let $\mathscr{P} = \{\Pi_n, X^{n-1}, k^{n+1}, f_n\}$ be a Postnikov system. Let $p_1 : Y^1 = \mathbf{I}^{f_1} \to X^0 = Y^0$ be the fibration constructed in §7 of Chapter I, and let $j_0 : X^0 \to Y^0$ be the identity map, $j_1 : X^1 \to Y^1$ the inclusion. Let $p_2 : Y^2 = \mathbf{I}^{j_1 \circ f_2} \to Y^1$ be the appropriate fibration, $j_2 : X^2 \hookrightarrow Y^2$. Proceeding in this way, we obtain a commutative diagram

(4.2)
$$\begin{array}{ccccccc} \cdots \to X^{n+1} & \xrightarrow{f_{n+1}} & X^n \to \cdots \to X^1 & \xrightarrow{f_1} & X^0 \\ \downarrow j_{n+1} & & \downarrow j_n \qquad\quad \downarrow j_1 & & \downarrow j_0 \\ \cdots \to Y^{n+1} & \xrightarrow[p_{n+1}]{} & Y^n \to \cdots \to Y^1 & \xrightarrow[p_1]{} & Y^0 \end{array}$$

in which

(4.3) $p_{n+1} : Y^{n+1} \to Y^n$ is a fibration $(n \geq 0)$,

(4.4) $j_n : X^n \to Y^n$ is an inclusion and a homotopy equivalence.

This construction suggests the notion of a *fibred Postnikov system*. This is a sequence $\mathfrak{Y} = \{\Pi_n, Y^{n-1}, p_n \mid n \geq 1\}$ wit the properties

(1) Π_n is an abelian group;
(2) Y^0 is contractible;
(3) $p_n : Y^n \to Y^{n-1}$ is a fibration whose fibre F_n is an Eilenberg–MacLane space $K(\Pi_n, n)$;
(4) the injection $\pi_n(F_n) \to \pi_n(Y^n)$ is an isomorphism.

We then have

(4.5) Theorem *Let $\mathscr{P} = \{\Pi_n, X^{n-1}, k^{n+1}, f_n\}$ be a Postnikov system. Then there is a fibred Postnikov system $\mathfrak{Y} = \{\Pi_n, Y^{n-1}, p_n\}$ and a commutative diagram* (4.2) *satisfying the conditions* (4.3), (4.4). *In particular,*

(1) *Y^{n-1} is connected and simple;*
(2) *$\pi_q(Y^n) \approx \Pi_q$ for $q = 1, \ldots, n$; $\pi_q(Y^n) = 0$ for $q > n$;*
(3) *$\pi_q(p_n) : \pi_q(Y^n) \to \pi_q(Y^{n-1})$ is an isomorphism for all $q \neq n$.* □

Remark. The reader may have observed the absence of the cohomology classes corresponding to the classes k^{n+1}. Of course, they are present implicitly, as the obstructions to cross-sections of the fibrations p_n (cf. Exercise 5).

Thus for many purposes we may replace a Postnikov system by a fibred one. This has certain disadvantages—the spaces Y^n are no longer CW-complexes—but these are compensated for by the many advantages possessed by fibre maps.

One of the advantages of dealing with a tower of fibrations is that it has a relatively well-behaved inverse limit (but see the remark below). Let $p_n : Y^n \to Y^{n-1}$ be a sequence of maps, and let Y_* be their inverse limit in the usual topology. Then Y_* is a Hausdorff space; let $Y = k(Y_*)$ be the associated compactly generated space. The projection of Y_* into Y_n defines a map $g_n : Y \to Y_n$ such that $p_n \circ g_n = g_{n-1}$. The following statements are easily verified, and are left to the reader.

(4.6) *The space Y is an inverse limit of the Y^n in the category $\mathscr{K}$; i.e., for any X in $\mathscr{K}$ and maps $f_n : X \to Y^n$, such that $p_n \circ f_n = f_{n-1}$, there is a unique map $f: X \to Y$ such that $g_n \circ f = f_n$.* □

(4.7) *If each of the maps p_n is a fibration, then each of the maps g_n is a fibration.* □

Let us consider the behavior of the homotopy groups. The commutative diagrams

$$\begin{array}{ccc} & \xrightarrow{\pi_q(g_n)} & \pi_q(Y^n) \\ \pi_q(Y) & & \downarrow \pi_q(p_n) \\ & \xrightarrow{\pi_q(g_{n-1})} & \pi_q(Y^{n-1}) \end{array}$$

define a homomorphism

$$\eta : \pi_q(Y) \to \varprojlim \pi_q(Y^n).$$

The homomorphism η is not, in general, an isomorphism. Instead, we have

(4.8) Theorem *There is a short exact sequence*

$$0 \longrightarrow \varprojlim_n{}^1 \pi_{q+1}(Y^n) \longrightarrow \pi_q(Y) \xrightarrow{\eta} \varprojlim_n \pi_q(Y^n) \longrightarrow 0.$$

To show that η is an epimorphism, let $\alpha_n \in \pi_q(Y^n)$ be elements such that

$$\pi_q(p_{n+1})\alpha_{n+1} = \alpha_n \quad (n = 0, 1, 2, \ldots).$$

We shall show that there are maps $f_n : \mathbf{S}^q \to Y^n$ such that f_n represents α_n and $p_{n+1} \circ f_{n+1} = f_n (n = 0, 1, 2, \ldots)$. Let $f_0 : \mathbf{S}^q \to Y^0$ be an arbitrary representative of α_0, and suppose that f_n has been found and has the above properties (whenever they make sense) for all $n \leq N$. If $f'_{N+1} : \mathbf{S}^q \to Y^{N+1}$ is an arbitrary representative of α_{N+1}, then $p_{N+1} \circ f'_{N+1}$ is a representative of $\pi_q(p_{N+1})\alpha_{N+1} = \alpha_N$, and therefore $p_{N+1} \circ f'_{N+1} \simeq f_N(\text{rel. } *)$. The homotopy lifting property then assures us that f'_{N+1} is homotopic (rel. $*$) to a map f_{N+1} such that $p_{N+1} \circ f_{N+1} = f_N$. This proves the existence of maps f_n with the required properties. They in turn define a map $\mathbf{F} : S^q \to Y$ representing an element $\alpha \in \pi_q(Y)$ such that $\pi_q(g_n)\alpha = \alpha_n$ for all n.

Let us next define a homomorphism $\lambda : \text{Ker } \eta \to \varprojlim^1 \pi_{q+1}(Y^n)$. Let $f : \mathbf{S}^q \to Y$ be a map representing an element α of Ker η, so that $g_n \circ f \simeq *$ for every n. For each n, choose an extension

$$h_n : \mathbf{E}^{q+1} \to Y^n$$

of $g_n \circ f$. The maps $h_n, p_{n+1} \circ h_{n+1} : \mathbf{E}^{q+1} \to Y^n$ agree on $\mathbf{S}^q$. Therefore, regarding $\mathbf{E}^{q+1}$ as a CW-complex with one 0-cell $*$, one q-cell $\mathbf{S}^q$, and one $(q + 1)$-cell, we may form the difference cochain

$$d^{q+1} = d^{q+1}(p_{n+1} \circ h_{n+1}, h_n);$$

since there is only one $(q + 1)$-cell, we may confuse d^{q+1} with its value on the positively oriented cell $\mathbf{E}^{q+1}$; thus

$$\beta_n = d^{q+1}(p_{n+1} \circ h_{n+1}, h_n) \in \pi_{q+1}(Y^n).$$

Let us consider the sequence

$$\beta = (\beta_1, \beta_2, \ldots);$$

it depends, of course, on the choices of the extensions h_n. For a different set of extensions h'_n, let

$$\gamma_n = d^{q+1}(h_n, h'_n);$$

then

$$\begin{aligned} \beta'_n &= d^{q+1}(p_{n+1} \circ h'_{n+1}, h'_n) \\ &= d^{q+1}(p_{n+1} \circ h'_{n+1}, p_{n+1} \circ h_{n+1}) + d^{q+1}(p_{n+1} \circ h_{n+1}, h_n) \\ &\quad + d^{q+1}(h_n, h'_n) \quad \text{by (5.13) of Chapter V} \\ &= \pi_{q+1}(p_{n+1})d^{q+1}(h'_{n+1}, h_{n+1}) + \beta_n + \gamma_n \quad \text{by (5.4') of Chapter V} \\ &= -\pi_{q+1}(p_{n+1})\gamma_{n+1} + \gamma_n + \beta_n \end{aligned}$$

and therefore β and β' determine the same element $\lambda(\alpha)$ of $\varprojlim_n{}^1 \pi_{q+1}(Y^n)$. It is not hard to see that λ is a well-defined homomorphism. We shall show that λ is an isomorphism.

To prove λ a monomorphism, suppose that $\lambda(\alpha) = 0$. Then the elements β_n satisfy the condition

$$\beta_n = \pi_{q+1}(p_{n+1})\gamma_{n+1} - \gamma_n$$

for some $\gamma_n \in \pi_{q+1}(X^n)$. By (5.12) of Chapter V, there exist extensions h'_n of $g_n \circ f$ such that $d^{q+1}(h_n, h'_n) = \gamma_n$. The above computation shows that $\beta'_n = 0$, and therefore

$$p_{n+1} \circ h'_{n+1} \simeq h'_n$$

for all n. Because the maps p_n are fibrations, it is easy to construct recursively new extensions h''_{n+1} of $g_n \circ f$ such that $p_{n+1} \circ h''_{n+1} = h''_n$. The maps h''_n then define a map $h : \mathbf{E}^{q+1} \to Y$ such that $g_n \circ h = h''_n$, and $h | \mathbf{S}^q = f$. Hence $f \simeq *$, $\alpha = 0$.

To prove λ an epimorphism, let $\beta = (\beta_0, \ldots, \beta_n, \ldots)$ be a sequence of elements $\beta_n \in \pi_{q+1}(Y^n)$. We shall define recursively a sequence of maps $h_n : \mathbf{E}^{q+1} \to Y^n$ with the property that $d^{q+1}(p_{n+1} \circ h_{n+1}, h_n)$ is defined and equal to β_n for every n. Let $h_0 : \mathbf{E}^{q+1} \to Y^n$ be arbitrary, and suppose that h_0, $\ldots, h_N$ have been defined and have the requisite properties. Then by (5.12) of Chapter V, there exists a map $h'_N : \mathbf{E}^{q+1} \to Y^N$ such that $d^{q+1}(h'_N, h_N) = \beta_N$. Since $\mathbf{E}^{q+1}$ is contractible, there exists a map $h_{N+1} : \mathbf{E}^{q+1} \to Y^{N+1}$ such that $p_{N+1} \circ h_{N+1} = h'_N$.

Define $f_n : \mathbf{S}^q \to Y^n$ by $f_n = h_n | \mathbf{S}^q$. Then

$$\begin{aligned} p_{n+1} \circ f_{n+1} &= p_{n+1} \circ h_{n+1} | \mathbf{S}^q \\ &= h_n | \mathbf{S}^q = f_n \end{aligned}$$

and therefore the maps f_n determine a map $f: \mathbf{S}^q \to Y$ such that $g_n \circ f = f_n$ for every n. The element $\alpha \in \pi_q(Y)$ represented by f then belongs to the kernel of η, and it is clear that $\lambda(\alpha) = \beta$. Hence λ is an epimorphism. □

(4.9) Corollary *If* $\mathfrak{Y} = \{\Pi_n, Y^n, p_n\}$ *is a fibred Postnikov system, then*

$$\pi_q(Y) \approx \varprojlim_n \pi_q(Y^n) \approx \Pi_q.$$

For in this case the $\lim^1$ term vanishes because $\pi_{q+1}(p_n)$ is an isomorphism for all large n. □

Theorem (4.8) and its Corollary (4.9) show that the inverse limit of a tower of fibrations is reasonably well-behaved. This observation, however, must not be taken too literally. For example, suppose that Y^n is the cartesian product $\mathbf{S}^1 \times \cdots \times \mathbf{S}^n$, while $p_n : Y^{n+1} \to Y^n$ is the projection on the product of the first n factors. Then Y is the cartesian product

$$\prod_{n=1}^{\infty} \mathbf{S}^n$$

(let us ignore for the moment the fact that we do not know how to topologize infinite products!), so that Y need not have the homotopy type of a CW-complex. Indeed, let Y_* be the (weak) product of the $\mathbf{S}^n$. It follows from Exercise 6 of Chapter IV and from Theorem (4.8) that the injection $\pi_q(Y_*) \to \pi_q(Y)$ is an isomorphism for all q, so that the inclusion map is a weak homotopy equivalence. Were Y to have the homotopy type of a CW-complex, the inclusion would be a genuine homotopy equivalence. Suppose that $f: Y \to Y_*$ is a homotopy inverse. Then Y is compact, so that $f(Y)$ is contained in $\prod_{i=1}^k \mathbf{S}^i$ for some k. But then

$$\operatorname{Im} \pi_{k+1}(f) \subset \bigoplus_{i=1}^{k} \pi_{k+1}(\mathbf{S}^i)$$

and so $\pi_{k+1}(f)$ cannot be an epimorphism.

We can now prove the main results of this section.

(4.10) Theorem *Let* $\mathfrak{Y} = \{\Pi_n, Y^n, p_n\}$ *be a fibred Postnikov system, and let* Y *be the inverse limit* $\varprojlim_n Y^n$. *Let* $h : X \to Y$ *be a* CW-*approximation, and let* $\{X^n, f_n\}$ *be a homotopy resolution for* X. *Then there are weak homotopy equivalences*

$$h_n : X^n \to Y^n \qquad (n = 0, 1, 2, \ldots)$$

such that the diagrams

$$\begin{array}{ccc} X & \xrightarrow{h} & Y \\ {\scriptstyle i_{n+1}}\downarrow & & \downarrow{\scriptstyle g_{n+1}} \\ X^{n+1} & \xrightarrow[h_{n+1}]{} & Y^{n+1} \\ {\scriptstyle f_{n+1}}\downarrow & & \downarrow{\scriptstyle p_{n+1}} \\ X^n & \xrightarrow[h_n]{} & Y^n \end{array}$$

are commutative.

Suppose that h_n has been defined and has the requisite properties for $n = 0, 1, \ldots, N$ $(N \geq 0)$. By Theorem (1.5), there is a map $h'_{N+1} : X^{N+1} \to Y^{N+1}$ such that $h_{N+1} \circ i_{N+1} = g_{N+1} \circ h$. The maps $p_{N+1} \circ h'_{N+1}$ and $h_N \circ f_{N+1}$ are extensions of $g_N \circ h$; again by Theorem (1.5),

$$p_{N+1} \circ h'_{N+1} \simeq h_N \circ f_{N+1} \qquad (\text{rel. } X).$$

Because p_{N+1} is a fibration, there is a map $h_{N+1} : X^{N+1} \to Y^{N+1}$ such that $p_{N+1} \circ h_{N+1} = h_N \circ f_{N+1}$ and $h_{N+1} \simeq h'_{N+1}(\text{rel. } X)$. In particular,

$$h_{N+1} \circ i_{N+1} = h_{N+1} | X = h'_{N+1} | X = h'_{N+1} \circ i_{N+1} = g_{N+1} \circ h. \qquad \square$$

(4.11) Theorem *Let X be a connected simple CW-complex, $\{X^n, f_n\}$ a homotopy resolution of X, and $\mathscr{P}$ the associated Postnikov system. Let $\mathfrak{Y}$ be a fibred Postnikov system satisfying the conclusion of Theorem* (4.5). *Then the maps*

$$j_n \circ i_n : X \to Y^n$$

determine a map $h : X \to Y = \varprojlim_n Y^n$, and h is a weak homotopy equivalence.

We have

$$\begin{aligned} p_{n+1} \circ j_{n+1} \circ i_{n+1} &= j_n \circ f_{n+1} \circ i_{n+1} \\ &= j_n \circ i_n; \end{aligned}$$

hence there is a uniquely determined map $h : X \to Y$ such that $g_n \circ h = j_n \circ i_n$. We have seen that

$$\pi_n(g_n) : \pi_n(Y) \approx \pi_n(Y^n).$$

Moreover,

$$\pi_n(i_n) : \pi_n(X) \approx \pi_n(X^n).$$

Since j_n is a homotopy equivalence,

$$\pi_n(j_n) : \pi_n(X^n) \approx \pi_n(Y^n).$$

It follows that

$$\pi_n(h) : \pi_n(X) \to \pi_n(Y)$$

is an isomorphism. □

Theorem (4.10) tells us how to associate a space X to each (fibred) Postnikov system $\mathfrak{Y}$. And Theorem (4.11) tells us that if $\mathfrak{Y}$ is the Postnikov system of a space X', then X and X' have the same homotopy type.

5 Some Examples

In this section, we shall work out a few examples to illustrate the theory.

EXAMPLE 1 (The 2-sphere). Let us recall that $\pi_2(\mathbf{S}^2)$ and $\pi_3(\mathbf{S}^2)$ are infinite cyclic groups, generated by the homotopy classes ι_2 and η_2 of the identity and the Hopf map, respectively. Thus the first non-trivial stage in the Postnikov system is the Eilenberg–MacLane space $K(\mathbf{Z}, 2) = \mathbf{P}^\infty(\mathbf{C})$. The Postnikov invariant k^4 belongs to $H^4(\mathbf{Z}, 2; \mathbf{Z})$, and so is a multiple of the cup square $\mathbf{b}_2^2$ of the fundamental class $\mathbf{b}_2 = \iota^2(K(\mathbf{Z}, 2))$. We shall show that $k^4 = \pm\mathbf{b}_2^2$.

In fact, let $u = m\mathbf{b}_2^2 \in H^4(\mathbf{Z}, 2; \mathbf{Z})$ and consider the fibration $p : W = W(u) \to K(\mathbf{Z}, 2)$, which is induced by a map $f : K(\mathbf{Z}, 2) \to K(\mathbf{Z}, 4)$ such that $f^*\mathbf{b}_4 = m\mathbf{b}_2^2$ from the path space fibration $p : \mathbf{P}K(\mathbf{Z}, 4) \to K(\mathbf{Z}, 4)$. There is a commutative diagram

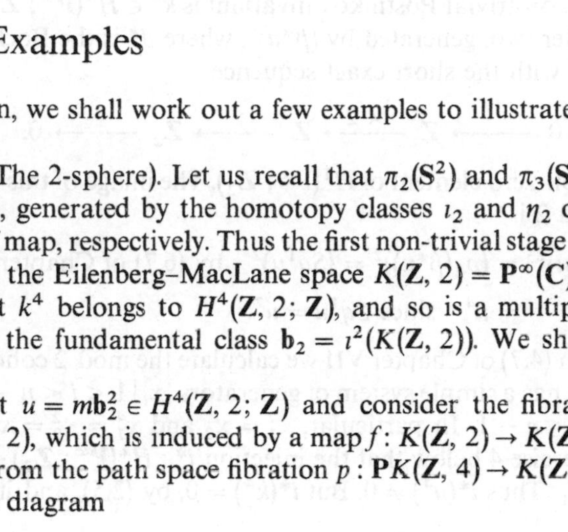

The diagram

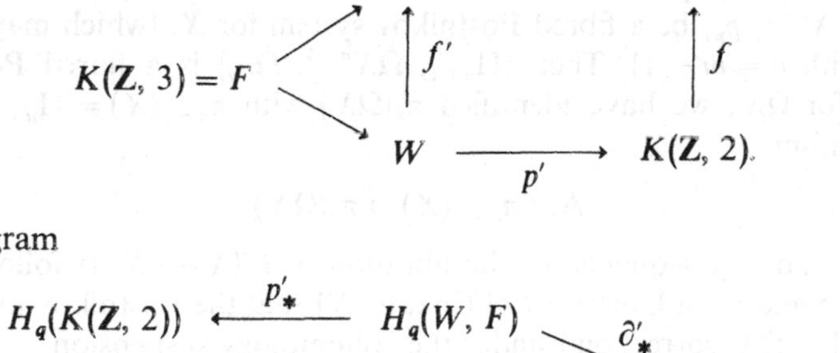

is commutative. By Theorem (6.1) of Chapter VII, p'_* and p_* are isomorphisms for $q \le 4$. Hence $H_3(W, F) = 0$ and $H_4(W, F)$ is an infinite cyclic group. Since $\mathbf{P}K(\mathbf{Z}, 4)$ is contractible, ∂_* is an isomorphism. But f'_* is a monomorphism with cokernel $\mathbf{Z}_m$, and it follows that the same is true of ∂'_*. By exactness of the homology sequence of (W, F), $H_3(W) \approx \mathbf{Z}_m$.

On the other hand, W has the same weak homotopy type as the space X^3 obtained from $X = \mathbf{S}^2$ by killing all the homotopy groups in dimensions ≥ 4. This can be accomplished by attaching only cells of dimensions ≥ 5. Thus $0 = H_3(\mathbf{S}^2) \approx H_3(W) \approx \mathbf{Z}_m$, which implies that $m = \pm 1$.

EXAMPLE 2 (The rotation group $\mathbf{O}_n^+$ $(n \geq 5)$). This time we have $\pi_1 = \mathbf{Z}_2$, so that $X^1 = K(\mathbf{Z}_2, 1)$ has the same homotopy type as $\mathbf{P}^\infty$, and $\pi_2 = 0, \pi_3 = \mathbf{Z}$. Thus the first non-trivial Postnikov invariant is $k^4 \in H^4(\mathbf{P}^\infty; \mathbf{Z})$. This group is cyclic of order two, generated by $(\beta^* u)^2$, where β^* is the Bockstein operator associated with the short exact sequence

$$0 \longrightarrow \mathbf{Z} \xrightarrow{\times 2} \mathbf{Z} \longrightarrow \mathbf{Z}_2 \longrightarrow 0,$$

and u is the non-zero element of $H^1(\mathbf{P}^\infty; \mathbf{Z}_2)$. The image of this element after reduction mod 2 is

$$\begin{aligned} \rho_2((\beta^* u)^2) &= (\rho_2(\beta^* u))^2 = (Sq^1 u)^2 \quad \text{by (6.7) of Chapter VIII} \\ &= u^4 \quad \text{since } Sq^1 u = u^2. \end{aligned}$$

In Theorem (4.7) of Chapter VII we calculate the mod 2 cohomology ring of $X = \mathbf{O}_n^+$. It has a simple system of generators $\{x_i \mid 1 \leq i \leq n-1\}$ such that $x_i^2 = x_{2i}$ if $2i \leq n-1$. In particular, $x_1^2 = x_2$ and $x_1^4 = x_2^2 = x_4 \neq 0$. It follows from Exercise 4 below that the injection $i^* : H^*(\mathbf{P}^\infty; \mathbf{Z}_2) \to H^*(X; \mathbf{Z}_2)$ maps u into x_1. Thus $i^*(u^4) \neq 0$. But $i^*(k^4) = 0$, by (2.5), and it follows that $k^4 = 0$.

EXAMPLE 3 (Loop spaces). Let X be an $(m-1)$-connected space $(m \geq 2)$, and let $\{\Pi_n, X^{n-1}, p_n\}$ be a fibred Postnikov system for X, (which may as well begin with $n = m - 1$). Then $\{\Pi_{n+1}, \Omega X^{n-1}, \Omega p_n\}$ is a fibred Postnikov system for ΩX; we have identified $\pi_n(\Omega X)$ with $\pi_{n+1}(X) = \Pi_{n+1}$ by the isomorphism

$$\Delta_* : \pi_{n+1}(X) \to \pi_n(\Omega X)$$

of the homotopy sequence of the fibration $p : \mathbf{P}'(X) \to X$. It follows from Exercise 5 and from Exercise 7 of Chapter VI that the Postnikov invariants of X and of ΩX correspond under the cohomology suspension

$$H^{n+2}(X^n; \Pi_{n+1}) \to H^{n+1}(\Omega X^n; \Pi_{n+1}).$$

In this way the Postnikov invariants of ΩX are determined by those of X.

EXAMPLE 4 (Suspensions). Let X be an $(m-1)$-connected space $(m \geq 2)$, and let $\{X^n, f_n\}$ be a homotopy resolution of X. Then (X^q, X) is $(q+1)$-connected; it follows from the Relative Hurewicz Theorem that

$H_r(X^q, X) = 0$ for $r \le q+1$. Since suspension is an isomorphism in homology, $H_r(\mathbf{S}X^q, \mathbf{S}X) = 0$ for $r \le q+2$. Since $\mathbf{S}X$ is 1-connected, we can apply the converse of the Relative Hurewicz Theorem to deduce that $(\mathbf{S}X^q, \mathbf{S}X)$ is $(q+2)$-connected. Therefore we can kill the homotopy groups of $\mathbf{S}X^q$ in dimensions $\ge q+2$ by attaching cells of dimension $\ge q+3$, obtaining in this way a resolving sequence $\{Y^{q+1}\}$ for $\mathbf{S}X$. It follows from Theorem (1.5) that the composite of $\mathbf{S}f_q : \mathbf{S}X^q \to \mathbf{S}X^{q-1}$ with the inclusion map $\mathbf{S}X^{q-1} \hookrightarrow Y^q$ has an extension $g_{q+1} : Y^{q+1} \to Y^q$, and the maps g_{q+1} are bonding maps. The suspension $\mathbf{S}\hat{X}^q$ of the relative mapping cylinder of f_{q+1} is naturally homeomorphic with the relative mapping cylinder $\widehat{\mathbf{S}X^q}$; moreover, $\mathbf{S}\hat{X}^q$ is a subspace of $\hat{Y}^{q+1}$. Routine diagram chasing shows that the diagram

$$\begin{array}{ccc}
H_{q+2}(\hat{X}^q, X^{q+1}) & \xrightarrow{\kappa_{q+2}(X)} & \pi_{q+1}(X) \\
\downarrow{\scriptstyle s_*} & & \downarrow{\scriptstyle \mathbf{S}_*} \\
H_{q+3}(\hat{Y}^{q+1}, Y^{q+2}) & \xrightarrow[\kappa_{q+3}(\mathbf{S}X)]{} & \pi_{q+2}(\mathbf{S}X)
\end{array}$$

is commutative; s_* is the composite of the suspension homomorphism

$$H_{q+2}(\hat{X}^q, X^{q+1}) \to H_{q+3}(\mathbf{S}\hat{X}^q, \mathbf{S}X^{q+1})$$

with the injection

$$H_{q+3}(\mathbf{S}\hat{X}^q, \mathbf{S}X^{q+1}) \to H_{q+3}(\hat{Y}^{q+1}, Y^{q+2}).$$

The relationship

$$s^*k^{n+3}(\mathbf{S}X) = \mathbf{S}_{**}k^{n+2}(X)$$

between the Postnikov invariants of X and $\mathbf{S}X$ is then elucidated by the diagram

$$\begin{array}{ccc}
 & & H^{q+3}(Y^{q+1}; \pi_{q+2}(\mathbf{S}X)) \\
 & & \downarrow{\scriptstyle s^*} \\
H^{q+2}(X^q; \pi_{q+1}(X)) & \xrightarrow{\mathbf{S}_{**}} & H^{q+2}(X^q; \pi_{q+2}(\mathbf{S}X))
\end{array}$$

in which s^* is the composite of the injection and suspension homomorphisms and $\mathbf{S}_{**}$ is induced by the homomorphism $\mathbf{S}_* : \pi_{q+1}(X) \to \pi_{q+2}(\mathbf{S}X)$ of coefficient groups.

Note that, if $q \le 2m-3$, then $\mathbf{S}_*$ is an isomorphism; in this range, then, the Postnikov invariants of X are determined by those of $\mathbf{S}X$.

EXAMPLE 5 (H-spaces). Let X be a connected CW-complex which is an H-space under a product $\mu : X \times X \to X$; we may assume that μ is cellular and $\mu | X \vee X$ is the folding map ∇. By (3.6) of Chapter IV, the space X is simple. Let X^* be an m-anticonnected extension of X; we may assume that X^* is a CW-complex having X as a subcomplex.

(5.1) Theorem *The space X^* is an H-space under a product $\mu^* : X^* \times X^* \to X^*$ extending that of X (so that (X^*, X) is an H-pair). The map μ^* is unique up to homotopy* (rel. $X \times X \cup X^* \vee X^*$).

The map μ has an extension $\mu_1 : X \times X \cup X^* \vee X^* \to X^*$ such that $\mu_1 | X^* \vee X^*$ is the folding map for X^*. Consider the diagram

$$\begin{array}{ccc} X \times X & \xrightarrow{\quad\mu\quad} & X \\ \downarrow & & \downarrow \\ X \times X \cup X^* \vee X^* & \overset{\mu_1}{\searrow} & \\ \downarrow & & \downarrow \\ X^* \times X^* & \xrightarrow[\mu^*]{\cdots\cdots} & X^* \end{array}$$

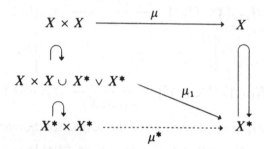

and the homology sequence

(5.2) $$\cdots \to H_q(X^* \times X^*, X \times X) \to H_q(X^* \times X^*, X \times X \cup X^* \vee X^*) \to H_{q-1}(X \times X \cup X^* \vee X^*, X \times X) \to \cdots$$

of the triple $(X^* \times X^*, X \times X \cup X^* \vee X^*, X \times X)$. Since the injection $\pi_q(X) \to \pi_q(X^*)$ is an isomorphism for all $q < m$, and an epimorphism for $q = m$, the same is true for the injection $\pi_q(X \times X) \to \pi_q(X^* \times X^*)$. Hence the pair $(X^* \times X^*, X \times X)$ is m-connected; by the Hurewicz Theorem, $H_q(X^* \times X^*, X \times X) = 0$ for all $q \leq m$. On the other hand, the injection

$$H_{q-1}(X^* \vee X^*, X \vee X) \to H_{q-1}(X \times X \cup X^* \vee X^*, X \times X)$$

is an isomorphism by the Excision Theorem. Again, since the injection $H_{q-1}(X) \to H_{q-1}(X^*)$ is an isomorphism for all $q \leq m$, the same is true for the injection $H_{q-1}(X \vee X) \to H_{q-1}(X^* \vee X^*)$. Hence $H_{q-1}(X^* \vee X^*, X \vee X)$, and therefore $H_{q-1}(X \times X \cup X^* \vee X^*, X \times X)$, vanishes for all $q \leq m$. By exactness of (5.2), $H_q(X^* \times X^*, X \times X \cup X^* \vee X^*) = 0$ for all $q \leq m$.

The obstructions to extending μ_1 lie in the groups

$$H^{q+1}(X^* \times X^*, X \times X \cup X^* \vee X^*; \pi_q(X^*)).$$

By the Künneth Theorem, these groups vanish for all $q \leq m - 1$. But $\pi_q(X^*) = 0$ for all $q \geq m$. Hence all obstructions vanish and the extension μ^* exists.

If μ_1^* and μ_2^* are extensions of μ_1, the obstructions to a homotopy between them lie in the groups

$$H^q(X^* \times X^*, X \times X \cup X^* \vee X^*; \pi_q(X^*)),$$

and these vanish for all q by a similar argument. □

Let X be a connected simple CW-complex and let $\{X^n, f_n\}$ be a homotopy resolution of X. By Theorem (5.1) each of the spaces X^n is an H-space under a product $\mu_n : X^n \times X^n \to X^n$.

(5.3) Theorem *The map* $f_n : X^n \to X^{n-1}$ *is an* H*-map for each* $n \geq 1$.

For $X^n \times X^n$ is a regular $(n+1)$-anticonnected extension of $X \times X$. By Theorem (1.5), the maps

$$\mu_{n-1} \circ (f_n \times f_0),\ f_n \circ \mu_n : X^n \times X^n \to X^{n-1}$$

are homotopic (rel. $X \times X$). □

In order to study the Postnikov invariants of an H-space, let us begin by examining the effect of the amplification process of §3 on the cohomology of an H-space X. Let $u \in H^n(X; \Pi)$; and let $q_u : W(u) \to X$ be an amplification of X by the class u. Let $g : W \to W(u)$ be a CW-approximation, and let $q = q_u \circ g : W \to X$. Let $\mu : X \times X \to X$ be the product in X, and recall that the class u is *primitive* if and only if $\mu^* u = p_1^* u + p_2^* u$, where p_1, $p_2 : X \times X \to X$ are the projections. Let $K = K(\Pi, n)$, and let $h : X \to K$ be a map such that $h^*\mathbf{b}_n = u$.

(5.4) Theorem *The following conditions are equivalent*:

(1) *the class* u *is primitive*;
(2) *the map* $h : X \to K$ *is an* H*-map*;
(3) W *is an* H*-space under a product* $\mu_1 : W \times W \to W$ *such that* $q : W \to X$ *is an* H*-map*.

Let $\mu_0 : K \times K \to K$ be the product in K (cf. §7 of Chapter V). Then h is an H-map if and only if $\mu_0 \circ (h \times h) \simeq h \circ \mu$, and this is true, by Corollary (6.20) of that Chapter, if and only if $\mu^* h^* \mathbf{b}_n = (h \times h)^* \mu_0^* \mathbf{b}_n$. But

$$\begin{aligned}(h \times h)^* \mu_0^* \mathbf{b}_n &= (h \times h)^*(\mathbf{b}_n \times 1 + 1 \times \mathbf{b}_n) \qquad (\mathbf{b}_n \text{ is primitive!})\\ &= h^*\mathbf{b}_n \times 1 + 1 \times h^*\mathbf{b}_n\end{aligned}$$

and this equals $\mu^* h^* \mathbf{b}_n$ if and only if $u = h^*\mathbf{b}_n$ is primitive. Hence (1) and (2) are equivalent.

To see that (2) implies (3), we observe that W is homotopically equivalent to the mapping fibre of h and that, by (6.18*) of Chapter III there is, for every space Y, an exact sequence

$$[Y, K'] \to [Y, W] \to [Y, X] \to [Y, K],$$

natural with respect to maps $Y' \to Y$ $(K' = \Omega K = K(\Pi, n-1))$. Thus there is a commutative diagram

(5.5)

$$\begin{array}{ccccccc}
[W \times W, K'] & \xrightarrow{j} & [W \times W, W] & \xrightarrow{\underline{q}} & [W \times W, X] & \xrightarrow{\underline{h}} & [W \times W, K] \\
\downarrow{\scriptstyle \bar{i}} & & \downarrow{\scriptstyle \bar{i}} & & \downarrow{\scriptstyle \bar{i}} & & \downarrow{\scriptstyle \bar{i}} \\
[W \vee W, K'] & \xrightarrow[j]{} & [W \vee W, W] & \xrightarrow[\underline{q}]{} & [W \vee W, X] & \xrightarrow[\underline{h}]{} & [W \vee W, K]
\end{array}$$

where $i: W \vee W \to W \times W$ and $j: K' \to W$ are inclusions. Let $\alpha = [\mu \circ (q \times q)] \in [W \times W, X]$. Then

$$\begin{aligned}
\underline{h}(\alpha) = [h \circ \mu \circ (q \times q)] &= [\mu_0 \circ (h \times h) \circ (q \times q)] \\
&= [\mu_0 \circ ((h \circ q) \times (h \circ q))] = 0
\end{aligned}$$

since $h \circ q \simeq 0$. By exactness of the top sequence of (5.5), there exists $\mu' : W \times W \to W$ such that $\underline{q}[\mu'] = [\mu \circ (q \times q)]$, i.e., $q \circ \mu' \simeq \mu \circ (q \times q)$. If $\nabla : W \vee W \to W$ is the folding map, then

$$\begin{aligned}
\underline{q}\bar{i}[\mu'] = [q \circ \mu' \circ i] &= [\mu \circ (q \times q) \circ i] = [\mu \circ i \circ (q \vee q)] \\
&= [\nabla \circ (q \vee q)] = [q \circ \nabla] = \underline{q}[\nabla].
\end{aligned}$$

By the extended exactness property (6.21*) of Chapter III, there exists $\alpha \in [W \vee W, K']$ such that $[\nabla] = \alpha \cdot \bar{i}[\mu']$. The homomorphism $\bar{i} : [W \times W, K'] \to [W \vee W, K']$ corresponds, under the isomorphisms of Corollary (6.20) of Chapter V, to the homomorphism

$$i^* : H^{n-1}(W \times W; \Pi) \to H^{n-1}(W \vee W; \Pi),$$

and the latter is an epimorphism. Hence there exists $\beta \in [W \times W, K']$ such that $\bar{i}(\beta) = \alpha$, and therefore

$$[\nabla] = \alpha \cdot \bar{i}[\mu'] = \bar{i}(\beta) \cdot \bar{i}[\mu'] = \bar{i}(\beta \cdot [\mu']).$$

We may take μ_1 to be any representative of $\beta \cdot [\mu']$.

Finally, assume that (3) holds. Then there is a commutative diagram

$$\begin{array}{ccc}
H^n(X \times X, X \vee X; \Pi) & \xrightarrow{q_2^*} & H^n(W \times W, W \vee W; \Pi) \\
\uparrow \beta^* & & \uparrow \beta_1^* \\
H^n(X \times X; \Pi) & \xrightarrow{q_1^*} & H^n(W \times W; \Pi) \\
\uparrow \mu^* & & \uparrow \mu_1^* \\
H^n(X; \Pi) & \xrightarrow[q^*]{} & H^n(W; \Pi)
\end{array}$$

in which the homomorphisms q_1^* and q_2^* are induced by the map $q \times q$, while β^* and β_1^* are the projections of Chapter III, §7. In §3 we saw that

$$\pi_r(q) : \pi_r(W) \to \pi_r(X)$$

is an isomorphism for $r \leq n - 2$ and an epimorphism for $r = n - 1$. By the Whitehead Theorem (7.13) of Chapter IV,

$$H_r(q): H_r(W) \to H_r(X)$$

has the same properties. It follows easily (since $n \geq 2$) that

$$H_r(q \times q): H_r(W \times W, W \vee W) \to H_r(X \times X, X \vee X)$$

is an isomorphism if $r = n - 1$ and an epimorphism if $r = n$. Hence $q_2^* : H^n(X \times X, X \vee X; \Pi) \to H^n(W \times W, W \vee W; \Pi)$ is a monomorphism. But $q^*u = 0$ and therefore $q_2^* \beta^* \mu^* u = 0$; hence $\beta^* \mu^* u = 0$, i.e., u is primitive. □

(5.6) Corollary *Let $\mathscr{P} = \{\Pi_n, X^{n-1}, k^{n+1}, f_n\}$ be a Postnikov system for the* H-*space X. Then*

(1) X^{n-1} *is an* H-*space;*
(2) $i_n : X \hookrightarrow X^{n-1}$ *is an* H-*map;*
(3) $f_n : X^n \to X^{n-1}$ *is an* H-*map;*
(4) $k^{n+1} \in H^{n+1}(X^{n-1}; \Pi_n)$ *is primitive.* □

6 Relative Postnikov Systems

The preceding discussions can be relativized, using Theorem (2.3) of Chapter V. The resulting theory is due to J. C. Moore [1]. We shall sketch the theory, leaving to the reader the task of filling in the sometimes irksome details.

Let X and B be 0-connected spaces. A map $f: X \to B$ is said to be *n-anticonnected* if and only if its mapping fibre $\mathbf{T}^f$ is n-anticonnected. This is the case if and only if the homomorphism $f_* : \pi_q(X) \to \pi_q(B)$ is an isomorphism for all $q > n$ and a monomorphism for $q = n$.

An *n-anticonnected extension* of a map $f: X \to B$ is a map $\bar{f}: \bar{X} \to B$ such that

(1) X is a subspace of $\bar{X}$ and $\bar{f} \mid X = f$;
(2) the pair $(\bar{X}, X)$ is n-connected;
(3) the map $\bar{f}$ is n-anticonnected.

If, instead of (2) we assume (2′) $(\bar{X}, X)$ is a relative CW-complex having no cells of dimension $\leq n$, we shall say that $\bar{f}$ is a *regular* n-anticonnected extension of f.

From Theorem (2.3) of Chapter V we deduce

(6.1) Theorem *If X and B are 0-connected spaces and n is a non-negative integer, then every map $f: X \to B$ has a regular n-anticonnected extension $\bar{f}: \bar{X} \to B$.* □

Let $\bar{f}: \bar{X} \to B$ be an n-anticonnected extension of $f: X \to B$ and let $\bar{F} = \mathbf{T}^{\bar{f}}$, $F = \mathbf{T}^f$ be their mapping fibres. Then F is a subspace of $\bar{F}$. Let $i: X \hookrightarrow \bar{X}$.

(6.2) Theorem *The injection $\pi_q(X) \to \pi_q(\bar{X})$ is an isomorphism for all $q < n$. The homomorphism $\pi_q(\bar{f}) : \pi_q(\bar{X}) \to \pi_q(B)$ is an isomorphism for all $q > n$. The diagram*

$$\begin{array}{ccc} \pi_n(X) & \xrightarrow{\pi_n(f)} & \pi_n(B) \\ & {\scriptstyle \pi_n(i)} \searrow \quad \nearrow {\scriptstyle \pi_n(\bar{f})} & \\ & \pi_n(\bar{X}) & \end{array}$$

is commutative, the injection $\pi_n(X) \to \pi_n(\bar{X})$ is an epimorphism, and $\pi_n(\bar{f}) : \pi_n(\bar{X}) \to \pi_n(B)$ is a monomorphism. The injection $\pi_q(F) \to \pi_q(\bar{F})$ is an isomorphism for all $q < n$, and $\pi_q(\bar{F}) = 0$ for all $q \geq n$.

Only the last sentence requires some argument. There is a commutative diagram

$$\begin{array}{ccccccccc} \pi_{q+1}(X) & \to & \pi_{q+1}(B) & \to & \pi_q(F) & \to & \pi_q(X) & \to & \pi_q(B) \\ \downarrow {\scriptstyle \pi_{q+1}(i)} & & \downarrow & & \downarrow & & \downarrow {\scriptstyle \pi_q(i)} & & \downarrow \\ \pi_{q+1}(\bar{X}) & \to & \pi_{q+1}(B) & \to & \pi_q(\bar{F}) & \to & \pi_q(\bar{X}) & \to & \pi_q(B) \end{array}$$

whose rows are the sequences (8.9) of Chapter IV for the maps f and $\bar{f}$; the vertical arrows denote injections. If $q < n$, then $\pi_{q+1}(i)$ is an epimorphism and $\pi_q(i)$ an isomorphism; by the Five-Lemma, the injection $\pi_q(F) \to \pi_q(\bar{F})$ is an isomorphism. □

(6.3) Theorem *Let $f_q : X^q \to B$ be a regular $(q+1)$-anticonnected extension of $f: X \to B$ $(q = n,\ n+1)$, and let $i_q : X \hookrightarrow X_q$. Then there is a map $g : X^{n+1} \to X^n$ such that $g \circ i_{n+1} = i_n$ and $f_n \circ g \simeq f_{n+1}$* (rel. X).

In the diagram

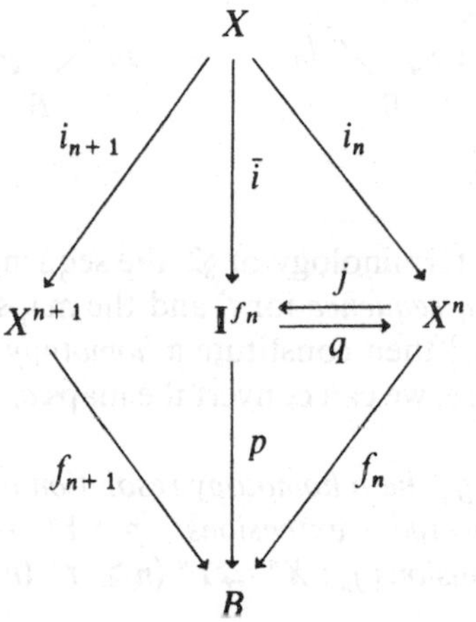

the maps p, q are the fibrations of $\mathbf{I}^{f_n}$ over B, X^n, respectively, while the maps j and $\bar{i} = j \circ i_n$ are inclusions, $q \circ j = 1$, $j \circ q \simeq 1$ (rel. X), $p \circ j = f_n$, and $f_n \circ q \simeq p$ (rel. X). Let F^n be the fibre of p. Then

$$p \circ \bar{i} = p \circ j \circ i_n = f_n \circ i_n = f = f_{n+1} \circ i_{n+1}.$$

Thus $\bar{i}: X \to \mathbf{I}^{f_n}$ is a partial lifting of f_{n+1}. The obstructions to extending $\bar{i}$ to a lifting of f_{n+1} lie in the groups

$$G_q = H^{q+1}(X^{n+1}, X; \pi_q(F^n))$$

(local coefficients!). Since (X^{n+1}, X) has no cells of dimension $\leq n+2$, the group G_q vanishes for all $q \leq n+1$. But F^n is $(n+1)$-anticonnected, so that $\pi_q(F^n)$, and therefore G_q, vanishes if $q \geq n+1$. Thus all obstructions vanish, and the lifting extension problem in question has a solution $g' : X^{n+1} \to \mathbf{I}^{f_n}$. Then $g = q \circ g'$ has the required properties. □

Now let X and B be 0-connected spaces, $f: X \to B$. Let $X^{-1} = B$, and, for every $n \geq 0$, let $f_n : X^n \to B$ be a regular $(n+1)$-anticonnected extension of f. Let $g_0 = f_0 : X^0 \to B$, and, for each $n \geq 0$, let $g_{n+1} : X^{n+1} \to X^n$ be a map satisfying the conclusion of Theorem (6.3). We then find, by induction, that the map $f'_n = g_0 \circ g_1 \circ \cdots \circ g_n$ is homotopic (rel. X) to f_n. Replacing f'_n by f_n, we have

(6.4) Theorem *Let X and B be 0-connected spaces, $f: X \to B$. Then there are regular $(n+1)$-anticonnected extensions $f_n : X^n \to B$ and maps $g_n : X^n \to X^{n-1}$ $(n \geq 0)$ such that the diagrams*

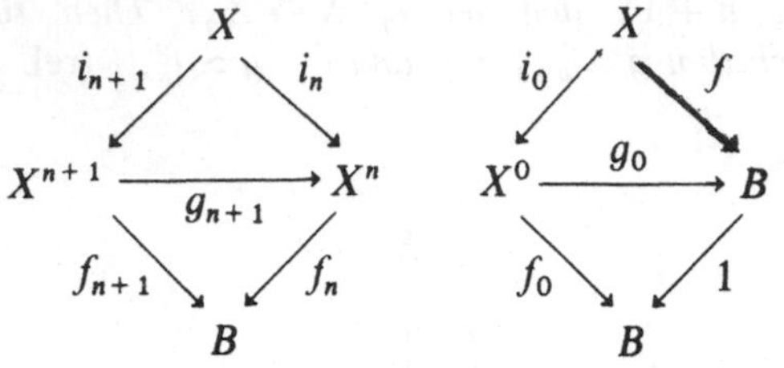

are strictly commutative. □

By analogy with the terminology of §2, the sequence of maps $f_n : X^n \to B$ will be called a *resolving sequence* for f, and the maps g_n *bonding maps*. The pair of sequences $\{f_n, g_n\}$ then constitute a *homotopy resolution* of f.

As in the absolute case, we can convert the maps g_n into fibrations, so that

(6.5) Theorem *Let $\{f_n, g_n\}$ be a homotopy resolution of $f: X \to B$. Then there exist $(n+1)$-anticonnected extensions $p_n : Y^n \to B$ of f, fibrations $q_n : Y^n \to Y^{n-1}$, and inclusions $j_n : X^n \hookrightarrow Y^n$ $(n \geq Y^n$ $(n \geq 0)$ such that*

(1) *j_n is a homotopy equivalence* (rel. X);
(2) *the octahedral diagram*

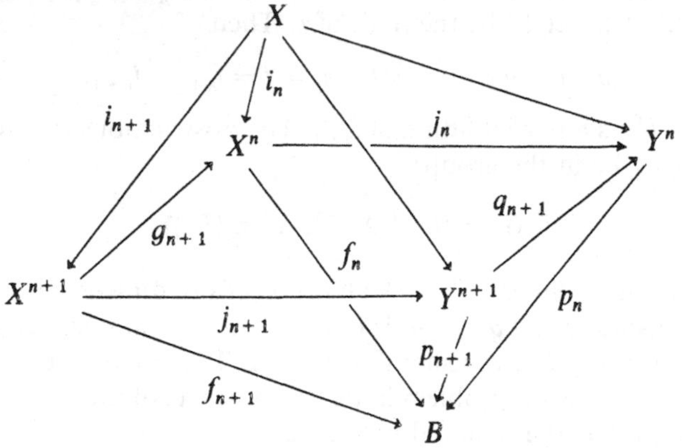

is commutative. □

Let $\{f_n, g_n\}$ be a resolution of $f: X \to B$. Let us now show, by analogy with the absolute case, how to construct X^{n+1} from X^n (or, rather, f_{n+1} from f_n).

As in the absolute case, we shall need some additional hypotheses. Specifically, we shall assume

(1) *X and B are* CW*-complexes;*
(2) *f is 1-connected;*
(3) *the group $\pi_1(B)$ operates trivially on the mapping fibre $\mathbf{T}^f$ of f.*

(A map satisfying conditions (2) and (3) is said to be *simple*).

Let F^n be the mapping fibre of f_n, and let $h : F \to X$, $h_n : F^n \to X^n$ the appropriate fibrations. There is an inclusion $i'_n : F \hookrightarrow F^n$, and the map $g_{n+1} : X^{n+1} \to X^n$ induces a map $g'_{n+1} : F^{n+1} \to F^n$ such that $g'_{n+1} \circ i'_{n+1} = g'_n$ and the diagram

(6.6)
$$\begin{array}{ccccc}
F & \xrightarrow{h} & X & & \\
{\scriptstyle i'_{n+1}}\downarrow & & \downarrow{\scriptstyle i_{n+1}} & \searrow{\scriptstyle f} & \\
F^{n+1} & \xrightarrow{h_{n+1}} & X^{n+1} & \xrightarrow{f_{n+1}} & B \\
{\scriptstyle g'_{n+1}}\downarrow & & \downarrow{\scriptstyle g_{n+1}} & \nearrow{\scriptstyle f_n} & \\
F^n & \xrightarrow[h_n]{} & X^n & &
\end{array}$$

is commutative. Let $\psi : K \to F$ be a CW-approximation.

(6.7) *There exist a homotopy resolution $\{K^n, \theta_n\}$ of K and weak homotopy equivalences $\psi_n : K^n \to F^n$ such that the diagram*

(6.8)
$$\begin{array}{ccc}
K & \xrightarrow{\psi} & F \\
{\scriptstyle j_{n+1}}\cap & & \downarrow{\scriptstyle i'_{n+1}} \\
K^{n+1} & \xrightarrow{\psi_{n+1}} & F^{n+1} \\
{\scriptstyle \theta_{n+1}}\downarrow & & \downarrow{\scriptstyle g'_{n+1}} \\
K^n & \xrightarrow[\psi_n]{} & F^n
\end{array}$$

is homotopy commutative (rel. K). □

Let us adjoin the diagram in (6.8) to that in (6.6) and then discard the column containing the F's. We obtain a new diagram

(6.9)

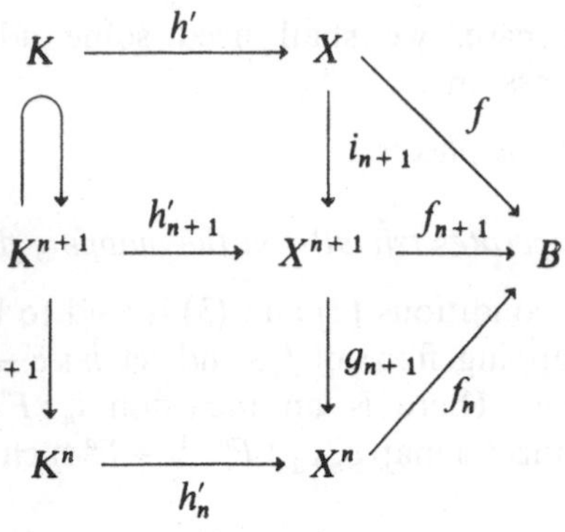

in which the two left-hand squares are homotopy commutative (rel. K). Let $\hat{K}^n$, $\hat{X}^n$ be the relative mapping cylinders of θ_{n+1}, g_{n+1}, respectively.

(6.10) Theorem *There is a map*

$$\phi_n : (\hat{K}^n; K^n, K^{n+1}) \to (\hat{X}^n; X^n, X^{n+1})$$

such that $\phi_n | K^n = h'_n$, $\phi_n | K^{n+1} = h'_{n+1}$, *and*

$$\pi_k(\phi_n) : \pi_k(\hat{K}^n, K^{n+1}) \to \pi_k(\hat{X}^n, X^{n+1})$$

is an isomorphism for every k. □

The groups $\pi_k(\hat{K}^n, K^{n+1})$ vanishing for $k \neq n+2$, the same is true for the groups $\pi_k(\hat{X}^n, X^{n+1})$. In particular, the latter pair is $(n+1)$-connected. It follows from the simplicity of the map f that $\pi_1(X^{n+1})$ operates trivially on $\pi_{n+2}(\hat{X}^n, X^{n+1})$, so that the Hurewicz map

$$\rho : \pi_{n+2}(\hat{X}^n, X^{n+1}) \to H_{n+2}(\hat{X}^n, X^{n+1})$$

is an isomorphism. Corresponding to the homomorphism

$$H_{n+2}(\hat{X}^n, X^{n+1}) \xrightarrow{\rho^{-1}} \pi_{n+2}(\hat{X}^n, X^{n+1}) \xrightarrow{\phi_*^{-1}} \pi_{n+2}(\hat{K}^n, K^{n+1}) \xrightarrow{\partial_*}$$
$$\pi_{n+1}(K^{n+1}) \xrightarrow{\pi_{n+1}(j_{n+1})^{-1}} \pi_{n+1}(K)$$

is a cohomology class $\kappa^{n+2}(f) \in H^{n+2}(\hat{X}^n, X^{n+1}; \pi_{n+1}(K))$. Its image under the injection is a class

$$k^{n+2}(f) \in H^{n+2}(X^n; \pi_{n+1}(K)),$$

called the $(n+2)$nd *Moore–Postnikov invariant of f*. Evidently the image of $k^{n+2}(f)$ under the injection

$$\psi_n^* : H^{n+2}(X^n; \pi_{n+1}(K)) \to H^{n+2}(K^n; \pi_{n+1}(K))$$

is the $(n+2)$nd Postnikov invariant of K.

Let $p_K : W(k^{n+2}(K)) \to K^n$, $p_f : W(k^{n+2}(f)) \to X^n$ be the amplifications of K^n, X^n by the appropriate cohomology classes. Then there is a commutative diagram

$$
\text{(6.11)} \qquad
\begin{array}{ccc}
W(k^{n+2}(K)) & \xrightarrow{\psi'_n} & W(k^{n+2}(f)) \\
{\scriptstyle p_K}\downarrow & & \downarrow{\scriptstyle p_f} \\
K^n & \xrightarrow[\psi_n]{} & X^n
\end{array}
$$

Moreover, the space X^{n+1} can be recovered from X^n with the aid of the class $k^{n+2}(f)$:

(6.12) Theorem *The diagram* (6.11) *can be enlarged to a commutative diagram*

$$
\begin{array}{ccccccc}
K^{n+1} & & & \xrightarrow{\psi_{n+1}} & & & X^{n+1} \\
 & \searrow^{s'_{n+1}} & & & & {}^{s_{n+1}}\swarrow & \\
{\scriptstyle g'_{n+1}}\downarrow & & W(k^{n+2}(K)) & \xrightarrow[\psi'_n]{} & W(k^{n+2}(f)) & & \downarrow{\scriptstyle g_{n+1}} \\
 & & {\scriptstyle p_K}\downarrow & & \downarrow{\scriptstyle p_f} & & \\
 & & K^n & \xrightarrow[\psi_n]{} & X^n & &
\end{array}
$$

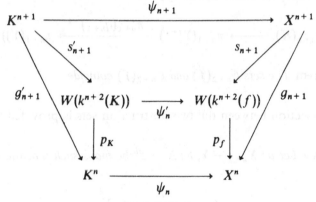

such that s'_{n+1} *and* s_{n+1} *are weak homotopy equivalences.* □

7 Postnikov Systems and Obstruction Theory

The notion of Postnikov system gives rise to an alternative approach to obstruction theory. Let (X, A) be a CW-pair, Y a simple space, $f: A \to Y$ a map, and let $\mathfrak{P} = \{\Pi_n, Y^n, p_n\}$ be a fibred Postnikov system for Y. Then there are maps $q_n : Y \to Y_n$ such that the diagrams

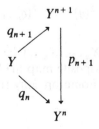

$$
\begin{array}{ccc}
 & & Y^{n+1} \\
 & {}^{q_{n+1}}\nearrow & \\
Y & & \downarrow{\scriptstyle p_{n+1}} \\
 & {}_{q_n}\searrow & \\
 & & Y^n
\end{array}
$$

are commutative. Since Y^0 is contractible, the map $q_0 \circ f : A \to Y^0$ has an extension $f_0 : X \to Y^0$. Suppose that $f_n : X \to Y^n$ is an extension of $q_n \circ f : A \to Y^n$. The map $p_{n+1} : Y^{n+1} \to Y^n$ is a fibration with fibre $K = K(\Pi_{n+1}, n+2)$, and the map $q_{n+1} \circ f$ is a partial lifting of f_n; the local coefficient system is simple. By the results of §6 of Chapter VI, the primary obstruction $\bar{\gamma}^{n+2}(q_{n+1} \circ f)$ to extending $q_{n+1} \circ f$ to a lifting of f_n lies in the group $H^{n+2}(X, A; \Pi_{n+1})$, and the vanishing of this obstruction is necessary and sufficient for the existence of such a lifting.

The set of these obstructions, as f_n ranges over all extensions of $q_n \circ f$, is a subset $\bar{\mathcal{O}}_{n+2}(f)$ of $H^{n+2}(X, A; \Pi_{n+1})$. On the other hand, in §5 of Chapter V we introduced the set $\mathcal{O}_{n+2}(f)$ consisting of the cohomology classes $c^{n+2}(g)$, as g ranges over all extensions $g : X_{n+1} \to Y$ of f. (N.B.: we are identifying the coefficient groups $\pi_{n+1}(K)$ and $\pi_{n+1}(Y)$ of these obstructions with Π_{n+1}; they are isomorphic under the composite

(7.1) $$\pi_{n+1}(K) \longrightarrow \pi_{n+1}(Y^{n+1}) \xrightarrow{\pi_{n+1}(q_{n+1})^{-1}} \pi_{n+1}(Y)).$$

(7.2) Theorem *The sets $\mathcal{O}_{n+2}(f)$ and $\bar{\mathcal{O}}_{n+2}(f)$ coincide.*

The connection between the two obstruction sets is provided by

(7.3) Lemma *Let $g : X_{n+1} \to Y$, $h : X \to Y^n$ be maps such that the diagram*

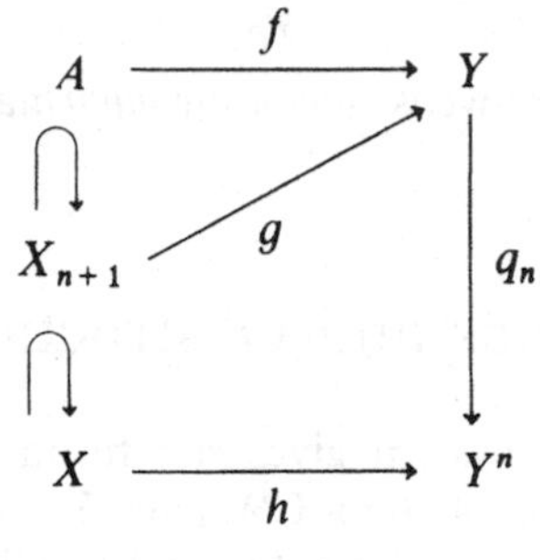

is commutative. Then

$$c^{n+2}(g) = \bar{c}^{n+2}(q_{n+1} \circ g).$$

Let $\phi_\alpha : (\Delta^{n+2}, \dot{\Delta}^{n+2}) \to (X_{n+2}, X_{n+1})$ be a characteristic map for an $(n+2)$-cell E_α. Then the composite map $(h | X_{n+2}) \circ \phi_\alpha$ induces a fibration $p' : Z \to \Delta^{n+2}$, which is fibre homotopically trivial. There is a commutative diagram

(7.4)

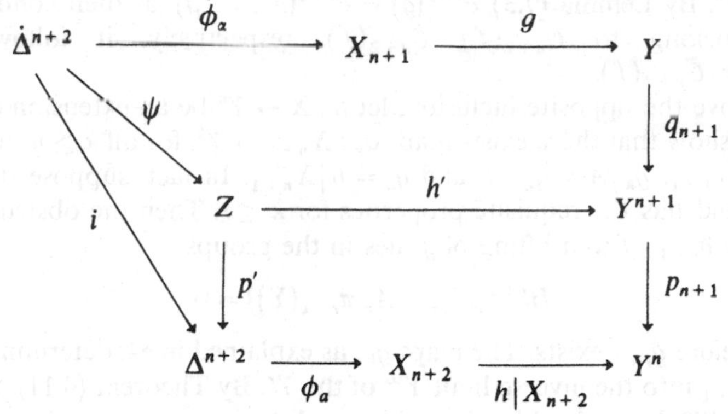

in which i is the inclusion map, $\dot{\phi}_\alpha = \phi_\alpha|\dot{\Delta}^{n+2}$, and ψ is defined by the conditions $p' \circ \psi = i$, $h' \circ \psi = q_{n+1} \circ g \circ \dot{\phi}_\alpha$ (this makes sense because $(h|X_{n+2}) \circ \phi_\alpha \circ i = q_n \circ g \circ \dot{\phi}_\alpha = p_{n+1} \circ q_{n+1} \circ g \circ \dot{\phi}_\alpha$). We may take $K = p_{n+1}^{-1}\, h\phi_\alpha(\mathbf{e}_0)$; let $K' = p'^{-1}(\mathbf{e}_0)$, so that h' maps K' homeomorphically upon K. Then there is a commutative diagram

$$\begin{array}{ccccc}
\pi_{n+1}(K') & \xrightarrow{i'} & \pi_{n+1}(Z) & & \\
\downarrow{\scriptstyle h_2'} & & \downarrow{\scriptstyle h_1'} & & \\
\pi_{n+1}(K) & \xrightarrow[i]{} & \pi_{n+1}(Y^{n+1}) & \xleftarrow[\pi_{n+1}(q_{n+1})]{} & \pi_{n+1}(Y)
\end{array}$$

in which i and i' are injections and h_1', h_2' are induced by the map h'. The value of $\bar{c}^{n+2}(q_{n+1} \circ g)$ on E_α is the element $h_2' i'^{-1}(\alpha)$, where α is the element of $\pi_{n+1}(Z)$ represented by ψ. But

$$h_2' i'^{-1}(\alpha) = i^{-1} h_1'(\alpha);$$

the image of this element under the isomorphism (7.1) is

$$\pi_{n+1}(q_{n+1})^{-1} i i^{-1} h_1'(\alpha) = \pi_{n+1}(q_{n+1})^{-1} h_1'(\alpha).$$

It follows from the commutativity of the diagram (7.4) that the latter element is represented by $g \circ \dot{\phi}_\alpha$. But this element is also the value of $c^{n+2}(g)$ on the cell E_α. □

We can now prove Theorem (7.2). Suppose that $g : X^{n+1} \to Y$ is an extension of f. The groups

$$H^{q+1}(X, X_{n+1}; \pi_q(Y^n))$$

vanish for all q; therefore the map $q_n \circ g : X_{n+1} \to Y^n$ has an extension

$h: X \to Y^n$. By Lemma (7.3), $c^{n+2}(g) = \bar{c}^{n+2}(q_{n+1} \circ g)$; as their cohomology classes belong to $\mathcal{O}_{n+2}(f)$, $\bar{\mathcal{O}}_{n+2}(f)$, respectively, it follows that $\mathcal{O}_{n+2}(f) \subset \bar{\mathcal{O}}_{n+2}(f)$.

To prove the opposite inclusion, let $h: X \to Y^n$ be an extension of $q_n \circ f$. We next show that there exists maps $g_k: X_{n+1} \to Y^k$, for all $k \geq n$, such that $p_k \circ g_k = g_{k-1}$, $g_k | A = q_k \circ f$, and $g_n = h | X_{n+1}$. In fact, suppose that g_k is defined and has the requisite properties for $k \leq r$. Then the obstruction to extending $q_{r+1} \circ f$ to a lifting of g_r lies in the groups

$$H^{r+2}(X_{n+1}, A; \pi_{r+1}(Y)) = 0.$$

and therefore g_{r+1} exists. The maps g_k, as explained in §4, determine a map g_∞ of X_{n+1} into the inverse limit Y^∞ of the Y^r. By Theorem (4.11), the map $q_\infty: Y \to Y^\infty$ determined by the q_r is a weak homotopy equivalence. Hence there is a map $g': X_{n+1} \to Y$ such that $q_\infty \circ g' \simeq g_\infty$. Now the map g_∞ is an extension of $q_\infty \circ f$; hence

$$q_\infty \circ g' | A \simeq g_\infty | A = q_\infty \circ f.$$

Again, since q_∞ is a weak homotopy equivalence, $g' | A \simeq f$, and therefore g' is homotopic to a map $g: X_{n+1} \to Y$ such that $g \circ | A = f$. Then $q_n \circ g \simeq g_n = h | X_{n+1}$; hence h is homotopic to a map $h': X \to Y^n$ such that $h' | X_{n+1} = q_n \circ g$. By Lemma (7.3), $c^{n+2}(g) = \bar{c}^{n+2}(q_n \circ g)$, and it follows as before that $\bar{\mathcal{O}}_{n+2}(f) \subset \mathcal{O}_{n+2}(f)$.

Let us apply these ideas to a special case of importance. Let us assume that the space Y is $(n-1)$-connected and also that $\pi_i(Y) = 0$ for $n < i < q$. Let $\mathfrak{P}$ be a fibred Postnikov system for Y; we may assume that $Y^i = Y^0 = \{*\}$ for all $i < n$, $Y^n = \cdots = Y^{q-1} = K(\Pi, n)$, where $\Pi = \pi_n(Y)$. The first non-trivial Postnikov invariant is $k^{q+1} \in H^{q+1}(\Pi, n; G)$, where $G = \pi_q(Y)$. According to §8 of Chapter V, k^{q+1} determines a cohomology operation

$$\theta: H^n(\ \ ; \Pi) \to H^{q+1}(\ \ ; G).$$

Let (X, A) be a CW-pair, $f: A \to Y$. The primary obstruction to extending f is

$$\gamma^{n+1}(f) = (-1)^n \delta^* f^* \iota^n(Y) \in H^{n+1}(X, A; \Pi),$$

and this vanishes if and only if there is a class $u \in H^n(X; \Pi)$ whose image under the injection $H^n(X; \Pi) \to H^n(A; \Pi)$ is $f^*\iota^n(Y)$. When this is so, f has an extension $f_u: X_{n+1} \to Y$ such that $f_u^*\iota^n(Y)$ is the image of u under the (isomorphic) injection $H^n(X; \Pi) \to H^n(X_{n+1}; \Pi)$. The map f_u, in turn, can be extended over X_q (because the relevant obstructions vanish), and, if g_0, g_1 are two such extensions, then $g_0 | X_{q-1} \simeq g_1 | X_{q-1}$ (rel. A). If $G: \mathbf{I} \times X_{q-1} \to Y$ is a homotopy between the latter two maps, we can form the difference cochain

$$d^q = d^q(g_0, G, g_1) \in \Gamma^q(X, A; \Pi),$$

as well as the obstructions $c^{q+1}(g_0)$, $c^{q+1}(g_1)$ to extending g_0, g_1, respectively, over X_{q+1}. By Theorem (5.6′) of Chapter V, these cochains are related by the formula

$$c^{q+1}(g_1) - c^{q+1}(g_0) = \delta d^n.$$

Hence the obstructions $c^{q+1}(g)$ to extending over X_{q+1} an extension $g : X_q \to Y$ of f_u lie in a single cohomology class

$$z^{q+1}(f_u) \in H^{q+1}(X, A; G),$$

whose vanishing is necessary and sufficient for f_u to be extendible over X_{q+1}. The class $z^{q+1}(f_u)$ is called the *secondary obstruction* to extending f_u.

In discussing the secondary obstruction, we shall first suppose that $A = \varnothing$.

(7.5) Theorem *Suppose $A = \varnothing$ and let $u \in H^n(X; \Pi)$. Then*

$$z^{q+1}(f_u) = \theta(u).$$

We shall apply Theorem (7.2), not for $A = \varnothing$, but rather for $A = X_{n+1}$, $f = f_u$. Then $\bar{\mathcal{O}}_{q+1}(f_u) = \mathcal{O}_{q+1}(f_u) = \{z^{q+1}(f_u)\}$. On the other hand, since the injection $H^n(X; \Pi) \to H^n(X_{n+1}; \Pi)$ is an isomorphism, so is the injection $[X, K(\Pi, n)] \to [X_{n+1}, K(\Pi, n)]$, so that there is a map $h : X \to K(\Pi, n)$, unique up to homotopy, such that $h \mid X_{n+1} \simeq q_n \circ f_u$. But

$$\begin{aligned}
\bar{\gamma}^{q+1}(q_{n+1} \circ f_u) &= h^*\bar{\gamma}^{q+1}(1) \quad \text{by Theorem (6.3) of Chapter VI} \\
&= h^*k^{q+1} \quad \text{by Exercise 5} \\
&= \theta(u) \quad \text{by definition of } \theta,
\end{aligned}$$

and therefore $\bar{\mathcal{O}}_{q+1}(f_u) = \{\theta(u)\}$. □

For the case of a general subcomplex A, we observe that, if f_u and f_v are extensions of f, then $w = (f_u, f_v)^*\iota^n(Y)$ is defined and belongs to the group $H^n(X, A; \Pi)$. The reader may therefore expect that

$$z^{q+1}(f_v) - z^{q+1}(f_u) = \theta(w). \tag{7.6}$$

This would, however, be too optimistic. In fact, Formula (7.6) holds whenever the operation θ is additive. When θ is not additive, the two sides of (7.6) differ by a term which is related to the deviation

$$\theta_2(x, y) = \theta_2(x + y) - \theta_2(x) - \theta_2(y)$$

of θ from additivity. We shall defer a discussion of these questions to Volume II, referring the reader, in the meantime to Steenrod [2] and Eilenberg and MacLane [2].

In order to discuss higher obstructions in an analogous way, we shall need the notion of higher-order cohomology operations. For example, let Y be a space with just two non-zero homotopy groups, $\pi_n(Y) = \Pi$ and $\pi_q(Y) = G$ $(n < q)$; thus the homotopy type of Y is determined by Π, G and the Postnikov invariant $k^{q+1} \in H^{q+1}(\Pi, n; G)$. Let $\theta : H^n(\ ; \Pi) \to H^{q+1}(\ ; G)$ be the corresponding cohomology operation. Furthermore, let $x \in H^{r+1}(Y; Q)$. If X is a CW-complex and $u \in H^n(X; \Pi)$, there is a map $f: X \to K(\Pi, n)$ such that $f^*\mathbf{b}^n = u$.

Now there is a fibration $p : Y \to K(\Pi, n)$ (induced from the path-space fibration by a map $g : K(\Pi, n) \to K(G, q+1)$ such that $f^*\mathbf{b}^{q+1} = k^{q+1}$); the fibre of p is $F = K(G, q)$. Consider the problem

(7.7)
$$\begin{array}{ccc} & & Y \\ & {\scriptstyle f'} \nearrow & \downarrow {\scriptstyle p} \\ X & \longrightarrow & K(\Pi, n) \end{array}$$

of lifting f; this has a solution if and only if

$$0 = f^*k^{q+1} = \theta(u).$$

If $f' : X \to Y$ is a lifting of f, we may form the cohomology class $f'^*x \in H^{r+1}(X; Q)$. But the map f' is not uniquely determined by the given data; the elements f'^*x, for all liftings f' of f, range over a subset of $H^{r+1}(X; Q)$. To handle this kind of problem, let us consider the subset $\Xi = \Xi(X)$ of $H^n(X; \Pi) \times H^{r+1}(X; Q)$ consisting of all pairs (u, y) such that $y = f'^*x$ for some $f' : X \to Y$ such that $f'^*\iota^n(Y) = u$. The set $\Xi(X)$ is a binary relation (not necessarily additive); it has the naturality property that, if $h : X' \to X$ is a map, then $h^* \times h^*$ maps $\Xi(X)$ into $\Xi(X')$. Note that the domain of Ξ is $\operatorname{Ker} \theta = \{u \in H^n(X; \Pi) \mid \theta(u) = 0\}$ and its image is the union of the above-mentioned subsets of $H^{r+1}(X; Q)$ for all $u \in \operatorname{Ker} \theta$. The relation Ξ is called a *secondary cohomology operation*. If Ξ happens to be additive, then it determines, as in Appendix B, a homomorphism of a subgroup of $H^n(X; \Pi)$ into a quotient group of $H^{r+1}(X; Q)$, and it is this homomorphism which is usually called a secondary cohomology operation.

The class $x \in H^{r+1}(Y; Q)$ may be used as the Postnikov invariant to construct a new space Y' with three non-trivial homotopy groups $\pi_n = \Pi$, $\pi_q = G$, $\pi_r = Q$; and a cohomology class of the new space Y' gives rise to a *tertiary operation*. This process can be continued to obtain more and more complicated operations; and the study of higher obstructions to extending a continuous map is intimately related to these notions.

Exercises

1. Let $p : \tilde{X} \to X$ be an n-connective fibration ($n \geq 1$). Then $H_q(\tilde{X})$ is isomorphic with the Eilenberg homology group $H_q^{(n)}(X)$ (Chapter IV, §5).
2. Let $p : X \to \mathbf{S}^3$ be a 3-connective fibration. Calculate the homology groups of X.
3. Prove that, if X is a CW-complex of dimension n with base vertex, then

$\mathbf{F}(X, K(\Pi, n))$ has the same weak homotopy type as $\Pi_{q=1}^{n} K(H^{n-q}(X; \Pi), q)$. What can you say if $\dim X \geq n$?

4. Let X be an $(n-1)$-connected CW-complex; then the first non-trivial stage in its Postnikov system is $X^n = K(\pi_n, n)$. The imbedding $X \hookrightarrow X^n$ corresponds, by obstruction theory, to a cohomology class $z \in H^n(X; \pi_n)$. Prove that $z = \iota^n(X)$ is the characteristic class of X.

5. Prove that the primary obstruction to a cross-section of the fibration $q_{n+1} : W^{n+1} \to X^n$ of 3 is k^{n+2}, under a suitable identification of Π with $\pi_{n+1}(\Omega K_n)$.

6. Prove (7.6) under the hypothesis that the operation θ is additive.

CHAPTER X

On Mappings into Group-like Spaces

This Chapter continues the discussion of H-spaces which was begun in Chapter III. If G is a group-like space, then $[X, G]$ is a group for any space X. This group need not be abelian (although it is if X is the suspension of a space Y). It is natural to ask whether it may be nilpotent. It turns out that the degree of nilpotence is closely related to the Lusternik–Schnirelmann category of the space X. Indeed, a mild change in the definition of the latter in order to relate it more closely to homotopy notions, results in the theorem that, if cat $X < c$ and G is a 0-connected group-like space, then $[X, G]$ is nilpotent of class $< c$.

The notion of category was proposed by Lusternik and Schnirelmann [L–S] in 1934, and was used by them to study the stationary points of real-valued functions on a manifold. In his thesis [1], written in 1941, Fox made a thorough study of the notion. The homotopy-theoretic modification was suggested by the present author [8] in 1956. Section 1 is devoted to some elementary properties of the new notion. Section 2 is devoted to some general properties of group-like spaces. In §3, the nilpotency question is raised and the above theorem proved.

In §4 the special case $X = X_1 \times \cdots \times X_k$ is considered. This space is filtered by subspaces $P_i = \{x \in X \mid x_j = * \text{ for at least } k - i \text{ values of } j\}$, and the corresponding subgroups $\Gamma_i = \{[f] \mid f \mid \Gamma_i \simeq *\}$ are studied. If, in particular, each of the X_i is a sphere, this leads to an explicit central chain of length k, in accordance with the fact that cat $(\mathbf{S}^{n_1} \times \cdots \times \mathbf{S}^{n_k}) = k$.

As the notion of nilpotence is based on the behavior of commutators, it is natural to consider the *commutator product*; if $f_i : X_i \to G$ $(i = 1, 2)$ is the homotopy class of $\alpha_i \in [X_i, G]$, and if $f : X_1 \times X_2 \to G$ is defined by

$$\begin{aligned} f(x_1, x_2) &= [f_1(x_1), f_2(x_2)] \\ &= (f_1(x_1)f_2(x_2))(f_1(x_1)^{-1}f_2(x_2)^{-1}), \end{aligned}$$

then $f|X_1 \vee X_2$ is nullhomotopic, therefore f defines a map $\bar{f}: X_1 \wedge X_2 \to G$, whose homotopy class is called the Samelson product $\langle\alpha_1, \alpha_2\rangle$ of α_1 and α_2. This product is studied in §5, and it is shown that, for maps of spheres of positive dimension, the Samelson product is biadditive and satisfies an anticommutative law and a kind of Jacobi identity. When G is the space of loops of a space X, the groups $\pi_q(G)$ and $\pi_{q+1}(X)$ are isomorphic. The Samelson product

$$\pi_p(G) \otimes \pi_q(G) \to \pi_{p+q}(G)$$

translates to a pairing

$$\pi_{p+1}(X) \otimes \pi_{q+1}(X) \to \pi_{p+q+1}(X),$$

the *Whitehead product.* And the Jacobi identity for the Samelson product gives rise to one for the Whitehead product, which is proved in §7.

The Whitehead product was introduced by J. H. C. Whitehead [2] in 1941, and it was conjectured at that time by Weil that it satisfied a Jacobi identity. Fox and I made some effort to prove it, but without success, and it was not until thirteen years later that independent proofs were found by Uehara and Massey [1], Nakaoka and Toda [1], Hilton [1] and myself [6].

We conclude this Chapter with a summary of elementary results relating the various operations which can be defined in homotopy groups: composition product, reduced join, Whitehead product, and suspension.

1 The Category of a Space

In 1934 Lusternik and Schnirelmann [L–S] introduced the notion of category of a manifold, and proved the important theorem that the category of a manifold M is a lower bound for the number of stationary points of a well-behaved real-valued function on M. This result was then used in order to prove the existence of closed geodesics under appropriate hypotheses.

According to Lusternik and Schnirelmann the category Cat M of M is the least number of elements in a covering of M by closed sets, each of which is contractible in M. The same definition applies without change to an arbitrary topological space X, and the category of X is not only a topological, but even a homotopy, invariant of X. In 1941 Fox proposed to alter the definition by replacing closed by open coverings; and he made a systematic study of the properties of this invariant, as well as the relations among the different notions of category.

In this section we shall present still another notion of category, which seems better adapted for the homotopy theory of spaces with nondegenerate base points. Accordingly, let X be such a space, n a positive integer. Let X_k^n be the subspace of the n-fold Cartesian power X^n of X consisting of all points, at least $n - k$ of whose coördinates are equal to the base point $*$. Thus X_0^n is the base point of X^n, $X_1^n = X \vee \cdots \vee X$, and $X_n^n = X^n$.

We shall say that X *has category less than n* if and only if the n-fold diagonal map

$$\Delta_n : (X, *) \to (X^n, X_{n-1}^n)$$

is compressible.

(1.1) Lemma *If X has category less than n, then X has category less than* $n + 1$.

For $X_{n-1}^n \times X \subset X_n^{n+1}$ and $\Delta_{n+1}(x) = (\Delta_n(x), x)$ for $x \in X$. Thus, if $h' : \mathbf{I} \times X \to X^n$ is a compression of Δ_n, then the map $h : \mathbf{I} \times X \to X^{n+1}$ defined by

$$h(t, x) = (h'(t, x), x)$$

is a compression of Δ_{n+1}. □

It follows from Lemma (1.1) that we can define the category cat X of the space X to be the least integer n such that X has category less than $n + 1$; cat $X = \infty$ if no such n exists. Thus cat $X \leq n$ if and only if X has category less than $n + 1$. In particular,

(1.2) cat $X = 0$ *if and only if X is contractible*; □

(1.3) cat $X \leq 1$ *if and only if X is an* H'*-space.* □

Suppose that cat $X < n$, and let $H : \mathbf{I} \times X \to X^n$ be a homotopy of Δ_n (rel. $*$) to a map $h : X \to X_{n-1}^n$. Then, if $p_i : X^n \to X$ is the projection on the ith factor, $p_i \circ H$ is a homotopy of the identity map to the map $h_i = p_i \circ h : X \to X$. Let $A_i = h_i^{-1}(*)$. Then A_i is closed, and the equality $X_{n-1}^n = \bigcup_{i=1}^n p_i^{-1}(*)$ and the inclusion $h(X) \subset X_{n-1}^n$ imply that $X = \bigcup_{i=1}^n A_i$. Moreover, $p_i \circ H$ is a deformation of X which contracts A_i to the base-point in X.

Conversely, suppose that $X = \bigcup_{i=1}^n A_i$, where A_i is a closed set which is contractible in X under a deformation $H_i : I \times X \to X$ of the whole space (rel. $*$). Then the H_i are the components of a homotopy $H : I \times X \to X^n$ of Δ_n to a map $h : X \to X^n$. Because the A_i cover X, $h(X) \subset X_{n-1}^n$, and it follows that cat $X < n$.

Thus our definition of cat X differs from that of Lusternik–Schnirelmann only by one and by the fact that the sets A_i are required to be contractible under a deformation of the whole space.

We now give a few properties of cat which will be useful later.

(1.4) Theorem *If X dominates Y, then* cat $X \geq$ cat Y.

For let $f: X \to Y$, $g: Y \to X$ be maps such that $f \circ g \simeq 1$, and let $H: \mathbf{I} \times X \to X^n$ be a compression of the n-fold diagonal map of X into X^n_{n-1}. Then $f^n \circ H \circ (1 \times g)$ is a compression of $f^n \circ \Delta_n \circ g = \Delta_n \circ f \circ g$ into Y^n_{n-1}. Combining this homotopy with the composite of Δ_n with a homotopy of the identity map of Y to $f \circ g$, we obtain a compression of the n-fold diagonal map of Y into Y^n_{n-1}. □

(1.5) Corollary *If X and Y have the same homotopy type, then* cat X = cat Y. □

(1.6) Theorem *For any map $f: X \to Y$,*

$$\text{cat } \mathbf{T}_f \leq 1 + \text{cat } Y.$$

We may assume cat $Y = n - 1 < \infty$. Because of Corollary (1.5), we may assume that f is an inclusion and replace $\mathbf{T}_f$ by the subspace $Z = Y \cup \mathbf{T}X$ of $\mathbf{T}Y$. Then the pairs $(Z, \mathbf{T}X)$ and (Z, Y) have the homotopy extension property. Since $\mathbf{T}X$ is contractible, there is a homotopy $f: \mathbf{I} \times Z \to Z$ of the identity to a map f_1 with $f_1(\mathbf{T}X) = *$. Let $g': \mathbf{I} \times Y \to Y^n$ be a homotopy of the diagonal map Δ'_n of Y to a map $g'_1: Y \to Y^n_{n-1}$; then g' can be extended to a homotopy $g: \mathbf{I} \times Z \to Z^n$ of the diagonal map Δ_n of Z to a map $g_1: Z \to Z^n$ such that $g_1(Y) \subset Y^n_{n-1}$. Then f, g define a map $h: \mathbf{I} \times Z \to Z^{n+1}$, and h is a homotopy of Δ_{n+1} to the map $h_1: Z \to Z^{n+1}$ defined by f_1 and g_1. Moreover,

$$h_1(\mathbf{T}X) \subset f_1(\mathbf{T}X) \times g_1(\mathbf{T}X) \subset * \times Z^n \subset Z^{n+1}_n$$
$$h_1(Y) \subset f_1(Y) \times g_1(Y) \subset Z \times Z^n_{n-1} \subset Z^{n+1}_n,$$

and therefore cat $Z < n + 1$. □

Let X be a CW-complex. A *stratification of X of height k* is a sequence of subcomplexes

$$\{*\} = P_0 \subset P_1 \subset \cdots \subset P_k = X$$

with the property that the boundary of each cell of P_i is contained in P_{i-1} $(i = 1, \ldots, k)$. Thus P_i is obtained from P_{i-1} by attaching a collection of cells (possibly of different dimensions); in fact, P_i is the mapping cone of a map $f: \Sigma \to P_{i-1}$, where Σ is a cluster of spheres. For example, the skeleta of a 0-connected CW-complex (whose 0-skeleton is a point) form a stratification of X. Again, if

$$\{*\} = P_0 \subset P_1 \subset \cdots \subset P_k = X$$
$$\{*\} = Q_0 \subset Q_1 \subset \cdots \subset Q_l = Y$$

are stratifications of X and Y, respectively, then the subcomplexes

$$R_j = \bigcup_{i=0}^{j} P_i \times Q_{j-i}$$

form a stratification of $X \times Y$.

(1.7) Corollary *If X has a stratification of height k, then* $\operatorname{cat} X \leq k$. □

(1.8) Corollary *If X is a CW-complex, then* $\operatorname{cat} X \leq \dim X$. □

The above results give useful upper bounds for the category of a space X. We can obtain lower bounds with the aid of cohomology theory.

(1.9) Theorem *Let X be a space of category $< n$, and let A be a ring. Then, if $u_i \in H^{p_i}(X; A)$ $(i = 1, \ldots, n)$ $(p_i > 0)$, we have*

$$u_1 \smile \cdots \smile u_n = 0.$$

For $u_1 \smile \cdots \smile u_n = \Delta_n^*(u)$, where u is the cross product of the u_i in $H^*(X^n; A)$. Since $u_i \in H^*(X, \{*\}; A)$, we may also form their cross product u' in $H^*((X, *)^n; A) = H^*(X^n, X_{n-1}^n; A)$, and $j^*u' = u$, where j^* is the injection. On the other hand, since $\operatorname{cat} X < n$, there is a map $f: X \to X_{n-1}^n$ such that $i \circ f \simeq \Delta_n$, where $i : X_{n-1}^n \hookrightarrow X^n$. Then

$$u_1 \smile \cdots \smile u_n = \Delta_n^* u = \Delta_n^* j^* u' = f^* i^* j^* u' = f^*(0) = 0. \qquad \square$$

(1.10) Corollary *The category of a product $X = \mathbf{S}^{p_1} \times \cdots \times \mathbf{S}^{p_n}$ of n spheres of positive dimensions is n.*

For $\mathbf{S}^{p_i}$ has a CW-decomposition consisting of one 0-cell $\{*\}$ and one p_i-cell. Therefore, if P_k is the set of all points of X, at least $n - k$ of whose coordinates are equal to $*$, then the P_k satisfy the hypotheses of Corollary (1.7), so that $\operatorname{cat} X \leq n$. On the other hand, $\mathbf{S}^{p_i}$ is an orientable manifold with fundamental class $\mathbf{s}_i$, and X is an orientable manifold with fundamental class

$$\mathbf{s} = \mathbf{s}_1 \times \cdots \times \mathbf{s}_n = p_1^* \mathbf{s}_1 \smile \cdots \smile p_n^* \mathbf{s}_n.$$

Since $\mathbf{s} \neq 0$, $\operatorname{cat} X \geq n$. □

(1.11) Corollary *The real projective space $\mathbf{P}^n$ has category n.*

For $\operatorname{cat} \mathbf{P}^n \leq \dim \mathbf{P}^n = n$, by Corollary (1.8). On the other hand, if u is the non-zero element of $H^1(\mathbf{P}^n; \mathbf{Z}_2)$, then $u^n \neq 0$, and so $\operatorname{cat} \mathbf{P}^n \geq n$. □

2 H_0-spaces

We have seen in Chapter III that, if G is an H-space, then the set $[X, G]$ of homotopy classes of maps of X into G admits a natural multiplication with unit element. Moreover, if G is homotopy associative, then the product in $[X, G]$ is associative, so that $[X, G]$ is a monoid. Of course, this monoid need not be a group. For example, suppose that G is a discrete monoid and that X is connected. Then $[X, G]$ is naturally isomorphic with G.

In §4 of Chapter III we have shown that G is group-like (and therefore $[X, G]$ is a group for every X) if and only if the shear map $\phi : G \times G \to G \times G$, defined by

$$\phi(x, y) = (x, xy)$$

is a homotopy equivalence. We now show that this is rather easy to ensure.

(2.1) Lemma *If G is 0-connected, the shear map $\phi : G \times G \to G \times G$ is a weak homotopy equivalence.*

To see this, recall first that, if $p_1, p_2 : G \times G \to G$ are the projections, then by Theorem (5.19) of Chapter III, the homomorphisms

$$p_{1*}, p_{2*} : \pi_n(G \times G) \to \pi_n(G)$$

represent the group $\pi_n(G \times G)$ as the direct product $\pi_n(G) \times \pi_n(G)$. Then the injections $i_{1*}, i_{2*} : \pi_n(G) \to \pi_n(G \times G)$ induce the dual representation as a direct sum. (N.B.: This is true even if $n = 1$, since $\pi_1(G)$ is abelian). Thus, if $\mu : G \times G \to G$ is the product in G, we have

$$\mu_* \circ i_{1*} = \mu_* \circ i_{2*} = 1,$$

and therefore

$$\mu_* = p_{1*} + p_{2*} : \pi_n(G \times G) \to \pi_n(G).$$

Now $p_1 \circ \phi = p_1$ and $p_2 \circ \phi = \mu$; thus

$$p_{1*} \circ \phi_* = p_{1*}, \; p_{2*} \circ \phi_* = p_{1*} + p_{2*},$$

and therefore ϕ corresponds, under the isomorphism of $\pi_n(G \times G)$ with $\pi_n(G) \times \pi_n(G)$, to the shear map of $\pi_n(G)$. But the latter map is an isomorphism. □

In this section we shall be concerned with 0-connected homotopy-associative H-spaces. Let us call such a space an H_0-*space.*

(2.2) Theorem *If G is a CW-complex which is an H_0-space, then G is group-like.*

For the shear map $\phi : G \times G \to G \times G$ is then a homotopy equivalence, and our result follows from (4.17) of Chapter III. □

In Chapter V we proved (Lemma (7.12)) that if $f: G' \to G$ is a CW-approximation and G is an H-space then G' admits an H-structure making f an H-map. We further have

(2.3) *If $f: G' \to G$ is a CW-approximation and G is an H_0-space, so is G'.*

Since G is 0-connected, so is G'. And the homotopy associativity of G and the fact that f is an H-map imply the homotopy associativity of G'. □

(2.4) Theorem *If G is an H_0-space, then $[X, G]$ is a group, under the natural multiplication, for every CW-complex X.*

For let $f: G' \to G$ be a CW-approximation, and apply (2.3) to find an H-structure on G' for which f is an H-map, and G' is an H_0-space. Then $\underline{f}: [X, G'] \to [X, G]$ is an isomorphism. But the shear map ϕ: $G' \times G' \to G' \times G'$ is a weak homotopy equivalence, by Lemma (2.1). Since $G' \times G'$ is a CW-complex, ϕ is a homotopy equivalence. By (4.17) of Chapter III, G' is group-like and therefore $[X, G']$ is a group. Hence $[X, G]$ is also. □

3 Nilpotency of $[X, G]$

While the groups $[S^n, G] = \pi_n(G)$ are abelian, even if $n = 1$, the group $[X, G]$ need not be. However, it turns out that $[X, G]$ is *nilpotent* under reasonable hypotheses. Let us begin by recalling some properties related to the concept of nilpotence. [Ha, Chapter 10], [Z, Chapter II, §6].

If Γ is a group, $x, y \in \Gamma$, their *commutator* is the element $[x, y] = xyx^{-1}y^{-1}$. If A, B are subsets of Γ, $[A, B]$ is the subgroup generated by $[a, b]$ for all $a \in A$, $b \in B$.

The *lower central series* of Γ is the chain of subgroups defined inductively by

$$Z_1(\Gamma) = \Gamma, \qquad Z_{i+1}(\Gamma) = [\Gamma, Z_i(\Gamma)] \quad \text{for } i \geq 0;$$

and Γ is *nilpotent* if and only if there is a non-negative integer c such that $Z_{c+1} = \{1\}$; the least such c is called the *class* of Γ. A *central chain of length* $i = 1, \ldots, k$ is a sequence

$$\Gamma = \Gamma_0 \supset \Gamma_1 \supset \cdots \supset \Gamma_k = \{1\}$$

of subgroups such that $[\Gamma, \Gamma_i] \subset \Gamma_{i+1}$ for $i = 0, 1, \ldots, k-1$. For example, the lower central series (of a nilpotent group) is a central chain; and Γ is nilpotent of class $\leq c$ if and only if Γ has a central chain of length c. Note that, if $\{\Gamma_i\}$ is an arbitrary central chain, then $Z_i(\Gamma) \subset \Gamma_{i-1}$ for all i.

We can also form *iterated commutators.* In fact, let us define the notion of an (iterated) *commutator of weight q* inductively as follows. Any element of Γ will be called a commutator of weight 1; and if x and y are commutators of weights r, s, respectively, we say that $[x, y]$ is a commutator of weight $r + s$. Of particular importance are the *special commutators* $[x_1, \ldots, x_q]$, defined inductively by

$$[x_1, \ldots, x_{q+1}] = [x_1, [x_2, \ldots, x_{q+1}]].$$

(3.1) *The following properties are equivalent*:

(i) Γ *is nilpotent of class* $< c$;
(ii) *All commutators of weight c vanish in* Γ;
(iii) *All special commutators* $[x_1, \ldots, x_c]$ *vanish in* Γ.

The following formulas will be useful later [Z, pp. 82, 84]:

(3.2) *If* $a, b, c \in \Gamma$, *then*

$$[a, bc] \equiv [a, b] \cdot [a, c] \pmod{Z_3(\Gamma)}.$$

(3.3) *If* $a, b, c \in \Gamma$, *then*

$$[a, [b, c]] \cdot [b, [c, a]] \cdot [c, [a, b]] \equiv 1 \pmod{Z_4(\Gamma)}.$$

Let G be a group-like space; it is convenient to assume that the base point e is a strict unit. The commutator map $\Phi : G \times G \to G$ is defined by

$$\Phi(x, y) = (xy)(x^{-1}y^{-1}).$$

Evidently

(3.4) *The map* $\Phi \mid G \vee G$ *is null-homotopic.*

We can also form iterated commutator maps

$$\Phi_n : G^n \to G$$

by

$$\Phi_1 \quad \text{is the identity map,}$$

$$\Phi_{n+1} = \Phi \circ (1 \times \Phi_n).$$

(in particular, $\Phi_2 = \Phi$). Then (3.4) generalizes to

(3.5) Theorem *The map* $\Phi_n \mid G_{n-1}^n$ *is nullhomotopic.*

This is trivial for $n = 1$, and is given by (3.4) for $n = 2$. Suppose $n \geq 2$ and $\Phi_n \,|\, G^n_{n-1}$ nullhomotopic. Now (G^n, G^n_{n-1}) is an NDR-pair, and so there is a homotopy $\Psi' : I \times G^n \to G$ of Φ_n such that $\Psi'(1 \times G^n_{n-1}) = e$. Then the map $\Psi : I \times G^{n+1} \to G \times G$ defined by

$$\Psi(t, x, y) = (x, \Psi'(t, y))$$

is a homotopy of $1 \times \Phi_n$ to a map $F : G^{n+1} \to G \times G$ such that $F(G \times G^n_{n-1}) \subset G \times e$. Since $F(e \times G^n) \subset e \times G$, we have

$$F(G^{n+1}_n) = F(G \times G^n_{n-1} \cup e \times G^n) \subset G \vee G.$$

But $\Phi_{n+1} = \Phi_2 \circ (1 \times \Phi_n)$ is homotopic to $\Phi_2 \circ F$, and therefore $\Phi_{n+1} \,|\, G^{n+1}_n$ is homotopic to $(\Phi_2 \,|\, G \vee G) \circ (F \,|\, G^{n+1}_n)$, which is nullhomotopic by (3.4). □

The next two theorems give sufficient conditions for $[X, G]$ to be nilpotent.

(3.6) Theorem *Let X have finite category $< c$, and let G be a 0-connected group-like space. Then $[X, G]$ is nilpotent of class $< c$.*

It suffices, by (3.1), to prove that all special commutators $[\alpha_1, \ldots, \alpha_c]$ vanish in $\Gamma = [X, G]$. Let $f_i : X \to G$ represent α_i. Then $[\alpha_1, \ldots, \alpha_n]$ is represented by the map

$$X \xrightarrow{\Delta_c} X^c \xrightarrow{f_1 \times \cdots \times f_c} G^c \xrightarrow{\Phi_c} G.$$

The map $f_1 \times \cdots \times f_c$ maps X^c_{c-1} into G^c_{c-1}, and Δ_c is compressible into X^c_{c-1}. Thus there is a homotopy commutative diagram

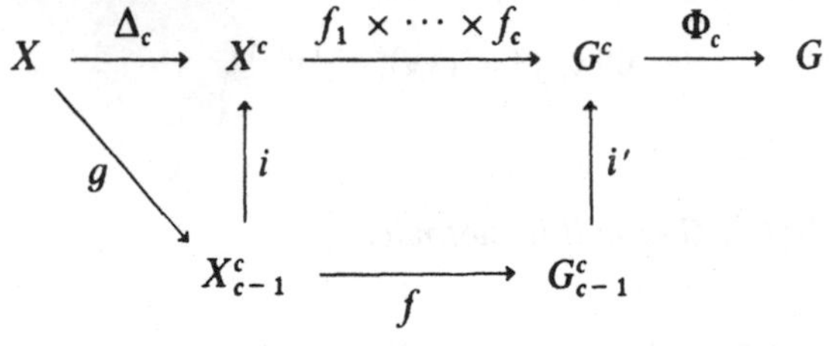

where i and i' are inclusions. Therefore

$$\Phi_c \circ (f_1 \times \cdots \times f_c) \circ \Delta_c \simeq \Phi_c \circ i' \circ f \circ g.$$

But $\Phi_c \circ i'$ is nullhomotopic, by Theorem (3.5). □

(3.7) Corollary *Let X be a* CW-*complex of finite category c, and let G be an* H_0-*space. Then the group $[X, G]$ is nilpotent of class $\leq c$.* □

(3.8) Corollary *If X is an n-dimensional* CW-*complex $(n < \infty)$ and G an* H_0-*space, then $[X, G]$ is nilpotent of class $\leq n$.* □

(3.9) Corollary *If X is a product of k spheres, and G is an H_0-space, then $[X, G]$ is nilpotent of class $\leq k$.* □

The nilpotency of $[X, G]$ in the next theorem also follows from Theorem (3.6), but Theorem (3.10) is stronger, since it gives an explicit central chain.

(3.10) Theorem *Let*

$$\{*\} = P_0 \subset P_1 \subset \cdots \subset P_c = X$$

be a stratification of the CW-complex X, and let G be an H_0-space. Let Γ_i be the set of all homotopy classes of maps $f: X \to G$ such that $f \mid P_i$ is nullhomotopic. Then $\Gamma_0, \ldots, \Gamma_c$ is a central chain for $\Gamma = [X, G]$.

We may assume that G is group-like. Let $f: X \to G$, $g: g: X \to G$ be representatives of $\alpha \in \Gamma$, $\beta \in \Gamma_i$, respectively. For each cell E_λ of the relative CW-complex (P_{i+1}, P_i), let x_λ be an interior point of E_λ. Then we can find closed cells $F_\lambda \subset \operatorname{Int} E_\lambda$ such that $x_\lambda \in \operatorname{Int} F_\lambda$ and $\dot{E}_\lambda$ is a deformation retract of $E_\lambda - F_\lambda$; it follows that P_i is a deformation retract of

$$Q = P_i \cup \bigcup_\lambda \overline{E_\lambda - F_\lambda},$$

and the set $R_0 = \{x_\lambda\}$ a deformation retract of $R = \bigcup_\lambda F_\lambda$. Since G is 0-connected, $f \mid R_0$, and therefore $f \mid R$, is nullhomotopic. Since $g \mid P_i$ is nullhomotopic, so is $g \mid Q$. By the homotopy extension property, f and g are homotopic to maps f', g' such that $f'(R) = g'(Q) = e$. But $[\alpha, \beta]$ is represented by the map

$$X \xrightarrow{\Delta} X \times X \xrightarrow{f' \times g'} G \times G \xrightarrow{\Phi} G$$

and $(f' \times g') \circ \Delta$ maps P_{i+1} into $G \vee G$, and it again follows from (3.4) that $[\alpha, \beta] \in \Gamma_{i+1}$. □

4 The Case $X = X_1 \times \cdots \times X_k$

We now examine this case in greater detail. Let P_i be the set of all points of X with at least $k - i$ coordinates equal to the base point. Then

$$\{*\} = P_0 \subset P_1 \subset \cdots \subset P_k = X.$$

We are going to examine the groups Γ_i of homotopy classes of maps $f: X \to G$ such that $f \mid P_i$ is nullhomotopic.

For each subset $\alpha \subset \{1, \ldots, k\}$, let $X_\alpha = \{x \in X \mid x_i = e_i \text{ for } i \notin \alpha\}$, $|\alpha|$ the cardinal of α, $X^\alpha = X_\alpha / X_\alpha \cap P_{|\alpha|-1}$. Thus X_α and X^α are homeomorphic with $\prod_{i \in \alpha} X_i$, $\bigwedge_{i \in \alpha} X_i$, respectively. Let $p_\alpha : X_\alpha \to X^\alpha$ be the identification map, and let $q_\alpha : X \to X_\alpha$ be the natural retraction; $q_\alpha(x)$ is the point y such that $y_i = x_i$ if $i \in \alpha$, $y_i = e_i$ if $i \notin \alpha$. Clearly $\alpha \neq \beta$ implies $q_\beta(X_\alpha) \subset X_\alpha$.

(4.1) Lemma *Let $(X; A_1, \ldots, A_n)$ be a CW $(n+1)$-ad, and let $A = A_1 \cup \cdots \cup A_n$. Suppose that there are retractions $\rho_i : X \to A_i$ such that $\rho_i(A_j) \subset A_j$ for all i, j. Let G be an H_0-space. Then the injection $\bar{j}: [X, A; G] \to [X, G]$ is a monomorphism.*

As usual, we may assume that G is group-like and has a strict unit. We must show that if $f_0, f_1 : X \to G$ are homotopic maps with $f_0(A) = f_1(A) = e$, then $f_0 \simeq f_1$ (rel. A). Let $B_r = A_1 \cup \cdots \cup A_r$; it suffices to show that $f_0 \simeq f_1$ (rel. B_r) implies $f_0 \simeq f_1$ (rel. B_{r+1}). Let $F : \mathbf{I} \times X \to G$ be a homotopy of f_0 to f_1 (rel. B_r), and define $F_1 : \mathbf{I} \times X \to G$ by

$$F_1(t, x) = F(t, x) \cdot F(t, \rho_{r+1}(x))^{-1}.$$

Then $x \in B_r$ implies $\rho_{r+1}(x) \in B_r$ and therefore $F_1(t, x) = e \cdot e^{-1} = e \cdot e = e$; and $x \in A_{r+1}$ implies $\rho_{r+1}(x) = x$. Thus for all $x \in B_{r+1}$, we have $F_1(t, x) = F(t, x) \cdot F(t, x)^{-1}$. Moreover, if $t = 0$ or 1, $F_1(t, x) = f_t(x)$.

Since the map $x \to x \cdot x^{-1}$ is nullhomotopic (rel. e), the map $F_1 | \mathbf{I} \times B_{r+1}$ is nullhomotopic (rel. $\dot{\mathbf{I}} \times B_{r+1}$). Let $C = \dot{\mathbf{I}} \times X \cup \mathbf{I} \times B_{r+1}$; then $F_1 | C$ is homotopic to the map $F_2 : C \to G$ such that

$$F_2(t, x) = f_t(x) \qquad (t \in \dot{\mathbf{I}}, x \in X),$$

$$F_2(t, x) = e \qquad (t \in \mathbf{I}, x \in B_{r+1}).$$

But $(\mathbf{I} \times X, C)$ is an NDR-pair, and therefore F_2 has an extension $F' : \mathbf{I} \times X \to G$, and F' is a homotopy of f_0 to f_1 (rel. B_{r+1}). □

Applying Lemma (4.1) to the space X and the subsets X_α with $|\alpha| = i$, we have

(4.2) Corollary *The injection $\bar{j}: [X, P_{i-1}; G] \to [X, G]$ maps the former group isomorphically upon Γ_{i-1}.* □

(4.3) Theorem *The group Γ_{i-1}/Γ_i is isomorphic with the direct product $\prod_{|\alpha|=i} [X^\alpha, G]$.*

For let $\bar{j}_\alpha : [X, A_{i-1}; G] \to [X_\alpha, X_\alpha \cap P_{i-1}; G]$ be the injection. Then the relative homeomorphism p_α induces an isomorphism $\bar{p}_\alpha : [X^\alpha, G] \approx [X_\alpha, X_\alpha \cap P_{i-1}; G]$. Let

$$\eta_\alpha = \bar{p}_\alpha^{-1} \circ \bar{j}_\alpha \circ \bar{j}^{-1} : \Gamma_{i-1} \to [X^\alpha, G].$$

The homomorphisms η_α define a homomorphism $\eta : \Gamma_{i-1} \to \Delta = \prod_{|\alpha|=i} [X^\alpha, G]$. Clearly $\Gamma_i \subset \operatorname{Ker} \eta$. Conversely, if $f \in \gamma \in \operatorname{Ker} \eta$, then, for each α, $f | X_\alpha$ is nullhomotopic (rel. $X_\alpha \cap P_{i-1}$), and therefore $f | P_i$ is nullhomotopic. Hence η induces a monomorphism of Γ_{i-1}/Γ_i into Δ. It remains only to show that η is an epimorphism.

Define $i_\alpha : X \to X^\alpha$ by $i_\alpha = p_\alpha \circ q_\alpha$. Since q_α can be factored through (X, P_{i-1}), it follows that $\bar{i}_\alpha[X^\alpha, G] \subset \Gamma_{i-1}$. Now

$$\eta_\alpha \circ \bar{i}_\alpha = \bar{p}_\alpha^{-1} \circ \bar{j}_\alpha \circ \bar{j}^{-1} \circ \bar{q}_\alpha \circ \bar{p}_\alpha;$$

since q_α is a retraction, $\bar{j}_\alpha \circ \bar{j}^{-1} \circ \bar{q}_\alpha$ is the identity, and therefore $\eta_\alpha \circ \bar{i}_\alpha$ is the identity. If $\alpha \neq \beta$, then $q_\alpha(X_\beta) \subset X_\alpha \cap X_\beta \subset P_{i-1}$, and therefore $\bar{j}_\beta \circ \bar{j}^{-1} \circ \bar{q}_\alpha = 0$; hence

$$\eta_\beta \circ \bar{i}_\alpha = \bar{p}_\beta^{-1} \circ \bar{j}_\beta \circ \bar{j}^{-1} \circ \bar{q}_\alpha \circ \bar{p}_\alpha = 0.$$

It follows that η is an epimorphism; moreover, we have seen that the $\bar{i}_\alpha$ are monomorphisms. □

Suppose further that X_i is the n-sphere $\mathbf{S}^{n_i}$; then X^α can be identified with $\mathbf{S}^{n(\alpha)}$, where $n(\alpha) = \sum_{i \in \alpha} n_i$. We consider $X_i = \mathbf{S}^{n_i}$ as a CW-complex with one 0-cell e_i and one n_i-cell; then X is a CW-complex and the filtration

$$\{*\} = P_0 \subset \cdots \subset P_k = X$$

a stratification of X. We then have, as a special case of Theorem (4.3),

(4.4) Theorem *The group* $\Gamma = [\mathbf{S}^{n_1} \times \cdots \times \mathbf{S}^{n_k}; G]$ *has a central chain* $\Gamma = \Gamma_0 \supset \Gamma_1 \supset \cdots \supset \Gamma_k = \{1\}$ *with*

$$\Gamma_{i-1}/\Gamma_i \approx \prod_{|\alpha| = i} \pi_{n(\alpha)}(G).$$ □

5 The Samelson Product

Let us consider the group $[X_1 \times X_2, G]$ and the monomorphisms

$$\bar{i}_1 : [X_1, G] \to [X_1 \times X_2, G],$$
$$\bar{i}_2 : [X_2, G] \to [X_1 \times X_2, G],$$
$$\bar{i}_{12} : [X_1 \wedge X_2, G] \to [X_1 \times X_2, G],$$

of §4. If $\alpha \in [X_1, G]$, $\beta \in [X_2, G]$, then

$$\eta_1[\bar{i}_1(\alpha), \bar{i}_2(\beta)] = [\alpha, 0] = 0,$$
$$\eta_2[\bar{i}_1(\alpha), \bar{i}_2(\beta)] = [0, \beta] = 0,$$

and therefore the element $[\bar{i}_1(\alpha), \bar{i}_2(\beta)]$ belongs to the kernel Γ_1 of the homomorphism η. As $\bar{i}_{12}$ maps the group $[X_1 \wedge X_2, G]$ isomorphically upon Γ_1, we may define the *Samelson product* $\langle\alpha, \beta\rangle \in [X_1 \wedge X_2, G]$ by

$$\langle\alpha, \beta\rangle = \bar{i}_{12}^{-1}[\bar{i}_1(\alpha), \bar{i}_2(\beta)].$$

The reduced join satisfies a commutative law up to natural homeomorphism. Therefore, if $\alpha \in [X_1, G]$, $\beta \in [X_2, G]$, we can identify $\langle \alpha, \beta \rangle \in [X_1 \wedge X_2, G]$ with $\langle \beta, \alpha \rangle^{-1} \in [X_2 \wedge X_1, G]$ under the natural isomorphism between the groups in which these elements lie. Thus the Samelson product satisfies a kind of "anti-commutative law". Moreover, the spaces $X_1 \wedge X_2 \wedge X_3$, $X_2 \wedge X_3 \wedge X_1$, and $X_3 \wedge X_1 \wedge X_2$ are homeomorphic. Since the Samelson product is defined in terms of commutators, we may expect some kind of "Jacobi identity" relating the products $\langle \alpha, \langle \beta, \gamma \rangle \rangle$, $\langle \beta, \langle \gamma, \alpha \rangle \rangle$ and $\langle \gamma, \langle \alpha, \beta \rangle \rangle$.

Suppose now that X_1 and X_2 are spheres $\mathbf{S}^p$, $\mathbf{S}^q$, respectively. Then $X_1 \wedge X_2$ can be identified with $\mathbf{S}^{p+q}$, and the Samelson product becomes a map of $\pi_p(G) \times \pi_q(G)$ into $\pi_{p+q}(G)$. In this case, $X_2 \wedge X_1$ can also be identified with $\mathbf{S}^{p+q}$. However, the map $\mathbf{S}^p \times \mathbf{S}^q \to \mathbf{S}^q \times \mathbf{S}^p$ which interchanges the coördinates induces a map of $\mathbf{S}^{p+q}$ into itself which has degree $(-1)^{pq}$. Thus we may expect the commutative law for the Samelson product to involve a non-trivial sign. In the same way, the map of $\mathbf{S}^{p+q+r}$ into itself, which corresponds to the map of $\mathbf{S}^p \times \mathbf{S}^q \times \mathbf{S}^r$ into $\mathbf{S}^q \times \mathbf{S}^r \times \mathbf{S}^p$ which cyclically permutes the coördinates, has degree $(-1)^{p(q+r)}$, and therefore we may expect a Jacobi identity with signs.

In the remainder of this section, we shall deal exclusively with the case that the X_i are all spheres. We shall assume that G is an H_0-space, and abbreviate $[\mathbf{S}^{n_1} \times \cdots \times \mathbf{S}^{n_k}, G]$ to $\pi_{n_1, \ldots, n_k}$. Let us first suppose that $k = 2$.

(5.1) Theorem *The map $(\alpha, \beta) \to \langle \alpha, \beta \rangle$ is bilinear, and so defines a pairing $\pi_p \otimes \pi_q \to \pi_{p+q}$. Moreover,*

$$\langle \beta, \alpha \rangle = (-1)^{pq+1} \langle \alpha, \beta \rangle.$$

To prove right linearity, let $\alpha \in \pi_p$, $\beta, \gamma \in \pi_q$. Then

$$\begin{aligned} [\bar{\imath}_1(\alpha), \bar{\imath}_2(\beta + \gamma)] &= [\bar{\imath}_1(\alpha), \bar{\imath}_2(\beta)\bar{\imath}_2(\gamma)] \\ &= [\bar{\imath}_1(\alpha), \bar{\imath}_2(\beta)] \cdot [\bar{\imath}_1(\alpha), \bar{\imath}_2(\gamma)] \quad (\mathrm{mod}\ \Gamma_2), \end{aligned}$$

by (3.2). But $\Gamma_2 = \{1\}$, by Theorem (3.10); applying $\bar{\imath}_{12}^{-1}$ to both sides of the resulting equality, we obtain

$$\langle \alpha, \beta + \gamma \rangle = \langle \alpha, \beta \rangle + \langle \alpha, \gamma \rangle.$$

Left linearity can be proved in a similar way. However, it follows from right linearity and the commutative law, which we now prove.

Let $t : \mathbf{S}^p \times \mathbf{S}^q \to \mathbf{S}^q \times \mathbf{S}^p$ be the map which interchanges the factors. In virtue of our orientation conventions, t has degree $(-1)^{pq}$. Furthermore, $t(\mathbf{S}^p \vee \mathbf{S}^q) \subset \mathbf{S}^q \vee \mathbf{S}^p$, so that t induces a map $t' : \mathbf{S}^{p+q} \to \mathbf{S}^{p+q}$ of the same degree. Clearly

$$i_1 \circ t = i_2, \qquad i_2 \circ t = i_1, \qquad i_{12} \circ t = t' \circ i_{12}.$$

Moreover, $\bar{t}'(u) = (-1)^{pq}u$ for $u \in \pi_{p+q}$. Hence

$$\begin{aligned}\langle \beta, \alpha \rangle &= \bar{i}_{12}^{-1}[\bar{i}_1(\beta), \bar{i}_2(\alpha)] = \bar{t}'^{-1}\bar{i}_{12}^{-1}\bar{t}[\bar{i}_1(\beta), \bar{i}_2(\alpha)]\\ &= (-1)^{pq}\bar{i}_{12}^{-1}[\bar{t}\bar{i}_1(\beta), \bar{t}\bar{i}_2(\alpha)]\\ &= (-1)^{pq}\bar{i}_{12}^{-1}[\bar{i}_2(\beta), \bar{i}_1(\alpha)]\\ &= (-1)^{pq+1}\bar{i}_{12}^{-1}[\bar{i}_1(\alpha), \bar{i}_2(\beta)] = (-1)^{pq+1}\langle \alpha, \beta \rangle. \qquad \square\end{aligned}$$

We can obtain information about iterated Samelson products like $\langle \alpha, \langle \beta, \gamma \rangle \rangle$ by taking $k = 3$. Consider the group $\pi_{p,q,r}$ and the monomorphisms $\bar{i}_1 : \pi_p \to \pi_{p,q,r}$, $\bar{i}_2 : \pi_q \to \pi_{p,q,r}$, $\bar{i}_3 : \pi_r \to \pi_{p,q,r}$ and

$$\bar{i}_{123} : \pi_{p+q+r} \to \pi_{p,q,r}.$$

The latter map is an isomorphism of π_{p+q+r} with Γ_3. Moreover, if $\alpha \in \pi_p$, $\beta \in \pi_q$, $\gamma \in \pi_r$, then $[\bar{i}_1(\alpha), \bar{i}_2(\beta), \bar{i}_3(\gamma)] \in \Gamma_2$, and we have

(5.2) Lemma *For all $\alpha \in \pi_p$, $\beta \in \pi_q$, $\gamma \in \pi_r$,*

$$\langle \alpha, \langle \beta, \gamma \rangle \rangle = \bar{i}_{123}^{-1}[\bar{i}_1(\alpha), \bar{i}_2(\beta), \bar{i}_3(\gamma)].$$

By definition,

$$\langle \alpha, \langle \beta, \gamma \rangle \rangle = \bar{i}_{12}^{-1}[\bar{i}_1(\alpha), \bar{i}_2\bar{i}_{12}^{-1}[\bar{i}_1(\beta), \bar{i}_2(\gamma)]].$$

Let $f = 1 \times p_{12} : \mathbf{S}^p \times \mathbf{S}^q \times \mathbf{S}^r \to \mathbf{S}^p \times \mathbf{S}^{q+r}$. Then $p_{12} \circ f$ and p_{123} are maps of $(\mathbf{S}^p \times \mathbf{S}^q \times \mathbf{S}^r, P_2)$ into $(\mathbf{S}^{p+q+r}, *)$ which preserve orientation, and therefore $p_{12} \circ f \simeq p_{123}$. Hence

$$\bar{f} \circ \bar{i}_{12} = \bar{f} \circ \bar{p}_{12} = \bar{p}_{123} = \bar{i}_{123}.$$

On the other hand, $i_2 \circ f = i_{23}$ and $i_1 \circ f = i_1$. Let h be the projection of $\mathbf{S}^p \times \mathbf{S}^q \times \mathbf{S}^r$ on $\mathbf{S}^q \times \mathbf{S}^r$. Then $i_{12} \circ h = i_{23}$, $i_1 \circ h = i_2$, $i_2 \circ h = i_3$. Thus

$$\begin{aligned}\bar{i}_{123}\langle \alpha, \langle \beta, \gamma \rangle \rangle &= \bar{f}\bar{i}_{12}\langle \alpha, \langle \beta, \gamma \rangle \rangle\\ &= \bar{f}[\bar{i}_1(\alpha), \bar{i}_2\bar{i}_{12}^{-1}[\bar{i}_1(\beta), \bar{i}_2(\gamma)]]\\ &= [\bar{f}\bar{i}_1(\alpha), \bar{f}\bar{i}_2\bar{i}_{12}^{-1}[\bar{i}_1(\beta), \bar{i}_2(\gamma)]]\\ &= [\bar{i}_1(\alpha), \bar{i}_{23}\bar{i}_{12}^{-1}[\bar{i}_1(\beta), \bar{i}_2(\gamma)]]\\ &= [\bar{i}_1(\alpha), \bar{h}[\bar{i}_1(\beta), \bar{i}_2(\gamma)]]\\ &= [\bar{i}_1(\alpha), [\bar{h}\bar{i}_1(\beta), \bar{h}\bar{i}_2(\gamma)]]\\ &= [\bar{i}_1(\alpha), [\bar{i}_2(\beta), \bar{i}_3(\gamma)]].\end{aligned}$$

(5.3) Lemma *If $\alpha \in \pi_p$, $\beta \in \pi_q$, $\gamma \in \pi_r$, then*

$$\bar{i}_{123}^{-1}[\bar{i}_2(\beta), \bar{i}_3(\gamma), \bar{i}_1(\alpha)] = (-1)^{p(q+r)}\langle \beta, \langle \gamma, \alpha \rangle \rangle,$$
$$\bar{i}_{123}^{-1}[\bar{i}_3(\gamma), \bar{i}_1(\alpha), \bar{i}_2(\beta)] = (-1)^{r(p+q)}\langle \gamma, \langle \alpha, \beta \rangle \rangle.$$

To prove the first relation, let $g : \mathbf{S}^p \times \mathbf{S}^q \times \mathbf{S}^r \to \mathbf{S}^q \times \mathbf{S}^r \times \mathbf{S}^p$ be the map given by

$$g(x, y, z) = (y, z, x)$$

Then g induces $g' : \mathbf{S}^{p+q+r} \to \mathbf{S}^{p+q+r}$; g and g' have the same degree $(-1)^{p(q+r)}$, and $i_{123} \circ g = g' \circ i_{123}$, $i_1 \circ g = i_2$, $i_2 \circ g = i_3$, $i_3 \circ g = i_1$. Hence, using Lemma (5.2), we find that

$$\begin{aligned} \bar{i}_{123}((-1)^{p(q+r)}\langle\beta, \langle\gamma, \alpha\rangle\rangle) &= \bar{g}\bar{i}_{123}\langle\beta, \langle\gamma, \alpha\rangle\rangle \\ &= \bar{g}[\bar{i}_1(\beta), \bar{i}_2(\gamma), \bar{i}_3(\alpha)] \\ &= [\bar{g}\bar{i}_1(\beta), \bar{g}\bar{i}_2(\gamma), \bar{g}\bar{i}_3(\alpha)] \\ &= [\bar{i}_2(\beta), \bar{i}_3(\gamma), \bar{i}_1(\alpha)]. \end{aligned}$$

The second relation follows in a similar way. □

The next theorem expresses a kind of "Jacobi identity."

(5.4) Theorem *If $\alpha \in \pi_p$, $\beta \in \pi_q$, $\gamma \in \pi_r$, then*

$$(-1)^{pr}\langle\alpha, \langle\beta, \gamma\rangle\rangle + (-1)^{pq}\langle\beta, \langle\gamma, \alpha\rangle\rangle + (-1)^{qr}\langle\gamma, \langle\alpha, \beta\rangle\rangle = 0.$$

For let $\alpha' = \bar{i}_1(\alpha)$, $\beta' = \bar{i}_2(\beta)$, $\gamma' = \bar{i}_3(\gamma)$. By (3.3),

(5.5) $$[\alpha', \beta', \gamma'] \cdot [\beta', \gamma', \alpha'] \cdot [\gamma', \alpha', \beta'] \equiv 1 \pmod{\Gamma_3}.$$

But $\Gamma_3 = \{1\}$, by Theorem (3.10), and therefore the congruence (5.5) becomes an equality. Applying $\bar{i}_{123}^{-1}$ and using Lemmas (5.2) and (5.3), we obtain the desired result. □

If X is an arbitrary 1-connected space, then ΩX is an H_0-space, and $\pi_q(\Omega X) \approx \pi_{q+1}(X)$. Thus the above pairing $(\alpha, \beta) \to \langle\alpha, \beta\rangle$ gives rise to a pairing $\pi_{p+1}(X) \otimes \pi_{q+1}(X) \to \pi_{p+q+1}(X)$. This pairing, due to J. H. C. Whitehead, will be studied in subsequent sections, without the hypothesis that X is 1-connected.

6 Commutators and Homology

Let G be a group-like space, $\Phi : G \times G \to G$ the commutator mapping of §3. The effect of Φ on the homology groups is by no means obvious. If G is homotopy commutative, then Φ is nullhomotopic and therefore $0 = \Phi_* : H_*(G \times G) \to H_*(G)$.

Let $\mu : G \times G \to G$ and $j : G \to G$ be the product and the inversion, respectively, and let us consider homology with coefficients in a commutative ring R.

(6.1) Lemma *If $x \in H_q(G)$ is primitive, then $j_*(x) = -x$.*

For the map

$$G \xrightarrow{\Delta} G \times G \xrightarrow{1 \times j} G \times G \xrightarrow{\mu} G$$

is nullhomotopic, and therefore

$$\begin{aligned} 0 &= \mu_*(1 \times j)_* \Delta_*(x) = \mu_*(1 \times j)_*(x \times 1 + 1 \times x) \\ &= \mu_*(x \times 1 + 1 \times j_* x) = x \cdot 1 + 1 \cdot j_* x = x + j_* x. \end{aligned}$$

□

It should be remarked that the above formula need not hold for non-primitive elements. Of course, $j_*(1) = 1$; and if x and y are primitive elements, then $j_*(xy) = j_*(y)j_*(x) = (-y)(-x) = yx$, which need not be equal to $-xy$.

(6.2) Lemma *If $x \in H_p(G)$, $y \in H_q(G)$ are primitive elements, then $\Phi_*(x \times y) = xy - (-1)^{pq}yx$.*

For Φ is the composite

$$G^2 \xrightarrow{\Delta \times \Delta} G^4 \xrightarrow{1 \times t \times 1} G^4 \xrightarrow{1 \times 1 \times j \times j}$$

$$G^4 \xrightarrow{\mu \times \mu} G^2 \xrightarrow{\mu} G,$$

where $t : G^2 \to G^2$ is the map which interchanges the factors; thus if $u \in H_p(G)$, $v \in H_q(G)$, then $t_*(u \times v) = (-1)^{pq} v \times u$. Hence

$$\begin{aligned} x \times y &\xrightarrow[(\Delta \times \Delta)_*]{} (x \times 1 + 1 \times x) \times (y \times 1 + 1 \times y) \\ &\qquad\qquad \text{(since } x, y \text{ are primitive)} \\ &= x \times 1 \times y \times 1 + x \times 1 \times 1 \times y + 1 \times x \times y \times 1 + 1 \times x \times 1 \times y \\ &\xrightarrow[(1 \times t \times 1)_*]{} x \times y \times 1 \times 1 + x \times 1 \times 1 \times y \\ &\qquad + (-1)^{pq} 1 \times y \times x \times 1 + 1 \times 1 \times x \times y \\ &\xrightarrow[(1 \times 1 \times j \times j)_*]{} x \times y \times 1 \times 1 - x \times 1 \times 1 \times y \\ &\qquad - (-1)^{pq} 1 \times y \times x \times 1 + 1 \times 1 \times x \times y \qquad \text{by Lemma (6.1)} \\ &\xrightarrow[(\mu \times \mu)_*]{} xy \times 1 - x \times y - (-1)^{pq} y \times x + 1 \times xy \\ &\xrightarrow[\mu_*]{} xy - xy - (-1)^{pq} yx + xy = xy - (-1)^{pq} yx. \end{aligned}$$

□

We can use Lemma (6.2) to determine the behavior of the product $\langle \alpha, \beta \rangle$ of §5 in homology.

(6.3) Theorem *If G is an* H_0*-space,* $\alpha \in \pi_k(G)$, $\beta \in \pi_l(G)$, *then*

$$\rho(\langle\alpha, \beta\rangle) = \rho(\alpha) \cdot \rho(\beta) - (-1)^{kl}\rho(\beta) \cdot \rho(\alpha).$$

Let $f: \mathbf{S}^k \to G$, $g: \mathbf{S}^l \to G$ be representatives of $\alpha \in \pi_k(G)$, $\beta \in \pi_q(G)$, respectively and let $h: \mathbf{S}^{k+l} \to G$ represent $\langle\alpha, \beta\rangle$. Let $p: \mathbf{S}^k \times \mathbf{S}^l \to \mathbf{S}^{k+l}$ be a map of degree 1 which collapses $\mathbf{S}^k \vee \mathbf{S}^l$ to the base point. Then it follows from the definition of $\langle\alpha, \beta\rangle$ in §5 that there is a homotopy commutative diagram

$$\begin{array}{ccc} \mathbf{S}^k \times \mathbf{S}^l & \xrightarrow{f \times g} & G \times G \\ \downarrow p & & \downarrow \Phi \\ \mathbf{S}^{k+l} & \xrightarrow[h]{} & G \end{array}$$

Now $p_*(\mathbf{s}_k \times \mathbf{s}_l) = \mathbf{s}_{k+l}$, and therefore

$$\begin{aligned} h_*(\mathbf{s}_{k+l}) &= h_* p_*(\mathbf{s}_k \times \mathbf{s}_l) \\ &= \Phi_*(f \times g)_*(\mathbf{s}_k \times \mathbf{s}_l) \\ &= \Phi_*(f_* \mathbf{s}_k \times g_* \mathbf{s}_l). \end{aligned}$$

But $f_* \mathbf{s}_k$ and $g_* \mathbf{s}_l$ are spherical and therefore primitive, homology classes, by (7.7) of Chapter III. Hence, by Lemma (6.2),

$$h_*(\mathbf{s}_{k+l}) = (f_* \mathbf{s}_k) \cdot (g_* \mathbf{s}_l) - (-1)^{kl}(g_* \mathbf{s}_l) \cdot (f_* \mathbf{s}_k).$$

But the elements $f_* \mathbf{s}_k$, $g_* \mathbf{s}_l$, and $h_*(\mathbf{s}_{k+l})$ are the images of α, β, and $\langle\alpha, \beta\rangle$, respectively, under the appropriate Hurewicz maps ρ. □

7 The Whitehead Product

Let $\alpha \in \pi_{p+1}(X)$, $\beta \in \pi_{q+1}(X)$ be elements, represented by maps $f: (E_1, \dot{E}_1) \to (X, *)$, $g: (E_2, \dot{E}_2) \to (X, *)$, where E_1 and E_2 are oriented cells of dimensions $p+1$, $q+1$, respectively. Then $E_1 \times E_2$ is a cell, oriented by the product of the given orientations of E_1 and of E_2; the base point of $E_1 \times E_2$ is the point $(*, *)$. The boundary $S = (E_1 \times E_2)^{\bullet} = E_1 \times \dot{E}_2 \cup \dot{E}_1 \times E_2$ is then an oriented $(p+q+1)$-sphere, and the map $h: (S, *) \to (X, *)$ defined by

(7.1)
$$h(x, y) = \begin{cases} f(x) & (x \in E_1, y \in \dot{E}_2), \\ g(y) & (x \in \dot{E}_1, y \in E_2), \end{cases}$$

represents an element $[\alpha, \beta] \in \pi_{p+q+1}(X)$ which, as the notation suggests, depends only on the homotopy classes α, β of f, g respectively. Moreover, the operation $(\alpha, \beta) \to [\alpha, \beta]$ is clearly natural; i.e.,

(7.2) *If* $\phi : X \to Y$, $\alpha \subset \pi_{p+1}(X)$, $\beta \in \pi_{q+1}(X)$, *then*

$$\phi_*[\alpha, \beta] = [\phi_* \alpha, \phi_* \beta] \in \pi_{p+q+1}(Y). \qquad \square$$

Suppose $p = q = 0$ (cf. Figure 10.1). Then $(\mathbf{I} \times \mathbf{I})^{\bullet}$ is the boundary of the unit square $\mathbf{I} \times \mathbf{I}$ in the plane $\mathbf{R}^2$, with the clockwise orientation, and with the origin as base point. The maps f, g are loops representing α, β and it is then clear that

(7.3) *If* $p = q = 0$, *then* $[\alpha, \beta] = \alpha\beta\alpha^{-1}\beta^{-1} \in \pi_1(X)$. $\square$

Thus the notation $[\alpha, \beta]$ for the Whitehead product is consistent with (and suggested by) our earlier notation for the commutator of two elements in a group.

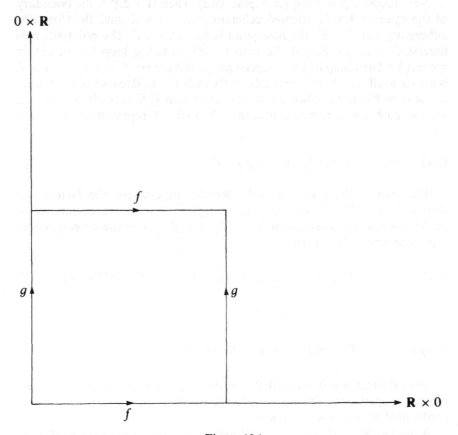

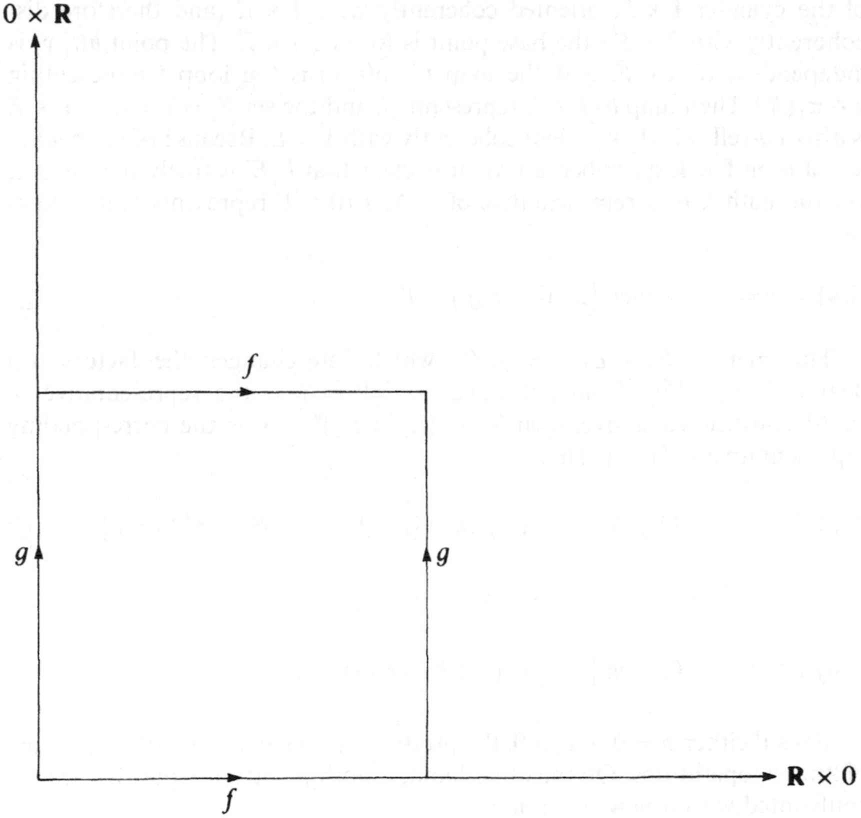

Figure 10.1

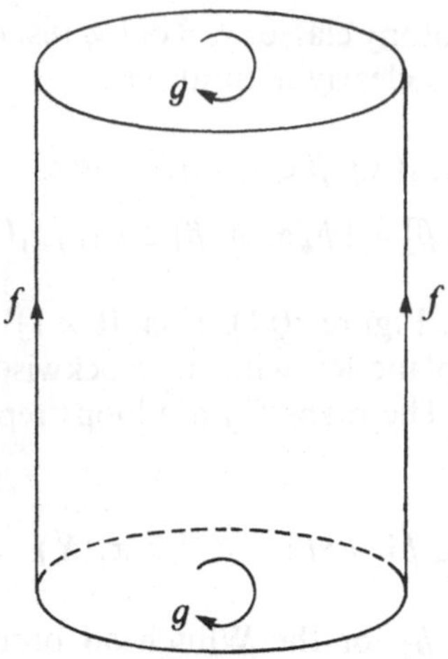

Figure 10.2

Next suppose $p = 0 < q$ (cf. Figure 10.2). Then $(\mathbf{I} \times E)^\bullet$ is the boundary of the cylinder $\mathbf{I} \times E$, oriented coherently with $1 \times E$ (and therefore discoherently with $0 \times E$); the base point is $(0, *) \in 0 \times \dot{E}$. The point $h(t, y)$ is independent of $y \in \dot{E}$, and the map $t \to h(t, y)$ is the loop f representing $\alpha \in \pi_1(X)$. Then map $h | 1 \times E$ represents β, and the set $E' = \mathbf{I} \times \dot{E} \cup 1 \times E$ is also a q-cell, which we orient coherently with $1 \times E$. Because of the behavior of h on $\mathbf{I} \times \dot{E}$ described above, it is clear that $h|E'$ is freely homotopic, via the path f, to a representative of β. As $h|0 \times E$ represents β, it follows that

(7.4) *If* $p = 0 < q$, *then* $[\alpha, \beta] = \tau_\alpha(\beta) - \beta$. □

The map $t : E_2 \times E_1 \to E_1 \times E_2$ which interchanges the factors has degree $(-1)^{(p+1)(q+1)}$, and, if $h : (E_1 \times E_2)^\bullet \to X$ is the representative of $[\alpha, \beta]$ constructed above, then $h \circ t : (E_2 \times E_1)^\bullet \to X$ is the corresponding representative of $[\beta, \alpha]$. Thus

(7.5) *If* $\alpha \in \pi_{p+1}(X)$, $\beta \in \pi_{q+1}(X)$, *then* $[\beta, \alpha] = (-1)^{(p+1)(q+1)}[\alpha, \beta]$. □

From (7.4) and (7.5) we deduce

(7.6) *If* $p > q = 0$, *then* $[\alpha, \beta] = (-1)^{p+1}(\tau_\beta(\alpha) - \alpha)$. □

Thus if either $p = 0$ or $q = 0$, the product $[\alpha, \beta]$ can be described in terms of known operations. On the other hand, if both p and q are positive, we are confronted with a new operation.

A map $h : \mathbf{S}^p \times \mathbf{S}^q \to X$ is said to have *type* (α, β) if and only if $h | \mathbf{S}^p \times \{*\}$

represents $\alpha \in \pi_p(X)$ and $h|\{*\} \times \mathbf{S}^q$ represents $\beta \in \pi_q(X)$. An important property of the Whitehead product is

(7.7) Theorem *Let $\alpha \in \pi_p(X)$, $\beta \in \pi_q(X)$. Then there exists a map of $\mathbf{S}^p \times \mathbf{S}^q$ into X of type (α, β) if and only if $[\alpha, \beta] = 0$.*

In fact, let us observe that $(\mathbf{S}^p \times \mathbf{S}^q, \mathbf{S}^p \vee \mathbf{S}^q)$ is a relative CW-complex with just one cell; a characteristic map for this cell is

$$\varpi_{p,q} = \varpi_p \times \varpi_q : (\mathbf{E}^p \times \mathbf{E}^q, (\mathbf{E}^p \times \mathbf{E}^q)^{\bullet}) \to (\mathbf{S}^p \times \mathbf{S}^q, \mathbf{S}^p \vee \mathbf{S}^q);$$

the attaching map for this cell is a representative of the Whitehead product $[\iota^1, \iota^2]$ of the homotopy classes of the inclusion maps $\mathbf{S}^p \to \mathbf{S}^p \vee \mathbf{S}^q$, $\mathbf{S}^q \to \mathbf{S}^p \vee \mathbf{S}^q$. If f, g are representatives of α, β respectively and $k = (f, g) : \mathbf{S}^p \vee \mathbf{S}^q \to X$ is the map determined by them, then there is a map of $\mathbf{S}^p \times \mathbf{S}^q$ into X of type (α, β) if and only if k can be extended over $\mathbf{S}^p \times \mathbf{S}^q$. Since $\varpi_{p,q}$ is a relative homeomorphism, this is so if and only if the map $k \circ \varpi_{p,q} | (\mathbf{E}^p \times \mathbf{E}^q)^{\bullet}$ can be extended over $\mathbf{E}^p \times \mathbf{E}^q$, i.e., is nullhomotopic. But the homotopy class of the latter map is $k_*[\iota^1, \iota^2] = [k_* \iota^1, k_* \iota^2] = [\alpha, \beta]$. □

(7.8) Corollary *If X is an* H-*space, then $[\alpha, \beta] = 0$ for every $\alpha \in \pi_p(X)$, $\beta \in \pi_q(X)$.*

For the map $\mu \circ (f \times g) : \mathbf{S}^p \times \mathbf{S}^q \to X$ has type (α, β) for any representatives f, g of α, β, respectively. □

In order to compare the Whitehead product in X with the Samelson product in ΩX, we shall make use of isomorphisms $\tau = \tau_p : \pi_{p+1}(X) \approx \pi_p(\Omega X)$. These can be defined in various ways, and the sign in the relation $\tau[\alpha, \beta] = \pm\langle \tau\alpha, \tau\beta \rangle$ will depend on the choices made for τ_p, τ_q, and τ_{p+q}. Let us, then, make the definition of τ_p very explicit. We shall make use of the adjointness relation

$$\mathbf{F}(\mathbf{S}X, Y) \approx \mathbf{F}(X, \Omega Y)$$

of §2, Chapter III, and the consequent relation

$$\pi_{p+1}(Y) = [\mathbf{S}^{p+1}, Y] = [\mathbf{S}\mathbf{S}^p, Y] \approx [\mathbf{S}^p, \mathbf{F}(\mathbf{S}, Y)] = \pi_p(\Omega Y).$$

Using the relative homeomorphism $\varpi_p : (\mathbf{I}^p, \dot{\mathbf{I}}^p) \to (\mathbf{S}^p, *)$ this yields

(7.9) *Let $f: (\mathbf{I}^{p+1}, \dot{\mathbf{I}}^{p+1}) \to (X, *)$ represent $\alpha \in \pi_{p+1}(X)$. Then the map $\tau f : (\mathbf{I}^p, \dot{\mathbf{I}}^p) \to (\Omega X, *)$ defined by*

$$\tau f(x_1, \ldots, x_p)(t) = f(t, x_1, \ldots, x_p)$$

represents $\tau(\alpha) \in \pi_p(\Omega X)$. □

We shall now prove

(7.10) Theorem *If* $\alpha \in \pi_{p+1}(X)$, $\beta \in \pi_{p+1}(X)$, *then*

$$\tau[\alpha, \beta] = (-1)^p \langle \tau\alpha, \tau\beta \rangle \in \pi_{p+q}(\Omega X).$$

We shall represent α and β by maps

$$f : (\mathbf{I}^{p+1}, \dot{\mathbf{I}}^{p+1}) \to (X, *),$$
$$g : (\mathbf{I}^{q+1}, \dot{\mathbf{I}}^{q+1}) \to (X, *);$$

then $[\alpha, \beta]$ is represented by the map

$$h : (S, *) \to (X, *),$$

defined by (7.1); recall that

$$S = \dot{\mathbf{I}}^{p+q+2} = (\mathbf{I}^{p+1} \times \mathbf{I}^{q+1})^{\bullet} = \dot{\mathbf{I}}^{p+1} \times \mathbf{I}^{q+1} \cup \mathbf{I}^{p+1} \times \dot{\mathbf{I}}^{q+1}.$$

Let us agree to identify $\mathbf{I}^{k+1}$ with $\mathbf{I} \times \mathbf{I}^k$ for every k. Let

$$K = 0 \times \mathbf{I}^p \times 0 \times \mathbf{I}^q \cup \mathbf{I} \times \dot{\mathbf{I}}^p \times \mathbf{I} \times \mathbf{I}^q \cup \mathbf{I} \times \mathbf{I}^p \times \mathbf{I} \times \dot{\mathbf{I}}^q \subset S;$$

the union of the second and third sets retracts by deformation into a subset of the contractible set $0 \times \mathbf{I}^p \times 0 \times \mathbf{I}^q$ (by contracting the first and third factors of each to the point 0); thus K is itself contractible. Let E be the product $\mathbf{I} \times \mathbf{I}^p \times \mathbf{I}^q$, and let $\phi : E \to S$ be the map defined as follows: subdivide $\mathbf{I}$ by inserting the partition points 1/4, 3/8, 1/2, 5/8, 3/4; then each of the sets $\mathbf{I} \times x \times y$ ($x \in \mathbf{I}^p$, $y \in \mathbf{I}^q$) is mapped in the obvious piecewise linear fashion into the closed polygon in S with vertices

$$*, \quad (0, x, 0, y), \quad (1, x, 0, y), \quad (1, x, 1, y),$$
$$(0, x, 1, y), \quad (0, x, 0, y), \quad *$$

(cf. Figure 10.3, for the case $p = 0$, $q = 1$; the sphere S is the boundary of the depicted cube and the x-, y-, and z-axes are those of the first, third and fourth coordinates respectively. The shaded area represents the image of ϕ, and K is the union of the two unshaded faces with the part of the z-axis lying between them). Then $\phi(\dot{E}) \subset K$, and $\phi : (E, \dot{E}) \to (S, K)$ is a relative homeomorphism. The cell $[1/4, 3/8] \times \mathbf{I}^p \times \mathbf{I}^q$ is mapped by ϕ upon the cell $\mathbf{I} \times \mathbf{I}^p \times 0 \times \mathbf{I}^q$, preserving the order of the coordinates, and the latter cell is coherent with $(-1)^p$ times the natural orientation of S. Thus, if $e \in \pi_n(E, \dot{E})$ and $s \in \pi_n(S, K)$ are the generators of these infinite cyclic groups determined by the natural orientations of E, S, respectively, we have $\phi_*(e) = (-1)^p s$.

Consider now the map $h \circ \phi : E \to X$. Let $F = \mathbf{I}^p \times \mathbf{I}^q$, so that $E = \mathbf{I} \times F$, and let $\psi = \tau(h \circ \phi) : F \to \Omega X$; thus

$$\psi(x, y)(t) = h\phi(t, x, y).$$

Inspection of the definitions of h and ϕ reveals that

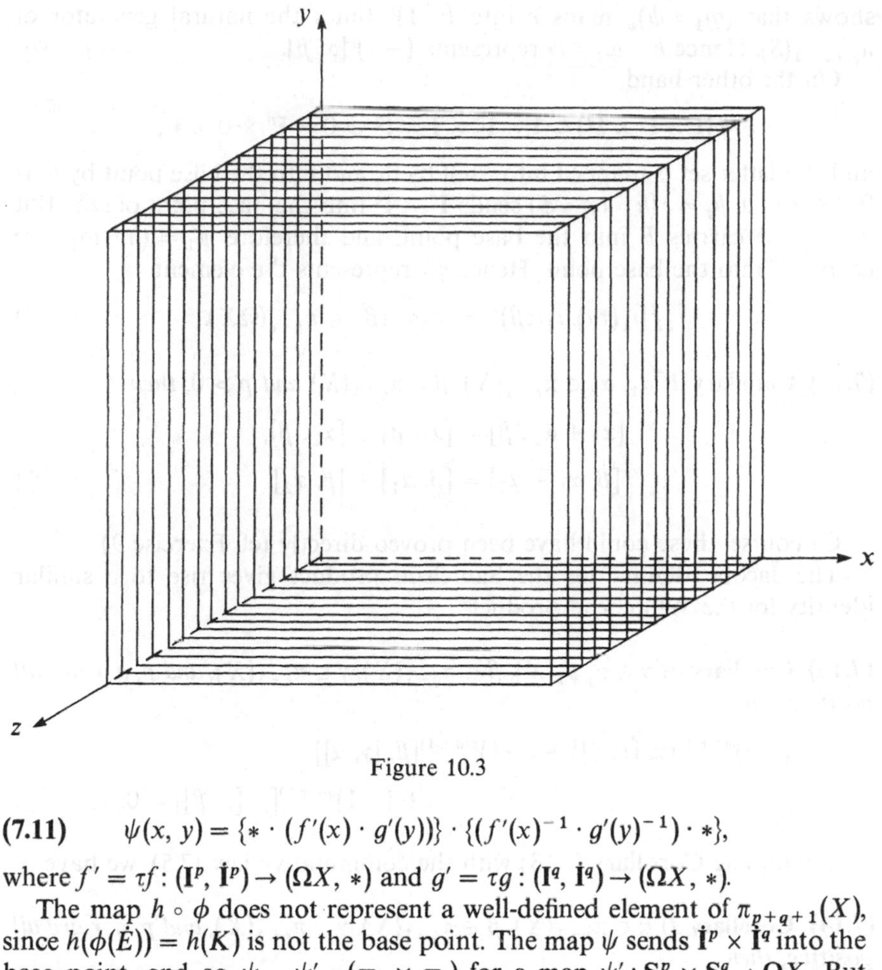

Figure 10.3

(7.11) $$\psi(x, y) = \{* \cdot (f'(x) \cdot g'(y))\} \cdot \{(f'(x)^{-1} \cdot g'(y)^{-1}) \cdot *\},$$

where $f' = \tau f : (\mathbf{I}^p, \dot{\mathbf{I}}^p) \to (\Omega X, *)$ and $g' = \tau g : (\mathbf{I}^q, \dot{\mathbf{I}}^q) \to (\Omega X, *)$.

The map $h \circ \phi$ does not represent a well-defined element of $\pi_{p+q+1}(X)$, since $h(\phi(\dot{E})) = h(K)$ is not the base point. The map ψ sends $\dot{\mathbf{I}}^p \times \dot{\mathbf{I}}^q$ into the base point, and so $\psi = \psi' \circ (\varpi_p \times \varpi_q)$ for a map $\psi' : \mathbf{S}^p \times \mathbf{S}^q \to \Omega X$. But $\psi'(\mathbf{S}^p \vee \mathbf{S}^q)$ is not the base point, so that ψ' does not factor through $\mathbf{S}^{p+q}$. However, ψ' is easily seen to represent the element $[\bar{i}_1(\tau\alpha), \bar{i}_2(\tau\beta)]$ of $\pi_{p,q}(\Omega X)$.

We have seen that K is contractible. The contraction of K constructed above can be extended to a homotopy $\omega_t : (S, K) \to (S, K)$ of the identity to a map ω_1 for which $\omega_1(K) = *$. Then $h \circ \omega_1 \circ \phi : (E, \dot{E}) \to (X, *)$, and the map $\omega_1 \circ \phi : (E, \dot{E}) \to (S, K)$ factors through $(S, *)$. But the injection $\pi_{p+q+1}(S, *) \to \pi_{p+q+1}(S, K)$ is an isomorphism, and commutativity of the diagram

$$\begin{array}{ccc}
\pi_{p+q+1}(E, \dot{E}) & \xrightarrow{(\omega_t \circ \phi)_*} & \pi_{p+q+1}(S, K) \\
 & \searrow^{(\omega_1 \circ \phi)_*} & \uparrow \\
 & & \pi_{p+q+1}(S, *)
\end{array}$$

shows that $(\omega_1 \circ \phi)_*$ maps e into $(-1)^p$ times the natural generator of $\pi_{p+q+1}(S)$. Hence $h \circ \omega_1 \circ \phi$ represents $(-1)^p[\alpha, \beta]$.

On the other hand,

$$\phi(\mathbf{I} \times \dot{\mathbf{I}}^p \times \dot{\mathbf{I}}^q) \subset \mathbf{I} \times \dot{\mathbf{I}}^p \times \mathbf{I} \times \dot{\mathbf{I}}^q \cup 0 \times \mathbf{I}^p \times 0 \times \mathbf{I}^q,$$

and the latter set is mapped into itself by ω_t and into the base point by h. It follows that $\psi_t = \tau(h \circ \omega_t \circ \phi)$ sends $\dot{\mathbf{I}}^p \times \dot{\mathbf{I}}^q$ into the base point of ΩX. But $h \circ \omega_1 \circ \phi$ carries $\dot{E}$ into the base point, and therefore $\psi'_1 = (h \circ \omega_1 \circ \phi)$ carries $\dot{F}$ into the base point. Hence ψ'_1 represents the element

$$\bar{i}_{12}^{-1}[\bar{i}_1(\tau\alpha), \bar{i}_2(\tau\beta)] = \langle\tau\alpha, \tau\beta\rangle \in \pi_{p+q}(\Omega X). \qquad \square$$

(7.12) Corollary *If* $\alpha_1, \alpha_2 \in \pi_{p+1}(X)$, $\beta \in \pi_{q+1}(X)$ *and* $p > 0$, *then*

$$[\alpha_1 + \alpha_2, \beta] = [\alpha_1, \beta] + [\alpha_2, \beta],$$
$$[\beta, \alpha_1 + \alpha_2] = [\beta, \alpha_1] + [\beta, \alpha_2]. \qquad \square$$

Of course, these could have been proved directly (cf. Exercise 9).

The Jacobi identity for the Samelson product gives rise to a similar identity for the Whitehead product:

(7.13) Corollary *If* $\alpha \in \pi_{p+1}(X)$, $\beta \in \pi_{q+1}(X)$, $\gamma \in \pi_{r+1}(X)$, *and* p, q, r *are all positive, then*

$$(-1)^{r(p+1)}[\alpha, [\beta, \gamma]] + (-1)^{p(q+1)}[\beta, [\gamma, \alpha]] + (-1)^{q(r+1)}[\gamma, [\alpha, \beta]] = 0. \qquad \square$$

Combining Corollary (7.13) with the commutative law (7.5), we have

(7.14) Corollary *If* $\alpha \in \pi_{p+1}(X)$, $\beta \in \pi_{q+1}(X)$, $\gamma \in \pi_{r+1}(X)$, *and* p, q, r *are all positive, then*

$$(-1)^{(p+1)(r+1)}[[\alpha, \beta], \gamma] + (-1)^{(q+1)(p+1)}[[\beta, \gamma], \alpha] + (-1)^{(q+1)(r+1)}[[\gamma, \alpha], \beta] = 0. \qquad \square$$

8 Operations in Homotopy Groups

In this section we shall discuss the Whitehead product and other operations in homotopy groups and explore some of their connections.

I. The composition product

This product, defined by composition of representative maps, associates to $\alpha \in \pi_r(X)$, $\beta \in \pi_n(\mathbf{S}^r)$, an element $\alpha \circ \beta \in \pi_n(X)$. If $f: \mathbf{S}^r \to X$ is a representative of α, then $f_*(\beta) = \alpha \circ \beta$. Since f_* is a homomorphism, we have

(8.1) Theorem *The composition operation is right additive, i.e., if* $\alpha \in \pi_r(X)$, $\beta_1, \beta_2 \in \pi_n(\mathbf{S}^r)$, *then* $\alpha \circ (\beta_1 + \beta_2) = \alpha \circ \beta_1 + \alpha \circ \beta_2$. □

On the other hand, composition is not, in general, left additive; an example will be given in Chapter XI. There are, however, two important special cases in which left additivity holds.

(8.2) Theorem *If* $\sigma \in \pi_{n-1}(\mathbf{S}^{r-1})$, $\alpha_1, \alpha_2 \in \pi_r(X)$, *then* $(\alpha_1 + \alpha_2) \circ (E\sigma) = \alpha_1 \circ E\sigma + \alpha_2 \circ E\sigma$.

Let $f: \mathbf{S}^{n-1} \to \mathbf{S}^{r-1}$ be a representative of σ, so that $1 \wedge f: \mathbf{S}^n \to \mathbf{S}^r$ is a representative of $E\sigma$. The standard coproduct in $\mathbf{S}^r$ is given by

$$\theta(\bar{t} \wedge y) = \begin{cases} (\overline{2t} \wedge y, *) & (0 \le t \le \frac{1}{2}), \\ (*, \overline{2t - 1} \wedge y) & (\frac{1}{2} \le t \le 1). \end{cases}$$

If $g_i : \mathbf{S}^r \to X$ is a representative of α_i $(i = 1, 2)$, then $\alpha_1 + \alpha_2$ is represented by $g \circ \theta : \mathbf{S}^r \to X$, where $g : \mathbf{S}^r \vee \mathbf{S}^r \to X$ is defined by

$$g(y, *) = g_1(y),$$
$$g(*, y) = g_2(y),$$

and $(\alpha_1 + \alpha_2) \circ E\sigma$ is represented by $h = g \circ \theta \circ (1 \wedge f)$. But (cf. Figure 10.4)

$$h(\bar{t} \wedge x) = \begin{cases} g_1(\overline{2t} \wedge f(x)) & (0 \le t \le \frac{1}{2}), \\ g_2(\overline{2t - 1} \wedge f(x)) & (\frac{1}{2} \le t \le 1). \end{cases}$$

This map is clearly a representative of $\alpha_1 \circ E\sigma + \alpha_2 \circ E\sigma$. □

(8.3) Theorem *If X is an* H-*space, then composition is biadditive, i.e., if* $\beta_1, \beta_2 \in \pi_r(X)$, $\alpha \in \pi_n(\mathbf{S}^r)$, *then*

$$(\beta_1 + \beta_2) \circ \alpha = \beta_1 \circ \alpha + \beta_2 \circ \alpha.$$

For if $f: \mathbf{S}^n \to \mathbf{S}^r$ represents α, then $\beta \circ \alpha = \bar{f}(\beta)$ for all $\beta \in \pi_r(X)$. But $\bar{f}: [\mathbf{S}^r, X] \to [\mathbf{S}^n, X]$ is a homomorphism. (This is just the naturality of the

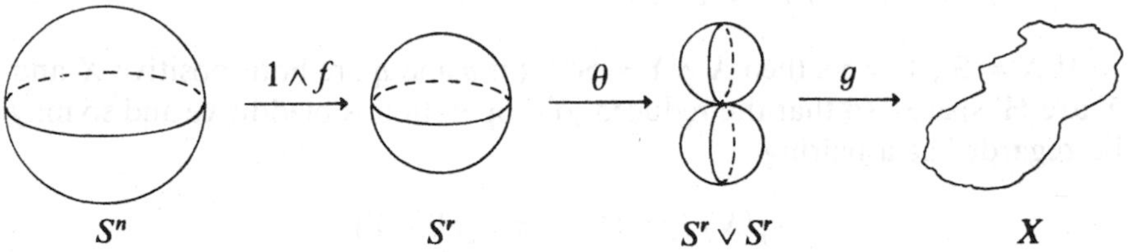

Figure 10.4

operation in $[Y, X]$ induced by the H-structure in X; by Theorem (5.21) of Chapter III, this agrees with the structure corresponding to the H′-structure on the H′-space Y, notably when Y is a sphere). □

(8.4) Corollary *If* $d = 1, 3,$ *or* 7, $\beta_1, \beta_2 \in \pi_r(\mathbf{S}^d)$, $\alpha \in \pi_n(\mathbf{S}^r)$, *then* $(\beta_1 + \beta_2) \circ \alpha = \beta_1 \circ \alpha + \beta_2 \circ \alpha$.

For then $\mathbf{S}^d$ is an H-space. □

Let $\iota_r \in \pi_r(\mathbf{S}^r)$ be the homotopy class of the identity map. Then $\alpha \circ (-\iota_n) = -(\alpha \circ \iota_n) = -\alpha$. On the other hand, $(-\iota_r) \circ \alpha$ is not, in general, equal to $-\alpha$, and the map $\alpha \to (-\iota_r) \circ \alpha$ is an automorphism $\nu = \nu_{n,r}$ of period two, of $\pi_n(\mathbf{S}^r)$. However,

(8.5) *If* $\sigma \in \pi_{n-1}(\mathbf{S}^{r-1})$, *then* $\nu(E\sigma) = -E\sigma$. □

(8.6) *If* $n = 1, 3,$ *or* 7, $\alpha \in \pi_n(\mathbf{S}^r)$, *then* $\nu(\alpha) = -\alpha$. □

II. The reduced join

The reduced join functor is compatible with homotopy and therefore induces an operation associating to each $\alpha \in [X, X']$, $\beta \in [Y, Y']$ an element $\alpha \wedge \beta \in [X \wedge Y, X' \wedge Y']$.

Suppose that Y is an H′-space. Then we have seen in Chapter III that $X \wedge Y$ is also an H′-space, and therefore there are natural operations defined in $[Y, Y']$, $[X \wedge Y, X' \wedge Y']$, respectively. It then makes sense to ask whether the operation of reduced join is right additive. A positive answer is furnished by

(8.7) Theorem *Let* Y *be an* H′-*space*, $\alpha \in [X, X']$, $\beta_1, \beta_2 \in [Y, Y']$. *Then*

$$\alpha \wedge (\beta_1 + \beta_2) = \alpha \wedge \beta_1 + \alpha \wedge \beta_2 \in [X \wedge Y, X' \wedge Y'].$$

The easy proof is left to the reader. □

(8.8) Corollary *If* X *is an* H′-*space*, $\alpha_1, \alpha_2 \in [X, X']$, $\beta \in [Y, Y']$, *then* $(\alpha_1 + \alpha_2) \wedge \beta = (\alpha_1 \wedge \beta) + (\alpha_2 \wedge \beta)$. □

If $X = \mathbf{S}^p$, $Y = \mathbf{S}^q$, then $X \wedge Y = \mathbf{S}^{p+q}$; if p and q are both positive X and Y are H′-spaces, so that the reduced join operation is biadditive and so may be regarded as a pairing

$$\pi_p(X) \otimes \pi_q(Y) \to \pi_{p+q}(X \wedge Y).$$

Suppose further that $X = \mathbf{S}^p$, $\alpha = \iota_p$ is the element represented by the identity map. If $p = 1$, $\alpha \wedge \beta$ is the suspension $E\beta$ of β, and it follows by induction (and the commutative and associative laws for the reduced join) that

(8.9) *If* $\alpha \in \pi_q(Y)$, *then* $\iota_p \wedge \alpha = E^p(\alpha)$. *In particular,* $\iota_p = \iota_1 \wedge \cdots \wedge \iota_1$ (p factors). □

(8.10) Lemma *If* $\alpha \in \pi_n(\mathbf{S}^r)$, *then* $\alpha \wedge \iota_1 = (-1)^{n-r} E\alpha$.

For there is a commutative diagram

$$\begin{array}{ccc} \mathbf{S}^n \wedge \mathbf{S}^1 & \xrightarrow{\alpha \wedge \iota_1} & \mathbf{S}^r \wedge \mathbf{S}^1 \\ \downarrow{\scriptstyle \tau_n} & & \downarrow{\scriptstyle \tau_r} \\ \mathbf{S}^1 \wedge \mathbf{S}^n & \xrightarrow[\iota_1 \wedge \alpha]{} & \mathbf{S}^1 \wedge \mathbf{S}^r \end{array}$$

where $\tau_k : \mathbf{S}^k \wedge \mathbf{S}^1 \to \mathbf{S}^1 \wedge \mathbf{S}^k$ is the map which interchanges the coördinates. Therefore τ_k has degree $(-1)^k$, and thus

$$\begin{aligned} \alpha \wedge \iota_1 &= ((-1)^r \iota_r) \circ (\iota_1 \wedge \alpha) \circ ((-1)^n \iota_n)) \\ &= (-1)^n(((-1)^r \iota_r) \circ E\alpha) \\ &= (-1)^{n-r} E\alpha \quad \text{by (8.5).} \end{aligned}$$ □

(8.11) Corollary *If* $\alpha \in \pi_{p+r}(\mathbf{S}^p)$, *then* $\alpha \wedge \iota_p = (-1)^{pr} E^p \alpha$. □

An interesting relation between the reduced join and composition operations has been given by Barratt and Hilton [1]:

(8.12) Theorem *If* $\alpha \in \pi_{p+k}(\mathbf{S}^p)$, $\beta \in \pi_{q+l}(\mathbf{S}^q)$, *then*

$$\alpha \wedge \beta = (-1)^{qk} E^q \alpha \circ E^{p+k} \beta = (-1)^{(q+l)k} E^p \beta \circ E^{q+l} \alpha.$$

For there is a commutative diagram

$$\begin{array}{ccc} \mathbf{S}^{p+k} \wedge \mathbf{S}^{q+l} & \xrightarrow{\alpha \wedge \iota_{q+l}} & \mathbf{S}^p \wedge \mathbf{S}^{q+l} \\ \downarrow{\scriptstyle \iota_{p+k} \wedge \beta} & \searrow{\scriptstyle \alpha \wedge \beta} & \downarrow{\scriptstyle \iota_p \wedge \beta} \\ \mathbf{S}^{p+k} \wedge \mathbf{S}^q & \xrightarrow[\alpha \wedge \iota_q]{} & \mathbf{S}^p \wedge \mathbf{S}^q \end{array}$$

and the formula follows from (8.9) and (8.11). □

(8.13) Corollary *If* $\alpha \in \pi_{p+k}(\mathbf{S}^p)$, $\beta \in \pi_{q+l}(\mathbf{S}^q)$, *then*

$$\beta \wedge \alpha = (-1)^{pl+qk+ql}\alpha \wedge \beta. \qquad \square$$

III. The Whitehead product

We have already defined this product and proved some of its properties in §7. One of them is a naturality property with respect to maps of its range. The effect of maps of the domain is much more subtle; while there is an explicit formula for $[\alpha \circ \gamma, \beta]$, it is quite complicated and involves notions (Hopf–Hilton invariants) which do not occur until Chapter XI. There are, however, some cases of special interest which can be stated quite simply.

The most elementary result in this direction is due to Hilton and J. H. C. Whitehead [1].

(8.14) Theorem *If* $\alpha \in \pi_p(X)$, $\beta \in \pi_q(X)$, $\gamma \in \pi_m(\mathbf{S}^p)$, $\delta \in \pi_n(\mathbf{S}^q)$, *and if* $[\alpha, \beta] = 0$, *then* $[\alpha \circ \gamma, \beta \circ \delta] = 0$.

We use the criterion of Theorem (7.7). Since $[\alpha, \beta] = 0$, there is a map $f : \mathbf{S}^p \times \mathbf{S}^q \to X$ of type (α, β). Let $g : \mathbf{S}^m \to \mathbf{S}^p$, $h : \mathbf{S}^n \to \mathbf{S}^q$ be representatives of γ, δ, respectively. Then $f \circ (g \times h) : \mathbf{S}^m \times \mathbf{S}^n \to X$ has type $(\alpha \circ \gamma, \beta \circ \delta)$, and therefore $[\alpha \circ \gamma, \beta \circ \delta] = 0$. $\square$

To formulate the next result, we need the notion of *join*. If X and Y are spaces in $\mathscr{K}_*$, their join is the subspace

$$X * Y = \mathbf{T}X \times Y \cup X \times \mathbf{T}Y$$

of $\mathbf{T}X \times \mathbf{T}Y$: the base point of $X * Y$ is $(*, *) \in X \times Y = (\mathbf{T}X \times Y) \cap (X \times \mathbf{T}Y)$. If $f : X \to X'$ and $g : Y \to Y'$ are maps, the map $\mathbf{T}f \times \mathbf{T}g : \mathbf{T}X \times \mathbf{T}Y \to \mathbf{T}X' \times \mathbf{T}Y'$ sends $X * Y$ into $X' * Y'$, inducing a map $f * g : X * Y \to X' * Y'$. Thus the join is a functor: $\mathscr{K}_* \times \mathscr{K}_* \to \mathscr{K}_*$.

It is more customary to define the rectilinear join of *free* spaces X and Y as the quotient space obtained from the disjoint union $X + (\mathbf{I} \times X \times Y) + Y$ by identifying each $x \in X$ with $0 \times x \times Y$ and each $y \in Y$ with $1 \times X \times y$; let $(1 - t)x + ty$ be the image of (t, x, y) under these identifications. The rectilinear join is then the union of the line segments $[x, y] = \{(1 - t)x + ty \mid t \in \mathbf{I}\}$, no two of which have a point in common except for possible end-points. The maps

$$x \to (x, *)$$

$$(t, x, y) \to \begin{cases} (x, 2t \wedge y) & (t \le \frac{1}{2}), \\ (2(1 - t) \wedge x, y) & (t \ge \frac{1}{2}), \end{cases}$$

$$y \to (*, y)$$

define a homeomorphism of the rectilinear join with $X * Y$. The inverse of this homeomorphism is the map defined by

$$(x, s \wedge y) \to \left(1 - \frac{s}{2}\right)x + \frac{s}{2}y,$$

$$(s \wedge x, y) \to \frac{s}{2}x + \left(1 - \frac{s}{2}\right)y.$$

If X and Y are spaces in $\mathscr{K}_*$, let X_*, Y_*, as usual, be the free spaces obtained from them by ignoring the base points. The space $\mathbf{T}X$ is obtained from $\mathbf{T}X_*$ by collapsing the contractible set $\mathbf{T}\{*\}$ to a point; the quotient map is therefore a homotopy equivalence $(\mathbf{T}X_*, X) \to (\mathbf{T}X, X)$ (rel. X). Similarly, the quotient map $(\mathbf{T}Y_*, Y) \to (\mathbf{T}Y, Y)$ is a homotopy equivalence (rel. Y). It follows that the product of these maps is a homotopy equivalence between the pairs

$$(\mathbf{T}X_* \times \mathbf{T}Y_*, X_* * Y_*),$$

$$(\mathbf{T}X \times \mathbf{T}Y, X * Y).$$

Thus the join $X * Y$ has the same homotopy type as the rectilinear join of X_* and Y_*.

The operation of join is compatible with homotopy, and so defines an operation, associating with $\alpha \in [X, X']$, $\beta \in [Y, Y']$, an element $\alpha * \beta \in [X * Y, X' * Y']$. That this operation is not essentially new follows from

(8.15) Lemma *There is a contractible subset $K(X, Y)$ of $X * Y$ such that $(X * Y, K(X, Y))$ is an* NDR-*pair and $X * Y/K(X, Y)$ is naturally homeomorphic with $\mathbf{S}(X \wedge Y)$.*

The subset in question is defined by

$$K(X, Y) = \mathbf{T}X \vee \mathbf{T}Y.$$

The map $p : X * Y \to \mathbf{S}(X \wedge Y)$ defined by

$$p(t \wedge x, y) = \frac{t}{2} \wedge x \wedge y,$$

$$p(x, t \wedge y) = \left(1 - \frac{t}{2}\right) \wedge x \wedge y,$$

sends $\mathbf{T}X \vee \mathbf{T}Y$ into the base point. Its restriction to $\mathbf{T}X \times Y$ is the identification map $(\mathbf{T}X \times Y, \mathbf{T}X \vee Y) \to (\mathbf{T}X \wedge Y, *)$, composed with the natural homeomorphism of the latter space with $\mathbf{T}_+(X \wedge Y)$; similarly, $p \mid X \times \mathbf{T}Y$ is the composite of the identification map $(X \times \mathbf{T}Y, X \vee \mathbf{T}Y) \to (X \wedge \mathbf{T}Y, *)$ with the natural homeomorphism of $X \wedge \mathbf{T}Y$ with $T_-(X \wedge Y)$. These maps fit together to define a relative homeomorphism of $(X * Y, \mathbf{T}X \vee \mathbf{T}Y)$ with $(\mathbf{S}(X \wedge Y), *)$. □

Trivial to prove is

(8.16) Lemma *If* $f: X \to X'$, $g: Y \to Y'$, *then* $f * g$ *maps* $\mathbf{T}X \vee \mathbf{T}Y$ *into* $\mathbf{T}X' \vee \mathbf{T}Y'$ *and the diagram*

$$\begin{array}{ccc} X * Y & \xrightarrow{f * g} & X' * Y' \\ \Big\downarrow p & & \Big\downarrow p' \\ \mathbf{S}(X \wedge Y) & \xrightarrow[\mathbf{S}(f \wedge g)]{} & \mathbf{S}(X' \wedge Y') \end{array}$$

is commutative. □

(8.17) Corollary *The operations* $(\alpha, \beta) \to \alpha * \beta$, $(\alpha, \beta) \to E(\alpha \wedge \beta)$, *are naturally equivalent.* □

We can now prove:

(8.18) Theorem *Let* $\alpha \in \pi_{p+1}(X)$, $\beta \in \pi_{q+1}(X)$, $\gamma \in \pi_m(\mathbf{S}^p)$, $\delta \in \pi_n(\mathbf{S}^q)$. *Then*

$$[\alpha \circ E\gamma, \beta \circ E\delta] = [\alpha, \beta] \circ E(\gamma \wedge \delta).$$

For let $a : (\mathbf{E}^{p+1}, \mathbf{S}^p) \to (X, *)$, $b : (\mathbf{E}^{q+1}, \mathbf{S}^q) \to (X, *)$ represent α, β, respectively. Let $c : \mathbf{S}^m \to \mathbf{S}^p$, $d : \mathbf{S}^n \to \mathbf{S}^q$ be representatives of γ, δ, respectively. Then $\mathbf{T}c \times \mathbf{T}d : \mathbf{E}^{m+1} \times \mathbf{E}^{n+1} \to \mathbf{E}^{p+1} \times \mathbf{E}^{q+1}$ sends $\mathbf{S}^m * \mathbf{S}^n$ into $\mathbf{S}^p * \mathbf{S}^q$, and its restriction $c * d$ to $\mathbf{S}^m * \mathbf{S}^n$ represents, according to (8.16), the element $E(\gamma \wedge \delta)$. If $h : \mathbf{S}^p * \mathbf{S}^q \to X$ is the representative of $[\alpha, \beta]$ given by (7.1), we have

$$\begin{aligned} h(c * d)(x, y)) &= h(\mathbf{T}c(x), \mathbf{T}d(y)) \\ &= \begin{cases} a(\mathbf{T}c(x)) & \text{if } x \in \mathbf{E}^{m+1},\ y \in \mathbf{S}^n, \\ b(\mathbf{T}d(y)) & \text{if } x \in \mathbf{S}^m,\ y \in \mathbf{E}^{n+1}. \end{cases} \end{aligned}$$

and therefore h represents $[\alpha', \beta']$, where α', β' are the elements of $\pi_{m+1}(X)$, $\pi_{n+1}(X)$ represented by $a \circ \mathbf{T}c : (\mathbf{E}^{m+1}, \mathbf{S}^m) \to (X, *)$, $b \circ \mathbf{T}d : (\mathbf{E}^{n+1}, \mathbf{S}^n) \to (X, *)$. It remains only to verify

(8.19) Lemma *Let* $f: (\mathbf{E}^{r+1}, \mathbf{S}^r) \to (X, *)$ *represent* $\theta \in \pi_{k+1}(X)$, *and let* $g : \mathbf{S}^t \to \mathbf{S}^r$ *represent* $\gamma \in \pi_t(\mathbf{S}^r)$. *Then the map* $f \circ \mathbf{T}g : (\mathbf{E}^{t+1}, \mathbf{S}^t) \to (X, *)$ *represents* $\theta \circ E\gamma$.

For there is a commutative diagram

$$\begin{array}{ccccc}
(\mathbf{E}^{t+1}, \mathbf{S}^t) & \xrightarrow{\mathbf{T}g} & (\mathbf{E}^{r+1}, \mathbf{S}^r) & \searrow^{f} & \\
\downarrow & & \downarrow & & X \\
\mathbf{S}^{t+1} & \xrightarrow[\mathbf{S}g]{} & \mathbf{S}^{r+1} & \nearrow_{f'} &
\end{array}$$

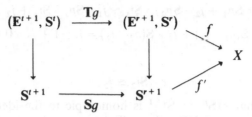

in which the vertical arrows denote the identification maps. The map f' represents θ and therefore $f' \circ \mathbf{S}g$ represents $\theta \circ E\gamma$. Hence $f \circ \mathbf{T}g$ represents $\theta \circ E\gamma$. □

Another useful property of the Whitehead product is

(8.20) Theorem *If* $\alpha \in \pi_p(X)$, $\beta \in \pi_q(X)$, *then* $E[\alpha, \beta] = 0$.

It suffices to prove this for the universal example, so that we may assume $\alpha = \iota_p$, $\beta = \iota_q$. Let us recall that $[\iota_p, \iota_q]$ is the homotopy class of the attaching map of the $(p+q)$-cell in the product complex $\mathbf{S}^p \times \mathbf{S}^q$, so that there is a commutative diagram

$$\begin{array}{ccc}
\mathbf{S}^{p+q-1} & \xrightarrow{f} & \mathbf{S}^p \vee \mathbf{S}^q \\
\downarrow & & \downarrow \\
\mathbf{E}^{p+q} & \xrightarrow[h]{} & \mathbf{S}^p \times \mathbf{S}^q
\end{array}$$

in which f represents $[\iota_p, \iota_q]$. Suspending this diagram, we obtain a commutative diagram

$$\begin{array}{ccc}
\mathbf{S}^{p+q} & \xrightarrow{\mathbf{S}f} & \mathbf{S}^{p+1} \vee \mathbf{S}^{q+1} \\
\downarrow & & \downarrow \\
\mathbf{SE}^{p+q} & \xrightarrow[\mathbf{S}h]{} & \mathbf{S}(\mathbf{S}^p \times \mathbf{S}^q).
\end{array}$$

We maintain that $\mathbf{S}^{p+1} \vee \mathbf{S}^{q+1}$ is a retract of $\mathbf{S}(\mathbf{S}^p \times \mathbf{S}^q)$. Given that this is so, let $r : \mathbf{S}(\mathbf{S}^p \times \mathbf{S}^q) \to \mathbf{S}^{p+1} \vee \mathbf{S}^{q+1}$ be a retraction. Then $r \circ \mathbf{S}h$ is an extension of $\mathbf{S}f$ over the $(p+q+1)$-cell $\mathbf{SE}^{p+q}$, and therefore $\mathbf{S}f$ is nullhomotopic.

Let $j_1 : \mathbf{S}^{p+1} \to \mathbf{S}^{p+1} \vee \mathbf{S}^{q+1}$, $j_2 : \mathbf{S}^{q+1} \to \mathbf{S}^{p+1} \vee \mathbf{S}^{q+1}$ be the usual inclusions. Then the maps

$$j_1 \circ \mathbf{S}p_1 : \mathbf{S}(\mathbf{S}^p \times \mathbf{S}^q) \to \mathbf{S}^{p+1} \vee \mathbf{S}^{q+1}, \; j_2 \circ \mathbf{S}p_2 : \mathbf{S}(\mathbf{S}^p \times \mathbf{S}^q) \to \mathbf{S}^{p+1} \vee \mathbf{S}^{q+1}$$

can be added to yield a map $r: \mathbf{S}(\mathbf{S}^p \times \mathbf{S}^q) \to \mathbf{S}^{p+1} \vee \mathbf{S}^{q+1}$. Then

$$r \circ \mathbf{S}i_1 = (j_1 \circ \mathbf{S}p_1 + j_2 \circ \mathbf{S}p_2) \circ \mathbf{S}i_1 \simeq j_1 \circ \mathbf{S}p_1 \circ \mathbf{S}i_1 + j_2 \circ \mathbf{S}p_2 \circ \mathbf{S}i_1$$
$$= j_1 \circ \mathbf{S}(p_1 \circ i_1) + j_2 \circ \mathbf{S}(p_2 \circ i_1) = j_1 \circ 1 + j_2 \circ 0 = j_1,$$

and similarly

$$r \circ \mathbf{S}i_2 \simeq j_2,$$

and it follows that $r | \mathbf{S}^{p+1} \vee \mathbf{S}^{q+1}$ is homotopic to the identity. Because, $(\mathbf{S}^p \times \mathbf{S}^q, \mathbf{S}^p \vee \mathbf{S}^q)$ is an NDR-pair, so is its suspension, and therefore r is homotopic to a retraction. □

Exercises

1. Prove that if G is an H_0 space, there is a short exact sequence

$$0 \to \pi_2(G)/2\pi_2(G) \to [\mathbf{P}^2(\mathbf{R}), G] \to {}_2\pi_1(G) \to 0$$

(for a group H, ${}_2H = \{x \in H | 2x = 0\}$). What can you say about the group extension?

2. (Spanier & Whitehead [2]). Prove that if $p: X \to B$ is a fibration with fibre F, and if F is a categorical subset of X, then F is an H-space.

3. (Hilton [2]). Prove that, if $\alpha \in \pi_1(X)$, $\beta \in \pi_{q+1}(X)$, $\gamma \in \pi_{r+1}(X)$, and q, r are positive, then

$$(-1)^r[\alpha, [\beta, \gamma]] + [\tau_\alpha(\beta), [\gamma, \alpha]] + (-1)^{q(r+1)}[\gamma, [\alpha, \beta]] = 0.$$

4. Prove that, if $\alpha \in \pi_1(X)$, $\beta \in \pi_1(X)$, $\gamma \in \pi_{r+1}(X)$, and $r > 0$, then

$$(-1)^r[\alpha, [\beta, \gamma]] + [\beta, [\gamma, \alpha]] + [\tau_\beta(\gamma), [\alpha, \beta]] = 0.$$

5. Prove that, if X and Y are 0-connected, then $\mathbf{S}(X \times Y)$ has the same weak homotopy type as $\mathbf{S}X \vee \mathbf{S}Y \vee \mathbf{S}(X \wedge Y)$.

6. (Copeland [1]). Let X be an $(n-1)$-connected CW-complex $(n \geq 1)$. Let $\iota^n(X) \in H^n(X, *; \pi_n(X))$ be the characteristic class. The Whitehead product is a pairing $\pi_n(X) \otimes \pi_n(X) \to \pi_{2n-1}(X)$. Let $\gamma^{2n} \in H^{2n}(X \times X, X \vee X; \pi_{2n-1}(X))$ be the cross product $\iota^n(X) \times \iota^n(X)$ defined by this pairing. Prove that the folding map $\nabla: X \vee X \to X$ can be extended over the $(2n-1)$-skeleton of the relative CW-complex $(X \times X, X \vee X)$, and that ∇ can be extended over the $2n$-skeleton if and only if $\gamma^{2n} = 0$.

7. Strengthen Theorem (2.1) of Chapter V as follows. Let π_1 be a group, and, for each $q \geq 2$, let π_q be an abelian group on which π_1 operates. Then there is a connected CW-complex X and a family of isomorphisms $\phi_q: \pi_q(X) \approx \pi_q$ such that,

(1) for all $\xi \in \pi_1(X)$, $\alpha \in \pi_q(X)$

$$\phi_q(\tau_\xi(\alpha)) = \phi_1(\xi) \cdot \phi_q(\alpha),$$

(2) for all $\alpha \in \pi_p(X)$, $\beta \in \pi_q(X)$, $p > 1, q > 1$, the Whitehead product $[\alpha, \beta]$ is zero.

8. (G. W. Whitehead [2]). Let $f: (\dot{\mathbf{I}}^{p+q+1}, \dot{\mathbf{I}}^{p+1} \times \dot{\mathbf{I}}^{q+1}) \to (X, *)$. Define maps f_+, $f_-: \dot{\mathbf{I}}^{p+q+1} \to X$ by

$$f_+ | \mathbf{I}^{p+1} \times \dot{\mathbf{I}}^{q+1} = f | \mathbf{I}^{p+1} \times \dot{\mathbf{I}}^{q+1},$$
$$f_+(\dot{\mathbf{I}}^{p+1} \times \mathbf{I}^{q+1}) = f_-(\mathbf{I}^{p+1} \times \dot{\mathbf{I}}^{q+1}) = *,$$
$$f_- | \dot{\mathbf{I}}^{p+1} \times \mathbf{I}^{q+1} = f | \dot{\mathbf{I}}^{p+1} \times \mathbf{I}^{q+1}.$$

Define maps $h_+ : (\mathbf{I}^{p+1}, \dot{\mathbf{I}}^{p+1}) \to (X, *)$, $h_- : (\mathbf{I}^{q+1}, \dot{\mathbf{I}}^{q+1}) \to (X, *)$ by

$$h_+(x) = f(x, *),$$
$$h_-(y) = f(*, y).$$

Let $\alpha, \alpha_+, \alpha_-, \beta_+, \beta_-$ be the elements of the appropriate homotopy groups of X represented by f, f_+, f_-, h_+, h_-, respectively. Prove that

$$\alpha = \alpha_+ + \alpha_- + [\beta_+, \beta_-].$$

9. Prove the biadditivity of the Whitehead product directly from the definition.

CHAPTER XI

Homotopy Operations

We have seen that the Eilenberg–MacLane spaces provide universal examples for cohomology operations. We shall also want to consider homotopy operations. The universal example for operations in one variable is a sphere; for operations in several variables it is a cluster

$$\Sigma = \mathbf{S}^{n_1} \vee \cdots \vee \mathbf{S}^{n_k}$$

of spheres. More precisely, let $\alpha \in \pi_n(\Sigma)$, and let $\beta_i \in \pi_{n_i}(X)$ $(i = 1, \ldots, k)$. Then there is a map $g : \Sigma \to X$ such that $g | \mathbf{S}^{n_i}$ represents β_i, and the correspondence

$$(\beta_1, \ldots, \beta_k) \to g_*(\alpha)$$

is a homotopy operation

$$\pi_{n_1}(X) \times \cdots \times \pi_{n_k}(X) \to \pi_n(X)$$

in k variables. Moreover, an easy naturality argument shows that every operation can be obtained in this way.

The homotopy groups of a cluster of spaces behave in a very complicated way, even if the spaces are spheres. A first result states that the group $\pi_n(X \vee Y)$ decomposes naturally into a direct sum $\pi_n(X) \oplus \pi_n(Y) \oplus \pi_{n+1}(X \times Y, X \vee Y)$. If X is $(p-1)$-connected and Y is $(q-1)$-connected then the summand $\pi_{n+1}(X \times Y, X \vee Y)$ vanishes if $n < p + q - 1$, and $\pi_{p+q}(X \times Y, X \vee Y) \approx \pi_p(X) \otimes \pi_q(Y)$ is imbedded in $\pi_{p+q-1}(X \vee Y)$ by the Whitehead product map.

Like cohomology operations, homotopy operations need not be additive. If $\alpha \in \pi_n(\mathbf{S}^r)$, and if $\iota^1, \iota^2 \in \pi_r(\mathbf{S}^r \vee \mathbf{S}^r)$ are the homotopy classes of the inclusions of S^r in the two summands of $\mathbf{S}^r \vee \mathbf{S}^r$, then

$$(\iota^1 + \iota^2) \circ \alpha = \iota^1 \circ \alpha + \iota^2 \circ \alpha + \partial_* \beta$$

for some $\beta \in \pi_{n+1}(\mathbf{S}^r \times \mathbf{S}^r, \mathbf{S}^r \vee \mathbf{S}^r)$, and the vanishing of the element β is necessary and sufficient for the operation defined by α to be additive. If $n = 2r - 1$, the group $\pi_{n+1}(\mathbf{S}^r \times \mathbf{S}^r, \mathbf{S}^r \vee \mathbf{S}^r)$ is an infinite cyclic group, generated by an element whose image under ∂_* is the Whitehead product $[\iota^1, \iota^2]$. Thus

$$(\iota^1 + \iota^2) \circ \alpha = \iota^1 \circ \alpha + \iota^2 \circ \alpha + H_0(\alpha)[\iota^1, \iota^2],$$

where $H_0(\alpha)$ is an integer, called the *Hopf invariant* of α. This notion was defined in a different way by Hopf, for $r = 2$ in [3] in 1931 and for general r in [5] in 1935. In the second paper he proved that the fibre maps which bear his name have Hopf invariant ± 1. Moreover there exist elements with Hopf invariant 2 for all even r. On the other hand, if r is odd, $H_0(\alpha)$ is always zero. These notions are discussed in §§1, 2.

The Hopf invariant cannot be described in terms of the usual concepts of homology theory, for if $f: \mathbf{S}^{2n-1} \to \mathbf{S}^r$, then the induced homomorphisms f_*, f^* of homology and cohomology groups are zero. In 1949 Steenrod [3] introduced new operations, the functional cup products, which suffice to describe H_0. Let $f: X \to Y$ and let $u \in H^p(Y)$, $v \in H^q(Y)$ be elements such that $f^*u = 0$ and $u \smile v = 0$. Then u and v determine an element $u \smile_f v$ of the quotient group $H^{p+q-1}(X)/f^*H^{p+q-1}(Y) + H^{p-1}(X) \smile f^*v$. If $X = \mathbf{S}^{2r-1}$, $Y = \mathbf{S}^r$, $u = v = \mathbf{s}^r$, then $f^* = 0$ and $u \smile_f v$ is a well-defined element of $H^{2r-1}(\mathbf{S}^{2r-1})$, which coincides up to sign with $H_0(\alpha)\mathbf{s}^{2r-1}$.

The Steenrod functional operations are introduced in §3 and the above application made in §4. Steenrod's construction of the functional operations did not come out of thin air, but was motivated by Hopf's original definition. Hopf considers a simplicial map of $\mathbf{S}^{2r-1}$ into $\mathbf{S}^r$. Then the counterimage of each well-chosen point x of $\mathbf{S}^r$ is a polyhedron of dimension $r - 1$, whose simplices can be oriented to obtain an $(r - 1)$-cycle $z(x)$. If $x \neq y$ then $z(x)$ and $z(y)$ are disjoint and their linking number $\lambda(z(x), z(y))$ is defined. Hopf proved that this number is independent of the choices made and depends only on the homotopy class of the map in question. Hopf's construction is presented in §5 and the connection with Steenrod's approach is established by making use of Poincaré duality in the manifolds $\mathbf{S}^{2r-1}$, $\mathbf{S}^r$.

In §6 we turn to the study of the homotopy groups of a cluster of spheres. These were determined in 1955 by Hilton [1] in terms of homotopy groups of spheres. Specifically, let $\Sigma = \mathbf{S}^{m_1+1} \vee \cdots \vee \mathbf{S}^{m_k+1}$, and let $\iota_j \in \pi_{m_j+1}(\Sigma)$ be the homotopy class of the inclusion of the jth summand. Then

$$\pi_{m+1}(\Sigma) \approx \bigoplus_{j=1}^{\infty} \pi_{m+1}(\mathbf{S}^{m_j+1})$$

the group $\pi_{m+1}(\mathbf{S}^{m_j+1})$ being imbedded in $\pi_{m+1}(\Sigma)$ by composition with a certain iterated Whitehead product $w_j(\iota_1, \ldots, \iota_k) \in \pi_{m_j+1}(\Sigma)$. A consequence is that all homotopy operations in several variables are generated by the composition and Whitehead product operations.

Hilton's proof was generalized by Milnor in 1956 by replacing the spheres $\mathbf{S}^{m_j+1}$ by suspensions $\mathbf{S}X_j$ $(j = 1, \ldots, k)$, and the remaining spheres $\mathbf{S}^{m_j+1}$ $(j > k)$ by iterated reduced joins $\mathbf{S}(X_{i_1} \wedge \cdots \wedge X_{i_r})$. Milnor's proof was never published, but has recently been reproduced by Adams in [4]. It is carried out in the semi-simplicial setting. A proof, inspired by Milnor's, in the topological category, has been given by Porter [1].

The algebraic notions needed to formulate the Hilton-Milnor Theorem are introduced in §6. The proof of the theorem itself is given in §7; its length is in great part due to the necessity of identifying the imbeddings which correspond to the Whitehead products in the case of spheres.

The Hilton decomposition of $\pi_{m+1}(\Sigma)$ has the following consequence. Let $\alpha \in \pi_{n+1}(\mathbf{S}^{r+1})$; then the element $(\iota_1 + \iota_2) \circ \alpha \in \pi_{n+1}(\mathbf{S}^{r+1} \vee \mathbf{S}^{r+1})$ has the decomposition

$$(\iota_1 + \iota_2) \circ \alpha = \iota_1 \circ \alpha + \iota_2 \circ \alpha + \sum_{j=3}^{\infty} w_j(\iota_1, \iota_2) \circ h_{j-3}(\alpha).$$

The elements $h_{j-3}(\alpha) \in \pi_{n+1}(\mathbf{S}^{m_j+1})$ are called *Hopf–Hilton invariants* of α. A few properties of the homomorphism h_j are worked out in §8. In particular, if $n = 2r$, then $h_j(\alpha) = 0$ for $j > 0$ and $h_0(\alpha) \in \pi_{2r+1}(\mathbf{S}^{2r+1})$ is determined by

$$h_0(\alpha) = H_0(\alpha)\iota.$$

Thus h_0 is a generalization of the Hopf invariant.

1 Homotopy Operations

Imitating the notion of cohomology operation, we may define a (primary) homotopy operation of type (n, r) to be a natural transformation $\theta : \pi_n \to \pi_r$. As these are not required to have any special algebraic properties (like additivity) it is convenient to take the range category $\bar{\mathscr{A}}$ to be that of abelian groups and arbitrary functions. The domain category may just as well be taken to be $\mathscr{K}_0$, the category of (compactly generated) spaces with base points. (There seems to be no particular advantage to start, as we did for cohomology, with the category $\mathscr{C}_0$ of CW-complexes with base points and then extend over all $\mathscr{K}_0$; however, the reader may verify that this program can be carried out if desired).

One example of such an operation comes immediately to mind. If $f : \mathbf{S}^r \to \mathbf{S}^n$, $g : \mathbf{S}^n \to X$ are maps, the homotopy class of their composition product $g \circ f$ depends only on the homotopy classes α, β of f, g respectively, and we may therefore define $\beta \circ \alpha$ to be that homotopy class. Then $\beta \circ \alpha = g_*(\alpha)$, and therefore

(1.1) *If* $\alpha_1, \alpha_2 \in \pi_r(\mathbf{S}^n)$, $\beta \in \pi_n(X)$, *then*

$$\beta \circ (\alpha_1 + \alpha_2) = \beta \circ \alpha_1 + \beta \circ \alpha_2 \in \pi_r(X). \qquad \square$$

On the other hand, it is not in general true that $\beta \circ \alpha$ depends linearly on β. (For an example, see §2).

If $\alpha \in \pi_r(\mathbf{S}^n)$ is a fixed element, then it is clear that the correspondence $\beta \to \beta \circ \alpha$ is a homotopy operation θ_α of type (n, r). Moreover,

(1.2) Theorem *If $\theta : \pi_n \to \pi_r$ is a homotopy operation, there is a unique element $\alpha \in \pi_r(\mathbf{S}^n)$ such that $\theta = \theta_\alpha$. Thus the set of all homotopy operations of type (n, r) is in one-to-one correspondence with $\pi_r(\mathbf{S}^n)$.*

For let $\iota_n \in \pi_n(\mathbf{S}^n)$ be the homotopy class of the identity map. If θ is an operation, then $\alpha = \theta(\iota_n) \in \pi_r(\mathbf{S}^n)$, and if $f : \mathbf{S}^n \to X$ represents $\beta \in \pi_n(X)$, we have

$$\theta_\alpha(\beta) = \beta \circ \alpha = f_*(\alpha) = f_*\theta(\iota_n) = \theta f_*(\iota_n) = \theta(\beta),$$

so that $\theta = \theta_\alpha$. On the other hand, if $\theta_\alpha = \theta_{\alpha'}$ then

$$\alpha = \theta_\alpha(\iota_n) = \theta_{\alpha'}(\iota_n) = \alpha'. \qquad \square$$

We shall need to consider operations in several variables. An *operation of type* $(n_1, \ldots, n_k; r)$ is a natural transformation $\theta : \pi_{n_1} \times \cdots \times \pi_{n_k} \to \pi_r$. Just as the n-sphere $\mathbf{S}^n$ is the "universal example" for operations in one variable, the cluster of spheres $\Sigma = \mathbf{S}^{n_1} \vee \cdots \vee \mathbf{S}^{n_k}$ is the universal example for operations in k variables, and we have the following generalization of Theorem (1.2).

(1.3) Theorem *The set of all operations of type $(n_1, \ldots, n_k; r)$ is in one-to-one correspondence with the group $\pi_r(\mathbf{S}^{n_1} \vee \cdots \vee \mathbf{S}^{n_k})$.*

For let $j_t : \mathbf{S}^{n_t} \to \Sigma$ be the inclusions, and let $\iota^t = j_{t*}(\iota_{n_t}) \in \pi_{n_t}(\Sigma)$. If $\beta_t \in \pi_{n_t}(X)$ is represented by $f_t : \mathbf{S}^{n_t} \to X$, the maps f_t define a map

$$f = (f_1, \ldots, f_k) : \Sigma \to X$$

such that $f \circ j_t = f_t$. Let $\alpha \in \pi_r(\Sigma)$, and define

$$\theta_\alpha(\beta_1, \ldots, \beta_k) = f_*(\alpha).$$

Then θ_α is an operation, and the proof of Theorem (1.2) extends easily to the present case. In particular, the element α is characterized by the formula

$$\alpha = \theta_\alpha(\iota^1, \ldots, \iota^k). \qquad \square$$

Thus it is of some importance to determine the homotopy groups $\pi_r(\Sigma)$, where Σ is a cluster of spheres, and the rest of this chapter will be devoted to the consideration of this problem.

We conclude this section by discussing the additivity problem for operations in one variable. In doing so, we shall need some information about the homotopy groups of a cluster of two spaces.

Consider the homotopy sequence

$$\textbf{(1.4)} \quad \cdots \to \pi_{r+1}(X \times Y) \xrightarrow{j_*} \pi_{r+1}(X \times Y, X \vee Y) \xrightarrow{\partial_*} \pi_r(X \vee Y) \xrightarrow{i_*} \pi_r(X \times Y) \to \cdots$$

and let us recall that the injections

$$i_{1*} : \pi_r(X) \to \pi_r(X \times Y),$$
$$i_{2*} : \pi_r(Y) \to \pi_r(X \times Y),$$

represent the group $\pi_r(X \times Y)$ as a direct sum; the dual representation as a direct product is given by

$$p_{1*} : \pi_r(X \times Y) \to \pi_r(X),$$
$$p_{2*} : \pi_r(X \times Y) \to \pi_r(Y).$$

Furthermore there are maps $j_1 : X \to X \vee Y, j_2 : Y \to X \vee Y, q_1 : X \vee Y \to X$, $q_2 : X \vee Y \to Y$ such that

$$i \circ j_t = i_t, \qquad p_t \circ i = q_t \qquad (t = 1, 2).$$

The homomorphism $\lambda : \pi_r(X \times Y) \to \pi_r(X \vee Y)$ defined by

$$\lambda = j_{1*} \circ p_{1*} + j_{2*} \circ p_{2*}$$

has the property that $p_{t*} \circ i_* \circ \lambda = p_{t*}$ $(t = 1, 2)$ and therefore $i_* \circ \lambda$ is the identity map. Hence

(1.5) Theorem *The injection*

$$j_* : \pi_r(X \times Y) \to \pi_r(X \times Y, X \vee Y)$$

is trivial, and the short exact sequence

(1.6) $$0 \to \pi_{r+1}(X \times Y, X \vee Y) \longrightarrow \pi_r(X \vee Y) \longrightarrow \pi_r(X \times Y) \to 0$$

splits. The injections

$$j_{1*} : \pi_r(X) \to \pi_r(X \vee Y),$$
$$j_{2*} : \pi_r(Y) \to \pi_r(X \vee Y),$$

together with the boundary operator

$$\partial_* : \pi_{r+1}(X \times Y, X \vee Y) \to \pi_r(X \vee Y)$$

represent the group $\pi_r(X \vee Y)$ as a direct sum:

$$\pi_r(X \vee Y) \approx \pi_r(X) \oplus \pi_r(Y) \oplus \pi_{r+1}(X \times Y, X \vee Y). \qquad \square$$

Remark. If $r = 1$, the above discussion must be modified slightly because $\pi_1(X)$ and $\pi_1(Y)$ are not necessarily abelian. In this case $\pi_1(X \vee Y)$ is the *free*, $\pi_1(X \times Y)$ the *direct* product of the groups $\pi_1(X)$, $\pi_1(Y)$. The sequence (1.6) remains exact, but does not, in general, split.

We next calculate the first non-vanishing relative homotopy group $\pi_{r+1}(X \times Y, X \vee Y)$.

(1.7) Theorem *Suppose that X is $(m-1)$-connected and Y is $(n-1)$-connected $(m, n \geq 2)$. Then $(X \times Y, X \vee Y)$ is $(m+n-1)$-connected and the map $\alpha \otimes \beta \to [\iota^1 \circ \alpha, \iota^2 \circ \beta]$ is an isomorphism of $\pi_m(X) \otimes \pi_n(Y)$ with the kernel of the injection $i_* : \pi_{m+n-1}(X \vee Y) \to \pi_{m+n-1}(X \times Y)$. Thus the injection $i_* : \pi_r(X \vee Y) \to \pi_r(X \times Y)$ is an isomorphism for $r < m+n-1$ and a split epimorphism with kernel $\pi_m(X) \otimes \pi_n(Y)$ for $r = m+n-1$.*

Since $m, n \geq 2$, X and Y are 1-connected, and so is $X \vee Y$, by the van-Kampen Theorem. By the Künneth Theorem, $H_r(X \times Y, X \vee Y) = 0$ for $r < m+n$, and

$$H_{m+n}(X \times Y, X \vee Y) \approx H_m(X, *) \otimes H_n(Y, *). \tag{1.8}$$

By the converses of the absolute and relative Hurewicz theorems, these groups are isomorphic with $\pi_{m+n}(X \times Y, X \vee Y)$ and $\pi_m(X) \otimes \pi_n(Y)$, respectively. It remains to make this isomorphism explicit. There is a commutative diagram

(1.9)

$$\begin{array}{ccccc}
\pi_m(X) \otimes \pi_n(Y) & \xrightarrow{\theta} & \pi_{m+n}(X \times Y, X \vee Y) & \xrightarrow{\partial_*} & \pi_{m+n-1}(X \vee Y) \\
\downarrow{\scriptstyle \rho \otimes \rho} & & \downarrow{\scriptstyle \rho} & & \downarrow{\scriptstyle \rho} \\
\mathrm{H}_m(X) \otimes H_n(Y) & \longrightarrow & H_{m+n}(X \times Y, X \vee Y) & \xrightarrow{\partial_*} & H_{m+n-1}(X \vee Y).
\end{array}$$

Let $f : (E_1, \dot{E}_1) \to (X, *)$, $g : (E_2, \dot{E}_2) \to (Y, *)$ be maps of oriented cells representing $\alpha \in \pi_m(X)$, $\beta \in \pi_n(Y)$, respectively. Then

$$f \times g : (E_1 \times E_2, (E_1 \times E_2)^{\bullet}) \to (X \times Y, X \vee Y)$$

represents an element $\alpha \times \beta \in \pi_{m+n}(X \times Y, X \vee Y)$ whose image under ρ is $(f \times g)_*(e_1 \times e_2) = f_* e_1 \times g_* e_2 = \rho(\alpha) \times \rho(\beta)$. Thus the correspondence $\alpha \otimes \beta \to \alpha \times \beta$ is the homomorphism θ of the diagram (1.9). Since $(f \times g) | (E_1 \times E_2)^{\bullet}$ is the representative h of $[\alpha, \beta]$ defined in §7, Chapter X, the truth of the rest of the theorem follows immediately. □

In the above discussion we assumed that X and Y were both 1-connected. If $m = n = 1$, then $\pi_2(X \times Y, X \vee Y)$ is the kernel N of the natural homomorphism of the free product $\pi_1(X) * \pi_1(Y)$ into the direct product $\pi_1(X) \times \pi_1(Y)$. By the theorem of Kurosh on subgroups of a free product, N is a free group, freely generated by the commutators $[j_{1*}(\alpha), j_{2*}(\beta)]$ where α, β range over all non-trivial elements of $\pi_1(X)$, $\pi_1(Y)$, respectively. The case $m = 1 < n$ we shall leave to the reader as an Exercise.

Let $\alpha \in \pi_r(\mathbf{S}^n)$. In order to study the additivity of the operation $\theta = \theta_\alpha$, let us consider the element $(\iota^1 + \iota^2) \circ \alpha \in \pi_r(\mathbf{S}^n \vee \mathbf{S}^n)$. By Theorem (1.5), this element can be represented in the form

$$(\iota^1 + \iota^2) \circ \alpha = \iota^1 \circ \alpha^1 + \iota^2 \circ \alpha^2 + \partial_* \beta;$$

projecting into the first and second factors, we find that $\alpha^1 = \alpha^2 = \alpha$. Thus

$$\textbf{(1.10)} \qquad (\iota^1 + \iota^2) \circ \alpha = \iota^1 \circ \alpha + \iota^2 \circ \alpha + \partial_* \beta.$$

If θ_α is additive, then $\iota^1 \circ \alpha + \iota^2 \circ \alpha = (\iota^1 + \iota^2) \circ \alpha$ and therefore $\beta = 0$. Conversely, suppose that $\beta = 0$. If $f_1, f_2 : \mathbf{S}^n \to X$ are maps representing $\beta_1, \beta_2 \in \pi_n(X)$, and if $f = (f_1, f_2) : \mathbf{S}^n \vee \mathbf{S}^n \to X$ is the map defined by them, then $f_* \iota^t = \beta_t$ for $t = 1, 2$, and therefore

$$(\beta_1 + \beta_2) \circ \alpha = f_*(\iota^1 + \iota^2) \circ \alpha = f_* \iota^1 \circ \alpha + f_* \iota^2 \circ \alpha = \beta_1 \circ \alpha + \beta_2 \circ \alpha,$$

so that θ_α is additive.

Let us say that α is *primitive* if and only if it satisfies (1.10) with $\beta = 0$. Thus we have proved

(1.11) Theorem *Let $\alpha \in \pi_r(\mathbf{S}^n)$. Then the operation $\theta_\alpha : \pi_n \to \pi_r$ defined by α is additive if and only if α is primitive.* □

(1.12) Corollary *If $r < 2n - 1$, then every homotopy operation of type (n, r) is additive.* □

Other sufficient conditions for additivity are given by Theorems (8.2) and (8.3) and Corollary (8.4) of Chapter X.

Suppose that $r = 2n - 1$. Then the image of $\partial_* : \pi_{2n}(\mathbf{S}^n \times \mathbf{S}^n, \mathbf{S}^n \vee \mathbf{S}^n) \to \pi_{2n-1}(\mathbf{S}^n \vee \mathbf{S}^n)$ is the infinite cyclic group generated by $[\iota^1, \iota^2]$. Thus (1.10) becomes

$$(\iota^1 + \iota^2) \circ \alpha = \iota^1 \circ \alpha + \iota^2 \circ \alpha + H_0(\alpha)[\iota^1, \iota^2],$$

where H_0 is a homomorphism of $\pi_{2n-1}(\mathbf{S}^n)$ into the additive group $\mathbf{Z}$ of integers. The integer $H_0(\alpha)$ is called the *Hopf invariant* of α.

It follows from the definition of H_0 that

(1.16) *If $\beta_1, \beta_2 \in \pi_n(X)$, $\alpha \in \pi_{2n-1}(\mathbf{S}^n)$, then*

$$(\beta_1 + \beta_2) \circ \alpha = \beta_1 \circ \alpha + \beta_2 \circ \alpha + H_0(\alpha)[\beta_1, \beta_2].$$ □

2 The Hopf Invariant

In this section we shall establish some elementary properties of the Hopf invariant. A more detailed discussion will appear in Volume II.

It is clear from the definition and the right-linearity of the composition operation that

(2.1) *The map $H_0 : \pi_{2n-1}(\mathbf{S}^n) \to \mathbf{Z}$ is a homomorphism, and $H_0(\alpha \circ (k\iota_{2n-1})) = kH_0(\alpha)$ for any integer k.* □

On the other hand,

(2.2) *For any integer k, $H_0((k\iota_n) \circ \alpha) = k^2 H_0(\alpha)$.*

For $\iota^t \circ (k\iota_n) \circ \alpha = (k\iota^t) \circ \alpha$, and therefore

$$\begin{aligned} H_0((k\iota_n) \circ \alpha)[\iota^1, \iota^2] &= (\iota^1 + \iota^2) \circ (k\iota_n) \circ \alpha \; - \iota^1 \circ (k\iota_n) \circ \alpha - \iota^2 \circ (k\iota_n) \circ \alpha \\ &= (k\iota^1 + k\iota^2) \circ \alpha - (k\iota^1) \circ \alpha - (k\iota^2) \circ \alpha \\ &= H_0(\alpha)[k\iota^1, k\iota^2] \\ &= k^2 H_0(\alpha)[\iota^1, \iota^2]; \end{aligned}$$

since $[\iota^1, \iota^2]$ has infinite order, the desired result follows. □

It follows from Theorem (1.13) that

(2.3) *If $\sigma \in \pi_{2n-2}(\mathbf{S}^{n-1})$, then $H_0(E\sigma) = 0$.* □

(2.4) Theorem *If n is odd then $H_0(\alpha) = 0$ for all $\alpha \in \pi_{2n-1}(\mathbf{S}^n)$.*

For

$$\begin{aligned} (\iota^1 + \iota^2) \circ \alpha - \iota^1 \circ \alpha - \iota^2 \circ \alpha &= H_0(\alpha)[\iota^1, \iota^2] \\ (\iota^2 + \iota^1) \circ \alpha - \iota^2 \circ \alpha - \iota^1 \circ \alpha &= H_0(\alpha)[\iota^2, \iota^1] \\ &= (-1)^n H_0(\alpha)[\iota^1, \iota^2]. \end{aligned}$$

But the left sides of these two relations are equal. Since $[\iota^1, \iota^2]$ has infinite order, we have $(1 + (-1)^{n+1})H_0(\alpha) = 0$, and therefore $H_0(\alpha) = 0$ if n is odd. □

On the other hand, if n is even, there always exist elements with Hopf invariant 2. In fact

(2.5) Theorem *If n is even, $H_0([\iota_n, \iota_n]) = 2$.*

For

$$\begin{aligned} H_0([\iota_n, \iota_n])[\iota^1, \iota^2] &= (\iota^1 + \iota^2) \circ [\iota_n, \iota_n] - \iota^1 \circ [\iota_n, \iota_n] - \iota^2 \circ [\iota_n, \iota_n] \\ &= [\iota^1 + \iota^2, \iota^1 + \iota^2] - [\iota^1, \iota^1] - [\iota^2, \iota^2] \\ &= [\iota^1, \iota^2] + [\iota^2, \iota^1] = 2[\iota^1, \iota^2] \end{aligned}$$

since n is even. □

We can now show that the composition operation fails to be additive in the first factor. We have seen that $\pi_3(\mathbf{S}^2)$ is an infinite cyclic group, generated by the homotopy class η of the Hopf fibration $p : \mathbf{S}^3 \to \mathbf{S}^2$. By Theorem (2.5), $H_0 : \pi_3(\mathbf{S}^2) \to \mathbf{Z}$ is non-zero, and it follows that H_0 is a monomorphism. By (2.2),

$$H_0((\iota_2 + \iota_2) \circ \eta) = H_0((2\iota_2) \circ \eta) = 4H_0(\eta).$$

Thus $(\iota_2 + \iota_2) \circ \eta = 4\eta$, while $\iota_2 \circ \eta + \iota_2 \circ \eta = 2\eta$.

Remark We shall see in Theorem (4.4) below that $H_0(\eta) = \pm 1$.

3 The Functional Cup Product

If $\alpha \in \pi_p(X)$, $\beta \in \pi_q(X)$, $(p, q > 1)$ and if $f : \mathbf{S}^{p+q-1} \to X$ is a representative of $[\alpha, \beta]$, then $f_* : H_{p+q-1}(\mathbf{S}^{p+q-1}) \to H_{p+q-1}(X)$ is trivial (because it can be factored through $H_{p+q-1}(\mathbf{S}^p \vee \mathbf{S}^q) = 0$). Thus the usual homological invariants associated with f must vanish. But this very triviality of f_* (and of f^*) can be used to construct new invariants, the Steenrod *functional cup products*, which we now describe.

We begin with some algebraic preliminaries (cf. Appendix B). Let

$$\Delta: \quad \begin{array}{ccccccccc} A_1 & \xrightarrow{\alpha_1} & A_2 & \xrightarrow{\alpha_2} & A_3 & \xrightarrow{\alpha_3} & A_4 & \xrightarrow{\alpha_4} & A_5 \\ \downarrow{\scriptstyle \lambda_1} & & \downarrow{\scriptstyle \lambda_2} & & \downarrow{\scriptstyle \lambda_3} & & \downarrow{\scriptstyle \lambda_4} & & \downarrow{\scriptstyle \lambda_5} \\ B_1 & \xrightarrow[\beta_1]{} & B_2 & \xrightarrow[\beta_2]{} & B_3 & \xrightarrow[\beta_3]{} & B_4 & \xrightarrow[\beta_4]{} & B_5 \end{array}$$

be a commutative diagram with exact rows in the category of abelian groups. Then $\beta_2^{-1} \circ \lambda_3 \circ \alpha_3^{-1} : A_4 \rightsquigarrow B_2$ is an additive relation Σ, and

$$\text{Dom } \Sigma = \text{Ker } \alpha_4 \cap \text{Ker } \lambda_4,$$
$$\text{Ker } \Sigma = \alpha_3(\text{Ker } \lambda_3),$$
$$\text{Im } \Sigma = \beta_2^{-1}(\text{Im } \lambda_3),$$
$$\text{Ind } \Sigma = \text{Im } \lambda_2 + \text{Im } \beta_1.$$

Thus Σ and Σ^{-1} induce homomorphisms

$$\sigma : \text{Ker } \alpha_4 \cap \text{Ker } \lambda_4 \to B_2/(\text{Im } \beta_1 + \text{Im } \lambda_2),$$
$$\tau : \beta_2^{-1}(\text{Im } \lambda_3) \to A_4/\alpha_3(\text{Ker } \lambda_3).$$

The relation Σ (as well as the associated homomorphism σ) is called the *suspension*, the relation Σ^{-1} (as well as the homomorphism τ) is the *transgression*, associated with the diagram Δ.

EXAMPLE 1. Let $p : X \to B$ be a fibration with fibre F. Then p^* maps the cohomology sequence of $(B, *)$ into that of (X, F), yielding a commutative diagram

$$\begin{array}{ccccccccc}
H^{q-1}(B) & \longrightarrow & H^{q-1}(*) & \longrightarrow & H^q(B, *) & \xrightarrow{j^*} & H^q(B) & \xrightarrow{i^*} & H^q(*) \\
\downarrow & & \downarrow & & \downarrow{\scriptstyle p_0^*} & & \downarrow{\scriptstyle p^*} & & \downarrow \\
H^{q-1}(X) & \xrightarrow[i^*]{} & H^{q-1}(F) & \xrightarrow[\delta^*]{} & H^q(X, F) & \longrightarrow & H^q(X) & \longrightarrow & H^q(F)
\end{array}$$

and the suspension and transgression are homomorphisms

$$\sigma^* : \text{Ker } p^* \to H^{q-1}(F)/\text{Im } i^*,$$

$$\tau^* : \delta^{*-1}(\text{Im } p_0^*) \to H^q(B)/\text{Ker } p^*,$$

respectively. Dually, the commutative diagram

$$\begin{array}{ccccccccc}
H_q(F) & \xrightarrow{i_*} & H_q(X) & \xrightarrow{j_*} & H_q(X, F) & \xrightarrow{\partial_*} & H_{q-1}(F) & \xrightarrow{i_*} & H_{q-1}(X) \\
\downarrow & & \downarrow{\scriptstyle p_*} & & \downarrow{\scriptstyle p_{0*}} & & \downarrow & & \downarrow \\
H_q(*) & \xrightarrow[i_*]{} & H_q(B) & \xrightarrow[j_*]{} & H_q(B, *) & \xrightarrow[\partial_*]{} & H_{q-1}(*) & \xrightarrow[i_*]{} & H_{q-1}(B)
\end{array}$$

yields suspension and transgression homomorphisms

$$\sigma : \text{Ker } i_* \to H_q(B)/p_* H_q(X)$$

$$\tau : j_*^{-1}(\text{Im } p_{0*}) \to H_{q-1}(F)/\partial_*(\text{Ker } p_{0*}).$$

EXAMPLE 2. Let $(X; A, B)$ be a proper triad. Then there is a commutative diagram

$$\begin{array}{ccccccccc}
H^{q-1}(X) & \to & H^{q-1}(B) & \to & H^q(X, B) & \to & H^q(X) & \to & H^q(B) \\
\downarrow & & \downarrow & & \downarrow{\scriptstyle \lambda^*} & & \downarrow & & \downarrow \\
H^{q-1}(A) & \to & H^{q-1}(A \cap B) & \to & H^q(A, A \cap B) & \to & H^q(A) & \to & H^q(A \cap B)
\end{array}$$

In this case λ^* is an isomorphism, and therefore the transgression

$$\Sigma^{-1} = \tau^* : H^{q-1}(A \cap B) \to H^q(X)$$

is a homomorphism, the coboundary operator or *seam homomorphism* of the Mayer–Vietoris cohomology sequence of the triad $(X; A, B)$. For the dual diagram, the suspension is a homomorphism

$$\Sigma = \sigma_* : H_q(X) \to H_{q-1}(A \cap B),$$

which is again the seam homomorphism of the Mayer–Vietoris homology sequence of $(X; A, B)$.

EXAMPLE 3. Let R be a commutative ring with unit; all cohomology groups will have coefficients in R. Let $f: X \to Y$ and let $v \in H^q(Y)$.

Consider the diagram

$$\begin{array}{ccccccccc} H^{p-1}(Y) & \xrightarrow{f^*} & H^{p-1}(X) & \xrightarrow{\delta^*} & H^p(\mathbf{I}_f, X) & \xrightarrow{k^*} & H^p(Y) & \xrightarrow{f^*} & H^p(X) \\ \Delta(f, v): \ \downarrow \lambda_1 & & \downarrow \lambda_2 & & \downarrow \lambda_3 & & \downarrow \lambda_1 & & \downarrow \lambda_2 \\ H^{p+q-1}(Y) & \xrightarrow[f^*]{} & H^{p+q-1}(X) & \xrightarrow[\delta^*]{} & H^{p+q}(\mathbf{I}_f, X) & \xrightarrow[k^*]{} & H^{p+q}(Y) & \longrightarrow & H^{p+q}(X) \end{array}$$

where k^* is the injection and the λ_i are determined by the cup product with v as follows: $\lambda_1(u) = u \smile v$, $\lambda_2(u) = u \smile f^*v$, and $\lambda_3(u) = u \smile p^*v$ where $p: \mathbf{I}_f \to Y$ is the projection and the last-mentioned cup product pairs $H^p(\mathbf{I}_f, X)$ with $H^q(\mathbf{I}_f)$ to $H^{p+q}(\mathbf{I}_f, X)$. The diagram is then commutative, and the suspension homomorphism

$$\sigma^* : \operatorname{Ker} f^* \cap \operatorname{Ker} \lambda_1 \to H^{p+q-1}(X)/f^*H^{p+q-1}(Y) + \lambda_2 H^{p-1}(X)$$

is defined. If $u \in \operatorname{Dom} \sigma^*$, so that $f^*u = 0$ and $u \smile v = 0$, let

$$u \smile_f v = \sigma^*(u) \in H^{p+q-1}(X)/f^*H^{p+q-1}(Y) + H^{p-1}(X) \smile f^*v;$$

the element $u \smile_f v$ is called the *functional cup product* of u and v. Since the homotopy type of the pair $(\mathbf{I}_f, X)$ depends only on the homotopy class of f, we have

(3.1) *If $f, g: X \to Y$ are homotopic maps, then $u \smile_f v = u \smile_g v$.* □

Therefore we may, and frequently shall, denote by $u \smile_\alpha v$ the element $u \smile_f v$ for any representative f of $\alpha \in [X, Y]$.

The functional cup product has certain naturality properties, which we now explain. Let $f: X \to Y$, $g: Y \to Z$, and let $u \in H^p(Z)$, $v \in H^q(Z)$, $u \smile v = 0$. If $f^*g^*u = 0$, then $u \smile_{g \circ f} v$ is defined and belongs to $H^{p+q-1}(X)/f^*g^*H^{p+q-1}(Z) + H^{p-1}(X) \smile f^*g^*v$. On the other hand, $g^*u \smile g^*v = 0$, and therefore $g^*u \smile_f g^*v$ is defined and belongs to $H^{p+q-1}(X)/f^*H^{p+q-1}(Y) + H^{p-1}(X) \smile f^*g^*v$. Thus the element $u \smile_{g \circ f} v$ is a coset of the subgroup

$$G = f^*g^*H^{p+q-1}(Z) + H^{p-1}(X) \smile f^*g^*v,$$

and $g^*u \smile_f g^*v$ is a coset of the subgroup

$$G' = f^*H^{p+q-1}(Y) + H^{p-1}(X) \smile f^*g^*v.$$

Since $G \subset G'$, the following statement is meaningful (and true):

(3.2) *Under the above conditions*

$$u \smile_{g \circ f} v \subset g^*u \smile_f g^*v.$$

In fact, the identity map of $\mathbf{I} \wedge X$, together with the map $g : Y \to Z$ induce a map of $(\mathbf{I} \wedge X) \vee Y$ into $(\mathbf{I} \wedge X) \vee Z$ which is compatible with the identification maps

$$(\mathbf{I} \wedge X) \vee Y \to \mathbf{I}_f$$
$$(\mathbf{I} \wedge X) \vee Z \to \mathbf{I}_{g \circ f}$$

and so induce a map $\bar{g} : (\mathbf{I}_f;\, X, Y) \to (\mathbf{I}_{g \circ f};\, X, Z)$. The map $\bar{g}$ induces, in turn, a map of the diagram $\Delta(g \circ f, v)$ into the diagram $\Delta(f, g^*v)$, which reduces to the identity on the cohomology groups of X, and our result follows from (2.3) of Appendix *B*, with β_{n+1} the identity map. □

Let X, Y, Z, u, v be as before, but let us make the stronger assumption that $g^*u = 0$. Then $u \smile_g v$ is defined and belongs to the group $H^{p+q-1}(Y)/g^*H^{p+q-1}(Z) + H^{p-1}(Y) \smile g^*v$. On the other hand, the element $u \smile_{g \circ f} v$ is also defined and belongs to the group $H^{p+q-1}(X)/f^*g^*H^{p+q-1}(Z) + H^{p-1}(X) \smile f^*g^*v$. The element $u \smile_g v$ is a coset of the subgroup

$$G = g^*H^{p+q-1}(Z) + H^{p-1}(Y) \smile g^*v$$

of $H^{p+q-1}(Y)$, which is mapped by f^* into the subgroup

$$G' = f^*g^*H^{p+q-1}(Z) + H^{p-1}(X) \smile f^*g^*v,$$

of $H^{p+q-1}(X)$. As $u \smile_{g \circ f} v$ is a coset of the latter subgroup, the following statement is meaningful (and true):

(3.3) *Under the above conditions*

$$f^*(u \smile_g v) \subset u \smile_{g \circ f} v.$$

This time the map $1 \wedge f : \mathbf{I} \wedge X \to \mathbf{I} \wedge Y$ and the identity map of Z are compatible with the identification maps

$$(\mathbf{I} \wedge X) \vee Z \to \mathbf{I}_{g \circ f}$$
$$(\mathbf{I} \wedge Y) \vee Z \to \mathbf{I}_g$$

and so induce a map $\bar{f} : (\mathbf{I}_{g \circ f};\, X, Z) \to (\mathbf{I}_g;\, Y, Z)$. This map, in turn, induces a map of the diagram $\Delta(g, v)$ into the diagram $\Delta(g \circ f, v)$ which reduces to the identity on the cohomology groups of Z. Our result now follows from (2.3) of Appendix B, with α_{n-1} the identity map. □

Now let X be an H′-space with coproduct $\theta : X \to X \vee X$. Let α, $\beta \in [X, Y]$, and let f, g, $h : X \to Y$ be representatives of α, β, and $\alpha + \beta$, respectively. Let $u \in H^p(Y)$, $v \in H^q(Y)$, and suppose that $f^*u = g^*u = 0$. By

Theorem (7.10*) of Chapter III, $h^*u = 0$. Then the elements $u \smile_f v$, $u \smile_g v$, and $u \smile_h v$ are defined and belong to the quotients of $H^{p+q-1}(X)$ by the subgroups

$$G_1 = f^*H^{p+q-1}(Y) + H^{p-1}(X) \smile f^*v,$$
$$G_2 = g^*H^{p+q-1}(Y) + H^{p-1}(X) \smile g^*v,$$
$$G = (f^* + g^*)H^{p+q-1}(Y) + H^{p-1}(X) \smile (f^*v + g^*v).$$

Evidently $G \subset G_1 + G_2$. Therefore the following statement is meaningful (and true):

(3.4) *Under the above conditions*

$$u \smile_{\alpha+\beta} v \subset u \smile_\alpha v + u \smile_\beta v.$$

For $h = k \circ \theta$, where $k = \nabla \circ (f \vee g)$. Now $k \circ j_1 = f$, $k \circ j_2 = g$, and therefore

$$j_1^* k^* u^* = f^*u = 0, \qquad j_2^* k^* u^* = g^*u = 0;$$

hence $k^*u = 0$. Therefore $u \smile_k v$ is defined and belongs to $H^{p+q-1}(X \vee X)/k^*H^{p+q-1}(Y) + H^{p-1}(X \vee X) \smile k^*v$; and hence $\theta^*(u \smile_k v)$ belongs to $H^{p+q-1}(X)/\theta^*k^*H^{p+q-1}(Y) + \theta^*(H^{p-1}(X \vee X) \smile k^*v)$; by (7.8*) of Chapter III, $\theta^*(H^{p-1}(X \vee X) \smile k^*v) = \theta^*H^{p-1}(X \vee X) \smile \theta^*k^*v = 0$. Since $\theta^* \circ k^* = h^*$, we see that $\theta^*(u \smile_k v)$ is a coset of the subgroup $h^*H^{p+q-1}(Y)$. On the other hand, $u \smile_h v$ is a coset of the same subgroup. Hence the inclusion guaranteed by (3.3) is an equality:

$$\theta^*(u \smile_k v) = u \smile_h v.$$

But $\theta^* = j_1^* + j_2^*$ and therefore

$$\begin{aligned}\theta^*(u \smile_k v) &= (j_1^* + j_2^*)(u \smile_k v)\\ &\subset j_1^*(u \smile_k v) + j_2^*(u \smile_k v)\\ &\subset u \smile_f v + u \smile_g v\end{aligned}$$

by another application of (3.3). □

We next show how $u \smile_f v$ can be computed in terms of singular cochains. Let U, V be cochains of Y representing u, v, respectively. Then $U \smile V = \delta A$ with $A \in C^{p+q-1}(Y)$ and $f^\# U = \delta B$ with $B \in C^{p-1}(X)$. Then

$$\begin{aligned}\delta(B \smile f^\# V) &= \delta B \smile f^\# V = f^\# U \smile f^\# V = f^\#(U \smile V)\\ &= f^\# \delta A = \delta f^\# A\end{aligned}$$

and therefore $Z = f^\# A - B \smile f^\# V$ is a cocycle of X.

(3.5) Lemma *The cohomology class of Z is a representative of $u \smile_f v$.*

Let $i : X \subsetneq \mathbf{I}_f$, so that $p \circ i = f$, and let $k : Y \subsetneq \mathbf{I}_f$, so that $p \circ k = 1$. Now

$$i^{\#}p^{\#}U = f^{\#}U = \delta B;$$

choose $C \in C^{p-1}(\mathbf{I}_f)$ with $i^{\#}C = B$. Then

$$i^{\#}(p^{\#}U - \delta C) = \delta B - \delta i^{\#}C = 0,$$

so that $p^{\#}U - \delta C$ is a cocycle of $\mathbf{I}_f$ mod X; moreover, $k^{\#}(p^{\#}U - \delta C) = U - \delta k^{\#}C \sim U$ in Y. Hence the cohomology class z of $p^{\#}U - \delta C$ is mapped by k^* into u. Then $(p^{\#}U - \delta C) \smile p^{\#}V$ is a cocycle of $\mathbf{I}_f$ mod X representing $\lambda_3(z)$. But

$$\begin{aligned}(p^{\#}U - \delta C) \smile p^{\#}V &= p^{\#}(U \smile V) - \delta(C \smile p^{\#}V)\\ &= p^{\#}(\delta A) - \delta(C \smile p^{\#}V)\\ &= \delta\{p^{\#}A - C \smile p^{\#}V\}\end{aligned}$$

and

$$i^{\#}(p^{\#}A - C \smile p^{\#}V) = f^{\#}A - B \smile f^{\#}V = Z,$$

so that δ^* maps the homology class of Z into $\lambda_3(z)$, and our result follows. □

With a little care, a similar result can be proved if X and Y are simplicial complexes and $f: X \to Y$ a simplicial map. In order to calculate cup products we need to order the vertices of X and of Y; and for the cochain map $f^{\#}$ to preserve cup products we need f to preserve the order of the vertices. But this is easily achieved; just order the vertices of Y arbitrarily by ordering $f^{-1}(y)$ arbitrarily for each vertex y of Y and then demanding that $f(x_1) < f(x_2)$ implies $x_1 < x_2$ for vertices x_1, x_2 of X. We then extend the ordering of X and Y to one of $\mathbf{I}_f$ by specifying that $x < y$ for each pair of vertices $x \in X$, $y \in Y$. Then the statement and proof of Lemma (3.5) go through just as in the singular case, and we have

(3.6) *The conclusion of Lemma* (3.5) *holds in the simplicial case.* □

Here is an interesting example of a functional cup product. Let $h : \mathbf{S}^{p+q-1} \to \mathbf{S}^p \vee \mathbf{S}^q$ be a representative of $[\iota^1, \iota^2]$, and let $u = q_1^* \mathbf{s}^p$, $v = q_2^* \mathbf{s}^q$. In this case $h^*u = 0$ and $u \smile v = 0$, so that $u \smile_h v$ is defined. Moreover, $h^*v = 0$ and $h^*H^{p+q-1}(\mathbf{S}^p \vee \mathbf{S}^q) = 0$. Therefore $u \smile_h v$ is a uniquely defined element of $H^{p+q-1}(\mathbf{S}^{p+q-1})$, and we have

(3.7) Theorem *In the group* $H^{p+q-1}(\mathbf{S}^{p+q-1})$, $u \smile_h v = -\mathbf{s}^{p+q-1}$.

Let $g : (\mathbf{E}^{p+q}, \mathbf{S}^{p+q-1}) \to (\mathbf{S}^p \times \mathbf{S}^q, \mathbf{S}^p \vee \mathbf{S}^q)$ be a simplicial map (with respect to appropriate subdivisions of the spaces involved) of degree $+1$; then $g|\mathbf{S}^{p+q-1} : \mathbf{S}^{p+q-1} \to \mathbf{S}^p \vee \mathbf{S}^q$ is a representative of $[\iota^1, \iota^2]$. Let U, V be cochains representing u, v, respectively. Then $h^{\#}U = \delta B$, and $U \smile V = 0$, so that we may take $A = 0$. Thus $-B \smile h^{\#}V$ represents $u \smile_h v$. Thus we must prove that $B \smile h^{\#}V$ represents $\mathbf{s}^{p+q-1}$. Now $\delta^*\mathbf{s}^{p+q-1} = \mathbf{e}^{p+q} = g^*(\mathbf{s}^p \times \mathbf{s}^q)$. Since $\delta^* : H^{p+q-1}(\mathbf{S}^{p+q-1}) \to H^{p+q}(\mathbf{E}^{p+q}, \mathbf{S}^{p+q-1})$ is an isomorphism it suffices to show that $g^*(\mathbf{s}^p \times \mathbf{s}^q) = \delta^*(u \smile_h v)$.

Let $i : \mathbf{S}^p \vee \mathbf{S}^q \subsetneq \mathbf{S}^p \times \mathbf{S}^q$, $j : \mathbf{S}^{p+q-1} \subsetneq \mathbf{E}^{p+q}$. Since $i^* : H^r(\mathbf{S}^p \times \mathbf{S}^q) \to H^r(\mathbf{S}^p \vee \mathbf{S}^q)$ is an isomorphism for $r = p, q$, there are cocycles U', V' such that $i^{\#}U' = U$, $i^{\#}V' = V$. Choose a cochain B' such that $j^{\#}B' = B$. Then $U' \smile V'$ represents $\mathbf{s}^p \times \mathbf{s}^q$, so that $g^{\#}(U' \smile V')$ represents $g^*(\mathbf{s}^p \times \mathbf{s}^q)$. But
sents $g^*(\mathbf{s}^p \times \mathbf{s}^q)$. But

$$
\begin{aligned}
j^{\#}g^{\#}(U' \smile V') &= j^{\#}(g^{\#}U' \smile g^{\#}V') = j^{\#}g^{\#}U' \smile j^{\#}g^{\#}V' \\
&= h^{\#}i^{\#}U' \smile h^{\#}i^{\#}V' = h^{\#}U \smile h^{\#}V \\
&= \delta B \smile h^{\#}V = \delta(B \smile h^{\#}V)
\end{aligned}
$$

and therefore $g^*(\mathbf{s}^p \times \mathbf{s}^q) = \delta^*(u \smile_h v)$. □

Remark. In the proof of Theorem (3.7) we used simplicial cochains rather than singular ones, because the cup product $U \smile V$ is actually zero, there being no simplicial chains of dimension $p + q$ in $\mathbf{S}^p \vee \mathbf{S}^q$.

4 The Hopf Construction

The join of two spaces X, Y is the subspace

$$X * Y = \mathbf{T}X \times Y \cup X \times \mathbf{T}Y$$

of $\mathbf{T}X \times \mathbf{T}Y$. The *Hopf construction* assigns to each map $f: X \times Y \to Z$ the map $g = \Gamma f: X * Y \to \mathbf{S}Z$ such that

$$
\begin{aligned}
g(x, t \wedge y) &= \varpi(t/2) \wedge f(x, y), \\
g(t \wedge x, y) &= \varpi((1 + t)/2) \wedge f(x, y).
\end{aligned}
$$

Thus g maps the triad $(X * Y; X \times \mathbf{T}Y, \mathbf{T}X \times Y)$ into the triad $(\mathbf{S}Z; \mathbf{T}_+ Z, \mathbf{T}_- Z)$, where $\mathbf{T}_+ Z$, $\mathbf{T}_- Z$ are copies of the cone $\mathbf{T}Z$ defined by

$$
\begin{aligned}
\mathbf{T}_+ Z &= \{\varpi(t) \wedge z \mid 0 \le t \le 1/2\}, \\
\mathbf{T}_- Z &= \{\varpi(t) \wedge z \mid 1/2 \le t \le 1\},
\end{aligned}
$$

and therefore induces a homomorphism of the cohomology Mayer–Vietoris sequence of the second triad into that of the first. In particular, there is a commutative diagram

$$\begin{array}{ccc} H^r(Z) & \xrightarrow{f^*} & H^r(X \times Y) \\ \Delta^* \downarrow & & \downarrow \Delta^* \\ H^{r+1}(\mathbf{S}Z) & \xrightarrow[g^*]{} & H^{r+1}(X * Y) \end{array}$$

where the Δ^* are the seam homomorphisms of the appropriate Mayer–Vietoris sequences.

The correspondence Γ is evidently compatible with homotopy and so induces a map $\Gamma : [X \times Y, Z] \to [X * Y, \mathbf{S}Z]$.

Remark 1. There are two reasonable ways to define Δ_*; they differ only in sign. We shall choose the sign as in Example 2 of §3; this agrees with Eilenberg–Steenrod [E–S] and Spanier [Sp], but disagrees with Greenberg [Gr]. The reason for this is that the seam homomorphism of the Mayer–Vietoris sequence of the triad $(\mathbf{S}^{n+1}; \mathbf{E}_+^{n+1}, \mathbf{E}_-^{n+1})$ preserves orientatation.

Remark 2. Let X and Y be CW-complexes and let $f : X \times Y \to Z$. If g' is any map of $(X * Y;\ X \times \mathbf{T}Y,\ \mathbf{T}X \times Y)$ into $(\mathbf{S}Z;\ \mathbf{T}_+ Z,\ \mathbf{T}_- Z)$ such that $g' | X \times Y = f$, then g' is homotopic to g. For $\mathbf{T}_+ Z$ is contractible and therefore $g | X \times \mathbf{T}Y$ and $g' | X \times \mathbf{T}Y$ are homotopic (rel. $X \times Y$); similarly $g | \mathbf{T}X \times Y \simeq g' | \mathbf{T}X \times Y$ (rel. $X \times Y$). These homotopies then fit together to yield a homotopy of g to g'.

Remark 3. Let X and Y be oriented spheres of dimension p, q, respectively. Then $X * Y$ is a sphere of dimension $p + q + 1$, which we may orient by requiring that $\Delta_* : H_{p+q+1}(X * Y) \to H_{p+q}(X \times Y)$ shall preserve orientation ($X \times Y$ being oriented by the cross-product of the orientations of X and Y).

EXAMPLE 1. Let D be one of the standard division algebras, viz. $\mathbf{R}$, $\mathbf{C}$, $\mathbf{Q}$, $\mathbf{K}$, and let d be the dimension of D over R. Let f: $\mathbf{S}^{d-1} \times \mathbf{S}^{d-1} \to \mathbf{S}^{d-1}$ be the map given by

$$f(x, y) = x^{-1}y;$$

then $g : \mathbf{S}^{2d-1} \to \mathbf{S}^d$ is the Hopf fibre map, up to homotopy. For we may represent $\mathbf{S}^{2d-1}$ as the set of all pairs $(x, y) \in D \times D$ such that $\|x\|^2 + \|y\|^2 = 1$; in this representation the sets $\mathbf{S}^{d-1} \times \mathbf{E}^d$ and $\mathbf{E}^d \times \mathbf{S}^{d-1}$ correspond to the subsets defined by the inequalities $\|x\| \le \|y\|$ and $\|x\| \ge \|y\|$, respectively; and the point $(x, y) \in \mathbf{S}^{d-1} \times \mathbf{S}^{d-1}$ corresponds to the point $(1/\sqrt{2}\,x, 1/\sqrt{2}\,y)$ of $\mathbf{S}^{2d-1}$. Similarly, we may represent $\mathbf{S}^d$ as the one-point compactification of $D = \mathbf{R}^d$; in this representation the sets $\mathbf{T}_+\mathbf{S}^{d-1}$ and $\mathbf{T}_-\mathbf{S}^{d-1}$ correspond respectively to the closure of the exterior and interior of the unit sphere. Finally, we may identify $\mathbf{P}^1(D)$ with $\mathbf{S}^d$ under the correspondence

$$[x, y] \to \begin{cases} x^{-1}y & \text{if } x \neq 0, \\ \infty & \text{if } x = 0. \end{cases}$$

The Hopf map sends $(x, y) \in \mathbf{S}^{2d-1}$ into the point $x^{-1}y$ or 1 of $\mathbf{S}^d$. With the identifications we have made, the triad $(\mathbf{S}^{2d-1}; \mathbf{S}^{d-1} \times \mathbf{E}^d, \mathbf{E}^d \times \mathbf{S}^{d-1})$ is sent into the triad $(\mathbf{S}^d; \mathbf{T}_+\mathbf{S}^{d-1}, \mathbf{T}_-\mathbf{S}^{d-1})$; and the map agrees with f on $\mathbf{S}^{d-1} \times \mathbf{S}^{d-1}$. By Remark 1, above, this map is homotopic to g.

Example 2. Let $\alpha \in \pi_r(\mathbf{O}_n)$ be represented by $h : \mathbf{S}^r \to \mathbf{O}_n$. We may consider $\mathbf{O}_n$ as a subspace of the function space $\mathbf{F}(\mathbf{S}^{n-1}, \mathbf{S}^{n-1})$. Therefore the adjoint of h is a map $f : \mathbf{S}^r \times \mathbf{S}^{n-1} \to \mathbf{S}^{n-1}$. Then $g = \Gamma f : \mathbf{S}^{r+n} \to \mathbf{S}^n$ and the correspondence $h \to g$ is compatible with homotopy and induces a function $J : \pi_r(\mathbf{O}_n) \to \pi_{r+n}(\mathbf{S}^n)$.

(4.1) Theorem *The map* $J : \pi_r(\mathbf{O}_n) \to \pi_{r+n}(\mathbf{S}^n)$ *is a homomorphism.*

For suppose that $h_i : \mathbf{S}^r \to \mathbf{O}_n$ represent $\alpha_i \in \pi_r(\mathbf{O}_n)$ $(i = 1, 2)$; we take the identity map as base point. The corresponding maps $f_i : \mathbf{S}^r \times \mathbf{S}^{n-1} \to \mathbf{S}^{n-1}$ have the property that $f_i(*, y) = y$ for all $y \in \mathbf{S}^{n-1}$. Let us assume, as we may, that $h_1(\mathbf{T}_-\mathbf{S}^{r-1}) = h_2(\mathbf{T}_+\mathbf{S}^{r-1}) = 1$. Then the map $h_0 : \mathbf{S}^r \to \mathbf{O}_n$ defined by $h_0 | \mathbf{T}_+\mathbf{S}^{r-1} = h_1 | \mathbf{T}_+\mathbf{S}^{r-1}$, $h_0 | \mathbf{T}_-\mathbf{S}^{r-1} = h_2 | \mathbf{T}_-\mathbf{S}^{r-1}$, represents $\alpha_1 + \alpha_2$. The corresponding maps $g_0, g_1, g_2 : \mathbf{S}^{n+r} \to \mathbf{S}^n$ are related by

$$g_0 | \mathbf{T}_+\mathbf{S}^{r-1} \times \mathbf{S}^{n-1} = g_1 | \mathbf{T}_+\mathbf{S}^{r-1} \times \mathbf{S}^{n-1},$$

$$g_0 | \mathbf{T}_-\mathbf{S}^{r-1} \times \mathbf{S}^{n-1} = g_2 | \mathbf{T}_-\mathbf{S}^{r-1} \times \mathbf{S}^{n-1},$$

$$g_1(\pi x, y) = g_2(x, y) = y \quad \text{for } x \in \mathbf{T}_+\mathbf{S}^{r-1} \times \mathbf{S}^{n-1},$$

where π is the obvious homeomorphism of $\mathbf{T}_+\mathbf{S}^{n-1}$ with $\mathbf{T}_-\mathbf{S}^{r-1}$. It follows from Lemma (5.11) of Chapter V that the elements γ_i represented by g_i $(i = 0, 1, 2)$, are related by

$$\gamma_0 = \gamma_1 + \gamma_2 .$$

But $\gamma_i = J(\alpha_i)$. □

Let $f : X \times Y \to Z$; an element $w \in H^r(Z)$ is said to be *f-primitive* if and only if $f^*w = u \times 1 + 1 \times v$ for some $u \in H^r(X)$, $v \in H^r(Y)$. Then there is a commutative diagram

$$\begin{array}{ccc} H^r(Z) & \xrightarrow{\ f^*\ } & H^r(X \times Y) \\ \Big\downarrow{\scriptstyle \Delta^*} & & \Big\downarrow{\scriptstyle \Delta^*} \\ H^{r+1}(\mathbf{S}Z) & \xrightarrow[\ g^*\ ]{} & H^{r+1}(X * Y) \end{array}$$

and it is clear that w is f-primitive if and only if $\Delta^* f^* w = 0$. Then $g^*\Delta^* w = 0$, and $\Delta^* w \smile \Delta^* w = 0$ because $\mathbf{S}Z$ is an H′-space, by (7.8*) of Chapter III. Therefore $\Delta^* w \smile_g \Delta^* w$ is defined and belongs to

$$H^{2r+1}(X * Y)/g^* H^{2r+1}(\mathbf{S}Z).$$

(4.2) Theorem *Let $f: X \times Y \to Z$, and let $w \in H^r(Z)$ be f-primitive: $f^*w = u \times 1 + 1 \times v \in H^r(X \times Y)$. Let $g = \Gamma f: X * Y \to \mathbf{S}Z$. Then*

$$(-1)^r \Delta^*(u \times v) \in \Delta^* w \smile_g \Delta^* w \in H^{2r+1}(X * Y)/g^* H^{2r+1}(\mathbf{S}Z).$$

Using the technique of CW-approximation, we may assume that the triads $(X * Y; X \times \mathbf{T}Y; \mathbf{T}X \times Y)$ and $(\mathbf{S}Z; \mathbf{T}_+ Z, \mathbf{T}_- Z)$ have been triangulated and that

$$g: (X * Y; X \times \mathbf{T}Y, \mathbf{T}X \times Y) \to (\mathbf{S}Z; \mathbf{T}_+ Z, \mathbf{T}_- Z)$$

is a simplicial map and therefore that $f \mid X \times Y: X \times Y \to Z$ is simplicial. It will be convenient to make the following convention: If Q is a subcomplex of P, $i: Q \subsetneq P$, and if $X \in C^r(P)$, then $X \mid Q$ is an abbreviation for $X \mid C_r(Q) = i^\# X \in C^r(Q)$.

Let $W \in Z^r(Z)$ and choose $C \in C^r(\mathbf{S}Z)$ such that $C \mid Z = W$. Define X_+, $X_- \in C^{r+1}(\mathbf{S}Z)$ by

$$X_+ \mid \mathbf{T}_+ Z = \delta C \mid \mathbf{T}_+ Z, \qquad X_+ \mid \mathbf{T}_- Z = 0,$$

$$X_- \mid \mathbf{T}_+ Z = 0, \qquad X_- \mid \mathbf{T}_- Z = \delta C \mid \mathbf{T}_- Z,$$

so that $\delta X_+ = \delta X_- = 0$, $X_+ + X_- = \delta C$. Then X_+ represents $\Delta^* w$, while X_- represents $-\Delta^* w$.

Let $U \in Z^r(X \times \mathbf{T}Y)$, $V \in Z^r(\mathbf{T}X \times Y)$ be cocycles such that $U \mid X \times \{*\}$ represents u and $V \mid \{*\} \times Y$ represents v. Then $U_0 = U \mid X \times Y$ and $V_0 = V \mid X \times Y$ represent $u \times 1$ and $1 \times v$, respectively. We may suppose U, V chosen so that

$$f^\# W = U_0 + V_0.$$

Define $E \in C^r(X * Y)$ by

$$E \mid \mathbf{T}X \times Y = V,$$

$$E \mid X \times \mathbf{T}Y = (g^\# C \mid X \times \mathbf{T}Y) - U;$$

then

$$\delta E \mid X \times \mathbf{T}Y = \delta g^\# C \mid X \times \mathbf{T}Y = g^\# \, \delta C \mid X \times \mathbf{T}Y = g^\# X_+ \mid X \times \mathbf{T}Y,$$

$$\delta E \mid \mathbf{T}X \times Y = 0 = g^\# X_+ \mid \mathbf{T}X \times Y,$$

so that $\delta E = g^\# X_+$.

Now $X_+ \mid \mathbf{T}_- Z = 0$ and $X_- \mid \mathbf{T}_+ Z = 0$, so that $X_- \smile X_+ = 0$. Hence

$$X_+ \smile X_+ = (\delta C - X_-) \smile X_+ = \delta C \smile X_+ = \delta(C \smile X_+).$$

Therefore by Lemma 3.5, $\Delta^* w \smile_g \Delta^* w$ is represented by the cocycle

$$T = g^*(C \smile X_+) - E \smile g^* X_+ = (g^* C - E) \smile g^* X_+.$$

But T vanishes on $\mathbf{T}X \times Y$, since $g^* X_+$ does. Also

$$\begin{aligned} T | X \times \mathbf{T}Y &= U \smile (g^* X_+ | X \times \mathbf{T}Y) = U \smile \delta(g^* C | X \times \mathbf{T}Y) \\ &= (-1)^r \, \delta(U \smile (g^* C | X \times \mathbf{T}Y)). \end{aligned}$$

The restriction of the cochain $U \smile (g^* C | X \times \mathbf{T}Y)$ to $X \times Y$ is

$$U_0 \smile f^* W = U_0 \smile (U_0 + V_0)$$

which represents $(u \times 1) \smile (u \times 1 + 1 \times v) = (u \smile u) \times 1 + u \times v$. Therefore T represents

$$(-1)^r \Delta^*((u \smile u) \times 1 + u \times v) = (-1)^r \Delta^*(u \times v). \qquad \square$$

We can now characterize the Hopf invariant in cohomological terms.

(4.3) Theorem *Let* $f: \mathbf{S}^{2n-1} \to \mathbf{S}^n$. *Then*

$$\mathbf{s}^n \smile_f \mathbf{s}^n = -H_0(f) \mathbf{s}^{2n-1}.$$

Let $\theta: \mathbf{S}^n \to \mathbf{S}^n \vee \mathbf{S}^n$ be the coproduct, so that θ is a representative of $\iota^1 + \iota^2$; moreover $j_1, j_2: \mathbf{S}^n \to \mathbf{S}^n \vee \mathbf{S}^n$ represent ι^1, ι^2, respectively. Then

$$\begin{aligned} \mathbf{s}^n \smile_f \mathbf{s}^n &= \theta^* u \smile_f \theta^* v = u \smile_{\theta \circ f} v, \quad \text{by (3.2)} \\ &= u \smile_{j_1 \circ f} v + u \smile_{j_2 \circ f} v + H_0(f) u \smile_h v, \quad \text{by (3.4)} \\ &= u \smile_{j_1 \circ f} v + u \smile_{j_2 \circ f} v - H_0(f) \mathbf{s}^{2n-1}, \quad \text{by Theorem (3.7)}. \end{aligned}$$

(The reader should check that the inclusions guaranteed by (3.2) and (3.4) are equalities in this special case). It remains to prove that $u \smile_{j_t \circ f} v = 0$ for $t = 1, 2$. But, by (3.2),

$$u \smile_{j_1 \circ f} v = j_1^* u \smile_f j_1^* v = j_1^* u \smile_f 0 = 0,$$

and similarly $u \smile_{j_2 \circ f} v = 0$. $\square$

Finally, we have

(4.4) Theorem *If* $g: \mathbf{S}^{2d-1} \to \mathbf{S}^d$ *is a Hopf fibre map then* $H_0(g) = \pm 1$.

(The ambiguity of sign is a reflection of the fact that we were not careful about orientations when we defined the Hopf maps, and also of the fact that if $g: \mathbf{S}^{2d-1} \to \mathbf{S}^d$ is a fibration, its composite with a homeomorphism of degree -1 of either domain or range is still a fibration).

For we have seen that there is a map $f: \mathbf{S}^{d-1} \times \mathbf{S}^{d-1} \to \mathbf{S}^{d-1}$ such that $g = \Gamma f$. The map f has type $(-\iota_{d-1}, \iota_{d-1})$, and therefore $f^*(\mathbf{s}^{d-1}) =$

$-\mathbf{s}^{d-1} \times 1 + 1 \times \mathbf{s}^{d-1}$. By Theorem (4.2),

$$\mathbf{s}^d \smile_g \mathbf{s}^d = \Delta^*\mathbf{s}^{d-1} \smile_g \Delta^*\mathbf{s}^{d-1} = -\Delta^*(-\mathbf{s}^{d-1} \times \mathbf{s}^{d-1})$$
$$= \mathbf{s}^{2d-1},$$

and our result follows from Theorem (4.3). □

Thus, while for every even d, there is a map of $\mathbf{S}^{2d-1}$ into $\mathbf{S}^d$ with Hopf invariant 2, we have only been able to prove existence of maps with Hopf invariant 1 when $d = 2, 4$, or 8. Adams [1] has proved that these are, in fact, the *only* values of d for which such a map exists.

5 Geometrical Interpretation of the Hopf Invariant

In this section we shall utilize Poincaré duality to give a homological description of the functional cup product in the case of a map between two manifolds. Applying Theorem (4.3), we then obtain a geometrical description of the Hopf invariant as a linking number; in fact, this description is the definition of his invariant given by Hopf at a time when cohomology was yet to be discovered.

Let us begin by recalling some notions connected with Poincaré duality. Suppose first that X is an oriented closed triangulated manifold with fundamental class $z \in H_n(X; \mathbf{Z})$. The Poincaré duality theorem then asserts that the homomorphism $z \frown : H^q(X; G) \to H_{n-q}(X; G)$ given by the cap product with z is an isomorphism for any coefficient group G.

Let K be a fixed triangulation of X, and let K' be its first barycentric subdivision. It is customary to agree that the vertices of K' are the barycentres b_σ of the simplices of K; however, it is convenient to modify this requirement by allowing the vertex of K' corresponding to a simplex σ of K to be *any* interior point a_σ of σ. The utility of this greater flexibility will be apparent in the following discussion.

If σ is a p-simplex of K, its *dual cell* $D(\sigma)$ is the sub-complex of K' consisting of all those simplices, each of whose vertices has the form a_τ for some simplex τ of K having σ as a face. The *boundary* $\dot{D}(\sigma)$ is defined similarly, by requiring that σ be a *proper* face of τ. The fact that X is a manifold implies that the pair $(D(\sigma), \dot{D}(\sigma))$ is a homology cell of dimension $n - p$; as $D(\sigma)$ is contractible, $\dot{D}(\sigma)$ is a homology sphere of dimension $n - p - 1$. The dual cells $D(\sigma)$, for all simplices σ of K, form a *block dissection* K^* of K' [H–W, p. 128]; accordingly, the homology groups of the three complexes K, K', and K^* are all isomorphic.

Let us recall that the subdivision operator is a chain map $Sd : C_*(K) \to C_*(K')$ and that there is a simplicial displacement $\theta : K' \to K$ such that, if

$\theta_{\#} : C_*(K') \to C_*(K)$ is the chain map induced by θ, then $\theta_{\#} \circ Sd$ is the identity map of $C_*(K)$, while $Sd \circ \theta_{\#}$ is chain homotopic to the identity map of $C_*(K')$. Dually, we have cochain maps $Sd^* : C^*(K') \to C^*(K)$ and $\theta^\# : C^*(K) \to C^*(K')$.

Let K be oriented, and let $z \in Z_n(K)$ be a cycle representing the given orientation; let $z' = Sdz$. For each oriented p-simplex σ of K, let u_σ be the elementary cochain such that

$$u_\sigma(\sigma) = +1,$$
$$u_\sigma(\tau) = 0 \qquad (\tau \neq \pm\sigma).$$

Let $u'_\sigma = \theta^* u_\sigma \in C^p(K')$. Then it can be verified that the $(n-p)$-chain $z' \frown u'_\sigma$ is a relative cycle of $(D(\sigma), \dot{D}(\sigma))$ whose homology class represents a generator of the infinite cyclic group $H_{n-p}(D(\sigma), \dot{D}(\sigma))$; thus $z' \frown u'_\sigma$ is an orientation of the dual cell of σ. Therefore the map $\sigma \to z' \frown u'_\sigma$ extends to an isomorphism $\phi : C^p(K) \approx C_{n-p}(K^*)$. As $\partial\phi(c) = \pm\phi(\delta c)$, the sign depending only on ϕ and on the particular convention used in the definition of the cap product, the map ϕ is an isomorphism of the cochain complex of K with the chain complex of K. From this fact the Poincaré duality theorem immediately follows.

Remark. For the cap product used by Greenberg [Gr] the relation in question is $\partial\phi(c) = (-1)^{p+1}\phi(\delta c)$.

Let $\mathscr{D} = \phi^{-1} : C_p(K^*) \to C^{n-p}(K)$.

The Poincaré duality theorem can then be exploited to give the intersection theory in X. At the homology level, this is the pairing

$$H_p(X) \otimes H_q(X) \to H_p(X) \otimes H^{n-q}(X) \to H_{p+q-n}(X),$$

where the first map is induced by the inverse of Poincaré duality and the second by the cap product. The calculation of this pairing can be effected by an appropriate operation at the chain level. This operation is a pairing

$$C_p(K) \otimes C_q(K^*) \to C_{p+q-n}(K'),$$

and it defined explicitly as the composite

$$C_p(K) \otimes C_q(K^*) \xrightarrow{1 \otimes \mathscr{D}} C_p(K) \otimes C^{n-q}(K) \xrightarrow{Sd \otimes 1} C_p(K') \otimes C^{n-q}(K)$$
$$\xrightarrow{1 \otimes \theta^\#} C_p(K') \otimes C^{n-q}(K') \xrightarrow{\frown} C_{p+q-n}(K').$$

Let $a \circ b$ be the value of this pairing at $a \otimes b$ for $a \in C_p(K)$, $b \in C_q(K^*)$. Then

(5.1) $$\partial(a \circ b) = \pm(\partial a) \circ b \pm a \circ (\partial b),$$

the signs depending only on p and q (and on the definition of the cap product; for the one used by Greenberg [Gr], we have

$$\partial(a \circ b) = (-1)^{n-q}(\partial a) \circ b + (-1)^{n+1} a \circ (\partial b)).$$

Let σ, τ be oriented simplices of K, of dimensions $p, n - q$ respectively. Then $\phi(u_\tau)$ is the oriented "cell" of K^* dual to τ, and

$$\sigma \circ \phi(u_\tau) = Sd\sigma \frown \theta^*(u_\tau).$$

Moreover, $\sigma \circ \phi(u_\tau)$ is a chain of K' lying on the intersection of σ' with $D(\tau)$. It then follows that, if $|x|$ is the carrier of the chain x (in K, K^*, or K' as the case may be), then $|a \circ b| \subset |a| \cap |b|$. This justifies, in part, the term "intersection" for the pairing we have just defined.

Let us now consider the special case $p + q = n$. Then $a \circ b$ is a 0-chain of K', whose Kronecker index $\varepsilon(a \circ b)$ is an integer, the *intersection number* $I(a, b)$ *of* a *and* b. Suppose instead that $p + q = n + 1$; then the boundary formula (5.1) yields

(5.2) $$0 = \varepsilon\partial(a \circ b) = \pm I(\partial a, b) \pm I(a, \partial b).$$

Finally, we can define linking numbers. Suppose that a and b are bounding cycles of dimensions p, q with $p + q = n$. Choose a chain c such that $\partial c = a$, and verify by means of (5.2) that the number $I(c, b)$ is independent of c; it is called the *linking number* $\lambda(a, b)$ of a and b.

Now let us suppose that M, N are oriented manifolds of dimensions $n + k$, n, respectively. Then f^* corresponds via Poincaré duality in the manifolds M, N to a homomorphism $f_! : H_p(N) \to H_{p+k}(M)$, called the *Hopf retromorphism*. Thus, if $\mu \in H_{n+k}(M)$, $\nu \in H_n(N)$ are the fundamental classes, we have

(5.3) $$f_!(\nu \frown x) = \mu \frown f^*(x)$$

for all $x \in H^*(N)$.

The retromorphism $f_!$ has the following useful properties [2]:

(5.4) *If* $u \in H_p(N)$, $v \in H_q(N)$, *then*

$$f_!(u \circ v) = f_! u \circ f_! v.$$

(5.5) *If* $u \in H_p(M)$, $v \in H_q(N)$, *then*

$$f_*(u \circ f_! v) = f_* u \circ v.$$

To calculate $f_!$, suppose that f is an order-preserving simplicial map of a triangulation K of M into a triangulation L of N. The vertices of L' are chosen to be the barycentres of the simplices of L. However, if σ is any simplex of K, we agree to take as the corresponding vertex of K' any point a_σ of $f^{-1}(b_\tau)$, where τ is the simplex of L which is the image of σ under f. (Note that b_σ need not lie in $f^{-1}(b_\tau)$). It follows that $f_\# \circ Sd = Sd \circ f_\#$ and therefore that $\theta'_\# \circ f_\# = f_\# \circ \theta_\#$ for the simplicial displacements $\theta : K \to K'$, $\theta' : L \to L'$. The map $f_\dagger = \mathscr{D}^{-1} \circ f^\# \circ \mathscr{D} : C_p(L^*) \to C_{p+k}(K^*)$ is then (up to sign) a chain map whose induced homomorphism is $f_!$. If τ is a simplex of K, σ a simplex of L, such that $f(\tau) \subset \sigma$, then $f(D(\tau)) \subset D(\sigma)$. It follows that, if σ is an oriented simplex of L, $\tau_1, \ldots, \tau_r$ the oriented simplices of K for which $f_{\#}(\tau_i) = \pm\sigma_i$, and z, w the fundamental classes of L, K, respectively, so that

$$f_{\dagger}(z' \frown u'_{\sigma}) = \sum_i \pm w' \frown u'_{\tau i},$$

then $f_{\dagger}$ maps the dual cell of σ into a chain of K^* lying on $f^{-1}(D(\sigma))$. Thus $|f_{\dagger}(c)| \subset f^{-1}|c|$ for every chain c of L^*.

If z is a *p-cycle* of L^*, then $f^{-1}(|z|)$ is a $(p+k)$-dimensional subcomplex of K^*, and the map $f_{\dagger}$ gives us a way of orienting the $(p+k)$-cells of $f^{-1}(|z|)$ so as to form the cycle $f_{\dagger}(z)$. In other words, $f_{\dagger}$ is, roughly speaking, a chain map carried by the relation f^{-1}.

Let us now suppose that M and N are oriented spheres of dimensions $2n-1$, n respectively $(n > 1)$. Let σ, τ be distinct n-simplices of L, oriented coherently with N; then b_{σ} and b_{τ} are 0-cycles of L^* and $f_{\dagger}(b_{\sigma})$, $f_{\dagger}(b_{\tau})$ are $(n-1)$-cycles of K^*, which bound because $H_{n-1}(M) = 0$. It is the linking number of these two cycles which was taken by Hopf as the definition of $H_0(f)$. In order to calculate this linking number using the machinery developed above, we encounter the difficulty that $f_{\dagger}(b_{\sigma})$ is a cycle of K^*, and not of K. However, this difficulty is merely technical; we merely have to perturb $f_{\dagger}(b_{\sigma})$ slightly to obtain a cycle of K, whose linking number with $f_{\dagger}(b_{\tau})$ is defined. The details follow in the proof of

(5.6) Theorem *The linking number of* $f_{\dagger}(b_{\sigma})$, $f_{\dagger}(b_{\tau})$ *is* $\pm H_0(f)$.

In fact, $\mathscr{D}(b_{\sigma}) = u_{\sigma}$ and $\mathscr{D}(b_{\tau}) = u_{\tau}$, and u_{σ}, u_{τ} are representatives of the cohomology class $\mathbf{s}^n$. Moreover,

$$f_{\dagger}(b_{\sigma}) = \mathscr{D}^{-1} f^{\#} \mathscr{D}(b_{\sigma}) = \mathscr{D}^{-1} f^{\#} u_{\sigma}.$$

Now $f^{\#} u_{\sigma} = \delta c$ for some $c \in C^{n-1}(K)$, and therefore

$$f_{\dagger}(b_{\sigma}) = \mathscr{D}^{-1}\,\delta c = \pm\partial\mathscr{D}^{-1}(c).$$

The group $C_*(K^*)$ is a subgroup of $C_*(K')$ and is mapped into $C_*(K)$ by $\theta_{\#}$; the perturbation mentioned above is just $\theta_{\#}\,|\,C_*(K^*)$. Thus the desired linking number is

$$\lambda(\theta_{\#}\, f_{\dagger}(b_{\sigma}), f_{\dagger}(b_{\tau})) = \pm\varepsilon(\theta_{\#}\mathscr{D}^{-1}(c) \circ f_{\dagger}(b_{\tau})).$$

But

$$\theta_{\#}\mathscr{D}^{-1}(c) = \theta_{\#}(Sdz \frown \theta^{\#} c) = \theta_{\#} Sdz \frown c = z \frown c,$$

and

$$\begin{aligned}\theta_{\#}\mathscr{D}^{-1}(c) \circ f_{\dagger}(b_{\tau}) &= \theta_{\#}\mathscr{D}^{-1}(c) \frown \mathscr{D} f_{\dagger}(b_{\tau})\\ &= \theta_{\#}\mathscr{D}^{-1}(c) \frown f^{\#}\mathscr{D}(b_{\tau})\\ &= (z \frown c) \frown f^{\#} u_{\tau}\\ &= z \frown (c \smile f^{\#} u_{\tau}).\end{aligned}$$

Thus

$$\begin{aligned}\lambda(\theta_{\#} f_{\dagger}(b_\sigma), f_{\dagger}(b_\tau)) &= \pm\varepsilon(z \frown (c \smile f^{\#}u_\tau))\\ &= \pm\langle c \smile f^{\#}u_\tau, z\rangle\\ &= \pm\langle \mathbf{s}^n \smile_f \mathbf{s}^n, \mathbf{s}_{2n-1}\rangle = \pm H_0(f),\end{aligned}$$

since $c \smile f^{\#}u_\tau$ is a representative of $\mathbf{s}^n \smile_f \mathbf{s}^n$. □

6 The Hilton–Milnor Theorem

Let X, Y be spaces with non-degenerate base points. Then the direct sum theorem insures that $H_n(X \vee Y) \approx H_n(X) \oplus H_n(Y)$ for all n. On the other hand, we have seen that this is no longer true for the homotopy groups. It is natural, then to seek to determine the homotopy groups of $X \vee Y$, if not in terms of the homotopy groups of X and Y, at least in terms of groups which we may consider known. This was first accomplished by Hilton, for the case that X and Y are spheres; his result states that

$$\pi_n(\mathbf{S}^{p+1} \vee \mathbf{S}^{q+1}) \approx \pi_n(\mathbf{S}^{p+1}) \oplus \pi_n(\mathbf{S}^{q+1}) \oplus \bigoplus_{i=1}^{\infty} \pi_n(\mathbf{S}^{r_i+1}),$$

where $\{r_i\}$ is a sequence of positive integers, depending on p and q, and tending to ∞ with i. The inclusion $\pi_n(\mathbf{S}^{r_i+1}) \to \pi_n(\mathbf{S}^{p+1} \vee \mathbf{S}^{q+1})$ is induced by composition with a suitable iterated Whitehead product. Later Hilton's result was generalized by Milnor to the case when X and Y are arbitrary suspensions. Milnor's proof was carried out in the category of semi-simplicial complexes; the proof we give here is carried out in the category $\mathscr{K}$, but is inspired by Milnor's. The idea in all of these proofs is to show that the loop space $\Omega\mathbf{S}(X \vee Y)$ has the same (weak) homotopy type as an infinite Cartesian product $\prod_{i=1}^{\infty} \Omega\mathbf{S}W_i$, where W_i is the reduced join of a number of copies of X and of Y.

The result for a cluster of two spaces extends, by induction, to a corresponding result for a cluster of any number of spaces. However, the result for $\Omega\mathbf{S}(X_1 \vee \cdots \vee X_k)$ is just as easy (i.e., no more difficult!) to state and gives more information. In order to formulate the result, we shall require some algebraic preliminaries [Ha, Chapter 11].

Let A be the free nonassociative ring with k generators $x_1, \ldots, x_k$. Thus A has an additive basis consisting of all parenthesized monomials in the x_i. We shall single out certain of these, referring to them as *basic products*.

Let us define the *weight* of a monomial to be the number of its factors. Thus $x_1, \ldots, x_k$ are the only monomials of weight 1; and any monomial of weight $r > 1$ has the unique form $m = m'm''$ with the weight of m equal to the sum of the weights of m' and m''.

We now define the basic products of weight r, by induction on r; and, for each such product m, a non-negative integer $r(m)$, called its *rank*. These are to be linearly ordered, in such a way that $m_1 < m_2$ if the weight of m_1 is less than the weight of m_2. The *serial number* $s(m)$ is the number of basic products $\leq m$ in terms of this ordering. The basic products of weight 1 are the generators $x_1, \ldots, x_k$; we agree that $x_i < x_j$ for $i < j$, and that $r(x_i) = 0$; of course, $s(x_i) = i$. Suppose that the basic products of weight less than n have been defined and linearly ordered, in such a way that $m_1 < m_2$ if the weight of m_1 is less than that of m_2; and suppose that the rank $r(m)$ of such a product has been defined. Then the basic products of weight n are all monomials of the form $m_1 m_2$, of weight n, for which m_1 and m_2 are basic products, $m_2 < m_1$, and $r(m_1) \leq s(m_2)$. Give these an arbitrary linear order, and define $r(m_1 m_2) = s(m_2)$.

For example, suppose that $k = 3$ and write x, y, z instead of x_1, x_2, x_3. Then the basic products of weight n are:

$n = 1$: x, y, z;

$n = 2$: yx, zx, zy;

$n = 3$: $(yx)x, (yx)y, (yx)z, (zx)x, (zx)y, (zx)z, (zy)y, (zy)z$.

We should observe here that the basic products are not uniquely defined, in the sense that the basic products of a given weight depend on the ordering of the basic products of lower weight.

Let us agree to enumerate the basic products by their serial numbers; i.e., for each positive integer m, w_m is the (unique) basic product whose serial number is m. Let T_m be the operation of right multiplication by w_m. If w is any basic product, then $r(w) = m$ if and only if $w = T_m(w')$ for a uniquely determined basic product w', and $r(w') \leq m$. If $r(w') = m$, we can split off another factor w_m, and so on. It follows that, if w is any basic product of rank $\leq m$, then w can be represented uniquely in the form $w = T_m^i(w_*)$ where i is a non-negative integer and w_* is a basic product of rank $< m$; moreover, $w_* > w_m$ unless $i = 0$, in which case w_* can be less than w_m, and, in fact, if $j < m$, then $r(w_j) < m$. Conversely, if w_* is a basic product such that $r(w_*) < m$ and $w_* > w_m$, then the elements $T_m^i w_*$ are basic products of rank $\leq m$. Thus

(6.1) *The basic products of rank $\leq m$ are given without repetition by $w = T_m^i(w_j)$ where*

(a) $i > 0, j > m$, *and* $r(w_j) < m$;
(b) $i = 0, j > m$, *and* $r(w_j) < m$;
(c) $i = 0, j < m$. □

Let Y be any multiplicative system. Given $y_1, \ldots, y_k \in Y$, there is a unique homomorphism of A into Y which carries x_i into y_i. If w is any monomial in

A, it is convenient to let $w(y_1, \ldots, y_k)$ be the image of w under this homomorphism. Thus $w(y_1, \ldots, y_k)$ is just the (parenthesized) monomial obtained from w by replacing x_i by y_i $(i = 1, \ldots, k)$. We shall use this device in several different ways.

Firstly we may take Y to be the class of all spaces with base point, the operation in Y being the reduced join. Then $w, (X_1, \ldots, X_k)$ is the reduced join of copies of the spaces $X_1, \ldots, X_k$, for each basic product w; moreover,

(6.2) $$w(X_1, \ldots, X_k) = X_1^{(a_1)} \wedge \cdots \wedge X_k^{(a_k)},$$

where, for any space X, $X^{(a)}$ is the reduced join of a copies of X; the integer a_i is just the number of occurrences of x_i in the word w.

Remark. The equality in (6.2) is really only a natural homeomorphism; in order to identify the two sides of the formula we need to make repeated use of the commutative and associative laws for the reduced join.

Secondly, we may take the product in question to be the Samelson product $[X_1, G] \times [X_2, G] \to [X_1 \wedge X_2, G]$, for any H_0-space G. If w is a basic product and $\alpha_i \in [X_i, G]$ $(i = 1, \ldots, k)$, then

$$w(\alpha_1, \ldots, \alpha_k) \in [w(X_1, \ldots, X_k), G].$$

Finally, let L be the free Lie algebra L generated by $y_1, \ldots, y_k$. Then a theorem of Marshall Hall [1] tells us that

(6.3) *The basic products $w(y_1, \ldots, y_k)$ form an additive basis for L.* □

We can now count the basic products in two different ways. The formulae are due to Witt. Let us begin by recalling some notions from number theory.

Let N be the set of positive integers, A a commutative ring, and let F be the set of all functions $f: N \to A$. Then F is itself a commutative ring under termwise addition and a product defined by

$$(f * g)(n) = \sum_{d \mid n} f\left(\frac{n}{d}\right) g(d).$$

Moreover, an element $f \in F$ is invertible if and only if $f(1)$ is an invertible element $u \in A$. When this is so, the inverse of f is the function g defined inductively by

$$g(1) = u^{-1}$$

$$g(n) = -u^{-1} \sum_{d}' f\left(\frac{n}{d}\right) g(d),$$

where the sum ranges over all proper divisors of n.

If s is the function such that $s(n) = 1$ for all n, the inverse of s is the *Möbius function* μ, defined by

$$\mu(1) = 1,$$
$$\mu(n) = 0 \quad \text{unless } n > 1 \text{ is square-free,}$$
$$\mu(p_1 \cdots p_k) = (-1)^k \quad \text{if } p_1, \ldots, p_k \text{ are distinct primes.}$$

The *Möbius inversion formula* is simply the statement that, if $f \in F$ and $g = f * s$, then $f = g * \mu$; i.e., if

$$g(n) = \sum_{d \mid n} f(d),$$

then

$$f(n) = \sum_{d \mid n} \mu(d) g\left(\frac{n}{d}\right)$$

We can now state the formulae of Witt [Ha, p. 169]:

(6.4) *The number of basic products involving x_i exactly n_i times is*

$$\frac{1}{n} \sum_{d \mid n_0} \mu(d) \frac{\left(\frac{n}{d}\right)!}{\left(\frac{n_1}{d}\right)! \cdots \left(\frac{n_k}{d}\right)!}$$

where n_0 is the greatest common divisor of $n_1, \ldots, n_k$, and $n = n_1 + \cdots + n_k$. □

(6.5) *The number of basic products of weight n is*

$$\frac{1}{n} \sum_{d \mid n} \mu(d) k^{n/d}.$$ □

With these preparations we can now state

(6.6) Theorem *Let $X_1, \ldots, X_k$ be connected* CW*-complexes with base vertices. Then the space $J(X_1 \vee \cdots \vee X_k)$ has the same homotopy type as the weak product*

$$\prod_w Jw(X_1, \ldots, X_k)$$

where w ranges over all admissible words in $x_1, \ldots, x_k$.

An explicit homotopy equivalence can be constructed in the following way. Let $j_t : X_t \to X = X_1 \vee \cdots \vee X_k$ be the usual inclusion ($t = 1, \ldots, k$), and let $i_t : X_t \to J(X)$ be the composite of j_t with the inclusion map $i_X : X \hookrightarrow J(X)$. For each basic product w, we can form the element

$$w(i_1, \ldots, i_k) : w(X_1, \ldots, X_k) \to J(X).$$

(Here, as elsewhere, it is convenient to ignore the distinction between a map and its homotopy class). By the universal property of the functor J (Theorem (2.5) of Chapter VII), w can be extended uniquely to a homomorphism

$$\hat{w}(i_1, \ldots, i_k) : Jw(X_1, \ldots, X_k) \to J(X).$$

Let us recall that the space $Y = \prod_{i=1}^{\infty} Jw_i(X_1, \ldots, X_k)$ is filtered by the subspaces

$$Y_r = \prod_{i=1}^{r} Jw_i(X_1, \ldots, X_k).$$

The external product of the maps $\hat{w}_i = \hat{w}_i(i_1, \ldots, i_k) : Jw_i(X_1, \ldots, X_k) \to J(X)$ is a map $h_r : Y_r \to J(X)$. Moreover, $h_{r+1} \mid Y_r = h_r$, and therefore there is a map $h : Y \to J(X)$ such that $h \mid Y_r = h_r$ for all r. Theorem (6.6) can now be strengthened to read

(6.7) Theorem (Hilton–Milnor Theorem). *The map $h : Y \to J(X)$ is a homotopy equivalence.*

7 Proof of the Hilton–Milnor Theorem

The proof is long and complicated. Had we been satisfied to prove Theorem (6.6), the proof would have been considerably shorter. But because of the applications we have in mind in §8, it is necessary to track down the isomorphism and prove that it can be expressed in terms of iterated Samelson products, i.e., prove Theorem (6.7).

The proof breaks up into five steps.

Throughout this Section, except when otherwise stated, we shall assume that the spaces mentioned are connected CW-complexes with base vertices. (However, certain spaces constructed in the course of the argument, e.g., ΩSX, need not be CW-complexes).

Step I *The spaces $J(X \vee Y)$ and $J(X) \times J(Y \vee (Y \wedge J(X)))$ have the same homotopy type.*

Step II *The spaces $J(Y \wedge J(X))$ and $J((Y \wedge X) \vee (Y \wedge X \wedge J(X))$ have the same homotopy type.*

Step III *The spaces $J(Y \wedge J(X))$ and $J(\bigvee_{i=1}^{\infty} Y \wedge X^{(i)})$ have the same homotopy type.*

Step IV *The spaces $J(X \vee Y)$ and $J(X) \times J(\bigvee_{i=0}^{\infty} Y \wedge X^{(i)})$ have the same homotopy type.*

Step V *The spaces $J(X_1 \vee \cdots \vee X_k)$ and $\prod_{i=1}^{\infty} Jw_i(X_1, \ldots, X_k)$ have the same homotopy type.*

In each case we shall show that a certain explicit map is a homotopy equivalence.

One feature of the proof is the interplay between the functors J and ΩS. Let us begin by recalling the connection between the functors.

The basic property of J is given by Theorem (2.5) of Chapter VII: if Q is a strictly associative H-space, then every map $f : Y \to Q$ can be uniquely extended to a homomorphism $\hat{f} : J(Y) \to Q$; $\hat{f}$ is called the *canonical extension* of f. Moreover, if $f_0 \simeq f_1$, then $\hat{f}_0 \simeq \hat{f}_1$. In particular, we may take $Q = \Omega^* W$. As we saw in §2 of Chapter III, there are maps $h : \Omega W \to \Omega^* W$ and $h' : \Omega^* W \to \Omega W$ such that $h' \circ h$ is the identity map, while $h \circ h'$ is homotopic to the identity. Moreover, h and h' are H-maps.

The adjoint of the identity map of $\mathbf{S}Y$ is a map $\lambda_0 : Y \to \Omega\mathbf{S}Y$, and $\lambda = h \circ \lambda_0 : Y \to \Omega^*\mathbf{S}Y$ has the canonical extension $\hat{\lambda} : J(Y) \to \Omega^*\mathbf{S}Y$. According to Theorem (2.6) of Chapter VII, $\hat{\lambda}$ is a weak homotopy equivalence. Let $\hat{\lambda}' = h' \circ \hat{\lambda} : J(Y) \to \Omega\mathbf{S}Y$. Then $\hat{\lambda}' | Y = \lambda_0$ and $\hat{\lambda}'$ is an H-map, as well as a weak homotopy equivalence.

The subspace Y of $J(Y)$ need not be a retract. However, it becomes one after suspension. In fact, the adjoint $r : \mathbf{S}J(Y) \to \mathbf{S}Y$ of the map $\hat{\lambda}' : J(Y) \to \Omega\mathbf{S}Y$ is a retraction. For $r|\mathbf{S}Y$ is the adjoint of $\lambda_0 = \hat{\lambda}' | Y$, and λ_0 is the adjoint of identity map of $\mathbf{S}Y$. The map r is the composite $\mathbf{e} \circ \hat{\lambda}'$, where $\mathbf{e} : \mathbf{S}\Omega\mathbf{S}Y \to \mathbf{S}Y$ is the evaluation map. Therefore we shall also call the map r the *evaluation map.*

The adjointness relation is an isomorphism $[\mathbf{S}Y, \mathbf{S}Z] \approx [Y, \Omega\mathbf{S}Z]$. The weak homotopy equivalence $\hat{\lambda}' : J(Z) \to \Omega\mathbf{S}Z$ induces an isomorphism $[Y, J(Z)] \approx [Y, \Omega\mathbf{S}Z]$. Therefore there is a canonical isomorphism $[\mathbf{S}Y, \mathbf{S}Z] \approx [Y, J(Z)]$. If $f : \mathbf{S}Y \to \mathbf{S}Z$, we shall also refer to any map $f' : Y \to J(Z)$ which corresponds to f under this isomorphism as a *J-adjoint* of f.

Let $f : \mathbf{S}Y \to \mathbf{S}Z$, and let $f' : Y \to J(Z)$ be a J-adjoint of f. Let $\hat{f}' : J(Y) \to J(Z)$ be the canonical extension of f'. We shall call $\hat{f}'$ a *J-extension* of f.

(7.1) *The diagram*

$$\begin{array}{ccc}
\Omega\mathbf{S}Y & \xrightarrow{\Omega f} & \Omega\mathbf{S}Z \\
\uparrow{\scriptstyle \hat{\lambda}'} & & \uparrow{\scriptstyle \hat{\lambda}'} \\
J(Y) & \xrightarrow[\hat{f}']{} & J(Z)
\end{array}$$

is homotopy commutative. If f is a homotopy equivalence, so is $\hat{f}'$.

The diagram in question can be enlarged to a diagram

$$\begin{array}{ccc}
\Omega\mathbf{S}Y & \xrightarrow{\Omega f} & \Omega\mathbf{S}Z \\
\uparrow{\scriptstyle h'} & & \uparrow{\scriptstyle h'} \\
\Omega^*\mathbf{S}Y & \xrightarrow{\Omega^* f} & \Omega^*\mathbf{S}Z \\
\uparrow{\scriptstyle \hat{\lambda}} & & \uparrow{\scriptstyle \hat{\lambda}} \\
J(Y) & \xrightarrow[\hat{f}']{} & J(Z)
\end{array}$$

and $h' \circ \hat{\lambda} = \hat{\lambda}'$ (for both Y and Z). The upper square is (strictly) commutative. Hence it suffices to prove the lower square homotopy commutative. The four maps involved in the lower square are homomorphisms and $\hat{\lambda} \circ \hat{f}' | Y = \lambda \circ f'$, $\Omega^* f \circ \hat{\lambda} | Y = \Omega^* f \circ \lambda = h \circ \bar{f}$. But

$$h' \circ \lambda \circ f' = \lambda' \circ f' \simeq \bar{f} = h' \circ h \circ \bar{f};$$

since h' is a homotopy equivalence, $\lambda \circ f' \simeq h \circ \bar{f}$. Thus $\hat{\lambda} \circ \hat{f}'$ and $\Omega^* f \circ \hat{\lambda}$ are the canonical extensions of the homotopic maps $\lambda \circ f'$, $h \circ \bar{f}$ and hence are themselves homotopic.

If f is a homotopy equivalence, so is Ωf. Since both maps $\hat{\lambda}'$ are weak homotopy equivalences, so is $\hat{f}'$. Since $J(Y)$ and $J(Z)$ are CW-complexes, $\hat{f}'$ is a homotopy equivalence.

With these preparations, we proceed to prove the Theorem.

Step I. Consider the projection $q = \mathbf{S}q_1 : \mathbf{S}(X \vee Y) \to \mathbf{S}X$, and let F be the mapping fibre of q. There is a fibration $p : F \to \mathbf{S}(X \vee Y)$ and the sequence

$$F \xrightarrow{p} \mathbf{S}(X \vee Y) \xrightarrow{q} \mathbf{S}X$$

is left exact. Moreover, q has the cross-section $\mathbf{S}j_1 : \mathbf{S}X \to \mathbf{S}(X \vee Y)$. Applying the functor Ω to this sequence yields a new left exact sequence

$$\text{(7.2)} \qquad \Omega F \xrightarrow{\Omega p} \Omega\mathbf{S}(X \vee Y) \xrightarrow{\Omega q} \Omega\mathbf{S}X,$$

and Ωq has the cross-section $\Omega\mathbf{S}j_1$.

(7.3) *For any space Q in $\mathscr{K}_*$, the homomorphism*

$$\underline{\Omega p} : [Q, \Omega F] \to [Q, \Omega\mathbf{S}(X \vee Y)]$$

is a monomorphism.

In §6 of Chapter III, we showed that the left exact sequence (7.2) can be extended to the left to obtain a left exact sequence

$$\Omega^2\mathbf{S}(X \vee Y) \xrightarrow{\Omega^2 q} \Omega^2\mathbf{S}X \xrightarrow{\Omega h} \Omega F \xrightarrow{\Omega p} \Omega\mathbf{S}(X \vee Y) \xrightarrow{\Omega q} \Omega\mathbf{S}X,$$

where $h : \Omega SX \to F$ is a connecting map. The sequence

$$[Q, \Omega^2\mathbf{S}(X \vee Y)] \xrightarrow{\underline{\Omega^2 q}} [Q, \Omega^2\mathbf{S}X] \xrightarrow{\underline{\Omega h}}$$
$$[Q, \Omega F] \xrightarrow{\underline{\Omega p}} [Q, \Omega\mathbf{S}(X \vee Y)].$$

is exact. But $\underline{\Omega^2 q}$ has the left inverse $\underline{\Omega^2 Sj_1}$, so that $\underline{\Omega^2 q}$ is an epimorphism. Hence $\underline{\Omega h} = 0$ and $\underline{\Omega p}$ is a monomorphism. □

We can use the external product of §4, Chapter III, to construct a map

$$\chi = \Omega j_1 \otimes \Omega p : \Omega\mathbf{S}X \times \Omega F \to \Omega\mathbf{S}(X \vee Y).$$

It is then easy to prove

(7.4) *The map χ is a weak homotopy equivalence.* □

It remains to determine the weak homotopy type of F. We shall construct a weak homotopy equivalence

$$\psi : \mathbf{S}Z \to F,$$

where $Z = Y \vee (Y \wedge W)$, $W = \Omega \mathbf{S}X$. It suffices to construct the adjoint

$$\tilde{\psi} : Z \to \Omega F$$

of ψ. By exactness of (7.2), it suffices to construct a map

$$\tilde{\omega} : Z \to \Omega \mathbf{S}(X \vee Y)$$

such that $\Omega q \circ \tilde{\omega}$ is nullhomotopic; for then $\tilde{\psi}$ is determined (uniquely, because of (7.3)) by $\Omega p \circ \tilde{\psi} \simeq \tilde{\omega}$. The map $\tilde{\omega}$ is determined by

$$\tilde{\omega} \mid Y = i_2' = \lambda_0 \circ j_2,$$

$$\tilde{\omega} \mid Y \wedge W = \langle i_2', \Omega \mathbf{S} j_1 \rangle.$$

The composite $(\Omega q) \circ i_2'$ is the constant map, and it follows that $\Omega q \circ \langle i_2', \Omega \mathbf{S} j_1 \rangle = \langle *, \Omega \mathbf{S} j_1 \rangle \simeq *$.

(7.5) Theorem *The map $\psi : \mathbf{S}(Y \vee (Y \wedge W)) \to F$ is a weak homotopy equivalence.*

This is not easy to see directly. To prove it, we shall exhibit another weak homotopy equivalence $\phi : F \to \mathbf{S}Z$, and show that $\phi \circ \psi : \mathbf{S}Z \to \mathbf{S}Z$ is a weak homotopy equivalence.

To construct ϕ, we begin with the observation

(7.6) *The space F is homeomorphic with the subspace $\mathbf{S}Y \times \Omega \mathbf{S}X \cup \{*\} \times \mathbf{P}'\mathbf{S}X$ of $\mathbf{S}Y \times \mathbf{P}'\mathbf{S}X$.*

Let us recall that F is the set of all pairs (z, u) such that $z \in \mathbf{S}X \vee \mathbf{S}Y$ and u is a path in $\mathbf{S}X$ with $u(0) = *$, $u(1) = q(z)$. If $z \in \mathbf{S}Y$, then $q(z) = *$ and $u \in \Omega \mathbf{S}X$; if $z \in \mathbf{S}X$, then z is determined by $u \in \mathbf{P}'\mathbf{S}X$. Thus a homeomorphism $F \to \mathbf{S}Y \times \Omega \mathbf{S}X \cup \{*\} \times \mathbf{P}'\mathbf{S}X$ is given by

$$(j_1(x), u) \to (*, u) \qquad (x \in \mathbf{S}X, u \in \mathbf{P}'\mathbf{S}X),$$

$$(j_2(y), u) \to (y, u) \qquad (y \in \mathbf{S}Y, u \in \Omega \mathbf{S}X).$$ □

The pair $(\mathbf{P}'\mathbf{S}X, \Omega \mathbf{S}X)$ is an NDR-pair, by Theorem (7.14) of Chapter I. Since $\mathbf{P}'\mathbf{S}X$ is contractible, we can apply Corollary (5.13) of the same Chapter to deduce

(7.7) *The space F has the same homotopy type as* $\mathbf{S}Y \times \Omega\mathbf{S}X/\{*\} \times \Omega\mathbf{S}X$. □

Our determination of the weak homotopy type of F will be completed once we have proved

(7.8) *If W and Y are 0-connected spaces in $\mathscr{K}_*$, then $\mathbf{S}Y \times W/\{*\} \times W$ has the same weak homotopy type as $\mathbf{S}Y \vee \mathbf{S}(Y \wedge W)$.*

We first observe that $\mathbf{S}Y \times W/\{*\} \times W$ is the suspension of the space $Y \times W/\{*\} \times W$. In fact, let us temporarily regard W as a free space and let W^+ be the corresponding space with base point. Then

$$\begin{aligned}\mathbf{S}Y \times W/\{*\} \times W = \mathbf{S}Y \wedge W^+ &= (\mathbf{S} \wedge Y) \wedge W^+ = \mathbf{S} \wedge (Y \wedge W^+) \\ &= \mathbf{S}(Y \times W/\{*\} \times W)\end{aligned}$$

(up to natural homeomorphism).

Let $f_1 = p_1 : \mathbf{S}Y \times W \to \mathbf{S}Y$ be the projection on the first factor, and let $f_2 : \mathbf{S}Y \times W \to \mathbf{S}Y \wedge W$ be the identification map. Then f_1 and f_2 annihilate the subspace $\{*\} \times W$, thereby inducing maps $f'_1 : \mathbf{S}Y \times W/\{*\} \times W \to \mathbf{S}Y$, $f'_2 : \mathbf{S}Y \times W/\{*\} \times W \to \mathbf{S}Y \wedge W$. Composing these with the respective inclusions $\mathbf{S}Y \hookrightarrow \mathbf{S}Y \vee (\mathbf{S}Y \wedge W)$, $\mathbf{S}Y \wedge W \hookrightarrow \mathbf{S}Y \vee (\mathbf{S}Y \wedge W)$, we obtain maps g_1, $g_2 : \mathbf{S}Y \times W/\{*\} \times W \to \mathbf{S}Y \vee (\mathbf{S}Y \wedge W)$. By the remark of the preceding paragraph, the latter can be added to obtain a map $g : \mathbf{S}Y \times W/\{*\} \times W \to \mathbf{S}Y \vee (\mathbf{S}Y \wedge W)$. It is easy to see, with the aid of Theorem (7.10) of Chapter III, that g is a homology equivalence. As the domain and range of g are the suspensions of 0-connected spaces, they are 1-connected, so that g is a weak homotopy equivalence by the converse of the Whitehead Theorem. □

Let us now consider the map $\phi \circ \psi$; as in the case of ϕ, it suffices to prove it a homology equivalence. We shall need to calculate ψ explicitly. For this, we observe that a map $\psi : Z \to F$ is determined by a map $\omega : Z \to \mathbf{S}(X \vee Y)$, together with a homotopy η of the constant map to $q \circ \omega_1$. The map ψ is then given by

$$\psi(z) = (\omega(z), \tilde{\eta}(z)),$$

where $\tilde{\eta} : Z \to \mathbf{P}'\mathbf{S}X$ is the adjoint of η.

The map $\psi_1 = \psi \,|\, \mathbf{S}Y$ is determined by the inclusion

$$\omega_1 = \mathbf{S}j_2 : \mathbf{S}Y \to \mathbf{S}X \vee \mathbf{S}Y,$$

together with the stationary homotopy of the constant map $q \circ \omega_1$.

The calculation of $\psi_2 = \psi \,|\, \mathbf{S}(Y \wedge W)$ is more difficult. There is a commutative diagram

$$\begin{array}{ccccccc}
[\mathbf{S}(Y \wedge W), F] & \xrightarrow{\alpha} & [Y \wedge W, \Omega F] & \xrightarrow{\underline{\Omega p}} & [Y \wedge W, \Omega\mathbf{S}(X \vee Y)] & \xrightarrow{\underline{q}} & [Y \wedge W, \Omega\mathbf{S}X] \\
\downarrow{\scriptstyle \overline{\mathbf{S}\pi}} & & \downarrow{\scriptstyle \bar{\pi}} & & \downarrow{\scriptstyle \bar{\pi}} & & \downarrow{\scriptstyle \bar{\pi}} \\
[\mathbf{S}(Y \times W), F] & \xrightarrow{\alpha} & [Y \times W, \Omega F] & \xrightarrow{\underline{\Omega p}} & [Y \times W, \Omega\mathbf{S}(X \vee Y)] & \xrightarrow{\underline{q}} & [Y \times W, \Omega\mathbf{S}X] \\
\downarrow{\scriptstyle \overline{\mathbf{S}i}} & & \downarrow{\scriptstyle \bar{i}} & & \downarrow{\scriptstyle \bar{i}} & & \downarrow{\scriptstyle \bar{i}} \\
[\mathbf{S}(Y \vee W), F] & \xrightarrow[\alpha]{} & [Y \vee W, \Omega F] & \xrightarrow[\underline{\Omega p}]{} & [Y \vee W, \Omega\mathbf{S}(X \vee Y)] & \xrightarrow[\underline{q}]{} & [Y \vee W, \Omega\mathbf{S}X]
\end{array}$$

where the α's are adjointness isomorphisms, $\pi : Y \times W \to Y \wedge W$ is the identification map, and i is the inclusion. The homomorphisms $\bar{\pi}$ (and therefore $\overline{S\pi}$) are monomorphisms, by Lemma (4.1) of Chapter X, and the columns of the diagram are exact. By (7.3), the homomorphisms $\underline{\Omega p}$ are monomorphisms.

The map $\psi_2 = \psi \,|\, \mathbf{S}(Y \wedge W)$ is determined by the property that $\underline{\Omega p}\alpha(\psi_2)$ is the Samelson product $\xi = \langle \lambda_1 \circ j_2, \Omega\mathbf{S}j_1 \rangle$. Let $\psi_3 \in [\mathbf{S}(Y \times W), F]$ be any element such that $\underline{\Omega p}\alpha(\psi_3) = \bar{\pi}(\xi)$. Then an easy diagram chase reveals that $\psi_3 = \overline{\mathbf{S}\pi}(\psi_2)$.

Recalling from §4 of Chapter X the definition of the Samelson product, we see that we are required to construct a map $\omega_3 : \mathbf{S}(Y \times W) \to \mathbf{S}(X \vee Y)$, together with a homotopy η_3 of the constant map to $q \circ \omega_3$. The map ω_3 corresponds under adjointness to the composite

$$Y \times W \xrightarrow{i'_2 \times \Omega\mathbf{S}j_1} \Omega\mathbf{S}(X \vee Y) \times \Omega\mathbf{S}(X \vee Y) \xrightarrow{\Phi} \Omega\mathbf{S}(X \vee Y),$$

where Φ is the commutator map. Thus

$$\omega_3(s \wedge (y, w)) = \begin{cases} j_2(4s \wedge y) & (0 \le s \le \frac{1}{4}), \\ j_1 w(4s - 1) & (\frac{1}{4} \le s \le \frac{1}{2}), \\ j_2((3 - 4s) \wedge y) & (\frac{1}{2} \le s \le \frac{3}{4}), \\ j_1 w(4 - 4s) & (\frac{3}{4} \le s \le 1). \end{cases}$$

The homotopy η_3 is easy to construct, if we realize that $q\omega_3(\ \ \wedge (y, w))$ is essentially the path $w \cdot w^{-1}$. In fact,

$$\eta_3(t, s \wedge (y, w)) = \begin{cases} * & (0 \le s \le \frac{1}{4}), \\ w\left(\dfrac{2t(4s - 1)}{3 - t}\right) & \left(\dfrac{1}{4} \le s \le \dfrac{5 - t}{8}\right), \\ w(t) & \left(\dfrac{5 - t}{8} \le s \le \dfrac{5 + t}{8}\right), \\ w\left(\dfrac{8t(1 - s)}{3 - t}\right) & \left(\dfrac{5 + t}{8} \le s \le 1\right). \end{cases}$$

This completes the definition of ψ.

The map ϕ is the composite

$$F \xrightarrow{\phi_0} \mathbf{S}Y \times W/\{*\} \times W \xrightarrow{g} \mathbf{S}Y \vee (\mathbf{S}Y \wedge W),$$

where g is the map constructed in the proof of (7.8); the map ϕ_0 sends (z, u) into the base point if $z \in \mathbf{S}X$, while

$$\phi_0(j_2(s \wedge y), w) = [s \wedge y, w],$$

where $[z, w]$ is the point of $\mathbf{S}Y \times W/\{*\} \times W$ corresponding to $(z, w) \in \mathbf{S}Y \times W$. Thus

$$\phi_0 \psi(s \wedge y) = \phi_0(j_2(s \wedge y), *) = [s \wedge y, *].$$

But $g_2[s \wedge y, *] = s \wedge y \wedge * = *$, so that $g|\mathbf{S}Y \times \{*\} \simeq g_1|\mathbf{S}Y \times \{*\}$. Hence $\phi \circ \psi|\mathbf{S}Y$ is homotopic to the identity map of $\mathbf{S}Y$.

On the other hand, $\phi_0 \circ \psi|\mathbf{S}Y \wedge W$ is determined by the map $\theta : \mathbf{S}(Y \times W) \to \mathbf{S}Y \times W/\{*\} \times W$ such that

$$\theta(s \wedge (y, w)) = \begin{cases} [4s \wedge y, *] & (0 \le s \le \frac{1}{4}), \\ * & (\frac{1}{4} \le s \le \frac{1}{2} \text{ or } \frac{3}{4} \le s \le 1), \\ [(3 - 4s) \wedge y, w'] & (\frac{1}{2} \le s \le \frac{3}{4}), \end{cases}$$

where

$$w'(t) = \begin{cases} w\left(\dfrac{2t(4s-1)}{3-t}\right) & (0 \le t \le 5 - 8s), \\ w(t) & (5 - 8s \le t \le 1) \end{cases}$$

if $s \le 5/8$ and

$$w'(t) = \begin{cases} w\left(\dfrac{8t(1-s)}{3-t}\right) & (0 \le t \le 8s - 5), \\ w(t) & (8s - 5 \le t \le 1) \end{cases}$$

if $s \ge 5/8$. Now

$$\textbf{(7.9)} \quad g\theta(s \wedge (y, w)) = \begin{cases} j_1(8s \wedge y) & (0 \le s \le \frac{1}{8}), \\ * & (\frac{1}{8} \le s \le \frac{1}{2} \text{ or } \frac{3}{4} \le s \le 1), \\ j_2((5 - 8s) \wedge y \wedge w') & (\frac{1}{2} \le s \le \frac{5}{8}), \\ j_1((6 - 8s) \wedge y) & (\frac{5}{8} \le s \le \frac{3}{4}). \end{cases}$$

To prove that $\psi \circ \phi$ is a homology equivalence, we make use of the inclusions

$$j_1 : \mathbf{S}Y \to \mathbf{S}Y \vee (\mathbf{S}Y \wedge W),$$
$$j_2 : \mathbf{S}Y \wedge W \to \mathbf{S}Y \vee (\mathbf{S}Y \wedge W),$$

and the projections

$$q_1 : \mathbf{S}Y \vee (\mathbf{S}Y \wedge W) \to \mathbf{S}Y,$$

$$q_2 : \mathbf{S}Y \vee (\mathbf{S}Y \wedge W) \to \mathbf{S}Y \wedge W.$$

Evidently, $q_1 \circ (\phi \circ \psi) \circ j_1 \simeq 1$, $q_2 \circ (\phi \circ \psi) \circ j_1 \simeq *$. If

$$\pi_1 = \mathbf{S}\pi : \mathbf{S}(Y \times W) \to \mathbf{S}(Y \wedge W)$$

is the identification map, it follows easily from (7.9) that $q_2 \circ (\phi \circ \psi) \circ j_1 \circ \pi_1$ is homotopic to the map

$$s \wedge (y, w) \to (1 - s) \wedge y \wedge w = \pi_1((1 - s) \wedge (y, w));$$

since $\bar{\pi}_1$ is a monomorphism, the map $q_2 \circ (\phi \circ \psi) \circ j_1$ is homotopic to the map $s \wedge y \wedge w \to (1 - s) \wedge y \wedge w$. From these facts, we see that $\phi \circ \psi$ is indeed a homology equivalence. □

(7.10) Corollary *The map $\Omega\psi : \Omega\mathbf{S}(Y \vee (Y \wedge \Omega\mathbf{S}X)) \to \Omega F$ is a weak homotopy equivalence.* □

(7.11) Corollary *The map*

$$\Omega\mathbf{S}j_1 \otimes \Omega(p \circ \psi) : \Omega\mathbf{S}X \times \Omega\mathbf{S}(Y \vee (Y \wedge \Omega\mathbf{S}X)) \to \Omega\mathbf{S}(X \vee Y)$$

is a weak homotopy equivalence.

For $\Omega\mathbf{S}j_1 \otimes \Omega(p \circ \psi) = (\Omega\mathbf{S}j_1 \otimes \Omega p) \circ (1 \times \Omega\psi) = \chi \circ (1 \times \Omega\psi)$. □

We can now make use of the relationship between the functors J and $\Omega\mathbf{S}$. In fact, let $\sigma : Y \vee (Y \wedge J(X)) \to J(X \vee Y)$ be the map such that

$$\sigma \mid Y = i_2,$$

$$\sigma \mid Y \wedge J(X) = \langle i_2, J(j_1)\rangle$$

($i_2 : Y \to J(X \vee Y)$ is the composite $Y \xrightarrow{j_2} X \vee Y \subset J(X \vee Y)$). Let

$$\gamma = J(j_1) \otimes \hat{\sigma} : J(X) \times J(Y \vee (Y \wedge J(X))) \to J(X \vee Y).$$

Step I will have been completed when we have proved

(7.12) Theorem *The map γ is a homotopy equivalence.*

We begin by observing that the diagram

$$\begin{array}{ccc}
Y \vee (Y \wedge J(X)) & \xrightarrow{\lambda_* = 1 \vee (1 \wedge \hat{\lambda}')} & Y \vee (Y \wedge \Omega\mathbf{S}X) \\
\sigma \downarrow & & \downarrow \tilde{\omega} \\
J(X \vee Y) & \xrightarrow[\hat{\lambda}']{} & \Omega\mathbf{S}(X \vee Y)
\end{array} \qquad (7.13)$$

is homotopy commutative. In fact,

$$\tilde{\lambda}' \circ \sigma \,|\, Y = \tilde{\lambda}' \circ i_2 = \lambda_0 \circ j_2 = \tilde{\omega} \,|\, Y = \tilde{\omega} \circ \lambda_* \,|\, Y,$$

$$\begin{aligned}
\tilde{\lambda}' \circ \sigma \,|\, Y \wedge J(X) &= \tilde{\lambda}' \circ \langle i_2, J(j_1)\rangle \\
&\simeq \langle \tilde{\lambda}' \circ i_2, \tilde{\lambda}' \circ J(j_1)\rangle \quad \text{since } \tilde{\lambda}' \text{ is an H-map} \\
&= \langle i_2', \Omega \mathbf{S} j_1 \circ \tilde{\lambda}'\rangle \\
&= \langle i_2', \Omega \mathbf{S} j_1\rangle \circ (1 \wedge \tilde{\lambda}') \\
&= (\tilde{\omega} \,|\, Y \wedge \Omega \mathbf{S} X) \circ (1 \wedge \tilde{\lambda}') \\
&= \tilde{\omega} \circ \lambda_* \,|\, Y \wedge J(X).
\end{aligned}$$

The map $\tilde{\omega}$ is the adjoint of a map $\omega : \mathbf{S}(Y \vee (Y \wedge \Omega \mathbf{S} X)) \to \mathbf{S}(X \vee Y)$, and $\tilde{\omega} \circ \lambda_*$ is the adjoint of $\omega \circ \mathbf{S}\lambda_*$. Thus σ is a J-adjoint of the latter map. It follows from (7.1) that the diagram

(7.14)
$$\begin{array}{ccc}
J(Y \vee (Y \wedge J(X))) & \xrightarrow{\hat{\sigma}} & J(X \vee Y) \\
\downarrow{\scriptstyle \tilde{\lambda}'} & & \\
\Omega \mathbf{S}(Y \vee (Y \wedge J(X))) & & \downarrow{\scriptstyle \tilde{\lambda}'} \\
\downarrow{\scriptstyle \Omega \mathbf{S}\lambda_*} & \searrow{\scriptstyle \Omega(\omega \circ \mathbf{S}\lambda_*)} & \\
\Omega \mathbf{S}(Y \vee (Y \wedge \Omega \mathbf{S} X)) & \xrightarrow[\Omega\omega]{} & \Omega \mathbf{S}(X \vee Y)
\end{array}$$

is homotopy commutative.

The map $\tilde{\lambda}' : J(X) \to \Omega \mathbf{S} X$ being a weak homotopy equivalence, it follows without difficulty that $\Omega \mathbf{S}\lambda_*$ is a weak homotopy equivalence, and therefore that

$$\tilde{\lambda}'' = \Omega \mathbf{S}\lambda_* \circ \tilde{\lambda}' : J(Y \vee (Y \wedge J(X))) \to \Omega \mathbf{S}(Y \vee (Y \wedge \Omega \mathbf{S} X))$$

is, too. The diagram

$$\begin{array}{ccc}
J(X) & \xrightarrow{J(j_1)} & J(X \vee Y) \\
\downarrow{\scriptstyle \tilde{\lambda}'} & & \downarrow{\scriptstyle \tilde{\lambda}'} \\
\Omega \mathbf{S} X & \xrightarrow[\Omega \mathbf{S}(j_1)]{} & \Omega \mathbf{S}(X \vee Y)
\end{array}$$

being commutative, and $\tilde{\lambda}' : J(X \vee Y) \to \Omega \mathbf{S}(X \vee Y)$ being an H-map, the diagram

$$
\begin{array}{ccc}
J(X) \times J(Y \vee (Y \wedge J(X))) & \xrightarrow{J(j_1) \otimes \hat{\sigma} = \gamma} & J(X \vee Y) \\
\Big\downarrow \hat{\lambda}' \times \hat{\lambda}'' & & \Big\downarrow \hat{\lambda}' \\
\Omega \mathbf{S} X \times \Omega \mathbf{S}(Y \vee (Y \wedge \Omega \mathbf{S} X)) & \xrightarrow[\Omega \mathbf{S} j_1 \otimes \Omega \omega]{} & \Omega \mathbf{S}(X \vee Y)
\end{array} \tag{7.15}
$$

is homotopy commutative. Since the remaining three maps in the diagram are weak homotopy equivalences, so is γ. □

Step II. Consider the three maps $p_1', \phi', p_2 : X \times J(X) \to J(X)$ defined by

$$p_1'(x, w) = x,$$
$$\phi'(x, w) = wx,$$
$$p_2(x, w) = w.$$

The map $\mathbf{S}p_1' - \mathbf{S}\phi' + \mathbf{S}p_2 : \mathbf{S}(X \times J(X)) \to \mathbf{S}J(X)$ is nullhomotopic on $\mathbf{S}(X \vee J(X))$ and therefore determines an element $v' \in [\mathbf{S}(X \wedge J(X)), \mathbf{S}J(X)]$, which is seen to be unique by taking adjoints and applying Lemma (4.1) of Chapter X.

In §2 of Chapter VII we saw that, if $\phi : X \times J(X) \to J(X)$ is the restriction to $X \times J(X)$ of the product in $J(X)$, then $\phi_* - p_{2*} : H_*(X \times J(X)) \to H_*(J(X))$ annihilates the image of the injection $H_*(\{*\} \times J(X)) \to H_*(X \times J(X))$ and induces an *isomorphism*

$$\bar{\phi}_* : H_*(X \times J(X)/\{*\} \times J(X)) \to H_*(J(X)).$$

Using this fact and Lemma (2.7) of Chapter VII, it is not difficult to see that the inclusion $\mathbf{S}X \subsetneq \mathbf{S}J(X)$ and the map v' induce a map

$$v_1 : \mathbf{S}X \vee \mathbf{S}(X \wedge J(X)) \to \mathbf{S}J(X)$$

which is a homotopy equivalence. Applying the functor $Y \wedge$ and using the natural homeomorphism $Y \wedge \mathbf{S}Q = \mathbf{S}(Y \wedge Q)$, we obtain a homotopy equivalence

$$v : \mathbf{S}((Y \wedge X) \vee (Y \wedge X \wedge J(X))) \to \mathbf{S}(Y \wedge J(X)).$$

Then

$$1 \vee v : \mathbf{S}(Y \vee (Y \wedge X) \vee (Y \wedge X \wedge J(X))) \to \mathbf{S}(Y \vee (Y \wedge J(X)))$$

is also a homotopy equivalence. The J-adjoint of $1 \vee v$ is a map

$$\beta : Y \vee (Y \wedge X) \vee (Y \wedge X \wedge J(X)) \to J(Y \vee (Y \wedge J(X)))$$

and its canonical extension

$$\hat{\beta} : J(Y \vee (Y \wedge X) \vee (Y \wedge X \wedge J(X))) \to J(Y \vee (Y \wedge J(X)))$$

is a homotopy equivalence, by (7.1).

(7.16) Corollary *The spaces* $J(X \vee Y)$ *and*

$$J(X) \times J(Y \vee (Y \wedge X) \vee (Y \wedge X \wedge J(X)))$$

have the same homotopy type. □

Let us calculate explicitly the isomorphism $\gamma \circ (1 \times \hat{\beta})$ obtained by combining Steps I and II. Evidently $\gamma \circ (1 \times \hat{\beta}) = J(j_1) \otimes \hat{\sigma} \circ \hat{\beta}$ and it behooves us to calculate $\hat{\sigma} \circ \hat{\beta}$. It suffices to calculate $\hat{\sigma} \circ \hat{\beta} | Q = \hat{\sigma} \circ \beta$, where $Q = Y \vee (Y \wedge X) \vee (Y \wedge X \wedge J(X))$.

The map $\beta | Y$ is the composite

$$Y \subsetneq Y \vee (Y \wedge J(X)) \subsetneq J(Y \vee (Y \wedge J(X)));$$

and $\hat{\sigma} \circ \beta | Y = \sigma | Y = i_2 : Y \to J(X \vee Y)$.

The map $\beta | Y \wedge X$ is the composite

$$Y \wedge X \subsetneq Y \wedge J(X) \subsetneq Y \vee (Y \wedge J(X)) \subsetneq J(Y \vee (Y \wedge J(X)));$$

and $\hat{\sigma} \circ \beta | Y \wedge X = \sigma | Y \wedge X = \langle i_2, J(j_1) | X \rangle = \langle i_2, i_1 \rangle$.

The map $\beta | Y \wedge X \wedge J(X)$ is induced by the map of $Y \wedge (X \times J(X))$ into $J(Y \wedge J(X))$ which sends $y \wedge (x, w)$ into the product (in $J(Y \wedge J(X))$)

$$(y \wedge x)(y \wedge wx)^{-1}(y \wedge w).$$

The map $\hat{\sigma}$ is a homomorphism; it must be remembered, however, that the relation $zz^{-1} = 1$ in $J(Q)$ holds only up to homotopy. Thus it is not necessarily true that $\sigma(z^{-1}) = \sigma(z)^{-1}$; however, the maps $z \to \sigma(z^{-1})$, $z \to \sigma(z)^{-1}$ are homotopic. Thus we may write

$$\hat{\sigma}((y \wedge x) \cdot (y \wedge wx)^{-1} \cdot (y \wedge w)) \simeq \sigma(y \wedge x)\sigma(y \wedge wx)^{-1}\sigma(y \wedge w),$$

where the symbol $\simeq$ means that the maps suggested by the two sides of the formula in question are homotopic. Now $\sigma | Y \wedge J(X)$ is the Samelson product $\langle i_2, J(j_1) \rangle$; this means that $\sigma | Y \wedge J(X)$ is induced by the map of $Y \times J(X)$ into $J(X \vee Y)$ which sends $y \wedge w$ into the commutator $[i_2(y), J(j_1)w]$, which we shall write, for short, simply as $[y, w]$. Thus $\hat{\sigma} \circ \beta | Y \wedge X \wedge J(X)$ is determined by the map of $Y \times X \times J(X)$ which sends $y \wedge x \wedge w$ into

$$\begin{aligned}[y, x][y, wx]^{-1}[y, w] &\simeq yxy^{-1}x^{-1}wxyx^{-1}w^{-1}y^{-1}ywy^{-1}w^{-1} \\ &\simeq yxy^{-1}x^{-1}wxyx^{-1}y^{-1}w^{-1}.\end{aligned}$$

On the other hand,

$$\begin{aligned}[[y, x], w] &= [y, x]w[y, x]^{-1}w^{-1} \\ &\simeq [y, x]w[x, y]w^{-1} \\ &= yxy^{-1}x^{-1}wxyx^{-1}y^{-1}w^{-1}.\end{aligned}$$

It follows that

$$\hat{\sigma} \circ \beta \,|\, Y \wedge X \wedge J(X) = \langle\langle i_2, i_1\rangle, J(j_1)\rangle.$$

Thus we have proved

(7.17) Theorem *A homotopy equivalence*

$$J(X) \times J(Y \vee (Y \wedge X) \vee (Y \wedge X \wedge J(X))) \to J(X \vee Y)$$

is given by $J(j_1) \otimes \hat{\sigma} \circ \hat{\beta}$. *The map* $\hat{\sigma} \circ \hat{\beta}$ *is the canonical extension of the map* $\tau : Y \vee (Y \wedge X) \vee (Y \wedge X \wedge J(X)) \to J(X \vee Y)$ *for which*

$$\tau \,|\, Y = i_2,$$

$$\tau \,|\, Y \wedge X = \langle i_2, i_1\rangle,$$

$$\tau \,|\, Y \wedge X \wedge J(X) = \langle\langle i_2, i_1\rangle, J(j_1)\rangle. \qquad \square$$

Step III. We first prove, by induction on n, that $\mathbf{S}(Y \wedge J(X))$ has the same homotopy type as

$$\mathbf{S}Z_n = \mathbf{S}(Z'_n \vee (Y \wedge X^{(n)} \wedge J(X)),$$

where $Z'_n = \bigvee_{i=1}^n Y \wedge X^{(i)}$. That this is true for $n = 1$ was proved in Step II. Suppose that $f_n : \mathbf{S}Z_n \to \mathbf{S}(Y \wedge J(X))$ is a homotopy equivalence. In Step II we constructed, for each space W, a homotopy equivalence

$$\nu_W : \mathbf{S}((W \wedge X) \vee (W \wedge X \wedge J(X))) \to \mathbf{S}(W \wedge J(X)).$$

The identity map of $\mathbf{S}Z'_n$, together with the map $\nu_{W(n)}$ for $W(n) = Y \wedge X^{(n)}$ define a map $1 \vee \nu_{W(n)} : \mathbf{S}Z_{n+1} \to \mathbf{S}Z_n$ which is a homotopy equivalence. Then $f_{n+1} = f_n \circ (1 \vee \nu_{W(n)})$ is also a homotopy equivalence. $\square$

Let $f'_n = f_n \,|\, \mathbf{S}Z'_n$; then $Z'_{n+1} = Z'_n \vee (Y \wedge X^{(n+1)}) \supset Z'_n$ and $f'_{n+1} \,|\, \mathbf{S}Z_n = f'_n$. Therefore the maps f'_n define a map

$$f' : \mathbf{S}Z' = \mathbf{S} \bigvee_{i=1}^{\infty} Y \wedge X^{(i)} = \bigcup_{n=1}^{\infty} \mathbf{S}Z'_n \to \mathbf{S}(Y \wedge J(X))$$

such that $f' \,|\, \mathbf{S}Z'_n = f'_n$ for all n.

Since X and Y are 0-connected, so is $J(X)$ and therefore the spaces $Y \wedge X^{(n)}$ and $Y \wedge X^{(n-1)} \wedge J(X)$ are n-connected. Hence $\mathbf{S}(Y \wedge X^{(n)})$ and $\mathbf{S}(Y \wedge X^{(n-1)} \wedge J(X))$ are $(n+1)$-connected. It follows that the spaces $\mathbf{S} \bigvee_{i=n+1}^{\infty} Y \wedge X^{(i)}$ is $(n+2)$-connected; by the Hurewicz Theorem, its reduced homology groups vanish in dimensions $\leq n+2$. Then the injection $\tilde{H}_q(Z'_n) \to \tilde{H}_q(Z')$ is an isomorphism for $q \leq n+2$, and it follows from the converse of the Whitehead Theorem that the injection $\pi_q(\mathbf{S}Z'_n) \to \pi_q(\mathbf{S}Z')$ is an isomorphism for $q \leq n+2$. Similarly, the injection $\pi_q(\mathbf{S}Z'_n) \to \pi_q(\mathbf{S}Z_n)$ is an isomorphism for $q \leq n+3$. The diagram

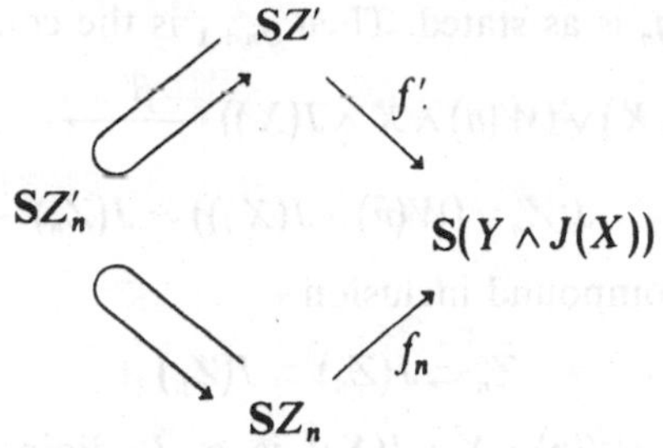

is commutative; since f_n is a homotopy equivalence, $\pi_q(f')$ is an isomorphism for all $q \le n+2$. Since n is arbitrary, $\pi_q(f')$ is a weak homotopy equivalence and therefore a homotopy equivalence.

The J-extension $\hat{g}_0 : J(Z') \to J(Y \wedge J(X))$ is, by (7.1), a homotopy equivalence. This completes Step III.

Step IV. The map $1 \vee f' : \mathbf{S}(Y \vee Z') \to \mathbf{S}(Y \vee (Y \wedge J(X)))$ is a homotopy equivalence. Therefore, if $g : Y \vee Z' \to J(Y \vee (Y \wedge J(X)))$ is a J-adjoint of $1 \vee f'$, then the map

$$(J(j_1) \otimes \hat{\sigma}) \circ (1 \times \hat{g}_1) = J(j_1) \otimes \hat{\sigma} \circ \hat{g}$$

is the desired homotopy equivalence. In order to make this map explicit, let us define a sequence of elements $v_n \in [Y \wedge X^{(n)}, J(X \vee Y)]$ by

$$v_0 = i_2,$$
$$v_{n+1} = \langle v_n, i_1 \rangle.$$

(7.18) Theorem *For each non-negative integer n,*

$$\hat{\sigma} \circ g \,|\, Y \wedge X^{(n)} = v_n .$$

The map $g \,|\, Y$ is the inclusion

$$Y \subsetneq J(Y) \subsetneq J(Y \vee (Y \wedge J(X))),$$

and $\hat{\sigma} \circ g \,|\, Y = \sigma \,|\, Y = i_2 = v_0$.

In order to prove the Theorem for $n > 0$, we shall need to study the behavior of the maps f_n. We shall need

(7.19) Lemma *Let $g_n : Z_n \to J(Y \wedge J(X))$ be a J-adjoint of f_n. Then*

$$\hat{\sigma} \circ g_n \,|\, Y \wedge X^{(i)} = v_i \qquad (i = 1, \ldots, n),$$
$$\hat{\sigma} \circ g_n \,|\, Y \wedge X^{(n)} \wedge J(X) = \langle v_n, J(j_1) \rangle.$$

The Lemma is proved by induction on n. If $n = 0$, $Z_n = Y \wedge J(X)$, f_0 is the identity map, so that g_0 is the inclusion map $Y \wedge J(X) \subsetneq J(Y \wedge J(X))$. Then $\hat{\sigma} \circ g_0 = \sigma \,|\, Y \wedge J(X) = \langle i_2, J(j_1) \rangle = \langle v_0, J(j_1) \rangle$.

Assume that $\hat{\sigma} \circ g_n$ is as stated. Then g_{n+1} is the composite

$$Z_{n+1} = Z'_n \vee (W(n) \wedge X) \vee (W(n) \wedge X \wedge J(X)) \xrightarrow{\beta'}$$
$$J(Z'_n \vee (W(n) \wedge J(X))) = J(Z_n) \xrightarrow{\hat{g}_n} J(Y \wedge J(X)),$$

where $\beta' \mid Z'_n$ is the compound inclusion

$$Z'_n \subsetneq J(Z'_n) \subsetneq J(Z_n)$$

and $\beta' \mid (W(n) \wedge X) \vee (W(n) \wedge X \wedge J(X))$ is a J-adjoint of $v_{W(n)}$. Thus, if $1 \leq i \leq n$,

$$\begin{aligned} \hat{\sigma} \circ g_{n+1} \mid Y \wedge X^{(i)} &= \hat{\sigma} \circ \hat{g}_n \mid Y \wedge X^{(i)} \\ &= \hat{\sigma} \circ g_n \mid Y \wedge X^{(i)} = v_i . \end{aligned}$$

The map $\beta' \mid W(n) \wedge X$ is the compound inclusion

$$W(n) \wedge X \subsetneq W(n) \wedge J(X) \subsetneq J(W(n) \wedge J(X)) \subsetneq J(Z_n);$$

hence

$$\begin{aligned} \hat{\sigma} \circ g_{n+1} \mid Y \wedge X^{(n+1)} &= \hat{\sigma} \circ g_{n+1} \mid W(n) \wedge X \\ &= \hat{\sigma} \circ g_n \mid W(n) \wedge X \\ &= \langle v_n, J(j_1) \rangle \mid W(n) \wedge X \\ &= \langle v_n, J(j_1) \mid X \rangle = \langle v_n, i_1 \rangle = v_{n+1}. \end{aligned}$$

To prove the last relation, consider the diagram

$$\begin{array}{ccccc} W(n) \wedge X \wedge J(X) & \xrightarrow{\beta'} & J(W(n) \wedge J(X)) & \xrightarrow{\hat{\sigma}_{W(n)}} & J(X \vee W(n)) \\ & & \Big\downarrow \hat{g}_n & & \Big\downarrow \hat{\phi} \\ & & J(Y \wedge J(X)) & \xrightarrow[\hat{\sigma}_Y]{} & J(X \vee Y) \end{array}$$

where $\hat{\phi}$ is the canonical extension of the map $\phi : X \vee W(n) \to J(X \vee Y)$ such that

$$\phi \circ i_1 = i_1,$$
$$\phi \circ i_2 = v_n .$$

Then

$$\begin{aligned} \hat{\phi} \circ \sigma \mid W(n) \wedge J(X) &= \hat{\phi} \circ \langle i_2, J(j_1) \rangle \\ &= \langle \hat{\phi} \circ i_2, \hat{\phi} \circ J(j_1) \rangle \\ &\qquad \text{(since } \hat{\phi} \text{ is a homomorphism)} \\ &= \langle v_n, J(j_1) \rangle \end{aligned}$$

while

$$\hat{\sigma} g_n | W(n) \wedge J(X) = \langle v_n, J(j_1) \rangle$$

by induction hypothesis. It follows that the diagram is commutative. But we have seen that

$$\hat{\sigma} \circ \beta' | W(n) \wedge X \wedge J(X) = \langle \langle i_2, i_1 \rangle, J(j_1) \rangle,$$

and therefore

$$\begin{aligned} \hat{\sigma} \circ \hat{g}_n \circ \beta' | W(n) \wedge X \wedge J(X) &= \hat{\phi} \circ \hat{\sigma} \circ \beta' | W(n) \wedge X \wedge J(X) \\ &= \hat{\phi} \langle \langle i_2, i_1 \rangle, J(j_1) \rangle \\ &= \langle \langle v_n, i_1 \rangle, J(j_1) \rangle \\ &= \langle v_{n+1}, J(j_1) \rangle. \qquad \square \end{aligned}$$

Theorem (7.18) now follows easily. In fact, $g | Y \wedge X^{(n)}$ is a J-adjoint of the map $f' | \mathbf{S}(Y \wedge X^{(n)}) = f'_n | \mathbf{S}(Y \wedge X^{(n)}) = f_n | \mathbf{S}(Y \wedge X^{(n)})$; thus we may assume $g | Y \wedge X^{(n)} = g_n | Y \wedge X^{(n)}$ and therefore

$$\hat{\sigma} \circ g | Y \wedge X^{(n)} = \hat{\sigma} \circ g_n | Y \wedge X^{(n)} = v_n . \qquad \square$$

Step V. We now commence the final assault. The idea of the proof is to make repeated use of Step IV. At the nth stage we obtain a decomposition of $J(X_1 \vee \cdots \vee X_k)$ as a product of n factors corresponding to the first n basic products and a remainder term of the form $J(X_{n+1} \vee \cdots)$, where X_{n+1} is the space corresponding to the $(n+1)^{\text{st}}$ basic product and the remaining terms correspond to certain higher basic products. What is needed to carry out the proof is a system of bookkeeping allowing us to keep track of which basic products occur, as well as a convergence argument allowing us to get a decomposition as an infinite product.

Let $X = X_1 \vee \cdots \vee X_k$, and if $m > k$, let

(7.20) $$X_m = w_m(X_1, \ldots, X_k),$$

where w_m is the mth basic product. (If $m \leq k$, then $w_m = x_m$, and therefore (7.20) holds for all $m \geq 1$.) Let

$$R_m = \bigvee_{\substack{i \geq m \\ r(i) < m}} X_i,$$

$$R'_m = \bigvee_{\substack{i > m \\ r(i) < m}} X_i,$$

so that $R_m = X_m \vee R'_m$ (we have abbreviated $r(w_i)$ to $r(i)$). Moreover,

$R_1 = X_1 \vee \cdots \vee X_k$. By Step IV, $J(R_m)$ has the same homotopy type as

$$J(X_m) \times J\left(\bigvee_{i \geq 0} R'_m \wedge X_m^{(i)}\right)$$
$$= J(X_m) \times J\left(\bigvee_{i \geq 0} \bigvee_{\substack{j > m \\ r(j) < m}} X_j \wedge X_m^{(i)}\right).$$

By (6.1), the second factor is equal (up to natural homeomorphism!) to

$$J\left(\bigvee_{\substack{q > m \\ r(q) \leq m}} X_q\right) = J(R_{m+1}).$$

Let $f_m : J(X_m) \times J(R_{m+1}) \to J(R_m)$ be a homotopy equivalence. Then

$$g_m = 1 \times f_m : \{J(X_1) \times \cdots \times J(X_{m-1})\} \times J(X_m) \times J(R_{m+1})$$
$$\to \{J(X_1) \times \cdots \times J(X_{m-1})\} \times J(R_m)$$

is also a homotopy equivalence. Let

$$h_m = g_1 \circ g_2 \circ \cdots \circ g_m | J(X_1) \times \cdots \times J(X_m).$$

Then $h_m | J(X_1) \times \cdots \times J(X_{m-1}) = h_{m-1}$, and therefore the maps h_m define a map $h : \prod_{m=1}^{\infty} J(X_m) \to J(R_1) = J(\bigvee_{i=1}^{k} X_i)$.

Each of the spaces $X_1, \ldots, X_k$ being 0-connected, it follows that, if w is any monomial of weight r, then $w(X_1, \ldots, X_k)$ is $(r-1)$-connected. As the number of admissible monomials of any given weight is finite, we conclude

(7.21) *The connectivity of X_m tends to ∞ with m.* □

Since the connectivity of R_m does not exceed the minimum of the connectivities of the X_i for all $i \geq m$, we further conclude

(7.22) *The connectivity of R_m tends to ∞ with m.* □

We can now prove

(7.23) *The map $h : \prod_{m=1}^{\infty} J(X_m) \to J(\bigvee_{i=1}^{k} X_i)$ is a homotopy equivalence.*

It suffices, because both spaces are CW-complexes, to prove that h is a weak homotopy equivalence. Let r be a positive integer, and choose an integer N such that X_m and R_{m+1} are r-connected for all $m \geq N$. The injections

$$\pi_r(Y_N) \to \pi_r(Y)$$
$$\pi_r(Y_N) \to \pi_r(Y_N \times J(R_{N+1}))$$

are then both isomorphisms. Commutativity of the diagram

$$\begin{array}{ccc} \pi_r(Y_N) & \longrightarrow & \pi_r(Y_N \times J(R_{N+1})) \\ \big\downarrow & & \big\downarrow g_{N*} \\ \pi_r(Y) & \xrightarrow[h_*]{} & \pi_r(X) \end{array}$$

and the fact that g_N is a homotopy equivalence allows us to deduce that h_* is an isomorphism. □

It remains to make explicit the homotopy equivalence h. The main problem is to devise a suitable nomenclature.

We first introduce certain products u_n in two variables x, y (in fact, these elements occur in any system of basic products in x, y). They are defined inductively by

$$u_0 = y,$$

$$u_{n+1} = u_n x.$$

The elements v_n introduced in Step IV can then be expressed in the form

$$v_n = u_n(i_1, i_2).$$

Let $j_m : X_m \to R_m$, $k_m : X_m \to J(R_m)$, $k'_m : R'_m \to J(R_m)$, be the inclusions; then $J(j_m) = \hat{k}_m : J(X_m) \to J(R_m)$. If $j > m$ and $r(j) < m$, so that X_j is a factor of R'_m, let $l'_j : X_j \to R'_m$ be the inclusion, and let $l_j = k'_m \circ l'_j : X_j \hookrightarrow J(R_m)$. It follows from Step IV that the homotopy equivalence

$$f_m : J(X_m) \times J(R_{m+1}) \to J(R_m)$$

has the form

(7.24) $$f_m = J(j_m) \otimes \hat{\theta}_m = \hat{k}_m \otimes \hat{\theta}_m$$

where

(7.25) $$\theta_m \mid R'_m \wedge X_m^{(i)} = u_i(k_m, k'_m),$$

and therefore

(7.26) $$\begin{aligned} \theta_m \mid X_j \wedge X_m^{(i)} &= u_i(k_m, k'_m) \circ (l'_j \wedge 1^{(i)}) \\ &= u_i(k_m, k'_m \circ l'_j) \\ &= u_i(k_m, l_j). \end{aligned}$$

We prove, by induction on m:

$$g_1 = f_1,$$

(7.27) $$g_1 \circ \cdots \circ g_m = \bigotimes_{i=1}^{m-1} \hat{\theta}_1 \circ \cdots \circ \hat{\theta}_{i-1} \circ \hat{k}_i \otimes \hat{\theta}_1 \circ \cdots \circ \hat{\theta}_{m-1} \circ f_m$$

$$(m \geq 2).$$

It suffices to prove

(7.28) $$\hat{\theta}_1 \circ \cdots \circ \hat{\theta}_{m-1} \circ f_m \circ (1 \times f_{m+1}) = \hat{\theta}_1 \circ \cdots \circ \hat{\theta}_{m-1} \circ \hat{k}_m \otimes \hat{\theta}_1 \circ \cdots \circ \hat{\theta}_m \circ f_{m+1}.$$

Composing (7.24) with $(1 \times f_{m+1})$, we find that

$$f_m \circ (1 \times f_{m+1}) = (\hat{k}_m \otimes \hat{\theta}_m) \circ (1 \times f_{m+1}) = \hat{k}_m \otimes \hat{\theta}_m \circ f_{m+1}.$$

Composing the latter relation with the homomorphism $\hat{\theta}_1 \circ \cdots \circ \hat{\theta}_{m-1}$, we obtain (7.28).

Since $f_m | J(X_m) = \hat{k}_m$, we find that

(7.29) $$h_m = g_1 \circ \cdots \circ g_m | J(X_1) \times \cdots \times J(X_m) = \bigotimes_{i=1}^{m} \hat{\theta}_1 \circ \cdots \circ \hat{\theta}_{i-1} \circ \hat{k}_i$$

Thus $h_m = \hat{\omega}_1 \otimes \cdots \otimes \hat{\omega}_m$, where $\hat{\omega}_t = \hat{\theta}_1 \circ \cdots \circ \hat{\theta}_{t-1} \circ \hat{k}_t$. It suffices to prove

(7.30) $$\hat{\omega}_t = \hat{w}_t(i_1, \ldots, i_k) \qquad (t = 1, 2, 3, \ldots).$$

Now both sides of (7.30) are homomorphisms $J(X_t) \to J(X)$, and therefore it suffices to prove that they agree on X_t, i.e.,

(7.31) $$\omega_t = \hat{\omega}_t | X_t = w_t(i_1, \ldots, i_k).$$

Suppose first that $1 \leq t \leq k$, so that $r(t) = 0$. If $m \leq t$, then X_t is one of the factors of R_m. Then k_t is the inclusion, and, if $m < t$, the formula (7.26), with $j = t$, $i = 0$, shows that $\hat{\theta}_m | X_t = u_0(k_m, l_t) = l_t$ is the inclusion. Hence ω_t is the inclusion $i_t : X_t \hookrightarrow J(R_1) = J(X)$; but

$$w_t(i_1, \ldots, i_k) = i_t.$$

Now let $n > k$, and assume that (7.31) holds for all $t < n$. Let $r = r(n)$; then $w_n = T_r^i(w_t) = u_i(w_r, w_t)$, where $i > 0$, $r(t) < r$, and $r < t < n$. Then X_n is a factor of R_q if $r < q \leq n$. Moreover, k_n is the inclusion, and it again follows from (7.26) with $j = n$, $i = 0$ that $\hat{\theta}_m | X_m$ is the inclusion map if $r < m < n$. Hence $\hat{\theta}_{r+1} \circ \cdots \circ \hat{\theta}_{n-1} \circ k_n$ is the inclusion map of X_n into $J(R_{r+1})$. Moreover, again by (7.26),

$$\hat{\theta}_r | X_n = \hat{\theta}_r | X_t \wedge X_r^{(i)} = u_i(k_r, l_t).$$

Thus $\omega_n = \hat{\theta}_1 \circ \cdots \circ \hat{\theta}_{r-1} \circ u_i(k_r, l_t)$; since the composite $\hat{\zeta} = \hat{\theta}_1 \circ \cdots \circ \hat{\theta}_{r-1}$ is a homomorphism, we have

$$\hat{\zeta} \circ u_i(k_r, l_t) = u_i(\hat{\zeta} \circ k_r, \hat{\zeta} \circ l_t)$$

By the induction hypothesis, $\hat{\zeta} \circ k_r = w_r(i_1, \ldots, i_k)$. Again by (7.26), we find that $\hat{\theta}_m | X_t$ is the inclusion provided that $r(t) < m < t$.

Hence $\hat{\theta}_r \circ \cdots \circ \hat{\theta}_{t-1} | X_t$ is the inclusion, and it follows that $\zeta \circ l_t = \hat{\theta}_1 \circ \cdots \circ \hat{\theta}_{t-1} \circ k_t = w_t(i_1, \ldots, i_k)$, again by the induction hypothesis. Thus

$$\begin{aligned}\omega_n &= u_i(w_r(i_1, \ldots, i_k), w_t(i_1, \ldots, i_k)) \\ &= w_n(i_1, \ldots, i_k).\end{aligned}$$

□

8 The Hopf–Hilton Invariants

Let us now suppose that each of the spaces X_i is a sphere $\mathbf{S}^{m_i}$ $(i = 1, \ldots, k)$. Then each of the spaces X_i $(i > k)$ is also a sphere $\mathbf{S}^{m_i}$. The element α_j belongs to $\pi_{m_j}(J(X)) \approx \pi_{m_j}(\Omega \mathbf{S}X) \approx \pi_{m_j+1}(\mathbf{S}X)$; let $\iota_j \in \pi_{m_j+1}(\mathbf{S}X)$ be the corresponding element. If $j > m$, the element $w_j(\alpha_1, \ldots, \alpha_k)$ is an iterated Samelson product of some of the elements $\alpha_1, \ldots, \alpha_k$. Because of Theorem (7.10) of Chapter X, the corresponding element $w_j(\iota_1, \ldots, \iota_k) \in \pi_{m_j+1}(SX)$ is, up to sign, an iterated Whitehead product. Therefore the Hilton–Milnor theorem can be recast to yield

(8.1) Theorem *There is an isomorphism*

$$\pi_{m+1}(\mathbf{S}^{m_1+1} \vee \cdots \vee \mathbf{S}^{m_k+1}) \approx \bigoplus_{j=1}^{\infty} \pi_{m+1}(\mathbf{S}^{m_j+1});$$

if $\beta_j \in \pi_{m+1}(\mathbf{S}^{m_j+1})$ is an arbitrary sequence of elements, the corresponding element of $\pi_{m+1}(\mathbf{S}^{m_1+1} \vee \cdots \vee \mathbf{S}^{m_k+1})$ is

$$\sum_{j=1}^{\infty} w_j(\iota_1, \ldots, \iota_k) \circ \beta_j. \tag{8.2}$$

□

(Note that, because m_j tends to ∞ with j, all but a finite number of the elements β_j are zero, so that the sum (8.2) is finite.).

Theorem (8.1) can be used to study the deviation of the composition operation from additivity in the second factor. In fact, suppose that $k = 2$, $m_1 = m_2 = r$. If $\alpha \in \pi_{n+1}(\mathbf{S}^{r+1})$, then $(\iota_1 + \iota_2) \circ \alpha$ belongs to $\pi_{n+1}(\mathbf{S}^{r+1} \vee \mathbf{S}^{r+1})$, and therefore can be expressed in the form

$$(\iota_1 + \iota_2) \circ \alpha = \iota_1 \circ \alpha' + \iota_2 \circ \alpha'' + \sum_{j=3}^{\infty} w_j(\iota_1, \iota_2) \circ h_{j-3}(\alpha).$$

The projection $q_1 : \mathbf{S}^{r+1} \vee \mathbf{S}^{r+1} \to \mathbf{S}^{r+1}$ has the property that $q_{1*}(\iota_1) = \iota$, the element of $\pi_{r+1}(\mathbf{S}^{r+1})$ corresponding to the identity map, while $q_{1*}(\iota_2) = 0$. Moreover, $q_{1*} w_j(\iota_1, \iota_2) = w_j(q_{1*} \iota_1, q_{1*} \iota_2) = w_j(\iota, 0)$, and this element is zero because of the biadditivity of the Whitehead product and the fact that $w_j(\iota_1, \iota_2)$ involves both ι_1 and ι_2 for all $j \geq 3$. It follows that $\alpha' = \alpha$, and, by a similar argument, $\alpha'' = \alpha$. Moreover, $h_j : \pi_{n+1}(\mathbf{S}^{r+1}) \to \pi_{n+1}(\mathbf{S}^{qr+1})$ is a homomorphism, where q is the weight of the basic product w_{j+3}. Thus

(8.3) Theorem *There are unique homomorphisms* $h_j : \pi_{r+1}(\mathbf{S}^{r+1}) \to \pi_{n+1}(\mathbf{S}^{qr+1})$ $(j = 0, 1, 2, \ldots)$ *such that, for any* $\alpha \in \pi_{n+1}(\mathbf{S}^{r+1})$,

(8.4) $$(\iota_1 + \iota_2) \circ \alpha = \iota_1 \circ \alpha + \iota_2 \circ \alpha + \sum_{j=0}^{\infty} w_{j+3}(\iota_1, \iota_2) \circ h_j(\alpha).$$

The method of the universal example allows us to conclude

(8.5) Theorem *If X is any space and* $\alpha \in \pi_{m+1}(\mathbf{S}^{r+1})$, $\beta_1, \beta_2 \in \pi_{r+1}(X)$, *then*

$$(\beta_1 + \beta_2) \circ \alpha = \beta_1 \circ \alpha + \beta_2 \circ \alpha + \sum_{j=0}^{\infty} w_{j+3}(\beta_1, \beta_2) \circ h_j(\alpha). \qquad \square$$

The homomorphisms $h_j : \pi_{n+1}(\mathbf{S}^{r+1}) \to \pi_{n+1}(\mathbf{S}^{qr+1})$ are called the *Hopf–Hilton homomorphisms*, and the element $h_j(\alpha)$ is the jth *Hopf–Hilton invariant of* α.

The commutative and associative laws for addition in the first factor allow us to deduce certain properties of the homomorphisms h_j. Unfortunately, these are difficult to work with, as the relevant calculations necessarily involve non-basic products, and the calculations needed to get rid of these become increasingly difficult as the dimension increases. We shall give a few of these calculations to illustrate the difficulties.

(8.6) Theorem *If* $n < 3r$ *and* r *is even, then* $2h_0(\alpha) = 0$.

In this case all basic products except w_1, w_2 and w_3 have weight at least 3 and therefore

$$h_j(\alpha) \in \pi_{n+1}(\mathbf{S}^{qr+1}) = 0$$

if $j \geq 1$. Thus (8.4) becomes

$$(\iota_1 + \iota_2) \circ \alpha = \iota_1 \circ \alpha + \iota_2 \circ \alpha + [\iota_2, \iota_1] \circ h_0(\alpha).$$

Interchanging ι_1 and ι_2, we obtain

$$(\iota_2 + \iota_1) \circ \alpha = \iota_2 \circ \alpha + \iota_1 \circ \alpha + [\iota_1, \iota_2] \circ h_0(\alpha);$$

but the product $[\iota_1, \iota_2]$ is not basic, and we must use the relation

$$\begin{aligned} [\iota_1, \iota_2] &= (-1)^{r+1}[\iota_2, \iota_1] \\ &= [\iota_2, \iota_1] \circ ((-1)^{r+1}\iota). \end{aligned}$$

Thus

$$(\iota_2 + \iota_1) \circ \alpha = \iota_2 \circ \alpha + \iota_1 \circ \alpha + [\iota_2, \iota_1] \circ ((-1)^{r+1}\iota) \circ h_0(\alpha).$$

Hence

$$h_0(\alpha) = ((-1)^{r+1}\iota) \circ h_0(\alpha) = (-\iota) \circ h_0(\alpha).$$

Since $h_0(\alpha) \in \pi_{n+1}(\mathbf{S}^{2r+1})$ and $n < 4r$, $h_0(\alpha)$ is a suspension element, we can use Theorem (8.5) of Chapter X to deduce that $(-\iota) \circ h_0(\alpha) = -h_0(\alpha)$, and our conclusion follows. □

A little more difficult is

(8.7) Theorem *If $n < 4r$, then*

$$h_1(\alpha) = h_2(\alpha) = 0 \quad \textit{if } r \textit{ is odd};$$

$$2h_0(\alpha) = 0, \qquad 3h_1(\alpha) = 0, \qquad h_2(\alpha) = -h_1(\alpha) \quad \textit{if } r \textit{ is even}.$$

First calculate $(\iota_2 + \iota_1) \circ \alpha$ as we did before. This time triple products will occur, and equating coefficients of $[[\iota_2, \iota_1], \iota_1]$ and $[[\iota_2, \iota_1], \iota_2]$ yields, in addition to the same relation $h_0(\alpha) = (-1)^{r+1}h_0(\alpha)$ as before, the new relations

$$h_1(\alpha) = (-1)^{r+1}h_2(\alpha), \qquad h_2(\alpha) = (-1)^{r+1}h_1(\alpha).$$

Next we calculate in $\pi_{n+1}(\mathbf{S}^{r+1} \vee \mathbf{S}^{r+1} \vee \mathbf{S}^{r+1})$. Expanding $(\iota_1 + (\iota_2 + \iota_3)) \circ \alpha$, we obtain a sum of terms, each of which has the form $w_i \circ h_j(\alpha)$, where w_i is a basic product. If, on the other hand, we expand $((\iota_1 + \iota_2) + \iota_3) \circ \alpha$, one of the terms we obtain is $[[\iota_3, \iota_2], \iota_1] \circ h_1(\alpha)$. The product $[[\iota_3, \iota_2], \iota_1]$ is not basic; however, if we use the Jacobi identity (Corollary 7.14 of Chapter X), we find that

$$[[\iota_3, \iota_2], \iota_1] = -[[\iota_2, \iota_1], \iota_3] + (-1)^r[[\iota_3, \iota_1], \iota_2],$$

and both products on the right-hand side are basic. We then find, by comparing coefficients of basic products as before, that

$$h_1(\alpha) + (-1)^r h_1(\alpha) = h_2(\alpha),$$

$$h_2(\alpha) = -h_1(\alpha).$$

Using these and the relations obtained earlier, we obtain the desired relations. □

Clearly we can continue this program, obtaining, for each k, a set R_k of relations which are valid whenever $n < kr$. It is tempting to conclude that R_{k+1} contains the old relations R_k, together with new relations among the h_i associated with basic products of weight k. However, this is a bit optimistic. For example, we have used the relation $(-\iota) \circ h_0(\alpha) = -h_0(\alpha)$, valid for $n < 4r$ because $h_0(\alpha)$ is a suspension element. However, if $n \geq 4r$, this is no longer true; for example,

$$(-\iota) \circ h_0(\alpha) = -h_0(\alpha) + [\iota, \iota] \circ h_0(h_0(\alpha)),$$

by (8.12) below.

But there is another kind of difficulty, which arises when we consider four-fold products. The product $[[[\iota_2, \iota_1], \iota_2], \iota_1]$ is not basic. Expanding it by the Jacobi identity, we get $0 = \pm[[[\iota_2, \iota_1], \iota_2], \iota_1] \pm [[[\iota_2, \iota_1], [\iota_2, \iota_1]] \pm [[[\iota_1, [\iota_2, \iota_1]], \iota_2]$ so that

$$[[[\iota_2, \iota_1], \iota_2], \iota_1] = \pm[[\iota_2, \iota_1], [\iota_2, \iota_1]] \pm [[[\iota_2, \iota_1], \iota_1], \iota_2].$$

The second term on the right is a basic product, but the first is not. However, by naturality of the Whitehead product,

$$\begin{aligned} [[\iota_2, \iota_1], [\iota_2, \iota_1]] &= [[\iota_2, \iota_1] \circ \iota, [\iota_2, \iota_1] \circ \iota] \\ &= [\iota_2, \iota_1] \circ [\iota, \iota]. \end{aligned}$$

Thus in our calculations involving products of a given weight, new terms involving products of lower weight may appear. Therefore we shall not continue this program further, leaving it to the interested reader to do so if he desires.

It is useful, however, to have formulas for $(k\beta) \circ \alpha$ for each integer k. These can be calculated inductively, using the "distributive law" (8.5). Thus we need information about the Whitehead products which involve only a single element $\beta \in \pi_{r+1}(X)$.

(8.8) Theorem *Let $\beta \in \pi_{r+1}(X)$. If r is even, $2[\beta, \beta] = 0$ and all Whitehead products in β of weight ≥ 3 are zero. If r is odd, $3[[\beta, \beta], \beta] = 0$ and all Whitehead products in β of weight ≥ 4 vanish.*

As usual, we may assume $\beta = \iota \in \pi_{r+1}(S^{r+1})$. Let p_k be the standard iterated product, defined by

$$p_2 = [\iota, \iota], \qquad p_{k+1} = [p_k, \iota].$$

Suppose first that r is even. Then, by (7.5) of Chapter X, $p_2 = (-1)^{r+1}p_2 = -p_2$, $2p_2 = 0$. Hence $2p_3 = 0$. But the Jacobi identity (7.14) of the same Chapter gives $3p_3 = 0$, and therefore $p_3 = 0$. It follows by induction that $p_k = 0$ for all $k \geq 3$.

Let $q_4 = [p_2, p_2]$, so that $2q_4 = 0$. Applying the Jacobi identity to the three elements p_2, ι, ι, we find that

$$\begin{aligned} 0 &= [[[\iota, \iota], \iota], \iota] + [[\iota, \iota], [\iota, \iota]] + [[\iota, [\iota, \iota]], \iota] \\ &= p_4 + q_4 + (-1)^{r+1}p_4 = q_4. \end{aligned}$$

It follows that all 3 and 4-fold products vanish.

Let p be a k-fold product ($k \geq 5$), and assume that all i-fold products vanish for $3 \leq i < k$. Then $p = [p^1, p^2]$ where p^j is an i_j-fold product and $0 < i_j < k$, $i_1 + i_2 = k$. Since $k \geq 5$, either $i_1 \geq 3$ or $i_2 \geq 3$ and our inductive hypothesis implies that $p = 0$.

The case r odd is a little more difficult. From the Jacobi identity we deduce that $3p_3 = 0$, and therefore $3p_k = 0$ for all $k \geq 3$. Applying the Jacobi identity to the elements p_2, ι, ι, we deduce that $q_4 + 2p_4 = 0$ and therefore $3q_4 = 0$. Let ι^* be the element of $\pi_{2r+1}(S^{2r+1})$ represented by the identity map. By naturality of the Whitehead product, we have

$$q_4 = [[\iota, \iota], [\iota, \iota]] = [p_2 \circ \iota^*, p_2 \circ \iota^*] = p_2 \circ [\iota^*, \iota^*].$$

By what we have already proved, $2[\iota^*, \iota^*] = 0$ and therefore $2q_4 = 0$. We have seen that $3q_4 = 0$, and therefore $q_4 = 0$. But then $2p_4 = 0$, and, since $3p_4 = 0$, we have $p_4 = 0$. Hence all 4-fold products vanish, and $p_k = 0$ for $k \geq 4$. Applying the Jacobi identity to the triple p_3, ι, ι, we find that

$$0 = \pm p_5 \pm p_5 \pm [p_3, p_2],$$

so that $[p_3, p_2] = 0$. Applying the Jacobi identity to p_2, ι, p_2, we obtain

$$0 = \pm[p_3, p_2] \pm [p_3, p_2] \pm [q_4, \iota],$$

so that $q_5 = [q_4, \iota] = 0$. It follows that all 5-fold products vanish. As for six-fold products, the only one to cause difficulty is $[p_3, p_3]$. This time we apply the Jacobi identity to the triple p_3, p_2, ι to obtain

$$\begin{aligned} 0 &= [[p_3, p_2], \iota] + [[p_2, \iota], p_3] + [[\iota, p_3], p_2] \\ &= [[p_3, p_2], \iota] + [p_3, p_3] \pm [p_4, p_2] \end{aligned}$$

But we have seen that $p_4 = 0$ and $[p_3, p_2] = 0$, and therefore $[p_3, p_3] = 0$.

Thus we have proved that all k-fold products vanish $(4 \leq k \leq 6)$. Let $k \geq 7$ and assume all i-fold products vanish for $4 \leq i < k$. If p is a k-fold product, $p = [p^1, p^2]$, where p^j is an i_j-fold product, $0 < i_j < k$, $i_1 + i_2 = k$. Since $k \geq 7$, either i_1 or i_2 is at least 4 and the corresponding product vanishes by the induction hypothesis. Hence $p = 0$ and the proof is complete. □

We can now give a formula for $(k\beta) \circ \alpha$.

(8.9) Theorem *For any integer k and any* $\alpha \in \pi_{n+1}(S^{r+1})$, $\beta \in \pi_{r+1}(X)$,

(8.10) $$(k\beta) \circ \alpha = k(\beta \circ \alpha) + \binom{k}{2}[\beta, \beta] \circ h_0(\alpha) - \binom{k+1}{3}[[\beta, \beta], \beta] \circ h_1(\alpha).$$

In particular, if r is even and $k \equiv 0$ *or* 1 (mod 4), *then* $(k\beta) \circ \alpha = k(\beta \circ \alpha)$; *and if r is odd and* $k \equiv 0$ *or* ± 1 (mod 9) *then*

(8.11) $$(k\beta) \circ \alpha = k(\beta \circ \alpha) + \binom{k}{2}[\beta, \beta] \circ h_0(\alpha).$$

Finally,

(8.12) $$(-\beta) \circ \alpha = -(\beta \circ \alpha) + [\beta, \beta] \circ h_0(\alpha).$$

Again, it suffices to prove the theorem in the special case $X = \mathbf{S}^{r+1}, \beta = \iota$. The theorem is proved by induction on r, it being trivial (and reducing to the equality $(k\iota) \circ \alpha = k\alpha$) when $n = r$. Assume that (8.10) holds for all $\alpha \in \pi_{n+1}(S^{q+1})$ with $m - q < n - r$. It follows that, for any integer l,

$$\begin{aligned}(l[\iota, \iota]) \circ h_0(\alpha) &= l\{[\iota, \iota] \circ h_0(\alpha)\} \\ &\quad + \binom{l}{2}[[\iota, \iota], [\iota, \iota]] \circ h_0(h_0(\alpha)) \\ &\quad - \binom{l+1}{3}[[[\iota, \iota], [\iota, \iota]], [\iota, \iota]] \circ h_1(h_0(\alpha)) \\ &= l([\iota, \iota] \circ h_0(\alpha))\end{aligned}$$

the last two terms vanishing by Theorem (8.8). Similarly,

$$(l[[\iota, \iota], \iota]) \circ h_1(\alpha) = l\{[[\iota, \iota], \iota] \circ h_1(\alpha)\}.$$

The formula (8.10) is patently true if $k = 0$ or 1. By Theorem (8.5),

(8.13) $$\begin{aligned}(2\iota) \circ \alpha &= 2\alpha + \sum_{j=0}^{\infty} w_{j+3}(\iota, \iota) \circ h_j(\alpha) \\ &= 2\alpha + [\iota, \iota] \circ h_0(\alpha) + [[\iota, \iota], \iota] \circ h_1(\alpha) + [[\iota, \iota], \iota] \circ h_2(\alpha),\end{aligned}$$

all higher terms vanishing by Theorem (8.8). Furthermore,

(8.14) $$\begin{aligned}(3\iota) \circ \alpha &= ((2\iota) + \iota) \circ \alpha \\ &= (2\iota) \circ \alpha + \alpha + [\iota, 2\iota] \circ h_0(\alpha) + [[\iota, 2\iota], 2\iota] \circ h_1(\alpha) \\ &\quad + [[\iota, 2\iota], \iota] \circ h_2(\alpha) \\ &= (2\iota) \circ \alpha + \alpha + 2[\iota, \iota] \circ h_0(\alpha) + 4[[\iota, \iota], \iota] \circ h_1(\alpha) \\ &\quad + 2[[\iota, \iota], \iota] \circ h_2(\alpha) \\ &= 3\alpha + 3[\iota, \iota] \circ h_0(\alpha) + 5[[\iota, \iota], \iota] \circ h_1(\alpha) \\ &\quad + 3[[\iota, \iota], \iota] \circ h_2(\alpha) \\ &= 3\alpha + 3[\iota, \iota] \circ h_0(\alpha) - [[\iota, \iota], \iota] \circ h_1(\alpha),\end{aligned}$$

since $3[[\iota, \iota], \iota] = 0$.

On the other hand,

$$\begin{aligned}(3\iota) \circ \alpha &= (\iota + 2\iota) \circ \alpha \\ &= \alpha + (2\iota) \circ \alpha + [2\iota, \iota] \circ h_0(\alpha) + [[2\iota, \iota], \iota] \circ h_1(\alpha) \\ &\quad + [[2\iota, \iota], 2\iota] \circ h_2(\alpha) \\ &= 3\alpha + 3[\iota, \iota] \circ h_0(\alpha) + 3[[\iota, \iota], \iota] \circ h_1(\alpha) \\ &\quad + 5[[\iota, \iota], \iota] \circ h_2(\alpha) \\ &= 3\alpha + 3[\iota, \iota] \circ h_0(\alpha) - [[\iota, \iota], \iota] \circ h_2(\alpha).\end{aligned}$$

Therefore, comparing the two formulae for $(3\iota) \circ \alpha$, we see that

(8.15) $$[[\iota, \iota], \iota] \circ h_1(\alpha) = [[\iota, \iota], \iota] \circ h_2(\alpha).$$

Thus (8.13) simplifies to

(8.16) $$(2\iota) \circ \alpha = 2\alpha + [\iota, \iota] \circ h_0(\alpha) - [[\iota, \iota], \iota] \circ h_1(\alpha)$$

(again we have made use of the fact that $3[[\iota, \iota], \iota] = 0$).

It follows from (8.14) and (8.16) that (8.10) holds for $k = 0, 1, 2, 3$. Assuming (8.10), we expand $((k+1)\iota) \circ \alpha$ to obtain

$$\begin{aligned}
((k+1)\iota) \circ \alpha &= (k\iota + \iota) \circ \alpha \\
&= (k\iota) \circ \alpha + \alpha + [\iota, k\iota] \circ h_0(\alpha) + [[\iota, k\iota], k\iota] \circ h_1(\alpha) \\
&\quad + [[\iota, k\iota], \iota] \circ h_2(\alpha) \\
&= (k\iota) \circ \alpha + \alpha + k[\iota, \iota] \circ h_0(\alpha) \\
&\quad + (k^2 + k)[[\iota, \iota], \iota] \circ h_1(\alpha) \\
&= k\alpha + \binom{k}{2}[\iota, \iota] \circ h_0(\alpha) - \binom{k+1}{3}[[\iota, \iota], \iota] \circ h_1(\alpha) \\
&\quad + \alpha + k[\iota, \iota] \circ h_0(\alpha) + (k^2 + k)[[\iota, \iota], \iota] \circ h_1(\alpha) \\
&= (k+1)\alpha + \binom{k+1}{2}[\iota, \iota] \circ h_0(\alpha) \\
&\quad + \frac{k(k+1)}{6}\{6 - (k-1)\}[[\iota, \iota], \iota] \circ h_1(\alpha).
\end{aligned}$$

The coefficient of the last term is

$$\begin{aligned}
\frac{k(k+1)}{6}(6 - k + 1) &= \frac{k(k+1)}{6}(-k - 2 + 9) \\
&= -\binom{k+2}{3} + 3\binom{k+1}{2}.
\end{aligned}$$

Since $3[[\iota, \iota], \iota] = 0$, we finally obtain

$$\begin{aligned}
((k+1)\iota) \circ \alpha &= (k+1)\alpha + \binom{k+1}{2}[\iota, \iota] \circ h_0(\alpha) \\
&\quad - \binom{k+2}{3}[[\iota, \iota], \iota] \circ h_1(\alpha).
\end{aligned}$$

This gives an inductive proof that (8.10) holds for all non-negative integers k.

If k is a positive integer, recall that

$$\binom{-k}{i} = (-1)^i\binom{k+i-1}{i};$$

in particular,

$$\binom{-k}{2} = \binom{k+1}{2}, \quad \binom{-k+1}{3} = -\binom{k+1}{3},$$

Thus (8.10) can be rewritten

$$(-k\iota) \circ \alpha = -k\alpha + \binom{k+1}{2}[\iota, \iota] \circ h_0(\alpha) + \binom{k+1}{3} \circ h_1(\alpha),$$

which is easily proved by expanding the equation

$$0 = ((k\iota) + (-k\iota)) \circ \alpha$$

and using (8.10).

This completes the proof of (8.10). The remainder of the theorem consists of easy consequences of (8.10) and the fact that $2[\iota, \iota] = 0$ and $[[\iota, \iota], \iota] = 0$ if r is even and $3[[\iota, \iota], \iota] = 0$ if r is odd. □

Now suppose that $n = 2r$. Then

$$h_0 : \pi_{2r+1}(\mathbf{S}^{r+1}) \to \pi_{2n+1}(\mathbf{S}^{2r+1}) \approx \mathbf{Z},$$

while, if $i > 0$, the range of h_i is $\pi_{2r+1}(\mathbf{S}^{kr+1}) = 0$ for some integer $k \geq 3$. In §1 we defined $H_0 : \pi_{2r+1}(\mathbf{S}^{r+1}) \to \mathbf{Z}$, and it is natural to ask whether these two homomorphisms are the same. In fact, let $\iota^* \in \pi_{2r+1}(\mathbf{S}^{2r+1})$ be the homotopy class of the identity map. Then

(8.17) Theorem *If* $\alpha \in \pi_{2r+1}(\mathbf{S}^{r+1})$, *then* $h_0(\alpha) = H_0(\alpha)\iota^*$.

This follows immediately by comparing (8.4) with the formula obtained from (1.16) by setting $\beta_1 = \iota_2$, $\beta_2 = \iota_1$. □

Thus h_0 is a true generalization of H_0, and the h_i can be thought of as higher generalized Hopf invariants.

Exercises

1. Determine $\pi_{n+1}(X \vee Y)$ when X is 0-connected and Y is $(n-1)$-connected.

2. Let $\theta^q : H^q(\ \ ; \Pi) \to H^{q+k}(\ \ ; G)$ be the components of a stable operation θ. If $f: X \to Y$, let

$$K^q = \{u \in H^q(Y; \Pi) \mid f^*u = 0 \text{ and } \theta^q(u) = 0\}$$
$$L^q = H^q(X; G)/\{f^*H^q(Y; G) + \theta^{q-k}H^{q-k}(X; \Pi)\}.$$

By analogy with the construction of §3, show how to define *functional operations* $\theta_f^q : K^q \to L^{q+k-1}$, and prove their principal properties.

3. Let $f: \mathbf{S}^{q+k-1} \to \mathbf{S}^q$ represent $\alpha \in \pi_{q+k-1}(\mathbf{S}^q)$ $(k > 1)$. Prove that α is detectable by a stable operation θ if and only if $\theta_f : H^q(\mathbf{S}^q; \Pi) \to H^{q+k-1}(\mathbf{S}^{q+k-1}; G)$ is non-zero.

4. Let $f: \mathbf{S}^p \times \mathbf{S}^p \to \mathbf{S}^p$ be a map of type (m, n) (i.e., $f \mid \mathbf{S}^p \times \{*\}$ has degree m and $f \mid \{*\} \times \mathbf{S}^p$ has degree n). Let $g: \mathbf{S}^{2p+1} \to \mathbf{S}^{p+1}$ be the map obtained from f by the Hopf construction. Prove that $H_0(g) = \pm mn$.
5. Prove that, if $\alpha \in \pi_{n+1}(\mathbf{S}^{r+1})$, $\beta \in \pi_m(\mathbf{S}^n)$, then
$$h_j(\alpha \circ E\beta) = h_j(\alpha) \circ E\beta.$$
6. Prove that, if $\alpha \in \pi_n(\mathbf{S}^r)$, $\beta \in \pi_{m+1}(\mathbf{S}^{n+1})$, then
$$h_j((E\alpha) \circ \beta) = E(\underbrace{\alpha \wedge \cdots \wedge \alpha}_{k \text{ factors}}) \circ h_j(\beta),$$
where k is the weight of the basic product w_{j+3}.

CHAPTER XII

Stable Homotopy and Homology

The adjoint of the identity map of $\mathbf{S}W$ is a map $\lambda_0 : W \to \Omega\mathbf{S}W$. For any CW-complex K, $[K, \Omega\mathbf{S}W] \approx [\mathbf{S}K, \mathbf{S}W]$; moreover, the injection $[K, W] \to [K, \Omega\mathbf{S}W]$ corresponds under this isomorphism to the suspension operator

$$\mathbf{S}_* : [K, W] \to [\mathbf{S}K, \mathbf{S}W].$$

If W is $(n-1)$-connected, then $(\Omega\mathbf{S}W, W)$ is $(2n-1)$-connected. Thus $\mathbf{S}_*$ is an isomorphism if $\dim K \leq 2n-2$ and an epimorphism if $\dim K = 2n-1$.

In particular, we may take K to be a sphere $\mathbf{S}^r$, so that

$$E = \mathbf{S}_* : \pi_r(W) \to \pi_{r+1}(\mathbf{S}W)$$

is an isomorphism for $r \leq 2n-2$ and an epimorphism if $r = 2n-1$. To study the behavior of E in higher dimensions we must investigate the relative groups $\pi_r(\Omega\mathbf{S}W, W)$. By the considerations of Chapter VII, we may replace $\Omega\mathbf{S}W$ by the reduced product $J(W)$. Now the groups $\pi_r(J(W), W)$ can be calculated in a range of dimensions, and this leads to an exact sequence

$$\pi_{3n-2}(W) \to \pi_{3n-1}(\mathbf{S}W) \to \cdots$$
$$\to \pi_q(W) \xrightarrow{E} \pi_{q+1}(\mathbf{S}W) \xrightarrow{H} \pi_{q+1}(W * W) \xrightarrow{P} \pi_{q-1}(W) \to \cdots$$

which is analogous, in a certain sense, to the sequence involving the homology suspension, which was studied in Chapter VIII. In particular, when $W = \mathbf{S}^n$, we obtain the sequence

$$\pi_{3n-2}(\mathbf{S}^n) \to \pi_{3n-1}(\mathbf{S}^{n+1}) \to \cdots$$
$$\to \pi_q(\mathbf{S}^n) \xrightarrow{E} \pi_{q+1}(\mathbf{S}^{n+1}) \xrightarrow{H} \pi_{q+1}(\mathbf{S}^{2n+1}) \xrightarrow{P} \pi_{q-1}(\mathbf{S}^n) \to \cdots$$

The homomorphism H is the Hopf–James invariant j_2 of Chapter VII, and the homomorphism P is determined by the observation that $P \circ E^2 : \pi_{q-1}(\mathbf{S}^{2n-1}) \to \pi_{q-1}(\mathbf{S}^n)$ is composition with the Whitehead product $[\iota_n, \iota_n] \in \pi_{2n-1}(\mathbf{S}^n)$. These results are developed in §§1, 2. As a consequence we calculate the homotopy groups $\pi_{n+1}(\mathbf{S}^n)$ and $\pi_{n+2}(\mathbf{S}^n)$ for every n.

In §3 we consider the suspension category $\mathscr{S}$. This was introduced by Spanier and J. H. C. Whitehead [1] in 1953 as a first approximation to homotopy theory. The objects of $\mathscr{S}$ are spaces in $\mathscr{K}_*$, but $\{X, Y\} = \mathscr{S}(X, Y)$ is the direct limit

$$[X, Y] \to [\mathbf{S}X, \mathbf{S}Y] \to [\mathbf{S}^2X, \mathbf{S}^2Y] \to \cdots$$

under iterated suspension. Thus, while $[X, Y]$ has no particular structure, $\{X, Y\}$ is an abelian group, and the operation $\{Y, Z\} \times \{X, Y\} \to \{X, Z\}$ of composition is biadditive.

Of particular importance are the stable homotopy groups $\sigma_q(X) = \{\mathbf{S}^q, X\}$. These form a graded group $\sigma_*(X)$; moreover, $\sigma_* = \sigma_*(\mathbf{S}^0)$ is a graded ring and $\sigma_*(X)$ is a graded σ_*-module. We prove a stable Hurewicz Theorem: if X is $(m-1)$-connected then $\sigma_m(X) \approx H_m(X)$. An exact sequence

$$\pi_{m+2}(X) \to H_{m+2}(X) \xrightarrow{\ b\ } \pi_m(X) \otimes \mathbf{Z}_2 \to \pi_{m+1}(X) \to H_{m+1}(X) \to 0$$

leads to a determination of $\sigma_{m+1}(X)$ up to a group extension. This sequence was studied by J. H. C. Whitehead [7] for $m = 2$ and by the present author [3] for $m > 2$.

Motivated by a desire to describe the group extension for $\sigma_{m+1}(X)$, we discuss group extensions in §5. If C is a free chain complex, then by the Universal Coefficient Theorem there is a short exact sequence

$$0 \to \operatorname{Ext}(H_{n-1}(C), G) \to H^n(C; G) \to \operatorname{Hom}(H_n(C), G) \to 0;$$

the sequence splits, but not naturally. Thus a class $u \in H^n(C; G)$ determines a homomorphism

$$u_* : H_n(C) \to G,$$

determined by the Kronecker index; but u does not determine in a natural way an element of $\operatorname{Ext}(H_{n-1}(C), G)$. Nevertheless, there is associated with u in a natural way an element $u_\dagger \in \operatorname{Ext}(H_{n-1}(C), \operatorname{Cok} u_*)$. And we further show that if X is an $(m-1)$-connected space, there is a canonically defined cohomology class $u = Sq^2\iota^m(X)$ such that the associated group extension $u_\dagger$ describes the extension for $\sigma_{m+1}(X)$.

Stable homotopy groups turn out to behave very much like homology groups. In fact, if the Eilenberg–Steenrod axioms are modified to take account of the fact that we are working in a category of spaces with base points, the stable homotopy functors satisfy the new axioms except for the Dimension Axiom. This leads in §6 to a consideration of homology theories in general; the two kinds of theories (those for pointed spaces and those for pairs) are compared and shown to be completely equivalent. In §7, the corresponding relationships for cohomology groups are established.

1 Homotopy Properties of the James Imbedding

Let W be a space in the category $\mathscr{K}_*$. In §2 of Chapter VII we constructed the James reduced product $J(W)$ and a weak homotopy equivalence $\hat{\lambda}: J(W) \to \Omega^*\mathbf{S}W$. Composing $\hat{\lambda}$ with the homotopy equivalence $h': \Omega^*\mathbf{S}W \to \Omega\mathbf{S}W$, we obtain a weak homotopy equivalence $\hat{\lambda}': J(W) \to \Omega\mathbf{S}W$. The restriction of $\hat{\lambda}'$ to the subspace W is the canonical imbedding $\lambda_0 : W \to \Omega\mathbf{S}W$, adjoint to the identity map of $\mathbf{S}W$.

It follows from the Five-Lemma that

(1.1) *The homomorphism*

$$\pi_q(\hat{\lambda}') : \pi_q(J(W), W) \to \pi_q(\Omega\mathbf{S}W, W)$$

is an isomorphism for every q. □

To see the significance of the groups $\pi_q(J(W), W)$, let us examine the injection

$$\pi_q(W) \to \pi_q(\Omega\mathbf{S}W).$$

More generally, if X is any space in $\mathscr{K}_*$, we have an injection

(1.2) $$[X, W] \to [X, \Omega\mathbf{S}W];$$

the latter set has a natural group structure, and is isomorphic, under the adjointness relation

(1.3) $$[X, \Omega\mathbf{S}W] \approx [\mathbf{S}X, \mathbf{S}W],$$

with the group $[\mathbf{S}X, \mathbf{S}W]$. Thus the injection (1.2) is equivalent, under the isomorphism (1.3), to a map

$$E : [X, W] \to [\mathbf{S}X, \mathbf{S}W].$$

On the other hand, the suspension functor $\mathbf{S}$ is compatible with the homotopy relation, and accordingly the map $f \to \mathbf{S}f$ induces a correspondence

$$\mathbf{S}_* : [X, W] \to [\mathbf{S}X, \mathbf{S}W],$$

which is again called the *suspension*.

(1.4) Lemma *The map* $E : [X, W] \to [\mathbf{S}X, \mathbf{S}W]$ *is the suspension.*

For let $f: X \to W$; then $\lambda_0 \circ f: X \to \Omega(\mathbf{S}W)$ is given by

$$(\lambda_0 \circ f)(x)(t) = t \wedge f(x)$$

for $t \in \mathbf{S}$, $x \in X$, and the corresponding map of $\mathbf{S}X$ into $\mathbf{S}W$ sends the point $t \wedge x$ into $t \wedge f(x)$. But the latter map is just the suspension $\mathbf{S}f$ of the map f. □

In particular, we are interested in the suspension

$$E : \pi_q(W) \to \pi_{q+1}(\mathbf{S}W);$$

and the groups $\pi_q(JW, W)$ measure, because of the exactness of the homotopy sequence of the pair (JW, W), the extent to which E fails to be an isomorphism.

That E is not always an isomorphism is shown by the example $W = \mathbf{S}^1$, $q = 2$. For we have seen in §8 of Chapter IV that $\pi_2(\mathbf{S}^1) = 0$, while $\pi_3(\mathbf{S}^2) \approx \mathbf{Z}$. This is contrary to the situation in homology theory, where $H_q(W) \approx H_{q+1}(\mathbf{S}W)$ for all q. The reason for this is that the excision axiom holds for the homology groups, but fails to hold for the homotopy groups (cf. Example 5, of Chapter IV, §7). In fact, consider the diagram

$$\pi_q(W) \xleftarrow{\partial_*} \pi_{q+1}(\mathbf{T}_+ W, W) \xrightarrow{k_*} \pi_{q+1}(\mathbf{S}W, \mathbf{T}_- W) \xleftarrow{j_*} \pi_{q+1}(\mathbf{S}W),$$

in which the homomorphisms j_* and k_* are injections. Note that ∂_* and j_* are isomorphisms because of the exactness of the appropriate homotopy sequences. The easy proof of

(1.5) *The composite* $j_*^{-1} \circ k_* \circ \partial_*^{-1}$ *is the suspension homomorphism* E

is left to the reader.

Thus E is equivalent to the homotopy excision k_*.

Let us recall the Freudenthal Suspension Theorem (Theorem (7.13) of Chapter VII):

If W is $(n-1)$-connected then $E : \pi_q(W) \to \pi_{q+1}(\mathbf{S}W)$ is an isomorphism for $q < 2n - 1$ and an epimorphism for $q = 2n - 1$. □

(1.6) Corollary *The suspension $E : \pi_q(\mathbf{S}^n) \to \pi_{q+1}(\mathbf{S}^{n+1})$ is an isomorphism for $q < 2n - 1$ and an epimorphism for $q = 2n - 1$.*

In fact, a more general result is true.

(1.7) Theorem *Let K be a* CW*-complex and let W be $(n-1)$-connected. Then $E : [K, W] \to [\mathbf{S}K, \mathbf{S}W]$ is an isomorphism if* $\dim K < 2n - 1$ *and an epimorphism if* $\dim K = 2n - 1$.

(We shall assume that $n > 1$, so that W is 1-connected. The case $n = 1$ is uninteresting and is relegated to the Exercises).

It suffices, by the definition of E, to verify that the injection $[K, W] \to [K, \Omega\mathbf{S}W]$ has the desired properties. Since $\bar{\lambda}' : (JW, W) \to (\Omega\mathbf{S}W, W)$ is a weak homotopy equivalence, it suffices to prove the same statement for the injection $[K, W] \to [K, JW]$. The latter follows in turn from the following two Lemmas.

(1.8) Lemma *If (P, Q) is an m-connected pair and X is a CW-complex of dimension r, then the injection $i_* : [X, Q] \to [X, P]$ is an isomorphism if $r < m$ and an epimorphism if $r = m$.*

(1.9) Lemma *If W is $(n - 1)$-connected, the pair (JW, W) is $(2n - 1)$-connected.*

PROOF OF LEMMA (1.8). Suppose first that $f: (X, *) \to (P, *)$. We may then regard f as a map of $(X, *)$ into (P, Q); if $r \leq m$, the latter map is compressible. This means that f is homotopic (rel. $*$) to a map of X into Q, and therefore i_* is an epimorphism.

Let $f_0, f_1 : (X, *) \to (Q, *)$ and suppose that $f_0 \simeq f_1$ (rel. $*$) in P. If $F : (\mathbf{I} \times X, \dot{\mathbf{I}} \times X \cup 1 \times \{*\}) \to (P, Q)$ is a homotopy of f_0 to f_1, and if $r < m$, then F is compressible to a map $G : \mathbf{I} \times X \to Q$. Then G is a homotopy of f_0 to f_1 in Q. Hence i_* is a monomorphism. □

PROOF OF LEMMA (1.9). In §2 of Chapter VII we saw that $J = J(W)$ is filtered by the subspaces $J^m = J^m(W)$, and that the identification maps $W^m \to J^m$ give rise to isomorphisms

$$\phi_m : H_*((W, e) \times (J^m, J^{m-1})) \approx H_*(J^{m+1}, J^m).$$

As $(J^1, J^0) = (W, e)$, it follows from the Künneth Theorem, by induction on m, that $H_q(J^{m+1}, J^m) = 0$ for all $q < (m + 1)n$. In order to use the converse of the relative Hurewicz Theorem, we need to know that J^m is 1-connected. Accordingly, we need to prove the following statement:

$$J^m \text{ is 1-connected and } (J^{m+1}, J^m) \text{ is } ((m + 1)n - 1)\text{-connected}.$$

As $(m + 1)n - 1 \geq 1$ for all $m \geq 1$, this is easily proved by induction on m.

It follows from Corollary (3.5) of Chapter II that

$$(J, J^m) \text{ is } ((m + 1)n - 1)\text{-connected}.$$

In particular, (J^2, J^1) is $(2n - 1)$-connected and (J, J^2) is $(3n - 1)$-connected and therefore $(2n - 1)$-connected. By Lemma (3.4) of Chapter II, $(J, W) = (J, J^1)$ is $(2n - 1)$-connected. □

2 Suspension and Whitehead Products

Let us see what can be said about the kernel and the cokernel of the suspension homomorphism E. Let W be an $(n - 1)$-connected space ($n \geq 2$). Then the injection $\pi_q(W) \to \pi_q(JW)$ can be composed with the standard isomorphism $\pi_q(JW) \approx \pi_q(\Omega SW) \approx \pi_{q+1}(SW)$; the composite is the suspension $E : \pi_q(W) \to \pi_{q+1}(SW)$. Therefore our task is to calculate the groups $\pi_q(JW, W)$.

We have seen that (JW, J^2W) is $(3n-1)$-connected, and therefore the injection $\pi_q(J^2W, W) \to \pi_q(JW, W)$ is an isomorphism for $q < 3n-1$ and an epimorphism for $q = 3n-1$. On the other hand, there is a relative homeomorphism $f: (W \times W, W \vee W) \to (J^2W, W)$. The spaces $W \times W$ and $W \vee W$ are 1-connected, and the pair $(W \times W, W \vee W)$ is $(2n-1)$-connected; moreover, the map $f \mid W \vee W : W \vee W \to W$ is n-connected. It follows from Theorem 7.14 of Chapter VII that

$$f_* : \pi_q(W \times W, W \vee W) \to \pi_q(J^2W, W)$$

is an isomorphism for $q \leq 3n-2$ and an epimorphism for $q = 3n-1$.

On the other hand, the identification map is a relative homeomorphism $g : (W \times W, W \vee W) \to (W \wedge W, *)$. The map $g \mid W \vee W : W \vee W \to \{*\}$ is still n-connected, and we deduce as before that $g_* : \pi_q(W \times W, W \vee W) \to \pi_q(W \wedge W)$ is an isomorphism for $q \leq 3n-2$ and an epimorphism if $q = 3n-1$.

The maps f and g are related by the property: if $u, v \in W \times W$ are points such that $f(u) = f(v)$, then $g(u) = g(v)$. Because f and g are identification maps, there is a unique map $h : (J^2W, W) \to (W \wedge W, *)$ such that $h \circ f = g$, and h is also a relative homeomorphism. As before, we conclude that $h_* : \pi_q(J^2W, W) \to \pi_q(W \wedge W)$ is an isomorphism for $q \leq 3n-2$ and an epimorphism for $q = 3n-1$. Let $\bar{h} : (JW, W) \to (J(W \wedge W), *)$ be the combinatorial extension (cf. §2 of Chapter VII) of h.

(2.1) Lemma *The diagram*

$$\begin{array}{ccc} \pi_q(J^2W, W) & \longleftarrow & \pi_q(JW, W) \\ \downarrow h_* & & \downarrow \bar{h}_* \\ \pi_q(W \wedge W) & \longrightarrow & \pi_q(J(W \wedge W)) \end{array}$$

in which the horizontal arrows denote injections, is commutative. The homomorphism $\bar{h}_* : \pi_q(JW, W) \to \pi_q(J(W \wedge W))$ *is an isomorphism for* $q \leq 3n-2$ *and an epimorphism for* $q = 3n-1$.

The commutativity of the diagram follows from the fact that $\bar{h}$ is an extension of h. Since $W \wedge W$ is $(2n-1)$-connected, $(J(W \wedge W), W \wedge W)$ is $(4n-1)$-connected, so that the injection $\pi_q(W \wedge W) \to \pi_q(J(W \wedge W))$ is an isomorphism for $q \leq 4n-2$, and the last statement follows from the properties we have already established. □

The following theorem is parallel, in an obvious sense, to the Homology Suspension Theorem (Theorem (1.5) of Chapter VIII).

(2.2) Theorem *If W is an $(n-1)$-connected space, there is an exact sequence*

(2.3) $$\pi_{3n-2}(W) \to \pi_{3n-1}(SW) \to \pi_{3n-1}(W * W) \to \cdots$$
$$\cdots \to \pi_q(W) \xrightarrow{E} \pi_{q+1}(SW) \xrightarrow{H} \pi_{q+1}(W * W) \xrightarrow{P} \pi_{q-1}(W)$$
$$\to \cdots$$

The sequence (2.3) is obtained from the appropriate segment of the homotopy sequence of the pair (JW, W) by replacing the groups $\pi_q(JW)$, $\pi_q(JW, W)$ by the isomorphic groups $\pi_{q+1}(SW)$, $\pi_{q+1}(W * W) \approx \pi_{q+1}(S(W \wedge W)) \approx \pi_q(J(W \wedge W))$. It is referred to, for lack of a better name, as the EHP-sequence.

The homomorphism E is the suspension; the homomorphism H is the second Hopf-James invariant j_2. The homomorphism P is more obscure; we shall identify it only in the special case $W = \mathbf{S}^n$. When this is the case, the EHP-sequence becomes

$$\cdots \to \pi_q(\mathbf{S}^n) \xrightarrow{E} \pi_{q+1}(\mathbf{S}^{n+1}) \xrightarrow{H} \pi_{q+1}(\mathbf{S}^{2n+1}) \xrightarrow{P} \pi_{q-1}(\mathbf{S}^n) \to \cdots$$

When $q \leq 3n - 2$, the double suspension

$$E^2 : \pi_{q-1}(\mathbf{S}^{2n-1}) \to \pi_{q+1}(\mathbf{S}^{2n+1})$$

is an isomorphism, and we have

(2.4) Theorem *The composite $P \circ E^2 : \pi_{q-1}(\mathbf{S}^{2n-1}) \to \pi_{q-1}(\mathbf{S}^n)$ is described by*

$$PE^2\alpha = [\iota_n, \iota_n] \circ \alpha.$$

There is a commutative diagram

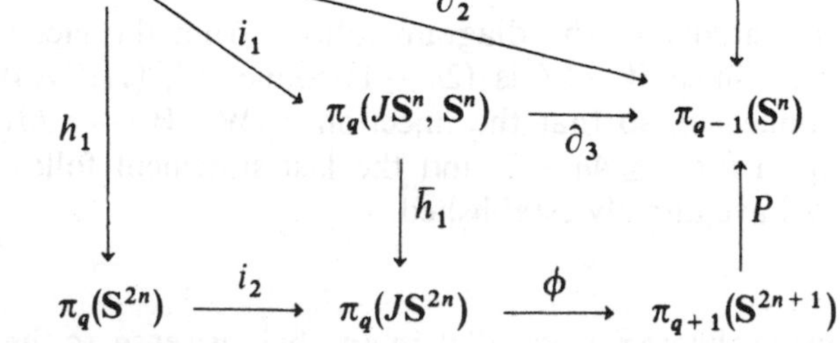

in which the homomorphisms k_1 and k_2 are induced by the characteristic map $k : (\mathbf{E}^{2n}, \mathbf{S}^{2n-1}) \to (J^2\mathbf{S}^n, \mathbf{S}^n)$ for the $2n$-cell in the CW-complex $J(\mathbf{S}^n)$, h_1

by the identification map $h : (J^2\mathbf{S}^n, \mathbf{S}^n) \to (\mathbf{S}^{2n}, *)$, $\bar{h}_1$ by the combinatorial extension $\bar{h} : (J\mathbf{S}^n, \mathbf{S}^n) \to (J\mathbf{S}^{2n}, *)$ of h, ϕ is the standard isomorphism, i_1 and i_2 are injections, and $\partial_1, \partial_2, \partial_3$ are the boundary operators of the appropriate homotopy sequences. The map $h \circ k$ has degree 1, and ∂_1 is an isomorphism; hence $h_1 \circ k_1 \circ \partial_1^{-1} = E$. We have seen that $\phi \circ i_2 = E$, and it follows that $P \circ E^2 = k_2$. Now $k_2(\alpha) = \beta \circ \alpha$, where $\beta \in \pi_{2n-1}(\mathbf{S}^n)$ is the homotopy class of the map $k | \mathbf{S}^{2n-1} : \mathbf{S}^{2n-1} \to \mathbf{S}^n$. The map k can be factored: $k = f \circ k'$, where $k' : (\mathbf{E}^{2n}, \mathbf{S}^{2n-1}) \to (\mathbf{S}^n \times \mathbf{S}^n, \mathbf{S}^n \vee \mathbf{S}^n)$ is the characteristic map for the $2n$-cell in the CW-complex $\mathbf{S}^n \times \mathbf{S}^n$. Thus $k' | \mathbf{S}^{2n-1}$ represents the Whitehead product $[\iota_n^1, \iota_n^2]$. But $f | \mathbf{S}^n \vee \mathbf{S}^n$ is the folding map ∇, and therefore $f \circ k' | \mathbf{S}^{2n-1}$ represents $[\iota_n, \iota_n]$. □

(2.5) Corollary *If* $\gamma \in \pi_{q-n-1}(\mathbf{S}^{n-1})$, *then*

$$P(E^{n+2}\gamma) = [\iota_n, E\gamma].$$

By Theorem (8.18) of Chapter X,

$$\begin{aligned}[\iota_n, E\gamma] &= [\iota_n \circ \iota_n, \iota_n \circ E\gamma] = [\iota_n \circ E\iota_{n-1}, \iota_n \circ E\gamma] \\ &= [\iota_n, \iota_n] \circ E(\iota_{n-1} \wedge \gamma) = [\iota_n, \iota_n] \circ E^n\gamma \\ &= PE^{n+2}\gamma.\end{aligned}$$ □

Since $q \le 3n - 2$, $q - n - 1 \le 2n - 3$, and therefore

$$E^{n+2} : \pi_{q-n-1}(\mathbf{S}^{n-1}) \to \pi_{q+1}(\mathbf{S}^{2n+1})$$

is an epimorphism. Hence every element in $\operatorname{Ker} E = \operatorname{Im} P$ has the form $[\iota_n, \xi]$ for some ξ. But we have seen (Theorem (8.20) of Chapter X) that the suspension of any Whitehead product is zero. Hence

(2.6) Corollary *If* $q \le 3n - 3$, *the kernel of* $E : \pi_q(\mathbf{S}^n) \to \pi_{q+1}(\mathbf{S}^{n+1})$ *is generated by all Whitehead products* $[\alpha, \beta]$ *with* $\alpha \in \pi_r(\mathbf{S}^n)$, $\beta \in \pi_s(\mathbf{S}^n)$, $r + s = q + 1$. □

(2.7) Corollary *If* $n \ge 3$, *the group* $\pi_{n+1}(\mathbf{S}^n)$ *is a cyclic group of order* 2, *generated by the* $(n-2)$*-fold suspension* η_n *of the homotopy class of the Hopf map.*

Let us recall that $\pi_3(\mathbf{S}^2)$ is the infinite cyclic group generated by η_2. The sequence

$$\pi_5(\mathbf{S}^5) \xrightarrow{P} \pi_3(\mathbf{S}^2) \xrightarrow{E} \pi_4(\mathbf{S}^3) \to 0$$

is exact, and $E : \pi_{n+1}(\mathbf{S}^n) \to \pi_{n+2}(\mathbf{S}^{n+1})$ is an isomorphism for $n \ge 3$. Moreover, by Corollary (2.5), $P(\iota_5) = [\iota_2, \iota_2]$. Now $H_0 : \pi_3(\mathbf{S}^2) \to \mathbf{Z}$ is an isomorphism and $H_0(\iota_5) = 2$ by Theorem (2.5) of Chapter XI. Hence $\operatorname{Im} P = 2\pi_3(\mathbf{S}^2)$, $\pi_4(\mathbf{S}^3) \approx \operatorname{Cok} P \approx \mathbf{Z}_2$. □

We can also calculate $\pi_{n+2}(\mathbf{S}^n)$, with the following result.

(2.8) Theorem *The group $\pi_{n+2}(\mathbf{S}^n)$ is a cyclic group of order two, generated by the composite $\eta_n \circ \eta_{n+1}$ $(n \geq 2)$.*

First recall that, by Corollary (8.13) of Chapter IV, the Hopf fibration $p : \mathbf{S}^3 \to \mathbf{S}^2$ induces isomorphisms $p_* : \pi_q(\mathbf{S}^3) \to \pi_q(\mathbf{S}^2)$ for all $q \geq 3$. In particular, $p_* : \pi_4(\mathbf{S}^3) \approx \pi_4(\mathbf{S}^2)$. But η_2 is the homotopy class of p, and $\pi_4(\mathbf{S}^3)$ is the cyclic group of order two generated by η_3. Hence $\pi_4(\mathbf{S}^2) \approx \mathbf{Z}_2$ is generated by $p_*(\eta_3) = \eta_2 \circ \eta_3$.

Next we apply Theorem (2.2) with $W = \mathbf{S}^n$, $n = 2$, to obtain the exact sequence

$$\pi_4(\mathbf{S}^2) \xrightarrow{E} \pi_5(\mathbf{S}^3) \xrightarrow{H} \pi_5(\mathbf{S}^5) \xrightarrow{P}$$
$$\pi_3(\mathbf{S}^2) \xrightarrow{E} \pi_4(\mathbf{S}^3) \longrightarrow 0$$

We have already calculated $E : \pi_3(\mathbf{S}^2) \to \pi_4(\mathbf{S}^3)$; its kernel is generated by $P(\iota_5) = \pm 2\eta_2$. Since $\pi_5(\mathbf{S}^5)$ and $\pi_3(\mathbf{S}^2)$ are infinite cyclic groups, P is a monomorphism and therefore $H = 0$ and $E : \pi_4(\mathbf{S}^2) \to \pi_5(\mathbf{S}^3)$ is an epimorphism. But we have seen (Theorem (8.9) of Chapter VIII) that $\eta_n \circ \eta_{n+1} \neq 0$ for all $n \geq 2$. Since $E(\eta_2 \circ \eta_3) = E\eta_2 \circ E\eta_3 = \eta_3 \circ \eta_4$, we deduce that E is an isomorphism, so that $\pi_5(\mathbf{S}^3) \approx \mathbf{Z}_2$ is generated by $\eta_3 \circ \eta_4$.

Again by Theorem (2.2), with $W = \mathbf{S}^3$, we have the exact sequence

$$\pi_7(\mathbf{S}^7) \xrightarrow{P} \pi_5(\mathbf{S}^3) \xrightarrow{E} \pi_6(\mathbf{S}^4) \xrightarrow{H} \pi_6(\mathbf{S}^7) = 0.$$

Thus E is an epimorphism whose kernel is generated by $P(\iota_7)$. By Theorem (2.4), $P(\iota_7) = [\iota_3, \iota_3]$; but $\mathbf{S}^3$ is an H-space, and therefore the Whitehead product $[\iota_3, \iota_3]$ vanishes, by Corollary (7.8) of Chapter X. Thus $E : \pi_5(\mathbf{S}^3) \approx \pi_6(\mathbf{S}^4)$. But $E : \pi_{n+2}(\mathbf{S}^n) \approx \pi_{n+3}(\mathbf{S}^{n+1})$ for $n \geq 4$, by the Freudenthal theorem. Hence Theorem (2.8) is true in all cases. □

The group $\pi_{n+1}(\mathbf{S}^n)$ was calculated by Freudenthal [1]. The results on $\pi_{n+2}(\mathbf{S}^n)$ were found independently by Pontryagin [2] and myself [4].

3 The Suspension Category

Let X, Y be spaces in the category $\mathscr{K}_*$. Then we can iterate the process of suspension to obtain a sequence

(3.1)
$$[X, Y] \xrightarrow{E} [\mathbf{S}X, \mathbf{S}Y] \to \cdots \to$$
$$[\mathbf{S}^n X, \mathbf{S}^n Y] \xrightarrow{E} [\mathbf{S}^{n+1} X, \mathbf{S}^{n+1} Y] \to \cdots$$

While $[X, Y]$ is just a set with distinguished element (the homotopy class of the constant map of X into the base point of Y), it follows from the discussion of Chapter III that $[\mathbf{S}^n X, \mathbf{S}^n Y]$ has a natural group structure for every $n \geq 1$, and this group is even abelian if $n \geq 2$. Moreover, the map $E : [\mathbf{S}^n X, \mathbf{S}^n Y] \to [\mathbf{S}^{n+1} X, \mathbf{S}^{n+1} Y]$ is a homomorphism if $n \geq 1$. Thus the sequence (3.1) is essentially a sequence of abelian groups and homomorphisms, and so has a direct limit group

$$\{X, Y\} = \varinjlim_n [\mathbf{S}^n X, \mathbf{S}^n Y].$$

The elements of $\{X, Y\}$ are called *S-maps of X into Y*.

Generalizing the above procedure slightly, we obtain a graded group $\{X, Y\}_*$; its component of degree m is defined by

$$\{X, Y\}_m = \varinjlim_n [\mathbf{S}^{m+n} X, \mathbf{S}^n Y];$$

thus, if $m \geq 0$,

$$\{X, Y\}_m = \{\mathbf{S}^m X, Y\},$$

while

$$\{X, Y\}_{-m} = \{X, \mathbf{S}^m Y\}.$$

The elements of $\{X, Y\}_m$ are called *S-maps of degree m.*

The above construction is well-behaved with respect to the composition product. In fact, it follows from Corollary (5.26) of Chapter III that the composition operator induces an operation

$$[Y, Z] \times [\mathbf{S}X, Y] \to [\mathbf{S}X, Z]$$

which is additive in the second factor, as well as an operation

$$[\mathbf{S}Y, Z] \times [X, Y] \xrightarrow{1 \times E} [\mathbf{S}Y, Z] \times [\mathbf{S}X, \mathbf{S}Y] \to [\mathbf{S}X, Z]$$

which is additive in the first. Moreover, the suspension operator commutes with composition, and therefore induces, for any spaces X, Y, Z an operation

$$\{Y, Z\} \times \{X, Y\} \to \{X, Z\}. \tag{3.2}$$

It follows by an easy argument from the above remarks that

(3.3) Theorem *The composition operation* (3.2) *is biadditive.* □

Thus the category $\mathscr{S}$ whose objects are spaces in $\mathscr{K}_*$ and for which $\mathscr{S}(X, Y) = \{X, Y\}$ is preadditive [MacL, p. 250]. The category $\mathscr{S}$ is called the (ungraded) *suspension category*. The full subcategory of $\mathscr{S}$ whose objects are the finite CW-complexes will be of particular importance later; it is the *Spanier–Whitehead category* $\mathscr{S}_0$.

Even more is true:

(3.4) Theorem *The categories $\mathscr{S}$ and $\mathscr{S}_0$ are additive.*

It suffices to show that these categories admit sums. For then a simple argument shows that the sums satisfy the axioms for a biproduct, and these are all that are needed in order that the category in question be additive [MacL, ibid.]. But $[X \vee Y, Z] = [X, Z] \times [Y, Z]$ and $\mathbf{S}(X \vee Y) = \mathbf{S}X \vee \mathbf{S}Y$; it follows that $\{X \vee Y, Z\} \approx \{X, Y\} \times \{Y, Z\}$ and therefore the operation $X \vee Y$ defines a sum in the category $\mathscr{S}$, as well as in the category $\mathscr{K}_*$. And if X and Y are finite CW-complexes, so is $X \vee Y$, so that the sum exists in $\mathscr{S}_0$ as well as in $\mathscr{S}$. □

A consequence of the above remarks is

(3.5) Theorem *For any space X, $\{X, X\}_*$ is a graded ring $\mathscr{X}_*$ and, for any space Y, $\{X, Y\}_*$ is a graded right $\mathscr{X}_*$-module $\mathscr{X}_*(Y)$ and $\{Y, X\}$ a graded left $\mathscr{X}_*$-module $\mathscr{X}^*(Y)$.* □

Of special importance is the case $X = \mathbf{S}^0$. The ring $\sigma_* = \{\mathbf{S}^0, \mathbf{S}^0\}_*$ is the *stable homotopy ring*, while $\sigma_*(X) = \{\mathbf{S}^0, X\}_*$ and $\sigma^*(X) = \{X, \mathbf{S}^0\}_*$ are the *stable homotopy and cohomotopy modules*, respectively, *of the space X*. Let us observe that the group σ_p is the direct limit of the sequence

$$\pi_{p+1}(\mathbf{S}^1) \to \cdots \to \pi_{p+k}(\mathbf{S}^k) \to \cdots,$$

while the q-th stable homotopy group $\sigma_q(X)$ is the direct limit of

$$\pi_q(X) \to \pi_{q+1}(\mathbf{S}X) \to \cdots \to \pi_{q+k}(\mathbf{S}^k X) \to \cdots.$$

If $\xi \in \sigma_p$ and $\alpha \in \sigma_q(X)$, then, for sufficiently large k, α has a representative $f: \mathbf{S}^{q+k} \to \mathbf{S}^k X$ and ξ a representative $g: \mathbf{S}^{p+q+k} \to \mathbf{S}^{q+k}$. The element $\alpha \cdot \xi \in \sigma_{p+q}(X)$ is then represented by the map $f \circ g: \mathbf{S}^{p+q+k} \to \mathbf{S}^k X$. Similarly, the q-th stable cohomotopy group $\sigma^q(X)$ of X is the direct limit of the sequence

$$[X, \mathbf{S}^q] \to [\mathbf{S}X, \mathbf{S}^{q+1}] \to \cdots \to [\mathbf{S}^k X, \mathbf{S}^{q+k}] \to \cdots;$$

if $\xi \in \sigma_p$, $\beta \in \sigma^q(X)$, there are representatives

$$h: \mathbf{S}^k X \to \mathbf{S}^{q+k},$$

$$g: \mathbf{S}^{q+k} \to \mathbf{S}^{q-p+k}$$

of β, ξ, respectively, and $\xi \cdot \beta \in \sigma^{q-p}(X)$ represented by the composite map $g \circ h: \mathbf{S}^k X \to \mathbf{S}^{q-p+k}$. (Remark: we have written the component of degree $-p$ of $\sigma^*(X)$ as $\sigma^p(X)$, in accordance with a general "sign-changing" convention which consists in changing signs when we convert subscripts to superscripts. This allows us to economize on notation; otherwise we would need a different symbol like $\tau_p(X)$ for the homotopy group $\{X, \mathbf{S}^0\}_p$ to avoid confusion with the homotopy group $\sigma_p(X) = \{\mathbf{S}^0, X\}_p$. A similar remark applies to the modules $\mathscr{X}_*(Y)$, $\mathscr{X}^*(Y)$).

The graded ring $\mathscr{X}_*$ is always associative (because the product is defined in terms of the operation of composition), but need not be commutative (in the graded sense; i.e., $\alpha\beta = (-1)^{pq}\beta\alpha$ for elements α, β of degrees p, q, respectively). However,

(3.6) Theorem *The graded ring σ_* is commutative.*

For if $\xi \in \sigma_k$, $\eta \in \sigma_l$ are represented by $\alpha \in \pi_{p+k}(\mathbf{S}^p)$, $\beta \in \pi_{q+l}(\mathbf{S}^q)$, respectively, then $\eta \circ \xi$ is represented by $E^p\beta \circ E^{q+l}\alpha$ and $\xi \circ \eta$ by $E^q\alpha \circ E^{p+k}\beta$. But the elements differ, according to Theorem 8.12 of Chapter X, by the sign $(-1)^{kl}$. □

At this moment, we do not have much explicit information about σ_*. In fact, with the machinery developed so far, we can only prove

(3.7) Theorem *The group σ_0 is an infinite cyclic group generated by the S-class ι of the identity map. The group σ_1 is a cyclic group of order 2 generated by the S-class η of the Hopf map $\mathbf{S}^3 \to \mathbf{S}^2$. The group σ_2 is a cyclic group of order two generated by η^2.*

The first statement follows from the fact that the group $\pi_n(\mathbf{S}^n)$ is the infinite cyclic group generated by the homotopy class ι_n of the identity map for all $n \geq 1$, the second from Corollary (2.7), and the last from Theorem (2.8). □

Remark. If $\alpha \in \{X, Y\}$ and $\beta \in \{Y, Z\}$, we may regard $\beta \circ \alpha$ as a function of α or of β; thus we may define

$$\alpha^*(\beta) = \beta_*(\alpha) = \beta \circ \alpha,$$

obtaining maps

$$\alpha^* : \{Y, Z\} \to \{X, Z\},$$
$$\beta_* : \{X, Y\} \to \{X, Z\}.$$

Associativity of ordinary composition implies that of composition of S-classes. Thus, for fixed Z, the correspondence $X \to \{X, Z\}$ defines a (contravariant) functor $T^Z : \mathscr{S} \to \mathscr{A}$: similarly, for fixed X, the correspondence $Y \to \{X, Y\}$ defines a (covariant) functor $T_X : \mathscr{S} \to \mathscr{A}$. In particular, the stable homotopy (cohomotopy) groups may be regarded as functors $\sigma_p : \mathscr{S} \to \mathscr{A}(\sigma^p : \mathscr{S} \to \mathscr{A})$.

Having calculated the groups $\sigma_q(S^0) = \sigma_q$ for $q = 0, 1, 2$, let us attempt to calculate these groups for an arbitrary space X.

(3.8) Lemma *If $q < 0$, then $\sigma_q(X) = 0$ for every space X.*

If $k \geq 0$, $\mathbf{S}^k X$ is $(k-1)$-connected, and therefore $\pi_{q+k}(\mathbf{S}^k X) = 0$. Hence $\sigma_q(X) = \underrightarrow{\lim}_k\, \pi_{q+k}(\mathbf{S}^k X) = 0$. □

For any space X, there is a commutative diagram

$$\begin{array}{ccc} \pi_n(X) & \xrightarrow{E} & \pi_{n+1}(\mathbf{S}X) \\ \downarrow{\scriptstyle \rho} & & \downarrow{\scriptstyle \rho} \\ H_n(X) & \xrightarrow[s]{} & H_{n+1}(\mathbf{S}X) \end{array}$$

in which E and s are the suspension operators in homology and homotopy, respectively, while ρ is the Hurewicz map. Hence there is a commutative diagram

$$\begin{array}{ccccccc} \pi_n(X) & \xrightarrow{E_0} & \pi_{n+1}(\mathbf{S}X) & \xrightarrow{E_1} & \pi_{n+2}(\mathbf{S}^2X) & \to & \cdots \\ \downarrow{\scriptstyle \rho_0} & & \downarrow{\scriptstyle \rho_1} & & \downarrow{\scriptstyle \rho_2} & & \\ H_n(X) & \xrightarrow[s_0]{} & H_{n+1}(\mathbf{S}X) & \xrightarrow[s_1]{} & H_{n+2}(\mathbf{S}^2X) & \to & \cdots \end{array} \tag{3.9}$$

The direct limit of the top line is the stable homotopy group $\sigma_n(X)$; as the homomorphisms s_k are all isomorphisms, we may identify the direct limit of the bottom line with $H_n(X)$. Thus the commutative diagram (3.9) induces a homomorphism $\tilde{\rho} : \sigma_n(X) \to H_n(X)$, which we may still term the Hurewicz map; and there is a commutative diagram

$$\begin{array}{ccc} \pi_n(X) & \xrightarrow{E_\infty} & \sigma_n(X) \\ & {\scriptstyle \rho}\searrow \quad \swarrow{\scriptstyle \tilde{\rho}} & \\ & H_n(X) & \end{array}$$

There is a stable analogue of the Hurewicz Theorem:

(3.10) Theorem *If X is an $(m-1)$-connected space, then $\sigma_n(X) = 0$ for $n < m$ and $\tilde{\rho} : \sigma_n(X) \to H_n(X)$ is an isomorphism for $n = m$ and an epimorphism for $n = m + 1$.*

For if X is $(m-1)$-connected, then $\mathbf{S}^k X$ is $(m+k-1)$-connected, so that $\pi_{n+k}(\mathbf{S}^k X) = 0$ for $n < m$. Hence the group $\sigma_n(X) = \underrightarrow{\lim}_k \pi_{n+k}(\mathbf{S}^k X)$ is also zero. If $n = m$, the map ρ_k is an isomorphism for k sufficiently large, and therefore $\tilde{\rho}$ is an isomorphism; and if $n = m + 1$, ρ_k is an epimorphism for k large, so that $\tilde{\rho}$ is an epimorphism. □

(3.11) Corollary *For any space X in $\mathscr{K}_0$, $\sigma_0(X) \approx H_0(X)$.* □

The determination of $\sigma_1(X)$ is more difficult. We first observe that, if η is the non-zero element of σ_1, then the map $\alpha \to \alpha \cdot \eta$ determined by the module structure of $\sigma_*(X)$ over σ_* is a homomorphism $\bar{\eta} : \sigma_m(X) \to \sigma_{m+1}(X)$. Since $2\eta = 0$, $\bar{\eta}(2\sigma_m(X)) = 0$ and therefore $\bar{\eta}$ in turn induces a homomorphism

$$\tilde{\eta} : \sigma_m(X) \otimes \mathbf{Z}_2 \to \sigma_{m+1}(X).$$

(3.12) Theorem *Let X be an $(m-1)$-connected space $(m \geq 3)$. Then there is an exact sequence*

(3.13)

$$\pi_{m+2}(X) \xrightarrow{\rho_2} H_{m+2}(X) \xrightarrow{b} \pi_m(X)/2\pi_m(X) \xrightarrow{\tilde{\eta}}$$
$$\pi_{m+1}(X) \xrightarrow{\rho_1} H_{m+1}(X) \to 0,$$

where ρ_1 and ρ_2 are the Hurewicz maps.

The sequence (3.13) is due to the present author [3]. A comparable sequence for $n = 2$, with the group $\pi_m(X) \otimes \mathbf{Z}_2$ replaced by the group $\Gamma(\pi_m(X))$ is due to J. H. C. Whitehead [7]. The homomorphism b is called the *secondary boundary operator.*

To prove Theorem (3.12), let us attach $(m+1)$-cells to the space X in order to kill the group $\Pi = \pi_m(X)$; specifically, let J be a system of generators for Π, let $f : \bigvee_{\alpha \in J} \mathbf{S}_\alpha^m \to X$ be a map such that $f \mid \mathbf{S}_\alpha^m$ represents α, and let $X^* = \mathbf{T}_f$ be the mapping cone of f. Let F be a free abelian group having a basis $\{x_\alpha \mid \alpha \in J\}$ and let $p : F \to \Pi$ be the homomorphism such that $p(x_\alpha) = \alpha$. Let R be the kernel of p, $i : R \subsetneq F$, so that there is an exact sequence

$$0 \longrightarrow R \xrightarrow{i} F \xrightarrow{p} \Pi \longrightarrow 0.$$

There is a commutative diagram (Figure 12.1) in which the top line is obtained from a portion of the homotopy sequence of the pair (X^*, X) by tensoring with $\mathbf{Z}_2$, the middle line is a portion of the homotopy sequence of (X^*, X), and the bottom line a portion of the homology sequence of (X^*, X). (Warning: the top line is not necessarily exact). Groups labelled with an asterisk are those of X^*, groups labelled with an obelisk are those of the pair (X^*, X), while the groups not otherwise labelled are those of X.

The homomorphisms ρ_k, ρ_k^*, $\sigma_k^\dagger$ are Hurewicz maps (absolute or relative); while the homomorphisms $\eta_k^\dagger$, η_k^*, are induced by composition with the generator $\eta_{m+k} \in \pi_{m+k+1}(\mathbf{S}^{n+k})$, and $\eta^\dagger$ by composition with the generator $\eta'_{m+k} \in \pi_{m+k+1}(\mathbf{E}^{m+k}, \mathbf{S}^{m+k-1}) \approx \pi_{m+k}(\mathbf{S}^{m+k-1})$.

The space X being $(m-1)$-connected, the space X^* is m-connected and the pair (X^*, X) m-connected. Since the relative CW-complex (X^*, X) has no cells in dimensions $\neq m+1$, the groups $H_q^\dagger$ are zero for $q \neq m+1$; hence i'_2 is an isomorphism and i'_1 a monomorphism. Since X and X^* are 1-connected, the Hurewicz maps ρ_0, ρ_1^*, and $\rho_1^\dagger$ are isomorphisms. Easy diagram-chasing shows that ρ_1 is an epimorphism.

We can use the Homotopy Map Excision Theorem (Theorem 7.14 of Chapter VII). In fact, the simultaneous attaching map is a relative homeomorphism $h : (E, S) \to (X^*, X)$, where $E = \bigvee_{\alpha \in J} \mathbf{E}_\alpha^{m+1}$, $S = \bigvee_{\alpha \in J} \mathbf{S}_\alpha^m$. The pair (E, S) is m-connected, and the map $f = h|S : S \to X$ is m-connected. Therefore $f_* : \pi_q(E, S) \to \pi_q(X^*, X)$ is an isomorphism for $q < 2m$. On the other hand,

$$\begin{aligned}\pi_q(E, S) &\approx \pi_{q-1}(S) \approx \bigoplus_{\alpha \in J} \pi_{q-1}(\mathbf{S}_\alpha^m) \\ &\approx \bigoplus_{\alpha \in J} \pi_q(\mathbf{E}_\alpha^{m+1}, \mathbf{S}_\alpha^m) \approx \pi_q(\mathbf{E}^m, \mathbf{S}^{m-1}) \otimes F,\end{aligned}$$

provided again that $q < 2m$, so that there are no Whitehead product terms. In particular, $\pi_{m+2}^\dagger \approx \pi_{m+2}(\mathbf{E}^{m+1}, \mathbf{S}^n) \otimes F \approx \pi_{m+1}^\dagger \otimes \mathbf{Z}_2$; clearly the isomorphism is given by composition with η'_{m+1}. Hence $\eta_1^\dagger$ is an isomorphism. Another diagram chase now allows us to conclude that $\operatorname{Ker} \rho_1 = \operatorname{Im} \eta_0$.

We next observe that there is an exact sequence

(3.14)

$$\pi_{m+1}^* \otimes \mathbf{Z}_2 \oplus \pi_{m+2} \xrightarrow{\alpha} \pi_{m+2}^* \xrightarrow{\beta} \pi_m \otimes \mathbf{Z}_2 \xrightarrow{\eta_0} \pi_{m+1} \xrightarrow{i_1} \pi_{m+1}^*$$

where

$$\alpha(y, z) = -\eta_1^* y + i_2 z, \qquad \beta = \partial''_2 \circ (\eta_1^\dagger)^{-1} \circ j_2$$

(cf. Exercise 1, Chapter V). The sequence (3.14) can be imbedded in a commutative diagram

(3.15)

$$\begin{array}{ccccccccc}
\pi_{m+1}^* \otimes \mathbf{Z}_2 \oplus \pi_{m+2} & \xrightarrow{\alpha} & \pi_{m+2}^* & \xrightarrow{\beta} & \pi_m \otimes \mathbf{Z}_2 & \xrightarrow{\eta_0} & \pi_{m+1} & \xrightarrow{i_1} & \pi_{m+1}^* \\
\Big\downarrow{\scriptstyle \gamma} & & \Big\downarrow{\scriptstyle \rho_2^*} & \overset{\beta'}{\nearrow} & & & & & \\
H_{m+2} & \xrightarrow[i'_2]{} & H_{m+2}^* & & & & & &
\end{array}$$

where $\gamma(y, z) = \rho_2(z)$. Applying what we have already proved to the m-

$$
\begin{array}{ccccccccccccccc}
\pi_{m+1}\otimes\mathbf{Z}_2 & \xrightarrow{i_2''} & \pi^*_{m+1}\otimes\mathbf{Z}_2 & \xrightarrow{j_2''} & \pi^\dagger_{m+1}\otimes\mathbf{Z}_2 & \xrightarrow{\partial_2''} & \pi_m\otimes\mathbf{Z}_2 & \longrightarrow & 0 & & & & & & \\
\downarrow{\scriptstyle \eta_1} & & \downarrow{\scriptstyle \eta_1^*} & & \downarrow{\scriptstyle \eta_1^\dagger} & & \downarrow{\scriptstyle \eta_0} & & \downarrow & & & & & & \\
\pi_{m+2} & \xrightarrow[i_2]{} & \pi^*_{m+2} & \xrightarrow[j_2]{} & \pi^\dagger_{m+2} & \xrightarrow[\partial_2]{} & \pi_{m+1} & \xrightarrow[i_1]{} & \pi^*_{m+1} & \xrightarrow[j_1]{} & \pi^\dagger_{m+1} & \xrightarrow[\partial_1]{} & \pi_m & \longrightarrow & 0 \\
\downarrow{\scriptstyle \rho_2} & & \downarrow{\scriptstyle \rho_2^*} & & \downarrow{\scriptstyle \rho_2^\dagger} & & \downarrow{\scriptstyle \rho_1} & & \downarrow{\scriptstyle \rho_1^*} & & \downarrow{\scriptstyle \rho_1^\dagger} & & \downarrow{\scriptstyle \rho_0} & & \\
H_{m+2} & \xrightarrow[i_2']{} & H^*_{m+2} & \xrightarrow[j_2']{} & H^\dagger_{m+2} & \xrightarrow[\partial_2']{} & H_{m+1} & \xrightarrow[i_1']{} & H^*_{m+1} & \xrightarrow[j_1']{} & H^\dagger_{m+1} & \xrightarrow[\partial_1']{} & H_m & \longrightarrow & 0
\end{array}
$$

Figure 12.1

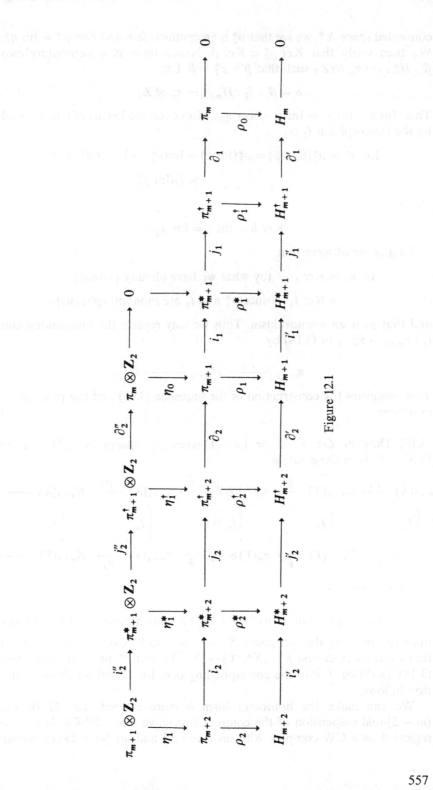

connected space X^*, we see that ρ_2^* is an epimorphism and Ker $\rho_2^* = \operatorname{Im} \eta_1^*$. We then verify that Ker $\rho_2^* \subset$ Ker β. Hence there is a homomorphism $\beta' : H_{m+2}^* \to \pi_m \otimes \mathbf{Z}_2$ such that $\beta' \circ \rho_2^* = \beta$. Let

$$b = \beta' \circ i_2' : H_{m+2} \to \pi_m \otimes \mathbf{Z}_2 .$$

Then Im b = Im β' = Im β = Ker η_0. Moreover, the kernel of b is mapped by the isomorphism i_2' on

$$\begin{aligned}\operatorname{Ker} \beta' = \rho_2^*(\operatorname{Ker} \beta) = \rho_2^*(\operatorname{Im} \alpha) = \operatorname{Im}(\rho_2^* \circ \alpha) &= \operatorname{Im}(i_2' \circ \gamma) \\ &= i_2'(\operatorname{Im} \gamma).\end{aligned}$$

Hence

$$\operatorname{Ker} b = \operatorname{Im} \gamma = \operatorname{Im} \rho_2 .$$

Finally, we observe that

$$\begin{aligned}\operatorname{Im} \eta_0 &= \operatorname{Ker} \rho_1 \quad \text{(by what we have already proved)} \\ &= \operatorname{Ker} i_1 \quad (\text{since } \rho_1^* \text{ and } i_1' \text{ are monomorphisms})\end{aligned}$$

and that ρ_1 is an epimorphism. Thus we may replace the homomorphism $i_1 : \pi_{m+1} \to \pi_{m+1}^*$ of (3.14) by

$$\pi_{m+1} \xrightarrow{\rho_1} H_{m+1} \longrightarrow 0.$$

This completes the construction of the sequence (3.13) and the proof of its exactness. □

(3.16) Theorem *Let X, Y be $(m-1)$-connected spaces $(m \geq 3)$, and let $f: X \to Y$. Then the diagram*

$$\begin{array}{ccccccccccc}\pi_{m+2}(X) & \xrightarrow{\rho_2} & H_{m+2}(X) & \xrightarrow{b} & \pi_m(X) \otimes \mathbf{Z}_2 & \xrightarrow{\tilde{\eta}} & \pi_{m+1}(X) & \xrightarrow{\rho_1} & H_{m+1}(X) & \longrightarrow & 0 \\ \downarrow f_* & & \downarrow f_* & & \downarrow f_* \otimes 1 & & \downarrow f_* & & \downarrow f_* & & \\ \pi_{m+2}(Y) & \xrightarrow[\rho_2]{} & H_{m+2}(Y) & \xrightarrow[b]{} & \pi_m(Y) \otimes \mathbf{Z}_2 & \xrightarrow[\tilde{\eta}]{} & \pi_{m+1}(Y) & \xrightarrow[\rho_1]{} & H_{m+1}(Y) & \longrightarrow & 0\end{array}$$

is commutative.

Only the second square from the left presents a problem. The obstructions to extending the composite $X \xrightarrow{f} Y \subsetneq Y^*$ over X^* all vanish, so that f has an extension $f^* : (X^*, X) \to (Y^*, Y)$. But f^* maps the diagrams (3.14), (3.15) for X into the corresponding ones for Y, and the desired relation follows.

We can make the homomorphism b more explicit. Let M be the $(m-2)$-fold suspension of the complex projective plane $\mathbf{P}^2(\mathbf{C})$; M can be regarded as a CW-complex with an n-cell $\mathbf{S}^m$ and an $(m+2)$-cell whose boundary is attached to the n-cell by the generator $\eta_m \in \pi_{m+1}(\mathbf{S}^m)$. If $m \geq 3$, M is a suspension and therefore $[M, X]$ is a group. Consider the sequence

$$\mathbf{S}^{m+1} \xrightarrow{\eta_m} \mathbf{S}^m \xrightarrow{i} M \xrightarrow{p} \mathbf{S}^{m+2}$$

where i is the inclusion and p collapses S^m to a point. Thus there is an exact sequence

(3.17) $$\pi_{m+2}(X) \to [M, X] \to \pi_m(X) \to \pi_{m+1}(X).$$

Let z be the generator of the infinite cyclic group $H_{m+2}(M)$ which is mapped by p_* into the positive orientation of $\mathbf{S}^{m+2}$. Then a kind of Hurewicz map $\tilde{\rho}$ is defined by

$$\tilde{\rho}([g]) = g_*(z)$$

for any $g : M \to X$; the easy verification of

(3.18) Lemma *The map $\tilde{\rho} : [M, X] \to H_{m+2}(X)$ is a homomorphism*

is left to the reader. □

(3.19) Lemma *The diagram*

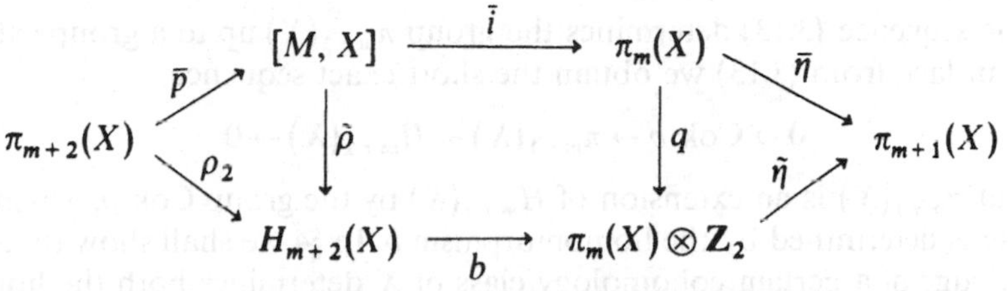

is commutative, and $\tilde{\rho}$ is an epimorphism (the homomorphism q is just reduction mod 2).

Only the middle square needs to be checked. The diagram is natural, and therefore we may use the method of the universal example. Thus we may assume $M = X$ and have only to prove that $q\bar{i}(\iota) = b\tilde{\rho}(\iota)$, where ι is the homotopy class of the identity map. Then $\tilde{\rho}(\iota) = z$. The space X^* has the homotopy type of $\mathbf{S}^{m+2}$. The homomorphism b is defined with the aid of the commutative diagram

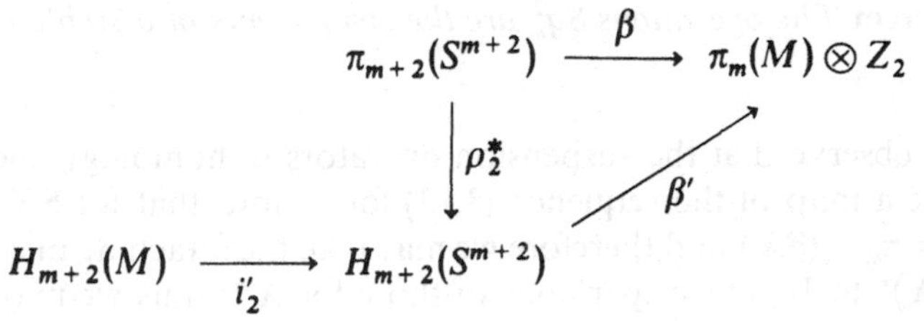

Consider the diagram of Figure 12.1, and note that $\pi_{m+1} = \pi^*_{m+1} = 0$. In this case j_2 is an epimorphism and therefore β maps the generator ι_{m+2} of π^*_{m+2} into the non-zero element ι_0 of $\pi_m \otimes \mathbf{Z}_2 \approx \mathbf{Z}_2$. It follows that $b(z) = \iota_0$. But clearly $q\bar{i}(\iota) = \iota_0$. □

The fact that $\tilde{\rho}$ is an epimorphism now follows from the Five-Lemma. Note that not every element of $H_{m+2}(X)$ need be spherical, but each element has the reasonably simple form $f_*(z)$ for some map $f: M \to X$.

Another consequence of Theorem (3.12) is

(3.20) Theorem *If* $m \geq 3$, $H_{m+2}(\Pi, m) \approx \Pi \otimes \mathbf{Z}_2$.

For if $X = K(\Pi, m)$, then $\pi_{m+1}(X) = \pi_{m+2}(X) = 0$; by exactness of (3.13), $b: H_{m+2}(X) \approx \Pi/2\Pi$. □

(3.21) Corollary *If* $m \geq 3$, $H^{m+2}(\Pi, m; G) \approx \mathrm{Hom}(\Pi \otimes \mathbf{Z}_2, G)$. *If, in addition,* Π *or* G *has exponent two, then* $H^{m+2}(\Pi, m; G) \approx \mathrm{Hom}(\Pi, G)$.

For $H_{m+1}(\Pi, m) = 0$ (Theorem 7.8 of Chapter V), and therefore the expected "Ext" term vanishes. □

The sequence (3.13) determines the group $\pi_{m+1}(X)$ up to a group extension; in fact, from (3.13) we obtain the short exact sequence

$$0 \to \mathrm{Cok}\, b \to \pi_{m+1}(X) \to H_{m+1}(X) \to 0$$

so that $\pi_{m+1}(X)$ is an extension of $H_{m+1}(X)$ by the group Cok b, which, of course is determined by the homomorphism b. In §4 we shall show that the knowledge of a certain cohomology class of X determines both the homomorphism b and the above extension.

Consider now the group $H^{m+2}(\Pi, m; \Pi \otimes \mathbf{Z}_2)$. The homomorphism $b: H_{m+2}(\Pi, m) \to \Pi \otimes \mathbf{Z}_2$ define an element $v \in H^{m+2}(\Pi, m; \Pi \otimes \mathbf{Z}_2)$; because b is an isomorphism, $v \neq 0$. The element v in turn determines a cohomology operation $\theta: H^m(\ \ ; \Pi) \to H^{m+2}(\ \ ; \Pi \otimes \mathbf{Z}_2)$. If $\Pi = \mathbf{Z}_2$, the group $H^{m+2}(\Pi, m; \Pi \otimes \mathbf{Z}_2) \approx \mathrm{Hom}(\mathbf{Z}_2, \mathbf{Z}_2)$ is cyclic of order two. But we have seen in §6 of Chapter VIII that the element $Sq^2 b_m \neq 0$, and it follows that when $\Pi = \mathbf{Z}_2$, $\theta = Sq^2$. Therefore it is natural to christen the above operation Sq^2_m for any group Π.

(3.22) Theorem *The operations* Sq^2_m *are the components of a stable operation* Sq^2.

We first observe that the suspension operators in homology and homotopy induce a map of the sequence (3.13) for X into that for $\mathbf{S}X$. Indeed, $E: \pi_m(X) \approx \pi_{m+1}(\mathbf{S}X)$, and therefore we may take the attaching maps for the cells of $(\mathbf{S}X)^*$ to be the suspensions of those for X^*; thus we may assume

$(\mathbf{S}X)^* = \mathbf{S}X^*$. Therefore the suspension operators map the homology and homotopy sequences for the pair (X^*, X) into the corresponding sequences for the pair $(\mathbf{S}X^*, \mathbf{S}X)$; because the Hurewicz map commutes with suspension and the homotopy operation defined by the η_n is stable, the whole diagram of Figure 12.1 for the pair (X^*, X) is mapped into that for the pair $(\mathbf{S}X^*, \mathbf{S}X)$. It follows that the diagram

(3.23)

$$\begin{array}{ccccccccccc}
\pi_{m+2}(X) & \longrightarrow & H_{m+2}(X) & \xrightarrow{b} & \pi_m(X)\otimes \mathbf{Z}_2 & \longrightarrow & \pi_{m+1}(X) & \longrightarrow & H_{m+1}(X) & \longrightarrow & 0 \\
\downarrow E & & \downarrow s_* & & \downarrow E\otimes 1 & & \downarrow E & & \downarrow s_* & & \\
\pi_{m+3}(\mathbf{S}X) & \longrightarrow & H_{m+3}(\mathbf{S}X) & \xrightarrow[b]{} & \pi_{m+1}(\mathbf{S}X)\otimes \mathbf{Z}_2 & \longrightarrow & \pi_{m+2}(\mathbf{S}X) & \longrightarrow & H_{m+2}(\mathbf{S}X) & \longrightarrow & 0
\end{array}$$

is commutative.

If $X = K(\Pi, m)$, then $H_{m+1}(X) = H_{m+2}(\mathbf{S}X) = 0$, so that the homomorphisms b of the diagram (3.23) determine unique elements

$$v_m \in H^{m+2}(X; \Pi/2\Pi)$$
$$v'_{m+1} \in H^{m+3}(\mathbf{S}X; \Pi/2\Pi),$$

and commutativity of (3.23) implies that $s^*v'_{m+1} = v_m$. If $k : \mathbf{S}K(\Pi, m) \to K(\Pi, m+1)$ is a transfer, then $h^*v_{m+1} = v'_{m+1}$, and it follows that $\sigma^*v_{m+1} = v_m$, as desired. □

4 Group Extensions and Homology

In the preceding section we determined the second non-vanishing stable homotopy group of a space X up to a group extension. This phenomenon occurs quite often in algebraic topology; the group C to be determined is placed in an exact sequence

$$A \xrightarrow{\alpha} B \longrightarrow C \longrightarrow D \xrightarrow{\beta} E,$$

and this determines C up to a group extension $\xi \in \operatorname{Ext}(\operatorname{Ker} \beta, \operatorname{Cok} \alpha)$. To ascertain the algebraic structure of C it remains to determine the extension ξ. Experience shows that in many cases the information necessary to determine ξ is carried by a certain cohomology class u. In this section we shall make this connection explicit; as an application we show how this machinery applies to the calculation of the group $\sigma_{m+1}(X)$ for an $(m-1)$-connected space X.

Let us recall some facts from elementary homological algebra. There are two customary ways of defining the functor Ext, one *via* short exact sequences, the other *via* resolutions. Let A, B be abelian groups; then two short exact sequences

$$\textbf{(4.1)}\qquad 0 \to B \xrightarrow{i} E \xrightarrow{p} A \to 0,$$

$$\textbf{(4.1')}\qquad 0 \to B \xrightarrow{i'} E' \xrightarrow{p'} A \to 0,$$

are said to be equivalent if and only if there is a homomorphism $h : E \to E'$ (necessarily an isomorphism) such that the diagram

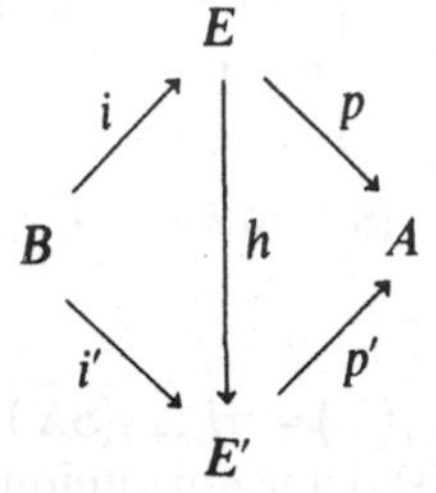

is commutative. Then $\operatorname{Ext}(A, B)$ is the set of equivalence classes of all short exact sequences (4.1). This defines $\operatorname{Ext}(A, B)$ as a set; it becomes an abelian group under the *Baer multiplication* (cf. Ex. 2, below).

The second definition makes use of the notion of *free resolution* of A; this is a short exact sequence

$$0 \to R \xrightarrow{j} F \xrightarrow{q} A \to 0$$

with F (and therefore R) free abelian. Then j induces a homomorphism

$$j^* = \operatorname{Hom}(j, 1) : \operatorname{Hom}(F, B) \to \operatorname{Hom}(R, B),$$

and we may define $\operatorname{Ext}(A, B)$ to be the cokernel of j^*.

The first definition has the advantages of being simple and conceptual, and the disadvantage that the group structure is not at all perspicuous. And, while the group structure is built into the second definition, it suffers from the apparent lack of naturality in the choice of a free resolution of A.

The two definitions, of course, are equivalent; a short exact sequence (4.1) and a homomorphism $g : R \to B$ determine the same element of $\operatorname{Ext}(A, B)$ if and only if there is a commutative diagram

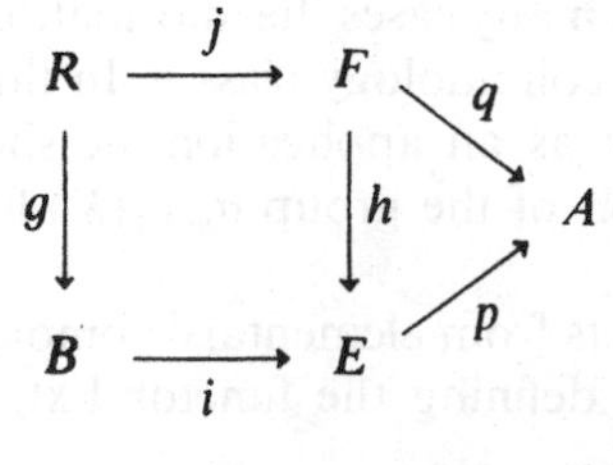

Each element $e \in \text{Ext}(A, B)$ and each positive integer m determines a homomorphism

$$e_m : {}_mA \to B/mB$$

as follows. Consider the diagram

$$A \xleftarrow{\;p\;} E \xrightarrow{\;m\cdot\;} E \xleftarrow{\;i\;} B,$$

where $m\cdot$ is the operation of multiplication by m. Then $i^{-1} \circ (m\cdot) \circ p^{-1} : A \rightsquigarrow B$ is an additive relation F, and one verifies that $\text{Dom}\, F = {}_mA$ and $\text{Ind}\, F = mB$. Thus F determines a homomorphism e_m as above.

The correspondence $e \to e_m$ is easily verified to be a homomorphism $r_m : \text{Ext}(A, B) \to \text{Hom}({}_mA, B/mB)$; the homomorphisms r_m determine a homomorphism

$$r : \text{Ext}(A, B) \to \prod_{m=2}^{\infty} \text{Hom}({}_mA, B/mB).$$

Evidently the kernel of r consists of all elements e for which B is a *pure subgroup* of E. Therefore r is a monomorphism (and hence e is determined by the homomorphisms e_m) if one of the following two conditions is satisfied [K, Vol. I, pp. 178–180]:

(1) A is a direct sum of cyclic groups;
(2) B is a torsion group of finite exponent.

In particular, r is a monomorphism whenever A is finitely generated.

Let C be a free graded chain complex. Then the Universal Coefficient Theorem asserts the existence of an isomorphism

$$H^n(C; G) \approx \text{Hom}(H_n(C), G) \oplus \text{Ext}(H_{n-1}(C), G).$$

However, this isomorphism is not natural, depending as it does on the choice of a subgroup P_n of C_n complementary to the group of cycles $Z_n(C)$. What one actually has is a splittable short exact sequence

(4.2)

$$0 \to \text{Ext}(H_{n-1}(C), G) \xrightarrow{\;\lambda\;} H^n(C; G) \xrightarrow{\;\mu\;} \text{Hom}(H_n(C), G) \to 0,$$

which is natural with respect both to chain maps $\phi : C \to C'$ and to coefficient group homomorphisms $f : G \to G'$.

Let $u \in H^n(C, G)$. If $\phi : C_n \to G$ is a cocycle representing u, then $\phi \mid Z_n(C) : Z_n(C) \to G$ maps the group $B_n(C)$ of bounding cycles into zero and thereby induces a homomorphism $u_* = \mu(u) : H_n(C) \to G$.

Let $e \in \operatorname{Ext}(H_{n-1}(C), G)$. The group $H_{n-1}(C)$ has the convenient resolution

$$0 \to B_{n-1}(C) \to Z_{n-1}(C) \to H_{n-1}(C) \to 0;$$

let $\phi : B_{n-1}(C) \to G$ be a homomorphism corresponding to e. Let $\partial' : C_n \to B_{n-1}(C)$ be the epimorphism defined by the boundary operator of C. Then $\phi \circ \partial' : C_n \to G$ is a cocycle, representing the element $\lambda(e) \in H^n(C; G)$.

Let $u \in H^n(C; G)$, and let $f : G \to G' = \operatorname{Cok} u_*$ be the projection. In view of the naturality of the sequence (4.2), there is a commutative diagram

$$\begin{array}{ccccc}
\operatorname{Ext}(H_{n-1}(C), G) & \xrightarrow{\lambda} & H^n(C; G) & \xrightarrow{\mu} & \operatorname{Hom}(H_n(C), G) \\
\downarrow f_2 & & \downarrow f_0 & & \downarrow f_1 \\
\operatorname{Ext}(H_{n-1}(C), G') & \xrightarrow[\lambda']{} & H^n(C; G') & \xrightarrow[\mu']{} & \operatorname{Hom}(H_n(C), G').
\end{array}$$

where the f_i are induced by f and the upper (lower) line is (4.2) for the group G (G'). Then

$$\mu' f_0(u) = f_1 \mu(u) = f \circ u_* = 0,$$

and therefore there exists $u_\dagger \in \operatorname{Ext}(H_{n-1}(C), G')$ such that $\lambda'(u_1) = f_0(u)$; the element $u_\dagger$ is unique because λ' is a monomorphism.

Thus each element $u \in H^n(C; G)$ determines canonically a pair of elements $(u_*, u_\dagger)$:

$$u_* : H_n(C) \to G,$$

$$u_\dagger \in \operatorname{Ext}(H_{n-1}(C), \operatorname{Cok} u_*).$$

The operations $u \to u_*$, $u \to u_\dagger$ have certain naturality properties, which we now explain. First, let $\phi : C \to C'$ be a chain map. Then there is a commutative diagram

$$\begin{array}{ccc}
H^n(C'; G) & \xrightarrow{\mu'} & \operatorname{Hom}(H_n(C'), G) \\
\phi^* \downarrow & & \downarrow \phi_1^* \\
H^n(C; G) & \xrightarrow[\mu]{} & \operatorname{Hom}(H_n(C), G).
\end{array}$$

Clearly,

(4.3) Theorem *If $u \in H^n(C'; G)$, then the element $(\phi^* u)_* = \mu(\phi^* u)$ is given by*

$$\mu(\phi^* u) = \phi_1^* \mu'(u) = u_* \circ \phi_*$$

where $\phi_* : H_n(C) \to H_n(C')$ is the homomorphism induced by ϕ.

On the other hand, $\operatorname{Im}(u_* \circ \phi_*) \subset \operatorname{Im} u_*$; let $\kappa : \operatorname{Cok}(u_* \circ f_*) \to \operatorname{Cok} u_*$ be the projection. By the naturality of the functor Ext, there is a commutative diagram

$$\begin{array}{ccc}
\operatorname{Ext}(H_{n-1}(C'), \operatorname{Cok}(u_* \circ f_*)) & \xrightarrow{\kappa'_*} & \operatorname{Ext}(H_{n-1}(C'), \operatorname{Cok} u_*) \\
\Big\downarrow \phi_2^* & & \Big\downarrow \phi_3^* \\
\operatorname{Ext}(H_{n-1}(C), \operatorname{Cok}(u_* \circ f_*)) & \xrightarrow[\kappa_*]{} & \operatorname{Ext}(H_{n-1}(C), \operatorname{Cok} u_*),
\end{array}$$

and we have

(4.4) Theorem *The elements*

$$u_\dagger \in \operatorname{Ext}(H_{n-1}(C'), \operatorname{Cok} u_*),$$
$$v_\dagger = (\phi^* u)_\dagger \in \operatorname{Ext}(H_{n-1}(C), \operatorname{Cok}(u_* \circ f_*))$$

are related by

$$\phi_3^*(u_\dagger) = \kappa_*(v_\dagger). \qquad \square$$

Let C be a free chain complex, m a positive integer. Then the short exact sequence

$$0 \to \mathbf{Z} \xrightarrow{m\,\cdot} \mathbf{Z} \xrightarrow{\rho_m} \mathbf{Z}_m \to 0$$

of coefficient groups gives rise to a long exact sequence

$$\cdots \to H_{n+1}(C; \mathbf{Z}_m) \xrightarrow{\beta_m} H_n(C) \xrightarrow{m\,\cdot} H_n(C) \xrightarrow{\rho'_m} H_n(C; \mathbf{Z}_m) \to \cdots;$$

the image of the Bockstein operator β_m is ${}_mH_n(C)$.

For any group G, the group G/mG is a $\mathbf{Z}_m$-module, and there is a natural homomorphism

$$\mu_m : H^n(C; G/mG) \to \operatorname{Hom}(H_n(C; \mathbf{Z}_m), G/mG).$$

Let $\rho_m : G \to G/mG$ be the projection, $\rho'_m : H^n(C; G) \to H^n(C; G/mG)$ the induced coefficient group homomorphism. If $u \in H^n(C; G)$, let $u_*^m = \mu_m \rho'_m(u) : H_n(C; \mathbf{Z}_m) \to G/mG$. Let $u_\dagger^m : {}_mH_{n-1}(C) \to G'/mG'$ be the homomorphism e_m determined by the extension $u_\dagger \in \operatorname{Ext}(H_{n-1}(C), G')$, and let $f_m : G/mG \to G'/mG'$ be the homomorphism determined by $f : G \to G'$.

(4.5) Theorem *For each $u \in H^n(C; G)$, the diagram*

$$
\begin{array}{ccccc}
H_n(C) & \xrightarrow{\rho'_m} & H_n(C; \mathbf{Z}_m) & \xrightarrow{\beta_m} & {}_mH_{n-1}(C) \\
\downarrow{\scriptstyle u_*} & & \downarrow{\scriptstyle u_*^m} & & \downarrow{\scriptstyle u_\dagger^m} \\
G & \xrightarrow[\rho_m]{} & G/mG & \xrightarrow[f_m]{} & G'/mG'
\end{array}
$$

is commutative.

Commutativity of the left-hand square is easy, and is left to the reader. To prove commutativity of the right-hand square, let $x \in C_n$ be a chain such that $\bar{x} = \rho_m(x)$ represents a given homology class $\xi \in H_n(C; Z_m)$; thus $\partial x = mc$ with $c \in Z_{m-1}(C)$; the cycle c represents $\beta_m(\xi) \in {}_mH_{n-1}(C)$. Let $\phi : C_n \to G$ be a representative of u; then $f \circ \phi : C_n \to G'$ annihilates the group of cycles, and therefore there is a unique homomorphism $\psi : B_{m-1}(C) \to G'$ such that $\psi \circ \partial' = f \circ \phi$. The homomorphism ψ corresponds to the extension $u_\dagger$, so that there is a commutative diagram

$$
\begin{array}{ccccccccc}
0 & \longrightarrow & G' & \longrightarrow & E & \longrightarrow & H_{n-1}(C) & \longrightarrow & 0 \\
 & & \uparrow{\scriptstyle \psi} & & \uparrow & & \uparrow{\scriptstyle 1} & & \\
0 & \longrightarrow & B_{n-1}(C) & \longrightarrow & Z_{n-1}(C) & \longrightarrow & H_{n-1}(C) & \longrightarrow & 0
\end{array}
$$

whose top line is the extension corresponding to $u_\dagger$. The element $\overline{\partial x} \in B_{n-1}(C)/mB_{n-1}(C)$ is the image of $\beta_m(\xi)$ under the homomorphism e_m corresponding to the bottom line of the diagram, and therefore $\psi(\overline{\partial x}) = f(\phi(x)) = f_m \rho_m(\phi(x))$. Thus

$$u_\dagger^m \beta_m(\xi) = f_m \rho_m(\phi(x)).$$

On the other hand, $\rho'_m(u)$ is represented by the cochain $\rho_m \circ \phi : C_n \to G/mG$, and therefore $u_*^m(\xi) = \rho_m(\phi(x))$; hence

$$f_m u_*^m(\xi) = f_m \rho_m(\phi(x)). \qquad \square$$

(4.6) Corollary *If G has no divisible subgroups, then u_* is determined by the homomorphisms u_*^m $(m = 2, 3, 4, \ldots)$. If $H_{n-1}(C)$ is a direct sum of cyclic groups (in particular, if $H_{n-1}(C)$ is finitely generated), then $u_\dagger$ is determined by the $u_\dagger^m$.* $\square$

Let C, D be free chain complexes, $\phi : C \to D$ a chain map; for simplicity we shall assume ϕ has degree 0. The *mapping cone* of ϕ is the (free) chain complex E such that

$$E_n = C_{n-1} \oplus D_n,$$

$$\partial(c, d) = (\partial c, f(c) - \partial d).$$

If $\iota : D \to E$, $\pi : E \to C$ are defined by

$$\iota(d) = (0, d),$$

$$\pi(c, d) = c,$$

then ι is a chain map of degree 0 and π a chain map of degree -1. The short exact sequence

$$0 \longrightarrow D \xrightarrow{\ \iota\ } E \xrightarrow{\ \pi\ } C \longrightarrow 0$$

gives rise to the exact homology sequence

(4.7)

$$\cdots \to H_{n+1}(D) \xrightarrow{\iota_{n+1}} H_{n+1}(E) \xrightarrow{\pi_n} H_n(C) \xrightarrow{\phi_n} H_n(D) \to \cdots$$

The sequence (4.7) in turn gives rise, for each n, to a short exact sequence

(4.8) $$0 \to \operatorname{Cok} \phi_n \to H_n(E) \to \operatorname{Ker} \phi_{n-1} \to 0,$$

representing an element of $\operatorname{Ext}(\operatorname{Ker} \phi_{n-1}, \operatorname{Cok} \phi_n)$. We proceed to show how the above considerations lead to a description of this extension.

A cohomology class $i^n \in H^n(D; H_n(D))$ is said to be *unitary* if and only if the homomorphism i^n_* is the identity. (Then $\operatorname{Cok} i^n_* = 0$, so that $i^n_\dagger = 0$ is uninteresting).

Let $i^n \in H^n(D; H_n(D))$ be a unitary class, and let $u^n = \phi^* i^n \in H^n(C; H_n(D))$. Then $u^n_* = \phi_n$, by (4.3), so that $u^n_\dagger \in \operatorname{Ext}(H_{n-1}(C), \operatorname{Cok} \phi_n)$. The inclusion $j : \operatorname{Ker} \phi_{n-1} \subsetneq H_{n-1}(C)$ induces a homomorphism $j^* : \operatorname{Ext}(H_{n-1}(C), G) \to \operatorname{Ext}(\operatorname{Ker} \phi_{n-1}, G)$.

(4.9) Theorem *The element of* $\operatorname{Ext}(\operatorname{Ker} \phi_{n-1}, \operatorname{Cok} \phi_n)$ *defined by the short exact sequence* (4.8) *is* $-j^*(u^n_\dagger)$.

Let $\omega : D_n \to H_n(D)$ be a cocycle representing the unitary class i^n; thus $\omega | Z_n(D)$ is the projection: $Z_n(D) \to H_n(D)$. The composite

$$C_n \xrightarrow{\ \phi\ } D_n \xrightarrow{\ \omega\ } H_n(D) \xrightarrow{\ f\ } \operatorname{Cok} \phi_n$$

annihilates the subgroup $Z_n(C)$, thereby inducing a homomorphism $\psi : B_{n-1}(C) \to \operatorname{Cok} \phi_n$ such that $\psi \circ \partial' = f \circ \omega \circ \phi$; since $\omega \circ \phi$ represents $u^n = \phi^* i^n$, the homomorphism ψ represents $u^n_\dagger$.

The free resolution

$$0 \to B_{n-1}(C) \to Z_{n-1}(C) \to H_{n-1}(C) \to 0$$

determines a free resolution

(4.10) $$0 \to B_{n-1}(C) \to Q_{n-1} \to \operatorname{Ker} \phi_{n-1} \to 0;$$

here $Q_{n-1} = \{z \in C_{n-1} \,|\, \partial z = 0 \text{ and } \phi(z) \in B_{n-1}(D)\}$. The free resolution (4.10), together with the homomorphism $-\psi : B_{n-1}(C) \to \operatorname{Cok} \phi_n$, represents the element $-j^*(u_\dagger^n)$. To prove Theorem (4.9), then, it behooves us to exhibit a commutative diagram

(4.11)

$$\begin{array}{ccccccccc}
0 & \longrightarrow & B_{n-1}(C) & \longrightarrow & Q_{n-1} & \longrightarrow & \operatorname{Ker}\phi_{n-1} & \longrightarrow & 0 \\
 & & \downarrow{\scriptstyle -\psi} & & \downarrow{\scriptstyle \theta} & & \downarrow{\scriptstyle 1} & & \\
0 & \longrightarrow & \operatorname{Cok}\phi_n & \longrightarrow & H_n(E) & \longrightarrow & \operatorname{Ker}\phi_{n-1} & \longrightarrow & 0
\end{array}$$

whose upper line is (4.10) and whose lower line is (4.8).

Let K be the kernel of ω; then

$$D_n = K + Z_n(D),$$
$$K \cap Z_n(D) = B_n(D).$$

Since $\partial' : D_n \to B_{n-1}(D)$ is an epimorphism and $\partial' Z_n(D) = 0$, the homomorphism ∂' maps K upon the free group $B_{n-1}(D)$, and therefore there is a homomorphism $\rho : B_{n-1}(D) \to D_n$ such that $\operatorname{Im}\rho \subset K$ and $\partial' \circ \rho = 1$.

If $z \in Q_{n-1}$, then $\partial z = 0$ and $\phi(z) \in B_{n-1}(D)$. Then $(z, \rho\phi(z))$ is a cycle of E; let

$$\theta(z) = [z, \rho\phi(z)]$$

be its homology class. The homomorphism $\iota_* : H_{n-1}(E) \to \operatorname{Ker}\phi_{n-1} \subset H_{n-1}(C)$ maps $\theta(z)$ into the homology class of z; hence the right-hand square of (4.11) is commutative. On the other hand, if $z = \partial b \in B_{n-1}(C)$, then $\psi(z) = f\omega\phi(b)$, and this element is mapped by the homomorphism induced by ι_* into $\iota_* \omega\phi(b)$. But

$$\begin{aligned}
\theta(z) &= [z, \rho\phi(z)] \\
&= [\partial b, \phi(b)] - [0, \phi(b) - \rho\phi\, \partial b] \\
&= -\iota_*\omega(\phi(b) - \rho\phi(\partial b)) \\
&= -\iota_*\omega\phi(b),
\end{aligned}$$

so that the left-hand square of (4.11) is commutative, too.

The reason that we have christened E the mapping cone of ϕ is that, if X and Y are spaces, $f : X \to Y$ a continuous map, and $\phi : \mathfrak{S}(X) \to \mathfrak{S}(Y)$ the chain map defined by f, then the mapping cone of ϕ is chain-equivalent to the singular complex of the mapping cone $\mathbf{T}_f$ of f [E–S, Chapter VII, Exercise C]. Hence our Theorem (4.9) translates into a topological theorem. The homology sequence

$$\cdots \to H_n(X) \xrightarrow{H_n(f)} H_n(Y) \longrightarrow H_n(\mathbf{T}_f) \longrightarrow H_{n-1}(X) \xrightarrow{H_{n-1}(f)} H_{n-1}(Y) \to \cdots$$

determines, as usual, a short exact sequence

(4.12) $$0 \to \operatorname{Cok} H_n(f) \to H_n(\mathbf{T}_f) \to \operatorname{Ker} H_{n-1}(f) \to 0$$

representing an element $e_n \in \operatorname{Ext}(\operatorname{Ker} H_{n-1}(f), \operatorname{Cok} H_n(f))$.

(4.13) Corollary *Let X, Y be spaces, $f: X \to Y$ a map, $\imath^n \in H^n(Y; H_n(Y))$ a unitary class, $u^n = f^*\imath^n \in H^n(X; H_n(Y))$. Then the element*

$$e_n \in \operatorname{Ext}(\operatorname{Ker} H_{n-1}(f), \operatorname{Cok} H_n(f))$$

defined by the short exact sequence (4.12) *is $-j^*(u_\dagger^n)$, where*

$$j^* : \operatorname{Ext}(H_{n-1}(X), \operatorname{Cok} H_n(f)) \to \operatorname{Ext}(\operatorname{Ker} H_{n-1}(f), \operatorname{Cok} H_n(f))$$

is the injection. □

We can now complete the program, promised in §3, of determining $\pi_{m+1}(X)$ as a group extension, when X is an $(m-1)$-connected space. In fact, let us kill the homotopy groups $\pi_i(X)$ for all $i > m$ by adjoining cells; thus we obtain a relative CW-complex (K, X), having no cells of dimension less than $n+2$, such that the injection $\pi_m(X) \to \pi_m(K)$ is an isomorphism and $\pi_i(K) = 0$ for all $i > m$. The space K is an Eilenberg–MacLane space $K(\Pi, m)$, where $\Pi = \pi_m(X)$, and the injection

$$H^m(K; \Pi) \to H^m(X; \Pi)$$

maps $\imath^m(K)$ into $\imath^m(X)$. Consider the diagram

(4.14)

$$\begin{array}{ccccccc}
H_{m+2}(X) & \xrightarrow{b} & \Pi \otimes \mathbf{Z}_2 & \xrightarrow{\tilde{\eta}} & \pi_{m+1}(X) & \xrightarrow{\rho} & H_{m+1}(X) \\
\uparrow 1 & & \uparrow b_0 & & \uparrow \partial & & \uparrow 1 \\
 & & & & \pi_{m+2}(K, X) & & \\
 & & & & \uparrow \rho_0^{-1} & & \\
H_{m+2}(X) & \xrightarrow[i]{} & H_{m+2}(K) & \xrightarrow[j]{} & H_{m+2}(K, X) & \xrightarrow[\partial_0]{} & H_{m+1}(X)
\end{array}$$

in which i and j are injections, ∂ and ∂_0 boundary operators, ρ and ρ_0 Hurewicz maps, b and b_0 secondary boundary operators; ρ_0 is an isomorphism, since (K, X) is $(m+1)$-connected.

(4.15) Lemma *The diagram* (4.14) *is commutative.*

Commutativity of the left-hand square follows from the naturality of the sequence (3.13) (Theorem (3.16)), that of the right-hand square from (4.8) of

Chapter IV. To prove commutativity of the central square, we use Lemma (3.19). Let $w \in H_{m+2}(K)$; then there is a map $f: M \to K$ such that f_* maps the generator z of $H_{m+2}(M)$ into w; we may assume that f is cellular, so that $f(\mathbf{S}^m) \subset X$. Then $b_0(w)$ is the image under reduction mod 2 of the element of $\Pi = \pi_m(K) = \pi_m(X)$ represented by $f|\mathbf{S}^m$, and $\tilde{\eta}b_0(w)$ is the element of $\pi_{m+1}(X)$ represented by the composite

$$\mathbf{S}^{m+1} \xrightarrow{\eta_m} \mathbf{S}^m \xrightarrow{f|\mathbf{S}^m} X.$$

Let $h: (\mathbf{E}^{m+2}, \mathbf{S}^{m+1}) \to (M, \mathbf{S}^m)$ be the characteristic map for the $(m+2)$-cell of M. Then $f \circ h$ represents an element $\alpha \in \pi_{m+1}(K, X)$ such that $\rho_0(\alpha) = j(w)$. Now $\partial\alpha$ is represented by $(f \circ h)|\mathbf{S}^{m+1}$; but $h|\mathbf{S}^{m+1}: \mathbf{S}^{m+1} \to \mathbf{S}^m$ represents η_m and therefore $\tilde{\eta}b_0(w) = \partial\alpha = \partial\rho_0^{-1}j(w)$. □

The diagram (4.14) represents an isomorphism between a part of the homology sequence of (K, X) and the sequence (3.13). If $\imath^{m+2} \in H^{m+2}(K; H_{m+2}(K))$, its image under the coefficient group homomorphism induced by b_0 is the class v described in §3 which determines the cohomology operation Sq_m^2. Hence the injection $H^{m+2}(K; \Pi \otimes Z_2) \to H^{m+2}(X; \Pi \otimes Z_2)$ maps $v = Sq_m^2 \imath^m(K)$ into $Sq_m^2 \imath^m(X)$. Therefore we conclude

(4.16) Theorem *If X is an $(m-1)$-connected space $(m \geq 3)$, then the extension*

$$0 \to \text{Cok } b \to \pi_{m+1}(X) \to H_{m+1}(X) \to 0$$

determined by the sequence (3.13) *corresponds to the element $u_\dagger$, where $u = Sq_m^2 \imath^m(X)$.* □

(N.B.: we have suppressed as unnecessary the minus sign with which $u_\dagger$ should be affected, since the coefficient group $\Pi \otimes \mathbf{Z}_2$ has exponent 2).

Since the operation Sq^2 is stable, we have

(4.17) Corollary *Let X be an $(m-1)$-connected space, $\Pi = \sigma_m(X)$, and let $u = Sq^2\imath^m(X) \in H^{m+2}(X; \Pi \otimes \mathbf{Z}_2)$. Then there is an exact sequence*

$$H_{m+2}(X) \xrightarrow{u_*} \Pi \otimes \mathbf{Z}_2 \longrightarrow \sigma_{m+1}(X) \longrightarrow H_{m+1}(X) \longrightarrow 0$$

and the short exact sequence

$$0 \to \text{Cok } u_* \to \sigma_{m+1}(X) \to H_{m+1}(X) \to 0$$

corresponds to the element $u_\dagger \in \text{Ext}(H_{m+1}(X), \text{Cok } u_*)$. □

5 Stable Homotopy as a Homology Theory

We have seen that homotopy and homology groups have many features in common. They differ in the fact that the former fail to have the excision property, and this is reflected in the fact that the suspension $E : \pi_n(X) \to \pi_{n+1}(\mathbf{S}X)$ is not an isomorphism. On the other hand, it is trivial that the suspension induces isomorphisms $E : \{X, Y\} \approx \{\mathbf{S}X, \mathbf{S}Y\}$, for both groups are direct limits of the same direct system. In particular, $E : \sigma_p(X) \approx \sigma_{p+1}(\mathbf{S}X)$ for every p and every space X. Thus it is reasonable to expect that the stable homotopy groups $\sigma_p(X)$ may behave more like homology groups than do the ordinary homotopy groups $\pi_p(X)$.

The stable homotopy groups are (covariant) functors $\sigma_p : \mathscr{K}_* \to \mathscr{A}$. Moreover, for each p, the suspension operators

$$E = E_p(X) : \sigma_p(X) \to \sigma_{p+1}(\mathbf{S}X)$$

are the components of a *natural transformation* E_p of the functor σ_p into the composite functor $\sigma_{p+1} \circ \mathbf{S}$. Their fundamental properties are given by

(5.1) Theorem *The system consisting of the functors σ_p and the natural transformations E_p has the following properties:*

(1) (Homotopy). *If $f_0, f_1 : X \to Y$ are homotopic maps, then $\sigma_n(f_0) = \sigma_n(f_1) : \sigma_n(X) \to \sigma_n(Y)$ for all n.*
(2) (Exactness). *If $f : X \to Y$ is a map, $j : Y \hookrightarrow \mathbf{T}_f$, then the sequence*

(5.2)
$$\sigma_n(X) \xrightarrow{\sigma_n(f)} \sigma_n(Y) \xrightarrow{\sigma_n(j)} \sigma_n(\mathbf{T}_f)$$

is exact for every n.
(3) (Suspension). *For every space X and integer n, the homomorphism*

$$E_n(X) : \sigma_n(X) \to \sigma_{n+1}(\mathbf{S}X)$$

is an isomorphism.

The homotopy property is immediate, and we have already observed that the suspension property holds. To prove exactness, observe that the sequence (4.2) is the direct limit of the sequences

(5.3)
$$\pi_{n+r}(\mathbf{S}^r X) \to \pi_{n+r}(\mathbf{S}^r Y) \to \pi_{n+r}(\mathbf{T}_{\mathbf{S}^r f}),$$

and it suffices to prove that the sequence (5.3) is exact for all sufficiently large r. Let $X' = \mathbf{S}^r X$, $Y' = \mathbf{S}^r Y$, $f' = \mathbf{S}^r f$; we may assume that f' is an inclusion and (Y', X') an NDR-pair. Then $\mathbf{T}_{f'}$ is the subspace $Y' \cup \mathbf{T}X'$ of $\mathbf{T}Y'$. There is a commutative diagram

$$\begin{array}{ccccc} \pi_{n+r}(X') & \xrightarrow{i} & \pi_{n+r}(Y') & \xrightarrow{l} & \pi_{n+r}(Y' \cup \mathbf{T}X') \\ & & \downarrow j & & \downarrow q \\ & & \pi_{n+r}(Y', X') & \xrightarrow[k]{} & \pi_{n+r}(Y' \cup \mathbf{T}X', \mathbf{T}X') \end{array}$$

where i, j, k, q and l are injections. Since $\mathbf{T}X$ is contractible, q is an isomorphism. The spaces X' and Y' are $(r-1)$-connected, and hence the pairs (Y', X') and $(\mathbf{T}X', X')$ are $(r-1)$-connected. By the Blakers–Massey Theorem, k is an isomorphism provided that $r \geq 2$ and $n + r \leq 2(r-1)$. Thus if $r \geq \max\{2, n+2\}$, Ker $l =$ Ker $j =$ Im i, and therefore the sequence (5.3) is exact. □

The properties (1)–(3) of Theorem (5.1) are reminiscent of the Eilenberg–Steenrod axioms. They differ in that our functors are defined on a category of spaces with base points, while the axioms are concerned with functors defined on a suitable category of free pairs. In order to compare our functors with homology functors, we need to formulate a notion of homology on a suitable pointed category and show how the two kinds of homology theory are related.

Let us recall that $\mathscr{K}_*$ is the category of spaces with non-degenerate base points. A *reduced homology theory* $\mathfrak{h}$ on $\mathscr{K}_*$ consists of

(1) a family of functors $h_n : \mathscr{K}_* \to \mathscr{A}$ $(n \in \mathbf{Z})$, together with
(2) a family of natural transformations

$$e_n : h_n \to h_{n+1} \circ \mathbf{S} \qquad (n \in \mathbf{Z}),$$

satisfying the following conditions, analogous to those of Theorem (5.1):

(1) (Homotopy). *If $f_0, f_1 : X \to Y$ are homotopic, then*

$$h_n(f_0) = h_n(f_1) : h_n(X) \to h_n(Y)$$

for all $n \in \mathbf{Z}$.

(2) (Exactness). *If $f : X \to Y$ and if $j : Y \subsetneq \mathbf{T}_f$, then the sequence*

$$h_n(X) \xrightarrow{h_n(f)} h_n(Y) \xrightarrow{h_n(j)} h_n(\mathbf{T}_f)$$

is exact for all $n \in \mathbf{Z}$.

(3) (Suspension). *The homomorphism*

$$e_n(X) : h_n(X) \to h_{n+1}(\mathbf{S}X)$$

is an isomorphism for all X and all $n \in \mathbf{Z}$.

Examples of homology theories come readily to mind.

EXAMPLE 1. Singular homology with coefficients in an arbitrary abelian group G is a homology theory $\mathfrak{h}(G)$.

EXAMPLE 2. Stable homotopy is a homology theory $\mathfrak{s}$, by Theorem (5.1).

EXAMPLE 3. If P is a space in $\mathscr{K}_*$ and $\mathfrak{h}$ is a homology theory, there is a homology theory $\mathfrak{h}^P$ for which

$$h_n^P(X) = h_n(P \wedge X),$$
$$h_n^P(f) = h_n(1 \wedge f) \quad \text{for } f: X \to Y,\ 1: P \subsetneq P,$$

while $e_n^P(X): h_n(P \wedge X) \to h_{n+1}(P \wedge \mathbf{S}X)$ coincides with

$$e_n(P \wedge X): h_n(P \wedge X) \to h_{n+1}(\mathbf{S}(P \wedge X))$$

under the natural homeomorphism of $P \wedge \mathbf{S}X$ with $\mathbf{S}(P \wedge X)$.

EXAMPLE 4. If m is a non-negative integer, we may take $P = \mathbf{S}^m$ in Example 3. It follows by iterated use of the Suspension Property that $h_n^P(X) \approx h_{n-m}(X)$. Similarly, $h_n^P(f)$ may be identified with $h_{n-m}(f)$ for any $f: X \to Y$. Finally, $e_n^P(X)$ may be identified with $e_{n-m}(X)$. Thus $\mathfrak{h}^P$ is obtained from $\mathfrak{h}$ by subtracting m from all indices. Similarly, we may reindex $\mathfrak{h}$ by adding m to all indices (though the resulting theory no longer has the form $\mathfrak{h}^P$). In either case, we shall say that the new theory is obtained from the old by *reindexing*.

EXAMPLE 5. If $\{\mathfrak{h}^\alpha \mid \alpha \in J\}$ is a family of homology theories, their direct sum $\mathfrak{h} = \bigoplus_\alpha \mathfrak{h}^\alpha$ is a homology theory, for which

$$h_n(X) = \bigoplus_\alpha h_n^\alpha(X),$$
$$h_n(f) = \bigoplus_\alpha h_n^\alpha(f) \quad \text{for } f: X \to Y,$$
$$e_n(X) = \bigoplus_\alpha e_n^\alpha(X).$$

Many more interesting examples will be studied in Volume II.

A homology theory $\mathfrak{h}$ is said to be *proper* if and only if it has the additional property

(4) (Dimension). If X is a 0-sphere, then $h_n(X) = 0$ for all $n \neq 0$.

Stable homotopy is not proper, for we have seen that $\sigma_1 = \sigma_1(\mathbf{S}^0) \approx \mathbf{Z}_2 \neq 0$. Nor can it be made proper by reindexing since $\sigma_0 = \sigma_0(\mathbf{S}^0) \approx \mathbf{Z}$ is also non-zero. Indeed, it can be proved (Exercise 4, below) that $\mathfrak{s}$ is not a direct sum of reindexed proper theories.

The *coefficient groups* of a homology theory are the groups $h_q(\mathbf{S}^0)$.

Two further properties are:

(5) (Additivity). *If* $\{X_\alpha \mid \alpha \in J\}$ *is a family of spaces,* $X = \bigvee_\alpha X_\alpha$, *then the injections* $h_n(X_\alpha) \to h_n(X)$ *represent the latter group as a direct sum.*

(6) (Isotropy). *If* $f : X \to Y$ *is a weak homotopy equivalence, then* $h_n(f) : h_n(X) \approx h_n(Y)$ *for every* n.

A homology theory $\mathfrak{h}$ is said to be *complete* if and only if it is additive and isotropic. For example, $\mathfrak{h}(G)$ and $\mathfrak{s}$ are complete.

Remark 1. The reader will have observed that we have not included the Dimension Property in the definition of a homology theory. Indeed, homology theories in our sense have been dubbed "generalized" or "extraordinary" in the literature. The reason for this is historical. When Eilenberg and Steenrod formulated their axioms for homology theory, very few examples of improper theories were known. In the intervening years, however, there has arisen a plethora of interesting new theories: stable homotopy, bordism, and the various K-theories. We shall study these new theories systematically in the second volume of this work.

Remark 2. Let $\mathscr{P}_*$ be the full subcategory of $\mathscr{K}_*$ whose objects are spaces having the homotopy type of CW-complexes. Let $h_n^0 : \mathscr{P}_* \to \mathscr{A}$, $e_n^0 : h_n^0 \to h_{n+1}^0 \circ S$ have the Homotopy, Exactness and Suspension Properties. We then say that $\mathfrak{h}^0 = \{h_n^0, e_n^0\}$ is a *reduced homology theory on* $\mathscr{P}_*$. The reader is invited (Exercise 3, below) to use standard techniques of CW-approximation to show that $\mathfrak{h}^0$ has an essentially unique extension to an isotropic theory $\mathfrak{h}$ on $\mathscr{K}_*$. Thus there is a one-to-one correspondence (up to natural isomorphism) between reduced homology theories on $\mathscr{P}_*$ and isotropic reduced theories on $\mathscr{K}_*$. Moreover, $\mathfrak{h}$ is additive (proper) if and only if $\mathfrak{h}^0$ is.

Let us derive some properties of a homology theory $\mathfrak{h}$ on $\mathscr{K}_*$.

(5.4) *If* (X, A) *is an* NDR-*pair in* $\mathscr{K}_*$, $i : A \subsetneq X$, *and* $p : X \to X/A$ *is the collapsing map, then the sequence*

(5.5)
$$h_n(A) \xrightarrow{h_n(i)} h_n(X) \xrightarrow{h_n(p)} h_n(X/A)$$

is exact.

For there is a commutative diagram

$$\begin{array}{ccccc} A & \xrightarrow{i} & X & \xrightarrow{j} & \mathbf{T}_i = X \cup \mathbf{T}A \\ & & & \searrow^{p} & \downarrow{q} \\ & & & & X/A \end{array}$$

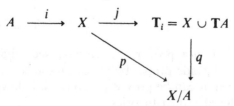

where q collapses $\mathbf{T}A$ to a point. The map q is a homotopy equivalence, by Corollary (5.13) of Chapter I. Apply the functor h_n to the diagram. Since

$h_n(q)$ is an isomorphism, exactness of (5.5) follows from the Exactness Property. □

(5.6) *If P is a space consisting of just one point, then $h_n(P) = 0$ for all n.*

Identifying P/P with P, we obtain from (5.4) an exact sequence

$$h_n(P) \xrightarrow{h_n(1)} h_n(P) \xrightarrow{h_n(1)} h_n(P).$$

But $h_n(1)$ is the identity map, so that

$$h_n(1) \circ h_n(1) = h_n(1) = 1.$$

But

$$h_n(1) \circ h_n(1) = 0$$

by exactness. This can only be true if $h_n(P) = 0$. □

(5.7) *If $f: X \to Y$ is a constant map, then $h_n(f) = 0$ for all n.*

For $h_n(f)$ factors

$$h_n(X) \to h_n(P) \to h_n(Y)$$

through $h_n(P) = 0$. □

(5.8) *If X, Y are spaces in $\mathscr{K}_*$, and $j_1 : X \to X \vee Y$, $j_2 : Y \to X \vee Y$ are the inclusions, then the injections*

$$j_{1*} = h_n(j_1) : h_n(X) \to h_n(X \vee Y),$$
$$j_{2*} = h_n(j_2) : h_n(Y) \to h_n(X \vee Y),$$

represent $h_n(X \vee Y)$ as a direct sum. The homomorphisms

$$q_{1*} : h_n(X \vee Y) \to h_n(X),$$
$$q_{2*} : h_n(X \vee Y) \to h_n(Y),$$

induced by the projections q_1, q_2, form the dual representation of $h_n(X \vee Y)$ as a direct product.

There is a commutative diagram

$$\begin{array}{ccc}
h_n(X) & \xrightarrow{1 = h_n(1)} & h_n(X) \\
& {\scriptstyle h_n(j_1)} \searrow \quad \nearrow {\scriptstyle h_n(q_1)} & \\
& h_n(X \vee Y) & \\
& {\scriptstyle h_n(j_2)} \nearrow \quad \searrow {\scriptstyle h_n(q_2)} & \\
h_n(Y) & \xrightarrow[1 = h_n(1)]{} & h_n(Y)
\end{array}$$

and the diagonal sequences are exact by (5.4). The result now follows from Lemma 13.1 of Chapter I of [E–S]. □

As in §7 of Chapter III (note that the arguments used there do not depend on the Dimension Property) we deduce

(5.9) *Let X be an* H′*-space with coproduct θ. Then*

$$\theta_*(x) = j_{1*}(x) + j_{2*}(x)$$

for all $x \in h_n(X)$. □

(5.10) *Let X be an* H′*-space, $f_1, f_2 : X \to Y$. Then*

$$(f_1 + f_2)_*(x) = f_{1*}(x) + f_{2*}(x)$$

for all $x \in h_n(X)$. □

(5.11) *Let $f: X \to Y$. Then*

$$(-\mathbf{S}f)_* = -(\mathbf{S}f)_* : h_n(\mathbf{S}X) \to h_n(\mathbf{S}Y).$$ □

The graded homology group $h_*(X)$ admits a module structure over the stable homotopy ring σ_*. To see this, let $u \in H_m(X)$, $\alpha \in \sigma_n$, and let $f: \mathbf{S}^{n+k} \to \mathbf{S}^k$ be a representative of α. Then $f \wedge 1 : \mathbf{S}^{n+k}X \to \mathbf{S}^k X$. Consider the composite

$$\begin{aligned} h_m(X) \xrightarrow{e^{n+k}} h_{m+n+k}(\mathbf{S}^{n+k}X) &\xrightarrow{h_{m+n+k}(f \wedge 1)} \\ &h_{m+n+k}(\mathbf{S}^k X) \xrightarrow{e^{-k}} h_{m+n}(X), \end{aligned}$$

where e^r is the rth iterate of the suspension operator e if $r \geq 0$, and the $(-r)$th iterate of e^{-1} if $r < 0$. The map $\mathbf{S}f: \mathbf{S}^{n+k+1} \to \mathbf{S}^{k+1}$ also represents α, and the diagram

$$\begin{array}{ccccccc} & & h_{m+n+k+1}(\mathbf{S}^{n+k+1}X) & \xrightarrow{h_{m+n+k+1}(\mathbf{S}f \wedge 1)} & h_{m+n+k+1}(\mathbf{S}^{k+1}X) & & \\ & \nearrow^{e^{n+k+1}} & \uparrow e & & \uparrow e & \searrow^{e^{-k-1}} & \\ h_m(X) & & & & & & h_{m+n}(X) \\ & \searrow_{e^{n+k}} & & & & \nearrow_{e^{-k}} & \\ & & h_{m+n+k}(\mathbf{S}^{n+k}X) & \xrightarrow[h_{m+n+k}(f \wedge 1)]{} & h_{m+n+k}(\mathbf{S}^k X) & & \end{array}$$

is commutative. It follows that the element

$$e^{-k}h_{m+n+k}(f \wedge 1)e^{n+k}(u)$$

depends only on u and the element α; let αu be this element.

(5.12) Theorem *The graded group $h_*(X)$ is a graded module over σ_* under the map $\alpha \otimes u \to \alpha u$. If $f: X \to Y$, then $h_*(f): h_*(X) \to h_*(Y)$ is a map of σ_*-modules.*

The verifications of the requisite properties

(1) $\alpha(u + v) = \alpha u + \alpha v$,
(2) $\alpha(\beta u) = (\alpha\beta)u$,
(3) $1u = u$,
(4) $f_*(\alpha u) = \alpha f_*(u)$,
for α, $\beta \in \sigma_*$, u, $v \in h_*(X)$, $f : X \to Y$ are verified by routine diagram chasing. It remains only to prove
(5) $(\alpha + \beta)u = \alpha u + \beta u$.

Let $f: \mathbf{S}^{n+k} \to \mathbf{S}^k$, $g : \mathbf{S}^{n+k} \to \mathbf{S}^k$ be representatives of α, $\beta \in \sigma_n$, respectively. Then

$$\begin{aligned}(\alpha + \beta)u &= e^{-k}(f + g)_* e^{n+k}u \\ &= e^{-k}(f_* + g_*)e^{n+k}u \quad \text{by (5.10)} \\ &= e^{-k}f_* e^{n+k}u + e^{-k}g_* e^{n+k}u \\ &= \alpha u + \beta u.\end{aligned}$$

□

(5.13) Corollary *Let $f: \mathbf{S}^n \to \mathbf{S}^n$ have degree d, $g : X \to Y$. Then*

$$(f \wedge g)_* e^n u = d e^n(g_* u)$$

for all $u \in h_q(X)$. □

We can now strengthen the Exactness Property. Let $f: X \to Y$. In §6 of Chapter III we exhibited a homotopy commutative diagram

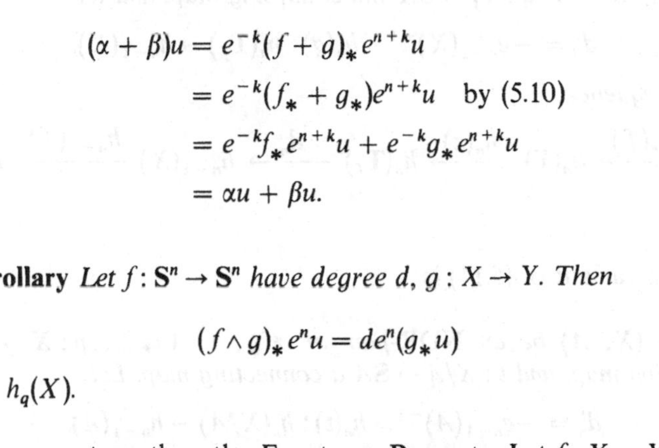

in which k, l, and m are inclusions, while q_1 and q_2 are homotopy equivalences. Applying the functor h_n to this diagram, we obtain a commutative diagram

$$\begin{array}{ccccccccc}
h_n(X) & \xrightarrow{h_n(f)} & h_n(Y) & \xrightarrow{h_n(k)} & h_n(\mathbf{T}_f) & \xrightarrow{h_n(l)} & h_n(\mathbf{T}_k) & \xrightarrow{h_n(m)} & h_n(\mathbf{T}_l) \\
 & & & & & \searrow^{h_n(q)} & \downarrow{\scriptstyle h_n(q_1)} & & \downarrow{\scriptstyle h_n(q_2)} \\
 & & & & & & h_n(\mathbf{S}X) & \xrightarrow{h_n(-\mathbf{S}f)} & h_n(\mathbf{S}Y)
\end{array}$$

Now $h_n(-\mathbf{S}f) = -h_n(\mathbf{S}f)$, by (5.11), so that the diagram

$$\begin{array}{ccc}
h_n(\mathbf{S}X) & \xrightarrow{h_n(-\mathbf{S}f)} & h_n(\mathbf{S}Y) \\
{\scriptstyle -e_{n-1}(X)}\uparrow & & \uparrow{\scriptstyle e_{n-1}(Y)} \\
h_{n-1}(X) & \xrightarrow[h_{n-1}(f)]{} & h_{n-1}(Y)
\end{array}$$

is commutative. We deduce

(5.14) *Let $f: X \to Y$, $q: \mathbf{T}_f \to \mathbf{S}X$ the collapsing map, and let*

$$d_n = -e_{n-1}(X)^{-1} \circ h_n(q): h_n(\mathbf{T}_f) \to h_{n-1}(X).$$

Then the sequence

$$h_n(X) \xrightarrow{h_n(f)} h_n(Y) \xrightarrow{h_n(k)} h_n(\mathbf{T}_f) \xrightarrow{d_n} h_{n-1}(X) \xrightarrow{h_{n-1}(f)} h_{n-1}(Y)$$

is exact. □

A useful variant of (5.14) is

(5.15) *Let (X, A) be an NDR-pair in $\mathscr{K}_*$, $i: A \subsetneq X$, $p: X \to X/A$ the identification map, and $t: X/A \to \mathbf{S}A$ a connecting map. Let*

$$d'_n = -e_{n-1}(A)^{-1} \circ h_n(t): h_n(X/A) \to h_{n-1}(A).$$

Then the sequence

$$h_n(A) \xrightarrow{h_n(i)} h_n(X) \xrightarrow{h_n(p)} h_n(X/A) \xrightarrow{d'_n} h_{n-1}(A) \xrightarrow{h_{n-1}(i)} h_{n-1}(X)$$

is exact. □

6 Comparison with the Eilenberg–Steenrod Axioms

Let $\mathfrak{h}$ be a reduced homology theory on $\mathscr{K}_*$, and let $\mathscr{K}_*^2$ be the category of free NDR-pairs. We shall show how $\mathfrak{h}$ determines an Eilenberg–Steenrod theory on $\mathscr{K}_*^2$ (i.e., one satisfying their axioms except for the Dimension

Axiom). Conversely, an Eilenberg–Steenrod theory on $\mathscr{K}_*^2$ determines a homology theory on $\mathscr{K}_*$, and there is a one-to-one correspondence (up to natural isomorphism) between the two kinds of theories.

A homology theory on $\mathscr{K}_*^2$ is one which satisfies the Eilenberg–Steenrod axioms, except for the Dimension Axiom. Such a theory consists of

(1) a family of functors $H_n : \mathscr{K}_*^2 \to \mathscr{A}$ ($n \in \mathbf{Z}$), together with
(2) a family of natural transformations $\partial_n : H_n \to H_{n-1} \circ R$, where $R : \mathscr{K}_*^2 \to \mathscr{K}_*^2$ is the shift functor, defined for pairs (X, A) by

$$R(X, A) = (A, \varnothing)$$

and for maps $f: (X, A) \to (Y, B)$ by

$$R(f) = f \,|\, A : (A, \varnothing) \to (B, \varnothing).$$

having the following properties:

(1) (Homotopy). *If $f_0, f_1 : (X, A) \to (Y, B)$ are homotopic, then*

$$H_n(f_0) = H_n(f_1) : H_n(X, A) \to H_n(Y, B)$$

for all $n \in \mathbf{Z}$.

(2) (Exactness). *If (X, A) is a pair in $\mathscr{K}_*^2$ and $i : A \subsetneq X, j : X \subsetneq (X, A)$, then the sequence*

$$\cdots \to H_{n+1}(X, A) \xrightarrow{\partial_{n+1}(X, A)} H_n(A) \xrightarrow{H_n(i)} H_n(X) \xrightarrow{H_n(j)}$$

$$H_n(X, A) \xrightarrow{\partial_n(X, A)} H_{n-1}(A) \to \cdots$$

is exact.

(3) (Excision). *Let X be the union of two closed sets A, B and suppose that $(A, A \cap B)$ is an* NDR-*pair. Then the injection*

$$H_n(A, A \cap B) \to H_n(X, B)$$

is an isomorphism for every n.

Remark. Our axioms differ slightly from those of Eilenberg–Steenrod in that the groups $H_n(X, A)$ are defined only for NDR-pairs and not for all pairs. For this reason the Excision Axiom takes a slightly different form.

The theory $\mathfrak{H}$ is said to be *proper* if and only if it satisfies

(4) (Dimension). *If P is a space consisting of just one point, then $H_n(P) = 0$ for all $n \neq 0$.*

The theory $\mathfrak{H}$ is said to be *additive* if and only if it satisfies

(5) (Additivity). *Let X be the topological sum $\sum X_\alpha$, and let A be a subspace of X, $A_\alpha = A \cap X_\alpha$. Then the injections $H_q(X_\alpha, A_\alpha) \to H_q(X, A)$ represent the group $H_q(X, A)$ as a direct sum.*

The theory $\mathfrak{H}$ is *isotropic* if and only if it satisfies

(6) (Isotropy). *Let* $f: X \to Y$ *be a weak homotopy equivalence. Then* $H_n(f): H_n(X) \approx H_n(Y)$ *for every* n.

A homology theory which is additive and isotropic is said to be *complete*.

Certain standard results of singular homology theory remain true for an arbitrary homology theory on $\mathscr{K}_*^2$, because their proofs do not make use of the Dimension Axiom. We enumerate some of these.

(6.1) *For any free space* X, $H_n(X, X) = 0$ *for all* n. □

(6.2) *If* A *is a deformation retract of* X, *then* $H_n(X, A) = 0$ *for all* n. □

(6.3) *If* (X, A) *and* (A, B) *are* NDR-*pairs, then the homology sequence*

$$\cdots \to H_{n+1}(X, A) \to H_n(A, B) \to H_n(X, B) \to H_n(X, A) \to \cdots$$

of the triple (X, A, B) *is exact.* □

Let P be a space consisting of a single point. For any space X, let $\tilde{H}_n(X)$ be the kernel of $H_n(f): H_n(X) \to H_n(P)$, where $f: X \to P$ is the unique map. Then $\tilde{H}_n$ is a functor: $\mathscr{K} \to \mathscr{A}$.

(6.4) *If* $x \in X$, *then the injection* $H_n(X) \to H_n(X, \{x\})$ *maps* $\tilde{H}_n(X)$ *isomorphically upon* $H_n(X, \{x\})$. □

(6.5) *If* $X \neq \varnothing$, *the sequence*

$$0 \to H_n(\tilde{X}) \to H_n(X) \to H_n(P) \to 0$$

is exact and splittable, so that

$$H_n(X) \approx H_n(\tilde{X}) \oplus H_n(P).$$ □

(6.6) *For any pair* (X, A), *the image of*

$$\partial_n(X, A): H_n(X, A) \to H_{n-1}(A)$$

is contained in $\tilde{H}_{n-1}(A)$. *Moreover, there is an exact sequence*

$$\cdots \to H_{n+1}(X, A) \to \tilde{H}_n(A) \to \tilde{H}_n(X) \to H_n(X, A) \to \tilde{H}_{n-1}(A) \to \cdots.$$ □

(6.7) (Map Excision Theorem). *If* $f: (X, A) \to (Y, B)$ *is a relative homeomorphism of* NDR-*pairs then*

$$H_n(f): H_n(X, A) \approx H_n(Y, B)$$

for all n. □

By induction from the Excision Property, we have

(6.8) (Finite Direct Sum Theorem). *Let $X = X_1 \cup \cdots \cup X_n \cup A$, where X_α and A are closed subsets of X ($\alpha = 1, \ldots, n$). Suppose that $X_\alpha \cap X_\beta \subset A$ for $\alpha \neq \beta$ and that $(X_\alpha, X_\alpha \cap A)$ is an* NDR*-pair. Let $X_\alpha^* = A \cup \bigcup_{\beta \neq \alpha} X_\beta$, $A_\alpha = X_\alpha \cap A$. Then*

(1) *the injections $i_\alpha : H_q(X_\alpha, A_\alpha) \to H_q(X, A)$ represent the group $H_q(X, A)$ as a direct sum;*
(2) *the injections $j_\alpha : H_q(X, A) \to H_q(X, X_\alpha^*)$ represent $H_q(X, A)$ as a direct product;*
(3) *the injection $k_\alpha = j_\alpha \circ i_\alpha : H_q(X_\alpha, A_\alpha) \to H_q(X, X_\alpha^*)$ is an isomorphism, and the representations $\{i_\alpha\}$, $\{j_\alpha\}$ are weakly dual.* □

We shall also need to consider the infinite case. It is then necessary to assume that the homology theory is additive.

(6.9) Theorem (General Direct Sum Theorem). *Let $X = A \cup \bigcup_{\alpha \in J} X_\alpha$, where A and X_α are closed in X, $X_\alpha \cap X_\beta \subset A$ for $\alpha \neq \beta$, and $(X_\alpha, X_\alpha \cap A)$ is an* NDR*-pair. Suppose, moreover, that X has the weak topology with respect to the sets A, X_α. Let $\mathfrak{H}$ be an additive homology theory. For each $\alpha \in J$, let $X_\alpha^* = A \cup \bigcup_{\beta \neq \alpha} X_\beta$, $A_\alpha = X_\alpha \cap A$. Then*

(1) *the injections $i_\alpha : H_q(X_\alpha, A_\alpha) \to H_q(X, A)$ represent the group $H_q(X, A)$ as a direct sum;*
(2) *the injections $j_\alpha : H_q(X, A) \to H_q(X, X_\alpha^*)$ represent $H_q(X, A)$ as a weak direct product;*
(3) *the injections $k_\alpha = j_\alpha \circ i_\alpha : H_q(X_\alpha, A_\alpha) \to H_q(X, X_\alpha^*)$ are isomorphisms, and the representations $\{i_\alpha\}$, $\{j_\alpha\}$ are weakly dual.*

Let $J_+ = J \cup \{\infty\}$, where $\infty \notin J$; we shall consider J_+ as a discrete space. Let $X_\infty = A$, and let $\bar{X}$ be the subspace

$$\bigcup_{\alpha \in J_+} \{\alpha\} \times X_\alpha$$

of $J_+ \times X$. If $\alpha \in J$, let

$$\begin{aligned} \bar{X}_\alpha &= \{\alpha\} \times X_\alpha, \\ \bar{A}_\alpha &= \{\alpha\} \times A_\alpha, \\ \bar{A} &= \{\infty\} \times A. \end{aligned}$$

Let $p : \bar{X} \to X$ be the restriction to $\bar{X}$ of the projection $J_+ \times X \to X$ on the second factor. Our hypothesis on the topology of X insures that p is a proclusion. Let

$$\bar{B} = p^{-1}(A) = \bar{A} \cup \bigcup_{\alpha \in J} \bar{A}_\alpha.$$

Then $(\bar{X}, \bar{B})$ is an NDR-pair and $p : (\bar{X}, \bar{B}) \to (X, A)$ a relative homeomorphism. Therefore

$$H_n(p) : H_n(\bar{X}, \bar{B}) \to H_n(X, A)$$

is an isomorphism, by (6.7).

Let $\bar{i}_\alpha : X_\alpha \to \bar{X}$ be the natural homeomorphism of X_α with $\{\alpha\} \times X_\alpha \subset \bar{X}$; then $\bar{i}_\alpha(X_\alpha \cap A) \subset \bar{B}$. Since $H_n(\bar{A}, \bar{A}) = 0$ for all n and $\mathfrak{H}$ is additive the homomorphisms $H_n(\bar{i}_\alpha) : H_n(X_\alpha, X_\alpha \cap A) \to H_n(\bar{X}, \bar{B})$ $(\alpha \in J)$ represent $H_n(\bar{X}, \bar{B})$ as a direct sum. But $p \circ \bar{i}_\alpha : (X_\alpha, X_\alpha \cap A) \subsetneq (X, A)$ and (1) follows.

That k_α is an isomorphism follows from the Excision Property. If $\alpha \neq \beta$, then $X_\alpha \subset X_\beta^*$ and therefore the injection $j_\beta \circ i_\alpha$ is trivial. Conclusions (2) and (3) follow from this fact. □

If Y is a fixed space and $\mathfrak{H}$ a homology theory on $\mathscr{K}_*^2$, then the functors H'_q and natural transformations ∂'_q defined by

$$H'_q(X, A) = H_q(X \times Y, A \times Y),$$
$$H'_q(f) = H_q(f \times 1) : H'_q(X, A) \to H'_q(X', A')$$
$$\text{for } f : (X, A) \to (X', A'),$$
$$\partial'_q(X, A) = \partial(X \times Y, A \times Y) : H'_q(X, A) \to H'_{q-1}(A),$$

constitute a homology theory $\mathfrak{H}'$ on $\mathscr{K}_*^2$.

With the aid of this remark we can use the results of §6 of Chapter II, insofar as they do not depend on the Dimension Property, to obtain some properties of a regular cell complex K, which will be useful later.

Let E be an n-cell of K, F an $(n-1)$-face of E, $F' = \overline{\dot{E} - F}$. Then

$$\partial : H_q(E \times Y, \dot{E} \times Y) \to H_{q-1}(\dot{E} \times Y, F' \times Y)$$

and

$$k : H_{q-1}(F \times Y, \dot{F} \times Y) \to H_{q-1}(\dot{E} \times Y, F' \times Y)$$

are isomorphisms. Hence

$$\sigma(E, F) = k^{-1} \circ \partial : H_q(E \times Y, \dot{E} \times Y) \to H_{q-1}(F \times Y, \dot{F} \times Y)$$

is an isomorphism, called the *general incidence isomorphism*. Evidently $\sigma(E, F)$ is natural with respect to maps $Y \to Y'$.

Let $\Gamma(K, Y)$ be the doubly graded group such that $\Gamma_{p,q}(K, Y) = H_{p+q}(K_p \times Y, K_{p-1} \times Y)$. Then $\Gamma(K, Y)$ is a chain complex with boundary operator

$$\partial : \Gamma_{p,q}(X, Y) \to \Gamma_{p,q-1}(X, Y).$$

Moreover, the injections

$$i_\alpha : H_{p+q}(E_\alpha \times Y, \dot{E}_\alpha \times Y) \to H_{p+q}(K_p \times Y, K_{p-1} \times Y)$$

for all p-cells E_α of K, represent the latter group as a direct sum. And we can use this remark to calculate the boundary operator in $\Gamma(K, Y)$. Specifically,

(6.10) Theorem *The boundary operator in $\Gamma(K, Y)$ is determined by the relation*

$$\partial \circ i_\alpha = \sum_\beta i_\beta \circ \sigma(E_\alpha, E_\beta),$$

where E_β ranges over all $(p-1)$-faces of E_α. □

(6.11) Theorem *Let F_1 and F_2 be $(n-1)$-faces of E which have a common $(n-2)$-face G. Then*

(1) if $n \geq 2$,

$$\sigma(F_1, G) \circ \sigma(E, F_1) + \sigma(F_2, G) \circ \sigma(E, F_2) = 0;$$

(2) if $n = 1$, let $p_i : F_i \times Y \to Y$ be the projection on the second factor. Then

$$H_{q-1}(p_1) \circ \sigma(E, F_1) + H_{q-1}(p_2) \circ \sigma(E, F_2) = 0.$$ □

By induction on n we can deduce from the above remarks that $H_q(E \times Y, \dot{E} \times Y) \approx H_{q-n}(Y)$ for every Y. However, we shall need to prove a somewhat more delicate relation.

It will be convenient in what follows to assume that K is oriented. We shall also make the convention that, if a cell of K is denoted by a capital letter, its preferred orientation will be denoted by the corresponding lower case letter.

(6.12) Theorem *For each n-cell E of K there is an isomorphism $e\times : H_k(Y) \to H_{n+k}(E \times Y, \dot{E} \times Y)$ such that, for every $(n-1)$-face F of E,*

$$\sigma(E, F)(e \times u) = [e : f] f \times u \tag{6.13}$$

for every $u \in H_k(Y)$.

If $n = 0$, $e\,\times$ is the isomorphism induced by the natural homeomorphism of Y with $E \times Y$. Suppose that $f\,\times$ has been defined for every cell F of dimension $q < n$ and satisfies the condition

$$\sigma(F, G)(f \times u) = [f : g] g \times u$$

for every $(q-1)$-face G of F.

Let E be an n-cell of K. For each $(n-1)$-face F of E and each $u \in H_k(Y)$, let $\phi(F)$ be the element

$$[e : f]\sigma(E, F)^{-1}(f \times u).$$

We shall show that, if F_1 and F_2 are faces of E, then $\phi(F_1) = \phi(F_2)$. Because of Property (3) of (6.2), Chapter II, we may assume that F_1 and F_2 have a common $(n-2)$-face G. When this is the case,

$$\begin{aligned}
\phi(F_2) &= [e:f_2]\sigma(E, F_2)^{-1}(f_2 \times u) \\
&= [e:f_2][f_2:g]\sigma(E, F_2)^{-1}\sigma(F_2, G)^{-1}(g \times u) \\
&\qquad \text{by induction hypothesis} \\
&= -[e:f_2][f_2:g]\sigma(E, F_1)^{-1}\sigma(F_1, G)^{-1}(g \times u) \\
&\qquad \text{by Theorem (6.11)} \\
&= [e:f_1][f_1:g]\sigma(E, F_1)^{-1}\sigma(F_1, G)^{-1}(g \times u) \\
&\qquad \text{by Theorem (6.6), Chapter II} \\
&= [e:f_1]\sigma(E, F_1)^{-1}(f_1 \times u) \quad \text{by induction hypothesis} \\
&= \phi(F_1).
\end{aligned}$$

We may therefore define $e \times u$ to be the common value of $\phi(F)$ for all $(n-1)$-faces F of E, and (6.13) then holds for the pair (E, F).

If $n = 1$, the proof has to be modified to make use of the second conclusion of Theorem (6.11) instead of the first. This is left to the reader. □

The map $e \times$ is natural in Y; i.e.,

(6.14) *Let $f\colon Y \to Z$, $u \in H_k(Y)$. Then*

$$(1 \times f)_*(e \times u) = e \times f_* u.$$ □

Let $\mathfrak{h}$ be a reduced homology theory on $\mathscr{K}_*$. We now show how to associate to $\mathfrak{h}$ a homology theory on the category $\mathscr{K}_*^2$. Let us recall that in §2 of Chapter III we imbedded the category of free pairs into $\mathscr{K}_*$ by the device of adjoining an external base point P; if X is a free space, $X^+ = X + P$ is a space with non-degenerate base point P, and if $f\colon X \to Y$ is a free map, then f extends to a map $f^+ : (X^+, P) \to (Y^+, P)$. Thus, if (X, A) is an object of $\mathscr{K}_*^2$, we shall define

$$H_n(X, A) = h_n(X^+/A^+),$$

and if $f\colon (X, A) \to (Y, B)$ is a free map, there is a uniquely determined map $f^\#$ making the diagram

$$\begin{array}{ccc}
X & \xrightarrow{\;f\;} & Y \\
\big\downarrow & & \big\downarrow \\
X^+ & \xrightarrow{\;f^+\;} & Y^+ \\
{\scriptstyle p_X}\big\downarrow & & \big\downarrow{\scriptstyle p_Y} \\
X^+/A^+ & \xrightarrow[\;f^\#\;]{} & Y^+/B^+
\end{array}$$

commutative (p_X and p_Y being the appropriate collapsing maps). We then define

$$H_n(f) = h_n(f^*) : H_n(X, A) \to H_n(Y, B).$$

Clearly, H_n is a functor: $\mathscr{K}^2_* \to \mathscr{A}$.

Next, if (X, A) is a pair in $\mathscr{K}^2_*$, let $\partial_n(X, A)$ be minus the composite

$$H_n(X, A) = h_n(X^+/A^+) \xrightarrow{h_n(t)} h_n(SA^+) \xrightarrow{e_{n-1}(A^+)} h_{n-1}(A^+) = H_n(A, \varnothing),$$

where $t : X^+/A^+ \to \mathbf{S}A^+$ is a connecting map. Evidently $\partial_n : H_n \to H_{n-1} \circ R$ is a natural transformation.

Our first result is

(6.15) Theorem *If $\mathfrak{h}$ is a boundary theory on $\mathscr{K}_*$ and $\mathfrak{H}$ is the theory defined above, then $\mathfrak{H}$ is a homology theory on $\mathscr{K}^2_*$.*

The Homotopy Property is immediate; for if $f_0, f_1 : (X, A) \to (Y, B)$ are homotopic then so are $f_0^\#$ and $f_1^\#$.

The Exactness Property follows immediately from the extended Exactness Property (5.15).

The Excision Property, too, is immediate. For if

$$k : (A, A \cap B) \subsetneq (X, B),$$

then $k^\# : (A^+/A^+ \cap B^+) \to (X^+/B^+)$ is a homeomorphism. □

Now let us instead suppose that we have a homology theory $\mathfrak{H}$ on $\mathscr{K}^2_*$. If X is a space in $\mathscr{K}_*$, let X^- be the free space obtained from X by ignoring the base point, and if $f: X \to Y$, let $f^- : X^- \to Y^-$ be the same function as f. (Thus we have defined a kind of "forgetful functor": $\mathscr{K}_* \to \mathscr{K}$). We now define

$$h_n(X) = H_n(X^-, \{*\}^-),$$
$$h_n(f) = H_n(f^-).$$

Remark. It will often be convenient to drop the minus signs and write

$$h_n(X) = H_n(X, \{*\}),$$
$$h_n(f) = H_n(f).$$

This will simplify the notation and should not cause undue confusion.

In order to define the suspension operator e_n, let X be a space in $\mathscr{K}_*$ and let $p : (\mathbf{T}X, X) \to (\mathbf{S}X, \{*\})$ be the collapsing map and

$$\partial_* : H_{n+1}(\mathbf{T}X, X) \to H_n(X, \{*\})$$

the boundary operator of the homology sequence of the triple $(\mathbf{T}X, X, \{*\})$.

Then $\mathbf{T}X$ is contractible and therefore ∂_* is an isomorphism by (6.2) and (6.3). Thus we may define $e_n(X)$ to be the composite

$$H_n(X, \{*\}) \xrightarrow{\partial_*^{-1}} H_{n+1}(\mathbf{T}X, X) \xrightarrow{H_{n+1}(p)} H_{n+1}(\mathbf{S}X, \{*\}).$$

Clearly, $e_n : h_n \to h_{n+1} \circ \mathbf{S}$ is a natural transformation of functors.

(6.16) Theorem *The theory $\mathfrak{h} = \{h_n, e_n\}$ is a homology theory on $\mathscr{K}_*$.*

The Homotopy Property is immediate. To prove exactness, let $f: X \to Y$ be a map in $\mathscr{K}_*$. Consider the triad $(\mathbf{T}_f; \mathbf{T}X, \mathbf{I}_f)$; by the Excision Property, the injection

$$k_1 : H_n(\mathbf{I}_f, X) \to H_n(\mathbf{T}_f, \mathbf{T}X)$$

is an isomorphism. Since $\mathbf{T}X$ is contractible, the injection

$$k_2 : H_n(\mathbf{T}_f, \{*\}) \to H_n(\mathbf{T}_f, \mathbf{T}X)$$

is an isomorphism. Since Y is a deformation retract of $\mathbf{I}_f$, the injection

$$k_3 : H_n(Y, \{*\}) \to H_n(\mathbf{I}_f, \{*\})$$

is an isomorphism. The composite $k_3 \circ f_*$ is the injection

$$i : H_n(X, \{*\}) \to H_n(\mathbf{I}_f, \{*\}).$$

Hence there is a commutative diagram

$$\begin{array}{ccccc}
H_n(X, \{*\}) & \xrightarrow{f_*} & H_n(Y, \{*\}) & \longrightarrow & H_n(\mathbf{T}_f, \{*\}) \\
& \searrow^{i} & \downarrow{\scriptstyle k_3} & & \downarrow{\scriptstyle k_1^{-1} \circ k_2} \\
& & H_n(\mathbf{I}_f, \{*\}) & \longrightarrow & H_n(\mathbf{I}_f, X)
\end{array}$$

and the Exactness Property now follows from the exactness of the homology sequence of the triple $(\mathbf{I}_f, X, \{*\})$.

The Suspension Property follows immediately by applying the Map Excision Theorem (6.7) to the map $p : (\mathbf{T}X, X) \to (\mathbf{S}X, \{*\})$. □

We have shown how each homology theory on $\mathscr{K}_*$ determines one on $\mathscr{K}_*^2$, and, conversely, to each homology theory on $\mathscr{K}_*^2$ there corresponds one on $\mathscr{K}_*$. We next show that these correspondences are essentially inverse to each other. Thus there is, up to natural isomorphism, a one-to-one correspondence between the two kinds of theories.

Let $\mathfrak{h}$ be a homology theory on $\mathscr{K}_*$, $\mathfrak{H}$ the corresponding theory on $\mathscr{K}_*^2$, and $\mathfrak{h}'$ the theory on $\mathscr{K}_*$ corresponding to $\mathfrak{H}$.

(6.17) Theorem *The theories $\mathfrak{h}$ and $\mathfrak{h}'$ coincide (up to natural isomorphism).*

Let X be a space in $\mathscr{K}_*$. Then

$$h_n'(X) = H_n(X^-, \{*\}^-) = h_n(X^{-+}/\{*\}^{-+}),$$

and the space $X^{-+}/\{*\}^{-+}$ is naturally homeomorphic with X. Similarly, if $f: X \to Y$, then

$$h_n'(f) = H_n(f^-) = h_n(f^{-\#}),$$

and $f^{-\#}$ coincides, in view of the above natural homeomorphisms, with f.

It remains to calculate the suspension homomorphism $e_n'(X)$. It will be convenient to abbreviate X^{-+} to $X^\flat$ for any space X and to identify $p^{-\#}$ with $p: (\mathbf{T}X)^\flat/X \to (\mathbf{S}X)^\flat/\{*\}^\flat$. Then the inverse of $e_n'(X)$ is the composite

$$\begin{aligned} h_{n+1}(\mathbf{S}X) = h_{n+1}((\mathbf{S}X)^\flat/\{*\}^\flat) &\xrightarrow{h_{n+1}(p)^{-1}} h_{n+1}((\mathbf{T}X)^\flat/X^\flat) \\ &\xrightarrow{h_{n+1}(t)} h_{n+1}(\mathbf{S}(X^\flat)) \xrightarrow{-e_n(X^\flat)^{-1}} \\ & h_n(X^\flat) \xrightarrow{h_n(i^\#)} h_n(X^\flat/\{*\}^\flat) = h_n(X) \end{aligned}$$

where $i: (X^-, \varnothing) \subsetneq (X^-, \{*\}^-)$, so that $i^\# : X^\flat \to X$ is the identity on X and sends the external base point P into the original base point $*$ of X. It will be helpful to examine Figure 12.2, which represents the union W of the space $(\mathbf{T}X)^\flat$ (the upper half of the figure, including the point P) with a copy of the cone $\mathbf{T}(X^\flat)$ (the lower half of the figure), whose intersection with $(\mathbf{T}X)^\flat$ is $X^\flat$. Thus $(\mathbf{T}X)^\flat$ is the disjoint union of P with the (reduced) cone over X, having the original base point $*$ as vertex, while $\mathbf{T}(X^\flat)$ is the (unreduced) cone over X, having P as vertex. If $s \in \mathbf{I}$, $x \in X$, let

$$[s, x] = \begin{cases} (2s) \wedge x \in (\mathbf{T}X)^\flat & (s \le \tfrac{1}{2}), \\ 2(1-s) \wedge x \in \mathbf{T}(X^\flat) & (s \ge \tfrac{1}{2}); \end{cases}$$

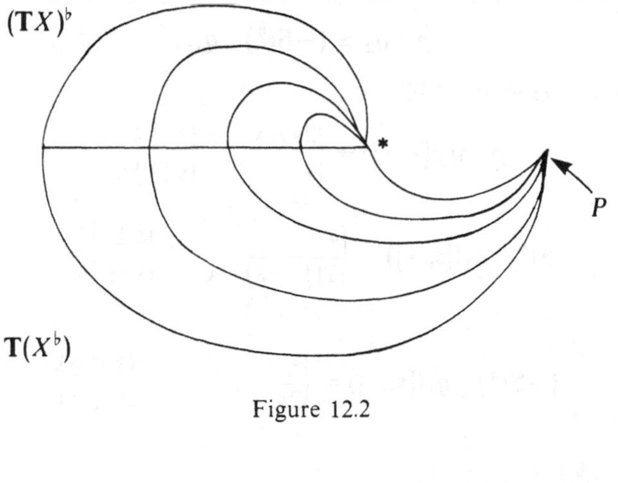

Figure 12.2

we may identify the point $[\frac{1}{2}, x]$ with x. Note that

$$
\begin{aligned}
&[0, x] = *,\\
\text{(6.18)}\qquad &[1, x] = P,\\
&[s, *] = \begin{cases} * & (s \le \frac{1}{2}),\\ 2(1-s) \wedge * & (s \ge \frac{1}{2}); \end{cases}
\end{aligned}
$$

as s increases from 0 to 1 the point $[s, *]$ remains stationary on the interval $[0, \frac{1}{2}]$, then moves along the line segment in $\mathbf{T}(X^\flat)$ from $*$ to P.

Let

$$q_1 : W \to W/(\mathbf{T}X)^\flat = \mathbf{S}(X^\flat),$$

$$q_2 : W \to W/\mathbf{T}(X^\flat) = (\mathbf{T}X)^\flat/X^\flat$$

be the collapsing maps. The connecting map $t : (\mathbf{T}X)^\flat/X \to \mathbf{S}(X^\flat)$ is a composite $q_1 \circ k$, where $k : (\mathbf{T}X)^\flat/X^\flat \to W$ is a homotopy inverse of q_2. Hence

$$t \circ q_2 \simeq q_1$$

Consider the diagram

$$
\text{(6.19)}\qquad
\begin{array}{ccc}
 & W & \\
{}^{q_1}\swarrow & & \searrow^{q_2} \\
\mathbf{S}(X^\flat) & \xleftarrow{\;\;t\;\;} & (\mathbf{T}X)^\flat/X^\flat \\
{}_{-\mathbf{S}i^\#}\searrow & & \swarrow_{p} \\
 & \mathbf{S}X &
\end{array}
$$

(6.20) Lemma *The diagram* (6.19) *is homotopy commutative.*

We have just seen that the upper triangle is homotopy commutative. Since q_2 is a homotopy equivalence, it suffices to prove that

$$p \circ q_2 \simeq (-\mathbf{S}i^*) \circ q_1.$$

The map $p \circ q_2$ is given by

$$p \circ q_2([s, x]) = \begin{cases} 2s \wedge x & (s \le \frac{1}{2}),\\ * & (s \ge \frac{1}{2}), \end{cases}$$

while

$$\mathbf{S}i^* \circ q_1([s, x]) = \begin{cases} * & (s \ge \frac{1}{2}),\\ \overline{2(1-s)} \wedge x & (s \ge \frac{1}{2}), \end{cases}$$

so that

$$(-\mathbf{S}i^*) \circ q_1([s, x]) = \begin{cases} * & (s \le \frac{1}{2}),\\ \overline{2s - 1} \wedge x & (s \ge \frac{1}{2}) \end{cases}$$

(the reader should check that these formulas are consistent with the relations (6.18)).

The map $h_1 : \mathbf{I} \times W \to \mathbf{S}X$ defined by

$$h_1(u, [s, x]) = \begin{cases} \overline{(2-u)s \wedge x} & \left(s \le \dfrac{1}{2-u}\right), \\ * & \left(s \ge \dfrac{1}{2-u}\right) \end{cases}$$

deforms $p \circ q_2$ into the map $k : W \to \mathbf{S}X$ such that

$$k([s, x]) = s \wedge x \qquad (s \in \mathbf{I},\, x \in X).$$

Similarly, the map $h_2 : \mathbf{I} \times W \to SX$ defined by

$$h_2(u, [s, x]) = \begin{cases} * & \left(s \le \dfrac{u}{1+u}\right), \\ \overline{s - u + su \wedge x} & \left(s \ge \dfrac{u}{1+u}\right) \end{cases}$$

deforms k into $(-\mathbf{S}i^*) \circ q_1$.

The diagram obtained from the lower half of (6.19) by applying the functor h_{n+1} can be enlarged to the commutative diagram

$$\begin{array}{ccc} & h_{n+1}((\mathbf{T}X)^\flat/X^\flat) & \\ {\scriptstyle h_{n+1}(t)} \swarrow & & \searrow {\scriptstyle h_{n+1}(p)} \\ h_{n+1}(\mathbf{S}(X^\flat)) & \xrightarrow{-h_{n+1}(\mathbf{S}i^*)} & h_{n+1}(\mathbf{S}X) \\ {\scriptstyle e_n(X^\flat)} \uparrow & & \uparrow {\scriptstyle e_n(X)} \\ h_n(X^\flat) & \xrightarrow[-h_n(i^*)]{} & h_n(X) \end{array}$$

We have seen that $e_n'(X)^{-1}$ is the composite

$$\begin{aligned} -h_n(i^*) &\circ e_n(X^\flat)^{-1} \circ h_{n+1}(t) \circ h_{n+1}(p)^{-1} \\ &= -e_n(X)^{-1} \circ h_{n+1}(\mathbf{S}i^*) \circ h_{n+1}(t) \circ h_{n+1}(p)^{-1} \\ &= e_n(X)^{-1} \circ h_{n+1}(p) \circ h_{n+1}(p)^{-1} \\ &= e_n(X)^{-1}. \end{aligned}$$

Hence the homology theories $\mathfrak{h}$ and $\mathfrak{h}'$ coincide. □

Finally, let $\mathfrak{H}$ be a homology theory on $\mathscr{K}^2_*$, let $\mathfrak{h}$ be the corresponding reduced theory on $\mathscr{K}_*$, and let $\mathfrak{H}'$ be the theory on $\mathscr{K}^2_*$ which corresponds to $\mathfrak{h}$.

(6.21) Theorem *The theories $\mathfrak{H}$ and $\mathfrak{H}'$ are naturally isomorphic.*

First observe that

$$H_n'(X, A) = H_n(X^+/A^+, \{*\});$$

if $i_X : (X, A) \subseteq (X^+, A^+)$ and $p_X : (X^+, A^+) \to (X^+/A^+, P)$, then

$$P_n(X, A) = H_n(p_X \circ i_X) : H_n(X, A) \approx H_n'(X, A).$$

Similarly, if $f: (X, A) \to (Y, B)$, there is a commutative diagram

$$\begin{array}{ccc}
H_n(X, A) & \xrightarrow{H_n(f)} & H_n(Y, B) \\
\Big\downarrow{\scriptstyle H_n(i_X)} & & \Big\downarrow{\scriptstyle H_n(i_Y)} \\
H_n(X^+, A^+) & \xrightarrow{H_n(f^+)} & H_n(Y^+, B^+) \\
\Big\downarrow{\scriptstyle H_n(p_X)} & & \Big\downarrow{\scriptstyle H_n(p_Y)} \\
H_n(X^+/A^+, P) & \xrightarrow[H_n(f^\#)]{} & H_n(Y^+/B^+, P)
\end{array}$$

and $H_n(f^\#) = H_n'(f)$. Therefore we have defined a natural isomorphism $P_n : H_n \approx H_n'$.

If (X, A) is a pair in $\mathscr{K}^2_*$, the homomorphism $\partial_n'(X, A)$ is minus the composite

$$H_n'(X, A) = H_n(X^+/A^+, P) \xrightarrow{H_n(t)} H_n(\mathbf{S}A^+, P)$$

$$\xrightarrow{H_n(p_1)^{-1}} H_n(\mathbf{T}A^+, A^+) \xrightarrow{\partial_1} H_{n-1}(A^+, P) = H_{n-1}'(A)$$

where $p_1 : (\mathbf{T}A^+, A^+) \to (\mathbf{S}A^+, *)$ is the collapsing map and ∂_1 is the boundary operator of the homology sequence of the triple $(\mathbf{T}A^+, A^+, P)$.

The diagram

(6.22)

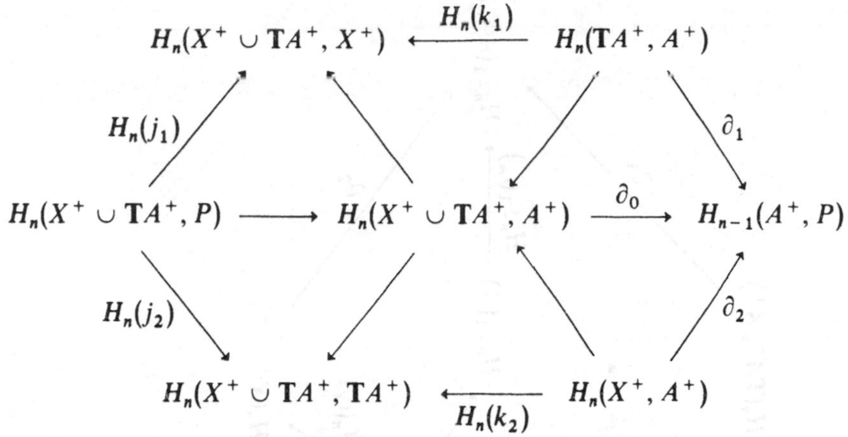

satisfies the hypotheses of the Hexagonal Lemma. Hence

(6.23) $$\partial_1 \circ H_n(k_1)^{-1} \circ H_n(j_1) = -\partial_2 \circ H_n(k_2)^{-1} \circ H_n(j_2).$$

From the diagram (6.22) let us remove the central group and the homomorphisms relating to it. From the resulting diagram we obtain a new diagram (Figure 12.3), in which the maps q_i are the appropriate collapsing maps. Observe that each region of the diagram is commutative except possibly for the large region labelled with an asterisk. Now $H_n(q_2)$ is an isomorphism, and

$$\begin{aligned}-\partial_n'(X, A) \circ H_n(q_2) &= \partial_1 \circ H_n(p_1)^{-1} \circ H_n(t) \circ H_n(q_2)\\ &= \partial_1 \circ H_n(k_1)^{-1} \circ H_n(q_1^{-1}) \circ H_n(q_4)\\ &= \partial_1 \circ H_n(k_1)^{-1} \circ H_n(j_1)\\ &= -\partial_2 \circ H_n(k_2)^{-1} \circ H_n(j_2).\end{aligned}$$

But $H_{n-1}(i_A) = P_{n-1}(A)$, so that

$$\begin{aligned}P_{n-1}(A) \circ \partial_n(X, A) \circ P_n(X, A)^{-1} \circ H_n(q_2)\\ &= \partial_2 \circ H_n(i_X) \circ P_n(X, A)^{-1} \circ H_n(q_2)\\ &= \partial_2 \circ H_n(p_X)^{-1} \circ H_n(q_2)\\ &= \partial_2 \circ H_n(k_2)^{-1} \circ H_n(q_3)^{-1} \circ H_n(q_2)\\ &= \partial_2 \circ H_n(k_2)^{-1} \circ H_n(j_2)\\ &= \partial_n'(X, A) \circ H_n(q_2).\end{aligned}$$

Since $H_n(q_2)$ is an isomorphism, we deduce that

$$\partial_n'(X, A) = P_{n-1}(A) \circ \partial_n(X, A) \circ P_n(X, A)^{-1},$$

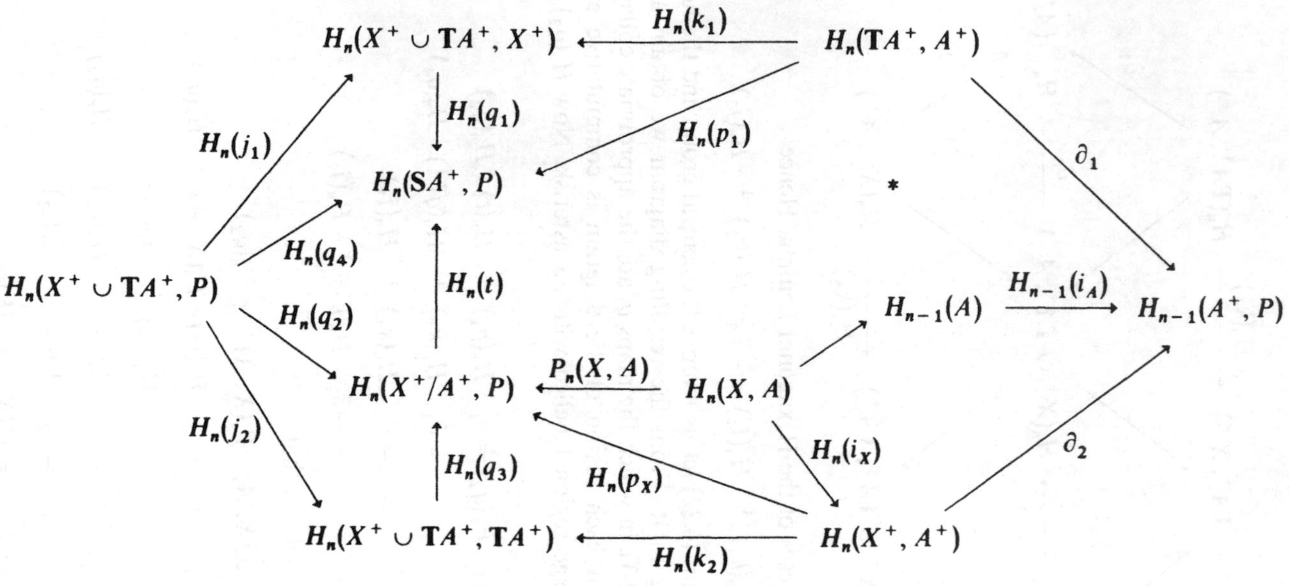
H_n(X⁺ ∪ TA⁺, X⁺)
H_n(k_1)
H_n(TA⁺, A⁺)
H_n(j_1)
H_n(q_1)
H_n(p_1)
∂_1
H_n(SA⁺, P)
*
H_n(q_4)
H_n(X⁺ ∪ TA⁺, P)
H_n(t)
H_n(q_2)
H_{n−1}(A)
H_{n−1}(i_A)
H_{n−1}(A⁺, P)
P_n(X, A)
H_n(X⁺/A⁺, P)
H_n(X, A)
H_n(j_2)
H_n(i_X)
∂_2
H_n(q_3)
H_n(p_X)
H_n(X⁺ ∪ TA⁺, TA⁺)
H_n(X⁺, A⁺)
H_n(k_2)

Figure 12.3

so that ∂_n' and ∂_n correspond under our natural isomorphism. Therefore the theories $\mathfrak{H}$ and $\mathfrak{H}'$ are isomorphic. □

We have now proved

(6.24) Theorem *There is a one-to-one correspondence (up to natural isomorphism) between homology theories on $\mathscr{K}_*^2$ and reduced homology theories on $\mathscr{K}_*$.* □

In proving the results of this section, we have been (perhaps unnecessarily) a little pedantic in our notation. Now that we have established Theorem (6.24), we can be more relaxed. Accordingly, and in view of (6.4), we shall use the following notation. If $\mathfrak{H} = \{H_n, \partial_n\}$ is a given homology theory on $\mathscr{K}_*^2$, the corresponding reduced theory will be denoted by $\tilde{\mathfrak{H}} = \{\tilde{H}_n, E_n\}$. And if $\tilde{\mathfrak{H}} = \{\tilde{H}_n, E_n\}$ is a given reduced homology theory, the corresponding theory on $\mathscr{K}_*^2$ will be christened $\mathfrak{H} = \{H_n, \partial_n\}$.

We conclude this section with

(6.25) Theorem *Let $\mathfrak{H}$ be a homology theory on $\mathscr{K}_*^2$. Then*

(1) *$\mathfrak{H}$ is proper if and only if $\tilde{\mathfrak{H}}$ is proper;*
(2) *$\mathfrak{H}$ is additive if and only if $\tilde{\mathfrak{H}}$ is additive;*
(3) *$\mathfrak{H}$ is isotropic if and only if $\tilde{\mathfrak{H}}$ is isotropic.*

The first statement follows from the fact that

$$\tilde{H}_n(S^0) \approx H_n(P).$$

Suppose that $\tilde{\mathfrak{H}}$ is additive. If the free space X is the topological sum $\sum X_\alpha$, $A \subset X$, $A_\alpha = X_\alpha \cap A$, then $X^+ = \bigvee_\alpha X_\alpha^+$, $A^+ = \bigvee_\alpha A_\alpha^+$, so that

$$X^+/A^+ = \bigvee_\alpha X_\alpha^+ / \bigvee_\alpha A_\alpha^+ = \bigvee_\alpha (X_\alpha^+/A_\alpha^+);$$

thus

$$H_n(X, A) = \tilde{H}_n(X^+/A^+) \approx \oplus \tilde{H}_n(X_\alpha^+/A_\alpha^+) = \oplus H_n(X_\alpha, A_\alpha)$$

under the injection homomorphisms.

Conversely, suppose that $\mathfrak{H}$ is additive. Let $\{X_\alpha \mid \alpha \in J\}$ be a family of spaces with non-degenerate base points x_α, $X = \bigvee_\alpha X_\alpha$, and let $\bar{X} = \sum_\alpha X_\alpha^-$ be the topological sum of the corresponding free spaces $\bar{A} = \sum_\alpha \{x_\alpha\}^-$. Then the natural map of $\bar{X}$ into X is a relative homeomorphism

$$p : (\bar{X}, \bar{A}) \to (X, *)$$

and therefore $H_n(p) : H_n(\bar{X}, \bar{A}) \approx H_n(X, *)$. If $\bar{i}_\alpha : (X_\alpha, \{x_\alpha\}) \to (\bar{X}, \bar{A})$ is the natural imbedding, then $p \circ \bar{i}_\alpha = i_\alpha : X_\alpha \subsetneq X$, and (2) now follows.

Suppose that $\mathfrak{H}$ is isotropic. If X and Y are spaces in $\mathscr{K}^*$, $f : X \to Y$ a weak homotopy equivalence, then $f : (X, *) \to (Y, *)$ is a weak homotopy equivalence of pairs, and therefore

$$\tilde{H}_n(f) = H_n(f) : H_n(X, *) \to H_n(Y, *)$$

is an isomorphism. Hence $\mathfrak{H}$ is isotropic.

Conversely, suppose that $\mathfrak{H}$ is isotropic and $f : (X, A) \to (Y, B)$ a weak homotopy equivalence of pairs.

Suppose first that $A = \varnothing$; it follows that $B = \varnothing$ and the extension $f^+ : X^+ \to Y^+$ is a weak homotopy equivalence. Hence $\tilde{H}_n(f^+) : \tilde{H}_n(X^+) \approx \tilde{H}_n(Y^+)$ for all n. But $H_n(X) = \tilde{H}_n(X^+)$ and $H_n(Y) = \tilde{H}_n(Y^+)$, so that $H_n(f) : H_n(X) \approx H_n(Y)$.

Suppose, on the contrary, that $A \neq \varnothing$, and choose a base point $x_0 \in A$. By Theorem (3.1) of Chapter V, f induces isomorphisms

$$f_1 : \pi_q(X, x_0) \approx \pi_q(Y, y_0),$$
$$f_2 : \pi_q(A, x_0) \approx \pi_q(B, y_0).$$

Since $\mathfrak{H}$ is isotropic, f induces isomorphisms

$$f'_1 : H_q(X, x_0) \approx H_q(Y, y_0),$$
$$f'_2 : H_q(A, x_0) \approx H_q(B, y_0).$$

By exactness and the Five-Lemma, f induces isomorphisms

$$H_q(f) : H_q(X, A) \approx H_q(Y, B).$$

Hence $\mathfrak{H}$ is isotropic. □

7 Cohomology Theories

In this section we discuss cohomology theories. Our discussion parallels that in §§5, 6 for homology theories. As the modifications are to a great extent merely formal, we shall for the most part eschew details.

A *reduced cohomology theory* $\mathfrak{H}^*$ on $\mathscr{K}_*$ consists of

(1) a family of contravariant functors $\tilde{H}^n : \mathscr{K}_* \to \mathscr{A}$ $(n \in \mathbf{Z})$, together with
(2) a family of natural transformations

$$E^n : \tilde{H}^{n+1} \circ \mathbf{S} \to \tilde{H}^n \qquad (n \in \mathbf{Z}),$$

satisfying the following conditions:

(1) (Homotopy). *If $f_0, f_1 : X \to Y$ are homotopic maps, then*

$$\tilde{H}^n(f_0) = \tilde{H}^n(f_1) : \tilde{H}^n(Y) \to \tilde{H}^n(X)$$

for all $n \in \mathbf{Z}$.

(2) (Exactness). *If $f : X \to Y$ and $j : Y \subsetneq \mathbf{T}_f$, then the sequence*

$$\tilde{H}^n(\mathbf{T}_f) \xrightarrow{\tilde{H}^n(j)} \tilde{H}^n(Y) \xrightarrow{\tilde{H}^n(f)} \tilde{H}^n(X)$$

is exact for all $n \in \mathbf{Z}$.

(3) (Suspension). *The homomorphism*

$$E^n(X) : \tilde{H}^{n+1}(\mathbf{S}X) \to \tilde{H}^n(X)$$

is an isomorphism for all X and all $n \in \mathbf{Z}$.

The theory $\mathfrak{H}^*$ is *proper* if and only if it satisfies

(4) (Dimension). *If X is a 0-sphere, then $\tilde{H}^n(X) = 0$ for all $n \neq 0$.*

The theory $\mathfrak{H}^*$ is *additive* if and only if it satisfies

(5) (Additivity). *If $\{X_\alpha \mid \alpha \in J\}$ is a family of spaces, $X = \bigvee_\alpha X_\alpha$, then the injections $\tilde{H}^n(X) \to \tilde{H}^n(X_\alpha)$ represent the group $\tilde{H}^n(X)$ as a direct product.*

The theory $\mathfrak{H}^*$ is *isotropic* if and only if it satisfies

(6) (Isotropy). *Let $f: X \to Y$ be a weak homotopy equivalence. Then $\tilde{H}^n(f) : \tilde{H}^n(Y) \approx \tilde{H}^n(X)$ for all n.*

A cohomology theory $\mathfrak{H}^*$ is *complete* if and only if it is additive and isotropic.

Let $\mathfrak{H}^*$ be a cohomology theory on $\mathscr{K}_*$. Then

(7.1) *If (X, A) is an NDR-pair in $\mathscr{K}_*$, $i : A \subsetneq X$, and if $p : X \to X/A$ is the collapsing map, then the sequence*

$$\tilde{H}^n(X/A) \xrightarrow{\tilde{H}^n(p)} \tilde{H}^n(X) \xrightarrow{\tilde{H}^n(i)} \tilde{H}^n(A) \tag{7.2}$$

is exact. □

(7.3) *If P is a space consisting of just one point, then $\tilde{H}^n(P) = 0$ for all n.* □

(7.4) *If $f: X \to Y$ is a constant map, then $\tilde{H}^n(f) = 0$ for all n.* □

(7.5) *If X, Y are spaces in $\mathscr{K}_*$, $j_1 : X \to X \vee Y$, $j_2 : Y \to X \vee Y$ the inclusions, then the injections*

$$j_1^* = \tilde{H}^n(j_1) : \tilde{H}^n(X \vee Y) \to \tilde{H}^n(X),$$
$$j_2^* = \tilde{H}^n(j_2) : \tilde{H}^n(X \vee Y) \to \tilde{H}_n(Y)$$

represent $\tilde{H}^n(X \vee Y)$ as a direct product. The homomorphisms

$$q_1^* : \tilde{H}^n(X) \to \tilde{H}^n(X \vee Y),$$
$$q_2^* : \tilde{H}^n(Y) \to \tilde{H}^n(X \vee Y),$$

induced by the projections, form the dual representation of $\tilde{H}^n(X \vee Y)$ *as a direct sum.* □

(7.6) *Let X be an* H′*-space with coproduct* θ. *Then*

$$\theta^* = j_1^* + j_2^* : \tilde{H}^n(X \vee X) \to \tilde{H}^n(X). \qquad \square$$

(7.7) *Let X be an* H′*-space,* $f_1, f_2 : X \to Y$. *Then*

$$(f_1 + f_2)^* = f_1^* + f_2^* : \tilde{H}^n(Y) \to \tilde{H}^n(X). \qquad \square$$

(7.8) *Let* $f: X \to Y$. *Then*

$$(-\mathbf{S}f)^* = -(\mathbf{S}f)^* : \tilde{H}^n(\mathbf{S}Y) \to \tilde{H}^n(\mathbf{S}X). \qquad \square$$

(7.9) *Let* $f: X \to Y$, *and let* $q : \mathbf{T}_f \to \mathbf{S}X$ *be the collapsing map. Let*

$$d^n = -\tilde{H}^n(q) \circ E^{n-1}(X)^{-1} : \tilde{H}^{n-1}(X) \to \tilde{H}^n(T_f).$$

Then the sequence

$$\cdots \to \tilde{H}^{n-1}(Y) \xrightarrow{\tilde{H}^{n-1}(f)} \tilde{H}^{n-1}(X) \xrightarrow{d^n} \tilde{H}^n(\mathbf{T}_f) \xrightarrow{\tilde{H}^n(k)} \tilde{H}^n(Y) \xrightarrow{\tilde{H}^n(f)} \tilde{H}^n(X) \to \cdots$$

is exact. □

(7.10) *Let* (X, A) *be an* NDR*-pair in* $\mathscr{K}_*$, $i : A \subsetneq X$, *and let* $p : X \to X/A$ *be the collapsing map and* $t : X/A \to \mathbf{S}A$ *a connecting map. Let*

$$d^n = -\tilde{H}^n(t) \circ E^{n-1}(A)^{-1} : \tilde{H}^{n-1}(A) \to \tilde{H}^n(X/A).$$

Then the sequence

$$\cdots \to \tilde{H}^{n-1}(X) \xrightarrow{\tilde{H}^{n-1}(i)} \tilde{H}^{n-1}(A) \xrightarrow{d^n} \tilde{H}^n(X/A) \xrightarrow{\tilde{H}^n(p)} \tilde{H}^n(X) \xrightarrow{\tilde{H}^n(l)} \tilde{H}^n(A) \to \cdots$$

is exact. □

As in the case of homology, we have to relate the above notion of cohomology theory to that of Eilenberg–Steenrod. Accordingly, a cohomology theory $\mathfrak{H}^*$ on $\mathscr{K}_*^2$ consists of

(1) a family of functors $H^n : \mathscr{K}_*^2 \to \mathscr{A}$ $(n \in \mathbf{Z})$ together with
(2) a family of natural transformations $\delta^n : H^{n-1} \circ R \to H^n$ $(n \in \mathbf{Z})$, having the following properties:

(1) (Homotopy). *If* $f_0, f_1 : (X, A) \to (Y, B)$ *are homotopic maps, then*

$$H^n(f_0) = H^n(f_1) : H^n(Y, B) \to H^n(X, A)$$

for all $n \in \mathbf{Z}$.

(2) (Exactness). *If (X, A) is a pair in $\mathscr{K}_*^2$ and $i : A \subsetneq X, j : X \subsetneq (X, A)$, then the sequence*

$$\cdots \to H^{n-1}(A) \xrightarrow{\delta^n(X, A)} H^n(X, A) \xrightarrow{H^n(j)} H^n(X) \xrightarrow{H^n(i)} H^n(A) \xrightarrow{\delta^{n+1}(X, A)} H^{n+1}(X, A) \to \cdots$$

is exact.

(3) (Excision). *Let X be the union of the closed sets A, B, and suppose that $(A, A \cap B)$ is an* NDR*-pair. Then the injection*

$$H^n(X, B) \to H^n(A, A \cap B)$$

is an isomorphism for every n.

The theory $\mathfrak{H}^*$ is *proper* if and only if it satisfies

(4) (Dimension). *If P is a space consisting of just one point, then $H^n(P) = 0$ for all $n \neq 0$,*

additive if and only if it satisfies

(5) (Additivity). *Let X be the topological sum $\sum_\alpha X_\alpha$ and let A be a subspace of X, $A_\alpha = A \cap X_\alpha$. Then the injections $H^n(X, A) \to H^n(X_\alpha, A_\alpha)$ represent the group $H^n(X, A)$ as a direct product,*

and *isotropic* if and only if

(6) (Isotropy). *Let $f: (X, A) \to (Y, B)$ be a weak homotopy equivalence. Then $H^n(f): H^n(Y, B) \approx H^n(X, A)$ for all n.*

An additive and isotropic theory is said to be *complete*.

Let $\mathfrak{H}^*$ be a cohomology theory on $\mathscr{K}_*^2$. Then

(7.11) *For any free space X, $H_n(X, X) = 0$ for all n.* □

(7.12) *If A is a deformation retract of X, then $H_n(X, A) = 0$ for all n.* □

(7.13) *If (X, A) and (A, B) are* NDR*-pairs, then the homology sequence*

$$\cdots \to H^{n-1}(A, B) \to H^n(X, A) \to H^n(X, B) \to H^n(A, B) \to H^{n+1}(X, A) \to \cdots$$

of the triple (X, A, B) is exact. □

Let P be a space consisting of just one point. For any space X, let $f: X \to P$ be the unique map, and let $\tilde{H}^n(X) = \operatorname{Cok} H^n(f): H^n(P) \to H^n(X)$. Then $\tilde{H}^n$ is a contravariant functor: $\mathscr{K} \to \mathscr{A}$.

(7.14) *If $x \in X$, the composite of the injection $H^n(X, \{x\}) \to H^n(X)$ with the projection $H^n(X) \to \tilde{H}^n(X)$ is an isomorphism.* □

(7.15) *If $X \neq \emptyset$, the sequence*

$$0 \to H^n(P) \to H^n(X) \to \tilde{H}^n(X) \to 0$$

is exact and splittable, so that

$$H^n(X) \approx \tilde{H}^n(X) \oplus H^n(P).$$ □

(7.16) *For any pair (X, A), the coboundary $\delta^n(X, A) : H^{n-1}(A) \to H^n(X, A)$ annihilates the image of $H^{n-1}(f) : H^{n-1}(P) \to H^{n-1}(A)$, inducing a homomorphism $\tilde{\delta}^n(X, A) : \tilde{H}^{n-1}(A) \to H^n(X, A)$. Moreover, the sequence*

$$\cdots \to \tilde{H}^{n-1}(A) \to H^n(X, A) \to \tilde{H}^n(X) \to \tilde{H}^n(A) \to H^{n+1}(X, A) \to \cdots$$

is exact. □

(7.17) (Map Excision Theorem). *If $f : (X, A) \to (Y, B)$ is a relative homeomorphism of* NDR*-pairs, then*

$$H^n(f) : H^n(Y, B) \to H^n(X, A)$$

is an isomorphism for all n. □

(7.18) (Finite Direct Product Theorem). *Let $X = A \cup X_1 \cup \cdots \cup X_n$, where A and X_α are closed ($\alpha = 1, \ldots, n$). Suppose that $X_\alpha \cap X_\beta \subset A$ for $\alpha \neq \beta$ and that $(X_\alpha, X_\alpha \cap A)$ is an* NDR*-pair. Let $X_\alpha^* = A \cup \bigcup_{\beta \neq \alpha} X_\beta$, $A_\alpha = X_\alpha \cap A$. Then*

(1) *the injections $i_\alpha^* : H^q(X, A) \to H^q(X_\alpha, A_\alpha)$ represent the group $H^q(X, A)$ as a direct product;*
(2) *the injections $j_\alpha^* : H^q(X, X_\alpha^*) \to H^q(X, A)$ represent $H^q(X, A)$ as a direct sum;*
(3) *the injection $k_\alpha^* = i_\alpha^* \circ j_\alpha^* : H^q(X, X_\alpha^*) \to H^q(X_\alpha, A_\alpha)$ is an isomorphism, and the representations $\{i_\alpha^*\}$, $\{j_\alpha^*\}$ are weakly dual.* □

(7.19) (General Direct Product Theorem). *Let $X = A \cup \bigcup_{\alpha \in J} X_\alpha$, where A and X_α are closed, $X_\alpha \cap X_\beta \subset A$ for $\alpha \neq \beta$, and $(X_\alpha, X_\alpha \cap A)$ is an* NDR*-pair. Suppose that $\mathfrak{H}^*$ is additive, and that X has the weak topology with respect to the subsets A, X_α. Let $X_\alpha^* = A \cup \bigcup_{\beta \neq \alpha} X_\beta$, $A_\alpha = X_\alpha \cap A$. Then*

(1) *the injections $i_\alpha^* : H^q(X, A) \to H^q(X_\alpha, A_\alpha)$ represent $H^q(X, A)$ as a direct product;*
(2) *the injections $j_\alpha^* : H^q(X, X_\alpha^*) \to H^q(X, A)$ represent $H^q(X, A)$ as a strong direct sum;*
(3) *the injection $k_\alpha^* = i_\alpha^* \circ j_\alpha^*$ is an isomorphism, and the representations $\{i_\alpha^*\}$, $\{j_\alpha^*\}$ are weakly dual.* □

(7.20) (General Incidence Isomorphism). *Let K be a regular cell-complex, E an n-cell of K, F an (n − 1)-face of E. Then there is an isomorphism*

$$\sigma^*(E, F): H^{q-1}(F \times Y, \dot{F} \times Y) \approx H^q(E \times Y, \dot{E} \times Y),$$

natural with respect to Y.

The isomorphism $\sigma^*(E, F)$ is the composite

$$H^{q-1}(F \times Y, \dot{F} \times Y) \xrightarrow{k^{*-1}} H^{q-1}(\dot{E} \times Y, F' \times Y) \xrightarrow{\delta^*} H^q(E \times Y, \dot{E} \times Y)$$

where k^* is the injection. □

Let $\Gamma^*(K, Y)$ be the doubly graded group such that $\Gamma^{p,q}(K, Y) = H^{p+q}(K_p \times Y, K_{p-1} \times Y)$. Then $\Gamma^*(K, Y)$ is a cochain complex with coboundary

$$\delta : \Gamma^{p,q}(K, Y) \to \Gamma^{p+1,q}(K, Y).$$

The injections

$$i^\alpha : \Gamma^{p,q}(K, Y) \to H^{p+q}(E_\alpha \times Y, \dot{E}_\alpha \times Y)$$

for all p-cells E_α of K, represent the former group as a direct product. And we can calculate δ in terms of this representation, as follows:

(7.21) *The coboundary operator in $\Gamma^*(K, Y)$ is determined by the relation*

$$i^\alpha \circ \delta = \sum_\beta \sigma^*(E_\alpha, E_\beta) \circ i^\beta,$$

where E_β ranges over all $(p - 1)$-faces of E_α.

(7.22) *Let F_1 and F_2 be $(n - 1)$-faces of E which have a common $(n - 2)$-face G. Then*

(1) *if $n \geq 2$,*

$$\sigma^*(E, F_1) \circ \sigma^*(F_1, G) + \sigma^*(E, F_2) \circ \sigma^*(F_2, G) = 0;$$

(2) *if $n = 1$, $p_i : F_i \times Y \to Y$ the projection on the second factor,*

$$\sigma^*(E, F_1) \circ H^{q-1}(p_1) + \sigma^*(E, F_2) \circ H^{q-1}(p_2) = 0.$$

(7.23) Theorem *For each n-cell E of K there is an isomorphism*

$$/e : H^{n+k}(E \times Y, \dot{E} \times Y) \to H^k(Y)$$

such that, for any $(n - 1)$-face F of E,

(7.24) $$\sigma^*(E, F)u/e = [e:f]u/f$$

for all $u \in H^{n+k-1}(F \times Y, \dot{F} \times Y)$.

The "slant product" isomorphism $/e$ is natural, in the sense

(7.25) *Let* $f: Y \to Z$, $u \in H^{n+k}(E \times Z, \dot{E} \times Z)$. *Then*

$$f^*(u/e) = (1 \times f)^* u/e.$$

We can now compare the two kinds of cohomology theories. If $\tilde{\mathfrak{H}}^*$ is a cohomology theory on $\mathscr{K}_*$ a theory $\mathfrak{H}^*$ on $\mathscr{K}_*^2$ is defined by

$$H^n(X, A) = \tilde{H}^n(X^+/A^+),$$
$$H^n(f) = \tilde{H}^n(f^\#),$$

where $f^\# : X^+/A^+ \to Y^+/B^+$ is induced by $f: (X, A) \to (Y, B)$; while $\delta^n(X, A)$ is minus the composite

$$\tilde{H}^{n-1}(A^+) \xrightarrow{E^{n-1}(A^+)} \tilde{H}^n(\mathbf{S}A^+) \xrightarrow{\tilde{H}^n(t)} \tilde{H}^n(X^+/A^+)$$

for a connecting map $t: X^+/A^+ \to \mathbf{S}A^+$. Conversely, if $\mathfrak{H}^*$ is a cohomology theory on $\mathscr{K}_*^2$, a theory $\tilde{\mathfrak{H}}^*$ on $\mathscr{K}_*$ is defined by

$$\tilde{H}^n(X) = H^n(X, \{*\}),$$
$$\tilde{H}^n(f) = H^n(f)$$

for $f: X \to Y$ in $\mathscr{K}_*$, while $E^n(X)$ is the composite

$$H^{n+1}(\mathbf{S}X, \{*\}) \xrightarrow{H^{n+1}(p)} H^{n+1}(\mathbf{T}X, X) \xrightarrow{\delta^{-1}} H^n(X, \{*\})$$

where $p: (\mathbf{T}X, X) \to (\mathbf{S}X, \{*\})$ is the collapsing map.

We then have:

(7.26) Theorem *There is a one-to-one correspondence (up to natural isomorphism)* $\tilde{\mathfrak{H}}^* \to \mathfrak{H}^*$ *between cohomology theories on* $\mathscr{K}_*$ *and on* $\mathscr{K}_*^2$. *This correspondence preserves propriety, additivity, and isotropy. In particular,* $\tilde{\mathfrak{H}}^*$ *is complete if and only if* $\mathfrak{H}^*$ *is.*

Exercises

1. Discuss Theorem (1.7) for the case $n = 1$.
2. Let

$$0 \longrightarrow B \xrightarrow{i_1} E_1 \xrightarrow{p_1} A \longrightarrow 0,$$
$$0 \longrightarrow B \xrightarrow{i_2} E_2 \xrightarrow{p_2} A \longrightarrow 0$$

be short exact sequences, corresponding to elements $\varepsilon_1, \varepsilon_2 \in \operatorname{Ext}(A, B)$. Let $E = P/Q$, where

$$P = \{(e_1, e_2) \mid p_1(e_1) = p_2(e_2)\} \subset E_1 \times E_2$$
$$Q = \{(i_1(b), -i_2(b) \mid b \in B\},$$

and define $i : B \to E$, $p : E \to A$ by

$$i(b) = i_1(b) + Q$$
$$p((e_1, e_2) + Q) = p_1(e_1).$$

Prove that

(1) the sequence $0 \to B \xrightarrow{i} E \xrightarrow{p} A \to 0$ is exact, thus determining an element $\varepsilon \in \mathrm{Ext}(A, B)$;

(2) $\varepsilon = \varepsilon_1 + \varepsilon_2$.

(The extension ε so defined is called the Baer product of ε_1 and ε_2).

3. Prove that every homology theory $\mathfrak{h}$ on $\mathscr{P}_*$ has a unique canonical extension $\hat{\mathfrak{h}}$ to an isotropic theory on $\mathscr{K}_*$. Show that $\mathfrak{h}$ is additive (proper) if and only if $\hat{\mathfrak{h}}$ is.

4. Prove that stable homotopy is not a direct sum of reindexed proper theories.

5. Determine the structure of (a) $H_*(X; G)$, (b) $\sigma_*(X)$ as a σ_*-module.

6. Prove that, if Y is a free space, $\delta_p \times : H_q(Y) \to H_{p+q}(\Delta^p \times Y, \dot{\Delta}^p \times Y)$ is the isomorphism of Theorem (6.12), $\pi : (\Delta^p \times Y, \dot{\Delta}^p \times Y) \to (\mathbf{S}^p Y^+, *)$ is the identification map, and e^p is the iterated suspension, then the diagram

$$\begin{array}{ccc} H_q(Y) & = & \tilde{H}_q(Y^+) \\ {\scriptstyle \delta_p \times}\downarrow & & \downarrow{\scriptstyle e^p} \\ H_{p+q}(\Delta^p \times Y, \dot{\Delta}^p \times Y) & \xrightarrow[\tilde{H}_{p+q}(\pi)]{} & \tilde{H}_{p+q}(\mathbf{S}^p Y^+) \end{array}$$

is commutative.

CHAPTER XIII

Homology of Fibre Spaces

We conclude this volume with an introduction to the method of spectral sequences for studying the homology of a fibration. In §1 we consider a filtered pair (X, A). The homology groups of the triples (X_p, X_{p-1}, A) are linked together in an intricate way. They can, however, be assembled into two graded groups which are connected by an exact triangle

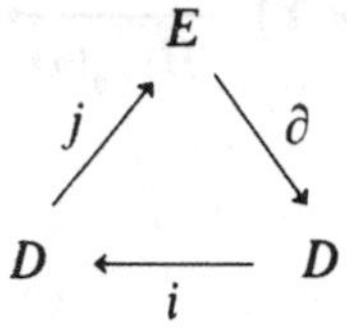

Such a diagram is called an exact couple; the notion is due to Massey [1; 2]. The basic operation on exact couples, that of *derivation*, gives rise to a new exact couple

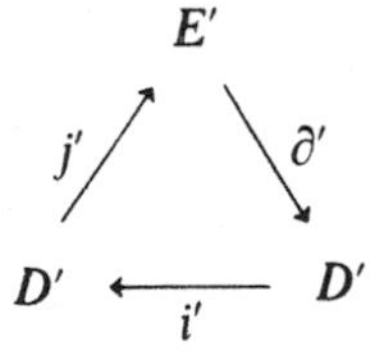

Hence the process can be iterated to obtain an infinite sequence of diagrams.

The composite $d = j \circ \partial$ has the property that $d \circ d = 0$. Hence E is a chain complex with respect to d. The group E' is just the homology group of the chain complex (E, d). Hence the above iteration process leads to a sequence $\{E^r\}$ of chain complexes, each of which is the homology group of the preceding, in other words, to a *spectral sequence*.

The algebra of exact couples is developed in §2. In §3 the resulting theory is applied to the exact couples associated with the homology and cohomology of a filtered pair. In the case of homology the spectral sequence converges to the associated graded group $\mathscr{G}H_*(X, A)$ associated with the filtration of $H_*(X, A)$ induced by the given filtration of the space. For cohomology, on the other hand, convergence does not always hold; its failure can be expressed with the aid of Milnor's functor $\lim^1$. Maps between filtered spaces are also considered, and a filtration preserving map induces a map between their couples. Homotopies are also considered, and it is shown that if two maps are connected by a filtration preserving homotopy, the induced maps of the associated couples are homotopic in the algebraic sense, from which it follows that the induced maps between the derived couples coincide.

If $f: (X, W) \to (B, A)$ is a fibration (i.e., $f: X \to B$ is a fibration and $W = f^{-1}(A)$) and (B, A) is a relative CW-complex, the filtration of (B, A) by skeleta induces a filtration of (X, W) by their counterimages. This situation is discussed in §4. Let $\mathfrak{G}$ be a homology theory (in the sense of Chapter XII; i.e., the Dimension Axiom is not assumed). The initial term $G_{p+q}(X_p, X_{p-1}) = E^1_{p,q}$ is identified *via* an isomorphism λ of $E^1_{p,q}$ with the chain complex of (B, A) with coefficients in the bundle of groups $G_q(\mathscr{F})$ formed by the homology groups of the fibres over the points of B. A fibre preserving map between two fibrations induces a filtration preserving map between the total spaces, as well as a chain map between the chain complexes of the base pairs. In §5 it is shown that λ is consistent with these maps; moreover, λ is an isomorphism of *chain complexes*. Thus the E^2 term is canonically isomorphic with $H_*(B, A; G_*(\mathscr{F}))$. Corresponding results are obtained for cohomology, and mild conditions on the fibration (and/or the homology theory $\mathfrak{G}$) which ensure convergence of the spectral sequence.

The next two Sections are devoted to applications. In §6 it is assumed that the fibre is a point; the spectral sequence then leads from the (ordinary) homology groups of (B, A) with coefficients in G_q (point) to the (extraordinary) homology groups of (B, A). Some applications are the stable forms of the exact sequence (3.13) of Chapter XII on the one hand and of the Steenrod classification theorem for maps of an n-complex into an $(n-1)$-sphere on the other. In §7 we consider an arbitrary fibration, but assume that we are dealing with ordinary homology theory. Applications include the Serre exact sequence which was proved under restrictive hypotheses in §6 of Chapter VII, as well as some qualitative results on homotopy groups; for example, if X is simple and $H_i(X)$ finitely generated (finite) for all $i > 0$, then $\pi_i(X)$ has the same properties.

In order to obtain deeper results, one needs to make use of the multiplicative structure of cohomology. This is done by introducing cross products into the spectral sequence, and then using the device of a diagonal map to convert these into cup products. This program is carried out in §8, and a few further applications bring our introductory treatment of spectral sequences, and with it, the first volume of the book, to a conclusion.

Leray [2, 3] inaugurated the use of spectral sequences in topology. His spectral sequence used cohomology theory of the Čech type and was defined for a more or less arbitrary continuous map (not necessarily a fibration). In 1951 Serre [1], using cubical singular theory, set up the spectral sequence for a map which satisfies conditions somewhat weaker than those for a fibration (the homotopy lifting property is assumed only for maps of finite complexes). Serre's paper was a landmark in homotopy theory; the applications in §§7, 9 and many more are due to him. In 1952 Massey [1, 2] introduced exact couples as a convenient formalism for handling the complicated algebra implicit in the use of spectral sequences.

Axioms for homology theory were announced by Eilenberg and Steenrod [1] in 1945, and by the time their book [E–S] appeared in 1952 their methods had already permeated algebraic topology. In their work the first six axioms have a general character, while the Dimension Axiom is specific; nevertheless it is given equal status with the others, no doubt because no very interesting examples of extraordinary theories were known. In spite of this, by 1955 the existence of the spectral sequence of §6 had become folklore. By the early '60's there had arisen a number of (extraordinary) homology theories of interest: stable homotopy, bordism, K-theory, etc. And it was Atiyah and Hirzebruch [1] who in 1961 made the first serious use of the said spectral sequence in their work on K-theory.

1 The Homology of a Filtered Space

Let $\{X_n \mid n \geq 0\}$ be an NDR-filtration of a space X, and let A be a subspace of X_0 such that (X_0, A) is an NDR-pair. Letting $X_n = A$ for all $n < 0$, we shall refer to $\{X_n \mid n \in \mathbf{Z}\}$ as an NDR-*filtration* of the pair (X, A).

Let $\mathfrak{H}$ be a homology theory (in the sense of Chapter XII). We shall be interested in the relationship among the homology groups of the various pairs which can be formed from X and the X_n. In particular, for each n, the groups $H_n(X_p, A)$ form a direct system under the injections $H_n(X_p, A) \to H_n(X_{p+1}, A)$, and we may consider their direct limit $\varinjlim_p H_n(X_p, A)$. On the other hand, the injections $H_n(X_p, A) \to H_n(X, A)$ induce a *canonical homomorphism* $\psi : \varinjlim_p H_n(X_p, A) \to H_n(X, A)$. It will be essential for us to assume that the homology theory in question is additive. For then

(1.1) Theorem (Milnor). *If $\mathfrak{H}$ is additive, the canonical homomorphism $\psi : \varinjlim_p H_n(X_p, A) \to H_n(X, A)$ is an isomorphism.* □

(1.2) Corollary *The group $H_n(X, A)$ is the union of the images of the injections $H_n(X_p, A) \to H_n(X, A)$.* □

We shall also need to consider additive cohomology theories $\mathfrak{H}^*$. For these the situation is not quite so simple. Indeed, one has, as before, an inverse system of groups $H^n(X_p, A)$ and a *canonical homomorphism* $\psi^* : H^n(X, A) \to \varprojlim_p H^n(X_p, A)$. Unfortunately, ψ^* is not always an isomorphism. Instead, one has

(1.3) Theorem (Milnor). *If $\mathfrak{H}^*$ is an additive cohomology theory, there is a short exact sequence*

$$0 \to \varprojlim_p{}^1 H^{n-1}(X_p, A) \to H^n(X, A) \xrightarrow{\psi^*} \varprojlim_p H^n(X_p, A) \to 0. \qquad \square$$

(1.4) Corollary *The intersection of the kernels of the injections $H^n(X, A) \to H^n(X_p, A)$ is the subgroup $\varprojlim_p{}^1 H^{n-1}(X_p, A)$.* $\square$

Throughout this chapter we shall assume that the homology and cohomology theories with which we are dealing are additive.

The homology sequences of the triples (X_p, X_{p-1}, A) are linked together in an intricate way (Figure 13.1). Let us see how we can make use of the information contained in this diagram to obtain results about the homology of the pair (X, A).

In Figure 13.1 the homology sequence of one particular triple (X_p, X_{p-1}, A) has been outlined. Moreover, the groups $H_n(X, A)$ have been adjoined to the bottom of the diagram. Let $J_{p,n-p}$ be the image of the injection $H_n(X_p, A) \to H_n(X, A)$; then there are inclusions

(1.5) $0 = J_{-1,n+1} \subset J_{0,n} \subset \cdots \subset J_{p,n-p} \subset J_{p+1,n-p-1} \subset \cdots \subset H_n(X, A),$

and, by (1.2)

$$H_n(X, A) = \bigcup_p J_{p,n-p}.$$

Thus $H_n(X, A)$ is filtered by the subgroups $J_{p,q}$ $(p + q = n)$, and we may form the doubly graded group $\mathscr{G}H_*(X, A)$ whose (p, q)th component is the quotient $J_{p,q}/J_{p-1,q+1}$. Thus we may hope to get information about $H_n(X, A)$ through a knowledge of the groups $H_n(X_p, A)$. And we may hope, in turn, to calculate the latter groups inductively, through a knowledge of the groups $H_n(X_p, X_{p-1})$.

As an example, suppose that (X, A) is a relative CW-complex with skeleta X_n, and that $\mathfrak{H}$ is singular homology theory with integral coefficients. By Theorem (2.11) of Chapter II and its Corollaries, the diagram simplifies enormously in this case (Figure 13.2). In fact, the groups above the row $p = n$ are all zero, the groups $H_n(X_p, X_{p-1})$ are zero for $n \neq p$, and the groups $H_n(X_p, A)$ are isomorphic with each other and with $H_n(X, A)$ for $p > n$. In this way we recapture the calculation of the homology groups of (X, A) from its chain complex $\Gamma(X, A)$ which was made in §2 of Chapter II.

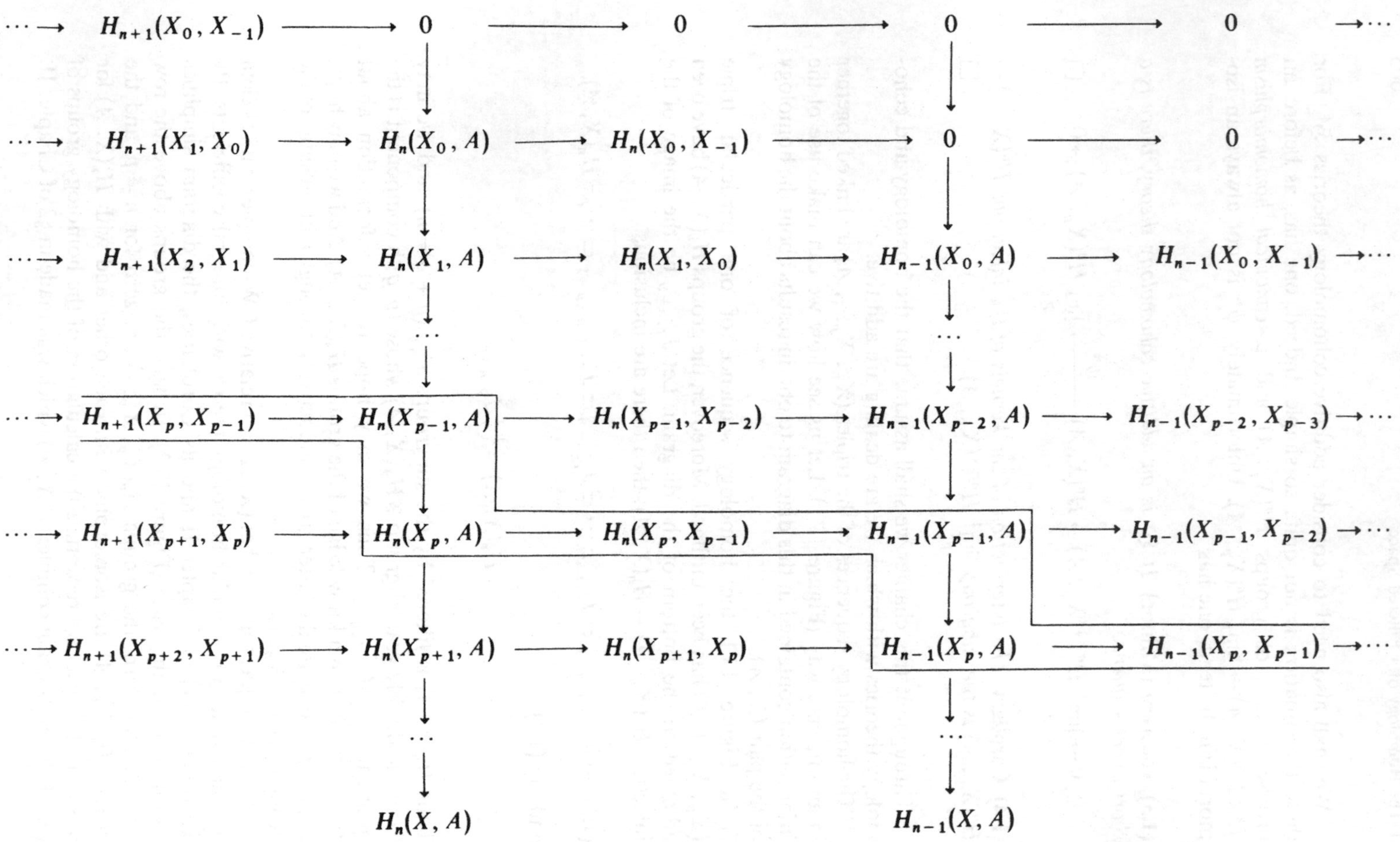
$\cdots \to H_{n+1}(X_0, X_{-1}) \to 0 \to 0 \to 0 \to 0 \to \cdots$
$\cdots \to H_{n+1}(X_1, X_0) \to H_n(X_0, A) \to H_n(X_0, X_{-1}) \to 0 \to 0 \to \cdots$
$\cdots \to H_{n+1}(X_2, X_1) \to H_n(X_1, A) \to H_n(X_1, X_0) \to H_{n-1}(X_0, A) \to H_{n-1}(X_0, X_{-1}) \to \cdots$
$\cdots \to H_{n+1}(X_p, X_{p-1}) \to H_n(X_{p-1}, A) \to H_n(X_{p-1}, X_{p-2}) \to H_{n-1}(X_{p-2}, A) \to H_{n-1}(X_{p-2}, X_{p-3}) \to \cdots$
$\cdots \to H_{n+1}(X_{p+1}, X_p) \to H_n(X_p, A) \to H_n(X_p, X_{p-1}) \to H_{n-1}(X_{p-1}, A) \to H_{n-1}(X_{p-1}, X_{p-2}) \to \cdots$
$\cdots \to H_{n+1}(X_{p+2}, X_{p+1}) \to H_n(X_{p+1}, A) \to H_n(X_{p+1}, X_p) \to H_{n-1}(X_p, A) \to H_{n-1}(X_p, X_{p-1}) \to \cdots$
$H_n(X, A)$
$H_{n-1}(X, A)$

Figure 13.1

$$
\begin{array}{ccccccccccc}
\cdots \to & 0 & \longrightarrow & 0 & \longrightarrow & 0 & \longrightarrow & 0 & \longrightarrow & 0 & \to \cdots \\
 & & & \downarrow & & & & \downarrow & & & \\
\cdots \to & H_{n+1}(X_{n+1}, X_n) & \longrightarrow & H_n(X_n, A) & \longrightarrow & H_n(X_n, X_{n-1}) & \longrightarrow & H_{n-1}(X_{n-1}, A) & \longrightarrow & H_{n-1}(X_{n-1}, X_{n-2}) & \to \cdots \\
 & & & \downarrow & & & & \downarrow & & & \\
\cdots \to & 0 & \longrightarrow & H_n(X_{n+1}, A) & \longrightarrow & 0 & \longrightarrow & H_{n-1}(X_n, A) & \longrightarrow & 0 & \to \cdots \\
 & & & \downarrow \approx & & & & \downarrow \approx & & & \\
 & & & \cdots & & & & \cdots & & & \\
 & & & \downarrow \approx & & & & \downarrow \approx & & & \\
 & & & H_n(X, A) & & & & H_{n-1}(X, A) & & &
\end{array}
$$

Figure 13.2

Let $E_{p,q} = H_{p+q}(X_p, X_{p-1})$, and let $d_{p,q} : E_{p,q} \to E_{p-1,q}$ be the boundary operator of the homology sequence of the triple (X_p, X_{p-1}, X_{p-2}). Then $d_{p,q} \circ d_{p+1,q}$ is the composite

(1.6) $$H_{p+q+1}(X_{p+1}, X_p) \to H_{p+q}(X_p, A) \to H_{p+q}(X_p, X_{p-1}) \to H_{p+q-1}(X_{p-1}, A) \to H_{p+q-1}(X_{p-1}, X_{p-2}),$$

and this is zero because the central segment is a portion of the homology sequence of the triple (X_p, X_{p-1}, A). Thus the groups $E_{p,q}$ are the components of a doubly graded chain complex E, whose boundary operator has degree $(-1, 0)$. The homology group of E is then a doubly graded group E', with

$$E'_{p,q} = \text{Ker } d_{p,q}/\text{Im } d_{p+1,q}.$$

The groups $D_{p,q} = H_{p+q}(X_p, A)$ are the components of a doubly graded group D; and the injections $i_{p,q}: D_{p,q} \to D_{p+1,q-1}$ the components of an endomorphism $i : D \to D$ of degree $(1, -1)$. The image of i is a doubly graded subgroup D' of D; it is convenient to take the (p, q)th component $D'_{p,q}$ of D' to be a subgroup of $D_{p,q}$, so that $D'_{p,q} = \text{Im } i_{p-1,q+1}$ (and *not* $\text{Im } i_{p,q}$). The injections $j_{p,q} : D_{p,q} \to E_{p,q}$ are the components of a homomorphism $j : D \to E$ of degree $(0, 0)$. And the boundary operators

$$\partial_{p,q} : E_{p,q} \to D_{p-1,q}$$

are the components of $\partial : E \to D$, of degree $(-1, 0)$. Thus the diagram of Figure 13.1 can be simplified to a diagram

(1.7)

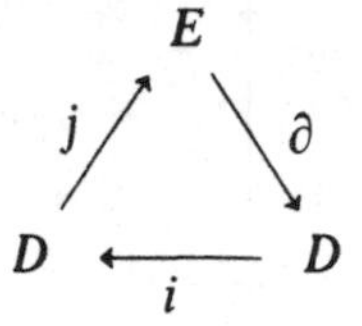

in the category of doubly graded groups.

The exactness of the homology sequences in Figure 13.1 is reflected in

(1.8) *The diagram* (1.7) *is exact, in the sense that*

$$\text{Ker } \partial = \text{Im } j,$$

$$\text{Ker } i = \text{Im } \partial,$$

$$\text{Ker } j = \text{Im } i. \qquad \square$$

Such a diagram is called an *exact couple*, and we shall study the algebra of exact couples in §2.

In the same way, the cohomology sequences of the triples (X, X_p, X_{p-1}) give rise to an exact couple

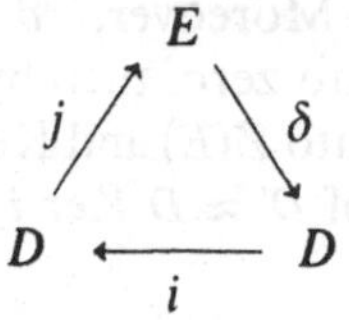

with

$$E^{p,q} = H^{p+q}(X_p, X_{p-1}),$$
$$D^{p,q} = H^{p+q}(X, X_{p-1});$$

the injections i, j have degrees $(-1, 1)$, $(0, 0)$, respectively, while the coboundary operator δ has degree $(1, 0)$.

Let $J^{p,n-p}$ be the kernel of the injection $H^n(X, A) \to H^n(X_{p-1}, A)$. Then

$$H^n(X, A) = J^{0,n} \supset \cdots \supset J^{p,n-p} \supset J^{p+1,n-p-1} \supset \cdots$$

However, the intersection of these subgroups is not, in general, zero. Instead, because of (1.4), we have

(1.9) *The intersection* $\bigcap_p J^{p,n-p}$ *is the subgroup* $\lim^1 H^{n-1}(X_p, A)$ *of* $H^n(X, A)$. □

Thus we have a descending filtration of the quotient

$$H^n(X, A)/\lim^1 H^{n-1}(X_p, A).$$

2 Exact Couples

In the algebraic discussion of this section we shall assume that we are working in a category of (possibly multiply) graded abelian groups. Each homomorphism thus has a well-defined degree (which may be a k-tuple of integers). The kernel of a homomorphism is then a graded group which inherits its graduation from its domain. Similarly, the image of a homomorphism is a graded group, whose graduation is inherited from the range.

Let

$\mathscr{C}$:

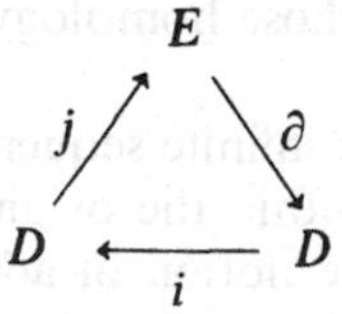

be an exact couple, and let $d = j \circ \partial : E \to E$. Then $d \circ d = j \circ \partial \circ j \circ \partial = 0$, since $\partial \circ j = 0$, so that E is a chain complex. Let $E' = H(E)$, and let $D' = \operatorname{Im} i$. Then $i|D$ maps D (and, *a fortiori*, D') into D', determining a homomorphism $i' : D' \to D'$. Moreover, ∂ maps $Z(E) = \operatorname{Ker} d$ into $\operatorname{Ker} j = D'$ and $B(E) = \operatorname{Im} d$ into zero, thereby inducing a homomorphism $\partial' : E' \to D'$. Finally, j maps D into $Z(E)$ and $\operatorname{Ker} i = \operatorname{Im} \partial$ into $B(E)$, thereby inducing a homomorphism j' of $D' \approx D/\operatorname{Ker} i$ into E'.

(2.1) Theorem *The diagram*

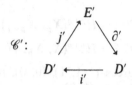

$$\mathscr{C}' : \quad \begin{array}{ccc} & E' & \\ {\scriptstyle j'} \nearrow & & \searrow {\scriptstyle \partial'} \\ D' & \xleftarrow[i']{} & D' \end{array}$$

is an exact couple.

For

$$\begin{aligned} \operatorname{Ker} \partial' &= \operatorname{Ker} \partial / B(E) = \operatorname{Im} j / B(E) = \operatorname{Im} j', \\ \operatorname{Ker} i' &= \operatorname{Ker} i \cap \operatorname{Im} i = \operatorname{Im} \partial \cap \operatorname{Ker} j, \\ \operatorname{Im} \partial' &= \partial(\operatorname{Ker} d) = \partial(\partial^{-1}(\operatorname{Ker} j)) = \operatorname{Ker} j \cap \operatorname{Im} \partial = \operatorname{Ker} i', \\ \operatorname{Ker} j' &= j^{-1}(\operatorname{Im} d)/\operatorname{Ker} i = j^{-1} j(\operatorname{Im} \partial)/\operatorname{Ker} i \\ &= \operatorname{Im} \partial + \operatorname{Ker} j/\operatorname{Ker} i = \operatorname{Ker} i + \operatorname{Ker} j/\operatorname{Ker} i \\ &= i(\operatorname{Ker} j) = i(D') = \operatorname{Im} i'. \end{aligned}$$

□

The exact couple $\mathscr{C}'$ is called the *derived couple* of the original exact couple $\mathscr{C}$. The process of derivation can be iterated indefinitely, yielding an infinite sequence of exact couples

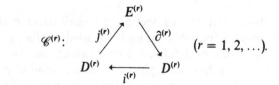

$$\mathscr{C}^{(r)} : \quad \begin{array}{ccc} & E^{(r)} & \\ {\scriptstyle j^{(r)}} \nearrow & & \searrow {\scriptstyle \partial^{(r)}} \\ D^{(r)} & \xleftarrow[i^{(r)}]{} & D^{(r)} \end{array} \qquad (r = 1, 2, \ldots).$$

such that $\mathscr{C}^{(1)} = \mathscr{C}$ and $\mathscr{C}^{(r+1)}$ is the derived couple of $\mathscr{C}^{(r)}$. The endomorphism $d^{(r)} = j^{(r)} \circ \partial^{(r)}$ has the property that $d^{(r)} \circ d^{(r)} = 0$, so that $E^{(r)}$ is a chain complex under $d^{(r)}$, whose homology group is $E^{(r+1)}$. In this way we obtain a *spectral sequence.*

Rather than deal with the infinite sequence of chain complexes $\{E^{(r)}\}$, it is often convenient to work within the original group E. The machinery for doing so is provided by the notion of additive relation; (cf. Appendix B, which the reader is encouraged to peruse if he is not already familiar with this notion).

Specifically, let i^k be the kth iterate of i, and let $\Delta^{(r)}$ be the additive relation $j \circ i^{-(r-1)} \circ \partial : E \rightsquigarrow E$.

(2.2) *The relation* $\Delta^{(r)}$ *has the following properties*:

(1) $\operatorname{Dom} \Delta^{(r)} = \partial^{-1} \operatorname{Im} i^{r-1}$,
(2) $\operatorname{Ker} \Delta^{(r)} = \partial^{-1} \operatorname{Im} i^r$,
(3) $\operatorname{Im} \Delta^{(r)} = j \operatorname{Ker} i^r$,
(4) $\operatorname{Ind} \Delta^{(r)} = j \operatorname{Ker} i^{r-1}$.

It follows by induction on r that

(2.3) *The group* $E^{(r)}$ *is the subquotient*

$$\operatorname{Dom} \Delta^{(r)}/\operatorname{Ind} \Delta^{(r)}$$

of E.

In order to make the inductive proof, we shall need to carry along a description of the whole couple $\mathscr{C}^{(r)}$. Thus we shall need to prove (2.3) and

(2.4) *The homomorphisms* $\partial^{(r)}, i^{(r)}, j^{(r)}$ *are given by*

$$\partial^{(r)}(x + j \operatorname{Ker} i^{r-1}) = \partial x, \qquad (x \in \partial^{-1} \operatorname{Im} i^{r-1}),$$

$$i^{(r)}(a) = i(a) \qquad (a \in \operatorname{Im} i^{r-1}),$$

$$j^{(r)}(i^{r-1}a) = j(a) + j \operatorname{Ker} i^{r-1} \qquad (a \in D),$$

by simultaneous induction. Both (2.3) and (2.4) are trivial for $r = 1$. Assume that the couple $\mathscr{C}^{(r)}$ is correctly described. Then, if $\partial x = i^{r-1}a$, we have

$$\textbf{(2.5)} \qquad d^{(r)}(x + j \operatorname{Ker} i^{r-1}) = j^{(r)} \partial x = j^{(r)} i^{r-1} a$$
$$= j(a) + j \operatorname{Ker} i^{r-1}.$$

If this is zero, we have $j(a) = j(a')$, $a' \in \operatorname{Ker} i^{r-1}$, whence

$$a = a' + i(a''),$$

$\partial x = i^{r-1}a = i^r a''$, so that $x \in \partial^{-1} \operatorname{Im} i^r$. Conversely, if $x \in \partial^{-1} \operatorname{Im} i^r$, $\partial x = i^r a$, then

$$d^{(r)}(x + j \operatorname{Ker} i^{r-1}) = ji(a) + j(\operatorname{Ker} i^{r-1}) = 0.$$

Thus $\operatorname{Ker} d^{(r)} = \partial^{-1} \operatorname{Im} i^r/\operatorname{Im} d^{(r-1)}$. Again, if (2.5) holds, then $0 = i\,\partial x = i^r a$, so that $\operatorname{Im} d^{(r)} \subset j \operatorname{Ker} i^r/j \operatorname{Ker} i^{r-1}$. Conversely, if $y = j(a)$, $i^r a = 0$, then $i^{r-1}a \in \operatorname{Ker} i = \operatorname{Im} \partial$, $i^{r-1}a = \partial x$ for some x; then $d^{(r)}(x + j \operatorname{Ker} i^{r-1}) = y + \operatorname{Ker} i^{r-1}$, $\operatorname{Im} d^{(r)} = j \operatorname{Ker} i^r/j \operatorname{Ker} i^{r-1}$, so that $E^{(r+1)} = \operatorname{Ker} d^{(r)}/\operatorname{Im} d^{(r)}$ is as described. The formulae in (2.4) follow without further difficulty. □

Let $Z^{(r)} = \text{Dom}\ \Delta^{(r)}$, $B^{(r)} = \text{Ind}\ \Delta^{(r)}$; then

(2.6) $E = Z^{(1)} \supset \cdots \supset Z^{(r)} \supset Z^{(r+1)} \supset \cdots \supset B^{(s+1)} \supset B^{(s)} \supset \cdots \supset B^{(1)} = 0.$

Letting $Z^{(\infty)} = \bigcap Z^{(r)}$, $B^{(\infty)} = \bigcup B^{(s)}$, we then have $Z^{(\infty)} \supset B^{(\infty)}$, and so we may define $E^{(\infty)} = Z^{(\infty)}/B^{(\infty)}$. In fact,

(2.7) *The group* $E^{(\infty)}$ *is isomorphic with each of the following double limits*:

$$\varinjlim_r \varprojlim_s Z^{(r)}/B^{(s)},$$

$$\varprojlim_s \varinjlim_r Z^{(r)}/B^{(s)}. \qquad \square$$

There are connections between the subquotients $E^{(\infty)}$ of E and certain subquotients of D. However, without additional hypotheses, these connections are rather tenuous. Therefore we shall formulate them only for certain special cases (for example, the homology and cohomology exact couple of a filtered pair).

A map (f, g) between two exact couples $\mathscr{C}_1$, $\mathscr{C}_2$, is *nullhomotopic* if and $g : D_1 \to D_2$ such that the diagram

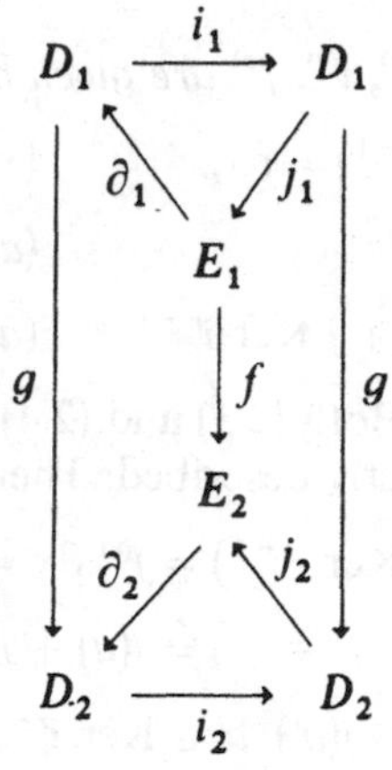

is commutative (except for the two triangles).

When this is so, $d_1 \circ f = f \circ d_2$, so that f is a chain map and induces a homomorphism $f' : E'_1 \to E'_2$. Moreover, g maps Im i_1 into Im i_2, inducing a map $g' : D'_1 \to D'_2$. And it is trivial to verify that f', g' define a map between the derived couples.

A map (f, g) between two exact couples $\mathscr{C}_1$, $\mathscr{C}_2$, is *nullhomotopic* if and only if there is a map $h : D_1 \to D_2$ such that

(1) $f = d_2 \circ h + h \circ d_1$,
(2) $g = \partial_2 \circ h \circ j_1$.

Two maps (f_0, g_0) and (f_1, g_1) are homotopic if and only if their difference

$(f_1 - f_0, g_1 - g_0)$ is nullhomotopic. When this is so, we see that f_0 and f_1 are chain homotopic, while g_0 and g_1 agree on $D' = \operatorname{Im} i$. Hence $f'_0 = f'_1$, $g'_0 = g'_1$, and the derived maps coincide.

Maps between couples can be composed: if $\phi = (f, g): \mathscr{C}_1 \to \mathscr{C}_2$ and $\phi' = (f', g'): \mathscr{C}_2 \to \mathscr{C}_3$, then $\phi' \circ \phi = (f' \circ f, g' \circ g): \mathscr{C}_1 \to \mathscr{C}_3$ is a map of couples. And if ϕ or ϕ' is nullhomotopic, so is $\phi' \circ \phi$. Hence

(2.8) *Let $\phi: \mathscr{C}_0 \to \mathscr{C}_1, \psi_0, \psi_1: \mathscr{C}_1 \to \mathscr{C}_2, \chi: \mathscr{C}_2 \to \mathscr{C}_3$ be maps of exact couples such that ψ_0 is homotopic to ψ_1. Then $\chi \circ \psi_0 \circ \phi$ and $\chi \circ \psi_1 \circ \phi$ are homotopic.* □

Remark. Since $E^{(k)}$ is a subquotient of $E = E^{(1)}$, there is a canonical additive relation $I_k: E \rightsquigarrow E^{(k)}$. If $r \geq k$, let $\Delta_k^{(r)} = I_k \circ \Delta^{(r)} \circ I_k^{-1}: E^{(k)} \rightsquigarrow E^{(k)}$. Then it is not hard to see (Exercise 5, below) that

$$\Delta_k^{(r)} = j^{(k)} \circ (i^{(k)})^{-(r-k)} \circ \partial^{(k)}.$$

In particular, the relation $\Delta_r^{(r)}$ is the boundary operator

$$d^{(r)} = j^{(r)} \circ \partial^{(r)}$$

of the couple $E^{(r)}$.

Thus the relation $\Delta^{(r)}$ does double duty. On the one hand, it serves to construct the complex $E^{(r)}$; on the other, it is the relation in $E^{(1)}$ corresponding to the operator $d^{(r)}$ in $E^{(r)}$.

3 The Exact Couples of a Filtered Space

Let us return to the special case of a filtration $\{X_p\}$ of a pair (X, A). We have seen that the group $H_n(X, A)$ is filtered by the subgroups $J_{p,q}$ with $p + q = n$. Moreover, if r is sufficiently large, then $H_{p+q}(X_{p-r}, A) = 0$, and therefore $Z_{p,q}^{(r)} = \operatorname{Ker} \partial_{p,q}$ is independent of r. It follows that $E_{p,q}^{(\infty)} = \operatorname{Ker} \partial_{p,q} / \bigcup B_{p,q}^{(r)}$ is the direct limit of the groups $E_{p,q}^{(r)}$ under the homomorphisms induced by the inclusions.

We can now make explicit the relations between the groups $E_{p,q}^{(r)}$ and the filtration (1.5) of $H_*(X, A)$. In fact, the composite of the injection

$$f = j_{p,q}: H_{p+q}(X_p, A) \to H_{p+q}(X_p, X_{p-1})$$

with the inverse of the injection

$$g: H_{p+q}(X_p, A) \to H_{p+q}(X, A)$$

is an additive relation $\Phi_{p,q}: H_{p+q}(X, A) \rightsquigarrow E_{p,q}$, and

(3.1) Theorem *The relation $\Phi_{p,q}$ induces an isomorphism*

$$\phi_{p,q}: J_{p,q}/J_{p-1,q+1} \approx E_{p,q}^{(\infty)}.$$

We must prove

(1) Dom $\Phi_{p,q} = J_{p,q}$;
(2) Ker $\Phi_{p,q} = J_{p-1,q+1}$;
(3) Im $\Phi_{p,q} = Z^{(\infty)}_{p,q}$;
(4) Ind $\Phi_{p,q} = B^{(\infty)}_{p,q}$.

The first statement is trivial, the second and third immediate consequences of the exactness of the homology sequence of the triple (X_p, X_{p-1}, A). The indeterminacy of $\Phi_{p,q}$ is $f(\text{Ker } g)$, by Example 1, §2 of Appendix B. But Ker g is the union of the kernels of the injections

$$g_r : H_{p+q}(X_p, A) \to H_{p+q}(X_{p+r-1}, A).$$

By exactness, the kernel of g_{r+1} is $g_r^{-1}(\text{Im } \partial_{p+r,q-r+1})$, and therefore $B^{(r)}_{p,q} = f(\text{Ker } g_{r+1})$. Thus

$$\begin{aligned}\text{Ind } \Phi_{p,q} &= f(\text{Ker } g) = f\Big(\bigcup_r \text{Ker } g_{r+1}\Big) \\ &= \bigcup_r f(\text{Ker } g_{r+1}) = \bigcup_r B^{(r)}_{p,q} = B^{(\infty)}_{p,q}.\end{aligned}$$ □

Let us summarize our results on the homology exact couple of the filtered pair (X, A).

(3.2) Theorem *Let* (X, A) *be a filtered pair. Then there is an ascending filtration*

(3.3) $$0 = J_{-1,n+1} \subset J_{0,n} \subset \cdots \subset J_{p,n-p} \subset \cdots \subset H_n(X, A)$$

of $H_n(X, A)$ *and a sequence of doubly graded chain complexes* $\{E^{(r)}_{p,q} \mid r \geq 1\}$ *such that*

(1) *the boundary operator of* $E^{(r)}$ *has degree* $(-r, r-1)$;
(2) *the homology group of* $E^{(r)}$ *is* $E^{(r+1)}$;
(3) *for each* p, q *and all sufficiently large* r, *there is an epimorphism*

$$E^{(r)}_{p,q} \twoheadrightarrow E^{(r+1)}_{p,q};$$

(4) *the graded group* $E^{(\infty)}$ *for which*

$$E^{(\infty)}_{p,q} = \varinjlim_r E^{(r)}_{p,q}$$

is the graded group $\mathscr{G}H_*(X, A)$ *associated with the filtration* (3.3) *of* $H_*(X, A)$. □

Let us turn our attention to the cohomology exact couple. In this case the operators $\{\delta_{(r)}, i_{(r)}, j_{(r)}\}$ of the $(r-1)$st derived couple have degrees $\{(r, 1-r), (-1, 1), (0, 0)\}$, respectively, and it follows that $d_{(r)} = j_{(r)} \circ \delta_{(r)}$ has degree $(r, 1-r)$. If r is sufficiently large, $X_{p-r} = A$ and the injection

$$H^{p+q}(X, X_{p-r}) \to H^{p+q}(X, A)$$

is the identity. For such r the kernel of the injection

$$H^{p+q}(X, X_{p-1}) \to H^{p+q}(X, X_{p-r})$$

is independent of r, and so is its image $B_{(r)}^{p,q}$ under the injection $j^{p,q}$. Thus

$$E_{(\infty)}^{p,q} = Z_{(\infty)}^{p,q}/B_{(\infty)}^{p,q} = \bigcap_r Z_{(r)}^{p,q}/B_{(\infty)}^{p,q}.$$

Again, we may define an additive relation $\Phi^{p,q}$. This time it is the composite of the injection $f = j^{p,q} : H^{p+q}(X, X_{p-1}) \to H^{p+q}(X_p, X_{p-1})$ with the inverse of the injection $g : H^{p+q}(X, X_{p-1}) \to H^{p+q}(X, A)$.

(3.4) Theorem *The relation $\Phi^{p,q}$ induces a monomorphism*

$$\phi^{p,q} : J^{p,q}/J^{p+1,q-1} \rightarrowtail E_{(\infty)}^{p,q}.$$

We must prove:

(1) Dom $\Phi^{p,q} = J^{p,q}$,
(2) Ker $\Phi^{p,q} = J^{p+1,q-1}$,
(3) Im $\Phi^{p,q} \subset Z_{(\infty)}^{p,q}$,
(4) Ind $\Phi^{p,q} = B_{(\infty)}^{p,q}$.

The first two statements are immediate consequences of the exactness of the appropriate cohomology sequences; the last follows from the identification of $B_{(\infty)}^{p,q}$ above. The image of $\Phi^{p,q}$, by exactness, is the kernel of $\delta^{p,q} : H^{p+q}(X_p, X_{p-1}) \to H^{p+q+1}(X, X_p)$; it is a subgroup of $Z_{(\infty)}^{p,q} = (\delta^{p,q})^{-1}(L^{p,q})$, where $L^{p,q}$ is the intersection of the images of the injections $H^{p+q+1}(X, X_{p+r}) \to H^{p+q+1}(X, X_p)$. □

Let us see what can be said about the cokernel of $\phi^{p,q}$. It follows from Milnor's result (1.3) that there is a commutative diagram

$$\begin{array}{ccccccccc}
0 \longrightarrow & \varprojlim_r^1 H^{p+q}(X_{p+r}, X_p) & \xrightarrow{\alpha_p} & H^{p+q+1}(X, X_p) & \xrightarrow{\beta_p} & \varprojlim_r H^{p+q+1}(X_{p+r}, X_p) & \longrightarrow 0 \\
& \downarrow{\scriptstyle \bar{\imath}^{p,q}} & & \downarrow{\scriptstyle i^{p+1,q}} & & \downarrow & \\
0 \longrightarrow & \varprojlim_r^1 H^{p+q}(X_{p+r}, X_{p-1}) & \xrightarrow[\alpha_{p-1}]{} & H^{p+q+1}(X, X_{p-1}) & \xrightarrow[\beta_{p-1}]{} & \varprojlim_r H^{p+q+1}(X_{p+r}, X_{p-1}) & \longrightarrow 0
\end{array}$$

induced by the relevant injections. Moreover,

$$\operatorname{Im} \alpha_p = \operatorname{Ker} \beta_p = L^{p,q},$$

$$\operatorname{Im} \alpha_{p-1} = \operatorname{Ker} \beta_{p-1} = L^{p-1,q+1}.$$

Thus

$$\begin{aligned}
\operatorname{Cok} \phi^{p,q} &\approx Z_{(\infty)}^{p,q}/\operatorname{Ker} \delta^{p,q} = (\delta^{p,q})^{-1}(L^{p,q})/\operatorname{Ker} \delta^{p,q} \\
&\approx L^{p,q} \cap \operatorname{Im} \delta^{p,q} = L^{p,q} \cap \operatorname{Ker} i^{p+1,q} \\
&\approx \operatorname{Ker} \bar{\imath}^{p,q}.
\end{aligned}$$

Thus

(3.5) Theorem *The cokernel of $\phi^{p,q}$ is isomorphic with the kernel of the homomorphism*

$$\bar{\imath}^{p,q} : \varprojlim_r{}^1 H^{p+q}(X_{p+r}, X_p) \to \varprojlim_r{}^1 H^{p+q}(X_{p+r}, X_{p-1})$$

induced by the appropriate injections. □

Summarizing, we have

(3.6) Theorem *Let (X, A) be a filtered pair. Then there is a descending filtration*

$$\bar{H}^n(X, A) = J^{0,n} \supset \cdots \supset J^{p,q} \supset J^{p+1,q-1} \supset \cdots, \quad \bigcap_{p+q=n} J^{p,q} = 0,$$

of the quotient

$$\bar{H}^n(X, A) = H^n(X, A)/\varprojlim_r{}^1 H^{n-1}(X_p, A) \approx \varprojlim_r H^n(X_p, A)$$

and a sequence of doubly graded cochain complexes $\{E_{(r)}^{p,q} \mid r \geq 1\}$ such that

(1) *the coboundary operator of $E_{(r)}$ has degree $(r, 1 - r)$;*
(2) *the cohomology group of $E_{(r)}$ is $E_{(r+1)}$;*
(3) *for each p, q and all sufficiently large r, there is a monomorphism*

$$E_{(r+1)}^{p,q} \rightarrowtail E_{(r)}^{p,q};$$

(4) *the graded group $\mathscr{G}\bar{H}^*(X, A)$ associated with the above filtration is a subgroup of the graded group $E_{(\infty)}$ for which*

$$E_{(\infty)}^{p,q} = \varprojlim_r E_{(r)}^{p,q}.$$ □

Let us now consider the effect of maps. Let (X, A) and (Y, B) be filtered by $\{X_p\}$, $\{Y_p\}$, respectively. Let $\mathscr{C}(X, A)$, $\mathscr{C}(Y, B)$ be the associated exact couples. Let $f: (X, A) \to (Y, B)$ be a map which respects the filtrations: $f(X_p) \subset Y_p$ for all p. Then f induces

$$f_\# : H_{p+q}(X_p, X_{p-1}) \to H_{p+q}(Y_p, Y_{p-1}),$$
$$f_\natural : H_{p+q}(X_p, A) \quad \to H_{p+q}(Y_p, B),$$

and it is clear that

(3.7) *The maps $f_\#, f_\natural$ constitute a map $f_* : \mathscr{C}(X, A) \to \mathscr{C}(Y, B)$.* □

More delicate is the effect of homotopies. If $\{X_p\}$ is a filtration of (X, A), then the spaces

$$\mathbf{I} \times X_{p-1} \cup \dot{\mathbf{I}} \times X_p$$

constitute a filtration of $\mathbf{I} \times (X, A) = (\mathbf{I} \times X, \mathbf{I} \times A)$. Let $f_0, f_1 : (X, A) \to (Y, B)$ be maps, and let

$$f : (\mathbf{I} \times X, \mathbf{I} \times A) \to (Y, B)$$

be a homotopy of f_0 to f_1 which respects filtration.

(3.8) Theorem *The maps $f_{0*}, f_{1*} : \mathscr{C}(X, A) \to \mathscr{C}(Y, B)$ induced by f_0 and f_1 are homotopic.*

The maps $p : X \to \mathbf{I} \times X$, $q : X \to \mathbf{I} \times X$ defined by

$$p(x) = (0, x), \qquad q(x) = (1, x)$$

respect filtration and therefore define maps $p_*, q_* : \mathscr{C}(X, A) \to \mathscr{C}(\mathbf{I} \times (X, A))$; moreover, $f_* \circ p_* = f_{0*}, f_* \circ q_* = f_{1*}$. By (2.8), it suffices to prove

(3.9) *The maps $p_*, q_* : \mathscr{C}(X, A) \to \mathscr{C}(\mathbf{I} \times X, \mathbf{I} \times A)$ are homotopic.*

It behooves us to define homomorphisms

$$\lambda_{r,s} : H_{r+s}(X_r, X_{r-1}) \to H_{r+s+1}(\mathbf{I} \times X_r \cup \dot{\mathbf{I}} \times X_{r+1}, \mathbf{I} \times X_{r-1} \cup \dot{\mathbf{I}} \times X_r)$$

such that

$$q_* - p_* = d \circ \lambda + \lambda \circ d, \tag{3.10}$$

$$q_\natural - p_\natural = \partial \circ \lambda \circ j. \tag{3.11}$$

The statements of (3.10) and (3.11), as well as their proofs, for a given degree (r, s) involve only a few subsets of X and their products with certain subsets of $\mathbf{I}$. It will simplify the notation if we agree that

$$\bar{P} = \mathbf{I} \times P, \qquad P_t = \{t\} \times P, \qquad \dot{P} = \dot{\mathbf{I}} \times P \qquad (t = 0, 1),$$

for any space P. The Fundamental Notational Convention of Chapter II continues in effect. In addition, the symbols p_i, q_i will denote homomorphisms induced by the maps p, q, respectively.

Let (R, Q, P) be an NDR-triple. We may then define a homomorphism

$$\lambda_m(R, Q, P) : H_n(Q, P) \to H_{n+1}(\bar{Q} \cup \dot{R}, \bar{P} \cup \dot{Q})$$

to be the composite of

(1) the homomorphism

$$H_n(Q, P) \to H_n(\bar{P} \cup \dot{Q}, \bar{P} \cup Q_0)$$

induced by q (which is easily seen, with the aid of the Excision Property, to be an isomorphism);

(2) the inverse of the boundary operator

$$H_{n+1}(\bar{Q}, \bar{P} \cup \dot{Q}) \to H_n(\bar{P} \cup \dot{Q}, \bar{P} \cup Q_0)$$

(which is an isomorphism because $(\bar{Q}, \bar{P} \cup Q_0) = (\mathbf{I}, 0) \times (Q, P)$ is a DR-pair);

(3) the injection

$$H_{n+1}(\bar{Q}, \bar{P} \cup \dot{Q}) \to H_{n+1}(\bar{Q} \cup \dot{R}, \bar{P} \cup \dot{Q}).$$

The homomorphism λ is then defined by

$$\lambda_{r,s} = \lambda_{r+s}(X_{r+1}, X_r, X_{r-1}).$$

We shall prove

(3.12) *Let* (S, R, Q, P) *be an* NDR-*quadruple. Then the relation*

$$\lambda_{n-1}(R, Q, P) \circ \partial_1 + \partial_2 \circ \lambda_n(S, R, Q) = q_1 - p_1$$

suggested by the diagram

(3.13)
$$\begin{array}{ccccc}
 & & H_{n-1}(Q, P) & & \\
 & \overset{\partial_1}{\nearrow} & & \searrow^{\lambda_{n-1}(R, Q, P)} & \\
H_n(R, Q) & & \xrightarrow{\;q_1 - p_1\;} & & H_n(\bar{Q} \cup \dot{R}, \bar{P} \cup \dot{Q}) \\
 & {}_{\lambda_n(S, R, Q)}\searrow & & \nearrow_{\partial_2} & \\
 & & H_{n+1}(\bar{R} \cup \dot{S}, \bar{Q} \cup \dot{R}) & &
\end{array}$$

holds.

The relation (3.10) follows immediately from (3.12) by taking $(S, R, Q, P) = (X_{r+1}, X_r, X_{r-1}, X_{r-2})$ and $n = r + s$.

The relation (3.11) also follows from (3.12). Let us first consider the quadruple (X_{r+1}, X_r, A, A) and observe that, since $H_{r+s-1}(A, A) = 0$, (3.12) asserts that

$$q_1 - p_1 = \partial_2 \circ \lambda_{r+s}(X_{r+1}, X_r, A).$$

Let us further consider the quadruple $(X_{r+1}, X_r, X_{r-1}, A)$ and the inclusion of the first quadruple in the second. This gives rise to a commutative diagram

$$\begin{array}{ccccc}
H_{r+s}(X_r, A) & \xrightarrow{\;\lambda'\;} & H_{r+s+1}(\bar{X}_r \cup \dot{X}_{r+1}, \bar{A} \cup \dot{X}_r) & \xrightarrow{\;\partial_2\;} & H_{r+s}(\bar{A} \cup \dot{X}_r, \bar{A}) \\
\downarrow{\scriptstyle j} & & \downarrow & & \downarrow{\scriptstyle j'} \\
H_{r+s}(X_r, X_{r-1}) & \xrightarrow[\lambda]{} & H_{r+s+1}(\bar{X}_r \cup \dot{X}_{r+1}, \bar{X}_{r-1} \cup \dot{X}_r) & \xrightarrow[\partial]{} & H_{r+s}(\bar{X}_{r-1} \cup \dot{X}_r, \bar{A})
\end{array}$$

where $\lambda' = \lambda_{r+s}(X_{r+1}, X_r, A)$, $\lambda = \lambda_{r+s}(X_{r+1}, X_r, X_{r-1})$. The homomorphism $j' \circ \partial_2 \circ \lambda' = j_2 \circ (q_1 - p_1)$ is the homomorphism $q_\natural - p_\natural$ of (3.11), while ∂, λ, and j have the same meaning as in (3.11) (for degree (r, s)). Thus (3.11) follows from commutativity of the above diagram.

To prove (3.12), we first establish a direct sum decomposition of the group $G = H_n(\bar{Q} \cup \dot{R}, \bar{P} \cup \dot{Q})$:

(3.14) *The homomorphisms*

$$\lambda = \lambda_{n-1}(R, Q, P) : H_{n-1}(Q, P) \to G,$$

$$p_1 : H_n(R, Q) \to G,$$

$$q_1 : H_n(R, Q) \to G$$

represent the group G as a direct sum.

Let us apply the Direct Sum Theorem to the pair $(\bar{Q} \cup \dot{R}, \bar{P} \cup \dot{Q})$ and the subspaces $\bar{Q}$, R_0, R_1. We find that the injections

$$i : H_n(\bar{Q}, \bar{P} \cup \dot{Q}) \to G,$$

$$i_0 : H_n(R_0, Q_0) \to G,$$

$$i_1 : H_n(R_1, Q_1) \to G$$

represent G as a direct sum. Since λ, p_1 and q_1 are the composites of i, i_0 and i_1 with the isomorphisms

$$H_{n-1}(Q, P) \xrightarrow{q_0} H_{n-1}(\bar{P} \cup \dot{Q}, \bar{P} \cup Q_0) \xrightarrow{\partial_0^{-1}} H_n(\bar{Q}, \bar{P} \cup \dot{Q}),$$

$$H_n(R, Q) \xrightarrow{p_2} H_n(R_0, Q_0),$$

$$H_n(R, Q) \xrightarrow{q_2} H_n(R_1, Q_1),$$

respectively, they also represent G as a direct sum. □

It also follows from the Direct Sum Theorem that

(3.15) *The injections*

$$j : G \to H_n(\bar{Q} \cup \dot{R}, \bar{P} \cup \dot{R}),$$

$$j_0 : G \to H_n(\bar{Q} \cup \dot{R}, \bar{Q} \cup R_1),$$

$$j_1 : G \to H_n(\bar{Q} \cup \dot{R}, \bar{Q} \cup R_0)$$

represent G as a direct product. The representations (λ, p_1, q_1) *and* (j, j_0, j_1) *are weakly dual.* □

We shall use the above representations to prove (3.12). Let $\lambda = \lambda_{n-1}(R, Q, P)$, $\bar{\lambda} = \lambda_n(S, R, Q)$. We must prove three relations:

(3.16) $$j \circ \partial_2 \circ \bar{\lambda} = -j \circ \lambda \circ \partial_1,$$

(3.17) $$j_0 \circ \partial_2 \circ \bar{\lambda} = -j_0 \circ p_1,$$

(3.18) $$j_1 \circ \partial_2 \circ \bar{\lambda} = j_1 \circ q_1.$$

As the reader may infer from the presence of the minus signs in (3.16) and (3.17), their proofs will involve the use of the Hexagonal Lemma. That of (3.18), on the other hand, is straight forward.

To prove (3.16), we first establish

(3.19) Lemma *The diagram*

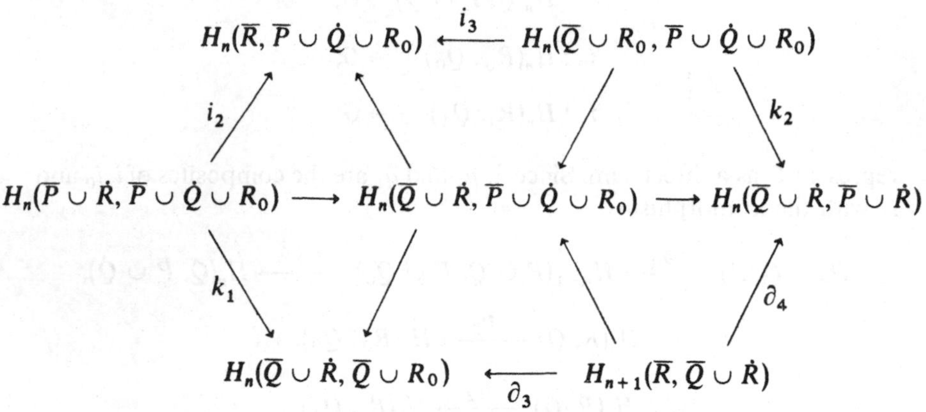

satisfies the hypotheses of the Hexagonal Lemma, and therefore

$$\partial_4 \circ \partial_3^{-1} \circ k_1 = -k_2 \circ i_3^{-1} \circ i_2.$$

The requisite exactness and commutativity properties are evident. Since $(\bar{R}, \bar{Q} \cup R_0) = (\mathbf{I}, 0) \times (R, Q)$ is a DR-pair, it follows from the exactness of the homology sequences of the appropriate triples that i_3 and ∂_3 are isomorphisms. □

(3.20) Lemma *The diagram*

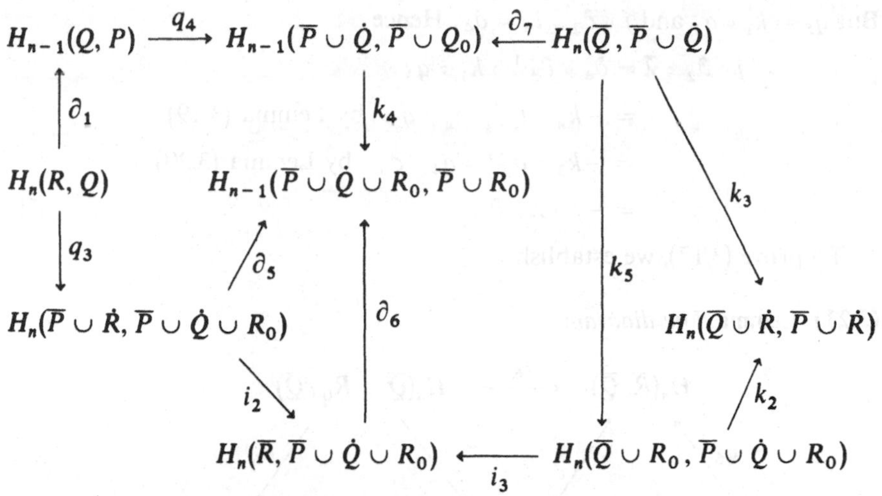

is commutative and all its homomorphisms except possibly ∂_1, ∂_5 and i_2 are isomorphisms. Hence

$$k_3 \circ \partial_7^{-1} \circ q_4 \circ \partial_1 = k_2 \circ i_3^{-1} \circ i_2 \circ q_3 .$$

The commutativity is evident. That q_3, q_4, k_3, k_4 and k_5 are isomorphisms follows from the Excision Property. That ∂_6, ∂_7 and i_3 are isomorphisms follows from the fact that $(\bar{R}, \bar{P} \cup R_0)$, $(\bar{Q}, \bar{P} \cup Q_0)$ and $(\bar{R}, \bar{Q} \cup R_0)$, respectively, are DR-pairs. □

We can now prove (3.16). The homomorphism $j \circ \lambda \circ \partial_1$ is the composite

$$H_n(R, Q) \xrightarrow{\partial_1} H_{n-1}(Q, P) \xrightarrow{q_4} H_{n-1}(\bar{P} \cup \dot{Q}, \bar{P} \cup Q_0)$$
$$\xrightarrow{\partial_7^{-1}} H_n(\bar{Q}, \bar{P} \cup \dot{Q}) \xrightarrow{i_4} H_n(\bar{Q} \cup \dot{R}, \bar{P} \cup \dot{Q})$$
$$\xrightarrow{j} H_n(\bar{Q} \cup \dot{R}, \bar{P} \cup \dot{R}).$$

But the composite of the last two injections is k_3. Hence

$$j \circ \lambda \circ \partial_1 = k_3 \circ \partial_7^{-1} \circ q_4 \circ \partial_1$$

On the other hand, $j \circ \partial_2 \circ \bar{\lambda}$ is the composite

$$H_n(R, Q) \xrightarrow{q_5} H_n(\bar{Q} \cup \dot{R}, \bar{Q} \cup R_0) \xrightarrow{\partial_3^{-1}} H_{n+1}(\bar{R}, \bar{Q} \cup \dot{R})$$
$$\xrightarrow{i_5} H_{n+1}(\bar{R} \cup \dot{S}, \bar{Q} \cup \dot{R}) \xrightarrow{\partial_2}$$
$$H_n(\bar{Q} \cup \dot{R}, \bar{P} \cup \dot{Q}) \xrightarrow{j} H_n(\bar{Q} \cup \dot{R}, \bar{P} \cup \dot{R}).$$

But $q_5 = k_1 \circ q_3$ and $j \circ \partial_2 \circ i_5 = \partial_4$. Hence

$$\begin{aligned} j \circ \partial_2 \circ \bar{\lambda} &= \partial_4 \circ \partial_3^{-1} \circ k_1 \circ q_3 \\ &= -k_2 \circ i_3^{-1} \circ i_2 \circ q_3 \quad \text{by Lemma (3.19)} \\ &= -k_3 \circ \partial_7^{-1} \circ q_4 \circ \partial_5 \quad \text{by Lemma (3.20)} \\ &= -j \circ \lambda \circ \partial_1. \end{aligned}$$

□

To prove (3.17), we establish

(3.21) Lemma *The diagram*

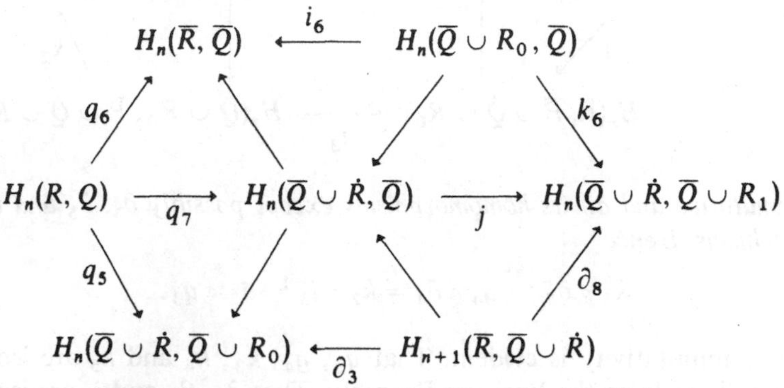

satisfies the hypotheses of the Hexagonal Lemma, and therefore

$$k_6 \circ i_6^{-1} \circ q_6 = -\partial_8 \circ \partial_3^{-1} \circ q_5.$$

Again, the requisite exactness and commutativity properties are clear, and we have seen that ∂_3 is an isomorphism. That i_6 is an isomorphism follows from the fact that $(\bar{R}, \bar{Q} \cup R_0)$ is a DR-pair. □

(3.22) Lemma *The diagram*

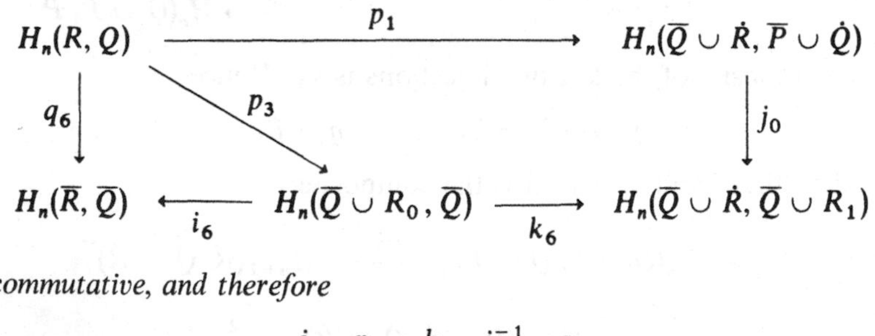

is commutative, and therefore

$$j_0 \circ p_1 = k_6 \circ i_6^{-1} \circ q_6.$$

The right-hand side of the diagram is trivially commutative. To show that the left-hand triangle is commutative, observe that the maps $p, q : (R, Q) \rightarrow$

$(\bar{R}, \bar{Q})$ are homotopic, and therefore their induced homomorphisms $p_4, q_6 : H_n(R, Q) \to H_n(\bar{R}, \bar{Q})$ coincide. Then $i_6 \circ p_3 = p_4 = q_6$. □

We can now prove (3.17). The homomorphism ∂_8 is the composite $j_0 \circ \partial_2 \circ i_5$; hence

$$\begin{aligned} j_0 \circ \partial_2 \circ \bar{\lambda} &= \partial_8 \circ \partial_3^{-1} \circ q_5 \\ &= -k_6 \circ i_6^{-1} \circ q_6 \quad \text{by Lemma (3.21)} \\ &= -j_0 \circ p_1 \quad \text{by Lemma (3.22).} \end{aligned}$$ □

Finally, we may observe that $j_1 \circ \partial_2 \circ i_5 = \partial_3$, and therefore

$$j_1 \circ \partial_2 \circ \bar{\lambda} = \partial_3 \circ \partial_3^{-1} \circ q_5 = q_5 = j_1 \circ q_1.$$

This completes the proof of (3.12). □

Similarly, a filtration preserving map f induces homomorphisms

$$\begin{aligned} f^{\#} &: H^{p+q}(Y_p, Y_{p-1}) \to H^{p+q}(X_p, X_{p-1}), \\ f^{\#} &: H^{p+q}(Y, Y_{p-1}) \to H^{p+q}(X, X_{p-1}) \end{aligned}$$

(3.7*) *The maps $f^{\natural}, f^{\natural}$ constitute a map $f^* : \mathscr{C}^*(Y, B) \to \mathscr{C}^*(X, A)$.* □

And if $f: (\mathbf{I} \times X, \mathbf{I} \times A) \to (Y, B)$ is a homotopy between two maps f_0, f_1 which respects filtration,

(3.8*) Theorem *The maps f_0^*, f_1^* induced by f_0 and f_1 are homotopic.* □

The proofs are dual to those of (3.7) and (3.8), respectively.

4 The Spectral Sequence of a Fibration

Let $f : X \to B$ be a fibration with fibre F, and let $\mathfrak{G} = \{G_n, \partial_n\}$ be an additive homology theory (we use the letter G, rather than H, because although we are interested in only one homology theory at a time, others may, and indeed do, crop up; in particular, ordinary homology groups with local coefficients appear in the case in hand). In this section we shall show how to associate with the fibre map f an exact couple whose spectral sequence leads from the ordinary homology of the base with (local) coefficients in the (extraordinary) homology of the fibre to the (extraordinary) homology of the total space.

Let (B, A) be a relative CW-complex, $f: X \to B$ a fibration, $W = f^{-1}(A)$, and let B_p be the p-skeleton of (B, A), $X_p = f^{-1}(B_p)$. We shall describe $f: (X, W) \to (B, A)$ as a *fibration of pairs*.

Throughout this section we shall assume that the fibrations $f: (X, W) \to (B, A)$ with which we are dealing have the property that the base space B is 0-connected. Then the fibres all have the same homotopy type, and we assume, in addition, that they, too are 0-connected.

(4.1) Theorem *The spaces X_p form an* NDR*-filtration of* (X, W).

By Theorem (7.14) of Chapter I, the pair (X_p, X_{p-1}) is an NDR-pair for every p. Therefore, by the results of §6 of Chapter I, it suffices to prove that X has the weak topology with respect to the family of subspaces X_p.

Let C be a subset of X whose intersection with X_p is closed for every p. If K is any compact subset of X, then $f(K)$ is compact and therefore contained in B_p for some p. Then $K \subset X_p$, and therefore

$$C \cap K = C \cap (X_p \cap K) = (C \cap X_p) \cap K$$

is closed. Since X is compactly generated, C is closed. □

The fibration $\{X_p\}$ gives rise to an exact couple $\mathscr{C}(f)$ as in §3, and our first objective is to calculate the group

$$E^1_{p,q} = G_{p+q}(X_p, X_{p-1}).$$

Let $\{E_\alpha\}$ be the p-cells of (B, A), and let $E^*_\alpha = B_p - \operatorname{Int} E_\alpha$. Let

$$\begin{aligned} Q_\alpha &= f^{-1}(E_\alpha), \\ \dot{Q}_\alpha &= f^{-1}(\dot{E}_\alpha), \\ Q^*_\alpha &= f^{-1}(E^*_\alpha). \end{aligned}$$

(4.2) Theorem *The injections*

$$i_\alpha : G_{p+q}(Q_\alpha, \dot{Q}_\alpha) \to G_{p+q}(X_p, X_{p-1})$$

represent the group $G_{p+q}(X_p, X_{p-1})$ as a direct sum. The injections

$$j_\alpha : G_{p+q}(X_p, X_{p-1}) \to G_{p+q}(X_p, Q^*_\alpha)$$

represent $G_{p+q}(X_p, X_{p-1})$ as a weak direct product. The representations $\{i_\alpha\}$, $\{j_\alpha\}$ are weakly dual.

This follows from the General Direct Sum Theorem (6.9) of Chapter XII, as soon as we have verified that X_p has the weak topology with respect to the subspaces X_{p-1}, Q^p_α. Let C be a subspace of X_p whose intersections with each Q^p_α and with X_{p-1} are closed. If K is a compact subset of X_p, then $f(K)$ is compact; by (1.1) of Chapter II, there is a finite set $J \subset J_p$ such that $f(K) \subset$

$B_{p-1} \cup \bigcup_{\alpha \in J} E_\alpha^p$. Then $K \subset X_{p-1} \cup \bigcup_{\alpha \in J} Q_\alpha$, so that

$$\begin{aligned} C \cap K &= C \cap \left(X_{p-1} \cup \bigcup_{\alpha \in J} Q_\alpha \right) \cap K \\ &= \left\{ (C \cap X_{p-1}) \cup \bigcup_{\alpha \in J} (C \cap Q_\alpha) \right\} \cap K \end{aligned}$$

is closed. Since X is compactly generated, C is closed. □

Let $\phi_\alpha : (\Delta^p, \dot{\Delta}^p) \to (E_\alpha, \dot{E}_\alpha)$ be a characteristic map. Let $z_\alpha = \phi_\alpha(\mathrm{e}_0)$, $F_\alpha = f^{-1}(z_\alpha)$. The map ϕ_α induces a fibration $f_\alpha : (T_\alpha, \dot{T}_\alpha) \to (\Delta^p, \dot{\Delta}^p)$, so that there is a commutative diagram

$$\begin{array}{ccc} T_\alpha & \xrightarrow{\ \psi_\alpha\ } & Q_\alpha \\ {\scriptstyle f_\alpha}\downarrow & & \downarrow{\scriptstyle f|Q_\alpha} \\ \Delta^p & \xrightarrow[\ \phi_\alpha\ ]{} & E_\alpha \end{array}$$

Since Δ^p is contractible, f_α is fibre homotopically trivial, so that there is a strong trivialization $\bar{h}_\alpha : (\Delta^p \times F_\alpha, \dot{\Delta}^p \times F_\alpha) \to (T_\alpha, \dot{T}_\alpha)$.

(4.3) *The map* $\psi_\alpha : (T_\alpha, \dot{T}_\alpha) \to (Q_\alpha, \dot{Q}_\alpha)$ *is a relative homeomorphism.*

The map $\phi_\alpha : (\Delta^p, \dot{\Delta}^p) \to (E_\alpha, \dot{E}_\alpha)$ is a relative homeomorphism; let $\omega_\alpha : E_\alpha - \dot{E}_\alpha \to \Delta^p - \dot{\Delta}^p$ be the inverse of $\phi_\alpha | \Delta^p - \dot{\Delta}^p$. Then the map $\tau_\alpha : Q_\alpha - \dot{Q}_\alpha \to T_\alpha - \dot{T}_\alpha$ defined by

$$\tau_\alpha(x) = (x, \omega_\alpha f(x))$$

is the inverse of $\psi_\alpha | T_\alpha - \dot{T}_\alpha$. It remains only to verify that ψ_α is a proclusion.

If K_0 is a compact subset of Q_α, $\psi_\alpha^{-1}(K_0)$ is a closed subset of the compact space $K_0 \times \Delta^p$, so that $\psi_\alpha^{-1}(K_0)$ is compact.

Let C be a subset of Q_α such that $\psi_\alpha^{-1}(C)$ is closed. Then $\psi_\alpha^{-1}(C) \cap \psi_\alpha^{-1}(K_0) = \psi_\alpha^{-1}(C \cap K_0)$ is closed, and therefore compact, for every compact subset K_0 of Q_α. Hence $C \cap K_0 = \psi_\alpha \psi_\alpha^{-1}(C \cap K_0)$ is compact and therefore closed. Since Q_α is compactly generated, C is closed. □

From Theorem (4.2) and the Map Excision Theorem ((6.7) of Chapter XII) we deduce

(4.4) Corollary *The homomorphisms*

$$v_\alpha = i_\alpha \circ G_{p+q}(\psi_\alpha) : G_{p+q}(T_\alpha, \dot{T}_\alpha) \to G_{p+q}(X_p, X_{p-1})$$

represent the latter group as a direct sum, and the representations $\{v_\alpha\}$, $\{j_\alpha\}$ *are weakly dual.* □

Let $h_\alpha = \psi_\alpha \circ \bar{h}_\alpha : (\Delta^p \times F_\alpha, \dot{\Delta}^p \times F_\alpha) \to (Q_\alpha, \dot{Q}_\alpha)$.

(4.5) *The homomorphism*

$$G_{p+q}(h_\alpha) : G_{p+q}(\Delta^p \times F_\alpha, \dot{\Delta}^p \times F_\alpha) \to G_{p+q}(Q_\alpha, \dot{Q}_\alpha)$$

is an isomorphism, independent of the strong trivialization $\bar{h}_\alpha$.

Since $\bar{h}_\alpha$ is a homotopy equivalence, $G_{p+q}(\bar{h}_\alpha)$ is an isomorphism. By the Map Excision Theorem $G_{p+q}(\psi_\alpha)$ is an isomorphism. The independence follows from Theorem (7.29) of Chapter I. □

Let δ_p be the canonical orientation of Δ^p. In Theorem (6.12) of Chapter XII we established an isomorphism

$$\delta_p \times : G_q(Y) \to G_{p+q}(\Delta^p \times Y, \dot{\Delta}^p \times Y)$$

for every space Y. Taking $Y = F_\alpha$ and composing the isomorphisms $\delta_p \times$, $G_{p+q}(h_\alpha)$, we obtain an isomorphism $G_q(F_\alpha) \approx G_{p+q}(Q_\alpha, \dot{Q}_\alpha)$. According to Theorem (4.2), we have

(4.6) Theorem *The homomorphisms*

$$i_\alpha \circ G_{p+q}(h_\alpha) \circ (\delta_p \times) : G_q(F_\alpha) \to G_{p+q}(X_p, X_{p-1})$$

represent the latter group as a direct sum. □

In §8 of Chapter IV we saw that if $u : \mathbf{I} \to B$ is a path from b_0 to b_1, there is a uniquely defined homotopy class of maps, called *admissible*, of $F_1 = f^{-1}(b_1)$ into $F_0 = f^{-1}(b_0)$; a map h is admissible if and only if there is a homotopy $H : \mathbf{I} \times F_1 \to X$ of h to the inclusion map $F_1 \hookrightarrow X$ such that $fH(t, y) = u(t)$ for all $(t, y) \in \mathbf{I} \times F_1$. As in Chapter VI, the homomorphisms $G_q(h) : G_q(F_1) \to G_q(F_0)$ give rise to a local coefficient system $G_q(\mathscr{F})$ in the space B. The homology groups of B with coefficients in $G_q(\mathscr{F})$ are those of a chain complex $\Gamma_*(B, A; G_*(\mathscr{F}))$. The elements of $\Gamma_p(B, A; G_q(\mathscr{F}))$ are finite formal sums

$$\sum_{\alpha \in J_p} g_\alpha e_\alpha$$

with $g_\alpha \in G_q(F_\alpha)$. Thus $\Gamma_p(B, A; G_q(\mathscr{F})) \approx \bigoplus_{\alpha \in J_p} G_q(F_\alpha)$, so that the doubly graded groups $\Gamma_*(B, A; G_*(\mathscr{F}))$ and E^1 are isomorphic. In fact, let e_α be the orientation of E_α determined by the characteristic map ϕ_α; then the map

$$\lambda : \Gamma_p(B, A; G_q(\mathscr{F})) \to G_{p+q}(X_p, X_{p-1})$$

given by

$$\lambda(xe_\alpha) = i_\alpha G_{p+q}(h_\alpha)(\delta_p \times x) \qquad (x \in G_q(F_\alpha))$$

is an isomorphism.

The isomorphism λ is natural in a sense which we now explain. Let $f : X \to B, f' : X' \to B'$ be fibrations. A *fibre preserving map* of f' into f is a pair of maps $g : B' \to B$, $h : X' \to X$ making the diagram

$$\begin{array}{ccc} X' & \xrightarrow{h} & X \\ {\scriptstyle f'}\downarrow & & \downarrow{\scriptstyle f} \\ B' & \xrightarrow[g]{} & B \end{array}$$

strictly commutative. If B and B' are 0-connected, as we are assuming, the map g is uniquely determined; in fact $g = f \circ h \circ (f')^{-1}$. Therefore we shall often say that h is a fibre preserving map.

EXAMPLE 1. Let $f: X \to B$ be a fibration and let $g : B' \to B$ be an arbitrary map. Let $f' : X' \to B'$ be the induced fibration. Then there is a uniquely defined map $h : X' \to X$ such that the pair (g, h) is fibre preserving.

EXAMPLE 2. Let $f' : X' \to B, f : X \to B$ be fibrations, and let $\lambda : X' \to X$ be a fibre homotopy equivalence. Then λ (or rather the pair $(\lambda, 1 : B \to B)$) is a fibre preserving map.

We shall be concerned, in particular, with the case that (B', A') and (B, A) are relative CW-complexes and $g : (B', A') \to (B, A)$ is cellular. When this is so, we shall say that (g, h) (and even h!) is *cellular*. Then $h(X'_p) \subset X_p$, so that h induces a map of the couple $\mathscr{C}(f')$ into the couple $\mathscr{C}(f)$. In particular, h induces $h_\# : G_{p+q}(X'_p, X'_{p-1}) \to G_{p+q}(X_p, X_{p-1})$.

On the other hand, the local coefficient systems $G_q(\mathscr{F}')$ and $G_q(\mathscr{F})$ are related by a homomorphism

$$\gamma_q = G_q(g, h) : G_q(\mathscr{F}') \to g^* G_q(\mathscr{F}).$$

Therefore g induces a chain map

$$g_\# : \Gamma_p(B', A'; G_q(\mathscr{F}')) \to \Gamma_p(B, A; G_q(\mathscr{F})).$$

Remark. Just as the isomorphisms $G_q(F_{b_0}) \approx G_q(F_{b_1})$ are induced by *admissible maps* $h : F_{b_0} \to F_{b_1}$, so the homomorphisms $G_q(F_{b'}) \to G_q(F_b)$ determined by the homomorphism γ_q are induced by certain maps, also called admissible, of $F' = F'_{b'}$ into $F_b = F$. Specifically, a map $h' : F' \to F$ is said to be *admissible* if and only if there is a path $u : \mathbf{I} \to B$ from $g(b')$ to b and a homotopy $G_t : F' \to X$ such that $G_0 = h | F'$, $G_1 = h'$, and $G_t(F')$ lies in the fibre over $u(t)$ for each $t \in \mathbf{I}$.

(4.7) Theorem *The diagram*

$$\begin{array}{ccc} \Gamma_p(B', A'; G_q(\mathscr{F}')) & \xrightarrow{g_\#} & \Gamma_p(B, A; G_q(\mathscr{F})) \\ \downarrow \lambda & & \downarrow \lambda \\ G_{p+q}(X'_p, X'_{p-1}) & \xrightarrow[h_\#]{} & G_{p+q}(X_p, X_{p-1}) \end{array}$$

is commutative.

The proof of Theorem (4.7) being somewhat lengthy, we defer it to §5.

The map λ being an isomorphism of doubly graded groups, it is natural to ask whether it is an isomorphism of chain complexes. That this is so is the burden of

(4.8) Theorem *The diagram*

$$\begin{array}{ccc} \Gamma_p(B, A; G_q(\mathscr{F})) & \xrightarrow{\partial} & \Gamma_{p-1}(B, A; G_q(\mathscr{F})) \\ \downarrow \lambda & & \downarrow \lambda \\ G_{p+q}(X_p, X_{p-1}) & \xrightarrow[\partial]{} & G_{p+q-1}(X_{p-1}, X_{p-2}) \end{array}$$

is commutative.

As this follows from Theorem (4.7), we shall defer its proof, too, to §5.

The upshot of our discussion can be summarized in

(4.9) Theorem *Let* $f: (X, W) \to (B, A)$ *be a fibration of pairs such that* (B, A) *is a relative* CW*-complex and the base* B *and fibre* F *are* 0*-connected. Let* $\mathfrak{G}$ *be an additive homology theory on* $\mathscr{K}_*^2$. *Then there is a filtration*

$$0 = J_{-1, n+1} \subset J_{0, n} \subset \cdots \subset J_{p, n-p} \subset J_{p+1, n-p-1} \subset \cdots \subset G_n(X, W),$$

a sequence of doubly graded chain complexes $\{E^r \mid r \geq 2\}$, *and an isomorphism* $\lambda_* : H_p(B, A; G_q(\mathscr{F})) \approx E^2_{p, q}$, *such that*

(1) *the boundary operator* $d^r : E^r \to E^r$ *has degree* $(-r, r-1)$;
(2) $E^{r+1} = H(E^r)$;
(3) *for each* p, q, *there is an epimorphism*

$$E^r_{p, q} \twoheadrightarrow E^{r+1}_{p, q}$$

for all sufficiently large r;
(4) $G_n(X, W) = \bigcup_{p=0}^{\infty} J_{p, n-p}$;

(5) *there is an isomorphism*

$$J_{p,q}/J_{p-1,q+1} \approx E^{\infty}_{p,q} = \varinjlim_{r} E^{r}_{p,q};$$

(6) *the isomorphism λ_* is natural with respect to cellular fibre preserving maps.* □

(We have written E^r instead of $E^{(r)}$ for typographical convenience).

There are analogous results for cohomology. There are minor differences because of the fact that we have to deal with direct product, rather than direct sum, decompositions. In this section we shall state without comment the results for cohomology that we shall need. The proofs of Theorem (4.7*) and Theorem (4.8*) will require minor modifications, which we shall discuss in §5.

Let $\mathfrak{G}^*$ be an additive cohomology theory on $\mathscr{K}^2_*$. Then

(4.2*) Theorem *The injections*

$$i^{\alpha} : G^{p+q}(X_p, X_{p-1}) \to G^{p+q}(Q_{\alpha}, \dot{Q}_{\alpha})$$

represent the former group as a direct product. The injections

$$j^{\alpha} : G^{p+q}(X_p, Q^*_{\alpha}) \to G^{p+q}(X_p, X_{p-1})$$

represent the latter group as a strong direct sum. The representations $\{i^{\alpha}\}$, $\{j^{\alpha}\}$ are weakly dual. □

(4.4*) Corollary *The homomorphisms*

$$v^{\alpha} = G^{p+q}(\psi_{\alpha}) \circ i^{\alpha} : G^{p+q}(X_p, X_{p-1}) \to G^{p+q}(T_{\alpha}, \dot{T}_{\alpha})$$

represent the former group as a direct product and the representations $\{v^{\alpha}\}$, $\{j^{\alpha}\}$ are weakly dual. □

(4.5*) *The homomorphism*

$$G^{p+q}(h_{\alpha}) : G^{p+q}(Q_{\alpha}, \dot{Q}_{\alpha}) \to G^{p+q}(\Delta^p \times F_{\alpha}, \dot{\Delta}^p \times F_{\alpha})$$

is an isomorphism, independent of the strong trivialization $\bar{h}_{\alpha}$. □

In Theorem (7.23) of Chapter XII we established an isomorphism

$$/\delta_p : G^{p+q}(\Delta^p \times Y, \dot{\Delta}^p \times Y) \to G^q(Y)$$

for every space Y.

(4.6*) Theorem *The homomorphisms*

$$(/\delta_p) \circ G^{p+q}(h_{\alpha}) \circ i^{\alpha} : G^{p+q}(X_p, X_{p-1}) \to G^q(F_{\alpha})$$

represent the former group as a direct product. □

An isomorphism $\lambda^* : G^{p+q}(X_p, X_{p-1}) \to \Gamma^p(B, A; G^q(\mathscr{F}))$ is then defined by

$$\lambda^*(u)(e_\alpha) = h_\alpha^* i^\alpha u/\delta_p.$$

(4.7*) Theorem *Let*

$$\begin{array}{ccc} (X', W') & \xrightarrow{h} & (X, W) \\ {\scriptstyle f'}\downarrow & & \downarrow{\scriptstyle f} \\ (B', A') & \xrightarrow[g]{} & (B, A) \end{array}$$

be a cellular fibre preserving map. Then the diagram

$$\begin{array}{ccc} G^{p+q}(X_p, X_{p-1}) & \xrightarrow{h^\#} & G^{p+q}(X'_p, X'_{p-1}) \\ {\scriptstyle \lambda^*}\downarrow & & \downarrow{\scriptstyle \lambda^*} \\ \Gamma^p(B, A; G^q(\mathscr{F})) & \xrightarrow[g^\#]{} & \Gamma^p(B', A'; G^q(\mathscr{F}')) \end{array}$$

is commutative. □

(4.8*) Theorem *The diagram*

$$\begin{array}{ccc} G^{p+q-1}(X_{p-1}, X_{p-2}) & \xrightarrow{\delta^*} & G^{p+q}(X_p, X_{p-1}) \\ {\scriptstyle \lambda^*}\downarrow & & \downarrow{\scriptstyle \lambda^*} \\ \Gamma^{p-1}(B, A; G^q(\mathscr{F})) & \xrightarrow[(-1)^{p-1}\delta^*]{} & \Gamma^p(B, A; G^q(\mathscr{F})) \end{array}$$

is commutative. □

In order to get good convergence properties, i.e., to avoid the complications due to the appearance of the unpleasant $\varprojlim^1$, we shall legislate it out of existence. Specifically, we shall impose additional conditions which will guarantee that the appropriate $\varprojlim^1$ terms vanish. These are subsumed in the hypothesis of

(4.9*) Theorem *Let $f: (X, W) \to (B, A)$ be a fibration of pairs such that (B, A) is a relative* CW*-complex and the base B and fibre F are* 0*-connected. Let $\mathfrak{G}^*$ be an additive cohomology theory. Assume that either*

(a) *the pair (B, A) is finite-dimensional, or*
(b) *there is an integer N such that $G^q(F) = 0$ for all $q < N$.*

Then there is a filtration

$$G^n(X, W) = J^{0,n} \supset \cdots \supset J^{p,n-p} \supset J^{p+1,n-p-1} \supset \cdots$$

of $G^n(X, W)$, *a sequence of doubly graded cochain complexes* $\{E_r \mid r \geq 2\}$, *and an isomorphism* $\lambda^* : E_2^{p,q} \approx H^p(B, A; G^q(\mathscr{F}))$, *such that*

(1) *the coboundary operator* $d_r : E_r \to E_r$ *has degree* $(r, 1-r)$;
(2) $E_{r+1} = H(E_r)$;
(3) *for each* p, q *there is a monomorphism*

$$E_{r+1}^{p,q} \rightarrowtail E_r^{p,q}$$

for all sufficiently large r;
(4) $\bigcap_{p=0}^{\infty} J^{p,n-p} = 0$;
(5) *there is an isomorphism*

$$J^{p,q}/J^{p+1,q-1} \approx E_\infty^{p,q};$$

(6) *the isomorphism* λ^* *is natural with respect to cellular fibre preserving maps.*

To verify (4), we must show that, for each n, $\underline{\lim}_p^1 G^{n-1}(X_p, W) = 0$. It is sufficient to show that the injection

$$G^{n-1}(X_{p+1}, W) \to G^{n-1}(X_p, W)$$

is an epimorphism for p large enough, and this is so provided that $G^n(X_{p+1}, X_p) \approx \Gamma_{p+1}(B, A; G^{n-p-1}(\mathscr{F})) = 0$. Our hypotheses guarantee this.

To verify (5) it suffices, by Theorem (3.5), to show that, for each p, q, the injection

$$G^{p+q}(X_{p+r}, X_p) \to G^{p+q}(X_{p+r-1}, X_p)$$

is an epimorphism for r large, and this is so provided that $G^{p+q+1}(X_{p+r}, X_{p+r-1}) = \Gamma_{p+r}(B, A; G^{q-r+1}(\mathscr{F})) = 0$ for large r. Again this is guaranteed by our hypotheses. □

We conclude this Section by remarking that our results can be extended to the case of a fibration over an arbitrary pair (B, A). In order to do this we need to assume that our homology (cohomology) theory satisfies the Isotropy Condition, and is therefore complete. For let $g : (K, L) \to (B, A)$ be a weak homotopy equivalence, and let $f' : (Q, P) \to (X, W)$ be the induced fibration. Then it follows by the Five Lemma that f' is a weak homotopy equivalence and by the isotropy condition that f induces isomorphisms of the homology (cohomology) groups. One then shows that the derived couples of the exact couples $\mathscr{C}(f')$ for all CW-approximations g form a transitive system over (X, W), and therefore determine an exact couple uniquely associated to the given fibration. The details are straightforward but tedious, and are left to the enterprising reader.

In the ensuing sections we shall study the spectral sequence in further detail. Two special cases are of importance:

I. The fibre is a point P. Then the coefficient systems are simple, and the spectral sequence relates the homology groups of (B, A) with coefficients in the groups $G_q = G_q(P)$ to the (extraordinary) homology groups $G_q(B, A)$.

II. The homology theory $\mathfrak{G}$ is ordinary homology theory with coefficients in a group G. The spectral sequence relates the homology groups of the base with local coefficients in the homology of the fibre to the homology of the total space.

5 Proofs of Theorems (4.7) and (4.8)

We begin with the proof of Theorem (4.7). In Corollary (4.4) we established the direct sum representation

$$\nu_\alpha : G_{p+q}(T_\alpha, \dot{T}_\alpha) \to G_{p+q}(X_p, X_{p-1})$$

and its weakly dual representation

$$j_\alpha : G_{p+q}(X_p, X_{p-1}) \to G_{p+q}(X_p, Q_\alpha^*)$$

as a weak direct product. Similarly we have representations

$$\begin{aligned} \nu'_\beta &: G_{p+q}(T'_\beta, \dot{T}'_\beta) \to G_{p+q}(X'_p, X'_{p-1}), \\ j'_\beta &: G_{p+q}(X'_p, X'_{p-1}) \to G_{p+q}(X'_p, Q'^*_\beta). \end{aligned}$$

These correspond to direct sum and weak direct product representations

$$\begin{aligned} \mu_\alpha &: H_p(\Delta^p, \dot{\Delta}^p; G_q(F_\alpha)) \to H_p(B_p, B_{p-1}; G_q(\mathscr{F})), \\ l_\alpha &: H_p(B_p, B_{p-1}; G_q(\mathscr{F})) \to H_p(B_p, E_\alpha^*; G_q(\mathscr{F})), \\ \mu'_\beta &: H_p(\Delta^p, \dot{\Delta}^p; G_q(F'_\beta)) \to H_p(B'_p, B'_{p-1}; G_q(\mathscr{F}')), \\ l'_\beta &: H_p(B'_p, B'_{p-1}; G_q(\mathscr{F}')) \to H_p(B'_p, E'^*_\beta; G_q(\mathscr{F}')). \end{aligned}$$

The homomorphism $g_\#$ is determined by the composites $g_{\alpha\beta} = l_\alpha \circ g_\# \circ \mu'_\beta$, just as the homomorphism $h_\#$ is determined by the composites $h_{\alpha\beta} = j_\alpha \circ h_\# \circ \nu'_\beta$. It is clear from the definitions that the homomorphisms λ respect the decompositions. Therefore, in order to prove Theorem (4.7), it suffices to prove

(5.1) *For each α, β, the diagram*

(5.2)
$$\begin{array}{ccc}
H_p(\Delta^p, \dot{\Delta}^p; G_q(F'_\beta)) & \xrightarrow{g_{\alpha\beta}} & H_p(B_p, E^*_\alpha; G_q(\mathscr{F})) \\
\Big\downarrow \lambda & & \Big\downarrow \lambda \\
G_{p+q}(T'_\beta, \dot{T}'_\beta) & \xrightarrow[h_{\alpha\beta}]{} & G_{p+q}(X_p, Q^*_\alpha)
\end{array}$$

is commutative.

To avoid verbiage, we shall say that the pair (E_α, E'_β) is *regular* when this is the case.

We first prove

(5.3) *If $E_\alpha \not\subset g(E'_\beta)$, then (E_α, E'_β) is regular.*

For let x be an interior point of E_α which does not belong to $g(E'_\beta)$. Then E^*_α is a deformation retract of $B_p - \{x\}$ and therefore $g \circ \phi'_\beta : (\Delta^p, \dot{\Delta}^p) \to (B_p, E^*_\alpha)$ is compressible, so that $g_{\alpha\beta} = 0$. On the other hand, by the homotopy lifting extension property, the map $h \circ \psi'_\beta : (T'_\beta, \dot{T}'_\beta) \to (X_p, Q^*_\alpha)$ is compressible, so that $h_{\alpha\beta} = 0$. □

From now on we shall assume that $E_\alpha \subset g(E'_\beta)$. We next prove

(5.4) *Suppose there is a map $k : (\Delta^p, \dot{\Delta}^p) \to (\Delta^p, \dot{\Delta}^p)$ such that $\phi_\alpha \circ k \simeq g \circ \phi'_\beta : (\Delta^p, \dot{\Delta}^p) \to (B_p, E^*_\alpha)$. Then the pair (E_α, E'_β) is regular.*

We may assume that $k(\mathbf{e}_0) = \mathbf{e}_0$. Consider the three-dimensional diagram

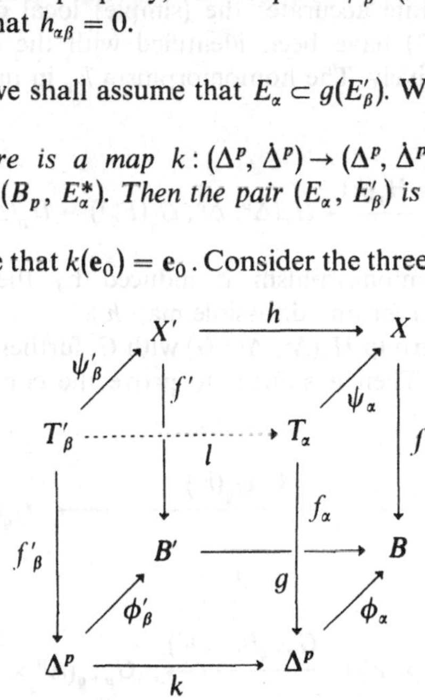

in which the bottom square is homotopy commutative and the back and two sides are strictly commutative. We shall find a map $l : T'_\beta \to T_\alpha$ making the top square homotopy commutative and the front square strictly commutative.

Let $F : \mathbf{I} \times \Delta^p \to B$ be a homotopy of $g \circ \phi'_\beta$ to $\phi_\alpha \circ k$. Then $F_1 = F \circ (1 \times f'_\beta) : \mathbf{I} \times T'_\beta \to B$ is a homotopy of $g \circ \phi'_\beta \circ f'_\beta = f \circ h \circ \psi'_\beta$ to

$\phi_\alpha \circ k \circ f'_\beta$. By the homotopy lifting property, the map $h \circ \psi'_\beta$ is homotopic to a map χ such that $f \circ \chi = \phi_\alpha \circ k \circ f'_\beta$. The maps $\chi : T'_\beta \to X$ and $k \circ f'_\beta : T'_\beta \to \Delta^p$ define a map $l : T'_\beta \to T_\alpha$ such that $f_\alpha \circ l = k \circ f'_\beta$ and $\psi_\alpha \circ l = \chi \simeq h \circ \psi'_\beta$. Since $F(\mathbf{I} \times \Delta_p) \subset B_p$ and $F(\mathbf{I} \times \dot{\Delta}_p) \subset E^*_\alpha$, the maps χ and $h \circ \psi'_\beta$ are homotopic maps of $(T'_\beta, \dot{T}'_\beta)$ into (X_p, Q^*_α).

Consider the diagram

$$\begin{array}{ccccccc}
H_p(\Delta^p, \dot{\Delta}^p; G_q(F'_\beta)) & \xrightarrow{k_*} & H_p(\Delta^p, \dot{\Delta}^p; G_q(F_\alpha)) & \xrightarrow{H_p(\phi_\alpha)} & H_p(B_p, B_{p-1}; G_q(\mathscr{F})) & \xrightarrow{l_\alpha} & H_p(B_p, E^*_\alpha; G_q(\mathscr{F})) \\
\downarrow \lambda & & \downarrow \lambda & & \downarrow \lambda & & \downarrow \lambda \\
G_{p+q}(T'_\beta, \dot{T}'_\beta) & \xrightarrow[G_{p+q}(l)]{} & G_{p+q}(T_\alpha, \dot{T}_\alpha) & \xrightarrow[G_{p+q}(\psi_\alpha)]{} & G_{p+q}(X_p, X_{p-1}) & \xrightarrow[j_\alpha]{} & G_{p+q}(X_p, Q^*_\alpha)
\end{array}$$

The composites of the homomorphisms in the top and bottom lines are $g_{\alpha\beta}$, $h_{\alpha\beta}$, respectively, by the given properties of the maps k, l. And we have seen that the center and right-hand squares are commutative. Therefore it remains to prove that the left-hand square is commutative. (The notation k_* in the top line is not quite accurate; the (simple) local coefficient systems $\phi^*_\alpha G_q(F)$ and $\phi'^*_\alpha G_q(F')$ have been identified with the coefficient groups $G_q(F_\alpha)$, $G_q(F'_\beta)$, respectively. The homomorphism k_* in question is then the composite

$$H_p(\Delta^p, \dot{\Delta}^p; G_q(F'_\beta)) \xrightarrow{H_p(k)} H_p(\Delta^p, \dot{\Delta}^p; G_q(F'_\beta)) \to H_p(\Delta^p, \dot{\Delta}^p; G_q(F_\alpha)),$$

where the second homomorphism is induced by the homomorphism $G_q(h') : G_q(F'_\beta) \to G_q(F_\alpha)$ for an admissible map h').

Let us identify the group $H_p(\Delta^p, \dot{\Delta}^p; G)$ with G; furthermore, let k_0 be the degree of the map k. Then it suffices to prove the commutativity of the diagram

$$\begin{array}{ccc}
G_q(F'_\beta) & \xrightarrow{k_0 G_q(h')} & G_q(F_\alpha) \\
\downarrow \delta_p \times & \text{I} & \downarrow \delta_p \times \\
G_{p+q}(\Delta^p \times F'_\beta, \dot{\Delta}^p \times F'_\beta) & \xrightarrow{G_{p+q}(k \times h')} & G_{p+q}(\Delta^p \times F_\alpha, \dot{\Delta}^p \times F_\alpha) \\
\downarrow G_{p+q}(\bar{h}'_\beta) & \text{II} & \downarrow G_{p+q}(\bar{h}_\alpha) \\
G_{p+q}(T'_\beta, \dot{T}'_\beta) & \xrightarrow[G_{p+q}(l)]{} & G_{p+q}(T_\alpha, \dot{T}_\alpha)
\end{array}$$

To prove the commutativity of I, let us collapse the subspaces $\dot{\Delta}^p \times F'_\beta$, $\dot{\Delta}^p \times F_\alpha$ to points. We obtain the diagram

$$\begin{array}{ccc}
G_q(F'_\beta) & \xrightarrow{k_0 G_q(h')} & G_q(F_\alpha) \\
\downarrow{\scriptstyle \delta_p \times} & & \downarrow{\scriptstyle \delta_p \times} \\
G_{p+q}(\Delta^p \times F'_\beta, \dot{\Delta}^p \times F'_\beta) & \xrightarrow{G_{p+q}(k \times h')} & G_{p+q}(\Delta^p \times F_\alpha, \dot{\Delta}^p \times F_\alpha) \\
\downarrow{\scriptstyle G_{p+q}(\pi')} & & \downarrow{\scriptstyle G_{p+q}(\pi)} \\
\tilde{G}_{p+q}(\mathbf{S}^p F'^{+}_\beta) & \xrightarrow[\tilde{G}_{p+q}(\bar{k} \wedge h')]{} & \tilde{G}_{p+q}(\mathbf{S}^p F^{+}_\alpha)
\end{array}$$

where π' and π are the identification maps and $\bar{k} : \mathbf{S}^p \to \mathbf{S}^p$ is induced by k. The bottom square being patently commutative, the commutativity of the top square is equivalent to that of the composite one. But $G_{p+q}(\pi') \circ \delta_p \times$ and $G_{p+q}(\pi) \circ \delta_p \times$ are the p-fold iterated suspensions, and the commutativity of the composite square follows from Corollary (5.13) of Chapter XII.

We now prove the commutativity of II. In order to do so, we must construct an admissible map $h' : F'_\beta \to F_\alpha$. Let us observe that ψ_α maps $f_\alpha^{-1}(\mathbf{e}_0)$ homeomorphically upon $F_\alpha = f^{-1}(z_\alpha) = f^{-1}\phi_\alpha(\mathbf{e}_0)$, that ψ'_β maps $f'^{-1}_\beta(\mathbf{e}_0)$ homeomorphically upon $F'_\beta = f'^{-1}(z'_\beta) = f'^{-1}\phi'_\beta(\mathbf{e}_0)$, and that g' maps F'_β into $f^{-1}g(z'_\beta)$. We shall therefore take the liberty of identifying $f_\alpha^{-1}(\mathbf{e}_0)$ with F_α and $f'^{-1}_\beta(\mathbf{e}_0)$, with F'_β. With these liberties of notation, we have

(5.5) Lemma *The map $l \mid F'_\beta : F'_\beta \to F_\alpha$ is admissible.*

We proved above that there is a homotopy $G_t : T'_\beta \to X$ of $h \circ \psi'_\beta$ to $\psi_\alpha \circ l$; this homotopy has the property that $f \circ G_t = F_t \circ f'_\beta$, where $F_t : \Delta^p \to B$ is a homotopy of $g \circ \phi'_\beta$ to $\phi_\alpha \circ k$. The map $t \to F_t(\mathbf{e}_0)$ is a path u in B from $g(z'_\beta)$ to z_α, and the homotopy $G_t \mid F'_\beta : F'_\beta \to X$ maps F'_β into $f^{-1}u(t)$ for every t. Moreover, $G_0 \mid F'_\beta = h \circ \psi'_\beta \mid F'_\beta$, $G_1 \mid F'_\beta = \psi_\alpha \circ l \mid F'_\beta$. Thus G_1 is admissible and therefore $l \mid F'$ is admissible.

Now consider the lifting extension problem

$$\begin{array}{ccc}
\dot{\mathbf{I}} \times \Delta^p \times F'_\beta \cup \mathbf{I} \times \{\mathbf{e}_0\} \times F'_\beta & \xrightarrow{\sigma} & T_\alpha \\
\cap & \nearrow & \downarrow{\scriptstyle f_\alpha} \\
\mathbf{I} \times \Delta^p \times F'_\beta & \xrightarrow[\tau]{} & \Delta^p
\end{array}$$

where

$$\sigma(0, u, y') = \bar{h}_\alpha(k(u), l(y')),$$
$$\sigma(1, u, y') = l\bar{h}'_\beta(u, y'),$$
$$\sigma(t, \mathbf{e}_0, y') = l(y');$$
$$\tau(t, u, y') = k(u).$$

Now $(\Delta^p, \{\mathbf{e}_0\})$ is a DR-pair, and therefore

$$(\mathbf{I} \times \Delta^p \times F'_\beta, \dot{\mathbf{I}} \times \Delta^p \times F'_\beta \cup \mathbf{I} \times \{\mathbf{e}_0\} \times F'_\beta) = (\mathbf{I}, \dot{\mathbf{I}}) \times (\Delta^p, \{\mathbf{e}_0\}) \times (F'_\beta, \varnothing)$$

is a DR-pair; by Lemma (7.15) of Chapter I, the problem has a solution, which is a homotopy between the maps $\bar{h}_\alpha \circ (k \times h')$ and $l \circ \bar{h}'_\beta$. □

(5.6) Corollary *Theorem* (4.7) *is true in the special case that* $(B', A') = (|K'|, |L'|)$, $(B, A) = (|K|, |L|)$, *where* (K', L') *and* (K, L) *are simplicial pairs, and* $g = |\phi|$ *for a simplicial map* $\phi : (K', L') \to (K, L)$. □

We now complete the proof of Theorem (4.7) by showing that (5.1) holds in general. By Lemma (1.4) of Chapter V, there is a subdivision $(K, \dot{K})$ of $(\Delta^p, \dot{\Delta}^p)$ and a map $g_1 : (\Delta^p, \dot{\Delta}^p) \to (B_p, E^*_\alpha)$ homotopic to $g \circ \phi'_\beta$ such that, for every simplex σ of K, either $g_1(\sigma) \subset E^*_\alpha$ or $g_1 | \sigma = \phi_\alpha \circ k_\sigma$ for some map $k_\sigma : (\sigma, \dot{\sigma}) \to (\Delta^p, \dot{\Delta}^p)$. Consider the diagram

$$\begin{array}{ccccc}
H_p(\Delta^p, \dot{\Delta}^p; G_q(F'_\beta)) & \xrightarrow{i} & H_p(|K_p|, |K_{p-1}|; G_q(F'_\beta)) & \xrightarrow{g_{1*}} & H_p(B_p, E^*_\alpha; G_q(\mathscr{F})) \\
\downarrow \lambda & & \downarrow \lambda & & \downarrow \lambda \\
G_{p+q}(T'_\beta, \dot{T}'_\beta) & \xrightarrow[j]{} & G_{p+q}(T'_\beta, T_0) & \xrightarrow[h_{1*}]{} & G_{p+q}(X_p, Q^*_\alpha)
\end{array}$$

where $T_0 = f'^{-1}_\beta(|K_{p-1}|)$, i and j are injections. Since $g \circ \phi'_\beta \simeq g_1$, there is a map h_1 such that (g_1, h_1) is a map of f'_β into f and $h_1 \simeq h \circ \psi'_\beta : (T'_\beta, \dot{T}'_\beta) \to (X_p, Q^*_\alpha)$. Evidently $g_{1*} \circ i = g_{\alpha\beta}$ and $h_{1*} \circ j = h_{\alpha\beta}$.

The characteristic maps $\tau_\sigma : (\Delta^p, \dot{\Delta}^p) \to (\sigma, \dot{\sigma})$ for the p-cells of K evidently satisfy the hypothesis of either (5.3) or (5.4). Therefore each of the pairs (σ, E_α) is regular. It follows that the right-hand square of the diagram is commutative. On the other hand, there is a simplicial map $\theta : (K_p, \dot{K}_p) \to (\Delta^p, \dot{\Delta}^p)$ which is homotopic to the identity and it follows that $|\theta|_* \circ i$ is the identity. There is a map $\theta' : T'_\beta \to T'_\beta$ such that $f'_\beta \circ \theta' = |\theta| \circ f'_\beta$ and $\theta' : (T'_\beta, \dot{T}'_\beta) \to (T'_\beta, \dot{T}'_\beta)$ is homotopic to the identity. Then $\theta'_* \circ j$ is the identity. It follows from Corollary (5.6) that $\lambda \circ \theta'_* = |\theta|_* \circ \lambda$, and from this fact the commutativity of the left-hand square follows. □

Let us now prove Theorem (4.8). It suffices to show that the homomorphisms $\partial \circ \lambda$ and $\lambda \circ \partial$ agree on every element of the form xe_α, where E_α is a p-cell of (B, A) and $x \in G(F_\alpha)$. Let $\phi_\alpha : (\Delta^p, \dot{\Delta}^p) \to (E_\alpha, \dot{E}_\alpha)$ be a characteristic map for the cell E_α. We shall regard $(\Delta^p, \dot{\Delta}^p)$ as a CW-pair under the obvious triangulation; and we may assume that $\dot{\phi}_\alpha = \phi_\alpha | \dot{\Delta}^p : \dot{\Delta}^p \to B$ is cellular.

With the notation as in the proof of Theorem (4.7), consider the three-dimensional diagram

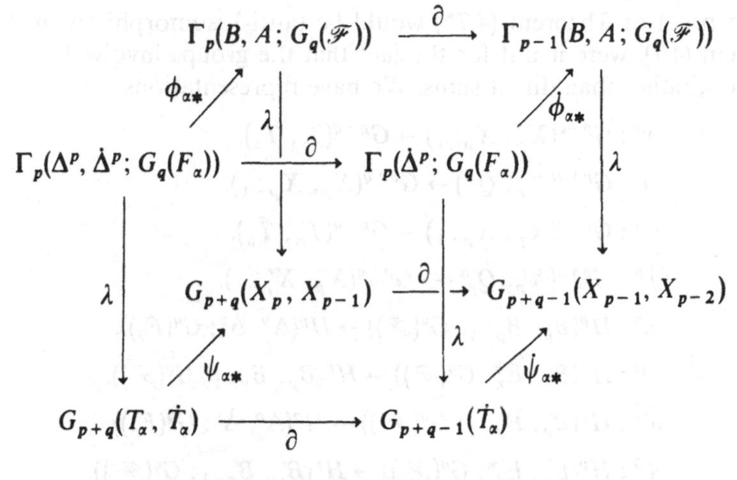

By Theorem (4.7), the two sides are commutative. The top and bottom are commutative because the boundary operators in question are natural (cf. §2 of Chapter VI and §6 of Chapter XII). We next show that the front face is commutative. Let $h : (\Delta^p \times F_\alpha, \dot{\Delta}^p \times F_\alpha) \to (T_\alpha, \dot{T}_\alpha)$ be a strong trivialization of f_α. Then $\dot{h} = h | \dot{\Delta}^p \times F_\alpha : \dot{\Delta}^p \times F_\alpha \to \dot{T}_\alpha$ is a strong trivialization of $\dot{f}_\alpha = f_\alpha | \dot{T}_\alpha$.

Now $\partial \delta_p = \sum_{i=0}^{p} (-1)^i \varepsilon_i$, where ε_i is the orientation of the ith face of δ_p, determined by the map $d_i : \Delta^{n-1} \to \Delta^n$; thus the incidence number $[\delta_p : \varepsilon_i] = (-1)^i$. Also $\lambda(x\delta_p) = h_*(\delta_p \times x)$, while

$$
\begin{aligned}
\lambda\, \partial(x\delta_p) &= \lambda \sum_{i=0}^{p} (-1)^i x\varepsilon_i = \sum_{i=0}^{p} (-1)^i \dot{h}_*(\varepsilon_i \times x) \\
&= \dot{h}_*\left(\sum_{i=0}^{p} [\delta_p : \varepsilon_i] \varepsilon_i \times x \right) \\
&= \dot{h}_*\left(\sum_{i=0}^{p} \sigma(\Delta^p, \Delta_i^p) \right)(\delta_p \times x)) \quad \text{by Theorem (6.12), Chapter XII,} \\
&= \dot{h}_*\, \partial(\delta_p \times x) \quad \text{by Theorem (6.10), Chapter XII,} \\
&= \partial h_*(\delta_p \times x) \\
&= \partial \lambda(x\delta_p).
\end{aligned}
$$

It now follows that

$$\begin{aligned}\lambda \circ \partial \circ \phi_{\alpha *} &= \lambda \circ \dot{\phi}_{\alpha *} \circ \partial = \dot{\psi}_{\alpha *} \circ \lambda \circ \partial \\ &= \dot{\psi}_{\alpha *} \circ \partial \circ \lambda = \partial \circ \psi_{\alpha *} \circ \lambda \\ &= \partial \circ \lambda \circ \phi_{\alpha *},\end{aligned}$$

as desired. □

The proof of Theorem (4.7*) would be (anti-) isomorphic with that of Theorem (4.7), were it not for the fact that the groups involved are direct products, rather than direct sums. We have representations

$$\begin{aligned}
v^{\alpha} &: G^{p+q}(X_p, X_{p-1}) \to G^{p+q}(T_\alpha, \dot{T}_\alpha),\\
j^{\alpha} &: G^{p+q}(X_p, Q_\alpha^*) \to G^{p+q}(X_p, X_{p-1}),\\
v'^{\beta} &: G^{p+q}(X'_p, X'_{p-1}) \to G^{p+q}(T'_\beta, \dot{T}'_\beta),\\
j'^{\beta} &: G^{p+q}(X'_p, Q'^*_\beta) \to G^{p+q}(X'_p, X'_{p-1}),\\
\mu^{\alpha} &: H^p(B_p, B_{p-1}; G^q(\mathscr{F})) \to H^p(\Delta^p, \dot{\Delta}^p; G^q(F_\alpha)),\\
l^{\alpha} &: H^p(B_p, E_\alpha^*; G^q(\mathscr{F})) \to H^p(B_p, B_{p-1}; G^q(\mathscr{F})),\\
\mu'^{\beta} &: H^p(B'_p, B'_{p-1}; G^q(\mathscr{F}')) \to H^p(\Delta^p, \dot{\Delta}^p; G^q(F'_\beta)),\\
l'^{\beta} &: H^p(B'_p, E'^*_\beta; G^q(\mathscr{F}')) \to H^p(B'_p, B'_{p-1}; G^q(\mathscr{F}')).
\end{aligned}$$

However, the homomorphism $g^\#$ is not determined by the homomorphisms

$$g^{\alpha\beta} = \mu'^{\beta} \circ g^\# \circ l^{\alpha},$$

nor is $h^\#$ determined by the

$$h^{\alpha\beta} = v'^{\beta} \circ h^\# \circ j^{\alpha}.$$

Let us fix a cell E'_β of (B', A'). Let $J_\beta = \{\alpha \mid E_\alpha \subset g(E'_\beta)\}$, and observe that the set J_β is finite. Let

$$B_p^* = B_{p-1} \cup \bigcup_{\alpha \in J_\beta} E_\alpha, \qquad B_p^\dagger = B_{p-1} \cup \bigcup_{\alpha \notin J_\beta} E_\alpha;$$

then

$$B_p = B_p^* \cup B_p^\dagger, \qquad B_{p-1} = B_p^* \cap B_p^\dagger.$$

Similarly, let $X_p^* = f^{-1}(B_p^*)$, $X_p^\dagger = f^{-1}(B_p^\dagger)$, so that

$$X_p = X_p^* \cup X_p^\dagger, \qquad X_{p-1} = X_p^* \cap X_p^\dagger.$$

(5.7) *The injections*

$$\begin{aligned}
\mu^* &: H^p(B_p, B_{p-1}; G^q(\mathscr{F})) \to H^p(B_p^*, B_{p-1}; G^q(\mathscr{F}^*)),\\
\mu^\dagger &: H^p(B_p, B_{p-1}; G^q(\mathscr{F})) \to H^p(B_p^\dagger, B_{p-1}; G^q(\mathscr{F}^\dagger))
\end{aligned}$$

represent the group $H^p(B_p, B_{p-1}; G^q(\mathscr{F}))$ as a direct product. The injections

$$l^* : H^p(B_p, B_p^\dagger; G^q(\mathscr{F})) \to H^p(B_p, B_{p-1}; G^q(\mathscr{F})),$$
$$l^\dagger : H^p(B_p, B_p^*; G^q(\mathscr{F})) \to H^p(B_p, B_{p-1}; G^q(\mathscr{F}))$$

represent the same group as a direct sum. The two representations are weakly dual. □

(5.8) *The injections*

$$v^* : G^{p+q}(X_p, X_{p-1}) \to G^{p+q}(X_p^*, X_{p-1}),$$
$$v^\dagger : G^{p+q}(X_p, X_{p-1}) \to G^{p+q}(X_p^\dagger, X_{p-1})$$

represent the group $G^{p+q}(X_p, X_{p-1})$ as a direct product. The injections

$$j^* : G^{p+q}(X_p, X_p^\dagger) \to G^{p+q}(X_p, X_{p-1}),$$
$$j^\dagger : G^{p+q}(X_p, X_p^*) \to G^{p+q}(X_p, X_{p-1})$$

represent the same group as a direct sum. The two representations are weakly dual. □

(5.3*) *The homomorphisms*

$$\mu'^\beta \circ g^\# \circ l^\dagger : H^p(B_p, B_p^*; G^q(\mathscr{F})) \to H^p(\Delta^p, \dot{\Delta}^p; G^q(F')),$$
$$v'^\beta \circ h^\# \circ j^\dagger : G^{p+q}(X_p, X_p^*) \to G^{p+q}(T'_\beta, \dot{T}'_\beta)$$

are zero.

For each $\alpha \notin J_\beta$, choose an interior point x_α of E_α which does not belong to $g(E_\beta)$. Let C be the totality of such points; then B_p^* is a deformation retract of $B_p - C$ and therefore the map $g \circ \phi'_\beta : (\Delta^p, \dot{\Delta}^p) \to (B_p - C, B_p^*)$ is compressible, and the first statement follows. A similar argument proves the second. □

For each $\alpha \in J$, let

$$\bar{l}^\alpha : H^p(B_p, E_\alpha^*; G^q(\mathscr{F})) \to H^p(B_p, B_p^\dagger; G^q(\mathscr{F}))$$

be the injection, so that $l^* \circ \bar{l}^\alpha = l^\alpha$. Similarly, let

$$\bar{j}^\alpha : G^{p+q}(X_p, Q_\alpha^*) \to G^{p+q}(X_p, X_p^\dagger)$$

be the injection, so that $j^* \circ \bar{j}^\alpha = j^\alpha$. The $\bar{l}^\alpha, \bar{j}^\alpha$ represent the relevant groups as finite direct sums. The homomorphism $g^\#$ is determined by the homomorphisms $\mu'^\beta \circ g^\#$ for every β. For each such β, the homomorphism $\mu'^\beta \circ g^\#$ is determined by the $g^{\alpha\beta}$ with $\alpha \in J_\beta$. Similarly, the homomorphism $h^\#$ is determined by the $h^{\alpha\beta}$ for each β and each $\alpha \in J_\beta$. Therefore it suffices to prove the commutativity of the diagram

$$
\text{(5.2*)}\qquad
\begin{array}{ccc}
G^{p+q}(X_p, Q_\alpha^*) & \xrightarrow{h^{\alpha\beta}} & G^{p+q}(T'_\beta, \dot{T}'_\beta) \\
\downarrow \lambda^* & & \downarrow \lambda^* \\
H^p(B_p, E_\alpha^*; G^q(\mathscr{F})) & \xrightarrow[g^{\alpha\beta}]{} & H^p(\Delta^p, \dot{\Delta}^p; G^q(F'_\beta))
\end{array}
$$

for all such pairs. And this we cheerfully leave to the reader. □

We conclude with one remark about the sign in Theorem (4.8*). The presence of this sign is due to our convention about the definition of the coboundary operator, and the fact that the proof of Theorem (4.8*) is carried out by calculating incidence numbers. *Verbum sap*!

6 The Atiyah–Hirzebruch Spectral Sequence

Let $\mathfrak{G}$ be a complete homology theory, and let $G_q = G_q(P)$, where P is a space consisting of just one point. The spectral sequence measures the extent to which the groups $G_n(B, A)$ are determined by the integral homology groups $H_p(B, A)$ and by the coefficient groups G_q.

(6.1) Theorem *Suppose that $G_q = 0$ for all $q < 0$. Then, for any pair (B, A), $G_q(B, A) = 0$ for all $q < 0$.*

In this and other arguments involving spectral sequences, the argument is best understood by a diagram in which each lattice point (p, q) is associated with the group $E^2_{p,q}$ (see Figure 13.3). The boundary operator d^2 is visualized by an arrow pointing from the point (p, q) to the point $(p - 2, q + 1)$. The group E^r for $r > 2$ may be visualized on a separate diagram; however, it is usually more enlightening to visualize $E^r_{p,q}$ as a subquotient of $E^2_{p,q}$ and d^r as an additive relation, visualized by an arrow from the point (p, q) to the point $(p - r, q + r - 1)$. Notice that each arrow runs from a point on the line $p + q = n + 1$, say, to a point on the parallel line $p + q = n$.

The groups $E^2_{p,q}$ with $p + q = n$ may be thought of as the building blocks out of which the groups $G_n(B, A)$ are to be constructed. For example, if it should happen that $E^2_{p,q} = 0$ for all p, q with $p + q = n$, then $G_n(B, A) = 0$.

In all cases with which we are concerned here, the diagram is confined to the right half-plane (because $H_p(B, A; G) = 0$ for any G if $p < 0$). The hypothesis of Theorem (6.1) ensures that the diagram is confined to the first quadrant, and this also happens in other situations of interest to us.

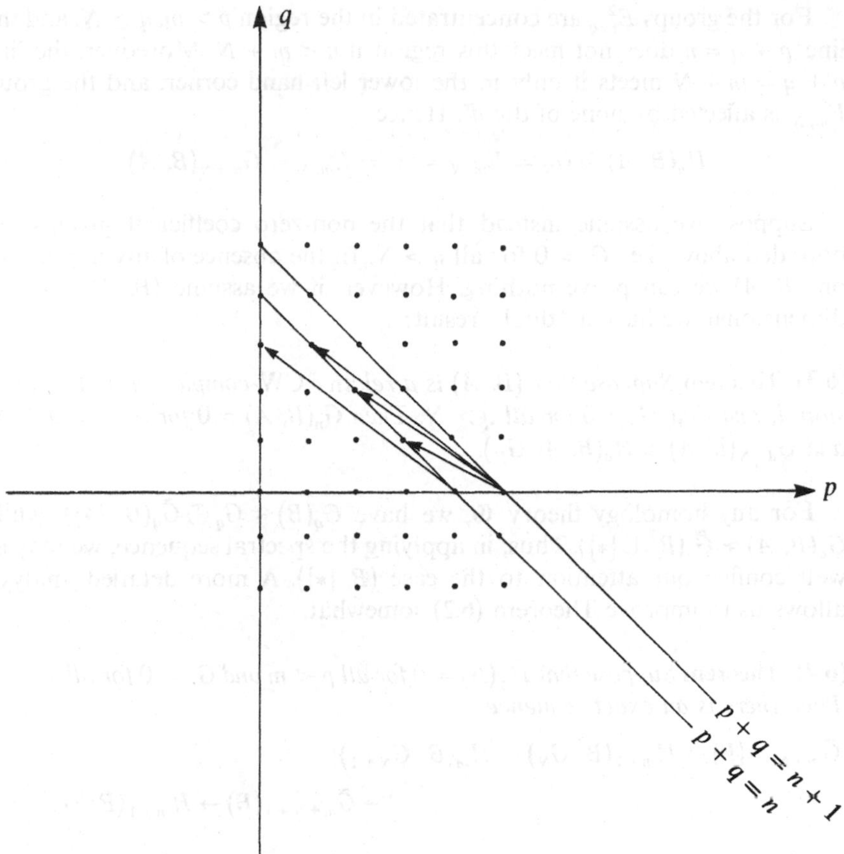

Figure 13.3

To prove Theorem (6.1), just observe that the line $p + q = n$ does not intersect the first quadrant when $n < 0$. Thus the remark at the end of the last paragraph but one applies. □

We can improve this somewhat. The blanket hypotheses we have made ensure that $H_0(B, A) = 0$ (except for the case $A = \varnothing$), and therefore the groups lying on the vertical axis vanish. If we assume that the integral homology groups $H_p(B, A)$ vanish for all $p < m$, then the same is true for the groups $E^2_{p,q}$. In particular, this will be the case if (B, A) is $(m - 1)$-connected.

(6.2) Theorem *Suppose that* $H_p(B, A) = 0$ *for all* $p < m$ *and that* $G_q = 0$ *for all* $q < N$. *Then* $G_n(B, A) = 0$ *for all* $n < m + N$, *and* $G_{m+N}(B, A) \approx H_m(B, A) \otimes G_N$.

For the groups $E^2_{p,q}$ are concentrated in the region $p \geq m$, $q \geq N$, and the line $p + q = n$ does not meet this region if $n < m + N$. Moreover, the line $p + q = m + N$ meets it only in the lower left hand corner, and the group $E^2_{m,N}$ is affected by none of the d^r. Hence

$$H_n(B, A) \otimes G_N = E^2_{m,N} = \cdots = E^\infty_{m,N} = G_{m+N}(B, A). \qquad \square$$

Suppose we assume instead that the non-zero coefficient groups are bounded above, i.e., $G_q = 0$ for all $q > N$. In the absence of any hypothesis on (B, A) we can prove nothing. However, if we assume (B, A) is finite-dimensional, we have a "dual" result:

(6.3) Theorem *Suppose that (B, A) is a relative* CW-*complex of finite dimension d, and that $G_q = 0$ for all $q > N$. Then $G_n(B, A) = 0$ for all $n > d + N$, and $G_{d+N}(B, A) \approx H_d(B, A; G_N)$.* $\square$

For any homology theory $\mathfrak{G}$, we have $G_q(B) \approx G_q \oplus \tilde{G}_q(B, \{*\})$, while $G_q(B, A) \approx \tilde{G}_q(B/A, \{*\})$. Thus, in applying the spectral sequence, we may as well confine our attention to the case $(B, \{*\})$. A more detailed analysis allows us to improve Theorem (6.2) somewhat.

(6.4) Theorem *Suppose that $\tilde{H}_p(B) = 0$ for all $p < m$ and $G_q = 0$ for all $q < N$. Then there is an exact sequence*

$$\tilde{G}_{m+N+2}(B) \to H_{m+2}(B; G_N) \to H_m(B; G_{N+1})$$
$$\to \tilde{G}_{m+N+1}(B) \to H_{m+1}(B; G_N) \to 0.$$

This is proved by examining Figure 13.4 (we may as well assume $N = 0$, as this can always be achieved by reindexing the theory $\mathfrak{G}$). The lines $p + q = m$, $m + 1$, $m + 2$ contain 1, 2, 3 lattice points, respectively, and the only relevant boundary operator is $d^2 : E^2_{m+2,0} \to E^2_{m,1}$. We have

$$\tilde{G}_{m+1}(B) = J_{m+1,0} \supset J_{m,1},$$

and

$$J_{m,1} = E^\infty_{m,1} = \text{Cok } d^2 : H_{m+2}(B; G_0) \to H_m(B; G_1),$$

while

$$J_{m+1,0}/J_{m,1} = E^\infty_{m+1,0} = E^2_{m+1,0} = H_{m+1}(B; G_0),$$

so that we have the exact sequence

$$0 \to \text{Cok } d^2 \to \tilde{G}_{m+1}(B) \to H_{m+1}(B; G_0) \to 0.$$

On the other hand, from

$$\tilde{G}_{m+2}(B) = J_{m+2,0} \supset J_{m+1,1} \supset \cdots$$

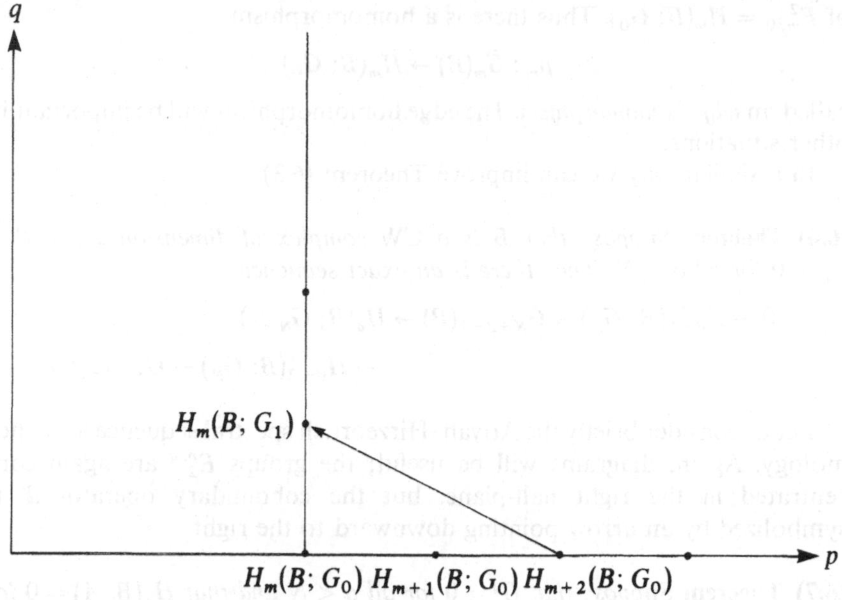

Figure 13.4

we have the short exact sequence

$$0 \to J_{m+1,1} \to \tilde{G}_{m+2}(B) \to E^{\infty}_{m+2,0} = \operatorname{Ker} d^2 \to 0.$$

Combining these two sequences with

$$0 \to \operatorname{Ker} d^2 \to H_{m+2}(B; G_0) \to H_m(B; G_1) \to \operatorname{Cok} d^2 \to 0,$$

we obtain the one desired. □

As stable homotopy satisfies the above condition and $\sigma_0 = \mathbf{Z}$, $\sigma_1 = \mathbf{Z}_2$, we have

(6.5) Corollary *Let B be* $(n-1)$*-connected. Then* $\sigma_n(B) \approx H_n(B)$ *and there is a short exact sequence*

$$\sigma_{n+2}(B) \to H_{n+2}(B) \to H_n(B) \otimes \mathbf{Z}_2 \to \sigma_{n+1}(B) \to H_{n+1}(B) \to 0. \quad \square$$

Of course, this is the conclusion of Theorem (3.12) of Chapter XII, except for the identification of the homomorphisms.

In general, if $G_q = 0$ for $q < 0$, the line $q = 0$ is the lower edge of the diagram. The group $E^{\infty}_{m,0}$ is the quotient $\tilde{G}_m(B)/J_{m-1,1}$; on the other hand, it is the intersection of the chain of subgroups

$$E^2_{m,0} \supset E^3_{m,0} \supset \cdots$$

of $E^2_{m,0} = H_m(B; G_0)$. Thus there is a homomorphism

$$\rho_m : \tilde{G}_m(B) \to H_m(B; G_0)$$

called an *edge homomorphism.* The edge homomorphism will be important in other situations.

In a similar way we can improve Theorem (6.3).

(6.6) Theorem *Suppose that B is a* CW*-complex of dimension d and that $G_q = 0$ for all $q > N$. Then there is an exact sequence*

$$0 \to H_{d-1}(B; G_N) \to \tilde{G}_{N+d-1}(B) \to H_d(B; G_{N-1})$$
$$\to H_{d-2}(B; G_N) \to \tilde{G}_{N+d-2}(B). \quad \square$$

Let us consider briefly the Atiyah–Hirzebruch spectral sequence for cohomology. Again, diagrams will be useful; the groups $E_2^{p,q}$ are again concentrated in the right half-plane, but the coboundary operator d^r is symbolized by an arrow pointing downward to the right.

(6.7) Theorem *Suppose that $G^q = 0$ for all $q < N$ and that $H_p(B, A) = 0$ for all $p < m$. Then*

(1) *$G^n(B, A) = 0$ for all $n < m + N$;*
(2) *$G^{m+N}(B, A) \approx H^m(B, A; G_N)$;*
(3) *there is an exact sequence*

$$0 \to H^{m+1}(B, A; G^N) \to G^{m+N+1}(B, A) \to H^m(B, A; G^{N+1}) \to$$
$$H^{m+2}(B, A; G^N) \to G^{m+N+2}(B, A). \quad \square$$

(6.8) Theorem *Suppose that $G^q = 0$ for all $q > N$ and that B is a* CW*-complex of dimension d. Then*

(1) *$\tilde{G}^n(B) = 0$ for all $n > d + N$;*
(2) *$\tilde{G}^{d+N}(B) \approx H^d(B; G^N)$;*
(3) *there is an exact sequence*

$$\tilde{G}^{d+N-2}(B) \to H^{d-2}(B; G^N) \to H^d(B; G^{N-1}) \to$$
$$\tilde{G}^{d+N-1}(B) \to H^{d-1}(B; G^N) \to 0. \quad \square$$

In theorem (6.8) we may take $\mathfrak{G}^*$ to be stable cohomotopy; note that $G^q = \sigma^q = \sigma_{-q} = 0$ for all $q > 0$. Thus

(6.9) Corollary *If B is a* CW*-complex of dimension d, then*

(1) *$\sigma^n(B) = 0$ for all $n > d$;*
(2) *$\sigma^d(B) \approx H^d(B; \mathbf{Z})$;*
(3) *there is an exact sequence*

$$\sigma^{d-2}(B) \to H^{d-2}(B; \mathbf{Z}) \to H^d(B; \mathbf{Z}_2) \to \sigma^{d-1}(B) \to H^{d-1}(B; \mathbf{Z}) \to 0. \quad \square$$

Part (2) of the conclusion is just the Hopf Classification Theorem (Corollary (6.19) of Chapter V) for the special case $Y = \mathbf{S}^d$. For $\sigma^d(B) = \{B, \mathbf{S}^d\} \approx [B, \mathbf{S}^d]$ provided that $d \geq 1$. Conclusion (3) gives, in a sense, the homotopy classification of maps of the d-dimensional complex B into $\mathbf{S}^{d-1}$; this time $\sigma^{d-1}(B) \approx [B, \mathbf{S}^{d-1}]$ provided that $d > 3$ (cf. Theorem (1.7) of Chapter XII).

7 The Leray–Serre Spectral Sequence

We return now to the case of a general fibration, but assume that the homology theory $\mathfrak{G}$ is ordinary homology theory $\mathfrak{H}(G)$ with coefficients in an abelian group G. In this case we have a *first quadrant* spectral sequence, since $E^2_{p,q} = H_p(B, A; H_q(\mathscr{F})) = 0$ unless $p \geq 0$ and $q \geq 0$. There are edge homomorphisms associated with both the lower and the left-hand edge.

We first discuss the former. In the first place, the local coefficient system $H_0(\mathscr{F}; G)$ is always simple. Thus $E^2_{p,0} = H_p(B, A; H_0(F; G)) \approx H_p(B, A; G)$. Moreover,

$$E^3_{p,0} = \operatorname{Ker} d^2 : E^2_{p,0} \to E^2_{p-2,1},\ E^4_{p,0} = \operatorname{Ker} d^3 : E^3_{p,0} \to E^3_{p-3,2}, \ldots.$$

Thus

$$H_p(B, A; G) = E^2_{p,0} \supset \cdots \supset E^r_{p,0} \supset E^{r+1}_{p,0} \supset \cdots \supset E^\infty_{p,0}$$

(in fact, the sequence eventually becomes stationary, since $d^r E^r_{p,0} = 0$ if $r > p$).

On the other hand,

$$H_p(X, W; G) = J_{p,0} \supset J_{p-1,1} \supset \cdots$$

and $J_{p,0}/J_{p-1,1} = E^\infty_{p,0}$. The composite

(7.1) $$H_p(X, W; G) \xrightarrow[\alpha_p]{} E^\infty_{p,0} \xrightarrow[\kappa_p]{} H_p(B, A; G)$$

is the edge homomorphism.

(7.2) Theorem *The edge homomorphism $\kappa_p \circ \alpha_p$ is the homomorphism $H_p(f): H_p(X, W; G) \to H_p(B, A; G)$ induced by the fibre map f.*

Let us consider the identity map of B as a fibration with a point P as fibre. Then the pair of maps $(1, f)$ is a fibre preserving map

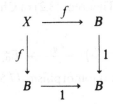

$$\begin{array}{ccc} X & \xrightarrow{f} & B \\ {\scriptstyle f}\downarrow & & \downarrow{\scriptstyle 1} \\ B & \xrightarrow[1]{} & B \end{array}$$

and therefore induces a map of exact couples. Thus we have a commutative diagram

$$\begin{array}{ccccc} H_p(B, A; G) = E^2_{p,0}(f) & \longleftarrow & E^\infty_{p,0}(f) & \twoheadleftarrow & H_p(X, W; G) \\ {\scriptstyle 1}\downarrow \quad\quad \downarrow & & \downarrow & & \downarrow {\scriptstyle f_*} \\ H_p(B, A; G) = E^2_{p,0}(1) & \longleftarrow & E^\infty_{p,0}(1) & \twoheadleftarrow & H_p(B, A; G) \end{array}$$

But $H_q(F; G) = 0$ for all G and therefore the second spectral sequence collapses and the edge homomorphism reduces to the identity map. □

(7.3) Corollary *The subgroup $J_{p-1,1}$ of $H_p(X, W; G)$ is the kernel of $f_* : H_p(X, W; G) \to H_p(B, A; G)$, and the subgroup $E^\infty_{p,0}$ of $H_p(B, A; G)$ is the image of f_*.* □

In particular, we may have $A = \{b_0\}$, $W = F$. The fibration $f: X \to B$ can also be regarded as a fibration $f_1 : (X, F) \to (B, \{b_0\})$; the identity map of X and the inclusion $B \subsetneq (B, \{b_0\})$ define a map of the first fibration into the second. Moreover, $E^2_{p,q}(f) = E^2_{p,q}(f_1)$ except that $E^2_{0,q}(f_1) = 0$. It follows that $E^r_{p,0}(f) = E^r_{p,0}(f_1)$ for all $r \le p$, while $E^\infty_{p,0}(f_1) = E^p_{p,0}(f_1) = E^p_{p,0}(f)$. Thus we have

(7.4) *The subgroup $E^p_{p,0}$ of $H_p(B; G)$ is the image of*

$$f_* : H_p(X, F; G) \to H_p(B; G) = H_p(B, b_0; G). \quad □$$

The other edge homomorphism is defined only in the absolute case $A = \varnothing$. In this case $d^r E^r_{0,q} = 0$ and therefore there are epimorphisms

$$E^2_{0,q} \twoheadrightarrow E^3_{0,q} \twoheadrightarrow \cdots \twoheadrightarrow E^\infty_{0,q}$$

(these, too, are eventually stationary since $d^r E^r_{r,q-r+1} = 0$ if $r > q + 1$), as well as an inclusion

(7.5) $$E^\infty_{0,q} = J_{0,q} \subsetneq H_q(X; G).$$

We have assumed that B is 0-connected, so that

$$E^2_{0,q} = H_0(B; H_q(\mathscr{F}; G))$$

is a quotient of $H_q(F; G)$, by Theorem (3.2) of Chapter VI. Thus we have an epimorphism

$$H_q(F; G) \xrightarrow{\lambda_q} E^\infty_{0,q}$$

whose composite with the monomorphism (7.5) is again called an edge homomorphism.

(7.6) Theorem *The edge homomorphism*

$$H_q(F; G) \to H_q(X; G)$$

is the injection.

Let us consider the restriction of f to $F = p^{-1}(\{b_0\})$ as a fibration f_0; then the inclusion map $F \subsetneq X$, $\{b_0\} \subsetneq B$ form a fibre preserving map of f_0 into f:

$$\begin{array}{ccc} F & \hookrightarrow & X \\ \downarrow{\scriptstyle f_0} & & \downarrow{\scriptstyle f} \\ \{b_0\} & \hookrightarrow & B \end{array}$$

and therefore induce a map of $\mathscr{C}(f_0)$ into $\mathscr{C}(f)$. Therefore there is a commutative diagram

$$\begin{array}{ccccc} H_q(F; G) = E^2_{0,q}(f_0) & \longrightarrow & E^\infty_{0,q}(f_0) & \longrightarrow & H_q(F; G) \\ {\scriptstyle 1}\downarrow \quad\quad \downarrow & & \downarrow & & \downarrow{\scriptstyle i_*} \\ H_q(F; G) \twoheadrightarrow E^2_{0,q}(f) & \longrightarrow & E^\infty_{0,q}(f) & \longrightarrow & H_q(X; G) \end{array}$$

But the spectral sequence for f_0 collapses, and the homomorphisms in the top line are identity maps. □

(7.7) Corollary *The subgroup* $E^\infty_{0,q} = J_{0,q}$ *of* $H_q(X; G)$ *is* $\operatorname{Im}\{i_* : H_q(F; G) \to H_q(X; G)\} = \operatorname{Ker}\{j_* : H_q(X; G) \to H_q(X, F; G)\}$. □

We have seen that there are epimorphisms $\alpha_n : H_n(X, F; G) \to E^n_{n,0}$, $\lambda_{n-1} : H_{n-1}(F; G) \to E^n_{0,n-1}$. The image groups are related by the operator d^n.

(7.8) Theorem *The diagram*

$$\begin{array}{ccc} H_n(X, F; G) & \xrightarrow{\partial_*} & H_{n-1}(F; G) \\ \downarrow{\scriptstyle \alpha_n} & & \downarrow{\scriptstyle \lambda_{n-1}} \\ E^n_{n,0} & \xrightarrow[d_n]{} & E^n_{0,n-1} \end{array}$$

is commutative.

We may assume that B has only one vertex b_0, so that $X_0 = F$ and $H_{n-1}(F; G) = H_{n-1}(X_0; G) = H_{n-1}(X_0, X_{-1}; G) = E^1_{0,n-1}$. Consider the commutative diagram

$$\begin{array}{ccccccc}
H_n(X, F; G) & & \xrightarrow{\quad\partial_*\quad} & & H_{n-1}(F; G) & & \\
\Big\uparrow{\scriptstyle i_2} & & & & \Big\downarrow{\scriptstyle i_1} & & \\
H_n(X_n, F; G) & \xrightarrow[j_1]{} & H_n(X_n, X_{n-1}; G) & \xrightarrow[\partial_1]{} & H_{n-1}(X_{n-1}; G) & \xrightarrow[j_2]{} & H_{n-1}(X_{n-1}, F; G)
\end{array}$$

The homomorphism i_2 is an epimorphism; let $z \in H_n(X, F; G)$ and choose z' such that $i_2 z' = z$. Then

$$i_1 \partial_* z = i_1 \partial_* i_2 z' = \partial_1 j_1 z'.$$

But z' represents an element $\zeta \in E^n_{n,0}$ and $\partial_* z$ represents an element $\xi \in E^n_{0,n-1}$ such that $d^n \zeta = \xi$. But $\alpha_n(z) = \zeta$ and $\xi = \lambda_{n-1} \partial_* z$. □

In §3 of Chapter XI we defined an additive relation, the transgression; it is the composite

$$H_n(B; G) \overset{f_*^{-1}}{\rightsquigarrow} H_n(X, F; G) \xrightarrow{\partial_*} H_{n-1}(F; G)$$

The corresponding homomorphism

$$\tau_* : \operatorname{Im} f_* \to H_{n-1}(F; G)/\partial_* \operatorname{Ker} f_*$$

is also called the transgression.

(7.9) Theorem *The homomorphism*

$$d^r : E^r_{r,0} \to E^r_{0,r-1}$$

is the transgression.

To prove this, chase the commutative diagram

$$\begin{array}{ccccc}
H_r(X; G) & \xrightarrow{j_*} & H_r(X, F; G) & \xrightarrow{\partial_*} & H_{r-1}(F; G) \\
\Big\downarrow{\scriptstyle \alpha_{r+1}} & & \Big\downarrow{\scriptstyle \alpha'_r} & & \Big\downarrow{\scriptstyle \lambda_{r-1}} \\
E^\infty_{r,0} = E^{r+1}_{r,0} & \longrightarrow & E^r_{r,0} & \xrightarrow{d^r} & E^r_{0,r-1} \\
& \searrow & \Big\downarrow & & \Big\downarrow \\
& & H_r(B; G) & & H_{r-1}(F; G)
\end{array}$$

and use (7.2), (7.4), (7.6), and (7.8).

The initial term of the Leray–Serre spectral sequence involves homology groups of the base with local coefficients in the homology groups of the fibre. As homology groups with local coefficients are relatively unfamiliar and accordingly difficult to work with, it is frequently necessary to assume that the local coefficient systems are simple. When this is so, we shall describe the fibration as *orientable*. More precisely, if $\mathfrak{G}$ is a homology theory, a fibration $f: X \to B$ is said to be $\mathfrak{G}$ *orientable* if and only if, for each q, the local coefficient system $G_q(\mathscr{F})$ is simple. If $\mathfrak{G} = \mathfrak{H}(G)$ is ordinary homology with coefficients in the abelian group G, we say that f is *G-orientable* if and only if it is $\mathfrak{H}(G)$-orientable.

As an application, let us derive the Serre exact sequence (6.3), Chapter VII, under less stringent hypotheses.

(7.10) Theorem *Let $f: X \to B$ be a G-orientable fibration, and assume that $H_i(B; \mathbf{Z}) = 0$ for $0 < i < m$, $H_j(F; G) = 0$ for $0 < i < n$. Then there is an exact sequence*

$$H_{m+n-1}(F; G) \to H_{m+n-1}(X; G) \to H_{m+n-1}(B; G) \to \cdots$$

$$\to H_{r+1}(B; G) \xrightarrow{\tau_*} H_r(F; G) \xrightarrow{i_*} H_r(X; G) \xrightarrow{f_*} H_r(B; G) \to \cdots.$$

Let us examine Figure 13.5. By the Universal Coefficient Theorem, $E^2_{p,q} = H_p(B; H_q(F; G)) \approx H_p(B; \mathbf{Z}) \otimes H_q(F; G) \oplus \operatorname{Tor}(H_{p-1}(B; \mathbf{Z}), H_q(F; G)) = 0$ if $0 < p < m$ or $0 < q < n$. Thus each line $p + q = r < m + n$ contains at most two non-zero groups $E^2_{r,0}$ and $E^2_{0,r}$. Hence the only possible non-zero boundary operator is $d^r: E^r_{r,0} \to E^r_{0,r-1}$. Thus we have exact sequences

$$0 \longrightarrow E^\infty_{r,0} \longrightarrow E^r_{r,0} \xrightarrow{d^r} E^r_{0,r-1} \longrightarrow E^\infty_{0,r-1}$$

and

$$0 \longrightarrow E^\infty_{0,r} \longrightarrow H_r(X; F) \longrightarrow E^\infty_{r,0}.$$

Combining these, we obtain the Serre sequence. The homomorphisms of the sequence are identified as τ_*, i_*, f_* with the aid of Theorems (7.9), (7.6), and (7.2), respectively. □

In a similar way we can derive the homology versions of the Gysin and Wang sequences. These are relegated to the Exercises.

We shall merely state the dual versions of the above results for cohomology.

(7.2*) Theorem *If $f: (X, W) \to (B, A)$ is a fibration, there is an epimorphism*

$$\kappa^p: H^p(B, A; G) \to E^{p,0}_\infty$$

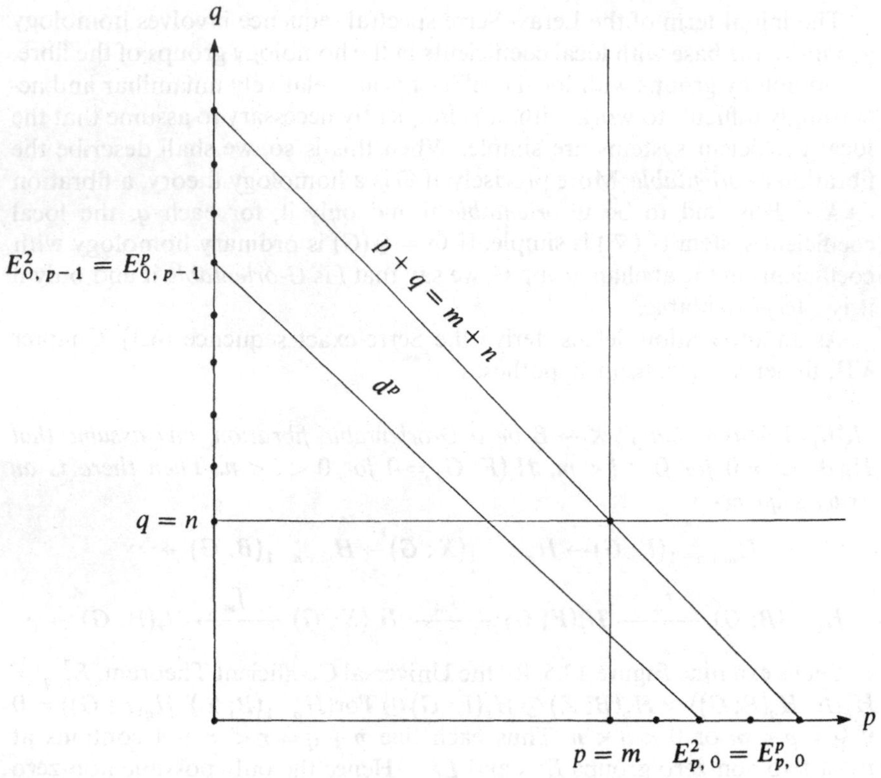

Figure 13.5

and a monomorphism

$$\alpha^p : E_\infty^{p,0} \to H^p(X, W; G)$$

whose composite is the projection

$$f^* : H^p(B, A; G) \to H^p(X, W; G). \qquad \square$$

(7.6*) Theorem *If $f: X \to B$ is a fibration, there is an epimorphism*

$$H^q(X; G) \to E_\infty^{0,q}$$

and a monomorphism

$$E_\infty^{0,q} \to H^q(F; G)$$

whose composite is the injection

$$i^* : H^q(X; G) \to H^q(F; G). \qquad \square$$

(7.9*) Theorem *The homomorphism*

$$d_r : E_r^{0,r-1} \to E_r^{r,0}$$

is the transgression τ^*. □

(7.10*) Theorem *Let $f : X \to B$ be a G-coörientable fibration, and suppose that $H_p(B) = 0$ for $0 < p < m$ and $H^q(F; G) = 0$ for $0 < q < n$. Then there is an exact sequence*

$$\cdots \to H^r(B; G) \xrightarrow{f^*} H^r(X; G) \xrightarrow{i^*} H^r(F; G) \xrightarrow{\tau^*} H^{r+1}(B; G)$$
$$\to \cdots \to H^{m+n-1}(X; G) \to H^{m+n-1}(F; G). \quad \square$$

(A fibration is G-coörientable if and only if the local coefficient system $H^q(\mathscr{F}; G)$ is simple for every q).

We now give some qualitative results which can be deduced from the spectral sequence.

(7.11) Theorem *Let $f: X \to B$ be a* **Z**-*orientable fibration, with fibre F. Then any two of the conditions*

(1) *for each p, $H_p(B)$ is finitely generated;*
(2) *for each q, $H_q(F)$ is finitely generated;*
(3) *for each n, $H_n(X)$ is finitely generated;*

imply the third.

Suppose first that (1) and (2) are satisfied. It then follows from the Universal Coefficient Theorem that $E_{p,q}^2$ is finitely generated for each p, q. Since $E_{p,q}^r$ is a subquotient of $E_{p,q}^2$, it, too, is finitely generated. Hence $E_{p,q}^\infty$ is finitely generated for all p, q.

The group $H_n(X)$ has a *finite* chain of subgroups

$$0 = J_{-1,n+1} \subset J_{0,n} \subset \cdots \subset J_{p-1,n-p+1} \subset J_{p,n-p} \subset \cdots \subset J_{n,0} = H_n(X)$$

such that the quotient groups

$$J_{p,n-p}/J_{p-1,n-p+1} \approx E_{p,n-p}^\infty$$

are finitely generated. It then follows by induction that each of the groups $J_{p,n-p}$ is finitely generated; in particular, $H_n(X)$ is.

Next, assume (2) and (3). Since B is 0-connected, the group $H_0(B) \approx \mathbf{Z}$ is finitely generated. Assume that $H_i(B)$ is finitely generated for all $i < p$. It follows from the Universal Coefficient Theorem that $E_{i,q}^2$ is finitely generated whenever $i < p$.

Now $H_p(B) = E_{p,0}^2$ has a finite chain of subgroups

$$E_{p,0}^2 \supset E_{p,0}^3 \supset \cdots \supset E_{p,0}^p \supset E_{p,0}^{p+1} = E_{p,0}^\infty;$$

the group $E^{\infty}_{p,0}$ is a quotient of $H_p(X)$ and therefore finitely generated. For each r $(2 \le r \le p)$, $E^{r+1}_{p,0}$ is the kernel of $d^r : E^r_{p,0} \to E^r_{p-r,r-1}$; thus d^r induces a monomorphism $E^r_{p,0}/E^{r+1}_{p,0} \rightarrowtail E^r_{p-r,r-1}$. The latter group being finitely generated, so is the quotient $E^r_{p,0}/E^{r+1}_{p,0}$. By downward induction on r we deduce that $E^r_{p,0}$ is finitely generated for each r; in particular, $H_p(B) = E^2_{p,0}$ is finitely generated.

The proof that (1) and (3) imply (2) involves no new ideas, and is left to the reader. □

(7.12) Theorem *Let Π be a finitely generated abelian group. Then $H_q(\Pi, n)$ is finitely generated for all q, n.*

This is proved by induction on n. The groups $H_q(\Pi, 1)$ were computed in principle in §7 of Chapter V; they turn out to be finitely generated. Assume that $H_q(\Pi, n)$ is finitely generated for all q $(n \ge 1)$. Then $\Omega K(\Pi, n+1)$ has the (weak) homotopy type of $K(\Pi, n)$, and there is a fibration

$$\Omega K(\Pi, n+1) \to \mathbf{P}K(\Pi, n+1) \to K(\Pi, n+1).$$

Since $n \ge 1$, $K(\Pi, n+1)$ is 1-connected, so that the fibration is orientable. But $\mathbf{P}K(\Pi, n+1)$ is acyclic, and *a fortiori*, its homology groups are finitely generated. By Theorem (7.11), so are those of $K(\Pi, n+1)$. □

(7.13) Theorem *Let X be a simple space and assume that $H_q(X)$ is finitely generated for each q. Let X_n be the terms of a Postnikov system for X. Then the groups $H_q(X_n)$ are finitely generated for all q, n.*

The space X_1 is an Eilenberg–Mac Lane space $K(\pi_1(X), 1)$. As $\pi_1(X)$ is abelian, it is isomorphic with the finitely generated group $H_1(X)$. By Theorem (7.12), $H_q(X_1)$ is finitely generated for all q.

Assume that $H_q(X_n)$ is finitely generated for all q $(n \ge 1)$. It follows that the relative homology groups $H_q(X_n, X)$ are finitely generated; in particular, $H_{n+2}(X_n, X)$ is. Now $\pi_{n+1}(X_n) = \pi_{n+2}(X_n) = 0$, so that $\partial : \pi_{n+2}(X_n, X) \to \pi_{n+1}(X)$ is an isomorphism. Since X is simple, $\pi_1(X)$ operates trivially on $\pi_{n+1}(X)$ and therefore on $\pi_{n+2}(X_n, X)$. By the Relative Hurewicz Theorem, $\rho : \pi_{n+2}(X_n, X) \to H_{n+2}(X_n, X)$ is an isomorphism. Thus $\pi_{n+1}(X) \approx H_{n+2}(X_n, X)$ is finitely generated.

As we have seen in (3.5) of Chapter IX, the space X_{n+1} has the weak homotopy type of the fibre W_{n+1} of a fibration $X_n \to K(\pi_{n+1}(X), n+2) = K$. By Theorem (7.12), the homology groups of K are finitely generated; by induction hypothesis, so are those of X_n. By Theorem (7.11), the homology groups of W_{n+1}, and therefore those of X_{n+1}, are finitely generated. □

(7.14) Corollary *If X is a simple space and $H_q(X)$ is finitely generated for all q, then $\pi_q(X)$ is finitely generated for all q.* □

(7.15) Theorem *Let* Λ *be a principal ideal domain,* $f\colon X \to B$ *a* Λ*-orientable fibration with fibre* F*. Then*

(1) F *is acyclic over* Λ *if and only if* $f_*\colon H_*(X;\Lambda) \to H_*(B;\Lambda)$ *is an isomorphism;*
(2) B *is acyclic over* Λ *if and only if the injection* $H_*(F;\Lambda) \to H_*(X;\Lambda)$ *is an isomorphism;*
(3) *if two of the three spaces* F, X, B *are acyclic over* Λ*, so is the third.*

Suppose that F is acyclic. By the Universal Coefficient Theorem, $E^2_{p,q} = 0$ for all $q > 0$. Then $E^\infty_{p,q} = 0$ for all $q > 0$ and $E^2_{p,0} = E^\infty_{p,0} = H_p(X;\Lambda)$. That f_* is an isomorphism follows from Theorem (7.2).

Suppose that F is not acyclic, so that there is an integer q such that $H_q(F;\Lambda) \neq 0$, $H_j(F;\Lambda) = 0$ for $0 < j < q$. Then $d^r E^r_{r,q-r+1} = 0$ and $d^r E^r_{q+1,0} = 0$ for all $r \leq q$. Hence $E^{q+1}_{q+1,0} = H_{q+1}(B;\Lambda)$, $E^{q+1}_{0,q} = H_q(F;\Lambda)$ and there is an exact sequence

$$H_{q+1}(X;\Lambda) \xrightarrow{f_*} H_{q+1}(B;\Lambda) \xrightarrow{d^{q+1}}$$
$$H_q(F;\Lambda) \longrightarrow H_q(X;\Lambda) \xrightarrow{f_*} H_q(B;\Lambda).$$

Since $H_q(F;\Lambda) \neq 0$, either f_* is not a monomorphism in dimension q or f_* is not an epimorphism in dimension $q+1$.

Suppose that B is acyclic. By the Universal Coefficient Theorem, $E^2_{p,q} = 0$ for all $p > 0$. Thus $E^\infty_{p,q} = 0$ for all $p > 0$ and $E^2_{0,q} = E^\infty_{0,q} = H_q(X;\Lambda)$. By Theorem (7.6), the injection is an isomorphism.

Suppose that B is not acyclic, so that there is an integer p such that $H_p(B;\Lambda) \neq 0$, but $H_i(B;\Lambda) = 0$ for $0 < i < p$. Then $d^r E^r_{p,0} = 0$ and $d^r E^r_{r,p-r-1} = 0$ for all $r < p$. Hence $E^p_{p,0} = H_p(B;\Lambda)$, $E^p_{0,p-1} = H_{p-1}(F;\Lambda)$ and there is an exact sequence

$$H_p(F;\Lambda) \xrightarrow{i_*} H_p(X;\Lambda) \longrightarrow H_p(B;\Lambda) \xrightarrow{d^p}$$
$$H_{p-1}(F;\Lambda) \xrightarrow{i_*} H_{p-1}(X;\Lambda).$$

Since $H_p(B;\Lambda) \neq 0$, either i_* is not a monomorphism in dimension $p-1$ or i_* is not an epimorphism in dimension p.

The third statement follows from the other two. □

There is no conclusion of comparable strength to be deduced from the hypothesis that X is acyclic. For example, the path space fibration $\Omega B \to \mathbf{P}B \to B$ has $\mathbf{P}B$ acyclic, but we can only conclude that $H_q(\Omega B) \approx H_{q+1}(B)$ in a range of dimensions (cf. Corollary (6.5), Chapter VII).

(7.16) Corollary *Let $f : X \to B$ be a* **Z**-*orientable fibration with fibre F. Then any two of the conditions*

(1) *for each $p > 0$, $H_p(B)$ is finite;*
(2) *for each $q > 0$, $H_q(F)$ is finite;*
(3) *for each $n > 0$, $H_n(X)$ is finite;*

imply the third.

Our hypotheses guarantee, by virtue of Theorem (7.11) that all three spaces have finitely generated homology groups. In the presence of this condition, the finiteness of the homology groups in positive dimensions is equivalent to rational acyclicity. Thus our result follows from Theorem (7.15).

(7.17) Corollary *Let Π be a finite abelian group. Then $H_q(\Pi, n)$ is finite for all $q > 0$ and all n.*

The groups $H_q(\Pi, 1)$ are finite, by the calculations of §7, Chapter V. The finiteness of $H_q(\Pi, n)$ follows, by induction on n, from Corollary (7.16).

□

(7.18) Theorem *Let X be a simple space whose integral homology groups $H_q(X)$ are finite for all $q > 0$. Let X_n be the terms of a Postnikov system for X. Then $H_q(X_n)$ is finite for all $q > 0$ and all n.*

This follows by essentially the same arguments as Theorem (7.13). □

(7.19) Corollary *If X is a simple space and $H_q(X)$ is finite for all $q > 0$, then $\pi_q(X)$ is finite for all q.* □

Examination of the above proofs reveals that relatively few properties of finitely generated and of finite groups (and these of a very general character) are involved in the above proofs. In his paper [3], Serre has formulated the notion of a "class" of groups and has found far-reaching generalizations of the above results. We shall not discuss these notions for lack of space, but hope to return to them in the second volume.

8 Multiplicative Properties of the Leray–Serre Spectral Sequence

Continuing our discussion of the Leray–Serre spectral sequence, we now assume that we are dealing with cohomology with coefficients in a commutative ring Λ. We shall first discuss the behavior of cross products; then we can study cup products with the aid of diagonal maps.

For simplicity, we shall consider the absolute case. The changes needed to handle the general case are minor, and are left to the reader.

Let $f' : X' \to B', f'' : X'' \to B''$ be fibrations with fibres F', F'', respectively. Then $f = f' \times f'' : X \to B$ is a fibration with fibre F, where

$$X = X' \times X'', \qquad B = B' \times B'', \qquad F = F' \times F''.$$

Let us first consider the behavior of the E_1 term of the spectral sequence. The cross product in cohomology is a pairing

$$\mu : H^{p+q}(X'_p, X'_{p-1}) \otimes H^{s+t}(X''_s X''_{s-1}) \to H^{p+q+s+t}(X_{p,s}, X_{p,s-1} \cup X_{p-1,s})$$

where $X_{p,s} = X'_p \times X''_s$; we also define

$$X_{\infty,s} = X' \times X''_s,$$
$$X_{p,\infty} = X'_p \times X'',$$
$$X^*_{p,s} = \bigcup_{\substack{a+b=p+s \\ a \neq p}} X_{a,b}$$

As we did with the cross product in the cohomology of a CW-complex in §2 of Chapter II, we can compose μ with the composite

$$H^{p+q+s+t}(X_{p,s}, X_{p,s-1} \cup X_{p-1,s}) \to H^{p+q+s+t}(X_{p+s}, X^*_{p,s})$$
$$\to H^{p+q+s+t}(X_{p+s}, X_{p+s-1}),$$

where the second homomorphism is an injection and the first the inverse of an isomorphic injection. In this way we obtain a pairing

$$\mu_1 : E_1^{p,q}(f') \otimes E_1^{s,t}(f'') \to E_1^{p+s,q+t}(f).$$

On the other hand, for each $b = (b', b'') \in B$, the cross product maps $H^q(F'_{b'}) \otimes H^t(F''_{b''})$ into $H^{q+t}(F_b)$. Let u' be a path in B' from b'_1 to b'_0 and u'' a path in B'' from b''_1 to b''_0, and let $u : \mathbf{I} \to B$ be the path such that

$$u(t) = (u'(t), u''(t)).$$

Let F'_i be the fibre of f' over b'_i, F''_i the fibre of f'' over b''_i; then $F_i = F'_i \times F''_i$ is the fibre of f over $b_i = (b'_i, b''_i)$. If $h' : F'_0 \to F'_1$ is u'-admissible, $h'' : F''_0 \to F''_1$ is u''-admissible, then $h = h' \times h'' : F_0 \to F_1$ is u-admissible, and the diagram

$$\begin{array}{ccc} H^q(F'_1) \otimes H^t(F''_1) & \longrightarrow & H^{q+t}(F_1) \\ \Big\downarrow {\scriptstyle h'^* \otimes h''^*} & & \Big\downarrow {\scriptstyle h^*} \\ H^q(F'_0) \otimes H^t(F''_0) & \longrightarrow & H^{q+t}(F_0) \end{array}$$

in which the horizontal arrows represent the cross product pairings, is commutative.

These pairings can be used to define cross products in the cohomology of the base spaces. Specifically, if $c' \in \Gamma^p(B'; H^q(\mathscr{F}'))$, $c'' \in \Gamma^s(B''; H^t(\mathscr{F}''))$, their cross product $c = c' \times c'' \in \Gamma^{p+s}(B; H^{q+t}(\mathscr{F}))$ is given by

$$c(e'_\alpha \times e''_\beta) = (-1)^{ps} c'(e'_\alpha) \times c''(e''_\beta).$$

The coboundary formula

$$\delta(c' \times c'') = \delta c' \times c'' + (-1)^p c' \times \delta c''$$

is readily verified. Thus the cross product of cochains induces a cross product mapping

$$\Gamma^p(B'; H^q(\mathscr{F}')) \otimes \Gamma^s(B''; H^t(\mathscr{F}'')) \to \Gamma^{p+s}(B; H^{q+t}(\mathscr{F})).$$

The isomorphisms λ of §4 can now be used to compare the two cross product pairings under discussion.

(8.1) Theorem *The diagram*

$$\begin{array}{ccc} E_1^{p,q}(f') \otimes E_1^{s,t}(f'') & \xrightarrow{\mu_1} & E_1^{p+s,q+t}(f) \\ \downarrow{\scriptstyle \lambda^*(f') \otimes \lambda^*(f'')} & & \downarrow{\scriptstyle \lambda^*(f)} \\ \Gamma^p(B'; H^q(\mathscr{F}')) \otimes \Gamma^s(B''; H^t(\mathscr{F}'')) & \longrightarrow & \Gamma^{p+s}(B; H^{q+t}(\mathscr{F})) \end{array}$$

is commutative except for the sign $(-1)^{(p+q)s}$.

Let E'_α, E''_β be cells of B', B'' of dimensions p, s, respectively; then $E_{\alpha,\beta} = E'_\alpha \times E''_\beta$ is a $(p+s)$-cell of B. It will be convenient to modify the discussion of §4 by using $\Delta^{p,s} = \Delta^p \times \Delta^s$, rather than Δ^{p+s}, as the domain of a characteristic map for $E_{\alpha,\beta}$. Consider the diagram (see Figure 13.6) where the c_i are cross product maps and $n = p+q+s+t$; in particular, $j \circ k^{-1} \circ c_1 = \mu_1$. We shall prove that the bottom square commutes except for the sign $(-1)^{qs}$. The remainder of the diagram is patently commutative. This being said, we have, for $u' \in E_1^{p,q}(f')$, $u'' \in E_1^{s,t}(f'')$,

$$(-1)^{qs} c_4(h_\alpha^* i^\alpha u'/\delta p \otimes h_\beta^* i^\beta u''/\delta_s) = h_{\alpha,\beta}^* i^{\alpha,\beta} \mu_1(u' \otimes u'')/\delta_p \times \delta_s.$$

Let

$$v' = h_\alpha^* i^\alpha(u'),$$
$$v'' = h_\beta^* i^\beta(u''),$$
$$v = h_{\alpha,\beta}^* i^{\alpha,\beta} \mu_1(u' \times u''),$$
$$\lambda' = \lambda^*(f'), \qquad \lambda'' = \lambda^*(f''), \qquad \lambda = \lambda^*(f).$$

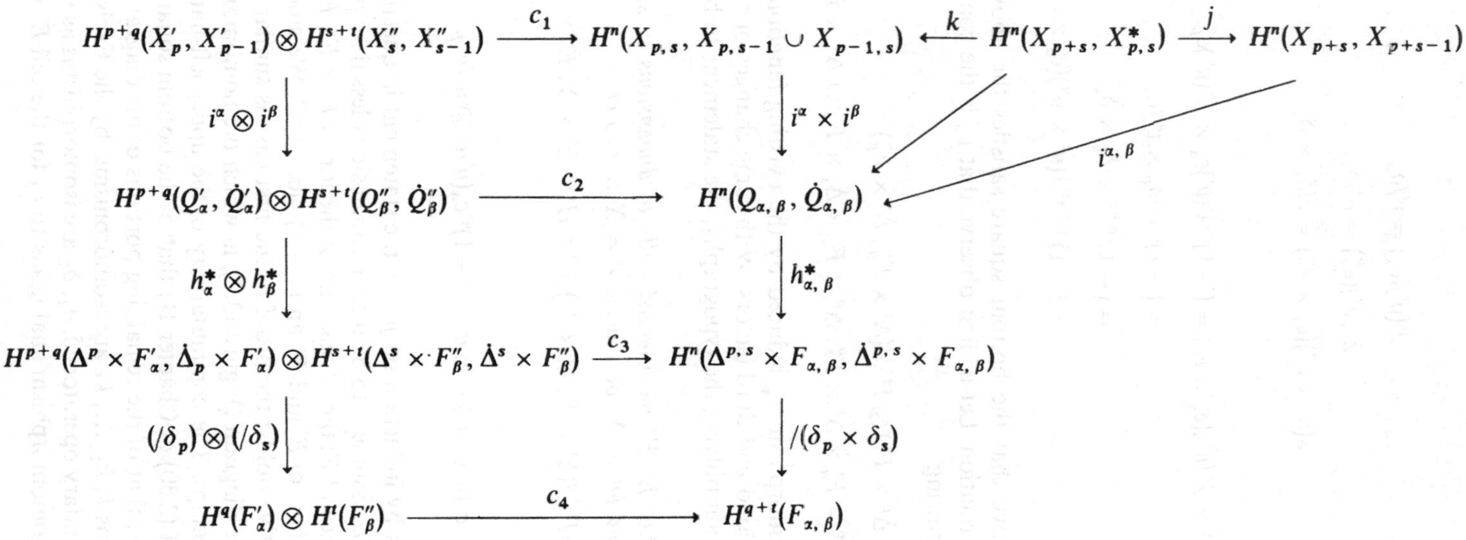
$H^{p+q}(X'_p, X'_{p-1}) \otimes H^{s+t}(X''_s, X''_{s-1})$
c_1
$H^n(X_{p,s}, X_{p,s-1} \cup X_{p-1,s})$
k
$H^n(X_{p+s}, X^*_{p,s})$
j
$H^n(X_{p+s}, X_{p+s-1})$
$i^\alpha \otimes i^\beta$
$i^\alpha \times i^\beta$
$i^{\alpha,\beta}$
$H^{p+q}(Q'_\alpha, \dot{Q}'_\alpha) \otimes H^{s+t}(Q''_\beta, \dot{Q}''_\beta)$
c_2
$H^n(Q_{\alpha,\beta}, \dot{Q}_{\alpha,\beta})$
$h^*_\alpha \otimes h^*_\beta$
$h^*_{\alpha,\beta}$
$H^{p+q}(\Delta^p \times F'_\alpha, \dot{\Delta}_p \times F'_\alpha) \otimes H^{s+t}(\Delta^s \times F''_\beta, \dot{\Delta}^s \times F''_\beta)$
c_3
$H^n(\Delta^{p,s} \times F_{\alpha,\beta}, \dot{\Delta}^{p,s} \times F_{\alpha,\beta})$
$(/\delta_p) \otimes (/\delta_s)$
$/(\delta_p \times \delta_s)$
$H^q(F'_\alpha) \otimes H^t(F''_\beta)$
c_4
$H^{q+t}(F_{\alpha,\beta})$

Figure 13.6

Then

$$\lambda'(u')(e'_\alpha) = v'/\delta_p,$$
$$\lambda''(y'')(e''_\beta) = v''/\delta_s,$$
$$\lambda(u' \times u'')(e'_\alpha \times e''_\beta) = v/\delta_p \times \delta_s$$

and therefore

$$\begin{aligned}\lambda'(u') \times \lambda''(u'')(e'_\alpha \times e''_\beta) &= (-1)^{ps}\lambda'(u')e'_\alpha \times \lambda''(u'')e''_\beta \\ &= (-1)^{ps}v'/\delta_p \times v''/\delta_s \\ &= (-1)^{(p+q)s}v/\delta_q \times \delta_s \\ &= (-1)^{(p+q)s}\lambda(u' \times u'')(e'_\alpha \times e''_\beta).\end{aligned}$$

We now prove that the bottom square satisfies the above-mentioned commutativity relation. Let us first observe that c_3 is the composite of the cross product pairing

$$\begin{aligned}&H^{p+q}(\Delta^p \times F'_\alpha, \dot{\Delta}^p \times F'_\alpha) \otimes H^{s+t}(\Delta^s \times F''_\beta, \dot{\Delta}^s \times F''_\beta) \\ &\quad \to H^n(\Delta^p \times F'_\alpha \times \Delta^s \times F''_\beta, \Delta^p \times F'_\alpha \times \dot{\Delta}_s \times F''_\beta \cup \dot{\Delta}^p \times F'_\alpha \times \Delta^s \times F''_\beta)\end{aligned}$$

with the homomorphism τ_* induced by the twisting function which interchanges the second and third factors. With some changes of notation made for the sake of simplifying the typography, the statement to be proved is

(8.2) Lemma *Let E', E'' be oriented cells of dimensions p, s, respectively, $E = E' \times E''$, and let X, Y, be spaces, $Z = X \times Y$. Let*

$$u' \in H^{p+q}(E' \times X, \dot{E}' \times X), u'' \in H^{s+t}(E'' \times Y, \dot{E}'' \times Y).$$

Then

$$c_3(u' \otimes u'')/e' \times e'' = (-1)^{qs}c_4(u'/e' \otimes u''/e'').$$

The proof is by induction on $p + s$, the statement in question being true when $p = s = 0$. Assume it to be true in dimensions less than $p + s$. Let F be a $(p + s - 1)$-face of E; then F has one of the forms $F' \times E''$, $E' \times F''$, where F' is a $(p - 1)$-face of E' and F'' an $(s - 1)$-face of E''. We shall assume the second alternative holds; the proof in the first case is similar.

Consider the diagram (Figure 13.7), in which the homomorphisms c'_i are cross product maps. The commutativity of the upper left-hand corner is a consequence of (2.30) of Chapter II; that of the bottom square is by induction hypothesis; that of the remaining portions of the diagram is clear. The homomorphisms $k, k_1, \ldots, k_5$ are isomorphisms, by the Excision Theorem. And the coboundary operators $\delta, \delta_1, \delta_2$ are isomorphisms as well. The map $\tau^* \circ c'_1$ is the homomorphism analogous to c_3 for the cell $E' \times E''$, the map

$$
\begin{array}{ccccc}
H^{p+q}(E' \times X, \dot{E}' \times X) \otimes H^{s+t}(E'' \times Y, \dot{E}'' \times Y) & \xrightarrow{c_1'} & H^n(E' \times X \times E'' \times Y, E' \times X \times \dot{E}'' \times Y \cup \dot{E}' \times X \times E'' \times Y) & \xrightarrow{\tau^*} & H^n(E \times Z, \dot{E} \times Z) \\
\Big\uparrow (-1)^{p+q}(1 \otimes \delta) & & \Big\uparrow \delta_1 & & \Big\uparrow \bar{\delta}_2 \\
 & & H^{n-1}(E' \times X \times \dot{E}'' \times Y \cup \dot{E}' \times X \times E'' \times Y, E' \times X \times F_0'' \times Y \cup \dot{E}' \times X \times E'' \times Y) & \xrightarrow{\tau^*} & H^{n-1}(\dot{E} \times Z, F_0 \times Z) \\
 & & \Big\downarrow k_3 & & \Big\downarrow k_4 \\
H^{p+q}(E' \times X, \dot{E}' \times X) \otimes H^{s+t-1}(\dot{E}'' \times Y, F_0'' \times Y) & \xrightarrow{c_2'} & H^{n-1}(E' \times X \times \dot{E}'' \times Y, E' \times X \times F_0'' \times Y \cup \dot{E}' \times X \times \dot{E}'' \times Y) & & \\
\Big\downarrow 1 \otimes k & & \Big\downarrow k_5 & & \\
H^{p+q}(E' \times X, \dot{E}' \times X) \otimes H^{s+t-1}(F'' \times Y, \dot{F}'' \times Y) & \xrightarrow{c_3'} & H^{n-1}(E' \times X \times F'' \times Y, E' \times X \times \dot{F}'' \times Y \cup \dot{E}' \times X \times F'' \times Y) & \xrightarrow{\tau^*} & H^{n-1}(F \times Z, \dot{F} \times Z) \\
\Big\downarrow (/e') \otimes (/f'') & & & & \Big\downarrow /f \\
H^q(X) \otimes H^t(Y) & & \xrightarrow{\quad (-1)^{q(s-1)} c_4 \quad} & & H^{q+t}(Z)
\end{array}
$$

Figure 13.7

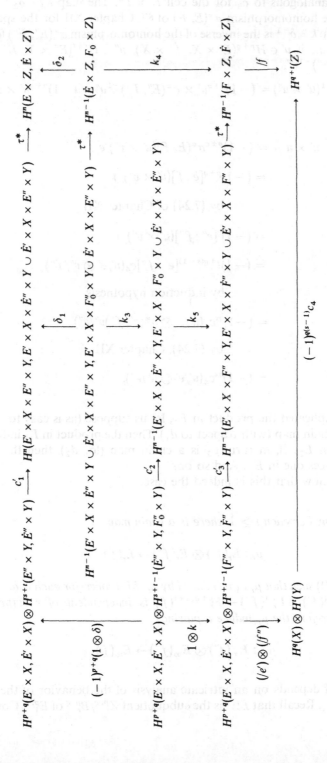

$\tau^* \circ c_3'$ that analogous to c_3 for the cell $E' \times F''$. The map $k_4 \circ \delta_2^{-1}$ is the inverse of the homomorphism $\sigma^*(E, F)$ of §7, Chapter XII for the space Z, while the map $k \circ \delta^{-1}$ is the inverse of the homomorphism $\sigma^*(E'', F'')$ for the space Y. Thus, if $u' \in H^{p+q}(E' \times X, \dot{E}' \times X)$, $u'' \in H^{s+t}(E'' \times Y, \dot{E}'' \times Y)$, $v'' = \sigma^*(E'', F'')^{-1}u''$,

$$\sigma^*(E, F)^{-1}(u' \times u'') = (-1)^{p+q}u' \times \sigma^*(E'', F'')^{-1}u'' = (-1)^{p+q}u' \times v''.$$

Hence

$$\begin{aligned} u' \times u''/e &= (-1)^{p+q}\sigma^*(E, F)(u' \times v'')/e \\ &= (-1)^{p+q}[e : f](u' \times v'')/f \\ &\qquad \text{by (7.24) of Chapter XII} \\ &= (-1)^{q}[e'' : f''](u' \times v'')/f \\ &= (-1)^{q+q(s-1)}[e'' : f'']c_4(u'/e' \otimes v''/f'') \\ &\qquad \text{by induction hypothesis} \\ &= (-1)^{qs}c_4(u'/e' \otimes \sigma^*(e'', f'')v''/e'') \\ &\qquad \text{by (7.24), Chapter XII} \\ &= (-1)^{qs}c_4(u'/e' \otimes u''/e''). \end{aligned}$$

□

Having explicated the product in E_1, let us suppose (as is easy to verify) that μ_1 is a chain map (with respect to d_1). Then the product in E_1 induces a product μ_2 in E_2. If, in turn, μ_2 is a chain map (for d_2), then the latter product induces one in E_3. And so on.

We now show that this is indeed the case.

(8.3) Theorem *For each* $r \geq 1$, *there is a chain map*

$$\mu_r : E_r(f') \otimes E_r(f'') \to E_r(f)$$

of degree (0, 0) *such that* μ_{r+1} *is induced by* μ_r. *Moreover, for each* p, q, s, t, *the map* $\mu_r : E_r^{p,q}(f') \otimes E_r^{s,t}(f'') \to E_r^{p+q,s+t}(f)$ *is independent of* r *sufficiently large and therefore the* μ_r *induce a pairing*

$$\mu_\infty : E_\infty(f') \otimes E_\infty(f'') \to E_\infty(f).$$

The proof depends on an intricate analysis of the behavior of the cross product in E_1. Recall that $E_r^{p,q}$ is the subquotient $Z_r^{p,q}/B_r^{p,q}$ of $E_1^{p,q}$. Consider the commutative diagram

$$
\begin{array}{ccccc}
H^{p+q}(X'_{p+r-1}, X'_{p-1}) & \xrightarrow{\delta'_1} & H^{p+q+1}(X', X'_{p+r-1}) & \xrightarrow{i'_2} & H^{p+q+1}(X'_{p+r}, X'_{p+r-1}) \\
\downarrow{\scriptstyle i'_1} & & \downarrow{\scriptstyle j'_1} & & \\
H^{p+q}(X'_p, X'_{p-1}) & \xrightarrow{\delta'_2} & H^{p+q+1}(X', X'_p) & & \\
& \searrow{\scriptstyle \delta'_3} & \downarrow{\scriptstyle i'_3} & & \\
& & H^{p+q+1}(X'_{p+r-1}, X'_p). & &
\end{array}
$$

An element $u' \in E_1^{p,q}(f')$ belongs to $Z_r^{p,q}$ if and only if $\delta'_2 u \in \operatorname{Im} j'_1 = \operatorname{Ker} i'_3$, i.e., $u \in \operatorname{Ker}(i'_3 \circ \delta'_2) = \operatorname{Ker} \delta'_3 = \operatorname{Im} i'_1$. If $v' \in H^{p+q}(X'_{p+r-1}, X'_{p-1})$, $i'_1(v') = u'$, and if $[u']$ is the element of E_r represented by u', then $d_r[u'] = [i'_2 \delta'_1 v']$. A comparable diagram reveals that, if $u'' \in E_1^{s,t}(f'')$ belongs to $Z_r^{s,t}$, then there is an element $v'' \in H^{s+t}(X''_{s+r-1}, X''_{s-1})$ such that $i''_1(v'') = u''$, and $d_r[u''] = [i''_2 \delta''_1 v'']$.

The reader will recall that the product μ_1 was defined by starting with the cross product $u' \times u''$ of two classes $u' \in E_1^{p,q}(f') = H^{p+q}(X'_p, X'_{p-1})$, $u'' \in E_1^{s,t}(f'') = H^{s+t}(X''_s, X''_{s-1})$. This product does not lie in the correct group; to place it there, we apply a number of injection homomorphisms and their inverses. And this is the idea underlying all the complications we encounter in proving Theorem (8.3).

Our first task is to show that, if $u' \in Z_r^{p,q}$ and $u'' \in Z_r^{s,t}$, then $u = u' \cdot u'' = \mu_1(u' \otimes u'') \in Z_r^{p+q,s+t}$. We must therefore exhibit an element $v \in H^{p+q+s+t}(X_{p+s+r-1}, X_{p+s-1})$ such that $i_1(v) = u' \cdot u''$. We may start with the element $v' \times v''$, which belongs to the group in the upper left hand corner of the commutative diagram

$$
\begin{array}{ccc}
H^n(X_{p+r-1,s+r-1}, X_{p+r-1,s-1} \cup X_{p-1,s+r-1}) & \xrightarrow{i_{1,1}} & H^n(X_{p,s}, X_{p,s-1} \cup X_{p-1,s}) \\
\uparrow{\scriptstyle k_1} & & \uparrow{\scriptstyle k_2} \\
H^n(X_{\infty,s-1} \cup X_{p+r-1,s+r-1} \cup X_{p-1,\infty}, X_{\infty,s-1} \cup X_{p-1,\infty}) & \longrightarrow & H^n(X_{p+s}, X^*_{p,s}) \\
\downarrow{\scriptstyle l_1} & & \downarrow{\scriptstyle l_2} \\
H^n(X_{p+s+r-1}, X_{p+s-1}) & \xrightarrow[i_1]{} & H^n(X_{p+s}, X_{p+s-1})
\end{array}
$$

in which the injections k_1 and k_2 are isomorphisms, $n = p + q + s + t$. The injection $i_{1,1}$ is induced by the cross product of the inclusion maps

$(X'_p, X'_{p-1}) \subseteq (X'_{p+r-1}, X'_{p-1})$, $(X''_s, X''_{s-1}) \subseteq (X''_{s+r-1}, X''_{s-1})$; by naturality of cross products, $i_{1,1}(v' \times v'') = u' \times u''$. Let $v = l_1 k_1^{-1}(v' \times v'')$; then it follows from the commutativity of the diagram that

(8.4) $$i_1(v) = u' \cdot u'' = u.$$

We must next calculate $d_r([u])$. As we are trying to show that μ_r is a chain map, we must prove that

(8.5) $$d_r([u]) = d_r([u']) \cdot [u''] + (-1)^{p+q}[u'] \cdot d_r([u'']).$$

By the remarks made at the beginning of the proof, the element $d_r([u])$ is represented by the image of v under the coboundary operator

$$\delta_3 : H^n(X_{p+s+r-1}, X_{p+s-1}) \to H^{n+1}(X_{p+s+r}, X_{p+s+r-1}).$$

There is a commutative diagram

$$\begin{array}{ccc}
H^n(X_{p+r-1,s+r-1}, X_{p+r-1,s-1} \cup X_{p-1,s+r-1}) & & \\
\uparrow k_1 & & \\
H^n(X_{\infty,s-1} \cup X_{p+r-1,s+r-1} \cup X_{p-1,\infty}, X_{\infty,s-1} \cup X_{p-1,\infty}) & & \\
 & \searrow l_1 & \\
 & & H^n(X_{p+s+r-1}, X_{p+s-1}) \\
\downarrow \delta_4 & & \downarrow \delta_3 \\
 & & H^{n+1}(X_{p+s+r}, X_{p+s+r-1}) \\
 & \nearrow l_3 & \\
H^{n+1}(X_{\infty,s+r-1} \cup X_{p+r-1,\infty}, X_{\infty,s-1} \cup X_{p+r-1,s+r-1} \cup X_{p-1,\infty}) & &
\end{array}$$

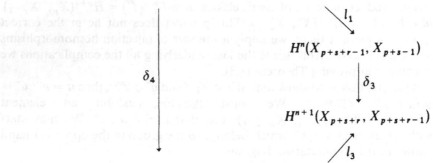

and $\delta_3(v) = \delta_3 l_1 k_1^{-1}(v' \times v'') = l_3 \delta_4 k_1^{-1}(v' \times v'')$.

Next consider the commutative diagram (Figure 13.8); the k_i are isomorphisms, and, according to our discussion in §2 of Chapter II, l_4 and l_5 represent the group

$$H^{n+1}(X_{\infty,\sigma} \cup X_{\pi,\infty}, X_{\infty,s-1} \cup X_{\pi,\sigma} \cup X_{p-1,\infty})$$

(where $\pi = p + r - 1$ and $\sigma = s + r - 1$) as a direct product, while j_2, j_3 form a weakly dual representation of the same group as a direct sum. Moreover,

$$\delta_5 k_3^{-1}(v' \times v'') = \delta'_1 v' \times v'',$$
$$\delta_6 k_5^{-1}(v' \times v'') = (-1)^{p+q} v' \times \delta''_1 v''.$$

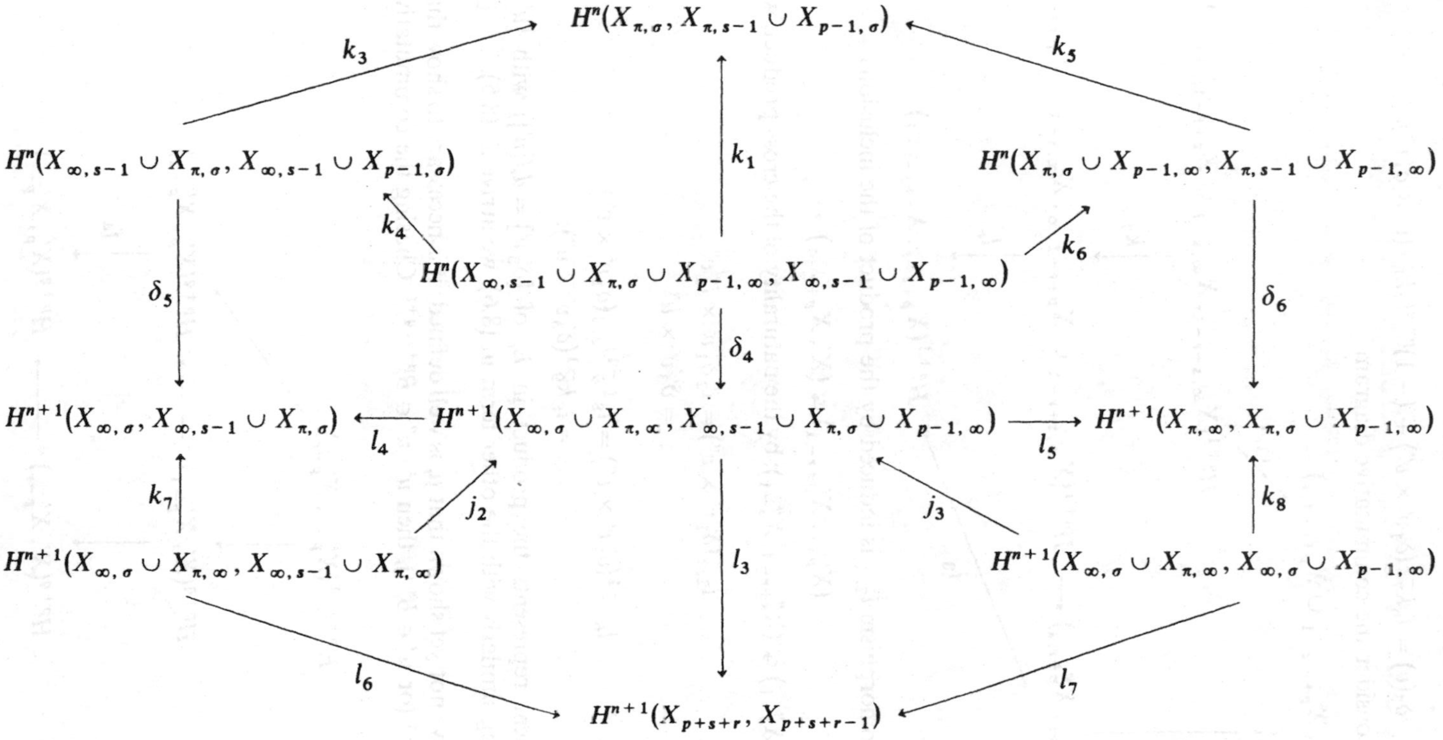

Figure 13.8

Thus

(8.6) $$\delta_3(v) = l_6 k_7^{-1}(\delta_1' v' \times v'') + (-1)^{p+q} l_7 k_8^{-1}(v' \times \delta_1'' v'').$$

Now consider the commutative diagram

$$\begin{array}{ccc}
H^{n+1}(X_{p+r,s}, X_{p+r,s-1} \cup X_{p+r-1,s}) & & \\
\uparrow k_9 & \nwarrow i_{2,1} & \\
 & & H^{n+1}(X_{\infty,s+r-1}, X_{\infty,s-1} \cup X_{p+r-1,s+r-1}) \\
 & & \downarrow k_7 \\
H^{n+1}(X_{p+s+r}, X^*_{p+r,s}) & \longleftarrow & H^{n+1}(X_{\infty,s+r-1} \cup X_{p+r-1,\infty}, X_{\infty,s-1} \cup X_{p+r-1,\infty}) \\
 & \searrow l_8 & \downarrow l_6 \\
 & & H^{n+1}(X_{p+s+r}, X_{p+s+r-1})
\end{array}$$

The homomorphism $i_{2,1}$ is induced by the product of the inclusions

$$(X'_{p+r}, X'_{p+r-1}) \subsetneq (X', X'_{p+r-1})$$

and $(X''_s, X''_{s-1}) \subsetneq (X''_{s+r-1}, X''_{s-1})$; by the naturality of the cross product, we have

$$\begin{aligned}
i_{2,1}(\delta_1' v' \times v'') &= i_2' \delta_1' v' \times i_1'' v'' \\
&= \delta_3' v' \times u''
\end{aligned}$$

Hence

$$\begin{aligned}
l_6 k_7^{-1}(\delta_1' v' \times v'') &= l_8 k_9^{-1} i_{2,1}(\delta_1' v' \times v'') \\
&= l_8 k_9^{-1}(\delta_3' v' \times u'');
\end{aligned}$$

this element represents the product in E_r of $[\delta_3' v'] = d_r([u'])$ with $[u'']$. Proceeding similarly with the other term in (8.6), we arrive at (8.5). □

We have not yet shown that μ_r is well defined; it is necessary to show that if $u' \in B_r^{p,q}$ (or $u'' \in B_r^{s,t}$), then $u' \cdot u'' \in B_r^{p+s,q+t}$. Chasing the commutative diagram

$$\begin{array}{ccc}
H^{p+q-1}(X'_{p-1}, X'_{p-r}) & & \\
\downarrow & \searrow & \\
H^{p+q}(X', X'_{p-1}) & \longrightarrow & H^{p+q}(X'_p, X'_{p-1}) \\
\downarrow & & \downarrow j_4' \\
H^{p+q}(X', X'_{p-r}) & \longrightarrow & H^{p+q}(X'_p, X'_{p-r})
\end{array}$$

we see that $B_r^{p,q}$ is the kernel of the injection j_4'. Let $u' \in B_r^{p,q}$, $u'' \in Z_r^{s,t}$. Then $u' \in \operatorname{Ker} j_4'$ and $u'' = i_1''(v'')$, $v'' \in H^{s+t}(X''_{s+r-1}, X''_{s-1})$. We shall prove that $u' \cdot u''$ belongs to the kernel of $j_4 : H^n(X_{p+s}, X_{p+s-1}) \to H^n(X_{p+s}, X_{p+s-r})$.

Once more we need a large commutative diagram (Figure 13.9). Again, the k_i are isomorphisms, the homomorphism i_2 is induced by the product of the inclusions

$$(X'_p, X'_{p-1}) \subsetneq (X'_p, X'_{p-1}), \qquad (X''_s, X''_{s-1}) \subsetneq (X''_{s+r-1}, X''_{s-1}),$$

and the homomorphism j_5 by the product of the inclusions

$$(X'_p, X'_{p-r}) \subsetneq (X'_p, X'_{p-1}), \qquad (X''_{s+r-1}, X''_{s-1}) \subsetneq (X''_{s+r-1}, X''_{s-1}).$$

The element $u' \times u''$ lies in the group in the upper right-hand corner; by naturality of the cross product, $u' \times u'' = i_2(u' \times v'')$, $j_5(u' \times v'') = 0$. Moreover, $u' \cdot u'' = l_2 k_2^{-1}(u' \times u'')$. Hence

$$\begin{aligned} j_4(u' \cdot u'') &= j_4 l_2 k_2^{-1}(u' \times u'') \\ &= j_4 l_2 k_2^{-1} i_2(u' \times v'') \\ &= j_4 l_{10} k_{11}^{-1}(u' \times v'') \\ &= l_9 k_{10}^{-1} j_5(u' \times v'') = 0. \end{aligned}$$

Similarly we show that, if $u' \in Z_r^{p,q}$, $u'' \in B_r^{s,t}$, then $u' \cdot u'' \in B_r^{p+s,q+t}$. Hence μ_r is well-defined. The remaining statements in Theorem (8.3) are obvious. □

The products we have defined have commutativity and associativity properties which we now formulate. Let $f' : X' \to B'$, $f'' : X'' \to B''$ be fibrations, and let $f = f' \times f'' : X' \times X'' \to B' \times B''$, $\check{f} = f'' \times f' : X'' \times X' \to B'' \times B'$. Let $t_X : X' \times X'' \to X'' \times X'$, $t_B : B' \times B'' \to B'' \times B'$ be the twisting maps which simply interchange the factors. Then t_X respects filtration and induces a map of the exact couple for f into that for $\check{f}$. In particular, t_X^* maps $E_r(\check{f})$ into $E_r(f)$ for every r.

(8.7) Theorem *The products of Theorem* (8.3) *are commutative in the sense that, if* $u' \in E_r^{p,q}(f')$, $u'' \in E_r^{s,t}(f'')$, *then*

$$t_X^*(u'' \cdot u') = (-1)^{(p+q)(s+t)} u' \cdot u''. \qquad \square$$

Similarly, let $f_i : X_i \to B_i \quad (i = 1, 2, 3)$, and let

$$f_{i,j} = f_i \times f_j : X_i \times X_j \to B_i \times B_j,$$

$$f = f_1 \times f_2 \times f_3 : X_1 \times X_2 \times X_3 \to B_1 \times B_2 \times B_3.$$

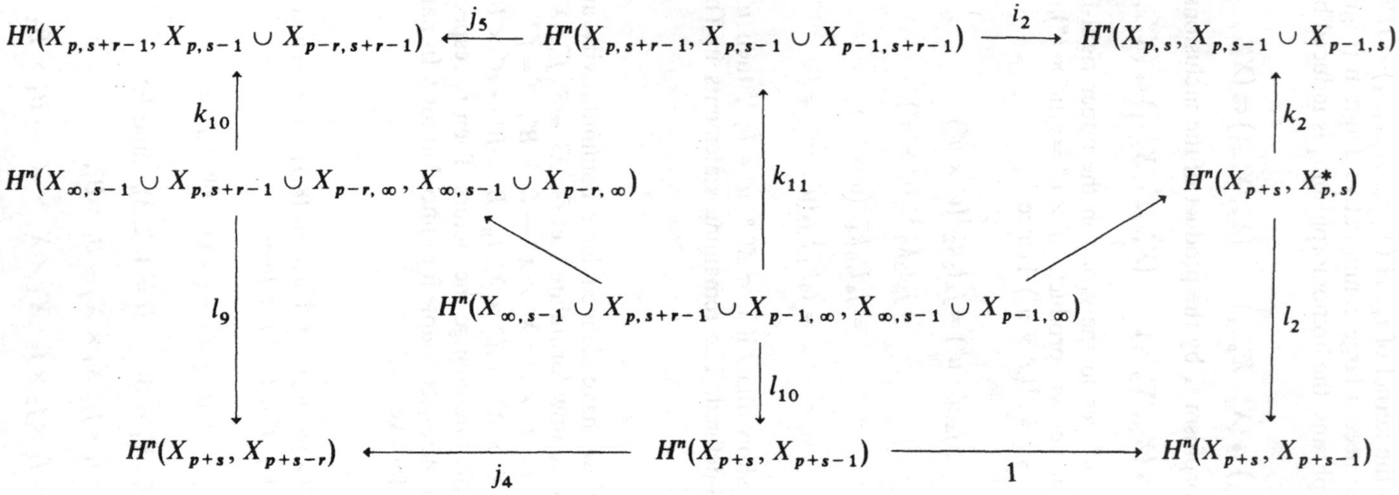
$H^n(X_{p,s+r-1}, X_{p,s-1} \cup X_{p-r,s+r-1})$
j_5
$H^n(X_{p,s+r-1}, X_{p,s-1} \cup X_{p-1,s+r-1})$
i_2
$H^n(X_{p,s}, X_{p,s-1} \cup X_{p-1,s})$
k_{10}
k_{11}
k_2
$H^n(X_{\infty,s-1} \cup X_{p,s+r-1} \cup X_{p-r,\infty}, X_{\infty,s-1} \cup X_{p-r,\infty})$
$H^n(X_{p+s}, X^*_{p,s})$
l_9
$H^n(X_{\infty,s-1} \cup X_{p,s+r-1} \cup X_{p-1,\infty}, X_{\infty,s-1} \cup X_{p-1,\infty})$
l_2
l_{10}
$H^n(X_{p+s}, X_{p+s-r})$
j_4
$H^n(X_{p+s}, X_{p+s-1})$
1
$H^n(X_{p+s}, X_{p+s-1})$
Figure 13.9

Then f coincides with $f_{1,2} \times f_3$ and with $f_1 \times f_{2,3}$, up to natural homeomorphism. For each r, the products of Theorem (8.3) are pairings

$$\alpha_r : E_r(f_1) \otimes E_r(f_2) \to E_r(f_{12}),$$

$$\beta_r : E_r(f_2) \otimes E_r(f_3) \to E_r(f_{23}),$$

$$\lambda_r : E_r(f_{12}) \otimes E_r(f_3) \to E_r(f),$$

$$\mu_r : E_r(f_1) \otimes E_r(f_{23}) \to E_r(f).$$

(8.8) Theorem *The products of Theorem* (8.3) *are associative in the sense that the diagrams*

$$\begin{array}{ccc} E_r(f_1) \otimes E_r(f_2) \otimes E_r(f_3) & \xrightarrow{\alpha_r \otimes 1} & E_r(f_{12}) \otimes E_r(f_3) \\ {\scriptstyle 1 \otimes \beta_r} \downarrow & & \downarrow {\scriptstyle \lambda_r} \\ E_r(f_1) \otimes E_r(f_{23}) & \xrightarrow[\mu_r]{} & E_r(f) \end{array}$$

are commutative. □

Theorems (8.7) and (8.8) are easy consequences of the known commutativity and associativity properties of the cohomology cross product.

The products of Theorem (8.3) have certain naturality properties, which we leave to the reader to formulate.

Finally, we discuss the relationship between the products in E_∞ and in the total spaces. The cohomology group $H^n(X)$ is filtered by the images $J^{p,n-p}(f)$ of the injection homomorphisms

$$H^n(X, X_{p-1}) \to H^n(X),$$

and

$$H^n(X) = J^{0,n}(f) \supset \cdots \supset J^{p,n-p}(f) \supset J^{p+1,n-p-1}(f) \supset \cdots,$$

$$\bigcap_{p+q=n} J^{p,q}(f) = 0.$$

Similarly, there are filtrations $\{J^{p,q}(f')\}$, $\{J^{s,t}(f'')\}$ of $H^*(X')$, $H^*(X'')$, respectively. The cross product maps $H^{p+q}(X', X'_{p-1}) \otimes H^{s+t}(X'', X''_{s-1})$ into $H^n(X, X_{\infty,s-1} \cup X_{p-1,\infty})$; as $X_{p+s-1} \subset X_{\infty,s-1} \cup X_{p-1,\infty}$, there is a commutative diagram

$$\begin{array}{ccc} H^{p+q}(X', X'_{p-1}) \otimes H^{s+t}(X'', X''_{s-1}) & \longrightarrow & H^n(X, X_{p+s-1}) \\ \downarrow & & \downarrow \\ H^{p+q}(X') \otimes H^{s+t}(X'') & \longrightarrow & H^n(X) \end{array}$$

Hence the cross product maps $J^{p,q}(f')\otimes J^{s,t}(f'')$ into $J^{p+s,q+t}(f)$. Moreover, the groups $J^{p+1,q-1}(f')\otimes J^{s,t}(f'')$ and $J^{p,q}(f')\otimes J^{s+1,t-1}(f'')$ are mapped into $J^{p+s+1,q+t-1}(f)$. Thus there is induced a pairing of the quotient groups

$$\bar{\mu}: E_{\infty}^{p,q}(f')\otimes E_{\infty}^{s,t}(f'')\to E_{\infty}^{p+s,q+t}(f).$$

(8.9) Theorem *The pairings $\bar{\mu}$ and μ_{∞} coincide.*

Let $u'\in Z_{\infty}^{p,q}(f')\subset H^{p+q}(X'_p, X'_{p-1})$, $u''\in Z_{\infty}^{s,t}(f'')\subset H^{s+t}(X''_s, X''_{s-1})$. Then there are classes $v'\in H^{p+q}(X', X'_{p-1})$, $v''\in H^{s+t}(X'', X''_{s-1})$ which map into u', u'', respectively, under the appropriate injections. Let $w'\in H^{p+q}(X')$, $w''\in H^{s+t}(X'')$ be the images of v', v'', respectively, under the relevant injections. Recalling the definition of the additive relation $\Phi^{p,q}$ of §3, we see that $(u', w')\in\Phi^{p,q}(f')$, $(u'', w'')\in\Phi^{s,t}(f'')$. Then the element $v'\times v''\in H^n(X, X_{\infty,s-1}\cup X_{p-1,\infty})$ is mapped by the injection into an element u of $H^n(X_{p+s}, X_{p+s-1})$; by an easy modification of the proof of (8.4), we see that $u=u'\cdot u''$. On the other hand, the naturality of the cross product map ensures that the injection $H^n(X, X_{\infty,s-1}\cup X_{p-1,\infty})\to H^n(X)$ maps $v'\times v''$ into $w'\times w''$. Thus $(u'\cdot u'', w'\times w'')\in\Phi^{p+s,q+t}(f)$. Hence

$$\bar{\mu}([u']\otimes[u''])=[u'\cdot u'']=\mu_{\infty}([u']\otimes[u'']).$$

We now turn to the consideration of cup products. Let $f: X\to B$ be a filtration with fibre F; then $f\times f: X\times X\to B\times B$ is a fibration with fibre $F\times F$. The diagonal maps of X and B define a fibre preserving map of the first fibration into the second; the induced map on the fibre F is again the diagonal map. In this way we obtain a map of $\mathscr{C}^*(f\times f)$ into $\mathscr{C}^*(f)$, and, in particular, for each $r\geq 2$, a chain map of $E_r(f\times f)$ into $E_r(f)$. Combining these chain maps with the operations μ_r, we obtain pairings $E_r(f)\otimes E_r(f)\to E_r(f)$. From our discussion of cross products we then deduce

(8.10) Theorem *Let $f: X\to B$ be a fibration with 0-connected base B and fibre F, and let Λ be a commutative ring, $\mathfrak{H}^*$ the cohomology theory with coefficients in Λ. Then, for each r, the group E_r is a commutative and associative algebra over Λ and $d_r: E_r\to E_r$ a derivation. Moreover, E_{r+1} is the cohomology algebra of E_r. The product in E_2 corresponds up to sign, under the isomorphism λ^* of §4, to the multiplication in $H^*(B; H^*(\mathscr{F}))$ determined by the cup products in $H^*(B)$ and $H^*(F)$. Finally, the products in E_r determine one in E_{∞}, and E_{∞} is isomorphic with the associated graded algebra $\mathscr{G}H^*(X)$.*

9 Further Applications of the Leray–Serre Spectral Sequence

In this section we take advantage of the extra information furnished by the multiplicative structure in the spectral sequence, in order to obtain further applications.

In our first applications, the coefficient group for the cohomology theory will be the field $\mathbf{Z}_0$ of rational numbers.

(9.1) Theorem *Let $p : X \to B$ be a fibration and assume B is 1-connected and X is acyclic over $\mathbf{Z}_0$. Let n be an odd integer. Then the following conditions are equivalent*:

(1) *the fibre F is a rational homology n-sphere*;
(2) *there is an element $u \in H^{n+1}(B)$ such that $H^*(B)$ is the polynomial algebra $\mathbf{Z}_0[u]$.*

(As usual, the name of the coefficient domain is omitted).

Assume first that (1) holds. Then

$$E_2^{p,q} = H^p(B; H^q(F))$$

so that $E_2^{p,q} = 0$ unless $q = 0$ or $q = n$. Thus there is only one possible non-trivial coboundary operator d_{n+1}, and $E_2 = E_{n+1}$; since X is acyclic, $E_\infty^{p,q} = 0$ unless $p = q = 0$. It follows that

$$d_{n+1} : E_{n+1}^{p-n-1,n} \to E_{n+1}^{p,0}$$

is an isomorphism for all $p > 0$. Now $E_2^{0,n} = H^0(B; H^n(F) \approx H^n(F)$ is one-dimensional, generated by $z \in H^n(F)$. Let $u = d_{n+1} z \in H^{n+1}(B; H^0(F)) \approx H^{n+1}(B)$. If $x \in H^{p-n-1}(B)$, then $xz \in H^{p-n-1}(B; H^n(F))$, and

$$d_{n+1}(xz) = xd_{n+1}(z) = xu.$$

An easy inductive argument now shows that $H^p(B) = 0$ unless $p \equiv 0 \pmod{n+1}$, and that $H^{k(n+1)}(B)$ is a one-dimensional space generated by u^k. Hence $H^*(B)$ is as stated.

Now assume that (2) holds. The first non-trivial operator is d_{n+1}, and it follows that $H^q(F) = 0$ for $0 < q < n$. Since $E_\infty^{p,q} = 0$ unless $p = q = 0$, there must exist $z \in H^n(F) \approx H^0(B; H^n(F))$ such that $d_{n+1} z = u$. It follows that $d_{n+1}(u^k z) = u^k d_{n+1}(z) = u^{k+1}$. Moreover, dim $H^n(F) = 1$; for otherwise there exists $u_1 \in H^n(F)$ with $d_{n+1} u_1 = 0$ and u_1 survives to a non-zero term in E_∞. Thus $E_\infty^{p,q} = 0$ for all $(p, q) \neq (0, 0)$, $q \leq n$. If F is not a homology sphere, there is a non-zero element $v \in H^q(F)$, $q > n$. We may assume that $H^k(F) = 0$ for $n < k < q$. But then $d_r v = 0$ for all r, so that v survives to a non-zero element of E_∞, a contradiction.

(9.2) Theorem *Let $p : X \to B$ be a fibration, and assume that B is 1-connected and X is acyclic over $\mathbf{Z}_0$. Let $n > 1$ be an odd integer. Then the following conditions are equivalent*:

(1) *B is a rational homology n-sphere*;
(2) *there is an element $u \in H^{n-1}(F)$ such that $H^*(F)$ is the polynomial algebra $\mathbf{Z}_0[u]$.*

Suppose first that (1) holds. Then $d_r = 0$ unless $r = n$, $E_2 = E_n$, and

$$d_n : E_n^{0,q} \to E_n^{n,q-n+1}$$

is an isomorphism for all $q > 0$. Let z generate the one-dimensional space $H^n(B) \approx H^n(B; H^0(F))$, and let $u = d_n^{-1}(z) \in H^{n-1}(F) \approx H^0(B; H^{n-1}(F))$. Then $d_n(u^k) = ku^{k-1}z$. Again an easy induction shows that $G^q(F) = 0$ unless $q \equiv 0 \pmod{n-1}$, while $H^{k(n-1)}(F)$ is a one-dimensional space generated by u^k. Hence $H^*(F)$ is as stated.

Suppose that (2) holds. Then the first non-trivial coboundary operator is d_n. Hence $H^q(B) = 0$ if $0 < q < n$ and $d_n(u)$ is a non-zero element of $H^n(B)$. If u_1 is an element of $H^n(B)$ not in the image of d_n, then u_1 survives to E_∞, a contradiction. Again $d_n(u^k) = ku^{k-1}z$, so that d_n maps $E_n^{0,k(n-1)}$ isomorphically on $E_n^{n,(k-1)(n-1)}$, and it follows that $E_\infty^{p,q} = 0$ for all $(p, q) \neq (0, 0)$ with $p \leq n$. As before, a non-zero element of $H^p(B)$ of lowest degree $> n$ would survive to E_∞, and this contradiction ensures that B is a homology n-sphere over $\mathbf{Z}_0$.

Remark. The argument of Theorem (9.1) is still valid if $\mathbf{Z}_0$ is replaced by an arbitrary field (or even an arbitrary principal ideal domain), but that of Theorem (9.2) does not.

(9.3) Corollary *If n is even, $H^*(\mathbf{Z}, n; \mathbf{Z}_0)$ is a polynomial algebra $\mathbf{Z}_0[u]$, $u \in H^n(\mathbf{Z}, n; \mathbf{Z}_0)$. If n is odd, $K(\mathbf{Z}, n)$ is a rational homology n-sphere.*

We have seen that $K(\mathbf{Z}, 1) = \mathbf{S}^1$. The rest follows by induction, using Theorems (9.1) and (9.2) alternatively.

(9.4) Corollary *If n is odd, $H_q(\mathbf{Z}, n)$ is a finite group for all $q > n$. If n is even, $H_q(\mathbf{Z}, n)$ is finite unless $q \equiv 0 \pmod n$, and $H_{kn}(\mathbf{Z}, n)$ is the direct sum of a finite group and an infinite cyclic group.*

By Theorem (7.12), $H_q(\mathbf{Z}, n)$ is finitely generated for all q, n. Hence $H_q(\mathbf{Z}, n)$ is a direct sum $F_q + T_q$, where F_q is free and T_q finite. The rank of F_q is the dimension of the rational vector space $H_q(\mathbf{Z}, n; \mathbf{Z}_0)$; hence $F_q = 0$ or $\mathbf{Z}$ as the case may be.

(9.5) Theorem *If n is odd, the homotopy groups $\pi_q(\mathbf{S}^n)$ are finite for all $q > n$.*

Let $f: \mathbf{S}^n \to K(\mathbf{Z}, n)$ be a map such that $f^*\mathbf{b}_n$ generates $H^n(\mathbf{S}^n)$, and let X be the mapping fibre of f. By Theorem (7.11), the homology groups of X are finitely generated. It follows from Corollary (9.4) that $f_* : H_*(\mathbf{S}^n; \mathbf{Z}_0) \to H_*(K(\mathbf{Z}, n); \mathbf{Z}_0)$ is an isomorphism. By Theorem (7.15), X is rationally acyclic and therefore $H_q(X)$ is finite for all $q > 0$. By Corollary (7.19) the homotopy groups of X are all finite. But $\pi_q(X) \approx \pi_q(\mathbf{S}^n)$ for all $q > n$.

(9.6) Theorem *If n is even, then $\pi_q(\mathbf{S}^n)$ is finite for all $q > n$, except that $\pi_{2n-1}(\mathbf{S}^n)$ is the direct sum of an infinite cyclic and a finite group.*

Let us consider the Stiefel manifold $\mathbf{V}_{2n+1,2}$. By Corollary (10.14) of Chapter IV, the homology groups of $\mathbf{V}_{2n+1,2}$ are finite, except that $H_0(\mathbf{V}_{2n+1,2}) \approx H_{4n-1}(\mathbf{V}_{2n+1,2}) \approx \mathbf{Z}$. By the Hopf Theorem, there is a map $f: \mathbf{V}_{2n+1,2} \to \mathbf{S}^{4n-1}$ such that $f^*: H^{4n-1}(\mathbf{S}^{4n-1}) \approx H^{4n-1}(\mathbf{V}_{2n+1,2})$. Let F be the mapping fibre of f. Since $f_*: H_*(\mathbf{V}_{2n+1,2}; \mathbf{Z}_0) \to H_*(\mathbf{S}^{4n-1}; \mathbf{Z}_0)$ is an isomorphism, it follows from Theorem (7.15) that F is rationally acyclic; by Corollary (7.19), the homotopy groups of F are finite. Hence $\pi_q(\mathbf{V}_{2n+1,2})$ is a finite group except that $\pi_{4n-1}(\mathbf{V}_{2n+1,2}) = \mathbf{Z} \oplus F$, where F is finite.

Now consider the homotopy sequence

$$\cdots \to \pi_q(\mathbf{S}^{2n-1}) \to \pi_q(\mathbf{V}_{2n+1,2}) \to \pi_q(\mathbf{S}^{2n}) \to \pi_{q-1}(\mathbf{S}^{2n-1}) \to \pi_{q-1}(\mathbf{V}_{2n+1,2})$$

of the fibration $\mathbf{S}^{2n-1} \to \mathbf{V}_{2n+1,2} \to \mathbf{S}^{2n}$. If $q \neq 2n, 4n-1$, then $\pi_q(\mathbf{V}_{2n+1,2})$ and $\pi_{q-1}(\mathbf{S}^{2n-1})$ are finite, and therefore $\pi_q(\mathbf{S}^{2n})$ is finite. If $q = 2n$, $\pi_q(\mathbf{S}^{2n}) = \mathbf{Z}$. Finally, we have the exact sequence

$$\pi_{4n-1}(\mathbf{S}^{2n-1}) \to \pi_{4n-1}(\mathbf{V}_{2n+1,2}) \xrightarrow{p_*} \pi_{4n-1}(\mathbf{S}^{2n}) \to \pi_{4n-2}(\mathbf{S}^{2n-1})$$

The homomorphism p_* has finite kernel and cokernel, and it follows that $\pi_{4n-1}(\mathbf{S}^{2n}) \approx \mathbf{Z} \oplus F'$ with F' finite.

Exercises

1. Let (X, A) be a pair, filtered by $\{X_n\}$. Let

$$\bar{D}_{p,q} = H_{p+q}(X, X_{p-1}),$$
$$E_{p,q} = H_{p+q}(X_p, X_{p-1}),$$

and form an exact couple

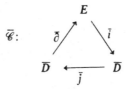

$$\bar{\mathscr{C}}: \quad \begin{array}{ccc} & E & \\ {\scriptstyle \bar{\partial}}\nearrow & & \searrow{\scriptstyle \bar{i}} \\ \bar{D} & \xleftarrow[\bar{j}]{} & \bar{D} \end{array}$$

Show that $\bar{\mathscr{C}}$ leads to the same spectral sequence as the exact couple of §3.

2. Derive the Wang sequence from the Leray–Serre spectral sequence of a fibration $F \to X \to \mathbf{S}^n$ (including the multiplicative properties we proved in Chapter VII).

3. Do the same for the Gysin sequence of an orientable fibration $\mathbf{S}^n \to X \to B$.

4. Complete Exercise 1 of Chapter VII calculating the homology groups of the space W^n of free loops in $\mathbf{S}^n$ for n even.

5. Prove the statement made in the Remark at the end of §2.

Appendix A Compact Lie Groups

The first two Sections outline the most important results on compact Lie groups and their classifying spaces.

In §3 we introduce the spinor groups **Spin** (n); these are the universal covering groups of the rotation groups. Section 4 introduces the Cayley algebra. The group $\mathbf{G}_2$ of automorphisms of $\mathbf{K}$ is examined in §5. The group $\mathbf{G}_2$ is seen to act transitively on the unit sphere $\mathbf{S}^6$ in $\mathbf{K}_0$, the subspace of pure imaginary Cayley numbers, and the isotropy group is seen to be the unimodular unitary group $\mathbf{U}_3^+$. In fact, $\mathbf{G}_2$ acts transitively on the space $\mathbf{V}_{7,2}$ of orthogonal pairs of unit vectors in $\mathbf{K}_0$, and this time the isotropy group is isomorphic with the multiplicative group $\mathbf{S}^3$ of unit quaternions.

In §6 we introduce the exceptional Jordan algebra $\mathfrak{J}$. The algebra $\mathfrak{J}$ is the set of all 3×3 Hermitian matrices with entries in the Cayley algebra $\mathbf{K}$, under the Jordan product

$$X \circ Y = \tfrac{1}{2}(XY + YX).$$

A study of its idempotents is made and the space P of primitive idempotents is identified as the Cayley projective plane. In §7 we study the automorphism group of $\mathfrak{J}$; it is the second exceptional Lie group $\mathbf{F}_4$. The group $\mathbf{F}_4$ acts transitively on P, and the isotropy group is **Spin** (9). Other subgroups of $\mathbf{F}_4$ are identified; of special interest is a transitive action of **Spin** (9) on $\mathbf{S}^{15}$ with isotropy group **Spin** (7). This action is not unrelated to the Hopf fibration $\mathbf{S}^{15} \to \mathbf{S}^8$.

1 Subgroups, Coset Spaces, Maximal Tori

We begin by sketching some of the most important properties of compact Lie groups. For details the reader is referred to Chevalley [C_1], Hochschild [Ho], or Helgason [He]. We also recommend the survey articles by Samelson [3] and Borel [4].

We first observe that the hypothesis of compactness is not restrictive from the point of view of homotopy theory. This follows from Theorem (1.1), below, which was proved by Élie Cartan [Ca] for 1-connected groups, and in full generality by Malcev and Iwasawa.

(1.1) Theorem *Let G be a connected Lie group. Then G has a maximal compact subgroup K, unique up to conjugacy, and G is homeomorphic with a Cartesian product $K \times \mathbf{R}^m$. In particular, G and K have the same homotopy type.* □

For example, let $G = \mathbf{L}^+(n)$, the group of linear operators on $\mathbf{R}^n$ with positive determinant. Then we may take $K = \mathbf{O}^+(n)$, and if $H(n)$ is the space of symmetric operators, $H^+(n)$ that of the positive definite ones, then $H(n)$ is a Euclidean space of dimension $\binom{n+1}{2}$ and the exponential map sends $H(n)$ homeomorphically upon $H^+(n)$. Moreover the map $(X, U) \to X \cdot U$ $(X \in H^+(n), U \in \mathbf{O}^+(n))$ is a homeomorphism of $H^+(n) \times \mathbf{O}^+(n)$ with $\mathbf{L}^+(n)$. The representation $A = X \cdot U$ of a matrix $A \in \mathbf{L}^+(n)$ is the *polar decomposition* of A.

The Structure Theorem

(1.2) Theorem *If G is a compact connected Lie group, then G has a finitely many sheeted covering group*

$$\tilde{G} = T^l \times G_1 \times \cdots \times G_r,$$

where T^l is a toral group and each of the groups G_i is a simply connected compact simple group. □

is due to Élie Cartan.

Remark. The word "simple" is used in the sense customary in dealing with topological groups; such a group G is said to be simple if and only if it is non-abelian and its only proper normal subgroups are discrete or open. For a Lie group G, this is equivalent to the assertion that the Lie algebra of G is simple.)

Cartan has also classified the compact simple groups, completing earlier work of Killing. Up to local isomorphism, these are given in

(1.3) Theorem *The compact simple groups are given without repetition (up to*

local isomorphism) by four infinite sequences:

$\mathbf{U}^+(n)$	$(n \geq 2)$,
$\mathbf{O}^+(2n+1)$	$(n \geq 2)$,
$\mathbf{Sp}(n)$	$(n \geq 3)$,
$\mathbf{O}^+(2n)$	$(n \geq 4)$,

as well as five exceptional groups

$$\mathbf{G}_2, \mathbf{F}_4, \mathbf{E}_6, \mathbf{E}_7, \mathbf{E}_8,$$

of dimensions

$$14, 52, 78, 133, 248,$$

respectively. The following sets consist of mutually locally isomorphic groups:

$$\{\mathbf{U}^+(2), \mathbf{O}^+(3), \mathbf{Sp}(1)\},$$
$$\{\mathbf{Sp}(2), \mathbf{O}^+(5)\},$$
$$\{\mathbf{U}^+(4), \mathbf{O}^+(6)\}. \qquad \square$$

(cf. Exercises 3, 4).

Let H be a closed subgroup of the compact Lie group G. Then H itself is a Lie group, and the natural map $p : G \to G/H$ of G into the space of left cosets of H in G is the projection of a fibre bundle [St_1, p. 30]. By (7.13) of Chapter I,

(1.4) *The map $p : G \to G/H$ is a fibration.* $\square$

More generally,

(1.5) *Let $p : X \to B$ be the projection of a principal bundle with structural group G, and let H be a closed subgroup of G, $C = X/H$ the orbit space. Then the natural map $p : C \to B$ is a fibration with fibre G/H.* $\square$

(1.6) Corollary *If $H \supset K$ are closed subgroups of G, then the natural map $p : G/K \to G/H$ is a fibration with fibre H/K.* $\square$

As particular applications of Corollary (1.6), we have the fibrations given in Table A.1 and their complex and quaternionic analogues. Of particular importance are the *Hopf fibrations.*

$$\mathbf{S}^{n-1} \to \mathbf{P}^{n-1}, \mathbf{S}^{2n-1} \to \mathbf{P}^{n-1}(\mathbf{C}), \mathbf{S}^{4n-1} \to \mathbf{P}^{n-1}(\mathbf{Q})$$

with fibres $\mathbf{S}^0, \mathbf{S}^1, \mathbf{S}^3$, respectively. These are obtained from the last line in Table A.1 and its analogues over $\mathbf{C}, \mathbf{Q}$ by taking $k = 1$. In particular, we may take $n = 2$; in this case the projective space is itself a sphere and we have the

Table A.1

Fibration	Fibre
$\mathbf{O}(n) \to \mathbf{V}_{n,k}$	$\hat{\mathbf{O}}(n-k) \approx \mathbf{O}(n-k)$
$\mathbf{O}(k+l) \to \mathbf{G}_{k,l}$	$\mathbf{O}(k,l) \approx \hat{\mathbf{O}}(k) \times \mathbf{O}(l)$
$\mathbf{V}_{n,k} \to \mathbf{G}_{n-k,k}$	$\mathbf{O}(k,l)/\hat{\mathbf{O}}(l) \approx \mathbf{O}(k)$

original Hopf fibrations

$$\mathbf{S}^1 \to \mathbf{S}^1,\ \mathbf{S}^3 \to \mathbf{S}^2,\ \mathbf{S}^7 \to \mathbf{S}^4.$$

Let us describe the Hopf fibrations more explicitly. To give a uniform description, let D be a normed associative real division algebra of dimension d. As is well known, there are, up to isomorphism, only three examples, viz. $\mathbf{R}$, $\mathbf{C}$, and $\mathbf{Q}$. Now $\mathbf{P}^{m-1}(D)$ is the space of one-dimensional subspaces of D^m; and the unit sphere in D^m is a sphere of dimension $dm - 1$. Each point of $\mathbf{S}^{dm-1}$ lies on a unique one-dimensional subspace; and the Hopf fibration attaches to each point of the sphere that subspace which contains it. Analytically, points of $\mathbf{S}^{dm-1}$ are described by coördinates $(x_1, \ldots, x_m)$ with $x_i \in D$ and $\sum_{i=1}^m \|x_i\|^2 = 1$. On the other hand, a one-dimensional subspace is determined by any non-zero vector in it; thus points of $\mathbf{P}^{m-1}$ can be considered as equivalences classes $[x_1, \ldots, x_m]$ of m-tuples $(x_1, \ldots, x_m)$, with $x_i \in D$ and not all $x_i = 0$, two m-tuples $(x_1, \ldots, x_m)$ and $(x'_1, \ldots, x'_m)$ being equivalent if and only if there is an element $x \in D$ (perforce non-zero) such that $x'_i = xx_i$ for $i = 1, \ldots, m$. The elements $(x_1, \ldots, x_m)$ are called *homogeneous coördinates* of the point $[x_1, \ldots, x_m]$. The Hopf map sends the point $(x_1, \ldots, x_m) \in \mathbf{S}^{dm-1}$ into the point $[x_1, \ldots, x_m] \in \mathbf{P}^{m-1}(D)$.

There is one other example of a normed division algebra; it is the non-associative algebra $\mathbf{K}$ of Cayley numbers. As a vector space $\mathbf{K} = \mathbf{R}^8$, and it has a two-sided unit element $\mathbf{e}_0$. To describe the multiplication table, we first single out certain ordered triples of $\{1, \ldots, 7\}$ as *admissible*. These are all triples which can be obtained from the triple (1, 2, 4) by applying the following operations:

(1) applying a cyclic permutation to the members of the triple;
(2) applying a cyclic permutation to the integers 1, ..., 7.

(These can be visualized with the aid of Figure A.1, in which each of the sides and medians of the triangle, oriented as in the figure, give rise to an admissible triple, and the dotted circle gives one more admissible triple). The multiplication table is now completely determined by the condition

(1.7) *For each admissible triple (p, q, r) the subspace spanned by $\mathbf{e}_0$, $\mathbf{e}_p$, $\mathbf{e}_q$ and $\mathbf{e}_r$ is a subalgebra, and the linear transformation of $\mathbf{Q}$ into this subspace which sends 1, i, j, k into $\mathbf{e}_0$, $\mathbf{e}_p$, $\mathbf{e}_q$, $\mathbf{e}_r$, respectively is an isomorphism.* □

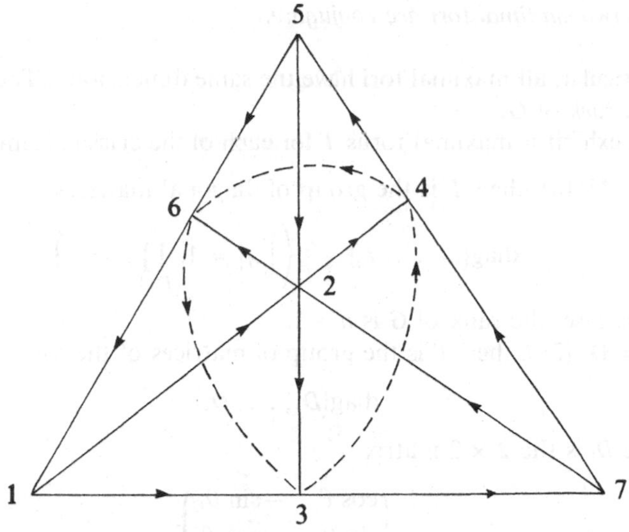

Figure A.1

For example,

$$\mathbf{e}_1\mathbf{e}_2 = \mathbf{e}_4 = -\mathbf{e}_2\mathbf{e}_1, \qquad \mathbf{e}_2\mathbf{e}_4 = \mathbf{e}_1 = -\mathbf{e}_4\mathbf{e}_2,$$

$$\mathbf{e}_4\mathbf{e}_1 = \mathbf{e}_2 = -\mathbf{e}_1\mathbf{e}_4, \qquad \mathbf{e}_1^2 = \mathbf{e}_2^2 = \mathbf{e}_4^2 = -\mathbf{e}_0.$$

As $\mathbf{K}$ is not associative, the notion of vector space over $\mathbf{K}$ is meaningless. And if we attempt to describe a "Cayley projective space" by homogeneous coördinates, we find that the proportionality relation between m-tuples of Cayley numbers is not an equivalence relation. There is, however, a Hopf map $\mathbf{S}^{15} \to \mathbf{S}^8 = \mathbf{P}^1(\mathbf{K})$. For let $\mathbf{S}^{15}$ be represented as the set of pairs (x, y) of Cayley numbers with $\|x\|^2 + \|y\|^2 = 1$, and let $\mathbf{S}^8$ be the one-point compactification of $\mathbf{K}$. Then the map $p : \mathbf{S}^{15} \to \mathbf{S}^8$ defined by

$$p(x, y) = \begin{cases} \bar{x}^{-1}y & \text{if } x \neq 0, \\ \infty & \text{if } x = 0, \end{cases}$$

is continuous, and can be verified to be a fibration. Moreover, the mapping cone of p is a 16-manifold, which is appropriately called the *Cayley projective plane.* (See also §7, below.)

Among the subgroups of a connected compact Lie group G, the toral subgroups play an important role. Especially important are the maximal tori. They have, *inter alia,* the following properties:

(1.8) *A maximal torus is a maximal abelian subgroup of G.* □

The converse is, however, false. For example, let A be the group of diagonal matrices in $\mathbf{O}_3^+$; then A is a finite group of order 4, but is easily seen to be maximal abelian.

(1.9) *Any two maximal tori are conjugate.* □

In particular, all maximal tori have the same dimension l. The integer l is called the *rank* of G.

Let us exhibit a maximal torus T for each of the classical simple groups.

(1) If $G = \mathbf{U}^+(n)$, then T is the group of diagonal matrices

$$\operatorname{diag}\{z_1, \ldots, z_n\} \qquad \left(|z_i| = 1, \prod_{i=1}^{n} z_i = 1\right).$$

In this case, the rank of G is $n - 1$.

(2) If $G = \mathbf{O}^+(2n)$, then T is the group of matrices of the form

$$\operatorname{diag}\{D_1, \ldots, D_n\},$$

where D_i is the 2×2 matrix

$$\begin{pmatrix} \cos\theta_i & -\sin\theta_i \\ \sin\theta_i & \cos\theta_i \end{pmatrix}.$$

In this case, the rank of G is n.

(3) If $G = \mathbf{O}^+(2n + 1)$, then T is the group of matrices of the form

$$\operatorname{diag}\{D_1, \ldots, D_n, 1\}$$

with D_i as above. In this case, G has rank n.

(4) If $G = \mathbf{Sp}(n)$, T is the group of matrices of the form

$$\operatorname{diag}\{z_1, \ldots, z_n\},$$

where each z_i is a complex number of absolute value 1 (we regard $\mathbf{C}$ as a subalgebra of $\mathbf{Q}$). Again, G has rank n.

2 Classifying Spaces

Let us recall some facts from the theory of fibre bundles. For further details and proofs the reader is referred to the treatises of Steenrod [St$_1$] and Milnor and Stasheff [M–S].

Let $p : X \to B$ be the projection of a fibre bundle with fibre F and structural group G; recall that p is a principal fibration if and only if $G = F$, acting on itself by left translation. For a general bundle, the bundle maps of F into X form a space $\tilde{X}$, the total space of the *principal associated bundle*. Moreover, the original bundle can be recovered from the principal associated bundle: the total space X can be identified as the quotient space of $\tilde{X} \times F$ under the equivalence relation

$$(\tilde{x}, gy) \sim (\tilde{x}g, y).$$

And two bundles over a space B are isomorphic if and only if their principal associated bundles are.

For example, let M be a compact differentiable n-manifold imbedded in $\mathbf{R}^{n+k}$. The tangent vectors to M form the total space of the tangent bundle of M; the principal associated bundle is the frame bundle of M, the points of whose total space are the orthonormal n-tuples $(v_1, \ldots, v_n)$ tangent to M at some point x. The map $\gamma : M \to \mathbf{G}_{n,k}$, which assigns to each $x \in M$ the n-plane through the origin parallel to the tangent plane to M at x, is called the *Gauss map*, and is of the highest importance in differential geometry.

The space $E_{k,n}$ of pairs (π, x), where π is an n-plane through the origin in $\mathbf{R}^{n+k}$ and $x \in \pi$, is the total space of a bundle $\mathscr{B}(k, n)$ over $\mathbf{G}_{n,k}$. The total space of the principal associated bundle is the Stiefel manifold $\mathbf{V}_{n+k,n}$. The map $\gamma : M \to \mathbf{G}_{n,k}$ induces a bundle $\gamma^*\mathscr{B}(k, n)$ over M, and this bundle turns out to be just the tangent bundle of the manifold M.

Motivated by these considerations, Whitney proved that every bundle with structural group $\mathbf{O}(n)$ over a complex K of dimension $\leq k$ is induced by a map of K into $\mathbf{G}_{k,n}$. Steenrod completed Whitney's analysis by observing that $\mathbf{G}_{k,n}$ is a *classifying space* for $\mathbf{O}(n)$-bundles over such a base, i.e., the isomorphism classes of bundles over K are in one-to-one correspondence with the homotopy classes of maps of K into $\mathbf{G}_{k,n}$. The crucial fact used in the proof is that the Stiefel manifold $\mathbf{V}_{n+k,n}$ is $(k-1)$-connected (Chapter IV, (10.12)).

The restriction on the dimension k can be removed. In fact, $\mathbf{V}_{n+k+1,n} = \mathbf{O}(n+k+1)/\hat{\mathbf{O}}(k+1)$, and

$$\mathbf{O}(n+k) \cap \hat{\mathbf{O}}(k+1) = \hat{\mathbf{O}}(k),$$

so that the inclusion map induces an imbedding

$$\mathbf{V}_{n+k,n} \to \mathbf{V}_{n+k+1,n}.$$

It is not difficult to see that the $\mathbf{V}_{n+k,n}$ form an expanding sequence of spaces forming an NDR-filtration of their direct limit $\mathbf{V}(n)$. Similarly we have imbeddings

$$\mathbf{G}_{k,n} \to \mathbf{G}_{k+1,n}$$

and their direct limit $\mathbf{G}(n)$. We may identify $\mathbf{V}(n)$ with the space of n-frames in $\mathbf{R}^\infty$ and $\mathbf{G}(n)$ with the space of n-dimensional vector subspaces of $\mathbf{R}^\infty$. The projection $p : \mathbf{V}(n) \to \mathbf{G}(n)$ which assigns to each n-frame in $\mathbf{R}^\infty$ the vector subspace spanned by it is easily seen to be the projection of a principal fibration with group $\mathbf{O}(n)$. Moreover, the space $\mathbf{V}(n)$ is ∞-connected. In fact, ([M–S], Theorems 5.6, 5.7):

(2.1) Theorem *The space* $\mathbf{G}(n)$ *is a classifying space for bundles having* $\mathbf{O}(n)$ *as structural group,* i.e., *the equivalence classes of such bundles over a paracompact base space B are in one-to-one correspondence with the homotopy classes of maps of B into* $\mathbf{G}(n)$. □

Let G be a compact Lie group. Then the Peter–Weyl Theorem asserts that G has a faithful orthogonal representation $\rho : G \to \mathbf{O}(n)$ for some positive integer n. Thus we may regard G as a subgroup of $\mathbf{O}(n)$, so that G acts on the space $\mathbf{V}(n)$. Let $B(G)$ be the orbit space $\mathbf{V}(n)/G$; then

(2.2) Theorem *The projection $p : \mathbf{V}(n) \to B(G)$ is a principal fibration with structural group G, and the space $B(G)$ is a classifying space for bundles over a paracompact base space with structural group G.* □

The notation $B(G)$ is somewhat misleading, for $B(G)$ depends on the faithful representation ρ. However, if $\rho' : B(G) \to \mathbf{O}(m)$ is another faithful representation, and $B'(G) = \mathbf{V}(m)/\mathrm{Im}\,\rho'$, we have

(2.3) Theorem *The spaces $B(G)$ and $B'(G)$ have the same homotopy type.*

Let $\mathscr{B}$, $\mathscr{B}'$ be the appropriate universal bundles. Since $B'(G)$ is classifying, there is a map $f : B(G) \to B'(G)$ such that $f^*\mathscr{B}' \approx \mathscr{B}$. Similarly, there is a map $g : B'(G) \to B(G)$ such that $g^*\mathscr{B} \approx \mathscr{B}'$. Then $(g \circ f)^*\mathscr{B} \approx f^*g^*\mathscr{B} \approx f^*\mathscr{B}' \approx \mathscr{B}$, and it follows that $g \circ f$ is homotopic to the identity map of $B(G)$. Similarly, $f \circ g$ is homotopic to the identity map of $B'(G)$. □

Let H be a closed subgroup of G and let $\rho : G \to \mathbf{O}(n)$ be a faithful representation. Then $\rho \,|\, H$ is also faithful, so that we may consider that we have inclusions

$$H \subset G \subset \mathbf{O}(n).$$

Then $B(G) = \mathbf{V}(n)/G$, $B(H) = \mathbf{V}(n)/H$, and, by (1.5), we have

(2.4) Theorem *If H is a closed subgroup of G, there is a fibration $B(H) \to B(G)$ with fibre G/H.* □

3 The Spinor Groups

The rotation group $\mathbf{O}^+(n)$ is not simply connected; it is shown in Chapter IV that $\pi_1(\mathbf{O}^+(n))$ is an infinite cyclic group if $n = 2$ and a cyclic group of order two if $n \geq 3$. The spinor group $\mathbf{Spin}(n)$ is a connected two-sheeted covering group of $\mathbf{O}^+(n)$; thus $\mathbf{Spin}(n)$ is the universal covering group of $\mathbf{O}^+(n)$ if $n \geq 3$.

An explicit construction of the group $\mathbf{Spin}(n)$ can be made with the aid of Clifford algebras. In this section we outline this theory. For complete details the reader is referred to Chevalley [C_2], Husemoller [Hu], or Jacobson [J].

Let V be a real vector space, with inner product $\langle\ ,\ \rangle$ and norm $\|\ \|$. Let us consider an associative algebra A with unit element 1, and a linear mapping $\lambda : V \to A$ such that $\lambda(x)^2 = \|x\|^2 \cdot 1$ for all $x \in V$. The Clifford algebra $C(V)$ is the universal example for such mappings; like all universal examples, it is uniquely determined up to isomorphism.

Let $v_1, \ldots, v_n$ be an orthonormal basis of V. We then construct $C(V)$ as a vector space (of dimension 2^n) having one basis element v_A for each subset A of $N = \{1, \ldots, n\}$. It is convenient to identify v_i with $v_{\{i\}}$, so that V is a subspace of C. Let $\varepsilon(A, B) = (-1)^m$, where m is the number of pairs (a, b) with $a \in A, b \in B, a > b$. The product in C is then defined by bilinearity and the condition

(3.1) $$v_A v_B = \varepsilon(A, B) v_{A+B},$$

where $A + B$ is the symmetric difference of the sets A, B. Then

(1) $v_\varnothing$ is the unit element 1;
(2) $v_i^2 = 1, (i = 1, \ldots, n)$;
(3) $v_i v_j = -v_j v_i, (i \neq j)$;
(4) if $A = \{a_1, \ldots, a_k\}, a_1 < \cdots < a_k$, then

$$v_A = v_{a_1} \cdots v_{a_k}.$$

The desired universal property is easily proved; if $\lambda : V \to A$ is a linear map, $\lambda(x)^2 = \|x\|^2 \cdot 1$, its extension $\Lambda : C \to A$ is defined by

$$\Lambda(v_A) = \lambda(v_{a_1}) \cdots \lambda(v_{a_k})$$

for A as in (4), above. In particular,

(3.2) $$x^2 = \|x\|^2 \cdot 1 \qquad (x \in V \subset C).$$

Let us apply the process of *polarization* to the equation (3.2); i.e., substitute $y + z$ for x and subtract from the resulting equation those obtained from (3.2) by substituting y and z for x. We obtain

(3.3) $$yz + zy = 2\langle y, z \rangle \cdot 1 \qquad (y, z \in V).$$

The structure of C is made explicit by

(3.4) Theorem *If n is even, C is a central simple algebra. If $n \equiv 3 \pmod 4$, then C is a simple algebra, whose center, spanned by the elements 1, $v_N = v_1 \cdots v_n$, is isomorphic with the field of complex numbers. If $n \equiv 1 \pmod 4$, C is the direct sum of the ideals generated by the central idempotents $u = \frac{1}{2}(1 + v_N)$, $v = \frac{1}{2}(1 - v_N)$; the algebras Cu and Cv are central simple.*

□

Let C^e be the subspace of C spanned by the v_A for which A has an even number of elements, and let C^o be the subspace spanned by the remaining basis elements. Then C^e is a subalgebra of C and $C = C^e \oplus C^o$.

(3.5) Theorem *If n is odd, C^e is a central simple algebra. If $n \equiv 0 \pmod 4$, C^e is a simple algebra whose center, spanned by the elements 1, v_N is isomorphic with the field of complex numbers. If $n \equiv 2 \pmod 4$, C^e is the direct sum of the ideals generated by the central idempotents u, v, and the algebras $C^e u$, $C^e v$ are central simple.* □

We now define a linear map $c \to \bar{c}$ of C into itself which sends each basis element v_A into $(-1)^{k(k-1)/2} v_A$, where k is the number of elements of A. It is not hard to verify

(3.6) *The map $c \to \bar{c}$ is an antiautomorphism of period 2 of the algebra C.* □

The above map will be called the *principal involution* of C.

The *Clifford group* $\Gamma = \Gamma(V)$ is the set of invertible elements $c \in C$ such that $cVc^{-1} \subset V$. For $c \in \Gamma$, the map $\chi(c): V \to V$ defined by $\chi(c)x = cxc^{-1}$ is orthogonal, and $\chi: \Gamma \to \mathbf{O}(V)$ is an orthogonal representation of the group Γ, called the *vector representation.*

(3.7) Lemma *If $0 \neq x \in V$, then $x \in \Gamma$, and $\chi(x) = -\rho_x$, where ρ_x is the reflection in the hyperplane orthogonal to x.*

For $x^2 = \|x\|^2 \cdot 1$, so that $x^{-1} = (1/\|x\|^2)x$. If $y \in V$, then, by (3.3),

$$xy = -yx + 2\langle x, y\rangle \cdot 1,$$

and therefore

$$\begin{aligned} xyx^{-1} &= -y + 2\langle x, y\rangle x^{-1} \\ &= -y + 2\frac{\langle x, y\rangle}{\langle x, x\rangle} x \in V. \end{aligned}$$

Moreover,

$$\begin{aligned} \chi(x)x &= xxx^{-1} = x, \\ \chi(x)y &= -y \quad \text{if } y \perp x, \end{aligned}$$

so that $\chi(x) = -\rho_x$. □

Let $\Gamma^e = \Gamma \cap C^e$. If $x_1, \ldots, x_{2r} \in V$, then $x = x_1 \cdots x_{2r} \in \Gamma^e$ and $\chi(x) = \rho_{x_1} \cdots \rho_{x_{2r}}$. Now every rotation $\sigma \in \mathbf{O}^+(V)$ is a product $\rho_{x_1} \cdots \rho_{x_k}$, and k is even since $\det \sigma = +1$, $\det \rho_{x_i} = -1$. Hence $\chi(\Gamma^e)$ contains $\mathbf{O}^+(V)$.

(3.8) Lemma *The vector representation χ maps Γ^e upon $\mathbf{O}^+(V)$ and* Ker $\chi \cap \Gamma^e$ *is the set $\mathbf{R}^* \cdot 1$ of all non-zero scalar multiples of the unit element.*

The kernel of χ consists of all $c \in \Gamma$ such that c commutes with every element of V; since V generates C, Ker χ is the intersection of Γ with the center Z of C.

Suppose that n is even. Then $Z = \mathbf{R} \cdot 1$ and therefore Ker $\chi = \mathbf{R}^* \cdot 1$. If $x \in V$, $x \neq 0$, then no scalar multiple of x belongs to Γ^e and therefore $\chi(x) \notin \chi(\Gamma^e)$. Thus

$$\mathbf{O}(V) \supset \chi(\Gamma) \underset{\neq}{\supset} \chi(\Gamma^e) \supset \mathbf{O}^+(V);$$

since $\mathbf{O}^+(V)$ has index two in $\mathbf{O}(V)$, we have

$$\mathbf{O}(V) = \chi(\Gamma), \qquad \mathbf{O}^+(V) = \chi(\Gamma^e).$$

Suppose, on the other hand that n is odd. If $\sigma \in \mathbf{O}(V)$, then $\sigma(x)^2 = \|\sigma(x)\|^2 \cdot 1 = \|x\|^2 \cdot 1$ for all $x \in V$; by the universal property of C, there is an automorphism $\bar{\sigma}: C \to C$ such that $\bar{\sigma}(x) = \sigma(x)$ for all $x \in V$. Moreover, if $\sigma, \tau \in \mathbf{O}(V)$, then $\overline{\sigma \circ \tau} = \bar{\sigma} \circ \bar{\tau}$. The automorphism $\bar{\sigma}$ must map the center Z of C into itself. Thus the map $\sigma \to \bar{\sigma} | Z$ is a representation θ of $\mathbf{O}(V)$ in the two-dimensional vector space Z. If $n > 1$, $\mathbf{O}^+(V)$ is simple, and therefore $\theta | \mathbf{O}^+(V)$ is trivial. Let σ_0 be the reflection in the subspace spanned by $v_2, \ldots, v_n$, so that $\sigma_0(v_1) = -v_1$, $\sigma_0(v_i) = v_i$ for $i = 2, \ldots, n$. Then

$$\bar{\sigma}_0(v_N) = \bar{\sigma}_0(v_1 \cdots v_n) = \sigma_0(v_1) \cdots \sigma_0(v_n) = -v_N,$$

and therefore $\bar{\sigma}(v_N) = -v_N$ for all $\sigma \in \mathbf{O}(V) - \mathbf{O}^+(V)$. Thus

$$\bar{\sigma}(v_N) = (\det \sigma) v_N \qquad (\sigma \in \mathbf{O}(V)). \tag{3.9}$$

If $n = 1$, then (3.9) holds trivially.

If $c \in \Gamma$, the restriction to V of the inner automorphism τ_c is $\chi(c)$, and therefore $\overline{\chi(c)} = \tau_c$; since e_N belongs to the center of C, $\tau_c(v_N) = v_N$, and it follows from (3.9) that det $\chi(c) = 1$. Thus $\chi(\Gamma) = \chi(\Gamma^e) = \mathbf{O}^+(V)$. Moreover, Ker $\chi \cap \Gamma^e = \Gamma^e \cap Z = \mathbf{R}^* \cdot 1$, since $v_N \notin \Gamma^e$. □

(3.10) Corollary *An element $c \in C$ belongs to Γ^e if and only if there exist non-zero elements $x_1, \ldots, x_{2r} \in V$ such that $c = x_1 \cdots x_{2r}$.* □

Let us apply the principal involution to the relation

$$cxc^{-1} = \chi(c)x \qquad (c \in \Gamma, x \in V),$$

to obtain

$$\bar{c}^{-1}x\bar{c} = \chi(c)x = cxc^{-1}.$$

Hence $\bar{c}c$ commutes with every element of V, and therefore $\bar{c}c \in Z$. If $c \in C^e$, then $\bar{c} \in C^e$, so that $\bar{c}c \in Z \cap C^e$; by Lemma (3.8),

$$\bar{c}c = \lambda(c) \cdot 1$$

with $0 \neq \lambda(c) \in R$. If $c_1, c_2 \in \Gamma^e$, then

$$(c_1 c_2)(\overline{c_1 c_2}) = c_1 c_2 \bar{c}_2 \bar{c}_1 = c_1 \lambda(c_2) \bar{c}_1$$
$$= \lambda(c_2) c_1 \bar{c}_1 = \lambda(c_2)\lambda(c_1) \cdot 1;$$

hence

(3.11) $$\lambda(c_1 c_2) = \lambda(c_1)\lambda(c_2),$$

so that $\lambda : \Gamma^e \to \mathbf{R}^*$ is a homomorphism. The kernel of λ is a subgroup of Γ^e, the *spinor group* $\mathbf{Spin}(V)$.

(3.12) Theorem *An element $c \in C$ belongs to* $\mathbf{Spin}(V)$ *if and only if there exist unit vectors $x_1, \ldots, x_{2r} \in V$ such that $c = x_1 \cdots x_{2r}$.*

If $x \in V$, then $\bar{x} = x$, and therefore $x\bar{x} = x^2 = \|x\|^2 \cdot 1$, so that $\lambda(x) = \|x\|^2$. If $c \in \Gamma^e$, then $c = y_1 \cdots y_{2r}$ with $0 \neq y_i \in V$, and $\lambda(c) = \prod_{i=1}^{2r} \lambda(y_i) = \prod_{i=1}^{2r} \|y_i\|^2$. Thus $c \in \mathbf{Spin}(V)$ if and only if $\Pi\|y_i\|^2 = 1$. When this is so, we also have $c = x_1 \cdots x_{2r}$, where $x_i = y_i/\|y_i\|$. □

(3.13) Theorem *The homomorphism χ maps* $\mathbf{Spin}(V)$ *upon* $\mathbf{O}^+(V)$, *and* $\operatorname{Ker} \chi \cap \mathbf{Spin}(V) = \{1, -1\}$.

For we have seen that $\chi(\Gamma^e) = \mathbf{O}^+(V)$, $\operatorname{Ker} \chi \cap \Gamma^e = \mathbf{R}^* \cdot 1$. If $\sigma \in \mathbf{O}^+(V)$, then there exist non-zero vectors $x_1, \ldots, x_{2r} \in V$ such that $\sigma = \chi(x_1 \cdots x_{2r}) = \chi(x_1) \cdots \chi(x_{2r})$. But $(1/\|x_i\|) \cdot 1 \in \operatorname{Ker} \chi \cap \Gamma^e$, and therefore

$$\chi(x_i) = \chi\left(\frac{x_i}{\|x_i\|}\right),$$
$$\sigma = \chi\left(\frac{x_1}{\|x_1\|} \cdots \frac{x_{2r}}{\|x_{2r}\|}\right) \in \chi(\mathbf{Spin}(V)).$$

Moreover, $c \in \operatorname{Ker} \chi \cap \Gamma^e$ if and only if $c = \alpha \cdot 1$, $\alpha \in \mathbf{R}^*$. Then

$$c\bar{c} = \alpha^2 \cdot 1,$$

so that $\lambda(c) = \alpha^2$; thus $c \in \mathbf{Spin} V \cap \operatorname{Ker} \chi$ if and only if $\alpha^2 = 1$, $\alpha = \pm 1$.

□

The invertible elements of C form an open set U (if $\lambda_x : C \to C$ is the operation of left multiplication by x, then λ_x is invertible if and only if $\det \lambda_x \neq 0$, and when this is so, $x^{-1} = \lambda_x^{-1}(1)$). The coördinates of xy (with respect to some basis for C) are polynomials in the coördinates of x and y, and the coördinates of xyx^{-1} have the form

$$P(x_1, \ldots, x_r, y_1, \ldots, y_r)/Q(x_1, \ldots, x_r)$$

where P is a polynomial and $Q(x_1, \ldots, x_r) = \det \lambda_x \neq 0$. Hence U is a Lie group and $\chi : U \to \mathbf{O}(V)$ a continuous homomorphism. Finally, $\mathbf{Spin}(V)$ is a closed subgroup of U and therefore a Lie group.

(3.14) Theorem *If $n \geq 2$, the group $\mathbf{Spin}(V)$ is a compact connected Lie group and $\chi|\mathbf{Spin}(V) : \mathbf{Spin}\,V \to \mathbf{O}^+(V)$ a two sheeted covering. If $n \geq 3$, $\chi|\mathbf{Spin}(V)$ is the universal covering of $\mathbf{O}^+(V)$.*

Since $\chi|\mathbf{Spin}(V)$ is an epimorphism and $\mathrm{Ker}(\chi|\mathbf{Spin}(V))$ has two elements, and since $\mathbf{O}^+(V)$ is compact, it follows that $\mathbf{Spin}(V)$ is compact and $\chi|\mathbf{Spin}(V)$ a covering map. To prove that $\mathbf{Spin}(V)$ is connected, it suffices, from general properties of covering spaces, to exhibit a path in $\mathbf{Spin}(V)$ from 1 to -1. Such a path $u : \mathbf{I} \to \mathbf{Spin}(V)$ is given by

$$\begin{aligned} u(t) &= (c(t)v_1 - s(t)v_2)(c(t)v_1 + s(t)v_2) \\ &= c(t)^2 v^2 + c(t)s(t)v_1 v_2 - s(t)c(t)v_2 v_1 - s(t)^2 v_2^2 \\ &= \{c(t)^2 - s(t)^2\} \cdot 1 + 2c(t)s(t)v_1 v_2 \\ &= c(2t) \cdot 1 + s(2t)v_1 v_2, \end{aligned}$$

where

$$c(t) = \cos \frac{\pi}{2} t, \qquad s(t) = \sin \frac{\pi}{2} t.$$

That $\chi|\mathbf{Spin}(V)$ is the universal covering of $\mathbf{O}^+(V)$ follows from the fact, proved in Theorem (10.7) of Chapter IV, that $\pi_1(\mathbf{O}^+(V)) = \mathbf{Z}_2$ if $\dim V \geq 3$. □

Suppose that V is a subspace of W (as an inner product space). If $i : V \to W$ is the inclusion map, then, in $C(W)$, we have the relation

$$i(x)^2 = \|i(x)\|^2 \cdot 1 = \|x\|^2 \cdot 1;$$

by the universal property of Clifford algebras, i has an extension to a homomorphism $\bar{\imath} : C(V) \to C(W)$ of algebras. It is easy to see that $\bar{\imath}$ is, in fact, a monomorphism, and that $\bar{\imath}(\mathbf{Spin}(V)) \subset \mathbf{Spin}(W)$. Thus we may regard $C(V)$ as a subalgebra of $C(W)$ and $\mathbf{Spin}(V)$ as a subgroup of $\mathbf{Spin}(W)$. We may identify $\mathbf{O}^+(V)$ with the subgroup of $\mathbf{O}^+(W)$ which leaves fixed each vector in the orthogonal complement of V. It is then easy to verify

(3.15) *The diagram*

$$\begin{array}{ccc} \mathbf{Spin}\,(V) & \xrightarrow{\bar{\imath}|\mathbf{Spin}\,(V)} & \mathbf{Spin}\,(W) \\ {\scriptstyle \chi|\mathbf{Spin}\,(V)}\downarrow & & \downarrow{\scriptstyle \chi|\mathbf{Spin}\,(W)} \\ \mathbf{O}^+(V) & \hookrightarrow & \mathbf{O}^+(W) \end{array}$$

is commutative. □

In particular, we may consider the sequence of subspaces $\{\mathbf{R}^n\}$ of $\mathbf{R}^\infty$. Let $\mathbf{Spin}(n) = \mathbf{Spin}(\mathbf{R}^n)$; then we have an expanding sequence of spaces

$$\mathbf{Spin}(1) \subset \mathbf{Spin}(2) \subset \cdots \subset \mathbf{Spin}(n) \subset \mathbf{Spin}(n+1) \subset \cdots$$

and we may put

$$\mathbf{Spin} = \bigcup_n \mathbf{Spin}(n);$$

the space **Spin** is filtered by its subspaces $\mathbf{Spin}(n)$.

4 The Cayley Algebra K

In this section we develop some further properties of the Cayley algebra. Let λ_x, ρ_x be the operations of left and right translations, respectively, defined by

$$\lambda_x(y) = \rho_y(x) = xy$$

for $x, y \in K$. The matrix of $\lambda_x(x = \sum_{i=0}^{7} x_i\mathbf{e}_i \in \mathbf{K})$ is

$$L_x = \begin{pmatrix} x_0 & -x_1 & -x_2 & -x_3 & -x_4 & -x_5 & -x_6 & -x_7 \\ x_1 & x_0 & -x_4 & -x_7 & x_2 & -x_6 & x_5 & x_3 \\ x_2 & x_4 & x_0 & -x_5 & -x_1 & x_3 & -x_7 & x_6 \\ x_3 & x_7 & x_5 & x_0 & -x_6 & -x_2 & x_4 & -x_1 \\ x_4 & -x_2 & x_1 & x_6 & x_0 & -x_7 & -x_3 & x_5 \\ x_5 & x_6 & -x_3 & x_2 & x_7 & x_0 & -x_1 & -x_4 \\ x_6 & -x_5 & x_7 & -x_4 & x_3 & x_1 & x_0 & -x_2 \\ x_7 & -x_3 & -x_6 & x_1 & -x_5 & x_4 & x_2 & x_0 \end{pmatrix},$$

while the matrix of ρ_y is

$$R_y = \begin{pmatrix} y_0 & -y_1 & -y_2 & -y_3 & -y_4 & -y_5 & -y_6 & -y_7 \\ y_1 & y_0 & y_4 & y_7 & -y_2 & y_6 & -y_5 & -y_3 \\ y_2 & -y_4 & y_0 & y_5 & y_1 & -y_3 & y_7 & -y_6 \\ y_3 & -y_7 & -y_5 & y_0 & y_6 & y_2 & -y_4 & y_1 \\ y_4 & y_2 & -y_1 & -y_6 & y_0 & y_7 & y_3 & -y_5 \\ y_5 & -y_6 & y_3 & -y_2 & -y_7 & y_0 & y_1 & y_4 \\ y_6 & y_5 & -y_7 & y_4 & -y_3 & -y_1 & y_0 & y_2 \\ y_7 & y_3 & y_6 & -y_1 & y_5 & -y_4 & -y_2 & y_0 \end{pmatrix}.$$

Let $\bar{x} = x_0 - \sum_{i=1}^{7} x_i\mathbf{e}_i$; then one verifies directly that

(4.1) $$x\bar{x} = \bar{x}x = \|x\|^2\mathbf{e}_0,$$

that $L_{\bar{x}}$ and $R_{\bar{y}}$ are the transposes L_x^*, R_y^* of L_x, R_y, respectively, and that

(4.2) $$L_x L_x^* = \|x\|^2 I = R_x R_x^*.$$

Thus if $\|x\| = 1$, L_x and R_x are orthogonal matrices, and therefore λ_x and ρ_x

are orthogonal linear transformations. In particular, they preserve the inner product, so that

(4.3) $$\langle xa, xb\rangle = \langle a, b\rangle = \langle ax, bx\rangle \quad \text{for } \|x\| = 1.$$

(We have written $\langle a, b\rangle$, rather than $a \cdot b$, to avoid confusion with the product in **K**). Then

$$\frac{1}{\|c\|^2}\langle ca, cb\rangle = \left\langle \frac{c}{\|c\|}a, \frac{c}{\|c\|}b\right\rangle = \langle a, b\rangle$$

so that

(4.4) $$\langle ca, cb\rangle = \langle c, c\rangle\langle a, b\rangle = \langle ac, bc\rangle,$$

which holds even if $c = 0$. Polarization of (4.4) (substitute $x + y$ for c and subtract the equations resulting from (4.4) by substituting each of x, y for c) yields

(4.5) $$\langle xa, yb\rangle + \langle ya, xb\rangle = 2\langle x, y\rangle\langle a, b\rangle.$$

Because $L_{\bar{x}}$ and $R_{\bar{x}}$ are the transposed matrices of L_x and R_x, respectively, we have

(4.6) $$\langle xa, b\rangle = \langle a, \bar{x}b\rangle,$$

(4.7) $$\langle ax, b\rangle = \langle a, b\bar{x}\rangle.$$

Put $a = \mathbf{e}_0$ in (4.6) to obtain

$$\langle x, b\rangle = \langle \mathbf{e}_0, \bar{x}b\rangle.$$

Put $a = \mathbf{e}_0$ and replace b, x by $\bar{x}$, $\bar{b}$ in (4.7) to obtain

$$\langle \mathbf{e}_0, \bar{x}b\rangle = \langle \bar{b}, \bar{x}\rangle.$$

By symmetry of the inner product, we have

(4.8) $$\langle x, b\rangle = \langle \bar{x}, \bar{b}\rangle.$$

From (4.4) and (4.6) we deduce

$$\langle a, (\bar{c}c)b\rangle = \|c\|^2\langle a, b\rangle = \langle ca, cb\rangle = \langle a, \bar{c}(cb)\rangle;$$

since this is true for every a, we have

(4.9) $$(\bar{c}c)b = \bar{c}(cb),$$

and similarly

(4.10) $$b(c\bar{c}) = (bc)\bar{c}.$$

From (4.9) and (4.10) we easily deduce

(4.11) $$c^2b = c(cb), \qquad bc^2 = (bc)c;$$

polarization yields

$$(4.12)\qquad \begin{aligned}(xy)b + (yx)b &= x(yb) + y(xb),\\ b(xy) + b(yx) &= (bx)y + (by)x.\end{aligned}$$

The *associator* of a triple (x, y, z) is

$$[x, y, z] = (xy)z - x(yz).$$

Equations (4.12) then assert that

$$(4.13)\qquad \begin{aligned}[x, y, b] + [y, x, b] &= 0,\\ [b, y, x] + [b, x, y] &= 0.\end{aligned}$$

It follows from (4.13) that, if σ is any permutation of $\{1, 2, 3\}$, then

$$(4.14)\qquad [x_{\sigma(1)}, x_{\sigma(2)}, x_{\sigma(3)}] = (\operatorname{sgn} \sigma)[x_1, x_2, x_3].$$

In particular,

(4.15) $[x, y, z] = 0$ whenever two of x, y, z are equal.

Therefore

(4.16) Theorem *The algebra* **K** *is alternative*; i.e., *the subalgebra* (*with unit*) *generated by any two elements of* **K** *is associative.* □

However, **K** is not associative; for

$$(\mathbf{e}_1\mathbf{e}_2)\mathbf{e}_3 = \mathbf{e}_4\,\mathbf{e}_3 = -\mathbf{e}_6,$$

while

$$\mathbf{e}_1(\mathbf{e}_2\,\mathbf{e}_3) = \mathbf{e}_1\mathbf{e}_5 = \mathbf{e}_6.$$

In fact, **K** is the best-known example of an alternative algebra which is not associative.

The relation $[a, x, y] = [x, y, a]$ implied by (4.14), when written out, yields

$$(4.17)\qquad (ax)y + x(ya) = a(xy) + (xy)a.$$

Replacing, firstly x by ax, and secondly, y by ya, and using (4.11), we obtain two relations

$$(4.18)\qquad \begin{aligned}(a^2x)y + (ax)(ya) &= a((ax)y) + ((ax)y)a,\\ (ax)(ya) + x(ya^2) &= a(x(ya)) + (x(ya))a;\end{aligned}$$

adding these, we obtain

$$\begin{aligned}(4.19)\quad (a^2x)y + 2(ax)(ya) + x(ya^2) &= a\{(ax)y + x(ya)\} + \{(ax)y + x(ya)\}a\\ &= a\{a(xy) + (xy)a\} + \{a(xy) + (xy)a\}a \qquad \text{by (4.17)}\\ &= a^2(xy) + 2a(xy)a + (xy)a^2 \qquad \text{by (4.11)}\end{aligned}$$

(note that $a((xy)a) = (a(xy))a$ by the alternative law, and we may write both expressions as $a(xy)a$). Substitute a^2 for a in (4.17) to obtain

(4.20) $$(a^2x)y + x(ya^2) = a^2(xy) + (xy)a^2;$$

subtracting (4.20) from (4.19), and dividing the result by 2, we obtain

(4.21) $$(ax)(ya) = a(xy)a.$$

Finally, let the *real* and *imaginary parts* of $x \in K$ be defined by

$$\mathscr{R}(x)\mathbf{e}_0 = \tfrac{1}{2}(x + \bar{x}),$$
$$\mathscr{I}(x) = \tfrac{1}{2}(x - \bar{x});$$

thus, if $x = \sum_{i=0}^{7} x_i \mathbf{e}_i$ then

$$x_0 = \mathscr{R}(x),$$
$$\mathscr{I}(x) = \sum_{i=1}^{7} x_i \mathbf{e}_i.$$

The elements $x \in \mathbf{K}$ with $\mathscr{R}x = 0$ form a subspace $\mathbf{K}_0$; the elements of $\mathbf{K}_0$ are said to be *pure imaginary*.

(4.22) Theorem *Every element* $x \in \mathbf{K}$ *satisfies the quadratic equation*

(4.23) $$x^2 - 2\mathscr{R}(x)x + \|x\|^2\mathbf{e}_0 = 0.$$

If x is not a scalar multiple of $\mathbf{e}_0$, (4.23) *is the only monic quadratic equation satisfied by* x.

For

$$\begin{aligned} x^2 &= x(2\mathscr{R}(x)\mathbf{e}_0 - \bar{x}) \\ &= 2\mathscr{R}(x)x - x\bar{x} \\ &= 2\mathscr{R}(x)x - \|x\|^2\mathbf{e}_0. \end{aligned}$$

Suppose that x satisfies two distinct monic quadratic equations

$$x^2 + \alpha x + \beta\mathbf{e}_0 = 0,$$
$$x^2 + \gamma x + \delta\mathbf{e}_0 = 0.$$

Then

$$(\alpha - \gamma)x + (\beta - \delta)\mathbf{e}_0 = 0$$

which implies that x and $\mathbf{e}_0$ are linearly dependent. □

5 Automorphisms of K

The object of this section is to determine the automorphism group $\mathbf{G}_2$ of $\mathbf{K}$. Let us begin with some observations about pure imaginary elements, i.e., elements of $\mathbf{K}_0$.

First, let $a \in \mathbf{K}_0$, so that $\bar{a} = -a$. Hence

$$a^2 = -a\bar{a} = -\|a\|^2\mathbf{e}_0.$$

In particular,

(5.1) *If* $a \in \mathbf{K}_0$, $\|a\| = 1$, *then* $a^2 = -\mathbf{e}_0$.

Next let a, b be an orthonormal pair in $\mathbf{K}_0$. Then

$$\begin{aligned}\langle ab, \mathbf{e}_0\rangle &= \langle a, \mathbf{e}_0\bar{b}\rangle \quad \text{by (4.6)}\\ &= \langle a, \bar{b}\rangle = -\langle a, b\rangle = 0,\end{aligned}$$

so that ab also belongs to $\mathbf{K}_0$. Moreover,

(5.2) $$ba = (-\bar{b})(-\bar{a}) = \bar{b}\bar{a} = \overline{ab} = -ab.$$

Furthermore,

$$\langle a, ab\rangle = \langle a\mathbf{e}_0, ab\rangle = \langle \mathbf{e}_0, b\rangle = 0 \quad \text{by (4.4)},$$

and similarly

$$\langle b, ab\rangle = 0.$$

Finally,

$$\|ab\| = \|a\| \cdot \|b\| = 1.$$

Let L be the subspace of $\mathbf{K}$ spanned by $\mathbf{e}_0$, a, b, and ab.

(5.3) Theorem *If* (a, b) *is an orthonormal pair in* $\mathbf{K}_0$, *and* L *is the subspace of* $\mathbf{K}$ *spanned by* $\mathbf{e}_0$, a, b *and* ab, *then* L *is a subalgebra of* $\mathbf{K}$, *isomorphic with the quaternion algebra* $\mathbf{Q}$. *Moreover, the linear map of* $\mathbf{Q}$ *into* L *which sends* $1, i, j, k$ *into* $\mathbf{e}_0$, a, b, ab, *respectively, is an isomorphism of algebras.*

It suffices to determine the multiplication table of the basis $\{\mathbf{e}_0, a, b, ab\}$ for L. Of course, $\mathbf{e}_0$ is the unit element; since a, b, ab belong to $\mathbf{K}_0$ and all have norm 1, we have

$$a^2 = b^2 = (ab)^2 = -\mathbf{e}_0,$$

by (5.1).

We have seen that $ba = -ab$. Moreover

$$\begin{aligned}a(ab) &= a^2b = -b \quad \text{by (4.11)}\\ &= -(ab)a \quad \text{by (5.2)},\end{aligned}$$

and similarly,

$$(ab)b = -a = -b(ab).$$

This completes the determination of the multiplication table. □

We next consider an orthonormal triple (a, b, c) of elements of $\mathbf{K}_0$. Let us call such a triple *special* if and only if c is orthogonal to ab.

If (a, b, c) is special, then

$$\begin{aligned}\langle a, bc\rangle &= \langle \bar{b}a, c\rangle \quad \text{by (4.6)}\\ &= \langle -ba, c\rangle = \langle ab, c\rangle = 0 \quad \text{by (5.2)}\end{aligned}$$

and similarly $\langle b, ca\rangle = 0$. Since $ab = -ba$, c is orthogonal to ab if and only if it is orthogonal to ba. Thus

(5.4) *Any permutation of the members of a special triple yields a special triple.*

We next prove

(5.5) *If (a, b, c) is a special triple, then*

(5.6)
$$(ab)c = -a(bc),$$

so that $[a, b, c] = 2(ab)c$.

For

$$\begin{aligned}a(bc) &= -[a, b, c] + (ab)c\\ &= -[b, c, a] + (ab)c \quad \text{by (4.14)}\\ &= -(bc)a + b(ca) + (ab)c\\ &= a(bc) + b(ca) + (ab)c \quad \text{by (5.2)}\end{aligned}$$

so that

$$(ab)c = -b(ca).$$

Hence

(5.7)
$$(ab)c = (ca)b.$$

Permuting a, b, c cyclically, we find

(5.8)
$$\begin{aligned}(ca)b &= (bc)a\\ &= -a(bc) \quad \text{by (5.2)}.\end{aligned}$$

Combining (5.7) and (5.8), we obtain (5.6). □

The main step in determining $\mathbf{G}_2$ is

(5.9) Theorem *If (a, b, c) is a special triple, there is an automorphism τ of* $\mathbf{K}$ *such that*

$$\tau(\mathbf{e}_1) = a,\ \tau(\mathbf{e}_2) = b,\ \tau(\mathbf{e}_7) = c.$$

We first observe that the elements

$$\mathbf{e}_0,\ a,\ b,\ ca,\ ab,\ c(ab),\ cb,\ c$$

form an orthonormal octuple. As these elements all have norm one, we have only to verify that they are mutually orthogonal. We have seen that $\mathbf{e}_0$, a, b, ab are mutually orthogonal. As λ_c is orthogonal, it follows that c, ca, cb, $c(ab)$ are also mutually orthogonal. It remains to prove that x and cy are orthogonal for all $x, y \in L$. We may assume $\|y\| = 1$, so that $y^2 = -\mathbf{e}_0$. Then

$$\langle cy, x\rangle = \langle cy, -xy^2\rangle = \langle cy, -(xy)y\rangle = -\langle c, xy\rangle = 0$$

by (4.11) and (4.3) and the fact that $xy \in L$.

Let τ be the linear transformation which carries $\mathbf{e}_0, \mathbf{e}_1, \ldots, \mathbf{e}_7$ into $\mathbf{e}_0$, a, b, ca, ab, $c(ab)$, cb, c, respectively. The elements $a, b, \ldots, c$ are mutually orthogonal unit vectors in $\mathbf{K}_0$, and it follows from (5.1) and (5.2) that they anticommute and each has square $-\mathbf{e}_0$. As $\tau(\mathbf{e}_0) = \mathbf{e}_0$ is the unit element of $\mathbf{K}$, it remains to prove that $\tau(\mathbf{e}_i\mathbf{e}_j) = \tau(\mathbf{e}_i)\tau(\mathbf{e}_j)$ for $1 \le i < j \le 7$. These are easily proved, using (4.11), (4.21), (5.1), (5.2), and (5.5). As a sample, consider the case $i = 3$, $j = 5$. We have

$$\begin{aligned}\tau(\mathbf{e}_3)\tau(\mathbf{e}_5) &= (ca)(c(ab))\\ &= -(ca)((ca)b) \quad \text{by (5.5)}\\ &= -(ca)^2 b \quad \text{by (4.11)}\\ &= b \quad \text{by (5.1)}\end{aligned}$$

while

$$\tau(\mathbf{e}_3\mathbf{e}_5) = \tau(\mathbf{e}_2) = b.$$

The remaining cases are similar and (mostly) easier, and the patient reader should have no trouble with them. □

(5.10) Theorem *Let τ be an automorphism of* $\mathbf{K}$. *Then τ is orthogonal, and the vectors $\tau(\mathbf{e}_1)$, $\tau(\mathbf{e}_2)$, $\tau(\mathbf{e}_7)$ form a special triple.*

Let $x \in \mathbf{K}$, and apply τ to both sides of (4.23). Since τ is an automorphism, $\tau(x^2) = \tau(x)^2$, $\tau(\mathbf{e}_0) = \mathbf{e}_0$, so that

(5.11) $$\tau(x)^2 - 2\mathscr{R}(x)\tau(x) + \|x\|^2\mathbf{e}_0 = 0.$$

If x is not a scalar multiple of $\mathbf{e}_0$, neither is $\tau(x)$, and we deduce from Theorem (4.22) that

(5.12) $$\mathscr{R}(\tau(x)) = \mathscr{R}(x), \ \|\tau(x)\|^2 = \|x\|^2.$$

If $x = \lambda\mathbf{e}_0$, then $\tau(x) = \lambda\tau(\mathbf{e}_0) = \lambda\mathbf{e}_0$, and we see that (5.12) is still satisfied.

The second relation of (5.12) assures us that τ is orthogonal. In particular, τ carries $\mathbf{K}_0$ into itself and it is clear that τ carries special triples into special triples. □

Let $\mathbf{G}_2$ be the group of automorphisms of $\mathbf{K}$. Each element of $\mathbf{G}_2$ is an orthogonal transformation leaving fixed the unit vector $\mathbf{e}_0$; thus $\mathbf{G}_2$ may be regarded as a subgroup of $\mathbf{O}_7$. Evidently $\mathbf{G}_2$ is closed, so that

(5.13) *The group* $\mathbf{G}_2$ *is a compact Lie group.* □

The group $\mathbf{G}_2$, as a subgroup of $\mathbf{O}_7$, acts on the manifold $\mathbf{V}_{7,2}$. By Theorems (5.9), (5.10) the action is transitive, and the map $\tau \to \tau(\mathbf{e}_7)$ is a one-to-one and continuous map of the isotropy group upon the set $\mathbf{S}_0$ of all unit vectors which are orthogonal to each of $\mathbf{e}_1$, $\mathbf{e}_2$ and $\mathbf{e}_4$. The set $\mathbf{S}_0$ is the unit sphere in the four-dimensional subspace of $\mathbf{K}_0$ orthogonal to $\mathbf{e}_1$, $\mathbf{e}_2$ and $\mathbf{e}_4$. Hence the map $\pi : \mathbf{G}_2 \to \mathbf{V}_{7,2}$ defined by

$$\pi(\tau) = (\tau(\mathbf{e}_1), \tau(\mathbf{e}_2))$$

is a fibration with fibre $\mathbf{S}^3$. Since $\mathbf{S}^3$ and $\mathbf{V}_{7,2}$ are connected, $\mathbf{G}_2$ is connected, and therefore $\mathbf{G}_2 \subset \mathbf{O}_7^+$.

The group $\mathbf{G}_2$ also acts on $\mathbf{S}^6$; the action is again transitive, and the isotropy group H is a subgroup of $\mathbf{O}_6^+$, which we may regard as the rotation group of the six-dimensional subspace $\mathbf{K}_1$ of $\mathbf{K}$ orthogonal to $\mathbf{e}_0$ and $\mathbf{e}_1$. The restriction of $\lambda_{\mathbf{e}_1}$ to $\mathbf{K}_1$ is an orthogonal transformation θ, and because of (4.11),

$$\theta(\theta(x)) = \mathbf{e}_1(\mathbf{e}_1 x) = \mathbf{e}_1^2 x = -x,$$

so that $-\theta^2$ is the identity. Hence $\mathbf{K}_1$ may be considered as a complex unitary space with

$$ix = \mathbf{e}_1 x$$

and inner product

$$\langle\langle x, y\rangle\rangle = \langle x, y\rangle + i\langle \mathbf{e}_1 x, y\rangle;$$

note that

$$\begin{aligned}\langle\langle x, x\rangle\rangle &= \langle x, x\rangle + i\langle \mathbf{e}_1 x, x\rangle \\ &= \langle x, x\rangle + i\langle \mathbf{e}_1 x, \mathbf{e}_0 x\rangle \\ &= \langle x, x\rangle + i\langle x, x\rangle\langle \mathbf{e}_1, \mathbf{e}_0\rangle = \|x\|^2.\end{aligned}$$

If $\tau \in H$, then $\tau(\mathbf{e}_1) = \mathbf{e}_1$ implies that $\tau \circ \lambda_{\mathbf{e}_1} = \lambda_{\mathbf{e}_1} \circ \tau$, and therefore τ is a linear transformation of the complex vector space $\mathbf{K}_1$. Moreover,

$$\begin{aligned}\langle\langle \tau x, \tau y\rangle\rangle &= \langle \tau x, \tau y\rangle + i\langle \mathbf{e}_1 \tau x, \tau y\rangle \\ &= \langle x, y\rangle + i\langle \tau(\mathbf{e}_1 x), \tau y\rangle \\ &= \langle x, y\rangle + i\langle \mathbf{e}_1 x, y\rangle = \langle\langle x, y\rangle\rangle,\end{aligned}$$

so that τ is unitary. Hence H is contained in the unitary group of $\mathbf{K}_1 \approx \mathbf{U}_3$. We claim that $H = \mathbf{U}_3^+$.

It suffices to show that $H \subset \mathbf{U}_3^+$; for it follows from the fibration

$$H \to \mathbf{G}_2 \to \mathbf{S}^6$$

that $\dim H - \dim \mathbf{G}_2 = 6$. But from the fibration

$$\mathbf{S}^3 \to \mathbf{G}^2 \to \mathbf{V}_{7,2}$$

we deduce that $\dim \mathbf{G}_2 = 3 + \dim \mathbf{V}_{7,2} = 3 + 11 = 14$. Hence $\dim H = 8$; since $\dim \mathbf{U}_3^+ = 8$, it follows from invariance of domain that H is open, as well as closed in $\mathbf{U}_3^+$, and therefore $H = \mathbf{U}_3^+$.

Let $\tau \in H$, and let $f_2, f_3 \in \mathbf{K}_1$ be an orthonormal pair of characteristic vectors for τ. Then

$$0 = \langle\langle f_2, f_3\rangle\rangle = \langle f_2, f_3\rangle + i\langle \mathbf{e}_1 f_2, f_3\rangle$$

and therefore $\langle f_2, f_3\rangle = \langle \mathbf{e}_1 f_2, f_3\rangle = 0$; thus $(\mathbf{e}_1, f_2, f_3)$ is a special triple. Therefore, by Theorem (5.9) there exists $\sigma \in \mathbf{G}_2$ such that $\sigma(\mathbf{e}_1) = \mathbf{e}_1$, $\sigma(\mathbf{e}_2) = f_2$, $\sigma(\mathbf{e}_3) = f_3$. Let $\tau' = \sigma^{-1}\tau\sigma$, so that

$$\tau'(\mathbf{e}_2) = e^{i\alpha}\mathbf{e}_2, \qquad \tau'(\mathbf{e}_3) = e^{i\beta}\mathbf{e}_3,$$

i.e.,

$$\begin{aligned}\tau'(\mathbf{e}_2) &= \mathbf{e}_2 \cos\alpha + \mathbf{e}_4 \sin\alpha, \\ \tau'(\mathbf{e}_3) &= \mathbf{e}_3 \cos\beta + \mathbf{e}_7 \sin\beta,\end{aligned}$$

and therefore

$$\begin{aligned}\tau'(\mathbf{e}_5) = \tau'(\mathbf{e}_2\mathbf{e}_3) &= \tau'(\mathbf{e}_2)\tau'(\mathbf{e}_3) \\ &= \mathbf{e}_5(\cos\alpha\cos\beta - \sin\alpha\sin\beta) \\ &\quad + \mathbf{e}_6(-\sin\alpha\cos\beta - \cos\alpha\sin\beta),\end{aligned}$$

so that

$$\tau'(\mathbf{e}_5) = e^{-i(\alpha+\beta)}\mathbf{e}_5.$$

Then $\det\tau = \det\tau' = e^{i\alpha}e^{i\beta}e^{-i(\alpha+\beta)} = 1$, $\tau \in \mathbf{U}_3^+$. □

6 The Exceptional Jordan Algebra $\mathfrak{J}$

Let $\mathfrak{J}$ be the set of all 3 × 3 "Hermitian" matrices with entries in the Cayley algebra **K**. An element $X \in \mathfrak{J}$ has the form

(6.1)
$$X = \begin{pmatrix} \xi_1 & x_3 & \bar{x}_2 \\ \bar{x}_3 & \xi_2 & x_1 \\ x_2 & \bar{x}_1 & \xi_3 \end{pmatrix}$$

with $x_i \in \mathbf{K}$, $\xi_i \in \mathbf{R}$. Thus $\mathfrak{J}$ is a vector space of dimension 27 over **R**. The space $\mathfrak{J}$ is not closed under matrix multiplication; however, it is closed under the *Jordan product*, defined by

$$X \circ Y = \tfrac{1}{2}(XY + YX).$$

Thus $\mathfrak{J}$ is a commutative algebra over **R**, but $\mathfrak{J}$ is not associative. For more information on $\mathfrak{J}$ the reader is referred to Jacobson [J].

Let E_{ij} be the 3 × 3 integral matrix whose sole non-zero entry is a one in the ith row and jth column ($i, j = 1, 2, 3$); thus

$$E_{ij}E_{kl} = \delta_{jk}E_{il},$$

where δ_{jk} is the Kronecker delta. Let $E_i = E_{ii}$, $F_i = I - E_{ii}$, where I is the identity matrix; and, for each $x \in K$, let

$$\alpha_1(x) = xE_{23} + \bar{x}E_{32},$$
$$\alpha_2(x) = xE_{31} + \bar{x}E_{13},$$
$$\alpha_3(x) = xE_{12} + \bar{x}E_{21}.$$

Thus α_i is a linear isomorphism of **K** with a subspace $\mathfrak{U}_i$ of $\mathfrak{J}$, and $\mathfrak{J}$ is the direct sum

(6.2)
$$\mathfrak{J} = \mathbf{R}E_1 \oplus \mathbf{R}E_2 \oplus \mathbf{R}E_3 \oplus \mathfrak{U}_1 \oplus \mathfrak{U}_2 \oplus \mathfrak{U}_3.$$

The matrix X of (6.1) is then given by

$$X = \sum_{i=1}^{3} \xi_i E_i + \sum_{i=1}^{3} \alpha_i(x_i).$$

The multiplication table of $\mathfrak{J}$ is then given by the commutative law and

(6.3)
$$E_i \circ E_j = \delta_{ij}E_i;$$
$$E_i \circ \alpha_j(x) = \begin{cases} 0 & (i = j), \\ \tfrac{1}{2}\alpha_j(x) & (i \neq j), \end{cases}$$
$$\alpha_i(x) \circ \alpha_i(y) = \langle x, y\rangle F_i,$$
$$\alpha_i(x) \circ \alpha_{i+1}(y) = \tfrac{1}{2}\alpha_{i+2}(\bar{y}\bar{x}) \quad \text{(indices mod 3)}.$$

Note that equations (6.3) are unchanged by cyclic permutation of the indices (1, 2, 3). Therefore the map $\gamma : \mathfrak{J} \to \mathfrak{J}$ defined by

$$\gamma(X) = \begin{pmatrix} \xi_2 & x_1 & \bar{x}_3 \\ \bar{x}_1 & \xi_3 & x_2 \\ x_3 & \bar{x}_2 & \xi_1 \end{pmatrix}$$

is an automorphism, called the *circulator*.

Let $t(X) = \sum \xi_i$ be the trace of the matrix X. If $X \in \mathfrak{J}$, then $X \circ X = \frac{1}{2}(X^2 + X^2) = X^2$; moreover, $X \circ X^2 = X \circ (X \circ X) = (X \circ X) \circ X$, so that we can write $X \circ X \circ X$ without ambiguity. It is convenient to abbreviate the latter expression to X^3 (*warning*: this may not coincide with either of the matrix products $X \cdot X^2$ or $X^2 \cdot X$). In fact, the algebra $\mathfrak{J}$ is *power-associative*, so that we may write $X^n = X \circ \cdots \circ X$ (n factors) for any parenthesization of the product.

Besides the trace function t, the quadratic and cubic forms defined by

$$q(X) = t(X^2)$$
$$c(X) = t(X^3)$$

are of importance for the structure of the algebra $\mathfrak{J}$. We can calculate these forms explicitly, with the results

$$\begin{aligned} q(X) &= \sum \xi_i^2 + 2 \sum \|x_i\|^2, \\ c(X) &= \sum \xi_i^3 + 3\{(\xi_2 + \xi_3)\|x_1\|^2 + (\xi_3 + \xi_1)\|x_2\|^2 + (\xi_1 + \xi_2)\|x_3\|^2\} \\ &\quad + 6\langle x_1, x_2, x_3 \rangle, \end{aligned}$$

where

$$\langle x_1, x_2, x_3 \rangle = \langle x_2, x_3, x_1 \rangle = \langle x_3, x_1, x_2 \rangle = \langle x_1 x_2, \bar{x}_3 \rangle.$$

In particular,

(6.4) *The quadratic form q is positive definite.*

The space $\mathfrak{J}$ is then an inner-product space with norm $\sqrt{q(X)}$. However, it will somewhat simplify subsequent calculations if we define instead

$$\|X\|^2 = \tfrac{1}{2} q(X) = \tfrac{1}{2} \sum \xi_i^2 + \sum \|x_i\|^2,$$

so that the associated inner product is

$$\begin{aligned} \langle X, Y \rangle &= \tfrac{1}{2}\{\|X + Y\|^2 - \|X\|^2 - \|Y\|^2\} \\ &= \tfrac{1}{2} t(X \circ Y). \end{aligned}$$

We then observe

(6.5) *The direct sum decomposition* (6.2) *is orthogonal. Moreover, the map $\alpha_i : \mathbf{K} \to \mathfrak{U}_i$ is an isometry.*

Let us introduce formal[1] characteristic roots $\rho_i (i = 1, 2, 3)$, so that

$$t(X) = \Sigma \rho_i,$$
$$q(X) = \Sigma \rho_i^2,$$
$$c(X) = \Sigma \rho_i^3.$$

Then the elementary symmetric functions $\sigma_i(X)$ of the ρ_i can be calculated by Newton's identities, and we have

$$\sigma_1(X) = t(X) = \Sigma \xi_i,$$
$$\sigma_2(X) = \tfrac{1}{2}\{t(X)^2 - q(X)\}$$
$$= \sum_{i<j} \xi_i \xi_j - \sum \|x_i\|^2$$
$$\sigma_3(X) = \xi_1 \xi_2 \xi_3 - \{(\xi_2 + \xi_3)\|x_1\|^2 + (\xi_1 + \xi_3)\|x_2\|^2 + (\xi_1 + \xi_2)\|x_3\|^2\} + 2\langle x_1, x_2, x_3 \rangle.$$

The *characteristic polynomial* of X is the polynomial

$$\phi(\lambda) = \lambda^3 - \sigma_1(X)\lambda^2 + \sigma_2(X)\lambda - \sigma_3(X),$$

and we have a kind of Cayley–Hamilton Theorem

(6.6) *The matrix $X \in \mathfrak{J}$ satisfies the relation*

$$X^3 - \sigma_1(X)X^2 + \sigma_2(X)X - \sigma_3(X)I = 0,$$

which is established by tedious but direct calculation.

(6.7) Theorem *If τ is an automorphism of $\mathfrak{J}$, then $\sigma_i(\tau(X)) = \sigma_i(X)$ $(i = 1, 2, 3)$. The forms t, q, c are invariant under τ. In particular, τ is an orthogonal transformation.*

The set K of elements X which satisfy some non-trivial quadratic equation $X^2 + qX + rI = 0$ is easily seen to be a proper algebraic subset of $\mathfrak{J}$; thus K is closed and nowhere dense, and its complement U is dense and invariant under τ. For $X \in U$, $X' = \tau(X)$, we have $\tau(X^2) = (X')^2$, $\tau(X^3) = (X')^3$. Hence

$$(X')^3 - \sigma_1(X)(X')^2 + \sigma_2(X)X' - \sigma_3(X)I = 0.$$

But

$$(X')^3 - \sigma_1(X')(X')^2 + \sigma_2(X')X' - \sigma_3(X')I = 0.$$

[1] It can be proved that, for each $X \in \mathfrak{J}$ there is an automorphism σ of $\mathfrak{J}$ such that $\sigma(X)$ is a diagonal matrix; the ρ_i are then the diagonal elements of $\sigma(X)$. In this sense, the ρ_i are characteristic roots of X.

Subtracting these two equations, we obtain a quadratic relation, whose triviality implies that $\sigma_i(X') = \sigma_i(X)$ for $i = 1, 2, 3$. Since these relations hold for $X \in U$, they hold by continuity for every $X \in \mathfrak{J}$. □

In order to study the automorphisms of $\mathfrak{J}$, we first examine the idempotents. If X is the matrix (6.1), then $X^2 = X$ if and only if the following conditions are satisfied:

$$\begin{aligned}
&\xi_1^2 + \|x_2\|^2 + \|x_3\|^2 = \xi_1; \qquad (\xi_2 + \xi_3)x_1 + \bar{x}_3\bar{x}_2 = x_1;\\
(6.8)\qquad &\xi_2^2 + \|x_1\|^2 + \|x_3\|^2 = \xi_2; \qquad (\xi_1 + \xi_3)x_2 + \bar{x}_1\bar{x}_3 = x_2;\\
&\xi_3^2 + \|x_1\|^2 + \|x_2\|^2 = \xi_3; \qquad (\xi_1 + \xi_2)x_3 + \bar{x}_2\bar{x}_1 = x_3.
\end{aligned}$$

Note that these conditions are unchanged under cyclic permutation of the integers 1, 2, 3.

(6.9) Lemma *If X is an idempotent, $0 \neq X \neq I$, then $t(X) = 1$ or $t(X) = 2$.*

For $X^2 = X$, $X^3 = X \circ X^2 = X \circ X = X^2 = X$, so that the Cayley-Hamilton equation (6.6) becomes

$$\{1 - \sigma_1(X) + \sigma_2(X)\}X - \sigma_3(X)I = 0.$$

If $X = \lambda I$ is a scalar multiple of the identity matrix I, then $\lambda^2 = \lambda$ and therefore $X = 0$ or $X = I$, and we have excluded this case. Hence X is not a scalar multiple of I, and we conclude

$$1 - \sigma_1(X) + \sigma_2(X) = 0,$$
$$\sigma_3(X) = 0.$$

By our calculation of the $\sigma_i(X)$, the first relation reduces to

$$\begin{aligned}
0 &= 1 - t(X) + \tfrac{1}{2}\{t(X)^2 - t(X^2)\}\\
&= 1 - t(X) + \tfrac{1}{2}\{t(X)^2 - t(X)\}\\
&= \tfrac{1}{2}\{t(X)^2 - 3t(X) + 2\},
\end{aligned}$$

which implies $t(X) = 1$ or $t(X) = 2$. □

An idempotent E is *primitive* if and only if $E \neq 0$ and there do not exist non-zero idempotents E', E'' such that

$$E = E' + E'', \qquad E' \circ E'' = 0.$$

Equivalently,

(6.10) *A non-zero idempotent E is primitive if and only if the only idempotents X such that $E \circ X = X$ are 0 and E.*

(6.11) Lemma *An idempotent E is primitive if and only if $t(E) = 1$.*

It follows from Lemma (6.9) that E is primitive if $t(E) = 1$. We have also seen, in Lemma (6.9), that $t(E) = 0, 3$ if and only if $E = 0, I$, respectively. Suppose that $t(E) = 2$. Let

$$E = \begin{pmatrix} \varepsilon_1 & e_3 & \bar{e}_2 \\ \bar{e}_3 & \varepsilon_2 & e_1 \\ e_2 & \bar{e}_1 & \varepsilon_3 \end{pmatrix}.$$

From the relations (6.8) and

(6.12) $$\varepsilon_1 + \varepsilon_2 + \varepsilon_3 = 2,$$

we deduce that

(6.13) $$\begin{aligned} \|e_1\|^2 &= (\varepsilon_2 - 1)(\varepsilon_3 - 1), \\ e_2 e_3 &= (\varepsilon_1 - 1)\bar{e}_1, \end{aligned}$$

together with the relations obtained from (6.13) by permuting the indices cyclically.

If $\varepsilon_1 \neq 0$, let

$$X = \frac{1}{\varepsilon_1} \begin{pmatrix} 0 & 0 & 0 \\ 0 & 1 - \varepsilon_3 & e_1 \\ 0 & \bar{e}_1 & 1 - \varepsilon_2 \end{pmatrix},$$

and verify that X is an idempotent $\neq 0, E$, and that $E \circ X = X$. Hence E is not primitive. If $\varepsilon_1 = 0$, deduce from (6.12) and (6.13) that $\varepsilon_2 = \varepsilon_3 = 1$, $e_1 = e_2 = e_3 = 0$, so that

$$E = E_2 + E_3$$

is not primitive. □

Let P be the set of all primitive idempotents. If $X \in P$, we have seen that $t(X) = \xi_1 + \xi_2 + \xi_3 = 1$. It follows from this fact and the relations (6.8) that

(6.14) *A matrix X belongs to P if and only if the following conditions are satisfied*:

$$\begin{aligned} x_2 x_3 &= \xi_1 \bar{x}_1, & \|x_1\|^2 &= \xi_2 \xi_3, & & \\ x_3 x_1 &= \xi_2 \bar{x}_2, & \|x_2\|^2 &= \xi_3 \xi_1, & \xi_1 + \xi_2 + \xi_3 &= 1. \\ x_1 x_2 &= \xi_3 \bar{x}_3, & \|x_3\|^2 &= \xi_1 \xi_2, & & \end{aligned}$$ □

Let P_0 be the subspace of P consisting of all matrices X such that $\xi_1 = 0$.

(6.15) Lemma *The space P_0 is homeomorphic with $\mathbf{S}^8$.*

We take $\mathbf{S}^8$ to be the set of all pairs (ζ, z) with $z \in \mathbf{K}$, $\zeta \in \mathbf{R}$, $\|z\|^2 + \zeta^2 = 1$.

The map $\phi : \mathbf{S}^8 \to P_0$ defined by

$$\phi(\zeta, z) = \tfrac{1}{2}\begin{pmatrix} 0 & 0 & 0 \\ 0 & 1+\zeta & z \\ 0 & \bar{z} & 1-\zeta \end{pmatrix}$$

is readily verified to be a homeomorphism of $\mathbf{S}^8$ with P_0. The inverse map is given by

$$\phi^{-1}(X) = (2\xi_2 - 1, 2x_1). \qquad \square$$

(6.16) Theorem *The space P is homeomorphic with the Cayley projective plane.*

The set of all pairs $y = (y_1, y_2)$ with $y_i \in \mathbf{K}$, $\|y\|^2 = \|y_1\|^2 + \|y_2\|^2 \le 1$ is a 16-cell E, and a relative homeomorphism $\psi : (E, \dot{E}) \to (P, P_0)$ is defined by

$$\psi(y) = \begin{pmatrix} 1 - \|y\|^2 & \sqrt{1-\|y\|^2}\,\bar{y}_1 & \sqrt{1-\|y\|^2}\,y_2 \\ \sqrt{1-\|y\|^2}\,y_1 & \|y_1\|^2 & y_1 y_2 \\ \sqrt{1-\|y\|^2}\,\bar{y}_2 & \bar{y}_2\bar{y}_1 & \|y_2\|^2 \end{pmatrix},$$

the inverse map is given by

$$\theta(X) = (\xi_1^{-1/2}\bar{x}_3, \xi_1^{-1/2}\bar{x}_2) \qquad (X \in P - P_0).$$

Therefore P is homeomorphic with the mapping cone of the map

$$\omega = \phi^{-1} \circ (\psi \,|\, \dot{E}) : \dot{E} \to \mathbf{S}^8,$$

and

$$\omega(y_1, y_2) = (2\|y_1\|^2 - 1, 2y_1 y_2).$$

Let $\sigma : \mathbf{S}^8 \to \mathbf{K} \cup \{\infty\}$ be stereographic projection from the south pole upon the equatorial plane, so that

$$\sigma(\zeta, z) = \begin{cases} \dfrac{z}{1+\zeta} & (\zeta \ne -1), \\ \infty & (\zeta = -1). \end{cases}$$

Then

$$\sigma\omega(y_1, y_2) = \begin{cases} \dfrac{2y_1 y_2}{2\|y_1\|^2} = \dfrac{y_1}{\|y_1\|^2} y_2 = \bar{y}_1^{-1} y_2 & (y_1 \ne 0), \\ \infty & (y_1 = 0), \end{cases}$$

so that $\sigma \circ \omega$ is the Hopf map of §1. $\square$

7 The Exceptional Lie Group $\mathbf{F}_4$

The group in question is the group of automorphisms of the algebra $\mathfrak{J}$. It is a closed subgroup of the full linear group $\mathbf{L}(\mathfrak{J})$; by Theorem (6.7), it is contained in the orthogonal group $\mathbf{O}(\mathfrak{J})$. Thus

(7.1) *The group $\mathbf{F}_4$ is a compact Lie group.* □

The group $\mathbf{F}_4$ acts on the space P of primitive idempotents. We shall see (Theorem 7.21, below) that the action is transitive. Thus there is a fibration

$$H \to \mathbf{F}_4 \to P,$$

where H is the subgroup of $\mathbf{F}_4$ consisting of all automorphisms leaving fixed a primitive idempotent E. Our first objective is to ascertain the structure of H.

Let $\varepsilon_i : \mathfrak{J} \to \mathfrak{J}$ be the operation of multiplication by the primitive idempotent E_i ($i = 1, 2, 3$). Then

$$\varepsilon_i(X) = \xi_i E_i + \tfrac{1}{2}\sum_{j \neq i} \alpha_j(x_j),$$

and it follows that ε_i is semi-simple and its characteristic values are $0, \frac{1}{2}$, and 1. Moreover, $\mathfrak{J}$ is the orthogonal direct sum

$$\mathfrak{J} = \mathfrak{J}_i(0) \oplus \mathfrak{J}_i(\tfrac{1}{2}) \oplus \mathfrak{J}_i(1), \tag{7.2}$$

where $\mathfrak{J}_i(\lambda)$ is the space of characteristic vectors for λ. Evidently

$$\mathfrak{J}_i(0) = \mathfrak{U}_i + \sum_{j \neq i} \mathbf{R}E_j,$$

$$\mathfrak{J}_i(\tfrac{1}{2}) = \sum_{j \neq i} \mathfrak{U}_j,$$

$$\mathfrak{J}_i(1) = \mathbf{R}E_i.$$

Let H_i be the set of all automorphisms of $\mathfrak{J}$ which map E_i into itself. Then each of the subspaces $\mathfrak{J}_i(\lambda)$ is invariant under H_i. Moreover, F_i is fixed by each element of H_i, so that the orthogonal complement $\mathfrak{V}_i$ of $\mathbf{R}F_i$ in $\mathfrak{J}_i(0)$ is invariant under H_i. Let $\mathfrak{W}_i = \mathfrak{J}_i(\frac{1}{2})$. Then

(7.3) Theorem *There is an orthogonal direct sum decomposition*

$$\mathfrak{J} = \mathbf{R}E_i \oplus \mathbf{R}F_i \oplus \mathfrak{V}_i \oplus \mathfrak{W}_i, \tag{7.4$_i$}$$

invariant under the action of H_i. □

We shall take the isotropy group H to be the group H_1. To simplify the notation, let us drop the subscript $i = 1$ in $(7.4)_i$. The subspaces $\mathbf{R}E$, $\mathbf{R}F$, $\mathfrak{V}$,

$\mathfrak{W}$ then consist of all matrices of the form

(7.5) $$\xi E = \begin{pmatrix} \xi & 0 & 0 \\ 0 & 0 & 0 \\ 0 & 0 & 0 \end{pmatrix},$$

(7.6) $$\eta F = \begin{pmatrix} 0 & 0 & 0 \\ 0 & \eta & 0 \\ 0 & 0 & \eta \end{pmatrix},$$

(7.7) $$V(\xi, x) = \begin{pmatrix} 0 & 0 & 0 \\ 0 & \xi & x \\ 0 & \bar{x} & -\xi \end{pmatrix},$$

(7.8) $$W(y, z) = \begin{pmatrix} 0 & z & \bar{y} \\ \bar{z} & 0 & 0 \\ y & 0 & 0 \end{pmatrix},$$

respectively. Moreover, $\mathfrak{V}$ and $\mathfrak{W}$ are the orthogonal direct sums

$$\mathfrak{V} = \mathbf{R}V_0 \oplus \mathfrak{U}_1,$$
$$\mathfrak{W} = \mathfrak{U}_2 \oplus \mathfrak{U}_3,$$

where V_0 is the matrix $-E_2 + E_3$.

Let us study the behavior of the product in $\mathfrak{J}$ under the decomposition (7.4). Our first observation is

(7.9) *If* $V = V(\xi, x)$, $W = W(y, z)$, *then*

$$V \circ W = \tfrac{1}{2}W(-\xi y + \bar{x}\bar{z}, \xi z + \bar{y}\bar{x}).$$

Moreover,

$$\|V \circ W\| = \tfrac{1}{2}\|V\| \cdot \|W\|.$$

The first statement is immediate. To prove the second, observe that

$$\|-\xi y + \bar{x}\bar{z}\|^2 = \xi^2\|y\|^2 - 2\xi\langle y, \bar{x}\bar{z}\rangle + \|x\|^2\|z\|^2,$$
$$\|\xi z + \bar{y}\bar{x}\|^2 = \xi^2\|z\|^2 + 2\xi\langle z, \bar{y}\bar{x}\rangle + \|x\|^2\|y\|^2.$$

But

$$\langle z, \bar{y}\bar{x}\rangle = \langle zx, \bar{y}\rangle = \langle y, \bar{x}\bar{z}\rangle$$

by (4.7) and (4.8), and therefore

$$\begin{aligned}\|V \circ W\|^2 &= \tfrac{1}{4}\{\|-\xi y + \bar{x}\bar{z}\|^2 + \|\xi z + \bar{y}\bar{x}\|^2\} \\ &= \tfrac{1}{4}(\xi^2 + \|x\|^2)(\|y\|^2 + \|z\|^2) \\ &= \tfrac{1}{4}\|V\|^2 \cdot \|W\|^2.\end{aligned}$$

□

We next calculate the product of two elements of $\mathfrak{W}$. Let $W_i = W(y_i, z_i)$ $(i = 1, 2)$. Then

$$W_1 \circ W_2 = \begin{pmatrix} \langle y_1, y_2\rangle + \langle z_1, z_2\rangle & 0 & 0 \\ 0 & \langle z_1, z_2\rangle & \frac{1}{2}(\bar{z}_1\bar{y}_2 + \bar{z}_2\bar{y}_1) \\ 0 & \frac{1}{2}(y_2 z_1 + y_1 z_2) & \langle y_1, y_2\rangle \end{pmatrix};$$

developing this product with respect to the decomposition (7.4), we find that

(7.10) $$W_1 \circ W_2 = \langle W_1, W_2\rangle(E + \tfrac{1}{2}F) + W_1 * W_2,$$

where

(7.11) $$W_1 * W_2 = \tfrac{1}{2}V(\langle z_1, z_2\rangle - \langle y_1, y_2\rangle, \bar{z}_1\bar{y}_2 + \bar{z}_2\bar{y}_1).$$

The products $(V, W) \to V \circ W$ and $(W_1, W_2) \to W_1 * W_2$ determine each other. In fact, if V, W_1, W_2 are as above, then

$$\begin{aligned}\langle V \circ W_1, W_2\rangle &= \tfrac{1}{2}\{\langle -\xi y_1 + x\bar{z}_1, y_2\rangle + \langle \xi z_1 + \bar{y}_1\bar{x}, z_2\rangle\} \\ &= \tfrac{1}{2}\{\xi(\langle z_1, z_2\rangle - \langle y_1, y_2\rangle) + \langle \bar{x}, y_2 z_1 + y_1 z_2\rangle\} \\ &= \tfrac{1}{2}\{\xi\langle z_1, z_2\rangle - \langle y_1, y_2\rangle + \langle x, \bar{z}_1\bar{y}_2 + \bar{z}_2\bar{y}_1\rangle\} \\ &= \langle V, W_1 * W_2\rangle.\end{aligned}$$

Thus

(7.12) *If $W_1, W_2 \in \mathfrak{W}$, then*

$$W_1 \circ W_2 = \langle W_1, W_2\rangle(E + \tfrac{1}{2}F) + W_1 * W_2,$$

where $W_1 * W_2 \in \mathfrak{V}$ is uniquely determined by the condition

$$\langle V \circ W_1, W_2\rangle = \langle V, W_1 * W_2\rangle$$

for all $V \in \mathfrak{V}$, $W_1, W_2 \in \mathfrak{W}$. □

The remaining calculations needed to make explicit the product in $\mathfrak{J}$ are easily carried out, and we have

(7.13) Theorem *The product in $\mathfrak{J}$ is determined by the commutative law,* (7.9), (7.12), *and the relations*

$$\begin{aligned} E^2 &= E, & E \circ F &= 0, & E \circ V &= 0, & E \circ W &= \tfrac{1}{2}W, \\ & & F^2 &= F, & F \circ V &= V, & F \circ W &= \tfrac{1}{2}W, \\ & & & & V_1 \circ V_2 &= \langle V_1, V_2\rangle F, & & \end{aligned}$$

whenever $V, V_1, V_2 \in \mathfrak{V}$, $W \in \mathfrak{W}$. □

An easy calculation gives

(7.14) *If $V \in \mathfrak{V}$, $W \in \mathfrak{W}$, then*

(7.15) $$V \circ (V \circ W) = \tfrac{1}{4}\|V\|^2 W,$$

and therefore

(7.16) $$V_1 \circ (V_2 \circ W) + V_2 \circ (V_1 \circ W) = \tfrac{1}{2}\langle V_1, V_2\rangle W$$

for all $V_1, V_2 \in \mathfrak{V}$, $W \in \mathfrak{W}$. □

It is clear that the structure of $\mathfrak{J}$ depends in a crucial way on the product $(V, W) \to V \circ W$. This is made more explicit by

(7.17) Lemma *Let* $h: \mathfrak{J} \to \mathfrak{J}$ *be an orthogonal linear transformation which respects the direct sum decomposition* (7.4). *Then h is an automorphism if and only if* $h(I) = I$ *and* $h(V \circ W) = h(V) \circ h(W)$ *for all* $V \in \mathfrak{V}$, $W \in \mathfrak{W}$.

These conditions are clearly necessary for h to be an automorphism. We prove their sufficiency.

Since h respects (7.4) we have $h(E) = \lambda E$, $h(F) = \mu F$, and therefore $I = h(I) = h(E + F) = \lambda E + \mu F$, which implies that $\lambda = \mu = 1$. In order to show that h preserves the multiplication table as given in Theorem (7.13), the only nontrivial verifications are:

(1) $h(V_1 \circ V_2) = h(V_1) \circ h(V_2)$,
(2) $h(W_1 \circ W_2) = h(W_1) \circ h(W_2)$.

For (1), we have

$$\begin{aligned} h(V_1) \circ h(V_2) &= \langle h(V_1), h(V_2)\rangle F \quad (\text{since } h(V_i) \in \mathfrak{V}) \\ &= \langle V_1, V_2\rangle F \quad (\text{since } h \text{ is orthogonal}) \\ &= \langle V_1, V_2\rangle h(F) \\ &= h(V_1 \circ V_2). \end{aligned}$$

For (2), observe that $h(W_1), h(W_2) \in \mathfrak{W}$, and therefore

$$\begin{aligned} h(W_1) \circ h(W_2) &= \langle h(W_1), h(W_2)\rangle (E + \tfrac{1}{2}F) + h(W_1) * h(W_2) \\ &= \langle W_1, W_2\rangle (E + \tfrac{1}{2}F) + h(W_1) * h(W_2) \\ &\qquad (\text{since } h \text{ is orthogonal}), \end{aligned}$$

while

$$\begin{aligned} h(W_1 \circ W_2) &= \langle W_1, W_2\rangle (h(E) + \tfrac{1}{2}h(F)) + h(W_1 * W_2) \\ &= \langle W_1, W_2\rangle (E + \tfrac{1}{2}F) + h(W_1 * W_2). \end{aligned}$$

Thus it suffices to prove that $h(W_1) * h(W_2) = H(W_1 * W_2)$.

Let $V_1 \in \mathfrak{B}$; then

$$\begin{aligned}\langle h(V_1), h(W_1) * h(W_2)\rangle &= \langle h(V_1) \circ h(W_1), h(W_2)\rangle \quad \text{(by (7.12))}\\ &= \langle h(V_1 \circ W_1), h(W_2)\rangle \quad \text{(by hypothesis)}\\ &= \langle V_1 \circ W_1, W_2\rangle \quad \text{(since } h \text{ is orthogonal)}\\ &= \langle V_1, W_1 * W_2\rangle \quad \text{(by (7.12))}\\ &= \langle h(V_1), h(W_1 * W_2)\rangle\\ &\qquad \text{(since } h \text{ is orthogonal).}\end{aligned}$$

Since h maps $\mathfrak{B}$ *upon* $\mathfrak{B}$, this implies that

$$h(W_1 * W_2) = h(W_1) * h(W_2).$$ □

We can now determine the structure of H. In fact, we shall prove

(7.18) Theorem *There is an isomorphism* $\eta : \mathbf{Spin}(\mathfrak{B}) \approx H$.

We shall define a homomorphism $\eta : \mathbf{Spin}(\mathfrak{B}) \to \mathbf{O}^+(\mathfrak{J})$ and prove that η is an isomorphism of $\mathbf{Spin}(\mathfrak{B})$ upon H. As H leaves invariant the decomposition (7.4), it behooves us to construct representations of $\mathbf{Spin}(\mathfrak{B})$ in each of the subspaces occurring there. Since $\mathbf{R}E$ and $\mathbf{R}F$ are one-dimensional, the corresponding representations must be trivial. Moreover, there is at hand the vector representation $\chi : \mathbf{Spin}(\mathfrak{B}) \to \mathbf{O}^+(\mathfrak{B})$ of §3. We proceed to construct a representation $\zeta : \mathbf{Spin}(\mathfrak{B}) \to \mathbf{O}^+(\mathfrak{W})$, called the *spin representation*.

Let $\mathfrak{E}$ be the algebra of endomorphisms of the vector space $\mathfrak{W}$, and define $\theta : \mathfrak{B} \to \mathfrak{E}$ by

$$\theta(X) = \lambda_{2X} | \mathfrak{W},$$

where λ_{2X} is the operation of left-(= right-)multiplication by $2X$ in the Jordan algebra $\mathfrak{J}$. Then

$$\begin{aligned}\theta(X)^2(W) &= 4X \circ (X \circ W)\\ &= \|X\|^2 W \quad \text{by (7.15).}\end{aligned}$$

Thus $\theta(X)^2 = \|X\|^2 I$; by the universal property of the Clifford algebra $C = C(\mathfrak{B})$, θ extends to a homomorphism $\bar{\theta} : C \to \mathfrak{E}$. Let $\bar{\theta}_e : C^e \to \mathfrak{E}$ be the restriction of $\bar{\theta}$. The dimension of $\mathfrak{B}$ being 9, that of $\mathfrak{W}$ being 16, we have $\dim C^e = 2^8$, $\dim \mathfrak{E} = 16^2 = 2^8$; moreover, by Theorem (3.5), the algebra C^e is simple. Since $\bar{\theta}_e$ is manifestly non-trivial, we have

(7.19) *The homomorphism* $\bar{\theta}_e : C^e \to \mathfrak{E}$ *is an isomorphism.* □

Let $\zeta : \mathbf{Spin}(\mathfrak{B}) \to \mathfrak{E}$ be the restriction of $\bar{\theta}_e$. If $X \in \mathfrak{B}$, $\|X\| = 1$, then

$$\|\zeta(X)(W)\| = \|\lambda_{2X}(W)\| = 2\|X \circ W\| = \|X\| \cdot \|W\| = \|W\| \quad \text{by (7.9)}$$

and therefore $\zeta(X)$ is orthogonal. Since $\mathbf{Spin}(\mathfrak{V})$ is connected, the map $X \to \det \zeta(X)$ is constant. If $u \in \mathbf{Spin}(\mathfrak{V})$ then, by Theorem (3.12), u is the product of an even number of unit vectors $X_i \in \mathfrak{V}$, and we conclude

(7.20) *The representation ζ is a monomorphism*

$$\zeta : \mathbf{Spin}(\mathfrak{V}) \to \mathbf{O}^+(\mathfrak{W}). \qquad \square$$

We can now prove Theorem (7.18). The representation $\eta : \mathbf{Spin}(\mathfrak{V}) \to \mathbf{O}^+(\mathfrak{J})$ is defined to be the direct sum of the trivial representation in $\mathbf{R}E \oplus \mathbf{R}F$, the vector representation χ in $\mathfrak{V}$ and the spin representation ζ in $\mathfrak{W}$. Since ζ is a monomorphism, so is η.

To prove that $\operatorname{Im} \eta \subset H$, it suffices, in view of Lemma (7.17), to prove that, if $u \in \mathbf{Spin}(\mathfrak{V})$, then $h = \eta(u)$ satisfies the relation $h(V \circ W) = h(V) \circ h(W)$ for all $V \in \mathfrak{V}$, $W \in \mathfrak{W}$. We may assume, because of Theorem (3.12), that u is the product (in $C(\mathfrak{V})$) of two unit vectors X, Y. In this case, it follows from Lemma (3.7) that $h = \rho_X \rho_Y$, so that

$$h(V) = V - 2\{\langle X, V\rangle - 2\langle X, Y\rangle\langle V, Y\rangle\}X - 2\langle V, Y\rangle Y.$$

Moreover,

$$h(W) = 4X \circ (Y \circ W),$$

and, by the same token,

$$h(V \circ W) = 4X \circ (Y \circ (V \circ W)).$$

On the other hand,

$$h(V) \circ h(W) = 4V \circ (X \circ (Y \circ W)) - 2\langle V, Y\rangle Y \circ (X \circ (Y \circ W))$$
$$- 2\{\langle X, V\rangle - 2\langle X, Y\rangle\langle V, Y\rangle\}X \circ (X \circ (Y \circ W)).$$

A little calculation, using (7.16) as necessary, reveals that the latter expression is equal to $4X \circ (Y \circ (V \circ W)) = h(V \circ W)$.

Let $h \in H$, so that $h(E) = E$. Since h is an automorphism, $h(I) = I$, and therefore $h(F) = F$. We have seen (Theorem (6.7)) that h is orthogonal. By Theorem (7.3), h respects the direct sum decomposition (7.4).

The vector representation χ maps $\mathbf{Spin}(\mathfrak{V})$ *upon* $\mathbf{O}^+(\mathfrak{V})$, and therefore there exists $u \in \mathbf{Spin}(\mathfrak{V})$ such that $\chi(u) = h \mid \mathfrak{V}$. Let $h' = \eta(u)$, $h_0 = h_0' = h \mid \mathfrak{V}$, $h_1 = h \mid \mathfrak{W}$, $h_1' = h' \mid \mathfrak{W}$. Then, for all $V \in \mathfrak{V}$, $W \in \mathfrak{W}$,

(1) $h_0(V) \circ h_1(W) = h_1(V \circ W)$,
(2) $h_0(V) \circ h_1'(W) = h_1'(V \circ W)$.

Moreover, by (1),

$$h_1(h_0^{-1}(V) \circ h_1^{-1}(W)) = h_0(h_0^{-1}(V)) \circ h_1 h_1^{-1}(W)) = V \circ W,$$

i.e.,

(3) $h_0^{-1}(V) \circ h_1^{-1}(W) = h_1^{-1}(V \circ W)$.

Apply h_1^{-1} to both sides of (2) to obtain

$$\begin{aligned} h_1^{-1}h_1'(V \circ W) &= h_1^{-1}(h_0(V) \circ h_1'(W)) \\ &= h_0^{-1}h_0(V) \circ h_1^{-1}h_1'(W) \text{ by (3)} \\ &= V \circ h_1^{-1}h_1'(W). \end{aligned}$$

Hence the operator $\bar{h} = h_1^{-1} \circ h_1'$ commutes with $\theta(\frac{1}{2}V) = \lambda_V \,|\, \mathfrak{W}$ for all $V \in \mathfrak{V}$. It follows that $\bar{h}$ commutes with $\bar{\theta}(u)$ for all $u \in C$. But $\bar{\theta}$ maps C *upon* $\mathfrak{E}$, so that $\bar{h}$ commutes with every element of $\mathfrak{E}$. Hence $\bar{h}$ is a scalar multiple of the identity; as $\bar{h}$ is orthogonal, $\bar{h} = \varepsilon I$, $\varepsilon = \pm 1$. Let $u' = \varepsilon u \in \mathbf{Spin}(\mathfrak{V})$. Then h and $\eta(u')$ agree on each of the subspaces in (7.4), and therefore $h = \eta(u') \in \operatorname{Im} \eta$. □

Let us now consider the full automorphism group $\mathbf{F}_4$. If $\sigma \in \mathbf{F}_4$, then $\pi(\sigma) = \sigma(E)$ is again a primitive idempotent, so that we have a map $\pi : \mathbf{F}_4 \to P$, and $\pi(\sigma_1) = \pi(\sigma_2)$ if and only if $\sigma_1^{-1}\sigma_2 \in H$. We shall prove

(7.21) Theorem *The group* $\mathbf{F}_4$ *acts transitively on* P, *so that* $\mathbf{F}_4/H$ *is homeomorphic with the Cayley projective plane.*

Let $X \in P$, so that X satisfies the conditions of (6.14). Decomposing X with respect to (7.4), we have

$$X = \xi_1 E + \tfrac{1}{2}(\xi_2 + \xi_3)F + V + W,$$

where

$$V = \begin{pmatrix} 0 & 0 & 0 \\ 0 & \frac{1}{2}(\xi_2 - \xi_3) & x_1 \\ 0 & \bar{x}_1 & -\frac{1}{2}(\xi_2 - \xi_3) \end{pmatrix} \in \mathfrak{V},$$

$$W = \begin{pmatrix} 0 & \bar{x}_3 & x_2 \\ x_3 & 0 & 0 \\ \bar{x}_2 & 0 & 0 \end{pmatrix} \in \mathfrak{W}.$$

Note that $\|V\|^2 = \frac{1}{2}(\xi_2 - \xi_3)^2 + \|x_1\|^2 = \frac{1}{2}(\xi_2 - \xi_3)^2 + \xi_2\xi_3 = \frac{1}{2}(\xi_2 + \xi_3)^2$. Since $\mathbf{O}^+(\mathfrak{V})$ is transitive on the unit sphere in $\mathfrak{V}$, so is H, and therefore there exists $h \in H$ such that

$$h(V) = \begin{pmatrix} 0 & 0 & 0 \\ 0 & \frac{1}{2}(\xi_2 + \xi_3) & 0 \\ 0 & 0 & -\frac{1}{2}(\xi_2 + \xi_3) \end{pmatrix}.$$

Thus $h(X) = \xi_1 E + \frac{1}{2}(\xi_2 + \xi_3)F + h(V) + W'$, where

$$W' = \begin{pmatrix} 0 & y_3 & \bar{y}_2 \\ \bar{y}_3 & 0 & 0 \\ y_2 & 0 & 0 \end{pmatrix},$$

so that

$$h(X) = \begin{pmatrix} \xi_1 & y_3 & \bar{y}_2 \\ \bar{y}_3 & \xi_2 + \xi_3 & 0 \\ y_2 & 0 & 0 \end{pmatrix}.$$

The matrix $h(X)$ is again in P, and so must satisfy the conditions (6.14). In particular,

$$\|y_2\|^2 = \xi_1 \cdot 0 = 0,$$

so that $y_2 = 0$.

Now consider the idempotent E_3 and the decomposition $(7.4)_3$:

$$\mathfrak{J} = \mathbf{R}E_3 \oplus \mathbf{R}F_3 \oplus \mathfrak{V}_3 \oplus \mathfrak{W}_3 .$$

Developing $h(X)$ with respect to the new direct sum decomposition, we have

$$\begin{aligned} h(X) &= \tfrac{1}{2}(\xi_1 + \xi_2 + \xi_3)F_3 + V_3 \\ &= \tfrac{1}{2}F_3 + V_3, \end{aligned}$$

where

$$V_3 = \begin{pmatrix} \xi_1 - \frac{1}{2} & y_3 & 0 \\ \bar{y}_3 & \frac{1}{2} - \xi_1 & 0 \\ 0 & 0 & 0 \end{pmatrix}$$

and $\|V_3\|^2 = (\xi_1 - \frac{1}{2})^2 + \|y_3\|^2 = \xi_1^2 - \xi_1 + \frac{1}{4} + \xi_1(\xi_2 + \xi_3) = \frac{1}{4}$. As before, there is an element h' belonging to the subgroup H_3 of F_4 fixing E_3 such that

$$h'(V_3) = \begin{pmatrix} \frac{1}{2} & 0 & 0 \\ 0 & -\frac{1}{2} & 0 \\ 0 & 0 & 0 \end{pmatrix}.$$

Then

$$h'h(X) = \tfrac{1}{2}F_3 + h'(V_3) = E. \qquad \square$$

(7.22) Corollary *The map* $\pi : \mathbf{F}_4 \to P$ *is a fibration with fibre* **Spin**(9). $\square$

There is an interesting connection between the representations χ, ζ of Spin($\mathfrak{V}$) and the Hopf map $h : \mathbf{S}^{15} \to \mathbf{S}^8$. In fact, define $h_1 : \mathfrak{W} \to \mathfrak{V}$ by

$$h_1(W) = 2W * W.$$

It is immediate that

(7.23) *If* $W = W(y, z)$, *then* $h_1(W) = V(\xi, x)$, *where*

$$\xi = \|z\|^2 - \|y\|^2,$$
$$x = 2\bar{z}y,$$

and $\|h_1(W)\| = \|W\|^2$. $\square$

Let $\mathbf{S}^{15}$, $\mathbf{S}^8$ be the unit spheres in the inner-product spaces $\mathfrak{W}$, $\mathfrak{V}$, respectively. Then $h = h_1 | \mathbf{S}^{15} : \mathbf{S}^{15} \to \mathbf{S}^8$, and it follows from the arguments of Theorem (6.16) that

(7.24) *The map* $h : \mathbf{S}^{15} \to \mathbf{S}^8$ *is a Hopf fibration.* $\square$

Let $\mu : \mathbf{O}^+(\mathfrak{V}) \times \mathfrak{V} \to \mathfrak{V}$, $\mu' : \mathbf{O}^+(\mathfrak{W}) \times \mathfrak{W} \to \mathfrak{W}$ be the actions of the appropriate groups.

(7.24) Theorem *The diagram*

$$\begin{array}{ccccc}
\mathbf{Spin}\,(\mathfrak{V}) \times \mathfrak{W} & \xrightarrow{\zeta \times 1} & \mathbf{O}^+(\mathfrak{W}) \times \mathfrak{W} & \xrightarrow{\mu'} & \mathfrak{W} \\
\Big\downarrow \chi \times h_1 & & & & \Big\downarrow h_1 \\
\mathbf{O}^+(\mathfrak{V}) \times \mathfrak{V} & & \xrightarrow[\mu]{\hspace{6em}} & & \mathfrak{V}
\end{array}$$

is commutative.

Commutativity says that

$$h_1(\zeta(u)W) = \chi(u)h_1(W) \tag{7.25}$$

for all $u \in \mathbf{Spin}(\mathfrak{V})$, $W \in \mathfrak{W}$. As χ and ζ are the restrictions to $\mathbf{Spin}(\mathfrak{V})$ of representations of the Clifford algebra $C = C(V)$, it makes sense to state that (7.25) holds for an element $u \in C$; and it is clear that, if (7.25) holds for $u = u_1$ and for $u = u_2$, where u_1, $u_2 \in C$, then it also holds for $u = u_1 u_2$. Therefore, because of (3.12), it suffices to prove that (7.25) holds whenever $u = V \in \mathfrak{V}$, $\|V\| = 1$.

In this case, $\zeta(V)W = 2V \circ W$, so that

$$h_1(\zeta(X)W) = 8(V \circ W) * (V \circ W),$$

while

$$\begin{aligned}\chi(V)h_1(W) &= -\rho_V(h_1(W)) \quad \text{(by Lemma (3.7))}\\ &= -2W * W + 4\langle V, W * W\rangle V\end{aligned}$$

Let $V = V(\xi, x)$, $W = W(y, z)$. Then $h_1(\zeta(V)W) = V(\xi', x')$, where

(7.26)
$$\begin{aligned}\xi' &= (\xi^2 - \|x\|^2)(\|z\|^2 - \|y\|^2 + 4\xi\langle xy, \bar{z}\rangle \\ &= (\|y\|^2 - \|z\|^2) + 2\xi[\xi(\|z\|^2 - \|y\|^2) + 2\langle x, \bar{z}\bar{y}\rangle] \\ &= -(\|z\|^2 - \|y\|^2) + 4\xi\langle V, W * W\rangle,\end{aligned}$$

(7.27)
$$x' = 2[-\xi^2\bar{z}\bar{y} + \xi(\|z\|^2 - \|y\|^2)x + (xy)(zx)].$$

But

$$\begin{aligned}(xy)(zx) &= x(yz)x \quad \text{by (4.21)} \\ &= [x(yz) + (\bar{z}\bar{y})\bar{x}]x - \bar{z}\bar{y}\|x\|^2 \quad \text{by Theorem (4.16)} \\ &= 2\langle x, \bar{z}\bar{y}\rangle x - \|x\|^2\bar{z}\bar{y}.\end{aligned}$$

Substituting this relation in (7.27) and simplifying the result, we find that

$$\begin{aligned}x' &= 2\{-\bar{z}\bar{y} + [\xi(\|z\|^2 - \|y\|^2) + 2\langle x, \bar{z}\bar{y}\rangle]x\} \\ &= -2\bar{z}\bar{y} + 4\langle V, W * W\rangle x,\end{aligned}$$

so that

$$\begin{aligned}h_1(\zeta(V)W) &= -2W * W + 4\langle V, W * W\rangle V \\ &= \chi(V)h_1(W).\end{aligned}$$
□

Let us take as base point of $\mathbf{S}^{15}$ the point $W_0 = \alpha_2(1)$, and as the base point of $\mathbf{S}^8$ the point $h(W_0) = V_0$. The maps $\sigma \to \sigma(W_0)$, $\tau \to \tau(V_0)$ are fibrations $p_1 : \mathbf{O}^+(\mathfrak{W}) \to \mathbf{S}^{15}$, $p_2 : \mathbf{O}^+(\mathfrak{V}) \to \mathbf{S}^8$.

(7.28) Corollary *The diagram*

$$\begin{array}{ccccc} \mathbf{Spin}\,(\mathfrak{V}) & \xrightarrow{\zeta} & \mathbf{O}^+(\mathfrak{W}) & \xrightarrow{p_1} & \mathbf{S}^{15} \\ \downarrow{\scriptstyle\chi} & & & & \downarrow{\scriptstyle h} \\ \mathbf{O}^+(\mathfrak{V}) & & \xrightarrow[p_2]{\hspace{4em}} & & \mathbf{S}^8 \end{array}$$

is commutative.

For if $u \in \mathbf{Spin}(\mathfrak{V})$, then

$$\begin{aligned}hp_1\zeta(u) &= h(\zeta(u)W_0) \\ &= \chi(u)h(W_0) \quad \text{by (7.24)} \\ &= \chi(y)V_0 = p_2\chi(u).\end{aligned}$$
□

The group H acts on $\mathbf{S}^{15}$ (through the representation ζ) and on $\mathbf{S}^8$ (through the representation χ). The isotropy groups

$$H_0 = \{\sigma \in H \mid \sigma(V_0) = V_0\},$$
$$H_* = \{\sigma \in H \mid \sigma(W_0) = W_0\}$$

are related, because of Corollary (7.28), by the inclusion $H_* \subset H_0$. Thus there is a fibration

(7.29) $$H_0/H_* \to H/H_* \to H/H_0.$$

It follows from the transitivity of the action ζ, proved in Theorem (7.31) below, that

(7.30) *The fibration* (7.29) *is the Hopf fibration*

$$\mathbf{S}^7 \longrightarrow \mathbf{S}^{15} \xrightarrow{\ h\ } \mathbf{S}^8. \qquad \square$$

Let us recall, from the beginning of §6, that there are isometries $\alpha_i : \mathbf{K} \to \mathfrak{U}_i$ $(i = 1, 2, 3)$. Moreover, let $\mathfrak{U}_0 = \alpha_1(\mathbf{K}_0) \subset \mathfrak{U}_1$.

(7.31) Theorem *The group H acts transitively on $\mathbf{S}^{15}$, and there are isomorphisms $H_0 \approx \mathbf{Spin}(\mathfrak{U}_1)$, $H_* \approx \mathbf{Spin}(\mathfrak{U}_0)$.*

Remark 1. The transitivity follows from the rest by a dimension count. For dim $H_* = \dim \mathbf{Spin}(\mathfrak{U}_0) = \dim \mathbf{Spin}(7) = 21$, and therefore

$$\dim H/H_* = \dim \mathbf{Spin}(9) - \dim \mathbf{Spin}(7) = 36 - 21 = 15.$$

Thus H/H_* is a submanifold of $\mathbf{S}^{15}$ which is closed (because H is compact) and open (by invariance of domain), and hence $H/H_* = \mathbf{S}^{15}$.

Remark 2. There are imbeddings $\mathfrak{U}_0 \subsetneq \mathfrak{U}_1 \subsetneq \mathfrak{V}$, which induce imbeddings $\mathbf{Spin}(\mathfrak{U}_0) \subsetneq \mathbf{Spin}(\mathfrak{U}_1) \subsetneq \mathbf{Spin}(\mathfrak{V})$. The second of these is, indeed, the inclusion $H_0 \subsetneq H$. However, it is impossible that the inclusion $H_* \subsetneq H$ should be induced by any inclusion $\mathfrak{U}_0 \subsetneq \mathfrak{V}$; for if this were the case, then the coset space would be $\mathbf{V}_{9,2}$, and not $\mathbf{S}^{15}$.

The map $\chi \circ \eta^{-1} : H \to \mathbf{O}^+(\mathfrak{V})$ is just the restriction map sending each automorphism σ into $\sigma \mid \mathfrak{V}$. Hence $\chi\eta^{-1}(H_0)$ is the subgroup of $\mathbf{O}^+(\mathfrak{V})$ fixing the vector V_0. But the latter subgroup can be identified with $\mathbf{O}^+(\mathfrak{U}_1)$, and therefore $\eta^{-1}(H_0) = \chi^{-1}\mathbf{O}^+(\mathfrak{U}_1) = \mathbf{Spin}(\mathfrak{U}_1)$, by (3.14).

In order to study the group H_*, let us observe that the circulator γ maps W_0 into the point $V_1 = \alpha_1(1)$. The inner automorphism γ_* defined by γ maps H_0 into itself and maps H_* upon the subgroup of H_0 consisting of all automorphisms which have V_1 fixed. The space $\mathfrak{U}_0$ being the orthogonal complement in $\mathfrak{U}_1$ of $\mathbf{R}V_1$, it follows by the argument above that η maps $\mathbf{Spin}(\mathfrak{U}_1)$ upon $\gamma H_* \gamma^{-1}$. $\square$

Let us scrutinize more carefully the action of H_0. Since H_0 maps each of the subspaces $\mathfrak{U}_i$ into itself, it follows that, for each $\sigma \in H_0$, there are uniquely determined linear maps $\sigma_i : \mathbf{K} \to \mathbf{K}$ such that

(7.32) $$\sigma(\alpha_i(x)) = \alpha_i(\sigma_i(x)) \qquad (i = 1, 2, 3;\ x \in \mathbf{K}).$$

Moreover, the map $\sigma \to \sigma_i$ is a linear representation ω_i of H_0.

(7.33) *The maps σ_i have the following properties*:

(1) $\langle \sigma_i(x), \sigma_i(y) \rangle = \langle x, y \rangle$,
(2) $\sigma_1(\bar{z}\bar{y}) = \overline{\sigma_3(z)}\,\overline{\sigma_2(y)}$,
(3) $\sigma_2(\bar{x}\bar{z}) = \overline{\sigma_1(x)}\,\overline{\sigma_3(z)}$,
(4) $\sigma_3(\bar{y}\bar{x}) = \overline{\sigma_2(y)}\,\overline{\sigma_1(x)}$.

In particular, the representations ω_i are orthogonal.

These properties are easily derived with the aid of (6.3), and are left to the reader. □

Conversely, it is easy to see that

(7.34) *If $\sigma_i : \mathbf{K} \to \mathbf{K}$ are linear maps satisfying the conditions of* (7.33), *there is a unique element $\sigma \in H_0$ such that* (7.32) *holds for all $x \in K$.* □

The representations $\omega_i : H_0 \to \mathbf{O}^+(K)$ define a representation $\omega : H_0 \to \mathbf{O}^+(K) \times \mathbf{O}^+(K) \times \mathbf{O}^+(K)$.

(7.35) *The representation ω is faithful.*

For if σ_i is the identity map for $i = 1, 2, 3$, then σ leaves each of the subspaces $\mathfrak{U}_i$ pointwise fixed. Since already $\sigma(E_i) = E_i$, it follows that σ is the identity map. □

With the aid of the above remarks, we can prove

(7.36) Theorem (Principle of Triality). *For each $\rho \in \mathbf{O}^+(\mathbf{K})$ there exist ρ', $\rho'' \in \mathbf{O}^+(\mathbf{K})$ such that, for all $u, v \in \mathbf{K}$,*

(7.37) $$\rho(uv) = \rho'(u)\rho''(v).$$

The elements ρ', ρ'' are unique up to a common scalar factor $\varepsilon = \pm 1$.

We first observe that the isometry $\alpha_1 : \mathbf{K} \to \mathfrak{U}_1$, induces an isomorphism $\bar{\alpha}_1 : \mathbf{O}^+(\mathbf{K}) \to \mathbf{O}^+(\mathfrak{U}_1)$ such that the diagram

(7.38) $$\begin{array}{ccc} \mathbf{Spin}\,(\mathfrak{U}_1) & \xrightarrow{\eta|\mathbf{Spin}\,(\mathfrak{U}_1)} & H_0 \\ \chi \big\downarrow & & \big\downarrow \omega_1 \\ \mathbf{O}^+(\mathfrak{U}_1) & \xleftarrow[\bar{\alpha}_1]{} & \mathbf{O}^+(\mathbf{K}) \end{array}$$

is commutative. Hence ω_1 is an epimorphism, and the kernel of ω_1 consists of the two elements $\eta(\pm 1)$, where 1 is the unit element of the Clifford algebra $C(\mathfrak{U}_1)$. Since $\zeta(\pm 1) = \pm I$, the kernel of ω_1 consists of the identity map and the automorphism σ_0 such that $\sigma_0 | \mathfrak{U}_1 = I$, $\sigma_0 | \mathfrak{U}_2 = -I$, $\sigma_0 | \mathfrak{U}_3 = -I$.

Let $\rho \in \mathbf{O}^+(\mathbf{K})$; since ω_1 is an epimorphism, there exists $\sigma \in H_0$ such that $\sigma_1 = \rho$. Define

$$\rho'(u) = \overline{\sigma_3(\bar{u})},$$
$$\rho''(v) = \overline{\sigma_2(\bar{v})}.$$

Then (7.37) follows from (2) of (7.33). Note that changing the sign of σ entails changing the signs of both ρ' and ρ''.

To prove the uniqueness part, suppose that ρ, ρ'_0, and $\rho''_0 \in \mathbf{O}^+(\mathbf{K})$, and that $\rho(uv) = \rho'_0(u)\rho''_0(v)$ for all $u, v \in \mathbf{K}$. Define $\sigma_i : \mathbf{K} \to \mathbf{K}$ by

$$\sigma_1 = \rho,$$
$$\sigma_2(y) = \overline{\rho''_0(\bar{y})},$$
$$\sigma_3(z) = \overline{\rho'_0(\bar{z})}.$$

Then the maps σ_i are orthogonal, and (2) of (7.33) is satisfied. We shall prove that (3) and (4) are also satisfied. It then follows from (7.34) that there is an element $\sigma \in H_0$ such that $\omega_i(\sigma) = \sigma_i$ $(i = 1, 2, 3)$. As $\omega_1(\sigma) = \rho$, it follows from what we have already proved that $\rho'_0 = \varepsilon\rho'$, $\rho''_0 = \varepsilon\rho''$, $\varepsilon = \pm 1$.

The proofs of (3) and (4) are similar; we prove only (3). We may assume that $z \neq 0$; for both sides of (3) vanish when $z = 0$. Then

$$\overline{\sigma_1(x)}\ \overline{\sigma_3(z)} = \overline{\rho(x)}\rho'_0(\bar{z}).$$

But

$$\rho(x) = \frac{1}{\|z\|^2}\rho(\|z\|^2 x) = \frac{1}{\|z\|^2}\rho(\bar{z}(zx)) \quad \text{by (4.9)}$$
$$= \frac{1}{\|z\|^2}\rho'_0(\bar{z})\rho''_0(zx) \quad \text{by hypothesis,}$$

so that

$$\overline{\sigma_1(x)}\ \overline{\sigma_3(z)} = \frac{1}{\|z\|^2}[\overline{\rho''_0(zx)}\ \overline{\rho'_0(\bar{z})}]\rho'_0(\bar{z})$$
$$= \frac{1}{\|z\|^2}\overline{\rho''_0(zx)}\|\rho'_0(\bar{z})\|^2 \quad \text{by (4.10)}$$
$$= \overline{\rho''_0(zx)} \quad \text{since } \rho'_0 \text{ is orthogonal.}$$

But

$$\sigma_2(\overline{x\bar{z}}) = \overline{\rho''_0(zx)}. \qquad \square$$

There is one more subgroup of $\mathbf{F}_4$ that is of interest to us. Let τ be an automorphism of $\mathbf{K}$, and recall that $\tau(\bar{x}) = \overline{\tau(x)}$ for all $x \in \mathbf{K}$. It follows that, if $X \in \mathfrak{J}$ and

$$\tau_*(X) = \begin{pmatrix} \xi_1 & \tau(x_3) & \tau(\bar{x}_2) \\ \tau(\bar{x}_3) & \xi_2 & \tau(x_1) \\ \tau(x_2) & \tau(\bar{x}_1) & \xi_3 \end{pmatrix}$$

is the matrix obtained from X by operating with τ on each of its entries, then $\tau_*(X) \in \mathfrak{J}$ and the map τ_* is an automorphism of $\mathfrak{J}$. Moreover, the map $\tau \to \tau_*$ is a monomorphism of $\mathbf{G}_2$ into $\mathbf{F}_4$, and its image consists of all $\sigma \in H_0$ such that $\sigma_1 = \sigma_2 = \sigma_3$. Evidently τ_* commutes with the circulator γ.

The group H_0 leaves each of the subspaces $\mathfrak{U}_i$ invariant; thus H_0 operates on the unit sphere $S_i^7 \subset \mathfrak{U}_i$. Accordingly, the subgroup H_* acts on S_i^7. By definition, H_* leaves fixed the vector $W_0 \in S_2^7$. On the other hand,

(7.39) Theorem *The group H_* acts transitively on S_1^7, and the subgroup of H_* leaving V_1 fixed is $\mathbf{G}_2$. Thus there is a fibration*

$$\mathbf{G}_2 \to \mathbf{Spin}(7) \to S_1^7.$$

Again transitivity follows by a dimension count once we have identified the isotropy group. An element $\sigma \in H_0$ belongs to H_* if and only if $\sigma_2(1) = 1$; and the isotropy group consists of all σ satisfying the further condition $\sigma_1(1) = 1$. By (4) of (7.33), $\sigma_3(1) = 1$. Since the σ_i are orthogonal, we deduce that $\sigma_i(\bar{u}) = \overline{\sigma_i(u)}$ $(i = 1, 2, 3)$. Putting $y = 1$ in (2), we obtain

$$\sigma_1(\bar{z}) = \overline{\sigma_3(z)} = \sigma_3(\bar{z}),$$

so that $\sigma_1 = \sigma_3$. Similarly $\sigma_2 = \sigma_3$. Finally, from (4) we deduce

$$\begin{aligned} \sigma_1(\bar{y}\bar{x}) &= \overline{\sigma_1(y)}\,\overline{\sigma_1(x)} \\ &= \sigma_1(\bar{y})\sigma_1(\bar{x}), \end{aligned}$$

from which it follows that σ_1 is an automorphism of $\mathbf{K}$, so that $\sigma \in \mathbf{G}_2$. Conversely, if $\sigma \in \mathbf{G}_2$, then $\sigma_1 = \sigma_2 = \sigma_3$ and $\sigma_1(1) = \sigma_2(1) = 1$, so that σ belongs to the isotropy group. □

To summarize our discussion of $\mathbf{F}_4$: there is a chain of subgroups with coset spaces and dimensions as indicated below

$$\begin{array}{ccccccccccccccc} 52 & & 36 & & 28 & & 21 & & 14 & & 8 & & 3 & & \\ \mathbf{F}_4 & \supset & \mathbf{Spin}\,(9) & \supset & \mathbf{Spin}\,(8) & \supset & \mathbf{Spin}\,(7) & \supset & \mathbf{G}_2 & \supset & \mathbf{U}^+(3) & \supset & \mathbf{S}^3 & \supset & \{1\} \\ & \mathbf{P}^2(\mathbf{K}) & & \mathbf{S}^8 & & \mathbf{S}^7 & & \mathbf{S}^7 & & \mathbf{S}^6 & & \mathbf{S}^5 & & \mathbf{S}^3 & \\ & & & & \mathbf{S}^{15} & & & & & & \mathbf{V}_{7,2} & & & & \end{array}$$

Exercises

1. For $a \in \mathbf{K}$, $\|a\| = 1$, the *inner transformation* $\tau_a : \mathbf{K} \to \mathbf{K}$ is well-defined by
$$\tau_a(x) = axa^{-1}.$$
Prove that τ_a is an automorphism of $\mathbf{K}$ if and only if $a^6 = 1$ (i.e., $a = \pm 1$ or $a = \pm\frac{1}{2} + (\sqrt{3}/2)u$, with $u \in \mathbf{K}_0$, $u^2 = -1$).

2. Describe maximal tori in $\mathbf{G}_2$ and $\mathbf{F}_4$.

3. Let Q be the set of all elements $\sigma \in H_*$ such that $\sigma_2(\mathbf{e}_1) = \mathbf{e}_1$. Show that $Q \approx \mathbf{Spin}(6)$ and that the representation ω_1 maps Q isomorphically upon a subgroup of $\mathbf{O}^+(\mathbf{K})$ isomorphic with $\mathbf{U}^+(4)$.

4. Let Q be as in #3, and let Q_0 be the set of all $\sigma \in \mathbf{K}$ such that $\sigma_2(\mathbf{e}_2) = \mathbf{e}_2$. Show that $Q_0 \approx \mathbf{Spin}(5)$ and that ω_1 maps Q_0 upon a subgroup of $\mathbf{O}^+(\mathbf{K})$ isomorphic with $\mathbf{Sp}(2)$.

Appendix B Additive Relations

An essential part of the algebraic machinery used in the study of algebraic topology is the theory of abelian groups and homomorphisms. On the other hand, many constructions needed there do not fit naturally into this framework. One of the simplest is concerned with the factorization problem

$$\begin{array}{ccc} A & \xrightarrow{\ f\ } & B \\ {\scriptstyle g}\downarrow & \overset{h}{\nearrow} & \\ C & & \end{array}$$

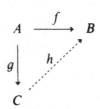

where g is an epimorphism, and g maps the kernel of f into zero. There is then known to be a unique solution $h : C \to B$ of the problem; h is usually defined by the requirement: $h(c) = f(a)$ for any a such that $g(a) = c$. One then has to verify that this defines h uniquely. It is tempting to write $h = f \circ g^{-1}$. The trouble with doing so is that, unless g is an *isomorphism*, g^{-1} is not a homomorphism. Thus, while the problem can be formulated within the category $\mathscr{A}$ of abelian groups and homomorphisms, and while its solution lies in $\mathscr{A}$ the above calculation takes us outside the category. In this Appendix, we shall, following MacLane [MacL], introduce a calculus which makes such constructions as the above legitimate.

In §1 we review some properties of direct sums and products, and additive relations themselves are treated in §2.

1 Direct Sums and Products

Let us begin by recalling some facts about finite direct sums and products of abelian groups (the category-minded reader may wish to replace $\mathscr{A}$ by an arbitrary abelian category). As all groups involved are abelian, we shall often omit the qualifying adjective.

Let $G_1, \ldots, G_r$ be a finite sequence of groups. A *direct product* of the G_k consists of a group P, together with homomorphisms $p_k : P \to G_k$ $(k = 1, \ldots, r)$ with the following universal property: for any group H and homomorphisms $f_k : H \to G_k$, there is a unique homomorphism $f: H \to P$ such that $p_k \circ f = f_k$ $(k = 1, \ldots, r)$. In other words, the problem typified by the diagrams

$$\text{(1.1)} \qquad \begin{array}{ccc} H & \dashrightarrow & P \\ & {}_{f_k}\searrow & \downarrow p_k \\ & & G_k \end{array} \qquad (k = 1, \ldots, r)$$

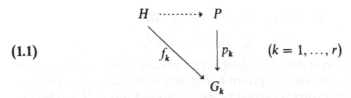

has a unique solution, so that $(P; p_1, \ldots, p_r)$ is a universal example for the problem in question.

(1.2) Theorem *Any finite sequence* $G_1, \ldots, G_r$ *has a direct product, which is unique up to isomorphism.*

Uniqueness is proved by "general nonsense," but existence, as is often the case, is proved by a specific construction. In fact, let $P = \prod_{k=1}^{r} G_k = G_1 \times \cdots \times G_r$ be the cartesian product of the G_k under componentwise addition: if $x = (x_1, \ldots, x_r)$, $y = (y_1, \ldots, y_r)$, then $x + y = (z_1, \ldots, z_r)$, where

$$z_i = x_i + y_i \qquad (i = 1, \ldots, r).$$

With p_k = projection on the kth factor, we see easily that $(P; p_1, \ldots, p_r)$ is a direct product of the G_k.

The direct product is a functor: $\mathscr{A} \times \cdots \times \mathscr{A} \to \mathscr{A}$. In fact, if $(P; p_1, \ldots, p_r)$ is a direct product of $G_1, \ldots, G_r$ and $(P'; p'_1, \ldots, p'_r)$ is a direct product of $G'_1, \ldots, G'_r$, and if $f_k : G_k \to G'_k$ are homomorphisms, there is a unique homomorphism $\bar{f}: P \to P'$ such that $p'_k \circ \bar{f} = f_k \circ p_k$ for $k = 1, \ldots, r$. If $P = \prod_{k=1}^{r} G_k$ and $P' = \prod_{k=1}^{r} G'_k$ are the specific products constructed above, then

$$\bar{f}(x_1, \ldots, x_r) = (f_1(x_1), \ldots, f_r(x_r))$$

for all $x_k \in G_k$, $k = 1, \ldots, r$. Defining $f_1 \times \cdots \times f_r = \bar{f}$, we see that the conditions for a functor are fulfilled.

The r-fold diagonal map

$$\Delta_r : G \to \underbrace{G \times \cdots \times G}_{r \text{ factors}}$$

is defined by $p_k \circ \Delta_r = 1$, the identity map of G $(k = 1, \ldots, r)$. If $f_k : H \to G_k$ are homomorphisms, the map

$$f = (f_1, \ldots, f_r) : H \to G_1 \times \cdots \times G_r$$

defined by

$$f = (f_1 \times \cdots \times f_r) \circ \Delta_r$$

is the solution of the problem (1.1); for

$$\begin{aligned} p_k \circ f &= p_k \circ (f_1 \times \cdots \times f_r) \circ \Delta_r \\ &= f_k \circ p_k \circ \Delta_r = f_k \circ 1 = f_k . \end{aligned}$$

Dual to the notion of direct product is that of direct sum. A *direct sum* of groups $G_1, \ldots, G_r$ consists of a group S, together with homomorphisms $j_1, \ldots, j_r$ having the following universal property: given a group H and homomorphisms $f_k : G_k \to H$, there is a unique homomorphism $f : S \to H$ such that $f \circ j_k = f_k$ $(k = 1, \ldots, r)$. In other words, $(S; j_1, \ldots, j_r)$ is a universal example for the problem typified by the diagrams

$$\begin{array}{ccc} S & \dashrightarrow & H \\ {\scriptstyle j_k}\uparrow & \nearrow{\scriptstyle f_k} & \\ G_k & & \end{array} \qquad (k = 1, \ldots, r).$$

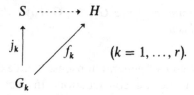

(1.3) Theorem *Any finite sequence of groups $G_1, \ldots, G_r$ has a direct sum, which is unique up to isomorphism.*

As before, uniqueness is proved by "general nonsense," existence by explicit construction.

Let S be a *direct sum* of $G_1, \ldots, G_r$. Then there are unique homomorphisms $q_k : S \to G_k$ such that

$$q_k \circ j_l = \begin{cases} 0 & \text{if } l \neq k, \\ 1_k & \text{if } l = k, \end{cases}$$

where 1_k is the identity map of G_k. Similarly, if P is a *direct product* of $G_1, \ldots, G_r$ there are unique homomorphisms $i_k : G_k \to P$ such that

$$p_l \circ i_k = \begin{cases} 0 & \text{if } l \neq k, \\ 1_k & \text{if } l = k. \end{cases}$$

(1.4) Theorem *The group S, together with the homomorphisms $q_k : S \to G_k$, is a direct product of $G_1, \ldots, G_r$. The group P, together with the homomorphisms $i_k : G_k \to P$, is a direct sum of the G_k. There is a unique isomorphism $h : S \to P$ such that*

$$h \circ j_k = i_k, \; p_k \circ h = q_k,$$

$$p_k \circ h \circ j_l = \begin{cases} 0 & \text{if } k \neq l, \\ 1_k & \text{if } k = l. \end{cases}$$

In particular, $P = \prod_{k=1}^{r} G_k$, with the homomorphism $i_k : G_k \to P$ defined by

$$i_1(x_1) = (x_1, 0, \ldots, 0)$$
$$i_2(x_2) = (0, x_2, \ldots, 0)$$

$$i_r(x_r) = (0, 0, \ldots, x_r)$$

is a direct sum, and, when we wish to consider it as a direct sum, we shall denote it by

$$G_1 \oplus \cdots \oplus G_r = \bigoplus_{k=1}^{r} G_k.$$

As in the case of the direct product, the direct sum is a functor. Indeed, let $f_k : G_k \to G'_k$ be homomorphisms ($k = 1, \ldots, r$). If $(S; j_1, \ldots, j_r)$ and $(S'; j'_1, \ldots, j'_r)$ are direct sums of the G_k, G'_k, respectively, there is a unique homomorphism $\underline{f} : S \to S'$ such that $\underline{f} \circ j_k = j'_k \circ f_k$ ($k = 1, \ldots, r$). If $S = \oplus G_k$, $S' = \oplus G'_k$, then

$$\underline{f}(x_1, \ldots, x_r) = (f_1(x_1), \ldots, f_r(x_r))$$

for all $x_k \in G_k$ ($k = 1, \ldots, r$). Defining $f_1 \oplus \cdots \oplus f_r = \underline{f}$, we see that the conditions for a functor are fulfilled.

Let

$$\nabla_r : \underbrace{G \oplus \cdots \oplus G}_{k \text{ factors}} \to G$$

be the homomorphism such that $\nabla_r \circ j_k$ is the identity map of G ($k = 1, \ldots, r$). If $f_k : G_k \to H$ are homomorphisms, then the map

$$f = \begin{pmatrix} f_1 \\ \cdots \\ f_k \end{pmatrix} : G_1 \oplus \cdots \oplus G_r \to H$$

defined by

$$f = \nabla_r \circ (f_1 \oplus \cdots \oplus f_r)$$

is the solution of the problem (1.3); for

$$f \circ j_k = \nabla_r \circ (f_1 \oplus \cdots \oplus f_r) \circ j_k = \nabla_r \circ j_k \circ f_k = f_k.$$

When $(S; j_1, \ldots, j_r)$ is a direct sum, we shall say that the homomorphisms $j_k : G_k \to S$ *represent S as a direct sum.* Similarly, if $(P; p_1, \ldots, p_r)$ is a direct product, we say that the homomorphisms $p_k : P \to G_k$ represent P as a *direct product.*

Remark. Our terminology differs from that of Eilenberg–Steenrod [E–S, p. 8]. These authors deal with only one notion, that of direct sum, and refer to $\{j_k\}$ as an *injective,* and to $\{p_k\}$ as a *projective,* representation as a direct sum.

Representations $\{i_k : A_k \to G\}$, $\{p_k : G \to B_k\}$ $(k = 1, \ldots, r)$ of the group G as a direct sum, product, respectively, are said to be *weakly dual* if and only if

$$p_k \circ i_l = 0 \quad \text{if } k \neq l,$$

while

$$p_k \circ i_k : A_k \to B_k$$

is an isomorphism. They are said to be *strongly dual* (or simply *dual*) if and only if they are weakly dual and, in addition, $A_k = B_k$, $p_k \circ i_k$ is the identity map. For example, the representations

$$\{j_k : G_k \to S\}, \qquad \{q_k : S \to G_k\},$$

as well as the representations

$$\{i_k : G_k \to P\}, \qquad \{p_k : P \to G_k\}$$

discussed above, are dual.

A group B, with dual representations

$$f_k : G_k \to B$$

$$g_k : B \to G_k$$

is called a *biproduct* of the G_k.

In view of these remarks, there is a unique homomorphism $\nabla_r : G \times \cdots \times G \to G$ such that $\nabla_r \circ i_k = 1$ $(i = 1, \ldots, r)$. The astute reader will already have verified that

$$\nabla_r(x_1, \ldots, x_r) = x_1 + \cdots + x_r.$$

In particular, $\nabla = \nabla_2$ is just the group operation in G.

It is trivial that

(1.5) *The operations of direct addition and multiplication are commutative and associative, up to natural isomorphism.*

We can now define addition in the set Hom(G, H) of all homomorphisms of G into H. In fact, if $f, g : G \to H$, then $f + g : G \to H$ is the composite

$$G \xrightarrow[\Delta]{} G \times G \xrightarrow[f \times g]{} H \times H \xrightarrow[\nabla]{} H.$$

(1.6) Theorem *The set* Hom(G, H) *is an abelian group under the above operation. The zero element of* Hom(G, H) *is the trivial homomorphisms which sends each element of G into* 0. *If $f: G \to H$, then $-f: G \to H$ is defined by*

$$(-f)(x) = -f(x)$$

for all $x \in G$.

We have forborn to define $f_1 + \cdots + f_r$ directly, as we could have done, as the composite

$$G \xrightarrow[\Delta_r]{} G \times \cdots \times G \xrightarrow[f_1 \times \cdots \times f_r]{} H \times \cdots \times H \xrightarrow[\nabla_r]{} H,$$

for this clearly coincides with the iterated sum of $f_1, \ldots, f_r$ for any parenthesization.

A useful criterion for a biproduct is given by

(1.7) Theorem *Let $B, G_1, \ldots, G_r$ be groups and let $f_k : G_k \to B$, $g_k : B \to G_k$ be homomorphisms ($k = 1, \ldots, r$). Then $(f_1, \ldots, f_r)$ and $(g_1, \ldots, g_r)$ are dual representations of B as a direct sum, product, respectively, if and only if*

(1) $g_k \circ f_l = 0$, $(k \neq l)$,
(2) $g_k \circ f_k = 1_k$,
(3) $f_1 \circ g_1 + \cdots + f_r \circ g_r = 1$, *the identity map of B.*

(of course, the addition in (3) takes place in the group Hom(B, B)).

We conclude this section with some remarks on infinite products. Let J be an arbitrary set; then there is no difficulty in defining the direct sum S and direct product P of a family of groups $\{G_\alpha \mid \alpha \in J\}$, just as we did in the finite case. If $i_\alpha : G_\alpha \to S$ represent S as a direct sum and $p_\alpha : P \to G_\alpha$ represent P as a direct product, then there is a homomorphism $k : S \to P$ such that

$$p_\beta \circ k \circ i_\alpha = \begin{cases} 0 & \text{if } \beta \neq \alpha, \\ 1, \text{ the identity map of } G_\alpha & \text{if } \beta = \alpha. \end{cases}$$

Moreover k is a *monomorphism*. The image of k in P is called the *weak direct product* of the G_α, and the homomorphisms $q_\alpha = p_\alpha \circ k$ are said to define a representation of S as a *weak direct product*, dual to its representation by the i_α as a direct sum. Similarly, it seems appropriate to say that the homomorphisms $j_\alpha = k \circ i_\alpha$ define a representation of P as a *strong direct sum*, dual to its representation by the p_α as a direct product.

2 Additive Relations

Motivated by the fact that, if $f: A \to B$ is a homomorphism of abelian groups, then the graph of f is a *subgroup* of $A \times B$ with special properties, we may define an *additive relation* $F: A \rightsquigarrow B$ to be an arbitrary subgroup of $A \times B$. For such a relation, we may further define the *domain* of $F = \mathrm{Dom}(F) = p_1(F)$; the *kernel* of $F = \mathrm{Ker}(F) = i_1^{-1}(F)$; the *image* of $F = \mathrm{Im}(F) = p_2(F)$; the *indeterminacy* of $F = \mathrm{Ind}(F) = i_2^{-1}(F)$. Thus F is (the graph of) a function from A to B (necessarily a homomorphism) if and only if $\mathrm{Dom}\, F = A$ and $\mathrm{Ind}\, F = 0$. Note that $\mathrm{Ker}\, F \subset \mathrm{Dom}\, F \subset A$ and $\mathrm{Ind}\, F \subset \mathrm{Im}\, F \subset B$.

We may also define certain quotient groups of A and B. These are the *coindeterminacy* of $F = \mathrm{Coin}\, F = A/\mathrm{Dom}\, F$; the *coimage* of $F = \mathrm{Coim}\, F = A/\mathrm{Ker}\, F$; the *cokernel* of $F = \mathrm{Cok}\, F = B/\mathrm{Im}\, F$; the *codomain* of $F = \mathrm{Cod}\, F = B/\mathrm{Ind}\, F$. These are, however, of lesser importance (except for the cokernel).

The *composite* of two additive relations is defined as is customary in the calculus of relations: if $F: A \rightsquigarrow B$ and $G: B \rightsquigarrow C$, then

$$G \circ F = \{(a, c) \in A \times C \mid (a, b) \in F \text{ and } (b, c) \in G \text{ for some } b \in B\}.$$

Clearly $G \circ F: A \rightsquigarrow C$ is again an additive relation.

The *converse* of an additive relation $F: A \rightsquigarrow B$ is defined by

$$F^{-1} = \{(b, a) \mid (a, b) \in F\};$$

clearly $F^{-1}: B \rightsquigarrow A$, $(F^{-1})^{-1} = F$, and $(G \circ F)^{-1} = F^{-1} \circ G^{-1}$ for $G: B \rightsquigarrow C$. Moreover,

$$\mathrm{Dom}\, F^{-1} = \mathrm{Im}\, F,$$

$$\mathrm{Ker}\, F^{-1} = \mathrm{Ind}\, F,$$

$$\mathrm{Im}\, F^{-1} = \mathrm{Dom}\, F,$$

$$\mathrm{Ind}\, F^{-1} = \mathrm{Ker}\, F.$$

However, it is not in general true that $F^{-1} \circ F$ is the identity map of A (nor is $F \circ F^{-1}$ the identity map of B).

Let $x \in \mathrm{Dom}\, F$, and let

(2.1) $$f(x) = \{y \in B \mid (x, y) \in F\}.$$

If $(x, y) \in F$ and $(x, y') \in F$, then $(0, y - y') \in F$, so that $y - y' \in \mathrm{Ind}\, F$. Conversely, if $y' \in \mathrm{Ind}\, F$, $(x, y) \in F$, then $(x, y + y') \in F$. Hence $f(x)$ is a coset of $\mathrm{Ind}\, F$, and $f: \mathrm{Dom}\, F \to \mathrm{Cod}\, F = B/\mathrm{Ind}\, F$ is easily seen to be a homomorphism. Conversely, let D be a subgroup of A, J a subgroup of B, $f: D \to B/J$ a homomorphism. Let $F = \{(x, y) \mid y \in f(x)\}$; then F is an additive relation. Moreover, it is easy to see that there is hereby established a one-to-one correspondence between the set of all additive relations $F: A \rightsquigarrow B$

on the one hand and the set of all homomorphisms of a subgroup of A into a quotient group of B on the other.

One can carry this a step farther. If F is an additive relation and f is defined by (2.1), then $\operatorname{Ker} f = \operatorname{Ker} F$ and $\operatorname{Im} f = \operatorname{Im} F/\operatorname{Ind} F$, so that f induces an isomorphism

$$\phi : \operatorname{Dom} F/\operatorname{Ker} F \approx \operatorname{Im} F/\operatorname{Ind} F$$

between subquotients of A and of B. Conversely, if $P = A_1/A_0$ and $Q = B_1/B_0$ are subquotients of A and B, respectively, and if $\phi : P \to Q$ is an isomorphism then the composite

$$A_1 \to A_1/A_0 = P \xrightarrow{\phi} Q = B_1/B_0 \to B/B_0$$

is a homomorphism of a subgroup of A into a quotient of B, which in turn gives rise to an additive relation F. Again, it is easy to see that there is a one-to-one correspondence between additive relations $F : A \rightsquigarrow B$ and isomorphisms $\phi : P \approx Q$ between subquotients of A, B, respectively.

Let $F, G : A \rightsquigarrow B$ be relations, and suppose that $F \subset G$. Then

$$\operatorname{Dom} F \subset \operatorname{Dom} G, \qquad \operatorname{Ker} F \subset \operatorname{Ker} G,$$
$$\operatorname{Im} F \subset \operatorname{Im} G, \qquad \operatorname{Ind} F \subset \operatorname{Ind} G,$$

and there is a natural map $\pi : B/\operatorname{Ind} F \to B/\operatorname{Ind} G$. Let $f : \operatorname{Dom} F \to B/\operatorname{Ind} F$, $g : \operatorname{Dom} G \to B/\operatorname{Ind} G$. Then we have the relation

$$g \,|\, \operatorname{Dom} F = \pi \circ f.$$

Therefore, if $x \in \operatorname{Dom} F$, then $f(x) \subset g(x)$, as subsets of B.

Example 1. Let $f : B \to A$, $g : B \to C$ be homomorphisms. Then $g \circ f^{-1} : A \rightsquigarrow C$, and

$$\operatorname{Dom}(g \circ f^{-1}) = \operatorname{Im} f,$$
$$\operatorname{Ker}(g \circ f^{-1}) = f(\operatorname{Ker} g),$$
$$\operatorname{Im}(g \circ f^{-1}) = \operatorname{Im} g,$$
$$\operatorname{Ind}(g \circ f^{-1}) = g(\operatorname{Ker} f).$$

Then $g \circ f^{-1}$ is a homomorphism $h : A \to C$ if and only if $\operatorname{Im} f = A$ and $G(\operatorname{Ker} f) = 0$, i.e., f is an epimorphism and $\operatorname{Ker} f \subset \operatorname{Ker} g$. When this is so,

$$\operatorname{Ker} h = f(\operatorname{Ker} g),$$
$$\operatorname{Im} h = \operatorname{Im} g.$$

EXAMPLE 2. Let $f: A \to B$, $g: C \to B$ be homomorphisms. Then $g^{-1} \circ f : A \rightsquigarrow C$, and

$$\begin{aligned}
\operatorname{Dom}(g^{-1} \circ f) &= f^{-1}(\operatorname{Im} g) \\
\operatorname{Ker}(g^{-1} \circ f) &= \operatorname{Ker} f \\
\operatorname{Im}(g^{-1} \circ f) &= g^{-1}(\operatorname{Im} f), \\
\operatorname{Ind}(g^{-1} \circ f) &= \operatorname{Ker} g
\end{aligned}$$

Thus $g^{-1} \circ f$ is a homomorphism $h: A \to C$ if and only if $\operatorname{Im} f \subset \operatorname{Im} g$ and $\operatorname{Ker} g = 0$, g is a monomorphism. When this is so,

$$\begin{aligned}
\operatorname{Ker} h &= \operatorname{Ker} f, \\
\operatorname{Im} h &= g^{-1}(\operatorname{Im} f).
\end{aligned}$$

EXAMPLE 3. A *ladder* is a commutative diagram

$$(\Delta) \qquad \begin{array}{ccccccccccc}
\cdots \to A_{n+2} & \xrightarrow{f_{n+2}} & A_{n+1} & \xrightarrow{f_{n+1}} & A_n & \xrightarrow{f_n} & A_{n-1} & \xrightarrow{f_{n-1}} & A_{n-2} \to \cdots \\
\downarrow h_{n+2} & & \downarrow h_{n+1} & & \downarrow h_n & & \downarrow h_{n-1} & & \downarrow h_{n-2} \\
\cdots \to B_{n+2} & \xrightarrow[g_{n+2}]{} & B_{n+1} & \xrightarrow[g_{n+1}]{} & B_n & \xrightarrow[g_n]{} & B_{n-1} & \xrightarrow[g_{n-1}]{} & B_{n-2} \to \cdots
\end{array}$$

with exact rows. Associated with such a ladder is the additive relation $\Sigma_n = g_{n+1}^{-1} \circ h_n \circ f_n^{-1} : A_{n-1} \rightsquigarrow B_{n+1}$. Evidently

(1) $\operatorname{Dom} \Sigma_n = \operatorname{Ker} f_{n-1} \cap \operatorname{Ker} h_{n-1}$,
(2) $\operatorname{Ker} \Sigma_n = f_n(\operatorname{Ker} h_n)$,
(3) $\operatorname{Im} \Sigma_n = g_{n+1}^{-1}(\operatorname{Im} h_n)$
(4) $\operatorname{Ind} \Sigma_n = \operatorname{Im} g_{n+2} + \operatorname{Im} h_{n+1}$,

so that Σ_n induces a homomorphism

$$\sigma_n : \operatorname{Ker} f_{n-1} \cap \operatorname{Ker} h_{n-1} \to B_{n+1}/\operatorname{Im} g_{n+2} + \operatorname{Im} h_{n+1},$$

called the *suspension*, and Σ_n^{-1} induces a homomorphism

$$\tau_n : g_{n+1}^{-1}(\operatorname{Im} h_n) \to A_{n-1}/f_n(\operatorname{Ker} h_n),$$

called the *transgression*, of the diagram (Δ). (The relations Σ_n, Σ_n^{-1} are also referred to as the suspension and the transgression, respectively).

Let (Δ') be another ladder whose entries are denoted by the corresponding primed letters, and let there be given a map of (Δ) into (Δ'), i.e., maps $\alpha_n : A_n \to A'_n$, $\beta_n : B_n \to B'_n$ making the appropriate 3-dimensional diagram commutative, i.e.,

$$f'_n \circ \alpha_n = \alpha_{n-1} \circ f_n,$$

$$g'_n \circ \beta_n = \beta_{n-1} \circ g_n,$$

$$\beta_n \circ h_n = h'_n \circ \alpha_n.$$

If Σ_n and Σ'_n are the additive relations associated with the diagrams (Δ), (Δ'), then we have an inclusion

(2.2)
$$\beta_{n+1} \circ \Sigma_n \subset \Sigma'_n \circ \alpha_{n-1}.$$

It follows that, if σ_n, σ'_n are the suspension homomorphisms defined by Σ_n, Σ'_n, respectively, and if $x \in \text{Dom } \Sigma_n = \text{Dom } \sigma_n$, then $\alpha_{n-1}(x) \in \text{Dom } \Sigma'_n = \text{Dom } \sigma'_n$, and

(2.3)
$$\beta_{n+1}(\sigma_n(x)) \subset \sigma'_n(\alpha_{n-1}(x)).$$

Example 4. Let (C, ∂) be a chain-complex, with homology group $H = H(C) = \text{Ker } \partial/\text{Im } \partial$. Then H is a subquotient of C, and may also be regarded as its own subquotient $H/\{0\}$. The identity map of H thus corresponds to an additive relation $U = C \rightsquigarrow H$. (In fact, $U = p \circ i^{-1}$, where $i : \text{Ker } \partial \subsetneq C$ and $p : \text{Ker } \partial \to H$ is the projection.) Evidently

$$\text{Dom } U = \text{Ker } \partial,$$

$$\text{Ker } U = \text{Im } \partial,$$

$$\text{Im } U = H,$$

$$\text{Ind } U = \{0\}.$$

Suppose that C_1, C_2 are chain-complexes, and let $U_i : C_i \rightsquigarrow H_i$ the relations defined above. Let $f: C_1 \to C_2$ be a chain-map. Then

$$U_2 \circ f \circ U_1^{-1} : H_1 \rightsquigarrow H_2;$$

in fact $U_2 \circ f \circ U_1^{-1}$ is the homomorphism

$$H(f) : H_1 \to H_2$$

induced by f.

Example 5. Let

$$(*) \qquad 0 \to A \xrightarrow{\ i\ } C \xrightarrow{\ p\ } B \to 0$$

be an exact sequence of chain-complexes. Then the boundary operator

$$\partial_* : H(B) \to H(A)$$

of the homology sequence of $(*)$ is the relation

$$U_A \circ i^{-1} \circ \partial \circ p^{-1} \circ U_B^{-1} : H(B) \rightsquigarrow H(A),$$

where ∂ is the boundary operator of the chain complex C.

EXAMPLE 6. Let A_1, B_1 be subgroups of A, B, respectively. We may then form the relation

$$B_1 | F | A_1 = F \cap (A_1 \times B_1) : A_1 \rightsquigarrow B_1.$$

If $A_1 = A$ or $B_1 = B$, we may abbreviate this to

$$B_1 | F = B_1 | F | A,$$
$$F | A_1 = B | F | A_1.$$

EXERCISES

1. Prove that $F \circ F^{-1} \circ F = F$, for any additive relation F.
2. Prove that every additive relation F can be factored in the form $F = g \circ f^{-1}$, where f and g are homomorphisms.
3. Prove that every additive relation can be factored in the form $F = g^{-1} \circ f$ where f and g are homomorphisms.
4. Let $F : A \rightsquigarrow B$. Prove that, whenever A_1 and A_2 are subgroups of A, B_1 and B_2 are subgroups of B, then

   (1) $F(A_1 \cap A_2) \subset F(A_1) \cap F(A_2)$,
   (2) $F(A_1 + A_2) \supset F(A_1) + F(A_2)$,
   (3) $F^{-1}(B_1 \cap B_2) \subset F^{-1}(B_1) \cap F^{-1}(B_2)$,
   (4) $F^{-1}(B_1 + B_2) \supset F^{-1}(B_1) + F^{-1}(B_2)$,
   (5) $F(A_1 \cap F^{-1}(B_1)) \supset F(A_1) \cap B_1$,
   (6) $F^{-1}(F(A_1) + B_1) \subset A_1 + F^{-1}(B_1)$.

   Equality holds in the following cases:

   (1) Ker $F = \{0\}$,
   (2) Dom $F = A$,
   (3) Ind $F = \{0\}$,
   (4) Im $F = B$,
   (5) $B_1 \supset$ Ind F,
   (6) $A_1 \subset$ Dom F.

5. Let

$$(\Delta) \qquad \begin{array}{ccccccccc} 0 & \longrightarrow & A & \xrightarrow{i} & B & \xrightarrow{p} & C & \longrightarrow & 0 \\ & & \downarrow f & & \downarrow g & & \downarrow h & & \\ 0 & \longrightarrow & A' & \xrightarrow[i']{} & B' & \xrightarrow[p']{} & C' & \longrightarrow & 0 \end{array}$$

be a *short* ladder. Prove that there is an exact sequence

$$0 \to \operatorname{Ker} f \to \operatorname{Ker} g \to \operatorname{Ker} h \xrightarrow{\ \sigma\ } \operatorname{Cok} f \to \operatorname{Cok} g \to \operatorname{Cok} h \to 0,$$

where σ is the suspension.

6. If $F : A \rightsquigarrow B$, show that $F \circ F^{-1}$ and $F^{-1} \circ F$ are the relations associated with the identity maps of the subquotients $\operatorname{Im} F/\operatorname{Ind} F$, $\operatorname{Dom} F/\operatorname{Ker} F$ of B, A respectively.

7. If $F : A \rightsquigarrow B$, $G : B \rightsquigarrow C$, show that

$$\operatorname{Dom}(G \circ F) = F^{-1}(\operatorname{Dom} G),$$
$$\operatorname{Ker}(G \circ F) = F^{-1}(\operatorname{Ker} G),$$
$$\operatorname{Im}(G \circ F) = G(\operatorname{Im} F),$$
$$\operatorname{Ind}(G \circ F) = G(\operatorname{Ind} F).$$

Then $\operatorname{Dom}(G \circ F) = A$ if and only if $\operatorname{Dom} F = A$ and $\operatorname{Im} F \subset \operatorname{Dom} G + \operatorname{Ind} F$, and $\operatorname{Ind}(G \circ F) = \{0\}$ if and only if $\operatorname{Ind} G = 0$ and $\operatorname{Dom} G \cap \operatorname{Ind} F \subset \operatorname{Ker} G$.

8. An additive relation $F : A \rightsquigarrow B$ is *null* if and only if the associated homomorphism $f : \operatorname{Dom} F \to \operatorname{Cod} F$ is zero. Prove that the following conditions are equivalent:

(1) F is null;
(2) $\operatorname{Ker} F = \operatorname{Dom} F$;
(3) $\operatorname{Ind} F = \operatorname{Im} F$;
(4) $F = A_0 \times B_0$ for some subgroups $A_0 \subset A$, $B_0 \subset B$.

9. Let $F : A \rightsquigarrow B, G : B \rightsquigarrow C$. Prove that $G \circ F$ is null if and only if $\operatorname{Im} F \cap \operatorname{Dom} G \subset \operatorname{Ind} F + \operatorname{Ker} G$. Deduce that $G \circ F$ is null if either F or G is null.

10. A *relational chain complex* is an abelian group C with an additive relation $F : C \rightsquigarrow C$ such that $F \circ F$ is null. Develop the homology theory of relational chain complexes.

11. Let $F : B$, $G : B \rightsquigarrow C$ be additive relations. Prove that there are exact sequences

$$0 \to \operatorname{Ker} F \to \operatorname{Ker}(G \circ F) \to \operatorname{Ker} G/\operatorname{Ker} G \cap \operatorname{Ind} F$$
$$\to \operatorname{Dom} G/\operatorname{Dom} G \cap \operatorname{Im} F \to \operatorname{Cok}(G \circ F) \to \operatorname{Cok} G \to 0,$$
$$0 \to \operatorname{Ind} G \to \operatorname{Ind}(G \circ F) \to \operatorname{Ind} F/\operatorname{Ind} F \cap \operatorname{Ker} G$$
$$\to \operatorname{Im} F/\operatorname{Im} F \cap \operatorname{Dom} G \to \operatorname{Coin}(G \circ F) \to \operatorname{Coin} F \to 0.$$

12. Let $A \supset B \supset C$, so that B/C is a subquotient of A. The canonical relation $J : A \rightsquigarrow B/C$ is the composite

$$A \xrightarrow{\ i^{-1}\ } B \xrightarrow{\ \pi\ } B/C$$

where i is the inclusion and π the natural projection. Prove that $J \circ J^{-1}$ is the identity map of B/C.

Bibliography

1 Books

[B] Bourbaki, N., (pseud.). *Éléments de Mathématique I. Les Structures Fondamentales de l'Analyse.* Livre II. Algèbre. Chapitre 3. Algèbre Multilinéaire. Actualités Scientifiques et Industrielles, No. 1044. Paris: Hermann, 1958.

[Ca] Cartan, Élie. *La Topologie des Groupes de Lie*, Actualités Scientifiques et Industrielles, No. 358. Paris: Hermann, 1936.

[C_1] Chevalley, Claude. *Theory of Lie Groups I.* Princeton Mathematical Series, No. 8. Princeton: Princeton University Press, 1946.

[C_2] Chevalley, Claude C. *The Algebraic Theory of Spinors.* New York: Columbia University Press, 1954.

[C–E] Cartan, Henri and Eilenberg, Samuel. *Homological Algebra.* Princeton Mathematical Series, No. 19. Princeton: Princeton University Press, 1956.

[C–F] Cooke, George E. and Finney, Ross L. *Homology of Cell Complexes* (*Based on Lectures by Norman E. Steenrod*). Mathematical Notes. Princeton: Princeton University Press, 1967.

[D] Dugundji, James. *Topology.* Boston: Allyn and Bacon, 1965.

[E–S] Eilenberg, Samuel and Steenrod, Norman. *Foundations of Algebraic Topology.* Princeton Mathematical Series, No. 15. Princeton: Princeton University Press, 1952.

[Gr] Greenberg, Marvin, *Lectures on Algebraic Topology.* Menlo Park, Calif.: W. A. Benjamin, 1966.

[Ha] Hall, Marshall, Jr. *The Theory of Groups.* New York: Macmillan, 1959.

[He] Helgason, Sigurdur. *Differential Geometry and Symmetric Spaces.* New York and London: Academic Press, 1962.

[Hi] Hilton, P. J. *Homotopy Theory and Duality.* New York: Gordon and Breach, 1965.

[H–W] Hilton, P. J. and Wylie, S. *Homology Theory. An Introduction to Algebraic Topology.* Cambridge, England: Cambridge University Press, 1960.

[Ho] Hochschild, G. *The Structure of Lie Groups.* San Francisco: Holden–Day, 1965.

[Hu] Husemoller, Dale. *Fibre Bundles*. New York: McGraw-Hill, 1966. Springer-Verlag, 1975.

[J] Jacobson, Nathan. *Structure and Representations of Jordan Algebras*. American Mathematical Society Colloquium Publications, No. 39. Providence: American Mathematical Society, 1968.

[K] Kurosh, A. G. *The Theory of Groups*. Translated from the Russian and edited by K. A. Hirsch. New York: Chelsea Publishing Company, 1955.

[L–S] Lusternik, L. and Schnirelmann, L. *Méthodes Topologiques dans les Problèmes Variationnels*. Actualités Scientifiques et Industrielles, No. 188. Paris, Hermann et Cie., 1934.

[L–W] Lundell, Albert T. and Weingram, Stephen. *Topology of* CW-*Complexes*. New York: Van Nostrand Reinhold, 1969.

[MacL] MacLane, Saunders. *Homology*. Die Grundlehren der Mathematischen Wissenschaften, Band 114. New York: Academic Press and Berlin-Heidelberg-New York: Springer-Verlag, 1963.

[M–S] Milnor, John W. and Stasheff, James D. *Characteristic Classes*. Annals of Mathematics Studies, No. 76. Princeton: Princeton University Press and University of Tokyo Press, 1974.

[Mo] Morse, Marston. *The Calculus of Variations in the Large*. American Mathematical Society Colloquium Publications, Vol. 18. New York: American Mathematical Society, 1934.

[Sp] Spanier, Edwin H. *Algebraic Topology*. New York: McGraw-Hill, 1966.

[St_1] Steenrod, Norman. *The Topology of Fibre Bundles*. Princeton Mathematical Series, No. 14. Princeton: Princeton University Press, 1951.

[St_2] Steenrod, N. E. *Cohomology Operations. Lectures by N. E. Steenrod written and revised by D. B. A. Epstein*. Annals of Mathematics Studies, No. 50. Princeton: Princeton University Press, 1962.

[V] Veblen, Oswald. *Analysis Situs*. American Mathematical Society Colloquium Publications, Volume V, Part II. New York: American Mathematical Society, 1921.

[W_1] Whitehead, George W. *Homotopy Theory*. Compiled by Robert J. Aumann. Cambridge, Mass. and London: The M.I.T. Press, 1966.

[W_2] Whitehead, George W. *Recent Advances in Homotopy Theory*. Conference Board of the Mathematical Sciences, Regional Conference Series in Mathematics, No. 5. Providence: American Mathematical Society, 1970.

[Z] Zassenhaus, Hans. *The Theory of Groups. Second English Edition*. New York: Chelsea Publishing Company, 1958.

2 Papers

Adams, J. F.

[1] On the non-existence of elements of Hopf invariant one. *Ann. of Math.* **(2) 72** (1960), 20–104.

[2] Vector fields on spheres, *Ann. of Math.* **(2) 75** (1962), 603–632.

Adems, José

[1] The relations on Steenrod powers of cohomology classes. In: *Algebraic Geometry and Topology. A Symposium in Honor of S. Lefschetz*. R. H. Fox, D. C. Spencer, A. W. Tucker (eds.), pp. 191–238. Princeton Mathematical Series, No. 12. Princeton: Princeton University Press, 1957.

[2] Un criterio cohomológico para determinar composciones esenciales de transformaciones, *Bol. Soc. Mat. Mexicana* **(2) 1** (1956), 38–48.

Alexander, J. W.
[1] On the deformation of the *n*-cell, *Proc. Nat. Acad. Sci. U.S.A.* **9** (1923), 406–7.

Atiyah, M. F. and Hirzebruch, F.
[1] Vector bundles and homogeneous spaces. In: *Proc. Sympos. Pure Math.*, **Vol. III**, pp. 7–38. Providence: American Mathematical Society, 1961.

Barcus, W. D. and Meyer, J.-P.
[1] The suspension of a loop space, *Amer. J. Math.* **80** (1958), 895–920.

Barratt, M. G.
[1] Track groups I. *Proc. London Math. Soc.* **(3) 5** (1955), 71–106.

Barratt, M. G. and Hilton, P. J.
[1] On join operations in homotopy groups, *Proc. London Math. Soc.* **(3) 3** (1953), 430–445.

Barratt, M. G. and Milnor, John
[1] An example of anomalous singular theory, *Proc. Amer. Math. Soc.* **13** (1962), 293–297.

Barratt, M. G. and Whitehead, J. H. C.
[1] The first non-vanishing group of an $(n + 1)$-ad, *Proc. London Math. Soc.* **(3) 6** (1956), 417–439.

Blakers, A. L.
[1] Some relations between homology and homotopy groups, *Ann. of Math.* **(2) 49** (1948), 428–461.

Blakers, A. L. and Massey, W. S.
[1] The homotopy groups of a triad I. *Ann. of Math.* **(2) 53** (1951), 161–205.

Borel, Armand
[1] Sur la cohomologie des espaces fibrés principaux et des espaces homogènes de groupes de Lie compacts, *Ann. of Math.* **(2) 57** (1953), 115–207.
[2] La cohomologie mod 2 de certains espaces homogènes, *Comment. Math. Helv.* **27** (1953), 165–197.
[3] Sur l'homologie et la cohomologie des groupes de Lie compacts et connexes, *Amer. J. Math.* **76** (1954), 273–342.
[4] Topology of Lie groups and characteristic classes, *Bull. Amer. Math. Soc.* **61** (1955), 397–432.

Borsuk, Karol
[1] Sur les prolongements des transformations continues, *Fund. Math.* **28** (1937), 99–110.

Brauer, Richard
[1] Sur les invariants intégraux des variétés des groupes de Lie simple clos, *C. R. Acad. Sci. Paris* **201** (1935), 419–421.

Brouwer, L. E. J.
[1] Über Abbildungen von Mannigfaltigkeiten, *Math. Ann.* **71** (1912), 97–115.

Browder, William
[1] Torsion in H-spaces, *Ann. of Math.* **(2) 74** (1961), 24–51.

Cartan, Élie
[1] Sur les invariants intégraux de certains espaces homogènes clos et les propriétés topologiques de ces espaces, *Ann. Soc. Pol. Math.* **8** (1929), 181–225.

Cartan, Henri
[1] Une théorie axiomatique des carrés de Steenrod, *C. R. Acad. Sci. Paris* **230** (1950), 425–427.

Cartan, Henri and Serre, Jean-Pierre
[1] Espaces fibrés et groupes d'homotopie I. Constructions générales, *C. R. Acad. Sci. Paris* **234** (1952), 288–290.

Čech, Edouard
[1] Höherdimensionale Homotopiegruppen. In: *Verhandlungen des Internationalen Mathematikerkongress*, Zürich, 1932, p. 203. Zürich and Leipzig: Orell Füssli, 1932.

Copeland, Arthur, Jr.
[1] On H-spaces with two non-trivial homotopy groups, *Proc. Amer. Math. Soc.* **8** (1957), 184–191.
[2] Binary operations on sets of mapping classes, *Michigan Math. J.* **6** (1959), 7–23.

Eilenberg, Samuel
[1] On the relation between the fundamental group of a space and the higher homotopy groups, *Fund. Math.* **32** (1939), 167–175.
[2] Cohomology and continuous mappings, *Ann. of Math.* **(2) 41** (1940), 231-251.
[3] Singular homology theory, *Ann. of Math.* **(2) 45** (1944), 407–447.
[4] Homology of spaces with operators I, *Trans. Amer. Math. Soc.* **61** (1947), 378–417; errata, ibid. **62** (1947), 548.

Eilenberg, Samuel and Mac Lane, Saunders
[1] Relations between homology and homotopy groups of spaces, *Ann. of Math.* **(2) 46** (1945), 480–509.
[2] On the groups $H(\Pi, n)$ III, *Ann. of Math.* **(2) 60** (1954), 513–557.

Eilenberg, Samuel and MacLane, Saunders
[1] Axiomatic approach to homology theory, *Proc. Nat. Acad. Sci. U.S.A.* **31** (1945), 117–120.

Fox, Ralph H.
[1] On the Lusternik-Schnirelmann category, *Ann. of Math.* **(2) 42** (1941), 333–370.
[2] On homotopy type and deformation retracts, *Ann. of Math.* **(2) 44** (1943), 40–50.
[3] Natural systems of homomorphisms. Preliminary report. *Bull. Amer. Math. Soc.* **49** (1943), 373. Abstract No. 172.

Freudenthal, Hans
[1] Über die Klassen der Sphärenabbildungen, *Compositio Math.* **5** (1937), 299–314.

Gysin, Werner
[1] Zur Homologietheorie der Abbildungen und Faserungen der Mannigfaltigkeiten, *Comment. Math. Helv.* **14** (1941), 61–122.

Hall, Marshall, Jr.
[1] A basis for free Lie rings and higher commutators in free groups, *Proc. Amer. Math. Soc.* **1** (1950), 575–581.

Hilton, P. J.
[1] On the homotopy groups of the union of spheres, *J. London Math. Soc.* **30** (1955), 154–172.
[2] Note on the Jacobi identity for Whitehead products, *Proc. Cambridge Philos. Soc.* **57** (1961), 180–182.

Hilton, P. J. and Whitehead, J. H. C.
[1] Note on the Whitehead product, *Ann. of Math.* **(2) 58** (1953), 429–442.

Hopf, Heinz
[1] Abbildungsklassen n-dimensionaler Mannigfaltigkeiten, *Math. Ann.* **96** (1926), 209–224.

[2] Zur Algebra der Abbildungen von Mannigfaltigkeiten, *J. Reine Angew. Math.* **103** (1930), 71–88.
[3] Über die Abbildungen der dreidimensionalen Sphäre auf die Kugelfläche, *Math. Ann.* **104** (1931), 639–665.
[4] Die Klassen der Abbildungen der *n*-dimensionalen Polyeder auf die *n*-dimensionalen Sphäre. *Comment. Math. Helv.* **5** (1933), 39–54.
[5] Über die Abbildungen von Sphären auf Sphären niedriger Dimension, *Fund. Math.* **25** (1935), 427–440.
[6] Über die Topologie der Gruppenmannigfaltigkeiten und ihre Verallgemeinerungen, *Ann. of Math.* **(2) 42** (1941), 22–52.
[7] Fundamentalgruppe und zweite Bettische Gruppe, *Comment. Math. Helv.* **14** (1942), 257–309.
[8] Beiträge zur Homotopietheorie, *Comment. Math. Helv.* **17** (1945) 307–326.

Hu, Sze-Tsen
[1] An exposition of the relative homotopy theory, *Duke Math. J.* **14** (1947), 991–1033.

Hurewicz, Witold
[1] Beiträge zur Topologie der Deformationen I–IV, *Nederl. Akad. Wetensch. Proc.*, **Ser. A 38** (1935), 112–119, 521–528; **39** (1936), 117–126, 215–224.
[2] On the concept of fiber space, *Proc. Nat. Acad. Sci. U.S.A.* **41** (1955), 956–961.

Hurewicz, Witold and Steenrod, Norman E.
[1] Homotopy relations in fibre spaces, *Proc. Nat. Acad. Sci. U.S.A.* **27** (1941), 60–64.

James, Ioan M.
[1] Reduced product spaces, *Ann. of Math.* **(2) 62** (1955), 170–197.

Leray, Jean
[1] Sur la forme des espaces topologiques et sur les points fixes des représentations, *J. Math. Pures Appl.* **(9) 24** (1945), 95–167.
[2] L'anneau spectral et l'anneau filtré d'homologie d'un espace localement compact et d'une application continue, *J. Math. Pures Appl.* **(9) 29** (1950), 1–139.
[3] L'homologie d'un espace fibré dont la fibre est connexe, *J. Math. Pures Appl.* **(9) 29** (1950), 169–213.

Massey, William S.
[1] Exact couples in algebraic topology I, II, *Ann. of Math.* **(2) 56** (1952), 363–396.
[2] Exact couples in algebraic topology III, IV, V, *Ann. of Math.* **(2) 57** (1953), 248–286.
[3] On the cohomology ring of a sphere bundle, *J. Math. Mech.* **7** (1958), 265–290.

Miller, Clair
[1] The second homology group of a group; relations among commutators, *Proc. Amer. Math. Soc.* **3** (1952), 588–595.

Milnor, John W.
[1] The geometric realization of a semi-simplicial complex, *Ann. of Math.* **(2) 65** (1957), 357–362.
[2] On spaces having the homotopy type of a CW-complex, *Trans. Amer. Math. Soc.* **90** (1959), 272–280.
[3] On axiomatic homology theory, *Pacific J. Math.* **12** (1962), 337–341.
[4] On the construction FK. In: *Algebraic Topology—a Student's Guide*, by J. F. Adams, pp. 119–136; *London Mathematical Society Lecture Note Series*, **No. 4**. Cambridge, England: Cambridge University Press, 1972.

Milnor, John W. and Moore, John C.
[1] On the structure of Hopf algebras, *Ann. of Math.* **(2) 81** (1965), 211–264.

Moore, John C.
[1] Semi-simplicial complexes and Postnikov systems. In: *Symposium Internacional de Topologia Algebraica*, pp. 232–247. Mexico City. Universidad Nacional Autónoma de México and UNESCO, 1958.

Nakaoka, Minoru and Toda, Hirosi
[1] On Jacobi identity for Whitehead products, *J. Inst. Polytech. Osaka City Univ.*, **Ser A 5** (1954), 1–13.

Namioka, Isaac
[1] Maps of pairs in homotopy theory. *Proc. London Math. Soc.* **(3) 12** (1962), 725–738.

Olum, Paul
[1] Obstructions to extensions and homotopies, *Ann. of Math.* **(2) 52** (1950), 1–50.

Poincaré, Henri
[1] Analysis situs, *J. Ecole Polytech.* **(2) 1** (1895), 1–121.

Pontrjagin, L. S.
[1] Homologies in compact Lie groups, *Rec. Math.*, **N.S. 6 (48)** (1939), 389–422.
[2] Homotopy classification of the mappings of an $(n + 2)$-dimensional sphere on an n-dimensional, *Doklady Akad. Nauk. SSSR* (*N.S.*) **70** (1950), 957–959. (Russian).

Porter, Gerald J.
[1] The homotopy groups of wedges of suspensions, *Amer. J. Math.* **88** (1966), 655–663.

Postnikov, M. M.
[1] Investigations in homotopy theory of continuous mappings. I. The algebraic theory of systems. II. The natural system and homotopy. In: *American Mathematical Society Translations*, **Series 2, Volume 7**, pp. 1–134. Providence: American Mathematical Society, 1957.
[2] Investigations in homotopy theory of continuous mappings. III. General theorems of extension and classification. In: *American Mathematical Society Translations*, **Series 2, Volume 11**, pp. 115–153. Providence: American Mathematical Society, 1959.

Puppe, Dieter
[1] Homotopiemengen und ihre induzierten Abbildungen, *Math. Z.* **69** (1958), 299–344.
[2] Some well known weak homotopy equivalences are genuine homotopy equivalences. In: *Symposia Mathematica*, **Vol. V** (INDAM, Rome, 1969/70), pp. 363–374. London: Academic Press, 1971.

Rothenberg, M. and Steenrod, Norman E.
[1] The cohomology of classifying spaces of H-spaces, *Bull. Amer. Math. Soc.* **71** (1965), 872–875.

Samelson, Hans
[1] Beiträge zur Topologie der Gruppenmannigfaltigkeiten, *Ann. of Math.* **(2) 42** (1941), 1091–1137.
[2] A connection between the Whitehead and the Pontryagin product, *Amer. J. Math.* **75** (1953), 744–752.
[3] Topology of Lie groups, *Bull. Amer. Math. Soc.* **58** (1952), 2–37.

Serre, Jean-Pierre
[1] Homologie singulière des espaces fibrés. Applications, *Ann. of Math.* **(2) 54** (1951), 425–505.

[2] Cohomologie modulo 2 des complexes d'Eilenberg–Mac Lane, *Comment. Math. Helv.* **27** (1953), 198–232.
[3] Groupes d'homotopie et classes de groupes abéliens, *Ann. of Math.* **(2) 58** (1953), 258–294.

Spanier, E. H. and Whitehead, J. H. C.
[1] A first approximation to homotopy theory, *Proc. Nat. Acad. Sci. U.S.A.* **39** (1953), 655–660.
[2] On fibre spaces in which the fibre is contractible, *Comment. Math. Helv.* **29** (1955), 1–8.

Steenrod, Norman E.
[1] Homology with local coefficients, *Ann. of Math.* **(2) 44** (1943), 610–627.
[2] Products of cocycles and extensions of mappings, *Ann. of Math.* **(2) 48** (1947), 290–320.
[3] Cohomology invariants of mappings, *Ann. of Math.* **(2) 50** (1949), 954–968.
[4] Cohomology operations. In: *Symposium Internacional de Topologia Algebraica*, pp. 165–185. Mexico City. Universidad Nacional Autonoma de Mexico and UNESCO, 1958.
[5] A convenient category of topological spaces, *Michigan Math. J.* **14** (1967), 133–152.

Steenrod, Norman E. and Whitehead, J. H. C.
[1] Vector fields on the *n*-sphere, *Proc. Nat. Acad. Sci. U.S.A.* **37** (1951), 58–63.

Stiefel, E.
[1] Richtungsfelder und Fernparallelismus in Mannigfaltigkeiten, *Comment. Math. Helv.* **8** (1936), 3–51.

Strøm, A.
[1] Note on cofibrations, *Math. Scand.* **19** (1966), 11–14.
[2] Note on cofibrations II, *Math. Scand.* **22** (1968), 130–142.

Thom, René
[1] Espaces fibrés en sphères et carrés de Steenrod, *Ann. Sci. École Norm. Sup.* **(3) 69** (1952), 109–182.

Toda, Hirosi
[1] Complex of the standard paths and *n*-ad homotopy groups, *J. Inst. Polytech. Osaka City Univ.*, **Ser. A 6** (1955), 101–120.

Uehara, Hirosi and Massey, W. S.
[1] The Jacobi identity for Whitehead products. In: *Algebraic Geometry and Topology. A Symposium in Honor of S. Lefschetz.* R. H. Fox, D. C. Spencer, A. W. Tucker, (eds.), pp. 361–377. Princeton Mathematical Series, No. 12. Princeton: Princeton University Press, 1957.

Wall, C. T. C.
[1] On the exactness of interlocking sequences, *L'Enseignement Mathématique* **12** (1966), 95–100.

Wang, H. C.
[1] The homology groups of the fibre bundles over a sphere, *Duke Math. J.* **16** (1949), 33–38.

Whitehead, George W.
[1] On the homotopy groups of spheres and rotation groups, *Ann. of Math.* **(2) 43** (1942), 634–640.
[2] On products in homotopy groups, *Ann. of Math.* **(2) 47** (1946), 460–475.
[3] On spaces with vanishing low-dimensional homotopy groups, *Proc. Nat. Acad. Sci. U.S.A.* **34** (1948), 207–211.

[4] The $(n+2)^d$ homotopy group of the n-sphere, *Ann. of Math.* **(2) 52** (1950), 245–247.
[5] Fibre spaces and the Eilenberg homology groups, *Proc. Nat. Acad. Sci. U.S.A.* **38** (1952), 426–430.
[6] On mappings into group-like spaces, *Comment. Math. Helv.* **28** (1954), 320–328.
[7] On the homology suspension, *Ann. of Math.* **(2) 62** (1955), 254–268.
[8] The homology suspension. In: *Colloque de Topologie*, Louvain, 1956, pp. 89–95. Liège: Georges Thone; Paris: Masson et Cie., 1957.
[9] Generalized homology theories, *Trans. Amer. Math. Soc.* **102** (1962), 227–283.

Whitehead, J. H. C.
[1] Simplicial spaces, nuclei and m-groups, *Proc. London Math. Soc.* **(2) 45** (1939), 243–327.
[2] On adding relations to homotopy groups, *Ann. of Math.* **(2) 42** (1941), 409–428.
[3] On the homotopy type of ANR's, *Bull. Amer. Math. Soc.* **54** (1948), 1133–1145.
[4] Combinatorial homotopy I, *Bull. Amer. Math. Soc.* **55** (1949), 213–245.
[5] On the realizability of homotopy groups, *Ann. of Math.* **(2) 50** (1949), 261–263.
[6] Combinatorial homotopy II, *Bull. Amer. Math. Soc.* **55** (1949), 453–496.
[7] A certain exact sequence, *Ann. of Math.* **(2) 52** (1950), 51–110.
[8] On the theory of obstructions, *Ann. of Math.* **(2) 54** (1951), 68–84.

Whitney, Hassler
[1] Sphere spaces, *Proc. Nat. Acad. Sci. U.S.A.* **21** (1935), 462–468.
[2] On maps of an n-sphere into another n-sphere, *Duke Math. J.* **3** (1937), 46–50.
[3] The maps of an n-complex into an n-sphere, *Duke Math. J.* **3** (1937), 51–55.
[4] Topological properties of differentiable manifolds, *Bull. Amer. Math. Soc.* **43** (1937), 785–805.
[5] On the theory of sphere bundles, *Proc. Nat. Acad. Sci. U.S.A.* **26** (1940), 148–153.

Index

D

E

F

R

S

Graduate Texts in Mathematics

continued from page ii

48 SACHS/WU. General Relativity for Mathematicians.
49 GRUENBERG/WEIR. Linear Geometry. 2nd ed.
50 EDWARDS. Fermat's Last Theorem.
51 KLINGENBERG. A Course in Differential Geometry.
52 HARTSHORNE. Algebraic Geometry.
53 MANIN. A Course in Mathematical Logic.
54 GRAVER/WATKINS. Combinatorics with Emphasis on the Theory of Graphs.
55 BROWN/PEARCY. Introduction to Operator Theory I: Elements of Functional Analysis.
56 MASSEY. Algebraic Topology: An Introduction.
57 CROWELL/FOX. Introduction to Knot Theory.
58 KOBLITZ. *p*-adic Numbers. *p*-adic Analysis, and Zeta-Functions. 2nd ed.
59 LANG. Cyclotomic Fields.
60 ARNOLD. Mathematical Methods in Classical Mechanics. Second Edition.
61 WHITEHEAD. Elements of Homotopy Theory.
62 KARGAPOLOV/MERLZJAKOV. Fundamentals of the Theory of Groups.
63 BOLLOBAS. Graph Theory.
64 EDWARDS. Fourier Series. Vol. I. 2nd ed.
65 WELLS. Differential Analysis on Complex Manifolds. 2nd ed.
66 WATERHOUSE. Introduction to Affine Group Schemes.
67 SERRE. Local Fields.
68 WEIDMANN. Linear Operators in Hilbert Spaces.
69 LANG. Cyclotomic Fields II.
70 MASSEY. Singular Homology Theory.
71 FARKAS/KRA. Riemann Surfaces.
72 STILLWELL. Classical Topology and Combinatorial Group Theory.
73 HUNGERFORD. Algebra.
74 DAVENPORT. Multiplicative Number Theory. 2nd ed.
75 HOCHSCHILD. Basic Theory of Algebraic Groups and Lie Algebras.
76 IITAKA. Algebraic Geometry.
77 HECKE. Lectures on the Theory of Algebraic Numbers.
78 BURRIS/SANKAPPANAVAR. A Course in Universal Algebra.
79 WALTERS. An Introduction to Ergodic Theory.
80 ROBINSON. A Course in the Theory of Groups.
81 FORSTER. Lectures on Riemann Surfaces.
82 BOTT/TU. Differential Forms in Algebraic Topology.
83 WASHINGTON. Introduction to Cyclotomic Fields.
84 IRELAND/ROSEN. A Classical Introduction to Modern Number Theory. Second Edition.
85 EDWARDS. Fourier Series. Vol. II. 2nd ed.
86 VAN LINT. Introduction to Coding Theory.
87 BROWN. Cohomology of Groups.
88 PIERCE. Associative Algebras.
89 LANG. Introduction to Algebraic and Abelian Functions. 2nd ed.
90 BRONDSTED. An Introduction to Convex Polytopes.
91 BEARDON. On the Geometry of Discrete Groups.
92 DIESTEL. Sequences and Series in Banach Spaces.
93 DUBROVIN/FOMENKO/NOVIKOV. Modern Geometry--Methods and Applications Vol. I.
94 WARNER. Foundations of Differentiable Manifolds and Lie Groups.
95 SHIRYAYEV. Probability, Statistics, and Random Processes.
96 CONWAY. A Course in Functional Analysis

97 KOBLITZ. Introduction to Elliptic Curves and Modular Forms.
98 BRÖCKER/TOM DIECK. Representations of compact Lie Groups.
99 GROVE/BENSON. Finite Reflection Groups. 2nd ed.
100 BERG/CHRISTENSEN/RESSEL. Harmonic Analysis on Semigroups: Theory of Positive Definite and Related Functions.
101 EDWARDS. Galois Theory.
102 VARDARAJAN. Lie Groups, Lie Algebras and Their Representations.
103 LANG. Complex Analysis. 2nd ed.
104 DUBROVIN/FOMENKO/NOVIKOV. Modern Geometry--Methods and Applications Vol. II.
105 LANG. $SL_2(\mathbf{R})$.
106 SILVERMAN. The Arithmetic of Elliptic Curves.
107 OLVER. Applications of Lie Groups to Differential Equations.
108 RANGE. Holomorphic Functions and Integral Representations in Several Complex Variables.
109 LEHTO. Univalent Functions and Teichmüller Spaces.
110 LANG. Algebraic Number Theory.
111 HUSEMÖLLER. Elliptic Cruves.
112 LANG. Elliptic Functions.
113 KARATZAS/SHREVE. Brownian Motion and Stochastic Calculus.
114 KOBLITZ. A Course in Number Theory and Cryptography.
115 BERGER/GOSTIAUX. Differential Geometry: Manifolds, Curves, and Surfaces.
116 KELLEY/SRINIVASAN. Measure and Integral, Volume 1.
117 SERRE. Algebraic Groups and Class Fields.
118 PEDERSEN. Analysis Now.
119 ROTMAN. An Introduction to Algebraic Topology.
120 ZIEMER. Weakly Differentiable Functions: Sobolev Spaces and Functions of Bounded Variation.
121 LANG. Cyclotomic Fields I and II. Combined 2nd ed.
122 REMMERT. Theory of Complex Functions.
Readings in Mathematics
123 EBBINGHAUS/HERMES et al. Numbers.
Readings in Mathematics
124 DUBROVIN/FOMENKO/NOVIKOV. Modern Geometry--Methods and Applications Vol. III.
125 BERENSTEIN/GAY. Complex Variables: An Introduction.
126 BOREL. Linear Algebraic Groups.
127 MASSEY. A Basic Course in Algebraic Topology.
128 RAUCH. Partial Differential Equations.
129 FULTON/HARRIS. Representation Theory: A First Course.
Readings in Mathematics.
130 DODSON/POSTON. Tensor Geometry.
131 LAM. A First Course in Noncommutative Rings.

Printed in the United States
By Bookmasters